MOLECULAR AND FUNCTIONAL DIVERSITY OF ION CHANNELS AND RECEPTORS

ANNALS OF THE NEW YORK ACADEMY OF SCIENCES

Volume 868

MOLECULAR AND FUNCTIONAL DIVERSITY OF ION CHANNELS AND RECEPTORS

Edited by Bernardo Rudy and Peter Seeburg

The New York Academy of Sciences
New York, New York
1999

♾ *The paper used in this publication meets the minimum requirements of American National Standard for Information Sciences—Permanence of Paper for Printed Library Materials. ANSI Z39.48-1984.*

Library of Congress Cataloging-in-Publication Data

Molecular and functional diversity of ion channels and receptors / editors Bernado Rudy, Peter Seeburg.
p. cm.—(Annals of the New York Academy of Sciences ; vol. 868)
Includes bibliographical references and index.
ISBN 1-57331-176-6 (alk. paper).—ISBN 1-57331-177-4 (pbk. : alk. paper)
1. Ion channels—Congresses. 2. Neurotransmitter receptors—Congresses. I. Rudy, Bernardo. II. Seeburg, P.H. '(Peter H.), 1944- . III. Series.
Q11.N5 vol. 868
[QH603.I54]
500 s—dc21
[571.6'4] 99-13744
CIP

ANNALS OF THE NEW YORK ACADEMY OF SCIENCES

Volume 868
April 30, 1999

MOLECULAR AND FUNCTIONAL DIVERSITY OF ION CHANNELS AND RECEPTORS[a]

Editors and Conference Chairs
BERNARDO RUDY AND PETER SEEBURG

CONTENTS

Introduction: Molecular Diversity of Ion Channels and Cell Function. *By* BERNARDO RUDY 1

The "Psychic" Neuron of the Cerebral Cortex. *By* PATRICIA S. GOLDMAN-RAKIC .. 13

Studies on Conditional Gene Expression in the Brain. *By* JASNA JERECIC, FRANK SINGLE, ULI KRÜTH, HEINZ KRESTEL, ROHINI KOLHEKAR, THORSTEN STORCK, KALEV KASK, MIYOKO HIGUCHI, ROLF SPRENGEL, and PETER H. SEEBURG 27

Part I. Na^+ Channels

Diversity of Mammalian Voltage-Gated Sodium Channels. *By* ALAN L. GOLDIN... 38

RNA Editing of a *Drosophila* Sodium Channel Gene. *By* CHRISTOPHER J. HANRAHAN, MICHAEL J. PALLADINO, LAURIE J. BONNEAU, and ROBERT A. REENAN 51

H^+-Gated Cation Channels. *By* Rainer Waldmann, Guy Champigny, ERIC LINGUEGLIA, JAN R. DE WEILLE, C. HEURTEAUX, and MICHEL LAZDUNSKI 67

Poster Papers

Identification of Voltage-Activated Na^+ and K^+ Channels in Human Steroid-Secreting Ovarian Cells. *By* A. BULLING, C. BRUCKER, U. BERG, M. GRATZL, and A. MAYERHOFER 77

[a]This volume is the result of a conference entitled **Molecular and Functional Diversity of Ion Channels and Receptors** held by the New York Academy of Sciences on May 14–17, 1998 in New York City.

Cloning and Functional Characterization of the Type III Na^+ Channel from Human Brain. *By* J. J. CLARE, T. J. DALE, X. XIE, T. C. PEAKMAN, and Y. CHEN 80

Slow Voltage-Dependent Inactivation of a Sustained Sodium Current in Stellate Cells of Rat Entorhinal Cortex Layer II. *By* JACOPO MAGISTRETTI and ANGEL ALONSO 84

Differential Distribution of Voltage-Gated Sodium Channel α- and β-Subunits in Human Brain. *By* W. R. J. WHITAKER, J. J. CLARE, and P. C. EMSON 88

Properties of Sodium Currents and Action Potential Firing in Isolated Cerebellar Purkinje Neurons. *By* INDIRA M. RAMAN and BRUCE P. BEAN 93

Slow Sodium Channel Inactivation in CA1 Pyramidal Cells. *By* TIMOTHY MICKUS, HAE-YOON JUNG, and NELSON SPRUSTON 97

Part II. Ca^{2+} Channels

Molecular and Functional Diversity of Voltage-Gated Calcium Channels. *By* HERMAN MORENO DAVILA 102

α_{1B} N-Type Calcium Channel Isoforms with Distinct Biophysical Properties. *By* ANTHONY STEA, STEFAN J. DUBEL, and TERRY P. SNUTCH 118

Molecular Characterization of Two Members of the T-Type Calcium Channel Family. *By* EDWARD PEREZ-REYES, JUNG-HA LEE, and LEANNE L. CRIBBS . . . 131

Interactions of Presynaptic Ca^{2+} Channels and Snare Proteins in Neurotransmitter Release. *By* WILLIAM A. CATTERALL 144

Dissection of the Calcium Channel Domains Responsible for Modulation of Neuronal Voltage-Dependent Calcium Channels by G Proteins. *By* ANNETTE C. DOLPHIN, KAREN M. PAGE, NICHOLAS S. BERROW, GARY J. STEPHENS, and CARLES CANTÍ 160

Neuronal Voltage-Activated Calcium Channels: On the Roles of the α_{1E} and β_3 Subunits. *By* STEPHEN M. SMITH, ERIKA S. PIEDRAS-RENTERÌA, YOON NAMKUNG, HEE-SUP SHIN, and RICHARD W. TSIEN 175

Voltage-Dependent Calcium Channel Mutations in Neurological Disease. *By* DANIEL L. BURGESS and JEFFREY L. NOEBELS 199

Poster Papers

Lambert-Eaton Antibodies Promote Activity-Dependent Enhancement of Exocytosis in Bovine Adrenal Chromaffin Cells. *By* KATHRIN L. ENGISCH, MARK M. RICH, NOAH COOK, and MARTHA C. NOWYCKY 213

L-Type Calcium Channel Regulation of Abnormal Tyrosine Hydroxylase Expression in Cerebella of Tottering Mice. *By* B. E. FUREMAN, D. B. CAMPBELL, and E. J. HESS 217

Okadaic Acid Antagonizes the Inhibitory Action of Internal GTP-γ-S on the Calcium Current of Dorsal Raphe Neurons but Not That of 5-HT_{1A} Receptor Activation. *By* JOHN S. KELLY and HRVOJE HEĆIMOVIĆ 220

Sensitivity to Conotoxin Block of Splice Variants of Rat α_{1B} (rbBII) Subunit of the N-Type Calcium Channel Coexpressed with Different β Subunits in *Xenopus* Oocytes. *By* H. J. MEADOWS and C. D. BENHAM 224

Quantitative Fluorescent RT-PCR Measurements of Postnatal Calcium Channel Gene Expression in Rat Hippocampal Subfields. *By* SHAN M. PRADHAN, SHUNDI GE, CHIEKO KURODA, JULIUS PETERS, and CHARLES E. NIESEN 228

Part III. K^+ Channels

Molecular Diversity of K^+ Channels. *By* WILLIAM A. COETZEE, YIMY AMARILLO, JOANNA CHIU, ALAN CHOW, DAVID LAU, TOM MCCORMACK, HERMAN MORENO, MARCELA S. NADAL, ANDER OZAITA, DAVID POUNTNEY, MICHAEL SAGANICH, ELEAZAR VEGA-SAENZ DE MIERA, and BERNARDO RUDY 233

Genomic Organization of Nematode 4TM K^+ Channels. *By* ZHAO-WEN WANG, MAYA T. KUNKEL, AGUAN WEI, ALICE BUTLER, and LAWRENCE SALKOFF . . . 286

Contributions of Kv3 Channels to Neuronal Excitability. *By* BERNARDO RUDY, ALAN CHOW, DAVID LAU, YIMY AMARILLO, ANDER OZAITA, MICHAEL SAGANICH, HERMAN MORENO, MARCELA S. NADAL, RICARDO HERNANDEZ-PINEDA, ARTURO HERNANDEZ-CRUZ, ALEV ERISIR, CHRISTOPHER LEONARD, and ELEAZAR VEGA-SAENZ DE MIERA 304

Functional and Molecular Aspects of Voltage-Gated K^+ Channel β Subunits. *By* OLAF PONGS, THORSTEN LEICHER, MICHAELA BERGER, JOCHEN ROEPER, ROBERT BÄHRING, DENNIS WRAY, KARL PETER GIESE, ALCINO J. SILVA, and JOHAN F. STORM 344

The Eag Family of K^+ Channels in *Drosophila* and Mammals. *By* BARRY GANETZKY, GAIL A. ROBERTSON, GISELA F. WILSON, MATTHEW C. TRUDEAU, and STEVEN A. TITUS 356

Small-Conductance Calcium-Activated Potassium Channels. *By* CHRIS T. BOND, JAMES MAYLIE, and JOHN P. ADELMAN 370

The Functional Role of Alternative Splicing of Ca^{2+}-Activated K^+ Channels in Auditory Hair Cells. *By* E. M. C. JONES, M. GRAY-KELLER, J. J. ART, and R. FETTIPLACE 379

Structure, G Protein Activation, and Functional Relevance of the Cardiac G Protein–Gated K^+ Channel, I_{KACh}. *By* KEVIN WICKMAN, GRIGORY KRAPIVINSKY, SHAWN COREY, MATT KENNEDY, JAN NEMEC, IGOR MEDINA, and DAVID E. CLAPHAM 386

Potassium Currents in Developing Neurons. *By* ANGELES B. RIBERA 399

Dysfunction of Delayed Rectifier Potassium Channels in an Inherited Cardiac Arrhythmia. *By* MICHAEL C. SANGUINETTI 406

Poster Papers

Topology of the Pore Region of an Inward Rectifier K^+ Channel, Kir2.1. *By* C. DART, M. L. LEYLAND, P. J. SPENCER, P. R. STANFIELD, and M. J. SUTCLIFFE 414

Virus-Mediated Modification of Cellular Excitability. *By* DAVID C. JOHNS, EDUARDO MARBAN, and H. BRADLEY NUSS 418

Functional Characterization of a Cloned Human Intermediate-Conductance Ca^{2+}-Activated K^+ Channel. *By* TINO DYHRING JØRGENSEN, BO SKAANING JENSEN, DORTE STRØBÆK, PALLE CHRISTOPHERSEN, SØREN-PETER OLESEN, and PHILIP KIÆR AHRING 423

Probing the Potassium Channel $K_V1/K_V1.1$ Interaction Using a Random Peptide Display Library. *By* STEPHEN J. LOMBARDI, AMY TRUONG, PAUL SPENCE, KENNETH J. RHODES, and PHILIP G. JONES 427

PKC Modulation of MinK Current Involves Multiple Phosphorylation Sites. *By* C. FREDERICK LO and RANDY NUMANN 431

The Role of Kir2.1 in the Genesis of Native Cardiac Inward-Rectifier K^+ Currents during Pre- and Postnatal Development. By TOMOE Y. NAKAMURA, KAREN LEE, MICHAEL ARTMAN, BERNARDO RUDY, and WILLIAM A. COETZEE 434

Kv2.1/Kv9.3, an ATP-Dependent Delayed-Rectifier K^+ Channel in Pulmonary Artery Myocytes. *By* AMANDA J. PATEL, MICHEL LAZDUNSKI, and ERIC HONORÉ 438

Functional Characterization of a Novel Mutation in KCNA1 in Episodic Ataxia Type 1 Associated with Epilepsy. *By* ALEXANDER SPAUSCHUS, LOUISE EUNSON, MICHAEL G. HANNA, and DIMITRI M. KULLMANN 442

Regulation of a Human Neuronal Voltage-Gated Potassium Channel (hKv1.1) by Protein Tyrosine Phosphorylation and Dephosphorylation. *By* QIANG WANG. . 447

Regulation of Firing Pattern through Modulation of Non-Sh K^+ Currents by Calcium/Calmodulin-Dependent Protein Kinase II in *Drosophila* Embryonic Neurons. *By* Wei-Dong Yao and Chun-Fang Wu 450

Expression of Kv1.2 Potassium Channels in Rat Sensory Ganglia: An Immunohistochemical Study. *By* SHIGERU YOKOYAMA, HISASHI TAKEDA, and HARUHIRO HIGASHIDA 454

Effects on Ion Permeation with Hydrophobic Substitutions at a Residue in Shaker S6 That Interacts with a Signature Sequence Amino Acid. *By* PAUL C. ZEI, EVA M. OGIELSKA, TOSHINORI HOSHI, and RICHARD W. ALDRICH. 458

Part IV. Glutamate Receptors

Recent Excitement in the Ionotropic Glutamate Receptor Field. *By* EDWARD B. ZIFF 465

The Arrangement of Glutamate Receptors in Excitatory Synapses. *By* YUTAKA TAKUMI, ATSUSHI MATSUBARA, ERIC RINVIK, and OLE P. OTTERSEN 474

Glutamate Receptor Anchoring Proteins and the Molecular Organization of Excitatory Synapses. *By* MORGAN SHENG and DANIEL T. PAK 483

Mice with Genetically Modified NMDA and AMPA Receptors. *By* ROLF SPRENGEL and FRANK N. SINGLE 494

GluRδ2 and the Development and Death of Cerebellar Purkinje Neurons in Lurcher Mice. *By* NATHANIEL HEINTZ and PHILIP L. DE JAGER 502

Expression Mechanisms Underlying NMDA Receptor–Dependent Long-Term Potentiation. *By* R. A. Nicoll and R. C. Malenka 515

Poster Papers

Activation of *N*-Methyl-D-Aspartate Receptors Reverses Desensitization of Metabotropic Glutamate Receptor, mGluR5, in Native and Recombinant Systems. *By* S. ALAGARSAMY, S. T. ROUSE, R. W. GEREAU IV, S. F. HEINEMANN, Y. SMITH, and P. J. CONN 526

Distribution of Group III mGluRs in Rat Basal Ganglia with Subtype-Specific Antibodies. *By* STEFANIA RISSO BRADLEY, DAVID G. STANDAERT, ALLAN I. LEVEY, and P. JEFFREY CONN 531

Characterization, Expression, and Distribution of GRIP Protein. *By* HUALING DONG, PEISU ZHANG, DEZHI LIAO, and RICHARD L. HUGANIR. 535

GluR2 Antisense Knockdown Produces Seizure Behavior and Hippocampal Neurodegeneration during a Critical Window. *By* LINDA K. FRIEDMAN and JANA VELÍSKOVÁ 541

Activation of Kainate Receptors on Rat Sensory Neurons Evokes Action Potential Firing and May Modulate Transmitter Release. *By* C. JUSTIN LEE, H. S. ENGELMAN, and A. B. MACDERMOTT 546

An Immunocytochemical Assay for Activity-Dependent Redistribution of Glutamate Receptors from the Postsynaptic Plasma Membrane. *By* DMITRI V. LISSIN, ROBERT C. MALENKA, and MARK VON ZASTROW 550

Activation of PKC Disrupts Presynaptic Inhibition by Group II and Group III Metabotropic Glutamate Receptors and Uncouples the Receptor from GTP-Binding Proteins. *By* THOMAS A. MACEK, HERVÉ SCHAFFHAUSER, and P. JEFFREY CONN 554

AMPA Receptor Forms a Biochemically Functional Complex with NSF and α- and β-SNAPS. *By* PAVEL OSTEN and EDWARD B. ZIFF 558

ABP: A Novel AMPA Receptor Binding Protein. *By* S. SRIVASTAVA and E. B. ZIFF 561

Part V. Nicotinic Receptors

Molecular Diversity of Neuronal Nicotinic Acetylcholine Receptors. *By* DANIEL S. MCGEHEE 565

Heteromeric Complexes of α5 and/or α7 Subunits: Effects of Calcium and Potential Role in Nicotine-Induced Presynaptic Facilitation. *By* R. GIROD, G. CRABTREE, G. ERNSTROM, J. RAMIREZ-LATORRE, D. MCGEHEE, J. TURNER, and L. ROLE 578

Nicotinic Modulation of Glutamate and GABA Synaptic Transmission in Hippocampal Neurons. *By* KRISTOFER A. RADCLIFFE, JANET L. FISHER, RICHARD GRAY, and JOHN A. DANI 591

The Role of β2-Subunit–Containing Nicotinic Acetylcholine Receptors in the Brain Explored with a Mutant Mouse. *By* CLÉMENT LÉNA and JEAN-PIERRE CHANGEUX 611

Poster Papers

Development of a Novel Class of Subtype-Selective Nicotinic Receptor Antagonist: Pyridine-*N*–Substituted Nicotine Analogs. *By* LINDA P. DWOSKIN, LINCOLN H. WILKINS, JAMES R. PAULY, and PETER A. CROOKS 617

Desensitization of Nicotinic Receptors in the Central Nervous System. *By* C. P. FENSTER, J. H. HICKS, M. L. BECKMAN, P. J. O. COVERNTON, M. W. QUICK, and R. A. J. LESTER 620

Methyllycaconitine-, α-Bungarotoxin–Sensitive Neuronal Nicotinic Receptor Operates Slow Calcium Signal in Skeletal Muscle End Plate. *By* IKUKO KIMURA and KATSUYA DEZAKI 624

Ultrastructural Immunolocalization of the α7 nAChR Subunit in Guinea Pig Medial Prefrontal Cortex. *By* MONA LUBIN, ALEV ERISIR, and CHIYE AOKI .. 628

Nicotinic Receptor Subunit mRNA Expression in Dopaminergic Neurons of the Rat Brain. *By* FRÉDÉRIC SGARD, ERIC CHARPANTIER, PASCAL BARNÉOUD, and FRANÇOIS BESNARD 633

Physostigmine and Atropine Potentiate and Inhibit Neuronal α4β4 Nicotinic Receptors. *By* R. ZWART, R. G. D. M. VAN KLEEF, and H. P. M. VIJVERBERG 636

The Long Cytoplasmic Loop of the α3 Subunit Targets Specific nAChR Subtypes to Synapses on Neurons *in Vivo*. *By* BRIAN M. WILLIAMS, MURALI KRISHNA TEMBURNI, SONIA BERTRAND, DANIEL BERTRAND, and MICHELE H. JACOB .. 640

Part VI. GABA and Glycine Receptors

Molecular and Functional Diversity of the Expanding GABA-A Receptor Gene Family. *By* PAUL J. WHITING, TIMOTHY P. BONNERT, RUTH M. MCKERNAN, SOPHIE FARRAR, BEATRICE LE BOURDELLÈS, ROBERT P. HEAVENS, DAVID W. SMITH, LOUISE HEWSON, MICHAEL R. RIGBY, DALIP J. S. SIRINATHSINGHJI, SALLY A. THOMPSON, and KEITH A. WAFFORD 645

Activity-Dependent Regulation of $GABA_A$ Receptors. *By* SILKE PENSCHUCK, JACQUES PAYSAN, OLIVIA GIORGETTA, and JEAN-MARC FRITSCHY 654

Structure and Functions of Inhibitory and Excitatory Glycine Receptors. *By* HEINRICH BETZ, JOCHEN KUHSE, VOLKER SCHMIEDEN, BODO LAUBE, JOACHIM KIRSCH, and ROBERT J. HARVEY 667

Poster Papers

Changes in $GABA_A$ Receptor–Mediated Synaptic Transmission in Oxytocin Neurons during Female Reproduction: Plasticity in a Neuroendocrine Context. *By* ARJEN B. BRUSSAARD and KAREL S. KITS 677

Structure-Function Relationships of the Human Glycine Receptor: Insights from Hyperekplexia Mutations. *By* TREVOR M. LEWIS and PETER R. SCHOFIELD ... 681

Regulation of Glycine Transport in Cultured Müller Cells by Ca^{2+}/Calmodulin-Dependent Enzymes. *By* ANA MARÍA LÓPEZ-COLOMÉ and ANA GADEA 685

Processing of $GABA_BR1$ in Heterologous Expression Systems. *By* J. MOSBACHER, K. KAUPMANN, V. SCHULER, D. RISTIG, K. STRUCKMEYER, T. PFAFF, A. KARSCHIN, M. F. POZZA, and B. BETTLER 689

Postsynaptic Colocalization of Gephyrin and $GABA_A$ Receptors. *By* MARCO SASSOÈ-POGNETTO, MAURIZIO GIUSTETTO, PATRIZIA PANZANELLI, DARIO CANTINO, JOACHIM KIRSCH, and JEAN-MARC FRITSCHY 693

Structural Requirements for the Interaction of Unsaturated Free Fatty Acids with Recombinant Human $GABA_A$ Receptor Complexes. *By* MICHAEL-ROBIN WITT, CLAUS FOG POULSEN, BIRTHE LÜKENSMEJER, SVEND ERIK WESTH-HANSEN, JUNICHI NABEKURA, NORIO AKAIKE, and MOGENS NIELSEN 697

Part VII. Serotonin and Purinergic Receptors Cyclic Nucleotide-Gated and Pacemaker Channels

5-HT Receptor Knockout Mice: Pharmacological Tools or Models of Psychiatric Disorders. *By* K. SCEARCE-LEVIE, J.-P. CHEN, E. GARDNER, and R. HEN 701

Functional and Molecular Diversity of Purinergic Ion Channel Receptors. *By* A. B. MACKENZIE, A. SURPRENANT, and R. A. NORTH. 716

Cyclic Nucleotide–Gated Channels: Molecular Mechanisms of Activation. *By* MARIE-CHRISTINE BROILLET and STUART FIRESTEIN 730

The HCN Gene Family: Molecular Basis of the Hyperpolarization-Activated Pacemaker Channels. *By* BINA SANTORO and GARETH R. TIBBS 741

Poster Paper

Ca^{2+}-Mediated Up-Regulation of I_h in the Thalamus: How Cell-Intrinsic Ionic Currents May Shape Network Activity. *By* ANITA LÜTHI and DAVID A. MCCORMICK. 765

* * *

Index of Contributors 771

Financial assistance was received from:

Supporters

- ICAGEN, INC.
- F. HOFFMANN-LA ROCHE, CNS DEPARTMENT
- MERCK RESEARCH LABORATORIES
- NATIONAL INSTITUTE OF NEUROLOGICAL DISORDERS AND STROKE, NATIONAL INSTITUTES OF HEALTH
- NATIONAL SCIENCE FOUNDATION

Contributors

- AMERICAN HEART ASSOCIATION
- AXON INSTRUMENTS, INC.
- BRISTOL-MYERS SQUIBB PHARMACEUTICAL INSTITUTE
- GLAXO WELLCOME
- HOECHST MARION ROUSSEL, INC.
- LILLY RESEARCH LABORATORIES
- NEUREX CORPORATION
- PARKE-DAVIS PHARMACEUTICAL RESEARCH DIVISION
- ROCHE BIOSCIENCE
- SIBIA, NEUROSCIENCES, INC.
- SMITHKLINE BEECHAM LABORATORES

Introduction

Molecular Diversity of Ion Channels and Cell Function

BERNARDO RUDY[a]

Department of Physiology and Neuroscience and Department of Biochemistry, New York University School of Medicine, 550 First Avenue, New York, New York 10016, USA

Diversity is an old and recurring theme in the study of the organization and function of the nervous system. The early neuroanatomists recognized that the variety of neuronal morphologies must be associated with functional specialization. A similar diversity of electrophysiological properties of neurons was revealed by the application of intracellular recording techniques to the nervous system several decades later. This theme is reemerging as we study the molecules responsible for the generation and transmission of signals in neurons.

The advances in molecular biology and genetics in the last decade allowed the identification and cloning of the protein components of ligand- and voltage-gated ion channels as well as metabotropic neurotransmitter and neuropeptide receptors (which modulate the activity of ion channels through intermediary second-messenger cascades), the molecules responsible for signal transmission in excitable systems such as the brain, the heart, and other muscles. This work has revealed the existence of an extraordinary diversity of molecular components, suggesting a degree of functional specificity well beyond that expected from previous functional and pharmacological studies. It is likely that these investigations will influence our conceptual view of the organization and function of the nervous system. Multiple types of many of the channels and receptors that are found in the nervous system are also seen in other tissues, suggesting novel functional specializations in other organs as well, and particularly in the cardiovascular and neuromuscular systems.

This volume of the *Annals of the New York Academy of Sciences* contains the proceedings of the conference "Molecular Diversity of Ion Channels and Receptors" sponsored by the New York Academy of Sciences and held at New York University School of Medicine, May 14–17, 1998. The conference, which discussed advances in the integration of the molecular biology and the physiology of ion channels and receptors, was organized by Peter Seeburg from the Max Planck Institute in Heidelberg and myself. The central question addressed at the meeting was how can the large amount of information obtained from the molecular cloning of components of ion channels and receptors be used to advance our understanding of cell and tissue function. The meeting brought together scientists from different disciplines, working with various neurotransmitter receptors and ion channels to share their experiences to gain an understanding of the physiological roles of these molecules. The conference consisted of six plenary sessions: Glutamate Receptors (chaired by Steve Heinemann), Ca^{2+} and Na^{+} Channels (chaired by Terry Snutch), Potassium Channels (chaired by Barry Ganetsky), Nicotinic and Serotonin Receptors (chaired by Lorna Role), GABA and Glycine Receptors (chaired by Heinrich Betz), and Pathophysiology of Ion Channels and Receptors (chaired by Peter Seeburg). The last session dealt with molecules

[a]Phone: 212-263-0431; fax: 212-689-9060; e-mail: rudyb01@mcrcr6.med.nyu.edu

discussed in the earlier sessions, but was designed explicitly to highlight the contributions that this field is making to the future of medicine. The conference also included a keynote presentation delivered by Patricia S. Goldman-Rakic entitled "From Channel Diversity to Higher Brain Function." The scope of the conference allowed the attendees to get a broad picture of the overall progress in the field. This was important because, although the problems that are being addressed in the study of different channels and receptors, and the approaches to solving them are clearly overlapping, individuals working with one type of channel rarely have the opportunity to listen to speakers from other disciplines.

A major task of current research in the field is to understand the physiological implications of the diversity that is being discovered through the molecular analysis. The conference presented work carried out in the last 10 years to characterize the functional roles of the cloned molecules, with an emphasis on the nervous system. Speakers represented the breadth of disciplines that are required to solve the problem: genetics, molecular biology and biochemistry, physiology and anatomy. In organizing the program for the conference, we tried to assemble a list of speakers that would address the variety of approaches being used to understand the physiological significance of the cloned components of ion channels and receptors, including biochemical characterization of native channels, analysis of patterns of expression, molecular and functional analysis of subunit composition, discovery of associated proteins, functional analysis of native and expressed channels, and physiological and pharmacological study of animal models in which specific channels are eliminated or modified. Mutations in genes encoding ion channels and receptors have been found to be a cause of human disease. Although the molecular analysis of ion channels is still a young field, the list of human diseases already found to be associated with defects of these molecules is truly impressive (TABLE 1). These findings illustrate the functional importance of these molecules and were an important topic of discussion at the conference. Moreover, the availability of the cDNAs encoding the many types of ion channels and receptors and the ability to study them functionally in heterologous expression systems is resulting in the discovery of new specific drugs and toxins. Not only will these be invaluable tools for experimental analysis, but the discoveries and novel methodologies that will emerge from this work are likely to contribute to the creation of a new pharmacology for pathologies of excitable systems.

This was a very timely meeting, as most of the molecular players have been cloned and the era of defining their physiological roles has begun. It was clear from the conference that knowledge of the function of ion channels and their role in cellular processes is quickly growing and reaching a state where important new insights can be gained on physiological and pathological situations in which ion channels are involved. Significant advances have taken place, as can be seen in this volume. About three years ago, L. Sivilotti and D. Colquhoun wrote a commentary in the American journal *Science* entitled "Acetylcholine Receptors: Too Many Channels, Too Few Functions."[1] This title could apply to many other families of channels and receptors. The chapters on nicotinic receptors show how much has been achieved in the three years since that commentary was written. The same is true of the other ion channels discussed in this volume. However, major challenges remain in our efforts to bridge the findings of molecular biology and physiology. We hope that this volume will have an impact on future research in this field and that scientists will find it useful. The problems addressed here are likely to become more acute as all the genes in the human and other species' genomes are identified

TABLE 1. Ion Channel Disorders[a]

Chloride channels
Myotonia congenita (CLC-1): dominant (Thompsen); recessive (Becker); Renal tubular disorders (CLC-5): hypercalciuric nephrolithiasis; Bartter's syndrome (CLC-KB); cystic fibrosis (epithelial chloride channel)
GABA receptors
Angleman or Prader-Willi: $GABA_A$ β3 receptor subunit
Alcohol nontolerant rat: $GABA_A$ α6 receptor subunit
Sodium channels
Skeletal muscle—α-subunit: SCN4A: Hyperkalemic periodic paralysis, Paramyotonia congenita, myotonia fluctuans, myotonia permanens, acetzolamide-responsive myotonia, malignant hyperthermia
Immune: Anti-GMI ganglioside antibodies, multifocal motor neuropathy; acute motor axonal neuropathy, ? Guillain-Barré & CIDP
Cardiac—α subunit (SCN5A): long QT syndrome (LQT3)
Epithelial, nonvoltage-gated Na^+ channels, α & β subunits
SCNNIA & SCNNIB: pseudohypoaldosteronism (Liddle's syndrome, hereditary hypertension)
Ca^{2+} channels
Hypokatemic periodic paralysis (CACNLIA3 αIS subunit), malignant hypertermia (CACNLIA3 αIS subunit), P-type Ca^{2+} channel, Lambert-Eaton myasthenic syndrome; CACNLIA4 α1A subunit: episodic ataxia type 2, familial hemiplegic migraine, tottering mice; αlE subunit: ? hemiplegic migraine; β4 subunit: lethargic mice; γ2 subunit: stargazer mice
Ryanodine receptor 1: malignant hyperthermia—central core disease
K^+ channels
Immune: neuromyotonia
Hereditary: Bartter syndrome: inward rectifing K^+ channel, subfamily J, member I; Long-QT syndromes: voltage-gated K^+ channels KVLQTI; HERG; Jervell & Lange-Nielsen syndrome: KCNEI; episodic ataxia/myokymia syndrome: voltage-gated K^+ channel (KCNAI); benign neonatal epilepsy: voltage-gated K^+ channel subunits KCNQ2 & KCNQ3 (also see ACHR α4 subunit); KCNN3 ? schizophrenia
Neural nicotinic receptors
α4 subunit dominant nocturnal frontal lobe epilepsy; α7 schizophrenia, attention disorder? lack of inhibition of P50 response to auditory stimulus
Glycine receptors
Hyperekplexia (startle disease): a-I subunit (strychnine binding)

[a]See: www.neuro.wustl.edu/neuromuscular/mother/chan.html

through the various genome projects; the task will be to place those discoveries in a physiological context.

There were nearly 500 attendees at this conference, and we all learned from the experiences with different molecules, from new techniques, and from the successes and failures of others. It was easy to appreciate the breadth of the truly explosive discoveries in the signaling molecules, which are sure to change fundamental physiological concepts and are already providing enormous opportunities for the design of new drugs for the treatment of human disease. The meeting brought together a representative spectrum of the work in the field and provided ample evidence that the molecular biology of signaling molecules is not merely an exercise in "stamp collecting," and that the placing of these discoveries into a physiological context is likely to have a major impact on physiology and medicine.

SPECIAL CHANNELS FOR SPECIAL FUNCTIONS

I first became interested in the diversity of ion channels during my Ph. D. thesis study at the University of Cambridge, where I examined the excitability of the giant axon of the marine worm *Myxicola*. This nerve is incapable of sustaining action potential firing when stimulated at frequencies higher than ~5 Hz (FIG. 1). I did voltage-clamp analysis of the voltage-dependent currents of this axon to understand the basis of this phenomenon. These studies demonstrated that the limited frequency response of this axon was due to the presence of novel sodium channels with unusual properties.[2–5] After opening, the Na^+ channels in this axon enter an inactivated state from which recovery is very slow. The rate of entry into this state is fast, and thus a fraction of the sodium channels enter is state after each action potential. This generates a cumulative inactivation of the sodium conductance, which explains the response of this axon to continuous stimulation. It is exciting that a sodium conductance with similar properties has now been found in mammalian neurons (see chapter by Nelson Spruston in the Na^+ CHANNELS section). Since then I have been fas-

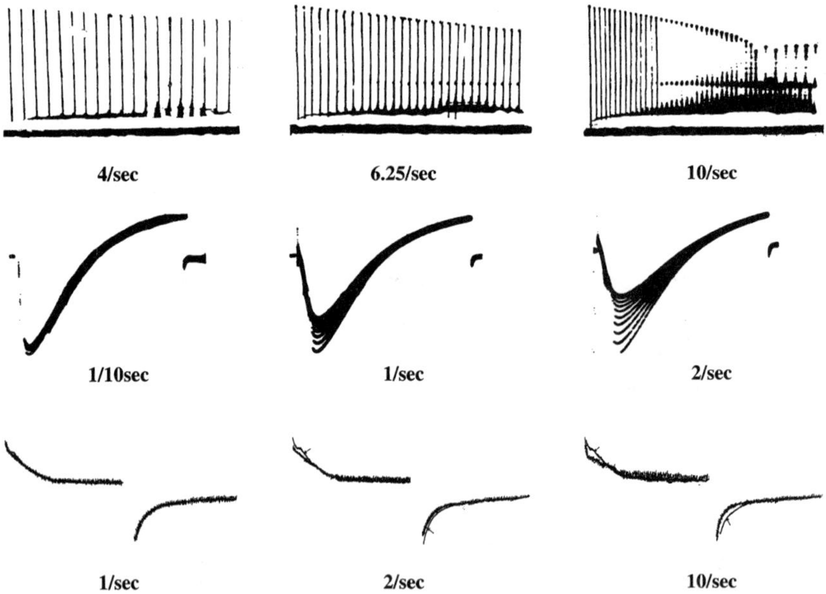

FIGURE 1. Na^+ channel slow inactivation limits the frequency response of *Myxicola* giant axons. **Upper records:** frequency sensitivity of action potential generation; **middle records:** frequency sensitivity of sodium currents; **lower records:** frequency sensitivity of Na^+ channel gating currents. (Modified from Refs. 2, 4, & 5.)

cinated by the idea that there might be "special" channels that participate in specialized functions in neuronal tissue, and that it might be possible to explain some of the complexity of signal transmission in the nervous system at the channel level. The NMDA receptor is perhaps one of the most impressive examples of this. The molecular work has revealed the potential existence of hundreds if not thousands of different channels, providing the opportunity of discovering many new such "special channels."

I believe that we are not dealing with merely a large amount of irrelevant redundancy; and that, although a few channel types might be sufficient for providing the fundamental properties of excitability as one may see in a simple unicellular system, the functional diversity achieved by more complex living systems depends on the fine tuning and increased abilities of modulating responses that can be obtained with many such molecules, supporting the idea that "redundancy" plays a key role in the evolution of complex biological systems.[6]

DIVERSITY OF ION CHANNELS AND RECEPTORS: SPECIFICITY VS. REDUNDANCY

We speculate that the power of generating variation by gene duplication and divergence may have played a role in the generation of complexity in the nervous system of animals with elaborate nervous systems. Nevertheless, the physiological significance of the molecular diversity that is being discovered remains to be elucidated. Unfortunately, it turns out to be quite difficult to determine a priori how much of the observed diversity is functionally important. General theory at this time turns out to be of little help here.

We function in the discipline of biology under certain basic assumptions that may turn out not to be correct. For example, we usually assume that genetic change is the result of random events, and therefore we infer that the observed molecular diversity is the result of selection among these random changes through the process of natural selection. However, there are increasing examples that many genetic changes are not all that random. In fact, only a few weeks after our meeting, a conference, also sponsored by the New York Academy of Sciences (Molecular Strategies in Biological Evolution) presented many examples of fast, nonrandom, and apparently directed mutagenesis and discussed the possibility that there are specific sites in the genome containing information to regulate the rate, extent, and location of genetic change.

We also usually assume that conservation of sequence throughout evolution indicates selective pressures acting at the level of the protein, and therefore that if a protein is conserved it must be doing something important. But this assumption may not be correct either, and alternative molecular strategies can be imagined. However, if the assumption is not always correct, then we have no a priori way to decide that a particular channel protein is likely to be "important" based on sequence conservation alone. Even if most sequence conservation is due to selective pressures on the protein, the reason why the sequence is conserved may be unrelated to the functions that one might find more interesting. It is possible that a protein is nol useful in particular organisms and that mutations are selected against because they could result in detrimental effects, for example the mutated protein could have dominant negative effects on other useful molecules or introduce other forms of deleterious gain of function (as in the case of the weaver mice).[7,8]

The reductionist approach involved in integrating the findings from molecular biology into cell and tissue physiology is not without its dangers, and needs to be used cautiously. Stephen Jay Gould and Richard C. Lewontin wrote a beautiful paper a few years ago analyzing the problems associated with following what they called an adaptationist program[9]: assuming that elements of a whole are there for a purpose, and that we can determine what that purpose is. They present in their paper wonderful examples of the errors and fallacies that this approach can lead to. Nevertheless, we often tend to fall into this "adaptationist program," and I believe I have done so in my work. It is difficult to do biological experimentation without some degree of reductionism, which—I am sure everybody will agree—can be fruitful if we are careful with our interpretations. The evidence of the successes in the field was clear at the conference and is obvious in the contents of this volume. I think there is no question that in addition to having learned a great deal about the molecular mechanisms of channel function, the molecular biology of channels and receptors has also contributed to our understanding of physiology and medicine.

THE CONTENTS OF THIS VOLUME

This volume of the *Annals* has been organized by channel or receptor type. In addition to chapters from the speakers at the plenary sessions, each section is preceded by a review chapter, contributed by a leading scientist, on the molecular biology of the channels or receptors discussed in that section. These review chapters summarize the diversity of molecular components of each type of channel and provide a background for the specific contributions dealing mainly with the integration of this diversity with cell and tissue function. The conference also included two poster sessions, with over 50 posters each. The posters were a key component of the meeting, contributing additional examples of work being done to understand the physiological significance of the molecular diversity, complementing the talks in achieving the goals of the conference. Emphasis was placed on encouraging and facilitating the attendance of students and postdoctoral researchers. The large attendance of young people was one of the most important achievements of the meeting. Poster presenters were invited to submit short chapters, "extended abstracts," on their presentations. Nearly 50 of these were selected for publication in this volume, and are presented at the end of each section.

There are several important groups of channels and receptors that we were not able to include in the plenary sessions at the conference, such as cyclic nucleotide–gated channels and purinergic receptors. To make this volume more comprehensive, we invited leading investigators to write review chapters on these channels. Literally just weeks before the conference took place, the group of Gareth Tibbs, Steve Siegelbaum, and Eric Kandel identified the first components of the pacemaker channels mediating the hyperpolarizing (I_h) or "funny" (I_f) current. The pacemaker channels were the last elusive group of channels whose molecular components had escaped identification. Gareth Tibbs was able to present a poster at the meeting and agreed to write a review chapter for this volume (see also Refs. 10–12). These chapters are included in Section VII, SEROTONIN AND PURINERGIC RECEPTORS, CYCLIC NUCLEOTIDE–GATED AND PACEMAKER CHANNELS.

This volume contains outstanding examples of the work performed to understand the physiological significance of the advances in the molecular analysis of ion channels and receptors. The chapters include several themes, which illustrate the many approaches that researchers are using.

Much of the cloning work has been done without prior biochemical analysis and therefore without the isolation and characterization of native proteins. Thus, when the molecular components identified by cloning are first discovered, we usually do not even know the exact relationship between them and native channels—that is, which subunits fit with which channels. As explained in the review chapters to each of the channels and receptors in this volume, the functional units are multimeric complexes of several subunits. In heterologous expression systems, many combinations of subunits can be formed having distinct functional properties. Some of these subunits can function by themselves, and others cannot, but can modify the channels expressed by the components that are functional alone. The problem is complex; however, as illustrated in many examples found in this volume, progress is being made, on questions such as which combinations of subunits may exist *in vivo*, where they might be present; and on establishing functional differences between receptors having different subunit combinations reconstituted in model expression systems (often called heterelogous expression systems), of which the *Xenopus* oocyte system, pioneered by Ricardo Miledi, has been the most important in the field.

The ability of the subunits of channels and receptors to coassemble in multiple combinations results in a potentially colossal number of distinct channel types. One notable example is the large number of subunits of $GABA_A$ receptors that have been cloned (see the chapter by Paul Whiting in this volume). Given that $GABA_A$ receptors are thought to be pentameric, the variety of receptors that could be generated from different combinations of these subunits is truly staggering. The studies described in Whiting's chapter suggest that many of these combinations may not be frequently found in the nervous system. Nevertheless, there are clearly many more subtypes of functional $GABA_A$ receptors in brain cells than were believed prior to the molecular work, suggesting variations in synaptic inhibition not previously expected (see also recent papers by Xiang *et al.* and McDonald *et al.*[13,14]). Moreover, as discussed by Jean-Marc Fritschy in his chapter, the subunit composition of these channels in neurons can be dynamic, changing with development and activity. These changes may play an important role in brain plasticity.

Biochemical analysis can provide direct information on subunit composition; however, the scarcity of channels and receptors and their enormous diversity make this approach extremely difficult. Nevertheless, this approach has been extremely fruitful in several fields (see chapters by Alan Goldin in the Na^+ CHANNELS section; by Herman Moreno and by William Caterall in the Ca^{2+} CHANNELS section; by Olaf Pongs and by Kevin Wickman and David Clapham in the K^+ CHANNEL section; and by Heinrich Betz in the GABA AND GLYCINE RECEPTORS section).

One important element in defining the functional roles of cloned components is the painstaking job of analyzing their patterns of expression in neurons and other cells. *In situ* hybridization histochemistry, even of nonabundant channel and receptor mRNAs, can be of high enough quality to provide the patterns of expression in various cellular elements of the CNS. Moreover, the availability of the primary sequence of the channel subunits allows the generation of site-specific antibodies that become invaluable tools in localizing the proteins, particularly at the subcellular level, as well as for biochemical analysis. These approaches have been very important in many of the families of channels and receptors discussed in this volume.

The nerve cell is highly compartmentalized, and this organization is fundamental to its function. In this context, the use of good-quality channel and receptor antibodies for high-resolution immunoelectron microscopy localization studies is of great functional importance. The chapter by Ole P. Ottersen on the use of postembedding immunogold techniques in the GLUTAMATE RECEPTORS section is an excellent example of this approach.

Many, if not most, molecules, including ion channels and receptors, function in the cell in microdomains, such that molecules that participate together in a given function often exist as macromolecular complexes. The discovery of proteins associated with the principal subunits of ion channels and receptors has already provided important information as to the function of these subunits. This is a powerful approach, which is contributing truly novel and exciting information for understanding how channels function in cells and promises to be a key element of future research, as illustrated by the molecular analysis of associated proteins of K^+ channels and glutamate receptors (see chapter by Morgan Sheng in the GLUTAMATE RECEPTORS section), of δ2 glutamate receptors (see chapter by Nathaniel Heintz in the GLUTAMATE RECEPTORS section), of Ca^{2+} channels (see chapter by William Catterall in the CA^{2+} CHANNEL section) and of glycine receptors (see chapter by Heinrich Betz in the GABA AND GLYCINE RECEPTORS section).

Genetic models, particularly in the fruitfly *Drosophila*, have been key to identifying and cloning novel channel components leading to the identification of new gene families. The explosion in our knowledge of the molecular components of K^+ channels is due, to a large extent, to the cloning of the Shaker gene in *Drosophila*.[15–18] This gene encodes subunits of voltage-gated K^+ channels, and the availability of nucleotide sequences for Shaker cDNAs opened the field for the molecular analysis of K^+ channels in many species, including mammals. Unfortunately, due to prior commitments Lily Jan, whose laboratory was one of the first to clone Shaker cDNAs and has remained a leading scientist in the molecular biology of K^+ channels, was unable to attend the conference. Her recent attempts at elucidating the subunit compositions that may exist in the inward rectifier channel family using several different methods are unraveling a complexity larger than that in voltage-gated channels (see chapter by Coetzee *et al.*). These studies will be key in understanding the molecular and functional diversity of inward rectifiers. Genetic models in *Drosophila* were used as well in the laboratory of Barry Ganetzky to clone the *eag*, and *slowpoke* genes (see chapter on the eag family of K^+ channels). The cloning of eag led to the identification of a new family of K^+ channels and the cloning of HERG, a human homologue, mutations of which are responsible for LQT syndrome, a lethal form of cardiac arrhythmia (see chapter by M. Sanguinetti). The cloning of the slowpoke gene allowed the characterization of the key molecular components of large-conductance Ca^{2+}-activated K^+ channels. These genetic models are also important tools for studying the physiological significance of the identified components.

The cloning of new components of important ion channels and receptors has continued in the last few years, providing additional important tools for this work. John Adelman succeeded in cloning the small-conductance Ca^{2+}-activated K^+ channels and has begun to elucidate the diversity of these channels; Olaf Pongs discovered a family of accessory (or β subunits) of Kvl voltage-gated K^+ channels (see chapters in the K+ CHANNEL section); and Eduardo Perez-Reyes recently cloned the primary subunits of T-type Ca^{2+} channels (see chapter in the CA^{2+} CHANNEL section).

The various "Genome Projects" and the sequencing of "expressed sequence tags" (EST's) are also contributing to the field by discovering new members of ion channel and receptor gene families that can be identified based on sequence similarities. This strategy, used by several of the authors of this volume, is quickly becoming the main method of identifying new sequences; we may expect that it will have an even more powerful influence in the field in the future, as more sequences are being deposited and as we improve our methods to identify them (see chapter by L. Salkoff in the K^+ CHANNELS section).

The study of the modulation of cloned ion channels and receptors expressed in heterologous expression systems by kinases and other postranslational modifications also provides important clues as to their native functional roles and can result in the discovery of novel channel properties and modulations.

In addition to the existence of large numbers of genes encoding many of the subtypes of channel or receptor subunits, diversification of components of these molecules also utilizes many of the other molecular mechanisms capable of generating variation. There are examples of alternative splicing in many of the families of channels and receptors discussed here. Alternative splicing as a means of generating diversity is particularly prominent in the primary subunits of large-conductance Ca^{2+}-activated K^+ channels. The gene encoding these proteins has at least 4–5 independent alternative splicing sites, with the possibility of generating a large number of isoforms, of which more than 10 have already been found. This alternative splicing serves to produce a large variety of Ca^{2+}-activated K^+ channels differing in Ca^{2+} sensitivity, and may be a key molecular mechanism used to tune individual auditory hair cells to respond to specific sounds (to illustrate this beautiful example of the significance of molecular diversity, we invited Robert Fettiplace to write a chapter, which can be found in the K^+ CHANNELS section). The alternative splicing of these molecules may be modulated under interesting physiological conditions.[19]

Another interesting mechanism to modify channel components posttranscriptionally is RNA editing. First discovered in AMPA glutamate receptors by Peter Seeburg (see chapters by Rolf Sprengel in the GLUTAMATE RECEPTORS section), it has now been observed in Na^+ channels (see chapter by Robert Reenan), K^+ channels,[20] and serotonin receptors.[21]

The information collected from the cellular and subcellular localization studies together with the functional analysis in heterologous expression systems can now be used to generate hypotheses and perform experiments that may allow the discovery of physiological roles for the various channels and receptors that have been identified. The availability of specific pharmacological tools is particularly useful. Significant for much of the progress have been the patch clamp technique developed by Erwin Neher and Bert Sakmann, together with the new methodologies that these scientists have been developing, which have allowed the application of electrophysiological analysis to an almost unlimited number of cell types and experimental preparations. Excellent examples of these efforts are already appearing in the literature and are demonstrating that, indeed, the diversity is associated with functional specialization, as illustrated in several chapters in this volume.

Often there are no adequate pharmacological tools with which the role of specific channels or receptors can be investigated. Gene elimination methods such as antisense hybrid arrest to eliminate specific mRNA transcripts and the use of dominant negative strategies to obliterate the function of the native protein are becoming increasingly used, and methods to apply these technologies both in isolated cells or tissue slices, as well as *in vivo*, are beginning to be developed; several chapters illustrate these approaches. Unfortunately, it is still difficult to introduce DNA into some types of cell, such as neurons; the use of viral vectors to introduce DNA agents to produce the specific elimination of components promises to expand and facilitate this approach (see abstract by David Johns *et al.*

in the K$^+$ Channels section). Another technique to "block" specific channels or receptors is the use of antibodies with functional effects, a method that has been used by Nakanishi's group (who, unfortunately due to prior commitments was also unable to attend the meeting) to analyze the function of specific types of metabotropic glutamate receptors.[22] These powerful approaches can have an impact on the future of medicine most directly by creating new types of pharmacological tools.

Animal models having mutations in specific channel or receptor components provide additional tools to investigate physiological roles of specific channels. The analysis of these models can become very complicated, and it is here where the cautionary ideas of Gould and Lewontin discussed earlier become particularly pertinent. The observed deficits reflect the consequences of the mutations in a complex system, and only indirectly may provide information on the "roles" of the mutated genes. Not only can there be compensatory gene expression that can obscure the role of the defective protein, but the molecular defect can initiate a complicated and nonlinear cascade of changes that can be difficult to reconstruct. This can be particularly serious if the mutated genes are expressed at early developmental stages or in plastic tissues, in which case development will take place in a new molecular context. Nevertheless, I believe that these models, used in conjunction with the other approaches described here, could be uniquely useful by allowing a correlation between molecular elements and function at different levels of organization. These animal models include natural mutations in mice and human. Moreover, the development of the homologous recombination technology in mouse embryonic stem cells (ES cells) allows the generation of such mutants by design. The methodology allows the introduction of different types of mutations, but the elimination of the gene to generate "knockout" mice has so far been the dominant form used in the field. These approaches have been applied extensively to the study of glutamate receptors (see chapters by Rolf Sprengel, Stephen Heinemann, and Nathaniel Heinz in the Glutamate Receptors section) and are beginning to be utilized with other channels and receptors (see chapters by Kevin Wickman and David Clapham, Olaf Pongs, Barry Ganetzky, and Bernardo Rudy in the K$^+$ channels section; by Heinrich Betz in the GABA and glycine receptors section; by Jeffrey Noebels in the Ca^{2+} channels section; by Clement Lena in the nicotinic receptors section; and by Renée Hen in the Serotonin and Purinergic Receptors, Cyclic Nucleotide–Gated and Pacemaker Channels section). Improved methods to prepare inducible transgenic animals and to target the mutations to specific cell populations are being developed and are particularly promising given the potential problems described earlier (see chapter by Seeburg in this volume). I expect the use of transgenic approaches to increase significantly in coming years, and the chapters in this volume start to illustrate the power as well as the problems of this approach.

ACKNOWLEDGMENTS

I wish to thank all the colleagues and friends that made the meeting such a successful gathering and all those who helped in ensuring the high quality of this volume. I specifically would like to thank the New York Academy of Sciences (NYAS) for sponsoring the conference, Peter Seeburg for his invaluable efforts in organizing this meeting; the NYAS Conference Committee and the reviewers of the program for their support and constructive suggestions; Rashid Shaikh, Director of the Meetings Program at the Academy, for his help and trust. I wish to thank the chairpersons of the various sessions at the meeting—Steven Heinemann, Terry Snutch, Barry Ganetsky, Lorna Role, and Heinrich Betz—for their help in preparing the program and other assistance in preparing the conference. I wish to thank all the speakers and poster presenters as well as all the colleagues who attended the meeting for enriching the discussion and for making precious contributions to the success of this effort. I wish to thank the staff at the NYAS and in particular Sue Davies, Renée Wilkerson, and Katie Schrader, who had to do a lot of hard work to make the meeting possible. I want to thank the staff at the *Annals*, including Bill Boland (Executive Editor) and Richard Stiefel (Associate Editor), who edited this volume, as well as Sheila Kane, Mary Hannigan, and Justine Cullinan for their patience and great efforts in producing it. I also want to thank the personnel at the New York University School of Medicine, who did an outstanding job to ensure the smooth running of the meeting, particularly during a weekend.

I wish to thank also the people in my laboratory—Alan Chow, Yimy Amarillo, Marcela Nadal, Michael Saganich, Eleazar Vega, David Lau, Tom McCormack, Herman Moreno, and Ander Ozaita—who did a lot of work helping to prepare the conference and this volume, as well as for their tolerance and loyalty.

And last, but certainly not least, I want to thank very specially all the financial contributors. Fundraising is the key to have a meeting like this in today's world; without the interest and support that we received from these generous contributors this meeting would not have taken place (see list of contributors on page xv).

REFERENCES

1. L. Sivilotti & D. Colquhoun. 1995. Acetylcholine receptors: Too many channels, too few functions. Science **269:** 1681–1682.
2. Rudy, B. 1976. Sodium gating currents in Myxicola giant axons. Proc. R. Soc. Lond. Ser. B Biol. Sci. **193:** 469–475.
3. Rudy, B. 1976. A kinetic model for slow inactivation in nerves. Biophys. J. **17:** 45a.
4. Rudy, B. 1981. Inactivation in Myxicola giant axons responsible for slow and accumulative adaptation phenomena. J. Physiol. (Lond.) **312:** 531–549.
5. Rudy, B. 1981. Slow inactivation of voltage-dependent channels. *In* Nerve Membrane. G. Matsumoto & W. Katani, Eds. 89–11. University of Tokyo Press. Tokyo.
6. Newman, 1994. J. Evol. Biol. **7:** 1467–1488.
7. Kofuji, P., M. Hoffer, K.J. Millen, J.H. Millonig, N. Davidson, H.A. Lester & M.e. Hatten. 1996. Functional analysis of the weaver mutant GIRK2 K^+ channel and rescue of weaver granule cells. Neuron **16:** 941–952.
8. Slesinger, P.A., N. Patil, Y.J. Liao, Y.N. Jan, L.Y. Jan & D.R. Cox. 1996. Functional effects of the mouse weaver mutation on G protein–gated inwardly rectifying K+ channels. Neuron **16:** 321–331.
9. Gould, S.J. & R.C. Lewontin. 1979. The spandrels of San Marco and the Panglossian paradigm: A critique of the adaptationist programme. Proc. R. Soc. Lond. Ser. B Biol. Sci. **205:** 581–598.
10. Santoro, B., D.T. Liu, H. Yao, D. Bartsch, E.R. Kandel, S.A. Siegelbaum & G.R. Tibbs. 1998. Identification of a gene encoding a hyperpolarization-activated pacemaker channel of brain. Cell **93:** 717–729.
11. Gauss, R., R. Seifert & U.B. Kaupp. 1998. Molecular identification of a hyperpolarization-activated channel in sea urchin sperm. Nature **393:** 583–587.
12. Ludwig, A., X. Zong, M. Jeglitsch, F. Hofmann & M. Biel. 1998. A family of hyperpolarization-activated mammalian cation channels. Nature **393:** 587–591.
13. Xiang Z., J.R. Huguenard & D.A. Prince. 1998. Cholinergic switching within neocortical inhibitory networks. Science **281:** 985–988.

14. McDonald, B., A. Amato, C.N. Connolly, D. Benke, S.J. Moss & T. Smart. 1998. Adjacent phosphorylation sites on GABAA receptor subunits determine regulation by cAMP-dependent protein kinase. Nature Neurosc. **1:** 23–28.
15. Papazian, D.M., T.L. Schwarz, B.L. Tempel & Y.N. Jan. 1987. Cloning of genomic and complimentary DNA from *Shaker*, a putative potassium channel gene from *Drosophila*. Science **237:** 749–753.
16. Tempel, B.L., D.M. Papazian, T.L. Schwarz, Y.N. Jan & L.Y. Jan. 1987. Sequence of a probable potassium channel component encoded at *Shaker* locus of *Drosophila*. Science **237:** 770–775.
17. Kamb, A., L. Iverson & M.A. Tanouye. 1987. Molecular characterization of *Shaker*, a *Drosophila* gene that encodes a potassium channel. Cell **50:** 405–413.
18. Baumann, A., I. Krah-jentgens, R. Muller, F. Muller-Holtkamp, R. Seidel, N. Kecskemethy, J. Casal, A. Ferrus & O. Pongs. 1987. Molecular organization of the maternal effect region of the *Shaker* complex of *Drosophila*: Characterization of an IA channel transcript with homology to vertebrate Na^+ channel. EMBO J. **6:** 3419–3429.
19. Xie, J. & D.P. McCobb. 1998. Control of alternative splicing of potassium channels by stress hormones. Science **280:** 443–446.
20. Patton, D.E., T. Silva & F. Banzanilla. 1997. RNA editing generates a diverse array of transcripts encoding squid Kv2 K+ channels with altered functional properties. Neuron **19:** 711–722.
21. Burns, C.M., H. Chu, S.M. Rueter, L.K. Hutchinson, H. Canton, E. Sanders-Bush & R.B. Emeson. 1997. Regulation of serotonin-2C receptor G-protein coupling by RNA editing. Nature **387:** 303–308.
22. Shigemoto, R., T. Abe, S. Nomura, S. Nakanishi & T. Hirano. 1994. Antibodies inactivating mGluR1 metabotropic glutamate receptor block long-term depression in cultured Purkinje cells. Neuron **12:** 2145–1255.

The "Psychic" Neuron of the Cerebral Cortex

PATRICIA S. GOLDMAN-RAKIC[a]

Section of Neurobiology, Yale University School of Medicine, 333 Cedar Street, New Haven, Connecticut 06520-8001, USA

ABSTRACT: Remarkable advances in the identification, cloning, and localization of ion channels and receptors in the central nervous system have opened up unprecedented possibilities for relating structure to physiological function at the subcellular level of analysis. A singularly advanced property of select central nervous system neurons is their ability to exhibit increases in firing rate in relation to the mnemonic trace of a preceding event, a property that has been referred to as "working memory." Single-cell recordings from the prefrontal cortex of nonhuman primates have revealed neurons in the prefrontal cortex that possess "memory fields" analogous to the receptive field properties of sensory neurons. The integrity of these neurons has been shown to be essential for accurate performance in memory tasks performed by trained monkeys (and humans). We can now show that the excitability and/or tuning of these prefrontal neurons are subject to modulatory influences by dopamine, serotonin, GABA, and glutamate among other peptides and conventional neurotransmitters. I will describe the dopaminergic, serotonergic, and GABAergic innervation of pyramidal neurons engaged in working memory and the localization of neurotransmitter receptors through which they exert their actions. The findings reveal a remarkable degree of diversity in the subcellular localization and functionality of the five cloned dopamine receptors (D1, D2, D3, D4, and D5) and two serotonin (5HT2A and 5HT3) receptors that have been examined to date. The potential now exists for linking systems neurobiology with molecular biophysics to comprehend the highest functions of information processing that distinguish our species.

Ramon y Cajal designated the pyramidal cell of the cerebral cortex the "psychic" neuron of the brain. When this designation was made, in the late nineteenth century, knowledge of this neuron class was limited to its morphological characteristics; its pyramid-shaped cell body; its long axon that penetrated white matter; its characteristic vertically extended apical dendrite studded with spines; and its bushy basiliar dendritic arbor. Only with the introduction of the silver impregnation and axonal transport methods by the middle of the twentieth century was it possible to define the destinations and synaptic targets of pyramidal cell axons and map the connections of populations of cortical neurons to distant cortical areas (e.g., cortico-cortical connections) and subcortical effector mechanisms. By extrapolation from estimates made on hippocampal pyramidal neurons, the cortical pyramidal neuron integrates literally thousands of afferent inputs and through its efferent connections regulates skeletal and smooth muscles involved in movement and affect. The appellation *psychic* is amply justified because such neurons, particularly those residing in the higher association cortices of mammals, are at the end of the "information highway"; i.e., by virtue of their sensory monosynaptic connections with the higher-order sensory cortices, they are privy to information from the current environment as well as from repositories of stored knowledge. A significant feature of the cortical pyramidal cell, clarified only within the last two decades, is that it processes information and directs actions via glutamate transmission, differentiating it from the other major class of cortical

[a]Phone 203-785-4808; fax, 203-785-5263; e-mail: patricia.goldman-rakic@yale.edu

neuron—the local circuit neuron, which is nonpyramidal in shape, possesses smooth dendrites, and for the most part utilizes the amino acid, gamma-amino-butyric acid (GABA), for neurotransmission. Unlike pyramidal neurons, the nonpyramidal cells do not innervate distant structures; rather their axonal terminations are confined to local targets, either nearby pyramidal neurons or other interneurons.

A major objective of systems neuroscience is to understand both the specific and generic functions of pyramidal neurons in different cortical areas and the cellular mechanisms by which they perform the operations that are essential to human cognition. A full understanding of the capacity of even a single pyramidal cell to integrate its myriad inputs and generate a decisive action requires knowledge not only of its biophysical properties but of its circuitry, signaling mechanisms, and contributions to information processing *in vivo*. In this chapter, I describe the efforts of my laboratory to understand the pyramidal cells of the prefrontal cortex, the area of the brain most closely and most often associated with the executive functions of the brain. We and others have been able to characterize prefrontal neurons functionally—i.e., "on line"—as they are actively engaged in a cognitive function for which the prefrontal cortex is specialized, working memory. We also address the circuit and receptor mechanisms that regulate pyramidal cell excitability *in vivo*. This work may seem far afield from membrane biophysics, but I hope to demonstrate that a connection may exist between the disposition of neurotransmitter receptors and the localization of channels that could illuminate the biological basis of cognition. Neuroscientists are in a position to integrate information from many levels of analysis in pursuit of the goal of understanding human cognition at a neurobiological level.

WORKING MEMORY—THE PSYCHIC FUNCTION OF THE PREFRONTAL NEURON

Studies in nonhuman primates dating back to the midthirties have established a role for the prefrontal cortex in higher cortical functions. Although the major primary sensory and primary motor domains of the cerebrum had been mapped, the functional map of the vast areas of the association cortex were provisional at best. However, a major discovery of this period was the critical role of the prefrontal cortex for performance on a particular kind of behavioral task—called the delayed-response task—which required animals to hold an item of information in mind for several seconds and to update a mental representation of that input on a moment-by-moment basis. In the modern era, we have come to recognize that the function tapped by these laboratory tasks is essentially that which cognitive psychologists refer to as "working memory." Working memory is the inferred ability to hold information transiently "on-line," essential for the temporal integration of ideas and action, for comprehension and thought.[1,2] A remarkable conjunction of physiology and behavior occurred when it was observed that prefrontal neurons could exhibit sustained tonic activity triggered by the brief presentation of a stimulus[3–5] as this distinguished neurons of the association cortex from sensory neurons that are time-locked to the stimulus. A neural mechanism had thus been isolated for the aphorism "out of sight—out of mind" that so often has been used to describe patients with prefrontal lesions. The sustained activity observed in many prefrontal cortical pyramidal cells has been shown to be content specific with individual neurons coded to specific items of information, such as the location of an object space mapped in egocentric coordinates (FIG. 1),[5,6] the direction of a prior response,[7] and the identity of an object such as a face or fruit (FIG. 2).[8,9] Further, the activities of prefrontal neurons are often polarized, exhibiting excitatory responses to targets in preferred directions and inhibitory responses to targets in nonpreferred directions.[5] In this way the neuron is endowed with both a "memory field" and an opponent memory field, while the averaged response across all other targets in space may not differ significantly from the background activity of the cell (FIG. 1). A major theme in cortical physiology is that the receptive field of a pyramidal neuron, including the memory field of the prefrontal neuron, is established by the nature of its afferent input, including lateral inhibitory input.

EXCITATORY/INHIBITORY INTERACTIONS IN CORTICAL CIRCUITS

Pyramidal and nonpyramidal cells constitute the major cellular constituents of the cerebral cortex, and it is widely believed that understanding the interaction of these two principal components of cortical architecture holds the key to a mechanistic understanding of cortical function. Based on an early observation, Mountcastle *et al.* were among the first to suggest that "regular" and "thin" spikes observed in extracellular recording studies of primate cortical neurons corresponded to pyramidal and "stellate" nonpyramidal neuronal morphologies, respectively.[10] Considerable progress has recently been made in deciphering the local circuits by which these major classes of neuron interact. *In vitro* intracellular work has confirmed that "regular-spiking" (RS) and "fast-spiking" (FS) neurons differ in their base widths and correspond to neurons with pyramidal and sparsely spiny stellate morphologies, respectively.[11] Furthermore, fast-spiking neurons are parvalbumin positive, indicating that these GABAergic interneurons are likely the basket and chandelier cells,[12] which provide a major source of inhibitory input to the soma and proximal dendritic regions of pyramidal cells (e.g., Ref. 13). As will be described below, the physiological distinction between fast-spiking and regular-spiking neurons can be used to extrapolate the functional properties of putative interneurons and pyramidal neurons in extracellular recordings obtained from trained monkeys as they perform working memory tasks.

Until recently, all major progress on the excitatory-inhibitory interactions has come from the study of *in vitro* preparations with the one significant limitation—that the activity of neurons examined in living slices cannot be time-locked to naturally occurring events. On the other hand, while single-unit studies in behaving animals have had the advantage of being able to time-lock neuronal activity to behaviorally relevant events, they have not generally been able to distinguish the neuronal type from which recordings were obtained or to perform intracellular recordings. A recent study in this laboratory has extended this line of investigation to the nonhuman primate. Wilson *et al.*, used wave form analysis to classify task-related neurons either as interneurons or pyramidal neurons in monkeys while they performed an oculomotor task that required them to make visually guided or memory-guided eye movements to directional targets in space.[8] Inverted patterns of activity between nearby putative pyramidal cells and nonpyramidal cells or interneurons were observed when the neurons, which constituted a pair of fast- and regular-spiking cells, were recorded sequentially within 400 microns of each other. This study showed that (1) interneurons, like pyramidal neurons, express directional preferences; and (2) the patterns of activity expressed by closely adjacent pyramidal and nonpyramidal neurons are often

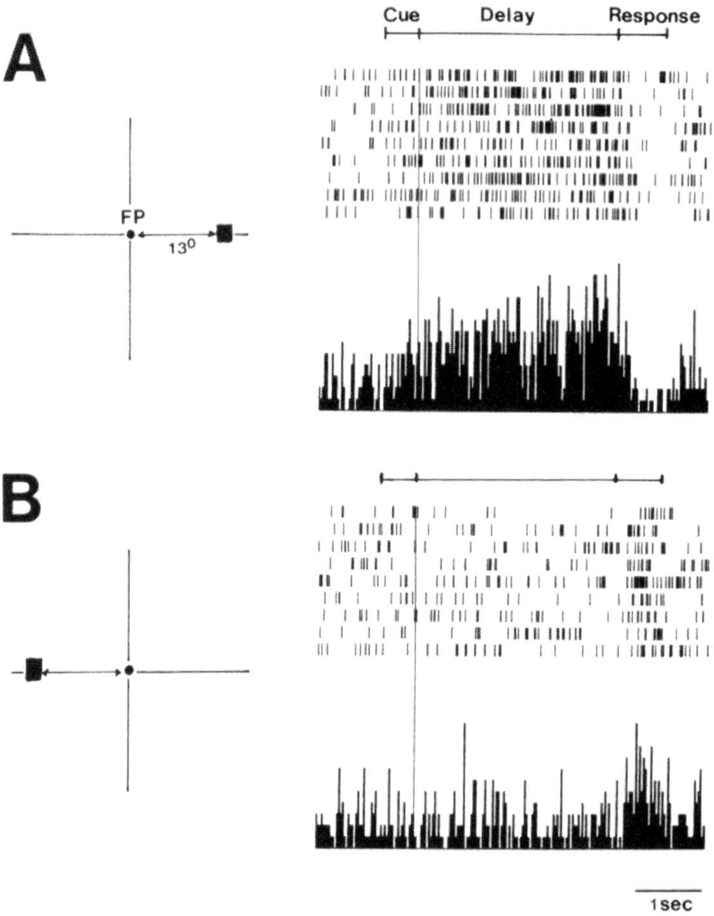

FIGURE 1. The functional specificity of prefrontal neuronal activity recorded *in vivo* as a monkey performed a delayed-response task. The neuron's activity is precisely time-locked to the events in the task: the presentation of a spatial cue (500 ms); the delay period interposed between the cue and the response (3000 ms); and the response period after the delay (1000 ms). The raster display above each histogram reveals the neuron's firing rate on numerous trials during which the animal was required to remember the target located 13° to the right of the fixation point (A); or during trials when the animal recalled the stimulus on the left (B). The firing rate was consistency enhanced during the delay period whenever the monkey recalled the rightward cue, its preferred target location; such enhancement was not observed for leftward targets (or other "nonpreferred" locations). The preferential activation of a prefrontal neuron during memory intervals (during which no stimulus is present and no response required) is termed the neuron's "memory field" in analogy with receptive field properties of neurons in sensory cortices. (Reprinted, with permission, from Goldman-Rakic *et al.*[22])

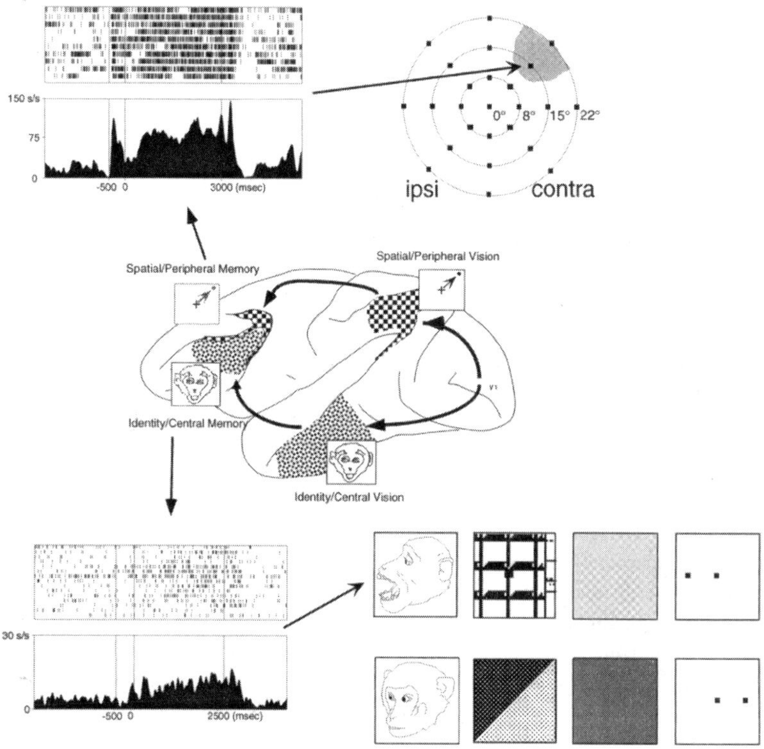

FIGURE 2. Diagram representing the topographic architecture of prefrontal cortex. Neurons with specificities for representations of visuospatial stimuli are located in a dorsal or superior region of the prefrontal cortex, which receives cortico-cortical afferents from the posterior parietal regions that process visual information originating in the magnocellular visual pathway. Neurons that respond to the identity of objects—e.g., faces—are located in the inferior convexity of the prefrontal cortex innervated by afferents from the inferotemporal cortex and parvocellular visual pathway.

complementary, such that as a nonpyramidal neuron increases its rate of discharge, a nearby pyramidal neuron decreases its rate. These findings provided suggestive evidence that a crossed directional form of feed-forward inhibition may play a role in the construction of a memory field in prefrontal neurons. Based on these findings, FIGURE 3 shows a GABAergic interneuron interconnecting two pyramidal cells with opposite "best" directions. According to this scheme, pyramidal cells with opposite best directions communicate via inhibitory interneurons such that a pyramidal neuron with a 90° memory field exhibits enhanced firing during the delay of trials in which the monkey is recalling a 90° target but is inhibited on trials when the memorandum is at the 270° location. A reciprocal pathway allows for a pyramidal neuron with a 270° memory field to inhibit one with a 90° memory field. This model suggests that nonpyramidal interneurons could provide essential feed-forward inhibitory influences between pyramidal cells with opponent memory fields. It is clear from electron-microscopic evidence that pyramidal cells innervate interneurons in the prefrontal cortex, and GABAergic interneurons innervate pyramidal cells[14] in the manner illustrated.

Most recently, we have been able to record from pyramidal and nonpyramidal neurons simultaneously at the same electrode site, and thereby to pick up neurons that are presumably within 50 microns of each other.[15] This difference in procedure resulted in a revision in the circuit basis of pyramidal-nonpyramidal interaction. We again confirmed that fast-spiking neurons possess tuned sensorimotor activity, and we revealed for the first time that they exhibit tuned delay period activity—i.e., "memory fields" similar to those found in regular-spiking neurons. Thus, as shown also by the earlier study by Wilson *et al.*,[8] it is clear that GABAergic interneurons are well-tuned cells and not simply regulators of threshold responses in their pyramidal neuronal targets. However, when RS and FS neurons were recorded simultaneously through the same electrode, it emerged that the pyramidal and nonpyramidal neuron pairs exhibited *isodirectional* rather than the cross-directional timing observed by Wilson *et al.* Further, cross-correlation analyses, described in full in Rao *et al.*[15] support the view that while interactions are presumably taking place between adjacent FS and RS neurons, each is the independent recipient of directional input from sensory centers. Altogether the results of these investigations revealed a key feature of the microcircuitry of working memory. As depicted in FIGURE 3, an interneuron in one cortical microcolumn bears a different relationship to its pyramidal target neurons, depending on the distance between them: being isodirectional when they lie within the same microcolumn and cross-directional when the RS neuron is in a distant cortical column. The striking functional arrangements between adjacent and separated FS and RS neurons support a microcolumnar functional architecture in the dorsolateral prefrontal cortex for spatial memory fields similar to that found in other areas of cortex for sensory receptive fields.

COMPARTMENTALIZATION OF NEUROTRANSMITTER RECEPTORS IN PYRAMIDAL NEURONS

The prefrontal cortex in primates is a major target of the dopamine afferents[16,17] and serotoninergic afferents[18] that originate in the brain stem. Experimental depletion of dopamine in prefrontal areas of rhesus monkeys has been shown to produce impairments in working memory performance.[19] Not unexpectedly, similar working memory deficits are present in Parkinson patients, due to the endogenous depletion of cortical dopamine (e.g., Refs. 20, 21). In pursuit of the anatomical and functional basis of dopamine actions in cortical circuits, we have learned that dopamine axons, originating in brain stem nuclei, form terminal specializations on the shafts and spines of cortical pyramidal neurons that are indistinguishable from traditional synaptic specializations.[22–24] Moreover, the spines of pyramidal neurons are often the targets of paired dopamine and excitatory terminals, the latter presumed to arise from other cortical or thalamic sources.[22] Synaptic triads, as we have termed these arrangements, are found in prefrontal, premotor, and motor cortex, suggesting that this anatomical architecture may be widespread and common to many cortical areas.

The distribution of dopamine afferents in the cortex is in register with the distribution of D1 and D2 receptor sites in this structure. In addition, we have learned that the D1 family of dopamine receptors are at least 20-fold more abundant in prefrontal cortex.[25,26] Most

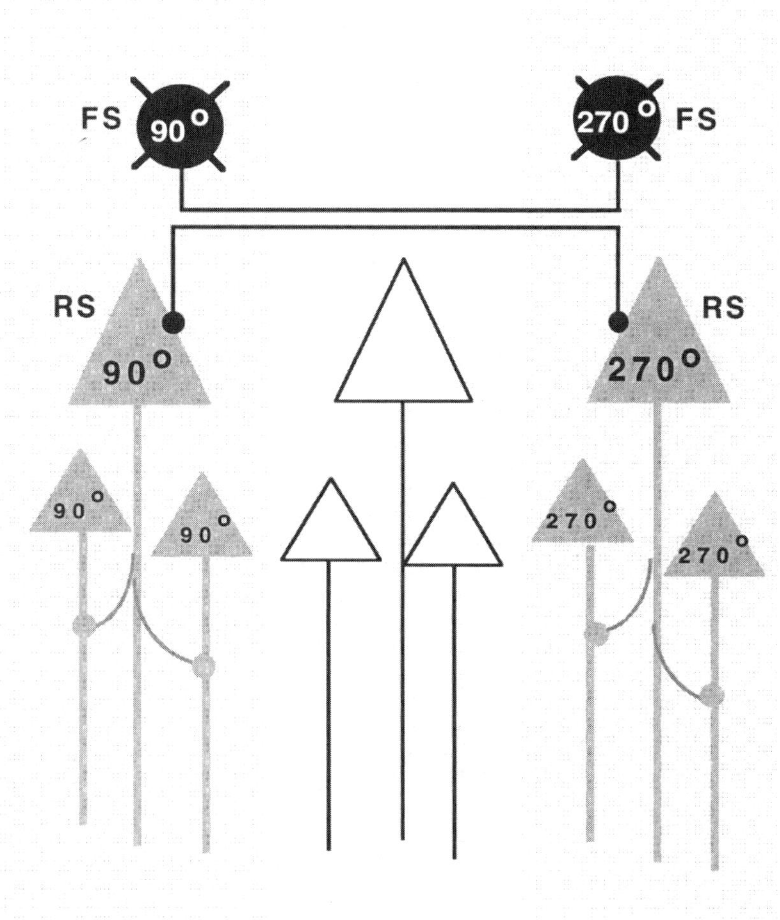

FIGURE 3. A hypothetical model of pyramidal-nonpyramidal interconnectivity in layer III of prefrontal cortex chat is in accord both with anatomical evidence and functional properties as revealed in Rao *et al.*[15] and Wilson *et al.*[8] In this model, nonpyramidal neurons and pyramidal neurons in the same microcolumn have isodirectional memory fields due to common afferents; nonpyramidal neurons and pyramidal neurons in distant columns have cross-directional relationships, due to lateral inhibition.

significant, ultrastructural analysis shows that the distal dendrites and spines of pyramidal cells are the most prominent cellular element labeled by antisera directed against the D1 receptor.[27] In sharp contrast, the 5-HT2A serotonin receptor, which is also prevalent in cerebral cortex, does not reside in the distal dendrites or spines, but rather is concentrated in the proximal apical dendritic portion of the pyramidal neuron[18] (FIG. 4). These findings indicate that the pyramidal neuron is highly compartmentalized with respect to the loca-

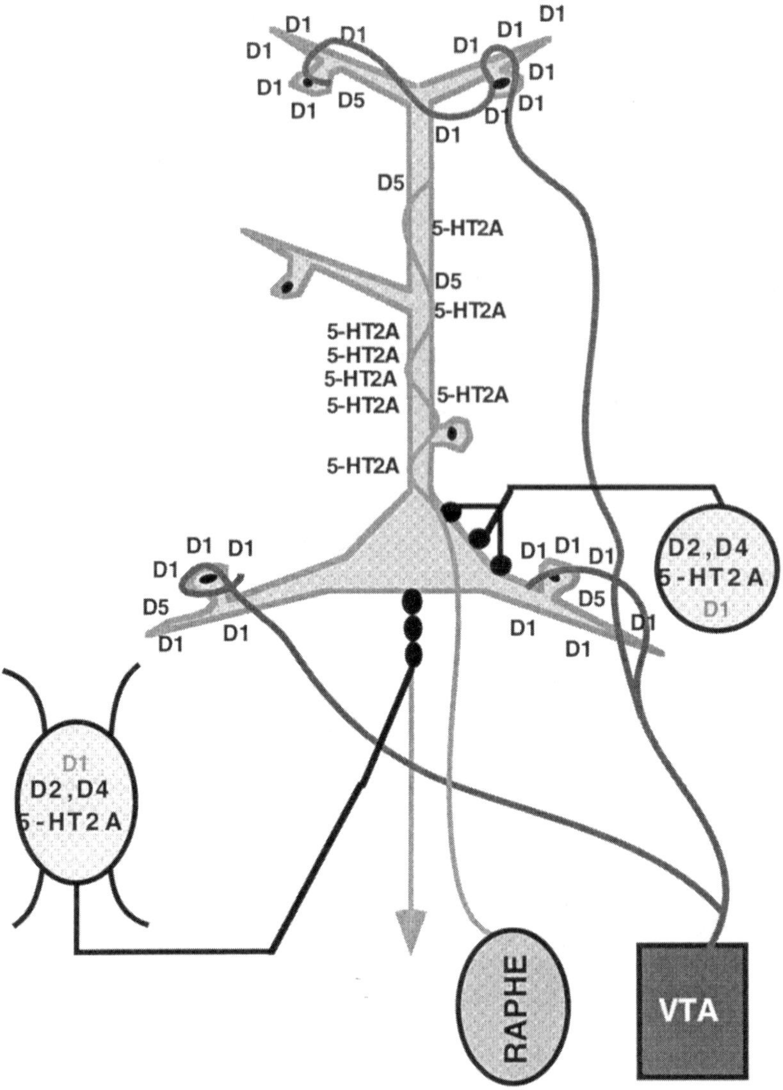

FIGURE 4. Diagram of a pyramidal neuron in layer III revealing a compartmentalization of neurotransmitter receptors. The dopamine D1 receptor is located primarily in distal portions of the cell—at distal dendrites and spines. The serotonin 5-HT2A receptor is preferentially located in the proximal portion of the cell—i.e., in the apical dendrite and somal region. L-type and N-type calcium channels have been reported to be similarly differentially distributed.

tion of subtype-specific ligand-gated receptors and presumably with respect to the signaling pathways by which these neurons are activated.

The ionic basis of neurotransmission in cortical circuits is not known. However, in striatal slices, dopamine can both inhibit[28,29] or excite[30] neurons. Both actions have been shown to be D1 mediated and to involve interactions with glutamate receptors.[31] There is evidence that D1 action in striatum enhances L-type calcium channels.[30,32,33] In the cerebral cortex, likewise, the physiological results have been contradictory. It is of interest in the context of this volume on ion channels that the disposition of D1 receptors along the distal portion of dendrites places them in close proximity to N-type calcium channels, which have been found to be preferentially localized along the dendritic surface and less in somal regions[34–36] and to contribute to voltage-gated calcium fluxes into dendrites. The serotonin 5-HT2A receptor, in contrast, may be more associated with L-type calcium channels, which do appear to be located in the somal region of the neuron.[34,37,38]

DOPAMINE MODULATION OF MNEMONIC FUNCTION IN PREFRONTAL NEURONS

The cellular basis of receptive field properties is among the most challenging issues in the study of higher cortical function. To date, neurotransmitter-specific actions on cortical neurons has been confined largely to examination of *in vitro* systems. Now, very recently, studies carried out by Graham Williams and Srini Rao in our laboratory have shown that the memory fields of prefrontal cortex are modulated by neurotransmitters such as dopamine, serotonin, and the inhibitory neurotransmitter GABA in highly differentiated ways. This is significant not only because many disorders of cognition involve dysfunction in dopamine, serotonergic, and/or GABAergic neurotransmission, but because they reveal the endogenous modulators of cognition on normal processes. These neurotransmitters act at specific receptors, which, because of availability of subtype-specific antibodies and molecular probes, can now be identified and localized in functionally specific synaptic circuits. To reveal these relationships, it is necessary to study neurotransmitter function *in vivo* at the cellular level. This has been achieved by the technical feat of iontophoretic application of drugs onto functionally active neurons while animals are engaged in processing information. Using this method, we have demonstrated that both dopamine and serotonin can modulate the tuning or sharpness of the prefrontal neuron's memory field "on line" by differential actions at the D1 and 5-HT2 receptors, respectively (Ref. 39 and G.V. Williams, S. Rao & P.S. Goldman-Rakic, unpublished observations). Both receptors modulate excitatory transmission in neurons that are involved directly in the mnemonic component of the task.[39] In the case of the D1 receptor, excessive dopamine levels destroy pyramidal cell tuning (not shown), while partial blockade (with a D1-specific antagonist, SCH 39166) *enhances* the memory fields of many prefrontal neurons without altering their general excitability (FIG. 5). The neuron's activity is enhanced *only* in the delay period and *only* for the neuron's preferred direction. In contrast, neuronal firing is often inhibited for the direction opposite to the preferred target and unaltered or inhibited for all other non-preferred directions (FIG. 5). We suggested that moderate D1 receptor block may set an optimal level of cAMP production and enhancement of NMDA receptor–mediated preferred inputs through regulation of persistent sodium conductances.[39] Attenuation of L- or N-type calcium conductances may account for the loss of tuning in pyramidal cells by excessive dopamine levels. The extraordinary specificity of D1 antagonist drugs on single-cell responses for preferred afferents can best be accounted for by the specific synaptic arrangement of D1 receptors in spines that receive a dopamine afferent and excitatory inputs, presumably from the visual pathway carrying topographically specified visuospa-

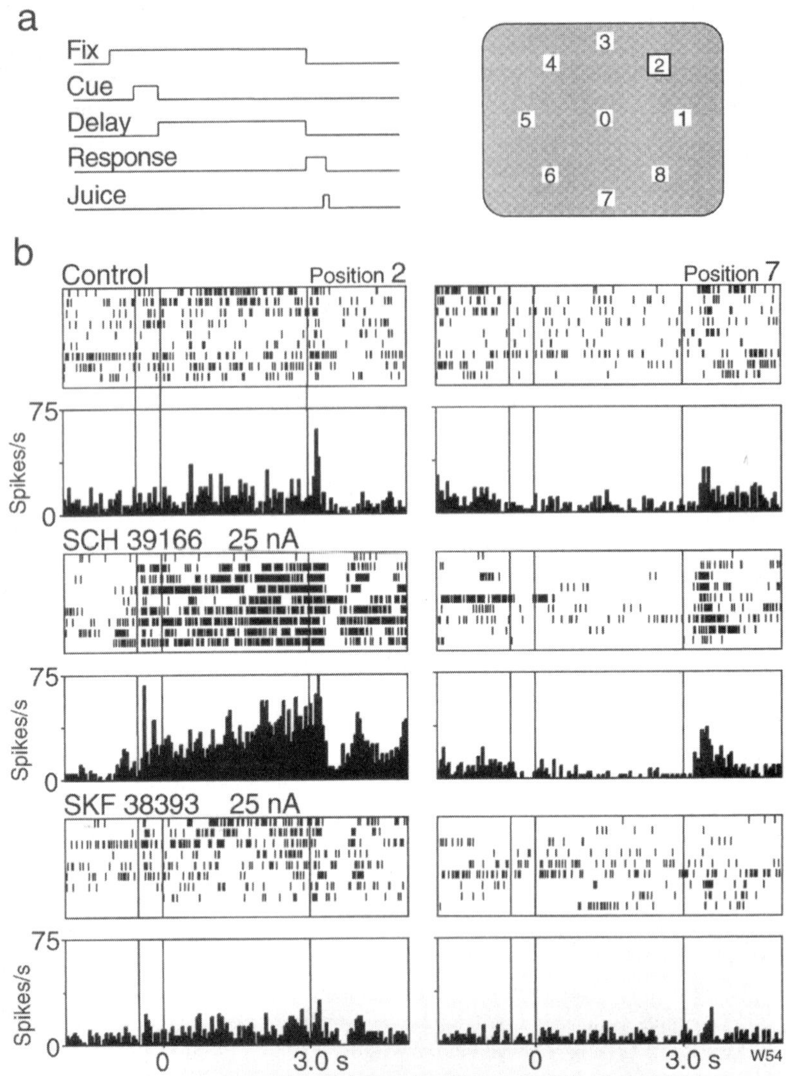

FIGURE 5. a, left, schematic sequence of events in each phase of the ODR task (aligned in time with rasters and histograms shown immediately below); **right,** depiction of the 8 positions of the target (plus the central fixation position, 0) to be remembered for guidance of oculomotor response. **b,** Effect of SCH 39166 on response of neuron W54 (rastergram above, histogram below; bin, 50 ms) for a target (position 2) in the memory field **(left)** and for a target (position 7) in nearly the opposite location in space **(right)**. **Top two rows,** control recording showing significant but weak delay activity. **Middle two rows,** SCH 39166 (25 nA) produces dramatic enhancement of activity during the delay period when the target is in the memory field but produces an inhibition of activity when the target is in position 7. **Bottom two rows,** SKF 38393, a D1 agonist, reverses the effect of SCH 39166 and reduces delay activity to "background" level.

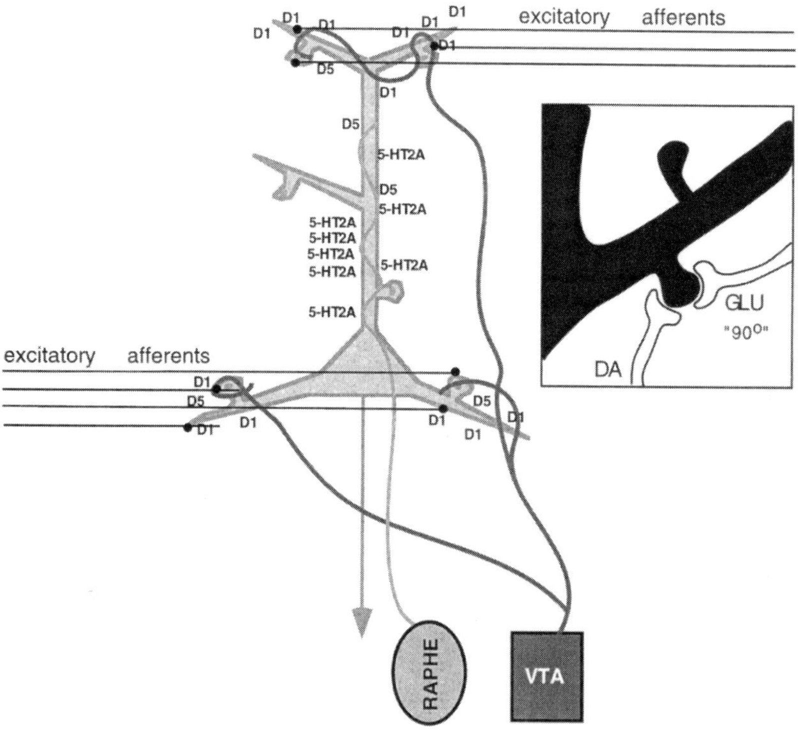

FIGURE 6. Schema of the dopaminergic (DA) and glutamatergic (GLU) afferent input on the pyramidal cell distal dendrites and spines based on electron-microscopic observations described in Goldman-Rakic *et al.*[22] The triadic synaptic arrangement shown can account for the specificity of physiological effects shown in FIGURE 5 and described in the text. The majority of excitatory afferents are assumed to be cortico-cortical afferents known to arise from visual areas in the posterior parietal cortex.

tial coordinate information (FIG. 6). This specificity of D1 action is presumed to occur by actions at the spines of pyramidal neurons where the D1 receptor is colocalized with glutamate receptors and where excitatory afferents terminate (FIG. 6).[22,23] Our finding of higher density of D1 receptors in prefrontal cortex draws attention to the potential functional significance of these receptors for cognitive processes in normal individuals and for deficits such as the thought fragmentation characteristic of diseases like schizophrenia.

As indicated, enhancement of the neuron's memory field or spatial tuning is observed only at low doses of antagonist; at higher doses, or in the presence of the dopamine agonist, neuronal firing is reduced or strongly inhibited. The latter results indicate that dopamine influences mnemonic function through D1 stimulation and that either too little dopamine[19,40] or excessive dopamine (e.g., Ref. 41) as in amphetamine psychosis and experimental situations, may result in cognitive dysfunction through excessive or insufficient D1 stimulation, respectively. It is not surprising that working memory is vulnerable to the influence of dopamine dysfunction, and a common process may underlie the cognitive deficits in Parkinson's disease, in aged individuals, and in schizophrenics. From this work, it appears that the effect of cortical dopamine, in part, is to modulate excitatory neurotransmission at the level of spines on distal dendrites, both basilar and apical, and thereby attentuate or enhance sensory signaling in the neurons that influence the direction and quality of cognitive operations and motor commands. The interaction of dopamine and glutamate pathways and the ionic changes in the membrane properties of cortical pyramidal neurons, generally, and working memory circuits, specifically, is presently an area of active investigation (Refs. 42–44).

SEROTONIN RECEPTORS AND CIRCUITS

The actions at D1 receptors described above are not only dose dependent but highly pharmacologically specific. For example, the dopamine D2 receptor antagonists have null or inhibitory effects on neuronal tuning in prefrontal neurons.[39] A particularly striking example of functional and anatomic compartmentalization of prefrontal circuitry comes from new studies of serotonin modulation of working memory neurophysiology. These studies show that, in contrast to dopamine, serotonin at low doses, enhances the memory fields of prefrontal neurons (Williams, Rao & Goldman-Rakic, PS, unpublished observations). This action is achieved by direct action at 5-HT2A receptors, which have recently been located in the proximal dendrites of pyramidal neurons in the primate prefrontal cortex.[18] However, the 5-HT2A receptor is also located in the parvalbumin subset of GABAergic neurons, which include the so-named "basket" and "chandelier" cells.[45] As indicated above, the inhibitory interneurons that utilize GABA as a neurotransmitter play a powerful role in regulating a pyramidal neuron's memory field by feed-forward disinhibitory mechanisms directed for the most part at the soma and axon hillock region of the cell. Our recent examination of the memory field of a neuron *in vivo* is beginning to suggest that the mnemonic process is sculpted by a combination of GABAergic actions at the soma of a pyramidal cell, serotonergic actions at the level of 5HT2 receptors on its proximal dendrite, and by D1 dopaminergic action on the spines, which are the portions of the cell triggered by excitatory input. Thus, studies of the primate prefrontal cortex are succeeding not only to decipher the circuit and synaptic basis of a complex cognitive process that establishes the human capacity for thought and reasoning but also is laying the groundwork for a rational approach to drug design for cognitive enhancement in aging and in disease.

REFERENCES

1. Baddeley, A. 1986. Working Memory. Oxford University Press. London.
2. Just, M.A. & P.A. Carpenter. 1992. A capacity theory of comprehension: Individual differences in working memory. Psych. Rev. **99:** 122–149.
3. Fuster, J.M. & G.E. Alexander. 1971. Neuron activity related to short-term memory. Science **173:** 652–654.
4. Kubota, K. & H. Niki. 1971. Prefrontal cortical unit activity and delayed alternation performance in monkeys. J. Neurophysiol. **34:** 337–347.
5. Funahashi, S., C.J. Bruce & P.S. Goldman-Rakic. 1989. Mnemonic coding of visual space in the monkey's dorsolateral prefrontal cortex. J. Neurophysiol. **61:** 331–349.
6. Fuster, J.M. 1973. Unit activity in prefrontal cortex during delayed-response performance: Neuronal correlates of transient memory. J. Neurophysiol. **36:** 61–78.

7. FUNAHASHI, S., M.V. CHAFEE & P.S. GOLDMAN-RAKIC. 1993. Prefrontal neuronal activity in rhesus monkeys performing a delayed anti-saccade task. Nature **365:** 753–756.
8. WILSON, F.A.W., S.P.O SCALAIDHE & P.S. GOLDMAN-RAKIC. 1993. Dissociation of object and spatial processing domains in primate prefrontal cortex. Science **260:** 1955–1958.
9. Ó SCALAIDHE, S.P., F.A.W. WILSON & P.S. GOLDMAN-RAKIC. 1997. Areal segregation of face-processing neurons in prefrontal cortex. Science **278:** 1135–1138.
10. MOUNTCASTLE, V.B., V.M. TALBOT, H. SAKATA & J. HYVARINEN. 1969. Cortical neuronal mechanisms in flutter-vibration studied in unanesthetized monkeys. Neuronal periodicity and frequency discrimination. J. Neurophysiol. **32:** 452–484.
11. MCCORMICK, D.A., B.W. CONNORS. J.W. LIGHTHALL & D.A. PRINCE. 1985. Comparative electrophysiology of pyramidal and sparsely spiny stellate neurons of the neocortex. J. Neurophysiol. **54:** 782–806.
12. KAWAGUCHI, Y. 1995. Physiological subgroups of nonpyramidal cells with specific morphological characteristics in layer II/III of rat frontal cortex. J. Neurosci. **15:** 2638–2655.
13. MARTIN, K.A., P. SOMOGYI & D. WHITTERIDGE. 1983. Physiological and morphological properties of identified basket cells in the cat's visual cortex. Exp. Brain Res. **50:** 193–200.
14. WILLIAMS, S.M., P.S. GOLDMAN-RAKIC & C. LERANTH. 1992. The synaptology of parvalbumin-immunoreactive neurons in the primate prefrontal cortex. J. Comp. Neurol. **320:** 353–369.
15. RAO, S., G.V. WILLIAMS & P.S. GOLDMAN-RAKIC. Iso-directional spatial tuning if adjacent putative intemourons and pyramidal neurons during working memory: Functional evidence for a microcolumnar organization in primate prefrontal cortex. J. Neurophysiol. In press.
16. WILLIAMS, S.M. & P.S. GOLDMAN-RAKIC. 1993. Characterization of the dopaminergic innervation of the primate frontal cortex using a dopamine-specific antibody. Cereb. Cortex **3:** 199–222.
17. WILLIAMS, S.M. & P.S. GOLDMAN-RAKIC. 1998. Widespread origin of the primate mesofrontal dopamine system. Cereb. Cortex **8:** 321–345.
18. JAKAB, R.L. & P.S. GOLDMAN-RAKIC. 1998. 5-hydroxytryptamine$_{2A}$ serotonin receptors in the primate cerebral cortex: Possible site of action of hallucinogenic and antipsychotic drugs in pyramidal cell apical dendrites. Proc. Natl. Acad. Sci. USA **95:** 735–740.
19. BROZOSKI, T., R.M. BROWN, H.E. ROSVOLD & P.S. GOLDMAN. 1979. Cognitive deficit caused by regional depletion of dopamine in prefrontal cortex of rhesus monkey. Science **205:** 929–932.
20. BROWN, R.G. & C.D. MARSDEN. 1988. Internal versus external cues and the control of attention in Parkinson's disease. Brain **111:** 323–345.
21. GOTHAM, A.M., R.G. BROWN & C.P. MARSDEN. 1988. 'Frontal' cognitive function in patients with Parkinson's disease 'on' and 'off' levodopa. Brain **111:** 299–321.
22. GOLDMAN-RAKIC, P.S., C. LERANTH, S.M. WILLIAMS, N. MONS & M. GEFFARD. 1989. Dopamine synaptic complex with pyramidal neurons in primate cerebral cortex. Proc. Natl. Acad. Sci. USA **86:** 9015–9019.
23. SMILEY, J.F., S.M. WILLIAMS, K. SZIGETI & P.S. GOLDMAN-RAKIC. 1992. Light and electron microscopic characterization of dopamine-immunoreactive processes in human cerebral cortex. J. Comp. Neurol. **321:** 325–335.
24. SMILEY, J.F. & P.S. GOLDMAN-RAKIC. 1993. Heterogeneous target of dopamine synapses in monkey prefrontal cortex demonstrated by serial section electron microscopy: A laminar analysis using the silver-enhanced diaminobenzidine sulfide (SEDS) immunolabeling technique. Cereb. Cortex **3:** 223–238.
25. GOLDMAN-RAKIC, P.S., M.S. LIDOW & D.W. GALLAGER. 1990. Overlap of dopaminergic, adrenergic, and serotoninergic receptors and complementarily of their subtypes in primate prefrontal cortex. J. Neurosci. **10:** 2125–2138.
26. LIDOW. M.S., P.S. GOLDMAN-RAKIC, D.W. GALLAGER & P. RAKIC. 1991. Distribution of dopaminergic receptors in the primate cerebral cortex: Quantitative autoradiographic analysis using [3^H]raclopride, [3^H]spiperone and [3^H]SCH23390. Neuroscience. **40:** 657–671.
27. SMILEY, J.F., A.I. LEVEY, B.J. CILIAX & P.S. GOLDMAN-RAKIC. 1994. D1 dopamine receptor immunoreactivity in human and monkey cerebral cortex: Predominant and extrasynaptic localization in dendritic spines. Proc. Natl. Acad. Sci. USA **91:** 5720–5724.
32. SURTNEIER, D.J., J. BARGAS, H.C. HEMMINGS, JR., A.C. NAIM & P. GREENGARD. 1995. Modulation of calcium currents by a D1 dopaminergic protein kinase/phosphatase cascade in rat neostriatal neurons. Neuron **14:** 385–397.

28. CALABRESI, P., N. MERCURI, P. STANZIONE, A. STEFANI & G. BERNARDI. 1987. Intracellular studies on the dopamine-induced firing inhibition of neostriatal neurons in vitro: Evidence for Dl receptor involvement. Neuroscience **20:** 757–771.
29. UCHIMURA. N., H. HIGASHI & S. NISHI. 1986. Hyperpolarizing and depolarizing actions of dopamine via D-1 and D-2 receptors on nucleus accumbens neurons. Brain Res. **375:** 368–372.
30. HEMANDEZ-LOPEZ, S., J. BARGAS, D.J. SURMEIER, A. REYES & E. GALARRAGA. 1997. Dl receptor activation enhances evoked discharge in neostriatal medium spiny neurons by modulating an L-type CA^{2+} conductance. J. Neurosci. **17:** 3334–3342.
31. CEPEDA, C., N.A. BUCHWALD & M.S. LEVINE. 1993. Neuromodulatory actions of dopamine in the neostriatum are dependent upon the excitatory amino acid receptor subtypes activated. Proc. Natl. Acad. Sci. USA **90:** 9576–9580.
33. CEPEDA, C., C.S. COLWELL, J.N. ITRI, S.H. CHANDLER & M.S. LEVINE. 1998. Dopaminergic modulation of NMDA-induced whole cell currents in neostriatal neurons in slices: Contribution of calcium conductances. J. Neurophysiol. **79:** 82–94.
34. WESTENBROEK, R.E., M.K. AHHJANIAN & W.A. CATTERALL. 1990. Clustering of L-type Ca2+ channels at the base of major dendrites in hippocampal pyramidal neurons. Nature **347:** 281–284.
35. WESTENBROEK, R.E., J.W. HELL, C. WARNER, S.J. DUBEL, T.P. SNUTCH & W.A. CATTERALL. 1992. Biochemical properties and subcellular distribution of an N-type calcium channel a_1 subunit. Neuron **9:** 1099–1115.
36. WESTENBROEK, R.E., T. SAKURAI, E.M. ELLIOTT, J.W. HELL, T.V.B. STARR, T.P. SNUTCH & W.A. CATTERALL. 1995. Immunochemical identification and subcellular distribution of the α_{1A} subunits of brain calcium channels. J. Neurosci. **15:** 6403–6418.
37. AHLIJANIAN, M.K., R.E. WESTENBROEK & W.A. CATTERALL. 1990. Subunit structure and localization of dihydropyridine-sensitive calcium channels in mammalian brain, spinal cord, and retina. Neuron **4:** 819–832.
38. HELL, J.W., R.E. WESTENBROEK, C. WARNER, M.K. AHLIJANIAN, W. PRYSTAY, M.M. GILBERT, T.P. SNUTCH & W.A. CATTERALL. 1993. Identification and differential subcellular localization of the neuronal class C and class D L-type calcium channel α_1 subunits. J. Cell Biol. **123:** 949–962.
39. WILLIAMS, G.V. & P.S. GOLDMAN-RAKIC. 1995. Modulation of memory fields by dopamine Dl receptors in prefrontal cortex. Nature **376:** 572–575.
40. LANGE, K.W., T.W. ROBBINS, C.D. MARSDEN, M. JAMES, A.M. OWEN & G.M. PAUL. 1992. L-dopa withdrawal in Parkinson's disease selectively impairs cognitive performance in tests sensitive to frontal lobe dysfunction. Psychopharmacology **107:** 394–404.
41. MURPHY, B.L., A.F.T. ARNSTEN, P.S. GOLDMAN-RAKIC & R.H. ROTH. 1996. Increased dopamine turnover in the prefrontal cortex impairs spatial working memory performance in rats and monkeys. Proc. Natl. Acad. Sci. USA **93:** 1325–1329.
42. CEPEDA, C., Z. RADISAVIJEVIC, W. PEACOCK, M.S. LEVINE & N.A. BUCHWALD. 1992. Differential modulation by dopamine of responses evoked by excitatory amino acids in human cortex. Synapse **11:** 330–341.
43. YANG, C.R. & J.K. SEAMANS. 1996. Dopamine D1 receptor actions in layers V-VI rat prefrontal cortex neurons in vitro: Modulation of dendritic-somatic signal integration. J. Neurosci. **16:** 1922–1935.
44. LISMAN, J.E., J-M FELLOUS & X-J WANG. 1998. A role for the NMDA-receptor channels in working memory. Nature Neuroscience **1:** 273–275.
45. JAKAB, R.L. & P.S. GOLDMAN-RAKIC. 1998. Segregation of 5-HT2A and 5-HT3 serotonin receptors in primate prefrontal cortex. Soc. Neurosci. Abstr. **24:** 308.

Studies on Conditional Gene Expression in the Brain

JASNA JERECIC, FRANK SINGLE, ULI KRÜTH, HEINZ KRESTEL, ROHINI KOLHEKAR,[a] THORSTEN STORCK,[a] KALEV KASK,[b] MIYOKO HIGUCHI, ROLF SPRENGEL, AND PETER H. SEEBURG[c]

Department of Molecular Neuroscience, Max-Planck Institute for Medical Research, Jahnstr. 29, 69120 Heidelberg, Germany

ABSTRACT: This manuscript summarizes our recent attempts to regulate *in vitro* and *in vivo* the expression of genes encoding components and regulators of the postsynaptic machinery along with marker genes such as lacZ and GFP. In particular, we studied tTA-dependent regulation and utilized Cre in combination with reversible silencing by intron engineering of dominant negative alleles. We further present a "knockin" approach for on-site artificial regulation of chromosomal genes.

In the neurosciences, as in other areas of modern biology, there is a need to manipulate and study gene function by conditional means. There are currently two main methodologies in use that permit conditional gene regulation. One of them renders genes responsive to tTA (tetracycline-controlled transactivator) or the reverse system, rtTA,[1,2] allowing for reversible changes in gene expression. tTA activates transcription in the absence of tetracycline derivatives—e.g., doxycycline (dox), whereas rtTA requires dox for transcriptional activation. The other utilizes Cre recombinase for generating the removal of loxP-flanked gene segments.[3,4] Although promising in many respects, and put to spectacular use in the recent past, both systems are in need of further improvement.

The group of Eric Kandel has been the most prolific in utilizing the tTA/rtTA system to alter synaptic function in the mouse brain. First, Mayford *et al.*[5] generated a transgenic line in which tTA expression is from the promoter for the α subunit of Ca-calmodulin–dependent protein kinase II (CaMKII). Introducing into this mutant an additional, tTA-responsive transgene for a CaMKII mutant allowed these researchers to investigate the frequency dependence of changes in the strength of excitatory synapses on the expression level of the mutant kinase. They further correlated the conditionally altered synaptic plasticity with consequences in the acquisition of spatial memory. Kandel's group has recently used the rtTA system, also driven by the CaMKII promoter, to induce in mice the conditional expression of a calcineurin (a Ca-dependent phosphatase) mutant introduced as a transgene, again with consequences on synaptic plasticity and spatial memory storage/retrieval.[6] These studies clearly show the stunning potentials of tTA/rtTA-mediated gene regulation. The current drawbacks of this system lie in the relatively slow time course by which genes can be turned on or off or can be reinduced after shut-off. For tTA, shut-off in

[a]Present address: BASF-LYNX Bioscience, AG, Im Neuenheimer Feld 515, 69120, Heidelberg, Germany.

[b]Present address: AGY Therapeutics, Inc., Two Corporate Drive, South San Francisco, California 94080, USA.

[c]Corresponding author. e-mail: seeburg@mpimf-heidelberg.mpg.de

presence of dox takes days and, naturally, will depend on the turnover of the gene product. Reinduction by dox withdrawal after shut-off may take weeks, depending on how much dox had been applied for gene inactivation. The reason is that dox is stored in some tissues, leading to slowly declining systemic dox levels. In case of rtTA, high dox levels for activation appear to be needed, and thus inactivation by dox withdrawal will also follow a relatively slow timecourse. Nevertheless, these problems can be addressed by improvements in the chemistry of dox antagonists.

Cre recombinase has been put to similarly spectacular use in the study of memory function by Susumu Tonegawa and his colleagues. Collaborating with Mayford and Kandel, they reported[7,8] transgenic mouse mutants that express Cre, driven by the same promoter fragment that Mayford *et al.* had used to express tTA and rtTA. The CaMKII promoter–controlled Cre transgene was crossed into a mouse that had its NMDA receptor (NMDAR) NR1 subunit (for review, see Ref. 9) alleles marked by two loxP sites in different introns. A functional knockout of the NR1 alleles was achieved, relatively specific for the hippocampus, by selecting an appropriate Cre expressor. The analysis of this spatially restricted functional NMDAR knockout revealed a loss in certain forms of synaptic plasticity in the hippocampus and a concomitant loss in the acquisition of spatial memory. Though powerful in its own right, the drawbacks with Cre lie in the irreversibility of Cre-mediated recombination and a stringent need for spatial and temporal control over its expression. The study by Tonegawa and his colleagues was possible because CaMKII gene expression commences relatively late in development, mainly after birth, and because one of many transgenic founder lines was selected in which, due to transgene integration, expression was confined mainly to hippocampal structures. Late, selective Cre expression was a (perhaps not fully met) prerequisite to avoid adaptive prenatal changes in brain wiring and synaptic alterations in other important brain structures, which would detract from a clean correlation of postnatal changes in hippocampal synapses with concomitant changes in memory functions.

Our own studies to regulate expression of several genes mediating and modulating postsynaptic function have also made use of tTA, rtTA, and Cre. We constructed bidirectional cassettes to monitor tTA-dependent gene expression along with simultaneously expressed marker genes (lacZ, GFP) and explored the functionality of a multimodal cassette designed for knockin into chromosomal genes.

tTA-RESPONSIVE GENES

Transgenic Approaches

We generated mice in which certain transgenes should be rendered tTA responsive in their expression. To achieve this, we constructed a cassette in which a heptamerized tet operator is positioned between two CMV minimal promoters for bidirectional tTA-controlled transcriptional activation (FIG. 1). We inserted behind the minimal promoters the *lacZ* gene for facile histochemical expression analysis and a chimeric mouse/rat (for ease of sequence distinction of transgene and endogenous alleles by *in situ* hybridization) minigene encoding the principal NMDAR subunit NR1 in its wild-type form. In another construct, *lacZ* was combined with a minigene for a Ca^{2+}-impermeable form of NR1,

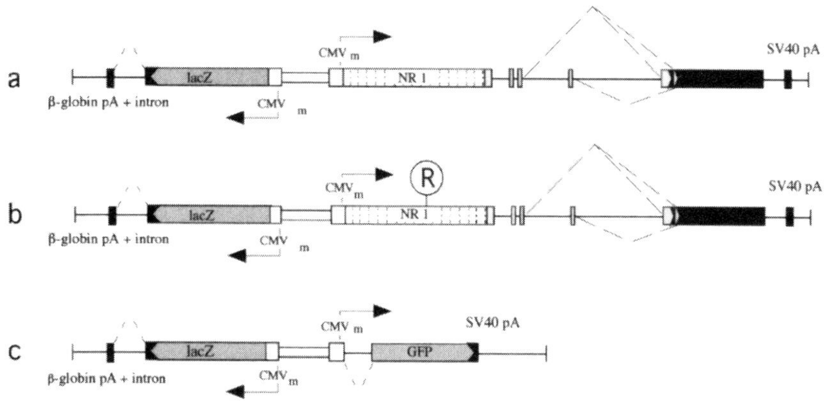

FIGURE 1. tTA-responsive minigenes. The three bidirectional cassettes for tTA-responsive minigenes—**(a)** *lacZ*/NR1; **(b)** *lacZ*/NR1(R); **(c)** *lacZ*/GFP—yielded founder lines, in which minigene expression was tTA sensitive in central neurons. The two CMV minimal promoters in the cassettes are separated by a heptamerized *lac* operator sequence (not indicated). The NR1 minigenes allow for different C-terminal splice forms, indicated by *thin broken lines*. The *black box* at the 3′ end of the NR1 genes is 3′ untranslated sequence.

NR1(R).[10] A third construct carried *lacZ* and the coding sequence for GFP. Founder lines were established having these constructs inserted in their genome. To test for tTA-dependent expression and regulation of the bidirectional cassettes in these lines we crossed the mutant mice with a transgenic line (kindly provided by M. Mayford and E. Kandel) in which tTA is well expressed from a CaMKII promoter, mainly in principal excitatory neurons of the postnatal forebrain.[5]

Three lines (M2 to M4), made double transgenic for tTA and the NR1/*lacZ* cassette, were analyzed.[11,12] All expressed β-galactosidase (beta-gal), with levels in transgenic lines M2 and M4 being approximately half of those in M3. *In situ* hybridization documented the tTA-mediated expression of the cassette-contained minigenes (not shown). Presumably due to the overexpression of NR1 (averaged in brain, 150% of wt), the mutant mice of line M3 appeared weaker and died prematurely before four weeks of age. Regulation of expression was attempted with this line by adding doxycycline to the drinking water (dox, 2 mg/ml) of pregnant females and double heterozygous mice. Expression of *lacZ* (by *in situ* hybridization) and beta-gal (by X-gal staining) was undetectable in postnatal day 20 (P20) mice kept on dox, demonstrating that the tTA-controlled expression from the bidirectional cassette can be silenced in the presence of the tet derivative (FIG. 2). To study if the down-regulated expression can be reactivated after dox withdrawal, we analyzed double-transgenic M3 animals that had been on dox until P20 and had then been kept for another three weeks free of dox. In these animals, beta-gal activity was readily visible, but was considerably reduced relative to M3 animals that had not received dox. Since dox can be absorbed by some tissues (bone, muscle) and can subsequently give rise to low systemic dox levels, we repeated this study using lower dox (200 and 50 μg/ml) levels. After 200 μg/ml of dox, reactivation of gene expression was not much better than after 2 mg/ml

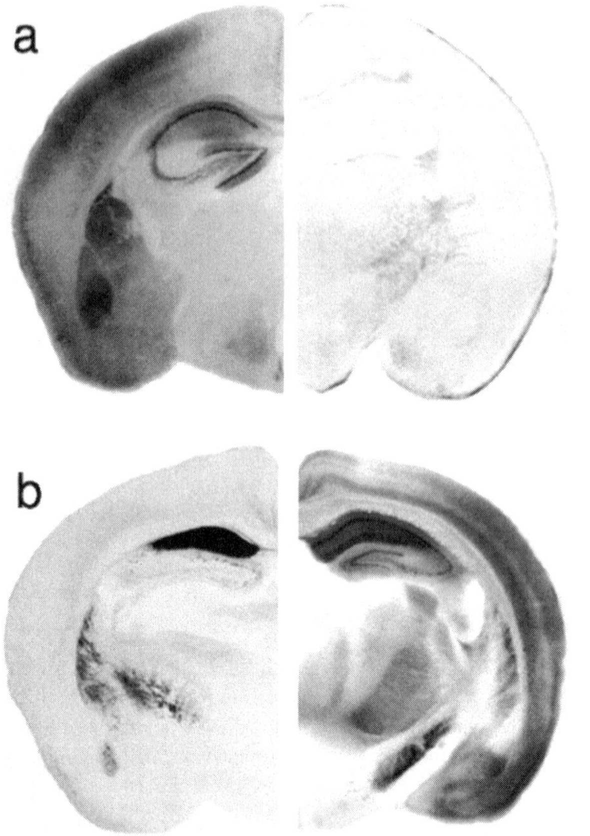

FIGURE 2. Coronal X-gal stained hemibrain sections of M3/CaMKII-tTA double heterozygotes demonstrate that beta-gal levels expressed from the bidirectional cassette are regulated by doxycycline (dox). **(a)** P18 animals. **Left**, no dox; **right,** raised in presence of dox. **(b) Left,** P20 brain of mouse raised for 10 days with dox and then 10 days without dox; **right,** P35 brain of mouse raised for 10 days with dox and then 25 days without dox. Transgenic NR1 expression from the bidirectional cassette is also regulated, as seen by *in situ* hybridization (not shown), and from the fact that the animals survive in presence of dox, but die in its absence.

(NR1, 120% of wt). The lowest dox levels also shut off bidirectional cassette expression, but permitted full reactivation of tTA-directed gene expression (FIG. 2). Interestingly, mice in which NR1 overexpression was fully reactivated did not develop the deficient, lethal phenotype, indicating that a critical developmental window of vulnerability to an excess of NR1 can be kept closed by dox.

Vulnerability was increased in transgenic mice overexpressing the Ca^{2+}-impermeable form NR1(R) rather than the wild-type NR1, in accord with the lethality of mice having the endogenous NR1 allele mutated to NR1(R) by homologous recombination.[12] In two double-heterozygous lines (R1,2) harboring the bidirectional *lacZ*/NR1(R) cassette and expressing tTA, *lacZ* expression and NR1 overexpression (115% of wild type) were considerably lower than in the M3 line, and yet these animals died by P16. Dox-mediated silencing of expression has not yet been attempted in the R1 and R2 lines.

We observed that nearly all founder lines transgenic for the bidirectional *lacZ*/NR1 cassettes showed tTA-dependent cassette expression in central neurons, although to different levels. Since tTA expression is unchanged, differences in cassette expression appear to reflect the influence of transgene position and copy number. The high success rate of transgenics expressing genes from bidirectional cassettes contrasts with poor functional expression in our lab from constructs having promoter fragments fused to the tTA gene (not shown). We tentatively attribute the failure of these attempts and the success with the bidirectional cassette to a degree of independence from the insertion locus of the cassette. As the cassette directs transcription in both directions, the central promoter region may be better shielded from outside interferences than in a unidirectional construct.

We presently analyze transgenic lines harboring a bidirectional *lacZ*/GFP expression cassette. In one such line, made double transgenic for tTA expression, the expression of the two marker genes driven by tTA was observed in many brain regions (FIG. 3). When "humanized" GFP[14] expression was monitored by laser scanning microscopy, it became apparent that not all pyramidal and granule cells fluoresce. A similar picture may emerge

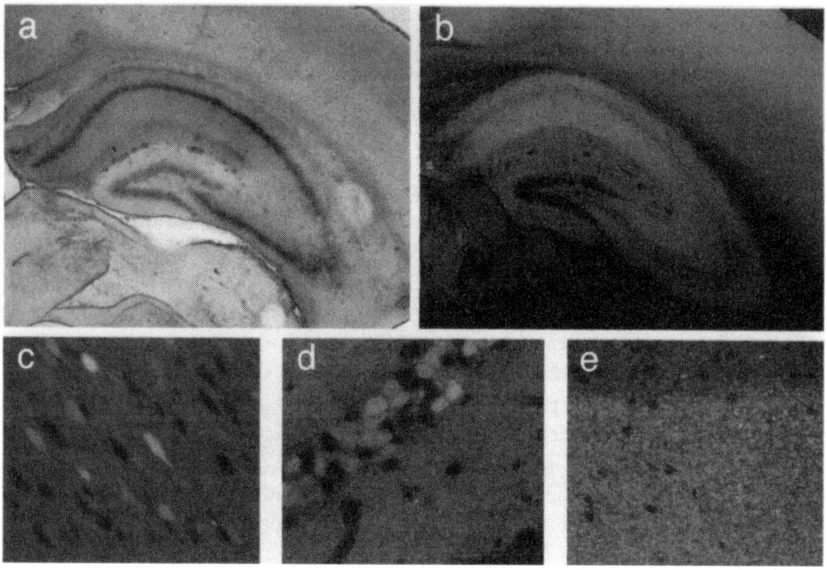

FIGURE 3. tTA-sensitive beta-gal (**a**) and GFP (**b–e**) expression in double transgenics, harboring a *lacZ*/GFP cassette and expressing tTA from a CaMKII promoter. **Upper panels** compare cortical and hippocampal *lacZ* and GFP expression in the mutant brain. **Lower panels** show higher magnification of cortex (**c**), hippocampal CA1 area (**d**), and striatum (**e**). Note the mosaic GFP expression in neocortical and hippocampal neurons.

for the tTA-driven beta-gal expression, although immunofluorescence for beta-gal and nuclear counterstaining has not yet been performed.

Knockin Approaches

We studied whether chromosomal genes can be put under the control of tTA for exogenous regulation of their expression by "knockin" of a multifunctional cassette into an exon for the 5′ untranslated region of their mRNAs.[15] We constructed such a cassette ("triTAUBi-AF") in a pBluescript backbone to contain, from 5′ to 3′, a tripartite leader sequence from adenovirus fused to the tTA coding region from which a potential splice donor site had been removed; an SV40 polyadenylation site; a "floxed" (flanked by loxP sites) tk-neo gene plus *URA* marker for the respective selection in murine embryonic stem (ES) cells (see below) and yeast; the transcriptional termination region of the human growth hormone gene; and, finally, a heptamerized tet operator preceding the CMV minimal promoter sequence (see FIGS. 4 and 5). To test the functionality of this cassette we inserted a plasmid between the full CMV promoter and a *lacZ* gene with a nuclear targeting signal (pnlacF).[16] This plasmid ("pTTT"; FIG. 4a) was transfected into HEK293 cells and after two days in presence or absence of dox (1 mg/ml), the cells were stained for beta-gal expression of the chromogenic substrate, X-gal. Only cells grown in absence of dox turned blue (approx. 20% of all cells), whereas untransfected cells and cells grown in presence of dox did not (FIG. 4b). In additional experiments the enzymatic activity of beta-gal was determined from lysed HEK293 cells that had been transiently transfected with several constructs. The constructs were pnlacF with and without the CMV promoter, pTTT, and pTTTCre, a pTTT derivative that had the *tk-neo* and *URA* markers removed by Cre recombinase in *E. coli*. To compare experimental results, the transfection efficiencies of the different constructs were monitored by nitrofecin, which reported beta-lactamase expression from a cotransfected eukaryotic beta-lactamase expression vector. The results (FIG. 4c), corrected for transfection efficiencies, showed that beta-gal expression from pTTT was 3.7- and from pTTTCre was 4.4-fold higher than from the CMVpnlacF construct, demonstrating a substantial boost of beta-gal expression by the CMV-driven tTA relative to the CMV-driven *lacZ* transcription. Furthermore, beta-gal levels from pTTT and pTTTCre were induced approximately 200-fold in the absence of dox, when compared to levels in the presence of the tet derivative.

We further selected via the neo gene on pTTT several stable HEK293 clones and analyzed these to assess the functionality of the cassette in different integration sites (FIG. 4). We found that all clones expressed the *lacZ* gene, and expression could be well suppressed by dox in approximately half of the clones. Together, the results obtained *in vitro* demonstrate the functionality of the multimodal cassette. Indeed, if similar results could be obtained *in vivo*, by cassette insertion into chromosomal genes, tet-controlled gene regulation over one to two orders of magnitude should suffice for most biological applications.

To test the "triTAUBi-AF" cassette in mice, we took murine ES cells and inserted the cassette by homologous recombination into an exonic position, 30 nucleotides upstream of the translational start codon ATG, of the genes for the AMPAR GluR-B subunit[9] and the NR1 subunit (FIG. 5a). To permit such a precise, restriction site–independent targeting of the regulatory cassette we utilized the homologous recombination machinery of yeast. We

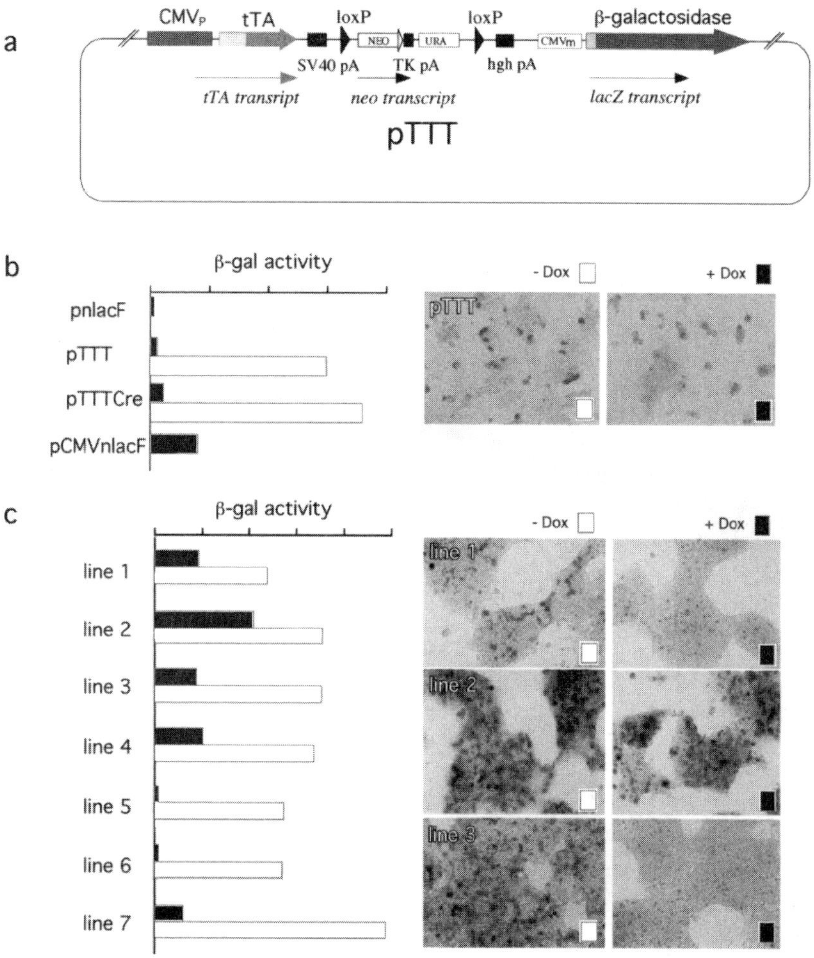

FIGURE 4. Composition and functionality of a cassette with conditional modules. The modules of the cassette (triTAUBi-AF; see also FIG. 5) carried on plasmid pTTT **(a)** are described in the text. pTTT was tested along with control plasmids by transient expression in HEK293 cells **(b).** Several stable HEK293 cell lines were also analyzed for regulatable *lacZ* expression **(c).** Beta-gal expression in absence (*white bars*) and presence (*black bars*) of dox is indicated in arbitrary units.

developed a method to construct the targeting vector for ES cells in yeast by cotransfecting into an appropriate strain the respective murine gene segment carried on a shuttle vector and the regulatory cassette harboring the *URA* marker (see above), and modified to carry at its ends recombinogenic arms of 500 bp, corresponding in sequence to the target in the mouse gene.[17]

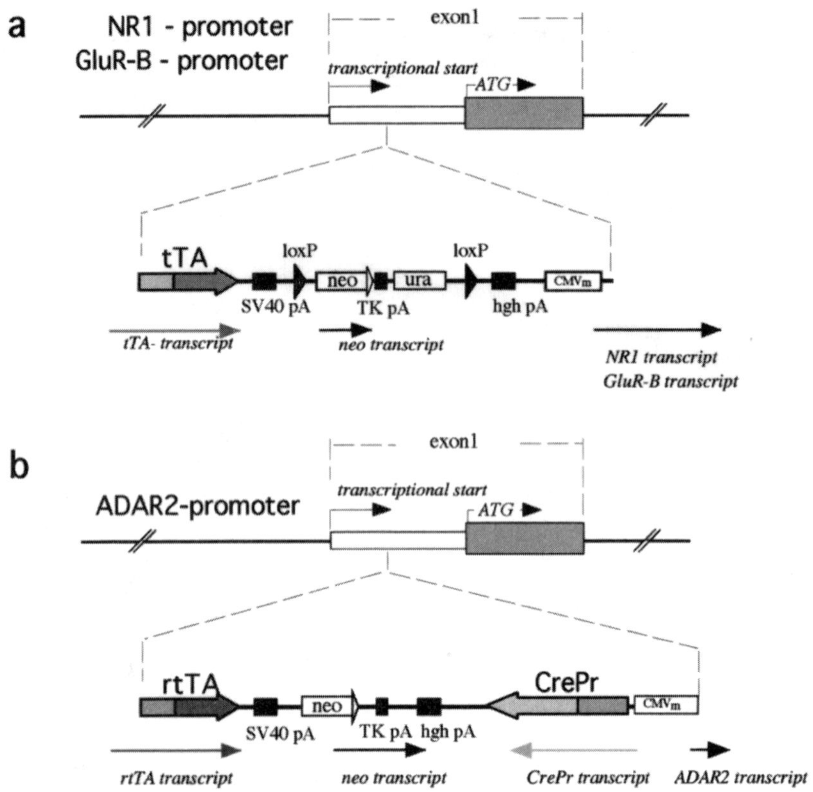

FIGURE 5. In murine ES cells, a multifunctional cassette ("tri-TAUBi-AF") was inserted by homologous recombination into exonic sequences for the 5′ untranslated transcript region of genes for NR1 and GluR-B **(a)**. Another, similar cassette for rtTA expression was recombined into an allele for the ADAR2 gene **(b)**.

ES cell clones successfully targeted with the regulatory cassette gave rise to chimeras, but these failed to transmit the targeted alleles through the germ line. To ascertain whether the modified alleles were expressed, and whether such expression was tet sensitive, we undertook several attempts to differentiate the NR1 gene-targeted ES cells into neurons, which are the only cell type expressing this gene. RT-PCR for tTA sequences from differentiated neurons remained inconclusive. NR1 gene expression was observed by RT-PCR because the targeted, differentiated ES cells harbored one wild-type NR1 allele. We currently repeat the cassette knockin into the NR1 gene.

We then targeted the ADAR2 gene (ADAR2, formerly termed RED1, is an RNA-dependent adenosine deaminase and a candidate enzyme for site-selective RNA editing of AMPAR GluR-B subunit pre-mRNA[18]) for multifunctional cassette insertion because ADAR2 is expressed in ES cells, and hence correctly targeted cells can be analyzed for conditional gene expression before generating the mutant mouse. We had modified the cassette (FIG. 5b) to incorporate rtTA instead of tTA and to place the coding sequence for Cre recombinase behind the operator/CMV minimal promoter. We targeted this construct into an exon for the 5′ untranslated region of ADAR2 mRNA, close to the translational initiator codon ATG; and we used an ES cell clone, transgenic for a *lacZ* gene that is silenced by insertion of a floxed sequence with in-frame translational stop codons such that silencing can be undone by Cre recombinase.[19] Correctly targeted transgenic ES cells did not express sufficient levels of Cre to activate the floxed *lacZ* gene. This may be due to the low endogenous ADAR2 expression and hence the low levels of tTA from the cassette-targeted ADAR2 gene, which, in turn, fail to activate the tTA-dependent Cre gene. As expected, control transfections into the targeted ES cells with plasmids for the transient expression of tTA and of Cre yielded beta-gal expression, as judged from the appearance of X-gal–stained cells.

A functionally successful knockin of the triTAUBi-AF cassette was recently obtained for the sk3 gene (sk genes encode small conductance, Ca^{2+}-activated K^+ channels)[20] in collaboration with John Adelman and his colleagues at the Vollum Institute.

Cre RECOMBINASE

Silencing and Unsilencing of Lethal Alleles

Previous work from our laboratories has shown that a single endogenous GluR-B allele rendered deficient for Q/R site editing leads to a seizure-prone, prematurely lethal phenotype in mice.[21] We silenced this dominant lethal allele by inserting into an intron a >3-kb sequence consisting of the genes for neo and tk. The presence of this bulky sequence interfered with expedient splicing and effectively attenuated the allele to express functional mRNA at approximately 5% of the wild-type level.[22] As a consequence, animals carrying this allele exhibit a milder phenotype, reach reproductive competence, and can be bred to homozygosity for the silenced allele. Since the intronic sequence insertion is flanked by loxP sites, the silenced allele can be unsilenced by Cre recombinase. If Cre-mediated unsilencing occurs widely in the brain, the allele will exert its dominant lethal effect. If unsilencing is restricted to particular neuronal populations, the effect should depend on cell type. Thus, these mice constitute an excellent model system in which the effect of a dominant lethal mutation can be studied in different cells and at different developmental and adult times, provided that mouse lines for regional and externally controlled Cre expression become available.

We have also attenuated the expression of a dominant lethal mutation in postsynaptic NMDAR channels by placing a floxed neo gene into an intron of the NR1 gene. The lethal mutation is at a critical position for the high Ca^{2+} permeability contributed by the NR1 subunit to the NMDAR channel.[10] The wild-type NR1 subunit carries an asparagine (N) residue in this channel position, whereas the mutant subunit contributes an arginine (R) residue. Carriers of an unsilenced NR1(R) allele die early, but carriers of a silenced NR1(R) allele survive.[13] Both NR1 alleles cannot be silenced since loss of most NMDARs is not compatible with (perinatal) life.[23,24]

CONCLUDING REMARKS

Existing methodologies for conditional gene regulation in the mouse—in particular, the approaches provided by use of tTA/rtTA and Cre—go a long way towards the desired goal. However, further improvements will be needed—for instance, in accelerating the time course of expression changes in tTA/rtTA-sensitive genes and in the temporal control of Cre activity. Our own studies concerning knockin of cassettes into genes for purposes of conditional regulation have not readily met with success. Unfortunately, the long time from construct to mouse precluded systematic studies. Certainly, the collective experience with the existing technology and the increasing momentum in its improvement (e.g., Refs. 25,26) will ultimately provide us with increasingly precise genetic means to control brain functions.

ACKNOWLEDGMENTS

We thank M. Mayford and E. Kandel for the tTA-expressing transgenic mouse, H. Bujard for tTA/rtTA plasmids and discussions, and N. Muzyczka for hGFP. This work was supported, in part, by HFSP, the Volkswagenstiftung, the German Chemical Industry, and the Bristol-Myers Squibb Company.

REFERENCES

1. Gossen, M. & H. Bujard. 1992. Tight control of gene expression in mammalian cells by tetracycline responsive promoters. Proc. Natl. Acad. Sci. USA **89:** 5547–5551.
2. Gossen, M. *et al.* 1995. Transcriptional activation by tetracyclines in mammalian cells. Science **268:** 1766–1796.
3. Sauer, B. & N. Henderson. 1990. Targeted insertion of exogenous DNA into the eukaryotic genome by the Cre recombinase. New Biol. **2:** 441–449.
4. Gu, H., Y.R. Zou & K. Rajewsky. 1993. Control of immunoglobulin switch recombination at individual switch recombination at individual switch regions evidenced through Cre-loxP mediated gene targeting. Cell **73:** 1155–1164.
5. Mayford, M. *et al.* 1996. Control of memory formation through regulated expression of a CaMKII transgene. Science **274:** 1678–1683.
6. Mansuy, I.M. *et al.* 1998. Inducible and reversible gene expression with the rtTA system for the study of memory. Cell **21:** 257–265.
7. Tsien, J.Z. *et al.* 1996. Subregion- and cell type-restricted gene knockout in mouse brain. Cell **87:** 1317–1326.
8. Tsien, J.Z., P.T. Huerta & S. Tonegawa. 1996. The essential role of hippocampal CA1 NMDA receptor-dependent synaptic plasticity in spatial memory. Cell **87:** 1327–1338.
9. Hollmann, M. & S. Heinemann. 1994. Cloned glutamate receptors. Annu. Rev. Neurosci. **17:** 31–108.
10. Burnashev, N. *et al.* 1992. Asparagine residue in the TM2 segment of NMDA receptor subunits controls Ca^{2+} permeability and Mg^{2+} block. Science **257:** 1415–1419.
11. Kistner, A. *et al.* 1996. Doxycycline-mediated quantitative and tissue-specific control of gene expression in transgenic mice. Proc. Natl. Acad. Sci. USA **93:** 10933–10938.
12. Jerecic J., A. Smith, M. Mayford, R. Sprengel & P.H. Seeburg. 1997. Conditional expression of NMDA receptor subunit NR1 in transgenic mice. Soc. Neurosci. Abstr. 451.6.
13. Single, F.N. *et al.* 1997. Effects of targeted point mutation at the N-site of the NMDA receptor channel M2 domain in mice. Soc. Neurosci. Abstr. 451.5.

14. Zolotukhin, S., M. Potter, W.W. Hauswirth, J. Guy & N.A. Muzyczka. 1996. 'Humanized' green fluorescent protein cDNA adapted for high-level expression in mammalian cells. J. Virol. **70:** 4646–4654.
15. Kolhekar, R., T. Storck, R. Sprengel & P.H. Seeburg. 1997. A single-step approach towards inducible gene expression. Soc. Neurosci. Abstr. 451.7.
16. Mercer, E.H., G.W. Hoyle, R.P. Kapur, R.L. Brinster & R.D. Palmiter. 1991. The dopamine beta-hydroxylase gene promoter directs expression of E. coli lacZ to sympathetic and other neurons in adult transgenic mice. Neuron **7:** 703–716.
17. Storck, T., U. Krüth, R. Kolhekar, R. Sprengel & P.H. Seeburg. 1996. Rapid construction in yeast of complex targeting vectors for gene manipulation in the mouse. Nucl. Acids Res. **24:** 4594–4596.
18. Melcher, T. *et al.* 1996. A candidate enzyme for RNA editing at the Q/R site in GluR-B pre-mRNA. Nature **379:** 460–464.
19. Akagi, K. *et al.* 1997. Cre-mediated somatic site-specific recombination in mice. Nucl. Acids Res. **25:** 1766–1773.
20. Koehler, M. *et al.* 1996. Small-conductance, calcium-activated potassium channels from mammalian brain. Science **273:** 1709–1714.
21. Brusa, R. *et al.* 1995. Early-onset epilepsy and postnatal lethality associated with an editing-deficient GluR-B allele in mice. Science **270:** 1677–1680.
22. Kask, K. *et al.* 1998. Neurological dysfunctions in mice expressing different levels of the Q/R site-unedited AMPA receptor subunit GluR-B. Submitted.
23. Forrest, D. 1994. Targeted disruption of NMDA receptor 1 gene abolishes NMDA response and results in neonatal death. Neuron **13:** 325–338.
24. Li, Y., R.S. Erzurumlu, C. Chen, S. Jhaveri & S. Tonegawa. 1994. Whisker-related neuronal patterns fail to develop in the trigeminal brainstem nuclei of NMDAR1 knockout mice. Cell **76:** 427–437.
25. Kellendonk, C. *et al.* 1996. Regulation of Cre recombinase activity by the synthetic steroid RU486. Nucl. Acids Res. **24:** 1404–1411.
26. Brocard, J., R. Feil, P. Chambon & D. Metzger. 1998. A chimeric Cre recombinase inducible by synthetic but not by natural ligands of the glucocorticoid receptor. Nucl. Acids Res. **26:** 4086–4090.

Diversity of Mammalian Voltage-Gated Sodium Channels

ALAN L. GOLDIN[a]

Department of Microbiology and Molecular Genetics, University of California, Irvine, California 92697-4025, USA

ABSTRACT: A variety of different isoforms of mammalian voltage-gated sodium channels have been identified. These channels can be classified into three different types. Eight type 1 isoforms have been identified in the CNS, PNS, skeletal muscle, and heart. All of these channels have been expressed in exogenous systems, and all of the genes have been mapped. Three type 2 isoforms have been identified in heart, uterus, and muscle. These channels diverge from the type 1 channels in critical regions, and have not been functionally expressed, so their significance is unknown. A single isoform identified in the PNS may represent a third class of channels, in that it diverges from both type 1 and 2 channels. The type 3 channel has not been functionally expressed.

Voltage-gated sodium channels consist of a highly processed subunit that is approximately 260 kDa and is associated with accessory β subunits in some tissues, such as brain and muscle.[1,2] A variety of α subunit isoforms have been detected by molecular cloning, biochemical purification, and electrophysiological recording. Unfortunately, the cDNA clones have been named in many different ways, with no consistent nomenclature for the various isoforms. To add to the confusion, many of the names include the tissue of origin for the clone, which does not always reflect the overall distribution of the specific isoform. For the purposes of this review, the isoforms have been assigned names based on sequence relatedness and approximate order of discovery. This convention is similar to that used for potassium channels,[3] and it has previously been used to designate some of the more recently identified sodium channel clones. The assigned names appear in TABLES 1 and 2, along with the original names, the gene names, Genbank accession numbers, and the references for the clones.

The mammalian sodium channel isoforms that have been identified thus far can be subdivided into three main groups. The type 1 channels (Nav1.x) share significant sequence similarity with each other,[4] and all but one have been functionally expressed in exogenous systems. The type 2 channels (Nav2.x) are approximately 50% identical to the type 1 channels, with significant differences in regions that are critical for channel function.[5] None of these channels have been expressed in an exogenous system yet, so that it is not possible to draw any conclusions about the properties of these channels. An additional cDNA clone has been isolated that is approximately 50% identical to either the type 1 or type 2 channels. This isoform may represent a third class, type 3 (Nav3.x).[6]

Four sodium channel isoforms are expressed primarily in the central nervous system (CNS), Nav1.1 (Scn1A, type I), Nav1.2 (Scn2A, type II and a splice variant termed Nav1.2A or type IIA), Nav1.3 (Scn3A, type III) and Nav1.6 (Scn8A, PN4a) (TABLE 1). Two isoforms have been detected in skeletal muscle, Nav1.4 (Scn4A, SkM1/μ1) in adult

[a]Phone: 949-824-5334; fax: 949-824-8598; e-mail: agoldin@uci.edu

TABLE 1. Mammalian Sodium Channel α Subunits

Channel Name	Gene Symbol	Original Name	Species	Function	Tissue	Size	Chromosome	Genbank Access #
Nav1.1	SCN1A	rat I[7]	Rat	yes[31]	CNS PNS	2009	Mouse 2[14] Human 2[87]	X03638
		HBSCI[88]	Human	no	CNS	partial		X65362
		GPBI	Guinea pig	no	CNS	partial		AF003372
Nav1.2	SCN2A	rat II[7]	Rat	yes[29]	CNS	2005	Mouse 2[14] Human 2[88,89]	X03639
		HBSCII[88]	Human	no	CNS	partial		X65361
		HBA[90]	Human	yes[90]	CNS	2005		M94055
Nav1.2A		rat IIA[30]	Rat	yes[30]	CNS	2005		X61149
Nav1.3	SCN3A	rat III[35,91]	Rat	yes[35,91]	CNS	1951	Mouse 2[14] Human 2[92]	Y00766
Nav1.4	SCN4A	SkM1, μ1[48]	Rat	yes[48]	skeletal muscle	1840	Mouse 11[93] Human 17[94,95]	M26643
		SkM1[95,96]	Human	yes[97]	skeletal muscle	1836		M81758
Nav1.5	SCN5A	SkM2[44] rH1[43]	Rat	yes[53]	denervated skeletal muscle, heart	2018	Mouse 9[98] Human 3[98]	M27902
		H1[99]	Human	yes[99]	heart	2016		M77235
Nav1.6	SCN8A	NaCh6[9]	Rat	no	CNS PNS	1976	Mouse 15[39] Human 12[20,39]	L39018
		PN4a[18]	Rat	yes[18]	CNS PNS	1976		AF049239A F049240
		Scn8a[32,39]	Mouse	yes[32]	CNS	1976		U26707 AF049617
		Scn8a[20]	Human	no	CNS	1980		AF050736
		CerIII[28]	Guinea pig	no	CNS	partial		AF003373
Nav1.7	SCN9A	PN1[66,69]	Rat	yes[66]	PNS	1984	Mouse 2[101,102]	U79568
		hNE-Na[68]	Human	yes[68]	medullary thyroid Ca	1977		X82835
		Nas[67]	Rabbit	no	Schwann cells	1984		U35238
Nav1.8	SCN10A	SNS[72]	Rat	yes[72]	PNS (DRG)	1957	Mouse 9[102]	X92184
		PN3[73]	Rat	yes[73]	PNS (DRG)	1956		U53833
		SNS[74]	Mouse	no	PNS	1958		Y09108
Nav2.1	SCN6A	Na 2.1[5]	Human	no	heart uterus muscle	1682	Human 2[100]	M91556
Nav2.2	SCN7A[a]	Na-G[103]	Rat	no	astrocytes	partial	Mouse 2[104]	M96578
		SCL11[76]	Rat	no	PNS (DRG)	1702		Y09164
Nav2.3		Na 2.3[75]	Mouse	no	heart uterus muscle	1681		L36179
Nav3.1		NaN[6]	Rat	no	PNS	1765		AF059030

[a]Scn7A may represent the same gene as Scn6A, since only one full-length cDNA has been isolated for these two genes from any single species.[105] SCL11 probably represents the same isoform as Na-G, based on 98% identity to the partial Na-G sequence.[76]

TABLE 2. Mammalian Sodium Channel β Subunits

Name	Gene Symbol	Original Name	Species	Tissue	Size	Chromosome	Genbank Access #
Naβ1.1	SCN1B	β1[33]	Rat	CNS	218	Mouse 7[106]	M91808
		β1[107]	Human	CNS	223	Human 19[108]	L10338
Naβ2.1	SCN2B	β2[34]	Rat	CNS	186	Mouse 9[109]	U37026 U37147
		β2[110]	Human	CNS	186	Human 11[110]	AF007783

tissue and Nav1.5 (Scn5A, SkM2/H1) in embryonic and denervated muscle. The Nav1.5 isoform is also present in heart muscle. Two type 1 isoforms are expressed primarily in the peripheral nervous system (PNS), Nav1.7 (Scn9A, PN1) and Nav1.8 (Scn10A, PN3). Three members of the type 2 family have been characterized, Nav2.1 (Scn6A) and Nav2.3 from heart tissue, and Nav2.2 (Scn7A, Na-G) from astrocytes. A recently identified isoform that was cloned from dorsal root ganglion (DRG) tissue, Nav3.1 (NaN), may represent a third family (type 3). Complementary DNA clones encoding the two accessory β subunits, Navβ1.1 (Scn1B) and Navβ1.2 (Scn2B), have also been isolated (TABLE 2).

TYPE 1 SODIUM CHANNEL α SUBUNITS

Central Nervous System Channels

Most of the sodium channel isoforms are expressed in multiple different tissues, but they will be discussed under the category of the tissue in which they are most abundant. Nav1.1 was originally identified in the CNS,[7] although it has also been found to be expressed at high levels in the PNS.[8] In contrast, the levels of Nav1.2 and Nav1.3 are significantly higher in the CNS than in the PNS.[8] Nav1.6 is the most abundantly expressed channel in the CNS, and it can also be detected in DRG cells.[9] Each of these isoforms is present in neurons[9,10] and glia,[9,11,12] although the function of the channels in glial cells is not well understood.[13]

The genes for the CNS sodium channels have been localized on mouse and human chromosomes (TABLE 1). Nav1.1, Nav1.2, and Nav1.3 are clustered on chromosome 2 in mice and humans. In the mouse, Nav1.2 and Nav1.3 are within 600 kb by physical mapping, and Nav1.1 and Nav1.2 are within 0.7 centimorgan by genetic linkage.[14] Alternative splicing of all four isoforms has been demonstrated.[15–20]

The isoforms in the CNS are present at different times in development, which has been studied most extensively in the rat. Three of the isoforms (Nav1.1, Nav1.2, and Nav1.6) are present at high levels in the adult CNS. Nav1.1 becomes detectable shortly after birth and increases until adulthood, whereas Nav1.2 becomes detectable during embryonic development and reaches maximal levels during adulthood.[21] Nav1.6 is the most abundantly expressed isoform in the CNS during adulthood,[9] although the levels of this channel are actually maximal during late embryonic and early postnatal periods.[22] Levels of Nav1.3 peak at birth, and this isoform becomes undetectable by adulthood.[21]

In the adult CNS, the different isoforms are present in different locations. Nav1.1 is the predominant channel in the caudal regions and the spinal cord, whereas levels of Nav1.2 are highest in the rostral regions.[21,23] There is no rostral-caudal gradient of Nav1.6 mRNA.[9] In the cerebellum, Nav1.1 is detectable in Purkinje cells but not in granule cells, Nav1.2 is expressed in both Purkinje[10] and granule cells,[24] and Nav1.6 is expressed predominantly in granule cells.[9] Nav1.1 is localized in the soma of neurons in a variety of CNS regions, including the hippocampus, cerebellum, and spinal cord, whereas Nav1.2 is axonal in distribution.[25]

Functional differences among the sodium channel isoforms have been inferred from correlations between electrophysiological recordings from native tissues and identification of the isoforms present in those tissues. For example, cerebellar Purkinje cells demonstrate unique persistent and resurgent currents.[26,27] Three different isoforms, Nav1.1, Nav1.2, and Nav1.6, have been detected in these cells,[10,28] and Vega-Saenz de Miera *et al.*[28] suggested that Nav1.1 mediates a transient current in Purkinje neurons, while Nav1.6 mediates a persistent current. Raman *et al.*[27] examined the persistent and resurgent currents in normal Purkinje cells and in cells from mice containing a null mutation for Nav1.6. Both currents were reduced in cells lacking Nav1.6, suggesting that this channel does contribute to persistent and resurgent currents.

One means of determining whether the different sodium channel isoforms mediate distinct conductances is to examine the properties of each isoform in isolation, which is most easily performed using an exogenous expression system. All four of the sodium channel isoforms that have been identified thus far in the CNS have been functionally expressed in exogenous systems (TABLE 1). The electrophysiological properties of the isoforms examined in *Xenopus* oocytes are generally similar, particularly when compared to the great variation observed for the voltage-gated potassium and calcium channels. All of the isoforms demonstrate fast inactivation, are blocked by nanomolar concentrations of TTX, and are modulated by the β1 and β2 subunits.[18,29–34]

There are subtle differences, however. In the absence of β subunits, the Nav1.6 channel inactivates more rapidly than any of the other isoforms,[32] and the Nav1.3 channel inactivates significantly more slowly.[35] Coexpression of the β subunits results in similar inactivation kinetics for Nav1.1, Nav1.2, and Nav1.6,[32] but Nav1.3 inactivates with biphasic kinetics, suggesting only partial modulation by the β_1 subunit.[36] The Nav1.6 isoform has a more positive voltage dependence of activation and a more negative voltage dependence of steady state inactivation compared to Nav1.1 and Nav1.2 in the absence of the β subunits.[32] However, coexpression of the β subunits causes a large hyperpolarizing shift in the voltage dependence of activation for Nav1.6, with no significant effect on the voltage dependence of steady state inactivation. Therefore, the voltage-dependent properties of Nav1.1, Nav1.2, and Nav1.6 are generally similar in the presence of the β subunits.

Nav1.1, Nav1.2, and Nav1.6 differ in the percentage of persistent current. Nav1.2 demonstrates the lowest percentage at all depolarizations, Nav1.1 shows a persistent current that is large at negative potentials and decreases with more positive membrane potentials, and Nav1.6 demonstrates a persistent current that increases linearly with more positive membrane potentials.[32] No resurgent current was detected for any of these three isoforms expressed in *Xenopus* oocytes.[32] The lack of resurgent current might indicate that the channels are processed differently in Purkinje cells, that accessory subunits other than the β subunits are necessary for resurgence, or that different splice variants may be responsible for resurgence.

No human diseases resulting from mutations in any of the CNS sodium channel α subunits have been identified yet, but several mutations in Nav1.6 have been characterized in mice. All of these mutations are recessive, and they cause a variety of symptoms ranging from mild ataxia to dystonia, paralysis and juvenile lethality.[37] These mutations include *med* and *med*^{tg}, both of which result in complete disruption of the Nav1.6 gene,[38–40] and *med*^{jo}, which is a single point mutation of Ala to Thr in the domain III S4-S5 linker.[41] This mutation produces an ataxic phenotype, which is most likely caused by changes in the voltage-dependent properties of the Nav1.6 channel.[41]

Skeletal and Heart Muscle Channels

There are two type 1 sodium channel isoforms present in skeletal and heart muscle, Nav1.4 and Nav1.5. Expression of these two channels has been characterized most extensively in the rat. The Nav1.4 isoform is expressed at high levels in adult rat skeletal muscle, at low levels in neonatal skeletal muscle, and not at all in brain or heart.[42] The Nav1.5 isoform is present at high levels in rat heart, but not in brain, liver, kidney, or uterus.[43,44] Nav1.5 is not observed in adult skeletal muscle, but it is detectable in neonatal skeletal muscle and after denervation of adult muscle.[44] Both isoforms are present in denervated muscle, although the increase in the level of sodium channel mRNA following denervation results from an induction of Nav1.5 expression.[45]

The two muscle sodium channel isoforms can be easily distinguished from one another and from the CNS isoforms on the basis of toxin sensitivity. Sodium channels present in adult skeletal muscle are sensitive to nanomolar concentrations of tetrodotoxin (TTX), like the CNS channels, and they are also sensitive to nanomolar concentrations of μ conotoxin, to which the CNS channels are resistant.[46,47] These sensitivities are observed when the Nav1.4 channel is expressed in an exogenous system.[48,49] Sodium channels expressed in cardiac muscle cells are resistant to nanomolar concentrations of TTX, requiring micromolar concentrations for inhibition.[50] These channels are more sensitive to inhibition by lidocaine than CNS channels.[51] Similar sensitivities are observed when the Nav1.5 channel is expressed in *Xenopus* oocytes.[52,53] The presence of a cysteine instead of an aromatic residue at one site in the pore region of domain I (the TTX resistance site) in Nav1.5 is primarily responsible for the relative resistance to TTX.[54–56] The same substitution is responsible for the greater sensitivity of Nav1.1 to block by cadmium and zinc.[54–57]

The electrophysiological properties of Nav1.4 and Nav1.5 are generally similar to those of the CNS channels, but with some important distinctions. Nav1.5 has a more negative voltage dependence of steady state inactivation than either Nav1.4 or any of the CNS isoforms.[58,59] Nav1.5 also inactivates more rapidly than either Nav1.4 or the CNS isoforms when the α subunit is expressed alone in *Xenopus* oocytes,[53,60] and co-expression of the β_1 subunit does not accelerate inactivation of Nav1.5.[61]

Mutations in Nav1.4 have been shown to cause three human neuromuscular diseases, hyperkalemic periodic paralysis (HYPP), paramyotonia congenita (PMC), and the potassium-aggravated myotonias.[62] HYPP is a disease in which increased levels of serum potassium lead to muscle hypoexcitability and paralysis. PMC patients experience cold-induced weakness and paralysis that is aggravated by increased muscle activity. These diseases are inherited in an autosomal-dominant manner, and result from mutations in many different regions of the channel, each of which causes defects in either voltage-dependent activation or inactivation.[62] Mutations in Nav1.5 have been shown to cause long QT syndrome, which is also inherited in a dominant manner.[63] The mutations that cause long QT syndrome are also located in multiple regions of the channel, and they all cause defects in sodium channel inactivation.[64,65]

Peripheral Nervous System Channels

There are two type 1 sodium channel isoforms that are expressed primarily in the PNS, Nav1.7 and Nav1.8. In addition, Nav1.1[8] and Nav1.6,[9,18] which were discussed previously with the other CNS channels, are present at lower levels in DRG neurons. Nav1.7 is widespread in the PNS, being present in all types of DRG neurons, in Schwann cells, and also in neuroendocrine cells.[66–68] Within neurons, Nav1.7 is localized to the neurite terminals, so that it is likely to have an important role in shaping the action potential.[69] It is expressed in PC12 cells, in which the level is induced by NGF.[70,71] The channel is sensitive to nanomolar concentration of TTX,[66,68] is slowly inactivating in oocytes, and is not modulated by the β_1 or β_2 subunits.[66]

The expression of Nav1.8 is more localized, being found primarily in small-diameter sensory neurons of the DRG and trigeminal ganglion, in which the channel has been observed during both neonatal and adult periods.[72–74] This limited localization may have important clinical significance, because the C fibers that transmit nociceptive impulses are small-diameter neurons, so that Nav1.8 may be involved in pathophysiological pain.[72,73] This channel is resistant to nanomolar concentrations of TTX, because of the presence of a serine rather than an aromatic residue at the TTX resistance site in domain I. It demonstrates slow inactivation in oocytes without modulation by the β subunits.[73]

TYPE 2 SODIUM CHANNEL α SUBUNITS

The type 2 sodium channels represent a distinctly different gene family, in that these sequences are less than 50% identical to those of the type 1 channels.[5,75,76] These isoforms are present at high levels in heart, skeletal muscle, and uterus; at low levels in brain, kidney, and spleen; and not at all in liver or smooth muscle. They have also been detected in astrocytes, suggesting that they may represent glial-specific channels. There are sequence differences in two major regions that have been shown to be critical for normal sodium channel function.[5] First, there are significantly fewer charges in the S4 regions, which are essential for voltage-sensitive gating.[77,78] Second, the interdomain III-IV linker, which is critical for fast inactivation,[79,80] is poorly conserved. However, it is not possible to evaluate the functional significance of these differences, because none of the type 2 channels has been functionally expressed in an exogenous system, despite numerous efforts.[75,76] There are a number of possible reasons for the inability to observe functional currents from any of these channels. The channels may require accessory subunits that have not yet been identified, or the full-length sequences may contain cloning artifacts. It is possible that these sequences represent pseudogenes, although the fact that all three clones contain uninterrupted reading frames makes this hypothesis less likely. It is also possible that the type 2 isoforms do not represent true voltage-gated sodium channels.[76] These three clones may be orthologs of a single type 2 gene, because they were isolated from different species.[105]

TYPE 3 SODIUM CHANNEL α SUBUNITS

The putative third family of sodium channel isoforms is represented by a single member, Nav3.1, which is less than 50% identical to any of the type 1 or type 2 channels.[6] The sequence of this channel is more similar to the type 1 than the type 2 family in the S4 segments and the domain III-IV linker. The number of positive charges in the S4 regions of domains I and IV are comparable to those present in type 1 channels, and the number in domains II and III are intermediate between types 1 and 2. The sequence in the domain III-IV linker includes the residues that are critical for normal inactivation. The Nav3.1 channel is expressed in small fibers (sensory neurons) of the DRG and trigeminal ganglion, and the level of expression is down-regulated after axotomy. The channel has not been expressed in an exogenous system yet, but it is predicted to be resistant to nanomolar concentrations of TTX based on the presence of a serine rather than an aromatic residue at the TTX resistant site in domain I.

SODIUM CHANNEL β SUBUNITS

Many of the mammalian sodium channel α subunits are associated with accessory β subunits *in vivo*. These include channels in the adult CNS, which are associated with both β_1 and β_2, and channels in adult skeletal muscle, which are associated with just β_1.[2] The β_2 subunit is covalently linked to the α subunit by disulfide bonds, whereas the β_1 subunit is noncovalently attached.[81] Complementary DNA clones encoding both of these subunits have been isolated from rats and humans (TABLE 2). The sequences of the two subunits are not homologous, but they both predict proteins with a single membrane-spanning region and an external amino terminus. The β_2 subunit sequence is notable in that it contains an immunoglobulin-like fold that is similar to contactin.[34]

Coexpression of the β_1 subunit with many of the α subunits in *Xenopus* oocytes modulates the electrophysiological properties of the channel, including accelerating inactivation and shifting the voltage dependence of steady state inactivation in the negative direction.[31–33] These effects require the extracellular domain of the β_1 subunit, but not the intracellular domain.[82,83] Coexpression of the β_2 subunit also modulates gating of the α subunit sodium channels, but to a lesser extent than the modulation observed for the β_1 subunit.[34] The β_2 subunit significantly increases membrane capacitance, however, suggesting that it is involved in insertion of the channels into the cellular membrane.[84,85] A mutation in the gene encoding the β_1 subunit (SCN1B) has been associated with generalized epilepsy with febrile seizures.[86]

CONCLUSIONS

The mammalian voltage-gated sodium channels represent an expanding collection of isoforms. The functional properties of the type 1 channels are relatively similar, at least when compared to the diversity of potassium and calcium channels, but they demonstrate important differences in localization and expression. The type 2 and 3 channels differ greatly from the type 1 family, suggesting that these channels have completely different roles *in vivo*, although any indication about their significance awaits functional expression. Because sodium channels have such an important function in the initiation and generation of action potentials, the differences among the type 1 channels are likely to have great physiological significance, and may provide opportunities to develop more specific pharmacological agents.

ACKNOWLEDGMENTS

I thank Dr. Miriam Meisler for assistance with TABLE 1. Work in the author's laboratory is supported by grants from the NIH, AHA, and NARSAD.

REFERENCES

1. CATTERALL, W.A. 1993. Structure and function of voltage-gated ion channels. Trends Neurosci. **16:** 500–506.
2. ISOM, L.L., K.S. DEJONGH & W.A. CATTERALL. 1994. Auxiliary subunits of voltage-gated ion channels. Neuron **12:** 1183–1194.
3. CHANDY, K.G. 1991. Simplified gene nomenclature. Nature **352:** 26
4. GOLDIN, A.L. 1995. Voltage-gated sodium channels. *In* Ligand- and Voltage-Gated Ion Channels, Vol. 2. R.A. North, Ed.: 73–112. CRC Press. Boca Raton, FL.
5. GEORGE, A.L., JR., T.J. KNITTLE & M.M. TAMKUN. 1992. Molecular cloning of an atypical voltage-gated sodium channel expressed in human heart and uterus: Evidence for a distinct gene family. Proc. Natl. Acad. Sci. USA **89:** 4893–4897.
6. DIB-HAJJ, S.D., L. TYRRELL, J.A. BLACK & S.G. WAXMAN. 1998. NaN, a novel voltage-gated Na channel, is expressed preferentially in peripheral sensory neurons and down-regulated after axotomy. Proc. Natl. Acad. Sci. USA **95:** 8963–8968.
7. NODA, M., T. IKEDA, T. KAYANO, H. SUZUKI, H. TAKESHIMA, M. KURASAKI, H. TAKAHASHI & S. NUMA. 1986. Existence of distinct sodium channel messenger RNAs in rat brain. Nature **320:** 188–192.
8. BECKH, S. 1990. Differential expression of sodium channel mRNAs in rat peripheral nervous system and innervated tissues. FEBS Lett. **262:** 317–322.
9. SCHALLER, K.L., D.M. KRZEMIEN, P.J. YAROWSKY, B.K. KRUEGER & J.H. CALDWELL. 1995. A novel, abundant sodium channel expressed in neurons and glia. J.Neurosci. **15:** 3231–3242.
10. BLACK, J.A., S. YOKOYAMA, H. HIGASHIDA, B.R. RANSOM & S.G. WAXMAN. 1994. Sodium channel mRNAs I, II and III in the CNS: Cell-specific expression. Mol. Brain Res. **22:** 275–289.
11. BLACK, J.A., S. YOKOYAMA, S.G. WAXMAN, Y. OH, K.B. ZUR, H. SONTHEIMER, H. HIGASHIDA & B.R. RANSOM. 1994. Sodium channel mRNAs in cultured spinal cord astrocytes: *In situ* hybridization in identified cell types. Mol. Brain Res. **23:** 235–245.
12. OH, Y., J.A. BLACK & S.G. WAXMAN. 1994. The expression of rat brain voltage-sensitive Na^+ channel mRNAs in astrocytes. Mol. Brain Res. **23:** 57–65.
13. SONTHEIMER, H., J.A. BLACK & S.G. WAXMAN. 1996. Voltage-gated Na^+ channels in glia: Properties and possible functions. Trends Neurosci. **19:** 325–331.
14. MALO, D., E. SCHURR, J. DORFMAN, V. CANFIELD, R. LEVENSON & P. GROS. 1991. Three brain sodium channel α-subunit genes are clustered on the proximal segment of mouse chromosome 2. Genomics **10:** 666–672.
15. SCHALLER, K.L., D.M. KRZEMIEN, N.M. MCKENNA & J.H. CALDWELL. 1992. Alternatively spliced sodium channel transcripts in brain and muscle. J. Neurosci. **12:** 1370–1381.
16. SARAO, R., S.K. GUPTA, V.J. AULD & R.J. DUNN. 1991. Developmentally regulated alternative RNA splicing of rat brain sodium channel mRNAs. Nucleic Acids Res. **19:** 5673–5679.
17. AHMED, C.M.I., V.J. AULD, H.A. LESTER, R. DUNN & N. DAVIDSON. 1990. Both sodium channel II and IIA α subunits are expressed in rat brain. Nucleic Acids Res. **18:** 5907.
18. DIETRICH, P.S., J.G. MCGIVERN, S.G. DELGADO, B.D. KOCH, R.M. EGLEN, J.C. HUNTER & L. SANGAMESWARAN. 1998. Functional analysis of a voltage-gated sodium channel and its splice variant from rat dorsal root ganglion. J. Neurochem. **70:** 2262–2272.

19. Plummer, N.W., M.W. McBurney & M.H. Meisler. 1997. Alternative splicing of the sodium channel *SCN8A* predicts a truncated two-domain protein in fetal brain and non-neuronal cells. J. Biol. Chem. **272:** 24008–24015.
20. Plummer, N.W., J. Galt, J.M. Jones, D.L. Burgess, L.K. Sprunger, D.C. Kohrman & M.H. Meisler. 1998. Exon organization, physical mapping, and polymorphic intragenic markers for the human neuronal sodium channel gene SCN8a. Genomics **54:** 287–296.
21. Beckh, S., M. Noda, H. Lübbert & S. Numa. 1989. Differential regulation of three sodium channel messenger RNAs in the rat central nervous system during development. EMBO J. **8:** 3611–3636.
22. Felts, P.A., S. Yokoyama, S. Dib-Hajj, J.A. Black & S.G. Waxman. 1997. Sodium channel α-subunit mRNAs I, II, III, NaG, Na6 and hNE (PN1): Different expression patterns in developing rat nervous system. Mol. Brain Res. **45:** 71–82.
23. Gordon, D., D. Merrick, V. Auld, R. Dunn, A.L. Goldin, N. Davidson & W.A. Catterall. 1987. Tissue-specific expression of the R_I and R_{II} sodium channel subtypes. Proc. Natl. Acad. Sci. USA **84:** 8682–8686.
24. Furuyama, T., Y. Morita, S. Inagaki & H. Takagi. 1993. Distribution of I, II and III subtypes of voltage-sensitive Na^+ channel mRNA in the rat brain. Mol. Brain Res. **17:** 169–173.
25. Westenbroek, R.E., D.K. Merrick & W.A. Catterall. 1989. Differential subcellular localization of the R_I and R_{II} Na^+ channel subtypes in central neurons. Neuron **3:** 695–704.
26. Raman, I.M. & B.P. Bean. 1997. Resurgent sodium current and action potential formation in dissociated cerebellar Purkinje neurons. J. Neurosci. **17:** 4517–4526.
27. Raman, I.M., L.K. Sprunger, M.H. Meisler & B.P. Bean. 1997. Altered subthreshold sodium currents and disrupted firing patterns in Purkinje neurons of *Scn8a* mutant mice. Neuron **19:** 881–891.
28. Vega-Saenz de Miera, E., B. Rudy, M. Sugimori & R. Llinas. 1997. Molecular characterization of the sodium channel subunits expressed in mammalian cerebellar Purkinje cells. Proc. Natl. Acad. Sci. USA **94:** 7059–7064.
29. Noda, M., T. Ikeda, H. Suzuki, H. Takeshima, T. Takahashi, M. Kuno & S. Numa. 1986. Expression of functional sodium channels from cloned cDNA. Nature **322:** 826–828.
30. Auld, V.J., A.L. Goldin, D.S. Krafte, J. Marshall, J.M. Dunn, W.A. Catterall, H.A. Lester, N. Davidson & R.J. Dunn. 1988. A rat brain Na^+ channel α subunit with novel gating properties. Neuron **1:** 449–461.
31. Smith, R.D. & A.L. Goldin. 1998. Functional analysis of the rat I sodium channel in *Xenopus* oocytes. J. Neurosci. **18:** 811–820.
32. Smith, M.R., R.D. Smith, N.W. Plummer, M.H. Meisler & A.L. Goldin. 1998. Functional analysis of the mouse Scn8a sodium channel. J. Neurosci. **18:** 6093–6102.
33. Isom, L.L., K.S. DeJongh, D.E. Patton, B.F.X. Reber, J. Offord, H. Charbonneau, K. Walsh, A.L. Goldin & W.A. Catterall. 1992. Primary structure and functional expression of the β_1 subunit of the rat brain sodium channel. Science **256:** 839–842.
34. Isom, L.L., D.S. Ragsdale, K.S. De Jongh, R.E. Westenbroek, B.F.X. Reber, T. Scheuer & W.A. Catterall. 1995. Structure and function of the β_2 subunit of brain sodium channels, a transmembrane glycoprotein with a CAM motif. Cell **83:** 433–442.
35. Joho, R.H., J.R. Moorman, A.M.J. VanDongen, G.E. Kirsch, H. Silberberg, G. Schuster & A.M. Brown. 1990. Toxin and kinetic profile of rat brain type III sodium channel expressed in *Xenopus* oocytes. Mol. Brain Res. **7:** 105–113.
36. Patton, D.E., L.L. Isom, W.A. Catterall & A.L. Goldin. 1994. The adult rat brain β_1 subunit modifies activation and inactivation gating of multiple sodium channel α subunits. J. Biol. Chem. **269:** 17649–17655.
37. Meisler, M.H., L.K. Sprunger, N.W. Plummer, A. Escayg & J.M. Jones. 1997. Ion channel mutations in mouse models of inherited neurological diseases. Ann. Med. (Helsinki) **29:** 569–574.
38. Kohrman, D.C., N.W. Plummer, T. Schuster, J.M. Jones, W. Jang, D.L. Burgess, J. Galt, B.T. Spear & M.H. Meisler. 1995. Insertional mutation of the motor endplate disease (*med*) locus on mouse chromosome 15. Genomics **26:** 171–177.
39. Burgess, D.L., D.C. Kohrman, J. Galt, N.W. Plummer, J.M. Jones, B. Spear & M.H. Meisler. 1995. Mutation of a new sodium channel gene, *Scn8a*, in the mouse mutant 'motor endplate disease.' Nat. Genet. **10:** 461–465.

40. KOHRMAN, D.C., J.B. HARRIS & M.H. MEISLER. 1996. Mutation detection in the *med* and *med*[J] alleles of the sodium channel *Scn8a*. J. Biol. Chem. **271:** 17576–17581.
41. KOHRMAN, D.C., M.R. SMITH, A.L. GOLDIN, J. HARRIS & M.H. MEISLER. 1996. A missense mutation in the sodium channel Scn8a is responsible for cerebellar ataxia in the mouse mutant *jolting*. J. Neurosci. **16:** 5993–5999.
42. TRIMMER, J.S., S.S. COOPERMAN, W.S. AGNEW & G. MANDEL. 1990. Regulation of muscle sodium channel transcripts during development and in response to denervation. Dev. Biol. **142:** 360–367.
43. ROGART, R.B., L.L. CRIBBS, L.K. MUGLIA, D.D. KEPHART & M.W. KAISER. 1989. Molecular cloning of a putative tetrodotoxin-resistant rat heart Na^+ channel isoform. Proc. Natl. Acad. Sci. USA **86:** 8170–8174.
44. KALLEN, R.G., Z.-H. SHENG, J. YANG, L. CHEN, R.B. ROGART & R.L. BARCHI. 1990. Primary structure and expression of a sodium channel characteristic of denervated and immature rat skeletal muscle. Neuron **4:** 233–242.
45. YANG, J.S.J., J.T. SLADKY, R.G. KALLEN & R.L. BARCHI. 1991. TTX-sensitive and TTX-insensitive sodium channel mRNA transcripts are independently regulated in adult skeletal muscle after denervation. Neuron **7:** 421–427.
46. CRUZ, L.J., W.R. GRAY, B.M. OLIVERA, R.D. ZEIKUS, L. KERR, D. YOSHIKAMI & E. MOCZYDLOWSKI. 1985. *Conus geographus* toxins that discriminate between neuronal and muscle sodium channels. J. Biol. Chem. **260:** 9280–9288.
47. MOCZYDLOWSKI, E., B.M. OLIVERA, W.R. GRAY & G.R. STRICHARTZ. 1986. Discrimination of muscle and neuronal Na-channel subtypes by binding competition between [^{3}H]saxitoxin and μ-conotoxins. Proc. Natl. Acad. Sci. USA **83:** 5321–5325.
48. TRIMMER, J.S., S.S. COOPERMAN, S.A. TOMIKO, J. ZHOU, S.M. CREAN, M.B. BOYLE, R.G. KALLEN, Z. SHENG, R.L. BARCHI, F.J. SIGWORTH, R.H. GOODMAN, W.S. AGNEW & G. MANDEL. 1989. Primary structure and functional expression of a mammalian skeletal muscle sodium channel. Neuron **3:** 33–49.
49. UKOMADU, C., J. ZHOU, F.J. SIGWORTH & W.S. AGNEW. 1992. μI Na^+ channels expressed transiently in human embryonic kidney cells: Biochemical and biophysical properties. Neuron **8:** 663–676.
50. BROWN, A.M., K.S. LEE & T. POWELL. 1981. Voltage clamp and internal perfusion of single rat heart muscle cells. J. Physiol. **318:** 455–477.
51. BEAN, B.P., C.J. COHEN & R.W. TSIEN. 1983. Lidocaine block of cardiac sodium channels. J. Gen. Physiol. **81:** 613–642.
52. CRIBBS, L.L., J. SATIN, H.A. FOZZARD & R.B. ROGART. 1990. Functional expression of the rat heart I Na^+ channel isoform. Demonstration of properties characteristic of native cardiac Na^+ channels. FEBS Lett. **275:** 195–200.
53. WHITE, M.M., L. CHEN, R. KLEINFIELD, R.G. KALLEN & R.L. BARCHI. 1991. SkM2, a Na^+ channel cDNA clone from denervated skeletal muscle, encodes a tetrodotoxin-insensitive Na^+ channel. Mol. Pharmacol. **39:** 604–608.
54. BACKX, P.H., D.T. YUE, J.H. LAWRENCE, E. MARBAN & G.F. TOMASELLI. 1992. Molecular localization of an ion-binding site within the pore of mammalian sodium channels. Science **257:** 248–251.
55. HEINEMANN, S.H., H. TERLAU & K. IMOTO. 1992. Molecular basis for pharmacological differences between brain and cardiac sodium channels. Pflügers Arch. **422:** 90–92.
56. SATIN, J., J.W. KYLE, M. CHEN, P. BELL, L.L. CRIBBS, H.A. FOZZARD & R.B. ROGART. 1992. A mutant of TTX-resistant cardiac sodium channels with TTX-sensitive properties. Science **256:** 1202–1205.
57. CHEN, L.-Q., M. CHAHINE, R.G. KALLEN & R. HORN. 1992. Chimeric study of sodium channels from rat skeletal and cardiac muscle. FEBS Lett. **309:** 253–257.
58. MAKIELSKI, J.C. 1996. The heart sodium channel phenotype for inactivation and lidocaine block. Jpn. Heart J. **37:** 733–739.
59. NUSS, H.B., G.F. TOMASELLI & E. MARBAN. 1995. Cardiac sodium channels (hH1) are intrinsically more sensitive to block by lidocaine than are skeletal muscle (μ1) channels. J. Gen. Physiol. **106:** 1193–1209.

60. SATIN, J., J.W. KYLE, M. CHEN, R.B. ROGART & H.A. FOZZARD. 1992. The cloned cardiac Na channel α-subunit expressed in *Xenopus* oocytes show gating and blocking properties of native channels. J. Membr. Biol. **130:** 11–22.

61. QU, Y., L.L. ISOM, R.E. WESTENBROEK, J.C. ROGERS, T.N. TANADA, K.A. MCCORMICK, T. SCHEUER & W.A. CATTERALL. 1995. Modulation of cardiac Na^+ channel expression in *Xenopus* oocytes by β1 subunits. J. Biol. Chem. **270:** 25696–25701.

62. CANNON, S.C. 1997. From mutation to myotonia in sodium channel disorders. Neuromusc. Disord. **7:** 241–249.

63. KASS, R.S. & M.P. DAVIES. 1996. The roles of ion channels in an inherited heart disease: Molecular genetics of the long QT syndrome. Cardiovasc. Res. **32:** 443–454.

64. WANG, Q., J. SHEN, Z. LI, K. TIMOTHY, G.M. VINCENT, S.G. PRIORI, P.J. SCHWARTZ & M.T. KEATING. 1995. Cardiac sodium channel mutations in patients with long QT syndrome, an inherited cardiac arrhythmia. Hum. Molec. Genet. **4:** 1603–1607.

65. WANG, D.W., K. YAZAWA, A.L. GEORGE, JR. & P.B. BENNETT. 1996. Characterization of human cardiac Na^+ channel mutations in the congenital long QT syndrome. Proc. Natl. Acad. Sci. USA **93:** 13200–13205.

66. SANGAMESWARAN, L., L.M. FISH, B.D. KOCH, D.K. RABERT, S.G. DELGADO, M. ILNIKCA, L.B. JAKEMAN, S. NOVAKOVIC, K. WONG, P. SZE, E. TZOUMAKA, G.R. STEWART, R.C. HERMAN, H. CHAN, R.M. EGLEN & J.C. HUNTER. 1997. A novel tetrodotoxin-sensitive, voltage-gated sodium channel expressed in rat and human dorsal root ganglia. J. Biol. Chem. **272:** 14805–14809.

67. BELCHER, S.M., C.A. ZERILLO, R. LEVENSON, J.M. RITCHIE & J.R. HOWE. 1995. Cloning of a sodium channel α subunit from rabbit Schwann cells. Proc. Natl. Acad. Sci. USA **92:** 11034–11038.

68. KLUGBAUER, N., L. LACINOVA, V. FLOCKERZI & F. HOFMANN. 1995. Structure and functional expression of a new member of the tetrodotoxin-sensitive voltage-activated sodium channel family from human neuroendocrine cells. EMBO J. **14:** 1084–1090.

69. TOLEDO-ARAL, J.J., B.L. MOSS, Z.-J. HE, A.G. KOSZOWSKI, T. WHISENAND, S.R. LEVINSON, J.J. WOLF, I. SILOS-SANTIAGO, S. HALEGOUA & G. MANDEL. 1997. Identification of PN1, a predominant voltage-dependent sodium channel expressed principally in peripheral neurons. Proc. Natl. Acad. Sci. USA **94:** 1527–1532.

70. D'ARCANGELO, G., K. PARADISO, D. SHEPHERD, P. BREHM, S. HALEGOUA & G. MANDEL. 1993. Neuronal growth factor regulation of two different sodium channel types through distinct signal transduction pathways. J. Cell Biol. **122:** 915–921.

71. TOLEDO-ARAL, J.J., P. BREHM, S. HALEGOUA & G. MANDEL. 1995. A single pulse of nerve growth factor triggers long-term neuronal excitability through sodium channel gene induction. Neuron **14:** 607–611.

72. AKOPIAN, A.N., L. SIVILOTTI & J.N. WOOD. 1996. A tetrodotoxin-resistant voltage-gated sodium channel expressed by sensory neurons. Nature **379:** 257–262.

73. SANGAMESWARAN, L., S.G. DELGADO, L.M. FISH, B.D. KOCH, L.B. JAKEMAN, G.R. STEWART, P. SZE, J.C. HUNTER, R.M. EGLEN & R.C. HERMAN. 1996. Structure and function of a novel voltage-gated, tetrodotoxin-resistant sodium channel specific to sensory neurons. J. Biol. Chem. **271:** 5953–5956.

74. SOUSLOVA, V.A., M. FOX, J.N. WOOD & A.N. AKOPIAN. 1997. Cloning and characterization of a mouse sensory neuron tetrodotoxin-resistant voltage-gated sodium channel gene, Scn10a. Genomics **41:** 201–209.

75. FELIPE, A., T.J. KNITTLE, K.L. DOYLE & M.M. TAMKUN. 1994. Primary structure and differential expression during development and pregnancy of a novel voltage-gated sodium channel in the mouse. J. Biol. Chem. **269:** 30125–30131.

76. AKOPIAN, A.N., V. SOUSLOVA, L. SIVILOTTI & J.N. WOOD. 1997. Structure and distribution of a broadly expressed atypical sodium channel. FEBS Lett. **400:** 183–187.

77. STÜHMER, W., F. CONTI, H. SUZUKI, X. WANG, M. NODA, N. YAHAGI, H. KUBO & S. NUMA. 1989. Structural parts involved in activation and inactivation of the sodium channel. Nature **339:** 597–603.

78. YANG, N. & R. HORN. 1995. Evidence for voltage-dependent S4 movement in sodium channels. Neuron **15:** 213–218.

79. PATTON, D.E., J.W. WEST, W.A. CATTERALL & A.L. GOLDIN. 1992. Amino acid residues required for fast sodium channel inactivation. Charge neutralizations and deletions in the III-IV linker. Proc. Natl. Acad. Sci. USA **89:** 10905–10909.
80. WEST, J.W., D.E. PATTON, T. SCHEUER, Y. WANG, A.L. GOLDIN & W.A. CATTERALL. 1992. A cluster of hydrophobic amino acid residues required for fast Na^+ channel inactivation. Proc. Natl. Acad. Sci. USA **89:** 10910–10914.
81. MESSNER, D.J. & W.A. CATTERALL. 1985. The sodium channel from rat brain—separation and characterization of subunits. J. Biol. Chem. **260:** 10597–10604.
82. CHEN, C. & S.C. CANNON. 1995. Modulation of Na^+ channel inactivation by the β_1 subunit: a deletion analysis. Pflügers Arch. **431:** 186–195.
83. MCCORMICK, K.A., L.L. ISOM, D. RAGSDALE, D. SMITH, T. SCHEUER & W.A. CATTERALL. 1998. Molecular determinants of Na^+ channel function in the extracellular domain of the β_1 subunit. J. Biol. Chem. **273:** 3954–3962.
84. SCHMIDT, J.W., S. ROSSIE & W.A. CATTERALL. 1985. A large intracellular pool of inactive Na channel α subunits in developing rat brain. Proc. Natl. Acad. Sci. USA **82:** 4847–4851.
85. SCHMIDT, J.W. & W.A. CATTERALL. 1986. Biosynthesis and processing of the α subunit of the voltage-sensitive sodium channel in rat brain. Cell **46:** 437–445.
86. WALLACE, R.H., D.W. WANG, R. SINGH, I.E. SCHEFFER, A.L. GEORGE, JR., H.A. PHILLIPS, K. SAAR, A. REIS, E.W. JOHNSON, G.R. SUTHERLAND, S.F. BERKOVIC & J.C. MULLEY. 1998. Febrile seizures and generalized epilepsy associated with a mutation in the Na^+-channel β1 subunit gene *SCN1B*. Nat. Genet. **19:** 366–370.
87. MALO, M.S., B.J. BLANCHARD, J.M. ANDRESEN, K. SRIVASTAVA, X.-N. CHEN, X. LI, E.W. JABS, J.R. KORENBERG & V.M. INGRAM. 1994. Localization of a putative human brain sodium channel gene (SCN1A) to chromosome band 2q24. Cytogenet. Cell Genet. **67:** 178–186.
88. LU, C.-M., J. HAN, T.A. RADO & G.B. BROWN. 1992. Differential expression of two sodium channel subtypes in human brain. FEBS Lett. **303:** 53–58.
89. LITT, M., J. LUTY, M. KWAK, L. ALLEN, R.E. MAGENIS & G. MANDEL. 1989. Localization of a human bain sodium channel gene (*SCN2A*) to chromosome 2. Genomics **5:** 204–208.
90. AHMED, C.M.I., D.H. WARE, S.C. LEE, C.D. PATTEN, A.V. FERRER-MONTIEL, A.F. SCHINDER, J.D. MCPHERSON, C.B. WAGNER-MCPHERSON, J.J. WASMUTH, G.A. EVANS & M. MONTAL. 1992. Primary structure, chromosomal localization, and functional expression of a voltage-gated sodium channel from human brain. Proc. Natl. Acad. Sci. USA **89:** 8220–8224.
91. KAYANO, T., M. NODA, V. FLOCKERZI, H. TAKAHASHI & S. NUMA. 1988. Primary structure of rat brain sodium channel III deduced from the cDNA sequence. FEBS Lett. **228:** 187–194.
92. MALO, M.S., K. SRIVASTAVA, J.M. ANDRESEN, X.-N. CHEN, J.R. KORENBERG & V.M. INGRAM. 1994. Targeted gene walking by low stringency polymerase chain reaction: Assignment of a putative human brain sodium channel gene (*SCN3A*) to chromosome 2q24-31. Proc. Natl. Acad. Sci. USA **91:** 2975–2979.
93. AMBROSE, C., S. CHENG, B. FONTAINE, J.H. NADEAU, M. MACDONALD & J.F. GUSELLA. 1992. The α-subunit of the skeletal muscle sodium channel is encoded proximal to Tk-1 on mouse chromosome 11. Mamm. Genome **3:** 151–155.
94. GEORGE, A.L., JR., D.H. LEDBETTER, R.G. KALLEN & R.L. BARCHI. 1991. Assignment of a human skeletal muscle sodium channel α-subunit gene (*SCN4A*) to 17q23.1-25.3. Genomics **9:** 555–556.
95. WANG, J., C.V. ROJAS, J. ZHOU, L.S. SCHWARTZ, H. NICHOLAS & E.P. HOFFMAN. 1992. Sequence and genomic structure of the human adult skeletal muscle sodium channel α subunit gene on 17q. Biochem. Biophys. Res. Commun. **182:** 794–801.
96. GEORGE, A.L., JR., J. KOMISAROF, R.G. KALLEN & R.L. BARCHI. 1992. Primary structure of the adult human skeletal muscle voltage-dependent sodium channel. Ann. Neurol. **31:** 131–137.
97. CHAHINE, M., P.B. BENNETT, A.L. GEORGE, JR. & R. HORN. 1994. Functional expression and properties of the human skeletal muscle sodium channel. Pflügers Arch. **427:** 136–142.
98. GEORGE, A.L., JR., T.A. VARKONY, H.A. DRABKIN, J. HAN, J.F. KNOPS, W.H. FINLEY, G.B. BROWN, D.C. WARD & M. HAAS. 1995. Assignment of the human heart tetrodotoxin-resistant voltage-gated Na^+ channel α-subunit gene (SCN5A) to band 3p21. Cytogenet. Cell Genet. **68:** 67–70.
99. GELLENS, M.E., A.L. GEORGE, JR., L. CHEN, M. CHAHINE, R. HORN, R.L. BARCHI & R.G. KALLEN. 1992. Primary structure and functional expression of the human cardiac tetrodotoxin-insensitive voltage-dependent sodium channel. Proc. Natl. Acad. Sci. USA **89:** 554–558.

100. George, A.L., Jr., J.F. Knops, J. Han, W.H. Finley, T.J. Knittle, M.M. Tamkun & G.B. Brown. 1994. Assignment of a human voltage-dependent sodium channel α-subunit gene (*SCN6A*) to 2q21-q23. Genomics **19:** 395–397.

101. Beckers, M.-C., E. Ernst, S. Belcher, J. Howe, R. Levenson & P. Gros. 1996. A new sodium channel α-subunit gene (*Scn9a*) from Schwann cells maps to the *Scn1a*, *Scn2a*, *Scn3a* cluster of mouse chromosome 2. Genomics **36:** 202–205.

102. Kozak, C.A. & L. Sangameswaran. 1996. Genetic mapping of the peripheral sodium channel genes, *Scn9a* and *Scn10a*, in the mouse. Mamm. Genome **7:** 787–792.

103. Gautron, S., G. Dos Santos, D. Pinto-Henrique, A. Koulakoff, F. Gros & Y. Berwald-Netter. 1992. The glial voltage-gated sodium channel: cell- and tissue-specific mRNA expression. Proc. Natl. Acad. Sci. USA **89:** 7272–7276.

104. Potts, J.F., M.R. Regan, J.M. Rochelle, M.F. Seldin & W.S. Agnew. 1993. A glial-specific voltage-sensitive Na channel gene maps close to clustered genes for neuronal isoforms on mouse chromosome 2. Biochem. Biophys. Res. Commun. **197:** 100–104.

105. Plummer, N.W. & M.H. Meisler. 1999. Evolution and diversity of the mammalian voltage-gated sodium channels. Genomics. In press.

106. Tong, J., J.F. Potts, J.M. Rochelle, M.F. Seldin & W.S. Agnew. 1993. A single β1 subunit mapped to mouse chromosome 7 may be a common component of Na channel isoforms from brain, skeletal muscle and heart. Biochem. Biophys. Res. Commun. **195:** 679–685.

107. McClatchey, A.I., S.C. Cannon, S.A. Slaugenhaupt & J.F. Gusella. 1993. The cloning and expression of a sodium channel β1-subunit cDNA from human brain. Hum. Molec. Genet. **2:** 745–749.

108. Makita, N., K. Sloan-Brown, D.O. Weghuis, H.H. Ropers & A.L. George, Jr. 1994. Genomic organization and chromosomal assignment of the human voltage-gated Na^+ channel β_1 subunit gene (*SCN1B*). Genomics **23:** 628–634.

109. Jones, J.M., M.H. Meisler & L.L. Isom. 1996. *Scn2b*, a voltage-gated sodium channel β2 gene on mouse chromosome 9. Genomics **34:** 258–259.

110. Eubanks, J., N. Srinivasan, M.B. Dinulos, C.M. Disteche & W.A. Catterall. 1997. Structure and chromosomal localization of the β2 subunit of the human brain sodium channel. NeuroReport **8:** 2775–2779.

RNA Editing of a *Drosophila* Sodium Channel Gene

CHRISTOPHER J. HANRAHAN, MICHAEL J. PALLADINO, LAURIE J. BONNEAU, AND ROBERT A. REENAN[a]

Department of Pharmacology, University of Connecticut Health Center, 263 Farmington Avenue, Farmington, Connecticut 06030-6125, USA

ABSTRACT: Extensive analysis of cDNAs from the *para* locus in *D. melanogaster* reveals posttranscriptional modifications indicative of adenosine-to-inosine RNA editing. Most of these edits occur in highly conserved regions of the Na^+ channel, and they occur in distant relatives of *D. melanogaster* as well. Sequence comparison between species has identified putative *cis*-acting elements important for each RNA editing site. Double-stranded RNA secondary structures with striking similarity to known RNA editing sites were generated based on these data. In addition, the RNA editing sites appear to be developmentally regulated. We have cloned a potential RNA editase, *DRED*, with a high degree of homology to the mammalian *RED1,2* genes. The *DRED* locus itself is highly regulated by transcription from alternative promoters and alternative splicings.

The discovery of adenosine-to-inosine (A to I) RNA editing in numerous systems has led to a revision in the concept of linear information flow from genomic DNA template to messenger RNA to final protein product. The biochemical realization of A to I RNA editing came with the discovery of an enzymatic activity that hydrolytically deaminates adenosine to inosine in double-stranded (ds) RNA substrates.[1,2] Such an activity has been demonstrated in a variety of tissues and cell types from both invertebrates and vertebrates (see Ref. 3 for review). Originally, the proposed biological significance of this activity was as a mechanism to regulate RNA:RNA interactions within the cell or as part of a cellular defense against dsRNA viruses. In fact, such modifications are seen in two examples: First, RNA viruses that replicate through a dsRNA intermediate are shown to undergo a relatively nonselective conversion of A to I (~50% A to I conversion), which has also been termed *hypermutation.* A to I RNA editing is seen as A to G transitions in cloned cDNAs synthesized from viral RNAs due to the base pairing preference of inosine, which is like that of guanosine. In synthesizing the first DNA strand, reverse transcriptase incorporates cytosine opposite an inosine, and the cytosine then directs the incorporation of guanosine during second-strand synthesis. A second example of this nonselective or extensive RNA editing is seen in a nonviral example in the *Drosophila 4f-RNP* gene, where extensive A to G (~50%) transitions are seen in both coding and noncoding regions of *4f-RNP* cDNAs.[4]

The role of this extensive mode of RNA editing is unclear. However, other activities have been identified that hint at a complex defense/regulatory role for RNA modification. A cytoplasmic factor has been identified from *Xenopus* oocytes that protects dsRNA from modification by A to I RNA editing, suggesting a mechanism for regulation of modifying

[a]Corresponding author. Phone: 860-679-7665; fax: 860-679-3693; e-mail: rreenan@neuron.uchc.edu

activities.[5] Conversely, a ribonuclease specific for inosine-containing RNAs (I-RNase) has been identified in mammalian nuclear extracts and is proposed to have a role in antiviral defense.[6] This is particularly cogent in light of the observation that an editing activity is inducible by interferons (see below). Another interesting result is the finding that viral antisense RNA along with extensive modification plays a role in the life cycle of polyoma virus.[7] Viral messages are modified during the viral life cycle due to sense-antisense ds-RNA production, and these heavily modified messages are stable and retained in the nucleus, never to be translated. This provides another potential role for dsRNA adenosine deaminases and inosine in regulation of gene expression.

SPECIFIC RNA EDITING OF NEURONAL GENES

In contrast to this ubiquitous nonspecific type of RNA editing are the examples of specific A to I RNA editing. Specific editing of a cellular mRNA was first demonstrated for the *GluR-B* gene of mammals. Initially, it was shown via site-directed mutagenesis that a particular arginine residue (R) in transmembrane domain II (TMII) of *GluR-B* channels was crucial in determining conductance and permeability properties of homo- and heteromeric *GluR* channels.[8] Subsequent studies revealed that genomic DNA encodes a glutamine (CAG) residue, while nearly all cDNAs from *GluR-B* encode arginine (CGG). Since inosine has base pairing properties like those of guanosine (discussed above), the authors proposed RNA editing as a mechanism to explain this nucleotide transition and hypothesized that the presence of inosine in the mRNA would direct the incorporation of a nontemplated amino acid at this position as well as direct the incorporation of G upon replication to generate cDNAs (FIG. 1). Subsequently, it was shown that the related *GluR-5, -6* genes utilize this editing process at the analogous residue as well, the so-called Q/R site.[9] Further diversity has been shown in the *GluR* genes at three other RNA editing sites.

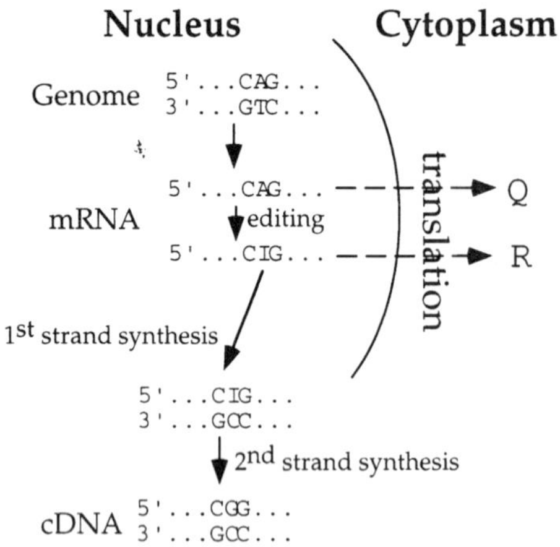

FIGURE 1. The effects of A to I RNA editing. A glutamine codon (CAG) is encoded by the genomic DNA, resulting in transcription of a CAG codon in mRNA. Editing converts the adenosine into inosine; the results of translation in the cytoplasm of the edited and unedited messages are shown. Also depicted are the conversion, *in vitro*, of mRNA into cDNA, showing the incorporation of cytosine opposite inosine and the eventual A to G transition, which results in the ds-cDNA.

The I/V and Y/C editing sites of the *GluR-6* gene in TMI alter the dependence of channel permeation on the Q/R site.[9] The R/G editing site, located near the flip/flop alternative exons of *GluR-B, -C, -D* genes, has been shown to alter the rate of recovery from desensitization.[10] The extent of RNA editing varies for analogous editing sites between the different *GluR* subunit genes. For instance, the Q/R site is edited at efficiencies of >99%, 39%, and 74% for the *GluR-B, -5, and -6* subunit genes, respectively. In addition to subunit differences in levels of *GluR* editing site utilization, editing has been shown to be developmentally regulated. The *GluR-B,C,D* R/G sites were shown to be strikingly regulated with low embryonic levels of editing rising steadily postnatally to adult levels, in the case of *GluR-C* more than an order of magnitude.[10] Other studies have shown comparable developmental changes for all of the *GluR* editing sites for *GluR-B, GluR-5,-6* with low levels early in development rising toward adult levels.[11,12] Spatial differences in editing frequencies have also been reported for the *GluR-5,-6* editing sites in rat brain.[13]

Another instance of specific RNA editing is seen in the hepatitis delta virus (HDV) anti-genomic RNA.[14] In this case, a stop codon (UAG) undergoes editing to generate a tryptophan codon (UGG) allowing readthrough of the delta-antigen gene. This transition is crucial for the life cycle of the virus via opposing roles for the short (unedited) and long (edited) forms of the delta-antigen in viral replication.[15] An activity that induces this modification was found to be present in extracts from *Drosophila* cells suggesting a universal mechanism for A to I editing that is conserved in many eukaryotes.[14]

Recently, two other examples of A to I RNA editing have been demonstrated in neuronally expressed proteins. First, the serotonin-2C G-protein–coupled receptor ($5\text{-HT}_{2C}R$) has been shown to produce seven major receptor isoforms by RNA editing.[16] In this case, four adenosines can be edited in a 13-nucleotide (nt) region (sites A–D) resulting in up to three amino acid changes. Editing in this case was shown to affect the coupling between the $5\text{-HT}_{2C}R$ and G-proteins with editing leading to a 10–15-fold decrease in receptor/G-protein interaction.

The second recent example of RNA editing is from the squid Kv2 potassium channel (*sqKv2*).[17] In *sqKv2*, there are 18 adenosines that are capable of being modified in a region spanning 320bp of the *sqKv2* single exon coding region. Of these sites, 12 result in amino acid substitutions. The editing frequencies for these various editing sites range from ~5%–98%. When two of these edits were engineered into a *sqKv2* genomic (unedited) expression construct, the resulting channels, expressed in *Xenopus* oocytes, were shown to differ in their rates of channel closure and inactivation with respect to the unedited version.

EDITING MECHANISM–dsRNA SECONDARY STRUCTURES

The mechanism by which specific RNA editing occurs was first elucidated for the *GluR-B* Q/R site.[18] Minigene constructs of the Q/R editing site were used to delineate *cis*-acting sequences required for RNA editing when transfected into mammalian cells. This approach revealed a region extending into the intron downstream of the RNA editing site that was essential for high level editing. This region contained an extended imperfect inverted repeat near the editing site and an exonic complementary sequence (ECS) further downstream (~300 bp from the edited adenosine) (FIG. 2). Mutational analysis of the ECS revealed that it must base-pair with the exonic sequences around the edited adenosine and that the sequences flanking the edited adenosine are also important. Similar analyses

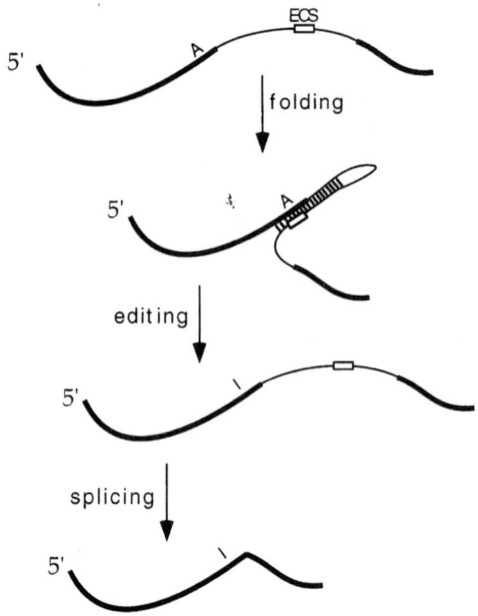

FIGURE 2. Mechanism of A to I RNA editing. Depicted is a primary transcript in the nucleus. *Bold lines* indicate exonic sequences, while *thin lines* represent introns. The primary transcript is thought to form a dsRNA secondary structure intermediate, which brings the exonic complementary sequence (ECS) into register with the editing site (shown by an adenosine nucleotide, A). Action of editases hydrolytically deaminate adenosine to inosine (I). Subsequently, the intron is removed, and the edited transcript leaves the nucleus.

revealed that the imperfect inverted repeat was also crucial for high levels of RNA editing and led to the proposal that the repeat sequences are necessary to juxtapose ECS and editing site sequences. Intron swapping experiments showed that the *GluR-B* Q/R site intron contained sufficient information to direct the editing of the *GluR-C* gene Q/R site.[19] The *GluR-C* gene does not undergo RNA editing even though its exonic sequences are similar to *GluR-B*. Secondary structure modeling of the *GluR-B* Q/R site predicts that the secondary structure formed has significant duplex dsRNA (contributed by the inverted repeat) interrupted in several locations by small stem-loop structures.[20] Though these small stem-loops appear not to be directly involved in the editing site/ECS interaction, deletion of several of these regions abolished or significantly reduced editing levels. Thus, the structural requirements for a functional RNA editing site appear to be complex, rather than simply based upon editing site/ECS hybridization. Another study, of the *GluR-5, -6* gene Q/R sites, revealed that the crucial ECS sequences can be much further away from the editing site (~1900 bp) than the *GluR-B* site (~300 bp).[21]

In contrast to the Q/R site, the *GluR-B* R/G site utilizes an ECS nearby the edited adenosine (~60 bp) in the downstream intron, and both the editing site and ECS are predicted to be contained within a long imperfect inverted repeat.[10] Mutations designed to destabilize this large stem-loop resulted in decreased levels of RNA editing, while mutations that stabilize the structure increased the editing level over that of the wild type.

The specific edit of HDV delta antigen has likewise been shown to require structural elements including dsRNA regions.[22,23] Though originally assumed to be a U to C editing mechanism, further experiments confirmed that HDV editing is of the A to I type. In fact, a mutation of G to A introduced directly 3′ to the edited adenosine gave rise to the recruitment of the introduced adenosine as a substrate for editing, and it was edited at a high frequency.[24]

Thus, all available data suggest that specific A to I RNA editing occurs through a dsRNA intermediate, with distant sequence elements being crucial for forming a secondary structure that encompasses the edited adenosine within a region of dsRNA. The edited adenosine can be base-paired, as in the case of the *GluR-B* Q/R site; mispaired with cytosine, as in the case of the *GluR-B* R/G site; or contained within a small bubble region, as with the *GluR-5,-6* Q/R sites.

EDITING ENZYMES

Several RNA editing enzymes have been cloned that act on dsRNA substrates and hydrolytically deaminate adenosine to inosine (see Ref. 25 for review). Collectively these enzymes are termed ADARs and historically are known as DRADA/dsRAD, RED1, and RED2. DRADA (dsRNA adenosine deaminase) clones were obtained by purifying an extensive A to I editing activity and using protein microsequence to design a degenerate PCR cloning strategy.[26,27] Sequence analysis revealed that DRADA contained a nuclear localization signal, three repeats of a dsRNA binding motif (DSRBM), and an adenosine deaminase domain with residues highly conserved amongst other deaminases. These studies also showed that DRADA is abundant in a variety of tissues including brain thus making it a candidate editase for the specific A to I editing of neuronally expressed genes. DRADA was also cloned and was characterized as an interferon-inducible gene product in keeping with the proposed role of nonspecific editing in antiviral defense.[28,29]

Using low-stringency hybridization to a DRADA catalytic domain probe, investigators obtained, by expression cloning, an enzyme capable of high-level specific editing of the *GluR-B* Q/R editing site and named the enzyme RED1 (RNA editase 1).[30] RED1 was shown to have a domain structure similar to that of DRADA except that RED1 has two DSRBMs rather than three. Enzymatically, RED1 is more efficient at editing the *GluR-B* R/G site and much more efficient at editing the *GluR-B* Q/R site. In addition, RED1 is expressed preferentially in the brain and, thus, represents a likely candidate for specific editing of the *GluR* gene products. Another enzyme, RED2 has been cloned using degenerate PCR based upon highly conserved regions of the DRADA and RED1 catalytic domains.[31] Structurally, RED2 appears highly related to RED1. Unlike the other editases, RED2 appears to be brain specific and contains a unique amino-terminal arginine-rich motif. Suprisingly, RED2 was not found to deaminate adenosine to inosine in either extended dsRNA duplex substrates or *GluR-B* minigene constructs containing the Q/R and R/G sites. However, a chimera composed of the DSRBMs of RED2 and the catalytic domain of RED1 was functional as a dsRNA-specific adenosine deaminase. These results suggest that RED2 is a brain-specific editase whose substrate has not yet been identified.

The editases DRADA and RED1 have been purified from several sources and characterized kinetically.[32,33] Most surprising is the observation that DRADA binds to Z-DNA, which has led to the proposal that DRADA activity or localization is affected by DNA topology near genes targeted for editing.[34] Several groups have shown RNA editing activity in cell-free extracts using *GluR-B* Q/R site synthetic substrates.[20,35–37] While these studies reveal strong similarities to *in vivo* editing, most results suggest that accessory factors play a role in increasing fidelity and extent of editing. It has also been shown that purified DRADA performs the HDV edit *in vitro* and that mutations affecting HDV editing *in vivo* have the same effect on the *in vitro* system.[38] Thus, DRADA is strongly implicated as the enzyme responsible for HDV editing.

RED1 has been purified from HeLa cell extracts based on high-efficiency editing of the *GluR-B* Q/R and R/G sites.[39,40] The activity *in vitro* suggests that RED1 alone is capable of editing both sites and suggests that RED1 may be the major editase acting on the *GluR* sites *in vivo*.

Mutagenic analyses have been carried out on DRADA in order to assess the significance of the various domain structures.[41,42] Not surprisingly, mutations in highly conserved residues within the catalytic domain, which are proposed to coordinate zinc, abolish deaminase activity. Deletion of the first and third DSRBM were found to greatly diminish enzyme activity; however, removal of the second DSRBM had little effect on enzymatic activity. Alternative splicing has been shown to generate different isoforms of both DRADA and RED1. DRADA undergoes alternative splicing near DSRBMs II and III.[43] The various isoforms of DRADA were shown to be equivalent in terms of enzymatic activity. However, the impact of deleting various DSRBMs on enzymatic function was shown to vary depending upon the splice-form into which the deletion mutations were introduced. Similarly, RED1 has been shown to produce alternatively spliced variants. One variant involves alternative splicing of an Alu cassette in the catalytic domain of hRED1.[44] The inclusion of this 120-bp alternative exon (hRED1-L) produces a protein that edits the *GluR-B* Q/R and R/G sites less efficiently than the shorter non–Alu-containing version (hRED1-S). It is proposed that these different forms may play a role in the temporal or cell-specific regulation of editing activity. Another study revealed additional alternative splicing that generates a carboxy-terminal truncated hRED1 protein.[45] The truncated versions of hRED1 were found to possess undetectable levels of editing activity on the known *GluR-B* editing sites. Since these truncated proteins are predicted to bind dsRNA substrates, it was proposed that they may play an inhibitory role in some cells.

The substrate requirements of dsRNA adenosine deaminases in general have been addressed. It has been shown that DRADA has a minimum substrate requirement of 15–20 bp of dsRNA but that efficient modification is seen only when substrate size exceeds 100 bp.[46] From this it was proposed that efficiently edited substrates may be bound by multiple copies of the enzyme. Another study utilized small substrates to probe the substrate specificity of DRADA.[47] In these experiments, DRADA was demonstrated to exhibit a strong 5′ neighbor preference. That is, the frequency of editing depended on the 5′ neighbor of the edited adenosine in the following manner: $A = U > C > G$. This rule is found to apply consistently in most cases of specific and nonspecific editing and agrees with the published *in vivo* data. Given this local substrate preference and the requirement for significant dsRNA secondary structure, the results with the serotonin-2C receptor editing site are quite interesting. It was found that the A and D sites, which are separated by only 11 bp, are efficiently edited by different enzymes.[16] The A site is most efficiently edited by DRADA, while the D site is efficiently edited by RED1. Both the A and D sites have preferred 5′ neighbors. Thus, the overall RNA secondary structure as well as the context of the targeted adenosine strongly influence enzyme-substrate interactions.

EDITING OF THE *para* LOCUS IN *DROSOPHILA*

The *para* locus in *Drosophila* has been shown to encode a voltage-gated Na^+ channel.[48] Coexpression of the para protein along with the product of the *tipE* locus in *Xenopus* oocytes leads to expression of a Na^+ conductance with properties predicted for an invertebrate Na^+ channel.[49] The *para* Na^+ channel locus is the object of intense regulatory control via alternative splicing. There are a minimum of 10 locations in the *para* transcription unit in which alternative splicing generates different channel isoforms (Ref. 50, unpublished results). Such extensive alternative splicing predicts a large number of Na^+ channel isoforms, many of which have been observed in *para* cDNAs. Additionally, this alternative splicing of *para* is developmentally regulated although the functional significance remains unknown.

We have obtained evidence of A to I–type RNA editing in *para* transcripts from *Drosophila melanogaster.* These editing events, like those described in other systems, were discovered as A to G transitions between independent cDNAs. We have discovered a total of four separate sites of RNA editing in the *para* transcript, each of which results in at least one amino acid substitution (FIG. 3). Fortunately, editing at each site generates, eradicates, or both generates and eradicates diagnostic restriction enzyme sites. Thus, isolating cDNAs by RT-PCR and characterizing them by restriction enzyme analysis reveals their editing status and provides an assay for quantifying relative levels of RNA editing at each site.

In order to demonstrate post-transcriptional modification of mRNAs, one must first rule out a number of alternative possibilities, including, polymerase error, cloning artifacts, duplicated genes, and polymorphism. We have cloned the genomic DNA from the region of each editing site by PCR, using the same primers used to amplify cDNAs, and have shown that in each case genomic DNAs encode adenosine. In characterizing editing of cDNAs, we have demonstrated that each editing site has an intrinsic level of modifica-

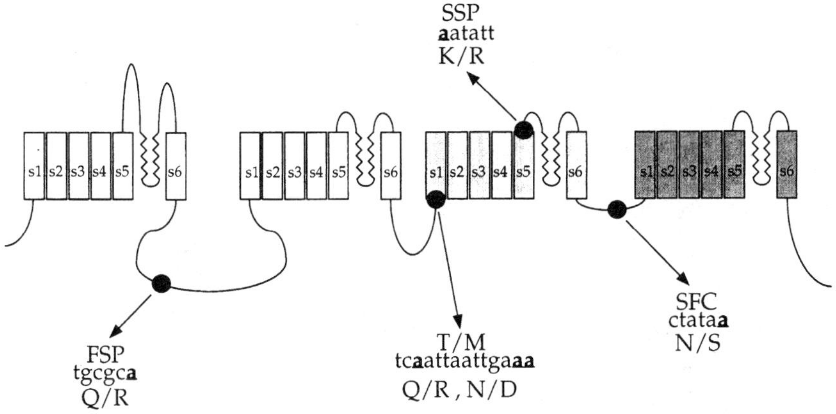

FIGURE 3. *para* RNA editing sites. Shown is the putative topology of the *para* Na^+ channel, with the positions of the identified RNA editing sites shown as *circles*. Each editing site is identified by a particular restriction enzyme site, FSPI, TaqI/MnlI, SSPI, or SFCI. The restriction enzyme site is listed below each site, with edited adenosines indicated in *bold* and *underlined*. Beneath the restriction enzyme recognition sites is given the amino acid change that results (unedited/edited).

tion in cDNAs derived from adult total RNA. For instance, the FSP editing site is edited at about 75% and the SSP site at about 22%. Several editing sites occur near characterized sites of alternative splicing, and we have observed that editing occurs in all splice forms and appears to occur independently of alternative splicing. Moreover, in the case of the SSP and SFC editing sites, we have isolated cDNAs that encompass both editing sites, and each site undergoes editing independently.

All but one of the RNA editing sites are maintained throughout evolution, being conserved in the species *Drosophila virilis*. The estimated time of divergence between *D. virilis* and *D. melanogaster* is placed at 60–80 million years ago.[51] Thus, editing at most sites of the *para* locus has been conserved for a significant evolutionary time. Last, the editing sites differ with respect to the developmental profile of RNA editing. For example, the T/M site is edited at 68% in embryo and 75% in adult cDNAs. In contrast, the SFC site is edited at 2% in embryo and 43% in adult, showing a striking increase in RNA editing between embryo and adult. These results are reminiscent of the developmental regulation of editing of the *GluR*s.

Thus, editing of the *para* locus cannot be explained by any of the above-mentioned artifactual mechanisms. The most likely explanation for these cDNA modifications is A to I–type RNA editing. Editing of the *para* Na^+ channel generates isoforms in addition to and in combination with those generated by alternative splicing and adds yet another level of complexity to *para* regulation.

PUTATIVE *CIS*-ACTING ELEMENTS AT *para* EDITING SITES: EVOLUTIONARY CONSERVATION

The fact that RNA editing is conserved between distant relatives provides a method for identifying *cis*-acting regions that may be important for the process of RNA editing. By analogy with *GluR* editing (Q/R and R/G sites), important *cis*-acting elements may be

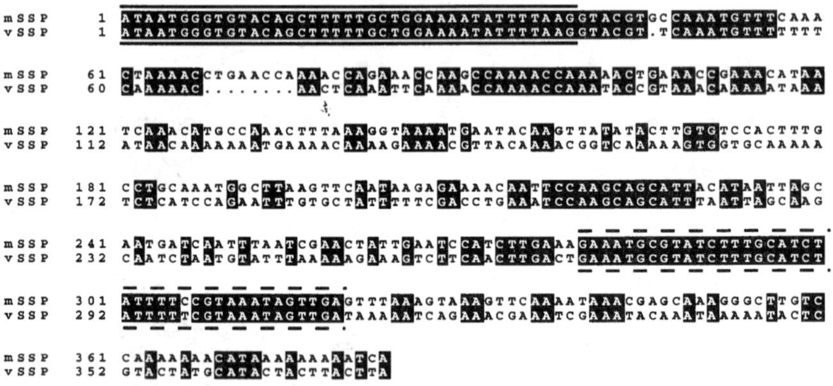

FIGURE 4. The SSP editing site exon and intronic sequences. The sequence alignment shown is for the genomic DNA comparison between *D. melanogaster* **(top)** and *D. virilis* **(bottom)** for the SSP editing site. *Boxing* indicates identity. The *solid lines* bracket exonic sequence. The edited adenosine occurs at nucleotide position 29. *Dashed lines* bracket the highly conserved intronic region, which is proposed to function as an ECS.

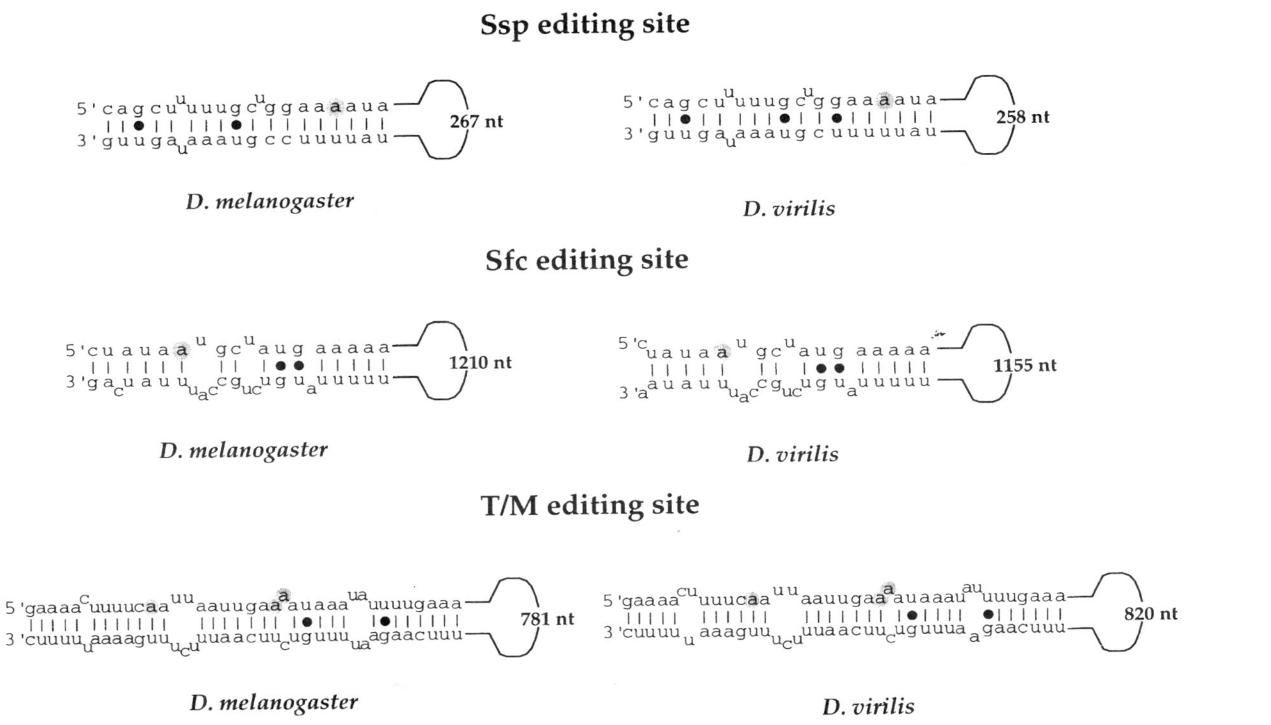

FIGURE 5. Local duplex structures at *para* editing sites. Secondary structures are shown with exonic sequences on the **top strand**, a loop (number indicates distance between edited adenosine and the residue opposite it in the duplex), and intronic sequences on the **bottom strand**. Edited adenosines are highlighted in *circles*. Editied adenosines are either base-paired or mispaired with cytosine.

located in introns near the RNA editing sites. The three conserved sites possess adjacent 3′ introns. So the genomic DNA encompassing the editing exon and at least one downstream intron were obtained, sequenced, and compared for regions of homology. Not surprisingly, exonic sequences were highly conserved (> 95%), while, in general, intronic sequences decreased to a level of about 55% identity at the nucleotide sequence level. However, for introns directly downstream of an exon containing an RNA editing site, we have found islands of high sequence homology between *D. melanogaster* and *D. virilis*. For instance, in the intron downstream of the SSP editing site, a region of 98% identity spanning 40 nt is seen (FIG. 4).

More importantly, when sequences encompassing this conserved region and the exonic editing site are used in predicting RNA secondary structures, a striking similarity to the *GluR* editing sites is seen. The conserved intronic region shows a high degree of complementarity to the exonic sequence surrounding the edited adenosine and is predicted to form an extended dsRNA duplex in this region (FIG. 5). Similar results are obtained with the T/M and SFC sites; namely, conserved regions within downstream introns are capable of base-pairing with the editing region, and these structures locally resemble the *GluR* Q/R and R/G sites as well as the HDV stop/W site. The coincidence of conservation of RNA editing and the presence of conserved intronic sequences capable of forming a base-paired structure at these *para* RNA editing sites is highly reminiscent of the *GluR* gene ECS/editing site mechanism. Therefore, it is likely that a similar mechanism is in place in *Drosophila* and explains specific A to I RNA editing.

THE FSP EDITING SITE: LOSS OR GAIN OF EDITING?

As mentioned earlier, the FSP editing site was not found to be conserved between *D. melanogaster* and *D. virilis*. Cloning of the *D. melanogaster* genomic DNA and subsequent sequence analysis revealed that a secondary structure could be formed in which the edited adenosine is contained within an extended imperfect inverted repeat and that the majority of this structure is exonic (FIG. 6). The local structure near the edited adenosine resembles the *GluR* sites. We cloned the *D.virilis* genomic DNA and found that the edited adenosine was conserved, as were the sequences surrounding the adenosine. However, the local structure formed by the identical base pairs in *D. virilis* has a more open bubble-like structure unlike the *GluR* sites. Moreover, if one uses the same extent of *D. virilis* genomic sequence as that used in forming the global RNA secondary structure, a structure is formed that is threefold less energetically stable in *D. virilis* than in *D. melanogaster*, and the edited adenosine is not contained within a region of dsRNA. We realized the novelty of the situation in the *Drosophilid* species—namely, that we could potentially determine whether the FSP editing site arose in melanogaster and sib-species or was lost from *D. virilis* and related species. Thus, we assayed available *Drosophila* species to determine the extent of editing in the genus *Drosophila* (FIG. 7). Our results indicate that the FSP editing site is present and edited in all members of the *Sophophora* subgenus and is either present but unedited or mutated in the *Drosophila* subgenus. In each case where we have obtained genomic DNA sequence from a member of the *Drosophila* subgenus, secondary structure calculations predict a destabilization of the secondary structure found in *D. melanogaster* (FIG. 6). We have not determined structures from other members of the *Sophophora* subgenus, but the extent of RNA editing at the *Sophophoran* FSP sites are similar to *D. melano-*

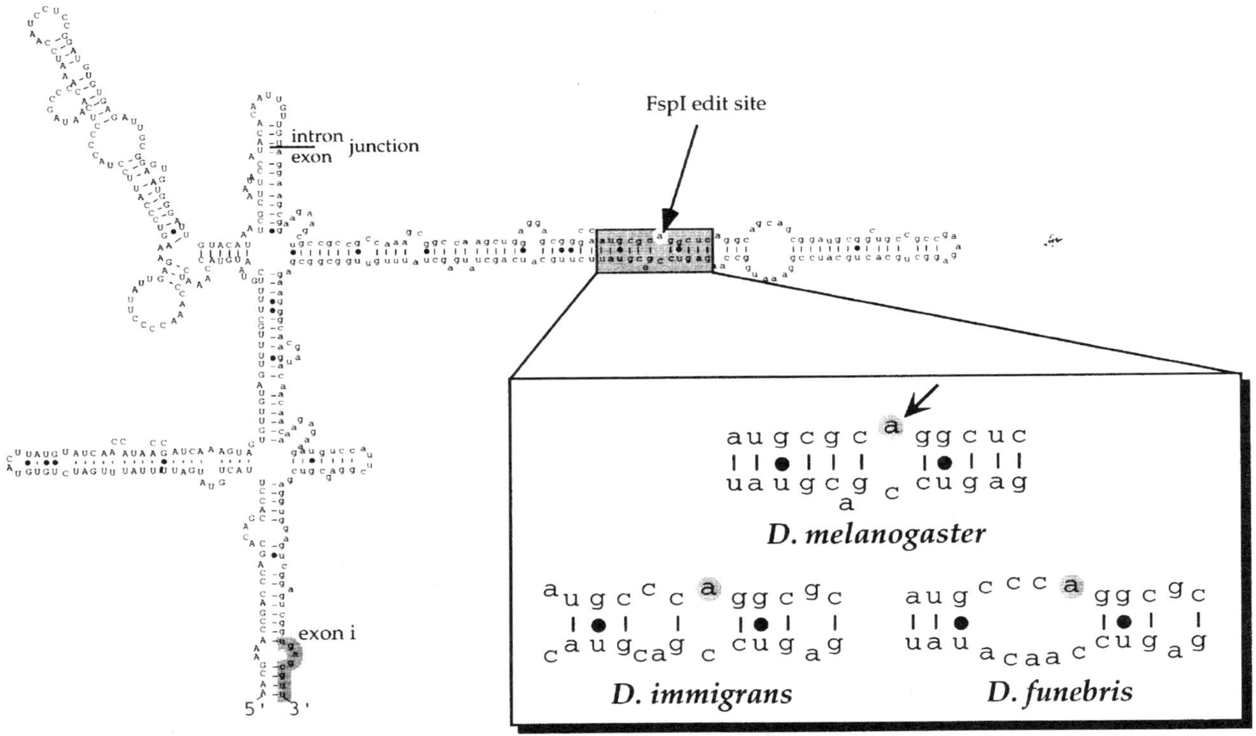

FIGURE 6. The FSP editing site secondary structure and lack of conservation. Shown is the predicted FSP site RNA secondary structure of *D. melanogaster* using the mfold program of the GCG software package. Exonic sequences are in *lower case*, while intron sequences are in *upper case*. **Inset:** the region of the edited adenosine in *D. melanogaster* is shown in comparison to two members of the *Drosophila* subgenus that do not show RNA editing.

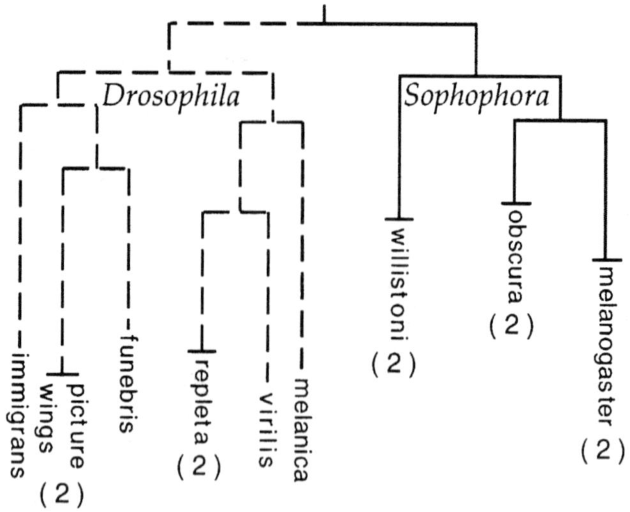

FIGURE 7. Phylogenetic relationships of members of the genus *Drosophila* and FSP site RNA editing. Shown is a phylogenetic tree of genus *Drosophila* branching into the two subgenera *Drosophila* and *Sophophora. Solid lines* indicate presence of RNA editing at the FSP site of *para*, while *dashed lines* indicate absence of editing of FSP. Numbers in *parentheses* indicate number of species tested if from a group of species. All other branches are single-species branches. As can be seen, we detect no editing in any member of the *Drosophila* subgenus. (Adapted from Powell & De Salle.[54])

gaster. Thus, we cannot determine whether the FSP site evolved *de novo* in the *Sophophora* branch or was lost from the *Drosophila* branch. Further studies in other insect genera may help to resolve this issue.

A *DROSOPHILA* EDITASE GENE: *DRED1*

Given the numerous specific RNA editing sites revealed in the *para* locus and the existence of an example of nonspecific A to I RNA editing of the *4f-RNP* gene in *Drosophila*, we proposed to identify a *Drosophila* editase gene. Comparison of the known editases reveals a highly conserved deaminase catalytic domain. Therefore, we designed degenerate PCR primers to several regions in the catalytic domain and obtained a PCR product of the appropriate size using adult polyA$^+$ mRNA as template in RT-PCR. Sequence analysis of the partial cDNA clone revealed a high degree of conservation to both DRADA and RED1 catalytic domains.

Gene-specific primers were designed to this cDNA and used in conjunction with degenerate primers designed to the DSRBMs of either DRADA or RED1. Despite repeated efforts to obtain *Drosophila* homologues of both DRADA and RED1, we were successful in obtaining only a RED1 homologue, which we named *DRED1* (*Drosophila* editase1). Full-length *DRED1* cDNAs were obtained by a modified RACE (rapid amplification of cDNA ends). The overall structure of *DRED1* resembles hRED1 in that it pos-

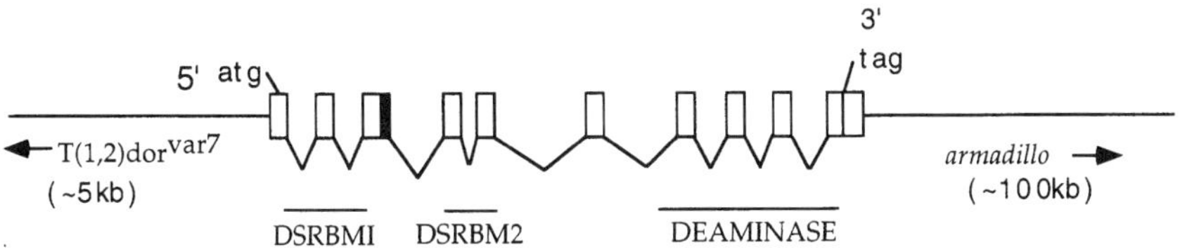

FIGURE 8. Cartoon of the genomic organization of the *DRED1* locus. *Open boxes* indicate constitutive exons. The *filled box* represents an alternative exon between DSRBM1 and DSRBM2. Starting atg and tag stop codon are shown. The direction of transcription is from telomere (T(1,2)dor^{var7}) towards the centromere (*armadillo* locus).

sesses two DSRBMs and a adenosine deaminase catalytic domain. The overall homology of *DRED1* with DRADA is 29% identity, while comparison to hRED1 reveals 41% identity and 55% similarity at the level of protein sequence. Sequence analysis reveals that the *DRED1* locus is distributed into at least 10 exons (Fig. 8), sharing at least several intron/exon boundaries with the hRED1. Interestingly, the *DRED1* locus undergoes alternative splicing between DSRBM1 and 2. Functional studies involving the mammalian enzymes indicate that the spacing between DSRBMs may affect differences in substrate specificity. In the short form of *DRED1*, the DSRBMs have a spacing similar to that of RED1, while in the long form the DSRBMs are spaced like those of DRADA. We have mapped the *DRED1* locus to the distal tip of the X chromosome via *in situ* hybridization to salivary polytene chromosomes. Using ordered P1 bacteriophage clones available from the Berkeley *Drosophila* Genome Project (BDGP), we have placed the *DRED1* transcription unit ~100 kb to the left of the *armadillo* locus and 6 kb to the right of a known translocation breakpoint (T(1,2)dor^{var7}). We are currently continuing our sequence effort in the region to determine the limits of the *DRED1* transcription unit and to identify flanking regulatory sequences.

CONCLUSIONS

We have identified four RNA editing sites in the *Drosophila para* gene, which encodes a voltage-gated Na^+ channel. The preliminary characterization of these editing sites indicates that they are most likely the result of A to I type RNA editing like that observed in mammalian *GluR*s, serotonin-2C receptor, and the squid potassium channel *sqKv2*. Like the mammalian *GluR* editing sites, certain *para* edit sites display a striking developmental regulation, while others appear to be edited at more or less constant levels throughout development. Three out of four editing sites have been conserved in the distant relative of *D. melanogaster, D. virilis.* We have used this conservation to identify *cis*-acting elements necessary for directing RNA editing. In each case, conservation of RNA editing is concomitant with conservation of intronic sequences capable of base-pairing with the exonic region around the editied adenosine. The edited adenosine can be base-paired, as is the case with the SSP, SFC, and positions 1,2 of the T/M sites; or mispaired with a cytosine opposite the edited adenosine, as in the FSP site and position 3 of the T/M site. Both of these local motifs are seen with editing sites in the *GluR*s, and both structures are edited efficiently by the RED1 editase. The 5′ neighbor preference at the *para* editing sites is consistent with the substrate specificity of DRADA and other RNA editing sites: 4/6 edited adenosines at *para* have adenosine as their 5′ neighbor, 2/6 have cytosine. None of the adenosines have guanosine as a 5′ neighbor; it is the least preferred 5′ neighbor. Modification at the FSP editing site was not conserved in the subgenus *Drosophila*. Consistent with this finding is the destabilization of the predicted local secondary structure with respect to *D. melanogaster*, a member of the *Sophophora* subgenus.

We have also cloned a *Drosophila* editase, *DRED1*. The *DRED1* locus encodes a protein with a high degree of homology to mammalian RED1. RED1 has been shown to efficiently and specifically edit the *GluR-B* editing sites. Thus, it is likely that *DRED1* is responsible for editing one or more of the *para* editing sites. Regulation of the *DRED1* protein through the production of alternative splice forms may play an important role in regulation of temporal and spatial aspects of *DRED1* activity as well as substrate specificity.

We are currently analyzing the *para* edit sites through the generation of minigene constructs and *Drosophila* transformation techniques. Through such experiments we hope to confirm the significance of the conserved *cis*-elements identified by evolutionary comparisons. Also, we are pursuing a reverse genetic approach to generating mutants of the *DRED1* locus in order to determine the role of this enzyme in the process of RNA editing in *Drosophila*.

The significance of RNA editing in gene regulation is currently appreciated in light of a large amount of information about a small number of targets and sites. The significance of a single site can best be addressed by the example of the early-onset epilepsy and postnatal lethality phenotypes seen in mice engineered with a Q/R site editing incompetent *GluR-B* gene.[52] However, our picture of the extent of A to I RNA editing in the genome may be predicted to be far from complete. A recent study showed that inosine is present in mRNAs from a variety of mammalian tissues reaching highest levels—one in 17,000 ribonucleotides—in the brain.[53] This suggests that there are many more RNA editing substrates. Thus, A to I RNA editing likely plays an important and central role in generating protein diversity and may well provide a heretofore unrecognized role in inherited disease.

ACKNOWLEDGMENTS

This work was supported by the University of Connecticut Health Center. We wish to thank members of the "fly group" at UCONN Health Center—namely, the laboratories of Steve Helfand and Jo Jack—for helpful discussions. We would also like to thank Barry Ganetzky, in whose laboratory these experiments were begun.

REFERENCES

1. Bass, B.L. & H. Weintraub. 1988. Cell **55:** 1089–1098.
2. Wagner, R.W., J.E. Smith, B.S. Cooperman & K. Nishikura. 1989. Proc. Natl. Acad. Sci. USA **86:** 2647–2651.
3. Bass, B.L. 1997. Trends Biochem. Sci. **22:** 157–162.
4. Petschek, J.P., M.R. Scheckelhoff, M.J. Mermer & J.C. Vaughn. 1997. Gene **240:** 267–276.
5. Saccomanno, L. & B.L. Bass. 1994. Mol. Cell. Biol. **14**(8): 5425–5432.
6. Scadden, A.D.J. & C.W.J. Smith. 1997. EMBO J. **16**(8): 2140–2149.

7. Kumar, M. & G.G. Carmichael. 1997. Proc. Natl. Acad. Sci. USA **94:** 3542–3547.
8. Sommer, B., M. Kohler, R. Sprengel & P.H. Seeburg. 1991. Cell **67:** 11–19.
9. Kohler, M., N. Burnashev, B. Sakmann & P.H. Seeburg. 1993. Neuron **10:** 491–500.
10. Lomeli, H., J. Mosbacher, T. Melcher, T. Hoger, J.R.P. Geiger, T. Kuner, H. Monyer, M. Higuchi, A. Bach & P.H. Seeburg. 1994. **266:** 1709–1713.
11. Lai, F.L., C. Chen, V.M.-Y. Lee & K. Nishikura. 1997. J. Neurochem. **69**(1): 43–52.
12. Paschen, W., J. Schmitt, C. Gissel & E. Dux. 1997. Dev. Brain Res. **98:** 271–280.
13. Belcher, S.M. & J.R. Howe. 1997. Mol. Brain Res. **52:** 130–138.
14. Casey, J.L. & J.L. Gerin. 1995. J. Virol. **69**(12): 7593–7600.
15. Zheng, H., T. Fu, D. Lazinski & J. Taylor. 1992. J. Virol. **66**(8): 4693–4697.
16. Burns, C.M., H. Chu, S.M. Rueter, L.K. Hutchinson, H. Canton, E. Sanders-Bush & R.B. Emeson. 1997. Nature **387:** 303–308.
17. Patton, D.E., T. Silva & F. Bezanilla. 1997. Neuron **19:** 711–722.
18. Higuchi, M., F.N. Single, M. Kohler, B. Sommer, R. Sprengel & P.H. Seeburg. 1993. Cell **75:** 1361–1370.
19. Egebjerg, J., V. Kukekov & S.F. Heinemann. 1994. Proc. Natl. Acad. Sci. USA **91:** 10270–10274.
20. Yang, J., P. Sklar, R. Axel & T. Maniatis. 1995. Nature **374:** 77–81.
21. Herb, A., M. Higuchi, R. Sprengel & P.H. Seeburg. 1996. Proc. Natl. Acad. Sci. USA **93:** 1875–1880.
22. Casey, J.L., K.F. Bergmann, T.L. Brown & J.L. Gerin. 1992. Proc. Natl. Acad. Sci. USA **89:** 7149–7153.
23. Greeve, J., D. Hartwig, E. Windler & H. Greten. 1994. Biochimie (Paris) **76:** 1209–1216.
24. Wu, T.-T., V.V. Bichko, W.-S. Ryu, S.M. Lemon & J.M. Taylor. 1995. J. Virol. **69**(11): 7226–7231.
25. Maas, S., T. Melcher & P.H. Seeburg. 1997. Curr. Opin. Cell. Biol. **9:** 343–349.
26. Kim, U., Y. Wang, T. Sanford, Y. Zeng & K. Nishikura. 1994. Proc. Natl. Acad. Sci. USA **91:** 11457–11461.
27. O'Connell, M.A., S. Krause, M. Higuchi, J.J. Hsuan, N.F. Totty, A. Jenny & W. Keller. 1995. Mol. Cell. Biol. **15**(3): 1389–1397.
28. Patterson, J.B. & C.E. Samuel. 1995. Mol. Cell. Biol. **15**(10): 5376–5388.
29. Patterson, J.B., D.C. Thomis, S.L. Hans & C.E. Samuel. 1995. Virology **210:** 508–511.
30. Melcher, T., S. Maas, A. Herb, R. Sprengel, P.H. Seeburg & M. Higuchi. 1996. Nature **379:** 460–464.
31. Melcher, T., S. Maas, A. Herb, R. Sprengel, M. Higuchi & P.H. Seeburg. 1996. J. Biol. Chem. **271**(50): 31795–31798.
32. O'Connell, M.A. & W. Keller. 1994. Proc. Natl. Acad. Sci. USA **91:** 10596–10600.
33. Kim, U., T.L. Garner, T. Sanford, D. Speicher, J.M. Murray & K. Nishikura. 1994. J. Biol. Chem. **269**(18): 13480–13489.
34. Herbert, A., K. Lowenhaupt, J. Spitzner & A. Rich. 1995. Proc. Natl. Acad. Sci. USA **92:** 7550–7554.
35. Hurst, S.R., R.F. Hough, P.J. Aruscavage & B.L. Bass. 1995. RNA **1:** 1051–1060.
36. Rueter, S.M., C.M. Burns, S.A. Coode, P. Mookherjee & R.B. Emeson. 1995. Science **267:** 1491–1494.
37. Dabiri, G.A., F. Lai, R.A. Drakas & K. Nishikura. 1996. EMBO J. **15**(1): 34–45.
38. Polson, A.G., B.L. Bass & J.L. Casey. 1996. Nature **380:** 454–456.
39. O'Connell, M.A., A. Gerber & W. Keller. 1997. J. Biol. Chem. **272**(1): 473–478.
40. Yang, J., P. Sklar, R. Axel & T. Maniatis. 1997. Proc. Natl. Acad. Sci. USA **94:** 4354–4359.
41. Lai, F., R. Drakas & K. Nishikura. 1995. J. Biol. Chem. **270**(29): 17098–17105.
42. Liu, Y. & C.E. Samuel. 1996. J. Virol. **70**(3): 1961–1968.
43. Liu, Y., C.X. George, J.B. Patterson & C.E. Samuel. 1997. J. Biol. Chem. **272**(7): 4419–4428.
44. Gerber, A., M.A. O'Connell & W. Keller. 1997. RNA **3:** 453–463.
45. Lai, F., C.X. Chen, K.C. Carter & K. Nishikura. 1997. Mol. Cell. Biol. **17**(5): 2413–2424.
46. Nishikura, K., C. Yoo, U. Kim, J.M. Murray, P.A. Estes, F.E. Cash & S.A. Liebhaber. 1991. EMBO J. **10**(11): 3523–3532.
47. Polson, A.G. & B.L. Bass. 1994. EMBO J. **13**(23): 5701–5711.
48. Loughney, K., R. Kreber & B. Ganetzky. 1989. Cell **58:** 1143–1154.

49. Warmke, J.W., R.A.G. Reenan, P. Wang, S. Qian, J.P. Arena, J. Wang, D. Wunderler, K. Liu, G.J. Kaczorowski, L.H.T. Van Der Ploeg, B. Ganetzky & C.J. Cohen. 1997. J. Gen. Physiol. **110:** 119–133.
50. Thackeray, J.R. & B. Ganetzky. 1994. J. Neurosci. **14**(5): 2569–2578.
51. Powell, J.R. 1997. Progress and Prospects in Evolutionary Biology: The *Drosophila* Model. Oxford University Press. New York.
52. Brusa, R., F. Zimmermann, D. Koh, D. Feldmeyer, P. Gass, P.H. Seeburg & R. Sprengel. 1995. Science **270:** 1677–1680.
53. Paul, M.S. & B.L. Bass. 1998. EMBO J. **17**(4): 1120–1127.
54. Powell, J.R. & R. DeSalle. 1995. Evol. Biol. **28:** 87–138.

H^+-Gated Cation Channels[a]

RAINER WALDMANN, GUY CHAMPIGNY, ERIC LINGUEGLIA, JAN R. DE WEILLE, C. HEURTEAUX, AND MICHEL LAZDUNSKI[b]

IPMC-CNRS, 660 route des Lucioles, Sophia Antipolis, 06560 Valbonne, France

ABSTRACT: H^+-gated cation channels are members of a new family of ionic channels, which includes the epithelial Na^+ channel and the FMRFamide-activated Na^+ channel. ASIC, the first member of the H^+-gated Na^+ channel subfamily, is expressed in brain and dorsal root ganglion cells (DRGs). It is activated by pHe variations below pH 7. The presence of this channel throughout the brain suggests that the H^+ might play an essential role as a neurotransmitter or neuromodulator. The ASIC channel is also present in dorsal root ganglion cells, as is its homolog DRASIC, which is specifically present in DRGs and absent in the brain. Since external acidification is a major factor in pain associated with inflammation, hematomas, cardiac or muscle ischemia, or cancer, these two channel proteins are potentially central players in pain perception. ASIC activates and inactivates rapidly, while DRASIC has both a fast and sustained component. Other members of this family such as MDEG1 and MDEG2 are either H^+-gated Na^+ channels by themselves (MDEG1) or modulators of H^+-gated channels formed by ASIC and DRASIC. MDEG1 is of particular interest because the same mutations that produce selective neurodegeneration in *C. elegans* mechanosensitive neurons, when introduced in MDEG1, also produce neurodegeneration. MDEG2 is selectively expressed in DRGs, where it assembles with DRASIC to radically change its biophysical properties, making it similar to the native H^+-gated channel, which is presently the best candidate for pain perception.

H^+-gated cation channels are ligand-gated ion channels activated by the simplest ligand possible. H^+-gated cation channels with different pH sensitivities and kinetics were reported in sensory neurons,[1–7] in neurons of the central nervous system (CNS),[7–9] and in oligodentrocytes.[10] The extracellular pH in a tissue can decrease by more than two pH units during tissue acidosis[11] that accompanies inflammation and many ischemic conditions, and there is very convincing evidence that the sensation of pain parallels the decrease in pH.[12] H^+-gated cation channels in sensory nerve endings were therefore proposed to be involved in the perception of pain that accompanies tissue acidosis.[1,6,11]

The recent cloning of four H^+-gated cation channel subunits showed that they are members of the NaC/DEG superfamily. This family includes (i) epithelial Na^+ channel (ENaC) subunits;[13–17] (ii) a family of proteins designated as degenerins (DEG); after mutations leading to a gain of function, these produce degeneration in mechanoreceptor neurons, which are required for touch sensation in the nematode[18–22] and which could be mechanosensitive channels; and (iii) a peptide-gated Na^+ channel, FaNaC, which is directly gated by the cardioexcitatory peptide FMRFamide in the snail *Helix aspersa*.[23]

The first H^+-gated channel subunit was cloned only recently and designated as ASIC1[24] (for *a*cid *s*ensing *i*onic *c*hannel *1*). Like other members of the NaC/DEG fam-

[a]This paper is dedicated to the memory of our friend and colleague Dr. Guy Champigny, who has contributed in a major way to our current understanding of the properties of this ion channel family.

[b]Corresponding author. Institut de Pharmacologie Moléculaire et Cellulaire, CNRS, 660 route des Lucioles, Sophia Antipolis, 06560 Valbonne, France. Phone: 33 (0)4 93 95 77 02 or 03; fax: 33 (0)4 93 95 77 04; e-mail: ipmc@ipmc.cnrs.fr

ily,[22,25] it has two transmembrane domains with a large extracellular protein component.[24] Like the FaNaC channel, it probably assembles as a tetramer.[26] The heterologously expressed channel activates when the extracellular pH is decreased rapidly from pH 7.4 to acidic pH values below pH 6.9 (FIG. 1). The ASIC1 channel has a conductance of ~14 pS. It is permeable to Na^+ and Li^+ as are other amiloride-sensitive channels of the NaC/DEG family; but it is also permeable to Ca^{2+} ($pNa^+/pLi^+ = 1.3$, $pNa^+/pCa^{2+} = 2.5$). ASIC1 desensitizes rapidly ($\tau_{inact} = 1.4$ s) with a single exponential time course. ASIC1 mRNA is present in both brain and sensory neurons. ASIC1 is blocked by 0.1 to 1-mM concentrations of amiloride ($K_D = 10$ μm) and of its derivatives benzamil and ethylisopropylamiloride (FIG. 2).

A mammalian neuronal degenerin homologue was in fact cloned before ASIC1 and named MDEG1[27] (for *m*ammalian *deg*enerin) or BNC1[28] (for *b*rain *N*a$^+$ *c*hannel *1*). MDEG1 (BNC1) shares 67% sequence identity with ASIC1, and it was demonstrated shortly after the cloning of ASIC1 that MDEG1 is also a H^+-gated cation channel[29,30] with biophysical properties distinct from ASIC1. The MDEG1 channel requires pH values below pH 5.5 for activation, desensitizes slower than ASIC1, and is selective for Na^+ over Ca^{2+}. The MDEG1 mRNA was detected in neurons of the central nervous system and is absent in sensory neurons.

Both ASIC1[24] and MDEG1[30] desensitize within a few seconds during prolonged application of extracellular acid. However, pain associated with tissue acidosis persists until the pH returns to neutral.[12] A biphasic H^+-gated cation current with a sustained component was described in sensory neurons[1] and was proposed to be responsible for the nonadapting pain that accompanies tissue acidosis.[1,11] Recently a new ASIC-type subunit designated as DRASIC (for *DRG a*cid *s*ensing *i*on *c*hannel) was cloned from dorsal root ganglia and displays such biphasic kinetics[31] with a sustained component (FIG. 3). The specific expression

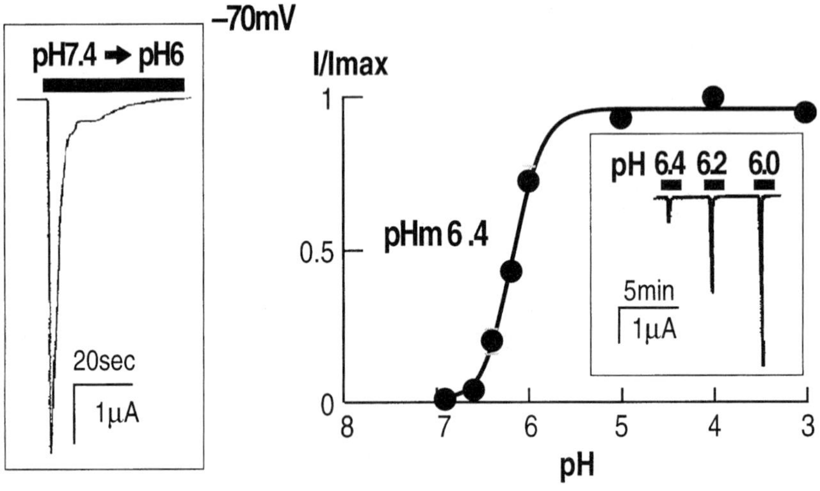

FIGURE 1. ASIC1 is a proton-gated channel involved in acid sensing. **Left:** Kinetics of the H^+-induced Na^+ current expressed in *Xenopus* oocytes. **Right:** pH dependence of the ASIC1 channel.[24]

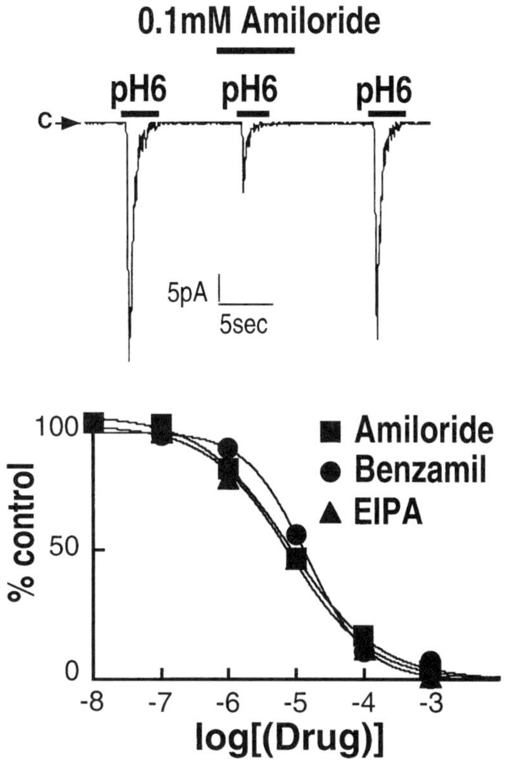

FIGURE 2. The ASIC1 current is blocked by amiloride and derivatives. EIPA = ethylisopropyl-amiloride.[24]

of DRASIC in sensory neurons and the kinetics of the DRASIC channel suggest that it is part of the sustained H^{+}-gated cation channel complex in sensory neurons. However, the sustained DRASIC current requires more acidic pH for activation (< pH 4) than the native H^{+}-gated current in sensory neurons[1] ($pH_{0.5}$ = 5.8), suggesting that a posttranslational modification or associated subunits are required to form the native H^{+}-gated cation channel.

Both ASIC1 and MDEG1 but also MDEG2 (a splice variant of MDEG1) show a widespread distribution pattern in brain.[24,29,32,33] The highest expression levels were found in the cerebellum in Purkinje and granule cells, the CA1-CA3 subfields of the hippocampal formation (dentate granule cells, Hiliar neurons, Pyramidal neurons), the neo- and allocortical regions (Pyramidal neurons), main olfactory bulb, habenula, and basolateral amygdaloid nuclei (see FIG. 4 for ASIC1). The ASIC-like subunits in sensory neurons (ASIC1, MDEG2, DRASIC) are predominantly expressed in small-diameter neurons (FIG. 4).[24,29,31] The different ASIC-like subunits in brain (ASIC1, MDEG1, MDEG2) as

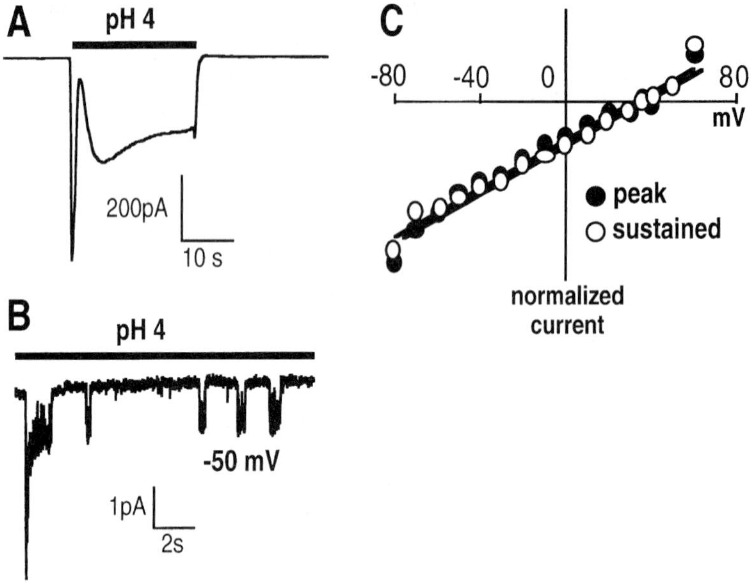

FIGURE 3. The expression of the DRASIC channel in COS cells. **A.** Biphasic kinetics of the channel activity after a pH jump from pH 7.3 to pH 4.[31] **B.** Unitary currents recorded from an outside-out patch held at −50mV.[31] **C.** Current-voltage relations indicating the Na^+ selectivity of both the peak and the sustained components of the DRASIC channel.[31]

well as the ASIC-like subunits in sensory neurons (ASIC1, MDEG2, DRASIC) are apparently coexpressed in the same neurons.[29,33]

Heteromultimeric assembly of subunits is commonly found in the DEG/NaC family of ion channels. The ENaC channel is a heteromultimer of three different types of subunit α, β, and γ,[15] and genetic evidence indicates that the association of the nematode degenerin MEC4 and MEC10 is essential for the mechanosensitive function.[20] Therefore, the apparent colocalization that we observed immediately suggested that ASIC/DRASIC/MDEG1 subunits also form heteromultimeric channels. This was indeed recently demonstrated for several subunit combinations in a variety of ways[29,33] (FIG. 5). For example, MDEG2, a splice variant of MDEG1 in which the first 236 amino acids of MDEG1 are replaced by a novel sequence turned out to be inactive when expressed alone. However coexpression of this splice variant MDEG2 modified the properties of both the DRASIC and MDEG1 channel subunits.[29] The homomultimeric MDEG1 channel has single exponential inactivation kinetics and is highly selective for Na^+. When coexpressed with MDEG2, the inactivation kinetics becomes biphasic, with a very slowly inactivating component that discriminates only poorly between Na^+ and K^+. When MDEG2 is coexpressed with DRASIC, similar selectivity changes were observed, and the sustained Na^+-selective DRASIC current (FIG. 3) become nonselective ($pNa^+ = pK^+$) (FIG. 5). Interestingly, the sustained component of the native biphasic H^+-gated cation channel recorded from sensory neurons

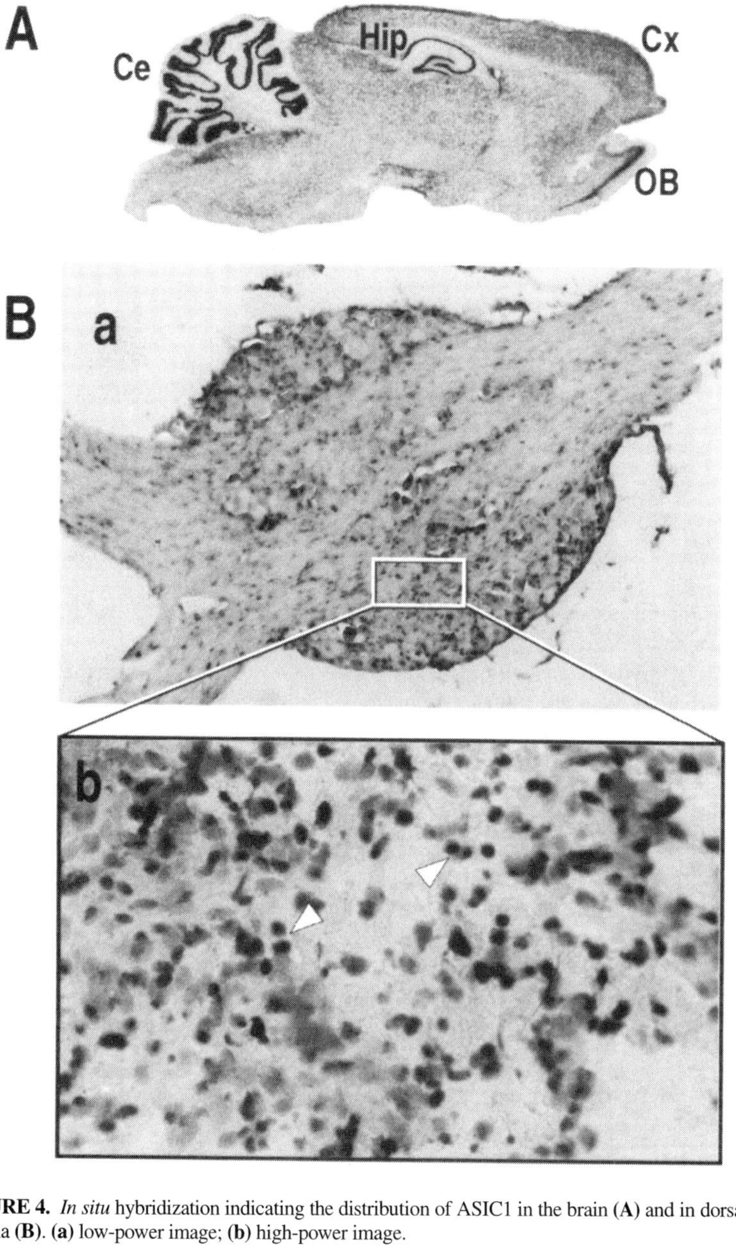

FIGURE 4. *In situ* hybridization indicating the distribution of ASIC1 in the brain **(A)** and in dorsal root ganglia **(B)**. **(a)** low-power image; **(b)** high-power image.

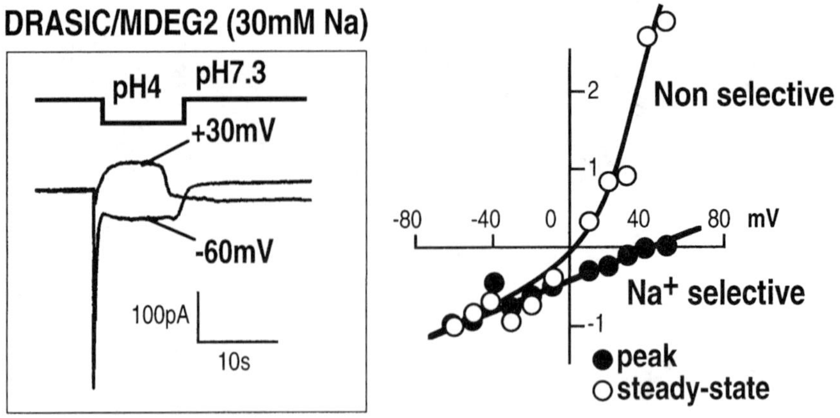

FIGURE 5. Assembly of MDEG2 with the DRASIC subunit changes the ionic selectivity of the sustained component of the current.

also does not discriminate between Na^+ and K^+,[1] suggesting that heteromultimeric DRASIC/MDEG2 channels are formed *in vivo*. Heteromultimeric assembly of ASIC subunits was directly demonstrated for ASIC1 and MDEG1 by coimmunoprecipitation after expression in SF9 cells.[33]

Mutations in the *C. elegans* degenerins cause late-onset neurodegeneration in the nematode.[18–20] Most of those mutations concern a substitution of an alanine situated just before the second transmembrane region for a bulkier amino acid, such as valine or phenylanaline. They were proposed to cause gain-of-function of the putative degenerin channel.[20] Interestingly, identical mutations introduced into ASIC and MDEG1 subunits can also cause constitutive channel activity (shown for ASIC1[33] and MDEG1[27]) as well as cell death.[27] Gain-of-function of the *C. elegans* degenerins causes neurodegeneration. Constitutively active ASIC/MDEG1 subunits have the potential to kill neurons, and it is possible that ASIC/MDEG1 mutations are involved in human forms of neurodegeneration just as mutations of their *C. elegans* homologues causes neuronal death in the nematode.

Beside creating constitutive Na^+ channel activity, mutations in MDEG1 corresponding to those in the nematode degenerins sequences (FIG. 6) also drastically change the gating properties of the MDEG1 channel. While the MDEG1 channel is activated only at relatively acidic pH ($pH_{0.5}$ = 4.1),[30] mutations replacing the key Gly residue situated just

→

FIGURE 6. Mutations that lead to neurodegeneration of mechanosensitive neurons in *C. elegans* lead to drastic changes of both the kinetics and pH dependence of the channel when introduced in MDEG1. **A. Higher part,** comparative sequences of the *C. elegans* degenerin MEC4 and MDEG1 just before the second transmembrane domain MII and identification of the Ala residue, which is mutated in nematode degenerins, and of the corresponding Gly_{430} residue in MDEG1. **Lower part,** kinetic changes observed when replacing Gly_{430} by other amino acids with bulky side chains. WT: wild type. **B.** Changes in the pH dependence of the MDEG1 channel after replacement of Gly430 by Ser, Cys, Phe, Thr, and Val residues.

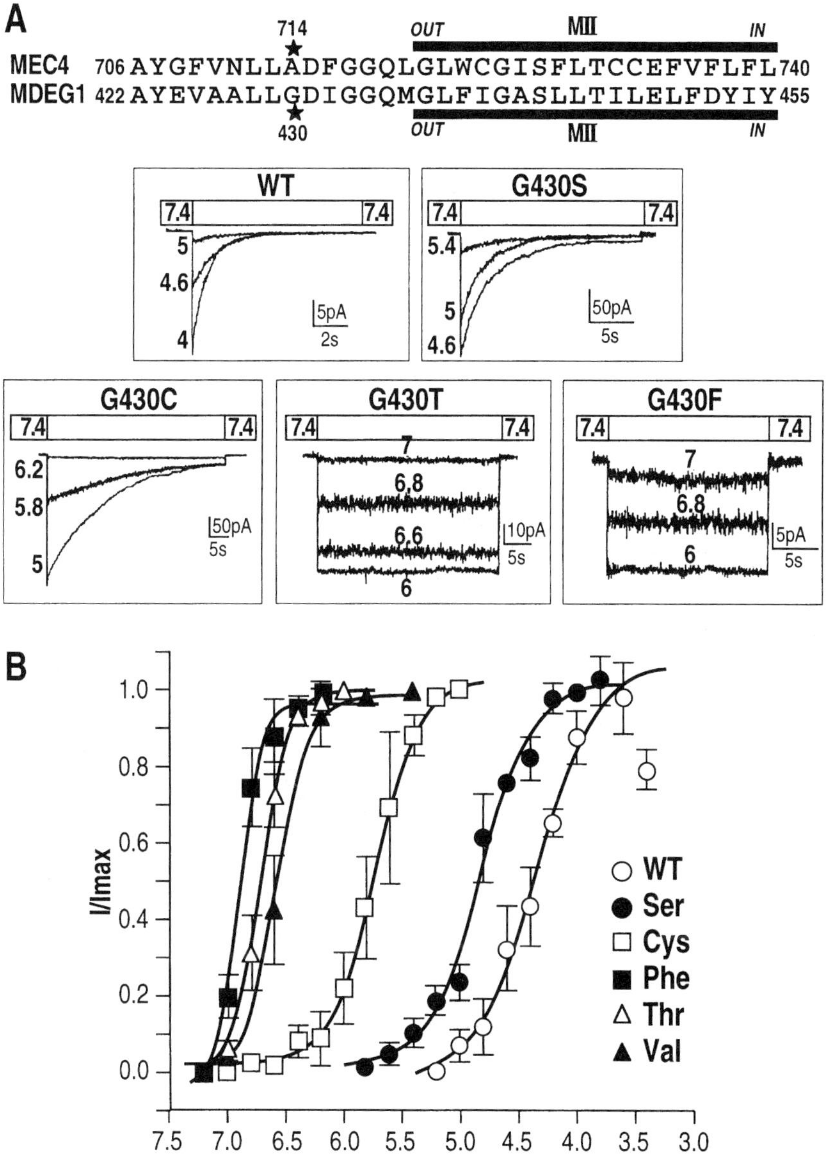
A
714
OUT
MII
IN
MEC4 706 AYGFVNLLADFGGQLGLWCGISFLTCCEFVFLFL 740
MDEG1 422 AYEVAALLGDIGGQMGLFIGASLLTILELFDYIY 455
430
OUT
MII
IN
WT
G430S
G430C
G430T
G430F
5pA
2s
50pA
5s
10pA
5pA
B
I/Imax
WT
Ser
Cys
Phe
Thr
Val

before the second transmembrane domain (FIG. 6) by a Val, a Phe, or any residue bulkier than Gly drastically alters the inactivation of the channel. It can even completely abolish it, and the channel then becomes permanently open after an acidification (FIG. 6). These mutations also drastically shift the pH dependence of the MDEG1 channel to much higher pH values (FIG. 6). The replacement of Gly_{430} by a Phe shifts the $pH_{0.5}$ value from pH 4.1 to pH 6.9. Clearly the protein domain situated just before the second transmembrane segment is essential for the gating of H^+-activated Na^+ channels.

H^+-gated cation channels are acid sensors, and this family of ion channels probably fulfills this role throughout the body. Tissue acidosis causes pain,[12] and an involvement of H^+-gated cation channels in nociception seems most likely. However, nociception cannot be the only role of H^+-gated cation channels, since most of the ASIC subunits cloned so far are also abundant in the CNS, and some of the subunits such as MDEG1 are only expressed there. The ASIC subunits in the CNS are highly conserved between species. ASIC2 protein sequences from rat and human are 99% identical, and this extreme conservation suggests a crucial role in neuronal function. ASIC channels are ligand-gated cation channels with biophysical properties (kinetics, ion selectivities) similar to those of other neuronal ligand-gated cation channels such as the ionotropic, purinergic, or NMDA receptors. The H^+-gated channels in the CNS require acidic pH fluctuations for activation; and some of them, such as the CNS-specific MDEG1 subunit ($pH_{0.5} = 4.1$)[29,30] or the native H^+-gated Na^+ channel described in hypothalamic neurons[8] ($pH_{0.5} = 4.9$), need a quite acidic pH for activity. This imposes the question whether the proton is the physiological activator in the CNS or whether there are other, yet-unidentified ligands. The proton is the only known activator so far, and it will indeed be a physiological activator if pH fluctuations that occur in the CNS are sufficiently rapid and acidic to activate the ASIC channels. ASIC1 starts to open below pH ~7,[24] a value that is very close to physiological pH. On the other hand, neuronal activity is known to be associated with pH fluctuations[34,35] that could probably activate ASIC1. It is not clear however whether pH fluctuations acidic enough to activate MDEG1 and the heteromultimeric ASIC1/MDEG1 channel are really taking place. Neurons dispose of mechanisms that could produce localized and rapid acidic pH fluctuations. Synaptic vesicles are acidic inside,[36,37] and a massive release of the content of those acidic vesicles associated with repetitive stimulation could probably provide the microenvironment for the activation of ASIC1 channels in the CNS. On the other hand, the synaptic degradative hydrolysis of neurotransmitters such as acetylcholine or ATP will also produce protons. Recordings with extracellular electrodes have clearly shown extracellular pH shifts (0.1–0.3 pH units) during sustained electrical activity in different regions of the nervous system.[34] However, the extent of the synaptic pH changes caused by synaptic activity is not yet known, but it is expected to be potentially larger. A closer examination of very local pH fluctuations in the CNS will be necessary to understand if there is enough acidity in specific regions of the CNS to activate the different ASIC-like channels.

Large changes in extracellular acidity are produced in the brain in the course of ischemia and epileptic seizures. Therefore, this class of ASIC-type channels will certainly be activated in these pathophysiological conditions. This activation would be expected to produce deleterious effects, including both cellular depolarization and a significant contribution to the well-known massive Na^+ entry, especially in ischemia, when the (Na^+, K^+) ATPase will be less active in pumping Na^+ out because of intracellular ATP depletion.

Blockers of the H^+-gated cation channels that are more specific than amiloride would be important in studying the role of those channels both in pain perception and in physiological and pathophysiological brain functions. Such specific inhibitors are not yet available, but the search for such blockers will be greatly facilitated with the availability of cDNA clones. Compared with the "classical" ligand-gated cation channels, the investigation of H^+-gated cation channels is still at an early stage. The idea that neurons might use the simplest ligand possible for cell-to-cell communication is attractive, and it is possible that the proton will have to be promoted into the role of an important neurotransmitter or neuromodulator.

ACKNOWLEDGMENTS

This work was supported by the Centre National de la Recherche Scientifique (CNRS), the Association Française contre les Myopathies (AFM) and the Association pour la Recherche sur le Cancer (ARC).

REFERENCES

1. BEVAN, S. & J. YEATS. 1991. Protons activate a cation conductance in a sub-population of rat dorsal root ganglion neurons. J. Physiol. **433:** 145–161.
2. KRISHTAL, O.A. & V.I. PIDOPLICHKO. 1981. Receptor for protons in the membrane of sensory neurons. Brain Res. **214:** 150–154.
3. AKAIKE, N., O.A. KRISHTAL & T. MARUYAMA. 1990. Proton-induced sodium current in frog isolated dorsal root ganglion cells. J. Neurophysiol. **63:** 805–813.
4. KOVALCHUK, YU.N., O.A. KRISHTAL & M.C. NOWYCKY. 1990. The proton-activated inward current of rat sensory neurons includes a calcium component. Neurosci. Lett. **115:** 237–242.
5. DAVIES, N.W., H.D. LUX & M. MORAD. 1988. Site and mechanism of activation of proton-induced sodium current in chick dorsal root ganglion neurons. J. Physiol. **400:** 159–187.
6. KRISHTAL, O.A. & V.I. PIDOPLICHKO. 1981. A receptor for protons in the membrane of sensory neurons may participate in nociception. Neuroscience **6:** 2599–2601.
7. AKAIKE, N. & S. UENO. 1994. Proton induced current in neuronal cells. Prog. Neurobiol. **43:** 73-83.
8. UENO, S., T. NAKAYE & N. AKAIKE. 1992. Proton-induced sodium current in freshly dissociated hypothalamic neurons of the rat. J. Physiol. **447:** 309–327.
9. GRANTYN, R. & H.D. LUX. 1988. Similarity and mutual exclusion of NMDA- and proton-activated transient Na^+-currents in rat tectal neurons. Neurosci. Lett. **89:** 198–203.
10. SONTHEIMER, H., M. PEROUANSKY, D. HOPPE, H.D. LUX, R. GRANTYN & H. KETTENMANN. 1989. Glial cells of the oligodentrocyte lineage express proton-activated Na^+ channels. J. Neurosci. Res. **24:** 496–500.
11. REEH, P.W. & K.H. STEEN. 1996. Tissue acidosis in nociception and pain. Prog. Brain Res. **113:** 143–151.
12. STEEN, K.H., U. ISSBERNER & P.W. REEH. 1995. Pain due to experimental acidosis in human skin: Evidence for non-adapting nociceptor excitation. Neurosci. Lett. **199:** 29–32.
13. CANESSA, C.M., J.D. HORISBERGER & B.C. ROSSIER. 1993. Epithelial sodium channel related to proteins involved in neurodegeneration. Nature **361:** 467–470.
14. LINGUEGLIA, E., N. VOILLEY, R. WALDMANN, M. LAZDUNSKI & P. BARBRY. 1993. Expression cloning of an epithelial amiloride-sensitive Na^+ channel. A new channel type with homologies to Caenorhabditis elegans degenerins. Febs Lett. **318:** 95–99.
15. CANESSA, C.M., L. SCHILD, G. BUELL, B. THORENS, I. GAUTSCHI, J.D. HORISBERGER & B.C. ROSSIER. 1994. Amiloride-sensitive epithelial Na^+ channel is made of three homologous subunits. Nature **367:** 463–467.

16. Lingueglia, E., S. Renard, R. Waldmann, N. Voilley, G. Champigny, H. Plass, M. Lazdunski & P. Barbry. 1994. Different homologous subunits of the amiloride-sensitive Na^+ channel are differently regulated by aldosterone. J. Biol. Chem. **269:** 13736–13739.
17. Waldmann, R., G. Champigny, F. Bassilana, N. Voilley & M. Lazdunski. 1995. Molecular cloning and functional expression of a novel amiloride-sensitive Na^+ channel. J. Biol. Chem. **270:** 27411–27414.
18. Chalfie, M. & E. Wolinsky. 1990. The identification and suppression of inherited neurodegeneration in Caenorhabditis elegans. Nature **345:** 410–416.
19. Driscoll, M. & M. Chalfie. 1991. The Mec-4 gene is a member of a family of *Caenorhabditis elegans* genes that can mutate to induce neuronal degeneration. Nature **349:** 588–593.
20. Huang, M. & M. Chalfie. 1994. Gene interactions affecting mechanosensory transduction in *Caenorhabditis elegans.* Nature **367:** 467–470.
21. Tavernarakis, N., W. Shreffler, S. Wang & M. Driscoll. 1997. unc-8, a Deg/EnaC family member, encodes a subunit of a candidate mechanically gated channel that modulates C. elegans locomotion. Neuron **18:** 107–119.
22. Lai, C.C., K. Hong, M. Kinnell, M. Chalfie & M. Driscoll. 1996. Sequence and transmembrane topology of MEC-4, an ion channel subunit required for mechanotransduction in Caenorhabditis elegans. J. Cell. Biol. **133:** 1071–1081.
23. Lingueglia, E., G. Champigny, M. Lazdunski & P. Barbry. 1995. Cloning of the amiloride-sensitive FMRFamide peptide-gated sodium channel. Nature **378:** 730–733.
24. Waldmann, R., G. Champigny, F. Bassilana, C. Heurteaux & M. Lazdunski. 1997. A proton gated cation channel involved in acid sensing. Nature **386:** 173–177.
25. Renard, S., E. Lingueglia, N. Voilley, M. Lazdunski & P. Barbry. 1994. Biochemical analysis of the membrane topology of the amiloride-sensitive Na^+ channel. J. Biol. Chem. **269:** 12981–12986.
26. Coscoy, S., E. Lingueglia, M. Lazdunski & P. Barbry. 1998. The Phe-Met-Arg-Phe-amide activated sodium channel is a tetramer. J. Biol. Chem. **273:** 8317–8322.
27. Waldmann, R., G. Champigny, N. Voilley, I. Lauritzen & M. Lazdunski. 1996. The Mammalian degenerin MDEG, an amiloride-sensitive cation channel activated by mutations causing neurodegeneration in *C. elegans.* J. Biol. Chem. **271:** 10433–10436.
28. Price, M.P., P.M. Snyder & M.J. Welsh. 1996. Cloning and expression of a novel human brain Na^+ channel. J. Biol. Chem. **271:** 7879–7882.
29. Lingueglia, E., J.R. De Weille, F. Bassilana, C. Heurteaux, H. Sakai, R. Waldmann & M. Lazdunski. 1997. A modulatory subunit of acid sensing ion channels in brain and dorsal root ganglion cells. J. Biol. Chem. **272:** 29778–29783.
30. Champigny, G., R. Voilley, R. Waldmann & M. Lazdunski. 1998. Mutations causing neurodegeneration in *Caenorhabditis elegans* drastically alter the pH sensitivity and inactivation of the mammalian H^+-gated Na^+ channel MDEG1. J. Biol. Chem. **273:** 15418–15422.
31. Waldmann, R., F. Bassilana, J. De Weille, G. Champigny, C. Heurteaux & M. Lazdunski. 1997. Molecular cloning of a non-inactivating proton-gated Na^+ channel specific for sensory neurons. J. Biol. Chem. **272:** 20975–20978.
32. Garcia-Anoveros, J., B. Derfler, J. Neville-golden, B.T. Hyman & D.P. Corey. 1997. BNaC1 and BNaC2 constitute a new family of human neuronal sodium channels related to degenerins and epithelial sodium channels. Proc. Natl. Acad. Sci. USA **94:** 1459–1464.
33. Bassilana, F., G. Champigny, R. Waldmann, J.R. De Weille, C. Heurteaux & M. Lazdunski. 1997. The acid-sensitive ionic channel subunit ASIC and the mammalian degenerin MDEG form a heteromultimeric H^+-gated Na^+ channel with novel properties. J. Biol. Chem. **272:** 28819–28822.
34. Chesler, M. & K. Kaila. 1992. Modulation of pH by neuronal activity. Trends Neurosci. **15:** 396–402.
35. Krishtal, O.A., Y.V. Osipchuk, T.N. Shelest & S.V. Smirnoff. 1987. Rapid extracellular pH transients related to synaptic transmission in rat hippocampal slices. Brain Res. **436:** 352–356.
36. Nguyen, M.L. & S.M. Parsons. 1995. Effects of internal pH on the acetylcholine transporter of synaptic vesicles. J. Neurochem. **64:** 1137–1142.
37. Wolosker, H., D.O. De Souza & L. De Meis. 1996. Regulation of glutamate transport into synaptic vesicles by chloride and proton gradient. J. Biol. Chem. **271:** 11726–11731.

Identification of Voltage-Activated Na^+ and K^+ Channels in Human Steroid-Secreting Ovarian Cells

A. BULLING,[a] C. BRUCKER,[b] U. BERG,[b] M. GRATZL,[a] AND A. MAYERHOFER[a,c]

[a]*Anatomisches Institut der Technischen Universität München, Biedersteiner Str. 29, D-80802 München, Germany*

[b]*Frauenklinik der Ludwig Maximilians-Universität, Maistr. 11, D-80333 München, Germany*

Cells communicate via extracellular molecules, including neurotransmitters, hormones, and growth factors. Depending on their nature, distinct signal transduction mechanisms exist, including receptor-mediated generation of second messengers (e.g., cAMP, cGMP), phosphorylation/dephosphorylation of receptors and channels, or changes in intracellular Ca^{2+} concentrations. Increased intracellular Ca^{2+} levels result from their release from intracellular Ca^{2+} stores or from the influx of extracellular Ca^{2+} through channels activated by the binding of a variety of ligands, or via voltage-activated Ca^{2+} channels.

Different types of cells have specific receptor types and signal transduction mechanisms tailored to their needs. In neurons, fast changes in membrane potential, carried along dendritic and axonal processes, are brought about mainly by activation of voltage-activated Na^+ and K^+ channels. In nerve terminals, arriving action potentials lead to subsequent activation of Ca^{2+} channels and finally to Ca^{2+} triggered release of neurotransmitters.

In aminergic and peptidergic endocrine cells, similar mechanisms underlies the exocytosis of hormones.[1] Thus their cell membranes contain voltage-activated Na^+, K^+, and Ca^{2+} channels. In contrast, channels, receptors, and transmembrane signaling in steroid-producing cells have not been studied very often so it is unknown whether these cells contain such a set of channels. There is evidence, however, that voltage-activated K^+ and Ca^{2+} channels are present in avian and porcine ovarian cells.[2,3] During the last few years we have focused our attention on a human steroidogenic ovarian cell type, namely luteinized granulosa cells (GCs).

Human GCs are obtained from *in vitro* fertilization patients, and can be cultivated for several days and used for functional studies. Like their counterparts in the corpus luteum, these cells produce progesterone, which is crucial for maintaining pregnancy. As shown previously, these cells possess receptors for various hormones and neurotransmitters (e.g., oxytocin, relaxin, catecholamines, acetylcholine[4–7]), and activation of these receptors is linked to increased Ca^{2+}, which is mainly, though not exclusively derived from intracellular stores. In an attempt to further characterize the signal transduction machinery of these cells, we have now started to examine their as yet unknown electrophysiological properties.

[c]Corresponding author. Phone: +49 89 4140 3150; fax: +49 89 397035; e-mail: Mayerhofer@lrz.tu-muenchen.de

RESULTS AND DISCUSSION

We studied luteinized human GC derived from *in vitro* fertilization 24 h after isolation and during the following days in culture using the patch-clamp technique. Stepwise depolarization elicited inward and outward currents (FIG. 1). The latter could be blocked by using a pipette solution containing CsCl and tetraethylammonium, thus indicating the presence of voltage-activated K^+ channels.

The observed inward currents were reversibly inhibited by the Na^+ channel blocker tetrodotoxin (TTX). The maximum Na^+ peak current density was –7.4 pA/pF. To identify the nature of the Na^+ channel, we performed RT-PCR using oligonucleotide primers homologous to regions present in the sequences of the α-subunits of Na^+ channels of the brain, muscle, and a peptidergic endocrine cell line. Analysis of the sequencing results of two PCR-derived cDNA clones showed that they are identical to a Na^+ channel, which has been found in the adrenal and in human C-cell carcinoma (FIG. 2).[8] Northern hybridization using a ^{32}P-labeled antisense cRNA probe transcribed from the cloned cDNA identified a transcript of approximately 7.5 kb. Using an antipan Na^+ channel α-subunit antibody, Western blotting showed a band at about 220 kD, and immunostaining indicated the presence of the protein in GC.

Our results show for the first time voltage-activated K^+ and Na^+ channels in typical steroidogenic endocrine cells. We identified a functional Na^+ channel, which was described earlier in peptidergic and aminergic endocrine cells. Its name "human neuroendocrine (hNE)" Na^+ channel type was coined after its first description in adrenal and C-cell carcinoma.[8] However, based on the present study of steroid-producing GCs, which are not of neuroendocrine origin, this name is misleading. We suggest that from now on this type of Na^+ channel be called "endocrine Na^+ channel."

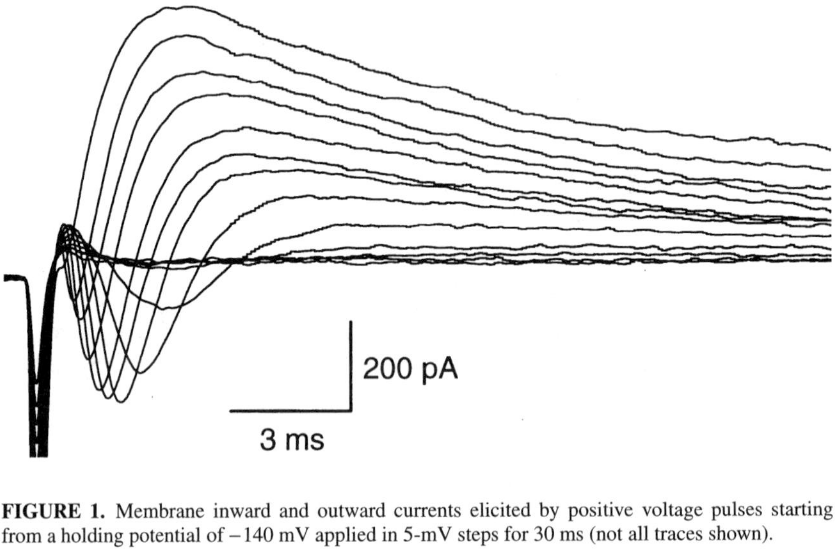

FIGURE 1. Membrane inward and outward currents elicited by positive voltage pulses starting from a holding potential of –140 mV applied in 5-mV steps for 30 ms (not all traces shown).

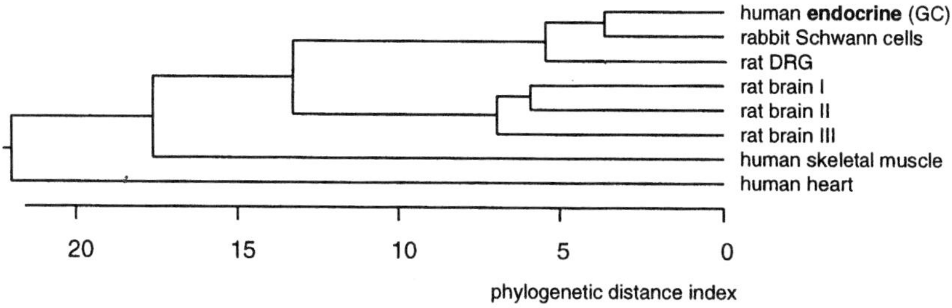

FIGURE 2. Phylogenetic tree of sodium channel sequences. The deduced sequence found in steroidogenic GC is identical to the one present in aminergic or peptidergic endocrine cells.

ACKNOWLEDGMENTS

This work was supported by Volkswagen-Stiftung DFG Ma 1080/10-1 and Graduiertenkolleg 333.

REFERENCES

1. Schlegel, W. & P. Mollard. 1995. Electrical activity and stimulus secretion coupling in neuroendocrine cells. *In* The Electrophysiology of Neuroendocrine Cells. H. Scherübl & J. Hescheler, Eds.: 23–38. CRC Press. Boca Raton.
2. Asem, E.K., J.L. Schwartz, G.A. Mealing, B.K. Tsang & J.F. Whitfield. 1988. Evidence for two distinct potassium channels in avian granulosa cells. Biochem. Biophys. Res. Commun. **155**(2): 761–766.
3. Mattioli, M., B. Barboni & L.J. DeFelice. 1993. Calcium and potassium currents in porcine granulosa cells maintained in follicular or monolayer tissue culture. J. Membrane Biol. **134**(1): 75–83.
4. Mayerhofer, A., K.J. Föhr, K. Sterzik & M. Gratzl. 1992. Carbachol increases intracellular free calcium concentrations in human granulosa-lutein cells in vitro. J. Endocrinol. **135:** 153–159.
5. Mayerhofer, A., K. Sterzik, H. Link, M. Wiemann & M. Gratzl. 1993. Oxytocin increases intracellular free calcium concentrations in human granulosa-lutein cells in vitro. J. Clin. Endocrinol. Metab. **77:** 1209–1214.
6. Mayerhofer, A., R. Engling, B. Stecher, A. Ecker, K. Sterzik & M. Gratzl. 1995. Relaxin triggers calcium transients in human granulosa-lutein cells. Eur. J. Endocrinol. **132:** 507–513.
7. Föhr, K.J., A. Mayerhofer, K., Sterzik, M. Rudolf, B. Rosenbusch & M. Gratzl. 1993. Concerted action of human chorionic gonadotropin and norepinephrine on intracellular free calcium in human granulosalutein cells: Evidence for the presence of a functional alpha-adrenergic receptor. J. Clin. Endocrinol. Metab. **76:** 367–373.
8. Klugbauer, N., L. Lacinova, V. Flockerzi & F. Hofmann. 1995. Structure and functional expression of a new member of the tetrodotoxin-sensitive voltage-activated sodium channel family from human neuroendocrine cells. EMBO J. **14:** 1084–1090.

Cloning and Functional Analysis of the Type III Na^+ Channel from Human Brain

J. J. CLARE,[a] T. J. DALE, X. XIE, T. C. PEAKMAN, AND Y. CHEN

Gene Function and Neuroscience Units, GlaxoWellcome, Medicines Research Centre, Stevenage, SG1 2NY, UK

Several different voltage-gated Na^+ channel α subunits have been identified in rat brain (types I, II, III, and VI). These have distinct expression patterns, both in different regions of the brain and at different stages of development, and presumably have specific physiological roles. Types I, II, and VI mRNA are all abundant in adult brain, whereas type III peaks at birth and becomes barely detectable in adults except in certain regions.[1,2] Each of these types has been cloned and expressed in *Xenopus laevis* oocytes. Although the α subunits alone are capable of forming functional channels, the β subunits also have important roles, for example, stabilizing the channel in a fast gating mode.[3] In contrast, when the rat type II channel is expressed in mammalian cells, a fast gating mode predominates, even in the absence of β subunits.[4] In the light of such differences, it is interesting to compare the properties of the other brain Na^+ channels when expressed in mammalian systems. Furthermore, since they are therapeutic targets in a number of significant diseases, it is important to determine if the human brain Na^+ channels have similar properties and share a similar distribution pattern to that of other species. Here, we describe the cloning of the type III α subunit from human brain and present a functional analysis when expressed in two different mammalian systems. A preliminary analysis of the distribution of this channel in human tissue is also presented.

Three overlapping clones encoding the human type III sequence were obtained from cDNA libraries derived from adult cerebellum and fetal total brain. Although these were extremely unstable, the full-length cDNA was assembled using an *E. coli* host engineered for reduced vector copy-number, propagating at reduced temperature (28°C), and by correcting spontaneous mutations by site-directed mutagenesis. To enable efficient production of expressing cell lines, and to avoid the selection of nonexpressors, this cDNA was inserted upstream of the encephalomyocarditis virus internal ribosome entry site (IRES), giving a vector encoding a polycistronic mRNA containing both the type III sequence and the G418 resistance determinant.

Following transient transfection of this vector into HEK293T cells, robust inward Na^+ currents were produced that were reversibly blocked by tetrodotoxin (100 nM) and that were not present in mock-transfected cells (FIG. 1A). The decay of current did not fit a single exponential function, and even after 100-ms depolarization a significant proportion of current persisted, suggesting the contribution of both transient and persistent components. The current–voltage relationship of the "persistent" component, measured 100 ms after depolarization, was similar to that of total current at the beginning of the pulse (FIG. 1B). The voltage dependence of activation, measured by plotting normalized conductance

[a]Corresponding author. Phone: 01438 763834; fax: 01438 764488; e-mail: jjc25153@glaxowellcome.co.uk

against membrane potential, fitted a single-order Boltzman curve (data not shown). The voltage dependence of inactivation after a 1-s conditioning pulse, as measured by plotting the normalized current during the test pulse against the prepulse potential, followed a biphasic curve also indicating two components (FIG. 1C). These data are consistent with the transient and persistent components being due to alternative states of the type III channel.

Biphasic inactivation kinetics were also observed in HEK293T cell lines stably expressing the type III channel but, in contrast, only fast-inactivating currents were produced in CHO cell lines (data not shown). The mechanism involved in generating these persistent currents, and the reason for the difference between cell types, is currently

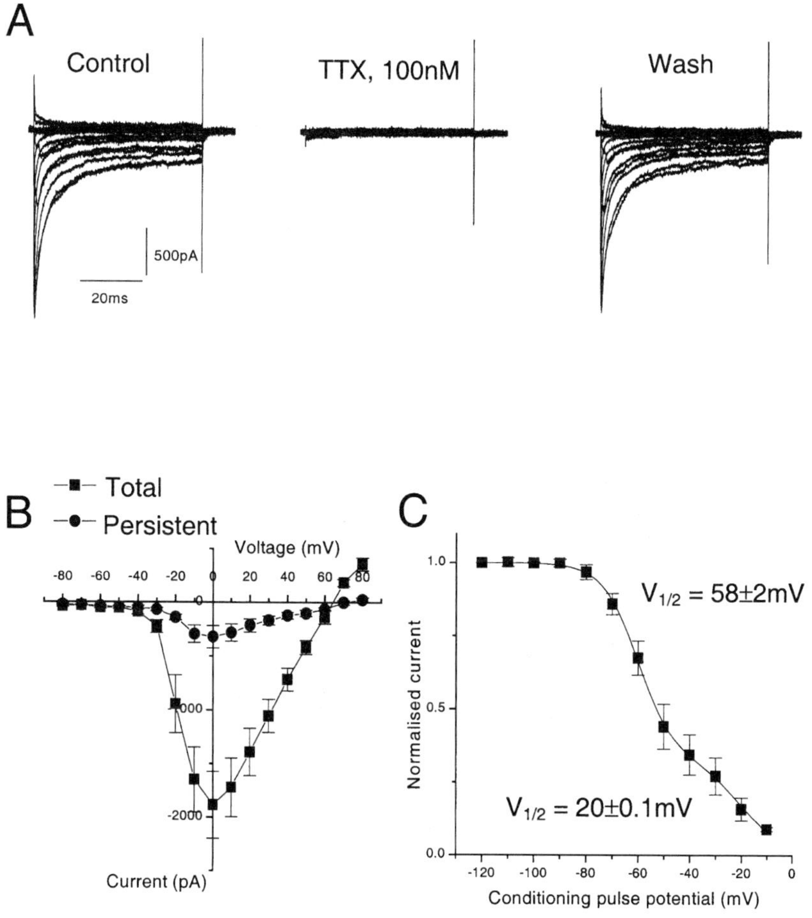

FIGURE 1. Biophysical properties of the human type III α subunit transiently expressed in HEK293T cells. **(A)** Inhibition of Na^+ currents by 100 nM TTX. **(B)** Current–voltage relationship ($n = 7$). **(C)** Voltage dependence of inactivation constructed using 1-s conditioning pulses ($n = 4$).

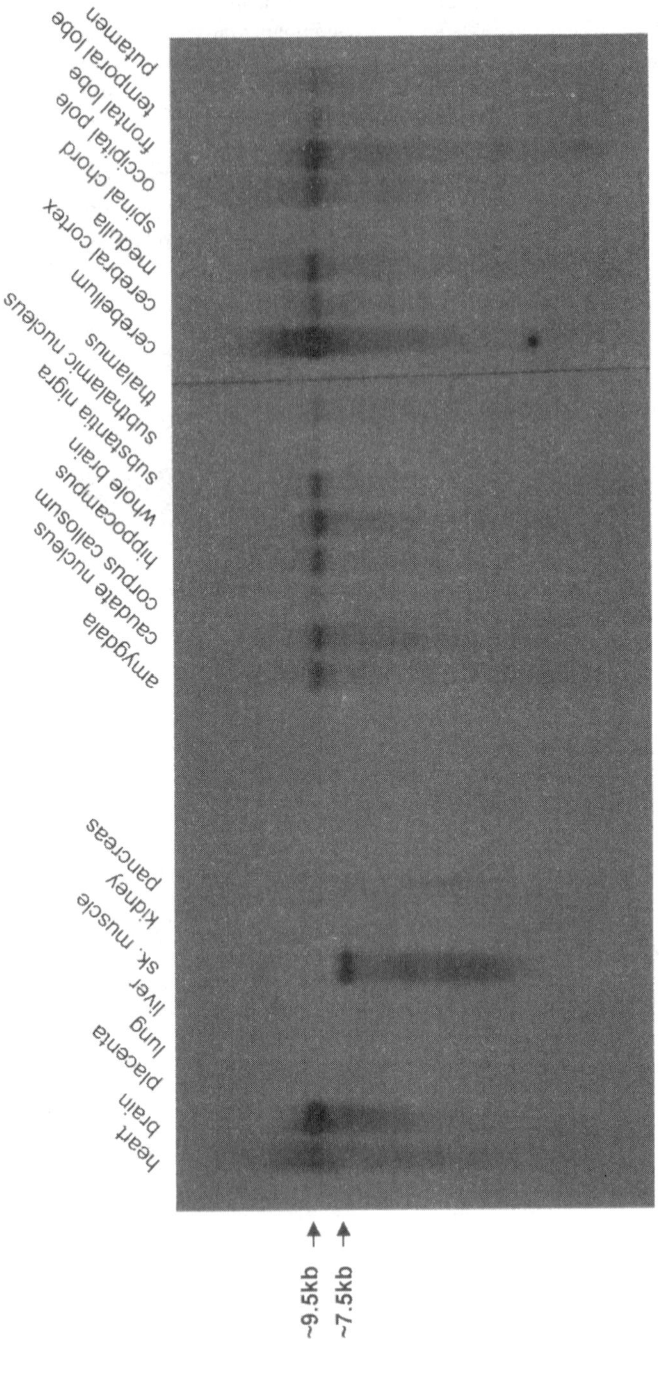

FIGURE 2. Northern blot showing distribution of type III mRNA of different human tissues.

unclear. Initial experiments have suggested that coexpression of β subunits does not substantially alter the proportion or kinetics of the persistent component. Persistent currents can be induced with the rat type II channel in HEK293 cells by overexpressing G-protein βγ subunits,[5] which may act by stabilizing the channel in a noninactivating state. A similar explanation could account for the phenomenon described here, except that, for the type III channel, this state may be more favored such that persistent currents occur even with low levels of free Gβγ. Alternatively, type III may be more susceptible to modulation by endogenous Gβγ than type II.

A hybridization probe derived from the 5′ untranslated region was used to determine the distribution and abundance of type III mRNA in different adult human tissues (FIG. 2). A ~9.5-kb band, which strongly hybridized in brain and also weakly in heart, was observed. In addition, a strongly hybridizing band of ~7.5 kb was found in skeletal muscle. It is not known if this putative splice variant encodes a functional channel, but this is currently under investigation. No expression was detected in placenta, lung, liver, kidney, and pancreas.

The distribution of type III mRNA in different adult brain regions was also examined. In contrast to rat, the human type III mRNA was found to be expressed in most brain regions examined. Type III mRNA was most abundant in cerebellum and frontal lobe, but was also found at moderate levels in amygdala, caudate nucleus, hippocampus, substantia nigra, medulla, occipital pole, and more weakly in subthalamic nucleus, thalamus, cerebral cortex, temporal lobe, and putamen. No expression was detected in corpus callosum and spinal chord.

The wide distribution and unusual biophysical properties of the type III Na^+ channel described here suggest it may have a more important physiological role in the adult human brain than has previously been thought. The biological significance of the type III persistent currents observed *in vitro* remains to be determined. Clearly, such currents would have a profound influence on neuronal firing patterns if they were also produced by type III channels in the brain. Although rarely observed, persistent Na^+ currents have indeed been described in neurons.[6] It is not known if particular channel subtypes or channel states are responsible.

Further characterization of type III and the other Na^+ channel subtypes, in both native and recombinant systems, is required to fully characterize the physiological and pathophysiological roles of these channels.

REFERENCES

1. BECK, S. *et al.* 1989. Differential regulation of three sodium channel mRNAs in the rat central nervous system during development. EMBO J. **8:** 3611–3616.
2. FELTS, P.A. *et al.* 1997. Sodium channel α-subunit mRNAs I, II, III, NaG, Na6 and hNE (PN1): Different expression patterns in developing rat nervous system. Mol. Brain Res. **45:** 71–82.
3. PATTON, D.E. *et al.* 1994. The adult rat brain β1 subunit modifies the activation and inactivation gating of multiple sodium channel α subunits. J. Biol. Chem. **269:** 17649–17655.
4. ISOM, L.L. *et al.* 1995. Functional co-expression of the β1 and type IIA α subunits of sodium channels in a mammalian cell line. J. Biol. Chem. **270:** 3306–3312.
5. MA, J.Y. *et al.* 1997. Persistent sodium currents through brain sodium channels induced by G protein βγ sub-units. Neuron **19:** 443–452.
6. TAYLOR, C.P. 1993. Na^+ currents that fail to inactivate. Trends Neurosci. **16:** 455–460.

Slow Voltage-Dependent Inactivation of a Sustained Sodium Current in Stellate Cells of Rat Entorhinal Cortex Layer II

JACOPO MAGISTRETTI[a] AND ANGEL ALONSO[b,c]

[a]*Laboratorio di Neurofisiologia Sperimentale, Istituto Nazionale Neurologico "Carlo Besta," Milano, Italy*

[b]*Department of Neurology and Neurosurgery, Montreal Neurological Institute and McGill University, Montréal, Québec, H3A 2B4, Canada*

The so-called persistent sodium current (I_{NaP}) has lately been devoted increasing attention for its functional role in important subthreshold and near-threshold neuronal activities. In a specific neuronal population of entorhinal cortex (EC), namely the stellate cells of layer II, which give rise to the main cortical projection to the hippocampus, I_{NaP} has been shown to critically participate in the generation of subthreshold membrane-potential oscillations in the theta-rhythm range.[1] This particular subthreshold activity is likely to contribute to the generation, at a population level, of synchronized oscillations that represent a well-known feature of EC and hippocampus, and are believed to play an important role in learning and memory processes.[2,3] In the present study we undertook a biophysical characterization of the I_{NaP} expressed by EC layer II stellate cells.

The study was performed both in acutely dissociated neurons (see reference 4 for the dissociation procedure) and in slices from young (P15-20) Long-Evans rats, by applying the patch-clamp technique in the whole-cell configuration. The ionic composition of extra- and intracellular recordings was suitable for isolating sodium currents (see the figure legends). In both isolated and *in situ* neurons the application of slow (50-mV/s) voltage ramps under voltage-clamp conditions revealed the presence of a prominent inwardly rectifying current that was suppressed by TTx (1 μM) (see for instance FIG. 1C), and was therefore identified as I_{NaP}. To establish whether this current could be simply accounted for by a classic Hodgkin-Huxley "window" current arising from the gating properties of the fast sodium current (I_{Na}), we applied protocols such as that shown in FIGURE 1. The voltage-dependence properties of I_{Na} were determined with standard activation and steady state inactivation protocols (FIG. 1A). Activation and steady state inactivation graphics of the conductance accounting for I_{Na} (G_{Na}) were constructed in single cells, and fitted with Boltzmann functions (FIG. 1B); the voltage dependence of the predicted window conductance (G_{NaW}) was thus derived (FIG. 1B, dotted line). In the same cells, I_{NaP} was also recorded and isolated via TTx subtraction (FIG. 1C), and the voltage dependence of the underlying sodium conductance (G_{NaP}) was derived. The comparison of the voltage dependences of G_{NaW} and G_{NaP} (FIG. 1D) revealed marked discrepancies in a voltage region positive to about −40 mV, where G_{NaW} would be predicted to rapidly decrease

[c]Address for correspondence: Dr. Angel Alonso, Department of Neurology and Neurosurgery, Montréal Neurological Institute, Room 753, 3801 University Street, Montréal, Québec, H3A 2B4, Canada. Phone: 514-398-6901; fax: 514-398-8106; e-mail: mdao@musica.mcgill.ca

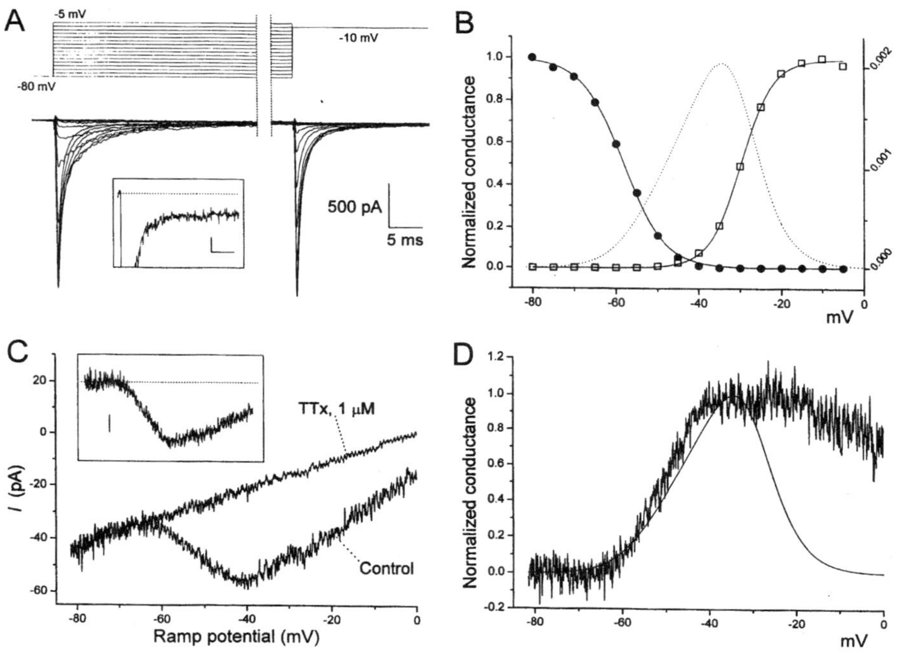

FIGURE 1. **(A)** Activation and steady state inactivation protocols for I_{Na} and currents recorded in a representative dissociated neuron. The extracellular recording solution contained (in mmol/L): 100 NaCl, 40 TEA-Cl, 10 Hepes (free acid), 2 $CaCl_2$, 3 $MgCl_2$, 0.2 $CdCl_2$, 5 4-aminopyridine, 25 glucose, pH 7.4. The intracellular recording solution contained 110 CsF, 10 Hepes-Na, 11 EGTA, 2 $MgCl_2$, pH 7.25. **Inset:** Detail of the current evoked at the test potential of −30 mV (*scale bars:* 20 pA, 25 ms). **(B)** Activation and steady state inactivation graphics for I_{Na} (same cell as in A), Boltzmann fittings and predicted window conductance (G_{NaW}), calculated as the product of the two Boltzmann functions (note the different *y*-axis scale). **(C)** isolation of I_{NaP} in the same cell as before. The currents were evoked with slow (50 mV/s) voltage ramps. **Inset:** TTx-subtracted trace (*scale bar:* 10 pA). **(D)** G_{NaP} (as derived from the current of panel C, **inset**) compared with the predicted G_{NaW} (both normalized to 1).

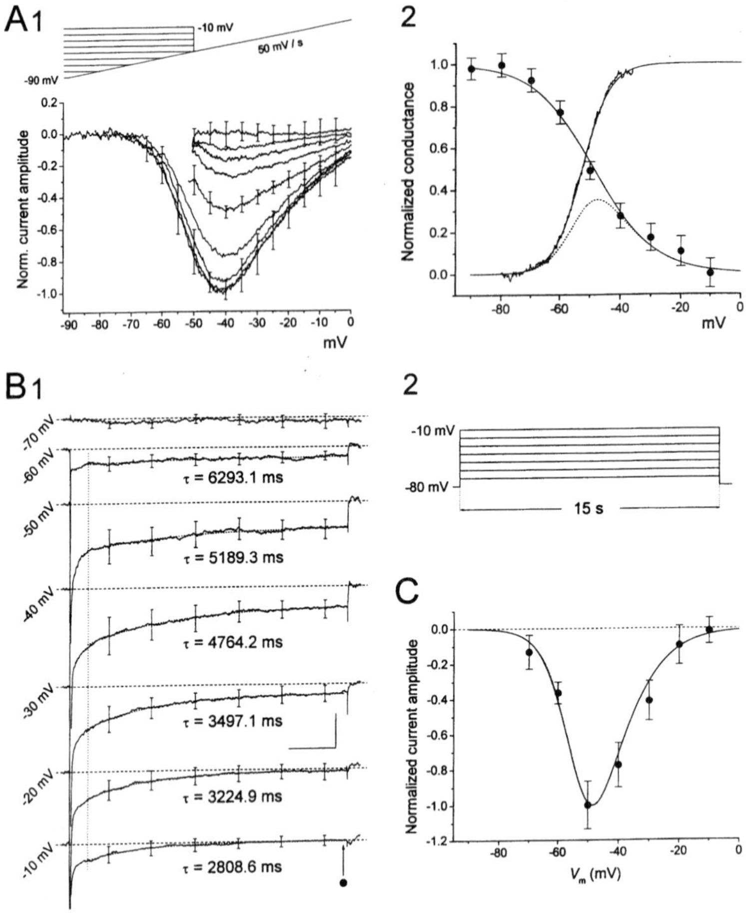

FIGURE 2. **(A1)** Steady state inactivation protocol for I_{NaP}, and average, normalized currents from 7 *in situ* neurons (mean ± SD). **(A2)** Normalized, average activation and steady state inactivation graphics derived from the currents depicted in (A1). Boltzmann fittings are also shown, along with the predicted G_{NaPW} (see text), which was calculated as the product of the two Boltzmann functions. **(B)** Long-lasting step protocol **(B2)** and average TTx-subtracted currents evoked with the same protocol in 8 *in situ* neurons (mean ± SD) **(B1)**. Single-exponential best fittings of the decay and time constants are also shown (the *dotted vertical line* indicates the starting time point for fittings). *Scale bars:* 50 pA, 2.5 s. **(C)** Graphic of the currents measured at the end of the 15-s pulses (see panel B1 *arrow* and *filled circle* at the bottom) as a function of the test potential (mean ± SEM). The predicted I_{NaPW} (*solid trace;* derived from G_{NaPW}, see panel A2) is also shown for comparison. Both quantities have been normalized to 1. The extracellular solution used for the recordings in (A) and (B) contained (in mmol/L): 34 NaCl, 26 $NaHCO_3$, 80 TEA-Cl, 5 KCl, 3 CsCl, 2 $CaCl_2$, 3 $MgCl_2$, 2 $BaCl_2$, 2 $CoCl_2$, 0.4 $CdCl_2$, 4 4-aminopyridine, 10 glucose, pH 7.4 when bubbled with 95% O_2/5% CO_2. The intracellular recording solution was the same as described in FIGURE 1 legend.

(with a nearly complete inactivation at about −10 mV), whereas G_{NaP} was still markedly sustained. Moreover, the size of the predicted G_{NaW} was much smaller than that of the G_{NaP} observed in the same cells. On average, the ratio of maximal G_{NaW} versus maximal G_{NaP} measured in the same cells was less than 0.1 (0.086 ± 0.034, mean ± SD, $n = 5$). These data demonstrate that by far most of the I_{NaP} expressed by EC layer II principal neurons cannot be accounted for by a classic Hodgkin-Huxley window current; rather, I_{NaP} must arise from some distinct channel species or channel gating modality.

A close inspection of the voltage dependence of G_{NaP} as obtained by ramp protocols (FIG. 1D) reveals a tendency to a decrease at relatively positive potentials, where a fully activated conductance would be expected to maintain a steady value. Such observations led us to hypothesize that a voltage- and time-dependent inactivation of I_{NaP} might be acting through the slow voltage ramp.[5] This idea was further confirmed by the observation that the amplitude of I_{NaP} decreased with decreasing slopes of the applied ramp (not shown). To characterize the possible voltage-dependent inactivation of I_{NaP} we applied the protocol shown in FIGURE 2A. Fifteen-second-long conditioning prepulses at increasingly positive potential levels preceded the voltage ramp (FIG. 2A1). I_{NaP} turned out to be progressively inactivated, and the inactivation was eventually nearly complete at about −10 mV. The average steady state inactivation curve (FIG. 2A2) showed a $V_{1/2}$ at −48.8 mV. When compared with the average activation curve of G_{NaP} as measured in the same cells, the inactivation curve allowed us to predict the existence of a window conductance arising from G_{NaP} (G_{NaPW}, dashed line in FIG. 2A2) and expected to generate a really persistent current (I_{NaPW}). This prediction was tested with experiments in which I_{NaP} was evoked with long-lasting step voltage protocols (FIG. 2B). These experiments confirmed the existence of a time- and voltage-dependent inactivation of I_{NaP}; the current decay took place with time constants of about 6.3 to 2.7 s, depending on the test potential. The residual currents observed at the end of the 15-s depolarizing pulses were measured, plotted as a function of the test potential, and compared with the values predicted for the window current, I_{NaPW}, arising from G_{NaPW} (FIG. 2C). This comparison returned a very good concordance.

We therefore conclude that the I_{NaP} expressed by EC layer II principal neurons is in fact a slowly decaying current. The wide overlapping between the activation and inactivation curves of the underlying conductance account for the generation of a really persistent window current (I_{NaPW}) in a voltage range that closely corresponds to that over which subthreshold membrane-potential oscillations are generated by the same neurons.[1]

REFERENCES

1. ALONSO, A. & R.R. LLINÁS.1989. Subthreshold Na$^+$-dependent theta-like rhythmicity in stellate cells of entorhinal cortex layer II. Nature **342:** 175–177.
2. LISMAN, J.E. & M.A.P. IDIART. 1995. Storage of 7 + 2 short term memories in oscillatory subcycles. Science **267:** 1512–1515.
3. WINSON, J. 1978. Loss of hippocampal theta rhythm results in spatial memory deficit in the rat. Science **201:** 160–163.
4. WHITE, J.A., A. ALONSO & A.R. KAY. 1993. A heart-like Na$^+$ current in the medial entorhinal cortex. Neuron **11:** 1037–1047.
5. FLEIDERVISH, I.A. & M.J. GUTNICK. 1996. Kinetics of slow inactivation of persistent sodium current in layer V neurons of mouse neocortical slices. J. Neurophysiol. **76**(3): 2125–2130.

Differential Distribution of Voltage-Gated Sodium Channel α- and β-Subunits in Human Brain

W. R. J. WHITAKER,[a,c] J. J. CLARE,[b] AND P. C. EMSON[a]

[a]*Department of Neurobiology, The Babraham Institute, Cambridge, Cambridgeshire, CB2 4AT, UK*

[b]*Gene Function Unit, GlaxoWellcome, Medicines Research Center, Stevenage, Hertfordshire, SG1 2NY, UK*

Neuronal voltage-gated sodium channels generate and propagate action potentials in the central nervous system (CNS). They are heteromultimeric proteins consisting of a heavily glycosylated α-subunit (M_r 260 kDa), which is both ion selective and voltage sensitive. The α-subunit is noncovalently linked to a β1(M_r 36 kDa) and to a β2 subunit (M_r 33 kDa) via a disulfide linkage.[1] Multiple brain-specific isoforms have been isolated and cloned from rat tissues. These include α-subunit subtypes: I and II,[2] IIA,[3] III,[4] and VI,[5] and two β-subunits, β1 and β2.[6] As with other ion channels, these isoforms are products of a multigene family.

Sodium channel α-subunits show a defined temporal and spatial distribution in the CNS of the rat, as revealed by Northern blot,[7] *in situ* mRNA hybridization,[8] and immunoprecipitation experiments.[9] Such heterogeneous expression indicates that each channel may perform a unique functional role in neuronal excitability. Distribution studies provide an insight into putative functional requirements, leading to an understanding of the mechanisms controlling neuronal excitability. To this end we have carried out the first detailed study of sodium channel α- and β-subunit distribution in human hippocampus, medial frontal gyrus, and cerebellum by *in situ* mRNA hybridization. Our results indicate that sodium channel α- and β-subunits show distinct regional expression patterns in human CNS.

MATERIALS AND METHODS

Oligonucleotide probes (36 mers) specific to human sodium channel α- and β-subunits were designed, synthesized, and purified by reverse-phase HPLC. A cocktail of two probes, targeted to coding and 3′ untranslated regions, was used in the hybridization reactions. Probes were 3′-end labeled with [^{35}S] dATP. Adjacent frozen cryostat sections were fixed briefly in 4% PFA and processed for standard radioactive *in situ* hybridization procedure. In brief, tissue sections were hybridized overnight at 37 °C and subsequently washed in 1×SSC at 55°C (×3). Initial autoradiography was carried out by exposing Kodak Bio-Max

[c]Corresponding author. Phone: 44 1223 832 312; fax: 44 1223 836 614; e-mail: william.whitaker @bbsrc.ac.uk

MR film to the slides for 2 weeks. Subsequently, slides were emulsion dipped at 37 °C in Ilford K5 emulsion and exposed for 6 to 8 weeks at 4 °C.

RESULTS

Individual oligonucleotide probes showed qualitatively similar hybridization patterns for each subtype. In all cases specific hybridization was displaced by incubation with excess unlabeled oligonucleotide (not shown). Specific cellular hybridization was seen with each probe (TABLE 1) in human hippocampal formation (FIG. 1, top panel), medial frontal gyrus (FIG. 1, middle panel) and cerebellum (FIG. 1, bottom panel).

DISCUSSION

The physiological and functional consequences of sodium channel diversity are as yet unknown. Evidence suggests that sodium channel subtypes may serve specialized roles in a global, integrated function reflected by their differential expression and localization in individual neurons. The observation that sodium channel expression is also modified extensively by mRNA splicing[10] and posttranslational modification, such as differential phosphorylation[11] adds a further level of complexity.

Here we report a detailed mRNA distribution study of neuronal sodium channel α-subunits and β-subunits in human brain. Types II, VI, and β1 are expressed at the highest level, with types I and III showing the lowest level of expression. The expression of types I

TABLE 1. Expression Comparisons Between Sodium Channel α I to VI and β1 and β2 Subunits in Hippocampus, Cortex, and Cerebellum

	Subtype					
	I	II	III	VI	β1	β2
Hippocampus						
CA3	+	+++	+	++++	+++	++
CA2	±	+++	±	+++	+++	++
CA1	±	++	±	++++	++	++
Dentate gyrus	++	+++	++	+++	+++	+++
Medial frontal gyrus						
Layers II/III	+	++	+	+++	++	++
Layers V/VI	+	++	+	++	++++	++
Cerebellum						
Granular layer	++	++++	+	+++	++++	+++
Purkinje cells	++	–	–	+++	+++	++
Molecular layer	–	–	–	–	±	–

SYMBOLS: ++++, very strong expression; +++, strong expression; ++, moderate expression; +, weak expression; ±, very weak/scattered expression; –, no expression.

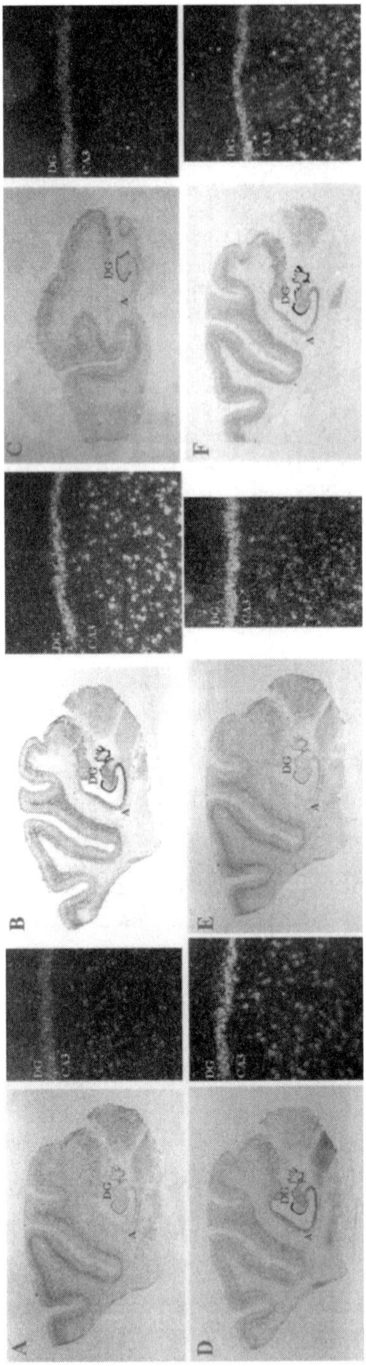
A
B
C
D
E
F

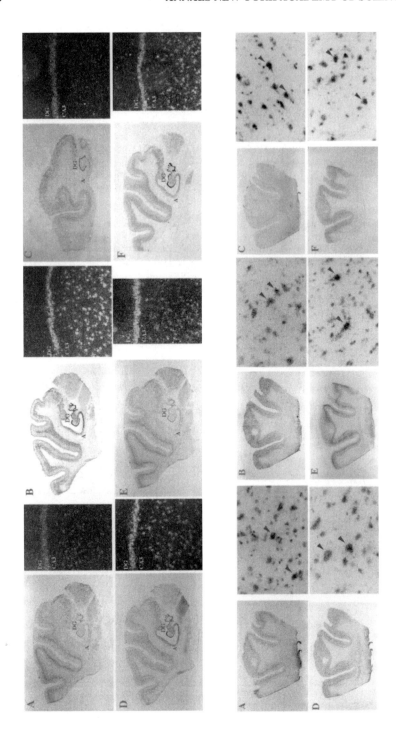

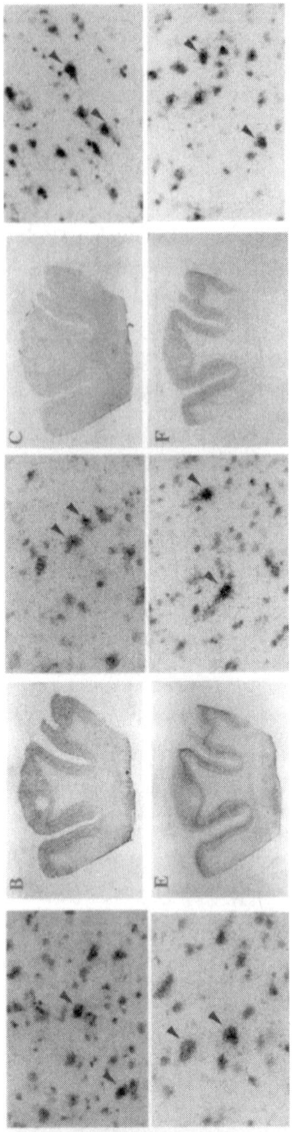
A
B
C
D
E
F

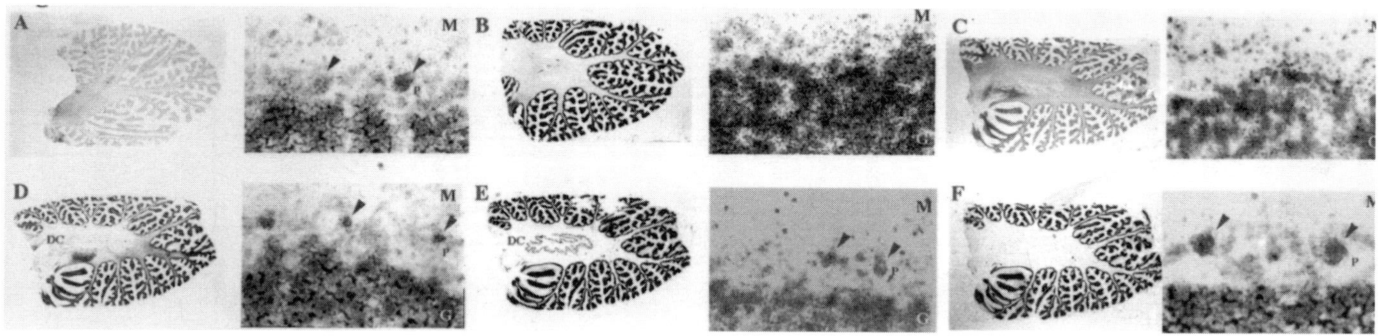

FIGURE 1. Top panel (*facing page*): Distribution of sodium channel type I **(A),** II **(B),** III **(C),** VI **(D),** β1 **(E),** and β2 **(F)** in human hippocampal formation and entorhinal cortex. In all cases, robust mRNA expression was seen in the granule cells of the dentate gyrus (DG). Differential expression was seen in the CA regions of the Ammon's Horn (A). Types I, III, and β1 showed moderate expression in CA3 pyramidal cells. Types II, VI, and β2 were strongly expressed in all the CA regions. Dark field microscopy images indicate varying expression levels at the dentate gyrus/ CA3 boundary. **Middle panel** (*facing page*): Distribution of sodium channel type I **(A),** II **(B),** III **(C),** VI **(D),** β1 **(E),** and β2 **(F)** in human medial frontal gyrus. Specific expression of sodium channel mRNA was seen in the different cortical layers (layers I–VI). Sodium channel type I showed expression in layers III, V, and VI. Sodium channel type II showed robust expression in the outer layers of the cortex and moderate expression in layers V and VI. This distribution was also seen with sodium channel VI. Similarly, sodium channel type III showed a low level of expression in the deeper cortical layers. β1 showed robust expression in the pyramidal cells of layer III compared to β2, which was expressed moderately in layer VI. Specific cellular hybridization is indicated by *arrows* on high-magnification photographs. **Bottom panel:** Distribution of sodium channel type I **(A),** II **(B),** III **(C),** VI **(D),** β1 **(E),** and β2 **(F)** in human cerebellum. In all cases expression was predominant in the granular layer (G). Types II and VI showed the most robust expression in this region. Additionally, types I, VI, and both β-subunits showed varying levels of mRNA expression (*arrows*) in Purkinje cells (P). Types VI and β1, respectively, showed moderate and strong expression in cells of the deep cerebellar nucleus (DC). Also, β1 showed scattered expression in cells of the molecular layer (M), assumed to be basket or stellate cells. No mRNA expression was seen in white matter.

and VI in human Purkinje cells supports their possible role in the generation of transient and persistent sodium conductances. Recent work in rat has revealed that type I[11] and type VI[12] may mediate the transient and persistent components, respectively, in Purkinje cells.

One striking exception is type III mRNA expression in human brain, which differs markedly from that observed in the rat brain. In the latter, mRNA levels reflect a predominantly embryonic/neonatal expression declining to negligible levels in the adult.[8] In the human brain we found a definite, albeit low-level, mRNA expression in cerebellum, hippocampus, and cerebral cortex. Observations that human type III sodium channel generates persistent currents (J. J. Clare, this volume) indicate that expression of this channel in adult neurons may have important functional implications.

REFERENCES

1. Catterall, W.A. 1992. Cellular and molecular biology of voltage-gated sodium channels. Physiol. Rev. **72**(4): S15–S48.
2. Noda, M. *et al.* 1986. Expression of functional sodium channels from cloned cDNA. Nature **322:** 826–828.
3. Auld, V.J. *et al.* 1988. A rat brain Na^+ channel α-subunit with novel gating properties. Neuron **1:** 449–461.
4. Kayano, T. *et al.* 1988. Primary structure of rat brain sodium channel type III deduced from the cDNA sequence. FEBS Lett. **228:** 187–194.
5. Schaller, K.L. *et al.* 1995. Alternatively spliced sodium channel transcripts in brain and muscle. J. Neurosci. **15**(5): 3231–3242.
6. Hartshorne. R.P. & W.A. Catterall. 1984. The sodium channel from rat brain: Purification and subunit composition. J. Biol. Chem. **259:** 1667–1675.
7. Gordon, D. *et al.* 1987. Tissue-specific expression of the R_I and R_{II} sodium channel subtypes. Proc. Natl. Acad. Sci. USA **84:** 8682–8686.
8. Felts, P.A. *et al.* 1997. Sodium channel α-subunit expression patterns in developing rat nervous system. Mol. Brain. Res. **45:** 71–82.
9. Oh, Y. *et al.* 1994. *In situ* hybridisation localisation of the Na^+ channel β1 subunit mRNA in rat CNS neurons. Neurosci. Lett. **176:** 119–122.
10. Schaller, K.L. *et al.* 1992. A novel sodium channel expressed in neurons and glia. J. Neurosci. **12**(4): 1370–1381.
11. Smith, R.D. & A.L. Goldin. 1998. Functional analysis of the rat I sodium channel in *Xenopus* oocytes. J. Neurosci. **18**(3): 811–820.
12. Raman, I.M. *et al.* 1997. Altered subthreshold sodium currents and disrupted firing patterns in Purkinje neurons of *Scn8a* mutant mice. Neuron **19:** 881–891.

Properties of Sodium Currents and Action Potential Firing in Isolated Cerebellar Purkinje Neurons

INDIRA M. RAMAN[a] AND BRUCE P. BEAN

Department of Neurobiology, Harvard Medical School, 220 Longwood Avenue, Boston, Massachusetts 02115, USA

UNUSUAL KINETICS OF SODIUM CURRENT IN PURKINJE NEURONS

We have characterized voltage-gated tetrodotoxin-sensitive sodium currents of Purkinje neurons of the cerebellum, with an interest in relating the properties of voltage-gated ion channels to the distinctive firing properties of those cells.[1] Step depolarizations from hyperpolarized potentials activate conventional sodium currents that decay by more than 99%. A family of sodium currents from a Purkinje cell isolated from a 2-week-old mouse and the corresponding current–voltage relation are shown in FIGURE 1A. Currents rose to a peak within a few hundred ms, and had a dominant decay time constant of 1 ms at −30 mV, which decreased with depolarization.[2]

An unusual component of sodium current, illustrated in FIGURE 1B, was evoked by step repolarizations from positive potentials after sodium currents had maximally inactivated.[2] This "resurgent" current is steeply and nonmonotonically voltage dependent, with the current being small but measurable at −70 mV, maximal at −30 mV, and undetectable at −10 mV. Although most recordings were done in reduced concentrations of sodium, in normal saline (150 mM sodium) the peak resurgent current was on the order of a nA. The rise and decay times of the currents were more than an order of magnitude slower than for the transient currents evoked by depolarization. Even brief (2-ms) steps to positive potentials could elicit resurgent current, suggesting that it may be evoked by spikelike voltage deflections. Resurgent current correlates with partial, rapid recovery from inactivation in the voltage range of −70 to −30 mV. This current was present in all Purkinje neurons, but not in CA3 neurons of the hippocampus.[2] Single-channel recordings as well as genetic studies suggest that the channels underlying resurgent current also contribute to conventional transient sodium current.[2,3]

CONTRIBUTION OF RESURGENT CURRENT TO SPIKING

Intact Purkinje neurons in brain slices are spontaneously active.[4–7] Isolated Purkinje cell bodies also fired spontaneously when studied under the current clamp. Firing was extremely regular, with a mean frequency of 50 Hz. FIGURE 2A shows that spontaneous firing can persist during blockade of voltage-gated calcium channels or hyperpolarization-gated cation channels (I_h), suggesting that the depolarizing drive between action potentials

[a]Corresponding author. Phone: 617-432-1768; fax: 617-432-3057; e-mail: iraman@warren.med.harvard.edu

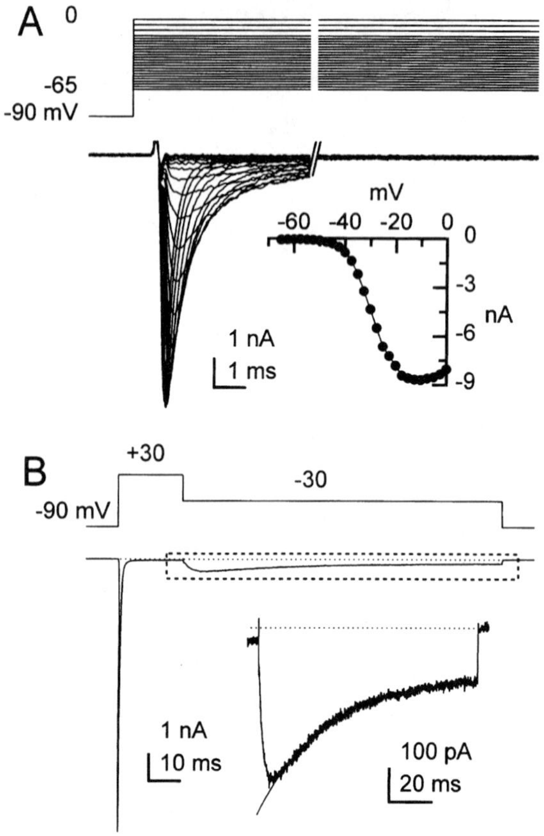

FIGURE 1. Transient and resurgent sodium currents in Purkinje neurons. Cells from 2–3-week-old mice were enzymatically isolated with protease 23.[2,3] Voltage-clamp recordings were made at room temperature in solutions containing (mM) 50 NaCl, 110 TEACl, 2 $BaCl_2$, 0.3 $CdCl_2$, and 10 HEPES (pH 7.4 with TEAOH), and were repeated in the same solution containing 300 nM tetrodotoxin. Subtraction allowed isolation of tetrodotoxin-sensitive current. **(A)** Currents were elicited with step depolarizations from −90 mV to potentials from −65 to 0 mV in 2.5-mV increments. The break in the trace represents 25 ms. For clarity, responses at −12.5, −7.5, and −2.5 mV are not illustrated. **Inset:** Peak current vs. voltage relation. **(B)** Currents were elicited with the voltage protocol shown. The resurgent current is evident upon repolarization to −30 mV. The area inside the *dashed lines* shown at higher gain in the **inset.** The decay phase of the current is fitted with a single exponential with a time constant of 31 ms. *Dotted lines* indicate 0 pA. Different cell from (A).

does not depend on calcium current and I_h. After firing was stopped by injection of hyperpolarizing current, a brief depolarization elicited bursts of spikes in an all-or-none fashion. This behavior is reminiscent of the all-or-none complex spikes seen in intact Purkinje cells. FIGURE 2B shows that bursts could occur during blockade of calcium and I_h currents.

We hypothesize that sodium channels that carry transient and resurgent current activate and inactivate with depolarization, much like conventional sodium channels, but, unlike

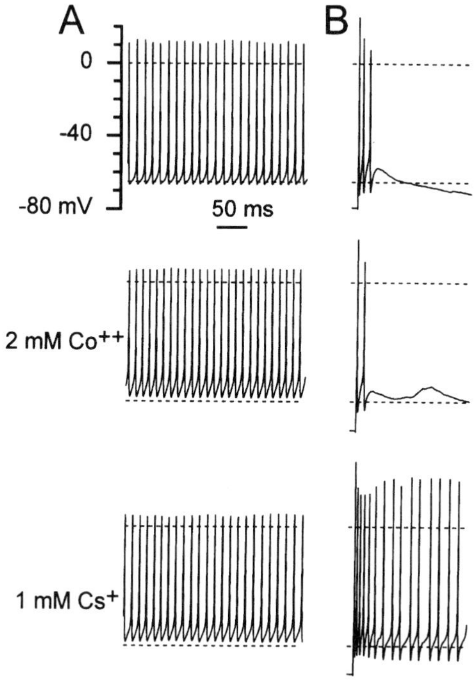

FIGURE 2. Action potentials elicited in Purkinje neurons. Current-clamp recordings were made at room temperature. Control solution **(top)** contained (mM) 150 NaCl, 4 KCl, 2 $CaCl_2$, 2 $MgCl_2$, 10 glucose, and 10 HEPES (pH 7.4 with NaOH). Recordings were repeated in the same solutions but with 2 mM $CoCl_2$ substituted for the $CaCl_2$ **(middle)** and with 1 mM CsCl added to the control solution **(bottom). (A)** Three hundred ms of spontaneous activity recorded in three ionic conditions. *Dotted lines* in (A) and (B) indicate 0 and −65 mV. *Scale bar* applies to all panels. **(B)** Spontaneous activity was silenced with 10 pA of hyperpolarizing current, and bursts were elicited with 1-ms depolarizations of 600 pA, in three ionic conditions. The current injection is indicated by the *bottom trace*. Different cell from (A).

conventional channels, they can reopen during recovery from inactivation. This reopening specifically occurs during a rapid phase of recovery at moderately negative potentials (> −70 mV) and enables channels to reactivate and provide the regenerative depolarization for the subsequent spike. When a relatively high availability of sodium channels exists, (e.g., following stronger hyperpolarizations), these kinetics can contribute to a briefer interval between spikes, promoting bursts of action potentials.

POSSIBLE MOLECULAR IDENTITY OF THE CHANNELS UNDERLYING RESURGENT CURRENT

Expression of the sodium channel alpha subunit Scn8a appears critical for normal production of resurgent current in Purkinje neurons.[3,8] Interestingly, however, *in situ* hybridizations indicate that this alpha subunit is present in many types of neurons,[9,10] at least some of which do not produce resurgent current (such as CA3 neurons). Also, expression of Scn8a in oocytes, with or without beta subunits, yields channels that carry transient but not resurgent current.[11] Possibly, some interaction or modification occurs in Purkinje neurons, allowing channels with this alpha subunit to produce resurgent sodium current.

REFERENCES

1. LLINÁS, R. 1988. The intrinsic electrophysiological properties of mammalian neurons: Insights into central nervous system function. Science **242:** 1654–1164.
2. RAMAN I.M. & B.P. BEAN. 1997. Resurgent sodium current and action potential formation in dissociated cerebellar Purkinje neurons. J. Neurosci. **17:** 4517–4526.
3. RAMAN I.M., L.K. SPRUNGER, M.H. MEISLER & B.P. BEAN. 1997. Altered subthreshold sodium currents and disrupted firing patterns in Purkinje neurons of *Scn8a* mutant mice. Neuron **19:** 881–891.
4. HOUNSGAARD, J. 1979. Pacemaker properties of mammalian Purkinje cells. Acta Physiol. Scand. **106:** 91–92.
5. LLINÁS R. & M. SUGIMORI. 1980. Electrophysiological properties of in vitro Purkinje cell somata in mammalian cerebellar slices. J. Physiol. **305:** 171–195.
6. LLINÁS R. & M. SUGIMORI. 1980. Electrophysiological properties of in vitro Purkinje cell dendrites in mammalian cerebellar slices. J. Physiol. **305:** 197–213.
7. HÄUSSER, M. & B. CLARK. 1997. Tonic synaptic inhibition modulates neuronal output pattern and spatiotemporal synaptic integration. Neuron **19:** 665–678.
8. BURGESS, D.L., D.C. KOHRMAN, J. GALT, N.W. PLUMMER, J.M. JONES, B. SPEAR & M.H. MEISLER. 1995. Mutation of a new sodium channel gene, *Scn8a*, in the mouse mutant 'motor endplate disease.' Nat. Genet. **10:** 461–465.
9. SCHALLER, K.L., D.M. KRZEMIEN, P.J. YAROWSKY, B.K. KRUEGER & J.H. CALDWELL. 1995. A novel, abundant sodium channel expressed in neurons and glia. J. Neurosci. **15:** 3231–3242.
10. VEGA-SAENZ DE MIERA, E., B. RUDY, M. SUGIMORI, & R. LLINÁS. 1997. Molecular characterization of the sodium channel subunits expressed in mammalian cerebellar Purkinje cells. Proc. Natl. Acad. Soc. USA **94:** 7059–7064.
11. SMITH, M.R., R.D. SMITH, N.W. PLUMMER, M.H. MEISLER & A.L. GOLDIN. 1998. Functional analysis of the mouse Scn8a sodium channel. J. Neurosci. **18:** 6093–6102.

Slow Sodium Channel Inactivation in CA1 Pyramidal Cells

TIMOTHY MICKUS, HAE-YOON JUNG, AND NELSON SPRUSTON[a]

Department of Neurobiology and Physiology, Institute for Neuroscience, Northwestern University, Evanston, Illinois 60208-3520, USA

Sodium channels in many types of excitable tissue often become less likely to open after long depolarizations or repetitive stimuli. This phenomenon has been named *slow inactivation*, and is distinct from the more common form of fast inactivation (first described by Hodgkin and Huxley) by its slower onset and slower time course of recovery. While the physiological importance and molecular mechanisms underlying this form of inactivation are unknown, it appears to be an important, general feature of most types of sodium channels that have been studied.

We and others have previously shown that sodium channels in the somata and dendrites of hippocampal CA1 pyramidal neurons undergo a cumulative form of slow inactivation.[1–3] This type of inactivation is partly responsible for the activity-dependent attenuation of action potentials as they propagate from the axon back into the dendrites of these neurons, and may also influence dendritic spike generation.[2,4] Gaining a more complete biophysical understanding of sodium channel function thus will give us more detailed insight into the activity dependence of dendritic excitability. It is also important to determine the relationship of the inactivation we have observed to that described by others.

To study this inactivation *in vitro*, hippocampal slices were prepared from 4–10-week-old Wistar rats as previously described.[2] Slices were bathed in a solution containing (in mM): 125 NaCl, 2.5 KCl, 25 $NaHCO_3$, 1.25 NaH_2PO_4, 1 $MgCl_2$, 2 $CaCl_2$, 25 dextrose, which was maintained at 35–37 °C and bubbled with a 95% O_2–5% CO_2 gas mixture to oxygenate and maintain a pH of 7.4. Neurons were visualized with infrared differential interference contrast microscopy.[5] Glass electrodes were coated with Sylgard, fire polished, and filled with a solution containing (in mM): 120 NaCl, 3 KCl, 10 HEPES, 2 $CaCl_2$, 1 $MgCl_2$, 30 tetraethylammonium chloride (TEA), 5 4-aminopyridine, pH 7.4 with NaOH. Cell-attached patch recordings were made on dendrites and somata of CA1 pyramidal neurons. Voltage-clamp recordings were made using a Dagan PC-ONE patch-clamp amplifier. Leak and capacitive currents were subtracted on-line by *P*/(-4) subtraction.

We focused our investigation on dendritic sodium channels, which undergo more slow inactivation than somatic sodium channels in CA1 neurons.[1,2] While characterizing the biophysical parameters of these channels, we found that the recovery time course from either a short or a long prepulse was nearly equal (τ_{recov} = 1.2 vs. 1.3 s, respectively). These results suggest that the "prolonged" inactivation of these sodium channels induced by short pulses and slow inactivation induced by longer pulses (as described by others) may be the same state.

In order to examine this issue more directly, we performed the experiments shown in FIGURE 1. We reasoned that a long depolarizing prepulse should induce slow inactivation of

[a]Corresponding author. Phone: 847-467-2734; fax: 847-467-4898; e-mail: spruston@nwu.edu

the sodium channels. In this way, if no further inactivation occurs during the train, then this would suggest that there is but one slow inactivated state (or several that are functionally indistinguishable). However, any further inactivation induced by a train would suggest that slow inactivation and prolonged inactivation are really separate phenomena, mediated by different states of the sodium channel. In FIGURE 1A and 1B, left, are control recordings in response to 20- and 50-Hz depolarizing trains without prepulses. FIGURE 1A, right, shows the results of the currents evoked during a 20-Hz train following a long depolarizing prepulse. As can be seen, no further inactivation occurred during the train; instead the currents actually recovered during the train to constant steady state level similar to that of the control steady state ($n = 4$ dendritic patches). The observed recovery was due to the interpulse interval, since with a higher frequency train (FIG. 1B), the currents remained at a smaller steady state amplitude with very little further recovery.

This result suggests that a simple model with a single slow inactivation state might be sufficient to explain the data. The model that best fit the data is shown in FIGURE 2. The model has four states: closed (C), open (O), fast inactivated (I_{fast}), and slow inactivated (I_{slow}). Transitions between all rates are possible, and the rates are voltage-dependent. This

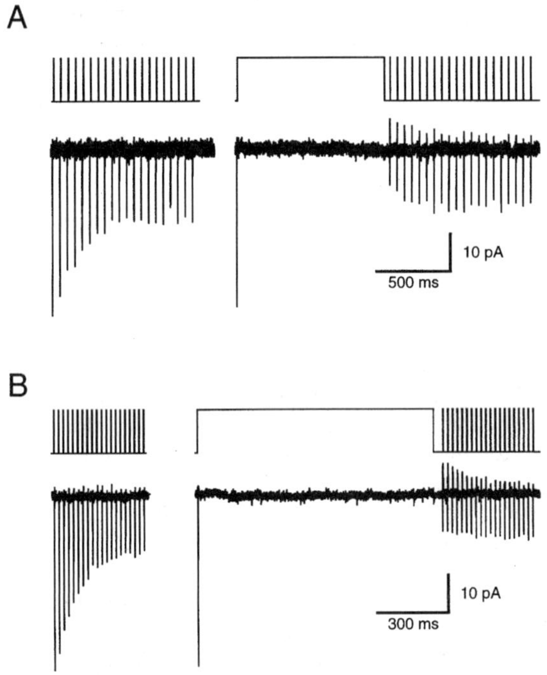

FIGURE 1. A long depolarizing prepulse prevents additional slow inactivation of dendritic sodium channels. Pulse protocols are shown above each trace. **(A)** Current responses to 20-Hz trains of 2 ms, 70-mV depolarizations without (*left*) and following a one-second, +70-mV prepulse (*right*). Each trace is the average of 3 responses. **(B)** Current responses from 50-Hz trains of depolarizations without (*left*) and following a one-second, +70-mV prepulse (*right*). Each trace is the average of 8 responses. Data from A and B are from the same dendritic recording, 160 µm from the soma. All commands were relative to the neuron's resting potential (about –65 mV).

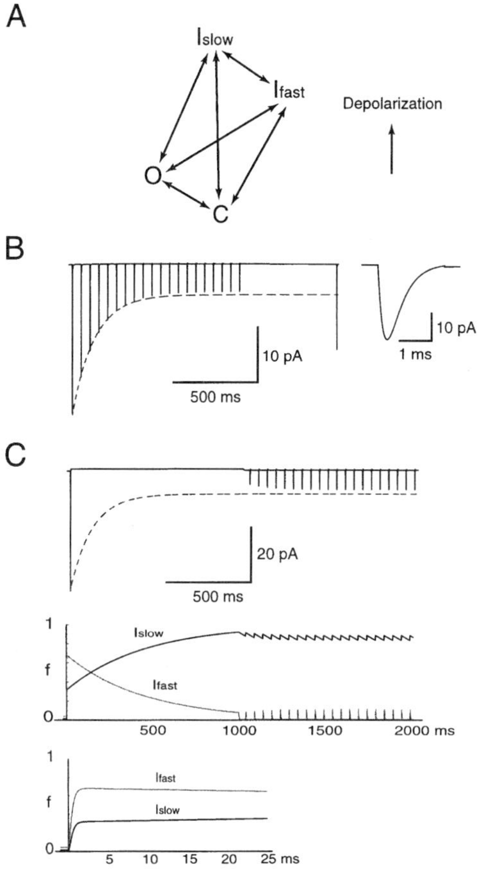

FIGURE 2. A simple gating model can reproduce many of the experimental results. **(A)** Our proposed gating scheme. C is closed, O is open, I_{fast} is fast inactivated, and I_{slow} is slow inactivated. Transitions toward the top of the figure are promoted by depolarization, while those toward the bottom are promoted by hyperpolarization. The model has been scaled to the change according to the free-energy difference between C and the other states at 0 mV. The length of the *vertical arrow* represents −5 kJ/mol. The scaling was done using the equation $\Delta G = -RT\ln(K)$, where ΔG is the free-energy difference relative to C, R is the gas constant, T is temperature, and K is the ratio of the forward and backward rate constants (at 0 mV). **(B)** Induction of and recovery from slow inactivation during simulations of a 20-Hz train of depolarizing pulses. The recovery pulse shown is 500 ms after the end of the train. At right, the first current from the train is shown on an enlarged time scale, to show the kinetics of the current. The *dashed lines* are the fit of three dendritic patches having the most extreme slow inactivation during a 20-Hz train. **(C)** Simulation of the experiment of FIGURE 1A. The model reproduces the same basic responses as the real data. The middle panel in C shows the fraction of channels in I_{fast} (*gray*) and I_{slow} (*black*) during the prepulse experiment. An expansion of the first 25 ms of the simulation is in the **bottom panel of C**, plotting the fraction of channels in both of I_{fast} and I_{slow}. The model initially has a rapid increase in I_{fast} and slower increase in I_{slow}, but the curves eventually cross over (see **middle panel**), signifying the tendency of this model to leave the I_{fast} state for the I_{slow} state at depolarized potentials.

gating scheme is presented in such a manner that transitions in the direction of the arrow (toward the top of the figure) are promoted by depolarizations, while transitions in the opposite direction are promoted by hyperpolarizations. We modeled data from three dendritic patches having the greatest amount of slow inactivation during a 20-Hz train of depolarizing pulses (indicated by the dashed line). The model is able to accurately follow the experimentally measured induction time course for slow inactivation (τ = 390 ms; data not shown) and has a very similar recovery time constant. A further test of this model is shown in FIGURE 2C, in which the experiment of FIGURE 1 is simulated. The model's responses (FIG. 2C, top) are qualitatively very similar to those of the sodium currents of FIGURE 1. The mechanism underlying entry into I_{slow} can be seen in the lower panels of FIGURE 2C, in which is plotted the fraction of channels (f) in the I_{fast} or I_{slow} states during the simulation. While initially many channels enter the I_{fast} state, during the continuous long depolarization $f(I_{fast})$ begins to decrease, while $f(I_{slow})$ increases and eventually surpasses $f(I_{fast})$, as indicated by the crossing of the two curves. This is a direct result of the I_{fast} to I_{slow} transition, which is promoted by long depolarizations. The model also suggests that there may be direct entry from O into the I_{slow} state. This can be seen more clearly in the bottom panel of FIGURE 2C, in which the rapid increase in $f(I_{slow})$ is more visible on an expanded time scale. This direct entry is likely to be the most important route of entry into slow inactivation during rapid depolarizations such as action potentials. This also suggests that some of the "fast" inactivation that is measured could be due to transitions from the open state to the slow inactivated state.

Although a simplified model of slow inactivation is useful for understanding how these channels could function, it is important to note what such a model cannot do. Other "ultraslow" forms of inactivation (e.g., Ref. 6) occur on a much longer time scale and therefore have different time constants than the model. In addition, we have found that the amount of recovery is also voltage dependent, with larger command potentials resulting in faster recovery from slow inactivation (data not shown). These results alone require that multiple slow inactivated states be added to sodium channel models. However, more work is needed to know accurately where the other slow inactivated states should be placed in such models. It will also be important to discover the molecular mechanisms involved in slow inactivation of sodium channels, as well as their function in a systems level context.

[**NOTE ADDED IN PROOF:** This work has recently been described in further detail (see Ref. 7).]

REFERENCES

1. COLBERT, C.M. *et al.* 1997. Slow recovery from inactivation of Na^+ channels underlies the activity-dependent attenuation of dendritic action potentials in hippocampal CA1 pyramidal neurons. J. Neurosci. **17**(17): 6512–6521.
2. JUNG, H.Y., T. MICKUS & N. SPRUSTON. 1997. Prolonged sodium channel inactivation contributes to dendritic action potential attenuation in hippocampal pyramidal neurons. J. Neurosci. **17**(17): 6639–6646.
3. MARTINA, M. & P. JONAS. 1997. Functional differences in Na^+ channel gating between fast-spiking interneurones and principal neurones of rat hippocampus. J. Physiol. (Lond.) **505**(Pt 3): 593–603.
4. GOLDING, N. & N. SPRUSTON. 1998. Dendritic sodium spikes are variable triggers of axonal action potentials in hippocampal CA1 pyramidal neurons. Neuron **21:** 1189–1200.

5. STUART, G., H.-U. DODT & B. SAKMANN. 1993. Patch-clamp recordings from the soma and dendrites of neurons in brain slices using infrared video microscopy. Pflüg. Arch. **423:** 511–518.
6. TOIB, A., V. LYAKHOV & S. MAROM. 1998. Interaction between duration of activity and time course of recovery from slow inactivation in mammalian brain Na^+ channels. J. Neurosci. **18**(5): 1893–1903.
7. MICKUS, T., H. JUNG & N. SPRUSTON. 1999. Properties of slow, cumulative sodium channel inactivation in rat hippocampal CA1 pyramidal neurons. Biophys. J. **76**(2): In press.

Molecular and Functional Diversity of Voltage-Gated Calcium Channels

HERMAN MORENO DAVILA[a]

Department of Physiology and Neuroscience, New York University Medical Center, 550 First Avenue, MSB 152, New York, New York 10016, USA

ABSTRACT: The contributing roles of voltage-gated calcium channels (VGCC) to the generation of electrical signaling are well documented. VGCCs open in response to depolarization of the plasma membrane and mediate the flux of calcium into excitable cells, which further depolarizes the membrane. But a more relevant role of VGCCs is to serve as highly regulated mechanisms to deliver calcium ions into specific intracellular locales for a variety of calcium-dependent processes including neurotransmitter release, hormone secretion, neuronal survival, and muscle contraction. Recent biochemical and molecular biological studies have demonstrated that the calcium channel pore-forming subunit (α1) is not an isolated entity, but in fact interacts physically with a variety of strategically localized proteins. The functional consequences of such interactions as well as other molecular aspects of VGCC will be discussed. Finally, although far from a final conclusion, what is currently known about the molecular composition of native calcium channels will be summarized.

Calcium ions play fundamental roles in the regulation of many cellular processes.[1,2] It is therefore essential that their intracellular levels be maintained under strict, yet dynamic control.[3,4] Voltage-gated calcium channels (VGCC) serve as one of the important mechanism for fast calcium flux into the cell, allowing entry of calcium ions across the plasma membrane. A diversity of VGCC have been described in both excitable and nonexcitable cells. Interestingly, a single cell can express a variety of VGCC. Different channel types are specifically localized in the cell and are therefore capable of generating distinct calcium compartments in the cell.[5] The principal subunits (α1) of VGCC belong to a gene family, whose members can form functional calcium channels by themselves when expressed in heterologous expression systems.[6] In native cells α1 subunits are expressed as multisubunit complexes with ancillary subunits, which modify the functional properties of α1 subunit.[7] Domains involved in different functional properties of VGCC have been mapped by mutagenesis, heterologous expression, and biophysical analysis.[8] The recently reported crystal structure of a K^+ channel[9] is in good agreement with many of the conclusions obtained from structure-function studies on calcium channels. In addition to the fundamental signaling role of calcium in cell physiology, consensus and controversies related to many aspects of these channels have provided an area of unique research interest—a fact reflected by many meetings and reviews that focus solely on these proteins. This review focuses mainly on the molecular and functional characterization of neuronal VGCC, as a way to integrate current molecular and biophysical information. Other topics on calcium channels such as subunit interactions and G protein modulation have been reviewed recently[10,11] (see also other chapters in this section).

[a]Phone: 212-263-5426; fax: 212-689-9060; e-mail: morenh01@mcrcr.med.nyu.edu

STRUCTURAL PROPERTIES OF VOLTAGE-GATED CALCIUM CHANNELS AND LOCALIZATION OF THE PORE-FORMING SUBUNITS

Calcium channels are hetero-oligomeric proteins consisting of a pore-forming subunit (termed α1), which is able to form functional channels on its own in heterologous expression systems, and a set of auxiliary or regulatory subunits. The molecular structure of calcium channels has been derived largely from the extensive work carried out with channels purified from the transverse tubule membranes of skeletal muscle,[7] which was facilitated by their high density in this tissue. The channels purified from skeletal muscle (termed L-type; see below) are known to contain four subunits: the α1 subunit (175 kDa), which carries the basic fingerprints of the channel (pseudotetrameric structure, voltage sensor, pore region, β subunit binding site, Ca^{++} binding site, dihydropiridines (DHP) and phenylalkylamide binding sites); a disulfide-linked subunit dimer α2/δ (143 kDa and 27 kDa); an intracellular β subunit (50 kDa); and the transmembrane γ subunit (33 kDa). The specific association of these subunits has been verified by coimmunoprecipitation and copurification experiments from both muscle and neuronal tissues.[12–15] Reconstitution of the purified muscle calcium channel complex in lipid bilayers results in the formation of functional calcium channels, sensitive to DHP and protein kinase A (PKA).[7] The primary structure of these proteins have been deduced by cloning their cDNAs from rabbit skeletal muscle;[16–20] these cDNAs were subsequently employed by many laboratories as probes to clone different α1, β, γ, and α2/δ subunits from a variety of tissues and species, including rat and human. In general, care must be exerted when the results obtained from studies with L-type muscle channels are extrapolated to other channels, as they may reflect peculiar properties of muscle channels that are not necessarily retained by other native channels.

Neuronal calcium channel α1 subunits are the product of at least seven different genes named α1 A–H (see TABLE 1). Two additional partial clones (putative calcium channels) termed α1F and α1I have been submitted to Genbank (U93305 and AL008716).

Inmunocytochemical studies have revealed a differential distribution of α1 calcium channel subunits. α1A and α1B are expressed mainly in dendrites and presynaptic terminals. In general α1A is concentrated in a larger number of nerve terminals when compared to α1B.[21–23] In both the rat and the human neuromuscular junction (NMJ), α1A is local-

TABLE 1. Characteristics of Voltage-Gated Calcium Channels

Voltage Dependence	Channel Type	α1 Subunit	Pharmacology of the Native Current
HVA	L	α1C, α1D, α1S	DHP(s), Phenylalkylamines, Benzodiazepines, ω-Aga IIIA, Mibefradil*, Cd^{++}
	N	α1B	ω-CTX GVIA, ω-Aga IIIA, Mibefradil*, Cd^{++}
	P	α1A?	sFTX, ω-Aga IIIA*, ω-CTXMVIIC, Cd^{++}
	Q	α1A??	ω-Aga IVA*, ω-CTX MVIIC
IVA	R	α1E?	Mibefradil* ω-Aga IIIA, Cd^{++}, Ni^{++}
LVA	T	α1G,α1H	Mibefradil, ω-Aga IIIA, Ni^{++}

ABBREVIATIONS AND SYMBOLS: ω-Aga IIIA, ω agatoxin IIIA; ω-CTX-GVIA, ω-conotoxin from *conus geographus* VIA; ω-CTX-MVIIC, ω-conotoxin from *Conus magus* VIIC; DHP, dihydropyridine; ?, debatable. * denotes intermediate sensitivity; otherwise only potent inhibitors were selected.

ized presynaptically, while α1B and α1A are both present in axon-associated Schwann cells.[24] Class E is localized mainly in cell bodies, in some cases in proximal dendrites, as well as in the distal dendritic branches of Purkinje cells. α1C and α1D are localized in cell bodies and proximal dendrites of central neurons.[25,26] The distribution and localization of the different α1 subunits are in good agreement with pharmacological and electrophysiological characterization of native calcium channels, as well as with calcium imaging both in the CNS and in NMJ.[27–32] Information on the protein localization of the recently cloned α1G and α1H subunits (see Perez Reyes, this volume) is not yet available, although it is known that they are expressed in the CNS, α1G being much more abundant than α1H.

TYPES OF NATIVE CALCIUM CHANNELS AND THEIR MOLECULAR COUNTERPART

Calcium channels have been classified based on their pharmacological and/or electrophysiological properties. The classification of voltage-dependent calcium channels divides them into three groups: high voltage–activated (HVA), which includes L-, N-, P-, and Q-types), intermediate (IVA, R-type), and low voltage–activated (LVA, T-type). It is worth noting that some authors classify R-type as HVA (see details below).[32,33]

HVA Channels

L-Type Calcium Channels

L-type calcium channels mediate DHP-sensitive currents, are HVA, and inactivate with a slow time course. The biophysical properties of L-type currents were originally described in chick dorsal root ganglion.[34] In sensory neurons,[35] the L-type current shows a slow inactivation during a 200-ms depolarization pulse, the decay time constant being higher than 500 ms. Single-channel analysis has showed that in 110 mM Ba^{2+} the slope conductance of this channel is 25 pS,[36] while its activation appears to adopt two gating modes, each of different duration. An important characteristic of the skeletal muscle calcium channel is the requirement of channel phosphorylation in order for the channel to undergo voltage-dependent potentiation.[37] The α1 subunit of these channel types can be α1C, α1D, or α1S.[7,12–14,17,37] Although a variety of DHP-sensitive currents have been described, they remain to be fully characterized.

L-type calcium channels are expressed in neuronal, endocrine, cardiac, smooth, and skeletal muscle, as well as in fibroblasts and kidney cells. In all these tissues, they serve many important cellular functions. For a while it was believed that L-type calcium channels were not involved in CNS neurotransmitter secretion. However, recent reports suggest a role for L-type calcium channels in this process.[38,39]

N-Type Currents

N-type currents are characterized by their irreversible blockade by ω-conotoxin (ω-CTX) GVIA. ω-conotoxin–sensitive currents with various biophysical characteristics have been described, with perhaps the greatest variety being in inactivation rates, which range from 100 ms in sympathetic neurons[40] to 1.5 s in supraoptic neurons,[41] when measured in equivalent divalent ion concentrations. Single-channel conductance measurements have also been used to identify these currents. The conductance is reported to be around 20 pS if these channels are activated above 0 mV in 110 mM Ba^{2+}.[42] However, the sensitivity of these microscopic currents to ω-conotoxin has not always been assessed, and this may lead to potential misinterpretations. The α1 subunit responsible for this current is proposed to be α1B. In fact, the high affinity of ω-conotoxin was used for the isolation and cloning of this subunit.[43] Nevertheless, there are still questions regarding observed differences in kinetics between expressed and native channels. These may be due to different native N-currents, cell-specific posttranslational modifications, and/or protein-protein interactions. Conversely, the idea of the existence of different currents gains support from the finding of biophysical differences between α1B splice variants.[44] Such variability could also be due to differential association with ancillary subunits, which are known to modify channel behavior.[45] The role of splice variants has not been fully addressed, since novel isoforms of the different subunits are still emerging. Although the purified N-type channel multisubunit complex from rabbit brain forms DHP-insensitive calcium currents when incorporated in lipid bilayers, these differ in several electrophysiological parameters from those of native channels, including different single-channel conductances.[43] These findings exemplify the importance of the intracellular milieu in the function of native channels.

P and Q Currents

After the initial description of P-type calcium channels in Purkinje cells[46,47] and the subsequent designations of the Q-type calcium channel as a separate channel category,[48] concern has arisen regarding the distinction between them. In fact, this is a case of high debate; a great deal of uncertainty still exists regarding the molecular counterpart of these currents. P-type calcium currents are DHP, ω-CTX GVIA–resistant currents; they activate above –50 mV, peaking at around +10 mV; in Purkinje cells they display very little if any inactivation over a period of one second.[46,49] Single-channel recordings (made in 110 mm Ba^{2+}) showed channels with conductance levels of 9, 14, and 19 ps.[49] It is not clear if each conductance represents a different type of channel or if the phenomenon represents subconductance levels of a single type of channel. P-type currents are characterized by their sensitivity to two fractions of the *Agelenopsis aperta* venom. The first fraction is a not-well-characterized funnel spider venom toxin (FTX), which is reported to be specific for DHP- and ω-CTX-insensitive HVA currents in both whole cell and in unitary P currents.[46,49] The second fraction, ω agatoxin IVA, is a polypeptide that inhibits macroscopic P currents with a K_d of 2–10 nM.[41,47,50] The ω-agatoxin IVA block can be reversed by a brief train of strong depolarization in different CNS areas.[47] Nevertheless, no single-channel blockade data is yet available for the classical P-type channel. Based on these pharmacological criteria, a cerebellar (granular neurons) P-type Ca^{++} channel with different biophysical properties from the one initially reported in Purkinje cells has been described. This novel current inactivates slowly.[51] Therefore, it is possible that there is more than one P-type calcium current. On the other hand, P-type current inactivation could be regulated by the local microenvironment or other conditions.

Q-type currents are HVA, insensitive to DHP and to ω-CTX-GVIA, and potently (but not specifically) blocked by ωCTX-MVIIC.[48] The Q-type current inactivates and is half blocked by 100 nM ω-Aga IVA, but requires prior elimination of P- and N-type Ca^{++} current components before it can be studied in isolation. This may be problematic because overlapping blocking effects of the toxins have been observed within the concentration range typically considered selective.[50]

The correlation between the α1 subunits and P- and Q-type currents has been controversial. After the initial cloning of α1A[52,53] (abundant protein in the cerebellum) and its expression in *Xenopus* oocytes by different groups,[52–55] it was proposed that α1A underlie Q- and to a lesser extent P- type currents due to the relative low sensitivity to ω-Aga IVA (half block 200 nM) and fast inactivation of α1A currents. However, recent experiments in mammalian cell lines strongly support the idea that the channels composed of α1A, α2/δ, and βIb are the most likely counterpart of native P-type channels,[56,57] since these channels are highly sensitive to ω-Aga IVA in both 293T and COS-7 cells. The currents inactivate slowly during depolarizing pulses of up to 600 ms COS-7 cells.[57] These findings are also in agreement with molecular dissection experiments demonstrating substantial reduction of native P-type currents by treatment of cerebellar neurons with antisense oligonucleotides against α1A subunit.[58] Nevertheless, none of the tested α1A subunit combinations (with βI, β2, or β3, plus α_2/δ) completely reproduce the pharmacological or electrophysiological properties of either P- or Q-types,[56,57] suggesting that other elements from the cell microenviroment may be necessary to reproduce the native properties of these channels. However, the consensus considers that the α1A subunit underlies the pore of both P- and Q-type channels. (See the chapter by Dr. T. Snutch in this section for a somewhat different view of the nature of P- and Q-type channels.)

IVA Channels

R currents are defined as the residual HVA calcium current observed after the application of a toxin cocktail that selectively blocks N-, L-, P-, and Q-type currents.[59] The biophysical properties of this current are difficult to distinguish from N- and Q-type currents in whole-cell mode. There are even some parameters that are reminiscent of T-type currents (τ_{decay} = 22 ms at 0 mV). Due to the way in which this current is defined, it shows relative insensitivity to known calcium channel blockers, with no selective inhibitors known (TABLE 1). Single-channel analysis in cerebellar granular cells revealed the existence of two populations of R-type currents, known as G1 and G2, with distinct voltage of activation (–40 mV and –25 mV).[51] Thus, as for other calcium channel types, it seems that multiple R-type channels can coexist in the same cell type. Perhaps the correlation between the R-type current and α1 subunit is the most controversial. Several groups[6,50,60] have proposed α1E as the pore-forming subunit of R-type currents, which when expressed in heterologous systems mediates currents with a pharmacological profile unlike that of N-, P-, or Q-type channels. Recent experiments correlating the time course of expression of α1E transcripts with that of R type currents and antisense studies[61,62] support this association. The situation is not completely clear, however, as there is a large variability in pharmacological and electrophysiological properties of expressed currents with α1E subunits from different species. For instance, depending on the species, the currents mediated by these subunits can be blocked by amiloride or ω-Aga IVA.[63,64] It is also possible that the portion of the current carried by channels containing any α1 subunit that is not completely blocked by the "toxin cocktail" could potentially be considered an R-type current. For instance, the inhibition produced by ω-Aga IVA and sFTX on α1A + α2/δ + βs currents in 293T cells reached saturation levels without complete block.[56] This leaves open the possibility that at least in certain conditions R-currents could be supported by any of these subunit combinations. Other ways of identifying this current include its sensitivity to Ni^{++}, although this has proved to be less informative, since the reported native R-type channel sensitivity lies between the IC_{50} for α1A and a α1E, α1E being more sensitive. This situation will remain difficult to resolve in the absence of specific blockers.

LVA Channels

The low-threshold calcium current was initially described in 1981 in guinea pig inferior olivary nucleus using intracellular sharp electrodes, and was localized to the soma and proximal dendrites by extracellular recordings of field potentials. In 1985 it was also found in thalamic neurons (reviewed in Ref. 65). Subsequently, the currents have been characterized both in whole-cell and single-channel studies in different cell types. Classical T channels start to open with weak depolarizations reaching voltages much more negative than those required to activate other voltage-gated calcium channels. The currents elicited are transient. Channel inactivation is prevented at very negative potentials, while channel opening is inhibited at a holding potential more positive than –60mV. Perhaps two of its most remarkable biophysical properties are its small conductance (8 pS in 110 mM Ba^{++})[66] and the slow kinetics of deactivation. In the CNS, the T-type channel is thought to be responsible for neuronal oscillatory activity, which is proposed to be involved in processes such as: sleep/wakefulness regulation, motor coordination, and neuronal circuit specification during ontogenesis.[65] In addition, T-type channels are involved in pacemaker activity, low-threshold calcium spikes, and rebound burst firing.[67,68] Investigation of the role of T-type channels in other systems (cardiovascular and endocrine in particular) has demonstrated that they are abundant in proliferating cells in both normal and pathological conditions.[68] The pharmacology of T-type channels is a theme of debate; the present consensus is that Ni^{++} sensitivity is not a reliable parameter by which to identify T-type currents. There is a high variability of sensitivity of different expressed α1 subunits to Ni^{++}, depending on the nature of the coexpressed auxiliary subunits as well as on the extracellular divalent concentrations. Furthermore, L-type channels can also be blocked by Ni^{++}.[68] New drugs, such as mibefradil, are being proposed as relatively specific T- and R-type calcium channel blockers (TABLE 1), although further experimentation is required to confirm this. One thing, however, seems apparent from the current literature; there exists more than one population of LVA channels. These different LVA currents have been described in mouse thalamic neurons, ND7-23 cells (rat DRG/neuroblastoma hybrid),[68] and rat hippocampal CA3 pyramidal neurons.[67] In thalamic and hippocampal neurons there is a T-type classical current (transient, Ni^{++} sensitive, DHP insensitive), but there are marked differences in a second component. In thalamic neurons this is proposed to be a new T-type channel, while in CA3 neurons it may correspond to an "anomalous" L-type current. In the hybrid cell line (DRG/neuroblastoma) the pharmacology of the two LVA currents is quite unusual, since they are sensitive to sFTX and ω-Aga IVA. These findings not only suggest the existence of various T-type currents but also reflect the difficulties involved in the pharmacological classification and dissection of currents.

Until recently, it was not known which α1 subunit(s) underlie(s) T-type channels. The α1G and α1H subunits isolated by Perez Reyes, Cribbs, and their groups[69,70] (see chapter by Perez-Reyes in this section) have the hallmark motifs of a VGCC, including a pseudo-tetrameric structure, the voltage sensor motif "S4," and the P region. Nevertheless, they lack other motifs conserved in the rest of the known α1 subunits such as the motifs involved in interactions with β subunits and calcium, a feature in good agreement with experiments that show the lack of effect of β subunit antisense oligos on native T-type calcium currents.[71] Expression of α1G (which is highly enriched in the CNS) or α1H (abundant in kidney and heart, but also present in the CNS) by themselves produced whole-cell calcium currents with biophysical characteristics very similar to native T-type currents. The pharmacological profile of these new α1 subunit is quite interesting, since both show high sensitivity to mibefradil (IC_{50} 1.4 μmol/L for α1G), whereas only α1H is highly sensitive to Ni^{++}.[69,70] The cloning of these new genes opens great possibilities for the characterization of new T-type–specific blockers, with potential use in a variety of pathologic entities, such as petite mal epilepsy and hypertension. Still, the identification of α1G and α1H does not close the loop on T-type channels, because it has been suggested that a specific isoform of α1E can form T-type channels.[63] More recently, it was also proposed that either α1C, α1B, or α1E could express T-type currents if devoid of auxiliary subunits.[72] Thus, not unlike the case with R-channels, cells may express channels mediating various types of T-like currents, which could be formed by different α1 subunits, depending on the presence or absence of auxiliary subunits.

ARE THE NON–PORE-FORMING SUBUNITS IMPORTANT IN DEFINING CALCIUM CHANNEL TYPE?

α1 subunits form functional calcium channels when expressed by themselves in heterologous expression systems, with the exception of the α1 subunit known as doe-1, an ortholog of α1E, which was isolated from the forebrain of the marine ray *Discopyge ommata.*[6] In the majority of cases tested so far, coexpression of auxiliary subunits affects the biophysical properties of the channels. Coexpression of α2/δ and γ subunits alters the time course and voltage dependence of activation and inactivation of cardiac α1C subunits and enhances macroscopic currents in *Xenopus* oocytes.[73] However, no effect was seen in the whole-cell currents in L cells coexpressing α1S or neuronal α1C and the same auxiliary subunits.[74]

In contrast with these variable results, coexpression of β subunits has important effects on every other α1 subunit tested so far. The coexpression of different β subunits alters different functional parameters. To date β subunits have been shown to change activation properties, steady state inactivation, inactivation kinetics (elicited with square pulses or trains of action potentials[75]), and peak current (which increases in most cases).[45] The functional physiological importance of β subunit on native neuronal VGCC has been explored with antisense treatment,[76] demonstrating a decrease in Bay K 8644–inducible current in native L-type channels. More recently it has been proposed that β subunits modify the interaction of calcium channel antagonist with the α1 subunit. For instance, channels formed by α1A + α2/δ + βIb are more sensitive to sFTX and ω-Aga IVA than α1 A + α2/δ + β2a or β3.[56] Similar results have been obtained in cells expressing α1Ca in which β3 increased the block by verapamil but not by other channel blockers.[68] Influence of β subunits on piperidine and nickel block of specific neuronal calcium channel has also been reported.[68,77] The amplitude of α1A, B, D, and E increased significantly with β5 and α2/δ.[45] In the case of α1Ca, based on biochemical and electrophysiological analysis, it was proposed that α2/δ produced an increase in the amount of α1 in the plasma membrane and an increase in open probability (Po), while the effect of β2a was due only to an increase in Po.[78] The effects of α2/δ subunits on the α1Ca subunit have recently been dissected, demonstrating that the δ domain of the protein is involved in changes of the gating properties of the α1 subunit, while the α2 is related to the number of channels in the membrane.[79] α2b/δ (the neuronal variant) has also been shown to affect the α1B-1 (N-type channel subunit) affinity to ω-CTX GVIA and the magnitude of barium currents, with no apparent changes in the kinetic properties of the current.[80] Novel effects of β subunits on calcium channel properties, such as modification of the sensitivity of the voltage sensor, involvement in channel regulation, and GTPase activation have also been reported.[81–84]

A well-defined interaction motif of α1, which specifically associates with β subunits, has been identified;[85] it is conserved in most but not all cloned α1 subunits.[69,70] Different regions believed to underlie interactions between α1E and βI,[68] as well as between α1A and β4,[86] are localized between loops II and III in α1E and in the carboxyl terminal in α1A.

ARE THE CONDITIONS OF CALCIUM CHANNEL RECORDING BETWEEN THE DIFFERENT GROUPS COMPARABLE?

Differences in pharmacological and electrophysiological data have been reported when identical combinations of calcium channel subunits are expressed in various systems.[55–57] One possible factor is the presence of endogenous calcium channel subunits in the model selected. For example, it has been reported that *Xenopus* oocytes (perhaps the most used heterologous expression system) possess at least two different β subunits (homologous to the β3 subunit) known to interact with expressed α1C subunits.[87] Therefore it is possible that experiments performed in oocytes are influenced by different parameters due to native β subunits, depending on the level of expression of a given batch of oocytes. It has also been reported that many of the commonly used immortalized mammalian cell lines have various native calcium currents.[87,88] The presence of native calcium currents has been reported in untransfected human epithelial kidney cells (HEK 293),[88] although no endogenous αIB-1, βI-2, or α2 (expression of other calcium channel subunits has yet to be done) was detected by mRNA measurements;[80] also, multiple recordings carried out in our laboratory have failed to demonstrate the presence of an endogenous calcium current under high extracellular barium (80 mM) conditions in the closely related HEK 293T cells. It has also been reported that COS-7 and mouse fibroblast (L cells) cells do not have native Ca^{++} currents. In the case of COS-7 it is known that they do not express any of the known calcium channel β subunits.[57,89]

Moreover, there are differences in the extracellular ionic strength used by different experimentalists. It is known that toxin affinity can vary depending on the extracellular divalent concentration. For instance, in two models, *Xenopus* oocytes and HEK 293T cells, the IC_{50} of sFTX for P-type channels is increased in high extracellular barium concentration.[56,90] Nickel sensitivity also changes with total extracellular divalent cation concentration.[68] In conclusion, the nature of the heterologous expression systems and the recording conditions must be taken into account when interpreting the results of heterologous expression experiments.

DOES ALTERNATIVE SPLICING OF CALCIUM CHANNEL SUBUNITS PRODUCE FUNCTIONAL DIFFERENCES?

Much of the molecular diversity of Ca^{++} channels is produced by the existence of multiple forms of α1 subunits.[89] For instance, the α1Ca and α1Cb isoforms have different sensitivities to DHPs,[91,92] but no clear differences in electrophysiological properties. The putative DHP binding site between these two isoforms are identical; therefore, this does not account for the observed differences.

Antibody studies have shown that α1C and α1A alternatively spliced isoforms are differentially expressed and have different sensitivities to phosphorylation by serine-threonine kinases.[93,94] Reports by Soong *et al.*[95] support the importance of alterative splicing in determining the inactivation kinetics of α1A currents. In addition, there have been reports of the differential phosphorylation, localization, and function of the splice variants of α1B, and α1D subunits.[93] Differences in the electrophysiological properties of the α1E have been identified between the mouse and human isoforms (where at least four isoforms have been described) as compared with the rat version (rbEII).[6,64] Moreover, differential sensitivity to β subunits of splice variants of α1E has been reported.[68] The α2 subunit has at least five different alternative splice variants (named α2a–e), which show tissue-specific distribution.[96] For instance, the brain expresses only α2b.

Calcium channel β subunits also further diversify by alternative splicing of primary transcripts,[97–99] resulting in isoforms with distinct patterns of tissue-specific expression. In the case of the β1 subunit, characterization of genomic sequences[97] has led to the conclusion that at least one skeletal muscle and two brain transcripts are the products of the same β1 gene. There is a common mechanism for the production of variants by the four subunit genes, which involves skipping exon 5 and/or 6, thus generating the noted forms (β1d, β2d, β3b, β3c, and β4d).[99,100] This type of splicing causes a frame shift and introduces early termination. The functional activity of these truncated β subunits remains to be determined, as does that of other splicing mechanisms in all four β subunit genes. In general, a large number of calcium channel properties are dictated by the specific isoform(s) of a given subunit.

COMPARTMENTALIZATION OF CELLULAR SIGNALING AND ITS RELATION TO CALCIUM CHANNELS

It is now clear that *in vivo* calcium channels do not just exist as isolated functional entities, but may be associated with proteins that regulate and/or define their cellular role. N-, P-, and Q-type channels are involved in calcium flux in presynaptic nerve terminals and synaptic transmission.[33,46] These channels are known to interact directly with multiple proteins of the synaptic vesicle release machinery in a complex and dynamic fashion [101–106] (see chapters by Tsien and by Catterall, this volume). In the case of the N-type α1 subunit (α1B) the structural motif involved in the interaction with syntaxin, synaptotagmin 1, and SNAP 25 has been identified.[101–103] The interaction between α1B with syntaxin and SNAP 25 is calcium regulated, while the interaction with synaptotagmin 1 is not. These interactions are modulated by serine/threonine kinases.[106] Native N-type–channel multisubunit complexes are also associated with a 95-kDa subunit of unknown function. P/Q channels also interact with the same set of synaptic proteins as N-type channels, including the 95-kDa protein, which has been proposed to be a short version of α1A.[107] Interestingly P/Q α1 subunit isoforms have different patterns of protein-protein interactions as compared to α1B. The α1A rbA isoform interacts with synaptotagmin and SNAP 25 but not with syntaxin, while the BI isoform interacts with the three proteins[104] and with a synaptic vesicle protein (part of the SNARE complex) known as cysteine string protein.[108] Some of these interactions are calcium regulated as well, although differently to what is seen for N-type channels. These series of interactions are known to have functional consequences, as coexpression of syntaxin 1A in *Xenous* oocytes with either α1A or α1B stabilizes the channel in the inactive state.[109] Alteratively, as could be expected, interruption of the interaction between presynaptic channels and syntaxin *in vivo* alters neurotransmission.[110] These series of events reflect the bidirectional modulation of signals between calcium channels and the exocytotic machinery. L-type calcium channels, which required phosphorylation by PKA to undergo voltage-dependent potentiation,[111] retain PKA activity when they are immunoprecipitated. Recently the physical link between the L-type calcium channel and PKA, through a PKA-anchoring protein named AKAP-15, has been identified.[111] There are several examples of the association between different channels and proteins involved in the formation of functional microdomains.[112]

All β calcium channel subunits identified so far share the Src homology 3 domains (SH3), found in a variety of intracellular proteins in organisms from yeast to man.[113] Their occurrence in several signaling molecules and cytoskeletal proteins suggests a role in mediating specific protein-protein interactions involved in signal transduction and/or specific targeting. For example, in K^+ channels it is known that there is direct *in vitro* association between SH3 domain of Src and Kv 1.3 channels.[114] This feature has not yet been explored in calcium channels, but it is interesting, since as reported in Moreno *et al.*[115] the cytosolic tyrosine kinase PYK2 induces modification of calcium channels containing the βIb subunit. In this case, PYK2 possesses two proline-rich regions, which could potentially bind the SH3 domain present in βIb.

CALCIUM CHANNELOPATHIES

The number of channelopathies is quickly growing, and their discovery and study are shedding new light on the pathophysiology of many neurological and nonneurological disorders. Information on the molecular genetics of calcium channels has led to the discovery that alterations in a variety of the calcium channel protein complexes are linked to diseases in rats, mice, and humans (see TABLE 2; see also chapter by J. Noebels in this section). This is perhaps not surprising given the central role of calcium as messenger in so many types of cellular processes, including cell death. Although different Ca^{++} channel subunits have been implicated in several CNS diseases, mutations in the α1A subunit are perhaps the most studied. It has been shown that missense mutation, insertions, and deletions are linked to different pathological states (TABLE 2).

TABLE 2. Calcium Channelopathies

Disease	Calcium Channel Subunit[a]	Model
Familial hemiplegic migraine[116]	α1A	Humans
Episodic ataxia type 2[117]	α1A	Humans
Spinocerebellar ataxia type 6[118]	α1A	Humans
Tottering and Learner phenotypes[119]	α1A	Mice
Lambert-Eaton syndrome[120]	α1A,[b] α1B,[b] and α1B[b]	Humans
Hypokalemic periodic paralysis[121]	α1S	Humans
Muscular dysgenesis[122]	α1S	Mice
Zucker diabetic fatty phenotypes[123]	α1C,[c] α1D[c]	Rats
Lethargic phenotype[124]	β4	Mice
Malignant hyperthermia[125]	α2/δ[d]	Humans
Stargazer[126]	γ	Mice

[a]Diseases are linked to mutation(s) of the gene unless otherwise indicated.

[b]This disease is characterized by antibodies against these subunits.

[c]Decrease in mRNA levels.

[d]Malignant hyperthermia is also linked to mutation in RYR1.

CONCLUDING REMARKS

Native calcium currents and their molecular counterparts have been studied by conjoint electrophysiological, biochemical, and molecular biological experimentation. Consensus exists on the pore-forming (α1) subunits of at least three types of calcium currents (L, N, and T). However, there is still much debate on the α1 subunit responsible for native P, Q, and R channels. Current molecular and pharmacological data reveals that in all VGCC types (L, N, P, Q, R, and T) there is more than one functional current, suggesting that in the near future new calcium channel types may be incorporated into this list. Although significant data is available for some VGCC, the exact subunit composition and stoichiometry of most native calcium channels remain to be elucidated. Future research on calcium channels may relate to such issues, as well as to the understanding of their role in a variety of physiological and pathological entities.

The identification of motifs that regulate the interaction of calcium channel subunits with partner molecules is also of high relevance. These interactions result in a highly compartmentalized set of proteins, involved in a variety of processes, from neurosecretion to muscle contraction.

ACKNOWLEDGMENTS

I thank Drs. B. Rudy and R. Llinás for their support and helpful comments on the manuscript. I also thank P. McIntosh, D. Palcantonakis, and Y. Amarillo for assistance and insightful commentary. Supported by Grants NS30989 and NS35215 to B. Rudy.

[NOTE ADDED IN PROOF: The complete cDNA sequence of α1I, as well as the mRNA localization of α1G, α1H, and α1I were recently published:

LEE, J.H *et al.* 1999. Cloning and expression of a novel member of the low voltage-activated T-type calcium channel family. J. Neurosci. **19**(6): 1912–1921.

TALLEY, E. *et al.* 1999. Differential distribution of three members of a gene family encoding low voltage activated (T-type) calcium channels. J. Neurosci. **19**(6): 1895–1911.]

REFERENCES

1. KATZ, B. 1969. The Release of Neuronal Transmitter Substances. C.C. Thomas, Ed. Liverpool University Press. Liverpool. Sherrington Lectures (**10**)1X: 2–5.
2. CHO, D.W. 1992. Excitotoxic cell death. J. Neurobiol. **23:** 1261–1276.
3. BAIMBRIDGE, K. *et al.* 1992. Calcium binding proteins in the nervous system. Trends Neurosci. **15:** 303–308.
4. PARK, Y. *et al.* 1996. Calcium clearance mechanism in isolated adrenal chromaffin cells. J. Physiol (Lond.) **492**(Pt. 2): 329–346.
5. LLINÁS, R. *et al.* 1992. Presynaptic calcium concentration microdomains and transmitter release. J. Physiol. (Lond.) **86:** 135–138.
6. ZHANG, J. *et al.* 1993. Distinctive pharmacology and kinetics of cloned neuronal calcium channels and their possible counterparts in mammalian CNS neurons. Neuropharmacology **32**(11): 1075–1088.
7. CATTRELL, W. 1995. Structure and function of voltage gated ion channels. Annu. Rev. Biochem. **64:** 493–531.
8. Sather, W. *et al.* 1994. Structural basis of ion channel permeation and selectivity. Curr. Opin. Neurobiol. **3:** 313–323.
9. Doyle, D. *et al.* 1998. The structure of the potassium channel: Molecular basis of K^+ conduction and selectivity. Science **280:** 69–77.
10. WALKER, D. & M. DE WAARD. 1998. Subunit interaction sites in voltage-dependent Ca^{2+} channels: Role in channel function. TINS **21**(4): 48–154.
11. DOLPHIN, A.C. 1998. Mechanisms of modulation of voltage-dependent calcium channels by G proteins. J. Physiol. (Lond.) **506**(Pt. 1): 3–11.
12. CURTIS, B. & W. CATTERALL. 1984. Purification of the calcium antagonist receptor of the voltage sensitive calcium channel from skeletal muscle transverse tubules. Biochemistry **23:** 2113–2118.
13. TAKAHASHI, M. *et al.* 1987. Subunit structure of dihydroyridine sensitive calcium channels from skeletal muscle. Proc. Natl. Acad. Sci. USA **84:** 5478–5482.
14. CURTIS, B. & W. CATTERALL. 1986. Reconstitution of the voltage sensitive calcium channel purified from skeletal muscle transverse tubules. Biochemistry **25:** 3077–3083.
15. FLOCKERZI, V. *et al.* 1986. Purified dihydropyridine binding site from skeletal muscle t-tubules is a functional calcium channel. Nature **323:** 66–68.
16. BOSSE, E. *et al.* 1990. The cDNA and deduced amino acid sequence of the gamma subunit of L type calcium channel from rabbit skeletal muscle. FEBS Lett. **267:** 153–156.
17. TANABE, T. *et al.* 1987. Primary structure of the receptor for calcium channel blockers from skeletal muscle. Nature **328:** 313–318.
18. ELLIS, S. *et al.* 1988. Sequence and expression of mRNAs encoding the alpha1 and alpha2 subunits of aDHP sensitive calcium channel. Science **241:** 1661–1664.
19. JAY, S. *et al.* 1990. Primary structure of the gamma subunit of the DHP sensitive calcium channel from skeletal muscle. Science **248:** 490–492.
20. RUTH, R. *et al.* 1989. Primary structure of the beta subunit of the DHP sensitive calcium channel from skeletal muscle. Science **245:** 1115–1118.
21. SAKURAI, T. *et al.* 1996. Biochemical properties and subcellular distribution of the BI and rbA isoforms of alpha 1A subunits of brain calcium channels. J. Cell Biol **134**(2): 511–528.

22. DAY, N. *et al.* 1996. Distribution of alpha 1A, alpha 1B, and alpha 1E voltage dependent calcium channel subunits in the human hippocampus and parahippocampal gyrus. Neuroscience **71**(4): 1013–1024.
23. LLINÁS, R. *et al.* 1992. Distribution and functional significance of the P type voltage dependent calcium channels in the mammalian central nervous system. TINS **15**(9): 351–353.
24. DAY, N. *et al.* 1997. Differential localization of voltage dependent calcium channel alpha 1 subunits at the human and rat neuromuscular junction. J. Neurosci. **17**(16): 6226–6235.
25. YOKOYAMA, C. *et al.* 1995. Biochemical properties and subcellular distribution of the neuronal class E calcium channel alpha 1 subunit. J. Neurosci. **15**(10): 6419–6432.
26. HELL, J. *et al.* 1993. Identification and differential subcellular localization of the neuronal class C and D L type calcium channel alpha 1 subunits. J. Cell Biol. **123**(4): 949–962.
27. WHEELER, D. *et al.* 1994. Roles of N type and Q type channels in supporting hippocampal synaptic transmission. Science **264**(5155): 107–111.
28. REID, C. *et al.* 1997. Nonuniform distribution of calcium channels subtypes on presynaptic terminals of excitatory synapses in hippocampal cultures. J. Neurosci. **17**(8): 2738–2745.
29. MINTZ, I. *et al.* 1995. Calcium control of transmitter release at a cerebellar synapse. Neuron **15**(3): 675–688.
30. REUTER, H. 1995. Measurements of exocytosis from single presynaptic nerve terminals reveal heterogeneous inhibition be calcium channel blockers. Neuron **14**(4): 773–779.
31. PROTTI, D. & O. UCHITEL. 1993. Transmitter release and presynaptic calcium currents blocked by the spider toxin omega-aga-iva. Neuroreport **5:** 333–336.
32. BIRNBAUMER, L. *et al.* 1994. The naming of voltage-gated calcium channels. Neuron **13:** 505–506.
33. REUTER, H. 1996. Diversity and function of presynaptic calcium channels in the brain. Curr. Opin. Neurobiol. **6:** 331–337.
34. NOWYCKY, M. *et al.* 1985. Three types of neuronal calcium channel with different calcium agonist sensitivity. Nature **316:** 440–443.
35. FOX, A.P. *et al.* 1987. Kinetic and pharmacological properties distinguishing three types of calcium currents in chick sensory neurones. J. Physiol. (Lond.) **394:** 149–172.
36. FOX, A.P. *et al.* 1987. Single-channel recordings of three types of calcium channels in chick sensory neurones. J. Physiol. (Lond.) **394:** 173–200.
37. CATTERALL, W. 1997. Modulation of sodium and calcium channels by protein phosphorylation and G proteins. Adv. Second Messenger Phosphoprotein Res. **31:** 159–181.
38. BONCI, A. *et al.* 1998. L-type calcium channels mediate a slow excitatory synaptic transmission in rat midbrain dopaminergic neurons. J. Neurosci. **8**(17): 6693–6703.
39. PROTTI, D.A. 1998. Calcium currents and calcium signaling in rod bipolar cells of rat retinal slices. J. Neurosci. **18**(10): 3715–3724.
40. PLUMMER, N. & P. Hess. 1991. Reversible uncoupling of inactivation in N type calcium channels. Nature **351:** 657–659.
41. FISHER, T. & C. BOURQUE. 1995. Distinct omega agatoxin sensitive calcium currents in somata and axon terminals of rat supraoptic neurons. J. Physiol. **489**(Pt. 2): 383–388.
42. ELMSLIE, K. 1997. Identification of the single channels that underlie the N type and L type calcium currents in bullfrog sympathetic neurons. J. Neurosci. **17:** 2658–2668.
43. DE WAARD, M. *et al.* 1994. Functional properties of the purified N type calcium channel from rabbit brain. J. Biol. Chem. **269**(9): 6716–6724.
44. LIN, Z. *et al.* 1997. Identification of functionally distinct isoforms of the N-type Ca^{2+} channel in rat sympathetic ganglia and brain. Neuron **18**(1): 153–166.
45. ISOM, L. *et al.* 1994. Auxiliary subunits of voltage gated ion channels. Neuron **12:** 1183–1194.
46. LLINÁS, R. *et al.* 1989. Blocking and isolation of a calcium channel from neurons in mammals and cephalopods utilizing a toxin fraction (FIX) from the funnel web spider poison. Proc. Natl. Acad. Sci. USA **86:** 1689–1693.
47. MINTZ, I. *et al.* 1992. P type calcium channels in rat central and peripheral neurons. Neuron **9:** 85–95.
48. RANDALL, A. & R. TSIEN. 1995. Pharmacological dissection of multiple types of calcium channel currents in rat cerebellar granule neurons. J. Neurosci. **15**(4): 2995–3012.
49. USOWICZ, M. *et al.* 1992. P type calcium channels in the somata and dendrites of adult cerebellar Purkinje cells. **9:** 1185–1199.

50. Pearson, H. *et al.* 1995. Characterization of calcium channel currents in cultured rat cerebellar granule neurones. J. Physiol. **482.3:** 493–509.
51. Tottene, A. *et al.* 1996. Functional diversity of P type and R type calcium channels in rat cerebellar neurons. J. Neurosci. **16**(20): 6353–6363.
52. Starr, T. *et al.* 1991. Primary structure of a calcium channel that is highly expressed in rat cerebellum. Proc. Natl. Acad. Sci. USA **88:** 5621–5625.
53. Mori, Y. *et al.* 1991. Primary structure and functional expression from complementary DNA of a brain calcium channel. Nature **350:** 398–402.
54. Sather, W. *et al.* 1993. Distinctive biophysical and pharmacological properties of class A (BI) calcium channel alpha 1A subunits. Neuron **11**(2): 291–303.
55. Stea, A. *et al.* 1994. Localization and functional properties of a rat brain alpha 1A calcium channel reflect similarities to neuronal Q and P type channels. Proc. Natl. Acad. Sci. USA **91:** 10576–10580.
56. Moreno, H. *et al.* 1997. B subunits influence the biophysical and pharmacological differences between P and Q type calcium currents expressed in a mammalian cell line. Proc. Natl. Acad. Sci. USA **94:** 14042–14047.
57. Berrow, N. *et al.* 1997. Properties of cloned rat alpha 1A calcium channels transiently expressed in the COS-7 cell line. Eur. J. Neurosci. **9:** 739–748.
58. Gillard, S. *et al.* 1997. Identification of a pore forming subunit of P type calcium channels: An antisense study on rat cerebellar Purkinje cells in culture. Neuropharmacology **36**(3): 405–409.
59. Randall, A. & R. Tsien. 1997. Contrasting biophysical and pharmacological properties of T type and R type calcium channels. Neuropharmacology **36**(7): 879–893.
60. Wakamori, M. *et al.* 1994. Distinctive functional properties of the neuronal BII (class E) calcium channel. Recept. Channels **2**(4): 303–314.
61. Pietrobon, A. *et al.* 1997. Correlation between native neuronal calcium channels and cloned calcium channel subunits. Soc. Neurosci. Abstr. 783.8a.
62. Piedras-Renteria, E.S. & R.W. Tsien. 1998. Antisense oligonucleotides against alpha1E reduce R-type calcium currents in cerebellar granule cells. Proc. Natl. Acad. Sci. USA **95**(13): 7760–7765.
63. Schneider, T. *et al.* 1994. Molecular analysis and functional expression of the human type E neuronal Ca2+ channel alpha 1 subunit. Recept. Channel **2:** 255–270.
64. Soong, T. *et al.* 1994. Structure and functional expression of a member of the low voltage activated calcium channel family. Science **260:** 1133–1136.
65. Llinás, R. 1988. The intrinsic electrophysiological properties of mammalian neurones: Insight into central nervous system function. Science **242:** 1654–1664.
66. Carbone, E. & H. Lux. 1985. Single low voltage activated calcium channels in chick and rat sensory neurons. Physiol. **386:** 571–601.
67. Avery, R. & D. Johnston. 1996. Multiple channel contribute to the low voltage activated calcium current in hippocampal CA3 pyramidal neurons. J. Neurosci. **16**(18): 5567–5582.
68. Ertel, S. & E. Ertel. 1997. Low voltage activated T type calcium channel. TIPS **18:** 37–42.
69. Perez-Reyes, E. *et al.* 1998. Molecular characterization of a neuronal low voltage activated T type calcium channel. Nature **391:** 896–900.
70. Cribbs, L.L. *et al.* 1998. Cloning and characterization of alpha1H from human heart, a member of the T-type Ca2+ channel gene family. Circ. Res. **83**(1): 103–109.
71. Lambert, R. *et al.* 1997. T type calcium current properties are not modified by calcium channel beta subunit depletion in nodosus ganglion neurons. J. Neurosci. **17:** 6621–6628.
72. Meir, A. & A. Dolphin. 1998. Known calcium channel alpha 1 subunits can form low threshold small conductance channels with similarities to native T type channels. Neuron **20:** 341–351.
73. Singer, D. M. Biel, I. Lotan, V. Flockerzi, F. Hofmann, N. Dascal *et al.* 1991. The roles of the subunits in the function of the calcium channel. Science **253:** 1553–1557.
74. Tomlinson, W. *et al.* 1993. Functional properties of a neuronal class C L type calcium channel. Neuropharmacology **32:** 1117–1126.
75. Patil, P.G. *et al.* 1998. Preferential closed-state inactivation of neuronal calcium channels. Neuron **5:** 1027–1038.
76. Berrow, N. *et al.* 1995. Antisense depletion of beta subunits modulates the biophysical and pharmacological properties of neuronal calcium channels. J. Physiol. (Lond.) **482**(Pt. 3): 481–491.

77. ZAMPONI, G. *et al.* 1996. Beta subunit coexpression and the alpha 1 subunit domain I-II linker affect piperidine block of neuronal calcium channels. J. Neurosci. **16**(8): 2430–2443.
78. SHISTIK, E. *et al.* 1995. Calcium current enhancement by alpha2/delta and beta subunits in Xenopus oocytes: Contribution of changes in channel gating and alpha1 protein level. J. Physiol. **489:** 1:55–62.
79. FELIX, R. *et al.* 1997. Dissection of functional domains of the voltage-dependent Ca2+ channel alpha2/delta subunit. J. Neurosci. **17**(18): 6884–6891.
80. BRUST, P. *et al.* 1993. Human neuronal voltage dependent calcium channels: studies on subunit structure and role in channel assembly. Neuropharmacology **32**(11)**:**1089–1102.
81. CENS, T. *et al.* 1998. Expression of beta subunit modulates surface potential sensing by calcium channels Pfluegers Arch. Eur. J. Physiol. **435**(6): 865–867.
82. CAMPBELL, V. *et al.* 1995. Inhibition of the interaction of G protein Go with calcium channels by the calcium channel beta subunit in rat neurones. J. Physiol. (Lond.) 485(Pt. 2): 365–372.
83. BOURON, A. *et al.* 1995. The beta 1 subunit is essential for modulation by PKC of an human and a non human L type calcium channel. FEBS Lett. **377**(2): 159–162.
84. CAMPBELL, V. *et al.* 1995. Voltage dependent calcium channel beta subunits in combination with alpha 1 subunits, have a GTPase activating effect to promote the hydrolysis of GTP by G alpha or in rat frontal cortex. FEBS Lett. **370**(1–2): 135–140.
85. PRAGNELL, M. *et al.* 1994. Calcium channel beta subunit binds to a conserved motif in the I-II cytoplasmic linker of the alpha 1-subunit. Nature **368**(6466): 66–70.
86. WALKER, D. *et al.* 1998. A beta 4 isoform–specific interaction site in the carboxyl-terminal region of the voltage-dependent Ca2+ channel alpha 1A subunit. J. Bio. Chem. **273**(4): 2361–2367.
87. TEREILUS, E. *et al.* 1997. A Xenopus oocyte beta subunit: Evidence for a role in the assembly/expression of voltage gated calcium channels that is separated from its role as a regulatory subunit. Proc. Natl.Acad. Sci: USA **94**(5): 1703–1708.
88. BERJUKOW, S. *et al.* 1996. Endogenous calcium channels in HEK 293. Br. J. Pharmacol. **118:** 748–754.
89. HOFMANN, F. *et al.* 1994. Molecular basis for calcium channel diversity. Annu. Rev. Neurosci. **17:** 399–418.
90. LIN, J., B. RUDY, & R. LLINÁS. 1990. Funnel-web spider venom and a toxin fraction block calcium current expressed from rat brain mRNA in Xenopus oocytes. Proc. Natl. Acad. Sci. USA **87:** 4538–4552.
91. WELLING, A. *et al.* 1992. Expression and regulation of cardiac and smooth muscle calcium channels. J. Pharmacol. **58**(Suppl. II) :258–262.
92. WELLING, A. *et al.* 1993. Stable co-expression of calcium channel alpha 1, beta and alpha2/delta subunits in a somatic cell line. J. Physiol. **471:** 749–765.
93. HELL, J. *et al.* 1994. Differential phosphorylation, localization and function of distinct alpha 1 subunits of neuronal calcium channels. Two size forms for class B, C and D 1 subunits with different COOH termini. Ann. N.Y. Acad. Sci. **747:** 282–293.
94. SAKURAI, T. *et al* 1995. Inmunochemical identification and differential phosphorylation of alternatively spliced forms of the alpha 1A subunit of brain calcium channels. J. Biol. Chem. **270**(36): 21234–21242.
95. SOONG, T. *et al.* 1994. Alternative splicing generates rat brain alpha 1A calcium channel isoforms with distinct eletrophysiological properties. Soc. Neurosci. Abstr. **20:** 70.
96. WILLIAMS, M.E. *et al.* 1992. Structure and functional expression of alpha 1, alpha 2, and beta subunits of a novel human neuronal calcium channel subtype. Neuron **1:** 71–84.
97. HULLIN, R. *et al.* 1992. Calcium channel beta subunit heterogeneity: Functional expression of cloned cDNA from heart, aorta and brain. EMBO J. **11:** 885–890.
98. POWERS, P. *et al.* 1992. Skeletal muscle and brain isoforms of a beta subunit of human voltage dependent calcium channels are encoded by a single gene. J. Biol. Chem. **167:** 22967–222972.
99. MURAKAMI, M. *et al.* 1996. Gene structure of the murine calcium channel beta3 subunit, cDNA and characterization of alternative splicing and transcription products. Eur. J. Biochem. **263**(1): 138–143.
100. PEREZ-REYES, E. *et al.* 1992. Cloning and expression of a cardiac/brain beta subunit of the L type calcium channel. J. Biol. Chem. **267:** 1792–1797.

101. SHENG, Z. *et al.* 1997. Interaction of the synprint site of N type calcium channels with the C2B domain of synaptotagmin I. Proc. Natl. Acad. Sci. USA **94**(10): 5405–5410.
102. SHENG, Z. *et al.* 1994. Identification of a syntaxin binding site on N type calcium channels. Neuron **13**(6)**:**1303–1313.
103. SHENG, Z. *et al.* 1996. Calcium dependent interaction of N type calcium channels with the synaptic core complex. Nature **379**(6564)**:**451–454.
104. RETTIG, J. *et al.* 1996. Isoform specific interaction of the alpha 1A subunits of brain calcium channels with the presynaptic proteins syntaxin and SNAP 25. Proc. Natl. Acad Sci. USA **93**(14): 7363–7368.
105. KIM, D. & W. CATTERALL. 1997. Calcium dependent and independent interactions of the isoforms of the alpha 1A subunit of brain calcium channels with presynaptic SNARE proteins. Proc. Natl. Acad. Sci. USA **94**(26): 14782–14786.
106. YOKOYAMA, C. *et al.* 1997. Phosphorylation of the synaptic protein interaction site on N type calcium channels inhibits interactions with SNARE proteins. J. Neurosci. **17**(18): 6929–6938.
107. SCOTT, V.E. *et al.* 1998. Evidence for a 95 kDa short form of the alpha1A subunit associated with the omega-conotoxin MVIIC receptor of the P/Q-type Ca2+ channels. J. Neurosci. **18**(2): 641–647.
108. LEVEQUE, C. *et al.* 1998. Interaction of cysteine string proteins with the alpha1A subunit of the P/Q-type calcium channel. J. Biol. **273**(22): 13488–13492.
109. BEZPROZVANNY, I. *et al.* 1995. Functional impact of syntaxin on gating of N type and Q type calcium channels. Nature **378**(6557): 623–626.
110. RETTIG, J. *et al.* 1997. Alteration of calcium dependence of neurotransmitter release by disruption of calcium channelsyntaxin interaction. J. Neurosci. **179**(17): 6647–6656.
111. GRAY, P.C. *et al.* 1998. Primary structure and function of an A kinase anchoring protein associated with calcium channels. Neuron **5:** 1017–1026.
112. SHENG, M. & E. KIM. 1996. Ion channel associated proteins. Curr. Opin. Neurobiol. **6:** 602–608.
113. SCHLESSINGER, J. 1992. SH2/SH3 signaling proteins. Curr. Opin. Genet. Dev. **4**(1): 20–25.
114. ESGUERRA, M. *et al.* 1994. Cloned calcium dependent K^+ channel modulated by a functionally associated protein kinase. Nature **369:** 563–565.
115. MORENO, H. *et al.* 1998. Calcium dependent activation of P type calcium channel by PYK2 mediated phosphorylation of an auxiliary subunit. Soc. Neurosci. Abstr. 621.19: 1577.
116. OPHOFF, R.A. *et al.* 1996. Familial hemiplegic migraine and episodic ataxia type-2 are caused by mutations in the Ca2+ channel gene CACNL1A4. Cell **87**(3): 543–552.
117. HESS, J.E. 1996. Migraines in mice? Cell **87:** 1149–1151.
118. ZHUCHENKO, O. *et al.* 1997. Autosomal dominant cerebellar ataxia (SCA6) associated with small polyglutamine expansions in the alpha 1A voltage dependent calcium channel. Nature Genet. **15:** 62–69.
119. FLETCHER, C. *et al.* 1996. Absence epilepsy in tottering mutant mice is associated with calcium channel defects. Cell **87:** 607–617.
120. LENNON, V. *et al.* 1995. Calcium channel antibodies in the Lambert-Eaton syndrome and other paraneoplastic syndromes. N. Eng. J. Med. **332:** 1467–1474.
121. PTACEK, L. *et al.* 1994. Dihydropyridine receptor mutations cause hypokalemic periodic paralysis. Cell **77:** 863–868.
122. CHAUDHARI, N. 1992. A single nucleotide deletion in the skeletal muscle specific calcium channel transcript of muscular dysgenesis (mdg) mice. J. Biol. Chem. **267:** 25636–25639.
123. ROE, M. *et al.* 1996. NIDDM is associated with loss of pancreatic beta cell L type calcium channel activity. Am. J. Physiol **270:** E133–E140.
124. BURGESS, D.L. *et al.* 1997. Mutation of the Ca2+ channel beta subunit gene Cchb4 is associated with ataxia and seizures in the lethargic (lh) mouse. Cell **88**(3): 385–392.
125. ILES, D. *et al.* 1994. Localization of the gene encoding the alpha2/delta subunits of the L type voltage dependent calcium channel to chromosome 7q and analysis of the segregation of flanking markers in malignant hyperthermia susceptible families. Hum. Molec. Genet. **3:** 969–975.
126. LETTS, V.A. *et al.* 1998. The mouse stargazer gene encodes a neuronal Ca2+-channel gamma subunit. Nat. Genet. **4:** 340–347.

α_{1B} N-Type Calcium Channel Isoforms with Distinct Biophysical Properties

ANTHONY STEA,[a] STEFAN J. DUBEL,[b] AND TERRY P. SNUTCH[b,c]

[a]*University-College of the Fraser Valley, 33844 King Road, Abbostford, B.C., Canada V2S 7M8*

[b]*Biotechnology Laboratory, Room 237-6174 University Boulevard, University of British Columbia, Vancouver, B.C. Canada V6T 1Z3*

ABSTRACT: N-type calcium channels both generate the initial calcium signal to trigger neurotransmitter release and also interact with synaptic release proteins at many mammalian central nervous system synapses. Two isoforms of the α_{1B} N-type channel from rat brain ($\alpha_{1B\text{-}I}$ and $\alpha_{1B\text{-}II}$) were found to differ in four regions: (1) a glutamate (Glu) to glycine (Gly) substitution in domain I S3; (2) a Gly to Glu substitution in the domain I-II linker; (3) the insertion or deletion of an alanine (Ala) in the domain I-II linker; and (4) the presence or absence of serine/phenylalanine/methionine/glycine (SFMG) in the linker between domain III S3-S4. Comparison of the electrophysiological properties of the $\alpha_{1B\text{-}I}$ and $\alpha_{1B\text{-}II}$ N-type channels shows that they exhibit distinct kinetics as well as altered current-voltage relations. Utilizing chimeric $\alpha_{1B\text{-}I}$ and $\alpha_{1B\text{-}II}$ cDNAs, we show that: (1) the Glu 177 to Gly substitution in domain I S3 increases the rate of activation by ~15-fold; (2) the presence or absence of Ala 415 in the domain I-II linker alters current-voltage relations by ~10 mV but does not affect channel kinetics; (3) the substitution of Gly 387 to Glu in the domain I-II linker also has no effect on kinetics; and (4) the presence or absence of SFMG (1236–1239) in domain III S3-S4 did not significantly affect channel current-voltage relations, kinetics, or steady state inactivation. We conclude that molecularly distinct α_{1B} isoforms are expressed in rat brain and may account for some of the functional diversity of N-type currents in native cells.

Voltage-gated calcium channels contribute to a number of physiological processes in the nervous system, including the regulation of calcium-dependent enzymes and gene transcription, the shaping of action potentials and firing patterns, and the initiation of neurotransmitter release. Of the four major classes of native neuronal calcium channels described to date (L-, N-, P/Q-, and T-types), N-type and P/Q-type channels are localized to presynaptic terminals and contribute to neurotransmitter release.[1–5] While both N-type and P/Q-type channels are inhibited by the neuropeptide toxin ω-conotoxin MVIIC (ω-CTX-MVIIC), N-type channels can be pharmacologically isolated by their selective and irreversible block by ω-conotoxin GVIA (ω-CTX-GVIA).[6–8] In contrast, P/Q-type channels are preferentially blocked by the spider toxin, ω-agatoxin IVA (ω-AGA-IVA).[9] Examination of central and peripheral neurons has revealed native N-type currents with distinct single-channel, kinetic, and voltage-dependent properties.[10–14] While a number of kinetic models have been proposed to account for the heterogeneous gating characteristics of N-type channels,[11,15–17] the exact molecular determinants involved remain to be precisely defined.

[c]Corresponding author: Dr. Terry P. Snutch, Biotechnology Laboratory, Rm. 237-6174 University Blvd., University of British Columbia, Vancouver, B.C. Canada V6T 1Z3. Phone: 604-822-6968; fax: 604-822-6470; e-mail: snutch@zoology.ubc.ca

High-threshold neuronal calcium channels (L-, N-, P/Q-types) are heterotrimeric complexes consisting of a pore-forming α_1 subunit containing four conserved structural domains, a β subunit that interacts cytoplasmically with the α_1 subunit domain I-II linker, and an $\alpha_2\delta$ subunit that contains a single transmembrane segment covalently linked to an extracellular component (reviewed in Ref. 18). Molecular cloning and expression studies have documented a larger degree of calcium channel diversity than predicted from pharmacological and electrophysiological studies. To date, four different α_1 subunit genes encoding L-type channels and two genes encoding T-type channels have been described. N-type, P/Q-type, and a novel type (α_{1E}) are each encoded by single genes[18–21] (see TABLE 1). With the exception of the T-type α_{1G} and α_{1H} subunits, the kinetic and current-voltage–dependent properties of currents induced by the α_1 subunits is modulated by coexpression any one of four different β subunits (β1–β4).[22–24]

TABLE 1. Pharmacological Profiles of Cloned and Native Calcium Currents

α_1 Subunit Gene	ω-CTX- GVIA	Dihydropyridines	ω-AGA-IVA	ω-CTX- MVIIC	Native Channel Type
α_{1A}	–	–	✓	✓	P/Q-type
α_{1B}	✓	–	–	✓	N-type
α_{1C}	–	✓	–	–	L-type
α_{1D}	–	✓	–	–	L-type
α_{1E}	–	–	–	–	Novel
α_{1F}	?	?	?	?	L-type[a]
α_{1G}	–	–	–	–	T-type
α_{1H}	–	–	–	–	T-type
α_{1S}	–	✓	–	–	L-type

[a] α_{1F} is predicted to be an L-type channel based on the conservation of residues that confer dihydropyridine sensitivity.[21]

Within most classes of α_1 subunits, closely related isoforms have been identified (reviewed in Ref. 18), although there are only a few conclusive reports demonstrating alternative splicing as the mechanism for generating calcium channel diversity.[25–27] We have previously reported the cloning and expression of α_{1B} N-type channel isoforms derived from rat brain,[28–31] and Lipscombe and coworkers[32] have recently reported further α_{1B} isoforms expressed in rat sympathetic ganglia. Together with a number of α_{1B} variants found in mouse, rabbit, and humans,[33–35] it is apparent that multiple α_{1B} N-type isoforms are expressed both across species and between tissues and cell types within a species. In the present study, we report that the rat brain $\alpha_{1B\text{-}I}$ and $\alpha_{1B\text{-}II}$ isoforms exhibit distinct biophysical properties which can be attributed to small amino acid alterations in defined regions of the N-type channel α_{1B} subunit.

MATERIAL AND METHODS

Expression of N-type Calcium Channels in Xenopus *Oocytes*

cDNAs encoding the rat brain calcium channel α_1 subunits, rbB-I ($\alpha_{1B\text{-}I}$ [28,29]) and rbB-II ($\alpha_{1B\text{-}II}$ [30,31]) were expressed in *Xenopus* oocytes as described previously.[31] Briefly, ovaries were surgically removed from anesthetized (0.17% 3-aminobenzoic acid ethyl ester; MS-222) mature female *Xenopus laevis* (Xenopus One, Ann Arbor, MI) and agitated for 2–3 h in 2 mg/ml collagenase (type IA; Sigma) dissolved in a Ca-free OR-2 solution containing (mM): NaCl, 82.5; KCl, 2; $MgCl_2$, 1; HEPES, 5 at a pH of 7.5. Prior to injection, oocytes were allowed to recover for 3–20 h at 18°C in standard oocyte saline (SOS) containing (mM): NaCl, 100; KCl, 2; $CaCl_2$, 1.8; $MgCl_2$, 1; HEPES, 5 at a pH of 7.5; supplemented with 2.5 mM sodium pyruvate and 10 µg/ml gentamycin sulphate (Sigma) or 100 U/ml penicillin-streptomycin. Nuclear injections were performed on stage V and VI oocytes using a Drummond Nanoject Automatic injector. Approximately 1–2 ng of each expression plasmid were injected into the nucleus (~10 nl total volume), and oocytes were maintained in supplemented SOS at 18°C for 2–5 days prior to electrophysiological recording. For this study, three full-length chimeric $\alpha_{1B\text{-}I}$ - $\alpha_{1B\text{-}II}$ cDNAs ($\alpha_{1B\text{-}7}$, $\alpha_{1B\text{-}39}$, $\alpha_{1B\text{-}52}$) were constructed into the vertebrate expression vector, pMT2[36] (see TABLE 2). Some experiments were also performed on another construct (30-14G) that is identical to $\alpha_{1B\text{-}II}$ except that it contains Gly 387 rather than Glu 387 (see Refs. 37,38). In all experiments the α_{1B} cDNAs were always coexpressed with the β_{1b} subunit.

Solutions and Data Analysis

Two-microelectrode voltage clamp experiments were performed on *Xenopus* oocytes using Axoclamp-2A or Geneclamp 500 amplifiers (Axon Instruments, Burlingame, CA) connected to an IBM-compatible computer with PCLAMP software (Axon Instruments). Microelectrodes were filled with 3 M KCl and typically had resistances of 0.5 to 1.5 MΩ. Ba currents were isolated by recording in a solution containing (mM): $BaCl_2$, 40; KCl, 2; tetraethylammonium chloride, 36; 4-amino-pyridine, 5; niflumic acid, 0.4; 5-nitro-2-(3-phenylpropylamino) benzoic acid (NPPB), 0.2; HEPES, 5 at pH 7.6. Many oocytes were injected with 10–30 nl of 100 mM BAPTA-free acid (10 mM HEPES, pH 7.2 with CsOH) to chelate intracellular calcium and prevent Ca-activated chloride current contamination.

Current recordings were capacitance and leak subtracted and filtered at 1000 Hz. Voltage-dependent properties were determined by fitting normalized current-voltage relations and steady-state inactivation data with smooth Boltzmann curves. Activation and inactivation rates were determined by fitting with single exponential curves. Oocytes expressing less than 100 nA of peak Ba current were not used in this study. Only cells with little or no endogenous chloride current (as indicated by an absence of chloride tail currents) were used in the analysis of current kinetics. Significant differences between various parameters were determined using a Student's *t*-test with the significance level set at $p \leq 0.01$. All values given in the text and figures are mean ± standard error of the mean.

RESULTS

The initial characterization of a full-length rat brain N-type channel α_1 subunit (called rbB-I or $\alpha_{1B\text{-}I}$) revealed slowly activating and inactivating whole-cell currents that were irreversibly blocked by 1–2 uM ω-CTX-GVIA (FIG. 1B and Refs. 28,29). The subsequent isolation of a second full-length α_{1B} cDNA (called rbB-II or $\alpha_{1B\text{-}II}$) revealed a fast-activating and -inactivating ω-CTX-GVIA–sensitive current that was more similar to that of native N-type currents (FIG. 1C and Refs. 30,31). Compared to $\alpha_{1B\text{-}I}$, the rat brain $\alpha_{1B\text{-}II}$ N-type channel isoforms differ in four regions: (1) a G to A transition resulting in a glutamate (Glu) to glycine (Gly) substitution at aa # 177 in domain I S3; (2) an A to G transition resulting in a Gly to Glu substitution at aa # 387 in the domain I-II linker; (3) the deletion of three base pairs (bp) in the domain I-II linker resulting in the absence of an alanine (Ala) at aa # 415; and (4) the deletion of 12 bp in the domain III S3-S4 loop resulting in the absence of four amino acids—serine, phenylalanine, methionine, glycine (SFMG) at aa # 1236–1239 (FIG. 1A).

In addition to the rat brain $\alpha_{1B\text{-}I}$ and $\alpha_{1B\text{-}II}$ isoforms, a number of α_{1B} variants have also been described in rat sympathetic ganglia[32] as well as in mouse, rabbit, and human.[33–35] To date, only the rat brain $\alpha_{1B\text{-}I}$ (rbB-I) isoform exhibits Glu 177 in domain I S3, while the rat brain $\alpha_{1B\text{-}II}$ and all other N-type channels possess Gly at this position. In contrast, only the rat brain $\alpha_{1B\text{-}II}$ isoform exhibits a Glu residue at position 387 in the domain I-II linker, while $\alpha_{1B\text{-}I}$ and all other N-type channels possess a Gly at this position. The presence or absence of Ala 415 in the domain I-II linker appears to be a prevalent N-type channel variant and has now been found in rat and mouse brain as well as in rat sympathetic ganglia. To date, the presence of the tetrapeptide SFMG in domain III S3-S4 has been detected in α_{1B} transcripts from rat brain and sympathetic ganglia and also in mouse brain (Genbank accession #2811218). Although not found in either of the rat brain $\alpha_{1B\text{-}I}$ or $\alpha_{1B\text{-}II}$ subtypes, rat sympathetic ganglia and mouse and human neuroblastomas express additional variants exhibiting the presence or absence of the dipeptide Glu/Thr in the domain IV S3-S4 linker.[32–35]

Comparison of the whole cell properties of $\alpha_{1B\text{-}I}$ and $\alpha_{1B\text{-}II}$ N-type channels shows that they exhibit striking differences in their kinetic properties (FIG. 1B and 1C). The $\alpha_{1B\text{-}II}$ currents exhibit fast activation (τ_{act} = 2.8 ± 0.2 ms) and inactivation (τ_{inact} = 112.1 ± 9.7 ms) compared to the slowly activating (τ_{act} = 41 ± 3.6 ms) and inactivating (τ_{inact} = 576.5 ± 97.9 ms) $\alpha_{1B\text{-}I}$ whole-cell currents (also see TABLE 2). This is more clearly observed with a 2 s depolarizing step (FIG. 2A).

In addition to the kinetic differences, a significant affect on current-voltage relation is also observed, with the $\alpha_{1B\text{-}II}$ current-voltage relation being shifted ~10 mV to the right of the $\alpha_{1B\text{-}I}$ (FIG. 2B and TABLE 2). There was no significant difference between the steady state inactivation curves of the two N-type channel isoforms (FIG. 2C and TABLE 1). Both $\alpha_{1B\text{-}I}$ and $\alpha_{1B\text{-}II}$ currents showed a range of expression of peak Ba currents from 100 nA to well over 2 μA ($\alpha_{1B\text{-}I}$ = 817 ± 312 nA, n = 9; $\alpha_{1B\text{-}II}$ = 1210 ± 182 nA, n = 32). The size of whole-cell currents did not significantly affect channel kinetics.

In order to determine the amino acid residues responsible for the kinetic and voltage-dependent differences between $\alpha_{1B\text{-}I}$ and $\alpha_{1B\text{-}II}$ N-type channels, several chimeric cDNAs were constructed and subsequently expressed in *Xenopus* oocytes. TABLE 2 summarizes the differences in the amino acid composition of the three chimeras ($\alpha_{1B\text{-}7}$, $\alpha_{1B\text{-}39}$, and $\alpha_{1B\text{-}52}$) in the four specific regions of variation between $\alpha_{1B\text{-}I}$ and $\alpha_{1B\text{-}II}$. The chimera $\alpha_{1B\text{-}39}$,

TABLE 2. Summary of the Electrophysiological Characteristics of Rat Brain α_{1B} Channels

α_{1B} cDNA	τ act[a] (ms)	τ inact[b] (ms)	V_{50} act[c] (mV)	V_{50} inact[d] (mV)
$\alpha_{1B\text{-I}}$[e] (E, G, +A, +SFMG)	41.0 ± 3.6 ($n = 9$)	576.5 ± 97.9 ($n = 6$)	-0.1 ± 1.9 ($n = 7$)	-63.5 ± 1.68 ($n = 6$)
$\alpha_{1B\text{-II}}$[f] (G, E, -A, -SFMG)	2.8 ± 0.2* ($n = 15$)	112.1 ± 9.7* ($n = 15$)	$+9.7 \pm 0.8$* ($n = 21$)	-67.5 ± 2.9 ($n = 5$)
$\alpha_{1B\text{-7}}$ (G, E, -A, +SFMG)	2.9 ± 0.2* ($n = 20$)	128.5 ± 8.5* ($n = 17$)	$+9.4 \pm 1.6$* ($n = 17$)	-67.1 ± 1.2 ($n = 14$)
$\alpha_{1B\text{-52}}$ (E, G, -A, +SFMG)	37.8 ± 3.1 ($n = 15$)	633.4 ± 64.1 ($n = 9$)	$+19.4 \pm 1.0$* ($n = 10$)	-57.5 ± 1.3 ($n = 7$)
$\alpha_{1B\text{-39}}$ (E, G, +A, -SFMG)	57.5 ± 14.3 ($n = 4$)	726.0 ± 88.6 ($n = 3$)	$+4.6 \pm 0.7$ ($n = 3$)	-57.7 ± 2.8 ($n = 3$)

[a]Single exponential curves were fitted to the initial activation phase of the inward Ba currents and τ values (ms) given in text.

[b]Single exponential curves were fitted to the decaying inactivation phase of the inward Ba currents and τ values (ms) given in text.

[c]The half-point of activation of the current-voltage relation was calculated by smooth curves fit to the raw I/V data.

[d]The half-point of the steady state inactivation curve was calculated by a smooth Boltzmann curve fit to data calculated from peak currents elicited from various holding potentials for 20 s.

[e]For a detailed survey of $\alpha_{1B\text{-I}}$ electrophysiology and pharmacology see Ref. 29.

[f]For a detailed survey of $\alpha_{1B\text{-II}}$ electrophysiology and pharmacology see Ref. 31.

* $p \leq 0.01$, Student's *t*-test, compared to $\alpha_{1B\text{-I}}$ as control.

which is identical to $\alpha_{1B\text{-I}}$ except that it is missing the SFMG in the putative extracellular loop between S3 and S4 of domain III, exhibited similar kinetics and current-voltage properties as $\alpha_{1B\text{-I}}$ (FIG. 3 and TABLE 2). Furthermore, the deletion of SFMG in $\alpha_{1B\text{-II}}$ did not affect sensitivity to ω-CTX-GVIA (FIG. 1; $n = 11$).

The presence or absence of Ala 415 in the domain I-II linker did not appear to be a determining factor of N-type channel kinetics since the $\alpha_{1B\text{-52}}$ channel differs from $\alpha_{1B\text{-I}}$ only by the Ala 415 deletion and has similar slow kinetics (FIG. 4A, TABLE 2). However, the current-voltage relation for the $\alpha_{1B\text{-52}}$ clone was significantly shifted ~19 mV to the right of the $\alpha_{1B\text{-I}}$ (FIG. 4B, TABLE 2). This rightward shift is similar to that seen for $\alpha_{1B\text{-II}}$ (TABLE 2), although even greater in magnitude. The steady state inactivation of $\alpha_{1B\text{-52}}$ was not significantly different than that of $\alpha_{1B\text{-I}}$ ($V_{50inact} = -58, -64$, respectively; FIG. 4C, TABLE 2).

The other two differences in $\alpha_{1B\text{-II}}$ compared to $\alpha_{1B\text{-I}}$ are amino acid substitutions with Gly at position 177 instead of Glu 177 in domain I S3, and Glu 387 instead of Gly 387 in the domain I-II cytoplasmic linker region. Gly 387 is highly conserved among cloned calcium channels.[18] Electrophysiological analyses of a full-length $\alpha_{1B\text{-II}}$ construct containing

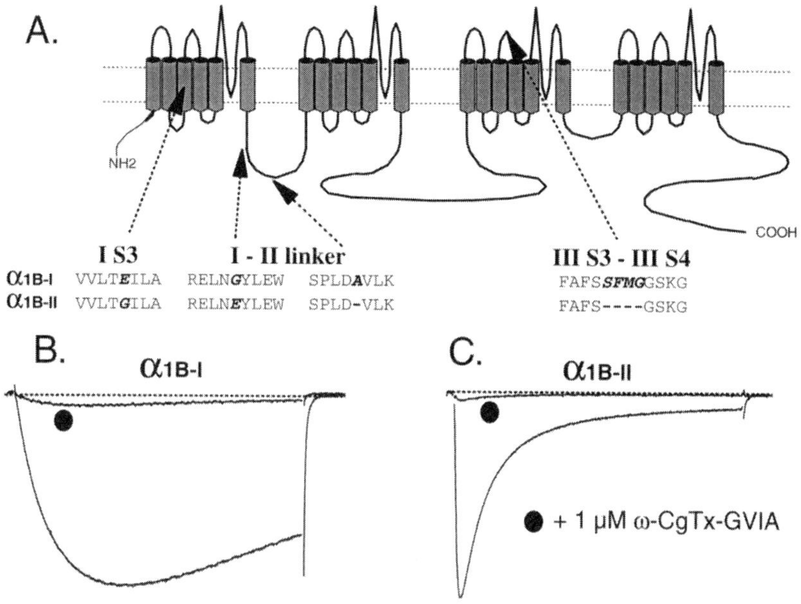

FIGURE 1. Amino acid sequence and functional differences between rat brain $\alpha_{1B\text{-}I}$ and $\alpha_{1B\text{-}II}$ N-type calcium channels. **A.** Schematic representation of sequence differences between $\alpha_{1B\text{-}I}$ and $\alpha_{1B\text{-}II}$. **B.** $\alpha_{1B\text{-}I}$ peak current traces in the presence and absence of 1 μM ω-CTX-GVIA (400 ms step from −100 mV to +20 mV). Note the slow kinetics of activation and inactivation. **C.** $\alpha_{1B\text{-}II}$ peak current traces in the presence and absence of 1 μM ω-CTX-GVIA (400 ms step from −100 mV to +20 mV).

Gly 387 (called 30-14G) showed fast kinetics ($n = 5$, not shown; and also see Refs. 37,38) indicating that Glu 387 is not a determining factor for channel kinetics. The most likely candidate for speeding the activation and inactivation kinetics of $\alpha_{1B\text{-}II}$ compared to $\alpha_{1B\text{-}I}$ is the Glu 177 to Gly 177 substitution in domain I S3. This is clearly seen in FIGURE 5 between constructs including the Gly 177, which display fast activation ($\alpha_{1B\text{-}II}$, $\alpha_{1B\text{-}7}$) and constructs with the Glu 177 in domain I S3, which show slowed kinetics ($\alpha_{1B\text{-}I}$, $\alpha_{1B\text{-}52}$, $\alpha_{1B\text{-}39}$; TABLE 2). N-type channel inactivation kinetics were similarly affected by the Glu 177 to Gly 177 substitution (FIG. 5B and TABLE 2).

DISCUSSION

Species-specific α_{1B} N-type channel isoforms have been cloned from mouse, rabbit, human, and rat.[28–35] Furthermore, native N-type currents from a variety of cell types and preparations exhibit distinct gating properties.[10–14] Attempts to correlate cloned α_1 subunits with native Ca currents are complicated by the fact that it is difficult to determine whether cross-species DNA sequence differences represent evolutionary divergence or

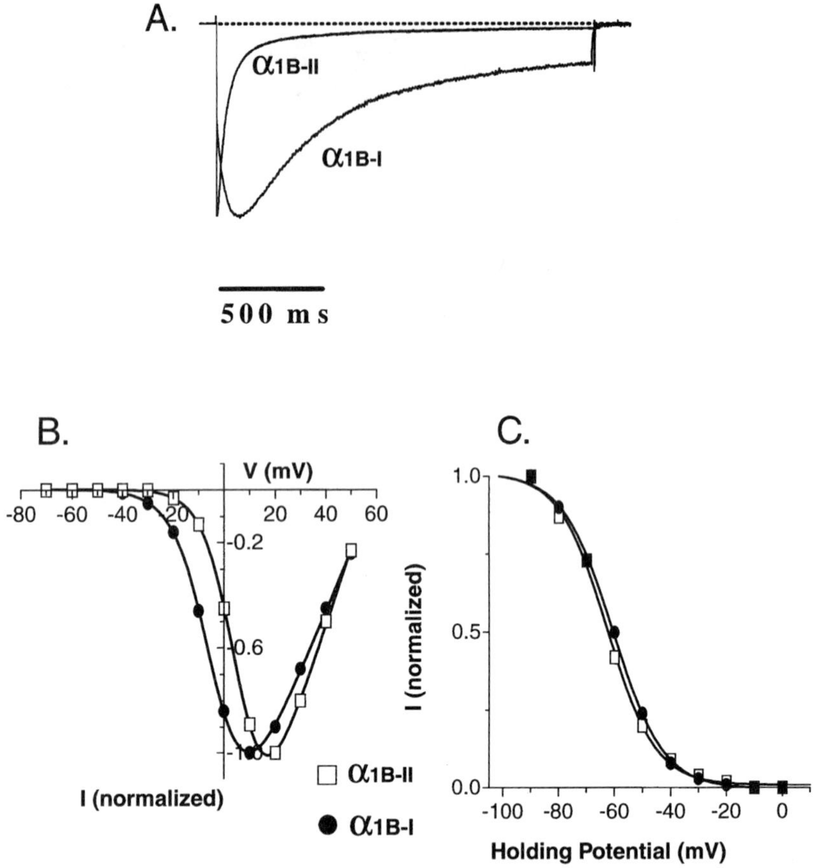

FIGURE 2. Kinetics and voltage-dependent differences between $\alpha_{1B\text{-}I}$ and $\alpha_{1B\text{-}II}$ N-type currents. **A.** Long test pulses (2 s steps from 100 mV to +20 mV) show increased activation and inactivation rates of the $\alpha_{1B\text{-}II}$ isoform. **B.** Current-voltage relation shows an ~10 mV depolarized shift in $\alpha_{1B\text{-}II}$. **C.** Steady state inactivation curves are not significantly different between $\alpha_{1B\text{-}I}$ and $\alpha_{1B\text{-}II}$.

rather that alternatively spliced variants are differentially expressed spatially and temporally. In the present paper we show that two rat brain α_{1B} N-type channel isoforms, $\alpha_{1B\text{-}I}$ and $\alpha_{1B\text{-}II}$ (or rbB-I and rbB-II), exhibit distinct electrophysiological properties. Together with the recent report of Lin and coworkers[32] describing α_{1B} isoforms expressed in rat sympathetic ganglia, the results suggest that multiple functionally distinct α_{1B} N-type calcium channels are expressed in mammalian neurons.

Examination of chimeric cDNAs constructed between different α_1 subunit genes together with *in vitro* mutagenesis studies have identified regions important for permeation, activation, excitation-contraction coupling, and β subunit binding.[39–42] In the present study, we examine chimeric channels derived from isoforms within a single class of α_1 subunit gene and find that modest amino acid alterations can have dramatic effects on N-type calcium channel functional properties. Of the four regions of amino acid difference between $\alpha_{1B\text{-}I}$ and $\alpha_{1B\text{-}II}$, two were found to significantly alter whole-cell N-type current characteristics. First, the single amino acid substitution of Glu 177 to Gly 177 in domain I S3 dramatically affected N-type channel activation and inactivation kinetics. Compared to the kinetically slower $\alpha_{1B\text{-}I}$ containing Glu 177, the $\alpha_{1B\text{-}II}$ subunit with Gly at this site exhibited an approximately 15-fold faster activation rate and an approximately 5-fold faster inactivation rate. The effect of substitutions in domain I S3 of N-type channel

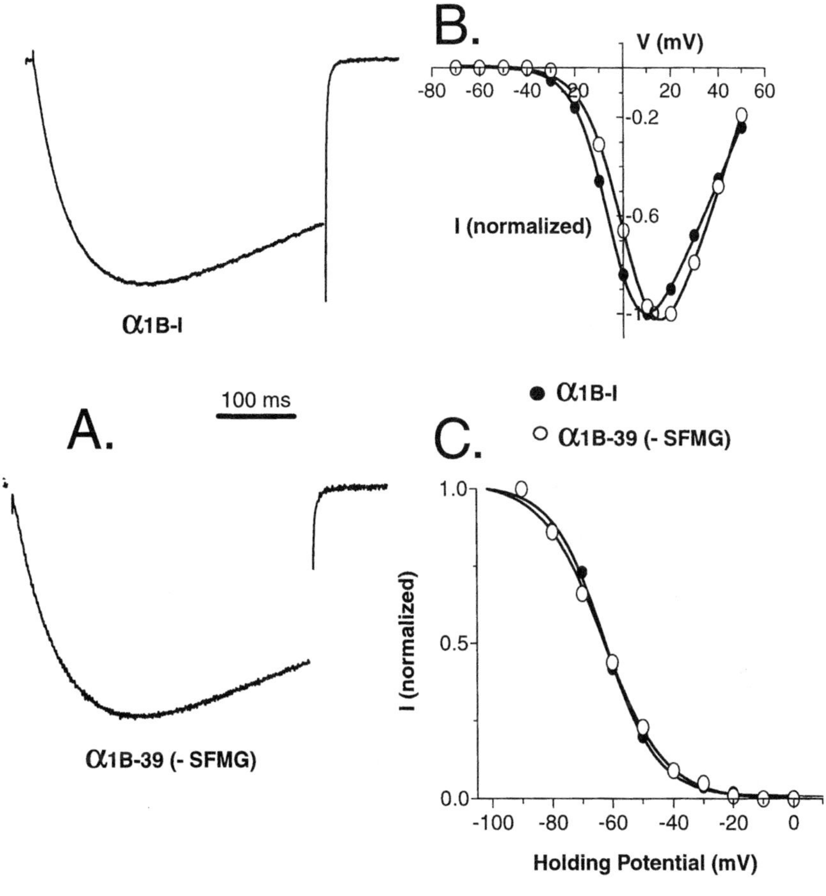

FIGURE 3. Deletion of the tetrapeptide SFMG in the extracellular loop between domain III S3 and III S4 does not alter kinetics or voltage dependence of α_{1B} N-type channels. **A.** Peak current traces of $\alpha_{1B\text{-}I}$ and $\alpha_{1B\text{-}39}$ show similar kinetics. **B.** Current-voltage relations of $\alpha_{1B\text{-}I}$ and $\alpha_{1B\text{-}39}$ show no significant differences. **C.** Steady state inactivation properties of $\alpha_{1B\text{-}I}$ and $\alpha_{1B\text{-}39}$ are similar to each other.

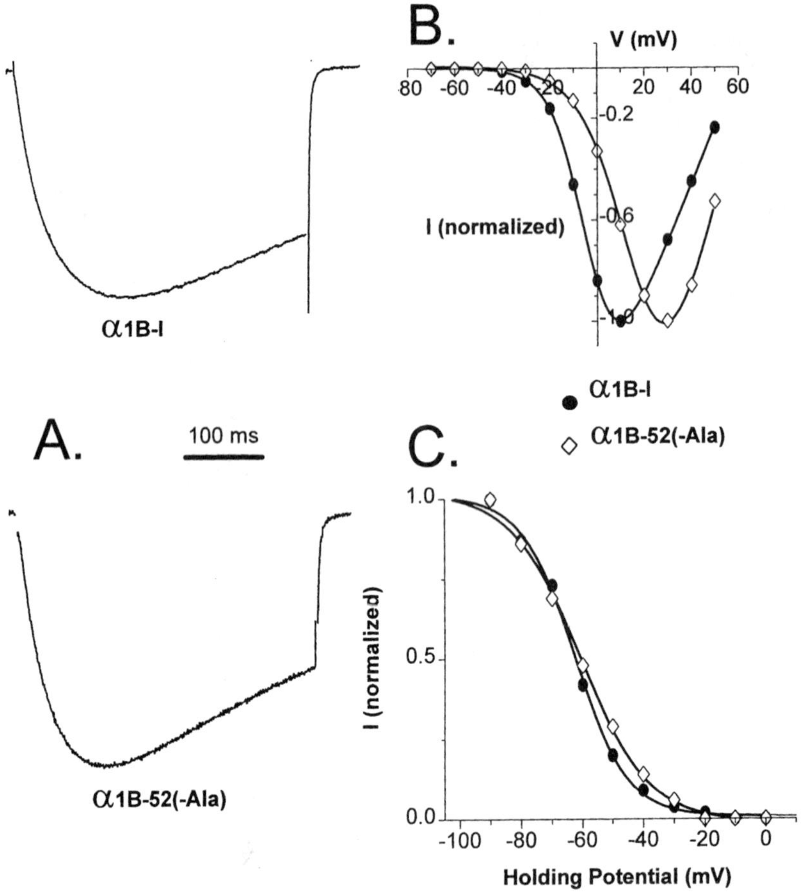

FIGURE 4. Deletion of a single alanine (Ala) residue in the domain I-II intracellular loop alters voltage dependence but not kinetics of α_{1B} currents. **A.** Peak current traces of $\alpha_{1B\text{-}I}$ and $\alpha_{1B\text{-}52}$ show similar kinetics. **B.** Current-voltage relations of $\alpha_{1B\text{-}52}$ are shifted 15–20 mV more depolarized compared to $\alpha_{1B\text{-}I}$. **C.** Steady state inactivation curves of $\alpha_{1B\text{-}I}$ and $\alpha_{1B\text{-}52}$ are not significantly different.

activation kinetics is in agreement with the work of Nakai, Beam, and coworkers examining L-type channel kinetics.[41]

In 11 RT-PCR products examined, Lin and coworkers[32] did not detect the brain Glu 177 variant in superior cervical ganglia (SCG) neurons. The authors suggest that the dramatically slowed activation kinetics observed between the rat brain $\alpha_{1B\text{-}I}$ and other N-type currents is less likely due to the Glu 177 in $\alpha_{1B\text{-}I}$ and more likely due to differences in expression levels, which might alter channel properties possibly through G-protein or calcium channel β subunit interactions. In the present study, the levels of rat brain $\alpha_{1B\text{-}I}$ and $\alpha_{1B\text{-}II}$–induced whole-cell currents were both similar to each other and to those reported by Lin *et al.*,[32] and current size did not affect the kinetics. Indeed, all constructs with the Gly

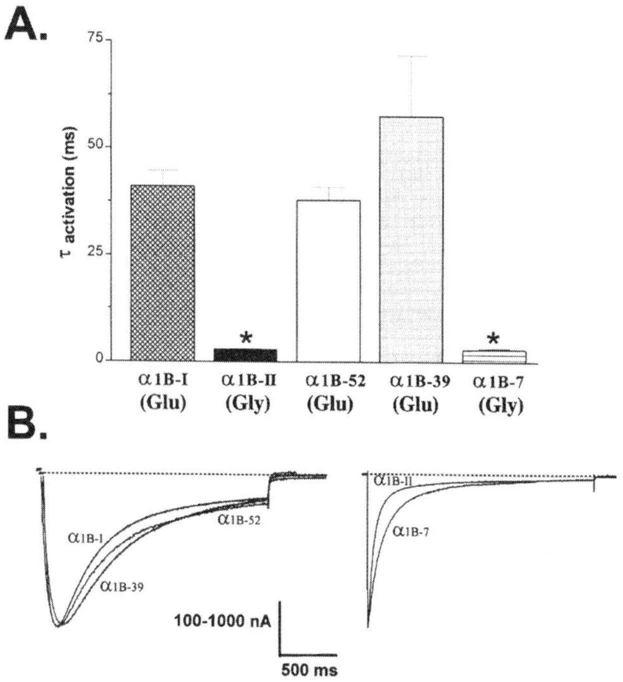

FIGURE 5. Substitution of a glutamate (E) to a glycine (G) in the domain I S3 region alter the kinetics of α_{1B} N-type calcium channels. **A.** The bar chart summarizes the activation rate of various α_{1B} constructs (see TABLE 2) showing the presence of either glutamate (E) or glycine (G) in the domain I S3 region at amino acid #177. Note the fast kinetics of constructs with glycine present. **B.** Long peak current traces (2 s) of the major constructs used in this study, grouping them with either glutamate (*left;* $\alpha_{1B\text{-I}}$, $\alpha_{1B\text{-}52}$, $\alpha_{1B\text{-}39}$) or glycine (*right;* $\alpha_{1B\text{-II}}$ and $\alpha_{1B\text{-}7}$) present in the domain I S3 region. The *asterisks* indicate statistical significance (student's *t*-test, $p < 0.01$) as compared to $\alpha_{1B\text{-I}}$.

177 substitution used in this study show that this site in domain I S3 can affect the activation rate by more than an order of magnitude (FIG. 5 and TABLE 2).

The second significant functional difference identified between the $\alpha_{1B\text{-I}}$ and $\alpha_{1B\text{-II}}$ affects N-type channel current-voltage relations. The presence or absence of Ala 415 in the domain I-II linker resulted in a 10-mV shift in $V_{50\ act}$ between the two N-type channel isoforms. Lin and coworkers[32] also found that the similar presence or absence of Ala 415 occurs in superior cervical ganglia neurons. In contrast to our results, the authors suggest that the Ala 415 variation has little influence on N-type channel gating, although they did not test this directly with constructs comparing the presence or absence of Ala 415.

Our results suggest that the absence of Ala 415 results in a depolarizing shift in N-type channel current-voltage relations and also that other structural elements are likely to affect current-voltage relations. For example, a comparison of the variants in SCG neurons showed that the presence or absence of SFMG in domain III S3-S4 and the presence or absence of the dipeptide Glu/Thr in domain IV S3-S4 also shifted N-type current-voltage relations.[32] In our study, the insertion or deletion of SFMG did not measurably affect any physiological property of N-type channels. These results underscore the importance of testing all possible chimeric combinations in order to help define sites that might functionally couple to each other.

It should also be noted that both Gly/Glu 387 and Ala 415 are within the β subunit binding site in the domain I-II linker.[42] Since coexpression of β subunits affects several calcium channel properties, including current-voltage relations and kinetics,[22–24] it is possible that alterations in this region could alter α_1 subunit–β subunit interactions, resulting in altered gating.

In order to confirm that the sequence differences between the rat brain $\alpha_{1B\text{-}I}$ and $\alpha_{1B\text{-}II}$ isoforms reflect alternative splicing of the single α_{1B} gene, it will be necessary to examine rat genomic DNA and intron/exon boundaries. However, genomic analyses of α_{1A} (P/Q-type), α_{1C} (L-type), and α_{1D} (L-type) subunit genes indicate that alternative splicing is a common mechanism to generate calcium channel variants.[25–27] Together with the data of Lin and coworkers[32] showing that α_{1B} isoforms are expressed in a tissue-specific manner, it is likely that most of the reported α_{1B} variants represent bona fide N-type channel variants. The exceptions to this may be Glu 177 in domain I S3 and Gly 387 in the domain I-II linker. Neither of these substitutions has been reported elsewhere, and it is possible that either could result from single base transitions (G to A or A to G) due to reverse transcription errors or to RNA editing. The exact nature of these and all other α_{1B} variants must await the examination of mammalian genomic DNA.

ACKNOWLEDGMENTS

We thank Dr. Mary M. Gilbert for helpful comments on the manuscript. The research was supported by a grant from the Medical Research Council (MRC) of Canada (to T.P.S.) and a postdoctoral fellowship from the MRC (to A.S.). T.P.S. is the recipient of an MRC Scientist Award.

REFERENCES

1. TAKAHASHI, T. & A. MOMIYAMA. 1993. Different types of calcium channels mediate central synaptic transmission. Nature **366:** 156–158.
2. WHEELER, D.B., A. RANDALL & R.W. TSIEN. 1994. Role of N-type and Q-type Ca channels in supporting hippocampal synaptic transmission. Science **264:** 107–111.
3. DUNLAP, K., J.I. LUEBKE & T.J. TURNER. 1995. Exocytotic Ca^{2+} channels in mammalian central neurons. Trends Neurosci. **18:** 89–98.
4. WESTENBROEK, R.E., J.W. HELL, C. WARNER, S.J. DUBEL, T.P. SNUTCH & W.A. CATTERALL. 1992. Biochemical properties and subcellular distribution of an N-type calcium channel α_1 subunit. Neuron **9:** 1–20.
5. WESTENBROEK, R.E., T. SAKURAI, E.M. ELLIOTT, J.W. HELL, T.V.B. STARR, T.P. SNUTCH & W.A. CATTERALL. 1995. Immunohistochemical identification and subcellular distribution of the α_{1A} subunits of brain calcium channels. J. Neurosci. **15:** 6403–6418.
6. OLIVERA, B.M., J.M. MCINTOSH, L.J. CRUZ, F.A. LUQUE & W.R. GRAY. 1984. Purification and sequence of a presynaptic peptide toxin from *Conus geographus* venom. Biochemistry **23:** 5087–5090.
7. BOLAND, L.M., J.A. MORRILL & B.P. BEAN. 1994. ω-Conotoxin block of N-type calcium channels in frog and rat sympathetic neurons. J. Neurosci. **14:** 5011–5027.

8. HILLYARD, D.R., V.D. MONJE, I.M. MINTZ, B.P. BEAN, L. NADASDI, J. RAMACHANDRAN, G. MILJANICH, A. AZIMI-ZOONOOZ, J.M. MCINTOSH, L.J. CRUZ, J.S. IMPERIAL & B.M. OLIVERA. 1992. A new *Conus* peptide ligand for mammalian presynaptic Ca^{2+} channels. Neuron **9:** 69–77.
9. MINTZ, I.M., V.J. VENEMA, K.M. SWIDEREK, T.D. LEE, B.P. BEAN & M.E. ADAMS. 1992. P-type channels blocked by the spider toxin ω-Aga-IVA. Nature **355:** 827–829.
10. LIPSCOMBE, D., S. KONGSAMUT & R.W. TSIEN. 1989. α-adrenergic inhibition of sympathetic neurotransmitter release mediated by selective modulation of N-type calcium channel gating. Nature **340:** 639–642.
11. BEAN, B.P. 1989. Neurotransmitter inhibition of neuronal calcium currents by changes in channel voltage dependence. Nature **340:** 153–156.
12. PLUMMER, M.R., D.E. LOGOTHETIS & P. HESS. 1989. Elementary properties and pharmacological sensitivities of calcium channels in mammalian peripheral neurons. Neuron **2:** 1453–1463.
13. JONES, S.W. & T.N. MARKS. 1989. Calcium currents in bullfrog sympathetic neurones: II inactivation. J. Gen. Physiol. **94:** 169–182.
14. PLUMMER, M.R. & P. HESS. 1991. Reversible uncoupling of inactivation in N-type calcium channels. Nature **351:** 657–659.
15. ELMSLIE, K.S., W. ZHOU & S.W. JONES. 1990. LHRH and GTP-γ-S modify calcium currents activation in bullfrog sympathetic neurons. Neuron **5:** 75–80.
16. BOLAND, L.M. & B.P. BEAN. 1993. Modulation of N-type calcium channels in bullfrog sympathetic neurons by luteinizing hormone–releasing hormone: Kinetics and voltage-dependence. J. Neurosci. **14:** 516–533.
17. PATIL, G., M. DE LEON, R.R. REED, S. DUBEL, T.P. SNUTCH & D.T. YUE. 1996. Elementary events underlying voltage-dependent G-protein inhibition of N-type calcium channels. Biophys. J. **71:** 2509–2521.
18. STEA, A., T.W. SOONG & T.P. SNUTCH. 1995. Voltage-gated calcium channels. *In* Handbook of Receptors and Channels; Ligand- and Voltage-Gated Ion Channels. R. Alan North, Ed.: 113–152. CRC Press. Boca Raton, Florida.
19. PEREZ-REYES E., L.L.CRIBBS, A. DAUD, A.E. LACERDA, J. BARCLAY, M.P. WILLIAMSON, M. FOX, M. REES & J.H. LEE. 1998. Molecular characterization of a neuronal low-voltage–activated T-type calcium channel. Nature **391:** 896–900.
20. CRIBBS L.L., J.H. LEE, J. YANG, J. SATIN, Y. ZHANG, A. DAUD, J. BARCLAY, M.P. WILLIAMSON, M. FOX, M. REES & E. PEREZ-REYES. 1998. Cloning and characterization of alpha1H from human heart, a member of the T-type Ca^{2+} channel gene family. Circ. Res. **83:** 103–109.
21. BECH-HANSEN, N.T., M.J. NAYLOR, T.A. MAYBAUM, W.G. PEARCE, B. KOOP, G.A. FISHMAN, M. METS, M.A. MUSARELLA & K.M. BOYCOTT. 1998. Loss-of-function mutations in a calcium-channel α_1-subunit gene in Xp11.23 cause incomplete X-linked congenital stationary night blindness. Nature Neurosci. **19:** 264–267.
22. LACERDA, A.E., H.S. KIM, P. RUTH, E. PEREZ-REYES, V. FLOCKERZI, F. HOFMAN, L. BIRNBAUMER & A.M. BROWN. 1991. Normalization of current kinetics by interaction between the α_1 and β subunits of the skeletal muscle dihydropyridine-sensitive Ca^{2+} channel. Nature **352:** 527–530.
23. PEREZ-REYES, E., A. CASTELLANO, H.S. KIM, P. BERTRAND, E. BAGGSTROM, A.E. LACERDA, X. WEI & L. BIRNBAUMER. (1992). Cloning and expression of a cardiac/brain β subunit of the L-type calcium channel. J. Biol. Chem. **267:** 1792–1797.
24. STEA, A., W.J. TOMLINSON, T.W. SOONG, E. BOURINET, S.J. DUBEL, S.R. VINCENT & T.P. SNUTCH. 1994. Localization and functional properties of a rat brain α_{1A} calcium channel reflect similarities to neuronal Q- and P-type channels. Proc. Natl. Acad. Sci. USA **91:** 10576–10580.
25. SNUTCH, T. P., W.J. TOMLINSON, J.P. LEONARD & M.M. GILBERT. 1991. Distinct calcium channels are generated by alternative splicing and are differentially expressed in the mammalian CNS. Neuron **7:** 45–57.
26. SOLDATOV, N.M. 1994. Genomic structure of human L-type Ca^{2+} channel. Genomics **22:** 77– 87.
27. IHARA, Y., Y. YAMADA, Y. FUJII, T. GONOI, H. YANO, K. YASUDA, N. INAGAKI, Y. SEINO & S. SEINO. 1995. Molecular diversity and functional characterization of voltage-dependent calcium channels (CACN4) expressed in pancreatic beta-cells. Mol. Endocrinol. **9:** 121–130.
28. DUBEL, S.J., T.V.B. STARR, J. HELL, M.K. AHLIJANIAN, J.J. ENYEART, W.A. CATTERALL., & T.P. SNUTCH. 1992. Molecular cloning of the α_1 subunit of an ω-conotoxin–sensitive calcium channel. Proc. Natl. Acad. Sci. USA **89:** 5058–5062.

29. Stea, A., S.J. Dubel, M. Pragnell, J.P. Leonard, K.P. Campbell & T.P. Snutch. 1993. A β subunit normalizes the electrophysiological properties of a cloned N-type Ca channel α_1 subunit. Neuropharmacology **32:** 1103–1116.
30. Dubel, S.J., A. Stea, & T.P Snutch. 1994. Two cloned rat brain N-type calcium channels have distinct kinetics. Soc. Neurosci. Abstr. 268.12.
31. Stea, A., T.W. Soong & T.P. Snutch. 1995. Determinants of PKC-dependent modulation of a family of neuronal calcium channels. Neuron **15:** 929–940.
32. Lin, Z., S. Haus, J. Edgerton & D. Lipscombe. 1997. Identification of functionally distinct isoforms of the N-type Ca^{2+} channel in rat sympathetic ganglia and brain. Neuron **18:** 153–166.
33. Williams, M.E., P.F. Brust, D.H. Feldman, S. Patthi, S. Simerson, A. Maroufi, A.F. McCue, G. Veliçelebi, S.B. Ellis & M. Harpold. 1992. Structure and functional expression of an ω-conotoxin–sensitive human N-type calcium channel. Science **257:** 389–395.
34. Fujita, Y., M. Mynlieff, R.T. Dirksen, M. Kim, T. Niidome, J. Nakai, T. Friedrich, N. Iwabe, T. Miyata, T. Furuichi, D. Furutama, K. Mikoshiba, Y. Mori & K.G. Beam. 1993. Primary structure and functional expression of the ω-conotoxin–sensitive N-type calcium channel from rabbit brain. Neuron **10:** 585–598.
35. Coppola, T., R. Waldmann, M. Borsotto, C. Heurteaux, G. Romey, M.-G. Mattei & M. Lazdunski. 1994. Molecular cloning of a murine N-type calcium channel α_1 subunit. Evidence for isoforms, brain distribution and chromosomal localization. FEBS Lett. **338:** 1–5.
36. Swick, A.G., M. Janicot, T. Cheneval-Kastelic, J.C. McLenithan & M.D. Lane. 1992. Promoter-cDNA–directed heterologous protein expression in *Xenopus laevis* oocytes. Proc. Natl. Acad. Sci. USA **89:** 1812–1816.
37. Bourinet, E., T.W. Soong, A. Stea & T.P. Snutch. 1996. Determinants of the G-protein dependent opioid modulation of neuronal calcium channels. Proc. Natl. Acad. Sci. USA **93:** 1486–1491.
38. Zamponi, G.W., E. Bourinet, D. Nelson, J. Nargeot & T.P. Snutch. 1997. Crosstalk between G proteins and protein kinase C mediated by the calcium channel α1 subunit. Nature **385:** 242–246.
39. Tanabe, T., K.G. Beam, B.A. Adams, T. Niidome & S. Numa. 1990. Regions of the skeletal muscle dihydropyridine receptor critical for excitation-contraction coupling. Nature **346:** 567–569.
40. Yang, J. P.T. Ellinor, W.A. Sather, J.-F. Zhang & R.W. Tsien. 1993. Molecular determinants of Ca^{2+} selectivity and ion permeation in L-type Ca^{2+} channels. Nature **366:** 158–161.
41. Nakai, J., B.A. Adams, K. Imoto & K.G. Beam. 1994. Critical roles of the S3-S4 linker of repeat I in activation of L-type calcium channels. Proc. Natl. Acad. Sci. USA **91:** 1014– 1018.
42. Pragnell, M., M. De Waard, Y. Mori, T. Tanabe, T.P. Snutch & K.P. Campbell. 1994. Calcium channel β subunit binds to a conserved motif in the I-II cytoplasmic linker of the α_1 subunit. Nature **368:** 68–70.

Molecular Characterization of Two Members of the T-Type Calcium Channel Family

EDWARD PEREZ-REYES,[a] JUNG-HA LEE, AND LEANNE L. CRIBBS

Department of Physiology, Loyola University Medical Center, 2160 South First Avenue, Maywood, Ilinois 60153, USA

ABSTRACT: In this chapter we review our recent studies on the cloning of two novel cDNAs (α1G and α1H), and present electrophysiological evidence that they encode low voltage-activated, T-type calcium channels ($Ca_vT.1$ and $Ca_vT.2$, respectively). The nucleotide sequences of these T channels are very different from high voltage–activated Ca^{2+} channels, which explains why they were not cloned earlier using homology-based strategies. We used a bioinformatic approach, cloning the first fragment *in silico*. We then used this fragment to screen human heart and rat brain λgt10 libraries, leading to the cloning of two full-length cDNAs derived from distinct genes (*CACNA1G* and *CACNA1H*). The deduced amino acid sequences of the T channels (α1G and α1H) are also very different from previously cloned Ca^{2+} and Na^+ channels; however, there are regions of structural similarity. For example, the T channels also contain four repeats, and within each repeat there are six putative membrane-spanning regions and a pore loop. Expression of these cloned channels in either *Xenopus* oocytes or HEK-293 cells leads to the formation of typical T-type currents. As observed for native T currents, these channels activate at potentials near the resting membrane potential, inactivate rapidly, deactivate slowly, and have a tiny single-channel conductance. The currents generated by α1G and α1H are nearly identical in terms of their voltage dependence and kinetics. We present preliminary evidence that nickel may serve as a valuable tool in discriminating between these subtypes.

The history of T-type Ca channels begins in the late 1950s, when it was realized that neurons could be triggered to fire after *hyperpolarizing* current injections (see references in Ref. 1). This phenomena was called *postanodal exaltation*, and is now more commonly referred to as *rebound burst firing*. Typically neurons fire action potentials after depolarizing current injections due to the activation of voltage-gated Na^+ and/or Ca^{2+} channels. The central role of low-threshold Ca^{2+} channels in rebound burst firing was later elucidated by Jahnsen and Llinás.[2,3] They showed that the hyperpolarization allowed a low-threshold Ca^{2+} channel to recover from inactivation. Once the hyperpolarization was removed, the membrane potential would return to its resting level, triggering the voltage-dependent activation of this channel.

Early classification schemes of voltage-gated Ca^{2+} channels were based on the voltage dependence of channel opening. Electrophysiological recordings from neurons, endocrine cells, and cardiac myocytes showed the presence of both low voltage–activated (LVA) and high voltage–activated Ca^{2+} channels (HVA). Tsien and colleagues[4] later classified LVA channels as T-type (transient), and subdivided neuronal HVA channels into L-type (long-lasting), and N-type (neither). These studies established that T-type channels could be defined by their voltage dependence of activation (−50 mV in 10 mM Ca^{2+}) and single channel conductance (8 pS in 110 mM Ba^{2+}). A third distinguishing feature is that T-type

[a]Corresponding author. Phone: 708-216-1240; fax: 708-216-6308; e-mail: eperez@luc.edu

channels close very slowly after a depolarizing pulse, giving rise to slow-deactivating (SD) tail currents.[5] These three properties define T-type channels as a subset of low voltage–activated channels. In most cases these terms are used interchangeably; however, there may be low voltage–activated channels that are not T-type.[6]

A more recent classification of voltage-gated Ca^{2+} channels based on pharmacology recognizes six types of channels, five of which are HVA channels—L-, N-, P-, Q-, and R-type. Molecular biology studies have revealed an even greater diversity of channels, as exemplified by the cloning of three genes that all encode L-type channels (α1S, α1C, and α1D). By combining cloning, expression, and pharmacology, the other three HVA genes cloned to date have been classified as α1A (P/Q-type), α1B (N-type), and α1E (R-type). Although the original report on the expression of α1E suggested that it was a member of the LVA family,[7] subsequent studies have demonstrated that it is a HVA channel.[8–10] Three reasons for the confusion are that both α1E and T-type channels inactivate at negative potentials, are blocked by low doses of Ni^{2+}, and lack a distinctive pharmacology. Two key properties that distinguish α1E from T-type channels are its fast deactivation and its two-fold larger single-channel conductance.[8,10,11]

Calcium influx causes two important effects: (1) it causes the membrane to depolarize further, thereby opening other voltage-activated channels; and (2) Ca^{2+} itself acts as a second messenger, triggering activation of a wide range of enzymes, muscle contraction, hormone and neurotransmitter release, and even gene transcription. Because low voltage–activated Ca^{2+} channels open so close to the resting membrane potential, they are thought to play an important role in the gating of other channels—that is, as pacemakers. Evidence for this pacemaker role has been described in brain and heart. In heart, Hagiwara *et al.*[12] showed that T-type currents could be selectively blocked by low doses of Ni^{2+}, sparing L-type currents. Application of Ni^{2+} to spontaneously beating sinoatrial (SA) nodal cells slowed their beating rate. Nickel had a more dramatic effect on pacemaker cycle length in latent atrial pacemaker cells, slowing the late slope of diastolic depolarization.[13]

In neurons, T-type channels mediate low-threshold Ca^{2+} spikes, which depolarize the membrane to the point where Na^+ or HVA Ca^{2+} channels begin to open. Dendrites have been found to contain considerable amounts of T-type channels, leading to the hypothesis that they play important roles in information processing.[14–16] In a well-polarized neuron, small excitatory postsynaptic potentials (EPSP) would trigger T-type channels directly.[17] As observed for rebound burst firing, in slightly depolarized neurons T channels could be activated by inhibitory postsynaptic potentials (IPSP). Reciprocal connections between any given neuron and an inhibitory interneuron allow for oscillatory responses to a single input. In many systems these oscillations are in the 5–10 Hz range. For example, prominent low-threshold Ca^{2+} spikes were recorded from inferior olivary neurons, which project climbing fibers to cerebellar Purkinje neurons. This oscillatory circuit may underlie tremor, which also shows a similar frequency.[3,18]

Low-threshold spikes are also thought to play an important role in thalamic signaling. Thalamic neurons display two distinct firing patterns. One, called *tonic firing*, occurs when the resting membrane is more positive than −60 mV; here, depolarizations cause repetitive firing due to activation of Na^+ channels (and repolarization due to K^+ channels). The second, called *burst firing*, occurs when the resting membrane potential is more negative; depolarizations trigger a Ca^{2+} spike, which further depolarizes the membrane to the Na^+ channel threshold. Considerable evidence supports the hypothesis that burst firing of thalamic neurons produces the 7–14-Hz spindle activity detected by electroencephalograms

(EEG).[19] Such sleep spindles are normally observed during non-REM sleep when the EEG is synchronized, but also during epileptic attacks (spike-wave discharges). These observations, coupled with the fact that many antiepileptics (ethosuxamide, valproate, phenytoin) can directly block T-type channels *in vitro,*[20] have led to the hypothesis that epilepsy may be caused by overactive T-type channels. Studies in animal models support this theory.[21] Specifically, increased T-current density was found in thalamic neurons in GAERS (*g*enetic *a*bsence *e*pileptic rats from *S*trasbourg). Now that we have identified the genes that encode T channels, this hypothesis can be tested directly by Northern blots to see if message levels are increased, and by sequencing to find the genetic basis of this mutation.

Due to their fast inactivation, T channels are not thought to play an important role in changing the concentration of intracellular Ca^{2+}. HVA channels that inactivate 10-fold more slowly are more ideally suited for this purpose. In cardiac muscle contraction, most of the Ca^{2+} influx is via L-type channels (which then triggers more release from intracellular stores). In neurotransmitter secretion, most of the Ca^{2+} influx is via N- and P/Q-type channels. An important exception are adrenal cortical cells, where T-type channels have been shown to be important in the secretion of both aldosterone[22] and cortisol.[23] Hormone secretion in these cells is unique in that steroid hormones are not packaged in vesicles for release, but rather Ca^{2+} triggers movement of the limiting substrate, cholesterol. Aldosterone secretion can be triggered by small elevations in serum K^+, which in turn slightly depolarize the glomerulosa cell.[24] Similarly, one can imagine that T-type channels may play a role in many cells to determine "basal" Ca^{2+} concentrations.

A unique property of T channels is that they deactivate slowly. This would allow for significant Ca^{2+} influx after short action potentials. Support for this hypothesis was provided by experiments using action potential waveforms to voltage clamp neurons.[25] This may be an important property in regulating gene expression in developing neurons and muscle, which express abundant amounts of T current. This property may also be important for the activation of Ca-dependent K^+ channels, which in turn play an important role in determining the electroresponsiveness of neurons by generating after-hyperpolarizing potentials (AHP).

Abnormal activity of T channels has been suggested in the following diseases: epilepsy,[21] resting tremor of Parkinson's patients,[18] hypertension,[26] restenosis after balloon angioplasty,[27] cardiac hypertrophy,[28] and diabetes.[29] The strongest evidence is for their role in hypertension and in some forms of epilepsy.

Overactivity of T channels may also produce the 3–6-Hz resting tremor found in Parkinson's disease.[18] Resting tremor has been correlated with the rhythmic firing of ventral lateral thalamic neurons, which sends projections to the premotor cortex. This rhythmic firing is due to the interplay of GABA-mediated IPSPs and rebound burst firing of T channels.[19] Tremor has been treated by chronic stimulation or surgical ablation of these thalamic neurons, or their inputs from the globus pallidus. It has been suggested that an effective T channel blocker would provide a less invasive means of therapy.[30]

One of the key tools in understanding the physiology and biochemistry of L-type channels was the development of selective antagonist drugs, clinically known as *calcium channel blockers*.[31] These drugs are useful in treating hypertension, coronary artery disease, and migraine. Until recently there were no useful blockers of T-type channels. Evidence that T-type calcium channel blockade may be clinically useful was provided by clinical trials with a new calcium antagonist, mibefradil (Ro 40-5967), showing relative selectivity for T-type calcium channels.[32,33] Mibefradil showed promising results in the treatment of hypertension and stable angina pectoris; however, the product was pulled from the market because of drug interactions. A role for T-type calcium channels in smooth muscle proliferation is suggested by a study in which mibefradil was shown effective in reducing neointima formation after vascular injury, while equivalent doses of L-type calcium channel blockers were ineffective, implicating mibefradil's ability to block T-type channels.[34] The therapeutic usefulness of some dihydropyridine compounds and the new nondihydropyridine compounds amiodarone and bepridil may be due to their effects on cardiac T-type channels.[35] Therefore, generation of stable cell lines that express T-type channels may provide a useful tool for high throughput screening and eventually lead to the development of new drugs.

RESULTS AND DISCUSSION

Cloning of the T-Type Ca Channel Family

After years of trying to clone T-type channels using PCR based strategies, we decided to try an *in silico* strategy, and clone using a computer chip to surf the GenBank DNA database. This strategy had been successfully used to identify novel ion channels.[36,37] These studies used the program BLAST, the *b*asic *l*ocal *a*lignment *s*earch *t*ool,[38] which can be accessed through the internet (www.ncbi.nlm.nih.gov). This program allows one to probe the database with either a nucleotide or protein sequence (tblastn). The limitation is that one must predict which region might be conserved in the novel gene. Our alternative cloning strategy was to use a text-based search (www2.ncbi.nlm.nih.gov/genbank/query_form.html) of the GenBank to find novel sequences that had homology to any region of cloned Ca^{2+} channels. In particular, we focused on the *e*xpressed *s*equence *t*agged (EST) division, which contains partial cDNA fragments cloned from normalized cDNA libraries,[39] and partially sequenced by the IMAGE Consortium.[40] One clone, H06096, was chosen based on its 30% identity to the carp α1S (GenBank #P22316) and sequenced in its entirety (#AF029228). The first 957 nucleotides of this clone appear to encode a voltage-gated Ca^{2+} channel, as the deduced amino acid sequence contains readily identifiable motifs including an S4 region and a pore loop. After residue 957, the open reading frame is lost, suggesting that this clone contains an intron.

Using the deduced amino acid sequence of AF029228, we searched the GenBank and found #U37548, which contains a putative protein found in the genomic DNA (cosmid C54D2.5) of *Caenorhabditis elegans.*[41] Hydropathy analysis of the *C. elegans* sequence suggested it had the four-domain structure typical of voltage-gated Na^+ and Ca^{2+} channels. AF029228 aligned best with the third domain (58% sequence identity). These results suggested the existence of a novel four-domain Ca^{2+} channel that was common in both nematodes and man.

Using the *BamH1* fragment of H06096 as probe (#162-977), we screened both rat brain and human heart λgt10 cDNA libraries. A full-length rat brain cDNA, referred to as either α1G or $Ca_vT.1a$ (GenBank #AF027984), was assembled from five overlapping clones.[42] Screening of the human heart library was performed at moderate stringency, allowing Dr. Cribbs to isolate a related sequence, α1H or $Ca_vT,2$ (GenBank #AF051947), derived from a distinct gene.[43] A full-length cDNA of α1H was assembled from four overlapping clones in the plasmid vector pSP72 (Promega). For expression in mammalian cells both full-length cDNAs were subcloned into pcDNA-3 (Invitrogen).

Sequence identity among the Ca^{2+} channel α1 subunits is highest in the putative membrane-spanning regions, with most changes being conservative with respect to structure (FIG. 1A). In FIGURE 1A the HVA channels have been reduced to two consensus sequences, one composed of the three L-type channels (S, C, and D) and the other the non-L HVA channels (A, B, and E). Similar alignments were used to generate evolutionary trees (FIG. 1B), which clearly show that the T channels form a distinct subfamily of the four-repeat superfamily. Charged residues are particularly conserved, with many charges being conserved across all domains and in voltage-gated Na^+ and K^+ channels.[44] The charged residues of the S4 regions are also conserved, consistent with their role as a voltage sensor.[45] The cation selectivity of Ca^{2+} channels requires a ring of negative charges provided by glutamate residues found at similar locations in each domain.[46,47] In both α1G and α1H, two of these glutamates are replaced by aspartates, suggesting altered selectivity. In contrast, there is little conservation of the sequences that link these regions within a domain, and even less between the intracellular loops that connect the domains. Notably missing are the motifs involved in binding the β subunit[48] and calcium ions.[49]

Northern analysis of α1G mRNA distribution indicated that it was expressed at high levels in the brain, in particular the thalamus, cerebellum, and amygdala. It is also expressed in the heart.[42] In contrast, α1H mRNA showed a broader distribution in peripheral tissues, with abundant expression in the kidney.[43]

Considerable evidence suggests a link between T-type Ca^{2+} channels and epilepsy.[21,50–52] To explore such a link our collaborators at the Rayne Institute mapped the genes encoding T-type channels. The α1G gene, *CACNA1G*, was mapped to the human chromosome 17q22 and the mouse chromosome 11.[42] A mouse neurological mutant with an ataxic phenotype, *teetering* (*tn*), maps close to the *Cacna1g* locus.[53,54] *Teetering* mice show difficulties in walking, sometimes having fits and upon falling continuing to pedal their feet.[54] Their brains show hindbrain dysgenesis—that is, atrophy of the cerebellum and midbrain. It is interesting to note that mutations in the genes of two calcium channel subunits have already been linked to epileptic and ataxic phenotypes.[55,56] The α1H gene, *CACNA1H*, was mapped to human chromosome 16p13.3 and mouse chromosome 17.[43] Both the polycystic kidney disease 1 (PKD1) gene and the tuberous scelerosis locus TSC2 have also been mapped to 16p13.3.[57] Interestingly the PKD1 (and PKD2) proteins have significant homology to one of the repeats of voltage-gated Ca^{2+} channels; however, they have not been reported to function as ion channels.[58]

Heterologous Expression of Cloned T-Type Channels

Xenopus laevis oocytes are one of the best expression systems for the preliminary characterization of ion channels. We injected oocytes with varying doses of rat α1G cRNA, then measured their activity with the two-microelectrode voltage clamp. In contrast to uninjected oocytes, which in some batches display small (<20 nA) endogenous high voltage–activated Ca^{2+} channel currents, oocytes injected with more than 1 ng α1G cRNA displayed distinctive low voltage–activated currents. These currents were characterized in terms of their voltage dependence of activation and inactivation, kinetics of activation, inactivation, and deactivation, and their single-channel conductance in 115 mM Ba^{2+}.[42]

A

```
                     S1                    S2                    S3                    S4
Repeat I
    GHcon   PWFE-VSMLVI-LNCVTLGMFRP FD-FIFAFFAVEMV-KMVALG- TWNRLDFFIV-AGM-EYSLD -SA-RTVRVLRPLRAINRVPSMRI
    SCDcon  KPFeiiILLtIFANCVALAiYIP vEY-FLiiFtvEaflKvIAYGl GWNlLDFiIV-vGlFsaILE VKALRAFRVLRPLRLVSGVPSLQV
    ABEcon  PPFEYMILATIIANCIVLALEQH TEPYFIGIFCFEAGIKIvALGF GWNVMDFiVVLsGILATaGT LRTLRAVRVLRPLKLVSGIPSLQi
Repeat II
    GHcon   KYF-RGIM-AILVNTLSMG-EYH SNIVFTS-FALEMLLKLLV-GP PYNIFDG-IVVISVWEIVGQ LRTFRL-RVLKLVRFLPAL-RQLV
    SCDcon  -vFYWLVI-lVfLNTLtIaSEHy ANkaLLaLFT-EMLlKMYSLGL lFNRFDCFiVCgGIlEtiLV lRCvRLLRIFKiTryWnSLsNLVA
    ABEcon  Q-FYW-VLslVALNTlCVAiVHy AEF-FLGLFl-EMslKMYGlGp SFNCFDfGVivGSiFEVvWA LRALRLLRIFKvTKYWaSLRNLVV
Repeat II
    GHcon   KMFDHVVLV-IFLNC-TIA-ERP SNYIFTA-F-AEM-VKVVALG- SWN-LDGLLV--S--DI-V- LRVLRLLRTLRPLRVISRA-GLKL
    SCDcon  tiFTNlIL-FIlLSSaaLAAEDP FDiaFT-iFTvEI-LKMTtyGA YFNiLDlLVV-VSLiSfGiq VKILRVLRVLRPLRAINRAKGLKH
    ABEcon  RYFEmcIL-VIA-SSIALAAEDP fDYvFTGVFTFEMVIKMIDlGL LWNILDFiVVsGALVAFAf- IKSLRVLRVLRPLKTIKRLPKLKA
Repeat IV
    GHcon   HYLDLFIT--I--NV-TM-MEHY CNY-FT--FVFE---KLVAF-F RWNQLDLAIVLLS-MGITLE IRIMRVLRIARVLKLLKMA-GMRA
    SCDcon  syFEYlMFvLImLNTiCLaMQHY LNm-FTg-FTvEMiLKliAFKp pWNvFDfLIVIGSIIDViLS itFFRLFRVMRLvKLLSRgEGiRT
    ABEcon  PpFEYtIMAMIALNTvVLMMKfY lNIvFTsvFSlECvLKviAFG- aWNiFDFvTVlGSITdIlvT lSFLrLFRAARLIKLLRQGYTIRI

                     S5                     P Loop                  S6
Repeat I                                              #
    GHcon   MLGNVLLLCFFVFFIFGIVGVQLWAGL FDN--YAWIAIFQVITLEGWVDIMY FYNFIYFILLIIVGSFFMINLCLVVIATQF
    SCDcon  PLlHIALLVLFviIIYAIIGLELF-GK FDNFaFaMLTVfQCITMEGWTDVLY ElPWvYFVsLvifGSFFvLNLVLGVLSGEF
    ABEcon  PLLQIGLLLFFAILmFAIIGLEFYmGK FDNILFAvLTVFQCITMEGWTd-LY TWNWLYFIPLIIIGSFFmLNLVLGVLSGEF
Repeat II                                             #
    GHcon   NVATFC-LLMLFIFIFSILGMHLFGCK FDSLLWAIVTVFQILTQEDWN-VLY SWAALYF-ALMTFGNYVLFNLLVAILVEGF
    SCDcon  SiASLLLLLFLFIIIFsLLGMQLFGGk FDNFPQaLltVFQiLTGEDWNsVMY mlVCIYFIILFiCGNYILLNVFLAIAVDNL
    ABEcon  SIISLLFLLFLFIVVFALLGMQLFGGq FDTFPAAImTVFQILTGEDWNeVMY M-ss-YFIVLTLFGNYTLLNVFLAIAVDNL
Repeat III                                            #
    GHcon   PIGNIV-ICCAFFIIFGILGVQLFKGK FDNLGQALMSLFVL-SKDGWV-IMY PWMLLYFISFLLIV-FFVLNMFVGVVVENF
    SCDcon  TIGNIviVTTLLQFMFACIGVQLFKGK FDNVLsAMMaLFTVSTFEGWP-LLY VEisIFFIIYIIIiAFFMMNIFVGFVIVTF
    ABEcon  NVFNILIVYmLFMFIFAVIAVQLFKGK YDNvlWALLTLFTVSTGEGWPqVLk MEmSIFYVVYFVVFPFFFVNIFVALIIITF
Repeat IV                                             #
    GHcon   QVGNLGLLFMLLFFI-AALGVELFG-L F-NFGMAFLTLFRVSTGDNWNGIMK --SP-YFV-FVL-AQFVLVNVV-AVLMKHL
    SCDcon  ALPYVALLIvMLFFIYAVIGMQmFGKi FQTFPQAVLLLFRCATGEAWQeImL nFA--YFISFYMLCAFLIINLFVAVIMDNF
    ABEcon  ALPYVCLLIAMLFFIYAIIGMQVFGNI FRtFfqaLMLLFRSATGEAWheIML dfAYfYFVSFIFlCSFLMLNLFVAVIMDNF
```

B

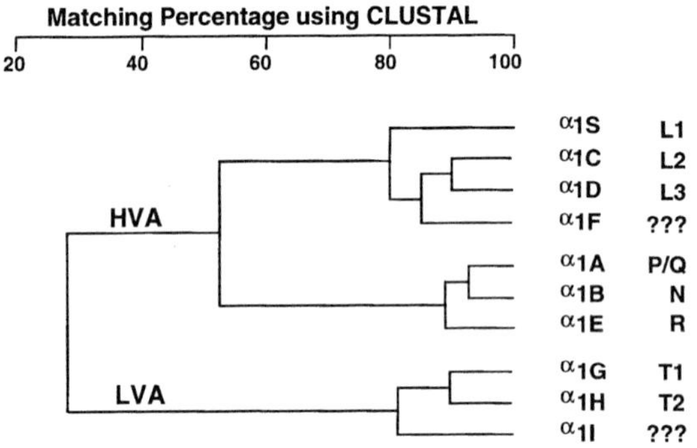

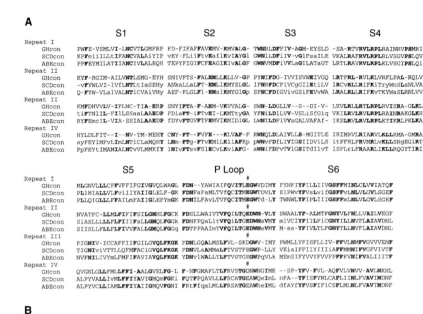

FIGURE 1. Alignment of all voltage-activated calcium channels. **(A)** The putative membrane-spanning regions and the pore loop sequences of each subfamily were reduced to a consensus (con). Conservation of each amino acid is indicated as follows: *uppercase* indicates it is conserved across all members of the each subfamily, *lowercase* indicates it is conserved in two out of three, and a *dash* indicates it is not conserved at all. Amino acids that are conserved across all eight calcium channels are shown in *bold*. The glutamate residues in the pore loops that determine Ca^{2+} selectivity are marked with a *hash* symbol. The GenBank accession numbers of the sequences are: α1G, AF027984; α1H, AF051947; α1S, L33798; α1C, L04569; α1D, M76558; α1A, X99897; α1B, M94172; and α1E, L27745. **(B)** An evolutionary tree was made using the alignment shown in **A**. Also shown are two new putative calcium channel sequences deduced from genomic DNA. The GenBank accession numbers of these partial sequences are: α1F, U93305; and α1I, AL008716.

These studies conclusively demonstrated that α1G encodes a T-type, low voltage–activated channel. It activated during test potentials that were 30 mV more negative than either α1C or α1E, it deactivated very slowly as observed for native T channels,[5] and it had a small single-channel conductance.[59]

The full-length cDNAs of both T channels have been subcloned into the mammalian expression vector pcDNA-3 (Invitrogen). This vector contains the cytomegalovirus pro-

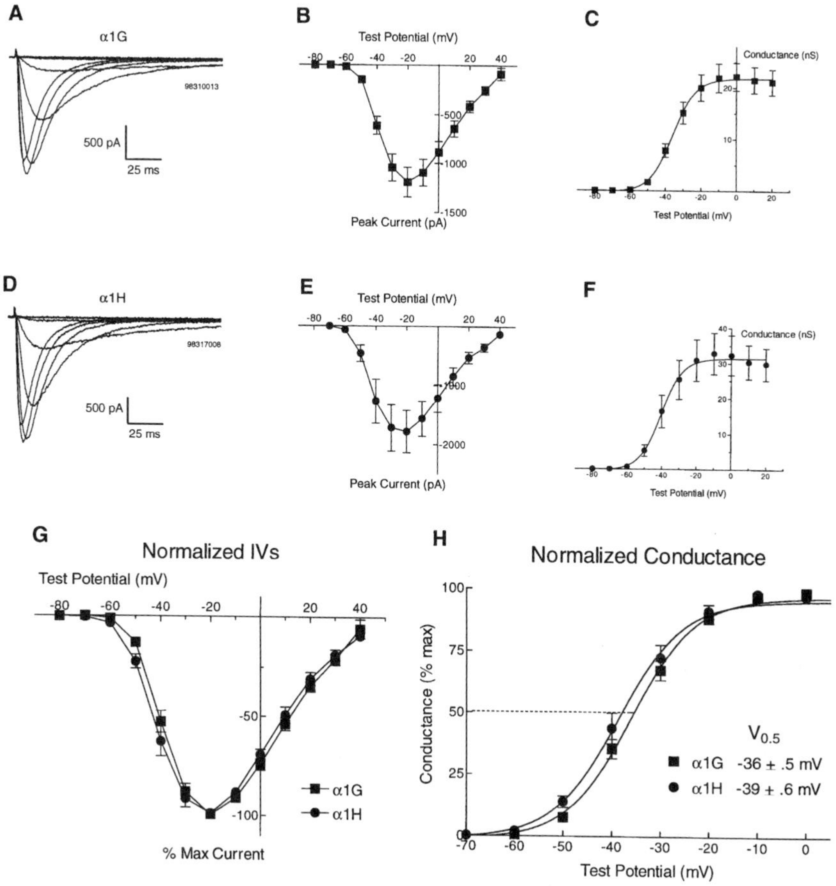

FIGURE 2. Voltage dependence of α1G and α1H activation. **(A, D)** Representative current traces elicited during step depolarizations from −90 mV to varying test potentials. The bath solution contained 10 mM Ba^{2+} (see Ref. 43 for details). Currents were recorded from stably transfected HEK-293 cell lines. **(B, E)** Peak currents were calculated and averaged over a number of cells (α1G, $n = 8$; α1H, $n = 6$). **(C, F)** Chord conductance was calculated by dividing the peak current measured by the driving force, then averaged. **(G, H)** The data from each cell were normalized to the observed maximum, then averaged. **(H)** The normalized conductance was fit with the Boltzmann equation (*smooth curves*). The midpoints of channel activation are shown in the *inset*. The slope of the Boltzmann fit was 6.5 for both channels.

moter for high expression of protein, and it contains the neomycin resistance gene for selection of a stable line. Stable cell lines for both α1G and α1H have been isolated and characterized. In contrast to our experience with α1C, we were able to make stable cell lines that continue to express T channels at high densities (40–50 pA/pF). Clearly the HEK-293 cells "like" to make T channels. These results were obtained by expression of the α1 subunit alone; however, as observed for K^+ channels, this does not rule out the existence of endogenous subunits. It is interesting to note that HEK-293 cells were made from human embryonic kidney, where α1H mRNA was abundantly expressed.

Preliminary comparisons show that α1G and α1H currents are very similar. FIGURE 2 shows unsubtracted current traces elicited from various test potentials, which are used to characterize the current-voltage (IV) relationship. Also shown are average IVs and data transformed into conductance. Typically α1H currents can be clearly observed during depolarizations to −60 mV, a voltage where little current can be detected in α1G-transfected cells. Plots of the normalized conductance show that α1H activates at slightly lower (3 mV) voltages (FIG. 2H).

Both α1G and α1H currents activate, deactivate, and inactivate with similar kinetics (FIG. 3). Similar kinetics have been observed for native T-type channels (reviewed in Refs. 11 and 60). Activation is typically slow near threshold, then accelerates to a limiting value of 1–2 ms (FIG. 3D). Similarly, inactivation is slow near threshold and accelerates to a limiting value of 14 ms. The combination of these kinetic properties leads to the formation of the criss-crossing pattern of successive current traces during the IV protocol. This pattern is typical of T-type currents.[61] In contrast, high voltage–activated channels show a distinct pattern because activation is uniformly fast. Another major difference between LVA and HVA currents is the rate at which they deactivate, which reflects the rate at which channels move from the open to the closed state.[5] These repolarization currents are called *tail currents*. The voltage protocol used to elicit these tail currents includes a short step to open the channels, followed by repolarization to various potentials (FIG. 3A). Typical currents are shown in FIGURE 3A and B. These tail currents can be well fit with a single exponential function to determine the rate constant of deactivation. Even at repolarization potentials of −120 mV the cloned channels deactivate relatively slowly (2.5 ms; FIG. 3C). This rate is even slower at more positive potentials. HVA currents deactivate at least 10-fold faster. For example, α1E and α1B currents were reported to deactivate with a tau of 0.25 ms.[10]

A defining characteristic of T-type Ca^{2+} channels is that their unitary conductance is tiny,[62,63] being two- to threefold lower than HVA channels in Ba^{2+}. Measurement of this conductance is complicated by the low probability of channel opening at negative potentials, where the driving force is larger. Therefore, we used tail current protocols to enhance channel opening at negative potentials. The average single-channel conductance for α1G was 7.5 pS.[42] An average of the values reported in the literature for neuronal T-type channels was 7.7 pS (n = 10 studies, reviewed in Ref. 11). A slightly smaller conductance (5.3 pS) was measured for α1H in HEK-293 cells.[43] We are currently investigating whether this difference in conductance can be ascribed to the different expression systems used, or to differences in the methods used to analyze the data (for α1G Dr. Antonio Lacerda and I used TRANSIT, while the α1H data was collected and analyzed with ChanAnal by Dr. Jon Satin, University of Kentucky). Preliminary results indicate the difference is due to analysis. The cloned T channels display a prominent subconductance state, which can be recognized and separated by TRANSIT.

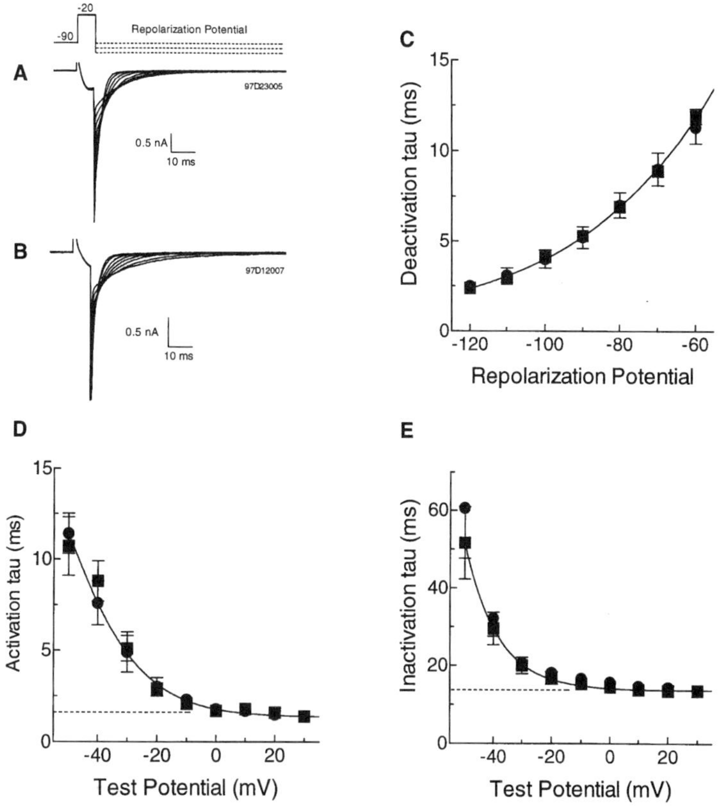

FIGURE 3. Kinetic analysis of α1G and α1H currents. **(A, B)** Representative current traces elicited during the tail voltage protocol. The data were fit with a single exponential to calculate the deactivation tau. **(C)** The deactivation tau is plotted as a function of repolarization. In all figures the α1G data (mean ± SEM) is plotted with *squares* (n = 4), and α1H with *circles* (n = 7). However, these symbols are difficult to see because the measured values are nearly identical. The kinetics of activation **(D)** and inactivation **(E)** were calculated by fitting the current traces obtained during the IV protocol (as shown in Fig. 2A) with two exponentials.

Sensitivity to pharmacological and toxicological agents has provided valuable tools for dissecting the contribution of channel subtypes to native currents. Mibefradil is a relatively new compound that has been shown to block LVA currents at lower doses than required for HVA currents.[33] We compared mibefradil block of α1G, α1H, and the L-type channel complex $\alpha 1C\alpha_2\beta_{2a}$ (Fig. 4A). Currents were elicited every 10 s with test pulses to the peak of their respective IV from a holding potential of −90 mV. Under these conditions, we observed that mibefradil blocked α1G and α1H with similar potency (IC_{50} = 1 μM). In contrast, 14-fold higher concentrations were required to block the cloned L-type currents. Similar results were obtained with native T- and L-type currents.[33] These results support the notion that mibefradil selectively blocks T-type channels, and provides indirect evi-

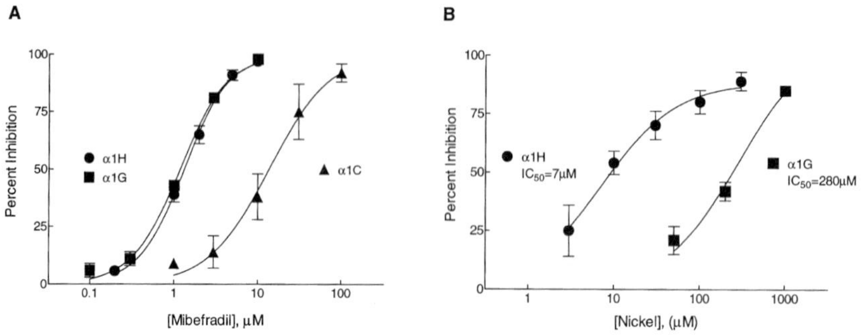

FIGURE 4. Pharmacological analysis of cloned channels. (**A**) mibefradil dose-response curves are shown for α1G, α1H, and α1C. The α1C subunit was coexpressed with rabbit α2a (gift from Dr. Tsutomu Tanabe) and rat β2a. Data were collected from stably transfected cells held at −90 mV and tested at the peak of their respective IVs every 10 s. The *smooth curves* were generated from a fit to the data with a sigmoidal dose-response equation. Mibefradil was a kind gift from Dr. Michel Manganel and Pierre Weber (F. Hofmann-La Roche, Ltd., Basel, Switzerland) (**B**) Nickel ($NiCl_2$) dose-response curves were generated using a similar protocol.

dence for a physiological role of T channels in controlling blood pressure and coronary blood flow.

Numerous studies have demonstrated that LVA currents can be selectively blocked by low concentrations of Ni^{2+}. Studies by Akaike and coworkers provided the first evidence that there were more than one T-type channel.[64,65] An overview of Ni^{2+} and Cd^{2+} block of neuronal LVA currents revealed the presence of at least two channels with very different sensitivities.[11] Therefore, we investigated the Ni^{2+} sensitivity of α1G and α1H (FIG. 4B). To our surprise, α1G was largely insensitive to Ni^{2+}, requiring 280 µM for half-maximal block. Similarly insensitive LVA currents have been recorded from neurons isolated from the hippocampus, frontal cortex, and dorsal horn (reviewed in Ref. 11). In contrast, α1H was 40-fold more sensitive to Ni^{2+} (FIG. 4B). Cardiac T currents are similarly Ni^{2+} sensitive,[12] suggesting that these currents are mediated by α1H. However, comparisons to native currents are premature at this time since there is a third gene on chromosome 22 (*CACNA1I*) that appears to encode a protein (α1I) similar to α1G and α1H (FIG. 1B).[66]

CONCLUSIONS

We have cloned and expressed the first two members of the T-type Ca^{2+} channel family. The biophysical properties of the cloned channels are nearly identical to those of T currents recorded from isolated cells. Three properties define T-type channels: activation at low voltages, slow deactivation, and small single-channel conductance. Nickel sensitivity and inactivation at negative potentials are not defining features, since some T channels are Ni^{2+} insensitive, and some HVA channels (e.g., α1E[7]) also inactivate at negative potentials. The cloning of the T-channel family opens up many new and exciting fields of research. Previously the distribution of these channels could be assayed only by using patch clamp recording of cells that survived the isolation procedure. Now cDNA probes can be used to determine the distribution and abundance of their mRNA transcripts. The cloned material can also be used to develop antibodies, which will provide more insights into the distribution of these channels. These tools will allow studies on the up- or down-regulation of these channels in disease states such as epilepsy, cardiac hypertrophy, and hypertension. Finally, cell lines expressing the cloned T channels may provide valuable tools for drug discovery.

ACKNOWLEDGMENTS

We thank Asif Daud and Jie Yang for valuable assistance with the cloning. We thank Qun Jiang for technical assistance. We thank Antonio Lacerda, Yi Zhang, and Jonathan Satin for help with the single-channel recordings. We thank Jane Barclay, Magali Williamson, Margaret Fox, and Michele Rees for mapping the T-channel genes. This work was supported by grants from the NHLBI and the Potts Foundation. Dr. Perez-Reyes is an Established Investigator of the American Heart Association.

REFERENCES

1. ANDERSEN, P., J.C. ECCLES & T.A. SEARS. 1964. The ventro-basal complex of the thalamus: types of cells, their responses and functional organization. J. Physiol. **174:** 370–399.
2. JAHNSEN, H. & R. LLINAS. 1984. Electrophysiological properties of guinea-pig thalamic neurones: An *in vitro* study. J. Physiol. **349:** 227–247.
3. JAHNSEN, H. & R. LLINAS. 1984. Ionic basis for the electroresponsiveness and oscillatory properties of guinea-pig thalamic neurones *in vitro*. J. Physiol. **349:** 227–247.
4. NOWYCKY, M.C., A.P. FOX & R.W. TSIEN. 1985. Three types of neuronal calcium channel with different calcium agonist sensitivity. Nature **316:** 440–443.
5. MATTESON, D.R. & C.M. ARMSTRONG. 1986. Properties of two types of calcium channels in clonal pituitary cells. J. Gen. Physiol. **87:** 161–182.
6. AVERY, R.B. & D. JOHNSTON. 1996. Multiple channel types contribute to the low-voltage–activated calcium current in hippocampal CA3 pyramidal neurons. J. Neurosci. **16:** 5567–5582.
7. SOONG, T.W. *et al.* 1993. Structure and functional expression of a member of the low voltage–activated calcium channel family. Science **260:** 1133–1136.
8. SCHNEIDER, T. *et al.* 1995. Molecular analysis and functional expression of the human type E $\alpha 1$ subunit. Recept. Channels **2:** 255–270.
9. WAKAMORI, M. *et al.* 1994. Distinctive functional properties of the neuronal BII (class E) calcium channel. Recept. Channels **2:** 303–314.
10. WILLIAMS, M.E. *et al.* 1994. Structure and functional characterization of neuronal alpha 1E calcium channel subtypes. J. Biol. Chem. **269:** 22347–22357.
11. HUGUENARD, J.R. 1996. Low threshold calcium currents in central nervous system neurons. Annu. Rev. Physiol. **58:** 329–348.
12. HAGIWARA, N., H. IRISAWA & M. KAMEYAMA. 1988. Contribution of two types of calcium currents to the pacemaker potentials of rabbit sino-atrial node cells. J. Physiol. (Lond.) **395:** 233–253.
13. ZHOU, Z. & S.L. LIPSIUS. 1994. T-type calcium current in latent pacemaker cells isolated from cat right atrium. J. Mol. Cell. Cardiol. **26:** 1211–1219.
14. MAGEE, J.C. & D. JOHNSTON. 1995. Synaptic activation of voltage-gated channels in the dendrites of hippocampal pyramidal neurons. Science **268:** 301–304.
15. KAVALALI, E.T. *et al.* 1997. Dendritic Ca^{2+} channels characterized by recordings from isolated hippocampal dendritic segments. Neuron **18:** 651–663.
16. PEDROARENA, C. & R. LLINAS. 1997. Dendritic calcium conductances generate high-frequency oscillation in thalamocortical neurons. Proc. Natl. Acad. Sci. USA **94:** 724–728.

17. Williams, S.R. *et al.* 1997. The "window" component of the low threshold Ca2+ current produces input signal amplification and bistability in cat and rat thalamocortical neurones. J. Physiol. **505:** 689–705.
18. Pare, D., R. Curro'Dossi & M. Steriade. 1990. Neuronal basis of the parkinsonian resting tremor: A hypothesis and its implications for treatment. Neuroscience **35:** 217–226.
19. Steriade, M. & R.R. Llinas. 1988. The functional states of the thalamus and the associated neuronal interplay. Physiol. Rev. **68:** 649–742.
20. MacDonald, R.L. & K.M. Kelly. 1995. Antiepileptic drug mechanisms of action. Epilepsia **36:** S2–12.
21. Tsakiridou, E. *et al.* 1995. Selective increase in T-type calcium conductance of reticular thalamic neurons in a rat model of absence epilepsy. J. Neurosci. **15:** 3110–3117.
22. Cohen, C.J. *et al.* 1988. Ca channels in adrenal glomerulosa cells: K^+ and angiotensin II increase T-type Ca channel current. Proc. Natl. Acad. Sci. USA **85:** 2412–2416.
23. Enyeart, J.J., B. Mlinar & J.A. Enyeart. 1993. T-type Ca2+ channels are required for adrenocorticotropin-stimulated cortisol production by bovine adrenal zona fasciculata cells. Mol. Endocrinol. **7:** 1031–1040.
24. Barrett, P.Q. *et al.* 1991. Ca2+ channels and aldosterone secretion: modulation by K+ and atrial natriuretic peptide. Am. J. Physiol. **261:** F706–719.
25. McCobb, D.P. & K.G. Beam. 1991. Action potential waveform voltage-clamp commands reveal striking differences in calcium entry via low and high voltage-activated calcium channels. Neuron **7:** 119–27.
26. Hermsmeyer, K. *et al.* 1997. Physiologic and pathophysiologic relevance of T-type calcium-ion channels: Potential indications for T-type calcium antagonists. Clin. Thera. **19:** 18–26.
27. Schmitt, R. *et al.* 1996. Prevention of neointima formation by mibefradil after vascular injury in rats: Comparison with ACE inhibition. Cardiovasc. Drugs Ther. **10:** 101– 105.
28. Nuss, H.B. & S.R. Houser. 1993. T-type Ca2+ current is expressed in hypertrophied adult feline left ventricular myocytes. Circ. Res. **73:** 777–782.
29. Wang, L. *et al.* 1996. Abnormally expressed low-voltage–activated calcium channels in beta-cells from NOD mice and a related clonal cell line. Diabetes **45:** 1678–1683.
30. Gomez-Mancilla, B. *et al.* 1992. Effect of ethosuximide on rest tremor in the MPTP monkey model. Movement Disorders **7:** 137–141.
31. Godfraind, T., R. Miller & M. Wibo. 1986. Calcium antagonism and calcium entry blockade. Pharmacol. Rev. **38:** 321–416.
32. Bezprozvanny, I. & R.W. Tsien. 1995. Voltage-dependent blockade of diverse types of voltage-gated Ca2+ channels expressed in Xenopus oocytes by the Ca2+ channel antagonist mibefradil (Ro 40-5967). Mol. Pharmacol. **48:** 540–549.
33. Mishra, S.K. & K. Hermsmeyer. 1994. Selective inhibition of T-type Ca2+ channels by Ro 40-5967. Circ. Res. **75:** 144–148.
34. Schmitt, R. *et al.* 1995. Mibefradil prevents neointima formation after vascular injury in rats. Possible role of the blockade of the T-type voltage-operated calcium channel. Arterioscler. Thromb. Vasc. Biol. **15:** 1161–1165.
35. Cohen, C.J., S. Spires & D. Van Skiver. 1992. Block of T-type Ca channels in guinea pig atrial cells by antiarrhythmic agents and Ca channel antagonists. J. Gen. Physiol. **100:** 703–728.
36. Ketchum, K.A. *et al.* 1996. A new family of outwardly rectifying potassium channel proteins with two pore domains in tandem. Nature **376:** 690–695.
37. Zhu, X. *et al.* 1995. Molecular cloning of a widely expressed human homologue for the *Drosophila trp gene.* FEBS Lett. **373:** 193–198.
38. Altschul, S.F. *et al.* 1990. Basic local alignment search tool. J. Mol. Biol. **215:** 403–410.
39. Soares, M.B. *et al.* 1994. Construction and characterization of a normalized cDNA library. Proc. Natl. Acad. Sci. USA **91:** 9228–9232.
40. Lennon, G. *et al.* 1996. The I.M.A.G.E. Consortium: An integrated molecular analysis of genomes and their expression. Genomics **33:** 151–152.
41. Wilson, R. *et al.* 1994. 2.2 Mb of contiguous nucleotide sequence from chromosome III of *C. elegans.* Nature **368:** 32–38.
42. Perez-Reyes, E. *et al.* 1998. Molecular characterization of a neuronal low voltage–activated T-type calcium channel. Nature **391:** 896–900.

43. Cribbs, L.L. *et al.* 1998. Cloning and characterization of α1H from human heart, a member of the T-type calcium channel gene family. Circ. Res. **83:** 103–109.
44. Jan, L.Y. & Y.N. Jan. 1990. A superfamily of ion channels. Nature **345:** 672.
45. Stuhmer, W. *et al.* 1989. Structural parts involved in activation and inactivation of the sodium channel. Nature **339:** 597–603.
46. Yang, J. *et al.* 1993. Molecular determinants of Ca2+ selectivity and ion permeation in L- type Ca2+ channels. Nature **366:** 158–161.
47. Tang, S. *et al.* 1993. Molecular localization of ion selectivity sites within the pore of a human L-type cardiac calcium channel. J. Biol. Chem. **268:** 13026– 13029.
48. Pragnell, M. *et al.* 1994. Calcium channel beta-subunit binds to a conserved motif in the I-II cytoplasmic linker of the alpha 1-subunit. Nature **368:** 67–70.
49. de Leon, M. *et al.* 1995. Essential Ca^{2+}-binding motif for Ca^{2+}-sensitive inactivation of L-type Ca^{2+} channels. Science **270:** 1502–1506.
50. Coulter, D.A., J.R. Huguenard & D.A. Prince. 1990. Differential effects of petit mal anticonvulsants and convulsants on thalamic neurones: Calcium current reduction. Br. J. Pharmacol. **100:** 800–806.
51. Kelly, K.M., R.A. Gross & R.L. Macdonald. 1990. Valproic acid selectively reduces the low-threshold (T) calcium current in rat nodose neurons. Neurosci. Lett. **116:** 233–238.
52. Todorovic, S.M. & C.J. Lingle. 1998. Pharmacological properties of T-type Ca2+ current in adult rat sensory neurons: Effects of anticonvulsant and anesthetic agents. J. Neurophysiol. **79:** 240–252.
53. Montgomery, J.C., K.A. Silverman & A.M. Buchberg. 1997. Chromosome 11. Mamm. Genome **7:** S190–208.
54. Meier, M. 1967. The neuropathy of teetering, a neurological mutation in the mouse. Arch. Neurol. **16:** 59–66.
55. Burgess, D.L. *et al.* 1997. Mutation of the Ca2+ channel subunit gene *Cchb4* is associated with ataxia and seizures in the lethargic (*lh*) mouse. Cell **88:** 385–392.
56. Ophoff, R.A. *et al.* 1996. Familial hemiplegic migraine and episodic ataxia type-2 are caused by mutations in the Ca^{2+} channel gene CACNL1A4. Cell **87:** 543–552.
57. T.E.P.K.D. Consortium. 1994. The polycystic kidney disease 1 gene encodes a 14 kb transcript and lies within a duplicated region on chromosome 16. Cell **77:** 881–894.
58. Mochizuki, T. *et al.* 1996. PKD2, a gene for polycystic kidney disease that encodes an integral membrane protein. Science **272:** 1339–1342.
59. Carbone, E. & H.D. Lux. 1987. Single low-voltage–activated calcium channels in chick and rat sensory neurones. J. Physiol. **386:** 571–601.
60. Chen, C.F. & P. Hess. 1990. Mechanism of gating of T-type calcium channels. J. Gen. Physiol. **96:** 603–630.
61. Randall, A.D. & R.W. Tsien. 1997. Contrasting biophysical and pharmacological properties of T- type and R-type calcium channels. Neuropharmacology **36:** 879–893.
62. Carbone, E. & H.D. Lux. 1984. A low voltage–activated, fully inactivating Ca channel in vertebrate sensory neurones. Nature **310:** 501–502.
63. Nilius, B. *et al.* 1985. A novel type of cardiac calcium channel in ventricular cells. Nature **316:** 443–446.
64. Akaike, N., P.G. Kostyuk & Y.V. Osipchuk. 1989. Dihydropyridine-sensitive low-threshold calcium channels in isolated rat hypothalamic neurones. J. Physiol. **412:** 181–195.
65. Ye, J.H. & N. Akaike. 1993. Calcium currents in pyramidal neurons acutely dissociated from the rat frontal cortex: A study by the nystatin perforated patch technique. Brain Res. **606:** 111–117.
66. Perez-Reyes, E. *et al.*, 1998. Molecular characterization of T-type calcium channels. *In* Low Voltage-Activated T-Type Calcium Channels. R.W. Tsien, J.-P. Clozel & J. Nargeot, Eds.: 290–305. Adis International. Chester, UK.

Interactions of Presynaptic Ca^{2+} Channels and Snare Proteins in Neurotransmitter Release

WILLIAM A. CATTERALL

Department of Pharmacology, Box 357280, University of Washington, Seattle, Washington 98195-7280, USA

ABSTRACT: N- and P/Q-type Ca^{2+} channels are localized in high density in presynaptic nerve terminals and are crucial elements in neuronal excitation-secretion coupling. In addition to mediating Ca^{2+} entry to initiate transmitter release, they are thought to interact directly with proteins of the synaptic vesicle docking/fusion machinery. These Ca^{2+} channels can be purified from brain as a complex with SNARE proteins, which are involved in exocytosis. In addition, N-type and P/Q-type Ca^{2+} channels are colocalized with syntaxin in high-density clusters in nerve terminals. The synaptic protein interaction (synprint) sites in the intracellular loop II-III (L_{II-III}) of both α_{1B} and α_{1A} subunits of N-type and P/Q-type Ca^{2+} channels bind to syntaxin, SNAP-25, and synaptotagmin. Ca^{2+} has a biphasic effect on the interactions of N-type Ca^{2+} channels with SNARE complexes, stimulating optimal binding in the range of 10–30 µM. PKC or CaM KII phosphorylation of the N-type synprint peptide inhibits interactions with SNARE complexes containing syntaxin and SNAP-25. Introduction of the synprint peptides into presynaptic superior cervical ganglion neurons reversibly inhibits EPSPs from synchronous transmitter release by 42%. At physiological Ca^{2+} concentrations, synprint peptides significantly reduce transmitter release in injected frog neuromuscular junctions in cell culture, consistent with detachment of 70% of the docked vesicles from Ca^{2+} channels as analyzed by a theoretical model. Together, these studies suggest that presynaptic Ca^{2+} channels not only provide the Ca^{2+} signal required by the exocytotic mechinery, but also contain structural elements that are integral to vesicle docking, priming, and fusion processes.

Neurotransmitter release is initiated by influx of Ca^{2+} through voltage-dependent Ca^{2+} channels within 200 µs of the action potential arriving at the synaptic terminal,[1,2] where clusters of presynaptic Ca^{2+} channels are thought to supply Ca^{2+} to initiate release.[3–5] Exocytosis of synaptic vesicles requires high Ca^{2+} concentration, with a threshold of 20–50µM and half-maximal activation at 190µM.[6,7] The brief rise in Ca^{2+} concentration to the level necessary for exocytosis likely occurs only in proximity to the Ca^{2+} channels,[8–10] since intracellular Ca^{2+} concentration falls off steeply as a function of distance away from the Ca^{2+} channels. This limits the distance from Ca^{2+} entry site to Ca^{2+} sensor sites. Increasing evidence supports the notion that the maintenance of such a critical intermolecular distance must involve a physical link between the release mechanism and the presynaptic Ca^{2+} channels.

Major progress has been made towards understanding the molecular mechanisms that underlie Ca^{2+}-dependent exocytosis by identifying proteins that are involved in the vesicle docking/fusion process at presynaptic nerve terminals and analyzing their interactions (reviewed in Refs.11,12). Vesicle docking and fusion are mediated by a stable core-complex of proteins, including the synaptic vesicle protein VAMP/synaptobrevin[13] and the plasmalemmal proteins syntaxin and SNAP-25.[14–21] Synaptotagmin,[22] a synaptic vesicle protein, binds Ca^{2+} [23,24] and interacts with syntaxin in a Ca^{2+}-dependent manner.[25,26] It is thought to serve as a Ca^{2+} sensor for fast, Ca^{2+}-dependent neurotransmitter release.[27–32]

Immunochemical studies in several laboratories have indicated a tight association of syntaxin and synaptotagmin with both N- and P/Q-type Ca^{2+} channels,[14,19,33–36] implicating both types of Ca^{2+} channels as components of synaptic vesicle docking/fusion machinery. Detailed structural characterization of the interactions among the neuronal Ca^{2+} channels and different components of the exocytotic apparatus and their regulation by Ca^{2+} and second messenger–activated phosphorylation pathways will shed light on the molecular mechanisms of neurotransmitter relasese. This article focuses on our recent results on the localization of specific Ca^{2+} channel types in nerve terminals and on the binding site in presynaptic Ca^{2+} channels that is responsible for their interaction with SNARE proteins.

LOCALIZATION OF Ca^{2+} CHANNELS IN PRESYNAPTIC TERMINALS

Ca^{2+} channels are a complex of five subunits: a principal pore-forming α_1 subunit of 190 to 250 kDa in association with a disulfide-linked $\alpha_2\delta$ dimer of 170 kDa, an intracellular β subunit of 55 to 72 kDa, and, in skeletal muscle, a transmembrane γ subunit of 33kDa[37,38] (FIG. 1). The primary structures of five distinct classes of high voltage–activated neuronal Ca^{2+} channels α_1 subunits have been cloned and characterized, termed classes A, B, C, D, and E. Cloned neuronal α_{1B} and α_{1A} encode N-type and P/Q-type Ca^{2+} channels, respectively, whereas the α_{1C} and α_{1D} subunits are components of L-type channels.[39–46] Both ω-conotoxin (CTx) GVIA-sensitive N-type and ω-agatoxin (Aga) IVA– and ω-conotoxin MVIIC–sensitive P/Q-type Ca^{2+} channels participate in neurotransmitter release at central and peripheral synapses.[47–49] Consistent with this idea, the α_{1B} subunits of N-type Ca^{2+} channels[44,50] and the α_{1A} subunits of P/Q-type Ca^{2+} channels[50] are localized at low density in dendrites and at high density in presynaptic nerve terminals of many neurons (FIG. 1). The plasma membrane SNARE protein syntaxin is colocalized in the same high-density clusters. Thus, the α_{1A} and α_{1B} subunits are specialized for neurotransmitter release and colocalized with the SNARE proteins, which carry out synaptic vesicle docking and exocytosis. What are the molecular specializations that allow these channels to bind SNARE proteins and efficiently initiate transmitter release?

THE SYNAPTIC PROTEIN INTERACTION (SYNPRINT) SITE ON N-TYPE Ca^{2+} CHANNELS

The amino acid sequences of all cloned α_1 subunits share overall structural features of four hydrophobic homologous transmembrane domains (I–IV) that are linked by intracellular hydrophilic loops of various lengths (FIG. 1). The homologous domains exhibit a high degree of sequence conservation among subtypes, but the cytoplasmic loops linking the domains are highly divergent. The most attractive candidates for synaptic protein interactions are the intracellular linkers and the cytoplasmic N- and C-terminal domains. The sequence variations of these cytoplasmic segments may reflect the capacity for distinct, functionally significant interactions of both N- and P/Q-type channels with cytoplasmic domains of synaptic proteins involved in vesicle docking/fusion processes.

To identify and characterize the cytoplasmic loops of the N-type channels interacting with synaptic proteins, a series of hexahistidine-tagged (His)-fusion proteins expressing various cytoplasmic segments of α_{1B} and recombinant syntaxin 1A, SNAP-25, or VAMP-2

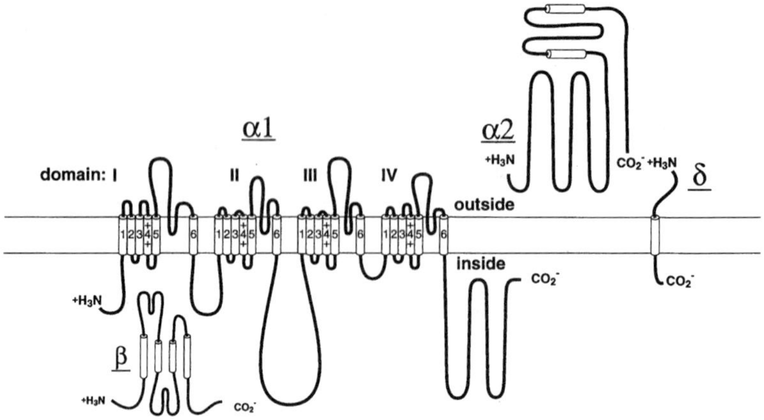

Type	Pore-forming Subunit	Location	Function
L	α_{1C}, α_{1D}	Cell bodies Dendrites	Gene Expression Synaptic Integration
N	α_{1B}	Nerve Terminals Dendrites Cell bodies	Transmitter Release
P/Q	α_{1A}	Nerve Terminals Dendrites Cell bodies	Transmitter Release
R	α_{1E}	Cell bodies Dendrites	Synaptic Integration Repetitive Firing

FIGURE 1. Subunit structure and localization of brain Ca^{2+} channels. **Top.** The subunit structure of brain Ca^{2+} channels is illustrated in the form of a transmembrane folding diagram. **Bottom.** The correspondence between Ca^{2+} channel type (L, N, P/Q, R) and their corresponding pore-forming α1 subunit (α_{1A}, α_{1B}, α_{1C}, α_{1D}, α_{1E}) is listed together with the main site of subcellular localization in major projection neurons in the brain and their main functional roles in those locations.

fused to glutathione S-transferase (GST) were generated. The recombinant GST-fusion proteins coupled to glutathione-Sepharose beads were used as an affinity matrix to screen His-fusion proteins for specific binding. The *in vitro* binding studies showed that both syntaxin 1A and SNAP-25, but not VAMP, specifically interact with the cytoplasmic loop ($L_{II\text{-}III}$) between homologous domains II and III of the α_{1B} subunit of the class B N-type Ca^{2+} channels from rat brain.[51,52] These interactions are mediated by a specific segment of amino acid sequence (residues 718–963). We use the term *synprint* to designate this *syn*aptic *pr*otein

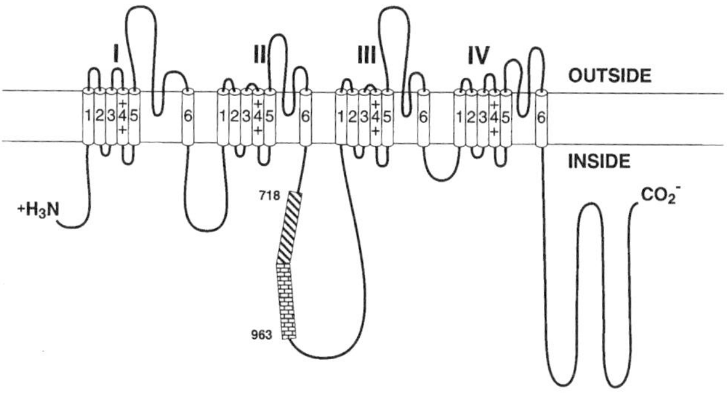

FIGURE 2. The synaptic protein interaction (synprint) sites on both N- and P/Q-type Ca^{2+} channels. Predicted topological structure of the α_1 subunits of class B N-type and class A P/Q-type Ca^{2+} channels with synprint site in the intracellular loop between homologous domains II and III ($L_{II\text{-}III}$) indicated by the *rectangle boxes*. Amino acid positions of N- and P/Q-type synprint sites are defined in the regions between 718–963, and 722–1036, respectively.

*int*eraction site on the Ca^{2+} channels (FIG. 2). The synprint site of N-type Ca^{2+} channels contains two adjacent binding regions that comprise two distinct subsites. The synprint peptide can specifically block coimmunoprecipitation of native N-type Ca^{2+} channels with syntaxin 1, indicating that this binding site is required for stable interaction of these two proteins.[51–53] This interaction takes place with the C-terminal one-third of syntaxin (residues 181–288), suggesting that neuronal Ca^{2+} channels bind to syntaxin 1A at a C-terminal site near the intracellular surface of the plasma membrane.[51–53]

Ca^{2+} DEPENDENCE OF INTERACTIONS WITH THE SYNPRINT SITE ON N-TYPE Ca^{2+} CHANNELS

An important unresolved issue in understanding neurotransmitter release is the mechanism of its Ca^{2+} dependence. Neurotransmitter release is initiated by influx of Ca^{2+} through voltage-dependent Ca^{2+} channels,[1,2] within 200 μs of the action potential arriving at the synaptic terminals, as the Ca^{2+} concentration increases from 100 nM to >200 μM.[9] Exocytosis requires high Ca^{2+} concentration, with a threshold of 20–50μM and half maximal activation at 190 μM.[6,7] Synaptotagmin may be the low-affinity Ca^{2+} sensor, since it binds to phospholipid and syntaxin in a Ca^{2+}-dependent manner in the range of 10 to 50 μM and 100 to 300 μM Ca^{2+}, respectively.[23,29,54–56] However, other Ca^{2+} responsive proteins may also be involved in the docking/fusion process, as residual neurotransmission persists in synaptotagmin-null mutants.[31,32,57]

To find additional Ca^{2+}-responsive interactions among synaptic proteins and Ca^{2+} channels, we measured the binding of these recombinant proteins *in vitro*.[52,58] We found that the interaction of the N-type synprint peptide with recombinant syntaxin, SNAP-25, the syntaxin–SNAP-25 dimer, or the synaptic core complex of syntaxin–SNAP-25–VAMP/synaptobrevin has a biphasic dependence on Ca^{2+} concentration, with maximal binding at approximately 20 μM free Ca^{2+}.[52] This Ca^{2+}-dependent interaction takes place in the same concentration range as the threshold for fast transmitter release. We also measured the effect of Ca^{2+} on binding of the native N-type Ca^{2+} channels specifically labeled with ω-CTx-GVIA to a complex of native syntaxin and SNAP-25 with GST-synaptobrevin bound to glutathione-Sepharose. Basal binding is observed in the absence of Ca^{2+}, maximal binding at 20 μM Ca^{2+}, and reduced binding at 100 μM Ca^{2+}.[52] Thus, the direct interaction of presynaptic Ca^{2+} channels with the synaptic fusion core complex is a Ca^{2+}-sensitive process and may play a key role in docking and fusion of synaptic vesicles.

SPECIFIC INTERACTIONS OF SYNPRINT SITES FROM THE ISOFORMS OF α_{1A} WITH SNARE PROTEINS

Based on studies of the effects of Ca^{2+} channel blockers on synaptic transmission, it is generally agreed that both ω-CTx–GVIA–sensitive N-type and ω-Aga–IVA–sensitive P/Q-type Ca^{2+} channels play a role in controlling synaptic transmission in the mammalian central nervous system.[47–49,59–65] In contrast to N-type Ca^{2+} channels, the corresponding synprint segment of $L_{II\text{-}III}$ from the rbA isoform of α_{1A}[42] does not bind to syntaxin.[51] This result raises the possibility that P/Q-type Ca^{2+} channels might have a distinct synprint site that is unable to interact with syntaxin in the transmitter release process.

The α_{1A} subunit of P/Q-type Ca^{2+} channels has multiple isoforms detected by cDNA sequencing and analysis with sequence-specific antibodies.[40,42,66] Examination of the amino acid sequence differences between the BI and rbA isoforms of α_{1A}[40,42] shows considerably lower identity (78%) in $L_{II\text{-}III}$ than in the remainder of the protein (>98%). The distinctly different levels of amino acid sequence identity in these regions suggest that this loop may be subject to alternative splicing. In support of this idea, Sakurai *et al.*[67] have found that α_{1A} isoforms with amino acid sequences characteristic of $L_{II\text{-}III}$ from rbA and BI are present in both rat and rabbit brain using site-directed antibodies. Thus, the synprint region of the α_{1A} subunit is subject to alternative splicing to yield at least two isoforms.

Binding data with fusion proteins containing the intracellular loop $L_{II\text{-}III}$ of these two α_{1A} isoforms demonstrates binding with different affinities to the presynaptic proteins syntaxin and SNAP-25.[53] The BI isoform has higher affinity for binding to both syntaxin and SNAP-25 than the rbA isoform has. Under *in vitro* binding conditions, binding of rbA to SNAP-25 is clearly detected, while binding to syntaxin is not.[53] If these binding interactions are required for efficient coupling of Ca^{2+} influx with synaptic vesicle fusion, these data may imply that a neuron could modulate the efficiency of synaptic transmission by regulating the expression of different isoforms of a single class A Ca^{2+} channel gene. Consistent with this idea, the BI and rbA isoforms of α_{1A} subunits are differentially distributed at synapses in rat brain.[67]

Similar to the synprint site of N-type channels, the synprint site from the BI isoform of α_{1A} involves two adjacent segments of the intracellular loop connecting domains II and III between amino acid residues 722 and 1036 (FIG. 2), and it binds specifically to the C-terminal one-third of syntaxin 1A (residues 181–288), whereas no binding to the N-terminal two-thirds of syntaxin 1A is observed.[53] These interactions of the BI synprint peptide with both syntaxin and SNAP-25 are competitively blocked by the corresponding synprint region of the N-type channels, indicating that these two channels bind to overlapping or identical regions of syntaxin as well as SNAP-25.[53] Immunoprecipitation studies show that an antibody against a fusion protein with the amino acid sequence of $L_{II\text{-}III}$ of the rbA isoform of α_{1A} coimmunoprecipitates SNAP-25 and syntaxin.[67] This immunoprecipitation was blocked by preincubation of the antibody with the corresponding fusion protein, supporting specific associations of syntaxin and SNAP-25 with class A Ca^{2+} channels by interactions with a site in L_{II-III} of α_{1A}.[67] Collectively, these results provide a molecular basis for a physical coupling of neuronal N-type and P/Q-type Ca^{2+} channels with synaptic vesicle docking/fusion complexes, enabling tight structural and functional association of Ca^{2+} entry sites and neurotransmitter release sites. Differences in the interactions of the synprint sites on α_{1B} and the rbA and BI isoforms of α_{1A} with SNARE proteins may alter the regulatory properties of synaptic transmission at different nerve terminals.

Ca^{2+}-INDEPENDENT INTERACTIONS WITH THE SYNPRINT PEPTIDES OF P/Q-TYPE Ca^{2+} CHANNELS

Binding of SNARE proteins to the synprint peptides from the rbA and BI isoforms of α_{1A} have different dependence on Ca^{2+} concentration from the synprint peptide of α_{1B}[68] (TABLE 1). The BI isoform of α_{1A} binds syntaxin and SNAP-25 in a Ca^{2+}-independent manner. The rbA isoform of α_{1A} does not bind to syntaxin appreciably *in vitro* and binds SNAP-25 in a Ca^{2+}-independent manner. The differences in Ca^{2+} dependence of interaction of these synprint peptides with different SNARE proteins suggests that the Ca^{2+} dependence of this interaction is not an essential element of the transmitter release pathway but serves a modulatory role that may confer different regulatory properties on transmitter release mediated by the different presynaptic Ca^{2+} channels.

INTERACTIONS OF THE SYNPRINT SITE WITH SYNAPTOTAGMIN

The vesicle SNARE protein synaptotagmin is thought to serve as the Ca^{2+} sensor for fast neurotransmitter release. Immunochemical studies show that it is associated with purified N-type and P/Q-type Ca^{2+} channels, similar to syntaxin and SNAP-25.[33,34] The interaction of synaptotagmin with the synprint sites of N-type and P/Q-type Ca^{2+} channels was measured using similar methods as described above.[69,70] Synaptotagmin forms a specific complex with synprint sites from both α_{1A} and α_{1B}. These two synprint peptides compete for binding to synaptotagmin, indicating that they bind to identical or overlapping sites in the helix-3 domain of syntaxin adjacent to the plasma membrane. Moreover, using both

TABLE 1. Summary of the Interactions of Ca^{2+} Channel Synprint Peptides with Synaptic Proteins.

	Syntaxin		SNAP-25		Synaptotagmin	
	Binding	Ca^{2+} dep	Binding	Ca^{2+} dep	Binding	Ca^{2+} dep
α_{1B}	+	+	+	+	+	–
$\alpha_{1A(rbA)}$	–	–	+	–	+	+
$\alpha_{1A(BI)}$	+	–	+	–	+	–

ABBREVIATION: Ca^{2+} dep, Ca^{2+} dependence.

immobilized recombinant proteins and native presynaptic membrane proteins, we found that the synprint peptide of N-type channels and synaptotagmin competitively interact with syntaxin.[69] These results predict that, in a nerve terminal, syntaxin molecules bound to Ca^{2+} channels cannot interact effectively with synaptotagmin, an interaction that is thought to be essential for transmitter release.

The competition between the synprint site of α_{1B} and synaptotagmin is Ca^{2+} dependent because of the Ca^{2+} dependence of the interactions between syntaxin and these two proteins. The affinity of N-type Ca^{2+} channels for binding to syntaxin is modulated by Ca^{2+} concentration, with maximal binding at a range of 10–30 μM near the threshold for neurotransmitter release.[52] In contrast, maximum binding of syntaxin to synaptotagmin I and II requires higher concentrations of Ca^{2+} in the range from 100 μM to 1 mM.[24–26,71] As the Ca^{2+} concentration increases beyond 30 μM, interaction of syntaxin with the synprint site of N-type Ca^{2+} channels will be weakened, and interaction with synaptotagmin will be strengthened. Thus, these studies provide potential biochemical correlates for the sequence of events during synaptic vesicle exocytosis: binding of syntaxin and SNAP-25 to N-type Ca^{2+} channels at low Ca^{2+} concentration, enhanced affinity of that interaction at Ca^{2+} concentrations in the range of 10 to 30 μM, and displacement of the synprint binding interaction on syntaxin by synaptotagmin at Ca^{2+} concentrations in the range of 100 μM and higher (FIG. 3). This sequence of protein–protein interactions with N-type Ca^{2+} channels may serve to control the triggering of exocytosis by regulating the interaction of syntaxin with synaptotagmin.

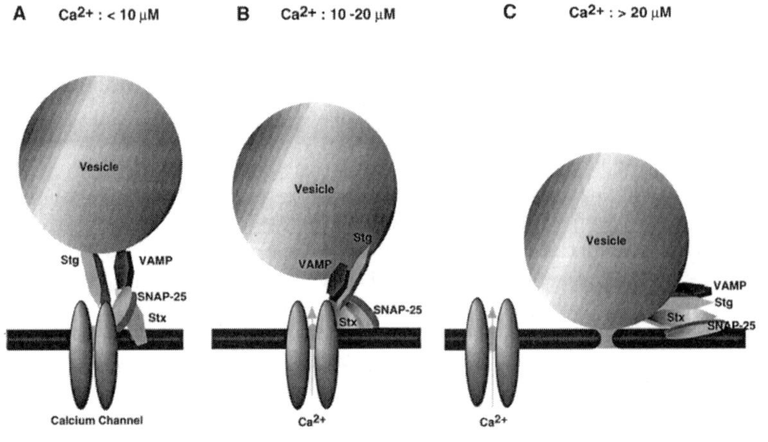

FIGURE 3. The proposed model for the sequential Ca^{2+}-dependent interactions of multiple proteins with syntaxin during synaptic vesicle docking/fusion process. **A**. The predocked vesicles form a low-affinity complex with the N-type Ca^{2+} channels through binding to syntaxin and SNAP-25 at resting [Ca^{2+}] level (<10 μM). **B**. The initial Ca^{2+} influx (10–20 μM) greatly increases the affinity of this coupling, so the binding energy of Ca^{2+} to this complex may contribute to the energetic driving force for the early priming steps of the fusion process. **C**. Finally, as the free Ca^{2+} reaches the threshold for release (>20 μM), the binding affinity of this coupling is reduced, and syntaxin and SNAP-25 dissociate from the channels. Higher levels of Ca^{2+} (above 30 μM) may be needed to enable displacement of syntaxin from the Ca^{2+} channels and efficient binding of synaptotagmin, in order for fusion to proceed. Thus, sequential Ca^{2+}-dependent interactions of multiple proteins with syntaxin may serve to order the biochemical events leading to membrane fusion.

Although binding of synaptotagmin to the α_{1B} subunit of N-type Ca^{2+} channels is Ca^{2+} independent, its binding to the rbA isoform of the α_{1A} subunit of P/Q-type Ca^{2+} channels is Ca^{2+} dependent, with maximum binding at 10 to 30 μM Ca^{2+}, similar to the Ca^{2+} dependence of binding of syntaxin and SNAP-25 to α_{1B}.[69] In contrast, synaptotagmin binding to the BI isoform of α_{1A} is Ca^{2+} independent[68] (TABLE 1). How might the differences in Ca^{2+} dependence of interactions with the SNARE proteins affect transmitter release initiated by N-type and P/Q-type Ca^{2+} channels? Because fast transmitter release is triggered by Ca^{2+} concentrations in the range of 100 μM or more, it is unlikely that the differences in the Ca^{2+} dependence of interaction of synprint sites with SNARE proteins at 10 to 30 μM Ca^{2+} are a key element in the basic transmitter release process. Instead, we propose that these Ca^{2+}-dependent interactions are important for Ca^{2+}-dependent modulation of transmitter release by post-tetanic potentiation and related short-term processes that modulate transmitter release in response to sustained changes in Ca^{2+} concentration in the 10 to 30 μM range (reviewed in Ref. 5). Our results predict differences in the Ca^{2+} dependence and molecular mechanisms of these Ca^{2+}-dependent forms of synaptic plasticity based on the differences in Ca^{2+} dependence of interaction of the synprint sites of Ca^{2+} channel subtypes with SNARE proteins.

PHYSIOLOGICAL SIGNIFICANCE OF THE INTERACTION OF N-TYPE Ca^{2+} CHANNELS WITH SNARE PROTEINS IN SYNAPTIC TRANSMISSION

The structural and functional coupling between Ca^{2+} entry sites and release sites of docked synaptic vesicles would ensure that neurotransmitter release is triggered rapidly when the action potential invades the nerve terminal.[3,4,10] The Ca^{2+} influx rapidly elevates the Ca^{2+} concentration to 200–300 μM in microdomains of approximately 0.3 μm^2 in the presynaptic terminal near release sites.[5,9] This transient Ca^{2+} influx triggers rapid vesicle fusion and neurotransmitter release through Ca^{2+} binding to receptor(s) on the exocytotic apparatus (FIG. 4A). Following Ca^{2+} channel closure, diffusion and sequestration rapidly decrease the Ca^{2+} concentration at the release site to subthreshold levels, terminating the exocytotic process. Although biochemical data support the hypothesis that there is a tight association between Ca^{2+} channels and exocytotic apparatus, which is consistent with current knowledge about synaptic proteins and Ca^{2+} channels, the functional roles of the synaptic protein–Ca^{2+} channel interactions in neuronal Ca^{2+}-activated exocytosis remain to be determined.

Our biochemical results predict that peptides containing synprint sites, if injected into neurons, would inhibit synaptic transmission by competitively binding to syntaxin and SNAP-25. This subsequently prevents their binding to presynaptic Ca^{2+} channels, increasing the distance between docked vesicles and Ca^{2+} channels and increasing the requirement for Ca^{2+} influx to initiate transmitter release. To test this hypothesis, we investigated the physiological relevance of this interaction by injecting competing peptides into the presynaptic cells of both sympathetic ganglion neuron synapses and *Xenopus* embryonic neuromuscular junctions in culture.[72,73]

Cultures of superior cervical ganglion neurons (SCGNs) are favorable for functional testing for several reasons. Peptides can be introduced into the relatively large (30–40 μm) presynaptic cell bodies by microinjection; the injected peptides can rapidly diffuse down short axons to nerve terminals forming synapses with adjacent neurons; the

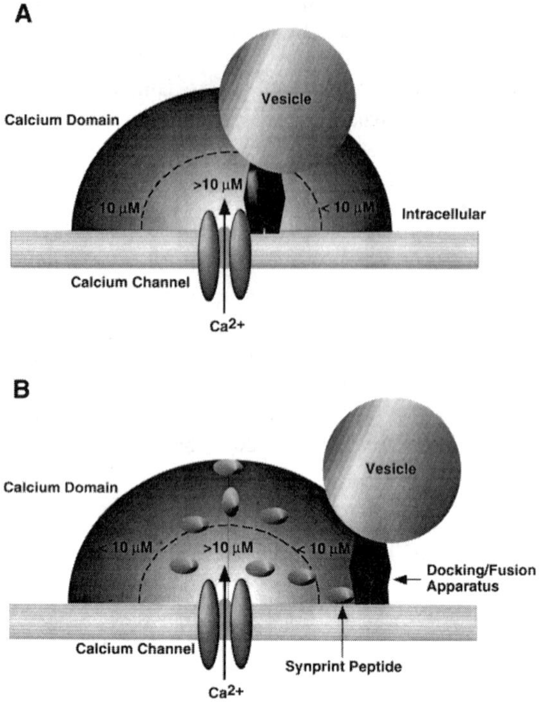

FIGURE 4. The proposed inhibitory role of injected peptides containing N-type synprint site on neurotransmission of SCG neurons and neuromuscular junctions. **A.** The physical link between N-type Ca^{2+} channels and the synaptic vesicle docking/fusion apparatus. Exocytosis of synaptic vesicles requires high Ca^{2+} concentration, with a threshold of 20–50 μM and half-maximal activation at 190 μM.[6,7] The brief rise in Ca^{2+} concentration to the level necessary for exocytosis likely occurs only in proximity to the Ca^{2+} channels,[8–10] since the intracellular Ca^{2+} concentration falls off steeply as a function of distance away from the Ca^{2+} channels. Thus, when binding to the synprint site, synaptic vesicles are docked in proximity to Ca^{2+} entry sites. Upon Ca^{2+} influx, the fusion apparatus is activated for rapid, Ca^{2+}-dependent, synchronous synaptic transmission. **B.** The peptides containing the synprint site competitively block the physical link between N-type Ca^{2+} channels and the synaptic vesicle docking/fusion apparatus, and subsequently remove predocked vesicles away from Ca^{2+} entry sites. This decreases the degree of efficiency by shifting Ca^{2+} dependence to higher values. Rapid, synchronous synaptic transmission is inhibited, while late, asynchronous EPSPs and paired-pulse facilitation are increased, consistent with the conclusion that synaptic vesicles are shifted from a pool primed for synchronous release to a pool not optimally primed or positioned for synchronous release.

effects on stimulated release of acetylcholine can be accurately monitored by recording the excitatory postsynaptic potentials (EPSPs) evoked by action potentials in presynaptic neurons; and only N-type Ca^{2+} channels control acetylcholine release at these synapses, so a homogeneous population of channels can be studied. Synaptic transmission was monitored between closely spaced pairs of neurons for 20–30 min, and then peptides containing the synprint site were allowed to diffuse into the presynaptic neurons from a suction pipette for 2–3 min. EPSPs were evoked by action potentials elicited by current pulses applied to the presynaptic cell through a recording microelectrode and were recorded with a second microelectrode in the nearby postsynaptic cell.[74,75] Peptides containing the synprint site from the α_{1B} subunit disrupt the interaction of native N-type channels with syntaxin and reduce synaptic transmission in SCGN synapses by 23% to 42% for different synprint peptides, without any effect on Ca^{2+} currents.[72] Rapid, synchronous synaptic transmission is inhibited, while late, asynchronous EPSPs and paired-pulse facilitation are increased, consistent with the conclusion that synaptic vesicles are shifted from a pool primed for synchronous release to a pool that is not optimally primed or positioned for synchronous release (FIG. 3B). The corresponding peptides from L-type Ca^{2+} channels have no effect on EPSPs. The relative efficiency for inhibition of transmitter release by three different peptides, $L_{II\text{-}III}$(718–963) > $L_{II\text{-}III}$(832–963) > $L_{II\text{-}III}$(718–859), was consistent with their rank order of affinity for *in vitro* binding to syntaxin.[53,72] These results provide direct evidence that binding of presynaptic Ca^{2+} channels to the synaptic docking/fusion complex is required for rapid, synchronous neurotransmitter release.

Early work at the frog neuromuscular junction revealed that fast synaptic transmission is steeply dependent on the external Ca^{2+} concentration $[Ca^{2+}]_e$, and the probability of acetylcholine release at the frog neuromuscular junction increases as the fourth power of $[Ca^{2+}]_e$.[76] To determine whether the inhibition of synaptic transmission by synprint peptides might be due to displacement of the docked vesicle away from the Ca^{2+} channels and resultant changes in the Ca^{2+} dependence of transmission, we used *Xenopus* nerve-muscle cocultures from developing embryos, in which synaptic transmission is mainly dependent on N-type Ca^{2+} channels.[77] This is an ideal preparation in which to study the effects of N-type synprint peptides on synaptic transmission because the Ca^{2+} transients in the presynaptic terminal can be imaged in parallel with measurements of synaptic transmission. During the first days of development, the embryos undergo cell divisions without substantial growth. Injection of synprint peptides into early blastomeres leads to loading of all progeny cells, including spinal cord neurons and muscle cells. Following cell culture, synaptic transmission of peptide-loaded and control cells were compared by measuring postsynaptic responses under different external Ca^{2+} concentrations. The dependence of synaptic transmission on Ca^{2+} concentration was shifted to higher concentrations so that, at physiological Ca^{2+} concentrations, approximately 50% reduction of transmitter release of injected neurons was observed. Analysis by a theoretical model indicated that at least 70% of the docked vesicles were detached from Ca^{2+} channels under these conditions. High Ca^{2+} concentrations can overcome this inhibition. Injection of the corresponding region of the L-type Ca^{2+} channels had virtually no effect.[73] These data suggest that disruption of the physical link between N-type Ca^{2+} channels and synaptic vesicle docking/fusion apparatus displaces predocked vesicles from Ca^{2+} entry sites, making neurotransmitter release less efficient by shifting its Ca^{2+} dependence to higher values (FIG. 4B). These findings are consistent with the functional data on rat superior cervical ganglion neurons,[72] where a maximum of 42% inhibition of synaptic transmission was observed following diffusion of synprint peptides.

MODULATION OF THE INTERACTIONS BETWEEN N-TYPE Ca^{2+} CHANNELS AND THE SYNAPTIC CORE COMPLEX BY PROTEIN PHOSPHORYLATION

The efficiency of synaptic transmission could be regulated by modulating the interactions of Ca^{2+} channels with synaptic protein complexes. Second messenger–activated regulation of neurotransmitter release via modulation of the interactions of proteins with the exocytotic apparatus has a potentially important role in synaptic plasticity. Several protein kinases including CaM KII, PKC, PKA, and PKG are expressed in presynaptic nerve terminals. Both presynaptic Ca^{2+} channels and SNARE proteins are phosphorylated by one or more of these protein kinases.[66,78–80] *In vitro* biochemical studies showed that phosphorylation of the synprint peptide with PKC and CaM KII, but not PKA or PKG, strongly inhibited binding of recombinant syntaxin or SNAP-25, and also inhibited the synprint peptide's interactions with native rat brain SNARE complexes containing syntaxin and SNAP-25.[58] These results suggest that phosphorylation of the synprint site by PKC or CaM KII may serve as a biochemical switch for interactions between N-type Ca^{2+} channels and SNARE protein complexes, and is a candidate presynaptic mechanism for the regulation of neurotransmission.

MODULATION OF Ca^{2+} CHANNEL FUNCTION BY INTERACTION WITH SNARE PROTEINS

Just as Ca^{2+} channel interactions with SNARE proteins are important for the anterograde Ca^{2+} signal that initiates fast transmitter release, retrograde signaling from the SNARE protein complex might also modulate Ca^{2+} channel function. In support of this idea, Bezprozvanny *et al.*[81] found that coexpression of syntaxin with N-type or P/Q-type Ca^{2+} channels in *Xenopus* oocytes caused a change in the voltage dependence of inactivation of the channels. This effect is most pronounced during long depolarizations, which would not normally occur in a nerve terminal. Similar experiments by Wiser *et al.*[82,83] detected both inhibition of Ca^{2+} channel expression and modulation of the voltage dependence of channel inactivation, and these effects are reversed by coexpression of synaptotagmin. Although the significance of these interactions in *Xenopus* oocytes for Ca^{2+} channel function in nerve terminals is unknown, these modulatory effects may reflect an important regulation of channel function *in vivo*. More direct evidence for functionally significant modulation of presynaptic Ca^{2+} channels by syntaxin comes from studies with botulinum toxins, which specifically cleave SNARE proteins.[84] Proteolytic cleavage of syntaxin prevents the normal G-protein modulation of presynaptic Ca^{2+} channels, consistent with direct effects of the SNARE protein on channel function.

Ca^{2+} CHANNEL INTERACTIONS WITH SNARE PROTEINS IN TRANSMITTER RELEASE

Understanding how presynaptic Ca^{2+} channels interact with SNARE proteins in a Ca^{2+}-dependent manner should not only help to clarify their functions in synaptic vesicle docking/fusion process but can also provide a model for how syntaxin-synaptotagmin interaction involving Ca^{2+} channels are regulated by Ca^{2+}. Several lines of evidence suggest that Ca^{2+} binding to synaptotagmin is part of the signal that initiates rapid exocytosis.[23,29,54–56] Our studies show that the interaction of N-type Ca^{2+} channels with the synaptic core complex is also dependent on changes in [Ca^{2+}] near the threshold level for initiation of transmitter release. We propose that docked vesicles form a low-affinity complex with the N-type Ca^{2+} channels through binding to syntaxin and SNAP-25 at resting [Ca^{2+}] level. Ca^{2+} influx greatly increases the affinity of this coupling, so the binding energy of Ca^{2+} to this complex may contribute to the energetic driving force for the early priming steps of the fusion process. Finally, as the free Ca^{2+} reaches the threshold for release (20–50 μM), the binding affinity of this coupling is reduced, and syntaxin and SNAP-25 dissociate from the channels. Higher levels of Ca^{2+} (above 30 μM) may be needed to enable displacement of syntaxin from the N-type Ca^{2+} channels and efficient binding of synaptotagmin in order for fusion to proceed. Thus, sequential Ca^{2+}-dependent interactions of multiple proteins with syntaxin may serve to order the biochemical events leading to membrane fusion (FIG. 3).

For presynaptic Ca^{2+} channels containing α_{1A} as their pore-forming subunit, different Ca^{2+}-dependent interactions are observed *in vitro*. We propose that the sequence of interactions of syntaxin first with the synprint site of presynaptic Ca^{2+} channels and then with synaptotagmin remains the same and is an essential element of the release pathway. For the BI isoform of α_{1A}, no Ca^{2+}-dependent interactions are observed *in vitro*, so it is likely that synaptotagmin displaces the synprint site from syntaxin without the aid of a Ca^{2+}-dependent decrease in binding affinity between the synprint site and syntaxin. This may require a higher concentration of Ca^{2+} to stimulate the interaction of synaptotagmin and syntaxin and therefore increase the steepness and cooperativity of the Ca^{2+} dependence of neurotransmitter release mediated by the BI isoform of α_{1A}. For the rbA isoform of α_{1A}, a biphasic dependence of binding to synaptotagmin is observed, but there is no Ca^{2+}-dependent binding to SNAP-25 or syntaxin. In this case, Ca^{2+}-dependent release of synaptotagmin from interaction with the synprint peptide may allow the interaction of synaptotagmin with syntaxin to occur more easily as Ca^{2+} concentrations increase and thereby enhance the release process. The differences in Ca^{2+}-dependent interactions of these synprint peptides may allow differential modulation of the release process by Ca^{2+} and perhaps by other influences such as protein phosphorylation.

REFERENCES

1. BARRETT, E.F. & C.F. STEVENS. 1972. The kinetics of transmitter release at the frog neuromuscular junction. J. Physiol. (Lond.) **227:** 691–708.
2. SMITH, S.J. & G.J. AUGUSTINE. 1988. Calcium ions, active zones and synaptic transmitter release. Trends Neurosci. **11:** 458–464.
3. PUMPLIN, D.W., T.S. REESE & R. LLINÁS. 1981. Are the presynaptic membrane particles the calcium channels? Proc. Natl. Acad. Sci. USA **78:** 7210–7214.
4. PUMPLIN, D.W. 1983. Normal variations in presynaptic active zones of frog neuromuscular junctions. J. Neurocytol. **12:** 317–323.
5. ZUCKER, R. 1993. Calcium and transmitter release. J. Physiol. (Paris) **87:** 25–36.
6. HEIDELBERGER, R., C. HEINEMANN, E. NEHER & G. MATTHEWS. 1994. Calcium dependence of the rate of exocytosis in a synaptic terminal. Nature **371:** 513–515.
7. VON GERSDORFF, H. & G. MATTHEWS. 1994. Dynamics of synaptic vesicle fusion and membrane retrieval in synaptic terminals. Nature **367:** 735–739.
8. COPE, T.C. & L.M. MENDELL. 1982. Distributions of EPSP latency at different group Ia-fiber-alpha-motoneuron connections. J. Neurophysiol. **47:** 469–478.

9. Llinás, R., M. Sugimori & R.B. Silver. 1992. Microdomains of high calcium concentration in a presynaptic terminal. Science **256:** 677–679.
10. Stanley, E.F. 1993. Single calcium channels and acetylcholine release at a presynaptic nerve terminal. Neuron **11:** 1007–1011.
11. Bajjalieh, S.M. & R.H. Scheller. 1995. The biochemistry of neurotransmitter secretion. J. Biol. Chem. **270:** 1971–1974.
12. Südhof, T.C. 1995. The synaptic vesicle cycle: A cascade of protein-protein interactions. Nature **375:** 645–653.
13. Trimble, W.S., D.M. Cowan & R.H. Scheller. 1988. VAMP-1: A synaptic vesicle-associated integral membrane protein. Proc. Natl. Acad. Sci. USA **85:** 4538–4542.
14. Bennett, M.K., N. Calakos & R.H. Scheller. 1992. Syntaxin: A synaptic protein implicated in docking of synaptic vesicles at presynaptic active zones. Science **257:** 255–259.
15. Bennett, M.K., J.E. Garcia-Arrars, L.A. Elferink, K. Peterson, A.M. Fleming, C.D. Hazuka & R.H. Scheller. 1993. The syntaxin family of vesicular transport receptors. **74:** 863–873.
16. Oyler, G.A., G.A. Higgins, R.A. Hart, E. Battenberg, M. Billingsley, F.E. Bloom & M.C. Wilson. 1989. The identification of a novel synaptosomal-associated protein, SNAP-25, differentially expressed by neuronal subpopulations. J. Cell. Biol. **109:** 3039–3052.
17. Söllner, T., S.W. Whiteheart, M. Brunner, H. Erdjument-Bromage, S. Geromanos, P. Tempst & J.E. Rothman. 1993. SNAP receptors implicated in vesicle targeting and fusion. Nature **362:** 318–324.
18. Calakos, N., M.K. Bennett, K. Peterson & R.H. Scheller. 1994. Protein-protein interactions contributing to the specificity of intracellular vesicular trafficking. Science **263:** 1146–1149.
19. O'Conner, V.M., O. Shamotienko, E. Grishin & H. Betz. 1993. On the structure of the 'synaptosecretosome.' Evidence for a neurexin/synaptotagmin/syntaxin/Ca^{2+} channel complex. FEBS Lett. **326:** 255–260.
20. Hayashi, T., H. McMahon, S. Yamasaki, T. Binz, Y. Hata, T.C. Südhof & H. Niemann. 1994. Synaptic vesicle membrane fusion complex: Action of clostridial neurotoxins on assembly. EMBO J. **13:** 5051–5061.
21. Chapman, E.R., S. An, N. Barton & R. Jahn. 1994. SNAP-25, a t-SNARE which binds to both syntaxin and synaptobrevin via domains that may form coiled coils. J. Biol. Chem. **269:** 27427–27432.
22. Matthew. W.D., L. Tsavaler & L.F. Reichardt. 1981. Identification of a synaptic vesicle-specific membrane protein with a wide distribution in neuronal and neurosecretory tissue. J. Cell Biol. **91:** 257–269.
23. Perin, M.S., V.A. Fried, G.A. Mignery, R. Jahn & T.C. Südhof. 1990. Phospholipid binding by a synaptic vesicle protein homologous to the regulatory region of protein kinase C. Nature **345:** 260–263.
24. Li, C., B.A. Davletov & T.C. Südhof. 1995. Distinct Ca^{2+} and Sr^{2+} binding properties of synaptotagmins. Definition of candidate Ca^{2+} sensors for the fast and slow components of neurotransmitter release. J. Biol. Chem. **270:** 24898–24902.
25. Li, C., B. Ullrich, J.Z. Zhang, R.G.W. Anderson, N. Brose & T.C. Südhof. 1995. $Ca(^{2+})$-dependent and -independent activities of neural and non-neural synaptotagmins. Nature **375:** 594–599.
26. Chapman, E.R., P.I. Hanson, S. An & R. Jahn. 1995. Ca^{2+} regulates the interaction between synaptotagmin and syntaxin 1. J. Biol. Chem. **270:** 23667–23671.
27. Elferink, L.A., M.R. Peterson & R.H. Scheller. 1993. A role for synaptotagmin (p65) in regulated exocytosis. Cell **72:** 153–159.
28. Bommert, K., M.P. Charlton, W.M. DeBello, G.J. Chin, H. Betz & G.J. Augustine. 1993. Inhibition of neurotransmitter release by C2-domain peptides implicates synaptotagmin in exocytosis. Nature **363:** 163–165.
29. Geppert, M., Y. Goda, R.E. Hammer, C. Li, T.W. Rosahl, C.F. Stevens & T.C. Südhof. 1994. Synaptotagmin I: A major Ca^{2+} sensor for transmitter release at a central synapse. Cell **79:** 717–727.
30. Broadie, K., H.J. Bellen, A. DiAntonio, J.T. Littleton & T.L Schwarz. 1994. Absence of synaptotagmin disrupts excitation-secretion coupling during synaptic transmission. Proc. Natl. Acad. Sci. USA **91:** 10727–10731.

31. LITTLETON, J.T., M. STERN, K. SCHULZE, M. PERIN & H.J. BELLEN. 1993. Mutational analysis of Drosophila synaptotagmin demonstrates its essential role in Ca($^{2+}$)-activated neurotransmitter release. Cell **74:** 1125–1134.
32. NONET, M.L., K. GRUNDAHL, B.J. MEYER & J.B. RAND. 1993. Synaptic function is impaired but not eliminated in C. elegans mutants lacking synaptotagmin. Cell **73:** 1291–1305.
33. LÉVÊQUE, C., T. HOSHINO, P. DAVID, Y. SHOJI-KASAI, K. LEYE, A. OMORI, B. LANG, O. EL FAR, K. SATO, N. MARTIN-MOUTOT, J. NEWSON-DAVIS, M. TAKAHASHI & M.J. SEAGAR. 1992. The synaptic vesicle protein synaptotagmin associates with calcium channels and is a putative Lambert-Eaton myasthenic syndrome antigen. Proc. Natl. Acad. Sci. USA **89:** 3625–3629.
34. LÉVÊQUE, C., O. EL FAR, N. MARTIN-MOUTOT, K. SATO, R. KATO, M. TAKAHASHI & M.J. SEAGAR. 1994. Purification of the N-type calcium channel associated with syntaxin and synaptotagmin. A complex implicated in synaptic vesicle exocytosis. J. Biol. Chem. **269:** 6306–6312.
35. YOSHIDA, A., C. OHO, A. OMORI, R. KUWAHARA, T. ITO & M. TAKAHASHI. 1992. HPC-1 is associated with synaptotagmin and omega-conotoxin receptor. J. Biol. Chem. **267:** 24925–24928.
36. EL FAR, O., N. CHARVIN, C. LEVEQUE, N. MARTIN-MOUTOT, M. TAKAHASHI & M.J. SEAGAR. 1995. Interaction of a synaptobrevin (VAMP)-syntaxin complex with presynaptic calcium channels. FEBS. Lett. **361:** 101–105.
37. TAKAHASHI, M., M.J. SEAGAR, J.F. JONES, B.F.X. REBER & W.A. CATTERALL. 1987. Subunit structure of dihydropyridine-sensitive calcium channels from skeletal muscle. Proc. Natl. Acad. Sci. USA **84:** 5478–5482.
38. CATTERALL, W.A. 1995. Structure and function of voltage-gated ion channels. Annu. Rev. Biochem. **65:** 493–531.
39. DUBEL, S.J., T.V.B. STARR, J. HELL, M.K. AHLIJANIAN, J.J. ENYEART, W.A. CATTERALL & T.P. SNUTCH. 1992. Molecular cloning of the α-1 subunit of an ω-conotoxin-sensitive calcium channel. Proc. Natl. Acad. Sci. USA **89:** 5058–5062.
40. MORI, Y., T. FRIEDRICH, M.-S. KIM, A. MIKAMI, J. NAKAI, P. RUTH, E. BOSSE, F. HOFMANN, V. FLOCKERZI, T. FURUICHI, K. MIKOSHIBA, K. IMOTO, T. TANABE & S. NUMA. 1991. Primary structure and functional expression from complementary DNA of a brain calcium channel. Nature **350:** 398–402.
41. SATHER, W.A., T. TANABE, J.-F. ZHANG, Y. MORI, M.E. ADAMS, & R.W. TSIEN. 1993. Distinctive biophysical and pharmacological properties of class A (BI) calcium channel alpha 1 subunits. Neuron **11:** 291–303.
42. STARR, T.V.B., W. PRYSTAY & T.P. SNUTCH. 1991. Primary structure of a calcium channel that is highly expressed in the rat cerebellum. Proc. Natl. Acad. Sci. USA **88:** 5621–5625.
43. STEA, A., S.J. DUBEL, M. PRAGNELL, J.P. LEONARD, K.P. CAMPBELL & T.P. SNUTCH. 1993. A beta-subunit normalizes the electrophysiological properties of a cloned N-type Ca^{2+} channel alpha 1-subunit. Neuropharmacology **32:** 1103–1116.
44. WESTENBROEK, R., J. HELL, C. WARNER, S. DUBEL, T. SNUTCH & W.A. CATTERALL. 1992. Biochemical properties and subcellular distribution of an N-type calcium channel α1 subunit. Neuron **9:** 1099–1115.
45. WILLIAMS, M.E., D.H. FELDMAN, A.F. MCCUE, R. BRENNER, G. VELICELEBI, S.B. ELLIS & M.M. HARPOLD. 1992. Structure and functional expression of alpha 1, alpha 2, and beta subunits of a novel human neuronal calcium channel. Neuron **8:** 71–84.
46. ZHANG, J.-F., A.D. RANDALL, P.T. ELLINOR, W.A. HORNE, W.A. SATHER, T. TANABE, T.L. SCHWARZ & R.W. TSIEN. 1993. Distinctive pharmacology and kinetics of cloned neuronal Ca^{2+} channels and their possible counterparts in mammalian CNS neurons. Neuropharmacology **32:** 1075–1088.
47. TSIEN, R.W., D. LIPSCOMBE, D.V. MADISON, K.R. BLEY & A.P. FOX. 1988. Multiple types of neuronal calcium channels and their selective modulation. Trends Neurosci. **11:** 431–438.
48. WU, L.-G. & P. SAGGAU. 1994. Pharmacological identification of two types of presynaptic voltage-dependent calcium channels at CA3-CA1 synapses of the hippocampus. J. Neurosci. **14:** 5613–5622.
49. MINTZ, I.M., B.L. SABATINI & W.G. REGEHR. 1995. Calcium control of transmitter release at a cerebellar synapse. Neuron **15:** 675–688.
50. WESTENBROEK, R., T. SAKURAI, E.M. ELLIOTT, J.W. HELL, T.V.B. STARR, T.P. SNUTCH & W.A. CATTERALL. 1995. Immunochemical identification and subcellular distribution of the α_{1A} subunits of brain calcium channels. J. Neurosci. **15:** 6403–6418.

51. SHENG, Z.-H., J. RETTIG, M. TAKAHASHI & W.A. CATTERALL. 1994. Identification of a syntaxin-binding site on N-type calcium channels. Neuron **13:** 1303–1313.
52. SHENG, Z.-H., J. RETTIG, T. COOK & W.A. CATTERALL. 1996. Calcium-dependent interaction of N-type calcium channels with the synaptic core-complex. Nature **379:** 451–454.
53. RETTIG, J., Z.-H. SHENG, D.K. KIM, C.D. HODSON, T.P. SNUTCH & W.A. CATTERALL. 1996. Isoform-specific interaction of the α_{1A} subunits of brain Ca^{2+} channels with the presynaptic proteins syntaxin and SNAP-25. Proc. Natl. Acad. Sci. USA **93:** 7363–7368.
54. BROSE, N., A.G. PETRENKO, T.C. SÜDHOF & R. JAHN. 1992. Synaptotagmin: A calcium sensor on the synaptic vesicle surface. Science **256:** 1021–1025.
55. DAVLETOV, B.A. & T.C. SÜDHOF. 1993. A single C2 domain from synaptotagmin I is sufficient for high affinity Ca^{2+}/phospholipid binding. J. Biol. Chem. **268:** 26386–26390.
56. CHAPMAN, E.R. & R. JAHN. 1994. Calcium-dependent interaction of the cytoplasmic region of synaptotagmin with membranes. Autonomous function of a single C2-homologous domain. J. Biol. Chem. **269:** 5735–5741.
57. DIANTONIO, A.K., D. PARFITT & T.L. SCHWARZ. 1993. Synaptic transmission persists in synaptotagmin mutants of Drosophila. Cell **72:** 1281–1290.
58. YOKOYAMA, C.T., Z.-H. SHENG & W.A. CATTERALL. 1997. Phosphorylation of the synaptic protein interaction site on N-type calcium channels inhibits interactions with SNARE proteins. J. Neurosci. **17:** 6929–6938.
59. HIRNING, L.D., A.P. FOX, E.W. MCCLESKEY, B.M. OLIVERA, S.A. THAYER, R.J. MILLER & R.W. TSIEN. 1988. Dominant role of N-type Ca^{2+} channels in evoked release of norepinephrine from sympathetic neurons. Science **239:** 57–60.
60. MINTZ, I.M., V.J. VENEMA, K.M. SWIDEREK, T.D. LEE, B.P. BEAN & M.E. ADAMS. 1992. P-type calcium channels blocked by the spider toxin omega-Aga-IVA. Nature **355:** 827–829.
61. UCHITEL, O.D., D.A. PROTTI, V. SANCHEZ, B.D. CHERKSEY, M. SUGIMORI & R. LLINÁS. 1992. P-type voltage-dependent calcium channel mediates presynaptic calcium influx and transmitter release in mammalian synapses. Proc. Natl. Acad. Sci. USA **89:** 3330–3333.
62. LUEBKE, J.I., K. DUNLAP & T.J. TURNER. 1993. Multiple calcium channel types control glutamatergic synaptic transmission in the hippocampus. Neuron **11:** 895–902.
63. TAKAHASHI, T. & A. MOMIYAMA. 1993. Different types of calcium channels mediate central synaptic transmission. Nature **366:** 156–158.
64. TURNER, T.J., M.E. ADAMS & K. DUNLAP. 1993. Multiple Ca^{2+} channel types coexist to regulate synaptosomal neurotransmitter release. Proc. Natl. Acad. Sci. USA **90:** 9518–9522.
65. WHEELER, D.B., A. RANDALL & R.W. TSIEN. 1994. Roles of N-type and Q-type Ca^{2+} channels in supporting hippocampal synaptic transmission. Science **264:** 107–111.
66. SAKURAI, T., J.W. HELL, A. WOPPMANN, G. MILJANICH & W.A. CATTERALL. 1995. Immunochemical identification and differential phosphorylation of alternatively spliced forms of the α_{1A} subunit of brain calcium channels. J. Biol. Chem. **270:** 21234–21242.
67. SAKURAI, T., R.E. WESTENBROEK, J. RETTIG, J. HELL & W.A. CATTERALL. 1996. Biochemical properties and subcellular distribution of the BI and rbA isoforms of α_{1A} subunits of brain calcium channels. J. Cell Biol. **134:** 511–528.
68. KIM, D.K & W.A. CATTERALL. 1997. Ca^{2+}-dependent and -independent interactions of the isoforms of the α_{1A} subunit of brain Ca^{2+} channels with presynaptic SNARE proteins. Proc. Natl. Acad. USA **94:** 14782–14786.
69. SHENG, Z.-H., C.T. YOKOYAMA & W.A. CATTERALL. 1997. Interaction of the synprint site of N-type Ca^{2+} channels with the C2B domain of synaptotagmin I. Proc. Natl. Acad. Sci. USA **94:** 5405–5410.
70. CHARVIN, N. & M.J. SEAGAR. 1997. Direct interaction of the calcium sensor protein synaptotagmin I with a cytoplasmic domain of the alpha 1A subunit of the P/Q-type calcium channel. EMBO J. **16:** 4591–4596.
71. KEE, Y. & R.H. SCHELLER. 1996. Localization of synaptotagmin-binding domains on syntaxin. J. Neurosci. **16:** 1975–1981.
72. MOCHIDA, S., Z.-H. SHENG, C. BAKER, H. KOBAYASHI & W.A. CATTERALL. 1996. Inhibition of neurotransmission by peptides containing the synaptic protein interaction site of N-type Ca^{2+} channels. Neuron **17:** 781–788.

73. RETTIG, J., C. HEINEMANN, U. ASHERY, Z.-H. SHENG, C.T. YOKOYAMA, W.A. CATTERALL & E. NEHER. 1997. Alteration of Ca^{2+} dependence of neurotransmitter release by disruption of Ca^{2+} channel/syntaxin interaction. J. Neurosci. **17:** 6647–6656.
74. MOCHIDA, S., H. KOBAYASHI, Y. MATSUDA, Y. YUDA, K. MURAMOTO & Y. NONOMURA. 1994. Myosin II is involved in transmitter release at synapses formed between rat sympathetic neurons in culture. Neuron **13:** 1131–1142.
75. MOCHIDA, S., H. SAISU, H. KOBAYASHI & T. ABE. 1995. Impairment of syntaxin by botulinum neurotoxin C1 or antibodies inhibits acetylcholine release but not Ca^{2+} channel activity. Neuroscience **65:** 905–915.
76. FATT, P. & B. KATZ. 1951. An analysis of the end-plate potential recorded with an intracellular electrode. J. Physiol. (Lond.) **115:** 320–370.
77. YAZEJIAN, B., D.A. DIGREGORIO, J.L. VERGARA, R.E. POAGE, S.D. MERINEY & A.D. GRINNELL. 1997. Direct measurements of presynaptic calcium and calcium-activated potassium currents regulating neurotransmitter release at cultured *Xenopus* nerve-muscle synapses. J. Neurosci. **17:** 2990–3001.
78. HELL, J.W., S.M. APPLEYARD, C.T. YOKOYAMA, C. WARNER & W.A. CATTERALL. 1994. Differential phosphorylation of two size forms of the N-type calcium channel $\alpha 1$ subunit which have different COOH-termini. J. Biol. Chem. **269:** 7390–7396.
79. HIRLING, L.D & R.H. SCHELLER. 1996. Phosphorylation of synaptic vesicle proteins: Modulation of the alpha SNAP interaction with the core complex. Proc. Natl. Acad. Sci. USA **93:** 11945–11949.
80. SHIMAZAKI, Y., T. NISHIKI, A. OMORI, M. SEKIGUCHI, Y. KAMATA, S. KOZAKI & M. TAKAHASHI, 1996. Phosphorylation of 25-kDa synaptosome-associated protein. Possible involvement in protein kinase C-mediated regulation of neurotransmitter release. J. Biol. Chem. **271:** 14548–14553.
81. BEZPROZVANNY, I., R.H. SCHELLER & R.W. TSIEN. 1995. Functional impact of syntaxin on gating of N-type and Q-type calcium channels. Nature **378:** 623–626.
82. WISER, O., M.K. BENNETT & D. ATLAS. 1996. Functional interaction of syntaxin and SNAP-25 with voltage-sensitive L- and N-type Ca^{2+} channels. EMBO J. **15:** 4100–4110.
83. WISER, O., D. TOBI, M. TRUS & D. ATLAS. 1997. Synaptotagmin restores kinetic properties of a syntaxin-associated N-type voltage sensitive calcium channel. FEBS Lett. **404:** 203–207.
84. STANLEY, E.F. & R.R. MIROTZNIK. 1997. Cleavage of syntaxin prevents G-protein regulation of presynaptic calcium channels. Nature **385:** 340–343.

Dissection of the Calcium Channel Domains Responsible for Modulation of Neuronal Voltage-Dependent Calcium Channels by G Proteins

ANNETTE C. DOLPHIN,[a] KAREN M. PAGE, NICHOLAS S. BERROW, GARY J. STEPHENS, AND CARLES CANTÍ

Department of Pharmacology, University College London, Gower Street, London WC1E6BT, United Kingdom

ABSTRACT: The molecular determinants for G-protein regulation of neuronal calcium channels remain controversial. We have generated a series of α1B/α1E chimeric channels, since rat brain α1E (rbEII), unlike human α1E, showed no G-protein modulation. The study, carried out in parallel using D2 receptor modulation of calcium currents in *Xenopus* oocytes of Gβγ modulation of calcium currents in COS-7 cells, consistently showed an essential role for domain I (from the N terminus to the end of the I-II loop) of the α1B Ca^{2+} channel in G-protein regulation, with no additional effect of the C terminal of α1B. The I-II loop alone of α1B, or the I-II loop together with the C-terminal tail, was insufficient to confer G-protein modulation of α1E (rbEII). We have further observed that the α1E clone rbEII is truncated at the N-terminus compared to other α1 subunits, and we isolated a PCR product from rat brain equivalent to a longer N-terminal isoform. The long N-terminal α1E, unlike the short form, showed G-protein modulation. Furthermore, the equivalent truncation of α1B (ΔN1-55) abolished G-protein modulation of α1B. Thus, we propose that the N terminus of α1B and α1E calcium channels contains essential molecular determinants for membrane-delimited G-protein inhibition, and that other regions, including the I-II loop and the C terminus, do not play a conclusive role alone.

Voltage-dependent calcium channels (VDCCs) play a major role in the function of neurons and other excitable cells. This includes control of transmitter release by Ca^{2+}, regulation of excitability by activation of calcium-dependent currents, and activation of other Ca^{2+}-dependent processes, including control of gene expression. Because of this unique property of Ca^{2+} entry to regulate so many cellular processes, the modulation of VDCCs is of great importance for the correct operation of cells. VDCCs consist of a transmembrane α1 subunit, which has four domains of similar structure, each containing six putative transmembrane segments. The α1 subunit is thought to form the pore of the channel. It copurifies with an intracellular β subunit and an extracellular α2 subunit, which is attached by S-S bonds to a transmembrane δ subunit[1,2] (FIG. 1). Several different α1 subunits have been cloned; α1C, D, and S all form 1,4-dihydropyridine (DHP)–sensitive L-type calcium channels, whereas α1A, B, and E, respectively, form P/Q,[3,4] N,[5] and possibly R[6] or a subtype of low voltage–activated T channels.[7] Furthermore, a novel α1 subunit

[a]Address for correspondence: Prof. A.C. Dolphin, Department of Pharmacology (Medawar Building), University College London, Gower Street, London WC1E6BT, United Kingdom. Phone: 0171 419 3054; fax: 0171 813 2808; e-mail: a.dolphin@ucl.ac.uk

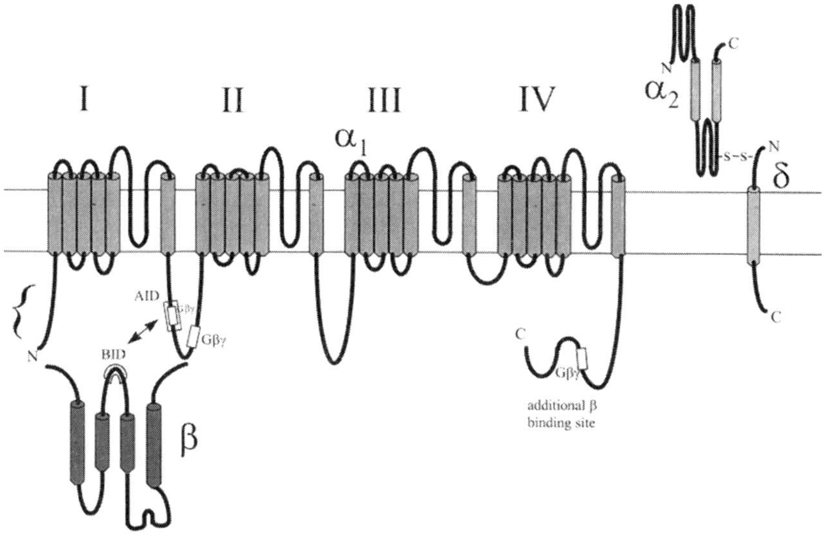

FIGURE 1. The VDCC oligomeric complex. Binding sites on the α1 I-II loop for VDCC β and Gβγ are shown. All VDCC α1 subunits have a binding site for β on this loop, but Gβγ binding has been shown only for α1A, α1B, and α1E. Regions of the C terminal tail that may be involved in VDCC β and Gβγ binding are also indicated. The region of the N-terminal tail that is essential for G-protein modulation is shown with a *brace*.

(α1G) has recently been cloned that codes for a subtype of rapidly inactivating T-type channel.[8]

Calcium channels may be modulated by a number of different means, including membrane-delimited and second-messenger pathways. For example, for native cardiac calcium channels, activation of the cyclic AMP–dependent phosphorylation pathway results in a marked increase in calcium current,[9] although the molecular substrate for the phosphorylation remains uncertain.[10] However, for neuronal channels, particularly N and P/Q, a major mechanism of modulation involves inhibition via the activation of heterotrimeric G proteins by seven transmembrane (7TM) receptors (for review see Refs. 11, 12). This inhibition is typified by a slowing of the current activation kinetics. Most evidence suggests that this represents a time-dependent recovery from the G protein inhibition that is present at hyperpolarized membrane potentials but can be overcome by depolarization.[13] At the single-channel level, it is manifested by a prolongation of the latency to first opening.[14,15] The voltage dependence is also evident in a shift to more depolarized potentials of the current activation-voltage relationship and the loss of inhibition at very large depolarizations.[13]

In neurons, the distribution of calcium channels is nonuniform, with α1A and α1B being particularly concentrated at synaptic terminals.[16,17] G protein–mediated modulation of these channels has been shown to occur at presynaptic terminals.[18] This mechanism is probably responsible for most of the presynaptic inhibition of synaptic transmission mediated by a wide variety of 7TM receptors in many areas of the nervous system.[11,18–20] Activation of such receptors will therefore reduce calcium entry into presynaptic terminals via VDCCs.

INHIBITION OF VDCCs BY G PROTEINS

The link between G protein activation and N or P/Q type calcium channel inhibition appears, from several lines of evidence, to be a direct one.[20,21] This led several groups to begin to study whether there is a direct link between activated G proteins and one of the subunits of the calcium channel complex. We initially examined whether the intracellular VDCC β subunit might be involved in G protein modulation. We developed an antisense strategy to deplete cultured dorsal root ganglion neurons of their VDCC β subunits by microinjection of an antisense oligonucleotide complementary to a region common to all β subunits.[22] In this study, we observed that depletion of the VDCC β subunit resulted in smaller calcium channel currents, with slowed activation kinetics, as would be expected for depletion of this accessory subunit (FIG. 2A).[22] It also resulted in an enhancement of the ability of the GABA-B receptor agonist (–)-baclofen to inhibit the residual currents[23] (FIG. 2A and B). Our hypothesis from these results was that there may be a competition between activated G protein and VDCC β subunit for binding to a site on the channel.[11,23] Thus, when the VDCC β subunit was depleted, the activated G protein was better able either to bind to its target and/or to have a functional inhibitory effect. Our result also indicated that the target of the activated G protein was not the VDCC β subunit itself, but more likely the VDCC α1 subunit.

INTERACTION BETWEEN VDCC α1 AND β SUBUNITS

The VDCC β subunit binds to a site (called the α interaction domain or AID) on the intracellular loop between transmembrane domains I and II of all VDCC α1 subunits.[24] There is also evidence for the presence of an additional binding site for the VDCC β subunit in the last 277 amino acids of α1E,[25] and a similar low-affinity binding site has been observed on the C terminal tail of α1A[26] (see FIG. 1). It is of interest that this binding site only showed affinity for β4 and β2a, but did not bind significantly to β1b.[26] If there is competition for binding between activated G protein and VDCC β subunit, the activated G protein moiety might bind to sites overlapping with those of VDCC β, either on the I-II loop or on the C terminal of α1. The hypothesis that there is competition for an overlapping binding site is supported by evidence that overexpression of VDCC β in *Xenopus* oocytes blocked agonist-mediated inhibition,[27] whereas expression of VDCCs in oocytes

FIGURE 2. Competition between VDCC β subunit and activated G protein for modulation of DRG calcium currents. Cultured dorsal root ganglion neurons were injected with an antisense oligonucleotide specific for the mRNA of all VDCC β subunits. The ability of the GABA-B agonist (–)-baclofen to inhibit the calcium channel currents was examined 60 h after microinjection. Baclofen was found consistently to be more effective in β subunit–depleted than in the control or nonsense-injected cells. **(A)** Examples of current traces before and during baclofen application in a β antisense–injected cell (**top traces**) or a nonsense-injected cell (**bottom traces**). **(B)** Bar graph of mean % inhibition by (–)-baclofen, for β antisense, nonsense, and control noninjected cells. Details of the work are to be found elsewhere.[22,23]

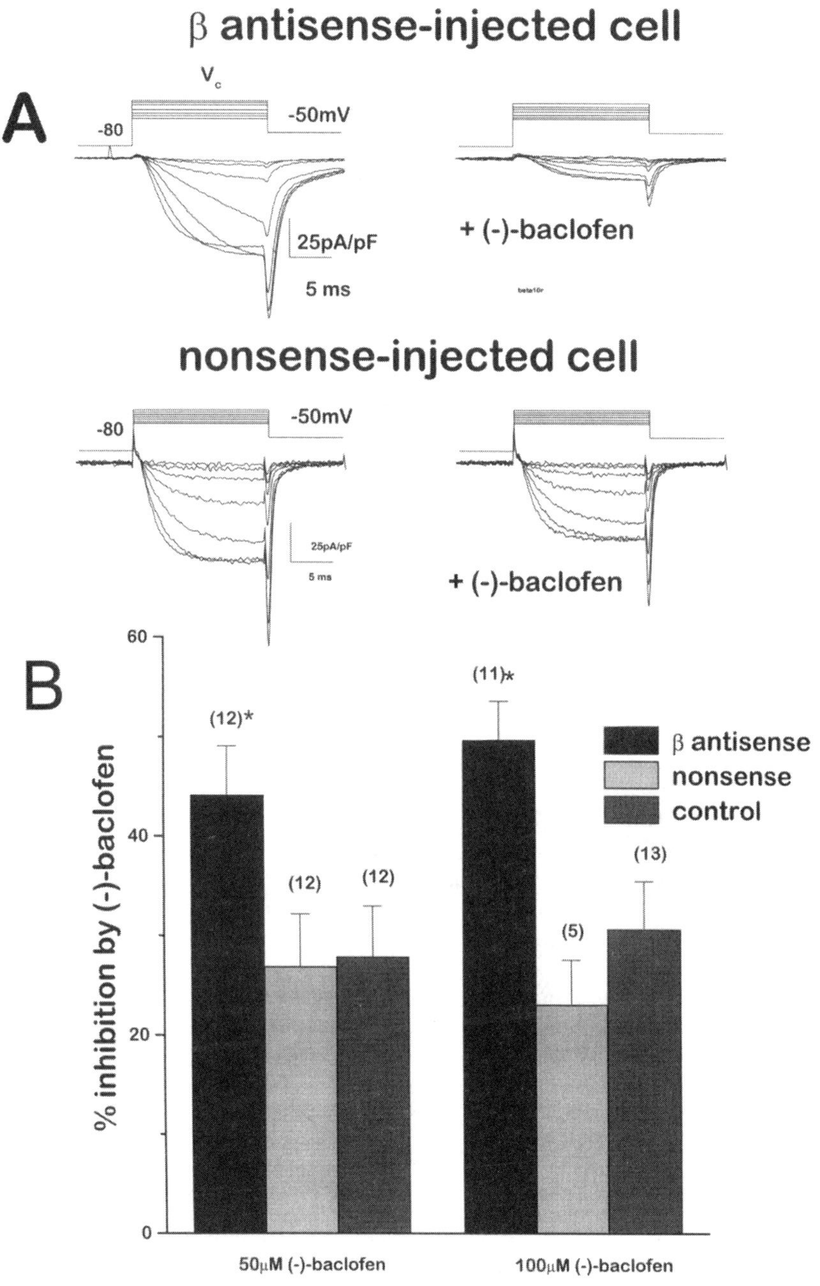
β antisense-injected cell
A
V_c
-80
-50mV
25pA/pF
5 ms
+ (-)-baclofen
nonsense-injected cell
-80
-50mV
25pA/pF
5 ms
+ (-)-baclofen
B
% inhibition by (-)-baclofen
60
40
20
0
(12)*
(12)
(12)
(11)*
(5)
(13)
β antisense
nonsense
control
50μM (-)-baclofen
100μM (-)-baclofen

in the absence of the VDCC β subunit led to an enhancement of receptor-mediated modulation.[28]

ROLE OF Gβγ SUBUNITS IN COUPLING TO VDCC α1 SUBUNITS

In most systems, the modulation of neuronal VDCCs is mediated by receptors coupled to pertussis toxin–sensitive G proteins,[29–31] and a number of groups showed that $G\alpha_o$ was primarily responsible for the effect.[32–34] However, in several systems, both $G\alpha_i$ and $G\alpha_o$ were involved,[35,36] and in a few systems, G_q- or G_s-coupled receptors produced similar modulation.[37–39] It was therefore hypothesized by two groups that Gβγ,[40,41] rather than any particular Gα, was the key factor involved in modulation. These studies showed that Gβγ overexpression led to the tonic inhibition of the calcium current, which could be reversed by a depolarizing prepulse, applied just before the test pulse, a hallmark of voltage-dependent inhibition of these channels.[40,41] Gβγ overexpression also occluded modulation by agonist. There is a clear precedent for this in the G protein–activated potassium channels (GIRKs), which are activated by Gβγs.[42–44] Furthermore, all Gβγ species tested in that system (except transducin Gβ1γ1) are similarly effective.[45–47] It appears likely that actual unbinding of Gβγ from the calcium channel occurs during a depolarizing prepulse, and rebinding occurs from the bulk phase. Of relevance to this, we have recently shown that the rate of reinhibition of α1B, following a depolarizing prepulse, can be decreased by expression of the β-adrenergic receptor kinase (βARK) βγ binding domain, and increased by overexpression of Gβγ subunits. These treatments will decrease and increase, respectively, the "free" concentration of Gβγ subunits available for binding.[48]

INVOLVEMENT OF THE α1 I-II LOOP IN G PROTEIN MODULATION

Initial results suggested that the intracellular I-II loop, which links domains I and II, may be of importance in G protein modulation. These included the finding that VDCC β subunits bind to the I-II loop,[49] and may therefore functionally compete with Gβγ subunits.[23] Gβγ subunits have been found previously to bind to sites on type 2 adenylyl cyclase and phospholipase Cβ2,[50] which have a characteristic central motif consisting of QQxER. While this motif is not necessarily indicative of a functional Gβγ binding site, it was also found to occur in the I-II loop of α1A, B, and E, intriguingly within the binding site described for the VDCC β subunit[51] (FIG. 1).

In reconstituted systems consisting of a VDCC α1/β combination and either an endogenous or an expressed G protein coupled receptor, classical modulation could be demonstrated for α1B (the N-type channel) and to a lesser extent for α1A, but not to any clear extent for rat α1E (rbEII clone),[28] despite the presence of a Gβγ binding site on the I-II loop (FIGS. 1, 3). Therefore we and others made chimeric channels between α1B, which shows the greatest G protein modulation, and those that showed no or less modulation, in an attempt to define the regions involved in this process.[51–54]

We have found, by making chimeras containing the I-II loop of either α1A or α1B in rat brain α1E (rbEII clone), that we can confer only a minor aspect of G protein–mediated inhibition onto this α1E clone, which is not significantly G protein modulated in our system,[53] or in others that have been tested.[28] In our first study, we substituted the I-II loop and part of the Is6 transmembrane segment from α1B into the α1E backbone, making an

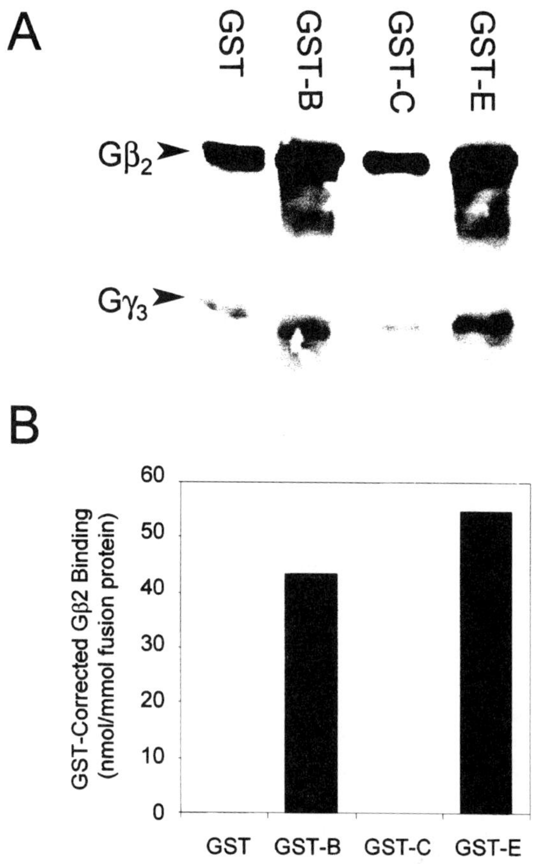

FIGURE 3. Binding of $G\beta_2\gamma_3$ to VDCC I-II loops. $G\beta_2$ and $G\gamma_3$ subunits were *in vitro* translated and transcribed from their respective cDNAs in the pCDNA3 vector in the presence of ^{35}S-methionine (with protease inhibitors and mevalonic acid lactone). Translation products were analyzed using SDS-PAGE, autoradiography, and scintillation counting. DNA fragments encoding the I-II loops of the VDCC α1B, C, and E were synthesized by PCR and ligated into the pGEX2T expression vector. After expression in *E. coli* Top10F′ the GST fusion proteins were first solubilized from the *E. coli* by sonication and the addition of 2% w/v Sarcosyl before immobilization on glutathione-sepharose beads. After three washes in binding buffer (1 mM HEPES pH 7.4, 150 mM NaCl, EDTA 1 mM, DTT 1 mM and 0.1% w/v CHAPS), the beads were resuspended, and the ^{35}S $G\beta2\gamma3$ subunits were added to a final concentration of 25 nM. After incubation for 16 hours at 4°C the beads were washed a further three times with binding buffer before resuspension in SDS sample buffer. **(A)** Samples were separated on 15% SDS-PAGE gels, transferred to PVDF membranes, and exposed to phosphor-imager screens, to visualize Gβ2 and Gγ3. **(B)** The level of Gβ2 subunits bound by the various fusion proteins was quantified by image analysis and scintillation counting. (N.S.B. and A.C.D., unpublished results).

α1E/BbEEE chimera (Chimera 3, FIG. 4). The two parameters that we have measured are kinetic slowing and inhibition of the calcium channel current amplitude. These may both be aspects of the same mechanism of G protein modulation. Kinetic slowing is probably due to the time- and voltage-dependent partial dissociation of Gβγ that occurs during the test pulse to activate inward calcium channel currents, whereas the inhibition of current amplitude, which is reversible by a very large depolarizing prepulse, may be due to inhibitory modulation by the Gβγ that remains bound during the small depolarization represented by the test pulse, but is driven off by depolarization to a large positive potential.

We examined modulation of the parental channels and chimera by GTPγS, following transient transfection of the relevant VDCC α1 subunits, together with β1b and α2-δ, in COS-7 cells. The replacement of the I-II loop and IS6 conferred on α1E the ability to exhibit some kinetic slowing in the presence of GTPγS, whereas α1E itself showed no modulation. In contrast, for currents resulting from α1B, GTPγS also produced a large decrease of the average maximum amplitude of the current, and this was reversed by a depolarizing prepulse. This property is not conferred on the α1E/BbEEE chimera. It is possible that the Gβγ unbinding from the α1E/BbEEE chimera shows a steeper voltage dependence than for α1B, so that all (rather than only some) of the Gβγ is removed by the test pulse. Our initial findings therefore suggest that the Is6 and/or the I-II linker region alone is not the most important region of the α1B channel in conferring full G protein modulation on rbEII. When these studies were repeated using β2a, rather than β1b, and modulation of the α1E/BbEEE chimera by overexpression of Gβ1γ2 was examined, even less modulation was observed, suggesting that the additional binding site of β2a may occlude G protein modulation in this chimera.[55]

In more recent studies using chimeras between α1B and rat brain α1E(rbEII), we have observed a significant role for the whole first domain of α1B ($\alpha 1B_{1\text{–}483}$ between the N terminus and the end of the I-II loop) in G protein modulation. Essentially, the chimera α1BbEEE (Chimera 7, FIG. 4) behaved very similarly to α1B, showing substantial modulation by Gβ1γ2 and by the dopamine D2 agonist, quinpirole, when this receptor is injected with the α1 subunit and the accessory subunits α2-δ and β2a in *Xenopus* oocytes. In contrast, chimeras containing the I-II loop of α1B alone in rat brain α1 E(rbEII), forming α1EbEEE (Chimera 4, FIG. 4) or the I-II loop plus the C terminus of α1B (α1EbEEEb, chimera 6) showed no G protein modulation.[56]

Several groups,[51,54] including ourselves (FIG. 3), have now shown that the I-II loops of several VDCCs bind Gβγ subunits. In FIGURE 3 an experiment is shown in which Gβ2 and Gγ3 subunits, *in vitro* transcribed and translated in the presence of ^{35}S methionine, were bound to immobilized GST fusion proteins of the I-II loops of α1B, α1C, and α1E, eluted, and separated by electrophoresis on a PAGE gel (FIG. 3A). Following subtraction of the background associated with the GST alone, binding of Gβ was observed only to α1B and α1E but not to the α1C I-II loop (FIG. 3B). In other studies, the residues critical for Gβγ binding to these I-II loops have been mapped.[51] In agreement with electrophysiological data, the AID part of the loop (the region involved in binding VDCC β) is one important domain, and this contains the QQIER sequence, some residues of which have been shown

FIGURE 4. Calcium channel α1 subunit chimeras and other constructs derived from α1B and rat brain α1E(rbEII). The α1B domains are shown with *dark shading* and the α1E(rbEII) domains with *light shading*. The origins of the N terminus, I-II loop, and C terminus are denoted by *small letters* (b or e), whereas the origins of the domains I-IV are given by *capital letters* (B or E).

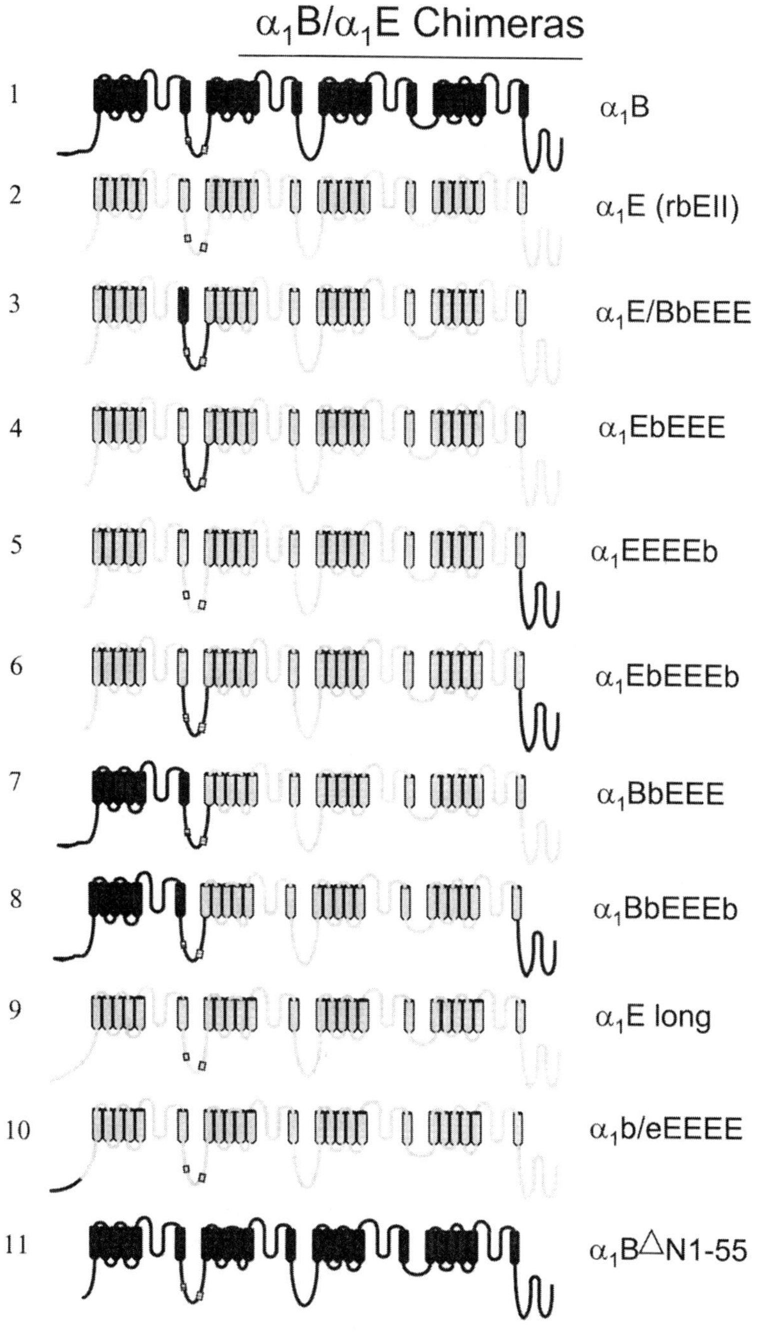
α1B/α1E Chimeras
1 α1B
2 α1E (rbEII)
3 α1E/BbEEE
4 α1EbEEE
5 α1EEEEb
6 α1EbEEEb
7 α1BbEEE
8 α1BbEEEb
9 α1E long
10 α1b/eEEEE
11 α1BΔN1-55

to be essential for Gβγ binding.[51] Catterall and colleagues provide additional data that the I-II loop is important in G protein modulation, first by using peptides.[57] Peptides alone do not prove that the α1 I-II loop is the site of modulation, but rather indicate whether the peptides bind to Gβγ and can therefore effectively compete for this mediator. However, in a second approach this group has mutated the QQIER sequence in α1A to that in α1C, which is QQLEE. This produced a reduction of modulation by GTPγS, using an HEK293 cell expression system.[57]

Zhang *et al.* (1996)[52] have made an extensive range of chimeras between α1B (strongly G protein modulated), α1A (less G protein modulated) and α1C (not G protein modulated), and examined modulation by somatostatin. Their conclusion was that the I-II linker had no role in conferring G protein modulatory properties to the channel, whereas the first transmembrane domain and the C terminal tail were of importance. Neither the I-II loop from α1A nor that from α1C reduced modulation in an α1B backbone. Nevertheless, it should be noted that in this study the chimera consisting of the first domain only of α1A in the α1B backbone (AaBBBb) showed greater modulation even than α1B. Furthermore, a chimera with the first domain and C terminal tail of α1C inserted into B (α1CbBBBc) still showed some G protein modulation,[52] whereas α1C itself showed no modulation.

G PROTEIN MODULATION OF α1E

It has been reported that the α1E subunit is either not (or only slightly) G protein modulated in most studies,[28,53,58] although its I-II loop contains a QQIER sequence, and the isolated I-II loop does bind to Gβγ *in vitro*[51] (FIG. 3). This has been taken by some as evidence that the I-II loop is not important for G protein modulation.[52] However, there are now two reports that, following expression without exogenous VDCC β subunits in *Xenopus* oocytes, human α1E does show some degree of G protein modulation, which is lost on coexpression of a β2a subunit and reduced by other VDCC β subunits.[59,60] This supports the notion that the VDCC β subunit inhibits receptor-mediated modulation,[23] possibly by competing with Gβγ for an overlapping binding site. It is conceivable that the reason that inhibition of α1E is either not seen at all or is much reduced in the presence of a VDCC β subunit is that binding of VDCC β to α1E may be particularly strong, although there is no direct evidence for this hypothesis. However, it must not be forgotten that *Xenopus* oocytes contain an endogenous β subunit, which is thought to be present in sufficient amount to promote surface expression of the α1E subunit, when heterologously expressed alone in this system, but may be insufficient to produce all the characteristic biophysical effects of β subunits.[25] It remains unclear whether the reason for the limited extent of modulation of α1E in the absence of exogenous β[60] is the presence of the endogenous β, or whether the ability of the α1E subunit to translate Gβγ binding into an effect on channel gating is intrinsically smaller than for α1B.

ROLE OF THE INTRACELLULAR C TERMINAL TAIL IN G PROTEIN MODULATION

Birnbaumer and colleagues[60] have further challenged the hypothesis that the I-II loop of human α1E is involved in G protein modulation, by making a chimera of human α1E containing the I-II loop of α1C. This remained G protein modulated (when expressed in the absence of β subunits), whereas a chimera containing part of the C terminal tail of α1C was not modulated.[60] This group therefore concluded that the C terminus rather than the I-II linker is involved in G protein modulation in α1E, and further showed that there is a binding site for Gβγ on the C terminal tail, corresponding to the region of the VDCC β binding site.[25] In contrast, from our own work, we found no significant enhancement of the modulation of the first-domain chimera between rat brain α1E(rbEII) and α1B (α1BbEEE, Chimera 7, FIG. 4) by the addition of the C terminal of α1B, forming α1BbEEEb (Chimera 8, FIG. 4).[56] Chimeras containing only the C terminus of α1B in α1E(rbEII) (forming α1EEEEb, Chimera 5, FIG. 4), or both the C terminal and the I-II loop (forming α1EbEEEb, Chimera 6, FIG. 4), show no G protein modulation.[56]

ROLE OF THE INTRACELLULAR N TERMINUS IN G PROTEIN MODULATION

The rat α1E (rbEII) subunit (L15453) has a shorter intracellular N terminal tail than the human, mouse, and rabbit clones. By RT-PCR of mRNA from rat cerebellar granule neurons, we cloned a cDNA fragment corresponding to a 50–amino acid extension of the N terminal sequence of α1E, and ligated this onto rbEII, forming $\alpha 1E_{long}$. The 5′ sequence of this clone (AF057029) showed very strong homology with the corresponding mouse, rabbit, and human α1E 5′ sequences. We have now observed that the $\alpha 1E_{long}$ clone is clearly G protein modulated, both by Gβ1γ2 in COS-7 cells, and by coexpression of the dopamine D2 receptor in *Xenopus* oocytes. However, the modulation is significantly less than for α1B[61] (FIG. 5A).

The α1E(rbEII) N terminal tail is 55 amino acids shorter than that of α1B, although the 40 amino acids that form the α1E(rbEII) N terminal tail do have a highly (82%) conserved counterpart in $\alpha 1B_{56-95}$ (FIG. 5B). Ligating the N terminal (1–55) tail of α1B onto rbEII produced α1b/eEEEE (Chimera 10, FIG. 4), a chimera whose modulation was intermediate between that of $\alpha 1E_{long}$ and α1B[61] (FIG. 5A). We have also obtained compelling evidence for the involvement of $\alpha 1B_{1-55}$ in its G protein modulation. Deletion of $\alpha 1B_{1-55}$ (forming the $\alpha 1B\Delta N_{1-55}$ construct, #11, FIG. 4) renders the α1B subunit, which exhibits the strongest degree of G protein sensitivity of all the α1 subunits, completely refractory to receptor-mediated inhibition and to the direct effect of Gβγ overexpression.[61] However, for both α1E and α1B, the biophysical properties of the truncated and N terminal extended forms are very similar,[61] suggesting that the truncation does not produce global structural changes, and that the N terminal 50 or 55 amino acids of $\alpha 1E_{long}$ and α1B, respectively, are playing a specific role in the process of calcium channel modulation by Gβγ subunits.

CONCLUSION

A number of experiments have indicated that the I-II loop of the α1 subunit is involved in the modulation of α1A and α1B calcium channels by G protein βγ subunits, and this loop clearly binds Gβγ subunits. However, several pieces of evidence suggest that this is not the only site involved, and alone is not sufficient for modulation in a channel that is otherwise not modulated.[52,53,56] There are now several pointers to a role for the C terminal tail, particularly in the small degree of modulation shown by α1E.[59,60] Furthermore the first domain of α1B also appears to play an essential role in its G protein modulation.[52,56]

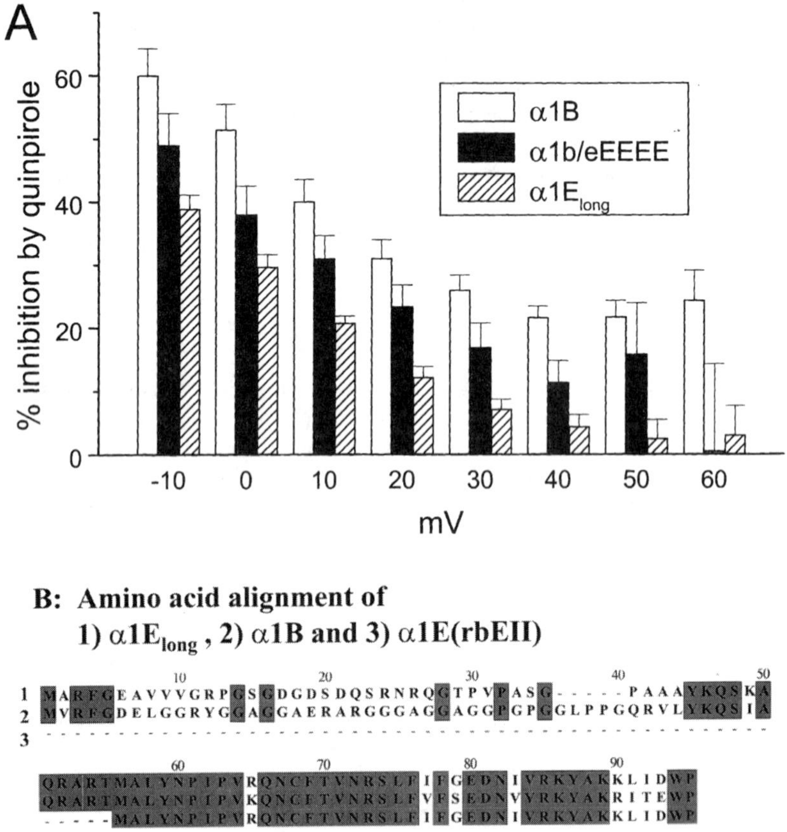

FIGURE 5. Sequence and function of $\alpha 1E_{long}$ and α1b/eEEEE. **(A)** The voltage dependence of the modulation of $\alpha 1E_{long}$, α1b/eEEEE (Chimera 10, FIG. 4), and α1B, by activation of exogenously expressed dopamine D2 receptors in *Xenopus* oocytes with 100nM quinpirole. The percent inhibition at each voltage is shown for α1B (*open bars*), α1b/eEEEE, (*shaded bars*), and $\alpha 1E_{long}$ (*hatched bars*). Experimental details have been described previously.[61] **(B)** The amino acid sequence of the N terminal tail of $\alpha 1E_{long}$ is compared to that of α1B and the truncated clone a1E(rbEII). Regions of amino acid identity are shown as *shaded sections.*

However, our recent results indicate that at least part of the importance of domain I lies in the first 50–55 amino acids of the N terminus.[61] For both α1B and α1E, the presence of this region is essential for their G protein modulation. Nevertheless, while our experiments rule out a role for the I-II loop alone in conferring G protein modulation on the truncated rbEII α1 subunit, which is otherwise not modulated, they have by no means excluded a role for the I-II loop in concert with other regions (the N terminus and possibly other parts of domain I). The consensus of evidence now suggests that the I-II loop is involved in the functional effects of Gβγs.

Goals for the future include determination of other sites of G protein binding and functional effects, elucidation of how these sites interact, and of the molecular mechanism of modulation by Gβγs. There is still little understanding of the way in which G protein binding is converted into an effect on latency of channel opening,[15] or additional inhibitory processes. It will also be of interest to evaluate whether the G protein α subunit plays a role in terminating the signal transduction process, as may be the case for GIRKs,[62] and to examine whether indeed there is competition between VDCC β and G protein βγ subunits for binding to calcium channels.

ACKNOWLEDGMENTS

We would like to acknowledge the contribution of all members of the group, past and present, on whose work we have built towards the studies from our laboratory that are described here. The work was supported by Wellcome Trust. C.C. is supported by a Marie Curie Fellowship from the European Community.

REFERENCES

1. Witcher, D.R., M. De Waard, J. Sakamoto, C. Franzini-Armstrong, M. Pragnell, S.D. Kahl & K.P. Campbell. 1993. Subunit identification and reconstitution of the N-type Ca^{2+} channel complex purified from brain. Science **261:** 486–489.
2. Liu, H., M. De Waard, V.E.S. Scott, C.A. Gurnett, V.A Lennon & K.P. Campbell. 1996. Identification of three subunits of the high affinity w-conotoxin MVIIC-sensitive Ca^{2+} channel. J. Biol. Chem. **271:** 13804–13810.
3. Sather, W.A., T. Tanabe, J.-F. Zhang, Y. Mori, M.E. Adams & R.W. Tsien. 1993. Distinctive biophysical and pharmacological properties of class A (BI) calcium channel α1 subunits. Neuron **11:** 291–303.
4. Gillard, S.E., S.G. Volsen, W. Smith, R.E. Beattie, D. Bleakman & D. Lodge. 1997. Identification of pore-forming subunit of P-type calcium channels: An antisense study on rat cerebellar Purkinje cells in culture. Neuropharmacology **36:** 405–409.
5. Fujita, Y., M. Mynlieff, R.T. Dirksen, M.-S. Kim, T. Niidome, J. Nakai, T. Friedrich, N. Iwabe, T. Miyata, T. Furuichi, D. Furutama, K. Mikoshiba, Y. Mori & K.G. Beam. 1993. Primary structure and functional expression of the w-conotoxin–sensitive N-type calcium channel from rabbit brain. Neuron **10:** 585–598.
6. Randall, A. & R.W. Tsien. 1995. Pharmacological dissection of multiple types of Ca^{2+} channel currents in rat cerebellar granule neurons. J. Neurosci. **15:** 2995–3012.
7. Soong, T.W., A. Stea, C.D. Hodson, S.J. Dubel, S.R. Vincent & T.P. Snutch. 1993. Structure and functional expression of a member of the low voltage-activated calcium channel family. Science **260:** 1133–1136.
8. Perez-Reyes, E., L.L. Cribbs, A. Daud, A.E. Lacerda, J. Barclay, M.P. Williamson, M. Fox, M. Rees & J. Lee. 1998. Molecular characterisation of a neuronal low-voltage-activated T type calcium channel. Nature **391:** 896–900.
9. Cachelin, A.B., J.E. De Peyer, S. Kokubun & H. Reuter. 1983. Ca^{2+} channel modulation by 8-bromocyclic AMP in cultured heart cells. Nature **304:** 462–464.
10. Klöckner, U., K. Itagaki, I. Bodi & A. Schwartz. 1992. β-Subunit expression is required for cAMP-dependent increase of cloned cardiac and vascular calcium channel currents. Pflügers Arch. **420:** 413–415.
11. Dolphin, A.C. 1995. Voltage-dependent calcium channels and their modulation by neurotransmitters and G proteins: G.L. Brown prize lecture. Exp. Physiol. **80:** 1–36.
12. Dolphin, A.C. 1998. Mechanisms of modulation of voltage-dependent calcium channels by G proteins. J. Physiol. (Lond.) **506:** 3–11.
13. Bean, B.P. 1989. Neurotransmitter inhibition of neuronal calcium currents by changes in channel voltage-dependence. Nature **340:** 153–155.

14. Carabelli, V., M. Lovallo, V. Magnelli, H. Zucker & E. Carbone. 1996. Voltage-dependent modulation of single N-type Ca^{2+} channel kinetics by receptor agonists in IMR32 cells. Biophys. J. **70:** 2144–2154.
15. Patil, P.G., M. De Leon, R.R. Reed, S. Dubel, T.P. Snutch & D.T. Yue. 1996. Elementary events underlying voltage-dependent G-protein inhibition of N-type calcium channels. Biophys. J. **71:** 2509–2521.
16. Westenbroek, R.E., J.W. Hell, C. Warner, S.J. Dubel, T.P. Snutch & W.A. Catterall. 1992. Biochemical properties and subcellular distribution of an N-type calcium channel $\alpha 1$ subunit. Neuron **9:** 1099–1115.
17. Westenbroek, R.E., T. Sakurai, E,M. Elliott, J.W. Hell, T.V.B. Starr, T.P. Snutch & W.A. Catterall. 1995. Immunochemical identification and subcellular distribution of the α_{1A} subunits of brain calcium channels. J. Neurosci. **15:** 6403–6418.
18. Toth, P.T., V.P. Bindokas, D. Bleakman, W.F. Colmers & R.J. Miller. 1993. Mechanism of presynaptic inhibition by neuropeptide Y at sympathetic nerve terminals. Nature **364:** 635–639.
19. Man-Son-Hing, H., M.J. Zoran, K. Lukowiak & P.G. Haydon. 1989. A neuromodulator of synaptic transmission acts on the secretory apparatus as well as on ion channels. Nature **341:** 237–239.
20. Hille, B. 1992. G protein-coupled mechanisms and nervous signaling. [Review]. Neuron **9:** 187–195.
21. Forscher, P., G.S. Oxford & D. Schulz. 1986. Noradrenaline modulates calcium channels in avian dorsal root ganglion cells through tight receptor-channel coupling. J. Physiol. **379:** 131–144.
22. Berro, N.S., V. Campbell, E.G. Fitzgerald, K. Brickley & A.C. Dolphin. 1995. Antisense depletion of β-subunits modulates the biophysical and pharmacological properties of neuronal calcium channels. J. Physiol. (Lond.) **482:** 481–491.
23. Campbell, V., N.S. Berrow, E.M. Fitzgerald, K. Brickley & A.C. Dolphin. 1995. Inhibition of the interaction of G protein G_o with calcium channels by the calcium channel β-subunit in rat neurones. J. Physiol. (Lond.) **485:** 365–372.
24. De Waard, M., M. Pragnell & K.P. Campbell. 1994. Ca^{2+} channel regulation by a conserved β subunit domain. Neuron **13:** 495–503.
25. Tareilus, E., M. Roux, N. Qin, R. Olcese, J.M. Zhou, E. Stefani & L. Birnbaumer. 1997. A *Xenopus* oocyte β subunit: Evidence for a role in the assembly/expression of voltage-gated calcium channels that is separate from its role as a regulatory subunit. Proc. Natl. Acad. Sci. USA **94:** 1703–1708.
26. Walker, D., D. Bichet, K.P. Campbell & M. De Waard. 1998. A β_4 isoform-specific interaction site in the carboxyl-terminal region of the voltage-dependent Ca^{2+} channel α_{1A} subunit. J. Biol. Chem. **273:** 2361–2367.
27. Roche, J.P., V. Anantharam & S.N. Treistman. 1995. Abolition of G protein inhibition of α_{1A} and α_{1B} calcium channels by co-expression of the β_3 subunit. FEBS Lett. **371:** 43–46.
28. Bourinet, E., T.W. Soong, A. Stea & T.P. Snutch. 1996. Determinants of the G protein-dependent opioid modulation of neuronal calcium channels. Proc. Natl. Acad. Sci. USA **93:** 1486–1491.
29. Marchetti, C., E. Carbone & H.D. Lux. 1986. Effects of dopmaine and noradrenaline on Ca channels of cultured sensory and sympathetic neurons of chick. Pflugers Arch. **406:** 104–111.
30. Dolphin, A.C., S.R. Forda & R.H. Scott. 1986. Calcium-dependent currents in cultured rat dorsal root ganglion neurones are inhibited by an adenosine analogue. J. Physiol. (Lond.) **373:** 47–61.
31. Holz, G.G.I., S.G. Rane & K. Dunlap. 1986. GTP-binding proteins mediate transmitter inhibition of voltage-dependent calcium channels. Nature **319:** 670–672.
32. McFadzean, I., I. Mullaney, D.A. Brown & G. Milligan. 1989. Antibodies to the GTP binding protein, G_o, antagonize noradrenaline-induced calcium current inhibition in NG108-15 hybrid cells. Neuron **3:** 177–182.
33. Baertschi, A.J., Y. Audigier, P.-M. Lledo, J.-M. Israel, J. Bockaert & J.-D. Vincent. 1992. Dialysis of lactotropes with antisense oligonucleotides assigns guanine nucleotide binding protein subtypes to their channel effectors. Mol. Endocrinol. **6:** 2257–2265.

34. Campbell, V., N. Berrow & A.C. Dolphin. 1993. $GABA_B$ receptor modulation of Ca^{2+} currents in rat sensory neurones by the G protein G_o: Antisense oligonucleotide studies. J. Physiol. (Lond.) **470:** 1–11.
35. Ewald, D.A., I.-H. Pang, P.C. Sternweis & R.J. Miller. 1989. Differential G protein-mediated coupling of neurotransmitter receptors to Ca^{2+} channels in rat dorsal root ganglion neurons in vitro. Neuron **2:** 1185–1193.
36. Toselli, M., J. Lang, T. Costa & H.D. Lux. 1989. Direct modulation of voltage-dependent calcium channels by muscarinic activation of a pertussis toxin–sensitive G-protein in hippocampal neurons. Pflügers Arch. **415:** 255–261.
37. Shapiro, M.S. & B. Hille. 1993. Substance P and somatostatin inhibit calcium channels in rat sympathetic neurons via different G protein pathways. Neuron **10:** 11–20.
38. Golard, A., L. Role & S.A. Siegelbaum. 1994. Substance P potentiates calcium channel modulation by somatostatin in chick sympathetic ganglia. J. Neurophysiol. **72:** 2683–2690.
39. Zhu, Y. & S.R. Ikeda. 1994. VIP inhibits N-type Ca^{2+} channels of sympathetic neurons via a pertussis toxin–insensitive but cholera toxin–sensitive pathway. Neuron **13:** 657–669.
40. Ikeda, S.R. 1996. Voltage-dependent modulation of N-type calcium channels by G protein βgamma subunits. Nature **380:** 255–258.
41. Herlitze, S., D.E. Garcia, K. Mackie, B. Hille, T. Scheuer & W.A. Catterall. 1996. Modulation of Ca^{2+} channels by G-protein βgamma subunits. Nature **380:** 258–262.
42. Logothetis, D.E., Y. Kurachi, J. Galper, E.J. Neer & D.E. Clapham. 1987. The β gamma subunits of GTP-binding proteins activate the muscarinic K^+ channel in heart. Nature **325:** 321–326.
43. Clapham, D.E. & E.J. Neer. 1993. New roles for G-protein β gamma dimers in transmembrane signalling. Nature **365:** 403–406.
44. Krapivinsky, G., L. Krapivinsky, K. Wickman & D.E. Clapham. 1995. Gβgamma binds directly to the G protein-gated k^+ channel, I_{KACh}. J. Biol. Chem. **270:** 29059–29062.
45. Wickman, K.D. & D.E. Clapham. 1995. G-protein regulation of ion channels. Curr. Opin. Neurobiol. **5:** 278–285.
46. Wickman, K., J.A. Inguinez-Liuhi, P.A. Davenport, R. Taussig, G.B. Krapivinsky, M.E. Linder, A.G. Gilman & D.E. Clapham. 1997. Recombinant G protein βgamma subunits activate the muscarinic-gated atrial potassium channel. Nature **368:** 255–257.
47. Yamada, M., Y. Ho, R.H. Lee, K. Kontani, T. Takahashi, T. Katada & Y. Kurachi. 1997. Muscarinic K^+ channels are activated by βgamma subunits and inhibited by the GDP bound form of the α subunit of transducin. Biochem. Biophys. Res. Commun. **200:** 1484–1490.
48. Stephens, G.J., N.L. Brice, N.S. Berrow & A.C. Dolphin. 1998. Facilitation of rabbit α1B calcium channels: Involvement of endogenous Gβgamma subunits. J. Physiol. (Lond.) **509:** 15–27.
49. Pragnell, M., M. De Waard, Y. Mori, T. Tanabe, T.P. Snutch & K.P. Campbell. 1994. Calcium channel β-subunit binds to a conserved motif in the I-II cytoplasmic linker of the α_1-subunit. Nature **368:** 67–70.
50. Chen, J., M. Devivo, J. Dingus, A. Harry, J. Li, J. Sui, D.J. Carty, J.L. Blank, J.H. Exton, R.H. Stoffel, J. Inglese, R.J. Lefkowitz, D.E. Logothetis, J.D. Hildebrandt & R. Iyengar. 1995. A region of adenylyl cyclase 2 critical for regulation by G protein βgamma subunits. Science **268:** 1166–1169.
51. De Waard, M., H.Y. Liu, D. Walker, V.E.S. Scott, C.A. Gurnett & K.P. Campbell. 1997. Direct binding of G-protein βgamma complex to voltage-dependent calcium channels. Nature **385:** 446–450.
52. Zhang, J.F., P.T. Ellinor, R.W. Aldrich & R.W. Tsien. 1996. Multiple structural elements in voltage-dependent Ca^{2+} channels support their inhibition by G proteins. Neuron **17:** 991–1003.
53. Page, K.M., G.J. Stephens, N.S. Berrow & A.C. Dolphin. 1997. The intracellular loop between domains I and II of the B type calcium channel confers aspects of G protein sensitivity to the E type calcium channel. J. Neurosci. **17:** 1330–1338.
54. Zamponi, G.W., E. Bourinet, D. Nelson, J. Nargeot & T.P. Snutch. 1997. Crosstalk between G proteins and protein kinase C mediated by the calcium channel α_1 subunit. Nature **385:** 442–446.

55. Stephens, G.J., N.S. Berrow, K.M. Page & A.C. Dolphin. 1997. Gβgamma subunits mediate aspects of G protein sensitivity of Ca^{2+} sensitivity of calcium channel α1 subunits by an action on the I-II loop. J. Physiol. (Lond.) **504:** 157P.
56. Stephens, G.J., C. Canti, K.M. Page & A.C. Dolphin. 1998. Role of domain I of neuronal Ca^{2+} channel α1 subunits in G protein modulation. J. Physiol. (Lond.) **509:** 163–169.
57. Herlitze, S., G.H. Hockerman, T. Scheuer & W.A. Catterall. 1997. Molecular determinants of inactivation and G protein modulation in the intracellular loop connecting domains I and II of the calcium channel α_{1A} subunit. Proc. Natl. Acad. Sci. USA **94:** 1512–1516.
58. Toth, P.T., L.R. Shekter, G.H. Ma, L.H. Philipson & R.J. Miller. 1996. Selective G-protein regulation of neuronal calcium channels. J. Neurosci. **16:** 4617–4624.
59. Mehrke, G., A. Pereverzev, H. Grabsch, J. Hescheler & T. Schneider. 1997. Receptor-mediated modulation of recombinant neuronal class E calcium channels. FEBS Lett. **408:** 261–270.
60. Qin, N., D. Platano, R. Olcese, E. Stefani & L. Birnbaumer. 1997. Direct interaction of Gβgamma with a C terminal Gβgamma binding domain of the calcium channel α1 subunit is responsible for channel inhibition by G protein coupled receptors. Proc. Natl. Acad. Sci. USA **94:** 8866–8871.
61. Page, K.M., C. Canti, G.J. Stephens, N.S. Berrow & A.C. Dolphin. 1998. Identification of the amino terminus of neuronal Ca^{2+} channel α1 subunits α1B and α1E as an essential determinant of G protein modulation. J. Neurosci. **18:** 4815–4824.
62. Schreibmayer, W., C.W. Dessauer, D. Vorobiov, A.G. Gilman, H.A. Lester, N. Davidson & N. Dascal. 1996. Inhibition of an inwardly rectifying K^+ channel by G-protein α-subunits. Nature **380:** 624–627.

Neuronal Voltage-Activated Calcium Channels: On the Roles of the α_{1E} and β_3 Subunits

STEPHEN M. SMITH,[a,b] ERIKA S. PIEDRAS-RENTERÌA,[a] YOON NAMKUNG,[c] HEE-SUP SHIN,[c,d] AND RICHARD W. TSIEN[a,e]

[a]Department of Molecular and Cellular Physiology, Howard Hughes Medical Institute,[b] The Beckman Center, Stanford University Medical Center, Stanford, California 94305, USA

Department of Life Science[c] and National CRI Center for Calcium and Learning[d] Pohang University of Science and Technology, Pohang, 790-784, Republic of Korea

ABSTRACT: Many neurons of the central and peripheral nervous systems display multiple high voltage–activated (HVA) Ca^{2+} currents, often classified as L-, N-, P-, Q, and R-type. The heterogeneous properties of these channels have been attributed to diversity in their pore-forming α_1, subunits, in association with various β subunits. However, there are large gaps in understanding how individual subunits contribute to Ca^{2+} channel diversity. Here we describe experiments to investigate the roles of α_{1E} and β_3 subunits in mammalian neurons. The α_{1E} subunit is the leading candidate to account for the R-type channel, the least understood of the various types of high voltage–activated Ca^{2+} channels. Incubation with α_{1E} antisense oligonucleotide caused a 53% decrease in the peak R-type current density, while no significant changes in the current expression were seen in sense oligonucleotide-treated cells. The specificity of the α_{1E} antisense oligonucleotides was supported by the lack of change in the amplitude of P/Q current. These results upheld the hypothesis that members of the E class of α_1 subunits support the high voltage–activated R-type current in cerebellar granule cells. We studied the role of the Ca^{2+} channel β_3 subunit using a gene targeting strategy. In sympathetic β_3-/- neurons, the L-type current was significantly reduced relative to wild type (wt). In addition, N-type Ca^{2+} channels made up a smaller proportion of the total Ca^{2+} current than in wt due to a lower N-type current density in a group of neurons with small total currents. Voltage-dependent activation of P/Q-type Ca^{2+} channels was described by two Boltzmann components with different voltage dependence. The absence of the β_3 subunit was associated with a shift in the more depolarized component of the activation along the voltage axis toward more negative potentials. The overall conclusion is that deletion of the β_3 subunit affects at least three distinct types of HVA Ca^{2+} channel, but no single type of channel is solely dependent on β_3.

Voltage-gated Ca^{2+} channels regulate Ca^{2+} entry in a potential-dependent manner and thereby contribute to Ca^{2+} signaling in a wide variety of cell types. A single opening of a Ca^{2+} channel can allow many hundreds or thousands of Ca^{2+} ions to flow into the cytoplasm, thus generating a rise in $[Ca^{2+}]_i$ that may control vital functions such as excitability, rhythmicity, transmitter or hormone release, contraction, metabolism, and gene expression.[1,2] To initiate such events effectively, Ca^{2+} channels have evolved as very efficient and highly regulated enzymes to catalyze the downhill flow of Ca^{2+} across membranes.[3] Activation of Ca^{2+} channels is steeply voltage dependent, similar to voltage-gated Na^+ and K^+ channels. Typically, Ca^{2+} channels open within one or a few milliseconds after

[e]Corresponding author: B-105 Beckman Center, Stanford University Medical Center, 300 Pasteur Drive, Stanford, California 94305. Phone 650-725-7557; fax: 650-725-2504; e-mail: rwtsien@stanford.edu

the membrane is depolarized from rest, and close (deactivate) within a fraction of a millisecond following repolarization. The basic features of Ca^{2+} channel selectivity and gating are intrinsic properties of all voltage-gated Ca^{2+} channels, evidently highly conserved in evolution. The powerful functional capabilities of Ca^{2+} channels are rooted in their molecular architecture. As far as we know, all voltage-gated Ca^{2+} channels are composed of multiple components, which come together to form a large macromolecular complex (~500 kDa). The generic structure contains five subunits, called α_1, $\alpha_2\delta$, β, and γ.[4–8] Each subunit has been cloned in one or more forms within the last dozen years.[9–11]

Multiple types of voltage-gated Ca^{2+} channels were first distinguished on the basis of their voltage and time dependence, single-channel conductance, and pharmacology.[12–14] These criteria have led to a widely accepted classification of Ca^{2+} channels as T-, L- N-, P-, Q-, and R-type.[15–17] This classification has proved useful in deciphering the varied functional roles of the channels in different organ systems, but the relationship between specific channel types and the underlying Ca^{2+} channel subunits remains incompletely understood. This paper reviews our recent efforts to resolve some remaining uncertainties about the relationship between the neuronal Ca^{2+} channel types and the underlying molecular subunits. Two areas of interest have been (1) the possible role of α_{1E} in supporting R-type current and (2) the contributions of the β_3 subunit.

ROLE OF α_{1E} IN SUPPORTING R-TYPE CALCIUM CHANNELS

Among the major categories of Ca^{2+} channels uncovered so far, R-type channels were the most recently defined and remain the least well understood. R-type Ca^{2+} channel currents were first identified in cerebellar granule cells as a current that remained in the presence of nimodipine, ω-CTx-GVIA, and ω-Aga-IVA, inhibitors of the L-, N-, and P/Q-type channels, respectively.[17–19] Recent experiments have demonstrated the importance of R-type channels for dendritic Ca^{2+} entry[20] and synaptic transmission.[21,22] In contrast to other high voltage–activated (HVA) Ca^{2+} channels, the molecular basis of R-type currents is not completely settled. The leading candidate for the molecular correlate of R-type currents is the α_{1E} subunit.[23–25] When expressed in *Xenopus* oocytes and HEK293 cells, α_{1E} subunits induced a prominently inactivating, fast-deactivating current that was highly sensitive to block by Ni^{2+}[23,24,26] and ω-Aga-IIIA,[27] similar to R-type current in cerebellar granule neurons.[17,18,28] However, it has also been suggested that α_{1E} might support a low voltage–activated Ca^{2+} channel instead of R-type currents.[23,29–32] We utilized an antisense strategy to test whether α_{1E} underlies the expression of the R-type current in cerebellar granule cells.[33]

α_{1A} Antisense Reduces P/Q-Type Current but Does Not Affect R-Type Current

As an initial test of the efficacy of antisense treatment in the cerebellar granule cells, we examined the effect of α_{1A} antisense that targeted nucleotides 145 to 162 of the rat α_{1A} sequence.[34] P/Q+R currents were measured in the presence of ω-CTx GVIA and nimodipine in the solution. We compared P/Q+R currents recorded from cells cultured in the presence of either antisense or sense α_{1A} oligonucleotides(ON) (FIG. 1A). Cells grown in the presence of α_{1A} antisense showed a peak current density of -13.93 ± 3.3 pA/pF ($n = 5$),

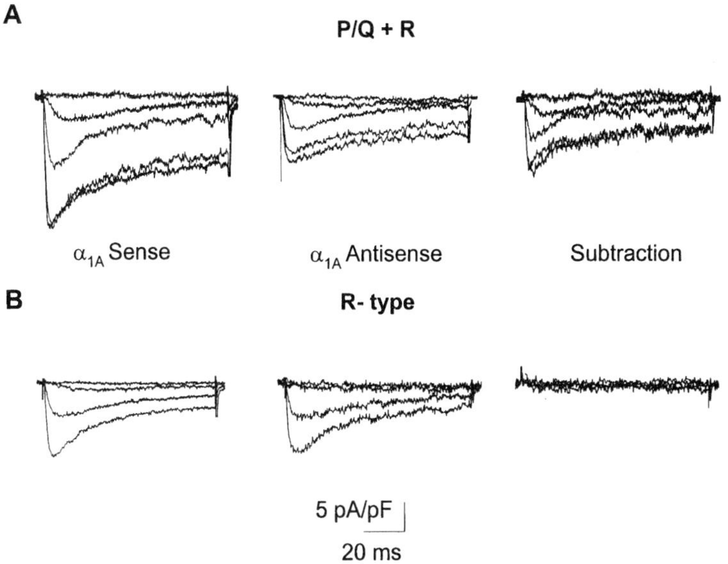

FIGURE 1. Antisense ONs against α_{1A} decrease the PQ+R components but do not affect the expression of R-type current. **(A)** Addition of α_{1A} antisense ON decreased the P/Q+R component (**middle trace**) compared to cells grown in the presence of α_{1A} sense ON (**left trace**). The subtracted traces (**right**) show the fast-inactivating component suppressed by α_{1A} antisense treatment. Ba^{2+} currents were elicited with depolarizing pulses from V_{hold} = –80 mV to V_{test} = –60, –40, –20, 0, and + 10 mV (data pooled from 5 cells). **(B)** α_{1A} antisense ON did not affect R-type currents (**middle trace**) compared to cells treated with α_{1A} sense (**left trace**). R type currents were elicited with depolarizing pulses to V_{test} = –60, –40, –20, and 0 mV from a V_{hold} = –80 mV (data pooled from 5 cells). (Reprinted in part by permission from Piedras-Renteria & Tsien.[33])

significantly smaller than the peak current density from sense-treated cells, –22.1 ± 2.1 pA/pF (n = 5) ($p < 0.036$). This confirms the generally accepted view that α_{1A} supports P/Q current.

The same oligonucleotides were tested against R-type currents, isolated pharmacologically by including ω-CTx MVIIC along with ω-CTx GVIA and nimodipine in the bath solution. FIGURE 1B shows averaged R-type currents from cells cultured in the presence of either antisense or sense α_{1A} ONs. In this case, current traces obtained in the presence of either ON were not different from each other. Peak current density was –11.5 ± 1.7 pA/pF (n = 5) in cells treated with α_{1A} antisense ON and –11.7 ± 1.7 pA/pF in cells treated with α_{1A} sense ON (n = 5). This finding was in line with the hypothesis that R-type currents arose from another α_1 subunit, possibly α_{1E}. Because the targeted region of α_{1A} (nucleotides 145 to 162) lacks any appreciable homology with the α_{1E} sequence, the ON treatment would not be expected to have an effect on the R-type current expression.

α_{1E} Antisense Specifically Decreased R-Type Current but Did Not Affect P/Q Current

As a direct test for involvement of the α_{1E} subunit, we tested an ON directed against nucleotides 582–599, a region located at repeat I between the S3 and S4 transmembrane domains.[23] This particular region of the protein was chosen for its lack of similarity to L-type Ca^{2+} channel sequences and its low homology with non–L-type channel subunits present in cerebellar granule cells, such as α_{1A} or α_{1B} (27% and 38%, respectively). Incubation of the neurons with α_{1E} antisense oligonucleotide caused a significant decrease in the expression of R-type currents, compared to untreated cells or cells grown in the presence of α_{1E} sense ON. FIGURE 2A illustrates averaged Ba^{2+} current traces obtained from untreated cells cultured in the presence of medium supplemented with 4 μM of α_{1E} antisense oligonucleotide and cells cultured with α_{1E} sense ON. As evident from the traces and the corresponding peak I–V curves (FIG. 2B), treatment with antisense ON in culture significantly reduced the peak amplitude of R-type current. In comparison to the mean peak current density for untreated controls, –10.0 ± 0.6 pA/pF ($n = 6$), peak current in cells treated with α_{1E} antisense averaged –4.8 ± 0.8 pA/pF ($n = 11$), a 53% decrease ($p < 0.01$). In contrast, the peak current density in granule cells treated with sense ON averaged –11.3 ± 3.3 pA/pF ($n = 5$), not significantly different from the untreated cells ($p = 0.84$), in support of the specificity of the antisense effect. We did not observe a total elimination of the residual current with the antisense treatment, which is not uncommon when using anti-

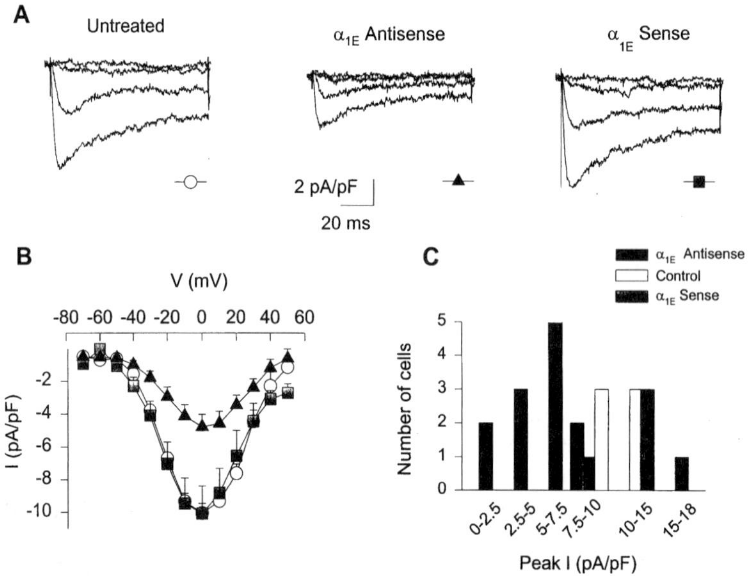

FIGURE 2. α_{1E} antisense ONs in the culture medium decrease the R-type current amplitude. **(A)** Activation of Ba^{2+} currents with various depolarizing pulses (V_{test} = –60, –40, –20, and 0 mV from a V_{hold} = –80 mV) in untreated cells, cells cultured in the presence of 4 mM α_{1E} antisense ON, and cells treated with 4 mM α_{1E} sense ON. Data pooled from 3 to 4 cells. R-type currents were measured in the presence of toxins to block L-, N-, and P/Q-type current components. **(B)** Current-voltage relationship averages for untreated cells (*circles*, $n = 6$), sense α_{1E} ON (*squares*, $n = 5$), and antisense α_{1E} ON (*triangles*, $n = 11$). Currents from cells treated with antisense ON were significantly smaller than the untreated ($p < 0.01$) and sense-treated controls ($p < 0.05$). **(C)** Peak current distribution in the untreated, antisense α_{1E} and sense α_{1E}-treated cell groups. (Reprinted in part by permission from Piedras-Renteria & Tsien.[33])

sense strategies.[30,35] This partial effect would be expected if the kinetics of α_{1E} turnover within the cell were slow, as is often found for membrane channel proteins. Using fluorescence microscopy we demonstrated considerable variation in the uptake of antisense ON tagged with fluorescein at the 3′ end of the sequence (data not shown). This was reflected by variation in the decrease in current induced by the antisense treatment across the entire granule cell population (FIG. 2C).

An additional test of the specificity of the α_{1E} antisense ON for R-type currents was to investigate its effect on P/Q-type currents. FIGURE 3A compares the peak current values recorded from untreated cells (C) and in antisense α_{1E}-treated neurons (A). The peak amplitude for P/Q+R current fell from 19.4 ± 1.7 pA/pF (n = 4) in control conditions to a value of 13.7 ± 1.1 pA/pF (n = 8) in neurons treated with α_{1E} antisense ON, a significant

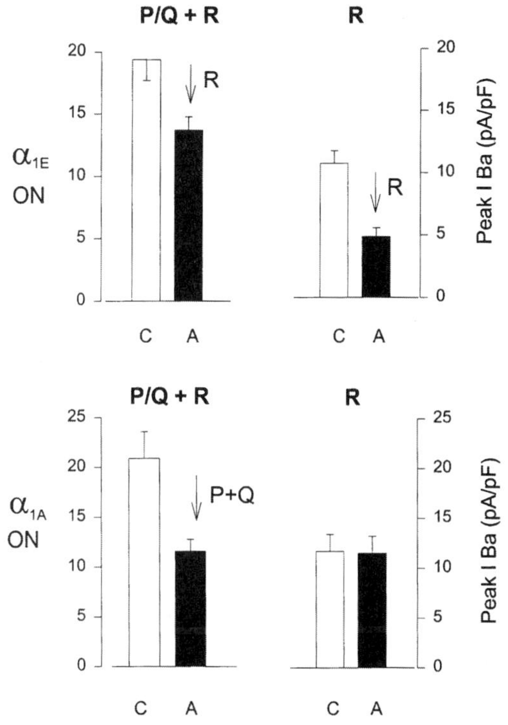

FIGURE 3. Specificity of antisense ON treatment. **(Top)** α_{1E} antisense treatment shows specific reduction of R-type current. The P/Q+R current components recorded in the presence and the absence of α_{1E} antisense were significantly different ($p < 0.05$) (**left panel**); this reduction α_{1E} antisense (~30%) can be accounted for by the decrease in R current alone (**right panel**). R-type currents were measured in the presence of 1 μM ω-CTx MVII C (n = 5 for the control group; n = 11 for antisense-treated cells). P/Q+R components were measured in the absence of the toxin (P/Q+R, n = 6 for controls, and n = 7 for antisense-treated cells). **(Bottom)** α_{1A} antisense treatment decreased the PQ+R currents compared to controls (**left panel**) ($p < 0.05$). Addition of α_{1A} antisense did not affect R-type current expression (**right panel**, n = 5 for controls and antisense groups). A, antisense; C, control (Reprinted in part by permission from Piedras-Renteria & Tsien.[33])

decrease ($p < 0.01$). The reduction was essentially the same as that found when peak R-type current was studied in isolation (left panel). Thus, the reduction in the aggregate P/Q+R current can be accounted for by a specific decrease in R-type current alone, without any change in the contribution P/Q current.

INFLUENCE OF THE β SUBUNIT

All Ca^{2+} channels in their native state appear to contain β subunits.[36–38] Unlike the pore-forming α_1 subunit, the contribution of the other subunits to Ca^{2+} channel function is less well understood. The β subunit was first identified in voltage-sensitive Ca^{2+} channels purified from skeletal muscle as an integral component of ~60 kDa molecular weight.[39,40] Four different types of β subunit are known to exist in mammals and are now known as β_1–β_4.[36,41] The cloning of the four β subunit types has permitted the investigation of their function in heterologous expression systems.[42–45] Such experiments suggest a complex role for the β subunit as it regulates the Ca^{2+} entry into the cell by increasing the peak Ca^{2+} current,[46–48] by shifting the voltage dependence of activation and inactivation,[44,48] and by modulating G-protein inhibition of the α_1 subunit.[49–51] We were interested in the physiological impact of β subunits in neurons and therefore investigated the effect of removal of the β_3 subunit, the most abundant β subunit in the brain.[52] The β_3 subunit is an intriguing candidate for investigation since it is associated with a majority of the N-type channel complexes in the brain,[53] and these Ca^{2+} channels play an important role in regulating Ca^{2+} entry into dendrites, soma, and synaptic terminals.[54–56] With these considerations in mind, we have used a gene targeting strategy to approach questions about the role of the β_3 subunit in neurons.[57]

Ca^{2+} Channel Currents Supported by N-, P/Q- and L-Type in Mouse and Rat SCG Neurons

Extensive studies of sympathetic neurons from rat and bullfrog have demonstrated the predominance.of N-type channels over the much smaller contributions of L-type[55] and R-type channels.[58] This is illustrated with a representative recording from a rat SCG neuron (FIG. 4A) in which the Ba^{2+} currents generated by N- and L-type channels constituted ~95% and ~4% of total. The prevalence of N-type current in well-studied SCG neurons and the common association of α_{1B} subunit and β_3 subunits encouraged us to examine effects of β_3 subunit deletion in mouse SCG neurons. We began by characterizing the relative contributions of various types of Ca^{2+} channels in SCG neurons from wild-type mice, since this had not been reported previously. Unexpectedly, we found significant components of global Ba^{2+} current (I_{Ba}) supported by N-, P/Q-, and L-type channels, as defined by an array of specific pharmacological inhibitors. FIGURE 4B shows representative records of I_{Ba} activated by depolarizing pulses from –80 mV to –10 mV. In the mouse SCG neuron (FIG. 4B), the total current was decreased by ~45% upon application of ω-conotoxin at 1 μM, a saturating concentration for inhibition of N-type current in other kinds of neurons. The subsequent application of 10 μM ω-conotoxin MVIIC in the continued presence of ω-conotoxin GVIA was used to define the contribution of P/Q-type current, ~26% of total in this example. Further application of nimodipine (10 μM) caused an additional inhibition

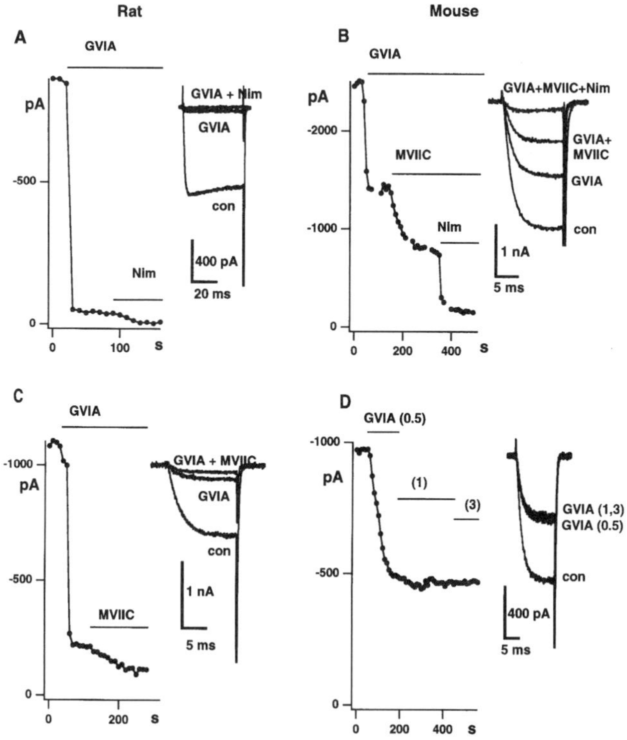

FIGURE 4. The pharmacological dissection of Ca^{2+} channel currents in rat and mouse SCG neurons. The response of peak I_{Ba} in a rat (**A,C**) and mouse (**B,D**) neurons to application of ω-conotoxin GVIA (0.5–3 μM), (ω-conotoxin MVIIC (5–10 μM), and nimodipine (10 μM). I_{Ba} was activated by depolarizations from –80 mV to –10 mV every 10 s. Representative current traces are shown on the **right**. The neuron in D is from a β_3-/- mouse. (Reprinted in part by permission from Namkung *et al.*[57])

of 23%, representing the contribution of L-type current, leaving a residuum of 6%. These findings for currents in the cell body are in line with previous data on the pharmacology of sympathetic transmission in mice, which shows a prominent contribution of P/Q-type channels (ω-Aga-IVA–sensitive) as well as N-type (ω-CTx-GIA–sensitive) channels.[59] The range of Ca^{2+} channel types in the mouse SCG neurons resembles the heterogeneity found in the great majority of CNS neurons such as cerebellar granule cells, which express an array of Ca^{2+} channel types, among them N-, P/Q, and L-type.[17]

Since a significant P/Q component has not previously been reported in SCG neurons of other species, we performed similar experiments in rat SCG neurons. A subgroup of the rat SCG neurons contained an appreciable P/Q current, averaging 15% of total current, along with N- and L-type currents (FIG. 4C). This conclusion rests on the presumption that the doses of the various inhibitors were sufficient to give maximal blockade. To verify this, additional controls were performed to rule out the possibility that N-type channels in mouse neurons might be unusually resistant to ω-conotoxin GVIA block,[60] leaving a residual N-type current to be blocked by ω-conotoxin MVIIC.[61] However, a threefold increase in concentration of ω-conotoxin GVIA produced no further blockade of I_{Ba} (FIG. 4D), confirming that saturating block of N-type channels had been obtained at 1 μM. Thus, the MVIIC-sensitive current in FIGURE 4B could be unambiguously identified as P/Q-type.

β_3 Deletion Reduces N- and L-Type Currents in Mouse SCG Neurons

Myotube L-type currents are reduced 10-fold by deletion of the β_1 subunit, leading to a failure of excitation-contraction coupling in these animal.[62] This lethal phenotype occurs because the absence of a β subunit causes a massive reduction in the insertion of α_{1S} in the cell membrane.[62] We hypothesized that deletion of the β_3 subunit might similarly affect the expression level of associated α_1 subunits, resulting in changes in the size of various components of the global I_{Ba}. We focused initially on N-type channels because of the well-described association between α_{1B} and β_3. FIGURE 5A shows pooled data for the fractional contribution of N-type current, plotted as a cumulative distribution. The proportion of N-

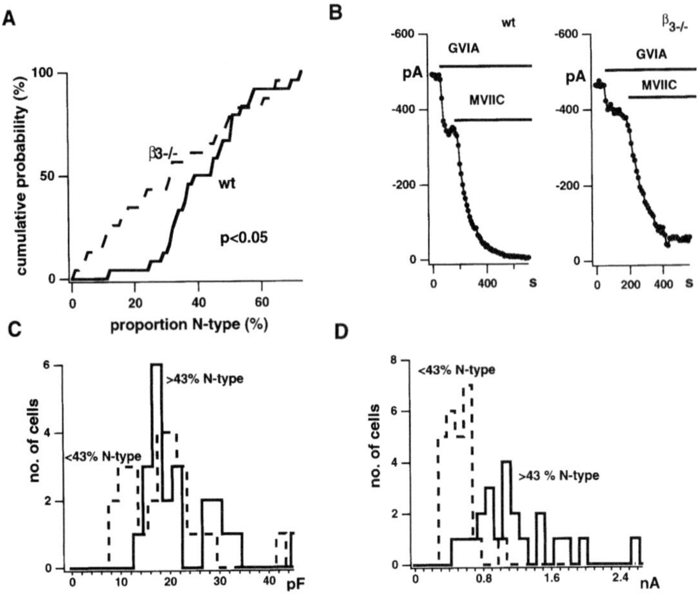

FIGURE 5. The N-type current is reduced in β_3-/- SCG neurons. **(A)** The cumulative probability plot for the proportion of I_{Ba} that is N-type for 24 wt (*solid*) and 23 β_3-/- (*broken*) SCG neurons. Plots were significantly different by Kolmogorov-Smirnov test ($p < 0.05$). **(B)** Effect of 1μM GVIA and 10 μM MVIIC on I_{Ba} in wt and β_3-/- neurons. **(C)** Histogram of cell size (pF) in which neurons have been divided according to whether the percentage of I_{Ba} carried by N-type channels was <43 (*dashed histogram*) or >43 (*solid histogram*). **(D)** Histogram of total I_{Ba} using same criteria to split data. The difference in average I_{Ba} between these groups was clearly significant, regardless of whether wt and β_3-/- cells were lumped together as shown or treated separately. Average values of I_{Ba} were 440 ± 55 vs. 1061 ± 153 pA for wt neurons ($p < 0.01$) and 425 ± 38 vs. 1013 ± 146 pA for β_3-/- cells ($p < 0.001$). (Reprinted in part by permission from Namkung *et al.*[57])

type current was significantly smaller in the absence of the β_3 subunit ($p < 0.05$ by Kolmogorov-Smirnoff test). The difference between the two groups of neurons reflects the fact that half of the β_3-/- SCG had less than 30% N-type current, whereas wt SCG neurons rarely showed such a small percentage (FIG. 5A,B).

The large variation in the relative proportion of N-type current (12–73% in wt, 1–67% in β_3-/- neurons) and its non-Gaussian distribution suggested some kind of cellular heterogeneity. This was consistent with our observations that cell membrane area and global I_{Ba} vary five- and eightfold, respectively (FIG. 5C,D) and with previous findings of functional heterogeneity among neurons in intact sympathetic ganglia.[63,64] Cell membrane area was considered as a possible basis for the heterogeneity in the population of neurons, based on previous work in sensory neurons that showed a strong correlation between cell size and the composition and magnitude of Ca^{2+} channel current.[65] FIGURE 5C shows the distributions of cell membrane areas in neurons with large or small contributions of N-type current, ignoring distinctions between wt and β_3-/- cells. Using a cut-off of 43%, near the median percentage of N-type current, to divide all neurons into roughly equal groups, we found very similar distributions of cell size in cells with a small proportion of N-type current (<43%, dashed histogram) and those with a large proportion (>43%, solid histogram).

We also considered the absolute size of I_{Ba} in SCG neurons as a possible basis or correlate of the heterogeneity (FIG. 5D). In this case, the overall distribution was bimodal, with peaks of low and high total I_{Ba}. The lower peak consisted almost entirely of neurons with a small proportion of N-type current (<43%, dashed line), and the higher peak was made up almost completely of cells with a large proportion of N-type current (>43%, solid line). This segregation was clear regardless of whether wt and β_3-/- cells were lumped together as in FIGURE 5D or treated separately (data not shown). Thus, the overall population of SCG could be divided into two groups according to I_{Ba} amplitude. Using 700 pA as a dividing line between them, we reanalyzed these groups separately (FIG. 6). The cells with large I_{Ba} by this criterion failed to show significant changes in either the proportion or absolute density of any of the individual currents. However, the neurons with small I_{Ba} displayed clear differences in the abundance of N-type current between wt and β_3-/-. In the small I_{Ba} group, elimination of the β_3 subunit caused a significant reduction in N-type current, expressed either as a proportion (24 ± 4% vs. 35 ± 3% in control, $p < 0.05$) or as current density (–6.0 ± 1.4 pA/pF vs. –11.0 ± 1.6 in control, $p < 0.05$). In contrast, the current density for global I_{Ba} and for contributions of other types of Ca^{2+} channel were not significantly affected.

The L-type current appeared to be reduced in size in association with deletion of the β_3 subunit. Pooling the data in the subgroups revealed a twofold difference in L-type current density between wt neurons (–4.1 ± 0.9 pA/pF, $n = 16$) and β_3-/-neurons (–1.9 ± 0.4 pA/pF; $n = 10$) ($p < 0.05$). The same trend was evident in the subgroups (FIG. 6A,B), although it did not reach statistical significance in the context of the smaller sample sizes.

Relative Importance of α_1 and β Subunits in G-Protein Mediated Inhibition

It has been proposed that Ca^{2+} channel β subunits play an important role in the regulation of Ca^{2+} channels by G protein–coupled receptors.[66–69] The β subunits appear to interact with at least two distinct domains of α_1 subunits, which also bind G protein $\beta\gamma$ subunits.[50,70–72] Thus, the β subunit may help regulate channel function by competing with

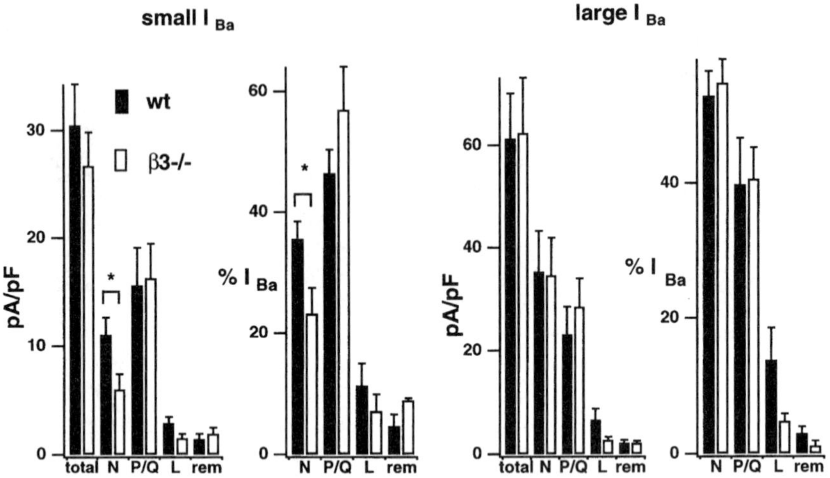

FIGURE 6. Histograms of the current density and proportion of small (**left**; $n = 6–16$) or large (**right**; $n = 3–11$) I_{Ba} carried by total, N, P/Q, L, or remainder channels types for wt (*solid*) and β_3-/- (*open*). N-type current density and proportion are significantly reduced in small I_{Ba} β_3-/- neurons. (Reprinted in part by permission from Namkung *et al.*[57])

$G_{\beta\gamma}$.[73] We investigated the possibility of a specific interaction between β subunits and G proteins by comparing the responses of wt and β_3-/- neurons to G protein–mediated signaling. Norepinephrine (NE) was a logical choice as a neuromodulator, since it is representative of a large class of agents that inhibit Ca^{2+} currents in a voltage-dependent manner[74–76] and can be released by sympathetic neurons themselves. FIGURE 7 illustrates the action of NE on N- and P/Q-type Ca^{2+} channel currents, recorded in the presence of nimodipine to eliminate L-type current. The current traces (FIG. 7A) and analysis (FIG. 7B,C) show data from a representative β_3-/- neuron. We used a voltage protocol (FIG. 7A) that distinguishes between voltage-dependent and voltage-independent components of inhibition.[74,77] In each trial, I_{Ba} was activated by two voltage pulses from –80 mV to –10 mV, the second pulse being preceded by a strong conditioning depolarization (+80 mV), sufficient to remove any voltage-dependent component of G-protein inhibition. Current amplitudes, measured 10 ms after the beginning of the first and second test pulses (I_1, and I_2), are plotted separately (filled and open symbols respectively, FIG. 7B) and as a ratio (FIG. 7C). In the absence of NE, I_1 and I_2 were almost identical ($I_2/I_1 = 0.96 \pm 0.01$ and 1.00 ± 0.02 for total current from wt and β_3-/- neurons). The application of NE caused a prompt decrease in peak I_{Ba} and a slowing of its activation during the first test pulse. The NE-induced inhibition was partially relieved by the strong conditioning depolarization, resulting in a sharp increase in the I_2/I_1 ratio (FIG. 7B). The effect of NE on the combination of N- and P/Q-type current (upper traces) was compared with the effect on P/Q-type current alone (in the presence of ω-CTx-GVIA, middle traces). The remaining P/Q-type current was also inhibited, but the voltage-dependent relief was considerably smaller than for total current, as seen in the smaller rise of I_2/I_1 (FIG. 7C). N-type current, obtained by subtracting records taken in the absence and presence of ω-CTx-GVIA (bottom traces), showed a correspond-

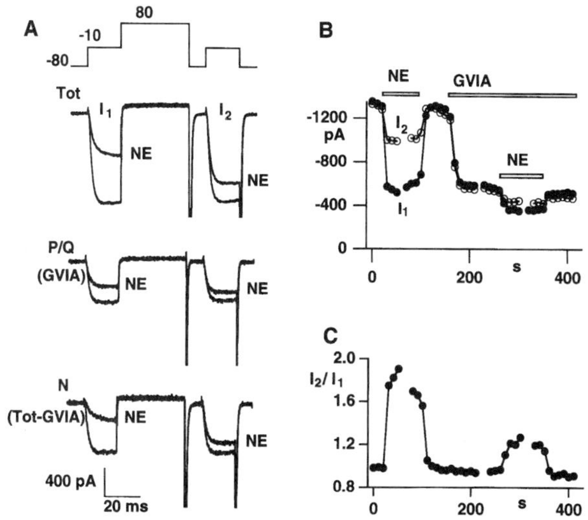

FIGURE 7. The voltage-dependent component of NE inhibition is greater for N-type currents than for P/Q-type. **(A)** Examples of "total" (**upper trace**), P/Q-type (**middle trace**), and N-type currents (**lower trace**), all in the presence of 10 µM nimodipine, in a β_3-/- neuron following the "double pulse" protocol (**inset**), before and after the application of 10 µM NE. **(B,C)** Time course of inhibition of currents (I_1 and I_2; B) and ratio (I_2/I_1), measured 10 ms after the step to –10 mV following application of NE (10 µM) and GVIA (1 µM). (Reprinted in part by permission from Namkung *et al.*[57])

ingly larger degree of voltage-dependent relief. This observation was confirmed in pooled results from experiments in wt and β_3-/- neurons where NE inhibition was broken down by channel type and according to whether dependent or independent of conditioning voltage (FIG. 8). The most notable finding is that voltage-dependent inhibition was significantly greater for N-type current than for P/Q current, not only in wt neurons ($p < 0.05$), but also in β_3-/- neurons ($p < 0.01$). In contrast, no significant difference was detected between wt and β_3-/- neurons in any of the other categories. Thus, these experiments suggest that the type of α_1 subunit is more important than the type of β subunit in determining the Ca^{2+} channel response to G-protein modulation.

The greater degree of voltage-dependent inhibition of N-type relative to P/Q-type channels is consistent with the behavior of native currents in chromaffin cells[78] and for α_{1A} and α_{1B} expressed in oocytes.[79,80] The differing sensitivity of N- and P/Q-type channels to the imposition and relief of G protein–mediated inhibition may provide a mechanism for fine control of synaptic strength. We could not confirm a recent report that coexpression of the β_3 subunit confers a decrease in sensitivity to G-protein modulation that was much greater for α_{1A} than α_{1B}. We did not detect such changes in sensitivity to NE when comparing Ca^{2+} channels in the absence and presence of β_3 (FIG. 8). One possible explanation is that in β_3-/- SCG neurons, in contrast to oocytes, β_1 or β_4 may compensate for the absence of β_3 subunit.

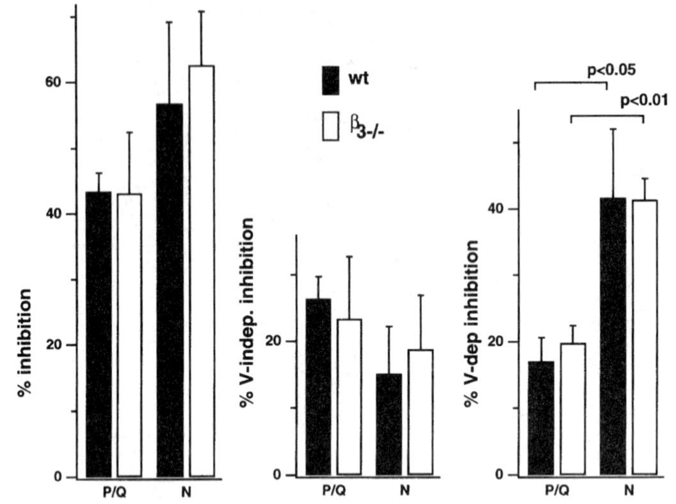

FIGURE 8. Histograms of the total, voltage-independent, and voltage-dependent components of NE inhibition of N-type and P/Q-type currents in wt (*solid*) and β_3-/- (*open*) SCG neurons (n = 5–9). (Reprinted in part by permission from Namkung *et al.*[57])

Shift in Preponderance of Modes of Gating of P/Q-Type Channels in β_3-/- Neurons

Single-channel analysis of P/Q-type channels generated by expression of α_{1A} in HEK293 cells has demonstrated three modes of gating and a relationship between the prevalence of each mode and the specific β subunit type.[81] In particular, the β_3 subunit was preferentially associated with the mode characterized by the lowest probability of opening and the most depolarized activation curve. To find out if β_3 subunits have an influence on gating of P/Q-type channels in neurons, we looked for differences in the voltage dependence of activation of P/Q currents in whole-cell recordings from wt and β_3-/- SCG cells. Tail currents were evoked by repolarizations to –80 mV following brief depolarizing pulses to various potentials (FIG. 9). Within 200–300 μs after the repolarizing step, tail currents could be fitted by a single exponential, with similar time constants for wt neurons (328 ± 51 μs, n = 6) and β_3-/- neurons (427 ± 170 μs, n = 3). Tail current amplitudes were measured 400 μs after the repolarization to reduce the series resistance error.[74,82] The dependence of tail current amplitude on depolarizing level could not be fitted with a single Boltzmann function, but required the sum of two standard Boltzmann functions (overall expression given in FIG. 9), with midpoint voltages separated by ~40 mV. This is in contrast to activation of P-type currents in Purkinje neurons, which can be well described by a single Boltzmann function.[83,84] The more complex activation characteristics in our cells have not been reported before for P- or Q-type channels, but they are reminiscent of the voltage-dependent activation of N-type channels, especially under the influence of G-protein modulation.[74,82,85]

The effect of β_3 deletion on P/Q channels in SCG neurons was mainly expressed as a prominent hyperpolarizing shift in the upper limb of the activation curve. The parameters describing each curve are half-activation voltages ($V'_{0.5}$,$V''_{0.5}$), current amplitudes normal-

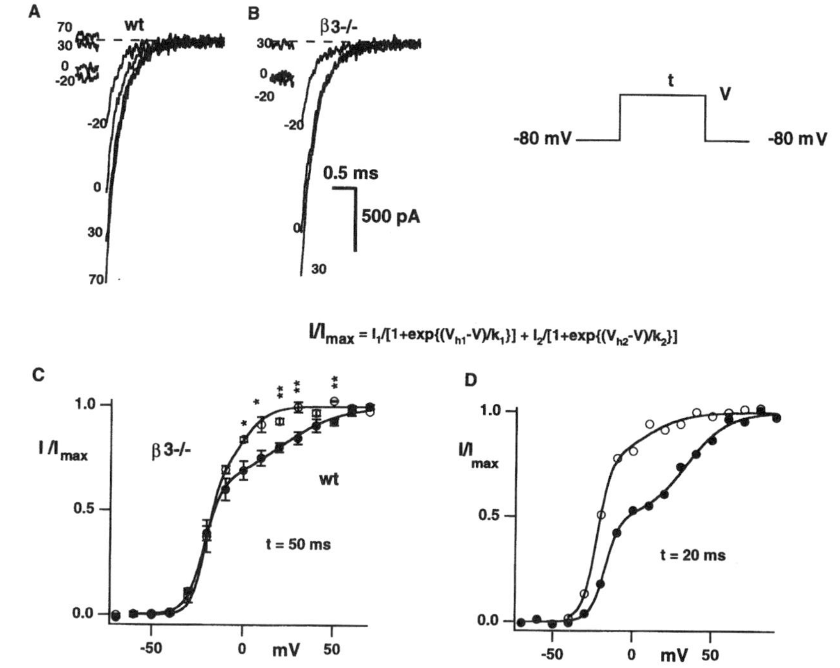

FIGURE 9. The activation curve for P/Q currents from β_3-/- neurons is shifted in a hyperpolarizing direction. Recordings performed in solutions containing 10 μM nimodipine and 1 μM ω-conotoxin GVIA to block L- and N-type channels. Examples of tail currents recorded at –80 mV from wt **(A)** and β_3-/- **(B)** neurons. **(C)** The plot of mean (± S.E.) normalized tail current amplitude for wt ($n = 6$) and β_3-/- ($n = 3$) neurons is different between 0 and 50 mV (* indicates $p < 0.05$ and ** $p < 0.01$ by Student's *t*-test). The curves are drawn using the mean values given in the text and the equation (see **inset**). **(D)** Shorter depolarizations of wt and β_3-/- neurons also produced different activation. Values for these fits were I′ (0.48 vs. 0.74), I″ (0.52 vs. 026), k′ (4.6 vs. 4.3 mV), k″ (12.8 vs. 3.7 MV), $V'_{0.5}$ (–17 vs. –22 mV) and $V''_{0.5}$ (34 vs. 5 mV). (Reprinted in part by permission from Namkung *et al.*[57])

ized to the maximum values (I′, I″), slope factors (k′, k) and midpoint voltages ($V'_{0.5}$,$V''_{0.5}$). Fits to tail-derived activation curves gave comparable values for wt and β_3-/- neurons for I′ (0.61 ± 0.05 vs. 0.67 ± 0.03), I″ (0.39 ± 0.05 vs. 0.33 ± 0.03), k′ (5.1 ± 0.7 vs. 3.9 ± 0.1 mV), k″ (16.8 ± 1.8 vs. 9.3 ± 2.7 mV), and $V'_{0.5}$ (–21 ± 1 vs. –20 ± 2 mV). However, the $V''_{0.5}$ for β_3-/- neurons (–1 ± 1 mV) was 21 mV more negative than $V'_{0.5}$ for wt neurons (20 ± 5 mV; $p < 0.05$). It is this difference that resulted in the leftward shift of the upper limb of the activation curve for the β_3-/- neurons (FIG. 9C). An even more striking displacement of the upper component of the activation curve was seen with 20-ms depolarizations (FIG. 9D). Thus, omission of the β_3 subunit exerts a significant effect on gating of the P/Q channels. Our data are in accord with single-channel recordings from expressed α_{1A} subunits,[81] showing that coexpression of β_3 subunits favors occupancy of low-probability gating modes.

Pharmacological and Kinetic Characterization of the P/Q Current

P/Q-type currents are supported by the α_{1A} subunit but display a considerable range of kinetic and pharmacological properties as expressed in various neurons. As originally defined in cerebellar Purkinje cells,[86,87] P-type currents are blocked by ω-AgaIVA with an IC_{50} of 2 nM or less.[88] In cerebellar granule neurons, a small P-type current appears along with a much larger Q-type current, a component inhibited by ω-AgaIVA with an IC_{50} of ~80 nM.[17] We characterized further the P/Q-type currents in mouse SCG neurons by testing ω-AgaIVA at a dose (20 nM) that would block almost all P-type channels while largely sparing Q-type channels. With ω-CTx-GVIA (1 μM) and nimodipine (10 μM) present to block N- and L-type channels, addition of 20 nM AgaIVA failed to block I_{Ba}. No signifi-

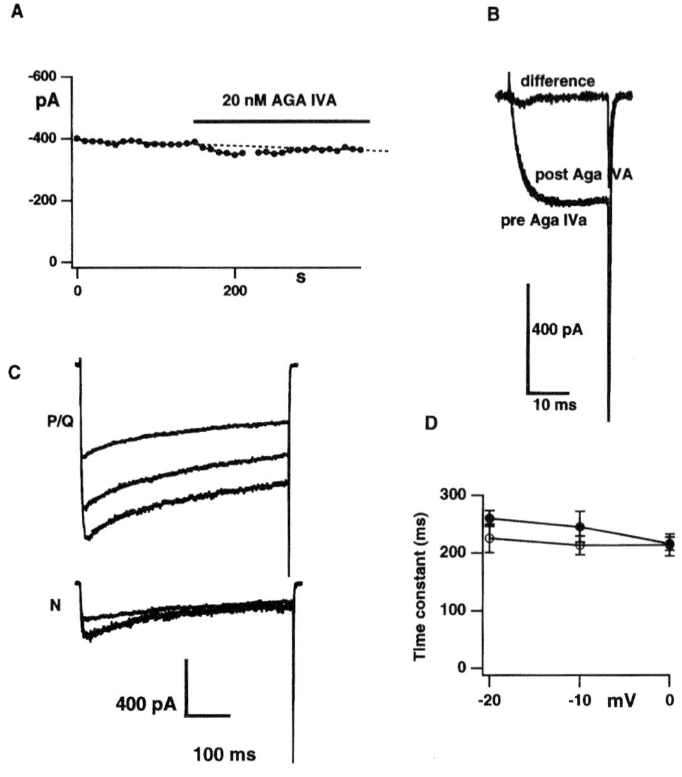

FIGURE 10. Further characterization of the P/Q-type current. **(A)** The response of the peak current to 20 nM Aga IVA. Currents were activated by depolarizations from –80 mV to –10 mV every 10 s. *Broken line* shows rundown predicted by least squares fit to data prior to application. **(B)** Difference current and mean currents pre- and post-Aga IVA application. Mean currents represent the five currents just before and at end of drug application. Post current has been amplified to take rundown into account. **(C)** P/Q- and N-type currents activated by 500-ms depolarizations from –90 mV to –20, –10, and 0 mV. Current decay is fit with a single exponential. **(D)** Time constant of decay is same for P/Q currents in wt (*closed circles*) and b3-/- (*open circles*) neurons at these voltages.

cant inhibition distinct from rundown could be detected in wt neurons (FIG. 10A,B). Likewise, I_{Ba} in β_3-/- SCG neurons was insensitive to ω-AgaIVA at this discriminatory concentration.

P- and Q-type currents have also been distinguished by their kinetics of inactivation. The classical P-type current in cerebellar Purkinje cells hardly decays with long depolarizations (time constants of ~2s[87]), whereas Q-type current decays with τ's of a few hundred milliseconds. We examined the decay of currents in the presence of GVIA (1 μM) and nimodipine (10 μM) (largely P/Q-type currents) during 500-ms-long depolarizations to –20, –10, and 0 mV from a holding potential of –90 mV (FIG. 10C). In wt neurons, the currents decayed exponentially with mean time constants of 210–260 ms, as expected for Q-type channels (filled symbols, FIG. 10D). Very similar τ's were obtained in experiments with β_3-/- neurons (open symbols, FIG. 10D). The overall conclusions about the nimodipine- and GVIA-insensitive current in mouse SCG neurons are as follows: (1) the pharmacological and kinetic characteristics of the nimodipine- and GVIA-insensitive current allow it to be identified as Q-type; (2) the Q-type current was recorded in the absence of any detectable P-type current; (3) the properties of the Q-type component were not detectably affected by deletion of the β_3 subunit.

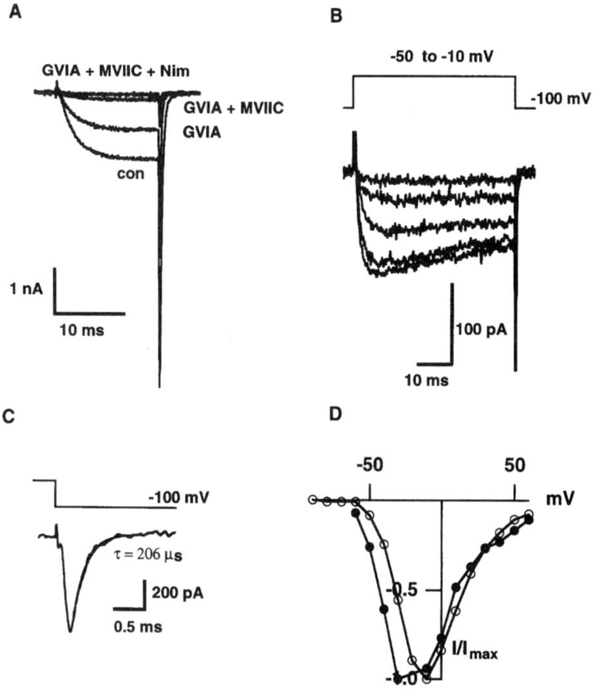

FIGURE 11. R-type currents in mouse SCG neurons. **(A)** Currents activated by stepping from –80 mV to –10 mV in the presence of saturating doses of GVIA, MVIIC, and nimodipine. **(B)** A family of curves activated after depolarizations from –100 mV in the same cell after application of blockers. **(C)** Deactivation of currents is rapid and fit well by a single exponential where τ = 206 μs. **(D)** Current voltage properties of peak currents for wt and β3-/- neurons.

R-Type Currents in Rat and Mouse SCG Neurons

R-type currents have been identified in the soma, dendrites, or presynaptic terminals of a wide variety of neurons.[17,54,89,90] Rat SCG neurons contain transcripts for the underlying α_{1E} subunit[58] and display a nimodipine- and ω-conotoxin GVIA-resistant current in whole-cell recordings.[91] We looked for R-type current in mouse SCG neurons by closely examining the Ba^{2+} currents that remained after application of saturating doses of nimodipine, GVIA, and MVIIC (FIG. 11A). FIGURE 11B shows a representative family of residual currents, recorded with the holding potential set at –100 mV to increase their size. As expected for R-type current, activation is voltage dependent and rapid, and the decay of current during the 50-ms test pulse was greater than observed for P/Q- or N-type currents (FIG. 10C). The time course of deactivation at –100mV was rapid (τ ~200 μs), typical of HVA channels and as expected for R-type channels. Furthermore, the R-type current activates over a more negative range of membrane voltage than the global currents, which are predominantly N- and P/Q-type channels (FIG. 11D; see ref. 57). Thus, the component of I_{Ba} described here shows both pharmacological and biophysical properties consistent with R-type current in other neuronal systems.[17]

DISCUSSION

α_{1E} Is Responsible for R-Type Current

Our findings with antisense oligonucleotides provided strong support for the hypothesis that α_{1E} supports R-type current. The antisense oligonucleotide designed against α_{1E} stood out in its ability to reduce R-type current. None of the control sequences tested—namely, a sense α_{1E} ON, an antisense α_{1A} ON, and a sense α_{1A} ON—affected the R-type current. The specific reduction of R-type current by α_{1E} antisense is consistent with previous studies showing that biophysical and pharmacological properties of R-type currents in neurons[17,92] are in close alignment with the characteristics of α_{1E} expressed in cell lines.[24,27,29] Examples of such similarity include single-channel conductances using Ba^{2+} as the charge carrier.[25,26,92] and sensitivity to the neurotoxin ω-Aga-IIIA.[27] Further study of unitary Ca^{2+} conductances of native R-type channels is needed to complete the comparisons with expressed α_{1E} subunits.[29] Pietrobon's group has provided evidence for two forms of unitary R-type channel activity, designated G2 and G3, differing by ~10 mV in their voltage dependence of activation.[89,92] We considered the possibility that our antisense sequence affects only one of these subtypes, but did not observe the expected changes in the voltage dependence of peak current, or in the rate of inactivation. Rather than invoking additional α_1 subunits, we prefer to hypothesize that multiple forms of R-type Ca^{2+} may arise from splice variations in α_{1E}[93,94] or from association of α_{1E} with diverse ancillary subunits. This kind of explanation may also apply to pharmacological studies with SNX-482, a new peptide neurotoxin that blocks α_{1E} currents in mammalian cell lines and R-type currents in nerve terminals of rat neurohypophysis, but fails to inhibit R-type current in rat cerebellar granule cells.[94] Interestingly, in cerebellar granule cells cultured under the conditions used by Tottene *et al.*,[92] SNX-482 appears able to block a subfraction of R-type current (D. Pietrobon, personal communication).

α_{1A} Subunits Support P/Q-Type Currents

Our study provided independent support for the generally accepted notion that α_{1A} subunits underlie P/Q-type currents. The most abundant voltage-gated Ca^{2+} channel currents in cerebellar granule cells, P/Q-type currents are blocked by ω-AgaIVA and ω-CTx MVIIC, like currents generated by α_{1A} cRNA in oocytes and cell lines.[95–98] We found that the antisense oligonucleotide designed against α_{1A} specifically reduced the peak amplitude of the P/Q-type components while leaving the R-type current unaffected. None of the control ONs (α_{1A} sense, α_{1E} antisense, and α_{1E} sense) had any effect on the P/Q components. Based on comparison of pooled data from sense- and from antisense-treated neurons (FIG. 1), the component suppressed by the α_{1A} antisense had a prominently decaying time course, as expected if Q-type current were predominant. These results may be compared with α_{1A} antisense experiments in cerebellar Purkinje cells,[35] in which P-type currents are strongly predominant.[16,87] α_{1A} antisense reduced P-type current in Purkinje neurons, consistent with previous findings of α_{1A} transcripts and immunoreactivity in these cells. Taken together, these studies leave little doubt that α_{1A} can support both Q- and P-type currents, whatever the explanation for how they differ in pharmacology and inactivation kinetics (see Ref. 98).

The β_3 Subunit Regulates Expression and/or Gating of Multiple Calcium Channels

An overall conclusion of this study is that deletion of a specific β subunit can cause deficits or alterations in multiple components of neuronal Ca^{2+} channel current. A reduction of N-type current density was prominent in sympathetic neurons that displayed a small total current. This finding is consistent with an independent study of β_3 subunit–deficient mice, which demonstrated a reduced level of N-type channel expression by measurements of Ca^{2+} channel currents and ω-conotoxin GVIA binding in dorsal root ganglion neurons.[99] We also demonstrated that the L-type current was markedly diminished in β_3-/- SCG neurons. In contrast, P/Q-type currents were not detectably reduced in overall amplitude, but nonetheless displayed a clear alteration in their voltage dependence of activation. Thus, the impact of deleting β_3 was found to extend across three different channel types, generated by distinct α_1 subunits. This is the most direct evidence to date that a single kind of β subunit can play a significant role in influencing the function of multiple pore-forming α_1 subunits. On the other hand, the persistence of each of the channel types supports the idea that individual α_1 subunits can rely on a variety of β subunits—none are strictly dependent on the β_3 subunit.

Comparison of β_3 Knockout with Functional Elimination of Other β Subunits

It is interesting to compare the functional consequences of β_3 with the effects of β_1 and β_4, the other β subunits that have been deleted or functionally eliminated to date. Elimination of the β_1 subunit was accomplished by conventional gene targeting and produced striking changes in skeletal myotubes,[62] including a marked diminution of L-type current, reduction of voltage-dependent charge movement, and elimination of Ca^{2+} transients and excitation-contraction coupling,[100] all effects being reversed upon reintroduction of β_1 by transfection.[101] The small L-type current that remained in β_1-/- myotubes may be accounted for by the residual expression of α_{1S}, the principal α_1 subunit of skeletal muscle.[102] Expression of the β_4 subunit was eliminated in the *lethargic* mouse, a strain originally isolated through observations of epileptic behavior,[103] and subsequently shown to bear a truncated mutation in the β_4 gene.[104] Further studies have indicated that forebrains of *lh/lh* mice contain no detectable truncated or full-length β_4 protein, and display a two-fold reduction in the number of ω-conotoxin GVIA binding sites.[105] These findings are broadly consistent with our experiments showing that N-type and L-type currents were both reduced in β_3-/- SCG neurons.[57] Taken together, these data support the idea that β subunits play an important role in regulating the level of expression of α_1 subunits in native neurons or myocytes, along the lines suggested by expression studies in oocytes or cell lines.[44,45,106]

Deletion of the β_3 subunit also produced striking changes in the voltage-dependent activation of P/Q-type channels, but without significant alteration of their level of expression. This illustrates the general principle that β subunits may exert independent effects on channel assembly/expression and channel gating.[70] In this context, it is interesting to consider recent evidence supporting the existence of multiple β subunit binding sites on the α_1 subunit.[70,107] Birnbaumer and colleagues proposed that a β subunit may bind with high affinity to one site on the α_1, subunit and influence the expression of the channel, while binding of a second β subunit (similar or different in type) to another site could then regulate channel gating.[70] Coronado and colleagues extended the idea of multiple interactions of β subunits with a single α subunit, in finding that both β_{1a} and β_{2a} subunits rescued the current density and channel gating properties in β_1-/- myotubes, while β_{1a} but not β_{2a} restored normal charge movement and calcium release (personal communication). Our own evidence supports the idea that assembly/expression can be well supported by a variety of β subunits, whereas the pattern of gating shows greater specificity for the particular β subunit type.

Varied Phenotypic Effects of Abolishing Specific β Subunits

The viability of the β_3 knockout stands in contrast with the embryonic lethal effect of deleting the β_1 subunit, caused by respiratory arrest,[62] or the epileptic effect of eliminating expression of the β_4 subunit.[104,105] Relevant considerations are the prevalence of individual subunits and their physiological roles. To detect the impact of β_3 deletion in the intact mouse, further tests of specific functions such as blood pressure or heart rate regulation may be needed. Cardiovascular changes might be anticipated, because of the role of sympathetic neurons in controlling cardiac activity and vascular tone, and because of the prominent expression of β_3 subunits in smooth muscle.[108] Altered nociception in β_3 knockout animals is another interesting possibility put forward by Flockerzi and colleagues,[99] consistent with the reduction of N-type currents and earlier evidence that these currents play an important role in pain transmission.[109] Meanwhile, our electrophysiological data may help explain the absence of overt changes in the development and behavior of the β_3-/- mice. Their neurological status will depend on changes in the functional properties of N- and P/Q-type Ca^{2+} channels, which often work together to regulate Ca^{2+} influx and exocytosis at presynaptic terminals in the CNS.[56] A reduction in N-type current like that found in SCG neurons would impair transmitter release, while the hyperpolarizing shift in the activation curve of P/Q-type channels would favor increased activation and enhanced synaptic transmission. The net effect in the intact animal may be much milder than either alteration alone.

ACKNOWLEDGMENTS

We thank all members of the Tsien laboratory for helpful discussion. This work was supported in part by a Creative Research Initiative program grant from the Ministry of Science and Technology, a medical science grant from the Ministry of Health and Welfare, a genetic engineering grant from the Ministry of Education of Korea, and grants from NIH and HHMI. S.M.S. is a Howard Hughes Medical Institute Physician Postdoctoral Fellow, and E.P.R. is an American Heart, Western States Affiliate.

REFERENCES

1. TSIEN, R.W. & R.Y. TSIEN. 1990. Calcium channels, stores, and oscillations. Annu. Rev. Cell Biol. **6:** 715–60.
2. HILLE, B. 1992. Ionic Channels of Excitable Membranes. Sinauer Associates. Sunderland, MA.
3. TSIEN, R.W., P. HESS, E.W. MCCLESKEY & R.L. ROSENBERG. 1987. Calcium channels: Mechanisms of selectivity, permeation, and block. Annu. Rev. Biophys. Biophys Chem. **16:** 265–190.
4. CATTERALL, W.A & B.M. CURTIS. 1987. Molecular properties of voltage-sensitive calcium channels. Soc. Gen. Physol. Ser. **41:** 201–213.
5. CAMPBELL, K.P., A.T. LEUNG & A.H. SHARP. 1988. The biochemistry and molecular biology of the dihydropyridine-sensitive calcium channel. Trends Neurosci. **11:** 425–430.
6. CATTERALL, W.A., M.J. SEAGAR & M. TAKAHASHI. 1988. Molecular properties of dihydropyridine-sensitive calcium channels in skeletal muscle. J. Biol. Chem. **263:** 3535–3538.
7. GLOSSMANN, H. & J. STRIESSNIG. 1990. Molecular properties of calcium channels. Rev. Physiol. Biochem. Pharmacol. **114:** 1–105.
8. LETTS, V.A., R. FELIX, G.H. BIDDLECOME, J. ARIKKATH, C.L. MAHAFFEY, A. VALENZUELA, F.S. BARTLETT, 2nd, Y. MORI, K. P. CAMPBELL & W. N. FRANKEL. 1998. The mouse stargazer gene encodes a neuronal Ca^{2+}-channel gamma subunit [see comments]. Nature Genet. **19:** 340–347.
9. TANABE, T., H. TAKESHIMA, A. MIKAMI, V. FLOCKERZI, H. TAKAHASHI, K. KANGAWA, M. KOJIMA, H. MATSUO, T. HIROSE & S. NUMA. 1987. Primary structure of the receptor for calcium channel blockers from skeletal muscle. Nature **328:** 313–318.
10. ELLIS, S.B., M.E. WILLIAMS, N.R. WAYS, R. BRENNER, A.H. SHARP, A.T. LEUNG, K.P. CAMPBELL, E. MCKENNA, W.J. KOCH & A. HUI. 1988. Sequence and expression of mRNAs encoding the α_1, and α_2 subunits of a DHP-sensitive calcium channel. Science **241:** 1661–1664.
11. RUTH, P., A. ROHRKASTEN, M. BIEL, E. BOSSE, S. REGULLA, H.E. MEYER, V. FLOCKERZI & F. HOFMANN. 1989. Primary structure of the β subunit of the DHP-sensitive calcium channel from skeletal muscle. Science **245:** 1115–1118.
12. CARBONE, E. & H. . LUX. 1984. A low voltage-activated, fully inactivating Ca^{2+} channel in vertebrate sensory neurones. Nature **310:** 501–502.
13. NOWYCKY M.C., A.P. FOX & R.W. TSIEN. 1985. Three types of neuronal calcium channel with different calcium agonist sensitivity. Nature **316:** 440–443.
14. MATTESON, D.R. & C.M. ARMSTRONG. 1986. Properties of two types, of calcium channels in clonal pituitary cells. J. Gen. Physiol. **87:** 161–182.
15. TSIEN, R.W., A.P. FOX, P. HESS, E.W. MCCLESKEY, B. NILIUS, M.C. NOWYCKY & R.L. ROSENBERG. 1987. Multiple types of calcium channel in excitable cells. Soc. Gen. Physiol. Ser. **41:** 167–187.

16. LLINAS R., M. SUGIMORI, D.E. HILLMAN & B. CHERKSEY. 1992. Distribution and functional significance of the P-type, voltage-dependent Ca^{2+} channels in the mammalian central nervous system. Trends Neurosci. **15:** 351–355.
17. RANDALL, A. & R.W. TSIEN. 1995. Pharmacological dissection of multiple types of Ca^{2+} channel currents in rat cerebellar granule neurons. J. Neurosci. **15:** 2995–3012.
18. ELLINOR, P.T., J.-F. ZHANG, A.D. RANDALL, M. ZHOU, T.L. SCHWARZ, R.W. TSIEN & W.A. HORNE. 1993. Functional expression of a rapidly inactivating neuronal calcium channel. Nature **363:** 455–458.
19. ZHANG, J.-F., A.D. RANDALL, P.T. ELLINOR W.A. HORNE, W.A. SATHER, T. TANABE, T.L. SCHWARZ & R.W. TSIEN. 1993. Distinctive pharmacology and kinetics of cloned neuronal Ca^{2+} channels and their possible counterparts in mammalian CNS neurons. Neuropharmacology **32:** 1075–1088.
20. KAVALALI, E.T., M. ZHUO, H. BITO & R.W. TSIEN. 1997. Dendritic Ca^{2+} channels characterized by recordings from isolated hippocampal dendritic segments. Neuron **18:** 651–663.
21. WU, L.G. & P. SAGGAU. 1994. Pharmacological identification of two types of presynaptic voltage-dependent calcium channels at CA3-CA1 synapses of the hippocampus. J. Neurosci. **14:** 5613–56122.
22. WANG G., G. DAYANITHI, S. KIM, D. HOM, L. NADASDI, R. KRISTIPATI, J. RAMACHANDRAN, E.L. STUENKEL, J.J. NORDMANN, R. NEWCOMB & J.R. LEMOS. 1997. Role of Q-type Ca^{2+} channels in vasopressin secretion from neurohypophysial terminals of the rat. J. Physiol. (Lond). **502:** 351–363.
23. SOONG, T.W., A. STEA, C.D. HODSON, S.J. DUBEL, S.R. VINCENT & T.P. SNUTCH. 1993. Structure and functional expression of a member of the low voltage-activated calcium channel family. Science **260:** 1133–1136.
24. WILLIAMS, M.E., L.M. MARUBIO, C.R. DEAL, M. HANS, P.F. BRUST, L.H. PHILIPSON, R.J. MILLER, E.C. JOHNSON, M.M. HARPOLD & S.B. ELLIS. 1994. Structure and functional characterization of neuronal α_{1E} calcium channel subtypes. J. Biol. Chem. **269:** 22347–22357.
25. SCHNEIDER, T., X. WEI, R. OLCESE, J.L. COSTANTIN, A. NEELY, P. PALADE, E. PEREZ-REYES, N. QIN, J. ZHOU, G.D. CRAWFORD *et al.* 1994. Molecular analysis and functional expression of the human type E neuronal Ca^{2+} channel alpha 1 subunit. Recept. Channels. **2:** 255–270.
26. WAKAMORI, M., T. NIIDOME, D. FURUTAMA, T. FURUICHI, K. MIKOSHIBA, Y. FUJITA, I. TANAKA, K. KATAYAMA, A. YATANI & A. SCHWARTZ. 1994. Distinctive functional properties of the neuronal BII (class E) calcium channel. Recept. Channels **2:** 303–314.
27. ROCK, D.M., W.A. HORNE, S. J. STOEHR, C. HASHIMOTO, R.Z. CONG, M.A. PALMA, D. HIDAYETOGLU & J. OFFORD. 1998. Does α_{1E} code for T-type Ca^{2+} channels? A comparison of recombinant E class Ca^{2+} channels with GH3 pituitary T-type and recombinant B class Ca^{2+} channels. *In* T-type Calcium Channels. R.W. Tsien, J.P. Clozel & J. Nargeot, Eds.: 279–289. Aidis Press.
28. RANDALL A.D. & R.W. TSIEN. 1997. Contrasting biophysical and pharmacological properties of T-type and R-type calcium channels. Neurophamacology. **36:** 879–93.
29. BOURINET, E., G. W. ZAMPONI, A. STEA T.W. SOONG, B.A. LEWIS, L.P. JONES, D.T. YUE & T.P. SNUTCH. 1996. The α_{1E} calcium channel exhibits permeation properties similar to low-voltage-activated calcium channels. J. Neurosci. **16:** 4983–4993.
30. PIEDRAS-RENTERÍA, E.S., C.C. CHEN & P.M. BEST. 1997. Antisense oligonucleotides against rat brain alpha1E DNA and it's atrial homologue decrease T-type calcium current in atrial myocytes. Proc. Natl. Acad. Sci. USA **94:** 14936–14941.
31. STEPHENS, G.J., K.M. PAGE, J.R. BURLEY, N.S. BERROW & A.C. DOLPHIN. 1997. Functional expression of rat brain cloned alpha1E calcium channels in COS-7 cells. Pfluegers Arch. **433:** 523–532.
32. MEIR, A. & A. C. DOLPHIN. 1998. Known calcium channel alphal subunits can form low threshold small conductance channels with similarities to native T-type channels. Neuron **20:** 341–351.
33. PIEDRAS-RENTERIA, E. S. & R. W. TSIEN. 1998. Antisense oligonucleotides against alpha1E reduce R-type calcium currents in cerebellar granule cells. Proc. Natl. Acad. Sci. USA **95:** 7760–7765.
34. STEA, A., W. J. TOMLINSON, T. W. SOONG, E. BOURINET, S. J. DUBEL, S. R. VINCENT & T. P. SNUTCH. 1994. Localization and functional properties of a rat brain α_{1A} calcium channel reflect similarities to neuronal Q- and P-type channels. Proc. Natl. Acad. Sci. USA **91:** 10576–10580.

35. Gillard, S.E., S.G. Volsen, W. Smith, R.E. Beattie, D. Bleakman & D. Lodge. 1997. Identification of pore-forming subunit of P-type calcium channels: An antisense study on rat cerebellar Purkinje cells in culture. Neuropharmacology. **36:** 405–409.
36. Hofmann, F., M. Biel & V. Flockerzi. 1994. Molecular basis for Ca^{2+} channel diversity. Annu. Rev. Neurosci. **17:** 399–418.
37. Isom, L.L., K.S. DeJongh & W.A. Catterall. 1994. Auxiliary subunits of voltage-gated ion channels. Neuron. **12:** 1183–194.
38. De Waard, M., C.A. Gurnett & K.P. Campbell. 1996. Structural and functional diversity of voltage-activated calcium channels. *In* Ion Channels. Plenum Press. New York. 41–87.
39. Glossmann, H., J. Striessnig, L. Hymel & H. Schindler. 1987. Purified L-type calcium channels: only one single polypeptide (α_1-subunit) carries the drug receptor domains and is regulated by protein kinases. Biomed Biochim Acta. **46:** S351–S356.
40. Takahashi, M., M.J. Seagar, J.F. Jones, B.F. Reber & W.A. Catterall. 1987. Subunit structure of dihydropyridine-sensitive calcium channels from skeletal muscle. Proc. Natl. Acad. Sci. USA **84:** 5478–5482.
41. Birnbaumer, L., K.P. Campbell, W.A. Catterall, M.M. Harpold, F. Hofmann, W.A. Horne, Y. Mori, A. Schwartz. T.P. Snutch, T. Tanabe & R.W. Tsien. 1994. The naming of voltage-gated calcium channels. Neuron. **13:** 505–506.
42. Pragnell, M., J. Sakamoto, S.D. Jay & K.P. Campbell. 1991. Cloning and tissue-specific expression of the brain calcium channel beta-subunit. FEBS Lett. **291:** 253–258.
43. Perez-Reyes, E., A. Castellano, H.S. Kim P. Bertrand, E. Baggstrom, A.E. Lacerda, X.Y. Wei & L. Birnbaumer. 1992. Cloning and expression of a cardiac/brain β subunit of the L-type calcium channel. J. Biol. Chem. **267:** 1792–1797
44. Castellano, A., X. Wei, L. Birnbaumer & E. Perez-Reyes. 1993. Cloning and expression of a third calcium channel beta subunit. J. Biol. Chem. **268:** 3450–3455.
45. Castellano, A., X. Wei, L. Birnbaumer & E. Perez-Reyes. 1993. Cloning and expression of a neuronal calcium channel β subunit. J. Biol. Chem. **268:** 12359–12366.
46. Stea, A., S.J. Dubel, M. Pragnell, J.P. Leonard, K.P. Campbell & T.P. Snutch. 1993. An α_1-subunit normalizes the electrophysiological properties of a cloned N-type Ca^{2+} channel α_1-subunit. Neuropharmacology **32:** 1103–1116.
47. Mori, Y., T. Friedrich, M.S. Kim, A. Mikami, J. Nakai, P. Ruth, E. Bosse, F. Hofmann, V. Flockerzi, T. Furuichi, K. Mikoshiba, K. Imoto, T. Tanabe & S. Numa. 1991. Primary structure and functional expression from complementary DNA of a brain calcium channel. Nature **350:** 398–402.
48. De Waard, M., D.R. Witcher, M. Pragnell, H. Liu & K.P. Campbell. 1995. Properties of the α_1-α anchoring site in voltage-dependent Ca^{2+} channels. J. Biol. Chem. **270:** 12056–12064.
49. Campbell, V., N.S. Berrow, E.M. Fitzgerald, K. Brickley & A.C. Dolphin. 1995. Inhibition of the interaction of G protein G(o) with calcium channels by the calcium channel beta-subunit in rat neurones. J. Physiol. **485:** 365–372.
50. Qin, N., D. Platano, R. Olcese E. Stefani & L. Birnbaumer. 1997. Direct interaction of gbetagamma with a C-terminal gbetagamma-binding domain of the Ca^{2+} channel alpha1 subunit is responsible for channel inhibition by G protein–coupled receptors. Proc. Natl. Acad. Sci. USA **94:**8866–8871.
51. Roche, J.P., V. Anantharam & S.N. Treistman. 1995. Abolition of G protein inhibition of alpha 1A and alpha 1B calcium channels by co-expression of the beta 3 subunit. FEBS Lett. **371:** 43–46.
52. Witcher, D.R., M. De Waard, S.D. Kahl & K.P. Campbell. 1994. Purification and reconstitution of N-type calcium channel complex from rabbit brain. Methods Enzymol. **238:** 335–348.
53. Scott, V.E., M. De Waard, H. Liu, C.A. Gurnett, D.P. Venzke, V.A. Lennon & K.P. Campbell. 1996. Beta subunit heterogeneity in N-type Ca^{2+} channels. J. Biol. Chem. **271:** 3207–3212.
54. Kavalali, E.T., M. Zhuo, H. Bito & R.W. Tsien. 1997. Dendritic Ca^{2+} channels characterized by recordings from isolated hippocampal dendritic segments. Neuron **18:** 651–663.
55. Plummer, M.R., D.E. Logothetis & P. Hess. 1989. Elementary properties and pharmacological sensitivities of calcium channels in mammalian peripheral neurons. Neuron **2:** 1453–1463.
56. Wheeler, D.B., A. Randall & R.W. Tsien. 1994. Roles of N-type and Q-type Ca^{2+} channels in supporting hippocampal synaptic transmission. Science **264:** 107–111.

57. NAMKUNG, Y., S. M. SMITH, S. B. LEE N. V. SKRYPNYK, H. L. KIM, H. CHIN, R. H. SCHELLER, R. W. TSIEN & H. S. SHIN. 1998. Targeted disruption of the Ca^{2+} channel beta3 subunit reduces N- and L-type Ca^{2+} channel activity and alters the voltage-dependent activation of P/Q-type Ca^{2+} channels in neurons. Proc. Natl. Acad. Sci. USA **95:** 12010–12015.
58. LIN, Z., C. HARRIS & D. LIPSCOMBE. 1996. The molecular identity of Ca channel alpha 1–subunits expressed in rat sympathetic neurons. J. Mol. Neurosci. **7**(4): 257–267.
59. WATERMAN, S. A. 1997. Role of N-, P- and Q-type voltage-gated calcium channels in transmitter release from sympathetic neurones in the mouse isolated vas deferens. Br. J. Pharmacol. **120:** 393–398.
60. HAWS, C. M., P. A. SLESINGER & J. B. LANSMAN. 1993. Dihydropyridine- and omega-conotoxin–sensitive Ca^{2+} currents in cerebellar neurons: Persistent block of L-type channels by a pertussis toxin–sensitive G-protein. J. Neurosci. **13:** 1148–1156.
61. MCDONOUQH, S. I, K. J. SWARTZ, I. M. MINTZ, L. M. BOLAND & B. P. BEAN. 1996. Inhibition of calcium channels in rat central and peripheral neurons by α-Conotoxin MVIIC. J. Neurosci. **16:** 2612–2623.
62. GREGG, R. G., A. MESSING, C. STRUBE, M. BEURG, R. MOSS, M. BEHAN, M. SUKHAREVA, S. HAYNES, J. A. POWELL, R. CORONADO & P. A. POWERS. 1996. Absence of the beta subunit (cchbl) of the skeletal muscle dihydropyridine receptor alters expression of the alpha 1 subunit and eliminates excitation-contraction coupling. Proc. Natl. Acad. Sci. Usa **93:** 13961–13966.
63. Eccles, J. C. 1935. The action potential of the superior cervical ganglion. J. Physiol. **85:** 179–206.
64. SKOK, V. L. 1973. Physiology of Autonomic Ganglia. Igaku Shoin Ltd. Tokyo.
65. SCROGGS, R. S. & A. P. FOX. 1992. Calcium current variation between acutely isolated adult rat dorsal root ganglion neurons of different size. J. Physiol. (Lond.) **445:** 639–658.
66. DOLPHIN, A. C. 1995. The G.L. Brown prize lecture. Voltage-dependent calcium channels and their modulation by neurotransmitters and G proteins. Exp. Physiol. **80:** 1–36.
67. ZAMPONI. G. W. , E. BOURINET, D. NELSON, J. NARGEOT & T. P. SNUTCH. 1997. Crosstalk between G proteins and protein kinase C mediated by the calcium channel alphal subunit [see comments]. Nature **385:** 442–446.
68. DE WAARD, M., M. PRAGNELL & K. P. CAMPBELL. 1994. Ca^{2+} channel regulation by a conserved beta subunit domain. Neuron **13:** 495–503.
69. HERLITZE, S., G. H. HOCKERMAN, T. SCHEUER & W. A. CATTERALL. 1997. Molecular determinants of inactivation and G protein modulation in the intracellular loop connecting domains I and II of the calcium channel alpha-1A subunit. Proc. Natl. Acad. Sci. USA **94:** 1512–1516.
70. TAREILUS, E., M. ROUX, N. QIN, R. OLCESE, J. ZHOU, E. STEFANI & L. BIRNBAUMER. 1997. A *Xenopus* oocyte beta subunit: Evidence for a role in the assembly/expression of voltage-gated calcium channels that is separate from its role as a regulatory subunit. Proc. Natl. Acad. Sci. USA **94:** 1703–1708.
71. DE WAARD, M., H. LIU, D. WALKER, V. E. SCOTT, C. A. GURNETT & K. P. CAMPBELL. 1997. Direct binding of G-protein betagamma complex to voltage-dependent calcium channels [see comments]. Nature **385:** 446–450.
72. PAGE, K. M., G. J. STEPHENS, N. S. BERROW & A. C. DOLPHIN. 1997. The intracellular loop between domains I and II of the B-type calcium channel confers aspects of G-protein sensitivity to the E-type calcium channel. J. Neurosci. **17:** 1330–1338.
73. CAMPBELL V., N. BERROW, K. BRICKLEY, K. PAGE, R. WADE & A. C. DOLPHIN. 1995. Voltage-dependent calcium channel beta-subunits in combination with alpha 1 subunits, have a GTPase activating effect to promote the hydrolysis of GTP by G alpha o in rat frontal cortex. FEBS Lett. **370:** 135–140.
74. BEAN, B. P. 1989. Neurotransmitter inhibition of neuronal calcium currents by changes in channel voltage dependence. Nature **340:** 153–156.
75. TSIEN, R. W., D. LIPSCOMBE, D. V. MADISON, K. R. BLEY & A. P. FOX. 1988. Multiple types of neuronal calcium channels and their selective modulation. Trends Neurosci. **11:** 431–438.
76. IKEDA, S. R. 1996. Voltage-dependent modulation of N-type calcium channels by G-protein beta gamma subunits. Nature **380:** 255–258.
77. ELMSLIE, K. S., P. J. KAMMERMEIER & S. W. JONES. 1994. Reevaluation of Ca^{2+} channel types and their modulation in bullfrog sympathetic neurons. Neuron **13:** 217–228.

78. Currie, K. P. & A. P. Fox. 1997. Comparison of N- and P/Q-type voltage-gated calcium channel current inhibition. J. Neurosci. **17:** 4570–4579.
79. Roche, J. P & S. N. Treistman. 1998. The Ca^{2+} channel beta3 subunit differentially modulates G-protein sensitivity of alpha1A and alpha1B Ca^{2+} channels. J. Neurosci. **18:** 878–886.
80. Zhang, J. F., P. T. Ellinor, R. W. Aldrich & R. W. Tsien. 1996. Multiple structural elements in voltage-dependent Ca^{2+} channels support their inhibition by G proteins. Neuron. **17:** 991–1003.
81. Luvisetto, S., B. Hivert, M. Spagnolo, P. Brust, M. Williams, K. Stauderman, M. Harpold & D. Pietrobon. 1998. Single channel properties of human $alpha_{1A}$ stably coexpressed with different beta subunits in HEK293 cells. Biophys. J. **74:** A120.
82. Ikeda, S. R. 1991. Double-pulse calcium channel current facilitation in adult rat sympathetic neurones. J. Physiol. (Lond.) **439:** 181–214.
83. Mintz, I. M. & B. P. Bean. 1993. GABAB receptor inhibition of P-type Ca^{2+} channels in central neurons. Neuron **10:** 889–898.
84. McDonough, S. I., R. A. Lampe, R. A. Keith & B. P. Bean. 1997. Voltage-dependent inhibition of N- and P-type calcium channels by the peptide toxin omega-grammotoxin-SIA. Mol. Pharmacol. **52:** 1095–1104.
85. Zhu, Y. & S. R. Ikeda. 1993. Adenosine modulates voltage-gated Ca^{2+} channels in adult rat sympathetic neurons. J. Neurophysiol. **70:** 610–620.
86. Llinás, R., M. Sugimori, D. E. Hillman & B. Cherksey. 1992. Distribution and functional significance of the P-type, voltage-dependent Ca^{2+} channels in the mammalian central nervous system. Trends Neurosci. **15:** 351–355.
87. Regan, L. J., D. W. Sah & B. P. Bean. 1991. Ca^{2+} channels in rat central and peripheral neurons: High-threshold current resistant to dihydropyridine blockers and α-Conotoxin. Neuron **6:** 269–280.
88. Mintz, I. M., M. E. Adams & B. P. Bean. 1992. P-type calcium channels in rat central and peripheral neurons. Neuron **9:** 85–95.
89. Forti, L., A. Tottene, A. Moretti & D. Pietrobon. 1994. Three novel types of voltage-dependent calcium channels in rat cerebellar neurons. J. Neurosci. **14:** 5243–5256.
90. Wu, L. G., J. G. Borst & B. Sakmann. 1998. R-type Ca^{2+} currents evoke transmitter release at a rat central synapse. Pro. Natl. Acad. Sci. USA **95:** 4720–4725.
91. Boland, L. M., J. A. Morrill & B. P. Bean. 1994. α-Conotoxin block of N-type calcium channels in frog and rat sympathetic neurons. J. Neurosci. **14:** 5011–5027.
92. Tottene, A., A. Moretti & D. Pietrobon. 1996. Functional diversity of P-type and R-type calcium channels in rat cerebellar neurons. J. Neurosci. **16:** 6353–6363.
93. Pereverzev, A., U. Klöckner, M. Henry, H. Grabsch, R. Vajna, S. Olyschläger, S. Viatchenko-Karpinski, R. Schröder, J. Hescheler & T. Schneider. 1998. Structural diversity of the voltage-dependent Ca^{2+} channel alpha1E-subunit. Eur. J. Neurosci. **10:** 916-925.
94. Newcomb, R., B. Szoke, A. Palma, R. Long, K. Tarczy-Hornoch, J. A. Loo, D. J. Dooley, W. Hopkins, R. Crea, J. Miljanich *et al.* 1997. Soc. Neurosci. Abstr. 856.
95. Sather, W. A., T. Tanabe, J.-F. Zhang, Y. Mori, M. E. Adams & R. W. Tsien. 1993. Distinctive biophysical and pharmacological properties of class A (BI) calcium channel α_1 subunits. Neuron **11:** 291–303.
96. De Waard, M. & K. P. Campbell. 1995. Subunit regulation of the neuronal alpha 1A Ca^{2+} channel expressed in Xenopus oocytes. J. Physiol. (Lond.) **485:** 619–634.
97. Berrow, N. S., N. L. Brice, I. Tedder, K. M. Page & A. C. Dolphin. 1997. Properties of cloned rat alpha1A calcium channels transiently expressed in the COS-7 cell line. Eur. J. Neurosci. **9:** 739–748.
98. Moreno, H., B. Rudy & R. Llinás. 1997. Beta subunits influence the biophysical and pharmacological differences between P- and Q-type calcium currents expressed in a mammalian cell line. Proc. Natl. Acad. Sci. USA **94:** 14042–14047.
99. Murakami, M., B. Fleischmann, A. Cavalie, C. Trost, A. Ludwig, F. Zimmermann, U. Wissenbach, H. Schwegler, F. Hofmann, J. Hescheler & V. Flockerzi. 1998. Altered pain perception due to reduced expression of n-type Ca channels in mice deficient in the Ca channel beta 3 subunit. In press.

100. STRUBE, C., M. BEURG, P. A. POWERS, R. G. GREGG & R. CORONADO. 1996. Reduced Ca^{2+} current, charge movement, and absence of Ca^{2+} transients in skeletal muscle deficient in dihydropyridine receptor beta 1 subunit. Biophys. J. **71:** 2531–2543.

101 BEURG. M., M. SUKHAREVA, C. STRUBE, P. A. POWERS, R. G. GREGG & R. CORONADO. 1997. Recovery of Ca^{2+} current, charge movements, and Ca^{2+} transients in myotubes deficient in dihydropyridine receptor beta 1 subunit transfected with beta 1 cDNA. Biophys. J. **73:** 807–818.

102. STRUBE, C., M. BEURG, M. SUKHAREVA, C. A. AHERN, J. A. POWELL, P. A. POWERS, R. G. GREGG & R. CORONADO. 1998. Molecular origin of the L-type Ca^{2+} current of skeletal muscle myotubes selectively deficient in dihydropyridine receptor beta11a subunit. Biophys. J. **75:** 207–217.

103. DUNG, H. C. & R. H. SWIGART. 1972. Histo-pathologic observations of the nervous and lymphoid tissues of "lethargic" mutant mice. Tex. Rep. Biol. Med. **30:** 23–39.

104. BURGESS, D. L., J. M. JONES, M. H. MEISLER & J. L. NOEBELS. 1997. Mutation of the Ca^{2+} channel beta subunit gene Cchb4 is associated with ataxia and seizures in the lethargic (1h) mouse. Cell **88:** 385–392.

105. MCENERY, M. W., T. D. COPELAND & C. L. VANCE. 1998. Altered expression and assembly of N-type calcium channel alpha1B and beta subunits in epileptic lethargic (lh/lh) mouse. J. Biol. Chem. **273:** 21435–21438.

106. DE WAARD, M., M. PRAGNELL & K. P. CAMPBELL. 1994. Ca^{2+} channel regulation by a conserved beta subunit domain. Neuron **13:** 495–503.

107. PRAGNELL, M., M. DE WAARD, Y. MORI, T. TANABE, T. P. SNUTCH & K. P. CAMPBELL. 1994. Calcium channel α-subunit binds to a conserved motif in the I-II cytoplasmic linker of the $\alpha 1$-subunit. Nature **368:** 67–70.

108. COLLIN, T., P. LORY, S. TAVIAUX, C. COURTIEU, P. GUILBAULT, P. BERTA & J. NARGEOT. 1994. Cloning, chromosomal location and functional expression of the human voltage-dependent calcium-channel beta 3 subunit. Eur. J. Biochem. **220:** 257–262.

109. MILJANICH, G. P. & J. RAMACHANDRAN. 1995. Antagonists of neuronal calcium channels: Structure, function, and therapeutic implications. Annu. Rev. Pharmacol. Toxicol. **35:** 707–734.

Voltage-Dependent Calcium Channel Mutations in Neurological Disease

DANIEL L. BURGESS AND JEFFREY L. NOEBELS[a]

Developmental Neurogenetics Laboratory, Department of Neurology, Baylor College of Medicine, Houston, Texas, 77030, USA

ABSTRACT: Calcium ion channel mutations disrupt channel function and create recognizable disease phenotypes in the nervous system. The broad array of underlying cellular alterations is commensurate with the expanding genetic diversity of the voltage-gated calcium ion channel complex and its critical role in regulating cell function. Currently, 16 calcium channel genes are known, and mutations in 7 of these are associated with distinct inherited neurological disorders. These mutations provide new insight into the structure and function of the channels, and link specific subunits to cellular disease processes, including altered excitability, synaptic signaling, and cell death. Studies of mutant channel behavior, subunit interactions, and the differentiation of neural networks demonstrate unique patterns of downstream rearrangement. Developmental analysis of molecular plasticity in these mutants is a critical step to define the intervening mechanisms that translate aberrant ion channel behavior into the diverse clinical phenotypes observed.

Voltage-dependent ion channels are literally defined by their ability to mediate ion flux across cellular membranes in response to transmembrane potential changes. We are accustomed to equate this function with control of membrane impulse activity and excitability; however, by influencing local cytoplasmic levels of the free ion, a ubiquitous second messenger, Ca^{2+} channels are unique among the superfamily of voltage-dependent ion channels in the diversity of additional cellular functions they regulate.[1] In the embryo, Ca^{2+} channels control proliferation, differentiation, and cell-cell interactions; in the developing and mature central nervous system, they regulate coupling of electrical excitation to gene expression, and modulate a wide variety of intracellular signaling pathways that lead in turn to neurite outgrowth, synaptogenesis, transmitter and hormone release, plasticity, and muscle contraction. This broad range of potential downstream alterations in cellular phenotypes and their effects on the developing brain has profound implications for those who seek to define the mechanisms by which Ca^{2+} channel mutations may result in clinical disorders.

SIMPLE AND COMPLEX PHENOTYPES OF INHERITED Ca^{2+} CHANNEL DISORDERS

Recently, several neurological disorders have been linked to mutations in calcium channel genes in humans and mice. The clinical phenotypes currently recognized range from relatively simple disruptions of function in a specific cell type—for example, the

[a]Address for correspondence: Jeffrey L. Noebels, M.D., Ph.D., Department of Neurology, Baylor College of Medicine, One Baylor Plaza, Houston, Texas 77030. Phone: 713-798-5860; fax: 713-798-7528; e-mail: jnoebels@bcm.tmc.edu

myopathy that follows altered skeletal muscle fiber calcium currents in human hypokalemic periodic paralysis and the muscular dysgenesis mouse;[2,3] and the retinal signaling defect linked to mutant L-type calcium channels in human congenital night blindness;[4,5] to the more complex signaling alterations within widespread brain circuits, including episodic ataxia, hemiplegia, dyskinesia, and epilepsy arising from disorders in various channel subunits.[6–11] These mutations reveal that the neurological phenotypes of inherited calcium channelopathies may arise at various ages of development, appear as intermittent or continuous deficits, and continue as stationary or progressive disorders. The syndromes may result directly from membrane excitability defects or from relatively unpredictable signaling disturbances that emerge from subtle alterations in cellular physiology. As in other hereditary brain disorders, the most clinically apparent elements of the phenotype may originate in only a small subset of cells expressing the mutant gene. Since abnormal activity itself can markedly disrupt subsequent brain development, the challenge in each of these disorders will be to unravel the natural history of the genetic lesion and identify mechanisms underlying each element of the neurological phenotype.

THE GENETICS OF Ca^{2+} CHANNELOPATHIES

Mammalian voltage-dependent Ca^{2+} channels are multimeric complexes of α_1, β, and $\alpha_2\delta$ subunits that copurify in a 1:1:1 molar ratio (see Ref. 12 for review). A fourth subunit, γ, is expressed in skeletal muscle, where the channel subunit ratio is 1:1:1:1. A brain-specific γ subunit has recently been identified,[11] but its exact interactions with the other channel subunits remain to be defined. The α_1 subunit consists of four homologous domains (I–IV), each with six transmembrane segments (S1–S6), which associate in the cytoplasmic membrane to form the ion-conducting pore. This subunit also contains the voltage sensor for the channel, which is believed to consist of several positively charged amino acid residues within the S4 transmembrane segment of each domain. The β subunit is a smaller, cytoplasmic protein that binds directly to the α_1 subunit through a single high-affinity binding motif on each subunit and one or more lower-affinity interaction sites.[13,14] The $\alpha_2\delta$ subunit is transcribed from a single gene, then translated and proteolytically processed into α_2 (N-terminus) and δ (C-terminus) subunits, which remain linked through disulfide bonds. A single transmembrane domain was identified within the δ subunit,[15] and both α_2 and δ subunits contain several potential glycosylation sites.[16] The brain and skeletal muscle γ subunits contain four transmembrane domains and are also glycosylated[11,17] (FIG. 1).

At least eight mammalian genes have been identified that encode α_1 subunits, four genes encoding β subunits, two genes for the γ subunit, and a single gene for $\alpha_2\delta$. This genetic diversity constitutes the primary basis for phenotypic variation arising from calcium channel mutations. The chromosomal locations of known human and mouse voltage-dependent Ca^{2+} channel subunit genes, and associated genetic disorders, are shown in TABLE 1.

Assembly of a mature Ca^{2+} channel complex, its biophysical properties, and ultimate subcellular destination are dictated in part by the specific subunit combinations expressed by each particular cell, and factors influencing their association. A second level of variation is therefore based on mutations that specifically alter the combinatorial possibilities afforded by the multiple members of each subunit. Loss or alteration of interaction sites contributes importantly to divergent calcium channel mutant phenotypes, as evidenced by the initial analysis of mouse mutants lacking specific subunits.[8,10,11,18,19] Further Ca^{2+} channel structural and functional diversity contributing to the phenotype of calcium channelopathies is generated by a variety of other mechanisms that are described in detail elsewhere.

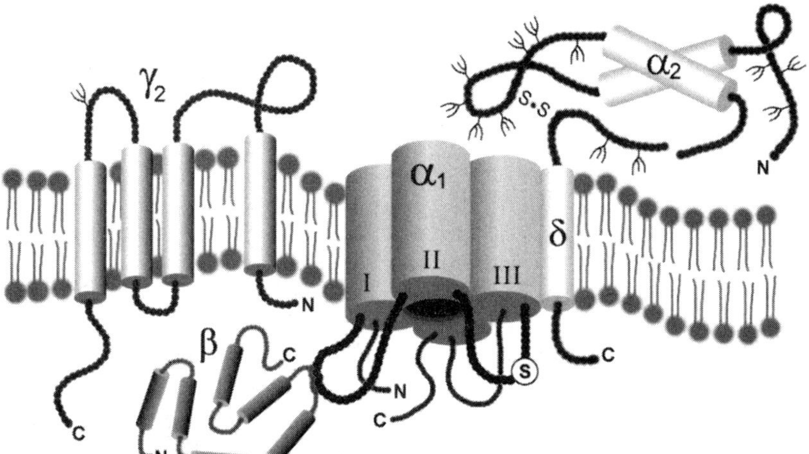

FIGURE 1. Subunit composition of neuronal voltage-dependent Ca^{2+} channels. The α_1 subunit consists of four homologous domains (I–IV), each comprising six transmembrane segments (S1–S6). Domains I–IV associate in the plasma membrane to form a Ca^{2+} conducting pore, and the S4 segment of each domain contains positively charged amino acids that are believed to act as the voltage sensor that initiates channel opening. The interaction between α_1 and the cytoplasmic β subunit is mediated by well-defined sequence motifs on β (BID, β subunit interaction domain) and on the I-II cytoplasmic linker of α_1 (AID). The BID sequence is highly conserved among all four β subunits, and AID is conserved among most α_1 subunits (with the possible exception of T-type channels), permitting formation of many distinct α_1-β combinations. The highly glycosylated $\alpha_2\delta$ subunit is transcribed from the *Cacna2* gene as a single mRNA that is posttranslationally cleaved into N-terminal α_2 and C-terminal δ subunits that remain attached through disulfide linkages. The transmembrane γ_2 subunit is able to modify the biophysical properties of α_{1A}-containing (P/Q-type) channels, but its linkage to other subunits in the channel complex is not yet defined. Several additional cellular proteins interact directly with the Ca^{2+} channel complex. The G-protein βγ complex modulates channel activity by binding to the I-II cytoplasmic loop of α_1. The synaptic proteins syntaxin-1A, synaptotagmin, SNAP-25, and synaptobrevin interact with the Ca^{2+} channel complex through a synaptic protein interaction, or "symprint" site (S) on the II-III linker of α_1, coupling excitation to neurotransmitter release at nerve terminals.

TABLE 1. Summary of Human and Mouse Voltage-Dependent Calcium Channel Genes

Subunit	Channel Type[a]	Human Gene	Location	Mouse Gene	Location	Associated Disease[b]
α_{1A}	P/Q	*CACNA1A*	19p13.1–p13.2	*Cacna1a*	8 (37.5)	EA2, FHM, SCA6, tottering (*tg*), leaner (*tg*[la])
α_{1B}	N	*CACNA1B*	9q34	*Cchn1a*	2 (17.4) A2	
α_{1C}	L	*CACNA1C*	12p13.3	*Cchl1a1*	[6 (54–64)]	
α_{1D}	L	*CACNA1D*	3p21.3–p21.2[C]	*Cacna1d*	14 (8.0)	
α_{1E}	R	*CACNA1E*	1q25–q31	*Cchra1*	1 (80.0)	
α_{1F}	L	*CACNA1F*[d]	Xp11.23	*Cacna1f*	[X (0–10 or 62–65)]	CSNB2
α_{1G}	T	*CACNA1G*	17q22[e]	*Cacna1g*	11 (55)	
α_{1H}	T	*CACNA1H*	16p13.3	*Cacna1h*	17 [10–11]	
α_{1S}	L	*CACNA1S*	1q31–q32	*Cacna1s*	1 (69.9) F	HOKPP, MHS1, muscular dysgenesis (*mdg*)
β_1	n.a.	*CACNB1*	17q21–q22	*Cacnb1*	11 (58.0)	t.i. (neonatal lethal)[f]
β_2	n.a.	*CACNB2*	10p12	*Cacnb2*	2 (11.5)	
β_3	n.a.	*CACNB3*	12q13	*Cacnb3*	15 (59.8)	t.i. (subclinical)[g]
β_4	n.a.	*CACNB4*	2q22–q23	*Cacnb4*	2 (33.9)	lethargic (*lh*)
$\alpha_2\delta$	n.a.	*CACNA2*	7q21-q22	*Cacna2*	5 (4.0)	
γ_1	L	*CACNG1*	17q24	*Cacng1*	[11 (56–66)]	
γ_2	n.a.	*CACNG2*	22q13	*Cacng2*	15 (45.2)	stargazer (*stg*), waggler (*stg*[wag])

NOTE: Nomenclature, genetic locations, and disease associations are according to Genome Database (GDB), Online Mendelian Inheritence in Man (OMIM), Mouse Genome Informatics (MGI), and the European Collaborative Interspecific Mouse BackCross (EUCIB) databases. Locations of mouse genes are given as: chromosome (centimorgan location) cytogenetic band. Locations in brackets are predictions based on mouse-human conserved linkage associations. n.a., not applicable; t.i., targeted inactivation.

[a] α_{1B}, α_{1C}, and α_{1E}, -subunits can exhibit properties of T-type channels under some assay conditions (Ref. 63).

[b] Uppercase symbols refer to human, lowercase to mouse: EA2, episodic ataxia type-2; FHM, familial hemiplegic migraine; SCA6, spinal cerebellar ataxia type-6; CSNB2, X-linked congenital stationary night blindness; HOKPP, hypokalemic periodic paralysis; MHS1, malignant hyperthermia susceptibility.

[c] GDB-approved location, CACNA1D was also mapped to 3p14.3 (Ref. 66).

[d] Ref. 64.

[e] Ref. 65.

[f] Ref. 18.

[g] Ref. 19.

ANALYSIS OF NEURONAL CALCIUM CHANNELOPATHIES IN MOUSE MUTANTS

α_1 Subunit Channelopathies

Recent efforts have identified mutations in the *CACNA1A* gene in three clinically distinct human neurological disorders: colon episodic ataxia type 2, familial hemiplegic migraine, and spinal cerebellar ataxia type 6.[6,7] These disorders illustrate the extensive phenotypic diversity that can arise from alterations in a single subunit isoform. The P/Q-type channels encoded by *CACNA1A* mediate a distinct set of Ca^{2+} currents with relatively high activation thresholds present on the dendrites, soma, and presynaptic terminals of many CNS neurons, predicting their involvement in impulse integration, plasticity, and the release of neurotransmitter. The analysis of cellular and neurological function in mice bearing mutations affecting this channel subtype provides valuable insight into the potential complexity of Ca^{2+} channelopathies, and highlights the challenges to understanding these neurological phenotypes in humans.

The Tottering (tg) *Mouse*

Mutations at the tottering locus disrupt the *Cacna1a gene*, which encodes the α_{1A} subunit of P/Q-type voltage-dependent Ca^{2+} channels.[8,9] To date, five alleles at the tottering locus have been reported, including *tg*, tg^{la}, tg^{rol}, tg^{4J}, and tg^{5J}.[20–23] Mice homozygous for the original allele, *tg*, exhibit three distinct, recessively inherited neurological phenotypes: epilepsy, ataxia, and episodic (or paroxysmal) dyskinesia. The epilepsy component is similar to human absence (petit mal) epilepsy and consists of brief (1–10 s), bilaterally synchronous 5–6 Hz cortical spike-wave discharges with concomitant behavioral arrest.[24] These seizures begin at about 3 weeks of age and continue throughout life. Ataxia becomes apparent after 4 weeks of age and is continuous. The third component, paroxysmal dyskinesia, is an intermittent disorder with an onset approximately concurrent with the ataxia. It consists of sudden attacks that last 20 to 40 minutes, occur a few times per day, and are characterized by a transient progression of focal dystonia and repetitive limb shaking associated with abnormal activation of several deep brainstem pathways.[24] During dyskinesia episodes the EEG is unremarkable, further supporting a subcortical origin for this component of the phenotype.

Phenotypic heterogeneity is observed among the tottering alleles tg^{la} (leaner), tg^{rol}, and tg^{5J}. Leaner mice have more severe ataxia, undergo early and progressive degeneration of cerebellar granule, Golgi, and Purkinje cells,[25] and do not exhibit episodes of dyskinesia. Mice homozygous for tg^{rol} are intermediate in phenotype between *tg* and tg^{la}. The tg^{5J} mutation results in a phenotype similar to *tg*, but unlike other *tg* alleles, inheritance is dominant.

Histological examination of *tg/tg* brains revealed a normal-sized cerebrum with an increase in the number of noradrenergic axons in regions innervated by the locus ceruleus, including hippocampus, cerebellum, and dorsal lateral geniculate;[26,27] excessive infra- and supragranular mossy fibers in the dentate gyrus;[28] and a decreased volume of the molecular layer of the cerebellum.[29] Aberrant gene expression, channel density, and intracellular signaling in adult *tg/tg* brain have also been reported.[30–36] The neuropathology of the other tottering alleles is equally complex. The anterior lobe of the tg^{rol} cerebellum is preferentially affected, while in tg^{la} the whole cerebellum is grossly reduced in size. A wide variety of functional alterations have been identified in tg^{rol} mutants.[37]

Cellular excitability has been analyzed in tottering mice, revealing a conditional hyperexcitability of hippocampal neurons during potassium-evoked network bursting.[38,39] Excessive bursting and reduced afterhyperpolarizations in cells that show normal excitability when studied at rest is an emergent property that could favor abnormal synchronization and epilepsy when expressed in specific brain circuits. Evidence of this abnormal network behavior is present at the earliest appearance of the clinical seizure phenotype, and may reflect the contribution of multiple excitability mechanisms.[40,41]

β Subunit Calcium Channelopathies

The Lethargic (lh) *Mouse*

The lethargic locus encodes the voltage-dependent Ca^{2+} channel β_4 subunit, and is the first example of a mammalian neurological disease caused by an inherited defect in a non–pore-forming subunit of a voltage-gated ion channel.[10] The mutation identified in the β_4 gene consists of a four–base pair insertion into a splice donor site within the *Cacnb4* gene that produces two abnormally spliced mutant mRNA isoforms. Both of these result in translational frameshifts and a predicted β_4 protein that is truncated prior to the essential α_1 subunit interaction domain. The predicted β_4 protein is likely to be completely nonfunctional, and the *lh* mutation can thus be considered a β_4 null mutant. A second allele, lh^{2J}, mutated in the promoter region, appears to eliminate β_4 mRNA expression and results in a mutant phenotype identical to *lh*.[23] Lethargic mice exhibit a neurological disorder strikingly similar to that seen in tottering, including spike-wave epilepsy, ataxia, and paroxysmal dyskinesia. The onset of ataxia, however, is at least two weeks earlier in *lh/lh* than in *tg/tg* mice. In addition to the triad of neurological disorders, lethargic homozygotes show a slightly reduced body weight and transient defects in nonneuronal tissues including the immune system.[42] Neuropathological observations in lethargic mouse brain are limited compared with tottering, but also reveal a reduction in the size of the cerebellar molecular layer (D.L. Burgess, unpublished observations). Regional increases in the levels of $GABA_B$ receptor binding have been reported,[43] providing an example of the downstream responses that may occur in a brain attempting to compensate for a persistent physiological abnormality.

Other β *Subunit Mutants*

The β_1 subunit is widely coexpressed with the other three β subunits in brain, but is the only β subunit expressed in skeletal muscle. The importance of this subunit is demonstrated by the neonatal lethality exhibited in mice homozygous for a targeted knockout of the β_1 gene,[18] although this phenotype prevents our understanding of its role in brain. Interestingly, targeted inactivation of the β_3 subunit fails to produce a clinical phenotype in spite of measurable decreases in N-type Ca^{2+} currents studied in superior cervical ganglion neurons.[19] It remains to be determined whether a more severe phenotype would be expressed on a different genetic background.

Gamma Subunit Channelopathies

The Stargazer (stg) *Mouse*

The stargazer mutant phenotype shows early onset ataxia, and spike-wave seizures.[44] Although *Cacng2* expression in brain closely resembles that of *Cacna1a* and *Cacnb4*, stargazer mice, unlike *tg* and *lh* homozygotes, do not exhibit paroxysmal dyskinesia, and show frequent head tossing and an inability to swim that is indicative of a vestibular defect. The waggler mutation arose on the MRL/MpJ mouse strain and is allelic to stargazer but more mildly affected.[45] The stargazer locus encodes the neuronal Ca^{2+} channel γ_2 subunit gene with a predicted protein structure consisting of four membrane-spanning domains and a cytosolic C-terminus 100 amino acids longer than the distantly related muscle γ_1 subunit.[11] An Etn retrotransposon inserted into the second intron of the *Cacng2* gene results in either aberrant transcriptional termination or inefficient splicing, leading to a striking reduction of the *Cacng2* transcript in brain, with the possibility of some residual γ subunit function. The exact mutation in the waggler allele, though not yet identified, also causes a severe decrease in *Cacng2* mRNA expression.

Histology of the central nervous system in stargazer mice reveals no obvious neuronal loss at symptomatic ages; however, a striking outgrowth of dentate granule cell axons in the hippocampus is observed shortly after the onset of epilepsy, despite the absence of molecular indicators of immediate early gene expression or cellular stress in these neurons.[46–48] The novel expression of a peptide neurotransmitter, neuropeptide Y, is also detected in these cells.[47] Analysis of growth factor expression as a possible component of the synaptic reorganization in stargazer brain reveals mild selective increases of brain-derived nerve growth factor (BDNF) in the region of mossy fiber sprouting that are present only at the onset of sprouting. Interestingly, in the cerebellum, there is a near total loss of BDNF expression within cerebellar granule cells, and defective synaptic plasticity in cerebellar circuits.[49,50] Hyperexcitability is present in stargazer neocortical slices studied *in vitro*, and an increase in an inward rectifying current similar to I_h in deep-layer cortical neurons has also been described;[51] however, the relationship of these electrophysiological abnormalities to the calcium channel defect requires further investigation.

MUTANT CHANNEL BEHAVIOR

The mechanisms involved in translating neuronal calcium channel mutations into the observed clinical phenotypes cannot be fully resolved until the basic nature of the altered channel behavior within the relevant circuits is understood. To date, the direct effects of the various mutations on the relevant calcium current subtype have only been studied in heterologous cell systems, in acutely dissociated Purkinje cells, or in slice preparations from mutant brain. Although these studies provide important initial insights into the effect of the mutation on the channel, it may be incorrect to assume that an altered kinetic parameter will behave similarly in other types of neurons containing different modulatory pathways, or be equally evident in all cells where the mutant gene product is expressed.

In the leaner allele of tottering, electrophysiological studies demonstrate marked decreases in P-type Ca^{2+} current density in acutely dissociated tg^{la}/tg^{la} cerebellar Purkinje cells, with little evidence of alterations in the voltage dependency of activation or inactivation. Cell-attached patch clamp recordings revealed no differences in single-channel conductance, but a threefold reduction in open probability of the channels, suggesting that the mutation reduces P-type current density through kinetic alterations rather than by a reduction in the number of functional channels.[52,53] Preliminary studies in *tg/tg* Purkinje cells indicate that the biophysical properties of tottering P-type currents are unaltered,[54] hinting that the decreased P-type current density observed in tg^{la}/tg^{la} Purkinje cells may not be an essential constituent of the ataxia phenotype, which is present in both mutants.

Interestingly, four human α_{1A} subunit mutations underlying familial hemiplegic migraine (FHM) have been expressed in oocytes, and three of these show altered kinetics of channel gating, including faster inactivation and slower recovery from inactivation. These changes might be predicted to reduce calcium entry during repetitive stimulation.[55] One of these mutations, T666M, lies within 19 amino acids of the tottering mutation in the S5–6 linker region of domain II, suggesting the functional alterations may be similar in the mouse model. In contrast, one of the human FHM mutations, R192Q, lacked an observable effect on calcium channel gating. Similarly, no apparent alterations in P-type calcium currents have been detected in either tottering or lethargic Purkinje cells studied *in vitro.*[54,56] Coexpression of the stargazer γ_2 subunit in BHK cells did not alter the peak calcium current amplitude or current-voltage relationship, but instead accentuated the inhibition of the current by prepulses at negative potentials,[11] similar to the negative voltage shift of the steady state inactivation curve reported for the muscle γ subunit.[57] The effects of the γ_2 subunit mutation in stargazer neurons remain to be determined.

THE COMPLEX CHANNELOPATHY AND FUNCTIONAL Ca^{2+} CHANNEL PROPERTIES

The nature of the molecular defect in the lethargic mouse predicts a novel "multiplex" channelopathy that is mechanistically distinct from those arising from intrinsic α_1 subunit mutations. The complexity arises from the ability of the β_4 subunit to interact with any one of at least four neuronal Ca^{2+} channel α_1 subunits in addition to its preferred partner, α_{1A}. While preferential α_1/β pairings ($\alpha_1 + \beta_4 > \beta_{2a} > \beta_{1b} >> \beta_3$) are inferred from binding studies of *in vitro* translated subunits,[13] coimmunoprecipitation experiments demonstrate that each of the four known β subunits can associate with native L (α_{1C}, α_{1D}), N (α_{1B}), and P/Q-type (α_{1A}), channels in brain,[58–60] and are thus likely to regulate more than a single α_1 subtype. Other experiments show that heterogeneous α_1-β pairings can occur within a presumably homogenous population of cultured PC12 cells.[61] Thus, while mutations in α_1 subunit genes affect the function of a single specific channel type, mutation of a β subunit can simultaneously modify multiple channel types, and independently alter calcium current properties at various sites and developmental stages in the CNS.

SUBUNIT RESHUFFLING IN THE LETHARGIC β_4 MUTATION

Not only can the loss of a single interacting β subunit alter function in more than one channel subtype, but the potential replacement of the missing β by alternative, coexpressed β subunits may also endow the affected channels with novel kinetic or neuromod-

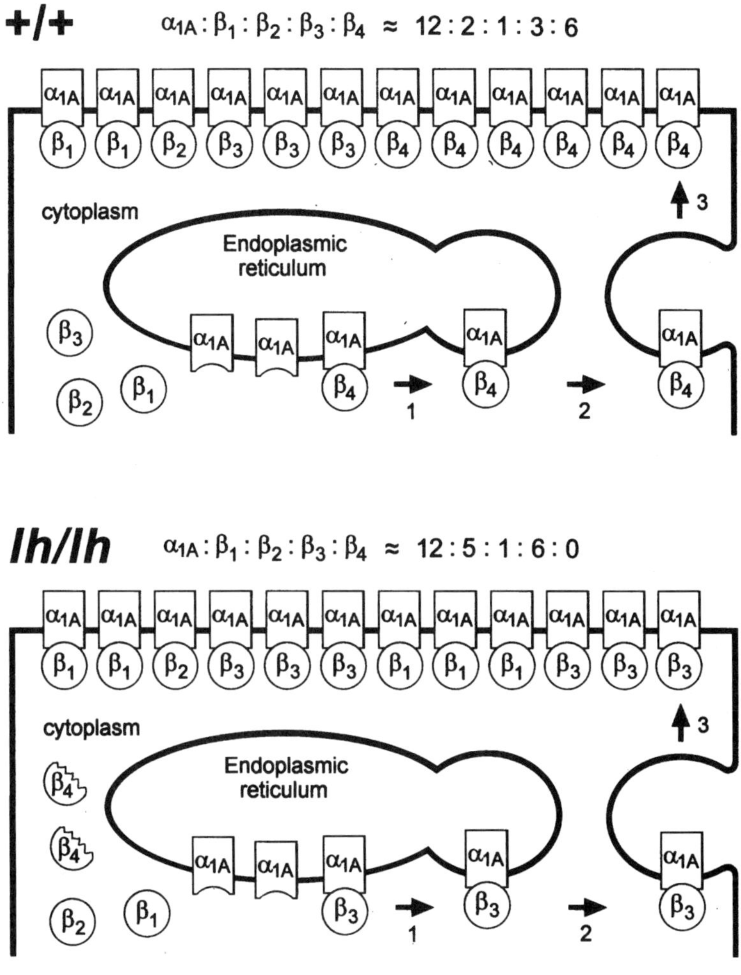

FIGURE 2. Ca^{2+} channel subunit reshuffling in lethargic brain. A simplified model is shown of hypothetical wild-type (+/+) and lethargic (*lh/lh*) central neurons expressing a single class of transmembrane α_1 subunit (α_{1A}) and four cytoplasmic β subunits (β_1, β_2, β_3, and β_4). In a wild-type neuron, α_{1A} associates in varying proportions with all four β subunits. In a lethargic neuron, the mutant β_4 protein cannot interact with α_{1A}, which is then available to form increased associations among the remaining β subunits. Although the total number of α_{1A}-containing channels remains constant, the ratio of specific α_{1A}-β subunit pairs is significantly altered. The α_{1A} and β subunit ratios shown are extrapolated from whole-brain coimmunoprecipitation analysis of wild-type and lethargic mice (D.L. Burgess, G.H. Biddlecome, S.I. McDonough, M.E. Diaz, C.A. Zilinski, B.P. Bean, K.P. Campbell, & J.L. Noebels, unpublished results). In an actual brain, the ratios in different neurons would be expected to vary from this average, depending on particular subunit expression profiles. The *arrows* indicate the essential role of β subunits in the subcellular trafficking of α_1 subunits. Association with the β subunit is believed to promote proper folding or stabilization of the α_1 protein (1), facilitating vesicle-mediated transport out of the endoplasmic reticulum (2) and expression of mature channels at the cell surface (3).

ulatory properties. Recent studies reveal that this reshuffling does indeed occur in brain missing a β subunit. Coimmunoprecipitation studies of lethargic mouse brain lacking the β_4 subunit reveal increased α_{1A} and α_{1B} steady state associations with the remaining β_1, β_2, and β_3 subunits, demonstrating the presence of novel calcium channel complexes.[56,62] Reshuffling is a posttranslational phenomena accomplished without measurable changes in mRNA transcription, and may explain the absence of detectable changes in resting P-type calcium currents in affected mutant cells[56] (FIG. 2). The reshuffling process is also predicted to be spatially and temporally nonuniform, since not all cells will provide the same ratio of available β subunits during brain development.

The general similarity of the tottering and lethargic phenotypes is consistent with a hypothesis that α_{1A} and β_4 are preferentially paired in wild-type mouse brain, as suggested by their overlapping brain mRNA expression profiles. The earlier onset of cerebellar ataxia in the lethargic mutant suggests that lack of β_4 from multiple channel subtypes in addition to the α_{1A}-containing P/Q-type, possibly in conjunction with its replacement by novel β_{1-3} combinations, may accelerate the appearance of at least one element of the neurological syndrome. Consequently, β channelopathies involving subunit reshuffling could contribute additional variation to an already complex phenotype.

CONCLUSION

Calcium ion channel mutations disrupt channel function and create recognizable disease phenotypes in the nervous system. The broad array of underlying cellular alterations is commensurate with the expanding genetic diversity of the voltage-gated calcium ion channel complex and its critical role in regulating cell function. Currently, 16 calcium channel genes are known, and mutations in 7 of these are associated with distinct inherited neurological disorders. These mutations provide new insight into the structure and function of the channels, and link specific subunits to cellular disease processes, including altered excitability, synaptic signaling, and cell death. Studies of mutant channel behavior, subunit interactions, and the differentiation of neural networks demonstrate unique patterns of downstream rearrangement. Developmental analysis of molecular plasticity in these mutants is a critical step to define the intervening mechanisms that translate aberrant ion channel behavior into the diverse clinical phenotypes observed. It remains to be determined which elements of each neurological phenotype are due to direct alterations in membrane excitability, synaptic integration, and neurotransmitter release in the mature nervous system; and which are secondary to complex downstream patterns of gene dysregulation during brain development. Site- and developmental stage–specific mutations of voltage-gated calcium channel subunits will help to identify the individual components of channel function that play a role in specific neurological phenotypes. Continued biological analysis of the nervous system in mutant mouse models of calcium channel disorders will increase our understanding of the cellular mechanisms underlying these neurological phenotypes in humans, and define critical molecular targets for therapeutic intervention.

ACKNOWLEDGMENTS

This work was supported by the American Epilepsy Society (D.L.B.); NIH Grant NS 29709 (J.L.N.), and The Blue Bird Foundation for Pediatric Neurology.

REFERENCES

1. Berridge, M.J. 1998. Neuronal calcium signaling. Neuron **21:** 13–26.
2. Ptacek, L.J., R. Tawil, R.C. Griggs, A.G. Engel, R.B. Layzer, H. Kwiecinski, P.G. McManis, L. Santiago, M. Moore & G. Fouad. 1994. Dihydropyridine receptor mutations cause hypokalemic periodic paralysis. Cell **77:** 863–868.
3. Beam, K.G., C.M. Knudson & J.A. Powell. 1986. A lethal mutation in mice eliminates the slow calcium current in skeletal muscle cells. Nature **320:** 168–170.
4. Bech-Hansen, N.T., M.J. Naylor, T.A. Maybaum, W.G. Pearce, B. Koop, G.A. Fishman, M. Mets, M.A. Musarella & K.M. Boycott. 1998. Loss-of-function mutations in a calcium-channel alpha1-subunit gene in Xp11.23 cause incomplete X-linked congenital stationary night blindness. Nat Genet **19:** 264–267.
5. Strom, T.M., G. Nyakatura, E. Apfelstedt-Sylla, H. Hellebrand, B. Lorenz, B.H. Weber, K. Wutz, N. Gutwillinger, K. Ruther, B. Drescher, C. Sauer, E. Zrenner, T. Meitinger, A. Rosenthal & A. Meindl 1998. An L-type calcium-channel gene mutated in incomplete X-linked congenital stationary night blindness Nat. Genet. **19**(3): 260–263.
6. Ophoff, R.A., G.M. Terwindt, M.N. Vergouwe, R. van Eijk, P.J. Oefner, S.M. Hoffman, J.E. Lamerdin, H.W. Mohrenweiser, D.E. Bulman, M. Ferrari, J. Haan, D. Lindhout, G.J. van Ommen, M.H. Hofker, M.D. Ferrari & R.R. Frants. 1996. Familial hemiplegic migraine and episodic ataxia type-2 are caused by mutations in the Ca^{2+} channel gene CACNL1A4. Cell **87**(3): 543–552.
7. Zhuchenko, O., J. Bailey, P. Bonnen, T. Ashizawa, D.W. Stockton, C. Amos, W.B. Dobyns, S.H. Subramony, H.Y. Zoghbi & C.C. Lee. 1997. Autosomal dominant cerebellar ataxia (SCA6) associated with small polyglutamine expansions in the alpha 1A voltage-dependent calcium channel. Nat. Genet. **15**(1): 62–69.
8. Fletcher, C.F., C.M. Lutz, T.N. O'Sullivan, J.D. Shaughnessy Jr., R, Hawkes, W.N. Frankel, N.G. Copeland & N.A. Jenkins. 1996. Absence epilepsy in tottering mutant mice is associated with calcium channel defects. Cell **87**(4): 607–617.
9. Doyle, J., X. Ren, G. Lennon & L. Stubbs. 1997. Mutations in the Cacnl1a4 calcium channel gene are associated with seizures, cerebellar degeneration, and ataxia in tottering and leaner mutant mice. Mamm. Genome **8:** 113–120.
10. Burgess, D.L., J.M. Jones, M.H. Meisler & J.L. Noebels. 1997. Mutation of the Ca^{2+} channel beta subunit gene *Cchb4* is associated with ataxia and seizures in the lethargic (*lh*) mouse. Cell **88:** 385–392.
11. Letts, V.A., R. Felix, G.H. Biddlecome, J. Arikkath, C.L. Mahaffey, A. Valenzuela, F.S. Bartlett 2nd., Y. Mori, K.P. Campbell & W.N. Frankel. 1998. The mouse stargazer gene encodes a neuronal Ca^{2+} channel gamma subunit Nat. Genet. **19:** 340–347.
12. Catterall, W.A. This volume.
13. De Waard, M., D.R. Witcher, M. Pragnell, H. Liu & K.P. Campbell. 1995. Properties of the α_1-β anchoring site in voltage-dependent Ca^{2+} channels. J. Biol. Chem. **270:** 12056–12064.
14. Witcher, D.R., M. De Waard, H. Liu, M. Pragnell & K.P. Campbell. 1995. Association of native Ca^{2+} channel β subunits with the α_1 subunit interaction domain. J. Biol. Chem. **270:** 18088–18093.
15. Jay, S.D., A.H. Sharp, S.D. Kahl, T.S. Vedvick, M.M. Harpold & K.P. Campbell. 1991. Structural characterization of the dihydropyridine-sensitive calcium channel alpha 2-subunit and the associated delta peptides. J. Biol. Chem. **266:** 3287–3293.
16. Ellis, S.B., M.E. Williams, N.R. Ways, R. Brenner, A.H. Sharp, A.T. Leung, K.P. Campbell, E. McKenna, W.J. Koch & A. Hui. 1988. Sequence and expression of mRNAs encoding the alpha 1 and alpha 2 subunits of a DHP-sensitive calcium channel. Science **241:** 1661–1664.
17. Jay, S.D., S.B. Ellis, A.F. McCue, M.E. Williams, T.S. Vedvick, M.M. Harpold & K.P. Campbell. 1990. Primary structure of the gamma subunit of the DHP-sensitive calcium channel from skeletal muscle. Science **248:** 490–492.
18. Gregg, R.G., A. Messing, C. Strube, M. Beurg, R. Moss, M. Behan, M. Sukhareva, S. Haynes, J.A. Powell, R. Coronado & P.A. Powers. 1996. Absence of the β subunit (*Cchb1*) of the skeletal muscle dihydropyridine receptor alters expression of the α_1 subunit and eliminates excitation-contraction coupling. Proc. Natl. Acad. Sci. USA **93:** 13961–13966.

19. SMITH, S.M., Y. NAMKUNG, R.H. SCHELLER, R.W. TSIEN & H.S. SHIN. 1998. Knock-out of the β_3 subunit reduces expression of N-type Ca^{2+} channel current in mouse dissociated sympathetic neurons. Biophys. J. **74:** 120.
20. GREEN, M.C. & R.L. SIDMAN. 1962. Tottering—A neuromuscular mutation in the mouse. J. Hered. **53:** 233–237.
21. TSUJI, S. & H. MEIER. 1971. Evidence for allelism of leaner and tottering in the mouse. Genet. Res. **17:** 83–88.
22. ODA, S.I. 1981. A new allele of the tottering locus, rolling mouse Nagoya, on Chromosome no. 8 in the mouse. Jpn. J. Genet. **56:** 295–299.
23. LUTZ, C.M., D.A. HOSFORD & W.N. FRANKEL. 1998. New mouse mutants harboring dominant and recessive mutations in the alpha-1A calcium channel and beta subunits. *In* collected abstracts of this conference, no. PI-28.
24. NOEBELS, J.L. & R.L. SIDMAN. 1979. Inherited epilepsy: Spike-wave and focal motor seizures in the mutant mouse tottering. Science **204:** 1334–1336.
25. HERRUP, K. & S.L. WILCZYNSKI. 1982. Cerebellar cell degeneration in the leaner mutant mouse. Neuroscience **7:** 2185–2196.
26. LEVITT, P. & J.L. NOEBELS. 1981. Mutant mouse tottering: Selective increase of locus ceruleus axons in a defined single-locus mutation. Proc. Natl. Acad. Sci. USA **78:** 4630–4634.
27. NOEBELS, J.L. 1984. A single gene error of noradrenergic axon growth synchronizes central neurones. Nature **310:** 409–411.
28. STANFIELD, B.B. 1989. Excessive intra- and supragranular mossy fibers in the dentate gyrus of tottering (tg/tg) mice. Brain Res. **480:** 294–299.
29. ISAACS, K.R. & L.C. ABBOTT. 1995. Cerebellar volume decreases in the tottering mouse are specific to the molecular layer. Brain Res. Bull. **36:** 309–314.
30. WILLOW, M., S.M. TAYLOR, W.A. CATTERALL & R.H. FINNELL. 1986. Down regulation of sodium channels in nerve terminals of spontaneously epileptic mice. Cell. Mol. Neurobiol. **6:** 213–220.
31. HESS, E.J. & M.C. WILSON. 1991. Tottering and leaner mutations perturb transient developmental expression of tyrosine hydroxylase in embryologically distinct Purkinje cells. Neuron **6:** 123–132.
32. PATEL, V.K., L.C. ABBOTT, A.K. RATTAN & G.A. TEJWANI. 1991. Increased methionine-enkephalin levels in genetically epileptic (tg/tg) mice. Brain Res. Bull. **27:** 849–852.
33. AUSTIN, M.C., M. SCHULTZBERG, L.C. ABBOTT, P. MONTPIED, J.R. EVERS, S.M. PAUL & J.N. CRAWLEY. 1992. Expression of tyrosine hydroxylase in cerebellar Purkinje neurons of the mutant tottering and leaner mouse. Brain Res. Mol. Brain Res. **15:** 227–240.
34. TEHRANI, M.H. & E.M. BARNES, JR. 1995. Reduced function of gamma-aminobutyric acidA receptors in tottering mouse brain: Role of cAMP-dependent protein kinase. Epilepsy Res. **22:** 13–21.
35. TEHRANI, M.H., B.J. BAUMGARTNER, S.C. LIU & E.M. BARNES, JR. 1997. Aberrant expression of GABAA receptor subunits in the tottering mouse: An animal model for absence seizures. Epilepsy Res. **28:** 213–223.
36. DE BARTOLOMEIS, A., V. KOPRIVICA, D. PICKAR, J.N. CRAWLEY & L.C. ABBOTT. 1997. Opioidergic and dopaminergic gene expression in the caudate-putamen and accumbens of the mutant mouse, tottering (tg/tg). Brain Res. Mol. Brain Res. **46:** 321–324.
37. MURAMOTO, O., I. KANAZAWA & K. ANDO. 1981. Neurotransmitter abnormality in rolling mouse Nagoya, an ataxic mutant mouse. Brain Res. **215:** 295–304.
38. HELEKAR, S. & J.L. NOEBELS. 1991. Synchronous hippocampal bursting unmasks latent network excitability alterations in an epileptic gene mutation. Proc. Natl. Acad. Sci. USA **88:** 4736–4740.
39. HELEKAR, S. & J.L. NOEBELS. 1994. Analysis of voltage-gated and synaptic conductances contributing to a gene-linked prolongation of depolarizing shifts in the epileptic mutant mouse tottering. J. Neurophysiol. **71:** 1–10.
40. HELEKAR, S.A. & J.L. NOEBELS. 1992. A burst-dependent excitability defect elicited by potassium at the developmental onset of spike-wave seizures in the tottering mutant. Dev. Brain Res. **65:** 205–210.
41. KOSTOPOULOS, G.K., & C.T. PSARROPOULOU. 1992. Possible mechanisms underlying hyperexcitability in the epileptic mutant mouse tottering. J. Neural Transm. Suppl. **35:** 109–124.

42. Dung, H.C. 1976. Relationship between the adrenal cortex and thymic involution in "lethargic" mutant mice. Am. J. Anat. **147:** 255–264.
43. Lin, F.H., Z. Cao & D.A. Hosford. 1993. Increased number of $GABA_B$ receptors in the lethargic (*lh/lh*) mouse model of absence epilepsy. Brain Res. **608:** 101–106.
44. Noebels, J.L., X. Qiao, R.T. Bronson, C. Spencer & M.T. Davisson. 1990. Stargazer: A new neurological mutant on chromosome 15 in the mouse with prolonged cortical seizures. Epilepsy Res. **7:** 129–135.
45. Sweet, H.O., R.T. Bronson, S. Cook, C.A. Spencer & M.T. Davisson. 1991. Waggler (wag). Mouse Genome **89:** 552.
46. Qiao, X. & J.L. Noebels. 1993. Developmental analysis of hippocampal mossy fiber outgrowth in a mutant mouse with inherited spike-wave seizures. J. Neurosci. **13:** 4622–4635.
47. Chafetz, R.S., W.K. Nahm & J.L. Noebels. 1995. Aberrant expression of Neuropeptide Y in hippocampal mossy fibers in the absence of local cell injury following the onset of spike-wave synchronization. Mol. Brain Res. **31:** 111–121.
48. Nahm, W. & J.L. Noebels. 1998. Non-obligate role of early or sustained expression of immediate-early gene proteins c-fos, c-jun and zif/268 in hippocampal mossy fiber sprouting. J. Neurosci. **18:** 9245–9255.
49. Qiao, X., F. Hefti, B. Knusel & J.L. Noebels. 1996. Selective failure of brain-derived neurotrophic factor mRNA expression in the cerebellum of stargazer, a mutant mouse with ataxia. J. Neurosci. **16:** 640–648.
50. Qiao, X., L. Chen, H. Gao, S. Bao, F. Hefti, R.F. Thompson & B. Knusel. 1998. Cerebellar brain-derived neurotrophic factor-TrkB defect associated with impairment of eyeblink conditioning in stargazer mutant mice. J. Neurosci. **18:** 6990–6999.
51. Di Pasquale, E., K.D. Keegan & J.L. Noebels. 1997. Increased excitability and inward rectification in layer V cortical pyramidal neurons in the epileptic mutant mouse Stargazer. J. Neurophysiol. **77:** 621–631.
52. Dove, L.S., L.C. Abbott & W.H. Griffith. 1998. Whole-cell and single-channel analysis of P-type calcium currents in cerebellar Purkinje cells of leaner mutant mice. J. Neurosci. **18:** 7687–7699.
53. Lorenzon, N.M., C.M. Lutz, W.N. Frankel & K.G. Beam. 1998. Altered calcium channel currents in Purkinje cells of the neurological mutant mouse leaner. J. Neurosci. **18:** 4482–4489.
54. Lorenzon, N.M., C.M. Lutz, W.N. Frankel, C.F. Fletcher, N.G. Copeland, N.A. Jenkins & K.G. Beam. 1997. Calcium currents in leaner and tottering mutant mice. Soc. Neurosci. Abstr. **23:** 783.1.
55. Kraus, R.L., M.J. Sinnegger, H. Glossmann, S. Hering & J. Streissnig. 1998. Familial Hemiplegic migraine mutations change α1A Ca^{2+} channel kinetics. J. Biol. Chem. **273:** 5586–5590
56. Burgess, D.L., G.H. Biddlecome, S.I. McDonough, M.E. Diaz, C.A. Zilinski, B.P. Bean, K.P. Campbell & J.L. Noebels. β Subunit reshuffling modifies N- and P/Q-type Ca^{2+} channel subunit compositions in lethargic mouse brain. Submitted.
57. Eberst, R., S. Dai, N. Klugbauer & F. Hofmann. 1997. Identification and functional characterization of a calcium channel gamma subunit. Pflugers Arch. **433:** 633–637.
58. Pichler, M., T.N. Cassidy, D. Reimer, H. Haase, R. Kraus, D. Ostler & J. Striessnig. 1997. Beta subunit heterogeneity in neuronal L-type Ca^{2+} channels. J. Biol. Chem. **272:** 13877–13882.
59. Scott, V.E., M. De Waard, H. Liu, C.A. Gurnett, D.P. Venzke, V.A. Lennon & K.P. Campbell. 1996. Beta subunit heterogeneity in N-type Ca^{2+} channels. J. Biol. Chem. **271:** 3207–3212.
60. Liu, H., M. De Waard, V.E.S. Scott, C.A. Gurnett, V.A. Lennon & K.P. Campbell. 1996. Identification of three subunits of the high affinity omega-conotoxin MVIIC-sensitive Ca^{2+} channel. J. Biol. Chem. **271:** 13804–13810.
61. Liu, H., R. Felix, C.A. Gurnett, M. De Waard, D.R. Witcher & K.P. Campbell. 1996. Expression and subunit interaction of voltage-dependent Ca^{2+} channels in PC12 cells. J. Neurosci. **16:** 7557–7565.
62. McEnery, M.W., T.D. Copeland & C.L. Vance. 1998. Altered expression and assembly of N-type calcium channel alpha1B and beta subunits in epileptic lethargic (*lh/lh*) mouse. J. Biol. Chem. **273:** 21435–21438.
63. Meir, A. & A.C. Dolphin. 1998. Known calcium channel alpha1 subunits can form low threshold small conductance channels with similarities to native T-type channels. Neuron. **20:** 341–351.

64. FISHER, S.E., A. CICCODICOLA, K. TANAKA, A. CURCI, S. DESICATO, M. D'URSO & I.W. CRAIG. 1997. Sequence-based exon prediction around the synaptophysin locus reveals a gene-rich area containing novel genes in human proximal Xp. Genomics **45:** 340–347.
65. PEREZ-REYES, E., L.L. CRIBBS, A. DAUD, A.E. LACERDA, J. BARCLAY, M.P. WILLIAMSON, M. FOX, M. REES & J.H. LEE. 1998. Molecular characterization of a neuronal low-voltage-activated T-type calcium channel. Nature **391:** 896–900.
66. SEINO, S., Y. YAMADA, R. ESPINOSA, B.M. LE & G.I. BELL. 1992. Assignment of the gene encoding the α_1 subunit of the neuroendocrine/brain-type calcium channel (CACNL1A2) to human chromosome 3, band p14.3. Genomics **13:** 1375–1377.

Lambert-Eaton Antibodies Promote Activity-Dependent Enhancement of Exocytosis in Bovine Adrenal Chromaffin Cells

KATHRIN L. ENGISCH,[a,b] MARK M. RICH,[c,d] NOAH COOK,[a] AND MARTHA C. NOWYCKY[a,e]

[a]*Department of Neurobiology and Anatomy, Allegheny University of the Health Sciences, MCP and Hahnemann University, 3200 Henry Avenue, Philadelphia, Pennsylvania, 19129, USA*

[c]*Department of Neuroscience, University of Pennsylvania School of Medicine, Philadephia, Pennsylvania 19129, USA*

Patients with the Lambert-Eaton myasthenic syndrome (LEMS) have an impairment in neuromuscular transmission that is thought to be caused by antibodies to presynaptic calcium channels.[1–4] Although the initial compound muscle action potential (a measure of transmitting neuromuscular junctions) is abnormally small, it is greatly increased by repetitive stimulation.[3,5] Accumulation of Ca^{2+} ions is often proposed to mediate activity-dependent facilitation. We have previously demonstrated, however, that bovine chromaffin cells display a form of short-term enhancement of secretion during stimulus trains that is not due to Ca^{2+} accumulation.[6] Therefore we used chromaffin cells as a model system to examine the mechanism by which Lambert-Eaton antibodies can affect exocytosis evoked by trains of depolarizing pulses.

Ca^{2+} currents in adult bovine adrenal chromaffin cells are primarily carried by P/Q- and N-type Ca^{2+} channels. We have operationally defined P/Q-type current and N-type current as the plateau and difference current, respectively, during a 320-ms depolarization (–90 to +20 mV). This characterization is based on the sensitivities of these components to the channel antagonists ω-agatoxin-IVa and ω-conotoxin GVIa. As reported previously IgGs from different patients had different effects.[7] The five LEMS IgGs inhibited the P/Q-type current by variable amounts: LEMS 1, 28%; LEMS 3, 43%, LEMS 4, 17%; LEMS 5, 53%, and LEMS 7, 25%. LEMS 1 and LEMS 3 also inhibited N-type current, each by 24%.

In bovine adrenal chromaffin cells exocytosis evoked by single-step depolarizations is a nonlinear function of the integral of the Ca^{2+} current:[6,8]

$$\text{Exocytosis} = 0.147 * (\text{Ca}^{2+}\text{ current integral})^{1.49}$$

(with exocytosis in fF of membrane capacitance and current integral in pC). Treatment of bovine chromaffin cells with LEMS IgG resulted in smaller exocytotic responses. The inhibition of exocytosis was proportional to the reduction in Ca^{2+} entry in LEMS-treated cells. The Ca^{2+} dependence of exocytosis described previously was not altered by any of

[b]Corresponding author. Present address: Kathrin L. Engisch, Ph.D., Department of Physiology, Emory University School of Medicine, Atlanta, Georgia 30322. Phone: 404-727-6710; fax: 404-727-2648; e-mail: kengisch@physio2.physio.emory.edu

[d]Present address: Mark M. Rich, M.D., Ph.D., Department of Neurology, Emory University School of Medicine, Atlanta, Georgia 30322.

[e]Present address: Martha C. Nowycky, Ph.D., Department of Physiology and Pharmacology, UMDNJ, Newark, New Jersey 07103-2714.

the five LEMS IgGs, whether the IgG inhibited both N- and P/Q-type current, or the P/Q-type current in isolation.

When bovine adrenal chromaffin cells are stimulated by trains of depolarizations there are short-term, activity-dependent changes in the Ca^{2+}–exocytosis relationship.[8] In FIGURE 1B the single-pulse input–output relationship is shown as a dotted curve; for some stimulus trains the Ca^{2+}–exocytosis relationship adheres to this curve (FIG. 1Bii). Often the Ca^{2+}-exocytosis relationship becomes enhanced during trains (FIG. 1Bi); this behavior is

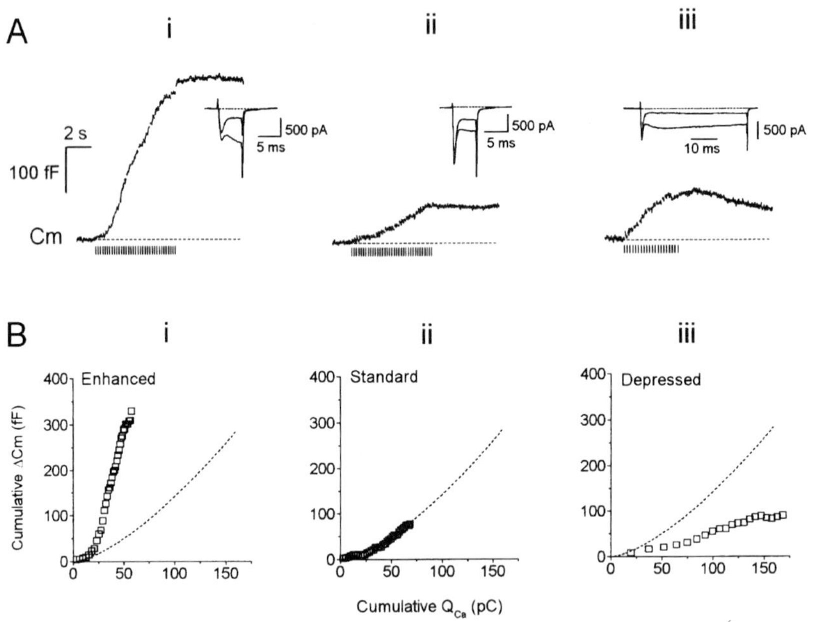

FIGURE 1. Examples of secretory behaviors observed during repetitive stimulation in control adrenal chromaffin cells. (**A**) Exocytosis was monitored as an increase in membrane capacitance (Cm) in perforated patch whole-cell voltage-clamp recordings. Stimulus trains of depolarizing pulses (from –90 mV to +20 mV) were applied at 200-ms intervals. The timing of the depolarizations are indicated by *gaps* and *vertical lines* beneath the traces. (**i**) A large increase in Cm evoked by a train of 5-ms pulses. (**ii**) A similar protocol of 5-ms depolarizations evokes a much smaller change in Cm during a train applied to a different cell. (**iii**) In a third cell a train of 40-ms depolarizations does not evoke substantially greater exocytosis than the train in (ii). **Insets:** Inward currents evoked by the first and last depolarization of the stimulus trains, illustrating inactivation of Ca^{2+} currents and the absence of a facilitation Ca^{2+} current. (**B**) For the traces shown in A, Cm increases summed over the stimulus train, plotted as a function of cumulative integrated Ca^{2+} entry. (**i**) The Ca^{2+}–exocytosis relationship for the response shown in Ai abruptly shifts after 7 pulses to an enhanced Ca^{2+}–exocytosis relationship, relative to the standard input–output relationship derived using single pulses[6,8] (*dashes*). (**ii**) The amount of exocytosis during the train in Aii has the same relationship to integrated Ca^{2+} entry as the average response evoked by single prolonged depolarizing pulses (*standard curve*; *dashes*). Note that a similar amount of total Ca^{2+} entry occurred in this and the cell in (Ai) (60 pC), but very different amounts of exocytosis were evoked. (**iii**) A depressed Ca^{2+}–exocytosis relationship was evoked by the train of 40-ms depolarizations shown in Aiii. As a result, little exocytosis is evoked despite total Ca^{2+} entry >150 pC.

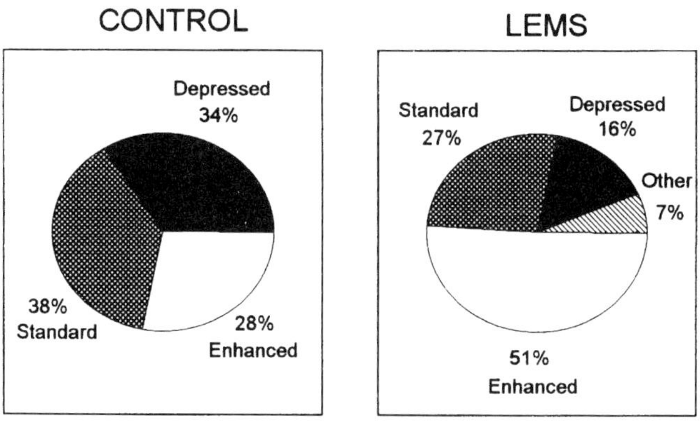

FIGURE 2. Enhancement during repetitive stimulation is more likely after treatment with LEMS IgGs. Pie graphs depicting the distribution of response behaviors evoked by trains of 10-ms depolarizing pulses applied at 200-ms intervals. Control data are plotted from Ref. 6, with $n = 42$ trains. LEMS data are from cells treated with any of the five LEMS IgGs; there are $n = 45$ trains. Enhanced (*white sections*), standard (*gray sections*), and depressed (*black sections*) Ca^{2+}–exocytosis relationships are defined as being above, on, or below the standard single-pulse relationship,[6,8] respectively. A novel category, other, is shown as a *striped section*, and includes trains that exhibited endocytosis, or docked secretion (exocytosis above standard curve initially followed by depression) or threshold secretion (exocytosis below standard curve initially, with exocytosis occurring late in the stimulus train).

usually associated with trains of very brief depolarizations (5–10 ms) or small amplitude currents. In the remaining trains exocytosis is depressed (FIG. 1Biii); this is most common for long-duration (40-ms) pulses. Enhancement is not the result of Ca^{2+} accumulation, because (1) greater Ca^{2+} accumulation occurs during trains of longer duration pulses, and (2) enhancement is prevented when the interpulse interval is shortened from 200 ms to 44 ms.[6]

We examined the secretory behaviors during stimulus trains in cells exposed to LEMS IgGs (FIG. 2). The percent of trains exhibiting enhancement was dramatically increased (white section) for trains of 10-ms pulses. The enhancement observed in LEMS-treated cells is not explained by Ca^{2+} accumulation, since the antibodies reduce Ca^{2+} entry per pulse. The increased frequency of observing enhancement is not due only to fewer trains showing depression, but also to a lower percent of trains with a standard Ca^{2+}–exocytosis relationship. These results suggest that the reduction in Ca^{2+} entry caused by treatment with LEMS antibodies may, via activation of enhancement, paradoxically lead to an overall increase in the amount of exocytosis observed during stimulus trains.

REFERENCES

1. KIM, Y.I. & E. NEHER. 1988. IgG from patients with Lambert-Eaton syndrome blocks voltage-dependent calcium channels. Science **239:** 405–408.
2. VINCENT, A., B. LANG & J. NEWSOM-DAVIS. 1989. Autoimmunity to the voltage-gated calcium channel underlies the Lambert-Eaton myasthenic syndrome, a paraneoplastic disorder. Trends Neurosci. **12:** 496–502.
3. SANDERS, D.B. 1994. Lambert-Eaton myasthenic syndrome: Pathogenisis and treatment. Semin. Neurol. **14:** 111–117.

4. LENNON, V.A., T.J. KRYZER, G.E. GRIESMANN *et al.* 1995. Calcium-channel antibodies in the Lambert-Eaton syndrome and other paraneoplastic syndromes. N. Engl. J. Med. **332:** 1467–1474.
5. ELMQVIST, D. & E.H. LAMBERT. 1968. Detailed analysis of neuromuscular transmission in a patient with the myasthenic syndrome sometimes associated with bronchogenic carcinoma. Mayo Clin. Proc. **43:** 689–713.
6. ENGISCH, K.L., N.I. CHERNEVSKAYA & M. C. NOWYCKY. 1997. Short-term changes in the Ca^{2+}-exocytosis relationship during repetitive pulse protocols in bovine adrenal chromaffin cells. J. Neurosci. **17:** 9010–9025.
7. GARCIA, K.D. & K.G. BEAM. 1996. Reduction of calcium currents by Lambert-Eaton syndrome sera: Motoneurons are preferentially affected, and L-type currents are spared. J. Neurosci. **16:** 4903–4913.
8. ENGISCH, K.L. & M.C. NOWYCKY. 1996. Calcium dependence of large dense-cored vesicle exocytosis evoked by calcium influx in bovine adrenal chromaffin cells. J. Neurosci. **16:** 1359–1369.

L-Type Calcium Channel Regulation of Abnormal Tyrosine Hydroxylase Expression in Cerebella of Tottering Mice

B. E. FUREMAN, D. B. CAMPBELL, AND E. J. HESS[a]

Department of Neuroscience and Anatomy, Pennsylvania State University College of Medicine, 500 University Drive, Hershey, Pennsylvania 17033, USA

Activity-dependent gene expression in the central nervous system depends on the proper regulation and function of voltage-dependent calcium channels (VDCCs). Calcium influx through VDCCs at the cell soma and proximal dendrites modifies a number of cellular processes via second messenger systems acting on genes and proteins with calcium-responsive properties. A role for VDCC-mediated changes in gene expression has been demonstrated *in vitro*.[1] By contrast, the *in vivo* regulation of gene expression by VDCCs has not been extensively studied.

We have used the neurological mouse mutant tottering (*tg*) as a model system for studying the effects of VDCC misregulation on calcium-responsive gene expression. The *tg* mutation alters the α_{1A} subunit gene, which encodes P/Q-type calcium channels.[2] Recently, we have demonstrated that *tg* mice exhibit an upregulation of L-type calcium channels in the cerebellum.[3] An abnormal cellular phenotype has also been identified in the cerebellum of *tg* mutant mice; the rate-limiting enzyme in catecholamine biosynthesis, tyrosine hydroxylase (TH), is ectopically expressed in cerebellar Purkinje cells of adult *tg* mice.[4] Here, we test the hypothesis that abnormal TH expression in *tg* mouse cerebellar Purkinje cells is regulated by L-type calcium channels. The results suggest that the upregulation of L-type channels in *tg* Purkinje cells is responsible for abnormal TH gene expression. In addition, the reduction in TH mRNA in response to L-type antagonist administration demonstrates that calcium responsive gene expression is amenable to pharmacologic manipulation *in vivo*.

MATERIAL AND METHODS

Adult *tg* mice aged 16–20 weeks were identified either by analysis of PCR-amplified simple sequence length polymorphisms in C57BL/6 +/*tg* × C57BL/6 +/*tg* cross-progeny,[5] or by the absence of oligosyndactylism in *Os* +/+ tg × *Os* +/+ *tg* cross-progeny. Age- and gender-matched C57BL/6 +/*tg* or *Os* +/+ *tg* mice were used as controls. Adult *tg* mice and heterozygous controls were injected with the specific L-type calcium channel antagonist nimodipine twice daily for a two-week period. Nimodipine (Research Biochemicals International) was prepared at 2 mg/mL in 14.5% ethanol/ 2.25% Tween80/ 83.25% 0.9% saline. The mice received a 5 mL/kg volume of nimodipine or vehicle twice daily at 12-hour intervals (20 mg/kg/day) for a total of 14 days. *In situ* hybridization using standard

[a]Corresponding author. Phone: 717-531-4440; fax: 717-531-5184; e-mail: ehess@psu.edu

TABLE 1. Effect of Chronic Nimodipine Administration on TH Expression in Cerebellar Purkinje Cells of *tg* Mice[a]

Treatment Group	Grain Density
Vehicle-treated *tg*	2757 ± 209
Nimodipine-treated *tg*	1877 ± 181[b]

[a]Data were quantified using NIH Image and analyzed using paired *t*-test. Grain densities are expressed as means ± SEM.

[b]Denotes $p < 0.0001$.

protocols with a cRNA probe transcribed from a 1.7-kb fragment of the mouse TH gene was used to assess TH gene expression in fresh frozen sagittal sections. Mean grain density was quantified using NIH Image software from dark-field images of emulsion-dipped slides after normalizing for threshold variability. Mean grain density was measured as square pixels remaining after the threshold was equalized. Counts of mean grain density were analyzed using a paired *t*-test.

RESULTS AND DISCUSSION

Chronic blockade of L-type calcium channels with nimodipine significantly decreased the abnormal expression of TH in *tg* mouse cerebellar Purkinje cells (TABLE 1). The *tg* mutation affects P/Q-type calcium channels, which are abundantly expressed in the cerebellum. Phenotypes resulting from the *tg* mutation include synchronous cortical spike-wave discharges, intermittent motor convulsions, and TH expression in cerebellar Purkinje cells. Although the effect of the *tg* mutation on P/Q-type function is unknown, we have previously demonstrated that an upregulation of L-type calcium channels in cerebellum likely underlies the convulsive phenotype displayed by *tg* mutant mice.[3] The reduction in cerebellar TH expression following nimodipine treatment suggests that ectopic gene expression in *tg* mice is a result of increased calcium influx due to L-type calcium channel upregulation. TH expression in Purkinje cells of *tg* mice represents a cellular phenotype resulting from secondary effects of the P/Q-type calcium channel mutation.

The ability to manipulate gene expression *in vivo* using calcium channel antagonists represents a new area of study in the regulation and function of voltage-gated calcium channels. Although calcium responsive gene expression has been demonstrated *in vitro*, this study is the first to demonstrate that TH gene expression can be manipulated *in vivo* using an L-type calcium channel antagonist. The abnormal TH expression in *tg* mice presents an excellent model system for parallel studies of voltage-gated calcium channel regulation and effects on calcium responsive gene expression.

REFERENCES

1. Brosenitsch, T.A., D. Salgado-Commissariat, D.L. Kunze & D.M. Katz. 1998. A role for L-type calcium channels in developmental regulation of transmitter phenotype in primary sensory neurons. J. Neurosci. **18:** 1047–1055.
2. Fletcher, C.F., C.M. Lutz, T.N. O'Sullivan, J.D. Shaughnessy, Jr., R. Hawkes, W.N. Frankel, N.G. Copeland & N.A. Jenkins. 1996. Absence epilepsy in tottering mutant mice is associated with calcium channel defects. Cell **87:** 607–617.
3. Campbell, D.B. & E.J. Hess. L-type channels contribute to the tottering mouse dystonic episodes. Mol. Pharmacol. In press.
4. Hess, E.J. & M.C. Wilson. 1991. Tottering and leaner mutations perturb transient developmental expression of tyrosine hydroxylase in embryologically distinct Purkinje cells. Neuron **6:** 123–132.
5. Campbell, D.B. & E.J. Hess. 1997. Rapid genotyping of mutant mice using dried blood spots for polymerase chain reaction (PCR) analysis. Brain Res. Prot. **1:** 117–123.

Okadaic Acid Antagonizes the Inhibitory Action of Internal GTP-γ-S on the Calcium Current of Dorsal Raphe Neurons but Not That of 5-HT$_{1A}$ Receptor Activation

JOHN S. KELLY[a] AND HRVOJE HEĆIMOVIĆ

Department of Pharmacology, The University of Edinburgh, 1 George Square, Edinburgh, EH8 9JZ, United Kingdom

A host of neurotransmitters have now been shown to inhibit voltage-dependent calcium currents (HVA) in a whole variety of neuronal preparations.[1] The inhibition is incomplete (usually around 50%) and associated with a substantial slowing of activation and a partial relaxation of current amplitude towards control levels at the end of a 100–150 ms depolarization. The inhibition is G-protein mediated and can be mimicked and made irreversible by an application of the nonhydrolyzable analogue of GTP, GTP-γ-S. For many cell types, it is believed the G-protein acts directly on the calcium channel proteins, rather than via a freely diffusible second messenger.[2] In particular, Dolphin and others have suggested that it is a specific interaction of the G-protein with the channel protein subunit that modulates the speed of channel opening.[1]

Other experiments have shown the transmitter-mediated inhibition of the HVA to be reduced by large depolarizations and towards the end of the test pulse. These findings have led to the suggestion that the modulation of opening by an interaction of the G-protein with the channel proteins is voltage dependent and that facilitation by a depolarizing prepulse and slowed activation by a transmitter or GTP-γ-S all reflect changes in the binding and unbinding of the G-protein with channel protein subunits. Recently, Zamponi *et al.*[3] suggested that endogenous phosphorylation of a G-βγ binding site on the α_1 subunit, I-II linker domain, can antagonize GTP-βγ–induced inhibition. Chen and Penington[4] have shown protein kinase C activation by 4β-phorbol 12-myristate, 13-acetate (PMA), to partially block the inhibitory effect of 5-HT$_{1A}$ receptor activation by serotonin on the calcium current of freshly isolated dorsal raphe neurons. Using an identical protocol, we have now shown phosphatase 1 inhibition, by internal perfusion with okadaic acid (OA), to have the opposite effect. OA substantially blocked the inhibitory action of internal perfusion with GTP-γ-S and left the inhibition evoked by activation of the 5-HT$_{1A}$ receptor by 8-OH-DPAT unaltered.

METHODS

Dorsal raphe neurons were acutely isolated, as described by Penington *et al.*[2] Coronal slices (400 μm) containing the dorsal raphe nucleus were prepared from young (3 to 6 week-old) male Wistar rats and incubated in PIPES buffer containing 0.05% (w/v) trypsin

[a]Phone: 44 131 650 3519; fax: 44 131 667 9381; e-mail: jskelly@ed.ac.uk

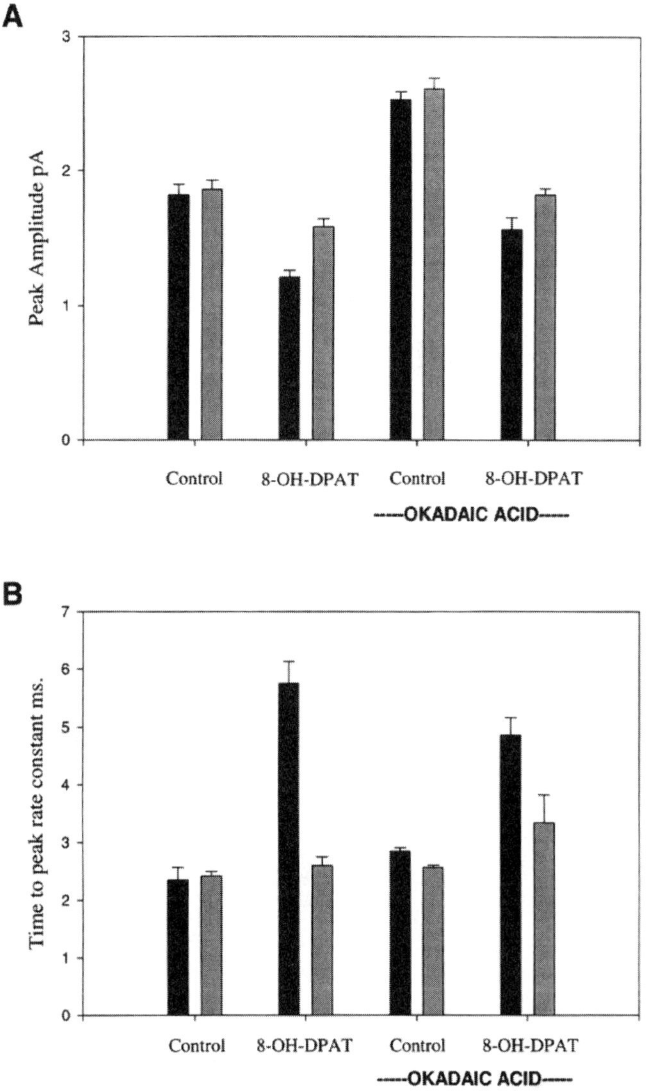

FIGURE 1. Histograms showing the lack of effect of okadaic acid on 8-OH–DPAT–mediated inhibition of I_{Ba} peak amplitude, the slowing of the rate of I_{Ba} activation, and the modest attenuating action of OA on depolarizing prepulse facilitation. In **A** and **B**, histograms show OA to increase I_{Ba} peak amplitude (control) without significantly modifying the rate of I_{Ba} activation. The second histogram of each paired histogram shows the result obtained following a depolarizing prepulse. In the absence and presence of OA, 8-OH-DPAT significantly reduced I_{Ba} peak amplitude and slowed the rate of I_{Ba} activation. However, in the presence of OA the depolarizing prepulse facilitation failed to reverse the 8-HO-DPAT–evoked reduction in peak amplitude and only partially restored the rate of I_{Ba} activation.

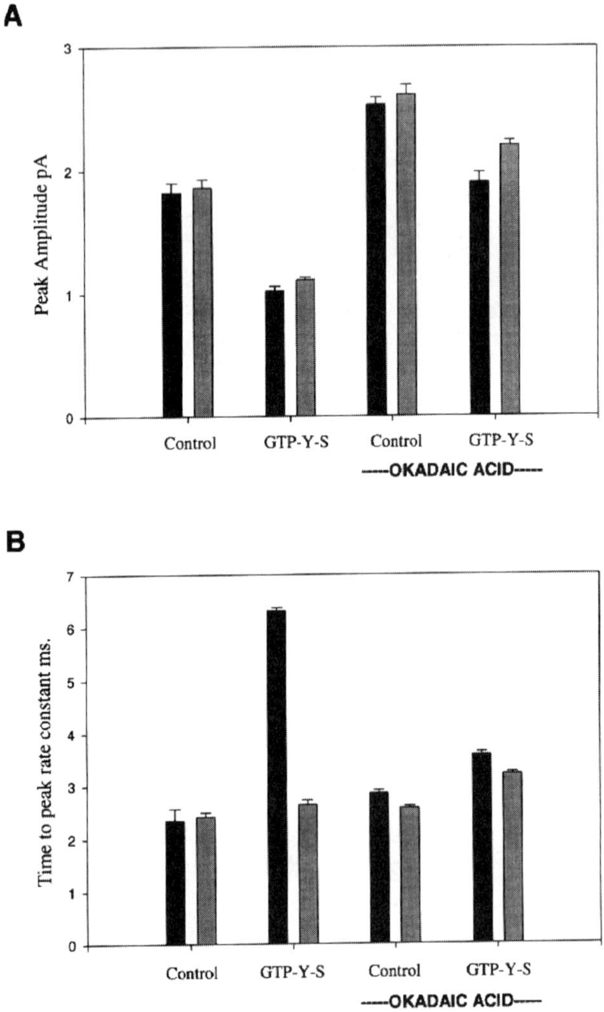

FIGURE 2. Histograms showing that OA greatly attenuated GTP-γ-S–mediated inhibition of I_{Ba} peak amplitude and completely blocked the GTP-γ-S–mediated slowing of the rate of I_{Ba} activation. In the absence of OA there was no depolarizing prepulse facilitation of the decreased I_{Ba} amplitude evoked by GTP-γ-S; in contrast the slowing of the rate of I_{Ba} activation was completely restored. The control histograms are from the same data as in FIG. 1. In the absence of OA, GTP-γ-S substantially reduced the peak amplitude of the I_{Ba} and greatly slowed the rate of I_{Ba} activation. However, in the presence of OA the inhibition of I_{Ba} peak amplitude was very much less than that seen with 8-OH-DPAT in FIG. 1, and there was no slowing of the rate of I_{Ba} activation as occurred with 8-OH-DPAT. In the absence of OA, the restorative effects of a depolarizing prepulse are striking; the slowed rate of I_{Ba} activation evoked by GTP-γ-S is totally reversed; however, no equivalent change occurs in I_{Ba} peak amplitude.

(Sigma Type XI) at 37 °C for 90 min, and then washed in fresh PIPES buffer before being resuspended in PIPES buffer. Individual dorsal raphe cells were then isolated by gentle trituration. After the whole-cell patch mode was established, voltage-dependent Ba^{2+} currents (I_{Ba}) were recorded by suppression of voltage-activated Na^+ and K^+ currents by changing the external bathing solution from Tyrode solution to one containing 138 mM TEACl, 5 mM $BaCl_2$, 5 mM 4-AP, 20 mM HEPES-Tris-OH, and 20 mM sucrose, pH 7.4. The pipette solution contained 30 mM TEACl, 28 mM Trizma-EGTA, 70 mM Trizma-PO_4, 2 mM MgATP, and 300 µM GTP at pH 7.2. In the whole-cell patch mode, isolated neurons were held voltage clamped at –110 mV, and peak I_{Ba} currents were elicited by a voltage step, 150 ms in duration, to –10 mV; 100 ms conditioning prepulses were applied 10 ms prior to the test pulse as a step from the holding potential of –100 mV to +40 mV. The individual experiments and the data analysis were conducted as described in McAllister-Williams and Kelly.[5] The currents were leak and capacity subtracted.

RESULTS AND CONCLUSIONS

Internal perfusion with 1 µM OA increased the amplitude of the peak I_{Ba} from 1.82 to 2.53 pA and marginally slowed the onset rate constant from 2.53 to 2.85 milliseconds. At 20 °C, 50 µM 8-OH-DPAT inhibited both the control and the OA-enhanced I_{Ba} by 34%, and in OA the onset rate constant was slowed 1.7-fold from 2.85 ms (n = 45) to 4.87 ms (n = 16) (FIG. 1). Perfusion of the cells with 200 µM GTP-γ-S produced a 44% inhibition and slowed the onset rate constant 2.7-fold from 2.35 ms (n = 34) to 6.32 ms (n = 10) (FIG. 2). However, when the cells were perfused with OA and GTP-γ-S, the OA-enhanced I_{Ba} was reduced by only 25%, and the onset rate constant increased only 1.3-fold from 2.85 (n = 45) to 3.58 ms (n = 7). One possible explanation of the paradox is that phosphorylation evoked by PMA and endogenous phosphorylation preserved by OA act differentially at multiple GTP-βγ binding sites.

ACKNOWLEDGMENT

This work was funded by the Wellcome Trust.

REFERENCES

1. DOLPHIN, A. C. 1998. Mechanisms of modulation of voltage-dependent calcium channels by G-proteins. J. Physiol. Lond. **506:** 3–11.
2. PENINGTON, N. J., J. S. KELLY & A. P. FOX. 1991. A study of the mechanism of Ca^{2+} current inhibition produced by serotonin in rat dorsal raphe neurons. J. Neurosci. **11:** 3594–3609.
3. ZAMPONI, G. W., E. BOURINET, D. NELSON, J. NARGEOT, & T. P. SNUTCH. 1997. Crosstalk between G proteins and protein kinase C mediated by calcium channel α_1 subunit. Nature **385:** 442–446.
4. CHEN, Y. & N. P. PENINGTON. 1996. Differential effects of PKC activation on 5-HT_{1A} receptor coupling to Ca^{2+} and K^+ currents in rat serotoninergic neurones. J. Physiol. Lond. **496:** 129–137.
5. MCALLISTER-WILLIAMS, R. H., & J. S. KELLY. 1995. The temperature dependence of high-threshold calcium currents recorded from adult rat dorsal raphe neurones. Neuropharmacology **34:** 1479–1490.

Sensitivity to Conotoxin Block of Splice Variants of Rat α_{1B} (rbBII) Subunit of the N-Type Calcium Channel Coexpressed with Different β Subunits in *Xenopus* Oocytes

H. J. MEADOWS[a] AND C. D. BENHAM

Neurosciences Research, SmithKline Beecham Pharmaceuticals, New Frontiers Science Park, Third Avenue, Harlow, Essex CM19 5AW, UK

Omega-conotoxins GVIA and MVIIC, which both effectively block N-type calcium channels, have been shown to have differential effects on sympathetic nerve–mediated contractions of guinea-pig ileum and rat vas deferens.[1] Omega-conotoxin MVIIC was 100-fold less active at vas deferens than at ileum, while conotoxin GVIA was equipotent on both preparations, indicating the possible existence of different subtypes of N-type calcium channels with different toxin sensitivities. A number of variants of the α_{1B} subunit of the N channel have been identified in both rat brain[2] and sympathetic ganglia with differing patterns of expression.[3] One of these variants of rbBII[4] had a 4–amino acid insert (SFMG) in domain III S3-S4, an extracellular site, and hence a good candidate for the modification of conotoxin binding. In this study we investigated whether this SFMG insert had any effect on toxin sensitivity.

In addition to variants of the α_{1B} subunit, N-type channels in rabbit brain have been shown to differ in their β subunit composition, and it has been suggested that this may account for some of the functional diversity of these channels.[5] Therefore, we also investigated whether altering the β subunit expressed in combination with α_{1B} and α_2 had any effect on conotoxin sensitivity.

Dissociated *Xenopus* oocytes were injected with cDNA (pMT2 vector) for the subunit combination of interest and incubated at 22°C. Calcium channel currents were recorded 5–7 days later using standard two-electrode voltage clamp techniques with 10 mM Ba^{2+} as the charge carrier. Omega-conotoxin MVIIA (which exhibits very similar pharmacology to GVIA) or MVIIC was bath applied and left in contact with the oocyte for 3 minutes. Percentage inhibition of peak inward current was measured at this time. Concentration-response curves were constructed to omega-conotoxins MVIIA and MVIIC on both isoforms of the α_{1B} subunit in combination with α_2 and β_3 subunits.

Omega-conotoxin MVIIA showed no significant difference in sensitivity between the two α_{1B} isoforms ($\alpha_{1B} + \alpha_2 + \beta_3$ IC50 18 nM; α_{1B}(+SFMG) + $\alpha_2 + \beta_3$ IC50 21 nM; FIG. 1A). A small change in MVIIC sensitivity was observed between the two isoforms ($\alpha_{1B} + \alpha_2 + \beta_3$ IC50 71 nM; α_{1B}(+SFMG) + $\alpha_2 + \beta_3$ IC50 42 nM; FIG. 1B), but only of the order of 2-fold, and therefore not great enough to account for the 100-fold difference in sensitivity observed by Boot[1] for MVIIC on sympathetic nerve–mediated contractions between guinea-pig ileum and rat vas deferens.

[a]Corresponding author. Phone: +44 1279 622000; e-mail: Helen_J_Meadows@sbphrd.com

A

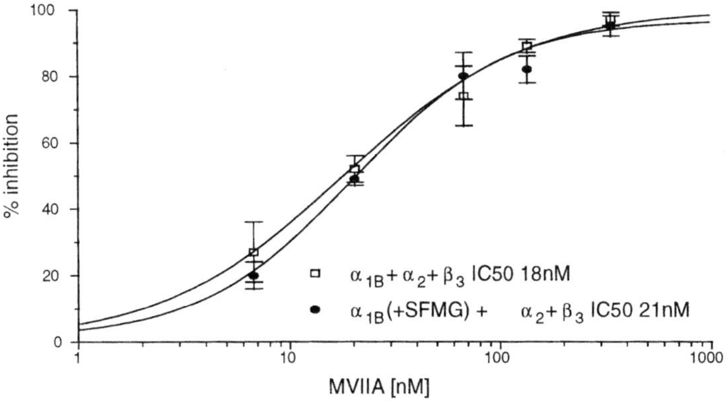

B

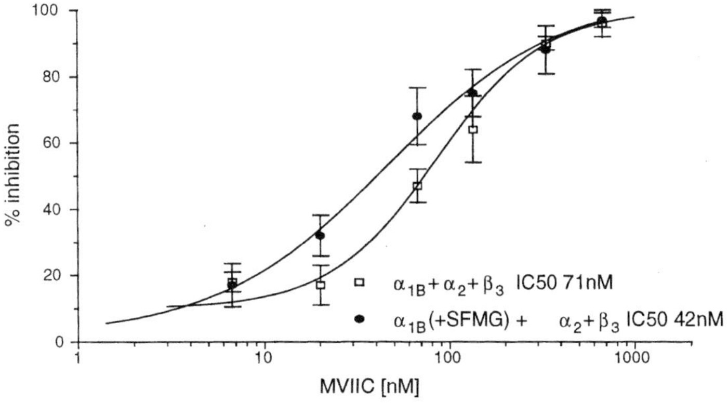

FIGURE 1. Concentration-dependent inhibitory effects of omega-conotoxins MVIIA (**A**) and MVIIC (**B**) on maximal calcium channel currents recorded from α_{1B} subunit variants in combination with α_2 and β_3 subunits. Holding potential −100 mV. Each point represents the mean ± SEM of 4 oocytes.

We next examined the dependency of block on the β subunit expressed. We chose a single concentration of each toxin giving >50% block on $\alpha_{1B}\alpha_2\ \beta_3$ to provide a suitable assay to identify any channel construct that was much less sensitive to conotoxin MVIIC. Coexpression of the β_{1b}, β_{2a}, β_3, or β_4 subunit had no effect on MVIIA or MVIIC sensitivity. All combinations were potently blocked by both toxins (FIG. 2A and 2B). This indicates that the β subunit does not have a dramatic effect on channel sensitivity to conotoxin MVIIC.

There are a number of other explanations for the differences in toxin sensitivity observed on nerve-mediated contractions from different tissues, including the possibility

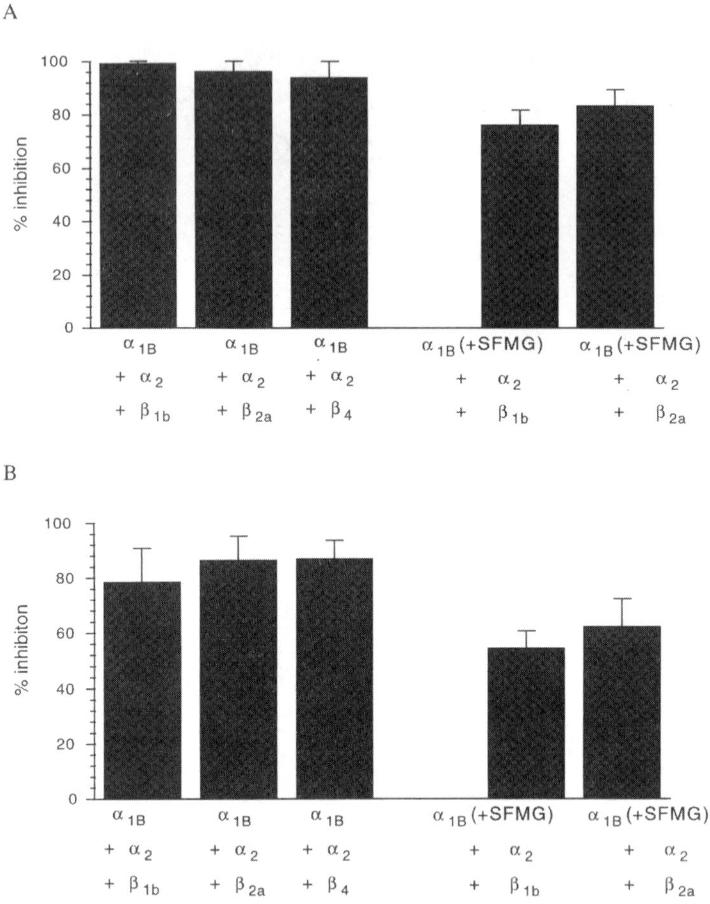

FIGURE 2. Inhibitory effects of omega-conotoxins MVIIA 100 nM **(A)** and MVIIC 100 nM **(B)** on maximal calcium channel currents recorded from different β subunits in combination with α_{1B} and α_2 subunits. Holding potential −100mV. Each point represents the mean ± SEM of at least 4 oocytes.

that the toxins have differential access to the channels in the whole-tissue preparations. It is still possible, however, that another untested splice variant of the α_{1B} subunit may be responsible for the differential toxin sensitivity observed by Boot,[1] and one potential candidate for this may be the ET insert in the domain IV S3-S4 extracellular linker.[3] Alpha$_{1B}$ variants containing these two amino acids (ET) were found to be differentially expressed in rat tissues. The major form of α_{1B} expressed in sympathetic ganglia contained these two amino acids (ET), whereas in the brain the majority of α_{1B} subunits lacked these two amino acids. The ET variant would therefore appear to be a good target for further investigation.

ACKNOWLEDGMENTS

We thank T. P. Snutch for providing the cDNA.

REFERENCES

1. Boot, J.R. 1994. Differential effects of ω-conotoxin GVIA and MVIIC on nerve stimulation induced contractions of guinea-pig ileum and rat vas deferens. Eur. J. Pharmacol. **258:** 155–158.
2. Dubel, S.J., A. Stea & T.P. Snutch. 1994. Two cloned rat brain N-type calcium channels have distinct kinetics. Soc. Neurosci. Abstr. **20:** 268.12.
3. Lin, Z., S. Haus, J. Edgerton & D. Lipscombe. 1997. Identification of functionally distinct isoforms of the N-type Ca^{2+} channel in rat sympathetic ganglia and brain. Neuron **18:** 153–166.
4. Stea, A., T.W. Soong & T.P. Snutch. 1995. Determinants of PKC-dependent modulation of a family of neuronal calcium channels. Neuron **15:** 929–940.
5. Scott, V.E.S., M. De Waard, H. Liu, C.A. Gurnett, D.P. Venzke, V.A. Lennon & K.P. Campbell. 1996. β Subunit heterogeneity in N-type Ca^{2+} channels. J. Biol. Chem. **271:** 3207–3212.

Quantitative Fluorescent RT-PCR Measurements of Postnatal Calcium Channel Gene Expression in Rat Hippocampal Subfields

SHAN M. PRADHAN,[a] SHUNDI GE,[a] CHIEKO KURODA,[b] JULIUS PETERS,[b] AND CHARLES E. NIESEN[a,c]

Divisions of [a]Neurology and [b]Pathology, Childrens Hospital Los Angeles, University of Southern California School of Medicine, Los Angeles, California 90027, USA

Mature function and activity of voltage-gated Ca^{2+} channels is attained over the first several weeks of postnatal life.[1,2] Previous work has described some aspects of these developmental changes. In hippocampal CA3 neurons, presynaptic Ca^{2+} channels were predominantly N-type early in development and switched to mostly Q-type later in life.[3] To learn more about the developmental regulation of Ca^{2+} channel genes, we studied the postnatal expression of the α_1 subunit, since this is the pore-containing portion of this heteromeric channel protein. We required a technique that afforded both regional localization and precise quantitation. While *in situ* hybridization allows regional localization of gene expression, it yields only semiquantitative results. We developed a novel quantitative fluorescent RT-PCR method that used a gene-specific deletion construct that coamplified with the cDNA for each channel gene. Thus, this gene construct could serve as an external standard and generate accurate copy numbers of the gene of interest.

We compared the regional expression of the α_1 subunit genes, rbA, rbB, rbC, and rbD, in three different hippocampal areas in the early postnatal period and in adulthood. Biweekly samples were taken in the first month of life since this is the period of most active growth and development of neurons and their connections.

Wistar rats from the first day of life (P0) through adulthood (ages P7, 13, 16, 25, and 80) were anesthesized with halothane and decapitated. Transverse dorsal hippocampal slices were isolated and then sectioned with a McIlwain chopper at 400 μm. Punch biopsies were taken from the somatic layers of the CA1, CA3, and dentate gyrus (DG) subfields using sterile 200-μL filter pipette tips (tip diameter 1 mm). Samples were transferred directly to 1.5-mL centrifuge tubes containing 1 mL TRIzol reagent (Gibco BRL) and RNA was isolated using standard protocols.

RNA samples were reverse-transcribed using MMLV-Reverse Transcriptase (GIBCO BRL), and published primer sequences for each of the four genes were used.[4] Deletion products were constructed for the four gene fragments and served as the external standards for quantitation in the PCR reactions. Each deletion product had priming sites identical to those of the full-size PCR product, which ensured coamplification at the same efficiency and differed only in containing the deletion of known length.

[c]Address for correspondence: Dr. Charles Niesen, Division of Neurology-#82, Childrens Hospital Los Angeles, 4650 Sunset Boulevard, Los Angeles, California 90027. Phone: 323-669-2525; e-mail: cniesen@hsc.usc.edu

Deletion products were made using the fusion primer method of Celi *et al.*[5] Fusion primers were constructed by directly linking the 3′ end of the sense primer to the 5′ end of a downstream sequence of 10–12 bases within the gene fragment or, similarly, linking the 3′ end of the antisense primer to such an upstream sequence. The deleted stretch of DNA was usually 10% of the size of the total gene fragment. A predetermined copy number of the deletion products was added to PCR reactions along with cDNA from the hippocampal slices. The PCR cycle was 32×, although this method has been shown to be PCR cycle number-independent.[6] The sense primer in each competitive PCR reaction was directly labeled at its 5′ end using substituted fluorescein phosphoramidites, yielding both a cDNA-derived fluorescent product and a deletion fragment-derived fluorescent product. Labeled fragments were resolved and sized, relative to a set of rhodamine-tagged molecular weight standards, with an ABI DNA Sequencer using Genescan software. Peak areas were measured in fluorescence units and standardized based on the input number of deletion product copies. cDNA copy numbers of the target genes were then calculated from the ratio of cDNA to deletion product areas under the curves.

FIGURE 1 shows the developmental gene expression for the four Ca^{2+} channel genes in the CA1 area and the dentate gyrus. Detectable expression of rbA was not observed until after the first week of life. Expression followed a sigmoidal curve. The adult level was attained near P30; 50% of maximum levels occurred at P16–17. This pattern was seen in both the CA1 and the dentate gyrus regions. All other genes showed overexpression in the early postnatal period compared to adult values. RbD levels were almost fourfold higher at P13 than at P80. Both rbB and rbC peaked at twice adult levels at the end of the second week of life and, contrary to what was expected, maximum values for these two genes occurred 4–13 days earlier in the dentate gyrus than in the CA1 region. Though not shown, gene expression in the CA3 area most closely followed the developmental pattern observed in the CA1 region.

One advantage of this technique is that absolute numbers of gene-specific cDNA copies can be measured. In FIGURE 2A, the genes associated with the Ca^{2+} channels that are involved in neurotransmission, rbA and rbB, showed a striking increase in expression from birth to adulthood and produced about one million cDNA copies each at P80. On the other hand, the transcript number for the genes associated with L-type channels, rbC and rbD, was 100–300-fold less at maturity (FIG. 2B).

This is the first report to give a detailed description of the developmental expression of the four major Ca^{2+} channel α_1 subunit genes. Each gene showed a different pattern of expression, suggestive of independent regulation. The rbA gene, which is associated with the P/Q channel, was the only one of the four genes whose expression began entirely in the early postnatal period. Adult levels were reached near P30, while the other three genes showed overexpression compared to adult values during the first month of life. Overexpression of neuronal genes during postnatal life has been observed for glutamate and GABA receptors, which, interestingly, occurred over the same time period.[7,8] Such phenomena have also been called *cytoplasmic pooling* and may represent a protective strategy to ensure that adequate numbers of the desired channel or receptor proteins reach their proper destinations.

Relative expression among the genes fits well with previous reports.[9] In our study, the absolute number of transcripts for rbA and rbB, the Ca^{2+} channel genes associated with neurotransmitter release, was 100–300-fold higher than expression for the L-type channel genes. We do not know the reason for this striking difference in expression and can only

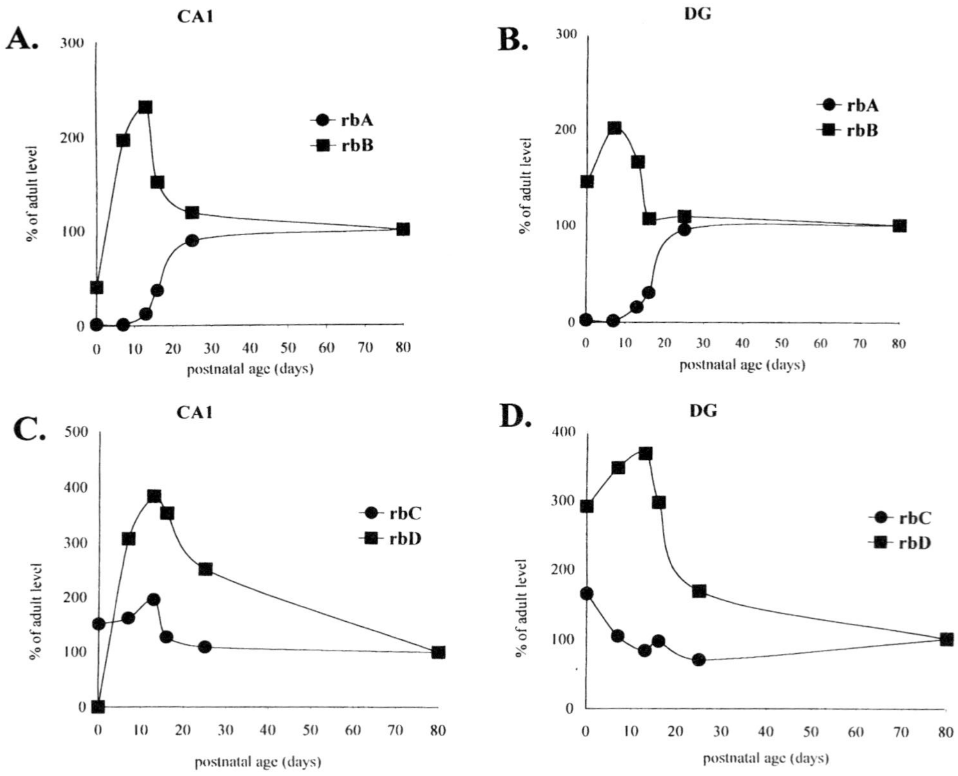

FIGURE 1. Developmental expression of the α_1 subunit genes in the hippocampal CA1 region **(A,C)** and dentate gyrus **(B,D)**. cDNA levels are compared relative to adult values.

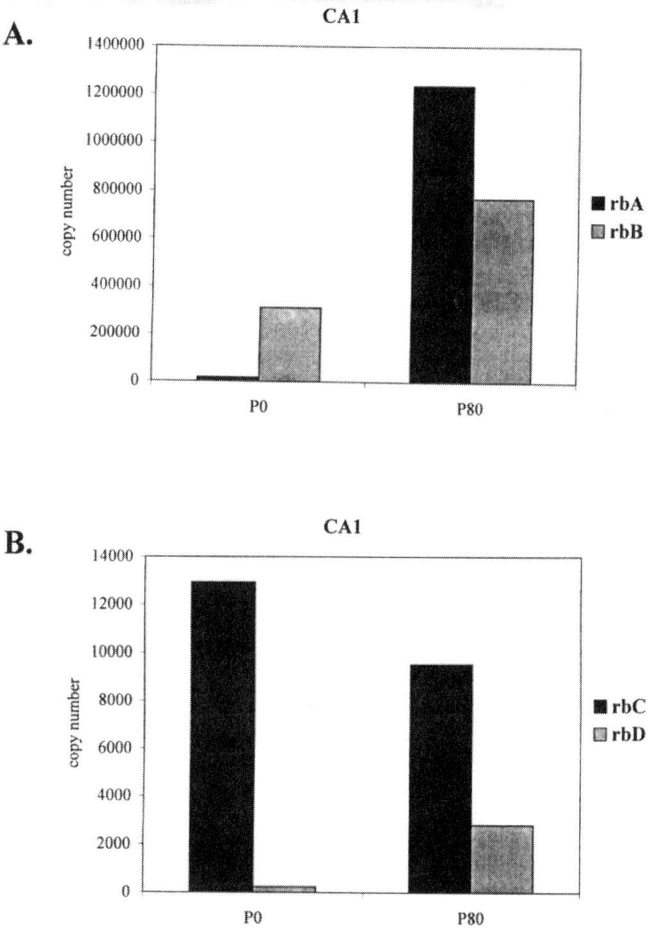

FIGURE 2. cDNA copy number for **(A)** rbA and rbB and **(B)** rbC and rbD at birth (P0) and in adulthood (P80) for the CA1 region of the hippocampus. Copy numbers are expressed per 62.5 ng RNA sample.

speculate on its significance. However, assigning importance to the magnitude of the copy number is not a simple matter. Subcellular localization as well as posttranscriptional factors may determine the number of mRNA transcripts. Despite the "low" copy number of the L-channel genes, their early expression in postnatal life implies a critical role in early neuronal development.

We feel this technique of fluorescent RT-PCR has provided an important first step in understanding the ontogeny of Ca^{2+} channels in hippocampal neurons and can be a useful tool in studying the control and regulation of these and other ion channels during neuronal maturation.

REFERENCES

1. THOMPSON, S.M. & R.K.S. WONG. 1991. Development of calcium current subtypes in isolated rat hippocampal pyramidal cells. J. Physiol. **439:** 671–689.
2. NIESEN, C.E., O.T. JONES, L.R. MILLS & P.L. CARLEN. 1993. Postnatal development of high threshold Ca^{2+} currents in rat hippocampal CA1 neurons. Soc. Neurosci. Abstr. **23:** 1719.
3. SCHOLZ, K.P. & R.J. MILLER. 1995. Developmental changes in presynaptic calcium channels coupled to glutamate release in cultured rat hippocampal neurons. J. Neurosci. **15:** 4612–4617.
4. LIU, H., R. FELIX, C.A. GURNETT, M. DE WAARD, D.R. WITCHER & K.P. CAMPBELL. 1996. Expression and subunit interaction of voltage-dependent Ca^{2+} channels in PC12 cells. J. Neurosci. **16:** 7557–7565.
5. CELI, F.S., M.E. ZENILMAN & A.R. SHULDINER. 1993. A rapid and versatile method to synthesize internal standards for competitive PCR. Nucleic Acids Res. **21:** 1047.
6. PETERS, J., H. WANG, S.C. SEEGER & C.P. REYNOLDS. 1998. Relative and absolute quantitation of MYCN mRNA in neuroblastoma cell lines by fluorescent RT-PCR. In preparation.
7. STANDLEY, S., G. TOCCO, M.F. TOURIGNY, G. MASSICOTTE, R.F. THOMPSON & M. BAUDRY. 1995. Developmental changes in α-amino-3-hydroxy-4-isoxazole proprionate receptor properties and expression in the rat hippocampal formation. J. Neurosci. **67:** 881–892.
8. WISDEN, W., D.J. LAURIE, H. MOYNER & P.H. SEEBURG. 1992. The distribution of 13 $GABA_A$ receptor subunit mRNAs in the rat brain. I. Telencephalon, diencephalon and mesencephalon. J. Neurosci. **12:** 1040–1062.
9. LUDWIG, A., V. FLOCKERZI & F. HOFMANN. 1997. Regional expression and cellular localization of the α_1 and β subunit of high voltage-activated calcium channels in rat brain. J. Neurosci. **17:** 1339–1349.

Molecular Diversity of K^+ Channels

WILLIAM A. COETZEE,[a,b,c] YIMY AMARILLO,[b] JOANNA CHIU,[c] ALAN CHOW,[b] DAVID LAU,[b] TOM McCORMACK,[b] HERMAN MORENO,[b] MARCELA S. NADAL,[b] ANDER OZAITA,[b] DAVID POUNTNEY,[c] MICHAEL SAGANICH,[b] ELEAZAR VEGA-SAENZ DE MIERA,[b] AND BERNARDO RUDY[b,d]

Departments of [b]Physiology and Neuroscience, [c]Pediatric Cardiology, and [d]Biochemistry, New York University School of Medicine, 550 First Avenue, New York, New York 10016, USA

ABSTRACT: K^+ channel principal subunits are by far the largest and most diverse of the ion channels. This diversity originates partly from the large number of genes coding for K^+ channel principal subunits, but also from other processes such as alternative splicing, generating multiple mRNA transcripts from a single gene, heteromeric assembly of different principal subunits, as well as possible RNA editing and post-translational modifications. In this chapter, we attempt to give an overview (mostly in tabular format) of the different genes coding for K^+ channel principal and accessory subunits and their genealogical relationships. We discuss the possible correlation of different principal subunits with native K^+ channels, the biophysical and pharmacological properties of channels formed when principal subunits are expressed in heterologous expression systems, and their patterns of tissue expression. In addition, we devote a section to describing how diversity of K^+ channels can be conferred by heteromultimer formation, accessory subunits, alternative splicing, RNA editing and posttranslational modifications. We trust that this collection of facts will be of use to those attempting to compare the properties of new subunits to the properties of others already known or to those interested in a comparison between native channels and cloned candidates.

The first molecular components of K^+ channels were identified only about a decade ago by molecular cloning methods.[1–5] However, the number of cloned and characterized components has grown so much that reviewing the molecular biology of K^+ channels has become a daunting, if not impossible, task in a chapter of these dimensions. Several excellent reviews, discussing specific aspects of this subject have appeared in recent years.[6–16] The present chapter presents a review as comprehensive as possible of all the K^+ channel subunits known to date, aimed mainly at scientists who might be interested in finding possible molecular correlates for their functional findings. We limited the scope by focusing primarily on mammalian K^+ channel principal and auxiliary subunits. Most of the data are presented in tabular format. It is possible that our tables have missing and even erroneous data. We wish to apologize to our colleagues for these errors and omissions and will appreciate receiving comments. Given space limitations we have not included specific references from which the data in the tables were extracted. To alleviate this constraint, we intend to publicize a web page[e] on which this information will be accessible along with additional data that similarly had to be omitted because of space considerations.

The availability of K^+ channel cDNAs has allowed enormous progress in the understanding of the structure and molecular mechanisms of function of K^+ channels. Important

[a]Corresponding authors: William A. Coetzee, D.Sc., Pediatric Cardiology, TH517, New York University School of Medicine, 550 First Avenue, New York, New York 10016. Phone: 212-263-8518; fax: 212-263-1393; e-mail: william.coetzee@med.nyu.edu and Bernardo Rudy, Department of Physiology and Neuroscience, New York University School of Medicine, 550 First Avenue, New York, New York 10016. Phone: 212-263-0431; fax: 212-689-9060; e-mail; rudyb01@ mcrcr6.med.nyu.edu

[e]http://k-channels.med.nyu.edu/

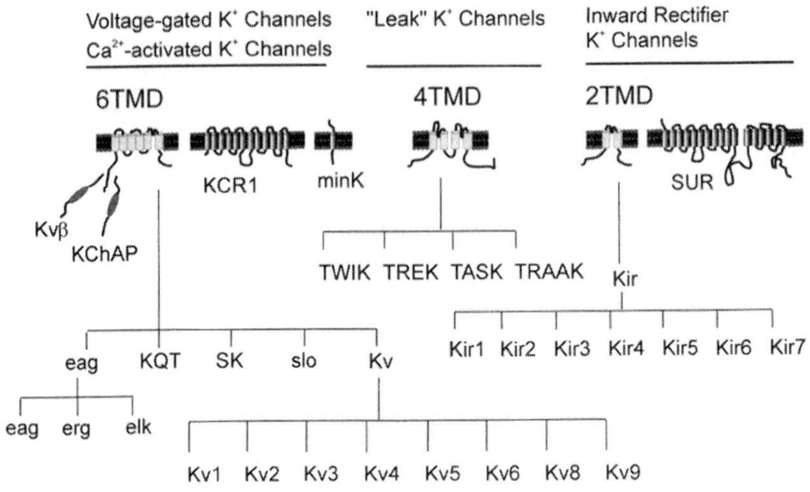

FIGURE 1. Schematic representation of the three groups of K^+ channel principal subunits. They are classified into three groups in terms of their predicted membrane topology—those that have six transmembrane domains (TMDs), those with four transmembrane domains and those with only two transmembrane domains. Each group of principal subunits is divided into discrete families on the basis of sequence similarity (see FIGS. 2 and 3). Each family can be further subdivided into several subfamilies, which often contain several closely related subfamily members. A functional classification places the voltage- and Ca^{2+}-regulated K^+ channels in the 6TMD group, the "leak" K^+ channels in the 4TMD group, and the inward rectifier K^+ (Kir) channels in the 2TMD group. Also shown in the figure are some of the auxiliary subunits that have been shown to alter expression levels and/or kinetics of K^+ channel principal subunits when expressed in heterologous expression systems. For clarity, they are grouped together with the principal subunits with which they have been shown to interact (see text for more details).

new insights into the mechanisms of ionic selectivity, voltage- and calcium-dependent gating, inactivation and blockade of these channels have been obtained. These efforts recently culminated with the crystallization and high-resolution structural analysis of a K^+ channel[17]—the first natural membrane channel for which high-resolution real structural information is now available. This work provided strong evidence in favor of a tetrameric K^+ channel structure. This structure was first suggested based on the similarities between six transmembrane K^+ channel subunits and each of the four internally homologous repeats of Na^+ and Ca^{2+} channels[18] and was supported experimentally.[17,19,20] In this structure, four independent subunits or homologous repeats of one or two subunits form the infrastructure of a channel with a fourfold symmetry around a central pore. We can expect that this breakthrough will bring a deeper and accelerated understanding of the structure and function of K^+ channels.

Less progress has been made in understanding the physiological significance of the enormous molecular diversity of K^+ channel protein subunits (TABLE 1[f]). Over 100 different proteins, subunits of distinct types of K^+ channels, have been identified to date, and the list is rapidly growing (FIG. 1). In addition to the pore-forming or principal subunits (often

[f] All tables appear at the end of the paper.

called α subunits), which determine the infrastructure of the channel, many K^+ channels (like Na^+ and Ca^{2+} channels) contain auxiliary proteins that can modify the properties of the channels, often significantly. Most of the known principal K^+ channel subunits express in heterologous expression systems as functional homomultimeric channel complexes. However, some principal subunits do not form functional homomultimeric channels, but must coassemble with other (similar) subunits for expression of functional channels (e.g., G-protein–activated K^+ channels; GIRKs). These subunits may be called coassembly principal subunits. All known principal subunits of K^+ channels show a certain amount of sequence and structural similarity, and they might all be related through evolution. There is sufficient primary sequence similarity between the different principal subunits that members of new families of K^+ channel proteins have been discovered by degenerate PCR or by screening databases of "expressed sequence tags" (ESTs) and sequences deposited by various genome projects. It remains possible, however, that some unknown K^+ channels may have principal subunits that are unrelated to those known today.

Auxiliary subunits (sometimes referred to as β subunits) have primary sequences not resembling principal subunits. They interact with channel complexes containing principal subunits and may alter their electrophysiological or biophysical properties, expression levels, or expression patterns. In addition, many K^+ channel molecular complexes interact with additional proteins such as regulatory enzymes and elements of the cytoskeleton (see chapters by Morgan Sheng and John Adelman, this volume). We term these *associated proteins*, with the understanding that the distinction between auxiliary subunits and associated proteins may not be simple on occasions. Since K^+ channels are thought to be multimers of principal subunits, which may form heteromeric channels with closely related principal subunits in various combinations, the number of possible distinct K^+ channels based on these different combinations may be in the order of hundreds, if not thousands (see DIVERSITY CONFERRED BY HETEROMULTIMER FORMATION..., below). However, it is not known how much of this potential diversity is actually used in native cells. A major task of future research is to identify physiological roles of the cloned proteins, starting with the identification of native channels containing specific types of cloned subunits. This is particularly important because most of the cloning work has been done in the absence of prior isolation of native proteins. Therefore, the exact relationship between the molecular components identified by cloning and native channels is, in most cases, not known and must be a priority for future research.

CLASSIFICATION OF K⁺ CHANNEL PRINCIPAL SUBUNITS

There are several types of K^+ channels, including voltage-gated and Ca^{2+}-activated K^+ channels, inward rectifiers, "leak" K^+ channels, and Na^+-activated K^+ channels.[21,22] Principal subunits of at least the first four types have already been identified, and they are divided into three groups based on structural properties (FIG. 1). The first group, consisting of six transmembrane domain (TMD) proteins, are components of voltage-gated (Kv) and Ca^{2+}-activated K^+ channels. The second group, consisting of proteins with two TMDs, are components of inward rectifier K^+ (Kir) channels. The third group, known as two-pore subunits, are components of "leak" K^+ channels. Each of these groups is further divided into families, which in turn are divided into subfamilies, with several closely related members within most of these subfamilies.

The first major group of K^+ channel principal subunits to be identified contain six transmembrane domains (TMDs) (S1-S6), with a conserved P (pore or H5) domain. Functionally, they form voltage- and/or Ca^{2+}-activated K^+ channels when expressed in heterologous expression systems. This group contains the Kv family (with eight subfamilies: Kv1–Kv6 and Kv8–Kv9) as well as members of the KQT, eag, SK, and slo families of principal subunits.

The second major group of pore-forming subunits are components of inward-rectifying K^+ (Kir) channels and the first members were first identified by expression cloning.[23–25] Kir principal subunits have a predicted membrane topology of two TMDs (M1–M2) and a pore domain, analogous to S5-P-S6 of the 6TMD K^+ channel subunits. There are currently seven subfamilies (Kir1–Kir7), most of which form K^+ channels with various degrees of inward rectification when expressed in heterologous expression systems.

A third group of mammalian K^+ channel principal subunits was recently described and contains four putative TMDs (M1–M4) and two P domains (P1 and P2).[12,26] Structurally, these principal subunits have a predicted membrane topology as if they consisted of two spliced Kir subunits. Whereas the 6TMD and 2TMD principal subunits are thought to assemble as tetrameric proteins to form functional channels (see above), the 4TMD subunits are thought to dimerize, thereby retaining the fourfold symmetry around the central pore.[12,27] There are currently four members in this novel family of K^+ channel principal subunits (FIG. 1 and TABLE 1), but it is possible that more members might be cloned in the near future (see later). Functionally, these principal subunits express K^+ selective channels that do not appear to gate in a manner as observed with channels formed by Kir principal subunits. Since the current responds to changes in extracellular K^+ concentration in a manner described by the Goldman-Hodgkin-Katz equation, these channels are also referred to as "leak" K^+ channels.[12] Recent reports [28] indicate that at least some of these channels can be extensively modulated (e.g., by arachidonic acid or pH; TABLE 1).

The P domain of K^+ channel principal subunits is critically important for channel function. Approaches using both mutagenesis [29,30] and X-ray crystallography[17] suggest a role for this domain in the formation of the K^+ selective pore of the channel. The consensus pore sequence[g] calculated from a simultaneous alignment of the P domains of K^+ channel principal subunits that are shown in TABLE 1 is: [TS]-[MLQ]-T-T-[IV]-G-Y-G[31] and appears to be hallmark of K^+ channel principal subunits.

Mammalian K⁺ Channel Principal and Auxiliary Subunits

TABLE 1 lists published sequences of mammalian (mostly from human, rat, and mouse) K^+ channel principal and auxiliary subunits cloned to date. As far as possible, we used standard nomenclature[32,33] to describe the various genes. In addition to Genbank or Swissprot accession numbers and trivial (author-assigned) names of known alternatively spliced variants, we also give the gene name as defined by the Human Genome Organization (HUGO) Nomenclature Committee,[h] chromosomal localization, and possible (or confirmed) associated diseases related to allelic variants. These data for principal subunits

[g] The consensus pore sequence was calculated from a simultaneous alignment of the pore regions of K^+ channel principal subunits using MEME (http://www.sdsc.edu/MEME). A residue was included if it had a probability of occurrence larger than 0.2.

[h] http://www.gene.ucl.ac.uk/nomenclature/

have been subdivided by functional classification, as described above. The auxiliary subunits are listed in terms of the main principal subunits with which they are thought to interact.

Genealogical Analysis of Genes Coding for Principal K^+ Channel Subunits

We performed a genealogical analysis of the K^+ channel principal subunits shown in TABLE 1 with the aim of examining relatedness between the various genes. For this analysis, we divided the K^+ channel principal subunits on the basis of their predicted transmembrane topology. FIGURE 2 is a phylogenetic tree of six TMD voltage-gated and Ca^{2+}-activated K^+ channels generated by parsimony analysis, and FIGURE 3 shows a similar analysis for the two TMD principal subunits representing Kir principal subunits.

For the six-TMD group, a strong primary node (bootstrap value of 99% in FIG. 2) exists that includes the members of all of the Kv subfamilies (Kv1–Kv6 and Kv8–Kv9). The question may arise whether the KQT family should be considered as part of the Kv family. However, the KQT family has a much weaker sequence identity to Kv genes (19–25%) than is found between members of the Kv subfamilies (33–50%) (TABLE 2A). It is therefore likely that the divergence of the KQT and the Kv gene families preceded the divergence of the different Kv subfamily members. Thus, for the purposes of this review, the KQT family will be regarded as a discrete family. The analysis shows that members of eag, erg, and elk subfamilies showed clustering with a strong node (bootstrap value of 100%), suggesting that they all originated from a single ancestral gene.

Interestingly, although there are large structural differences (e.g., the length of the C-terminus) between SK and slo, these families (which are all principal subunits of Ca^{2+}-activated K^+ channels) grouped together with a strong bootstrap value (73%; FIG. 2). Thus, within the regions used to perform the parsimony analysis (which included most of the six transmembrane domains), characters exist that diagnosed SK and Slo as close relatives.

It is often difficult to determine whether a gene belongs to a certain subfamily. For the purposes of this review, we used identity scores as the only criterion to subdivide genes into different subfamilies. Other criteria may exist, such as the presence of regulatory sequences (e.g., ATP-binding sequences, etc.), similar electrophysiological or pharmacological phenotypes in expressed channels, or similar regulation by metabolic pathways. Concentrating only on similarity of the primary sequences, it is interesting to note that an identity score of >55%[i] exists among members of individual Kv, eag, KQT, and slo subfamilies (TABLE 2A). In fact, this threshold value of ~55% holds true within each of the 6TMD K^+ channel *subfamilies* except for the SK genes. Within the SK gene family, SK4 appears to diverge from other SK family members (TABLE 2B), suggesting that SK4 might represent a member of a new emerging subfamily of SK genes.

Within the 2TMD family, members of the Kir2, Kir3, Kir5 (the latter having only a single member), and Kir6 subfamilies each fall into their own expected groups (FIG. 3). The classification of the remainder of the genes (members of Kir1, Kir4, and Kir7 subfamilies)

[i] Note that this number was derived from multiple sequence alignments where portions of unstable sequences were removed for the generation of trees. When whole sequences are aligned, this number may be smaller. The threshold value of 55% identity can be extracted from the percent identity table for the 2TMD potassium channels as well as the 6TMD channels; i.e., in both identity tables, genes that have identity values of less than 55% belong to different subfamilies.

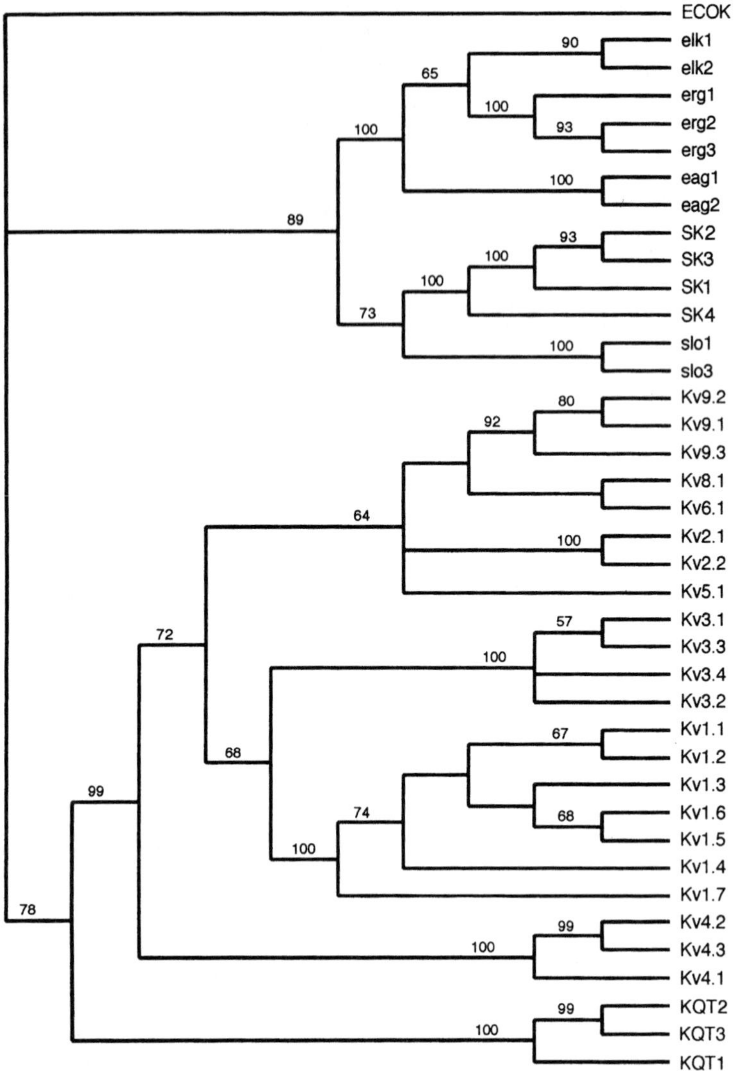

FIGURE 2. Strict consensus phylogenetic tree of 6TMD voltage-gated and calcium-activated K^+ channels generated by parsimony. The parsimony tree is generated by heuristic search using PAUP 3.1.1.[138] Ten random addition searches with TBR branch swapping were performed. The *E. coli* K^+ channel homologue ECOKCH (accession #L12044) was used as an outgroup). ClustalX was used to generate all alignments of amino acid sequences. By varying the alignment-parameters, we were able to identify regions that lack alignment stability. We proceeded to remove alignment ambiguous regions by implementing a "culling" procedure.[139] In this analysis, only regions that span the six transmembrane domains were used. All characters were equally weighted in the analysis. Bootstrap values generated using PAUP 3.1.1[138] are shown on the tree when available, and they represent a measure of node robustness. Sequences of eag2, elk1, and elk2 used in this analysis are from B. Ganetzky (see this volume).

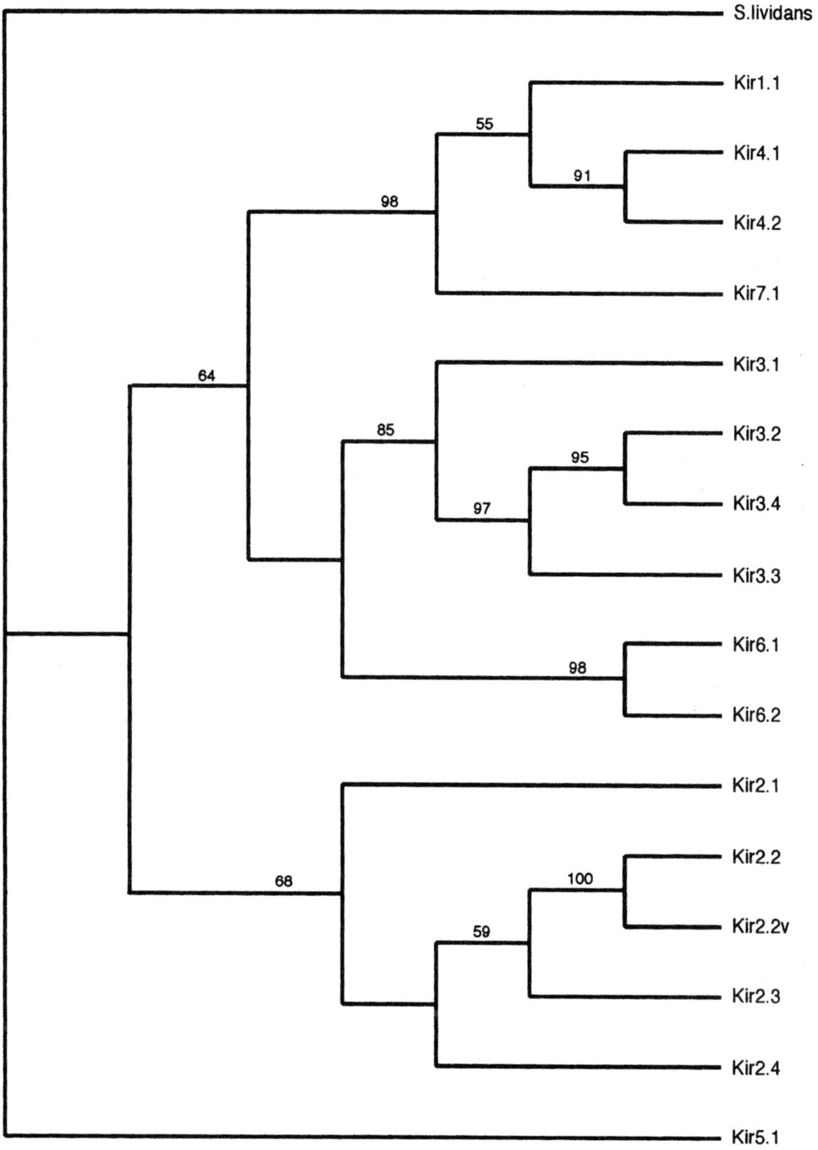

FIGURE 3. One of two most parsimonious trees of 2TMD inward rectifier K^+ channels that are generated by heuristic search. For methods used, please refer to the legend of FIG. 2. The *S. lividans* skc1 K^+ channel principal subunit (accession Z37969) was used as an outgroup. Except for small portions of the NH_2 and COOH terminals, most of the amino acid sequences were used in this analysis. To measure node robustness, we generated bootstrap values (shown on tree when available) using PAUP 3.1.1.

deserves comment. Following the cloning of Kir1.1 and its splice variants (TABLE 1), the primary sequences of closely related genes were published. Ambiguity arose in the naming of these genes. Whereas BIR10[34] was proposed to be named Kir4.1,[33] others have been referring to this gene as Kir1.2[35] because of its apparent sequence similarity to Kir1.1. Similarly, there has been equal ambiguity in the naming of Kir1.3/Kir4.2.[36,37] More recently, a member of a potentially new subfamily was cloned and was called Kir7.1,[38] but was also named Kir1.4 by others (Genbank accession number AB013890). As seen in TABLE 2C, a threshold identity value of >55% also exists among members of individual Kir subfamilies, except for the subfamily that includes Kir1.1, Kir1.2/Kir4.1, Kir1.3/Kir4.2, and Kir1.4/Kir7.1. When examining this particular subfamily (Kir1/Kir4/Kir7), an identity score as low as 36% is found, suggesting that (at least some of) these genes might indeed belong to different subfamilies. A more detailed analysis of identity scores of these genes (TABLE 2D) shows that Kir1.4/Kir7.1 is only 36–39% identical to the other three genes. Similarly, while Kir1.2/Kir4.1 and Kir1.3/Kir4.2 are 62% identical to each other, they are only 47% identical to Kir1.1. Thus, the possibility that discrete Kir4 and Kir7 subfamilies exist (as originally proposed)[33,38] must be considered. These considerations led us to adopt this nomenclature for the remainder of this chapter. Although our analysis supports the idea that members of the Kir1, Kir4, and Kir7 subfamilies are not sufficiently similar in terms of their primary sequences to be grouped as a single subfamily, the analysis shown in FIGURE 3 illustrates that these genes are more closely related to each other than to other Kir subfamily members.

In mammals, the 4TMD gene family is currently the smallest among the three main groups of K^+ channels, with only four representatives published to date. Among them, TRAAK and TREK are more closely related to each other than to either of the other two genes, TWIK and TASK, as indicated by alignment and phylogenetic analysis (data not shown). A more detailed description of 4TMD principal subunits in *C. elegans* appears elsewhere in this volume (see chapter by Salkoff).

CORRELATION OF K^+ CHANNEL PRINCIPAL SUBUNITS WITH NATIVE CHANNELS

The molecular diversity of K^+ channel-forming proteins (TABLE 1) is far greater than that found for native K^+ currents,[21] which adds additional complexity when one attempts to correlate the molecular candidate with a particular native current expressed in a specific tissue. In order to infer a relationship between heterologously expressed and native K^+ channels, there are several criteria that may be used.

1. A close resemblance between the biophysical properties of native channels and those of channels expressed in heterologous expression systems may signify a possible relationship. Further support for such a relationship will be provided if they are similarly affected by pharmacological compounds, toxins, and other interventions. In order to assist in these comparisons, we compiled the data presented in TABLE 4.
2. The tissue expression of mRNA (and protein) of the candidate principal subunits should correspond with the tissues in which the native current is expressed. Data in TABLE 5 should assist in this comparison.

3. The functional consequences of processes such as heteromultimeric assembly by different principal subunits, their regulation by auxiliary subunits, and posttranslational modification should be considered. We present examples where these processes have been shown to influence channel properties. Because of the amount of detail involved, this information is supplied mainly in textual format (DIVERSITY CONFERRED BY HETEROMULTIMER FORMATION..., below).
4. Strong support for a causal relationship between native and candidate cloned K^+ channel principal subunits can be obtained from experiments involving deletion or overexpression of the target channel principal subunits (transgenic animals or overexpression of wild-type, antisense, or dominant negative constructs in isolated cells). Since this is a relatively new area of investigation, there are only a few examples where these methodologies have been applied (see TABLE 3).

It is important to note, however, that these criteria do not exclusively define a relationship between cloned and native proteins. For example, negative results might well be caused by unknown events occurring at the molecular level, such as interactions of K^+ channel principal subunits with other cytosolic or membrane-bound proteins or their regulation by unknown endogenous compounds and peptides or poorly understood posttranslational modifications.

FUNCTIONAL PROPERTIES OF K+ CHANNELS IN HETEROLOGOUS EXPRESSION SYSTEMS

We have listed the functional and pharmacological properties of the currents expressed by K^+ channel principal subunits in heterologous expression systems in TABLE 4. Where possible, we report data obtained from expression studies using mammalian cells under patch-clamp conditions. Otherwise, we report data obtained from the *Xenopus* expression system. In cases in which there is close agreement between several reports on the same protein, we report an averaged value. When large discrepancies exist in published values to make it impossible to find a consensus value, we give a range of values. In come cases, we favored reports in which a more extensive functional analysis was performed.

In using these tables it is important to remember that experimental variables can have important effects on published parameters. In addition, a native channel composed of a particular subunit may have properties different than those in heterologous expression systems due to factors such as those listed in DIVERSITY CONFERRED BY HETEROMULTIMER FORMATION..., below, or other factors that are sometimes difficult to predict. We would like to mention three such examples of problems encountered using heterologous expression systems: First, it has been observed that Kir3.1 expresses G-protein–activated K^+ channels in *Xenopus* oocytes, but not in mammalian cultured cells. It is now clear that this is due to the presence of an intrinsic Kir3.4 subunit (called XIR) in *Xenopus* oocytes[39] (see chapter by Wickman in this volume) that is not found in mammalian cells and is required for expression of G-protein–activated K^+ channels. Second, artifacts can be caused by the unusually large amounts of channels expressed in heterologous expression systems. The flow of large currents may lead to effects such as K^+ accumulation in the extracellular spaces, which in turn can modify the behavior of expressed channels. Last, some channels can be very sensitive to particular elements in the extracellular or intracellular solution, such as blockade of several voltage-gated K^+ channels by Mg^{2+}. Care should therefore be exercised when comparing the electrophysiological phenotype of native K^+ currents with those presented in TABLE 4.

TISSUE EXPRESSION OF K^+ CHANNEL SUBUNITS

Studies on the cell- and tissue-specific expression of K^+ channel genes vary widely. Some genes such as Kv3 (see chapter by Rudy *et al.*, this volume), Kir3 (see chapter by Kevin Wickman *et al.*, this volume), and several Kv1's[16] have been studied in great detail in several laboratories. Others have been studied to a much lesser extent. We attempted to give a representative overview of expression patterns of known K^+ channel genes in TABLE 5. However, the nature of available information will be reflected in our compilation.

Different laboratories use different methods to analyze expression levels of various K^+ channel transcripts. Some of these methods are more quantitative in nature (Northern blot analysis and RNAse protection assays) but have a limited spatial resolution. In contrast, *in situ* hybridization techniques can have excellent spatial resolution, but quantification is more difficult. RT-PCR is very sensitive, and results obtained with this technique are difficult to quantify. Given this sensitivity, it is often difficult to interpret a result when no attempt at quantification is made. Antibodies have now been raised to a number of K^+ channel proteins, and they can be great tools for the analysis of protein products, but problems of specificity can arise. We present a table on the tissue and brain distribution of K^+ channel gene products (TABLE 5). However, caution should be exercised in using this table. For example, results for some genes may come from many studies using different methods, whereas for others data may have been derived from a single study. Great care should be taken when making a quantitative comparison of expression levels between different genes because of the variations in methodologies used in different laboratories as well as the subjectivity of different investigators when grading the intensities of signals.

DIVERSITY CONFERRED BY HETEROMULTIMER FORMATION, ACCESSORY SUBUNITS, ALTERNATIVE SPLICING, RNA EDITING AND POSTTRANSLATIONAL MODIFICATION

Over 50 mammalian genes encoding principal subunits of K^+ channels are listed in TABLE 1. The total number of different subunits, however, is even larger since many of these genes undergo RNA processing, such as alternative splicing resulting in multiple protein products from each gene. Considering these factors, this results in a large number (>100) of different mammalian principal subunits (TABLE 1).

However, the total number of different functional types of K^+ channels is probably significantly larger. In addition to the already large number of subunits, the diversity can be magnified by virtue of the oligomeric structure of the functional channel complex (see introductory section of this paper). Many of the principal subunits can form functional homomultimeric as well as heteromultimeric channels. Depending on the number of possible combinations, this can dramatically increase the number of possible functionally distinct channels obtained from these subunits.

Finally, channel function can also be influenced by auxiliary subunits and posttranslational modifications. If the channel complex exists with the auxiliary proteins in some cells and without them in other cells, or if it is differentially modified by posttranslational modification, these factors can also increase the diversity of functional units. This section highlights the contributions of these important mechanisms to K^+ channel function and diversity.

Heteromultimeric Assembly of Principal Subunits

Heteromultimeric assembly of K^+ channel proteins can theoretically provide the cell with a "modular" system for increasing K^+ channel diversity while reducing the requirement for more genes. For example, 210 functionally distinct tetrameric channels of different subunit combinations might be formed from seven different subunits.[j] Channel diversity can be increased even more if the possibility of heteromultimerization of alternative splice variants is considered.

When considering the large overlap in the expression of different K^+ channel subunits (see TABLE 5), it is quite possible that many native channels are the result of heteromeric association of primary subunits. However, coexpression of two subunits within a given cell does not necessarily warrant the occurrence of coassembly (see below). It is thus of great importance to elucidate the mechanisms that govern subunit recognition and assembly in order to understand the potential for heteromultimer formation and the composition of native channels. Here we discuss what is known about the rules governing subunit interactions and evidence for their formation in native tissue in the different families of K^+ channels.

Heteromultimeric Assembly of Kv and KQT Principal Subunits

For Kv1–Kv4 subfamilies, coexpression of different Kv cRNAs of the same, *but not of different*, subfamilies in *Xenopus* oocytes leads to the expression of channel properties that can not be the result of the sum of two independent channels. These currents have been interpreted as resulting from novel K^+ channel proteins formed as heteromultimers from different principal subunits of the same Kv subfamily.[40–44] These heteromultimeric channels tend to have properties that are intermediary between those of the two homomultimeric channels, but in some cases certain properties dominate. For example, a channel containing three noninactivating and one N-type inactivating subunit produces an inactivating channel.[20,41–43] In another example, Kv1.6 subunits have been shown to have a specific N-type inactivation-prevention (NIP) domain, which produces noninactivating currents when heteromultimerized.[45] Subfamily selectivity among Kv subunits is the result of recognition domains located in the N-terminal region of the protein and referred to as the NAB or T1 domains.[46,47] Interestingly, these domains are also involved in the interaction of Kv1 proteins with auxiliary (Kvβ) subunits.[48,49]

Evidence for the heteromultimerization of Kv proteins from the same subfamily *in vivo* has come from coimmunoprecipitation experiments in rat and mouse brain.[50–53] (also see

[j] [p+(n-1)]!/[p!(n-1)!], where p is the aggregation number and n is the number of subunits.

Rudy *et al.*, volume). For example, antibodies specific to Kv1.1 immunoprecipitate Kv1.2 and Kv1.4 proteins in nondenatured brain membrane extracts.[50,51] The coexpression of Kv1.4 and Kv1.2 in axons and terminals of many cells suggests the native A-type K^+ current may result from Kv1.4/Kv1.2 heteromultimers within these compartments.[50] Conversely, subunits of the same subfamily have also been shown to localize to separate compartments within the same cell.[54] The cellular mechanisms of subunit targeting and assembly could be important as well. Whether or not heteromultimers are formed will depend on whether channel assembly occurs before or after the subunits have been sorted and shipped to their appropriate destinations.

The two Kv2 subfamily members (Kv2.1 and Kv2.2) are able to function as homomultimeric or heteromultimeric channels. There are reports that their kinetics and voltage dependence are altered when they are coexpressed with members of other, closely related, Kv subfamilies (Kv5, Kv6, Kv8, Kv9; see FIG. 1).[55–57] An enormous potential for functional diversity is apparent when considering all of the possible combinations of these principal subunits. Interestingly, members of the Kv5, Kv6, Kv8, and Kv9 subfamilies do not express currents by themselves in heterologous expression systems. Since the primary sequence similarity of Kv5, Kv6, Kv8, and Kv9 principal subunits with the Kv2 subfamily members are particularly high in the T1 domain, one can speculate that these coassembly principal subunits may have evolved to regulate the function of Kv2-related proteins.

The KQT family currently has three members. The first to be described, KQT1 (KvLQT1), is found predominantly in the heart and liver, whereas KQT2 and KQT3 expression is restricted to the brain. Coexpression of KCNQ2 and KCNQ3 in *Xenopus* oocytes results in a 15-fold increase in current amplitude compared to expression of each subunit alone.[58] The overlapping distribution of KQT2 and KQT3 suggest they may function only as heteromultimers *in vivo*. In support of this argument, it is interesting to note that mutations of these two genes both give rise to the same genetic disorder (TABLE 1).

Heteromultimeric Assembly of eag, erg, and elk Subunits

There is currently a single published mammalian homologue of the *Drosophila* eag that produces currents in heterologous expression systems (TABLE 1). Although more eag family members have been described, the full-length cDNA coding sequences remain to be cloned and expressed (see chapter by Ganetzky, this volume). There are three members of the Erg subfamily with diverse physiological properties. However, their interactions have not been characterized. A mammalian elk homologue has been cloned and was found to produce functional currents. At least two new elk family members have been discovered.[59]

Heteromultimers between the two splice variants of erg (erg1a and erg1b; TABLE 1) have been studied *in vitro*, resulting in currents with properties more similar to those of the native current (I_{Kr}) in heart.[60] It is not yet known whether eag, erg, and elk principal subunits can coassemble as heteromultimers. With the cloning and expression of more members of each subfamily, it will be interesting to see if they follow the same rules for heteromultimerization as Kv channels. There is already evidence that subunit interactions within this family will be complicated. For example, it has been suggested that members of the eag family can interact with members of the Kv family in *Xenopus* oocytes.[61] However, whether such interactions will take place *in vivo* remains to be determined.

Heteromultimeric Assembly of Kir Subunits

There does not appear to be a consistent manner in which Kir subunits coassemble within or between subfamilies. The four members of the Kir3 subfamily serve as an example of where heteromultimerization of subunits from the same subfamily appears to be a necessary requirement for the formation of functional channels. Thus, the channels responsible for the acetylcholine-activated K^+ current in atrial muscle is a complex of Kir3.1 and Kir3.4 proteins (see chapter by Wickman *et al.*, this volume). Neurons probably contain functional heteromultimers of Kir3.1 with Kir3.2 or Kir3.3 (see chapter by Kevin Wickman, this volume; see also Ref. 62). In contrast to the Kir3 subunits, the evidence for heteromultimeric assembly within other Kir subfamilies members is less strong or there are even indications to the contrary. For example, despite the colocalization of Kir2.1 and Kir2.3 in neural tissue,[63] biochemical and electrophysiological experiments examining protein-protein interactions suggest that heteromultimeric coassembly of subunits within the Kir2 subfamily may not occur.[64] Recent experiments using antisense oligonucleotides directed against Kir2.1 transcripts demonstrated a specific inhibition of a 21-pS conductance channel (which is equivalent to the unitary conductance of Kir2.1 in heterologous expression systems) in cardiac myocytes.[65] Since the occurrence of native inward rectifier K^+ channels with different unitary conductances was unchanged, one could argue that heteromultimeric assemblies of Kir2.1 with other K^+ channel principal subunits did not occur (or more unlikely that, if it did, the single-channel conductance remained unchanged). The regions responsible for determining compatibility within members of the Kir2 subfamilies have been elucidated using a combination of deletion mutants and chimeric channels.[64] The results of such experiments reveal interactions conferred by domains found within the proximal C-terminus and second transmembrane segments. However, it is possible that other regions may also play a role.[66]

There are also published examples of heteromultimerization between subunits from different Kir subfamilies. For example, Kir5.1 (which by itself expresses no current in oocytes[34]) significantly alters expression levels and the single-channel conductance of Kir4.1, a member of a different subfamily.[67] The finding that Kir5.1 altered neither current amplitude nor the macroscopic phenotype of Kir1.1, Kir2.1, Kir2.3, Kir3.1, Kir3.2, or Kir3.4 suggests that a specific interaction between Kir4.1 and Kir5.1 may occur *in vivo*.

Heteromultimeric Assembly of slo and SK Subunits

The slo family currently has two members (TABLE 1). Heteromultimerization between the two slo principal subunits has not been examined. Despite the small number of family members, slo1 (the principal subunit of BK Ca^{2+}-activitated K^+ channels) has a very large number of possible alternative spliced versions. Tested splice variants show functional variability.[68–71] Heteromultimerization between splice variants could thus allow an enormous number of possible functionally diverse channels. Single-cell RT-PCR studies of individual hair cells of the chick cochlea demonstrate that multiple splice variants of slo are found within a single cell[72,73] (see also chapter by Robert Fettiplace, this volume), suggesting that at least a proportion of this potential diversity might indeed occur *in vivo*.

The SK family now has four members (TABLE 1). SK1–3 each express small-conductance Ca^{2+}-activated K^+ channels with similar properties but differing in their pharmacology (see chapter by Adelman, this volume). The fourth (SK4) expresses an intermediary-conductance Ca^{2+}-activated K^+ channel. Coexpression of these genes has not yet been tested, but their overlapping tissue distribution suggests that heteromultimer formation might be possible *in vivo*.[74] Although the SK1-SK3 subunits have very similar properties, the NH_2 and COOH terminal amino acid sequences vary significantly which could potentially provide additional diversity through differential modulation through associating proteins. Conversely, divergence in these regions may also prevent heteromultimer formation.

Heteromultimeric Assembly of "Leak" K^+ Channel Subunits

Heteromultimer formation within the 4TMD family of K^+ channels has not yet been explored. It can be predicted from the number of similar genes expressed in *C. elegans* that a multitude of "two-pore" subunits may exist in mammals (see chapter by Larry Salkoff, this volume). Whether or not the diversity of this family of K^+ channels is increased by heteromeric subunit interactions remains to be determined.

Auxiliary Subunits

There is increasing evidence for the existence of auxiliary subunits, some peripheral and some integral membrane proteins, that regulate the expression levels and functional properties of K^+ channel proteins (TABLE 1). Several types of K^+ channels are affected in this manner, including those consisting of Kv, KQT, eag, slo, and Kir principal subunits.

To date, the auxiliary subunits that have been characterized in most detail are the members of the Kvβ auxiliary subunit family. They are products of three genes (TABLE 1). We adopt the nomenclature shown in TABLE 1 to conform to convention and to signify their sequence relatedness. Affinity purification of brain K^+ channels using dendrotoxin led to the isolation of an auxiliary "β" subunit (β2 or Kvβ2) that binds noncovalently with a 1:1 stoichiometry to Kv1 principal subunits.[75,76] Identification of this sequence soon led to the isolation of related cDNAs. Three genes have been identified, and each has been shown to produce several isoforms by alternative splicing (TABLE 1). These subunits lack putative transmembrane domains, potential glycolsylation sites, or leader sequences, suggesting that they are cytoplasmic proteins.[76] Although they have some sequence similarities with aldo-keto reductase enzymes, Kvβ auxiliary subunits probably lack such enzymatic activity.[77] A detailed review of Kvβ subunits is present in this volume (chapter by Pongs *et al.*). Briefly, one function of the three Kvβ1 isoforms (which contain a variable inactivating ball domain) is to induce inactivation in otherwise noninactivating Kv1 channels by providing an extrinsic N-type inactivating domain.[78] In contrast, Kvβ2 accelerates inactivation only when Kv1.4 subunits form part of expressed (inactivating) Kv1 channels, probably by interacting with the intrinsic Kv1.4 inactivating ball.[79–81] For some Kv1 channels, Kvβ1 and Kvβ2 may also shift the voltage dependence of activation in heterologous expression systems.[82] A second role assigned to these auxiliary subunits is to act as chaperones during channel biosynthesis[83,84] and thus to increase expression levels, an effect first described for the interaction of Kvβ2 with Kv1.4.[79] While Kvβ1 and Kvβ2 proteins appear to interact exclusively with Kv1 principal subunits, Kvβ3 and a recently discovered K^+ channel auxiliary subunit, KChAP,[85] appear to interact also with Kv2 principal subunits and to enhance current levels without an effect on channel kinetics or gating.[85,86]

Although Kvβ3 and KChAP appear to have similar roles, their mechanisms of action might be different since chimeras between Kv2.1 and Kv2.2 indicate that the COOH-terminal end of the Kv2.2 protein is essential for its Kvβ3 sensitivity.[86] In contrast, KChAP appears to bind to the NH_2 termini of Kv1 and Kv2 principal subunits.[85]

Since assembly of Kv1 principal subunits occurs mainly in a subfamily-specific manner, it has been suggested that this subfamily specificity may also apply to their association with auxiliary subunits.[49,87] However, this may not strictly be the case, since there is evidence that the *Drosophila* auxiliary subunit homologue Hyperkinetic (*Hk*) associates with members of the *eag* and *Shaker* families and alters their expression levels and/or kinetics.[88] Recently, yet another auxiliary subunit has been identified, KCR1, which accelerates the activation of rat *eag* expressed in *Xenopus* oocytes or in COS-7 cells.[89]

MinK is a 15-kDa single-transmembrane protein that is present in cardiac and auditory cells.[5,90–93] that coassembles with KvLQT1 to form the slow cardiac repolarization current (I_{Ks}).[94–96] (see chapter by Sanguinetti, this volume) and with HERG to regulate the rapidly activating cardiac delayed rectifier (I_{Kr}).[97] Hence, minK contributes to two important outward currents that determine repolarization from the plateau phase of the action potential in ventricular myocytes.[98,99] Although it does not contain a pore domain characteristic of K^+ channels' principal subunits,[100,101] mutagenesis experiments suggest that minK may contribute to the formation of the channel pore,[102] implying that K^+-selective pores may include structures other than P domains or structures having a strict P-loop geometry.

The auxiliary subunit of high-conductance Ca^{2+}-activated K^+ channels (composed of principal subunits of the slo family) is a two-transmembrane protein of 191 amino acids and bears little sequence homology to any other known K^+ channel auxiliary subunit.[103] It has been shown that this subunit contributes to the high-affinity receptor for charybdotoxin[104] but apparently does not affect sensitivity to this drug. Mutagenesis has revealed that the large extracellular loop of the maxi K^+ channel auxiliary subunit has a restricted conformation with two important disulfide bridges. Specifically, four amino acids are critical in conferring high-affinity 251-Chtx binding to the complex. Functionally, it has been demonstrated that this subunit also confers higher Ca^{2+} sensitivity to Ca^{2+}-activated K^+ channels containing slo subunits (see TABLE 4).

Slob is a novel protein isolated by a yeast two-hybrid screen based on its interaction with the COOH-terminal domain of the *Drosophila* slowpoke (dSlo) Ca^{2+}-dependent K^+ channel.[105] Native Slob and dSlo coimmunoprecipitate together from fly head lysates, and they redistribute and colocalize in discrete intracellular structures when coexpressed in heterologous host cells. Although direct application of Slob to excised inside-out membrane patches can strongly activate dSlo (but not human slowpoke channels),[105] the *in vivo* function of Slob remains to be elucidated.

The Kir channel accessory subunit Kv2.2v (TABLE 1) acts as a negative regulator of the inward rectifier Kv2.2 through heteromeric assembly.

Finally, although Kir6 principal subunits (or C-terminal truncated Kir6 subunits) can express independently in heterologous expression systems,[106,107] the functional phenotype of native K_{ATP} channels is conferred only when Kir6 principal subunits are coexpressed with auxiliary subunits (sulfonylurea receptors, SUR1 or SUR2; see TABLE 1). SUR1 and SUR2 are members of the 12 transmembrane-spanning domain proteins, called ATP-binding cassette (ABC) proteins. SUR1 or SUR2 associate with Kir6 subunits as an octameric assembly.[108] Prevailing evidence suggests that the inhibitory effects of nucleotides on K_{ATP} channels are mediated via the principal Kir6 subunits, whereas the potentiation by

ADP is conferred by the sulfonylurea receptor subunit, SUR.[106] SUR auxiliary subunits are also responsible for conferring properties such as sensitivity to pharmacological compounds that block (e.g., glibenclamide) or increase (e.g., pinacidil) opening of the multimeric channel complex. The specific assembly between the principal and auxiliary subunits that takes place seems to be at least partly responsible for the functional diversity of native K_{ATP} channels (Table 3; reviewed by Babenko[109]).

Alternative Splicing

TABLE 1 lists alternative splice versions of principle and accessory K^+ channel subunits known to date. In many cases the use of alternative exons results in channels with diverse properties. TABLE 6 lists only the K^+ channel principal subunits studied so far where alternative splicing leads to diversity in channel function. The functional consequences of K^+ channel gene splicing have been grouped into four categories: (1) effects on electrophysiological properties, where splicing changes channel properties such as kinetics; (2) effects on expression, where splicing results in changes in gene expression and/or tissue distribution; (3) effects on subcellular localization, where alternative splicing results in changes in channel targeting to different cellular compartments; (4) effects on modulation, where splicing alters the effects of modulators (e.g., by protein kinases). Blank spaces indicate that no changes are observed or they remain to be tested. The purpose of TABLE 6 is to emphasize how K^+ channel diversity could be enhanced by mechanisms other than gene duplication and divergence. Furthermore, it should be emphasized that native channels may result from the heteromultimerization between different splice variants, thus increasing diversity even further.

RNA Editing

RNA editing by adenosine deamination has evolved as a mechanism to produce functionally diverse proteins from the same gene. The best-characterized example in the brain is the RNA editing of glutamate-activated receptor channel (GluR) mRNA[110,111] (see chapter by Sprengel in the Glutamate Receptor section, this volume). RNA editing of mammalian K^+ channels has not been described. However, in squid, five purine transitions found in cDNA clones encoding *sqKv2* K^+ channel are also generated by RNA editing.[112] The conductance-voltage relationships determined for the two most frequently edited sites of *sqKv2*, Y576C (pore region) and I597V (S6 segment), did not differ. However, the rate of channel closure upon repolarization was significantly affected by both substitutions.[112]

Posttranslational Modifications

Posttranslational modifications, particularly protein phosphorylation and dephosphorylation, are known to underlie modulation of the activity of ion channels, and hence, modulation of neuronal excitability.[113–115] Posttranslational processing or the distinct intracellular microenvironment of the channel could contribute to the diversity of K^+ channels in native cells. The modulation of Kv3.4 channels by protein kinase C (PKC) is a good example of how channel function and phenotype can be altered in this manner.[116] Phosphorylation of the amino-terminal activation domain of Kv3.4 by PKC suppresses N-type inactivation, converting these channels from rapidly inactivating A type to noninactivating delayed rectifier type.[116,117] Similarly, Roeper and coworkers found that the balance between phosphorylated and dephosphorylated Kv1.4 channels is regulated by changes in intracellular Ca^{2+} concentration, rendering Kv1.4 inactivation gating Ca^{2+} sensitive.[118] They showed that Ca^{2+}-calmodulin–dependent protein kinase (CaMKII) phosphorylation of a single amino-terminal residue of Kv1.4 slows inactivation gating and accelerates recovery from N-type inactivated states; while on the contrary, dephosphorylation of this residue induces a 5 to 10 times faster inactivation of Kv1.4.[118]

Finally, the dynamic regulation of ion channel interactions with the cytoskeleton may contribute to the diversity of K^+ channel properties and mediate aspects of synaptic plasticity (see chapter by Sheng, this volume). For example, Kir2.3 bind to PSD-95, a cytoskeletal protein of postsynaptic densities that clusters NMDA receptors and voltage-dependent K^+ channels: Kir2.3 colocalizes with PSD-95 in neuronal populations in forebrain, and a PSD-95/Kir2.3 complex occurs in hippocampus.[119] Within the C-terminal tail of Kir2.3, a serine residue critical for interaction with PSD-95, is also a substrate for phosphorylation by protein kinase A (PKA).[119] Thus, ion channel interactions with the postsynaptic density are regulated by a physiological mechanism, since stimulation of PKA in intact cells causes rapid dissociation of the channel from PSD-95.[119,120] It also follows from this work that any posttranslational modification that accounts, directly or indirectly, for changes in the localization or clustering of K^+ channels, may contribute to the functional diversity of this family of ion channels.

SUMMARY AND PERSPECTIVE

We attempted to provide a comprehensive, yet succinct, overview of the molecular diversity of K^+ channel subunits. However, due to the rapidly evolving nature of this field of research, it is very likely that this chapter will be outdated by the time of publication. With the recent sequencing of bacterial and yeast genomes, it became clear that many genes exist for which no definite function could be assigned. Given the relatively simple genome of these organisms compared to that of humans (~100,000 genes) and the fact that only a small proportion of the human genome has currently been sequenced, it will be no surprise if many more genes coding for K^+ channel principal and auxiliary subunits are uncovered in the near future. Importantly, sequences might become available to describe a variety of other protein subunits interacting with channels and to modify their function and expression patterns. The incredible diversity of subunits and subunit interaction at the molecular level is at present hard to reconcile with the more limited (although diverse in its own right) complement of native K^+ channels in cells. The correlation between molecular subunits and native K^+ channels in normal and pathophysiological states must remain to be a priority for future research.

These advances also have the promise of providing a wealth of information that will be of benefit to those studying regulation of K^+ channel expression. Promoters, enhancers, and silencers provide yet another level of diversity by virtue of controlling expression levels in response to environmental influences, as well as tissue-specific expression patterns. The recent progress in the determination of the 3D structure of a bacterial K^+ channel protein[17] will pave the way for similar studies on K^+ channel principal and auxiliary subunits. These structural analyses should provide better tools with which to study protein-protein interactions as well as drug-protein interaction. Ultimately, this new knowledge can be used to develop more effective protein-specific therapeutic approaches for pathophysiological states involving K^+ channel dysfunction. This, and the possibility of gene-specific treatment of diseases related to defects in channel subunit proteins, are increasingly areas of active research and will remain fertile areas for future studies.

TABLES

TABLE 1. Mammalian K^+ Channel Principal and Auxiliary Subunits

A. Principal Subunits of Voltage-Activated K^+ Channels

Subunit	Gene Locus Designation	Species	Trivial Name	Chromosome Localization	Alternative Splicing	Accession Number	Associated Diseases
Kv1.1	KCNA1	Human	HUKI	12p13		L02750	Episodic ataxia, myokymia syndrome
		Rat	RK1			X12589	
			RCK1				
			RBK1				
		Mouse	MK1	6		M30439	
			MBK1				
Kv1.2	KCNA2	Human	HUKIV	12		L02752	
		Rat	RBK2			J04731	
			RK2				
			RCK5				
			NGK1				
		Mouse	MK2	3		M30440	
Kv1.3	KCNA3	Human	HUKIII	1p21		M55515	
			HPCN3				
		Rat	RCK3			M31744	
			RGK5				
			KV3				
		Mouse	MK3	3		M30441	
Kv1.4	KCNA4	Human	HUKII	11p14		M55514	
			HPCN2				
		Rat	RCK4			M32867	
			RHK1				
			RIC3				
		Mouse		2		U03723	

A. Principal Subunits of Voltage-Activated K^+ Channels (*continued*)

Subunit	Gene Locus Designation	Species	Trivial Name	Chromosome Localization	Alternative Splicing	Accession Number	Associated Diseases
Kv1.5	KCNA5	Human	HK2	12p13		M55513	
			HPCN1				
		Rat	KV1			M27158	
			RK4				
			RCK7				
		Mouse		6		L22218	
		Mouse	Kv1.5_5′		Kv1.5a	C49507	
			Kv1.5_3′		Kv1.5b	C49507	
Kv1.6	KCNA6	Human	HBK2			X17622	
		Rat	KV2			M27159	
			RCK2				
		Mouse	MK1.6	6		M96688	
Kv1.7	KCNA7	Human		19q13.3			
		Mouse		7		AF032099	
Kv2.1	KCNB1	Human		20q13.2		L02840	
		Rat	DRK1			X16476	
		Mouse		2		M64228	
Kv2.2	KCNB2	Human				U69962	
		Rat	CDRK			M77482	
Kv3.1	KCNC1	Human	NGK2-KV4	11p15	Kv3.1a	S56770	
			KShIIIB				
		Mouse	NGK2	7		Y07521	
			Mshaw22				
		Rat	KV4		Kv3.1b	M37845	
			Raw2				
Kv3.2	KCNC2	Human		12			
		Rat	RKShIIIA		Kv3.2a	M34052	
			Rshaw12				
		Mouse	Mshaw12	10			
		Rat			Kv3.2b	M59211	
		Rat			Kv3.2c	M59213	
		Rat	Raw1		Kv3.2d	M84202	
Kv3.3	KCNC3	Human		19q13.3–13.4	Kv3.3[a]	AF055989	
		Rat	RKShIIID		Kv3.3a	M84210	
		Rat			Kv3.3b	M84211	
		Mouse	MShaw19	7	Kv3.3c	S69381	

A. Principal Subunits of Voltage-Activated K^+ Channels (*continued*)

Subunit	Gene Locus Designation	Species	Trivial Name	Chromosome Localization	Alternative Splicing	Accession Number	Associated Diseases
Kv3.4	KCNC4	Human	HKShIIIC	1p21	Kv3.4b	M64676	
		Rat	Raw3		Kv3.4a	X62841	
		Rat		10, 7	Kv3.4c		
Kv4.1	KCND1	Human	Kv4.1	Xp11.23	Kv4.1	AJ005898	
		Mouse	MShal			M64226	
Kv4.2	KCND2	Human		7q?			
		Rat	RK5			S64320	
		Mouse		6			
Kv4.3	KCND3	Human	Kv4.3M		Kv4.3a	AF048712	
		Rat	KShIVB			U42975	
		Rat	Kv4.3S		Kv4.3b	L48619	
		Human	Kv4.3L		Kv4.3c	AF048713	
		Rat				AB003587	
Kv5.1	KCNH1	Human	KH1	2p25		AF033382	
		Rat	IK8			M81783	
Kv6.1	KCNH2	Human	KH2	20q13		AF033383	
		Rat	K13			M81784	
Kv8.1		Human		8q22.3–8q24.1			
		Rat	Kv2.3r			X98564	
		Mouse				U62810	
Kv9.1	KCNS1	Human				AF043473	
		Mouse				AF008573	
Kv9.2	KCNS2	Human		8q22.4			Cohen syndrome?
		Mouse				AF008574	
Kv9.3	KCNS3	Human				AF043472	
		Rat				AF029056	
Eag1		Rat	Reag-1		Eag1a	Z34264	
		Mouse	Meag-1			U04294	
		Bull	BTeag1			Y13430	
		Bull	BTeag2		Eag1b	Y13431	
Eag2[b]		Rat				AF073891	
Elk1[c]		Rat				AF61957	
Elk2		Human	Helk-2[d]				
		Rat				AF073892	
		Mouse	Melk-2[e]				

A. Principal Subunits of Voltage-Activated K^+ Channels (*continued*)

Subunit	Gene Locus Designation	Species	Trivial Name	Chromosome Localization	Alternative Splicing	Accession Number	Associated Diseases
Erg1	LQT	Human	h-erg	7q35–7q36	Erg1a	U04270	LQT2 syndrome
		Rat	r-erg			Z96109	
		Mouse	Merg1a	5		AF012868	
		Mouse	Merg1a'		Erg1b	AF012871	
		Human[f]	HERGb		Erg1c		
		Mouse	Merg1b			AF012869	
Erg2		Rat				AF016192	
Erg3		Rat				AF016191	
KQT1	KCNQ1	Human	KvLQT1	11p15.5	KCNQ1a	U40990	Long QT syndrome, type I
		Mouse				U70068	
		Human	tKvLQT1		KCNQ1b	AF051426	
KQT2[g]	KCNQ2	Human		20q13.3	KCNQ2a	Y15065	
						AF033348	Benign familial neonatal convulsions (BFNC)
		Human	KQT2.1		KCNQ2b	AF074247	
KQT3	KCNQ3	Human		8q24		AF033347 (partial clone)	BFNC

[a] This sequence differs from Kv3.3a, but this could be a sequencing artifact.

[b] B. Ganetzky also described a partial rat Eag2 sequence in this volume (Reag-2). There are some differences between his sequence and the sequence submitted to Genbank (AF073891).

[c] B. Ganetzy also described a partial Elk1 sequence in this volume (Relk-1). There are clear differences between his sequence and the rat Elk1 sequence submitted to Genbank (AF61957).

[d] B. Ganetzky called this sequence Helk-2. According to the alignment on partial sequences, it seems to be the human homolog of rat Elk2 submitted to Genbank.

[e] Same case as in Helk-2.

[f] Refs. 60 and 140.

[g] The existence of 11 splice variants have been reported (no data available in databases).[141]

B. Principal Subunits of Ca^{2+}-Activated K^+ Channels

Subunit	Gene Locus Designation	Species	Trivial Name	Chromosome Localization	Alternative Splicing	Accession Number	Associated Diseases
SK1	KCNN1	Human	hSK1			U69883	
		Rat	rSK1			U69885	
SK2	KCNN2	Rat	rSK2			U69882	
SK3	KCNN3	Human	hKCa3	22q11–22q13.1		AF031815	Schizo-phrenia? Bipolar disease?
			hSK3				
		Rat	rSK3			U69884	
SK4	KCNN4	Human	hKCa4	19q13.2		AF000972	
			hSK4				
			hIKCa1				
			hIK1				
		Mouse	mIK1			AF042487	
Slo1	KCNMA1	Human	hslo1	10q22.2–10q23.1		U23767	
		Rat	rslo1			U55995	
		Mouse	mslo1[a]			L16912	
Slo3		Human	hslo3			n/a	
		Mouse	mslo3			AF039213	

[a]There is alternative splicing.

C. Principal Subunits of Inward Rectifier K^+ Channels

Subunit	Gene Locus Designation	Species	Trivial Name	Chromosome Localization	Alternative Splicing	Accession Number	Associated Diseases
Kir1.1	KCNJ1	Human	ROMK1	11q24	Kir1.1a	U12541	Barter's syndrome
		Rat	KAB-1			X72341	
		Human	ROMK2 ROMK1B		Kir1.1b	U12542	
		Rat				S69385	
		Mouse				AF012834	
		Human	ROMK3 ROMK1A		Kir1.1c[a]	U12543	
		Rat				S78155	
		Human	ROMK4		Kir1.1d	U12544	
		Human	ROMK5		Kir1.1e	U12545	
		Human	ROMK6		Kir1.1f[b]	U65406	
		Rat[c]				n/a	
		Human[d]	ROMK1C		Kir1.1g	n/a	
		Rat	ROMK6.1		Kir1.1h	AF081368	
Kir2.1	KCNJ2	Human	IRK1 HH-IRK1	17		U12507	
		Rat	RBL-IRK1			Q64273	
		Mouse	MMIRK1 MB-IRK1			X73052	
Kir2.2	KCNJ12	Human	IRK2 HIRK	17p11.1		L36069	
		Rat	RB-IRK2			X78461	
		Mouse	MB-IRK2			X80417	
Kir2.2v	KCNJN1	Human	HKIR2.2v	17p11.2–17p11.1		U53143	
Kir2.3	KCNJ4	Human	HIR HRK1 IRK3	22q13.1		U07364	
		Rat	BIR11			X87635	
		Mouse	MB-IRK3			U11075	
Kir2.4	KCNJ14	Rat	IRK4			AJ003065	

C. Principal Subunits of Inward Rectifier K^+ Channels (*continued*)

Subunit	Gene Locus Designation	Species	Trivial Name	Chromosome Localization	Alternative Splicing	Accession Number	Associated Diseases
Kir3.1	KCNJ3	Human	HGIRK1	2q24.1	Kir3.1a	U50964	
		Rat	GIRK1			L25264	
			KGA				
			KGB1				
			$Kir3.1_{10}$				
		Mouse	MBGIRK1			1582163	
		Rat	$Kir3.1_{01}$		Kir3.1b	U60025	
		Rat	$Kir3.1_{00}$		Kir3.1c	U42423	
			Kir3.1delta				
		Rat	$Kir3.1_{11}$		Kir3.1d	U72410	
			Kir3.1-deltaB				
Kir3.2	KCNJ7	Human	GIRK2	21q22.1	Kir3.2a[e]	L78480	Mapped to Down syndrome chromosome region 1
	KCNJ6		KATP2				
			BIR1				
		Rat	KATP-2			U21087	
		Mouse	GIRK2–1	16		U51122	Weaver mouse mutation
			MBGIRK2				
		Mouse	GIRK2A-1		Kir3.2b	U51123	
		Mouse	GIRK2A-2		Kir3.2c	U51124	
		Mouse	GIRK2B		Kir3.2d	U51125	
		Mouse	GIRK2C		Kir3.2e	U51126	
		Mouse	GIRK2D		Kir3.2f[f]	n/a	
Kir3.3	KCNJ9	Human	GIRK3	1q21–1q23		U52152	
		Rat	RBGIRK3			L77929	
		Mouse	MBGIRK3			U11860	
Kir3.4	KCNJ5	Human	GIRK4	11q24		U52154	
			CIR				
			hc-KATP1				
		Rat	rc-KATP1			L35771	
		Mouse				U72061	
Kir4.1 (Kir1.2)	KCNJ10	Human		1q		U52155	
		Rat	BIR10			X83585	
			KAB-2				

C. Principal Subunits of Inward Rectifier K$^+$ Channels (*continued*)

Subunit	Gene Locus Designation	Species	Trivial Name	Chromosome Localization	Alternative Splicing	Accession Number	Associated Diseases
Kir4.2 (Kir1.3)	KCNJ15	Human		21q22.2		U73191	Mapped to Down syndrome chromosome region 1
Kir5.1	KCNJ16	Rat	BIR9			P52191	
Kir6.1	KCNJ8	Human[g]	uKATP–1	12p12	Kir6.1a[h]	D50312	
						AF015605	
		Rat				D42145	
		Mouse	muKATP–1	6		D88159	
		Human			Kir6.1b	AF015606	
		Human			Kir6.1c	AF015607	
Kir6.2	KCNJ11	Human	BIR hBIR IKATP	11p15.1		D50582	Familial persistent hyper-insulinemic hypo-glycemia of infancy
		Rat	rBIR			D86039	
		Mouse	mBIR			D50581	
Kir7.1 (Kir1.4)	KCNJ13	Human				AF061118	
		Rat				AJ006129	

[a] The human and the rat ROMK3 transcripts are not true orthologs in terms of alternative splicing and exons usage.

[b] The human and the rat ROMK6 transcripts do not correspond in terms of alternative splicing and exons usage.

[c] Ref. 142.

[d] Ref. 143.

[e] The three mammalian transcripts (human, rat, and mouse) do not correspond to each other in terms of alternative splicing.

[f] Ref. 144.

[g] Two accession numbers are given here. D50312 represents the coding sequence, and AF015605 represents splice variant A for the 5′UTR.

[h] The three mammalian transcripts do not correspond to each other in terms of alternative splicing.

D. Principal Subunits of Two-Pore K^+ Channels

Subunit	Gene Locus Designation	Species	Trivial Name	Chromosome Localization	Alternative Splicing	Accession Number	Associated Diseases
TWIK	KCNK1	Human	TWIK-1	1q42–1q43		U33632	
		Mouse		8		U86009	
TREK	KCNK2	Mouse				U73488	
TASK		Human				AF006823	
		Rat				AF031384	
		Mouse				AF006824	
TRAAK		Mouse				AF056492	

E. Auxiliary Subunits of Kv and eag Channels

Subunit	Gene Locus Designation	Species	Trivial Name	Chromosome Localization	Alternative Splicing	Accession Number	Associated Diseases
Kvβ1	KCNA1B	Human	HKvβ1a	3q26.1	Kvβ1.1	U33428	
		Rat	Kvβ1			X70662	
		Mouse				X97281	
		Human	hKvβ3		Kvβ1.2	U16953	
		Human	hKvβ1.3		Kvβ1.3	L47665	
Kvβ2	KCNA2B	Human		1p36.3	Kvβ2.1	U33429	Charcot-Marie-Tooth disease, type 2A?
		Rat	RCKβ2 Kvβ2.1			X76724	
		Mouse		4		U31908	
		Human			Kvβ2.2	AF044253	
Kvβ3	KCNA3B	Human	KCNA3.1B	17q13	Kvβ3.1	AF016411	
		Rat	Kvβ3 RCKβ3			X76723	
		Mouse	mKvβ4		Kvβ3.2	U65593	
KChAP		Rat				AF032872	

F. Auxiliary Subunits of Ca^{2+}-Activated K^+ Channels

Subunit	Gene Locus Designation	Species	Trivial Name	Chromosome Localization	Alternative Splicing	Accession Number	Associated Diseases
Slo Beta		Human				U38907	
		Rat				U40602	
		Mouse				AJ001291	

G. Auxiliary Subunits of Inward Rectifying K^+ Channels

Subunit	Gene Locus Designation	Species	Trivial Name	Chromosome Localization	Alternative Splicing	Accession Number	Associated Diseases
SUR1	SUR	Human		11p15.1		Q09428	Familial persistent hyper-insulinemic hypo-glycemia of infancy
		Rat				X97279	
		Rat			SUR1B	AF039595	
SUR2		Human		12p12.1	SUR2A	AF061324	
		Mouse	mSUR	6		D80637	
		Human			SUR2B	AF061324	
		Rat				AF019628	
		Mouse				D086038	

TABLE 2. Sequence Identity (%)[a] between Genes Coding for K^+ Channel Principal Subunits

A. 6TMD K^+ Channel Principal Subunits

	Kv1	Kv2	Kv3	Kv4	Kv5	Kv6	Kv8	Kv9	KQT	SK	eag	erg	elk	slo
Kv1	**74-88**	46-49	44-48	41-45	38-40	37-39	35-36	36-42	19-24	13-18	7-13	11-14	11-13	11-15
Kv2		**99**	42-43	40-41	52	42	46	46-50	22-23	13-14	9-14	13	11-14	14-16
Kv3			**87-92**	37-38	38-40	34-35	34-35	33-39	19-23	11-14	7-11	12-14	11-13	11-18
Kv4				**87-94**	38-39	35-37	35-37	33-35	19-25	12-13	10-13	12-14	11-15	11-16
Kv5					**100**	38	41	39-45	22-25	12-14	7-11	10-11	12-13	12-15
Kv6						**100**	41	39-40	20-22	12-14	6-12	10-11	12	10-16
Kv8							**100**	46-49	20-24	13-16	9-15	13-14	14-16	11-12
Kv9								**59-63**	20-23	12-15	8-14	13-14	12-17	11-17
KQT									**55-71**	10-15	9-14	12-14	11-15	8-13
SK										**43-91**	5-12	8-13	10-13	12-15
eag											**70**	42-51	37-48	11-12
erg												**88-89**	51-54	9-14
elk													**67**	12
slo														**59**

B. SK Family Principal Subunits

	SK2	SK3	SK4
SK1	**91**	**85**	43
SK2		**84**	43
SK3			44

C. Kir Principal Subunits

	Kir1/Kir4/Kir7	Kir2	Kir3	Kir5	Kir6
Kir1/Kir4/Kir7	36-**62**	36-50	29-42	27-35	32-46
Kir2		**61-94**	45-52	44-48	39-46
Kir3			**58-77**	36-38	42-46
Kir5				**100**	37
Kir6					**69**

D. Kir1.1, Kir4.1 and Kir4.2 Principal Subunits

	Kir4.1 (Kir1.2)	Kir4.2 (Kir1.3)	Kir7.1 (Kir1.4)
Kir1.1	47	47	36
Kir4.1		**62**	39
Kir4.2			37

[a]The percent identity values were generated by importing alignment files created by Clustal X into the program Genedoc.

TABLE 3. Native Channels and Physiological Significance of the Molecular Diversity of K^+ Channels

Channel Proteins	Native Channels
Kv1.x (probably in heteromultimeric channels also containing Kvβ subunits)	4-AP–sensitive voltage-gated K^+ channels of both delayed rectifier and "A" types, including the "D" current (a dendrotoxin–sensitive voltage-dependent current with variable kinetics and voltage dependence; also sensitive to very low 4-AP concentrations).[121,122] "D" channels are probably various combinations of Kv1.2 (or Kv1.1 or Kv1.6) with other Kv1 proteins and Kvβ subunits. Kv1.4 (possibly in homomultimeric and heteromultimeric channels) might be responsible for fast "A" type K^+ currents in terminals and axons.[50,123,124] Use of antisense oligonucleotides suggests that Kv1.5 is responsible for the ultrarapid delayed rectifier current (I_{Kur}) in human atrium.[125]
Kv2.x, possibly in combination with Kv5.1, Kv6.1, Kv8.1, and Kv9.1–9.3	4-AP–sensitive delayed rectifiers with variable kinetics, slow inactivation and voltage dependence. Oxygen-sensitive currents in pulmonary artery myocytes.[126]
Kv3.1–Kv3.4 (may exist as homo- and heteromultimeric channels)	High voltage–activating, fast deactivating voltage-dependent K^+ currents, which may contain a fast inactivating component if they include Kv3.4 proteins. Channels containing mainly Kv3.1 and Kv3.2 proteins (and perhaps Kv3.3) are most likely delayed rectifiers with slow inactivation. Currents blocked by 1 mM TEA which are key in the repolarization of short action potentials in fast spiking neurons (see chapter by Rudy *et al.*, this volume).
Kv4.x (possibly in association with unidentified auxiliary subunits)	Classical low voltage– or subthreshold-activating A-type currents and the I_{to} in heart. Evidence for a relationship with I_{to} includes use of antisense oligonucleotides.[127,128] Blocked by mM concentrations of 4-AP.
KQT	KQT1 is responsible for the slowly activating delayed rectifier (I_{Ks}) in heart.[94,95] Mutations in this gene are responsible for a form of LQT syndrome (see chapter by Sanguinetti, this volume). KQT2 in association with KQT3 may form delayed rectifier type currents in neurons (see text). Mutations in these two genes cause a form of epilepsy (see TABLE 1).
eag	"M" current?[129,130]
elk	?
erg	Erg1 is responsible for the rapidly activating delayed rectifier current (I_{Kr}) in heart (in combination with minK).[97] Evidence includes use of antisense oligonucleotides.[131] LQT syndrome mapped to mutations of herg and minK (see TABLE 1 and chapter by Sanguinetti, this volume).
slo (different alternatively spliced versions, with and without a β subunit)	Maxi-K or BK Ca^{2+}-activated channels. Evidence includes biochemical purification from native tissue.[132,133]
SK 1–3	Small-conductance apamin-sensitive and -insensitive Ca^{2+}-activated K^+ channels (see chapter by Adelman, this volume).
SK4	Medium-conductance Ca^{2+}-activated K^+ channels.[134–136]
Two-pore K^+ channels	"Leak" K^+ channels, which help regulate the resting potential.[12]
Kir1.1–Kir1.3	Weak inward rectifiers.

TABLE 3. Native Channels and Physiological Significance of the Molecular Diversity of K^+ Channels (*continued*)

Channel Proteins	Native Channels
Kir2.1–2.4	Kir2.1: I_{K1} in cardiac ventricular myocytes, 21 pS channel.[65] Common strongly rectifying inward rectifier channel.
Kir3.1 + Kir3.4	G-protein–activated K^+ channels in neurons and heart (particularly in atrium).[137] Kir3.1 + Kir3.4 acetylcholine-activated K^+ current ($I_{K,Ach}$) in heart (see chapter by Kevin Wickman, this volume).
Kir6.1 plus SUR2B	K_{NDP} (33pS nucleotide–sensitive K^+ channel in vascular smooth muscle). Activated by ADP, GDP; blocked by glibenclamide (IC_{50} = 25 nM)
Kir6.2 plus SUR1	ATP-sensitive K^+ current ($I_{K,ATP}$) in pancreatic β-cells. $\gamma = 70$ pS; K_i (ATP) = 10 μM; KCOs (diazoxide > pinacidil), glibenclamide (IC_{50} < 10 nM)
Kir6.2 plus SUR2A	ATP-sensitive K^+ current ($I_{K,ATP}$) in heart and skeletal muscle. $\gamma = 80$ pS; K_i (ATP) = 175 μM; KCOs (pinacidil > diazoxide), glibenclamide (IC_{50} = 1.2 μM)
Kir6.2 plus SUR2B	ATP-sensitive K^+ current ($I_{K,ATP}$) in vascular smooth muscle. $\gamma =$ 80 pS; K_i (ATP) = 53 μM; KCOs (pinacidil > diazoxide), glibenclamide (IC_{50} = 1 μM)

NOTE: γ = unitary conductance.

TABLE 4. Functional Properties of K^+ Channel Proteins Expressed in Heterologous Expression Systems

A. Functional Properties of Voltage-Activated K^+ Channels

	Electrophysiological Properties										Pharmacological Properties (IC_{50})				
	Activation				Deactivation	Inactivation				Single Channel g					
	V_{on} (mV)	$V_{1/2}$ (mV)	k (mV)	τ^a	τ (ms)^b	$V_{1/2}$ (mV)	k (mV)	τ (ms)^c	τ_{rec}	(pS)	TEA_o (mM)	4-AP (mM)	DTX (nM)	CTX (nM)	Other Properties
Kv1.1	−60 to −50	−30	6 to 9	+++	14			Very slow^d		≈ 10	0.5	0.16 to 1.1	12 to 21	NB	TEA (human) IC_{50} 20 mM? HgTX IC_{50} 31 pM
Kv1.2	−40	−5 to 5	13	+	23			Very slow		9.2 to 17	> 10	0.2 to 0.8	2.8 to 24	1.7 to 17	HgTX IC_{50} 170 pM
Kv1.3	−50	−30	5 to 7	++	38	−44.7		250 to 600		9.6 to 14	10 to 50	0.2 to 1.5	250?	0.5 to 2.6	MgTX IC_{50} 230 pM HgTX IC_{50} 86 pM Shows cumulative inactivation
Kv1.4	−50	−22	12	+++	15 to 55	−65 to −45	+2.3	$\tau_1 \approx 20$ $\tau_2 \approx 40$	2.2 to 3.8s (−80 mV)	4.7	NB	0.7 to 13	> 200	NB	
Kv1.5	−50	−10	7	+++	23			Very slow		8	NB	< 0.1 to 0.4	NB	NB	
Kv1.6	−50	−20	6 to 8	+++				Very slow		9	1.7 to 7	0.3 to 1.5	20 to 25	1	MgTX IC_{50} 3 nM HgTX IC_{50} 6 nM
Kv1.7	−40	−20	8	+++	5.1 to 5.3	NA	NA	14	Slow	21	NB	0.25	NA	NB	MgTX IC_{50} 116 nM NxTX IC_{50} 18 nM ShK TX IC_{50} 13 nM
Kv2.1	−20 to −30	≈ 10	5 to 19	+	15 to 20	−20	5 to 13	Very slow^e	1.6 s (−90mV)	8 to 9	4 to 10	0.5 to 4.5	NB	NB	HaTX K_D 100 nM (see note *f*)

A. Functional Properties of Voltage-Activated K^+ Channels (*continued*)

	Electrophysiological Properties										Pharmacological Properties (IC_{50})				
	Activation				Deactivation	Inactivation				Single Channel g					
	V_{on} (mV)	$V_{1/2}$ (mV)	k (mV)	τ^a	τ (ms)^b	$V_{1/2}$ (mV)	k (mV)	τ (ms)^c	τ_{rec}	(pS)	TEA_o (mM)	4-AP (mM)	DTX (nM)	CTX (nM)	Other Properties
Kv2.2	−20 to −30	≈ 10	≈ 18	+	NA	−30	13	Very slow[e]	NA	≈ 14	≈ 8	1.5	NA	NA	Note *f*
Kv3.1	−10	+10 to 18	8 to 11	++	1.4			Very slow		16 to 27	0.15 to 0.2	0.02 to 0.6	NB	NB	Chromakalin K_D 0.237 mM
Kv3.2	−10	+10 to 13	7 to 9	++	2			Very slow		16 to 20	0.15	0.6 to 0.9	NB	NB	
Kv3.3	−10	7	6	++	2	5.2	6.1	240	NA	14	0.14	1.2			
Kv3.4	−10	+13 to 19	7 to 11	+++	NA	−20 to −32	4.8 to 8.3	10 to 20	NA	12 to 14	0.09 to 0.3	0.5 to 0.6	NB	NB	BDS-I IC_{50} 47 nM BDS-II IC_{50} 56 nM
Kv4.1	−60 to −50	≈ −10	= 13	++	NA	−69 to −50	4.7 to 8	$\tau_1$22.6 $\tau_2$86.4 $\tau_3$368[h]	200 ms (−100 mV)[g]	6 to 8	NB	9	NA	NA	
Kv4.2	≈ −50	−4 to −15	13 to 20	++	≈ 40	−41 to −66	6 to 8	τ_1 15 to 20 [≈ 80%] τ_2 61 to 120	150 to 300 ms (−80 mV)[g]	4 to 5	NB	2 to 5[i]	NB	NA	HpTX1 and HpTX2 IC_{50} 100 nM, HpTX3 IC_{50} 67 nM[j]
Kv4.3	−60 to −50	= −20	≈ 13	++	20 to 40	≈ −60	4.5 to 7	$\tau_1$30 [≈ 70%] τ_2 160	100 to 200 ms[g] (−80 to −100 mV)	NA	NB	5 to 10[i]	NA	NA	

A. Functional Properties of Voltage-Activated K^+ Channels (*continued*)

	Electrophysiological Properties										Pharmacological Properties (IC_{50})				
	Activation				Deactivation	Inactivation				Single Channel g					
	V_{on} (mV)	$V_{1/2}$ (mV)	k (mV)	τ^a	τ (ms)^b	$V_{1/2}$ (mV)	k (mV)	τ (ms)^c	τ_{rec}	(pS)	TEA_o (mM)	4-AP (mM)	DTX (nM)	CTX (nM)	Other Properties
Kv5.1, Kv6.1, Kv8.1, Kv9.x	No detectable current when expressed along (see text for heteromultimer formation with Kv2.x)														
KCNQ1	−50 to −40	−11.6	12.6	>100^k	Incomplete prepulse inactivation that does not produce current decay					NA	40% blocked by 96 mM	NB by 2 mM	NA	NB	Blocked by 10 μM Clofilium
KCNQ1 + minK	−50 to −20	7.5	16.5	>500	No inactivation					NA			NA		
KCNQ2	−60	NA	NA	>100						NA	90% blocked by 1 mM	NB by 2 mM	NA	NB	Not blocked by Clofilium
KCXNQ3	−70	NA	NA	>100	Inactivation might be similar to KCNQ1					NA	NB by 5 mM	NB by 2 mM	NA	NB	30% blocked by 10 μM clofilium
KCNQ2 +KCNQ3	−60	NA	NA							NA	20% blocked by 5 mM	NB by 2 mM	NA	NB	Low sensitivity to clofilium

A. Functional Properties of Voltage-Activated K^+ Channels (*continued*)

	Electrophysiological Properties										Pharmacological Properties (IC_{50})				
	Activation				Deactivation	Inactivation				Single Channel g					
	V_{on} (mV)	$V_{1/2}$ (mV)	k (mV)	τ[a]	τ (ms)[b]	$V_{1/2}$ (mV)	k (mV)	τ (ms)[c]	τ_{rec}	(pS)	TEA_o (mM)	4-AP (mM)	DTX (nM)	CTX (nM)	Other Properties
Eag1	–40	–7.5	19.6	++[l]	τ_1 4.6 $\tau_2$76 (–90 mV)			Slow		NA	28	> 100	NA	NA	M current? IKx in rod outer segment?
Erg1	–60 to –50	–21	6 to 8	Hundreds of ms	Erg1 and 2 behave as slowly activating delayed rectifiers (τ~100 ms), but at positive voltages (> 0 mV) there is a reduction in steady state (ss) current producing a negative slope I–V and inward rectification. Erg3 currents have a prominent transient component and activate 5–10 times faster and also have a negative slope I–V. Upon repolarization from positive voltages the 3 channels produce large currents that decay slowly. This behavior is due to the presence of a voltage-dependent "C" inactivation process, which is 2 to 10 times faster than activation at most voltages. Recovery from inactivation is very fast for all 3 channels (< 10 ms).					5 to 12 in symmetric $[K^+]$	50 NA	NA NA	NA NA	NA NA	Blocked by dofetilide E4031 K_D99 nM Blocked by 2 µM MK-499
Erg2	–40	–3.5	8.3							NA	NA	NA	NA	NA	E4031 K_D 116 nM
Erg3	–80 to –70	–44	7.2							NA		NA	NA	NA	E4031 K_D 193 nM
Elk1	–40[m]	9.3	13.1			69				NA	NB by 10 mM	NB by 10 mM			NB by 10 µM E4031

A. Functional Properties of Voltage-Activated K^+ Channels (*continued*)

NOTE: Data from mammalian cells used when available, oocytes if not. If considered in several papers and there is a consensus, we have placed an average. Sometimes we have selected papers in which there is a very detailed and careful analysis. If not consensus, we give the range. All values reported at ~20 °C.

ABBREVIATIONS: IC_{50}, Concentration at which 50% of inhibition is reached; V_{on}, activation voltage; $V_{1/2}$, voltage at which 50% of activation or inactivation is reached; k, slope; τ, time constant; TEA_o, extracellular tetraethyl ammonium; 4-AP, 4 aminopiridine; DTX, dendrotoxin; CTX, charibdotoxin; HgTX, hongotoxin; MgTX, margatoxin; HpTX, heteropodatoxin; NxTX, noxiustoxin; HaTX, hanatoxin, K_D, dissociation constant; NB, not blocked at high pharmacological concentrations; NA, no data available.

[a] τ of activation for fast activating currents: +++ < 10 ms; ++ 10–20 ms; + > 20 ms at + 20 to + 40 mV.

[b] At –60 mV unless otherwise stated.

[c] At 40 mV unless otherwise stated.

[d] Very slow = seconds.

[e] Incomplete inactivation.

[f] Several properties are modified by coexpression with Kv5.1, Kv6.1, Kv8.1, and Kv9.x (see section on heteromultimer formation).

[g] Recovery from inactivation calculated from fits to a single exponential but may not be monoexponential. Recovery from inactivation significantly accelerated by coexpression with 2 to 4 Kb brain mRNA.

[h] τ of inactivation for Kv4 channels from pulses at 20 mV. [%] = fractional amplitude of τ.

[i] Incomplete block. Complete block when coexpressed with 2 to 4 Kb brain mRNA.

[j] Blocked in a voltage dependent manner.

[k] Slow deactivation (>100 ms) with an initial hook. Deactivation is slowed down by coexpression with minK.

[l] Depends on V_H. Slowed down by hyperpolarization.

[m] Voltage dependence is highly sensitive to pH.

B. Functional Properties of Ca^{2+}-Activated K^+ Channels

	Voltage Dependence	Ca^{2+} Sensitivity	Single Channel γ (pS)	Pharmacology (IC_{50} or K_D)	Other Properties
Slo1	Conductance is voltage dependent. $V_{1/2}$ shifts with Ca^{2+} concentration from ≈50 mV at 4 μM [Ca^{2+}] to −30 mV at 100 μM [Ca^{2+}]. Voltage dependence is also shifted by coexpression with β subunit.	Ca^{2+} sensitivity is lower than channels from SK family and changes significantly depending on the alternative spliced version and on the presence of the β subunit.	> 200 in symmetric [K^+]	TEA 80 to 330 μM	CTX IC_{50} 2 to 40 nM. Iberiotoxin (IbTX) IC_{50} 6 to 11 nM. β subunit decreases sensitivity to IbTX (IC_{50} 102 to 160 nM) and confers sensitivity to the agonist DHS-1, and does not affect CTX sensitivity.
Slo3	Conductance is sensitive to voltage and pH but apparently not to Ca^{2+}		106 in symmetric [K^+]	TEA 49 mM	Expresses only in testis
SK1	Voltage independent	$K_{0.5}$ 0.71 μM Hill coefficient 3.9	9.2 in symmetric [K^+]	Apamin insensitive d-Tubocurarine 76.2 μM	
SK2	Voltage independent	$K_{0.5}$ 0.63 μM Hill coefficient 4.8	9.9 in symmetric [K^+]	Apamin 63 pM d-Tubocurarine 2.4 μM	
SK3	Voltage independent	Similar to SK1 and SK2	≈ 10	Apamin 1–2 nM	
SK4	Voltage independent	K_D 270 nM Hill coefficient 2.7	33 in symmetric [K^+]	Apamin insensitive	Not blocked by IbTX and MgTX; inwardly rectifying
			9 in asymmetric [K^+]	TEA 30 mM; CTX 10 nM Cortrimazole 387 nM	

C. Functional Properties of Inward-Rectifier K^+ Channels

					Blockers				
	γ (pS)[b]	Mean Open Time (ms)	Mean Closed Time (ms)	Rectification[c]	Spermidine[d]	Spermine[d]	$Ba^{2+}{}_o$ (μM)	$Cs^+{}_o$ (μM)	Special Properties
Kir1.1	31–39	24 ms (−80 mV)	1.1 & 35 ms (−80 mV)	Weak	2.16 mM	ND	70 (−100 mV)	1.2 mM (−120 mV)	Activated by ATP_i (blocked by Mg^{2+}; IC_{50} = 1.7–4.2 mM)[e]
Kir2.1	21–23	117–185 ms (−100 mV)	0.8 & 26ms (−100 mV)	Strong	8 nM & 2.9 μM	0.9 nM & 0.6 μM	0.15 (−140 mV)	14 (−140 mV); 54 mM (0 mV)	Blocked by Mg^{2+}; IC_{50} = 17 & 2170 μM[e]
Kir2.2	34–41	71 ms (−100 mV)	0.7 ms & 11.9 ms (−100 mV[f])	Strong	ND	ND	6 (−80 mV)	ND	
Kir2.3	13–16	ND	ND	Strong	ND	ND	93–183 (−130 mV)	30–47 (−130 mV)	Sensitive to pH_o and ATP_i
Kir2.4	15	ND	ND	Strong	ND	ND	390 (−80 mV)	8.06 mM (−80 mV)	
Kir3.1	27–42 (39[g])	0.26, 1.2 & 7.2 ms (−60 mV)	ND	Strong	ND	ND	10–94	94	Expresses as a heteromultimer with Kir3.2 or Kir3.4
Kir3.2	30 (35[h])	0.1 & 0.5 ms (0.5 & 3.5 ms[h]) (−80 mV[i])	ND	Strong	ND	10 μM	94 (−120 mV)	94 (−120 mV)	Expresses as a heteromultimer with Kir3.1 or Kir3.4
Kir3.3	—	—	—	—	—	—	—	—	No expression alone Coassembly subunit of Kir3.1 & Kir3.2

C. Functional Properties of Inward-Rectifier K^+ Channels (*continued*)

					Blockers				
	γ (pS)[b]	Mean Open Time (ms)	Mean Closed Time (ms)	Rectification[c]	Spermidine[d]	Spermine[d]	Ba^{2+}_o (μM)	Cs^+_o (μM)	Special Properties
Kir3.4	31–33 (35[h])	1.3 ms (−80 mV[h])		Strong	ND	ND	92 (−60 mV)	61 (−60 mV)	Coassembly subunit of Kir3.1 Inhibits expression of Kir3.2 May form stretch-activated K^+ channels
Kir4.1	22 & 36[j]	100–200 ms (−100 mV)	1 & 30 ms (−100 mV)	Weak	40 nM	ND	99 (−120 mV)	112 (0 mV)	Activated by ATP_i; inhibited by pH_i ($pK_a = 6.2$)
Kir4.2	—	—	—	—	—	—	—	—	No expression as homomultimeric subunits; inhibits expression of Kir1.1 or Kir1.2 currents
Kir5.1	—	—	—	—	—	—	—	—	No expression as homomultimeric subunits[k]
Kir7.1	0.05	ND	ND	Strong	ND	ND	1000	10,000	

[a] Excluding K_{ATP} channels, for which see TABLE 4D.
[b] Single-channel conductance (γ) was measured at high (140–150 mM) symmetrical K^+ concentrations.
[c] Rectification is conferred by block of the open channel by intracellular Mg^{2+} and polyamines.
[d] Measured at +40 mV.
[e] All Kir channels are blocked to some extent by Mg^{2+} (one of the prime reasons for rectification). IC_{50} values are not available in all cases.
[f] Channel activity of Kir2.2 occurred in bursts, separated by closings of >200 ms. The gap duration increases with hyperpolarization.
[g] When coexpressed with Kir3.4.
[h] When coexpressed with Kir3.1.
[i] Openings occur in bursts; the mean open time or burst kinetics were not analyzed.
[j] The 36 pS channel was only observed in *Xenopus* oocytes and not in mammalian cells, suggesting coassembly with endogenous *Xenopus* subunits.
[k] Has been reported to form functional heteromultimeric assemblies with Kir4.1

D. Functional Properties of ATP-Sensitive K^+ Channels

Principal Subunit with/without Auxiliary Subunit	Single-Channel Conductance γ (pS)	Mean Open Time (ms)	Mean Closed Time (ms)	Mean Burst Duration (ms)	Interburst Duration (ms)	Nucleotide Block (μM)	Blockers (nM)	Openers (μM)
Kir6.1 alone	70	3.31 (−60 mV)	0.9 (−60 mV)	ND	ND	ND	Glib (no block)	No effect
Kir6.1+SUR1	ND	ND	ND	ND	ND	ND	Tol (17.7 μM)	ND
Kir6.1+SUR2B	32.9	ND‘	ND	ND	ND	ATP = GTP (3)[a]	Glib (< 3 μM)	Pinacidil (< 100 μM)
Kir6.2 alone	57.6[b] 70	0.8–1.9 (−60 mV[b])	0.31&12.6 (−60 mV[b])	2.4–8.3 (−60 mV[b])	ND	Mg-ATP (106) Mg-ATP (115–145)[b]	No block	No effect
Kir6.2+SUR1	58–76 (73)[b]	2.1 (−80 mV)	0.56 (−80 mV)	17.9 (−60 mV)	24 & 243 (−60 mV)	ATP (6–34); Mg-ATP (28); ATP (13–18)[b]	Glib (1.8–8.6) > Tol (4 −32 μM)	Diazoxide (20–100 μM) > pinacidil
Kir6.2+SUR2A	79–80	1.3 (−80 mV)	0.16 (−80 mV)	ND	ND	ATP (100–148) > ADP>AMP Mg-ATP (172)	Glib (160) Glib (350)[c] Glib (630)[d] Tol (120 μM)	Pinacidil (10) > nicorandil >> diazoxide
Kir6.2+SUR2B	80.3	ND	ND	ND	ND	ATP (67.9) > Mg-ATP (300)	Glib (< 1 μM)	Pinacidil > diazoxide

[a] Activated by nucleotides.
[b] C-terminal truncated form of Kir6.2.
[c] Rb^+ efflux measurements.
[d] Binding assays.

E. Functional Properties of Two-Pore K^+ Channels

				Pharmacology	
	γ (pS)[a]	Rectification	Special Properties	Ba_0^{2+}	Cs_0^{2+}
TWIK	35	Inward (weak)		35–100 µM	Not blocked
TREK	14	Outward[b]		100 µM	Not blocked
TASK	16	Outward[b]	Highly sensitive to extracellular pH pKa = 6.3 Hill = 1.6	29% blocked by 100 µM	30% blocked by 100 µM
TRAAK	45	Outward[b]	Regulated by arachidonic acid and polyunsaturated fatty acids	1 mM at 30 mV	Not blocked

[a] Single-channel conductance (γ) was measured at high (140–150 mM) symmetrical K^+ concentrations.
[b] Obeys Goldman-Hodgkin-Katz current equation for changes in $[K]_o$.

TABLE 5. Tissue Expression of K^+ Channel Principal Subunits

	Tissue Distribution[a]							Brain Region							
	sp	He	Ki	SK	Lu	Br	Other	sp	CX	Th	CB	OB	HC	Other	Protein Distribution
Kv1.1	r	–	–	–	–	+		r	+	+	+	+	+		Axons and terminals.
Kv1.2	r	+[i]	–	ND	ND	+		r	+	+	+	+	+	Hb	Axons and nerve terminals Paranodal in nodes of Ranvier
Kv1.3	r	–	±	–	+	+	Re, spleen, lymph	r	ND	ND	+	+	+		Somatodendritic
Kv1.4	r	+	–	–	ND	+		r	+	+	+	+	+	CP	Axons and terminals
Kv1.5	r	+	+	+	+	+	Pituitary, aorta	r	ND	ND	ND	ND	+		Somas and proximal dendrites
Kv1.6	r/h	–	–	–	–	+		r	ND	ND	±	±	+		Somatodendritic
Kv1.7	r	+	±	+	+	±		ND							
Kv2.1	r	+	–	±	–	±		r	+	±	+	+	+	Hb	Somatic and proximal dendrites
Kv2.2	r	ND	ND	ND	ND	+		r	+	+	+	+	+		Somatic and neuropil
Kv3.1	r	±	–	±	–	+	T-lymph	r	+	+	+	+	+	RT	Somatic and axonal, terminals
Kv3.2	r	±	–	–	–	+		r	+	+	±	±	+		Somas and axon collaterals
Kv3.3	r	±	–	–	ND	+	Thy	r	+	±	+	+	+		
Kv3.4	r	±	–	+	–	+		r	±	–	±	–	±		
Kv4.1	r	±	ND	ND	ND	±		r	–	–	–	+	+		
Kv4.2	r	+	ND	ND	ND	+	Vas deferens[e]	r	+	+	+	+	+	Hb, CP	Somatodendritic
Kv4.3	r	+	+[e]	+[e]	+[e]	+	Vas deferens[e]	r	+	+	+	+	+	Hb, SNPC, CP	
Kv5.1	r	±	±	±	–	+		r	±	±	–	+	±		
Kv6.1	r	–	±	±	+	±	Li	r	+	+	+	+	+	A, CP	

TABLE 5. Tissue Expression of K^+ Channel Principal Subunits (*continued*)

	Tissue Distribution[a]							Brain Region							
	sp	He	Ki	SK	Lu	Br	Other	sp	CX	Th	CB	OB	HC	Other	Protein Distribution
KQT1	h	+	+	–	+	–	Pla, Pa	ND							
KQT2	h	–	–	–	–	+		h[a]	+	+	+	ND	+	A, C. nucl.	
KQT3	h	–	–	–	–	+		h	+	+	±	ND	+	A, C. nucl.	
Kv8.1[b]	r	–	–	–	–	+		r	+	–	+	+	+	CP	
Kv9.1	m/r	–	–	–	–	+		m	+	–	+	+	+		
Kv9.2	m	–	–	–	–	+		m	+	–	+	+	+	Hb, BAN	
Kv9.3	r	±	±	–	+	+		ND							
Kir1.1	r	–	+	–	±	+	Spleen	r	–[a]	±?[a]	–[a]	ND	–[a]	Hypothalamus	
Kir2.1	m	+	–	+	ND	+		r	+	+	+	+	+	Purkinje cells, CP	
Kir2.2	h/r/m	+/+/+	ND/+/+	ND/+/+	ND/+/+	ND/+/+		r	+	+	+	±	±	CP, Strong in Th and CB	
Kir2.3	h/r/m	+/–/–	–?/–/–	±/–/–?	–/–/ND	+		r	+	±	±?	+	+	CP, Th ret. nucleous, amygdala[a]	
Kir2.4	h/r	–/+	ND/+	ND/–	–/±	+		ND						Midbrain, brainstem	
Kir3.1	r/m	+	ND/–	±?	+/ND	+	Atrium > ventricle	r	+	+	+	+	+	Not in Purkinje cells	
Kir3.2	r/m	–	–	–	–	+	High in rat pancreas	r	+	+	–	+	+	C3 pyramidal cells	
Kir3.3	m	–	–	–	–	+		r	+	+	+	+	+	Brainstem, midbrain	
Kir3.4	r/m	+	+[e]/ND	–[e]/ND	–[e]/ND	+/–		r	+	+	±	+	±		
Kir4.1 (Kir1.2)	h/r	–/–	±/±	–/–	–/ND	+/+		r	ND	±	+	ND	±	A, C. callosum, S. nigra	

TABLE 5. Tissue Expression of K^+ Channel Principal Subunits (*continued*)

	Tissue Distribution[a]							Brain Region							
	sp	He	Ki	SK	Lu	Br	Other	sp	CX	Th	CB	OB	HC	Other	Prot. Dist.
Kir4.2 (Kir1.3)	h	–	+	–	±	–	Pa	ND							
Kir5.1	r[e]	+	±	+	±	+		r[e]	+[f]	±	+	ND	+[g]		
Kir6	r	+	+[h]	+	ND	?	Li	ND							
Kir6.2	r	+	ND	+	–	+	Pituitary, pa	r	+	+	+	+	+	Midbrain, CP, brainstem	
Kir7.1	h	–[e]	+[e]	–[e]	–[e]	+[e]	Prostate, testis, small intestine	ND						Purkinje and pyramidal cell layer	
Eag1	r	–	–	–	–	+		r	+	–	+	+	+		
Erg1	h[i]	+	ND	ND	ND	+	SCG, CG, SMG, Re, adr, thy	ND							
Erg2	r[h]	–	ND	ND	ND	–	CG, SMG, Re	ND							
Erg3	r[h]	–	ND	ND	ND	+	SCG, CG, SMG	ND							
SK1	r	+	–	–	–	+	Adr	r	+	–	+	+	+		
SK2	r	+	–	–	–	+	Adr	r	+	+	+	+	+	Pontine nucleus	
SK3	r	+	ND	ND	ND	+	Adr	r	–	+	–	+	–	Supraoptic nucleus	
SK4	h	–	–	–	+	–	Adr, pla	NA							
SKCa3	h	ND	ND	ND	ND	ND		ND							
SKCa4	h	+	-	+	+	-	Pla, thy, co, T-lymph., Li	NA							

TABLE 5. Tissue Expression of K⁺ Channel Principal Subunits (*continued*)

	Tissue Distribution[a]							Brain Region							
	sp	He	Ki	SK	Lu	Br	Other	sp	CX	Th	CB	OB	HC	Other	Prot. Dist.
IK1	h	-	-	-	+	ND	Thy, pla, stomach, co, bla, prostate	h	-	ND	-	ND	ND		
Slo	h	ND	-	+	+	+	Pancreatic islets, Li	d						Optic lobes	
Slo3	m/h	-	-	-	-	-	Testis	NA							
TWIK	m	+	+	±	+	+	Li	m	+	-	+	-	+		
TREK	m	+	+	±	+	+	Mainly in brain	m	±	-	+	+	+		
TASK	h/m/r	+/+/+	+/±/±	±?/–/±	+/+/+	+/±/±	Pla, pa, atria, not ventricle	m/r	+/+	+/ND	+/ND	ND/ND	+/+	CP	
TRAAK	m	-	-	-	-	+		m	+	±	+	+	+	CP	

ABBREVIATIONS: sp, specie; He, heart; Ki, kidney; Pa, pancreas; Lu, lung; Sk, skeletal muscle; S.M., vascular smooth muscle; Li, Liver; T-lymph, T-lymphocytes; Br, brain; thy, thymus; Adr, adrenal gland; co, colon; bla, bladder; pla, placenta; SCG, superior cervical ganglia; CG, celiac ganglia; SMG, superior mesenteric ganglia; re, retina; CX, cerebral cortex; HC, hippocampus; Th, dorsal thalamus; CB, cerebellum; OT, olfactory bulb; H, hypothalamus; CP, caudate putamen; Th ret. Nucleus, thalamic reticulate nucleolus; BAN, basolateral amygdaloid nucleus, C. nucl., caudate nucleus; SNPC, substantia nigra pars compacta; Rt, reticular thalamus; A, amygdala; Hb, medial hebenula; ND: not determined; NA: not applicable; d, *drosophila*; m, mouse; r, rat; h, human.

[a] Determined by Northern (RNA) analysis.

[b] Kv8.1/Kv2.3r differ in thalamus by species.

[c] Detected only in human, not in rat or mouse.

[d] Detected only in human and rat and not in mouse.

[e] Determined by RT-PCR of total RNA.

[f] Only BIR 10 isoform.

[g] Only BIR 10 and BIR 11 isoforms.

[h] Neonate.

[i] Determined by RNase protection analysis.

TABLE 6. Functional Diversity Conferred by Alternative Splicing

Splice Version	Comments	Effects on Functional Properties	Effects on Expression	Effects on Subcellular Localization or Modulation
Kv1.5Δ5′	Unusual splicing results in truncated 5′ end			
Kv1.5Δ3′	Truncated 3′ end	Nonfunctional, possible dominant negative	Predominant splice version	
Kv3.1a	(Alternative C-termini in all isoforms)		Predominant isoform during early development	Axons
Kv3.1b			Predominant in adults	Prox. dendrites, soma, axons
Kv3.2a–d	(Alternative C-termini in all isoforms)			Kv3.2a in basolateral membrane, Kv3.2b in apical membrane when expressed in MDCK cells. Kv3.2b but not Kv3.2a modulated by PKC
Kv4.3S	(Alternative C-termini in all isoforms)		Weak in heart, strong in brain	
Kv4.3M			Most abundant in brain, skeletal muscle, pancreas	
Kv4.3L			Weak in brain, major product in smooth muscle, heart, lung & kidney	
LQT1 isoform 2	N-terminal truncation	Dominant negative	Heart ventricle	
KQT2.1–2.11			Some isoforms demonstrate different temporal expression patterns	
Kir1.1a–f	Splicing of Kir1.1b,d–f results in alternative 5′ UTR's; Kir1.1a&c have alternative N-termini		Alternative exon usage results in changes in tissue expression	

TABLE 6. Functional Diversity Conferred by Alternative Splicing (*continued*)

Splice Version	Comments	Effects on Functional Properties	Effects on Expression	Effects on Subcellular Localization or Modulation
Kir3.1a–d	(Alternative C-termini in all isoforms)	Changes in G protein activation in Kir3.1c/3.1a heteromultimers	Alternative exon usage results in changes in tissue expression	
Kir3.2a–e	In Kir3.2a–e splicing results in alternative C-termini; Kir3.2e also has truncated N-terminus.		Kir3.2b&c expression in testes is restricted to a subset of tubules	
Kir6.1a	(Alternative 5′ UTR's in all isoforms)		Heart and pancreas	
Kir6.1b			Heart and skeletal muscle	
Kir6.1c			Heart	
slo	(slo channels have the ability to generate numerous splice isoforms. Human slo contains 5 splicing cassettes within the C-terminus)	(Tested isoforms demonstrate variations in kinetics and calcium sensitivity)	(Tested isoforms demonstrate unique expression patterns. Expression of isoforms in adrenal chromaffin tissue controlled by ACTH)	
Eag1a		Faster activation kinetics	Predominant isoform	
Eag1b	27aa insertion between S3 and S4			
Erg1a		Slower deactivation kinetics	Heart, brain, testes	
Erg1a′	N-terminal truncation	Deactivation kinetics more rapid	Not expressed abundantly	
Erg1b	Shorter/divergent N terminus	Deactivation kinetics more rapid than Erg1a, heteromultimerization with Erg1a increases deactivation kinetics	Heart	

REFERENCES

1. Papazian, D.M., T.L. Schwarz, B.L. Tempel, Y.N. Jan & L.Y. Jan. 1987. Cloning of genomic and complementary DNA from *Shaker*, a putative potassium channel gene from Drosophila. Science **237:** 749–753.
2. Tempel, B.L., D.M. Papazian, T.L. Schwarz, Y.N. Jan & L.Y. Jan. 1987. Sequence of a probable potassium channel component encoded at Shaker locus of Drosophila. Science **237:** 770–775.
3. Kamb, A., L.E. Iverson & M.A. Tanouye. 1987. Molecular characterization of Shaker, a Drosophila gene that encodes a potassium channel. Cell **50:** 405–413.
4. Stuhmer, W., M. Stocker, B. Sakmann, P. Seeburg, A. Baumann, A. Grupe & O. Pongs. 1988. Potassium channels expressed from rat brain cDNA have delayed rectifier properties. FEBS Lett. **242:** 199–206.
5. Takumi, T., H. Ohkubo & S. Nakanishi. 1988. Cloning of a membrane protein that induces a slow voltage-gated potassium current. Science **242:** 1042–1045.
6. Jan, L.Y. & Y.N. Jan. 1997. Voltage-gated and inwardly rectifying potassium channels. J. Physiol. (Lond.) **505:** 267–282.
7. Aguilar-bryan, L., J.P. Clement, G. Gonzalez, K. Kunjilwar, A. Babenko & J. Bryan. 1998. Toward understanding the assembly and structure of K_{ATP} channels. Physiol. Rev. **78:** 227–245.
8. Isomoto, S., Kondo, C. & Kurachi, Y. 1997. Inwardly rectifying potassium channels: Their molecular heterogeneity and function. Jpn. J. Physiol. **47:** 11–39.
9. Nichols, C.G. & A.N. Lopatin. 1997. Inward rectifier potassium channels. Ann. Rev. Physiol. **59:** 171–191.
10. Barry, D.M. & J.M. Nerbonne. 1996. Myocardial potassium channels: Electrophysiological and molecular diversity. Ann. Rev. Physiol. **58:** 363–394.
11. Quayle, J.M., M.T. Nelson & N.B. Standen. 1997. ATP-sensitive and inwardly rectifying potassium channels in smooth muscle. Physiol. Rev. **77:** 1165–1232.
12. Goldstein, S.A.N., K.W. Wang, N. Ilan & M.H. Pausch. 1998. Sequence and function of the two P domain potassium channels: Implication of an emerging superfamily. J. Mol. Med. **76:** 13–20.
13. Vega-Saenz De Miera, E., M. Weiser, C. Kentros, D. Lau, H. Moreno, P. Serodio & B. Rudy. 1994. Shaw-related K^+ channels in mammals. *In* Handbook of Membrane Channels. 41–78. Academic Press. New York.
14. Lehmann-Horn, F. & R. Rudel. 1996. Molecular pathophysiology of voltage-gated ion channels. Rev. Physiol. Biochem. Pharmacol. **128:** 195–268.
15. Dolly, J.O. & D.N. Parcej. 1996. Molecular properties of voltage-gated K^+ channels. J. Bioenerg. Biomembr. **28:** 231–253.
16. Chandy, K.G. & G.A. Gutman. 1995. Voltage-gated potassium channel genes. *In* Handbook of Receptors and Channels: Ligand and Voltage-Gated Ion Channels. ed. R.A. North, Ed.: 1–71. CRC Press. Boca Raton, FL.
17. Doyle, D.A., J.M. Cabral, R.A. Pfuetzner, A. Kuo, J.M. Gulbis, S.L. Cohen, B.T. Chait & R. Mackinnon. 1998. The structure of the potassium channel: Molecular basis of K^+ conduction and selectivity. Science **280:** 69–77.
18. Catterall, W.A. 1993. Structure and modulation of Na^+ and Ca^{2+} channels. Ann. N.Y. Acad. Sci. **707:** 1–19.
19. Mackinnon, R. 1991. Determination of the subunit stoichiometry of a voltage- activated potassium channel. Nature **350:** 232–235.
20. Mackinnon, R., R.W. Aldrich & A.W. Lee. 1993. Functional stoichiometry of Shaker potassium channel inactivation. Science **262:** 757–759.
21. Rudy, B. 1988. Diversity and ubiquity of K channels. Neuroscience **25:** 729–749.
22. Hille, B. 1992. Ionic Channels of Excitable Membranes. Sinauer Associates, Inc. Sunderland, MA.
23. Ho, K., C.G. Nichols, W.J. Lederer, J. Lytton, P.M. Vassilev, M.V. Kanazirska & S.C. Hebert. 1993. Cloning and expression of an inwardly rectifying ATP-regulated potassium channel. Nature **362:** 31–38.
24. Kubo, Y., T.J. Baldwin, Y.N. Jan & L.Y. Jan. 1993. Primary structure and functional expression of a mouse inward rectifier potassium channel. Nature **362:** 127–133.

25. Dascal, N., W. Schreibmayer, N.F. Lim *et al.* 1993. Atrial G protein–activated K+ channel: Expression cloning and molecular properties. Proc. Natl. Acad. Sci. USA. **90:** 10235–10239.
26. Lesage, F., E. Guillemare, M. Fink, F. Duprat, M. Lazdunski, G. Romey & J. Barhanin. 1996. TWIK-1, a ubiquitous human weakly inward rectifying K^+ channel with a novel structure. EMBO J. **15:** 1004–1011.
27. Lesage, F., R. Reyes, M. Fink, F. Duprat, E. Guillemare & M. Lazdunski. 1996. Dimerization of TWIK-1 K^+ channel subunits via a disulfide bridge. EMBO J. **15:** 6400–6407.
28. Fink, M., F. Lesage, F. Duprat, C. Heurteaux, R. Reyes, M. Fosset & M. Lazdunski. 1998. A neuronal two P domain K^+ channel stimulated by arachidonic acid and polyunsaturated fatty acids. Embo J. **17:** 3297–3308.
29. Taglialatela, M., M.S. Champagne, J.A. Drewe & A.M. Brown. 1994. Comparison of H5, S6, and H5-S6 exchanges on pore properties of voltage-dependent K^+ channels. J. Biol. Chem. **269:** 13867–13873.
30. Heginbotham, L., Z. Lu, T. Abramson & R. MacKinnon. 1994. Mutations in the K^+ channel signature sequence. Biophys. J. **66:** 1061–1067.
31. Bailie, T.L. & C. Elkan 1994. Fitting a mixture model by expectation maximization to discover motifs in biopolymers. AAAI Press. Menlo Park, CA.
32. Chandy, K.G. 1991. Simplified gene nomenclature. Nature **352:** 26.
33. Doupnik, C.A., N. Davidson & H.A. Lester. 1995. The inward rectifier potassium channel family. Curr. Opin. Neurobiol. **5:** 268–277.
34. Bond, C.T., M. Pessia, X.M. Xia, A. Lagrutta, M.P. Kavanaugh & J.P. Adelman. 1994. Cloning and expression of a family of inward rectifier potassium channels. Recept. Channels **2:** 183–191.
35. Shuck, M.E., T.M. Piser, J.H. Bock, J.L. Slightom, K.S. Lee & M.J. Bienkowski. 1997. Cloning and characterization of two K+ inward rectifier (Kir) 1.1 potassium channel homologs from human kidney (Kir1.2 and Kir1.3). J. Biol. Chem. **272:** 586–593.
36. Gosset, P., G.A. Ghezala, B. Korn, M.L. Yaspo, A. Poutska, H. Lehrach, P.M. Sinet & N. Créau. 1997. A new inward rectifier potassium channel gene (KCNJ15) localized on chromosome 21 in the Down syndrome chromosome region 1 (DCR1). Genomics **44:** 237–241.
37. Shuck, M.E., T.M. Piser, J.H. Bock, J.L. Slightom, K.S. Lee & M.J. Bienkowski. 1997. Cloning and characterization of two K^+ inward rectifier (K_{ir}) 1.1 potassium channel homologs from human kidney ($K_{ir}1.2$ and $K_{ir}1.3$). J. Biol. Chem. **272:** 586–593.
38. Krapivinsky, G., I. Medina, L. Eng, L. Krapivinsky, Y.H. Yang & D.E. Clapham. 1998. A novel inward rectifier K^+ channel with unique pore properties. Neuron **20:** 995–1005.
39. Hedin, K.E., N.F. Lim & D.E. Clapham. 1996. Cloning of a Xenopus laevis inwardly rectifying K^+ channel subunit that permits GIRK1 expression of I_{KACh} currents in oocytes. Neuron **16:** 423–429.
40. Christie, M.J., R.A. North, P.B. Osborne, J. Douglass & J. Adelman. 1990. Heteropolymeric potassium channels expressed in Xenopus oocytes from cloned subunits. Neuron **4:** 405–411.
41. Isacoff, E.Y., Y.N. Jan & L.Y. Jan. 1990. Evidence for the formation of heteromultimeric potassium channels in Xenopus oocytes. Nature **345:** 530–534.
42. McCormack, K., J.W. Lin, L.E. Iverson & B. Rudy. 1990. Shaker K^+ channel subunits from heteromultimeric channels with novel functional properties. Biochem. Biophys. Res. Commun. **171:** 1361–1371.
43. Ruppersberg, J.P., K.H. Schroter, B. Sakmann, M. Stocker, S. Sewing & O. Pongs. 1990. Heteromultimeric channels formed by rat brain potassium-channel proteins. Nature **345:** 535–537.
44. Weiser, M., E. Vega-Saenz De Miera, C. Kentros, H. Moreno, L. Franzen, D. Hillman, H. Baker & B. Rudy. 1994. Differential expression of Shaw-related K^+ channels in the rat central nervous system. J. Neurosci. **14:** 949–972.
45. Roeper, J., S. Sewing, Y. Zhang, T. Sommer, S.G. Wanner & O. Pongs. 1998. NIP domain prevents N-type inactivation in voltage-gated potassium channels. Nature **391:** 390–393.
46. Li, M., Y.N. Jan & L.Y. Jan. 1992. Specification of subunit assembly by the hydrophilic aminoterminal domain of the Shaker potassium channel. Science **257:** 1225–1230.
47. Shen, N.V. & P.J. Pfaffinger. 1995. Molecular recognition and assembly sequences involved in the subfamily-specific assembly of voltage-gated K^+ channel subunit proteins. Neuron **14:** 625–633.

48. YU, W., J. XU & M. LI. 1996. NAB domain is essential for the subunit assembly of both alpha-alpha and alpha-beta complexes of shaker-like potassium channels. Neuron **16:** 441–453.
49. SEWING, S., J. ROEPER & O. PONGS. 1996. Kv β1 subunit binding specific for shaker-related potassium channel α subunits. Neuron **16:** 455–463.
50. SHENG, M., Y.J. LIAO, Y.N. JAN & L.Y. JAN. 1993. Presynaptic A-current based on heteromultimeric K^+ channels detected in vivo. Nature **365:** 72–75.
51. WANG, H., D.D. KUNKEL, T.M. MARTIN, P.A. SCHWARTZKROIN & B.L. TEMPEL. 1993. Heteromultimeric K^+ channels in terminal and juxtaparanodal regions of neurons. Nature **365:** 75–79.
52. CHOW, A., C. FARB, A. ERISIR & B. RUDY. 1998. K^+ channel expression distinguishes between two subpopulations of parvalbumin-containing cortical interneurons. Submitted for publication.
53. HERNÁNDEZ-PINEDA, R., A. CHOW, Y. AMARILLO, H. MORENO, M. SAGANICH & B. RUDY. 1999. Kv3.1-Kv3.2 heteromultimeric channels underlie a high voltage-activating component of the delayed rectifier K^+ current in projecting neurons from the Globus Pallidus. J. Neurophysiol. In press.
54. MI, H., T.J. DEERINCK, M.H. ELLISMAN & T.L. SCHWARZ. 1995. Differential distribution of closely related potassium channels in rat Schwann cells. J. Neurosci. **15:** 3761–3774.
55. KRAMER, J.W., M.A. POST, A.M. BROWN & G.E. KIRSCH. 1998. Modulation of potassium channel gating by coexpression of Kv2.1 with regulatory Kv5.1 or Kv6.1 alpha-subunits. Am. J. Physiol. **274:** C1501–C1510
56. SALINAS, M., F. DUPRAT, C. HEURTEAUX, J.P. HUGNOT & M. LAZDUNSKI. 1997. New modulatory alpha subunits for mammalian Shab K^+ channels. J. Biol. Chem. 272: 24371–24379.
57. PATEL, A.J., M. LAZDUNSKI & E. HONORÉ. 1997. Kv2.1/Kv9.3, a novel ATP-dependent delayed-rectifier K^+ channel in oxygen-sensitive pulmonary artery myocytes. EMBO J. **16:** 6615–6625.
58. YANG, W.P., P.C. LEVESQUE, W.A. LITTLE, M.L. CONDER, P. RAMAKRISHNAN, M.G. NEUBAUER & M.A. BLANAR. 1998. Functional expression of two KvLQT1-related potassium channels responsible for an inherited idiopathic epilepsy. J. Biol. Chem. **273:** 19419–19423.
59. SHI, W., H.S. WANG, Z. PAN, R.S. WYMORE, I.S. COHEN, D. MCKINNON & J.E. DIXON. 1998. Cloning of a mammalian elk potassium channel gene and EAG mRNA distribution in rat sympathetic ganglia. J. Physiol. (Lond.) **511:** 675–682.
60. LONDON, B., M.C. TRUDEAU, K.P. NEWTON, A.K. BEYER, N.G. COPELAND, D.J. GILBERT, N.A. JENKINS, C.A. SATLER & G.A. ROBERTSON. 1997. Two isoforms of the mouse *ether-a-go-go–* related gene coassemble to form channels with properties similar to the rapidly activating component of the cardiac delayed rectifier K^+ current. Circ. Res. **81:** 870–878.
61. CHEN, M.L., T. HOSHI & C.F. WU. 1996. Heteromultimeric interactions among K^+ channel subunits from *Shaker* and *eag* families in Xenopus oocytes. Neuron **17:** 535–542.
62. LIAO, Y.J., Y.N. JAN & L.Y. JAN. 1996. Heteromultimerization of G-protein–gated inwardly rectifying K^+ channel proteins GIRK1 and GIRK2 and their altered expression in weaver brain. J. Neurosci. **16:** 7137–7150.
63. FINK, M., F. DUPRAT, C. HEURTEAUX, F. LESAGE, G. ROMEY, J. BARHANIN & M. LAZDUNSKI. 1996. Dominant negative chimeras provide evidence for homo and heteromultimeric assembly of inward rectifier K^+ channel proteins via their N-terminal end. FEBS Lett. **378:** 64–68.
64. TINKER, A., Y.N. JAN & L.Y. JAN. 1996. Regions responsible for the assembly of inwardly retifying potassium channels. Cell **87:** 857–868.
65. NAKAMURA, T.Y., M. ARTMAN, B. RUDY & W.A. COETZEE. 1998. Inihibition of rat ventricular I_{K1} with antisense oligonucleotides targeted to Kir2.1 mRNA. Am. J. Physiol. **274:** H892–H900
66. FINK, M., F. DUPRAT, C. HEURTEAUX, F. LESAGE, G. ROMEY, J. BARHANIN & M. LAZDUNSKI. 1996. Dominant negative chimeras provide evidence for homo and heteromultimeric assembly of inward rectifier K^+ channel proteins via their N-terminal end. FEBS Lett. **378:** 64–68.
67. PESSIA, M., S.J. TUCKER, K. LEE, C.T. BOND & J.P. ADELMAN. 1996. Subunit positional effects revealed by novel heteromeric inwardly rectifying K^+ channels. EMBO J. **15:** 2980–2987.
68. ADELMAN, J.P., K.Z. SHEN, M.P. KAVANAUGH, R.A. WARREN, Y.N. WU, A. LAGRUTTA, C.T. BOND & R.A. NORTH. 1992. Calcium-activated potassium channels expressed from cloned complementary DNAs. Neuron **9:** 209–216.

69. Tseng-Crank, J., C.D. Foster, J.D. Krause, R. Mertz, N. Godinot, T.J. Dichiara & P.H. Reinhart. 1994. Cloning, expression, and distribution of functionally distinct Ca^{2+}-activated K^+ channel isoforms from human brain. Neuron **13:** 1315–1330.
70. Xie, J. & D.P. McCobb. 1998. Control of alternative splicing of potassium channels by stress hormones. Science **280:** 443–446.
71. Butler, A., S. Tsunoda, D.P. McCobb, A. Wei & L. Salkoff. 1993. mSlo, a complex mouse gene encoding "maxi" calcium-activated potassium channels. Science **261:** 221–224.
72. Navaratnam, D.S., T.J. Bell, T.D. Tu, E.L. Cohen & J.C. Oberholtzer. 1997. Differential distribution of Ca^{2+}-activated K^+ channel splice variants among hair cells along the tonotopic axis of the chick cochlea. Neuron **19:** 1077–1085.
73. Rosenblatt, K.P., Z.P. Sun, S. Heller & A.J. Hudspeth. 1997. Distribution of Ca^{2+}-activated K^+ channel isoforms along the tonotopic gradient of the chicken's cochlea. Neuron **19:** 1061–1075.
74. Kohler, M., B. Hirschberg, C.T. Bond, J.M. Kinzie, N.V. Marrion, J. Maylie & J.P. Adelman. 1996. Small-conductance, calcium-activated potassium channels from mammalian brain. Science **273:** 1709–1714.
75. Rehm, H. & M. Lazdunski. 1988. Purification and subunit structure of a putative K^+ channel protein identified by its binding properties to dendrotoxin I. Proc. Natl. Acad. Sci. USA **85:** 4919–4923.
76. Scott, V.E.S., J. Rettig, D.N. Parcej, J.N. Keen, J.B.C. Findlay, O. Pongs & J.O. Dolly. 1994. Primary structure of a beta subunit of alpha-dendrotoxin-sensitive K^+ channels from bovine brain. Proc. Natl. Acad. Sci. USA **91:** 1637–1641.
77. McCormack, T. & K. McCormack. 1994. Shaker K^+ channel beta subunits belong to an NAD(P)H-dependent oxidoreductase superfamily [letter]. Cell **79:** 1133–1135.
78. Rettig, J., S.H. Heinemann, F. Wunder, C. Lorra, R. Wittka, O. Dolly & O. Pongs. 1994. Inactivation properties of voltage gated K^+ channels altered by the presence of β-subunit. Nature **369:** 289–294.
79. McCormack, K., T. Mccormack, M. Tanouye, B. Rudy & W. Stühmer. 1995. Alternative splicing of the human Shaker K^+ channel β1 gene and functional expression of the β2 gene product. FEBS Lett. **370:** 32–36.
80. Heinemann, S.H., J. Rettig, F. Wunder & O. Pongs. 1995. Molecular and functional characterization of a rat brain Kv beta 3 potassium channel subunit. FEBS Lett. **377:** 383–389.
81. Wang, Z., J. Kiehn, Q. Yang, A.M. Brown & B.A. Wible. 1996. Comparison of binding and block produced by alternatively spliced Kvbeta1 subunits. J. Biol. Chem. **271:** 28311–28317.
82. England, S.K., V.N. Uebele, J. Kodali, P.B. Bennett & M.M. Tamkun. 1995. A novel K^+ channel β-subunit (hKvβ1.3) is produced via alternative mRNA splicings. J. Biol. Chem. **270:** 28531–28534.
83. Shi, G.Y., K. Nakahira, S. Hammond, K.J. Rhodes, L.E. Schechter & J.S. Trimmer. 1996. β Subunits promote K^+ channel surface expression through effects early in biosynthesis. Neuron **16:** 843–852.
84. Nagaya, N. & D.M. Papazian. 1997. Potassium channel α and β subunits assemble in the endoplasmic reticulum. J. Biol. Chem. **272:** 3022–3027.
85. Wible, B.A., Q. Yang, Y.A. Kuryshev, E.A. Accili & A.M. Brown. 1998. Cloning and expression of a novel K^+ channel regulatory protein, KChAP. J. Biol. Chem. **273:** 11745–11751.
86. Fink, M., F. Duprat, F. Lesage, C. Heurteaux, G. Romey, J. Barhanin & M. Lazdunski. 1996. A new K^+ channel β subunit to specifically enhance Kv2.2 (CDRK) expression. J. Biol. Chem. **271:** 26341–26348.
87. Rhodes, K.J., B.W. Strassle, M.M. Monaghan, Z. Bekele-Arcuri, M.F. Matos & J.S. Trimmer. 1997. Association and colocalization of the Kvβ1 and Kvβ2 β-subunits with Kv1 α-subunits in mammalian brain K^+ channel complexes. J. Neurosci. **17:** 8246–8258.
88. Wilson, G.F., Z. Wang, S.W. Chouinard, L.C. Griffith & B. Ganetzky. 1998. Interaction of the K channel beta subunit, Hyperkinetic, with eag family members. J. Biol. Chem. **273:** 6389–6394.
89. Hoshi, N., H. Takahashi, M. Shahidullah, S. Yokoyama & H. Higashida. 1998. KCR1, a membrane protein that facilitates functional expression of non-inactivating K+ currents associates with rat EAG voltage-dependent K+ channels [In Process Citation]. J. Biol. Chem. **273:** 23080–23085.

90. Folander, K., J.S. Smith, J. Antanavage, C. Bennett, R.B. Stein & R. Swanson. 1990. Cloning and expression of the delayed-rectifier Isk channel from neonatal rat heart and diethylstilbestrol-rrimed rat uterus. Proc. Natl. Acad. Sci. USA **87:** 2975–2979.
91. Sakagami, M., K. Fukazawa, T. Matsunaga, H. Fujita, N. Mori, T. Takumi, H. Ohkubo & S. Nakanishi. 1991. Cellular localization of rat Isk protein in the stria vascularis by immunohistochemical observation. Hear. Res. **56:** 168–172.
92. Freeman, L.C. & R.S. Kass. 1993. Expression of a minimal K+ channel protein in mammalian cells and immunolocalization in guinea pig heart. Circ. Res. **73:** 968–973.
93. Varnum, M.D., A.E. Busch, C.T. Bond, J. Maylie & J.P. Adelman. 1993. The min K channel underlies the cardiac potassium current IKs and mediates species-specific responses to protein kinase C. Proc. Natl. Acad. Sci. USA **90:** 11528–11532.
94. Barhanin, J., F. Lesage, E. Guillemare, M. Fink, M. Lazdunski & G. Romey. 1996. K_vLQT1 and IsK (minK) proteins associate to form the I_{Ks} cardiac potassium current. Nature **384:** 78–80.
95. Sanguinetti, M.C., M.E. Curran, A. Zou, J. Shen, P.S. Spector, D.L. Atkinson & M.T. Keating. 1996. Coassembly of K_vLQT1 and minK (IsK) proteins to form cardiac I_{Ks} potassium channel. Nature **384:** 80–83.
96. Salata, J.J., N.K. Jurkiewicz, J. Wang, B.E. Evans, H.T. Orme & M.C. SanGuinetti. 1998. A novel benzodiazepine that activates cardiac slow delayed rectifier K+ currents. Mol. Pharmacol. **54:** 220–230.
97. McDonald, T.V., Z.H. Yu, Z. Ming, E. Palma, M.B. Meyers, K.W. Wang, S.A.N. Goldstein & G.I. Fishman. 1997. A minK-HERG complex regulates the cardiac potassium current I_{Kr}. Nature **388:** 289–292.
98. Sanguinetti, M.C. & N.K. Jurkiewicz. 1990. Two components of cardiac delayed rectifier K^+ current: Differential sensitivity to block by class III antiarrhythmic agents. J. Gen. Physiol. **96:** 194–214.
99. Sanguinetti, M.C. & N.K. Jurkiewicz. 1991. Delayed rectifier outward K+ current is composed of two currents in guinea pig atrial cells. Am. J. Physiol. **260:** H393–399.
100. Busch, A.E., M.P. Kavanaugh, M.D. Varnum, J.P. Adelman & R.A. North. 1992. Regulation by second messengers of the slowly activating, voltage-dependent potassium current expressed in Xenopus oocytes. J. Physiol. **450:** 491–502.
101. Blumenthal, E.M. & L.K. Kaczmarek. 1994. The minK potassium channel exists in functional and nonfunctional forms when expressed in the plasma membrane of Xenopus oocytes. J. Neurosci. **14:** 3097–3105.
102. Tai, K.K. & S.A.N. Goldstein. 1998. The conduction pore of a cardiac potassium channel. Nature **391:** 605–608.
103. Knaus, H.G., K. Folander, M. Garcia-Calvo, M.L. Garcia, G.J. Kaczorowski, M. Smith & R. Swanson. 1994. Primary sequence and immunological characterization of beta-subunit of high conductance Ca(2+)-activated K+ channel from smooth muscle. J. Biol. Chem. **269:** 17274–17278.
104. Hanner, M., R. Vianna-Jorge, A. Kamassah, W.A. Schmalhofer, H.G. Knaus, G.J. Kaczorowski & M.L. Garcia. 1998. The beta subunit of the high conductance calcium-activated potassium channel. Identification of residues involved in charybdotoxin binding. J. Biol. Chem. **273:** 16289–16296.
105. Schopperle, W.M., M.H. Holmqvist, Y. Zhou, J. Wang, Z. Wang, L.C. Griffith, I. Keselman, F. Kusinitz, D. Dagan & I.B. Levitan. 1998. Slob, a novel protein that interacts with the slowpoke calcium-dependent potassium channel. Neuron **20:** 565–573.
106. Tucker, S.J., F.M. Gribble, C. Zhao, S. Trapp & F.M. Ashcroft. 1997. Truncation of Kir6.2 produces ATP-sensitive K+ channels in the absence of the sulphonylurea receptor. Nature **387:** 179–183.
107. John, S.A., J.R. Monck, J.N. Weiss & B. Ribalet. 1998. The sulphonylurea receptor SUR1 regulates ATP-sensitive mouse Kir6.2 K^+ channels linked to the green fluorescent protein in human embryonic kidney cells. J. Physiol. (Lond.) **510:** 333–346.
108. Shyng, S. & C.G. Nichols. 1997. Octameric stoichiometry of the K_{ATP} channel complex. J. Gen. Physiol. **110:** 655–664.
109. Babenko, A.P., L. Aguilar-Bryan & J. Bryan. 1998. A view of sur/KIR6.X, KATP channels. Ann. Rev. Physiol. **60:** 667–687.

110. Seeburg, P.H., M. Higuchi & R. Sprengel. 1998. RNA editing of brain glutamate receptor channels: Mechanism and physiology. Brain Res. Brain Res. Rev. **26:** 217–229.
111. Maas, S., T. Melcher & P.H. Seeburg. 1997. Mammalian RNA-dependent deaminases and edited mRNAs. Curr. Opin. Cell Biol. **9:** 343–349.
112. Patton, D.E., T. Silva & F. Bezanilla. 1997. RNA editing generates a diverse array of transcripts encoding squid Kv2 K+ channels with altered functional properties. Neuron **19:** 711–722.
113. Jonas, E.A. & L.K. Kaczmarek. 1996. Regulation of potassium channels by protein kinases. Curr. Opin. Neurobiol. **6:** 318–323.
114. Levitan, I.B. 1994. Modulation of ion channels by protein phosphorylation and dephosphorylation. Ann. Rev. Physiol. **56:** 193–212.
115. Kaczmereck, K.L. & I.B. Levitan. 1987. Neuromodulation: The Biochemical Control of Neuronal Excitability. Oxford University Press. Oxford.
116. Covarrubias, M., A. Wei, L. Salkoff & T.B. Vyas. 1994. Elimination of rapid potassium channel inactivation by phosphorylation of the inactivation gate. Neuron **13:** 1403–1412.
117. Beck, E.J., R.G. Sorensen, S.J. Slater & M. Covarrubias. 1998. Interactions between multiple phosphorylation sites in the inactivation particle of a K+ channel. Insights into the molecular mechanism of protein kinase C action. J. Gen. Physiol. **112:** 71–84.
118. Roeper, J., C. Lorra & O. Pongs. 1997. Frequency-dependent inactivation of mammalian A-type K+ channel KV1.4 regulated by Ca2+/calmodulin-dependent protein kinase. J. Neurosci. **17:** 3379–3391.
119. Cohen, N.A., J.E. Brenman, S.H. Snyder & D.S. Bredt. 1996. Binding of the inward rectifier K^+ channel Kir 2.3 to PSD-95 is regulated by protein kinase A phosphorylation. Neuron **17:** 759–767.
120. Cohen, N.A., Q. Sha, E.N. Makhina, A.N. Lopatin, M.E. Linder, S.H. Snyder & C.G. Nichols. 1996. Inhibition of an inward rectifier potassium channel (Kim2.3) by G-protein βgamma subunits. J. Biol. Chem. **271:** 32301–32305.
121. Storm, J.F. 1988. Temporal integration by a slowly inactivating K+ current in hippocampal neurons. Nature **336:** 379–381.
122. Wu, R.L. & M.E. Barish. 1992. Two pharmacologically and kinetically distinct transient potassium currents in cultured embryonic mouse hippocampal neurons. J. Neurosci. **12:** 2235–2246.
123. Sheng, M., M.L. Tsaur, Y.N. Jan & L.Y. Jan. 1992. Subcellular segregation of two A-type K^+ channel proteins in rat central neurons. Neuron **9:** 271–284.
124. Debanne, D., N.C., Guerineau, B.H. Gahwiler & S.M. Thompson. 1997. Action-potential propagation gated by an axonal I(A)-like K^+ conductance in hippocampus. Nature **389:** 286–289.
125. Feng, J.L., B. Wible, G.R. Li, Z.G. Wang & S. Nattel. 1997. Antisense oligodeoxynucleotides directed against Kv1.5 mRNA specifically inhibit ultrarapid delayed rectifier K^+ current in cultured adult human atrial myocytes. Circ. Res. **80:** 572–579.
126. Patel, A.J., M. Lazdunski & E. Honore. 1997. Kv2.1/Kv9.3, a novel ATP-dependent delayed-rectifier K^+ channel in oxygen-sensitive pulmonary artery myocytes. EMBO J. **16:** 6615–6625.
127. Fiset, C., R.B. Clark, Y. Shimoni & W. Giles. 1997. Shal-type channels contribute to the Ca^{2+}-independent transient outward K^+ current in rat ventricle. J. Physiol. (Lond.) **500:** 51–63.
128. Nakamura, T.Y., W.A. Coetzee, E. Vega-Saenz De Miera, M. Artman & B. Rudy. 1997. Modulation of Kv4 channels, key components of rat ventricular transient outward K^+ current by PKC. Am. J. Physiol. **273:** H1775–H1786.
129. Marrion, N.V. 1997. Does r-EAG contribute to the M-current? Trends Neurosci. **20:** 243–244.
130. Mathie, A. & C.S. Watkins. 1997. Is EAG the answer to the M-current? Trends Neurosci. **20:** 14.
131. Feng, J.L., B. Wible, G.R. Li, Z. Wang & S. Nattel. 1996. Antisense oligonucleotides directed against Kv1.5 mRNA specifically inhibit the ultrarapid delayed rectifier in cultured adult human atrial myocytes. Circulation **94:** I-528(Abstr.).
132. Garcia-Calvo, M., H.G. Knaus, O.B. McManus, K.M. Giangiacomo, G.J. Kaczorowski & M.L. Garcia. 1994. Purification and reconstitution of the high-conductance, calcium-activated potassium channel from tracheal smooth muscle. J. Biol. Chem. **269:** 676–682.

133. Knaus, H.G., A. Eberhart, R.O. Koch, P. Munujos, W.A. Schmalhofer, J.W. Warmke, G.J. Kaczorowski & M.L. Garcia. 1995. Characterization of tissue-expressed alpha subunits of the high conductance Ca^{2+}-activated K^+ channel. J. Biol. Chem. **270:** 22434–22439.
134. Ishii, T.M., C. Silvia, B. Hirschberg, C.T. Bond, J.P. Adelman & J. Maylie. 1997. A human intermediate conductance calcium-activated potassium channel. Proc. Natl. Acad. Sci. USA **94:** 11651–11656.
135. Logsdon, N.J., J. Kang, J.A. Togo, E.P. Christian & J. Aiyar. 1997. A novel gene, hKCa4, encodes the calcium-activated potassium channel in human T lymphocytes. J. Biol. Chem. **272:** 32723–32726.
136. Joiner, W.J., L.Y. Wang, M.D. Tang & L.K. Kaczmarek. 1997. hSk4, a member of a novel subfamily of calcium-activated potassium channels. Proc. Natl. Acad. Sci. USA **94:** 11013–11018.
137. Corey, S.G. Krapivinsky, L. Krapivinsky & D.E. Clapham. 1998. Number and stoichiometry of subunits in the native atrial G-protein–gated K^+ channel, I_{KACh}. J. Biol. Chem. **273:** 5271–5278.
138. Swofford, D.L. 1993. PAUP. (3.1.1). (Computer Program). Smithsonian Institution. Washington DC, Sinauer Associates. Sunderland, MA.
139. Gatesy, J., R. DeSalle & W. Wheeler. 1993. Alignment-ambiguous nucleotide sites and the exclusion of systematic data. Mol. Phylogenet. Evol. **2:** 152–157.
140. Lees-Miller, J.P., C. Kondo, L. Wang & H.J. Duff. 1997. Electrophysiological characterization of an alternatively processed ERG K^+ channel in mouse and human hearts. Circ. Res. **81:** 719–726.
141. Nakamura, M., H. Watanabe, Y. Kubo, M. Yokoyama, T. Matsumoto, H. Sasai & Y. Nishi. 1998. KQT2, a new putative potassium channel family produced by alternative splicing. Isolation, genomic structure, and alternative splicing of the putative potassium channels. Recep. Channels **5:** 255–271.
142. Kondo, C., S. Isomoto, S. Matsumoto, M. Yamada, Y. Horio, S. Yamashita, K. Takemura-Kameda, Y. Matsuzawa & Y. Kurachi. 1996. Cloning and functional expression of a novel isoform of ROMK inwardly rectifying ATP-dependent K^+ channel, ROMK6 (Kir1.1f). FEBS Lett. **399:** 122–126.
143. Yano, H., L.H. Philipson, J.L. Kugler, Y. Tokuyama, E.M. Davis, M.M. Le Beau, D.J. Nelson, G.I. Bell & J. Takeda. 1994. Alternative splicing of human inwardly rectifying K+ channel ROMK1 mRNA. Molecular Pharmacology **45:** 854–860.
144. Wei, J., M.E. Hodes, R. Piva, Y. Feng, Y. Wang, B. Ghetti & S.R. Dlouhy. 1998. Characterization of murine girk2 transcript isoforms: Structure and differential expression. Genomics **51:** 379–390.

[Note added in proof: After this paper was submitted, Wang *et al.* (1998, Science **282:** 1890–1893) have presented strong evidence that KCNQ2 and KCNQ3 subunits form channels mediating the "M-current."]

Genomic Organization of Nematode 4TM K$^+$ Channels[a]

ZHAO-WEN WANG,[b] MAYA T. KUNKEL,[b] AGUAN WEI,[b] ALICE BUTLER,[b] AND LAWRENCE SALKOFF[b,c,d]

[b]*Department of Anatomy and Neurobiology and* [c]*Department of Genetics, Washington University School of Medicine, 660 South Euclid Avenue, St. Louis, Missouri 63110, USA*

ABSTRACT: As many as 50 genes in the *C. elegans* genome may encode K$^+$ channels belonging to the novel structural class of two-pore (4TM) channels. Many 4TM channels can be grouped into channel subfamilies. We analyzed 4TM channels in *C. elegans* using methods made possible by having complete genomic sequence. Two genes were chosen for comprehensive analysis, n2P16 and n2P17. By comparing the pattern of conservation in genomic DNA sequences between *C. elegans* and a closely related species, *C. briggsae*, we were able to identify all coding regions and predict the gene structure for these two genes. Given the extent of the 4TM channel family, we were surprised to discover that n2P17 produced at least six alternative transcripts encoding a constant central region and variable amino- and carboxyl-termini. Blocks of highly conserved DNA sequences in noncoding regions were also apparent and most likely confer important regulatory functions. The interspecies comparison of the deduced channel proteins revealed that the extracellular loop between M1 and P1 is an apparent hot spot for evolutionary change in both channels. This contrasts with the membrane-spanning domains that are highly conserved. Analysis of intron positions for 36 channels revealed that introns are frequently present at an identical position within the pore region, but very few are located in membrane-spanning domains.

C. elegans is a small, free-living soil nematode with a genome size of approximately one hundred million base pairs (about 2.5% of the human genome). Currently, greater than 80% of the *C. elegans* genome has been sequenced. Genomic DNA has been routinely analyzed by Genefinder,[1] a program that produces a tentative map of gene organization. From these data, we have compiled a list of approximately 70 genes encoding putative K$^+$ channels that can be divided into three structural classes based on the number of membrane-spanning domains: 2TM, 4TM, and 6TM. These structural classes may be further divided into family and subfamily groupings.[2] Almost all K$^+$ channels known from vertebrates fit within this organization. One surprise from this study[2] is the extent of the family of 4TM channels. Genes of this family encode channel subunits that have four transmembrane domains and two pore-forming regions per subunit. Many questions remain to be answered regarding the need for so many 4TM channels in *C. elegans*, their structure, function, and tissue distribution. The question of whether a similarly large family of 4TM channels exists in vertebrates is also intriguing.

[a]DNA sequences reported here as well as additional *C. elegans* potassium channel sequences are available at http://nt-salkoff.wustl.edu

[d]Corresponding author: Lawrence Salkoff, Ph.D., Department of Anatomy and Neurobiology, Washington University School of Medicine, Box 8108, 660 South Euclid Avenue, St. Louis, MO 63110. Phone: 314-362-3644; fax: 314-362-3446; e-mail: salkoffl@thalamus.wustl.edu

Several 4TM channels have been recently discovered in other species, including mammals,[3,4] *Drosophila*,[5] and plants;[6] and their functional properties have been examined by heterologous expression in *Xenopus* oocytes[3–5] or baculovirus-infected insect cells.[6] As predicted for a K^+ channel lacking a voltage sensing region, these channels are voltage insensitive, and the reversal potentials of currents from 4TM channels shift with a change in the equilibrium potential of K^+. Despite their similar topology, these channels differ somewhat in their observed properties, with reported examples of both inward[3] and outward[6] rectification. Various cellular factors affect their gating. For example, the activity of KCO1 is dependent on intracellular calcium,[6] while those of TWIK-1 and TASK are sensitive to internal[3] and external[4] pH, respectively.

With the genomic data of *C. elegans* and *C. briggsae* as the main resource, we performed a number of analyses and experiments aimed at addressing some basic questions related to 4TM channel biology. We found that (1) the coding DNA of a gene, including alternatively spliced forms, can be predicted by the pattern of genomic DNA mismatches between *C. elegans* and *C. briggsae*. Prediction of gene structures using this approach appears more accurate than that achieved with other available methods that are based on genomic data from a single species. (2) Segments of perfectly conserved DNA sequences that are not translated are present in 5′ and 3′ flanking DNA and in introns; such sequences apparently serve regulatory functions. (3) Areas of channel proteins that evolve the most rapidly are limited to small domains; one domain in particular is the extracellular loop following M1. (4) Intron insertion sites are not random but may be related to functional domains, as revealed by an analysis of intron sites for many members of the 4TM channel family of *C. elegans*. (5) Many 4TM channels appear to fall into channel subfamilies, and there is evidence that they may have mammalian orthologs.

MATERIALS AND METHODS

The genomic DNA sequences of 4TM K^+ channels were mostly compiled from annotated *C. elegans* sequence entries in the Genbank database at the National Center for Biotechnology Information (NCBI) (http://www3.ncbi.nlm.nih.gov/Entrez/nucleotide.html). In addition, finished *C. elegans* cosmids were screened by BLAST searches using the facilities of the Genome Sequencing Center (GSC) at Washington University School of Medicine. The consensus amino acid sequences derived from the K^+ channel pore domains were used as a query. Positive cosmids were further analyzed for exons containing four membrane-spanning domains (M1 to M4) and two pore regions (P1 and P2) in the correct order of topology.

The deduced amino acid sequence of each 4TM channel of *C. elegans* was used as a query for BLAST searching of the *C. briggsae* genome database at the BLAST server of GSC (http://genome.wustl.edu/gsc/blast/blast_servers.html) to look for orthologous genes in that species. Genomic DNA sequences of orthologous genes were aligned between the two species for comparative analysis using either *MicroGenie Sequence Analysis Program* (Beckman Instruments, Inc., Palo Alto, CA, USA) or *DNASTAR* (DNASTAR, Inc., Madison, Wisconsin, USA). Due to the presence of occasional large gaps in the genomic DNA sequence of one species, manual adjustments were necessary to achieve appropriate alignment.

The predicted coding DNA sequences of nine genes, including n2P2, n2P16, n2P17, n2P18, n2P20, n2P24, n2P26, n2P29, and n2P38, were verified experimentally by PCR using first strand cDNA and cDNA libraries (Stratagene, La Jolla, CA, USA) prepared from mixed-stage *C. elegans*. Primers used for RT-PCR were based on predicted gene sequence. The amplification of each gene was achieved by two consecutive rounds of RT-PCR (~30cycles/round) using nested primer pairs and *EXPAND™High Fidelity PCR System* (Boehringer Mannheim Corp., Indianapolis, IN, USA). Products were subcloned into a vector for automated DNA sequencing. The partial sequence of one gene, n2P31, was identified in the *C. elegans* EST database at http://www.ddbj.nig.ac.jp/c-elegans/html/CE_CLONE.html. Further sequence information of n2P31 was obtained from a cDNA clone that was kindly provided by Yuji Kohara, National Institute of Genetics, Mishima Shizuoka-Ken 411, Japan.

Intron positions were compiled for 36 putative 4TM channel genes of *C. elegans*. For genes with confirmed cDNA sequences, intron sites were determined by comparing cDNA sequences with genomic DNA sequences. Intron sites of the other genes were based on predicted gene structures in the NCBI and GSC database.

4TM channels were grouped into subfamilies by first aligning the domains (M1, P1-M2, M3-M4) of 36 predicted 4TM channel sequences with the neighbor-joining method within J. Felsenstein's package *PHYLIP* (http://evolution.genetics.washington.edu/phylip.html). Confidence limits of the tree were determined by the bootstrap method within the *PHYLIP* package. Channels that grouped with confidence intervals greater than 75% were used to generate a phylogenetic tree using the Clustal method with a PAM250 residue weight table within *DNASTAR*. The mammalian 4TM channel TASK-1 is a homolog of the subfamily consisting of the *C. elegans* channels n2P20 and n2P38. Other 4TM channels (e.g., TWIK-1, TREK-1, KCO1, ORK1) were not identified as homologs to any *C. elegans* channels by this analysis.

RESULTS

The Coding and Non-Coding Regions Show Different Patterns of Conservation

n2P16 and n2P17 were chosen for detailed analysis because complete genomic DNA sequence was available in both *C. elegans* and *C. briggsae*. Alignment of genomic DNA sequences between the two species is shown in FIGURES 1A and 1B. This alignment revealed different patterns of conservation in coding and noncoding regions. In exons, the majority of nucleotide mismatches occurred at the third nucleotide position of codons (third base wobble position). The proportion of mismatches at the third positions of four-fold degenerate codons was approximately 60–70%. This should be a good estimate for the rate of mutation in DNA not under selection pressure. Indeed, many regions within introns, where alignment was possible, appear to have drifted by a similar amount. Substitutions at the third position of codons generally do not alter the identities of coded amino acids, and are therefore silent. When mismatches occurred at the first position of codons, they generally belonged to leucine or arginine codons, and were also silent. In FIGURES 1A and 1B, a "ladder" pattern of mismatches can be seen in the coding regions. Thus, in spite of the fact that 40–50% of codons have nucleotide mismatches between species, few amino acid residues differed between *C. elegans* and *C. briggsae* (6.3% in n2P16, and 3.1% in n2P17). In contrast to coding DNA, introns and the noncoding DNA flanking the genes had a very different pattern of conservation characterized by length variation, a higher frequency of mismatches, and occasional blocks of identity that lacked the "ladder" pattern characteristic of coding DNA (FIG. 1).

Prediction of Gene Structure by the Pattern of Conservation

Prediction of gene structure solely from genomic DNA presents a challenge. Furthermore, the determination can be especially difficult when genes are alternatively spliced. We explored the possibility of predicting gene structures for n2P16 and n2P17 simply by aligning the genomic sequences from *C. elegans* and *C. briggsae*, and identifying exons by their characteristic pattern of conservation (as described above). The boundaries of exons were identified by splice junction consensus sequences (see the legend of FIG. 1) and a change in the pattern of conservation in the aligned genomic DNA sequences.

Using the above technique, all exons of n2P16 became quickly apparent. A putative translational start site (ATG) was identified at the beginning of the first predicted exon, and a stop codon (TAG in *C. elegans* and TAA in *C. briggsae*) was present at the end of the last predicted exon. Most corresponding exons for n2P16 had identical lengths in the two species. However, due to the presence of an additional intron in *C. briggsae*, the coding potential of exons 6 and 7 in *C. briggsae* was contained in a single exon in *C. elegans*. In

FIGURE 1 *(following five pages)*. Alignment of genomic DNA sequences between *C. elegans* and *C. briggsae* for n2P16 (**A**) and n2P17 (**B**). NCBI/GSC designations for the cosmids are indicated at the top of each sequence. Numbers on the left of the sequences indicate nucleotide positions in cosmids. Areas of nucleotide matches are *shaded in black* for exons and *dark gray* for noncoding regions. All mismatches are shown in *light gray*. Most of the mismatches in exons were silent, occurring at the third or first nucleotide position of degenerate codons. Except for a single intron starting with GC, all introns followed the GT-AG rule. Confirmed initiator start sites (ATG) and stop codons are *circled*. (**A**) The n2P16 gene had 12 confirmed exons in *C. elegans* versus 13 predicted exons in *C. briggsae*. Only a single form was predicted by comparative analysis, and only that single form was detected by RT-PCR (see FIG. 2A). (**B**) n2P17 is very complex, with three initiator methionines and three possible alternative stop sites. All the three predicted initiator methionines and two of the three predicted stops were experimentally confirmed (see FIG. 2B). The predicted but unconfirmed stop codon in exon 9 is *underlined*. The stop codon in exon 10 is out of frame in *C. briggsae*. Since polyT precedes the stop, it is possible that a T was omitted due to a sequencing error. The intervening amino acid sequence between the indicated alternative donor splice site of exon 10 and the nearby alternative stop codon is: VNQFEIRYRV in *C. elegans*. Note that a nonconventional GC (versus GT) is used in the donor splice site at the end of exon 9. We predicted that the intron following exon 9 begins with GC; this choice of splice junction would allow exon 9 to end with AG (A and G are known be common at the respective –2 and –1 positions) and maintain the reading frame of exon 9 coherent with that of exon 10. The use of this nonconventional donor splice site has been confirmed by RT-PCR.

The consensus sequences for *C. elegans* intron splice junctions are[17]:

5′ Splice Site

–3	–2	–1	+1	+2	+3	+4	+5	+6	+7	+8
A/C	A	G	G	T	A/G	A	G	T	T	T
	56	64	100	100		67	76	62	54	52

3′ Splice Site

–7	–6	–5	–4	–3	–2	–1	+1
T	T	T	T	C	A	G	A/G
53	89	98	70	83	100	100	

Numbers with plus or minus signs indicate nucleotide positions relative to splice junctions; numbers at the bottom show the percentage of occurrence of indicated nucleotides at each position.

A. n2pore16 : *C. elegans* cosmid F52E4 / *C. briggsae* cosmid G45M03

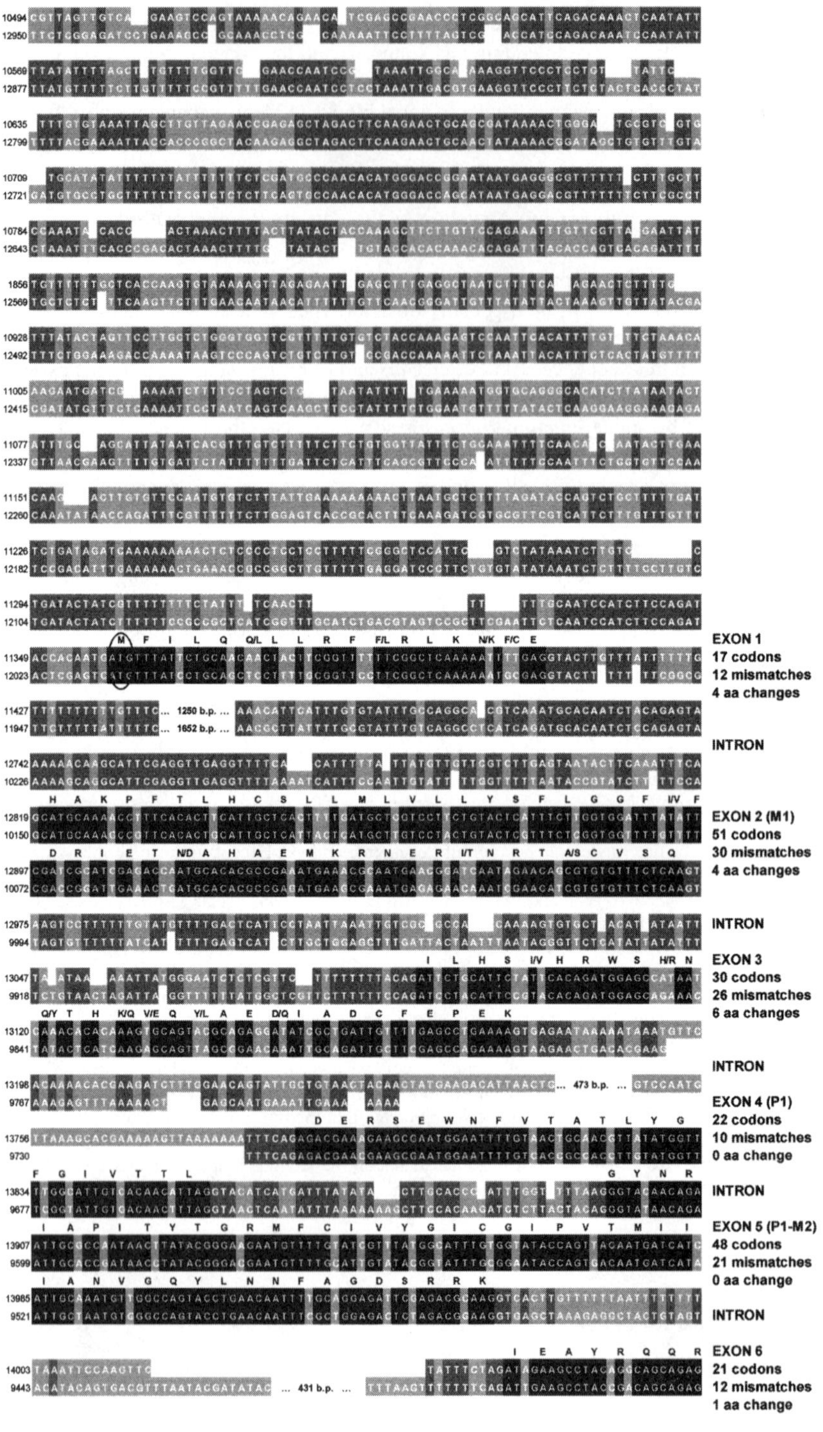

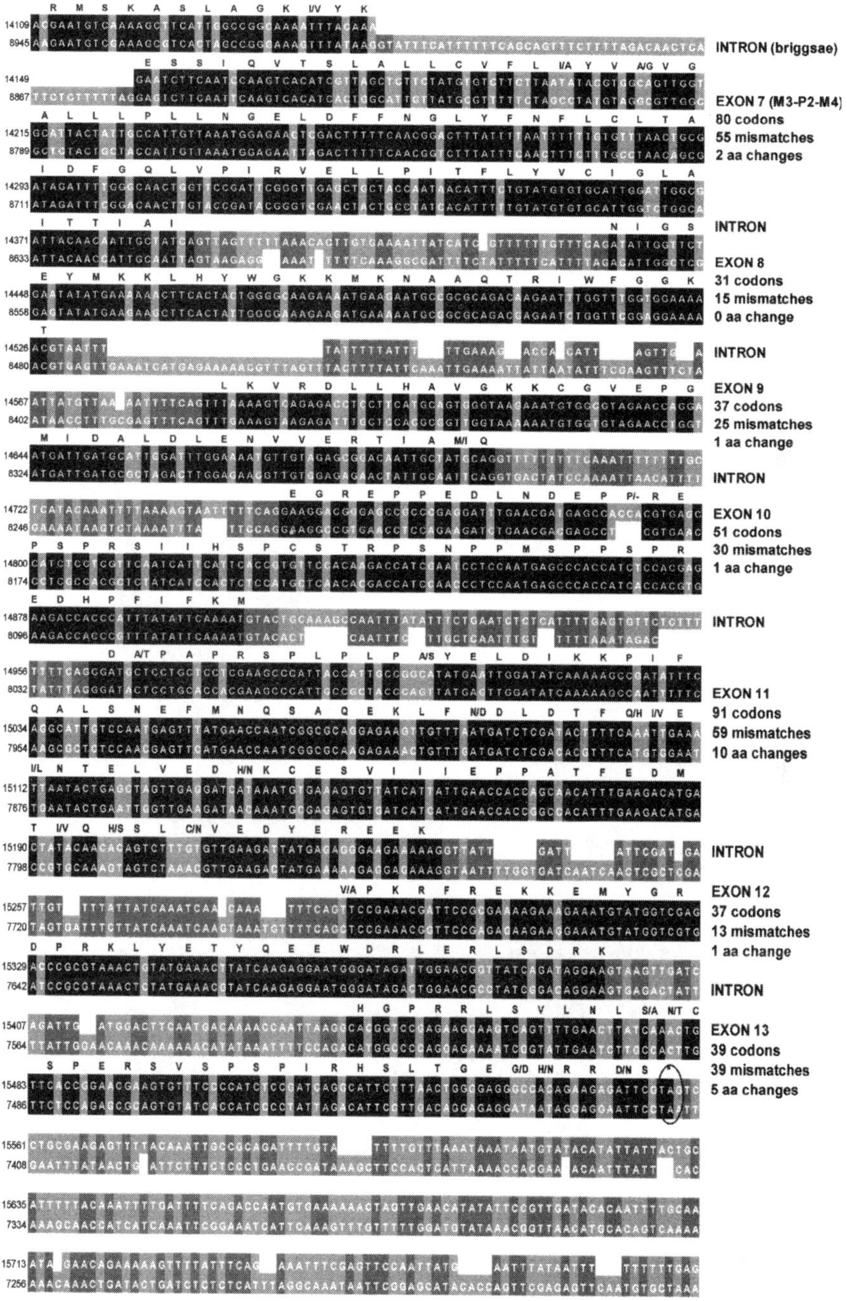
INTRON (briggsae)
EXON 7 (M3-P2-M4)
80 codons
55 mismatches
2 aa changes
INTRON
EXON 8
31 codons
15 mismatches
0 aa change
INTRON
EXON 9
37 codons
25 mismatches
1 aa change
INTRON
EXON 10
51 codons
30 mismatches
1 aa change
INTRON
EXON 11
91 codons
59 mismatches
10 aa changes
INTRON
EXON 12
37 codons
13 mismatches
1 aa change
INTRON
EXON 13
39 codons
39 mismatches
5 aa changes

B. n2pore17 : *C. elegans* cosmid C44E12 / *C. briggsae* cosmid G44E08

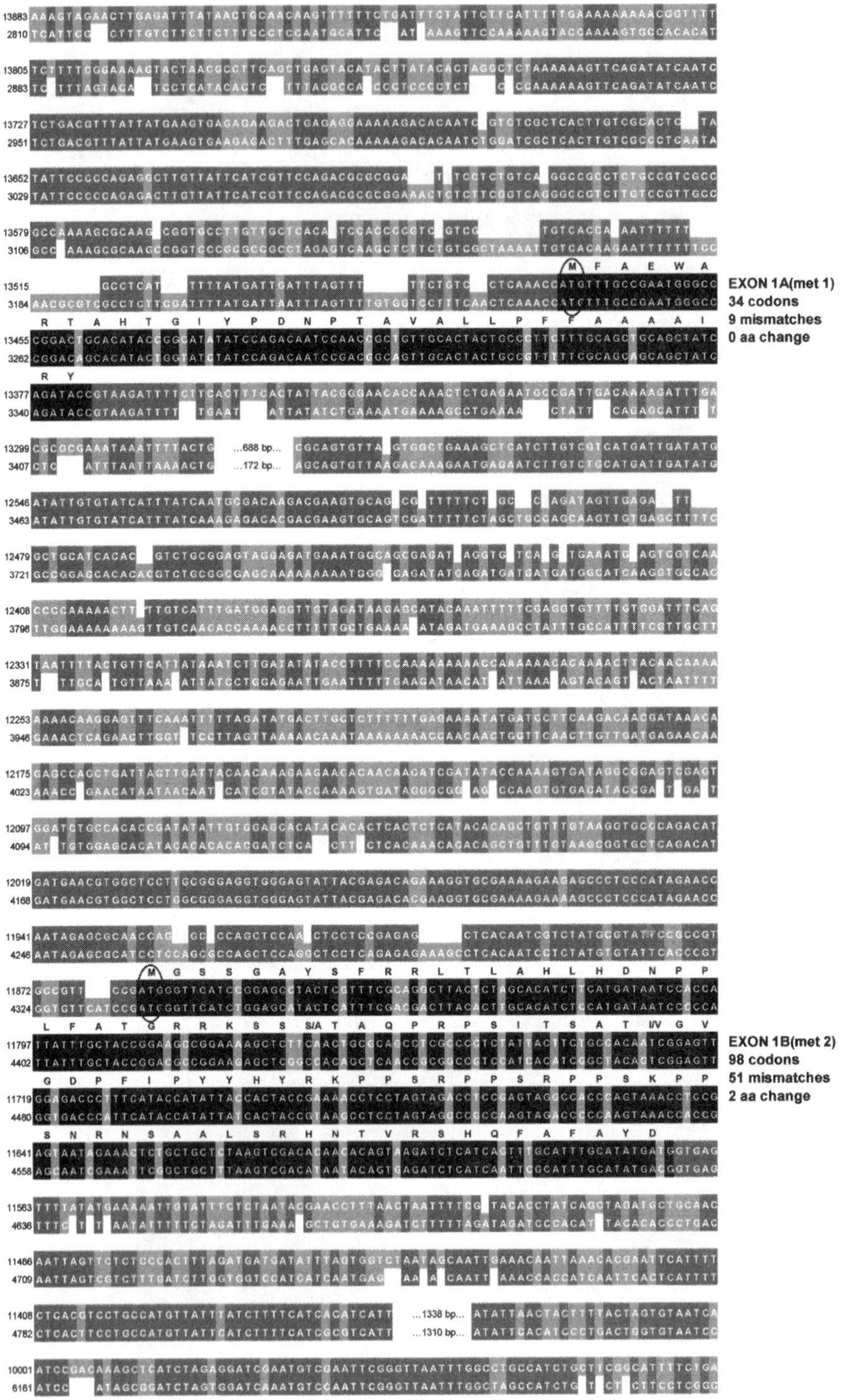

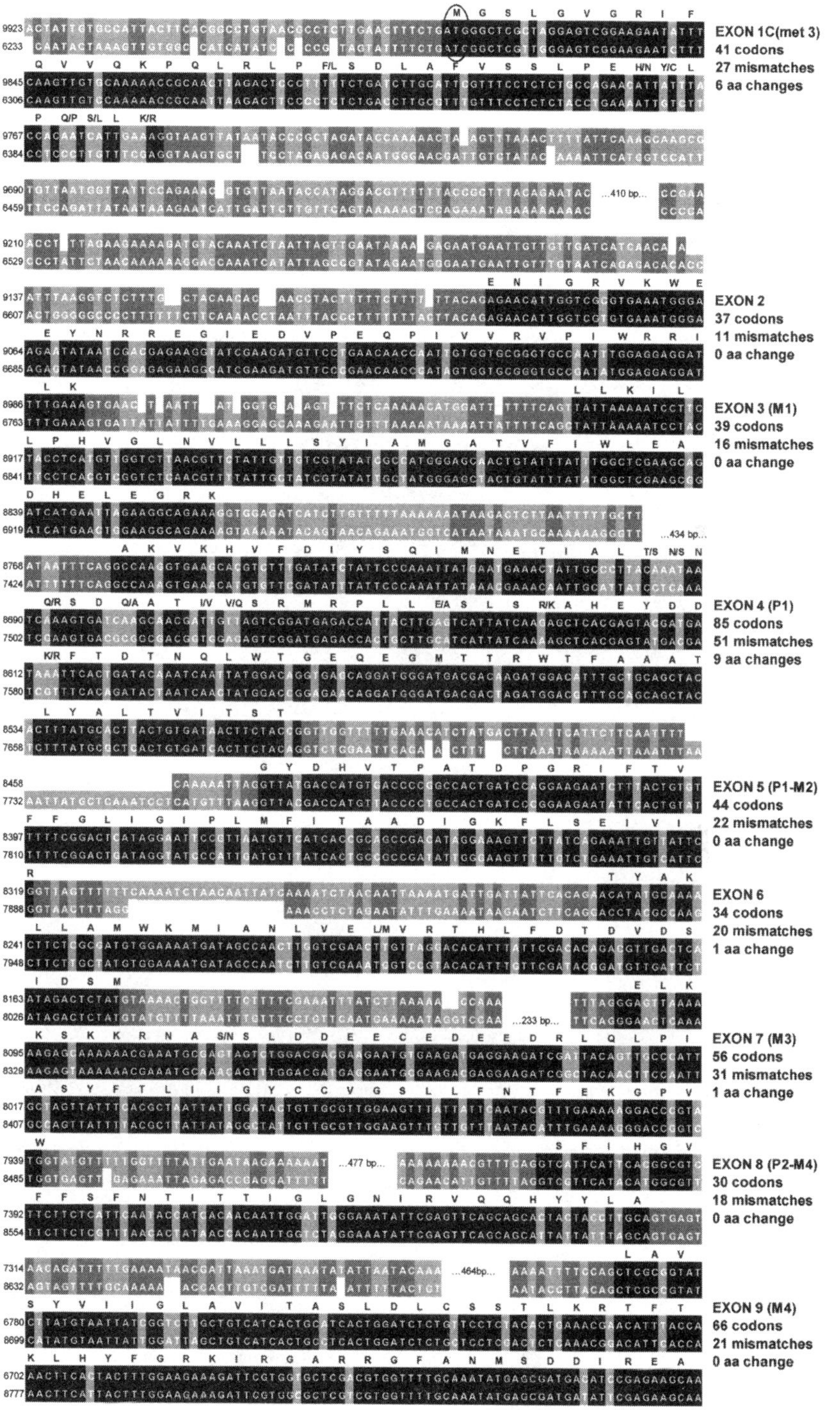
EXON 1C(met 3)
41 codons
27 mismatches
6 aa changes
EXON 2
37 codons
11 mismatches
0 aa change
EXON 3 (M1)
39 codons
16 mismatches
0 aa change
EXON 4 (P1)
85 codons
51 mismatches
9 aa changes
EXON 5 (P1-M2)
44 codons
22 mismatches
0 aa change
EXON 6
34 codons
20 mismatches
1 aa change
EXON 7 (M3)
56 codons
31 mismatches
1 aa change
EXON 8 (P2-M4)
30 codons
18 mismatches
0 aa change
EXON 9 (M4)
66 codons
21 mismatches
0 aa change

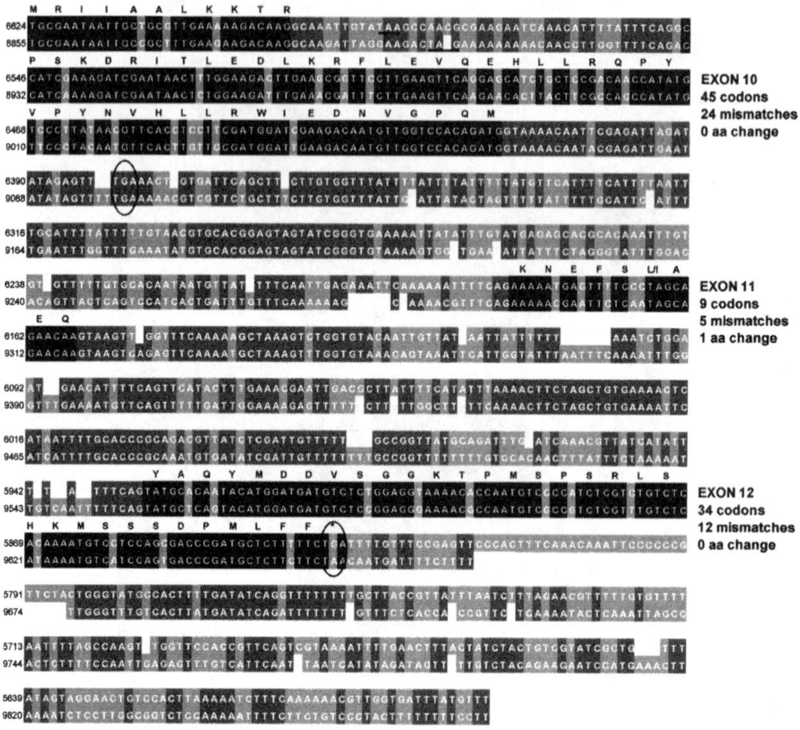

addition, *C. elegans* had an additional codon in exon 10 (FIG. 1A). A single open reading frame encoding the primary structure of the protein resulted when all exons were joined following the predicted splicing pattern (FIG. 2A-I).

In contrast, our analysis of n2P17 revealed a very complex structure (FIGS. 1B and 2B-I). Three upstream exons were identified having putative initiator ATGs. These exons (1A, 1B, and 1C, respectively) are separated by noncoding DNA, and appear to serve as alternative translational start sites. An examination of the potential splice junction sites of these and subsequent exons suggested that exons 1A, 1B, and 1C could all be independently and appropriately spliced to exon 2, the first putative common exon, to form a continuous open reading frame. Almost all of the remaining exons could be easily identified based on the described pattern of conservation, and splice junction consensus sequences. However, determination of the 3′-end of exon 9 was ambiguous because it was not followed by "GT," suggesting the use of a nonconventional splice site. We predicted that the intron following exon 9 started with "GC" (see the legend of FIG. 1). We also found three possible alternative stop sites located at the ends of exons 9, 10, and 12, respectively (FIG. 1B). The predicted gene structure of n2P17 is shown schematically in FIGURE 2B-I.

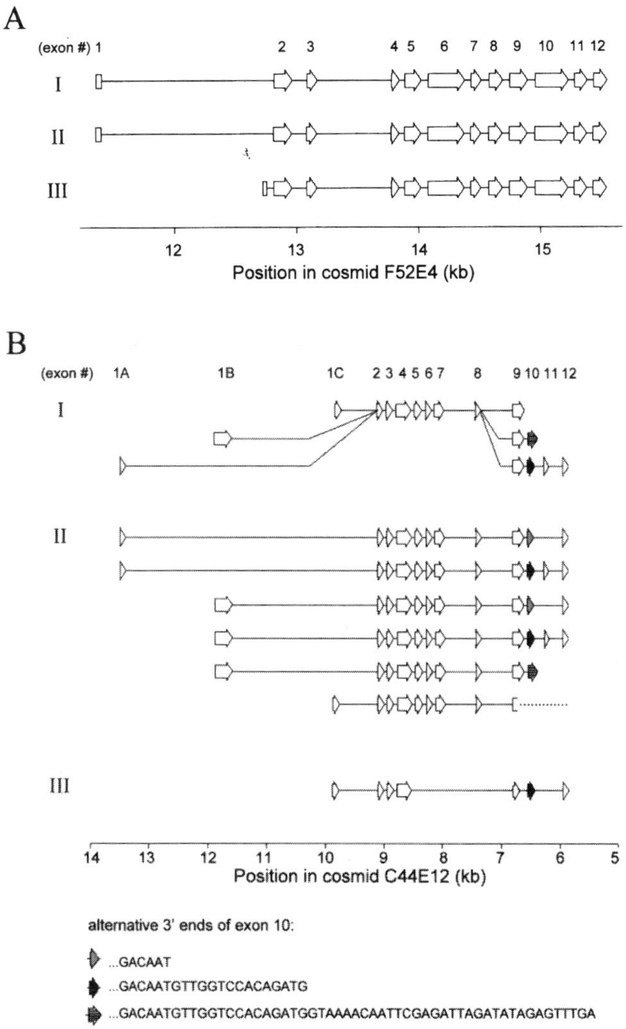

FIGURE 2. Gene organization shown diagrammatically. **(A)** n2P16. **I.** Prediction based on aligned genomic DNA sequences of *C. elegans* and *C. briggase*. **II.** Actual gene organization verified by RT-PCR. **III.** Gene structure predicted by *Genefinder*. *Genefinder* correctly predicted the structure of this simple gene except for chosing incorrect sequences for the first exon. **(B)** n2P17. **I.** Prediction based on aligned genomic DNA sequences of *C. elegans* and *C. briggase*. Three alternative initiator methionines and three alternative stop sites were predicted. All the alternative forms share common exons in the central part of the gene. **II.** Alternative forms experimentally verified. All the three predicted initiator methionines are used, and two of the three predicted stop sites were confirmed. The 3′ end of the cDNA starting from the third alternative methionine was undetermined. **III.** Gene structure predicted by *Genefinder*. *Genefinder* omitted two of the three alternative initiator methionines. In addition, it failed to correctly predict five exons in the central and 3′ regions of the gene, and the alternative stop in exon 10.

Verification of Predicted Gene Structures by RT-PCR

To test the accuracy of our predicted gene structures, we performed RT-PCR using nested primer pairs. To verify full-length transcripts, primers were chosen to correspond to the predicted 5′ start sites and 3′ stop sites. Partial transcripts were also amplified using primers matching internal sites. Only one cDNA form was amplified for n2P16 (FIG. 2A-II). The sequence of the single cDNA exactly matched our predicted sequence (FIG. 2A-I).

Due to the complexity of the predicted n2P17 gene structure, many RT-PCR experiments were performed. Five cDNA's encoding an entire open reading frame were verified. Two products were amplified from the first predicted alternative methionine, and three from the second predicted alternative methionine. We were able to verify that the third predicted methionine was also used as an alternative start site but the site of 3′ termination was undetermined. The 3′ end of n2P17 had complex alternative splicing. Two of three putative stop sites were confirmed. In addition, exon 10 terminated at alternative sites, and exon 11 was present only in some alternatively spliced forms (FIG. 2B-II).

All n2P17 cDNA's that terminated in exon 12 shared the same 3′ untranslated region. A polyA tail was identified 558 bases downstream of the stop codon in *C. elegans*. The sequence between the stop codon and polyA tail in the isolated cDNA perfectly matched

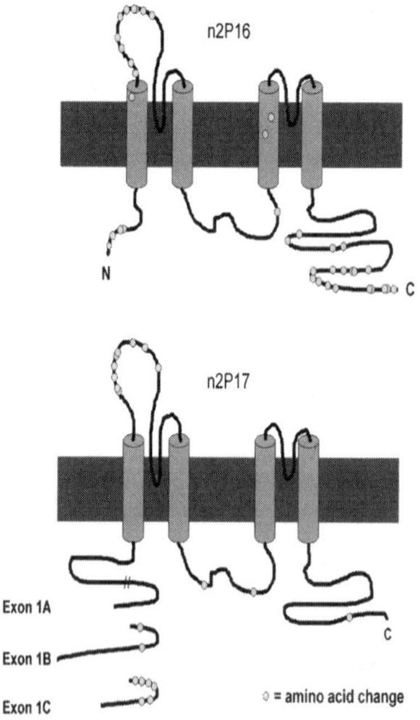

FIGURE 3. Illustration of amino acid differences between *C. elegans* and *C. briggsae* for n2P16 **(top)** and n2P17 **(bottom)**. Amino acid differences are indicated as *light gray circles*. In general, the amino acid sequence of each channel is highly conserved (~93% identical for n2P16, ~97% identical for n2P17). The extracellular loop following M1 is a conspicuous location for amino acid differences in both n2P16 and n2P17. In addition, the N-terminals of both channels and the C-terminus of n2P16 also show some amino acid differences.

the genomic DNA sequence without gaps. Because the polyadenylation site is only 25 bases upstream of the stop codon of a gene going in the opposite genomic DNA strand, the n2P17 gene most likely overlaps with the 3′ end of the downstream gene. The 3′ untranslated sequence following the stop codon in exon 10 also perfectly matched the genomic DNA sequence without gaps. Therefore, exons 11 and 12 were included in the 3′ untranslated region of the early terminating cDNA form (data not shown).

The experimentally determined gene structures are in remarkable agreement with our predictions based on the comparative analyses of *C. elegans* and *C. briggsae* genomic DNA sequences. There was perfect agreement between our prediction and the actual structure of n2P16. Even the structure of the complex gene n2P17 was accurately predicted with regard to alternative splicing at the 5′ end. The major complexities that could not be predicted were that exon 11 was not present in all forms, and that exon 10 was spliced at alternative 3′ sites. Thus, it appears that comparative analysis of genomic DNA sequences of *C. elegans* and *C. briggsae* is a more precise method of analysis to predict gene structure than *Genefinder.*[1] *Genefinder* failed to predict the first exon of n2P16 (FIG. 2A-III). For n2P17, *Genefinder* failed to predict two of the three alternative exons at the 5′ end, and a total of five additional exons in the center and 3′ region of the gene. An alternative stop site in exon 10 was also missed (FIG. 2B-III).

Some Areas within the Channel Protein Evolve More Rapidly Than Others

To view the evolutionary changes in 4TM channel structure between *C. elegans* and *C. briggsae*, we examined differences between the deduced amino acid sequences for n2P16 and n2P17. As shown in FIGURE 3, most of the amino acid sequence was highly conserved, with the exception of a conspicuous cluster of amino acid changes in the extracellular loop between M1 and P1. Mismatches were also found in the amino- and

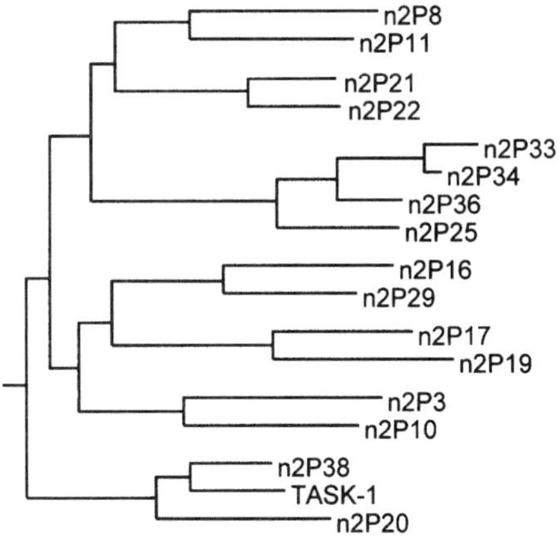

FIGURE 4. Phylogenetic tree of 4TM channels that group into subfamilies. The 16 *C. elegans* channels depicted above were grouped into 7 subfamilies by comparing the primary sequence between P1 and M4 using the Clustal method with a PAM250 residue weight table. The length of each branch reflects the relative number of substitutions between channels. n2P33, n2P34, n2P36, and n2P25 appear to belong to one subfamily, while each of the other six subfamilies consists of only two *C. elegans* 4TM channels. TASK-1 is the mammalian ortholog of n2P38.

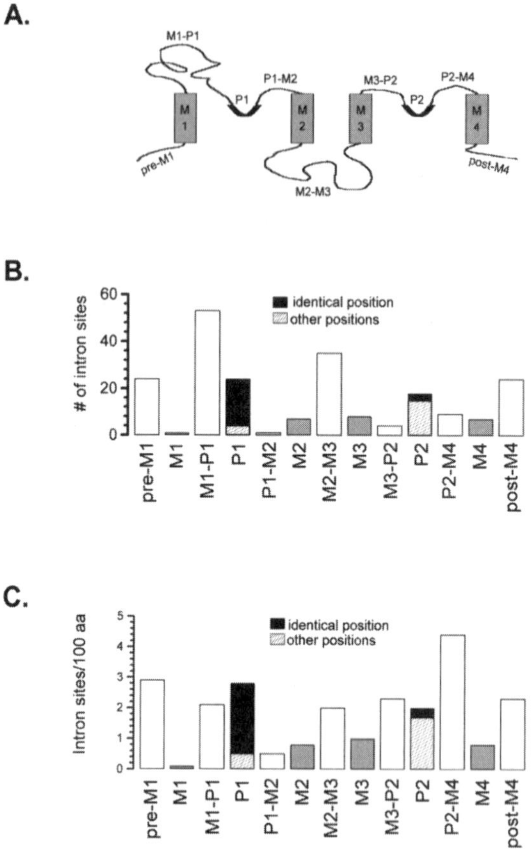

FIGURE 5. Intron positions within the genomic sequence encoding 36 4TM channels. **(A)** Diagram of 4TM channel structure. The P regions and membrane-spanning domains were assigned as in a previous study.[2] Twenty-four amino acid residues were assigned to each P region, with 17 preceding and 4 following the pore signature sequence "GYG." Membrane-spanning domains (M1 to M4) were determined according to Kite-Doolittle hydrophobicity profile. For regions preceding M1 (pre-M1) or following M4 (post-M4), only the sequence of the neighboring 30 (<30 in some channels) residues was considered. Note that the extracellular loop following M1 and the intracellular loop between M2 and M3 are conspicuously long. **(B)** Total number of intron sites within each region. Note that intron positions are not at random locations. They seldom occur in the four membrane-spanning domains, but unexpectedly are common in the two P regions. Intron sites occurring at the identical position (between the first and second nucleotides of the codon for the first glycine of the pore signature sequence "GYG") of the P region are shown by *black columns*, whereas those occurring elsewhere in the P region are represented by *hatched columns*. **(C)** The number of intron sites normalized by amino acid number in each region. Note the low frequency of introns in membrane-spanning domains and the high frequency in P regions. The significance of *black* and *hatched columns* is as described in (B).

carboxyl-termini. These observations suggest that some areas of the channel, in particular the extracellular loop between M1 and P1, evolve more rapidly than other regions. Notably, this hydrophilic extracellular loop is longer in 4TM channels than in any other K^+ channel type. Excluding this loop, the high level of conservation (greater than 90% identity) between orthologs of *C. elegans* and *C. briggsae* in the core region (M1 to M4) of 4TM channels is in striking contrast to the great differences seen in the same region among the different paralogs of 4TM channel members (which can be lower than 25% identity). This seems to indicate that the 4TM channel structural class has an ancient evolutionary origin.

Classification of 4TM Channels

During the original identification and classification of the large number of 4TM channels in the *C. elegans* genome, members of this structural class did not appear to group into subfamilies.[2] However, with the sequence project at 80% completion, the number of 4TM channel genes has nearly doubled. To analyze possible subfamily relationships, we first aligned the predicted primary sequence of the core of 36 4TM channels, correcting for length polymorphisms by removing the intracellular loop and the extracellular loop between M1 and P1. This aided in the alignment using the Clustal method. The aligned sequences were subsequently subjected to bootstrap analysis using J. Felsenstein's *PHYLIP* package (http://evolution.genetics.washington.edu/phylip.html). Through this analysis, 16 4TM channels grouped into seven different subfamilies (FIG. 4). The remaining 4TM channels did not appear to group by this analysis and thus may reflect unique genes. Interestingly, the mammalian 4TM channel TASK-1 clearly grouped with n2P20 and n2P38, and may be an ortholog of n2P38. Other 4TM channels identified in other species that were included in our analysis did not appear to be homologs of any presently known *C. elegans* channels. However, as more 4TM channels are revealed through genome sequencing projects, additional homologs to *C. elegans* 4TM channels may be identified.

Intron Sites Are Not Random

To determine if intron sites delimit functional domains, we compiled data of intron positions for 36 4TM channels. For some channel genes, including n2P2, n2P16, n2P17, n2P18, n2P20, n2P24, n2P26, n2P29, n2P31, and n2P38, the cDNA sequence had been experimentally determined, and therefore the deduced intron sites are accurate. Intron sites of the other channels were mostly compiled based on *Genefinder*-predicted gene structures. As shown in FIGURE 5, intron sites did not occur randomly; they were seldom present within genomic DNA encoding the four membrane-spanning domains. Among the 36 channels, introns occurred only 1, 7, 8, and 7 times within M1, M2, M3, and M4, respectively. Surprisingly, genomic DNA encoding the two P regions frequently contained introns. An intron site was found in the P1 region of 24 channels and in the P2 region of 18 channels. Most (20 out of 24) of the intron sites in P1 occurred between the first and second nucleotide of the codon for the first amino acid residue of the pore signature sequence "GYG," whereas only 3 of the 18 intron sites in P2 were in this identical location.

DISCUSSION

Over two-thirds of the predicted K^+ channel genes identified in the *C. elegans* genome are members of the novel structural class possessing two pore regions and four transmembrane domains. Analysis of 36 of these 4TM channel genes has revealed that many of these channels can be grouped into different subfamilies. There is preliminary data that mammalian orthologs exist in *C. elegans*, as suggested by the identification of a TASK-1 ortholog, n2P38. The finding of mammalian orthologs in *C. elegans* suggests that these channels are involved in fundamental, but as yet undetermined, roles in physiology. The fact that 4TM channels are present in plants[6] also suggests a very ancient origin in evolution.

C. elegans and *C. briggsae* diverged in evolution tens of millions of years ago.[7,8] DNA sequences undergo spontaneous mutations over time. The rate of spontaneous mutation varies between species and among nuclear genes of the same species.[9,10] However, DNA sequences important to the survival or propagation of the animal are conserved under selection pressure. Conserved sequences may be in exons and therefore encode proteins, or may be in noncoding regions and have essential functions. It has been noted in previous studies[7,11] that, in alignments of DNA encoding orthologous proteins in related species, there is commonly a high incidence of nucleotide mismatches at silent positions of codons (usually the third position). We envisaged that such a distinctive pattern of mismatches in genomic DNA sequence might be a major resource in predicting gene structure. Indeed, when comparing the genomic DNA sequences of two 4TM channel orthologs from *C. elegans* and *C. briggsae*, the coding regions became apparent as segments showing a "ladder" of third base mismatches bounded by potential splice donor and acceptor sites. By this analysis, we were able to predict not only all exons common to all transcripts, but also all alternative exons. In addition to identification of coding regions, this distinctive pattern of exon conservation also helped in determining splice junction site choice because correct splice sites must allow the pattern of third base mismatches to follow the open reading frame throughout the gene.

The n2P16 gene structure predicted by comparative analysis of *C. elegans* and *C. briggsae* genomic DNA sequences yielded 12 exons that could form a single open reading frame; this structure was verified by isolating and sequencing a cDNA clone. By contrast, n2P17 was found to be a very complex gene of at least six alternative forms that have a constant central region with variable 5′ and 3′ ends. In spite of its complexity, we were able to predict the overall gene structure for n2P17 with good accuracy. However, our method encountered limitations in predicting the complex alternative splicing near the 3′ end of the gene, and it was necessary to experimentally determine the complex pattern of alternative splicing of exons 10 and 11. In addition, the use of a nonconventional splice site made our prediction of the 3′ end of exon 9 difficult.

The noncoding regions were, in general, poorly conserved with respect to both length and sequence. However, we observed blocks of highly conserved DNA sequence as long as 50 nucleotides within introns and in the noncoding DNA flanking the genes (FIG. 1). These blocks of highly conserved sequence are likely serving important regulatory functions. However, they do not represent recognition sites for regulatory proteins that are widely used in other genes because BLAST searching of the entire *C. elegans* genomic database indicates that they are unique to these individual genes. The function of these highly conserved blocks of noncoding DNA remains to be determined.

The comparative analysis of *C. elegans* and *C. briggsae* proteins has allowed us to identify regions of the n2P16 and n2P17 channels that are highly conserved versus those that evolve rapidly. We found that the overall lengths of channel peptides were well conserved. Amino acid sequence in most areas of the channel, including the two pore regions, the four membrane-spanning domains, and the intracellular loop between M2 and M3, were particularly well conserved. However, a relatively large number of amino acid differences were clustered in the extracellular loop between M1 and P1. This extracellular loop is unusually long in most 4TM channels relative to the analogous region in other K^+ channel families. Indeed, the M1-to-P1 distance is often more than 10 times that of the M3-to-P2 distance in 4TM channels. This extracellular loop has been implicated in the dimerization of subunits of TWIK-1 channels via a disulfide bridge.[12] Located adjacent to the first pore, this loop could be the site of toxin binding or binding by regulatory factors, but no data presently exist on its function.

There have been various reports suggesting that intron sites delimit functional domains.[13,14] We examined the frequency of intron sites in different regions of the 4TM channels to determine if their locations were random or fell into a pattern. We found that introns were seldom present in DNA encoding the membrane-spanning domains. However, they were frequently present in the two pore regions. Among the 36 channels that we examined, 24 had an intron site within P1. Surprisingly, 20 out of these 24 intron sites were found at an identical position (between the first and second nucleotides of the codon for the first residue of the pore signature sequence "GYG"). It is remarkable that all K^+ channels of the voltage-gated *Shal* subclass that have been examined also have an intron at this identical site. This has been observed for mammals, *Drosophila*, and jellyfish, the latter being a surviving member of a truly ancient order of the *Metazoa*.[15] The significance of this conserved intron position is not known. Could there be regulatory information carried in the intron sequence? Examination of the length and sequence of introns within this site revealed no conservation. By this analysis, an obvious role for these introns in developmental or tissue-specific splicing regulation cannot be determined. On the other hand, it is possible that exons with a perfectly conserved splice site within the P region may serve a role in evolution to diversify K^+ channels; intergenic recombination across this intron would maintain the reading frame. Indeed, in two genes (n2P13 and n2P38), it is even possible to convert a 4TM channel into a 2TM channel by alternative splicing; both have an intron site at the same location in P1 as well as P2. This raises the interesting possibility of the production of alternatively spliced 2TM channels from a gene encoding a 4TM channel. Such a channel would include M1 and M4 plus one hybrid pore.

With the *C. elegans* genome project at 80% completion, we have identified about 40 4TM channel genes. Upon completion of the genome project, this total number may be as many as 50, excluding channels formed through alternative splicing. Why should such a "simple" animal as *C. elegans* need so many genes encoding 4TM channels? The pattern of tissue distribution of 4TM channels may be a key element in understanding the need for such an extensive family of 4TM channels and their function. GFP-promoter-reporter transformation experiments provide an efficient method for an initial determination of the cellular expression patterns of genes in *C. elegans*. Our preliminary experiments with promoter-GFP constructs for several of the 4TM channels[16] indicate that the family of 4TM channels is expressed in most tissue types, and that expression of a particular gene is not widespread but limited to a particular tissue type. In the nervous system, expression of a single 4TM channel is usually limited to a small subset of neurons. The highly tissue-

and cell-specific expression patterns of 4TM channels suggest that diverse requirements of each cell or tissue type to define their unique electrical properties may underlie the need for the existence of such an extensive family of 4TM channels in *C. elegans.*

Our results clearly indicate that analysis of patterns of genomic DNA mismatches between *C. elegans* and *C. briggsae* is a powerful method to determine gene structures and to identify important regulatory elements in the noncoding regions. We believe that the human genome project could benefit tremendously if the genome of a related mammalian species was also sequenced for comparative analysis. Since the mammalian genome is much more extensive and introns tend to be much larger than in *C. elegans*, it is more difficult for computer programs to predict gene structures using genomic data alone in humans. In addition, it is virtually impossible to identify most regulatory elements in noncoding regions using conventional procedures. In contrast, we have found that "reading" DNA sequences by aligning the genomic DNA sequences of two closely related species would enable the prediction of exons as well as conserved regulatory elements in the noncoding regions. The ancient Egyptian hieroglyphics had been a mystery before an inscribed stone (the Rosetta stone) was discovered in 1799. The Rosetta stone contained the same message in three different languages, including hieroglyphics, demotic, and Greek. Hieroglyphics was finally deciphered by comparing the same passage written in hieroglyphics and Greek. Aligned genomic DNA sequences of related mammalian species could well be the "Rosetta stone" for deciphering the human genome.

ACKNOWLEDGMENTS

This work was supported by a grant from the NIH to L.S., an NRSA to Z.W.W., and a fellowship from the McDonnell Center for Cellular and Molecular Neurobiology to M.T.K.

We would like to thank Matthew Schreiber, Alex Yuan, Tanya Mathur, Linda Lutfiyya, and Sean Eddy for many helpful comments and discussions.

REFERENCES

1. WILSON, R. *et al.* 1994. 2.2 Mb of contiguous nucleotide sequence from chromosome III of *C. elegans*. Nature **368:** 32–38.
2. WEI, A., T. JEGLA & L. SALKOFF. 1996. Eight potassium channel families revealed by the *C. elegans* genome project. Neuropharmacology **35:** 805–829.
3. LESAGE, F. *et al.* 1996. TWIK-1, a ubiquitous human weakly inward rectifying K^+ channel with a novel structure. EMBO J. **15:** 1004–1011.
4. DUPRAT, F. *et al.* 1997. TASK, a human background K^+ channel to sense external pH variations near physiological pH. EMBO J. **16:** 5464–5471.
5. GOLDSTEIN, S.A. *et al.* 1996. ORK1, a potassium-selective leak channel with two pore domains cloned from *Drosophila melanogaster* by expression in *Saccharomyces cerevisiae*. Proc. Natl. Acad. Sci. USA **93:** 13256–13261.
6. CZEMPINSKI, K. *et al.* 1997. New structure and function in plant K^+ channels: KCO1, an outward rectifier with a steep Ca^{2+} dependency. EMBO J. **16:** 2565–2575.
7. HESCHL, M.F. *et al.* 1990. Functional elements and domains inferred from sequence comparisons of a heat shock gene in two nematodes. J. Mol. Evol. **31:** 3–9
8. BUTLER, M.H. *et al.* 1981. Molecular relationships between closely related strains and species of nematodes. J. Mol. Evol. **18:** 18–23.

9. SHARP, P.M. & W.H. LI. 1989. On the rate of DNA sequence evolution in *Drosophila*. J. Mol. Evol. **28:** 398–402.
10. STRACHAN, T. & A.P. READ. 1996. Human Molecular Genetics. 241–273. Wiley-Liss. New York.
11. BURRIS, P.A. *et al.* 1998. The pore-forming and cytoplasmic domains of the neurogenic gene product, BIG BRAIN, are conserved between *Drosophila virilis* and *Drosophila melanogaster.* Gene **206:** 69–76.
12. LESAGE, F. *et al.* 1996. Dimerization of TWIK-1 K^+ channel subunits via a disulfide bridge. EMBO J. **15:** 6400–6407.
13. DURKIN, M.E., U.M. WEWER & A.E. CHUNG. 1995. Exon organization of the mouse entactin gene corresponds to the structural domains of the polypeptide and has regional homology to the low-density lipoprotein receptor gene. Genomics **26:** 219–228.
14. SUDHOF, T.C. *et al.* 1985. The LDL receptor gene: A mosaic of exons shared with different proteins. Science **228:** 815–822.
15. JEGLA, T. & L. SALKOFF. 1997. A novel subunit for *Shal* K^+ channels radically alters activation and inactivation. J. Neurosci. **17:** 32–44.
16. SALKOFF, L. *et al.* 1998. The impact of the *C. elegans* genome project on potassium channel biology. 1998. *In* Potassium Ion Channels: Molecular Structure, Function and Diseases. Y. Kurachi, L. Jan & M. Lazdunski, Eds. Academic Press. San Diego, CA. In press.
17. BLUMENTHAL, T. & K. STEWARD. 1997. RNA processing and gene structure. *In C. elegans*. D.L. Riddle, T. Blumenthal, B.J. Meyer & J.R. Priess, Eds.: 117–145. Cold Spring Harbor Laboratory Press. Plainview, NY.

Contributions of Kv3 Channels to Neuronal Excitability

BERNARDO RUDY,[a,b] ALAN CHOW,[a] DAVID LAU,[a] YIMY AMARILLO,[a] ANDER OZAITA,[a] MICHAEL SAGANICH,[a] HERMAN MORENO,[a] MARCELA S. NADAL,[a] RICARDO HERNANDEZ-PINEDA,[a,c] ARTURO HERNANDEZ-CRUZ,[c] ALEV ERISIR,[d] CHRISTOPHER LEONARD,[d] AND ELEAZAR VEGA-SAENZ DE MIERA[a]

[a]*Department of Physiology and Neuroscience, and Department of Biochemistry, New York University of Medicine, New York, New York 10016, USA*
[c]*Instituto de Fisiología Celular, Universidad Nacional Autónoma de México, México.*
[d]*Department of Physiology, New York Medical College, Valhalla, New York, USA*

ABSTRACT: Four mammalian Kv3 genes have been identified, each of which generates, by alternative splicing, multiple protein products differing in their C-terminal sequence. Products of the Kv3.1 and Kv3.2 genes express similar delayed-rectifier type currents in heterologous expression systems, while Kv3.3 and Kv3.4 proteins express A-type currents. All Kv3 currents activate relatively fast at voltages more positive than –10 mV, and deactivate very fast. The distribution of Kv3 mRNAs in the rodent CNS was studied by *in situ* hybridization, and the localization of Kv3.1 and Kv3.2 proteins has been studied by immunohistochemistry. Most Kv3.2 mRNAs (~90%) are present in thalamic-relay neurons throughout the dorsal thalamus. The protein is expressed mainly in the axons and terminals of these neurons. Kv3.2 channels are thought to be important for thalamocortical signal transmission. Kv3.1 and Kv3.2 proteins are coexpressed in some neuronal populations such as in fast-spiking interneurons of the cortex and hippocampus, and neurons in the globus pallidus. Coprecipitation studies suggest that in these cells the two types of protein form heteromeric channels. Kv3 proteins appear to mediate, in native neurons, similar currents to those seen in heterologous expression systems. The activation voltage and fast deactivation rates are believed to allow these channels to help repolarize action potentials fast without affecting the threshold for action potential generation. The fast deactivating current generates a quickly recovering afterhyperpolarization, thus maximizing the rate of recovery of Na^+ channel inactivation without contributing to an increase in the duration of the refractory period. These properties are believed to contribute to the ability of neurons to fire at high frequencies and to help regulate the fidelity of synaptic transmission. Experimental evidence has now become available showing that Kv3.1-Kv3.2 channels play critical roles in the generation of fast-spiking properties in cortical GABAergic interneurons.

The four known Kv3 genes were identified nearly a decade ago.[1–13] Injection of Kv3 cRNAs into *Xenopus* oocytes induced expression of voltage-gated K^+ channels that had unusual properties, which, as far as we could tell at the time, were unlike any voltage-dependent K^+ currents described in native neurons. In particular, the currents did not activate significantly until the membrane was depolarized to membrane potentials more positive than −10 mV. Yet Northern blot analysis showed that transcripts of at least three of the four genes were abundantly expressed in brain, some of them as abundantly as the most

[b]Address for correspondence: Department of Physiology and Neuroscience, New York University School of Medicine, 550 First Avenue, New York 10016. e-mail: Rudyb01@mcrcr6.med.nyu.edu

abundant K+ channel transcripts known then.[14,15] We considered three main possibilities to explain the apparent lack of Kv3-like currents in native cells:

1. that native Kv3 channels have different properties from those seen in the oocyte due to posttranslational modifications or interactions of Kv3 subunits with auxiliary proteins;
2. that Kv3 proteins are targeted to neuronal compartments, such as axons and terminals, with restricted accessibility to electrophysiological methods; and
3. that native Kv3 currents have properties such as those in the oocyte but they had not been described in neurons due to the problems of separating components of the total K+ current.

Moreover, we reasoned that native Kv3 currents may play very specific roles in neurons if their properties are similar to those in the oocyte.

We decided that in either case Kv3 proteins were a good point to start investigating the relationship between cloned channel components and native channels. We also found Kv3 proteins interesting to pursue because the evolution of Kv3 genes appeared to be different from the evolution of Kv genes of other subfamilies. Mammalian products of a given Kv subfamily are more similar to one of four Kv-like genes in *Drosophila* than to a mammalian product of a different subfamily, indicating that precursor genes to each subfamily existed prior to the divergence of chordates and arthropods. Kv3 proteins are more similar to the products of the *Drosophila* Shaw gene than to other *Drosophila* or mammalian Kv proteins.[15] However, the percentage of amino acid identity seen between mammalian Kv3 proteins and Shaw (49–56%) is significantly less than that seen between mammalian and fly homologues in the other Kv subfamilies (70–80%). Even perhaps more significant, while Kv3 proteins are more similar to Shaw in certain regions of the protein, they are more similar to members of other Kv subfamilies in the S4 domain and the sequences near this domain (see TABLE 6 in Ref. 15). Functionally, mammalian Kv3 channels differ considerably from Shaw channels, which lack the unique properties that distinguish Kv3 currents from other voltage-dependent currents (see sections on functional expression and neuronal roles, below). Gene conversion events, or other mechanisms have been suggested to explain the evolution of mammalian Kv3 genes.[15] Whatever the mechanism, mammalian-like Kv3 genes appear to be a recent evolutionary acquisition, specific to species in the evolutionary line leading to mammals, suggesting they might be used in neuronal properties that are specific to mammals and other vertebrates.

A several-year-long effort by our laboratory and others has shown that native Kv3 currents in neurons are similar to those in heterologous expression systems, and is beginning to provide evidence that they indeed play special roles in neuronal excitability.

A review on Kv3 genes was published in 1994.[15] This review discussed in detail several aspects of the work with these genes that will not be considered extensively in this article (such as the patterns of alternative splicing and genomic structure of Kv3 genes, the tissue expression and brain distribution of Kv3 mRNA transcripts, the properties of homomultimeric and heteromultimeric Kv3 channels in *Xenopus* oocytes, and the analysis of structural and functional domains of Kv3 proteins). Instead we will focus here on the work demonstrating that Kv3 channels in neurons have properties similar to those in heterologous expression systems and describe the studies that are particularly relevant to understanding their neuronal roles.

THE Kv3 K+ CHANNEL GENE SUBFAMILY

The Kv3 or Shaw-related subfamily in mammals is similar to the other Kv subfamilies in that several genes have been formed by gene duplication throughout animal evolution. There are four Kv3 genes known both in rodents and humans,[9,15] and their chromosomal locations have been determined in mouse and human (see chapter by Coetzee et al. in this volume).[16] In the last few years Kv3 genes have been identified in other vertebrate species.[17,18]

All four Kv3 genes known in mammals generate multiple products by alternative splicing. The analysis of cDNAs predicts the existence of twelve different Kv3 proteins in mammals (FIG. 1). The alternatively spliced transcripts of each gene have identical nucleotide sequences from the assumed starting ATG up to a point of divergence near, but prior to, the in-frame stop codon, predicting protein products with different carboxyl ends. The point of divergence conforms to the consensus sequence for donor splice junctions. These sequence relationships suggest that the different transcripts arise by alternative splicing of a primary transcript rather than by transcription from separate but highly homologous genes. Other data supporting this conclusion have been discussed previously (see Ref. 15).

Two sites of alternative splicing have been observed. In some Kv3 genes (such as Kv3.1) the divergence follows the second protein-coding exon, which encodes the transmembrane portion of the protein. In Kv3.2, the point of divergence occurs at the end of the third protein-coding exon. In Kv3.3 and Kv3.4 alternative splicing at both sites has been observed. In all cases, the alternative splicing changes *only* the C-terminal portion of the protein; the point of divergence starts at least 50 residues after the last membrane-spanning domain. In some instances, the alternatively spliced C-termini are very short: 10 amino acids in Kv3.1a. The longest occurs in Kv3.3a, where it is over 200 residues (FIGS. 1 and 3). The variable C-termini do not affect the electrophysiological properties of the currents expressed by the products of a particular Kv3 gene. Possible roles of this alternative splicing are discussed later in this chapter. In addition to the alternative-splicing generating proteins with divergent C-termini, alternative splicing at the 5′ end, generating transcripts with different 5′ UTRs, but not affecting the sequence of the protein, has been observed in some Kv3 genes.[14,15,19] In some genes encoding inward rectifying K+ channels, alternative splicing of the 5′ UTR results in the expression of transcripts having different tissue expression[20] (see also Coetzee *et al.,* in this volume). Perhaps the alternative splicing of the 5′ UTR of the Kv3.2 gene is associated with the differential developmental expression in the thalamus as compared to other brain areas (see below). The predicted amino acid sequences of the constant region (prior to the point of divergence of the alternative spliced isoforms) of the products of the four Kv3 genes is shown in FIGURE. 2, and the amino acid sequences of the alternatively spliced C-termini in FIGURE 3.

Kv3 CHANNELS IN MAMMALIAN HETEROLOGOUS-EXPRESSION SYSTEMS

The properties of Kv3 currents in *Xenopus* oocytes were previously reviewed.[15,21] Kv3 cDNAs have been transfected in several types of mammalian cells that have negligible intrinsic K+ currents. Both stable and transient transfection methods have been used. Our laboratory has studied at least one alternatively spliced isoform of each of the four Kv3

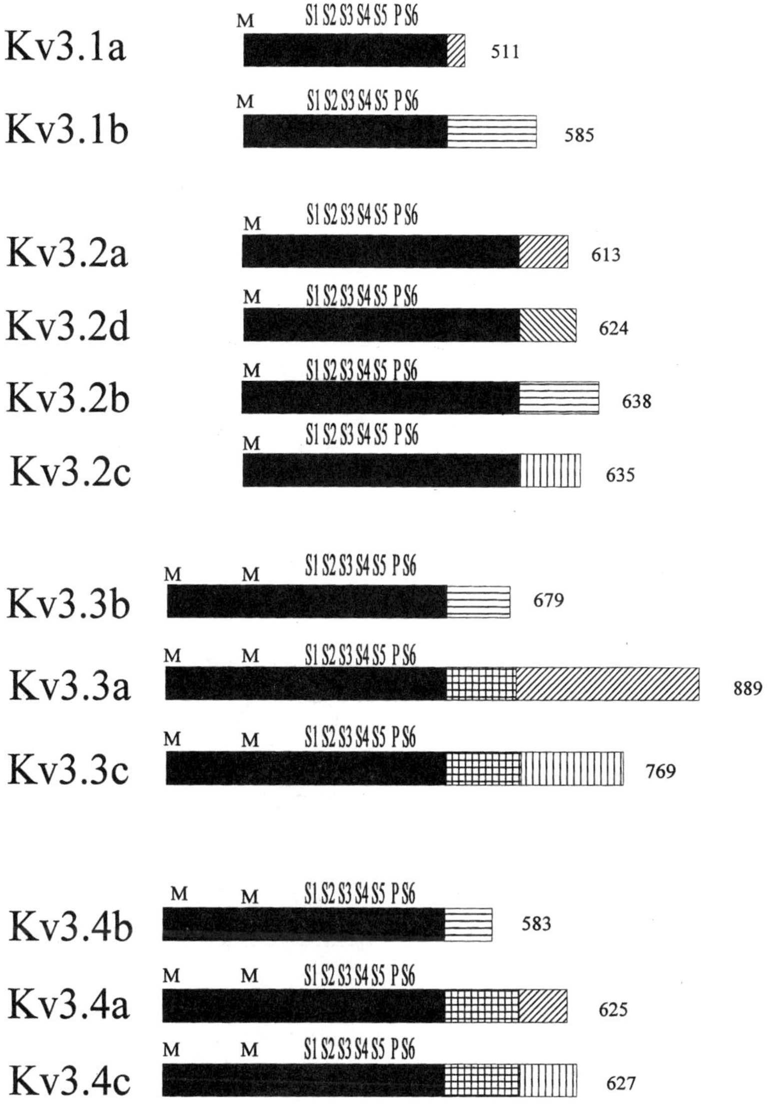

FIGURE 1. Mammalian Kv3 proteins. Four known Kv3 genes encode 12 proteins by alternative splicing. Kv3 isoforms from a given gene differ only in their C-terminal sequence. Proteins expressing inactivating channels (Kv3.3x and Kv3.4x) have an NH-terminal insert preceding the methionine (M) residue, which aligns with the starting methionine of Kv3.1 and Kv3.2 and is responsible for channel inactivation. S1–S6, trasmembrane domains; P, pore domain (also known as H5 domain).

```
Kv3.3   MLSSVCVWSFSGRQGTRKQHSQPAPTPQPPESSPPPLLPPPQQQCAQPGTAASPAGAPLSCGPGGRRAEP   70
Kv3.4   MISSVCVSSYRGRKSGNKPPSKTCLKEE                                             28

Kv3.1           MGQ     GDESERIVINVGGTRHQTYRSTLRTLPGTRLAWLAEPDAHSHFDYDPRA        52
Kv3.2           --K     IENN--VIL-------E------K--------L--SSEPQGDCLTAAGDKLQPLP   58
Kv3.3   CSGLPAVA--RHGGG-GD-GK-------V--E---------------G-T--E-AAR-----GT        134
Kv3.4           -AK     -EA--K-I--------E------------------D--GGGRPES-GGG        80

Kv3.1                                          DEFFFDRHPGVFAHILNYYRTGKLHCPAD   81
Kv3.2   PPLSPPPRPPPLSPVPSGCFEGGAGNCSSHGGNGSDHPGGGR------------YV-------------- 128
Kv3.3                                          -------------YV--------------  163
Kv3.4                                AGSSGSSGGGGGC------------YV--------------  121

Kv3.1   VCGPLYEEELAFWGIDETDVEPCCWMTYRQHRDAEEALDSFGGAPLDNSADDADADGPGDSGDGEDELEM 151
Kv3.2   -----F-------------------------------I-ETPD-IGGDPGD-E-LG-KRLGI--AAGL   198
Kv3.3   -----F----G----------A-------------------EAPDSSGN-NANAGGAHDAGL-D-AGAGG 233
Kv3.4   -----F----T--------------------------I-ESPDGGGGGAGPGDEAGD-ERELALQRLG  191

                                                              ↓     S1
Kv3.1   TKRLALSDSPDGRP               GGFWRRWQPRIWALFEDPYSSRYARYVAFASLFFILV    202
Kv3.2   GGPDGK                       S-R--KL---M-----------A--FI----------    241
Kv3.3   GGLDGAGGELKRLCFQDAGGGAGGPAGGPGGAG-TW-------V-----------A-------------I 303
Kv3.4   PHEGGSGPGAG                  S-GC-G----M-----------A--V-----------    239

                    ▲                  ▲                        S2
Kv3.1   SITTFCLETHERFNPIVNKTEIE        NVRNGTQVRYYREAETEAFLTYIEGVCVVWFTFEFLMRVV 265
Kv3.2   -----------A--IVK----          P-I---SAVLQY-I--DPA---V-------------V-I- 302
Kv3.3   -----------G-IH-S---VTQASPIPGAPPE-I-N-    -V---P----V----------------T 369
Kv3.4   -----------A--IDR-V---H        R-G-I-S--FR--V---PI---------M---L---V-I- 302

                          S3                               S4 •     •
Kv3.1   FCPNKVEFIKNSLNIIDFVAILPFYLEVGLSGLSSKAAKDVLGFLRVVRFVRILRIFKLTRHFVGLRVLG 335
Kv3.2   -S---L-----L---------------------------------------------------------- 372
Kv3.3   ---D----L-S------C---------------------------------------------------- 439
Kv3.4   C--DTLD-V--L-------------------------R-------------------------------- 372

         •     •     •   S5                                          H5
Kv3.1   HTLRASTNEFLLLIIFLALGVLIFATMIYYAERIGAQPNDPSASEHTHFKNIPIGFWWAVVTMTTLGYGD 405
Kv3.2   ----------------------------------V---------▲--Q---------------------- 442
Kv3.3   -------------------------------------D-D-ILG-N--Y--------------------- 509
Kv3.4   -------------------------------------R-S--RGND--D--------------------- 442

                      S6                           ✱
Kv3.1   MKPQTWSGMLVGALCALAGVLTIAMPVPVIVNNFGMYYSLAMAKQKLPKKKKKHIPRPPQLGSPNYCK   473
Kv3.2   ----------------------------------------------R-R-----PA-LAS--TF--   510
Kv3.3   ---K-------------------------------------------N---------P-------PD  579
Kv3.4   ---K-------------------------------------------R---V------E--I---   510

Kv3.1                                                         SVVNSPHHSTQSDTCPL  490
Kv3.2                                                         TEL-MACN-------LG  527
Kv3.3   PPPPPPPHPHHGSGGISPPPPITPPSMGVTVAGAYPPGPHTHPGLLRGGAGGLGIMGL-PLPAPGEP---  649
Kv3.4                                                         -EET--RD--Y---S-P  527

                ↓
Kv3.1   AQEE ILEINRA                                                             501
Kv3.2   KENR L--H--SVLSGDDSTGSEPPLSPPERLPIRRSSTRDKNRRGETCFLLTTGDYTCASDGGIRK      593
Kv3.3   ---- VI-T---                                                             660
Kv3.4   -R--GMV-RK--                                                             539
```

FIGURE 2. Amino acid sequence of the constant region of Kv3 proteins. Amino acids identical to those in Kv3.1 are shown with a *dash*. Gaps required to optimize the alignment are shown as *blanks*. Note that the second methionine of Kv3.3 and Kv3.4 aligns with the starting methionine of Kv3.1 and Kv3.2. The sequences between the first and the second methionine of Kv3.3 and Kv3.4 are required for channel inactivation. Exon boundaries are shown with arrows. The S1–S6 and H5 (or P) domains are *overlined*. Proline-rich sequences in Kv3.2 and Kv3 are *underlined*. Triangles indicate putative N-glycosylation sites. The positively charged residues in S4 have been *boxed*. The leucines in the Shaker leucine heptad repeat adjacent to the S4 domain are indicated with a *solid circle*. In Kv3 proteins the fourth position in the repeat is phenylalanine, but is a leucine in other Kv proteins. Residues in the P domain found by mutagenesis in Kv3.l to be important in determining pore properties are *boxed*. A cluster of positive charges after the S6 domain is indicated with a *star*. (Modified from Vega-Saenz de Miera *et al.*[15])

genes in Chinese Hamster Ovary (CHO) or HEK293T cells. Studies with mammalian cells have several advantages over the *Xenopus* oocyte expression system. The most important for the purpose of comparison with native neuronal channels is that currents can be recorded with the same methods (patch clamp in the whole cell and various patch configurations) and identical solutions that are used to record the currents in the native cells. Moreover, whole cell clamp of small mammalian cells (as compared to nearly 1-mm-in-diameter frog eggs) allows for better space clamp control and increased resolution. Furthermore, since we are studying mammalian proteins, it is more likely that the protein will be processed correctly in the mammalian cell than in the frog oocyte.

Untransfected CHO and HEK293T cells have negligible outward currents under the pulse protocols used to characterize Kv3 currents (<100 pA for the largest depolarizations). On the other hand, CHO cells transfected with each of the four Kv3 cDNAs (Kv3.1b, Kv3.2a, Kv3.3a, and Kv3.4a) have large voltage-dependent K^+ currents with a similar voltage-dependence, as observed in *Xenopus* oocytes (FIG. 4). The currents become apparent when the membrane is depolarized to potentials more positive than −10 mV. As in oocytes, Kv3.1 and Kv3.2 currents are of the delayed-rectifier type. Upon depolarization, they rise relatively fast, with a similar time course, to a maximum level that is maintained for the duration of pulses lasting a few hundred milliseconds. A slow inactivation becomes evident with pulses of longer duration, which is faster for Kv3.1 than for Kv3.2 channels (data not shown). Kv3.4 currents are of the A-type. They activate and inactivate very fast, as in oocytes. However, Kv3.3 currents, which are also transient when expressed in oocytes, resembled Kv3.1 and Kv3.2 currents when expressed in CHO or HEK-293T cells.

We do not yet understand why Kv3.3 currents are transient in oocytes but sustained in mammalian cells. The differences are not the result of the different methods of expressing the protein in the two expression systems, since CHO cells microinjected with the same Kv3.3a cRNA used in oocytes express identical currents to those seen in Kv3.3a-transfected cells. Inactivation of Kv3.3 and Kv3.4 channels is of the N-type[22] and can be removed by deleting NH-terminal inserts present in Kv3.3 and Kv3.4 proteins, but not in Kv3.1 and Kv3.2 proteins (FIGS. 1 and 2; see also Ref. 15). It is possible that the differences are the result of the different methods used to record whole cell currents in the two preparations: two-microelectrode voltage-clamp of *intact* cells in the case of oocytes and whole cell patch-clamp in the case of mammalian cells. Inactivation may be lost in the mammalian cells because of dialysis into the patch pipette of a cytoplasmic component

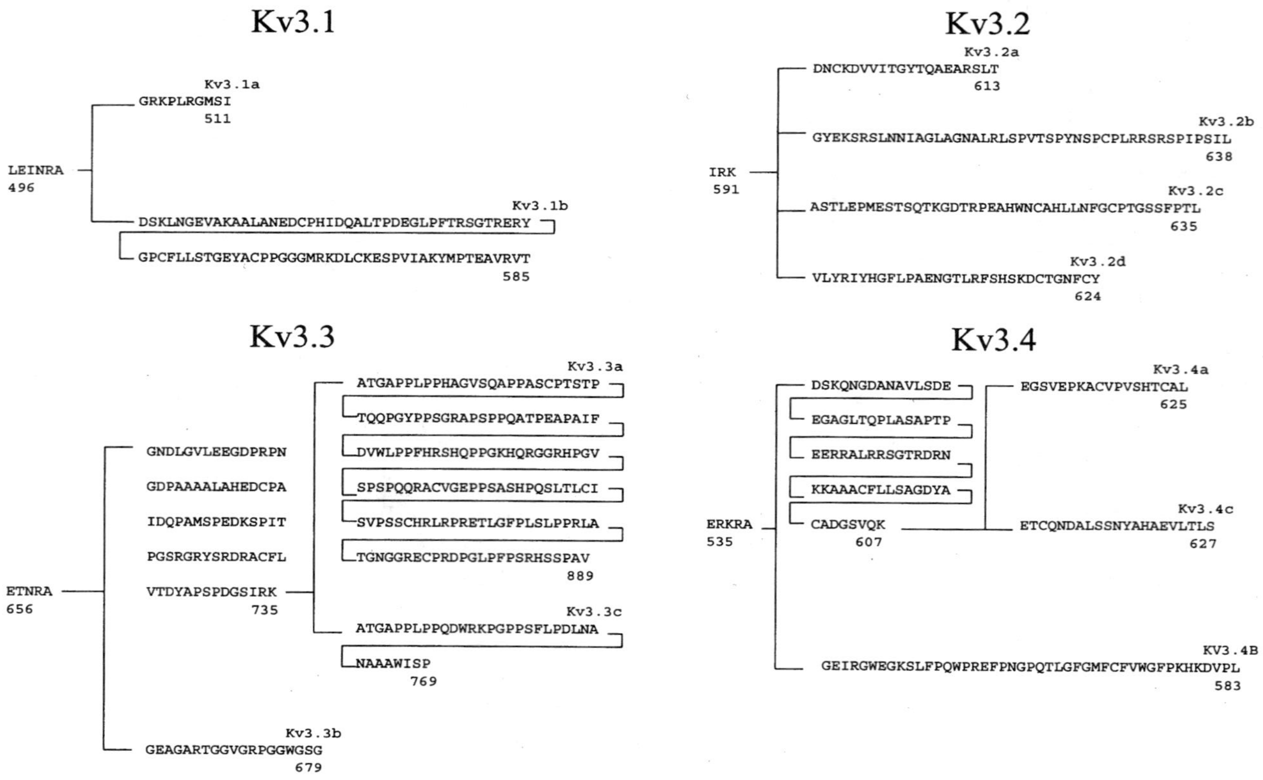

FIGURE 3. Amino acid sequences of the alternatively-spliced C-terminal region of 12 Kv3 proteins.

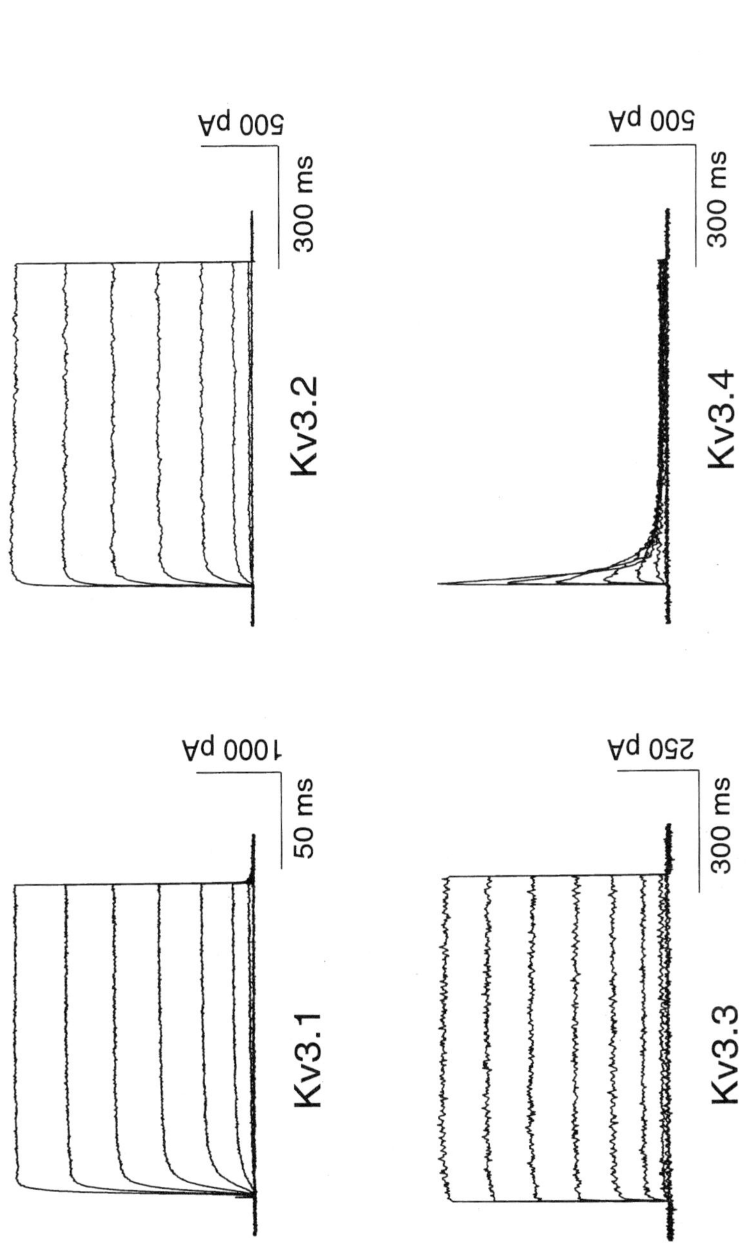

FIGURE 4. Kv3 currents in transfected HEK293 cells. Families of K^+ currents recorded from a holding potential of −80 mV in a Kv3.1, Kv3.2, Kv3.3, or Kv3.4 transfected cell during depolarizing pulses from −40 mV to +40 mV in 10-mV increments.

important for the inactivation process. Inactivation could also be distinct due to differences in the posttranslational processing of the protein in the two expression systems. However, the differences may not be as interesting as these possibilities would suggest. Kv3.3 proteins have a methionine in frame with the predicted starting methionine at the end of the NH2-terminal insert encoding the N-inactivation gate (FIGS. 1 and 2). It is possible that in the mammalian cell, translation of the protein starts at this second methionine. We should point out that if this turns out to be the explanation for the differences of Kv3.3 currents in oocytes and mammalian expression systems, this does not tell us how the protein will be translated in native neurons. The cDNAs that are available lack elements in their 5′ UTR that might be required for the proper selection of translation start site in mammalian cells.

Kv3 currents have several unusual properties that distinguish them from those of other delayed rectifier K^+ channels known (TABLE 1). These include their activation voltage range that is more positive than that of other voltage-gated K^+ channels. The channels with the nearest activation voltage (Kv2.1 and Kv2.2) show significant activation at 10–20 mV more negative potentials. Although Kv3 channel opening becomes significant at high potentials (more positive than −10 mV), the probability of channel opening increases steeply with voltage and >80% of the channels are opened between +30 and +40 mV. As a result of the steep voltage-dependence, the midpoints of the conductance-voltage relationships of Kv3 channels are not that different from those of other voltage-dependent K^+ currents (see Coetzee *et al.* in this volume). This distinguishes mammalian Kv3 currents from the currents expressed by the *Drosophila* Shaw protein, which also start activating at high voltages but have a very weak voltage-dependence, producing a midpoint of activation above +70 mV.[23,24]

The rate of rise of Kv3.1–Kv3.3 currents is relatively fast; faster than many other voltage-gated K^+ channels (e.g., Kv2.x; Kv1.2) but slower than that of other voltage-gated K^+ channels such as several Kv1 channels like Kv1.1, Kv1.4 and Kv1.5 (see Coetzee *et al.* in this volume). Kv3.4 currents rise faster than Kv3.1–Kv3.3 currents. In contrast to other delayed rectifiers, Kv3.1 and Kv3.2 currents are not significantly inactivated by depolarizing prepulses[25] and do not show cumulative inactivation (Ref. 26 and unpublished observations). Records of Kv3.1 and Kv3.2 currents in mammalian cells (but not in oocytes) often show an initial fast transient component. This is usually very small (<5% of the current) and is seen mainly in response to large depolarizing pulses (to >+30 mV). The origin of this component is not clear yet. In our experience its presence and magnitude are very variable; it is more readily seen in HEK293T than in CHO cells; and it is bigger in cells expressing large currents. Critz *et al.*[27] have suggested that it might be associated with K^+ accumulation in the extracellular space. This would be consistent with the variability we have observed. Moreover, other *fast-activating* K^+ channels, lacking a transient component in oocytes, also have a small transient component when expressed in some mammalian cells (see Ref. 26).

Another unusual feature of Kv3 currents is their fast rate of deactivation upon repolarization, first described for Kv3.1 currents expressed in NIH-3T3 and L929 cells by Grissmer *et al.*[26] These authors found that Kv3.1 currents deactivated about 10 times faster than other cloned voltage-gated K^+ channels known at the time. This feature is shared by non-inactivating Kv3.1–Kv3.3 currents (FIG. 5). Kv3.4 channels deactivate slowly because they can not close until inactivation is removed (see below). Inactivating Kv3.3 currents may behave in the same fashion. Since then, many new voltage-gated channels have been identified; however only one of them, Kv1.7, a non-neuronal member of the Kv1 family

TABLE 1. Properties of Kv3 Currents *in Vitro* and *in Vivo* (*abbreviations and notes overleaf*)

A. Electrophysiological Properties

	Activation					Inactivation			
	V_{on} (mV)	$V_{1/2}$ (mV)	k (mV)	t_{on} (mV)	τ_{off} (ms)	$V_{1/2}$ (mV)	k (mV)	τ (ms)	γ (pS)[a]
Kv3.1	–10	14	11	3.4	1.4	—	—	Slow	16 to 27
Kv3.2	–10	12.1	8.4	4	2.9	—	—	Very slow	16 to 20
Kv3.3	–10	12	10 to 14	4.4	1.9[b]	—	—	[c]	14
Kv3.4	–10	14	8.5	4	[d]	–20 to –32	4.8 to 8.3	10–20	12 to 14
$I_{GP\text{-}TEA}$ ("A")	–10	16.9	10.6	5	3.6	—	—	NA	NA
$I_{GP\text{-}TEA}$ ("B")	–10	15.5	14.3	3.42	1.4	—	—	14	NA

B. Pharmacological Properties and Modulation

	Pharmacological Properties (IC_{50})				Protein Kinase Modulation			
	TEA_0 (mM)	4-AP[e] (mM)	DTX (nM)	CTX (nM)	PKA	PKC	PKG	Other Properties
Kv3.1	0.15 to 0.2	0.02 to 0.6	NB	NB	No effect	Inhibits	Inhibits	Blocked by Chromakalin K_D 0.237 mM
Kv3.2	0.15	0.6 to 0.9	NB	NB	Inhibits	Inhibits Kv3.2b but not Kv3.2a	Inhibits	
Kv3.3	0.14	1.2			NA	Stimulates currents and suppresses inactivation	Inhibits	
Kv3.4	0.09 to 0.3	0.5 to 0.6	NB	NB	NA	Suppresses inactivation	Inhibits	Blocked by BDS-I IC_{50} 47nM & BDS-II IC_{50} 56nM

TABLE 1. Properties of Kv3 Currents *in Vitro* and *in Vivo* (*continued*)

ABBREVIATIONS: V_{on}: minimum voltage at which there is significant activation of the current; $V_{1/2}$: membrane potential at which the conductance is half maximal; k: slope of normalized g–V curve; t_{on}: time for the current to rise from 10 to 90% of its final value for non-inactivating currents and time to peak for inactivating currents, both at +40 mv; τ_{off}: time constant of deactivation at –60 mV; Inactivation τ: time constant of inactivation at +40 mV; NB: not blocked; NA: not available.

[a]Single-channel conductance from *Xenopus* oocytes in physiological potassium concentrations.

[b]Deactivation of the noninactivating Kv3.3 currents obtained in HEK 293 cells.

[c]Currents are noninactivating in CHO and HEK 293 cells, but inactivating in *Xenopus* oocytes; see text.

[d]See discussion on deactivation of Kv3.4 currents and Ruppersberg *et al.*[33]

[e]Block depends on pattern of stimulation.

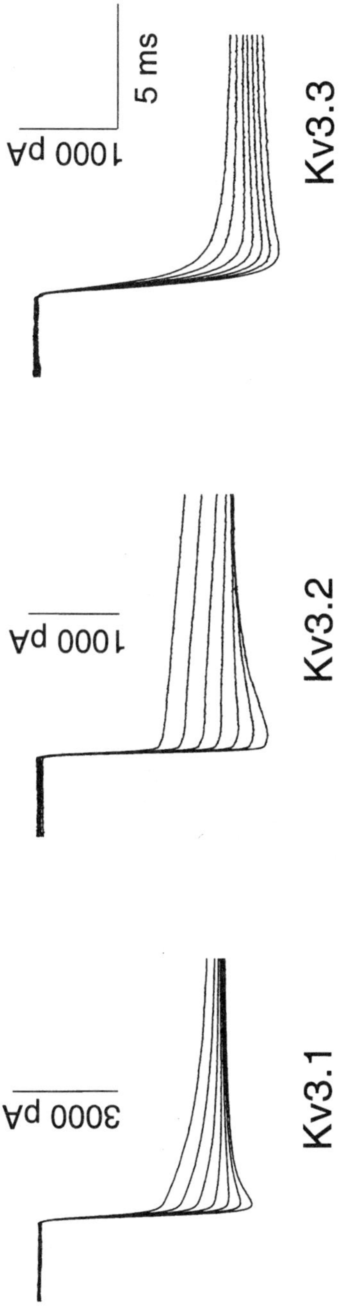

FIGURE 5. Deactivation kinetics of Kv3.1, Kv3.2, and Kv3.3 currents in HEK293 cells. Records of tail currents obtained by repolarizing cells expressing Kv3.1, Kv3.2, or Kv3.3 channels to a series of membrane potentials from −100 mV to −40 mV in 10 mV increments following a 20-ms depolarization to +20 mV.

deactivates fast (closing rates are about 2–3 times the deactivation rates of Kv3 channels.[28] The distinctive properties of Kv3 channels are likely to endow neurons with special electrophysiological properties (see below).

Kv3.1 and Kv3.2 channels (and perhaps also Kv3.3–Kv3.4) are blocked by intracellular Mg^{2+} in a voltage-dependent manner, producing a negative slope conductance in the conductance-voltage curve.[15] Variations in the conductance-voltage relationship between different studies on mammalian cells might be associated with differences in intracellular Mg^{2+} concentrations. Lower Mg^{2+} concentrations will tend to produce apparent shifts in voltage-dependence to the right. It was recently shown that the special sensitivity of Kv3.1 channels (as compared to other Kv channels) to intracellular Mg^{2+} could be transferred to a Shaker channel by transplanting the Kv3.1 P domain.[29]

TABLE 1 also includes the values of the single channel conductance of Kv3 channels, most of which have been derived from measurements in *Xenopus* oocytes. In general, Kv3 channels, particularly Kv3.1 and Kv3.2 have single channel conductances that are larger than most other voltage-gated K^+ channels (see TABLE 4 in Coetzee *et al.* in this volume).

While so far Kv3.3 and Kv3.4 channels have been given less attention than Kv3.1 and Kv3.2 channels, they show very interesting features. The NH-terminal inserts of both Kv3.3 and Kv3.4 responsible for the inactivation of these channels contain a cysteine residue surrounded by a very conserved sequence. A similar sequence is found in Kv1.4, a fast inactivating Shaker-related mammalian channel. The conservation of this sequence is even more notable given that there is otherwise very little amino acid sequence conservation in the amino ends of these proteins.[15] Ruppersberg *et al.*[30] showed that exposure of the cytoplasmic surface of Kv1.4 or Kv3.4 channels to air, presumably to oxygen (in inside-out excised membrane patches), resulted in removal of the inactivation process. In contrast to the effects of amino end deletions and other treatments, this inactivation removal was reversible. Inactivation was restored if the patch was reinserted into the cell or by exposure to reduced glutathione or DTT.[30] The effects of oxidation were not seen in channels expressed from a Kv1.4 protein where the cysteine was replaced with serine.[30] A model by which the redox state of this cysteine may modulate channel inactivation was proposed by these investigators. Similarly, exposure of oocytes expressing Kv1.4, Kv3.3, or Kv3.4 channels to high micromolar concentrations of external H_2O_2 resulted in a reversible removal of inactivation.[31] Concomitant with inactivation removal there was a significant increase in current magnitude, particularly large for Kv3.4 channels. The effects of H_2O_2 were specific to these three channels, and was not observed on channels expressed from a Kv3.3 cDNA in which the cysteine residue was mutated (Vega-Saenz de Miera and Rudy, unpublished observations). Moreover, there was no effect in channels which lack the conserved cysteine-containing sequence, including noninactivating Kv3.2 channels, inactivating *Drosophila* Shaker channels,[31] or inactivating Kv4.1 and Kv4.2 channels.[32] Effects similar to those of H_2O_2 were seen upon exposure to *N*-ethyl maleimide, a reagent which alkylates free sulfhydryls, but in this case the effects were, as expected, irreversible.

Another interesting feature of Kv3.4 channels expressed in heterologous expression systems was described by Ruppersberg *et al.*,[32] who found that inactivated Kv3.4a channels are able to pass current during repolarization. Apparently, these channels are unable to enter the resting closed state, when the membrane is hyperpolarized, until their inactivation is first removed, resulting in a long-lasting current (at voltages much more negative than those required for activation) while they are recovering from inactivation. It remains to be seen whether Kv3.3 channels, which inactivate more slowly, behave in a similar fashion. There is also no evidence that this behavior is seen in native channels containing Kv3.4 proteins. Such behavior could drastically affect the role played by these channels. If inactivated Kv3.4 channels are indeed able to pass current at negative potentials while they recover from inactivation, they could have subthreshold effects on cell excitability, unlike Kv3 channels lacking this property. Specifically, the first spike in a train could open and then quickly inactivate these channels. Upon repolarization of the membrane, the current through Kv3.4 channels recovering from inactivation could prevent the next spike or delay its onset.

Regulation of Kv3 Currents by Phosphorylation and Dephosphorylation

Several sites that fit the consensus sequence for protein-kinase phosphorylation are found in regions of Kv3 proteins that are thought to be intracellular. Some of these sites are found in products of all of the genes, some are found only in proteins encoded by a single gene, and some are specific to alternatively-spliced isoforms.[15] Three kinase systems, cAMP-dependent protein kinase (PKA), protein kinase C (PKC), and cGMP-dependent protein kinase (PKG) have now been shown to modulate Kv3 channel function in heterologous expression systems (TABLE 1).

PKA

This kinase inhibits Kv3.2a and Kv3.2b channels by phosphorylating the unique PKA-consensus site present in the C-terminal area of the protein, before the point of divergence of the alternatively spliced isoforms. Kv3.1b channels, which lack a consensus site for PKA phosphorylation, are not affected by PKA. Other Kv3 proteins, all of which also lack consensus PKA sites, have not been tested.[34,35]

PKC

This kinase modulates products of several Kv3 genes, but the functional effects vary depending on the subunit. PKC inhibits channels containing some alternatively spliced Kv3.2 isoforms, but not others (see the section entitled Functional Roles of Alternative Splicing below). PKC also inhibits Kv3.1 channels.[27,34,36] It has not yet been demonstrated that these modulations are the result of channel protein phosphorylation. On Kv3.4 channels, PKC slows down N-inactivation by phosphorylating several PKC sites present in the NH-terminal sequence forming the structure responsible for inactivation.[37,38] A similar effect of PKC activation is seen on Kv3.3a and Kv3.3b channels, presumably also due to phosphorylation of the inactivating NH-terminal sequence. However, in addition, PKC produces large (200–400%) increases in Kv3.3 current levels which are independent of the effects on inactivation, and can be induced with lower doses of phorbol esters than those required to slow down inactivation, and are also blocked by PKC inhibitors.[39,40]

PKG

Intracellular increases in cGMP concentrations, produced by the addition of permeable cGMP analogues or by stimulation of guanylyl cyclase with nitric oxide (NO) inhibit

Kv3.1b and Kv3.2a channels in a voltage-dependent manner. These effects are blocked by PKG inhibitors, but appear not to be the result of channel phosphorylation.[41,42] The data suggest that the channel inhibition produced by the activation of PKG is mediated by a specific phosphatase(s). We suggest that these enzymes dephosphorylate the channel proteins at sites that need to be phosphorylated for voltage-dependent activation to proceed normally. This hypothesis is consistent with the observation that exposure of the intracellular side of the membrane to alkaline phosphatase inhibits these channels as well as Kv3.3 and Kv3.4 channels.[40] The effects of PKG activation on Kv3.3 and Kv3.4 currents have not been tested.

Pharmacological Properties

At present there are no specific blockers for Kv3.1–Kv3.3 channels. However, in both *Xenopus* oocytes and in transfected mammalian cells, all Kv3 currents (including Kv3.4) are very sensitive to external tetraethylammonium (TEA) or 4-aminopyridine (4-AP) (TABLE 1). The sensitivity to TEA is particularly important and can be exploited to discriminate between known K^+ channels. Kv3 currents are > 80% blocked by 1 mM TEA. This concentration of TEA produces significant inhibition of only a few other known K^+ channels (see TABLE 4 in Coetzee *et al.* in this volume). These include the large-conductance Ca^{2+} activated K^+ channels containing proteins of the slo family (K_d 80–330 μM), Kv1.1 channels (K_d ~ 0.3 mM), and KCNQ2 (90% blocked by 1 mM). These channel types can be distinguished from Kv3 by other properties (see TABLE 4 in Coetzee *et al.* in this volume). Ca^{2+}-activated K^+ channels can be suppressed by using Ca^{2+} channel blockers in the extracellular solution, and by using BAPTA in the intracellular solution to chelate intracellular Ca^{2+}. There are also many peptide toxins that block these channels (e.g., iberotoxin, IbTx; charybdotoxin, CTX; see Coetzee *et al.* in this volume). Kv1.1 channels are blocked by dendrotoxin (DTX) (K_d ~ 10–20 nM). Kv3 currents are not affected by these or any of the other scorpion- or snake-venom-derived toxins that block other voltage-sensitive channels (TABLE 1). KCNQX are very slowly activating and deactivating K^+ channels (time constants of the order of hundred of ms to sec). In contrast to their high sensitivity to extracellular TEA, Kv3 channels are less sensitive to intracellular TEA than are many other K^+ channels.[43]

Kv3 channels are also very sensitive to 4-AP (TABLE 1). However, the values reported for the IC_{50} for this channel blocker in different studies vary widely (e.g., IC_{50}s ranging from 20 to 600 μM have been reported for Kv3.1). These variations probably reflect, at least in part, the complex kinetics of binding and unbinding of this drug to different states of the channel.[43,44] The degree of channel block by 4-AP depends both on time and on the immediate history of opening and closing of the channel, and therefore on the pattern of stimulation of the cell during the measurements. We therefore believe that TEA is a better blocker than 4-AP to discriminate between Kv3 channels and other K^+ channels. Moreover, many other K^+ channels are as sensitive or more sensitive than Kv3 channels to this drug[45] see also Coetzee *et al.* in this volume), including the ubiquitous "D" current.[46,47]

A new set of peptides obtained from the venom of the sea anemone *Anemonia sulcata,* known as blood-depressing substance I and II (BDS-I and BDS-II) were recently shown to block specifically, reversibly, and at low concentrations (IC_{50} values in the low nanomolar range) Kv3.4 channels in transfected COS cells. The toxins share no sequence similarity with other K^+ channel toxins derived from sea anemone and did not block other Kv or inward rectifying channels.[48] These toxins are likely to become important tools to investigate the roles of Kv3.4 channels in native tissue.

Properties of Kv3 Heteromeric Channels

Like other members of the Kv family, Kv3 proteins can form heteromultimeric channels with novel properties with other subunits of the same subfamily but not with proteins of other Kv subfamilies.[15,21,49] Heteromultimeric Kv channels have properties that are intermediary between those of the corresponding homomultimers, although some properties might be closer to those of one or the other homomultimer.[15,21,49–53] Since all Kv3 channels have similar voltage dependencies, one would expect that this parameter would not change in heteromultimers and this expectation has been confirmed experimentally. When two different Kv3 cRNAs are injected into *Xenopus* oocytes, the total current has a voltage-dependence similar to that of Kv3 homomultimers. Moreover, and perhaps not surprisingly, the currents recorded in cells co-expressing Kv3.1 and Kv3.2, which produce similar currents in a homomultimeric channel, are similar to Kv3.1 or Kv3.2 currents alone. The most interesting combinations are those of inactivating Kv3.3 or Kv3.4 proteins with those that produce sustained currents (Kv3.1 and Kv3.2). Oocytes injected with such combinations had fast inactivating currents that were several-fold larger than those seen in oocytes injected with the same amount of Kv3.4 or Kv3.3 cRNA alone. This is contrary to what might be expected from the algebraic sum of two independent currents and is consistent with the formation of heteromultimeric channels containing one or more Kv3.3 or Kv3.4 subunits plus several Kv3.1 or Kv3.2 subunits. The rate of inactivation increases with the number of inactivating subunits.[21] There is amplification of the transient current because a single inactivating subunit is sufficient to impart fast inactivating properties to the channel complex.[52]

TISSUE-SPECIFIC EXPRESSION OF Kv3 GENES

It is crucial to know where Kv3 proteins are found in order to identify cells where the properties of native Kv3 channels could be investigated and to be able to generate hypotheses about their physiological roles. We, in our laboratory, and others have therefore dedicated a large effort to determining the distribution of Kv3 products. In rodents, three of the four known Kv3 genes (Kv3.1–Kv3.3) are expressed mainly in brain. Kv3.4 is weakly expressed in brain, but strongly in skeletal muscle. Low levels of Kv3.1 are also seen in skeletal muscle, while low levels of Kv3.3 mRNAs are found in kidney and lung. Kv3.1, Kv3.3, and Kv3.4 products have also been identified in PC12 pheochromocytoma cells.[8,15,21] Kv3 genes are weakly expressed, if at all, in the heart. We and others did not detect any Kv3 transcripts in rat heart by Northern blot analysis.[5,14,15,21] Using the highly sensitive RNAse-protection method, Dixon and McKinnon[54] found extremely low levels for all four Kv3 genes in rat heart (faint signals were observed after very long [7–10-day] exposure as compared to robust signals produced by Kv3 probes in brain or by probes for other Kv subfamily mRNAs in heart after overnight exposure).

The presence of these weak signals might be of functional interest if they are derived from a small subpopulation of cells expressing significant levels of Kv3 transcripts. However, if such a subpopulation of cells exists, it remains to be identified. Kv3 transcripts might be more abundant in cardiac tissue in other species. In a study in enzymatically isolated ferret cardiac myocytes, Brahmajothi *et al.*[55] reported the presence of Kv3 transcripts in several myocytes and Kv3.3 transcripts have been amplified by PCR from mouse heart RNA.

By means of RNAse-protection, Kv3.4 transcripts were found to be abundantly expressed in sympathetic ganglia, at levels 273% higher than in brain. Kv3.3 and Kv3.1 produced very weak signals that could only be detected after 1 week's exposure of the autoradiograms (as compared to robust signals seen after overnight exposure for many other Kv transcripts). Kv3.2 transcripts were not detected.[56] Kv3.1 transcripts are also found in T lymphocytes, where they are thought to encode the principal subunits of the *l*-type K^+ channel[57] (see below). According to one report, Kv3.2 transcripts could be amplified by RT-PCR from transgenically derived betaTC3-neo insulinoma cells, but not from purified pancreatic β cells.[58] Except for this report, and the minimal expression in heart, Kv3.2 transcripts have not been detected outside the CNS. The tissue distribution of Kv3 mRNAs is summarized in TABLE 2.

Quantitative analysis[21] demonstrated that in the adult rat brain, Kv3.1 (Kv3.1b) and Kv3.3 transcripts were the most abundant Kv3 mRNAs. They appeared to be present in amounts similar to Kv1.2 mRNAs, an abundant Shaker-related transcript. Kv3.1 (Kv3.1b) and Kv3.3 mRNAs were ~3 times more abundant than Kv3.2 mRNAs, ~6–8 times more abundant than Kv3.4 transcripts, and about 1/5 as abundant as Na^+ channels α subunit mRNAs.

mRNA Distribution in Brain

In situ hybridization histochemistry has been used to study the distribution of Kv3 mRNAs in the central nervous system (CNS) of adult rats.[14,21,59,60] The work by Perney *et*

TABLE 2. Tissue Expression of Kv3 Transcripts in Rodents

	Kv3.1	Kv3.2	Kv3.3	Kv3.4
Brain	+++	+++	+++	+
Spinal cord	+++	-	+++	+
Sympathetic ganglia	±	-	±	+++
Skeletal muscle	+	-	-	+++
Kidney	-	-	+	-
Lung	-	-	+	-
Heart	±	±	±	±
Lymphocytes	++	NA	NA	NA

SYMBOLS: +++, abundant; ++, moderate; +, low levels; ± barely detectable levels; –, nondetectable; NA, data not available.

al.[59] on Kv3.1 and the work by Weiser *et al.*[21] on Kv3.1–Kv3.4 include, in addition to low-resolution X-ray autoradiography (which provides regional localization), a higher, cellular-resolution microscopic analysis of emulsion-dipped sections. These data have also been extensively reviewed[15] and are summarized in FIGURE 6. More recent studies in mice have largely confirmed the observations in rat. Kv3 genes appear to be expressed in neurons and not in glia.[21,59,61]

In the CNS each Kv3 gene shows a unique pattern of expression; however, transcripts of two or more Kv3 genes overlap in many neuronal populations. Kv3 genes are expressed in many, but not all, and not exclusively, GABAergic and glycinergic neurons in the CNS (FIG. 6). Kv3.1 and Kv3.3 transcripts overlap in many areas, particularly in the posterior part of the brain and in the spinal cord. Kv3.1 and Kv3.2 overlap in some neuronal populations, particularly in the anterior part of the brain, including the cortex, hippocampus, globus pallidus, deep cerebellar nuclei, and certain thalamic nuclei. However, in other areas, Kv3.1 and Kv3.2 transcripts actually show a reciprocal expression pattern. For example, Kv3.2 is strongly expressed in dorsal thalamic nuclei, but extremely weakly in the ventral thalamus (including the reticular thalamic nucleus and the ventral lateral geniculate), while Kv3.1 has the opposite pattern. Similarly, Kv3.2 is strongly expressed in the dorsal cochlear nucleus and the ventral lateral lemniscus nucleus, while Kv3.1 transcripts predominate in the ventral cochlear and dorsal lateral lemniscus nuclei.[21] Kv3.4 transcripts are weakly expressed in brain and only in a few neuronal types, usually in neurons also expressing other Kv3 genes. Kv3.1 and Kv3.3 are coexpressed in many neuronal populations, including most auditory central processing neurons, while Kv3.2 transcripts show the most restricted pattern of expression, with ~90% of Kv3.2 mRNAs being present in thalamic relay neurons *throughout* the dorsal thalamus. In fact, Kv3.2 gene expression is among the most specific of all known K^+ channels.

Protein Localization

Since we last reviewed the patterns of expression of Kv3 mRNAs,[15] specific antibodies against the protein products of two genes (Kv3.1 and Kv3.2) have been raised in our laboratory, allowing localization of the protein product. The antibodies against Kv3.2 are directed to the amino terminal area of the protein and recognize all alternatively spliced isoforms.[34,62,63] In the case of Kv3.1, antibodies have been raised to the C-termini of both the Kv3.1a[64,65] and Kv3.1b[66–68] (see also Refs. 69 and 70) proteins which are specific to each isoform.

Both Kv3.1 and Kv3.2 proteins are present in somas, axons, and terminals. The protein is concentrated in somatic, axonal, and presynaptic-terminal membranes and their underlying cytoplasm. The protein is not detected in dendrites, except in portions of primary dendrites close to the cell body. While most cells expressing Kv3.1 and Kv3.2 have protein in both the somatic and the axonal-terminal compartments (see section entitled Functional Roles of Alternative Splicing for the differential subcellular localization of the Kv3.1a and Kv3.1b isoforms), in thalamic relay neurons, the neuronal population containing the large majority of Kv3.2 transcripts in rodent brain, the protein is localized mainly in axons and terminals.[34] Kv3.2 antibodies stain the barrels in layer IV of somatosensory cortex,[34] and recent immunoelectron microscopy has confirmed the

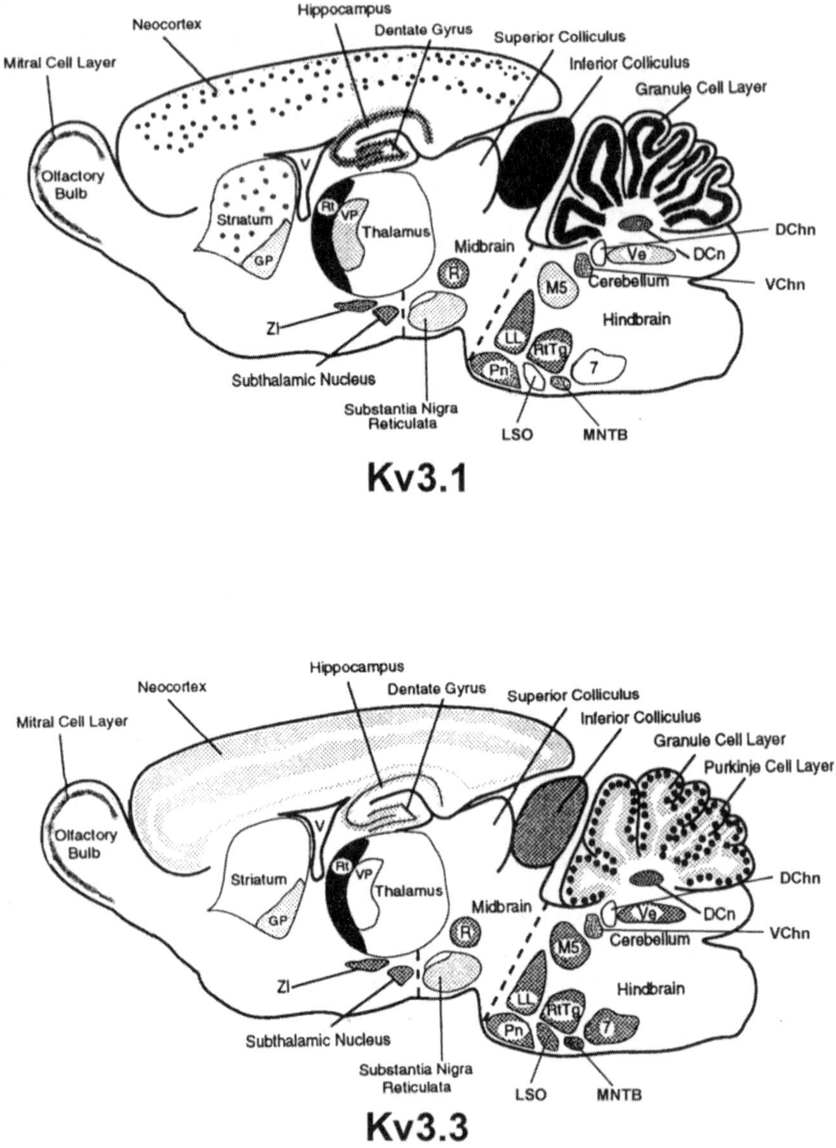

FIGURE 6 *(above and facing page).* Distribution of Kv3 mRNAs in the rat brain (similar patterns have been observed for some Kv3 transcripts in mice). The levels of expression of Kv3 genes based on *in situ* hybridization studies by Weiser *et al.*[21] are represented in these diagrams by different grades of *shading*. The position of some of the structures, particularly in the brain stem, is only approximate, and structures are shown in these diagrams that may not exist in the same sagittal plane. Some important structures expressing one or more Kv3 genes are not illustrated, such as several areas of the basal forebrain.

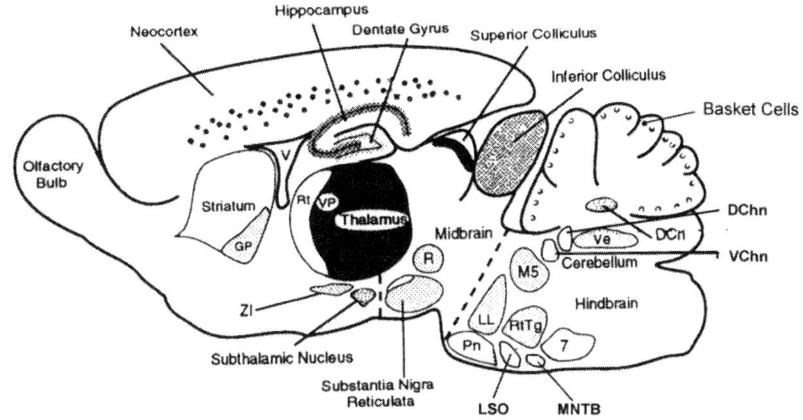

Kv3.2

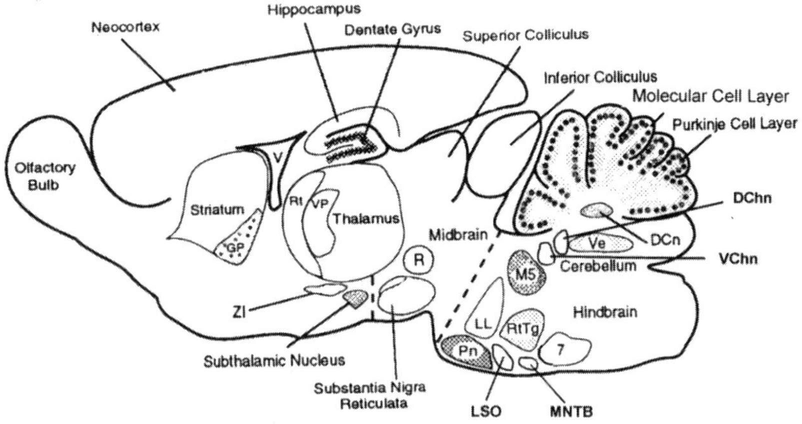

Kv3.4

FIGURE 6 *(continued).* Abbreviations: 7, facial nucleus; DChn, dorsal cochlear nucleus; DCn, deep cerebellar nuclei; GP, globus pallidus; LL, lateral lemniscus nuclei; LSO, superior olive; M5, trigeminal motor nucleus; MNTB, medial nucleus of the trapezoid body; Pn, pontine nuclei; R, red nucleus; Rt, reticular thalamic nucleus; RtTg, reticulo tegmental nucleus of the pons; VChn, ventral cochlear nucleus; Ve, vestibular nucleus; VP, ventral-posterior complex of the dorsal thalamus; ZI, zona incerta.

localization of Kv3.2 proteins throughout the axon and terminals of thalamocortical projections in the cortex. Apparently there is Kv3.2 protein in thalamocortical terminals with both spiny stellate neurons and with GABAergic cells.[62,63]

Kv3.1 and Kv3.2 proteins are strongly expressed in a small subpopulation of cortical (FIG. 7) and hippocampal neurons. Double-labeling immunofluorescence was used to identify the neuronal populations containing these proteins. In the cortex, Kv3.1 proteins are present in parvalbumin (PV)-containing GABAergic interneurons in rat and mouse. In fact there is a nearly one to one correlation between the two markers: virtually all cells containing PV express Kv3.1 and vice versa.[62,63,66,67] In the hippocampus, Kv3.1 is also expressed in many PV-containing interneurons but the relationship between the two markers is not as strict as in the neocortex.[66–68] Kv3.1 and parvalbumin are also co-localized in interneurons in the caudate nucleus.[66,69]

Kv3.2 proteins are expressed in cortical and hippocampal GABAergic interneurons as well. In the cortex strongly labeled cells are concentrated in deep layers (V–VI), while very weakly stained cells are seen in superficial layers (I–IV) (FIG. 7), in accordance with the intensity of *in situ* hybridization signals. Double-labeling shows that the weakly stained cells in superficial layers are positive for PV. In deep layers, about 70–80% of the cells expressing Kv3.2 are also positive for PV, and the remaining 20–30% are also GABAergic interneurons which may correspond to the calbindin- and somatostatin-containing Martinotti cells.[62,63] These data show that there are two types of PV-containing interneurons in the cortex with respect to their expression of Kv3 proteins. PV neurons in superficial layers contain mainly Kv3.1 protein, and PV neurons in deep layers contain both, Kv3.1 and Kv3.2.

Since all PV neurons are positive for Kv3.1, and since all PV neurons in deep cortical layers, also express Kv3.2 proteins, it is possible that the two subunits are part of the same heteromeric channels. To test this hypothesis we used co-immunoprecipitation assays from cortical membranes solubilized with nondenaturing detergents (as in Refs. 71 and 72). Antibodies against Kv3.1 immunoprecipitated both Kv3.1 and Kv3.2 proteins, and the antibodies against Kv3.2 also precipitated both subunits.[62,63] It is therefore likely that PV-containing interneurons in the cortex (particularly those in layers V–VI) have heteromeric Kv3.1–Kv3.2 channels. The channels in PV neurons in superficial layers might be Kv3.1 homomultimers or Kv3.1–Kv3.2 heteromultimers in which Kv3.1 proteins dominate. Kv3.1 and Kv3.2 proteins are also co-localized in the globus pallidus, where they are present in the major population of pallidal neurons, the PV-containing projecting cells. Co-immunoprecipitation and functional studies also suggest that in these neurons the channels are heteromeric Kv3.1–Kv3.2 proteins.[25]

In the hippocampus Kv3.2 proteins are expressed in several types of interneurons, including most Kv3.1 containing PV-positive basket cells, as well as interneurons containing somatostatin in the stratum oriens and in the dentate hilus which do not express Kv3.1 (McBain and Rudy, unpublished observations).

We have recently began to characterize the brain distribution of Kv3.3 proteins, which like Kv3.1 and Kv3.2 are expressed in somas and axons. The group of Olaf Pongs has raised antibodies against Kv3.4 proteins. In the CNS Kv3.4 expression was lower than several Kv1 proteins analyzed in the same study, and was seen in areas consistent with the localization of Kv3.4 transcripts.[73,74] In most of these neurons, staining with anti-Kv3.4 antibodies was predominantly seen in axons and terminals.

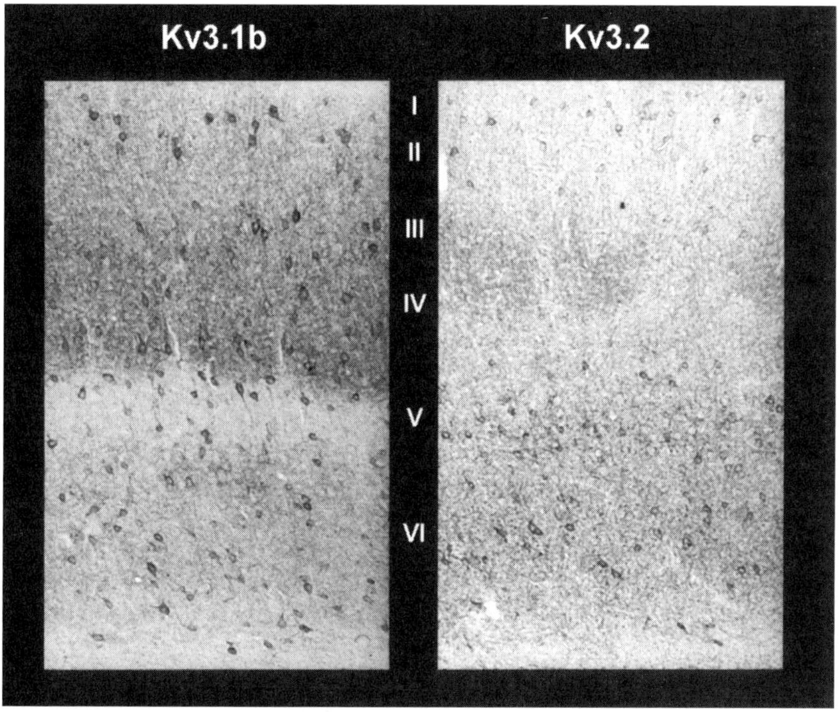

FIGURE 7. Kv3.1 and Kv3.2 proteins in neocortical interneurons. Immunoperoxidase staining of coronal sections of mouse somatosensory cortex for Kv3.1b and Kv3.2 proteins with specific polyclonal antibodies. There is strong somatic staining for Kv3.1b in a population of neurons distributed in layers II–VI. In the section stained with Kv3.2 antibodies there are strong labeled neurons in deep cortical layers (V–VI) and faintly stained cells in upper layers. There is strong staining of the neuropil with both antibodies, most prominent in areas around labeled cells. In addition, there is distinct staining of processes in the cortical barrels with Kv3.2-, but not Kv3.1b-antibodies.

Developmental Expression

Little is known about the expression of Kv3 genes during development. Although Perney *et al.*[59] reported *detectable* levels of expression of Kv3.1 mRNA transcripts in prenatal rats, all available studies agree that most of Kv3.1–Kv3.3 mRNA and protein expression develops after the first postnatal week. Perney *et al.*[59] found a large increase in expression of Kv3.1 transcripts between the 7th and 14th day after birth. By postnatal day 14, rats exhibited a pattern of expression in brain similar to that seen in the adult rat, although adult mRNA levels were somewhat higher than those at p14. In the hippocampus, Du *et al.*[68] could not detect Kv3.1b protein until postnatal day 8. There was a large increase between p8 and p20, but protein levels at p40 were still higher than those at p20. Goldman-Wohl *et al.*[13] studied the developmental expression of Kv3.3c transcripts in cerebellar Purkinje cells. Expression was first seen between postnatal day 8 and 10, and the levels increased as the cerebellum matured. Kv3.2 mRNAs in whole brain are barely detectable before postnatal day 8–10 (faint transcripts are seen after long exposures), but the levels start increasing rapidly after this age. Interestingly, the developmental expression of the Kv3.2 gene appears to be regulated independently in thalamic relay neurons, where the majority of Kv3.2 mRNAs are found in adult animals, and outside the thalamus. While near adult levels outside the thalamus are achieved by postnatal day 14, in thalamic relay neurons Kv3.2 expression is still low at this age and increases to near adult levels during the third postnatal week.[75,76] In the globus pallidus the levels of both Kv3.1 and Kv3.2 *proteins* were not detectable earlier than postnatal days 5–6, and increased sharply after postnatal day 8. The expression of Kv3.1 developed faster than the expression of Kv3.2: Maximum levels of Kv3.1 protein were seen at p15, while for Kv3.2, maximum levels were not achieved until p20. Once protein levels reached a maximum, they decreased slightly in older animals.[25]

The expression of Kv3.4 transcripts in skeletal muscle is also developmentally regulated.[77] Utilizing semi-quantitative RT-PCR, detectable levels of Kv3.4 mRNAs were found at embryonic day 17, which increased to adult levels during the first 2 weeks after birth. Interestingly, Kv3.4 mRNA expression is muscle fiber type-dependent, with higher levels found in fast muscles. Moreover, the expression levels decreased in myotonic mice or after denervation in wild-type animals, suggesting that excitation of the muscle cell regulates Kv3.4 gene expression.[77]

NATIVE Kv3 CURRENTS RESEMBLE HETEROLOGOUSLY EXPRESSED Kv3 CURRENTS

There is now evidence from several cell types showing that at least for Kv3.1, Kv3.2, and Kv3.4, native channels have properties similar to those in heterologous expression systems. In an initial study combining immunohistochemical and electrophysiological analysis in slice preparations, we found that drugs that block Kv3.1 currents in heterologous expression systems blocked a fraction of the K^+ current from hippocampal interneurons expressing Kv3.1b proteins that resembled Kv3.1 currents in heterologous expression systems.[68] Moreover, concentrations of 4-AP that block Kv3 currents shortened the duration of the action potential of interneurons expressing Kv3.1b proteins, but not in interneurons that did not stain with anti-Kv3.1b antibodies. However, difficulties associated with voltage-clamping neurons in slices limited the extent of the comparisons that could be made between native and heterologously expressed currents in that study.

More recently we have been able to do a more detailed comparison of native and expressed Kv3 currents in freshly dissociated neurons from the globus pallidus. Acutely dissociated and short-term cultured neurons are an excellent system for this kind of study, allowing better conditions for space clamp and pharmacological analysis. Identical recording conditions to those used to study Kv3 currents in CHO cells were used to eliminate differences produced by components of the intracellular or extracellular solutions or other experimental variations. We utilized extracellular Cd^{2+} and intracellular BAPTA, as well as tetrodotoxin (TTX) to suppress Ca^{2+}, Ca^{2+}-activated K^+, and Na^+ currents. 1 mM TEA was used to isolate Kv3-like currents in these cells.

As described earlier, the rat and mouse globus pallidus express both Kv3.1 and Kv3.2 proteins in somatic membrane, probably in heteromeric channels containing both subunits, present in the major neuronal population in this nucleus (the PV-containing projecting neurons) and at developmental stages in which dissociations were possible.[25] The GP plays key roles in the circuitry involved in movement control, and might also be involved in cognitive functions. There is therefore great interest in the anatomical and physiological characterization of this brain area.[78–80] The studies on Kv3 channels contribute novel information on the classification and cellular properties of pallidal neurons.

We found three morphological subtypes of neurons as a product of the enzymatic dissociation of the GP. The most common were medium to large cells with fusiform or triangular somas resembling the major population of pallidal neurons. When these cells were depolarized from a holding potential of –80 mV, the total current consisted of transient and sustained components. The transient component (which was variable) could be nearly completely inactivated by holding the cell at –40 mV. This holding potential also inactivated a significant portion of the sustained component. We assumed that Kv3.1–Kv3.2 currents contributed little to the suppressed sustained component since a VH of –40 mV has little effects on Kv3.1 and Kv3.2 currents expressed in CHO cells.[25] The currents recorded during depolarizing test pulses from a VH of –40 mV were of the delayed rectifier type. They could be separated into two components, which differed in activation and deactivation kinetics by 1 mM TEA (FIG. 8). The TEA-resistant current activated and deactivated more slowly than the TEA-sensitive component. The differences in kinetics were particularly large for the deactivation process (τ_{deact} = 25.64 ± 3.38 ms, for $I_{GP,R}$ and 2.27 ± 0.24 ms for $I_{GP,TEA}$).

A detailed comparison of the voltage-dependence and kinetics of the TEA-sensitive component showed that it has properties nearly identical to those of Kv3.1 and Kv3.2 currents in CHO cells (FIG. 8 and TABLE 1). In both cases the current starts activating between −10 and −20 mV, and the normalized conductance-voltage curves have similar slopes and midpoints of activation. When scaled to each other, the cloned and native currents have nearly superimposable activation and deactivation time courses when compared at the same voltages (FIG. 8; see also Refs. 25 and 81).

Previous electrophysiological analysis of the K^+ currents of pallidal neurons, in the same species, revealed a low voltage-activating fast inactivating current (I_A), a component with slower inactivation and slow recovery from inactivation that is blocked by micromolar concentrations of 4-AP (I_{As}), and two maintained components, one blocked (I_K) and one not blocked by 10 mM TEA.[82,83] None of these components resembles Kv3 currents, because the Kv3 channel-mediated current was buried in the I_K. These studies demonstrate that experimental conditions and methods to isolate individual components of the K^+ current tailored to search for specific current components are required before it is possible to determine whether native currents resemble those in heterologous expression systems.

The characteristics of the currents obtained from GP neurons described earlier were typical of the majority of the cells studied. However, we found that in a small group of cells (type "B" cells), 1 mM TEA blocked a fast inactivating current. Depolarizing pulses from a holding potential of –40 mV produced currents of the delayed rectifier type, similar to those seen in the cells described earlier (type-A cells), but had faster activation kinetics (rise time between 10 and 90% of ~14.0 ms for type "B" cells and ~22 ms for type "A" cells). Application of 1 mM TEA inhibited approximately 10–15% of the current. The TEA-resistant current had slower activation kinetics than the total current (10–90% rise time: 22.4 ± 2.5 ms). The TEA-sensitive component obtained by digital subtraction was composed predominantly of a current that activated rapidly, starting at voltages more positive than −10 mV, and inactivated quickly. The transient currents recorded from type "B"

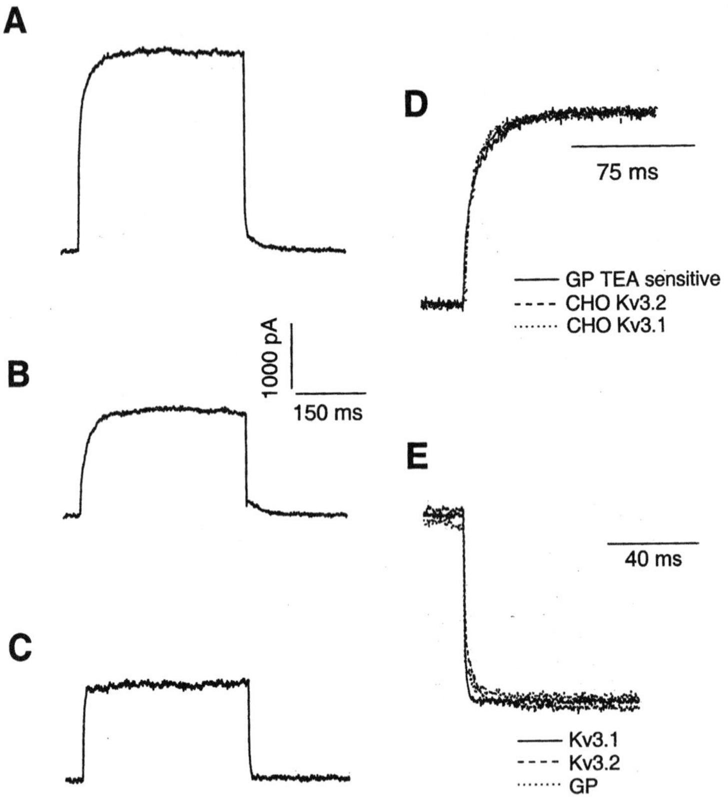

FIGURE 8. Voltage-dependent K+ currents in type "A" pallidal neurons. (**A-C**) Currents obtained from a GP neuron held at –40 mV during a depolarizing pulse to +40 mV before (**A**) and after (**B**) application of 1 mM TEA. The trace in **C** shows the TEA-sensitive component obtained by digital subtraction of the trace in **B** from the trace in **A**. (**D**) Superimposition of the first 150 ms of scaled records obtained from CHO cells transfected with Kv3.1 or Kv3.2 cDNAs and the TEA-sensitive current from a type "A" pallidal neuron. Records were obtained during voltage pulses to +30 mV applied from a holding potential of –40 mV. (**E**) Superimposition of scaled traces of tail currents obtained from the same cells as in **D** during repolarization to –40 mV following a voltage step to +40 mV.

neurons of the GP resembled closely the currents expressed by Kv3.4 proteins in voltage-dependence as well as in activation and inactivation kinetics. This result was surprising since *in situ* hybridization studies reported that signals for Kv3.4 mRNAs were "weakly above background."[21] Since we do not have antibodies against Kv3.4 proteins we used single-cell RT-PCR to investigate whether type-B cells in the GP contain Kv3.4 transcripts. These studies showed that Kv3.2 transcripts are found in most type A and some type B cells, Kv3.1 is found mainly in type A cells, and Kv3.4 only in type B cells.

As described earlier, heteromultimer formation between inactivating Kv3.4 subunits and Kv3.1 or Kv3.2 proteins results in amplification of the transient current. This phenomenon may account for the presence of the levels of Kv3.4-like currents seen in type "B" pallidal neurons in spite of the low-level expression of Kv3.4 transcripts. Nevertheless, relative to the other outward currents, the Kv3.4-like current in these neurons contributes a very small proportion of the total outward current. However, the current produces a large effect on the rise time of the total current, since the currents remaining after 1 mM TEA are much slower. This could be an important role for Kv3.4 currents in neurons, to accelerate the rate of rise of the repolarizing currents, without significantly increasing the steady-state levels of total outward current.

Kv3.1-like currents have also been described in central auditory-processing neurons[84] (see below), as well as in cells expressing Kv3.1 transcripts outside the nervous system. Grissmer *et al.*[57] showed that the current mediated by *l*-type channels in T lymphocytes was very similar to Kv3.1 currents in *Xenopus* oocytes when recordings in the two preparations were obtained with the same solutions.

It appears from these results that the main biophysical properties of native channels containing Kv3.1, Kv3.2, and Kv3.4 proteins, in both neurons and other cell types, are not significantly affected by factors such as associated subunits or posttranslational modifications as might be the case for other cloned subunits, at least in the cells studied until now.

Kv3.1 AND Kv3.2 CHANNELS PLAY UNIQUE ROLES IN THE GENERATION OF SUSTAINED HIGH FREQUENCY FIRING

The unusual properties of Kv3 channels, and in particular their voltage-dependence and their fast rate of deactivation, are likely to provide these channels with unique effects on neuronal excitability. Since Kv3 channels open significantly only when the membrane potential is depolarized beyond −10 mV, we expect that physiologically they will be activated mainly during action potentials. They may be activated also by receptor-mediated depolarizations that depolarize the membrane to potentials more positive than −10 mV, as might be encountered in the end bulb synapses in the ventral cochlear nucleus.[70]

It has been suggested that these channels are activated during the peak of the action potential and that, when present in sufficient amounts, they influence the rate of repolarization of the action potential and thus help dictate action potential duration.[15,34,66,67,69] Indeed, when HEK293 cells expressing Kv3.1 or Kv3.2 channels are voltage-clamped to a waveform in the shape of a brief action potential, current is not seen until after the action potential has reached its peak, achieving a maximum value during the repolarizing phase of the spike, in contrast to what is observed when cells expressing Kv1.1 are clamped to the same waveform (FIG. 9). These experiments also illustrate the effect of the fast deactivation of Kv3.1 and Kv3.2 channels. The K^+ current is quickly eliminated during the fast hyperpolarization and thus little Kv3.1/Kv3.2 currents remain during the interspike interval. The predominant somato-axonal localization of Kv3 proteins in CNS neurons (see above) is consistent with a role in the shaping and transmission of action potentials. The low levels of expression in fine dendrites suggests that these channels play little role in the local integration of postsynaptic signals at axo-dendritic synapses.

Although K^+ channels that activate at more negative voltages could also repolarize spikes and reduce their duration, high voltage-activating K^+ channels would modulate firing properties more selectively. By virtue of the fact that they are not opened until the spike has been generated, they are less likely to influence the initial threshold of an action

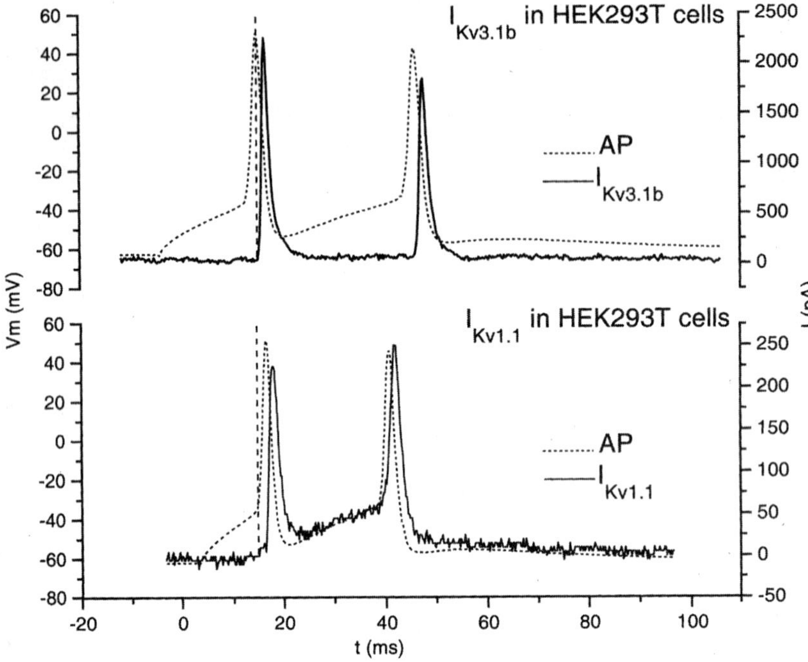

FIGURE 9. Currents evoked in cells expressing Kv3.1b or Kv1.1 when depolarized with a fast action potential. HEK293 cells were transfected with Kv3.1b cDNA **(upper traces)** or Kv1.1 cDNA **(lower traces)** and voltage-clamped to a waveform corresponding to the action potential (AP) recorded from a neocortical neuron. The currents, shown with *continuous black lines* have been superimposed on the command voltage trace. A *perpendicular broken line* has been drawn at the time that the current begins to rise (shown only during the first spike) to illustrate the temporal relationship between current activation and the action potential. The Kv1.1 current begins to rise early during the rising phase of the action potential, while the Kv3.1 current rises much later (similar results were obtained with Kv3.2). Note also that the Kv3.1 current is completely deactivated during the interspike interval, while the Kv1.1 channels close slowly and are activated by the inter-spike depolarization. After the second spike the Kv1.1 current deactivates slowly.

potential.[34,66] Given their fast deactivation, Kv3.1–Kv3.2 channels are also less likely to contribute to increase the duration of refractory periods. In neurons containing additional K^+ channels, their differential modulation by neurotransmitters and neuromodulators could independently regulate different aspects of the cell's electrical behavior.

Many of the neuronal populations expressing Kv3 gene products fire trains of action potentials at very high rates, such as neurons in the reticular thalamic nucleus (RT) and fast-spiking (FS) interneurons in the neocortex and the hippocampus.[14,21,59,66,67,68] It has been proposed that Kv3.1–Kv3.2 channels play key roles in the firing properties of these cells. An emphasis of recent investigations by us and others[68,70,84,85] has been on exploring the role of Kv3.1–Kv3.2 channels on high-frequency firing.

FS cells of the cerebral cortex are GABAergic inhibitory interneurons characterized by brief action potentials, a large and fast after-hyperpolarization (fAHP) and the ability to fire at high frequency with little accommodation. Cortical GABAergic interneurons play multiple central roles in the processing of cortical information. They are thought to be important in the synchronization of cortical circuits, in the establishment and reorganization of cortical representation maps, in the establishment of cortical columns during development, in the generation of some forms of cortical rhythms, and in the pathogenesis of seizures.[86–92]

In a recent study, we examined the effects of low TEA concentrations (<1 mM) on the electrophysiological properties of visually identified nonpyramidal cells in mouse cortical slices using whole-cell patch clamp methods.[93,94] We found that TEA had dramatic, reversible effects on the shape of the action potential and the repetitive firing characteristics of fast-spiking (FS) neurons. TEA nearly abolished the fast-afterhyperpolarization (fAHP), increased the spike duration, and decreased the maximum rate of spike repolarization (FIG. 10A; TABLE 3). Moreover, for a given current strength, TEA also produced a reduction in the steady firing rate (FIG. 10B). This effect of TEA was manifest over a large range of injected currents and reduced the maximum steady firing rate that could be achieved by direct-current injection into FS neurons (TABLE 3).

As described in the subsection in this paper entitled Pharmacological Properties (see also Coetzee *et al.* in this volume) at the concentrations of TEA used in the studies by Erisir *et al.,* the only known voltage-gated K^+ channels that are significantly blocked by this drug are, in addition to Kv3.1–Kv3.4 channels, those containing the K^+ channel proteins Kv1.1 and KCNQ2 as well as the BK (also called maxi-K) Ca^{2+}-activated K^+ channels containing proteins of the *slo* family. To test the possible involvement of these channels in the TEA-effect on FS neurons, Erisir *et al.* applied toxins specific for Kv1 and BK channels. 100 nM Dendrotoxin, which selectively blocks Kv1.1 and other Kv1 channels (see Coetzee *et al.* in this volume and Ref. 45), had few measurable effects on the shape of individual action potentials. However, and in contrast to the effects of TEA, it produced a significant *increase* in the maximum firing frequency achieved with current injection into FS neurons. Application of Charybdotoxin (100 nM) or Iberiotoxin (50 nM), which block Ca^{2+}-activated BK channels, produced minimal effects on spike shape or firing frequency.[93,94] The recently discovered KCNQ2 subunits express very slowly activating and deactivating (time constants of hundreds of milliseconds to seconds) voltage-gated K^+ channels that are unlikely to be activated during short action potentials and have not been shown to be expressed in cortical interneurons[95] (see also Coetzee *et al.* in this volume).

These results support the hypothesis that Kv3.1–Kv3.2 potassium channels are important for the elaboration of fast-spiking properties in cortical interneurons. Moreover, the fact that blockade of Kv1 channels had the opposite effect of low TEA concentrations supports the notion that the high voltage activation and rapid deactivation kinetics of Kv3.1–Kv3.2 channels are necessary for sustained high-frequency firing. This idea was further supported by results from voltage-clamp analysis of voltage-dependent K^+ currents recorded from macropatches excised from *somata* of FS cells. These currents showed two main components, the larger of which (>75% of the total) was blocked by 1 mM TEA. This component had a voltage-dependence and activation and deactivation kinetics that were similar to those of Kv3.1 and Kv3.2 currents in heterologous expression systems and the Kv3.1–Kv3.2-like current isolated from pallidal neurons described above. Thus, the native TEA-sensitive channels which were necessary for sustained high-frequency firing

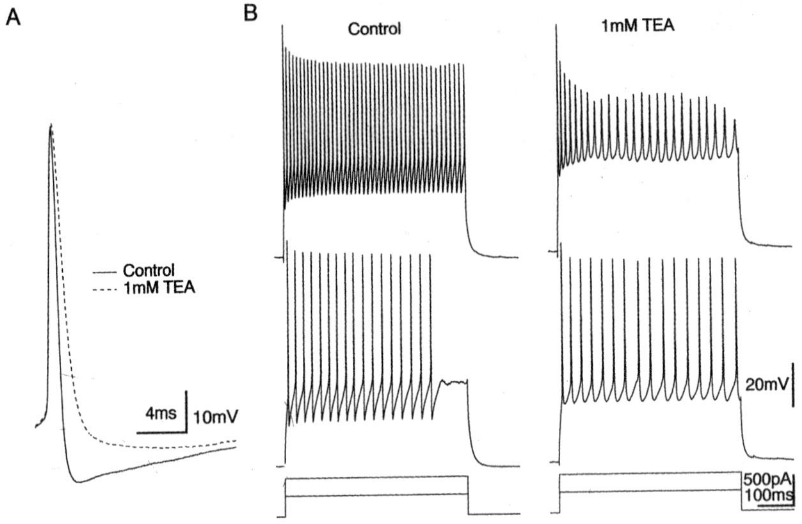

FIGURE 10. Low concentrations of TEA impair spike repolarization and slow high-frequency firing of fast-spiking (FS) layer II/III neocortical interneurons. (**A**) 1 mM TEA caused spike broadening by reducing the maximum rate of spike repolarization and suppressed the fast-AHP of single action potentials evoked by near-threshold depolarizations. (**B**) **left panel** shows repetitive firing of the same FS neuron in response to 350-pA **(bottom)** and 650pA currents under control conditions. **Right panel** shows the responses to identical currents in the presence of 1 mM TEA. TEA reduced the steady firing rate from 35 to 29 spikes/s and from 75 to 33 spikes/s for the 350- and 650pA currents, respectively.

in FS cells had similar properties to Kv3.1–Kv3.2 channels in heterologous expression systems.

Kv3 transcripts are also abundantly expressed in other sensory neurons in the brain, including in neurons of most of the midbrain and hindbrain auditory nuclei.[14,21,59,66,70] There is co-expression of Kv3.1 and Kv3.3 in many of these nuclei, including the ventral cochlear nucleus (VCN), the medial nucleus of the trapezoid body (MNTB), the lateral lemniscus nuclei and the inferior colliculus.[21] It remains to be shown whether the protein products are co-localized in the same neurons and whether they can be co-precipitated. There are also Kv3.2 mRNAs and proteins in the inferior colliculus (Ref. 21 and unpublished observations), but it has not yet been shown that Kv3.1 and Kv3.2 proteins are in the same neuronal populations or that they are present in the same channel complex. In other auditory nuclei there is actually a reciprocal expression between Kv3.1–Kv3.3 and Kv3.2 transcripts, as is seen in the thalamus. Kv3.2 is more abundantly expressed in the dorsal cochlear nucleus, while Kv3.1 and Kv3.3 are more abundant in the ventral nuclei. Similarly Kv3.2 mRNA transcripts and proteins are more abundant in the ventral lateral lemniscus nuclei and Kv3.1 and Kv3.3 in the dorsal nuclei.

These auditory neurons do not respond with high-frequency trains of action potentials when depolarized by steady current injection. However, many of these cells, includ-

TABLE 3. Effects of 1 mM TEA on the Firing Properties of Fast-Spiking Neocortical Interneurons

	FS Neurons	FS Neuron + 1 mm TEA
Spike duration (ms) (width at half-amplitude)	0.64	1.16
Maximal repolarization rate (mV/ms)	148.7	62.8
Fast after-potential (mV) (relative to threshold)	–12.5	0.8
Maximal firing frequency (Hz)	104.6	65.6

ing many of those expressing Kv3 products, are capable of entraining at very high frequencies when stimulated with high-frequency inputs (>600 Hz). The ability to follow high-frequency stimulation allows these neurons to preserve the timing information contained in auditory signals. The temporal precision of auditory signals is important for the faithful transmission and localization of sound.[84,96,97] Len Kaczmarek and coworkers have been studying the role of Kv3 channels in central auditory neurons. In the MNTB, application of low concentrations of TEA did not alter the ability of the cells to follow stimulation up to 200 Hz, but reduced their ability to follow higher frequency stimuli.[84]

Together, the pharmacological studies provide very strong circumstantial evidence for a role of Kv3.1–Kv3.2 channels in sustained high-frequency firing, but, of course, we cannot exclude the possibility that there are unknown K^+ channels that are also blocked by low TEA concentrations. However, the observations with different K^+ channel blockers and with modeling studies indicate that the relatively positive activation range and the deactivation rates of Kv3.1 and Kv3.2 channels are important for these apparent roles.[93,94]

How do the special properties Kv3.1 and Kv3.2 channels facilitate high-frequency firing? In trains of Na^+ spikes, the inter-spike interval is established in part by the amount of Na^+ channel inactivation that accumulates during the train. By increasing the rate of spike repolarization and by keeping action potentials short, Kv3.1–Kv3.2 currents would reduce the amount of Na^+ channel inactivation that occurs during the action potential. Moreover, the large, fast after-hyperpolarization produced by the Kv3.1–Kv3.2 current would function to speed up recovery from Na^+-channel inactivation and thus increase the amount of recovery that occurred following a spike. Indeed, examination of the effects of low TEA concentration on cortical interneurons revealed that 1 mM TEA produced a significant reduction in the maximum rate of rise of the second, but not the first, spike in a train. This suggested that there was significantly greater Na^+-channel inactivation at the onset of the second spike in the presence of TEA. This interpretation of the TEA results was strongly supported by results from computer simulations of a fast-spiking neuron.[94] In that model, blocking 95% of the Kv3.1–Kv3.2-like channels resulted in spike-broadening, a reduction in the after-hyperpolarization and a slowing of the frequency of repetitive firing for a given current injection as observed experimentally. Accompanying this slowdown in firing was a decrease in both the amount of recovery and speed of recovery from Na^+-channel inactivation. Thus, one function of the Kv3.1–Kv3.2 current is to reduce the impact of Na^+-channel inactivation on repetitive firing.

Another unique feature of Kv3.1–Kv3.2 currents is their rapid rates of deactivation. The importance of this for high-frequency firing was also apparent from the simulations. The large AHP produced by activation of Kv3.1–Kv3.2-channels also serves to rapidly terminate the Kv3.1–Kv3.2 current, which, because of its rapid deactivation rates, does not contribute increases in the duration of the refractory period. The situation is completely different for K^+ channels with slower deactivation kinetics. For example, a Kv1.3-like current deactivates slowly enough that its conductance decays relatively little during the interspike interval and therefore, it contributes to delaying the onset of the next spike. Hence, blocking such a current in the model produced an *increase* in spike frequency for a given current injection. These modeling studies strongly suggest that the particular activation range and fast deactivation kinetics of Kv3.1–Kv3.2 channels function specifically to enable sustained high-frequency firing.

The presence of Kv3.1–Kv3.2 channels alone is not sufficient for the generation of sustained high-frequency firing. The firing properties of the cell depend on the interplay between passive electrical properties of the cell and the various ionic conductances that are active when the cell is stimulated. Pallidal and brainstem auditory neurons may be a case in point. Two main types of neurons from the point of view of their intrinsic electrophysiological properties have been described in pallidal neurons recorded in guinea pig brain slices. Type I neurons were not spontaneously active and fired spike bursts with strong accommodation upon depolarization. Type II cells fired spontaneously at the resting level (~−60 mV) and displayed fast repetitive firing (≤200 Hz) with some accommodation when depolarized.[99,100] Type II cells resemble the "repetitively firing cells" recorded in an *in vivo* intracellular study in rats by Kita and Kitai,[101] while type I neurons resemble the periodic burst firing cells in this study. The majority of the cells in the rat study (73%) were of the repetitive firing type and most likely correspond to the neurons we classified as type "A" in the study described earlier.

The firing frequency of the repetitive firing cells in the globus pallidus is not as high or sustained as that of fast-spiking cortical neurons. The analysis made by Erisir *et al.*[93,94] of the currents in the latter cells suggests that they have a higher proportion of Kv3.1–Kv3.2-like currents and less Ca^{2+}-activated K^+ currents than do pallidal neurons. These differences in channel composition may explain the differences in firing frequency adaptation of the two cell types. Moreover, owing to the presence of a large component of inactivating current in PV+ pallidal neurons (see above), changes in the resting potential of the cell will change the contribution of Kv3.1–Kv3.2 channels to the total current. This may, in turn, change the maximum firing frequencies of these neurons.

The central auditory neurons described earlier can follow spikes at very high frequencies, but they respond with one or few spikes to steady depolarizations, unlike fast-spiking cortical interneurons. These different firing properties of the two cell types, both of which express abundant Kv3.1–Kv3.2 channels, could be explained by the different levels of low-voltage-activating dendrotoxin-sensitive K^+ channels.[84]

Kv3 CHANNELS IN AXONS AND TERMINALS

Kv3 proteins are prominently expressed in nonmyelinated axons and in presynaptic terminals in the CNS (see above); however not much is known about their function in these neuronal compartments. For that matter, little is known about the role of any specific

K^+ channels in axons and terminals in the CNS. This is not surprising, since until recently it was very difficult to study axonal and presynaptic function in central neurons. Nevertheless, there is clearly a large diversity of K^+ channels in axons and terminals (see Coetzee *et al.* in this volume), and we can expect a richness of functions as that seen in the generation of firing properties in the somatic compartment. It is also becoming clear that, as in the soma, several types of K^+ channels may coexist in the same terminal and/or in axons. Any mechanism affecting Ca^{2+} entry into the presynaptic terminal is likely to have strong effects on neurotransmitter release. K^+ channels are likely to modulate in complex and subtle ways the invasion, magnitude, and shape, and temporal relationships of action potentials of presynaptic terminals, which in turn will regulate the magnitude and kinetics of Ca^{2+} entry. Because of our ignorance of necessary details on action potential propagation and synaptic transmission in central neurons it is more difficult to generate predictable hypotheses about the role of specific channels. Therefore the investigation of these roles becomes more urgent.

Nevertheless, that presynaptic K^+ channels affect Ca^{2+} entry and/or neurotransmitter release is well established (in the CNS as well), and two K^+ channel blockers, 4-AP and DTX, are well known to increase transmitter release in several CNS synapses.[46,102–112]

We can expect that with the development of methods to study axonal and synaptic function in the CNS and of tools to specifically control the activity of specific K^+ channels, our knowledge of this area will grow enormously. A very recent study, taking advantage of the ability to record simultaneously from two connected cells, showing that an A-type K^+ channel, probably Kv1.4, could gate action potential propagation in hippocampal axons, illustrates this point.[113]

We expect that, as in the soma, Kv3 channels and their modulation will play special roles in axons and presynaptic terminals. These roles might be related to the ability of Kv3 channels to keep action potentials short and facilitate recovery of processes that accumulate during repetitive action potentials and can limit the frequency of action potential transmission. By minimizing these changes, Kv3 channels could facilitate the reliable transmission of impulses at high frequency. In synapses, by increasing action potential repolarization and helping to generate large fast AHPs without increasing the duration of the refractory period, Kv3 channels could help increase the temporal resolving abilities of the terminal. The presence of Kv3 channels together with other K^+ channels in terminals could allow the cell to regulate independently the amount of transmitter release per spike and the frequency of release.[34,66,70] Wang and Kaczmarek[98] recently showed that high-frequency firing speeds up the recovery of synaptic depression in auditory neurons by helping replenish the pool of releasable vesicles. Kv3 channels may thus also modulate the rates of recovery of terminals that undergo synaptic depression. The availability of pharmacological tools and mouse models (see below) to study Kv3 channels will be particularly useful to investigate the roles of these channels in axonal conduction and synaptic transmission.

FUNCTIONAL ROLES OF ALTERNATIVE SPLICING

The significance of the alternative splicing of Kv3 genes has remained unknown until recently. None of the alternatively spliced transcripts yet tested in heterologous expression systems (Kv3.1a;[1] Kv3.1b;[5] Kv3.2a;[2] Kv3.2b;[35] Kv3.2d, Moreno and Rudy, unpublished observations; Kv3.3a;[9] Kv3.3b, Vega-Saenz de Miera, unpublished observations; Kv3.4a;[4] and Kv3.4b[3]) expresses currents noticeably different from those of any other transcript from the same gene. However, a more in-depth analysis of the channels expressed by these transcripts that may reveal overlooked subtle differences is yet to be done. Since alternatively spliced C-termini have different putative-phosphorylation sequences (see TABLE 4 in Ref. 15), it was suggested that distinct alternatively spliced isoforms may respond differently to various second-messenger systems. Alternative splicing may thus allow for isoform-specific differences in the modulation of channel function by neurotransmitters and neuropeptides. We have also suggested that the divergent carboxy termini could be used in protein–protein interactions, conferring isoform-specific channel localization or mobility.

Evidence in support of these proposals is now becoming available. In a study comparing the modulation of Kv3.2a and Kv3.2b channels expressed in HEK-293T cells by PKA and PKC, we found that while both currents are similarly suppressed by PKA, only Kv3.2b channels are affected by phorbol esters activating PKC[35] (see TABLE 1). The similarity of the effects of PKA are consistent with the conclusion, mentioned earlier, that the effects of this kinase are due to phosphorylation at the single PKA consensus site, which is present in the Kv3.2 C-terminal sequence *prior* to the site of divergence of alternatively spliced isoforms. Several putative PKC phosphorylation sites are present in the Kv3.2b alternatively spliced C-terminus which are lacking in the Kv3.2a C-terminus. Phosphorylation of one or more of these sites may mediate the isoform-specific modulation by this kinase.

Data supporting the view that the alternatively spliced C-termini may participate in targeting Kv3 channels to different compartments of the neuronal membrane is also forthcoming. In a study in monolayers of polarized epithelial MDCK cells, Ponce *et al.*[114] showed that in Kv3.2a-transfected cells the Kv3.2 channels are expressed mainly in the basolateral membrane, while Kv3.2b is expressed mainly in the apical membrane. Interestingly a sequence analysis of the alternatively spliced C-termini of all four Kv3 genes showed that the sequences fall into two groups: those that resemble the C-terminus of Kv3.2a and those that resemble the C-terminus of Kv3.2b.[114]

More recently, we have been able to raise antibodies that distinguish the two Kv3.1 isoforms, and have begun a study of the cellular and subcellular distribution of the two proteins in CNS neurons. The results strongly support the notion that the divergent C-termini play a role in targeting the channels to distinct neuronal compartments. While Kv3.1b is expressed strongly in somatic, axonal, and terminal membranes (see above), Kv3.1a is expressed mainly in axons and terminals, including in many neuronal populations expressing Kv3.1b somatically, as in basket cells in the hippocampus.[64,65] We hypothesize that the Kv3.1a C-terminus (which resembles the C-terminus of Kv3.2b) contains signals for transport to axons and terminals, and that the Kv3.1b protein is carried to these compartments by forming heteromeric channels with Kv3.1a and other Kv3 proteins having the proper signals.[64,65]

CONCLUSIONS AND PERSPECTIVES

This chapter summarizes the results of a six-year effort by ourselves and others to discover the physiological roles of Kv3 proteins. Studies in heterologous expression systems, including *Xenopus* oocytes and transfected mammalian cell lines, showed that these proteins form homomultimeric and heteromultimeric delayed-rectifier and A-type voltage-gated K^+ channels mediating currents with several unusual properties. All Kv3 currents studied thus far have, compared to other voltage-dependent K^+ currents, a relatively positive activation voltage. Significant Kv3 currents are not seen until the membrane is depolarized beyond −10 mV.

Studies on the localization of Kv3 mRNA transcripts and proteins have been key to select neurons and other cells where the properties of native channels containing Kv3 proteins could be analyzed, and to generate hypotheses as to their physiological roles. In the cells studied so far, Kv3.1, Kv3.2, and Kv3.4 channels mediate currents which resembled those in heterologous expression systems. Native Kv3.3 channels have not been pursued yet.

Kv3.1 and Kv3.2 proteins are expressed in somatic and/or axonal-terminal membranes of selected neuronal populations in the CNS. These include many GABAergic neurons that are capable of firing action potentials at very high frequencies. Kv3.1 and/or Kv3.2 channels may play interesting roles in several central midbrain and brain stem sensory processing systems, and in the thalamocortical circuit. In some neurons, Kv3.1 and Kv3.2 proteins may be part of the same heteromeric channels. A careful utilization of the limited pharmacological tools available to differentiate Kv3 currents from other K^+ currents, and in particular the use of low concentrations of the K^+ channel blocker TEA (<1 mM), in electrophysiological recordings from cortical fast-spiking interneurons, support the suggestion that Kv3.1–Kv3.2 channels play *unique* roles in maintaining sustained high-frequency firing. These roles may critically depend on two of the unusual properties of the currents mediated by these channels: (1) Their voltage-dependence and (2) their fast closing rate upon membrane repolarization. The role of Kv3.1–Kv3.2 channels in fast-spiking neurons might not be easily replaceable by other K^+ channels.

In our laboratory[115,116] and others'[117,118] mouse lines with disrupted Kv3 genes have been generated, utilizing homologous recombination in embryonic stem cells. Mice that do not produce Kv3.1, Kv3.2, and/or Kv3.3 proteins are now available. Mice lacking Kv3.1 proteins have several behavioral deficits.[117,119] Kv3.2 knockout mice are very interesting: they have abnormal sleep cycles, epileptic seizures, and abnormal EEG patterns. These defects could well be associated with disruption of thalamocortical circuits and/or cortical inhibition, systems in which Kv3.2 proteins are so prominent. Bearing in mind all the precautions necessary to interpret results from this experimental paradigm (see the Introduction to this volume), we believe that these mice are likely to become important tools, in conjunction with the other experimental tools described in this chapter, to deepen our understanding of the physiological roles of Kv3 channels. In particular, if there have not been compensatory changes, and if the primary result of the genetic disruption has not initiated a cascade of secondary changes (as appears to be the case so far, and is probably minimized by the late appearance of these channels during development), the mice will allow us to confirm the roles of Kv3.1 and Kv3.2 channels on fast spiking and will be particularly helpful to analyze roles of Kv3 channels in signal transmission through axons and terminals. Moreover, these mice might be important experimental models to investigate the roles of the signaling patterns of the affected neuronal systems on brain function.

ACKNOWLEDGMENTS

The work in the Rudy lab described here was supported by NIH Grants NS30989, NS35215 and NSF Grant IBN9630832. Work in the C. L. lab is supported by NIH Grant NS27881.

REFERENCES

1. Yokoyama, S., K. Imoto, T. Kawamura, H. Higashida, N. Iwabe, T. Miyata & S. Numa. 1989. Potassium channels from NG108-15 neuroblastoma-glioma hybrid cells. Primary structure and functional expression from cDNAs. FEBS Lett. **259:** 37
2. McCormack, T., E.C. Vega-Saenz de Miera & B. Rudy. 1990. Molecular cloning of a member of a third class of Shaker-family K^+ channel genes in mammals [published erratum appears in Proc. Natl. Acad. Sci. U.S.A. 1991 May 1:88(9):4060]. Proc. Natl. Acad. Sci. USA **87:** 5227.
3. Rudy, B., K. Sen, E. Vega-Saenz de Miera, D. Lau, T. Ried & D.C. Ward. 1991. Cloning of a human cDNA expressing a high voltage-activating, TEA-sensitive, type-A K^+ channel which maps to chromosome 1 band p21. J. Neuroscie. Res. **29:** 401.
4. Schroter, K.H., J.P. Ruppersberg, F. Wunder, J. Rettig, M. Stocker & O. Pongs. 1991. Cloning and functional expression of a TEA-sensitive A-type potassium channel from rat brain. FEBS Lett. **278:** 211.
5. Luneau, C.J., J.B. Williams, J. Marshall, E.S. Levitan, C. Oliva, J.S. Smith, J. Antanavage, K. Folander, R.B. Stein, R. Swanson, *et al.* 1991. Alternative splicing contributes to K^+ channel diversity in the mammalian central nervous system. Proc. Natl. Acad. Sci. U.S.A. **88:** 3932.
6. Luneau, C., R. Wiedmann, J.S. Smith & J.B. Williams. 1991. Shaw-like rat brain potassium channel cDNAs with divergent 3′ ends. FEBS Lett. **288:** 163.
7. Vega-Saenz de Miera, E.C., K. Sen, P. Serodio, T. McCormack & B. Rudy. 1990. Description of a new class of potassium channel genes. Soc. Neurosci. Abst. **16:** 4.
8. Vega-Saenz de Miera, E.C., N. Chiu, K. Sen, D. Lau, J.W. Lin & B. Rudy. 1991. Toward an understanding of the molecular composition of K^+ channels: Products of at least nine distinct Shaker family K^+ channel genes are expressed in a single cell. Biophys. J. **59:** 197a.
9. Vega-Saenz de Miera, E., H. Moreno, D. Fruhling, C. Kentros & B. Rudy. 1992. Cloning of ShIII (Shaw-like) cDNAs encoding a novel high-voltage-activating, TEA-sensitive, type-A K^+ channel. Proc. R. Soc. Lond. B Biol. Sci. **248:** 9.
10. Ghanshani, S., M. Pak, J.D. McPherson, M. Strong, B. Dethlefs, J.J. Wasmuth, L. Salkoff, G.A. Gutman & K.G. Chandy. 1992. Genomic organization, nucleotide sequence, and cellular distribution of a Shaw-related potassium channel gene, Kv3.3, and mapping of Kv3.3 and Kv3.4 to human chromosomes 19 and 1. Genomics **12:** 190.
11. Ried, T., B. Rudy, E. Vega-Saenz de Miera, D. Lau, D.C. Ward & K. Sen. 1993. Localization of a highly conserved human potassium channel gene (NGK2-KV4; KCNC1) to chromosome 11p15. Genomics **15:** 405.
12. Haas, M., D.C. Ward, J. Lee, A.D. Roses, V. Clarke, P, D.E., D. Lau, E. Vega-Saenz de Miera & B. Rudy. 1993. Localization of Shaw-related K^+ channel genes on mouse and human chromosomes. Mamm. Genome **4:** 711.
13. Goldman-Wohl, D.S., E. Chan, D. Baird & N. Heintz. 1994. Kv3.3b: A novel Shaw type potassium channel expressed in terminally differentiated cerebellar Purkinji cells and deep cerebellar nuclei. J. Neurosci. **14:** 511.
14. Rudy, B., C. Kentros, M. Weiser, D. Fruhling, P. Serodio, E. Vega-Saenz de Miera, M.H. Ellisman, J.A. Pollock & H. Baker. 1992. Region-specific expression of a K^+ channel gene in brain. Proc. Natl. Acad. Sci. USA **89:** 4603.
15. Vega-Saenz de Miera, M. Weiser, C. Kentros, D. Lau, H. Moreno, P. Serodio & B. Rudy. 1994. Shaw-related K^+ channels in mammals. *In* Handbook of Membrane Channels. C. Peracchia, Ed. 41. Academic Press, Inc. Orlando, FL.

16. Coetzee, W.A., Y., J. Chiu, A. Chow, D. Lau, T. McCormack, T. Moreno, M. Nadal, A. Ozaita, D. Pountney, M. Saganich, E. Vega-Saenz de Miera & B. Rudy. 1999. Molecular diversity of K^+ channels. Ann. N.Y. Acad. Sci. This volume.
17. Rashid, A.J. & R.J. Dunn. 1998. Sequence diversity of voltage-gated potassium channels in an electric fish. Brain Res. Mol. Brain Res. **54:** 101.
18. Spitzer, N.C. & D. Gurantz. 1996 Xenopus Kv3.3 potassium channel transcripts are developmentally upregulated in the embryonic spinal cord. Soc. Neurosci. Abst. **22:** 1753.
19. Kentros, C., M. Weiser, E. Vega-Saenz de Miera, K. Morel, H. Baker & B. Rudy. 1992. Alternative splicing of the 5′ untranslated region of a gene encoding K channel components. Soc. Neurosci. Abst. **18**.
20. Erginel-Unaltuna, N., W.P. Yang & M.A. Blanar. 1998. Genomic organization and expression of KCNJ8/Kir6.1, a gene encoding a subunit of an ATP-sensitive potassium channel. Gene **211:** 71.
21. Weiser, M., E. Vega-Saenz de Miera, C. Kentros, H. Moreno, L. Franzen, D. Hillman, H. Baker & B. Rudy. 1994. Differential expression of Shaw-related K^+ channels in the rat central nervous system, J. Neurosci. **14:** 949.
22. Hoshi, T., W.N. Zagotta & R.W. Aldrich. 1990. Biophysical and molecular mechanisms of Shaker potassium channel inactivation [see comments]. Science **250:** 533.
23. Smith-Maxwell, C.J., J.L. Ledwell & R.W. Aldrich. 1998. Role of the S4 in cooperativity of voltage-dependent potassium channel activation. J. Gen Physiol. **111:** 399.
24. Johnstone, D.B., A. Wei, A. Butler, L. Salkoff & J.H. Thomas. 1997. Behavioral defects in C. elegans egl-36 mutants result from potassium channels shifted in voltage-dependence of activation, Neuron **19:** 151.
25. Hernández-Pineda, R., A. Chow, Y. Amarillo, H. Moreno, M. Saganich, E. Vega-Saenz de Miera, A. Hernández-Cruz & B. Rudy. 1999. Kv3.1–Kv3.2 heteromultimeric channels underlie a high voltage-activating component of the delayed rectifier K^+ current in projecting neurons from the Globus Pallidus. J. Neurophysiol. In press.
26. Grissmer, S., A.N. Nguyen, J. Aiyar, D.C. Hanson, R.J. Mather, G.A. Gutman, M.J. Karmilowicz, D.D. Auperin & K.G. Chandy. 1994. Pharmacological characterization of five cloned voltage-gated K^+ channels, types Kv1.1, 1.2, 1.3, 1.5, and 3.1, stably expressed in mammalian cell lines. Mol. Pharmacol. **45:** 1227.
27. Critz, S.D., B.A. Wible, H.S. Lopez & A.M. Brown. 1993. Stable expression and regulation of a rat brain K^+ channel, J. Neurochem. **60:** 1175.
28. Kalman, K., A. Nguyen, J. Tseng-Crank, I.D. Dukes, G. Chandy, C.M. Hustad, N.G. Copeland, N.A. Jenkins, H. Mohrenweiser, B. Brandriff, M. Cahalan, G.A. Gutman & K.G. Chandy. 1998. Genomic organization, chromosomal localization, tissue distribution, and biophysical characterization of a novel mammalian Shaker-related voltage-gated potassium channel, Kv1.7. J. Biol. Chem. **273:** 5851.
29. Harris, R.E. & E.Y. Isacoff. 1996. Hydrophobic mutations alter the movement of Mg^{2+} in the pore of voltage-gated potassium channels. Biophys. J. **71:** 209.
30. Ruppersberg, J.P., M. Stocker, O. Pongs, S.H. Heinemann, R. Frank & M. Koenen. 1991. Regulation of fast inactivation of cloned mammalian IK(A) channels by cysteine oxidation. Nature **352:** 711.
31. Vega-Saenz de Miera, E. & B. Rudy. 1992. Modulation of K^+ channels by hydrogen peroxide. Biochem. Biophys. Res. Commun. **186:** 1681.
32. Serodio, P., C. Kentros & B. Rudy. 1994. Identification of molecular components of A-type channels activating at subthreshold potentials. J. Neurophysiol. **72:** 1516.
33. Ruppersberg, J.P., R. Frank, O. Pongs & M. Stocker. 1991. Cloned neuronal IK(A) channels reopen during recovery from inactivation [see comments]. Nature **353:** 657.
34. Moreno, H., C. Kentros, E. Bueno, M. Weiser, A. Hernandez, E. Vega-Saenz de Miera, A. Ponce, W. Thornhill & B. Rudy. 1995. Thalamocortical projections have a K^+ channel that is phosphorylated and modulated by cAMP-dependent protein kinase. J. Neurosci. **15:** 5486.
35. McIntosh, P., H. Moreno, B. Robertson, & B. Rudy. 1998. Isoform-specific modulation of rat Kv3 potassium channel splice variants. J. Physiol. **511P:** 147.
36. Kanemasa, T., L. Gan, T.M. Perney, L.Y. Wang & L.K. Kaczmarek. 1995. Electrophysiological and pharmacological characterization of a mammalian Shaw channel expressed in NIH 3T3 fibroblasts. J. Neurophysiol. **74:** 207.

37. Covarrubias, M., A. Wei, L. Salkoff & T.B. Vyas. 1994. Elimination of rapid potassium channel inactivation by phosphorylation of the inactivation gate. Neuron **13:** 1403.
38. Beck, E.J., R.G. Sorensen, S.J. Slater & M. Covarrubias. 1998. Interactions between multiple phosphorylation sites in the inactivation particle of a K^+ channel. Insights into the molecular mechanism of protein kinase C action. J. Gen. Physiol. **112:** 71.
39. Vega-Saenz de Miera, E., H. Moreno & B. Rudy. 1994. Modulation of Kv3.3 K^+ channels by oxidation and phosphorylation, in Soc. Neurosci. Abst. **20:** 725.
40. Vega-Saenz de Miera, E., H. Moreno & B. Rudy. 1995. Phosphorylation may be required to activate Shaw related K^+ channels. Soc. Neurosci. Abst. **21:** 505.
41. Moreno, H., E. Bueno, A. Hernandez Cruz, A. Ponce & B. Rudy. 1995. Nitric oxide and cGMP modulate a presynaptic K^+ channel *in vitro*. Soc. Neurosci. Abst. **21:** 506.
42. Moreno, H., M.S. Nadal, E. Vega-Saenz de Miera & B. Rudy. 1999. Modulation of Kv3 potassium channels by a nitric oxide-activated phosphatase. J. Neurosci. Submitted for publication.
43. Shieh, C.C. & G.E. Kirsch. 1994. Mutational analysis of ion conduction and drug binding sites in the inner mouth of voltage-gated K^+ channels. Biophys. J. **67:** 2316.
44. Kirsch, G.E. & J.A. Drewe. 1993. Gating-dependent mechanism of 4-aminopyridine block in two related potassium channels. J. Gen. Physiol. **102:** 797.
45. Chandy, K.G. & G.A. Gutman. 1995. Voltage gated channels. *In* Handbook of Receptors and Channels: Ligand-gated and Voltage-gated Ion Channels. R.A. North, Ed.: 1–71. Boca Raton, FL.
46. Barish, M.E., M. Ichikawa, T. Tominaga, G. Matsumoto & T. Iijima. 1996. Enhanced fast synaptic transmission and a delayed depolarization induced by transient potassium current blockade in rat hippocampal slice as studied by optical recording. J. Neurosci. **16:** 5672.
47. Wu, R.L. & M.E. Barish. 1992. Two pharmacologically and kinetically distinct transient potassium currents in cultured embryonic mouse hippocampal neurons. J. Neurosci. **12:** 2235.
48. Diochot, S., H. Schweitz, L. Beress & M. Lazdunski. 1998. Sea anemone peptides with a specific blocking activity against the fast inactivating potassium channel Kv3.4. J. Biol. Chem. **273:** 6744.
49. McCormach, K., J.W. Lin, L.E. Iverson & B. Rudy. 1990. Shaker K^+ channel subunits from heteromultimeric channels with novel functional properties. Biochem. Biophys. Res. Commun. **171:** 1361.
50. Isacoff, E.Y., Y.N. Jan & L.Y. Jan. 1990. Evidence for the formation of heteromultimeric potassium channels in Xenopus oocytes [see comments]. Nature **345:** 530.
51. Ruppersberg, J.P., K.H. Schroter, B. Sakmann, M. Stocker, S. Sewing & O. Pongs. 1990. Heteromultimeric channels formed by rat brain potassium-channel proteins [see comments]. Nature **345:** 535.
52. MacKinnon, R. 1991. Determination of the subunit stoichiometry of a voltage-activated potassium channel. Nature **350:** 232.
53. Christie, M.J., R.A. North, P.B. Osborne, J. Douglass & J.P. Adelman. 1990. Heteropolymeric potassium channels expressed in Xenopus oocytes from cloned subunits. Neuron **4:** 405.
54. Dixon, J.E. & D. McKinnon. 1994. Quantitative analysis of potassium channel mRNA expression in atrial and ventricular muscle of rats. Circ. Res. **75:** 252.
55. Brahmajothi, M.V., M.J. Morales, R.L. Rasmusson, D.L. Campbell & H.C. Strauss. 1997. Heterogeneity in K^+ channel transcript expression detected in isolated ferret cardiac myocytes. Pacing Clin. Electrophysiol. **20:** 388.
56. Dixon, J.E. & D. McKinnon. 1996. Potassium channel mRNA expression in prevertebral and paravertebral sympathetic neurons. Eur. J. Neurosci. **8:** 183.
57. Grissmer, S., S. Ghanshani, B. Dethlefs, J.D. McPherson, J.J. Wasmuth, G.A. Gutman, M.D. Cahalan & K.G. Chandy. 1992. The Shaw-related potassium channel gene, Kv3.1, on human chromosome 11, encodes the type 1 K^+ channel in T cells. J. Biol. Chem. **267:** 20971.
58. Roe, M.W., J.F. Worley, III, A.A. Mittal, A. Kuznetsov, S. DasGupta, R.J. Mertz, S.M. Witherspoon, III, N. Blair, M.E. Lancaster, M.S. McIntyre, W.R. Shehee, I.D. Dukes & L.H. Philipson. 1996. Expression and function of pancreatic beta-cell delayed rectifier K^+ channels. Role in stimulus-secretion coupling. J. Biol. Chem. **271:** 32241.
59. Perney, T.M., J. Marshall, K.A. Martin, S. Hockfield & L.K. Kaczmarek. 1992. Expression of the mRNAs for the Kv3.1 potassium channel gene in the adult and developing rat brain, J. Neurophysiol. **68:** 756.

60. DREWE, J.A., S. VERMA, G. FRECH & R.H. JOHO. 1992. Distinct spatial and temporal expression patterns of K^+ channel mRNAs from different subfamilies. J. Neurosci. **12:** 538.
61. NGUYEN, T.D. & G. JESERICH. 1998. Molecular structure and expression of shaker type potassium channels in glial cells of trout CNS. Neurosci. Res. **51:** 284.
62. CHOW, A., A. ERISIR, C. FARB, D.H.P. LAU & B. RUDY. 1998. Kv3.1 and Kv3.2 proteins distingish three subpopulations of GABA-ergic interneurons in the mouse cortex. Soc. Neurosci. Abst. **24:** 1579.
63. CHOW, A., C. FARB, A. ERISIR, D. LAU & B. RUDY. 1999. K^+ channel expression distinguishes between two subpopulations of parvalbumin-containing cortical interneurons. J. Neurosci. Submitted for publication.
64. OZAITA, A., E. VEGA-SAENZ DE MIERA, A. CHOW, T.R. MUTH, M.J. CAPLAN & B. RUDY. 1998. Differential targeting of Kv3.1–Kv3.2 containing potassium channels produced by alternatively-spliced C-termini. Soc. Neurosci. Abst. **24:** 1580.
65. OZAITA, A., A. CHOW, M. MARTONE, M. ELLISMAN, E. VEGA-SAENZ DE MIERA & B. RUDY. 1999. Differential subcellular localization of the two Kv3.1 K^+ channel alternatively-spliced isoforms in brain neurons. J. Neurosci. Submitted for publication.
66. WEISER, M., E. BUENO, C. SEKIRNJAK, M.E. MARTONE, H. BAKER, D. HILLMAN, S. CHEN, W. THORNHILL, M. ELLISMAN & B. RUDY. 1995. THe potassium channel subunit Kv3.1b is localized to somatic and axonal membranes of specific populations of CNS neurons. J. Neurosci. **15:** 4298.
67. SEKIRNJAK, C., M.E. MARTONE, M. WEISER, T. DEERINCK, E. BUENO, B. RUDY & M. ELLISMAN. 1997. Subcellular localization of the K^+ channel subunit Kv3.1b in selected rat CNS neurons. Brain Res. **766:** 173.
68. DU J., L. ZHANG, M. WEISER, B. RUDY & C.J. MCBAIN. 1996. Developmental expression and functional characterization of the potassium-channel subunit Kv3.1b in parvalbumin-containing interneurons of the rat hippocampus. J. Neurosci. **16:** 506.
69. LENZ, S., T.M. PERNEY, Y. QIN, E. ROBBINS & M.F. CHESSELET. 1994. GABA-ergic interneurons of the striatum express the Shaw-like potassium channel Kv3.1. Synapse **18:** 55.
70. PERNEY, T.M. & L.K. KACZMAREK. 1997. Localization of a high threshold potassium channel in the rat cochlear nucleus. J. Comp. Neurol. **386:** 178.
71. SHENG, M., Y.J. LIAO, Y.N. JAN & L.Y. JAN. 1993. Presynaptic A-current based on heteromultimeric K^+ channels detected *in vivo*. Nature **365:** 72.
72. WANG, H., D.D. KUNKEL, T.M. MARTIN, P.A. SCHWARTZKROIN & B.L. TEMPEL. 1993. Heteromultimeric K^+ channels in terminal and juxtaparanodal regions of neurons. Nature **365:** 75.
73. VEH, R. W., R. LICHTINGHAGEN, S. SEWING, F. WUNDER, I.M. GRUMBACH & O. PONGS. 1995. Immunohistochemical localization of five members of the Kv1 channel subunits: Contrasting subcellular locations and neuron-specific co-localizations in rat brain. Eur. J. Neurosci. **7:** 2189.
74. LAUBE, G., J. ROPER, J.C. PITT, S. SEWING, U. KISTNER, C.C. GARNER, O. PONGS & R.W. VEH. 1996. Ultrastructural localization of Shaker-related potassium channel subunits and synapse-associated protein 90 to septate-like junctions in rat cerebellar Pinceaux. Mol. Brain Res. **42:** 51.
75. BUENO, E., D.H.P. LAU, A. CHOW, S. CHEN, G. RAMEAU, C. SEKIRNJAK, M.E. MARTONE, M. ELLISMAN, D. HILLMAN, B. RUDY & W. THORNHILL. 1995. Developmental expression of Kv3.2, Kv3.1, and GIRK K^+ channel proteins in the mammalian CNS. Soc. Neurosci. Abst. **21:** 1329.
76. KENTROS, C. 1996. The expression of the Kv3.2 gene. Ph.D. thesis. New York University Medical School, New York.
77. VULLHORST, D., R. KLOCKE, J.W. BARTSCH & H. JOCKUSCH. 1998. Expression of the potassium channel Kv3.4 in mouse skeletal muscle parallels fiber type maturation and depends on excitation pattern. FEBS Lett. **421:** 259.
78. DELONG, M.R. 1971. Activity of pallidal neurons during movement. J. Neurophysiol. **34:** 414.
79. CHUDLER, E.H. & W.K. DONG. 1995. The role of the basal ganglia in nociception and pain. Pain **60:** 3.
80. HAUBER, W., S. LUTZ, & M. MUNKLE. 1998. The effects of globus pallidus lesions on dopamine-dependent motor behaviour in rats. Neuroscience **86:** 147.
81. HERNANDEZ-PINEDA, R., H.-C. A., H. MORENO, A. CHOW & B. RUDY. 1996. Identification of voltage-gated K^+ channels containing Kv3 subunits in neurons from the globus pallidus. Soc. Neurosci. Abst. **22:** 1754.

82. Stefani, A., P. Calabresi, N.B. Mercuri & G. Bernardi. 1992. A-current in rat globus pallidus: A whole-cell voltage clamp study on acutely dissociated neurons. Neurosci. Lett. **144:** 4.
83. Stefani, A., A. Pisani, A. Bonci, F. Stratta & G. Bernardi. 1995. Outward potassium currents activated by depolarization in rat globus pallidus. Synapse **20:** 131.
84. Wang, L.Y., L. Gan, I.D. Forsythe & L.K. Kaczmarek. 1998. Contribution of the Kv3.1 potassium channel to high-frequency firing in mouse auditory neurones. J. Physiol. (London) **509:** 183.
85. Massengill, J.L., M.A. Smith, D.I. Son & D.K. O'Dowd. 1997. Differential expression of K4-AP currents and Kv3.1 potassium channel transcripts in cortical neurons that develop distinct firing phenotypes. J. Neurosci. **17:** 3136.
86. Sillito, A.M. 1984. Functional consideration of the operation of GABAergic inhibitory process in the visual cortex. *In* Cerebral Cortex. Vol. 2, A.P. Jones, E.G., Ed.: 107. Plenum Press. New York and London.
87. Gilbert, C.D. 1993. Circuitry, architecture, and functional dynamics of visual cortex. Cereb. Cortex **3:** 373.
88. Jones, E.G. 1993. GABAergic neurons and their role in cortical plasticity in primates. Cereb. Cortex **3:** 361.
89. Steriade, M., A. Nunez & F. Amzica. 1993. A novel slow (1 Hz) oscillation of neocortical neurons in vivo: Depolarizing and hyperpolarizing components. J. Neurosci. **13:** 3252.
90. Jefferys, J.G., R.D. Traub & M.A. Whittington. 1996. Neuronal networks for induced '40 Hz' rhythms [see comments]. Trends Neurosci. **19:** 202.
91. Jacobs, K.M. & J.P. Donoghue. 1991. Reshaping the cortical motor map by unmasking latent intracortical connections. Science **251:** 944.
92. Steriade, M. 1997. Synchronized activities of coupled oscillators in the cerebral cortex and thalamus at different levels of vigilance [published erratum appears in Cereb. Cortex 1997 Dec; **7(8):** 779], Cereb. Cortex **7:** 583.
93. Erisir, A., D. Lau, B. Rudy & C.S. Leonard. 1998. Low TEA concentration disrupts high frequency firing of fast spiking cells in mouse somatosensory cortex. Soc. Neurosci. Abst. **24:** 632.
94. Erisir, A., D. Lau, B. Rudy & C.S. Leonard. 1999. Contrasting effects on high frequency firing of fast spiking cortical interneurons produced by differential K^+ channel blockade, J. Neuroscience. Submitted for publication.
95. Yang, W.P., P.C. Levesque, W.A. Little, M.L. Conder, P. Ramakrishnan, M.G. Neubauer & M.A. Blanar. 1998. Functional expression of two KvLQT1-related potassium channels responsible for an inherited idiopathic epilepsy. J. Biol. Chem. **273:** 19419.
96. Raleigh, R. 1907. On our perception of sound direction. Phylosophical Magazine **13:** 214–232.
97. Brew, H.M. & I.D. Forsythe. 1995. Two voltage-dependent K^+ conductances with complementary functions in postsynaptic integration at a central auditory synapse. J. Neurosci. **15:** 8011.
98. Wang, L.Y. & L.K. Kaczmarek. 1998. High-frequency firing helps replenish the readily releasable pool of synaptic vesicles. Nature **394:** 384.
99. Nambu, A. & R. Llinas. 1994. Electrophysiology of globus pallidus neurons in vitro. J. Neurophysiol. **72:** 1127.
100. Nambu, A. & R. Llinas. 1997. Morphology of globus pallidus neurons: its correlation with electrophysiology in guinea pig brain slices [published erratum appears in J. Comp. Neurol. 1997 Mar 31; **380(1):** 154], J. Comp. Neurol. **377:** 85.
101. Kita, H. & S.T. Kitai. 1991. Intracellular study of rat globus pallidus neurons: Membrane properties and responses to neostriatal, subthalamic and nigral stimulation. Brain Res. **564:** 296.
102. Klein, M., J. Camardo & E.R. Kandel. 1982. Serotonin modulates a specific potassium current in the sensory neurons that show presynaptic facilitation in Aplysia. Proc. Natl. Acad. Sci. USA **79:** 5713.
103. Augustine, G.J. 1990. Regulation of transmitter release at the squid giant synapse by presynaptic delayed rectifier potassium current. J. Physiol. (London) **431:** 343.
104. Jackson, M.B., A. Konnerth & G.J. Augustine. 1991. Action potential broadening and frequency-dependent facilitation of calcium signals in pituitary nerve terminals. Proc. Natl. Acad. Sci. USA **88:** 380.

105. ROBERTS, W.M., R.A. JACOBS & A.J. HUDSPETH. 1990. Colocalization of ion channels involved in frequency selectivity and synaptic transmission at presynaptic active zones of hair cells. J. Neurosci. **10:** 3664.
106. ANDERSON, A.J. & A.L. HARVEY. 1988. Effects of the facilitatory compounds catechol, guanidine, noradrenaline and phencyclidine on presynaptic currents of mouse motor nerve terminals. Naunyn Schmiedebergs Arch. Pharmacol. **338:** 133.
107. VAUGHAN, C.W., S.L. INGRAM, M.A. CONNOR & M.J. CHRISTIE. 1997. How opioids inhibit GABA-mediated neurotransmission [see comments]. Nature **390:** 611.
108. COLMERS, W.F., K. LUKOWIAK & Q.J. PITTMAN. 1988. Neuropeptide Y action in the rat hippocampal slice: Site and mechanism of presynaptic inhibition. J. Neurosci. **8:** 3827.
109. SOUTHAN, A.P. & D.G. OWEN. 1997. The contrasting effects of dendrotoxins and other potassium channel blockers in the CA1 and dentate gyrus regions of rat hippocampal slices. Br. J. Pharmacol. **122:** 335.
110. HARVEY, A.L. & A.J. ANDERSON. 1985. Dendrotoxins: Snake toxins that block potassium channels and facilitate neurotransmitter release. Pharmacol. Ther. **31:** 33.
111. ROBITAILLE, R. & M.P. CHARLTON. 1992. Presynaptic calcium signals and transmitter release are modulated by calcium-activated potassium channels. J. Neurosci. **12:** 297.
112. WHEELER, D.B., A. RANDALL & R.W. TSIEN. 1996. Changes in action potential duration alter reliance of excitatory synaptic transmission on multiple types of Ca^{2+} channels in rat hippocampus. J. Neurosci. **16:** 2226.
113. DEBANNE, D., N.C. GUERINEAU, B.H. GAHWILER & S.M. THOMPSON. 1997. Action-potential propagation gated by an axonal I(A)-like K^+ conductance in hippocampus [published erratum appears in Nature 1997 Dec 4; **390:** 536]. Nature **389:** 286.
114. PONCE, A., E. VEGA-SAENZ DE MIERA, C. KENTROS, H. MORENO, B. THORNHILL & B. RUDY. 1997. K^+ channel subunit isoforms with divergent carboxy-terminal sequences carry distinct membrane targeting signals. J. Membr. Biol. **159:** 149.
115. LAU, D., M. CASTRO-ALAMANCOS, A. CHOW, A. OZAITA, E. VEGA-SAENZ DE MEIRA, S. MATHEW, J. GIBSON, B. CONNORS & B. RUDY. 1998. Targeted disruption in mouse of a voltage-gated potassium channel gene that is expressed predominantly in the thalamocortical system. J. Physiol. **511P:** 147.
116. LAU, D., M. CASTRO-ALAMANCOS, A. CHOW, A. OZAITA, E. VEGA-SAENZ DE MIERA, S. MATHEW, J. GIBSON, B.W. CONNORS & B. RUDY. 1998. Targeted disruption of a K^+ channel gene that is principally expressed in the presynaptic terminals of thalamic relay neurons in mice. Soc. Neurosci. Abst. **24:** 128.
117. HO, C.S., R.W. GRANE & R.H. JOHO. 1997. Pleiotropic effects of a disrupted K^+ channel gene: Reduced body weight, impaired motor skill and muscle contraction, but no seizures. Proc. Natl. Acad. Sci. USA **94:** 1533.
118. CHAN, E. 1997. Regulation and function of Kv3.3. Ph.D. thesis. Rockefeller University, New York.
119. JOHO, R.H., C.S. HO & G.A. MARKS. 1998. Increase in gamma oscillations and altered learning performance in a mouse deficient for the K^+ channel Kv3.1. Soc. Neurosci. **24:** 828.

Functional and Molecular Aspects of Voltage-Gated K$^+$ Channel β Subunits

OLAF PONGS,[a,b] THORSTEN LEICHER,[b] MICHAELA BERGER,[b]
JOCHEN ROEPER,[b] ROBERT BÄHRING,[b] DENNIS WRAY,[c] KARL PETER GIESE,[d]
ALCINO J. SILVA,[e] AND JOHAN F. STORM[f]

[b]*Institut für Neurale Signalverarbeitung, Center for Molecular Neurobiology, Martinistrasse 52, 20246 Hamburg, Germany*

[c]*University of Leeds, Department of Pharmacology, Leeds LS2 9JT, Great Britain*

[d]*Cold Spring Harbor Laboratory, Cold Spring Harbor, NY 11724, USA*

[e]*KPG, Department of Anatomy and Developmental Biology, University College London, Gower Street, London, WC1E 6BT, Great Britain*

[f]*Institute of Neurophysiology, P.O. Box 1104, Blindern, Oslo N 0317, Norway*

ABSTRACT: Voltage-gated potassium channels (Kv) of the *Shaker*-related superfamily are assembled from membrane-integrated α subunits and auxiliary β subunits. The β subunits may increase Kv channel surface expression and/or confer A-type behavior to noninactivating Kv channels in heterologous expression systems. The interaction of Kvα and Kvβ subunits depends on the presence or absence of several domains including the amino-terminal N-type inactivating and NIP domains and the Kvα and Kvβ binding domains. Loss of function of Kvβ1.1 subunits leads to a reduction of A-type Kv channel activity in hippocampal and striatal neurons of knock-out mice. This reduction may be correlated with altered cognition and motor control in the knock-out mice.

The activity of voltage-gated potassium (Kv) channels contributes to nervous excitability. Kv channels may be involved in determining the resting potential of cell membranes, in controlling thresholds of excitation in modulating wave forms and frequencies of action potentials, and in repolarizing depolarized membranes, for example.[1] Two main types of Kv channels have been described: (i) slowly inactivating, delayed rectifier–type Kv channels and (ii) rapidly inactivating (A-type) Kv channels. The former can make important contributions to action potential repolarization and to attenuation of cell excitability. The latter frequently operate in the subthreshold range of action potentials and may influence firing thresholds and frequencies.[2] Diverse A-type Kv channel subunits are expressed in neuronal pre- and postsynaptic compartments.[3–5] This may indicate that A-type Kv channels may have specialized functions for encoding pre- and postsynaptic nervous signals, which eventually contribute to learning, memory, and behavioral traits.[6,7]

Shaker-related Kv channels are assembled as tetramers from ion-conducting Kvα subunits, which are integral membrane proteins, and auxiliary Kvβ subunits.[8,9] It has been suggested that homo- and heteromultimeric assembly of α-subunits may contribute to the diversity of Kv channels found in the mammalian nervous system.[10,11] In addition, the coassembly with Kvβ subunits may also contribute to Kv channel diversity, particularly in conferring A-type behavior to Kv channels that otherwise behave like delayed rectifiers. A

[a]Corresponding author. e-mail: pointuri@uke.uni-hamburg.de

hallmark of Kv channel diversity is the wide range of time courses for inactivation that is observed *in vivo* and *in vitro* in heterologous expression systems. Two principal inactivation mechanisms have been recognized[8]: C-type inactivation correlated with carboxy-terminal Kvα-subunit structures,[12] and N-type inactivation conferred by "ball" domains in the amino termini of certain Kvα and Kvβ subunits.[13] Possibly, assembly of heteromultimers with one or more Kvα and/or Kvβ ball domains comprises one important principle in the generation of A-type Kv channel diversity.

It has been shown that biochemically purified Kv channels represent heterooligomers that contain Kvα and Kvβ subunits in a 1:1 ratio.[14,15] Cloning and analysis of the Kvβ subunit sequences suggested that they lack typical membrane-spanning segments and that potential N-glycosylation sites are absent, consistent with the lack of evidence for attached carbohydrate on the native protein.[15] The structural properties suggested that Kvβ subunits are peripheral proteins that may associate with a cytoplasmic domain of Kvα subunits. The notion was subsequently confirmed by delineating this domain to a particular sequence of the cytoplasmic amino-terminus of Kv1α subunits.[16,17] Here we describe the characterization of further domains on Kvα and Kvβ subunits which may be involved in the formation of A-type Kv channels in heterologous *in vitro* expression systems. Also, we show that Kvβ subunits contribute to A-type Kv channel activity *in vivo*.

STRUCTURE OF Kvβ GENES

The superfamily of *Shaker*-related Kvα subunits comprises to date nine subfamilies, Kv1 to Kv9,[18] while each subfamily may consist of several (three to nine) members. Altogether, ~30 distinct genes are known presently to encode *Shaker*-type Kvα subunits in the mammalian genome. In contrast, only three Kvβ genes have been cloned so far.[9,19] As indicated in FIGURE 1, the Kvβ1.1, Kvβ1.2, and Kvβ1.3 polypeptides have alternative amino-termini varying in length from 73 (Kvβ1.1) to 91 (Kvβ1.3) amino acids and a common core sequence of 328 amino acids.[20] This core sequence is highly conserved between Kvβ1, Kvβ2, and Kvβ3 type subunits. The Kvβ1 core sequence is ~85% identical to the corresponding Kvβ2 and Kvβ3 core sequences. The expression of Kvβ2 and Kvβ3 genes may also give rise to splice variants—that is, to Kvβ polypeptides with variant amino-termini.[19] Recently, a Kvβ4 subunit has been described in mouse.[21] Sequence comparison with rat and human Kvβ3 gene and cDNA sequences revealed that Kvβ4 is in fact a Kvβ3 splice variant and, therefore, has been renamed Kvβ3.2 (FIGURE 1).

COEXPRESSION OF Kvα AND Kvβ SUBUNITS

It has been shown that Kvβ1 and Kvβ2 subunits bind to Kv1α subunits.[16,17] The binding is apparently specific for Kv1α subunits. They contain in the amino-terminus a conserved sequence motif that does not occur in other Kvα subfamilies. It has been shown that this sequence motif is important for both biochemical and functional interaction of Kv1.5 and Kvβ1 subunits.[17] Therefore, it was proposed that the sequence motif represents a cytoplasmic amino-terminal domain of Kv1α subunits for specifically binding Kvβ1 (and Kvβ2) subunits. The binding of Kvβ subunits to Kv1α subunits may have several functional consequences. It may increase the cell surface expression of Kvα subunits.[22,23]

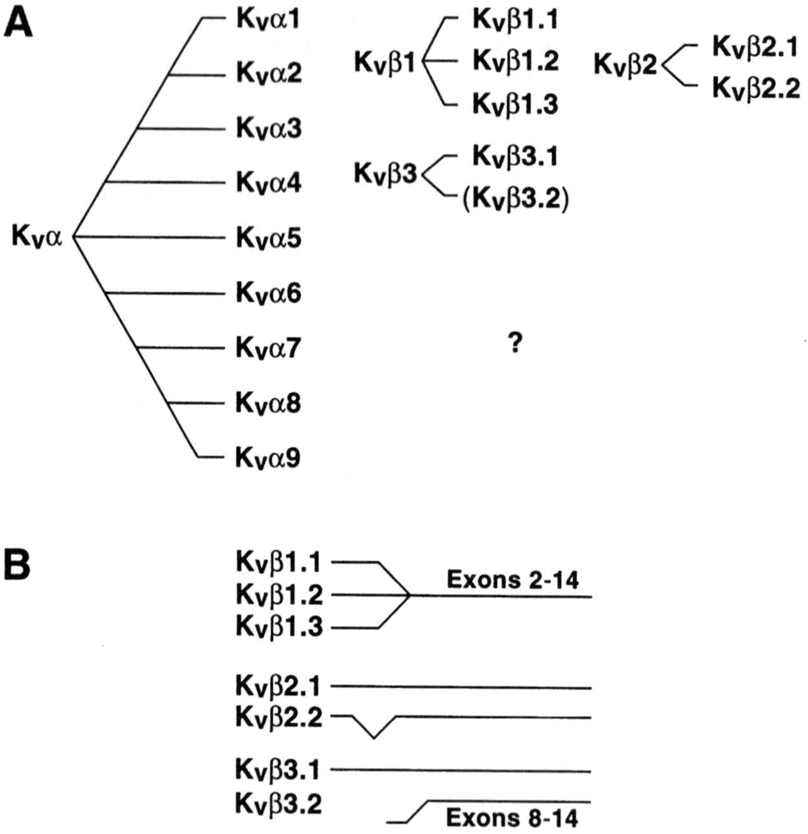

FIGURE 1. **A.** The *Shaker*-related voltage-gated K (Kv) channels are assembled from membrane-integrated ion-conducting Kvα subunits[8,18] and auxiliary Kvβ subunits.[13] The Kvα superfamily may be divided into subfamilies as indicated. Each subfamily consists of several members. The Kvβ family consists of three genes, giving rise to various splice variants.[19,20] The *question mark* indicates that it is not known whether all Kvα subunits are assembled in heteromultimeric Kv channels with Kvβ subunits. **B.** Exon compositions found in Kvβ cDNAs.[19–21] Kvβ1.1 to Kvβ31 share a conserved region defined by exons 2 to 14 and differ in their 5′ exons—i.e., the amino termini. Mouse Kvβ3.2 does not have Kvβ core exons 2 to 7, only exons 8–14 and a variant 5′ exon.[19,21]

In addition, one may observe a shift in the voltage dependence of activation of Kvα channels.[24] The shift usually occurs in the hyperpolarized direction. An important difference between Kvβ1 and Kvβ2 subunits is that the former contain an inactivating ball domain.[24] Thus, the coexpression of Kvβ1 subunits with certain Kv1α subunits may also confer rapid inactivation to the Kv channel.[13]

Previously, we have shown that the amino-terminus of Kvβ3 subunits may function as an inactivating ball domain like the Kvβ1.1 amino terminus. Yet, the coexpression of

Kv1.5 and Kvβ3.1 subunits in the *Xenopus* oocyte expression system did not yield Kv channels mediating rapidly inactivating currents unlike the coexpression of Kv1.5 and Kvβ1.1 subunits (FIG. 2). We investigated whether a particular domain in the Kv3.1 polypeptide sequence was responsible for the dysfunction of Kvβ3 subunits in the *Xenopus* oocyte expression system. For this purpose, we have constructed chimeras of Kvβ1.1 and Kvβ3.1 (FIG. 2). Kvβ31M is a chimeric polypeptide in which the first 229 amino acids stem from the Kvβ3.1 polypeptide sequence and the last 179 amino acids from the Kvβ1.1 polypeptide sequence. Coexpression of Kv1.5 and Kvβ31M subunits produced in *Xenopus*

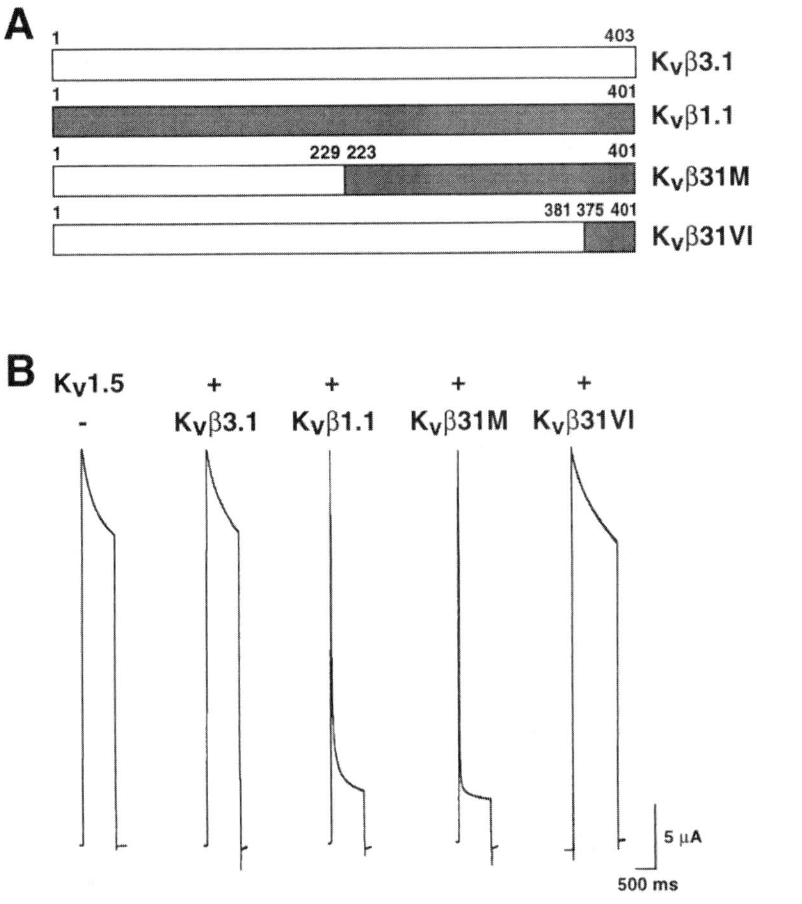

FIGURE 2. A. Diagram of Kvβ constructs used in coexpression studies with Kv1.5α subunits. Numbers on top refer to the first and last amino acid residues of the derived human Kvβ1.1[20] and Kvβ3.1[19] polypeptide sequences present in each construct. *Empty boxes* represent Kvβ31 wild-type sequences and *filled boxes* Kvβ1.1 wild-type sequences. **B.** Outward currents elicited in *Xenopus* oocytes two days after injection of Kv1.5 ± Kvβ mRNAs as indicated on top of each trace. Currents were measured in the two-electrode voltage-clamp configuration as described.[24]

oocytes Kv channels mediating rapidly inactivating potassium outward currents (τ_i = 45 ± 5 ms, n = 6). The time constants of inactivation were similar to the ones of inactivating currents mediated by Kv1.5/Kvβ1.1 channels (τ_i = 25 ± 4 ms, n = 6). Then we constructed a chimeric Kvβ31 polypeptide VI, in which the first 381 amino acids corresponded to those of Kvβ31 and the last 27 amino acids to those of Kvβ1.1. Coexpression of Kv1.5 and Kvβ31 VI subunits produced in *Xenopus* oocytes Kv channels which did not rapidly inactivate, (FIG. 2), but mediated currents like Kv1.5 in the absence of Kvβ subunits or Kv1.5 in the presence of Kvβ3.1. The results indicated that a Kvβ3.1 sequence stretch defined by amino acids 229 to 381 was responsible for the loss of Kvβ3.1 function. Replacement of this sequence by the equivalent Kvβ1.1 function resulted in a gain of function—that is, in the generation of A-type Kv channels. This relatively large Kvβ polypeptide region could be narrowed down in further experiments. As shown in FIGURE 3, a replacement of 27 amino acids of the Kvβ3 sequence (residues 354 to 381) by the equivalent Kvβ1.1 sequence (residues 347 to 374) was sufficient to generate Kvβ3.1 subunits (Kvβ31 VII) capable of conferring rapid inactivation to Kv1.5 subunits (τ_i = 30 ± 4 ms, n = 6). A sequence comparison showed that the respective Kvβ1.1 and Kvβ3.1 polypeptide sequences only differ by eight amino acid residues. Some or all of these residues should be responsible for the dysfunction of Kvβ3.1 subunits in the *Xenopus* oocyte expression system.

Combining our present understanding of Kvβ subunits and their possible role in conferring N-type inactivation, three possibilities may be considered for an explanation of Kvβ3.1 dysfunction. First, Kvβ3.1 subunits may not bind to Kv1.5 subunits. Second, the inactivating ball domain may not occlude the open Kv channel because it does not bind to its temporary docking site (the ball receptor) at the cytoplasmic side of the Kv1.5 pore.[25,26] Third, the Kvβ3.1 subunit may contain an inhibitory domain like the N-type inactivation-prevention (NIP) domain, recently discovered in Kv1.6 subunits.[27] Yet another alternative may be that Kvβ3.1 subunits require for activity some additional factor reminiscent of the KChAP-auxiliary proteins.[28] For example, Kv1.5 and mouse Kvβ3.2 subunits have been expressed in insect SF9 cells.[21] The possible association of Kv1.5 and mouse Kvβ3.2 was tested by immunoprecipitation from cells infected with constructs expressing the two subunits. Anti-Kv1.5 antibodies coprecipitated Kv1.5 and mouse Kvβ3.2 subunits. In control experiments, mouse Kvβ3.2 could not be precipitated by anti-Kv1.5 antibodies in the absence of Kv1.5. The data suggested the formation of Kv1.5-Kvβ3.2 complexes. As Kvβ3.2 is a splice variant of Kvβ3.1, it may be assumed that Kvβ3.1 is also able to bind to Kv1.5. In preliminary experiments, we have shown that Kvβ3.1 may, indeed, also bind to Kv1.5 (O. Pongs, unpublished experiment). The activity of the inactivating ball domain of Kvβ3.1 is evident from the experiments with chimeric Kvβ3/Kvβ1 constructs shown in FIGURES 2 and 3. Also, transfer of the Kvβ3.1 inactivating domain to Kvβ2 showed that the Kvβ3.1 inactivating domain can behave like the comparable Kvβ1.1 inactivating domain.[24] The data suggested that it binds to the Kv1α ball receptor and thereby induces the observed rapid Kv channel inactivation. In conclusion, these data suggest that Kvβ3.1 has the potential to bind to Kv1.5 subunits and contains a functional N-type inactivating domain.

Previously, we have reported the discovery of a NIP-domain in Kv1.6 subunits.[27] The NIP domain prevents inactivating ball domains to bind to the ball receptor by an as-yet-unknown mechanism.[27] Apparently, the NIP domain dominates the gating phenotype of *Shaker*-type Kv channels and renders the N-type inactivating ball domains dysfunctional. Heteromultimeric Kv channels containing ball and NIP domains mediate noninactivating delayed-rectifier type currents because of the dominant-negative activity of NIP domains.

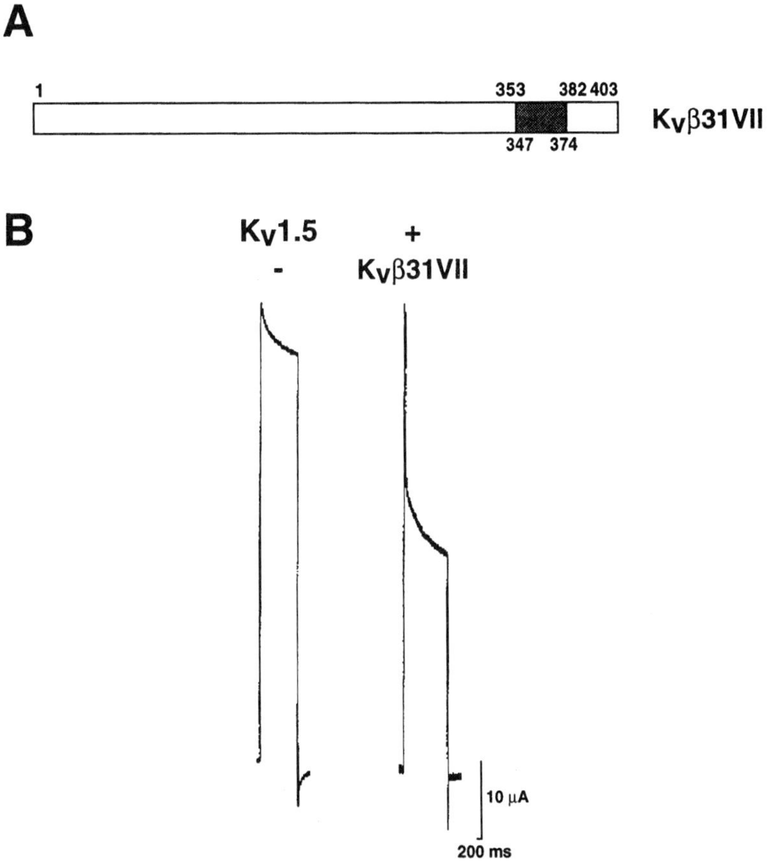

FIGURE 3. **A.** Diagram of Kvβ31VII construct used in coexpression study with Kv1.5α subunits. Numbers on top refer to Kvβ3.1 amino acid residues; numbers at bottom to the ones of Kvβ1.1. *Shaded area* indicates Kvβ1.1 sequence inserted into Kvβ31. **B.** Outward currents elicited in *Xenopus* oocytes two days after injection of Kv1.5 ± Kvβ31VII mRNAs as indicated on top of the traces. Currents were measured in the two-electrode voltage-clamp configuration as described.[24]

The NIP domain may consist of only a few amino acids.[27] Mutations in the NIP domain lead to loss of function, and its transfer to another Kv1α subunit leads to gain of function. Comparably, the Kvβ3.1 C-terminal domain, which is responsible for Kvβ3.1 dysfunction (see FIG. 3), may prevent the Kvβ3.1 inactivating domain from functioning. Alternatively, Kvβ3.1 may require, because of the C-terminal domain, an additional protein for function. In this context, a recent report[28] on cloning and expression of a regulatory K channel–associated protein (KChAP) is interesting. KChAP may bind to Kvα subunits like Kv1.5 as well as to the Kvβ C-terminus. Its coexpression with Kv2.1 and Kv2.2 may increase Kv channel surface expression. Although the precise function of KChAP is not known, we speculate that proteins like KChAP may aid Kvβ3.1 in generating A-type Kv channel gating behavior in heteromultimeric Kv channels. In summary, the structure-function analyses of Kvα/Kvβ heteromultimers in heterologous expression systems have indicated a complex set of domains regulating the biochemical and functional interaction of Kvα and Kvβ subunits (FIG. 4). In addition to the interaction domains, the presence of ball domains, NIP-domains, and ones like the Kvβ3.1 C-terminal domain described in this paper may affect the gating behavior of Kv channels.

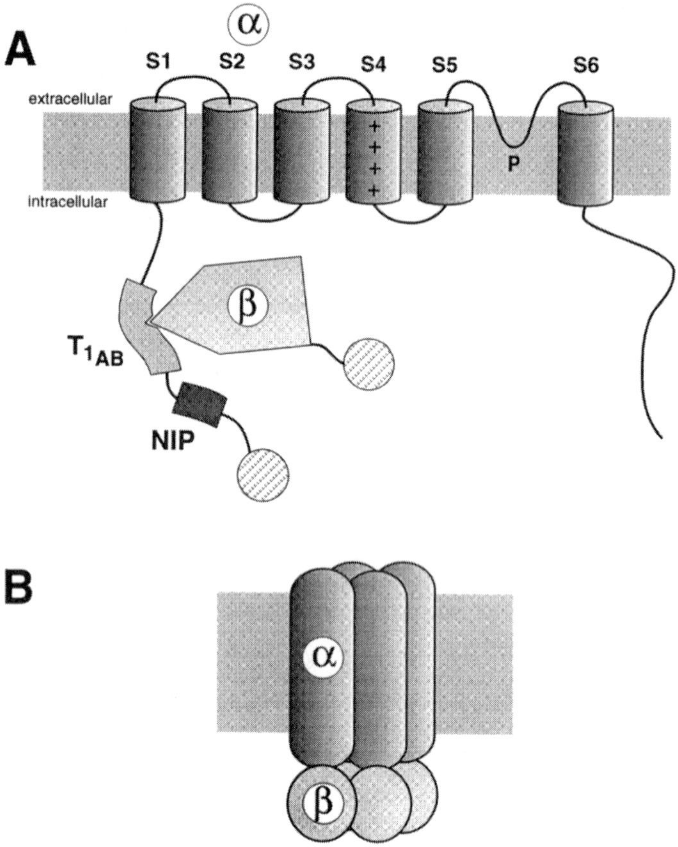

FIGURE 4. **A.** Diagram of location of domains on amino terminus of *Shaker*-related Kvα subunit and Kvβ subunit important for the gating behavior of heteromultimeric Kvα/Kvβ channels. The topological model of Kvα subunit[8] indicates membrane-spanning domains S1–S6 (*cylinders*). Membrane is *shaded*. A/P loop contributes to pore formation and S4 to voltage sensing (+). *Gray ribbon* corresponds to $T1_{AB}$ domain[34] implicated in the assembly of *Shaker* α subunits. *Black ribbon* indicates NIP domain of Kv1.6α subunits,[27] which prevents N-type inactivation. *Striped circles* illustrate amino-terminal localization of inactivation "ball" domains[25] in certain Kvα and Kvβ subunits. *Pointed shape* denotes location of Kvβ (β) interaction site. **B.** Tetrameric assembly of Kvα subunits as integral membrane proteins. They may be associated with cytoplasmic Kvβ subunits in a 1:1 stoichiometry.[15]

CONSEQUENCES OF Kvβ1.1 LOSS OF FUNCTION *IN VIVO*

To test the possible role of Kvβ1.1 subunits for the assembly of A-type Kv channels *in vivo*, we decided to target the Kvβ1.1 gene in mice and to produce Kvβ1.1 knock-out mice. The construction of the Kvβ1.1 targeting construct and the generation of homozygous Kvβ1.1 knock-out mice have been described elsewhere.[29] The loss of Kvβ1.1 protein did not markedly affect the expression of other Kvβ proteins in mouse brain nor that of Kv1α subunits that might interact with Kvβ1.1. Also, gross alterations in adult brain morphology were not apparent at the light microscope level. In the CNS of wild-type mice, Kvβ1.1 mRNA is prominently expressed in neurons of hippocampus and corpus striatum.[30] Therefore, neurons in these two areas of mouse brain might be expected to express Kvβ1.1 containing heteromultimeric Kv channels. Acute hippocampal slice preparations from adult wild-type and Kvβ1.1 knock-out mice were prepared, and somata of CA1 pyramidal neurons were patch clamped for investigating A-type Kv channel activity in these neurons. Typically the neurons expressed upon depolarization under voltage clamp rapidly inactivating outward currents (I_A) together with noninactivating sustained currents (I_{SO}) (FIG. 5A). Inactivation time constants (τ_i) for I_A in wild-type were like the ones in Kvβ1.1 knock-out mice (26 ± 3 ms, n = 6 mice, 14 cells; 27 ± 4 ms, n = 6 mice, 13 cells). However, the amplitudes of the elicited I_A currents in the CA1 pyramidal neurons of the knock-out mice were significantly smaller than the ones recorded from wild-type controls (FIG. 5C). Accordingly, the ratio of I_A and I_{SO} was smaller in the CA1 pyramidal neurons (0.36 ± 0.07) than the ones in wild type control littermates (0.61 ± 0.03; $p < 0.05$) (FIG. 5C). The results showed that the loss of Kvβ1.1 expression in mice was accompanied by a reduction of A-type Kv channel activity in CA1 pyramidal neurons. At present, we cannot decide whether the observed I_A amplitude reduction is due to a reduction of the cell surface expression of hippocampal A-type channels or to the loss of Kvβ1.1 containing heteromultimeric Kv channels or both. The results, do, however, clearly show that some A-type Kv channel activity depends on Kvβ1.1 expression in mouse hippocampus.

If A-type Kv channels contribute to action potential repolarization, it might be expected that repetitive firing of action potentials may induce an accumulation of inactivated A-type Kv channels and may concomitantly reduce the neuronal repolarizing capacity of action potentials. This situation may lead in spike trains to a frequency-dependent spike broadening of action potentials relative to the first spike. In acute hippocampal slice preparations of adult mouse brains, spike trains consisting of five spikes were elicited by current injection in CA1 pyramidal neurons of Kvβ1.1 knock-out mice and wild-type control littermates using sharp microelectrodes (FIG. 5B). Bursts 50 ms long consisting of five action potentials were compared. This comparison revealed that the fourth spike showed 38.6 ± 4.8% broadening in wild-type controls and 23.1 ± 3.9% broadening in the Kvβ1.1 knock-out mice (FIG. 5C). Although it is unlikely that there is a simple correlation between I_A amplitude and broadening in spike trains, it is interesting to note similar relative reductions of I_A amplitude and spike broadening in Kvβ1.1 knock-out mice. It is consistent with the notion that an accumulation of inactivated A-type Kv channels may contribute to spike broadening observed in the CA1 pyramidal neurons. The termination of spike trains in neurons may be followed by hypolarization of the membrane, the so-called afterhyperpolarization (AHP).[1] It has been shown that the slow AHP component is mediated in hippocampal CA1 pyramidal neurons by Ca^{2+}-activated K channels.[31,32] The activity of these K channels critically depends on intracellular Ca^{2+} concentrations. Somewhat unexpectedly,

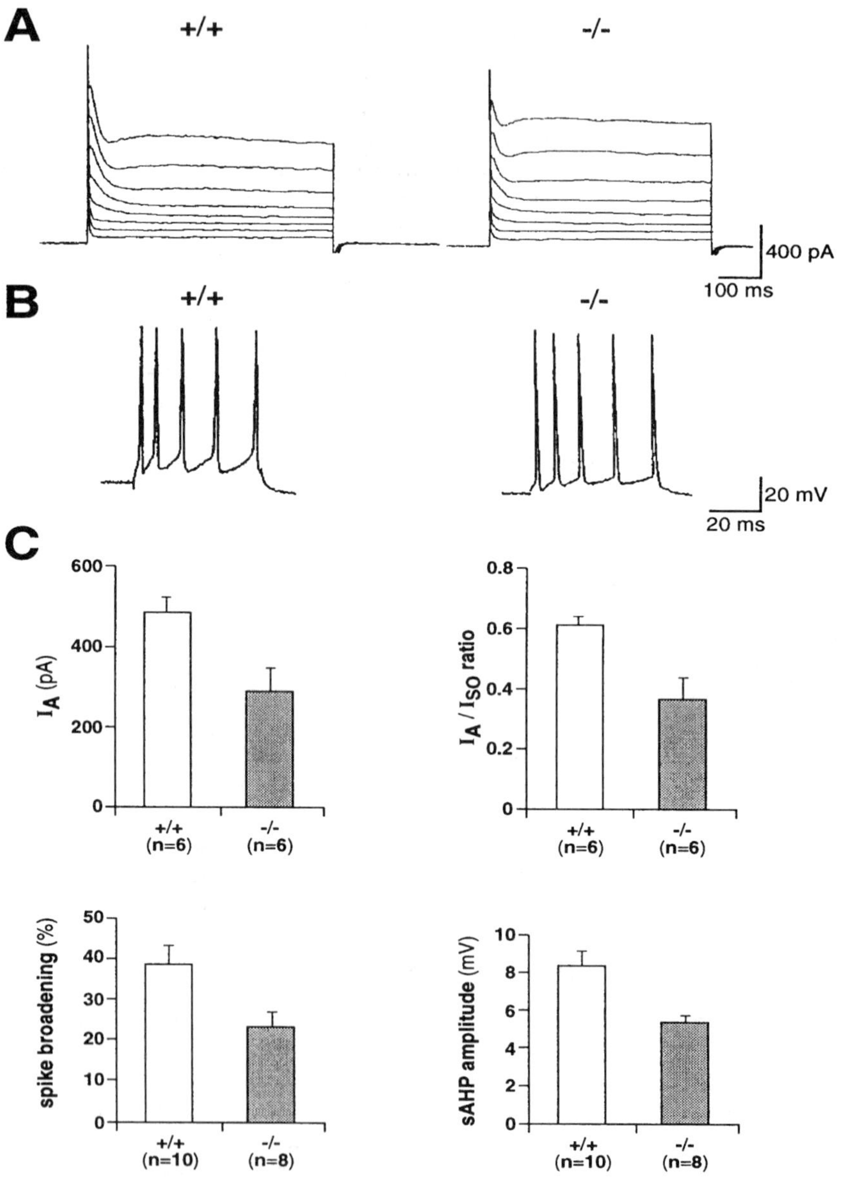
A
+/+
-/-
400 pA
100 ms
B
+/+
-/-
20 mV
20 ms
C
600
400
200
0
I_A (pA)
+/+
(n=6)
-/-
(n=6)
0.8
0.6
0.4
0.2
0
I_A / I_{SO} ratio
+/+
(n=6)
-/-
(n=6)
50
40
30
20
10
0
spike broadening (%)
+/+
(n=10)
-/-
(n=8)
10
8
6
4
2
0
sAHP amplitude (mV)
+/+
(n=10)
-/-
(n=8)

FIGURE 5. **A.** Outward currents elicited by 500-ms–long voltage commands (–70 to –10mV, in 10-mV increments) from a holding potential of –80mV in CA1 pyramidal neurons in acute brain slices of adult wild-type mice (+/+) and homozygous Kvβ1 knock-out mice (–/–). **B.** Representative spike trains in wild type (+/+) and mutant (–/–) mice elicited once every 10 s by injecting 50-ms-long depolarizing current pulses. **C.** Bar diagrams illustrate means ± SEM. *Empty bars* correspond to wild-type (+/+) and *filled bars* to mutant mice (–/–). A-type current amplitudes (I_A) were estimated from the peak current, which inactivated within 100 ms. The sustained current amplitudes (I_{SO}) were measured at the end of the 500–ms steps. Spike duration of fourth spikes was normalized with respect to the first-spike duration and expressed as spike broadening. SAHP amplitudes following 50-ms-long five-spike bursts are expressed in mV. Numbers below bars indicate the numbers of mice used in the experiments. On average 12 to 17 cells ($p < 0.05$) were recorded from. Original data are from Ref. 29.

we observed that the amplitudes of slow AHPs following a spike train were reduced in the Kvβ1.1 knock-out mice in comparison to wild-type controls (Fig. 5C). Apparently, the reduced spike broadening in Kvβ1.1 knock-out mice brings about a reduced activation of Ca^{2+}-activated K channels. Reduced action potential durations may lead to a reduced the influx of Ca^{2+} during action potential firing.[33] Thus, the intracellular rise in Ca^{2+}-activated K channels may be reduced in the Kvβ1.1-deficient mice. Alternatively, the loss of Kvβ1.1 expression may have caused a reduction in Ca^{2+}-activated K channel surface expression comparable to the observed reduction of A-type channel activity.

CONCLUSIONS

Shaker-related Kv channels are assembled from ion-conducting Kvα subunits, which are integral membrane proteins, and auxiliary Kvβ subunits. This leads to the formation of highly diverse heteromultimeric Kv channels that mediate outward currents with a wide spectrum of time courses for inactivation. The currents range from rapidly inactivating A-type (I_A) to noninactivating (sustained) delayed-rectifier type currents (I_{SO}). Rapid N-type inactivation of A-type Kv channels may be conferred by "ball" domains in the amino termini of certain Kvα subunits and Kvβ1 and Kvβ3 subunits. The presence of the "ball" domains appears to be important for the generation of A-type Kv channel diversity. Ball domain activity may be suppressed by other Kv subunit domains that prevent N-type inactivation in a dominant-negative manner. In adult mice, loss of Kvβ1.1 expression is correlated with a reduction of I_A amplitudes, spike broadening, and sAHP amplitudes in hippocampal CA1 pyramidal neurons.

ACKNOWLEDGMENT

This work was supported by the Deutsche Forschungsgemeinschaft.

REFERENCES

1. Hille, B. 1992. Ionic Channels of Excitable Membranes. Sinauer Associates. Sunderland, MA.
2. Connor, J.A. & C.F. Stevens. 1970. Voltage clamp studies of a transient outward membrane current in gastropod neural somata. J. Physiol. (Lond.) **213:** 21–30.

3. SHENG M., M.L. TSAUR, Y.N. JAN & L.Y. JAN. 1992. Subcellular segregation of two A-type K^+ channel proteins in rat central nervous system. Neuron **9:** 271–284.
4. VEH, R.W., R. LICHTINGHAGEN, S. SEWING, F. WUNDER, I.M. GRUMBACH & O. PONGS. 1995. Immunohistochemical localization of five members of the Kv1 channel subunits: Contrasting subcellular locations and neuron-specific colocalisations in rat brain. Eur. J. Neurosci. **7:** 2189–2205.
5. PONCE, A., E. VEGA-SAENZ DE MIERA, C. KENTROS, H. MORENO, B. THORNHILL & B. RUDY. 1997. K^+ channel subunit isoforms with divergent carboxy-terminal sequences carry distinct membrane targeting signals. J. Memb. Biol. **159:** 149–159.
6. ALKON, D.L. 1984. Calcium-mediated reduction of ionic currents: A biophysical memory trace. Science **30:** 1037–1045.
7. DUBNAU, J. & T. TULLY. 1998. Gene discovery in drosophila: new insights for learning and memory. Annu. Rev. Neurosci. **21:** 407–444.
8. JAN, L.Y. & Y.N. JAN. 1997. Cloned potassium from eukaryotes and prokaryotes. Annu. Rev. Neurosci. **20:** 21–123.
9. PONGS, O. 1995. Regulation of the activity of voltage-gated potassium channels by β-subunits. Neurosciences **7:** 137–146.
10. ISACOFF, E.Y., Y.N. JAN & L.Y. JAN. 1990. Evidence for the formation of heteromultimeric potassium channels in Xenopus oocytes. Nature **345:** 535–537.
11. RUPPERSBERG, J.P., K.H. SCHRÖTER, B. SAKMANN, M. STOCKER, S. SEWING & O. PONGS. 1990. Heteromultimeric channels formed by rat brain potassium-channel protein. Nature **345:** 535–537.
12. CHOI, K.L, R.W. ALDRICH & G. YELLEN. 1991. Tetraethyoammonium blockade distinguishes two inactivation mechanisms in voltage-gated K^+ channels. Proc. Nat. Acad. Sci. USA **88:** 5092–5095.
13. RETTIG, R., R.H. HEINEMANN, F. WUNDER, C. LORRA, D.N. PARCEJ, J.O. DOLLY & O. PONGS. 1994. Inactivation properties of voltage-gated K^+ channels altered by presence of β-subunit. Nature **369:** 289–294.
14. REHM, H. & M. LAZDUNSKI. 1998. Purification and subunit structure of a putative K^+ channel protein identified by its binding properties for dendrotoxin I. Proc. Natl. Acad. Sci. USA **85:** 4919–4923.
15. PARCEJ, DN. & J.O. DOLLY. 1989. Dendrotoxin receptor from bovine synaptic plasma membranes. Binding properties, purification and subunit composition of a putative constituent of certain voltage-activated K^+ channels. Biochem. J. **257:** 899–903.
16. YU, W., J. XU & M. LI. 1996. NAB domain is essential for the subunit assembly of both α-α and α-β complexes of *Shaker*-like potassium channels. Neuron **16:** 441–453.
17. SEWING, S., J. ROEPER & O. PONGS. 1996. Kvβ1 subunit binding specific for Shaker-related potassium channel α-subunits. Neuron **16:** 455–463.
18. CHANDY, K.G. & G.A. GUTMAN. 1995. CRC Handbook of Receptors and Channels, Vol. 2: 1–75. CRC Press. Boca Raton, FL.
19. LEICHER, T. & O. PONGS. 1998. Kvβ1, Kvβ2 and Kvβ3 subunits of human voltage-gated potassium channels encoded in genes with similar exon/intron structure, but greatly varying size. In preparation.
20. LEICHER, T., J. ROEPER, K. WEBER, X. WANG & O. PONGS. 1996. Structural and functional characterization of human potassium channel subunits β1 (KCNA1B). Neuropharmacol. **35:** 787–795.
21. FINK, M., F. DUPRAT, F. LESAG, C. HEURTEAUX, G. ROMEY, J. BARHANIN & M. LAZDUNSKI. 1996. A new K^+ channel β subunit to specifically enhance Kv2.2 (CDRK) expression. J. Biol. Chem. **271:** 26341–26348.
22. SHI, G., K. NAKAHIRA, H. SCOTT, K.J. RHODES, L.E. SCHECHTER & J.S. TRIMMER. 1996. β Subunits promote K^+ channel surface expression through effects early in biosynthesis. Neuron **16:** 843–852.
23 CHOUINARD, S.W., G.F. WILSON, A.K. SCHLIMGERN & B. GANETZKY. 1995. A potassium channel β subunit related to the aldo-keto reductase superfamily is encoded by the Drosophila hyperkinetic locus. Proc. Natl. Acad. Sci. USA **92:** 6763–6767.
24. HEINEMANN, S.H., J. RETTIG, H.R. GRAACK & O. PONGS. 1996. Functional characterization of Kvβ-subunits from rat brain. J. Physiol. (Lond.) **493:** 625–633.

25. ZAGOTTA, W., T. HOSHI & R.W. ALDRICH. 1990. Restoration of inactivation in mutants of *Shaker* potassium channels by a peptide derived from *Sh*B. Science **250:** 568–571.
26. ISACOFF, E.Y., Y.N. JAN & L.Y. JAN. 1991. Putative receptor for the cytoplasmic inactivation gate in the *Shaker* K^+ channel. Nature **353:** 86–90.
27. ROEPER, J., S. SEWING, Y. ZHANG, T. SOMMER, S.G. WANNER & O. PONGS. 1998. NIP domain prevents N-type inactivation in voltage-gated potassium channels. Nature **39:** 390–393.
28. WIBLE, B.A., Q. YANG, Y.A. KURYSHEV, E.A. ACCILI & A.M. BROWN. 1998. Cloning and Expression of a novel K^+ channel regulatory protein, KChAP. J. Biol. Chem. **273:** 11745–11751.
29. GIESE, K.P., J.F. STORM, D. REUTER, N.B. FEDOROV, L.R. SHAO, T. LEICHER, O. PONGS & A.J. SILVA. 1998. Reduced K^+ channel inactivation, spike broadening and afterhyperpolarization in Kvβ1.1-deficient mice with impaired learning. Submitted for publication.
30. BUTLER, D.M., J.K. ONO, T. CHANG, R.E. MCCAMAN & M.E. BARISH. 1998. Mouse brain potassium channel β1 subunit mRNA: Cloning and distribution during development. J. Neurobiol. **34:** 135–150.
31. STORM, J.F. 1990. Potassium currents in hippocampal pyramidal cells. Prog. Brain Res. **83:** 161–187.
32. SAH, P. 1996. Ca^{2+}-activated K^+ currents in neurons: Types, physiological roles and modulation. Trends Neurosci. **19:** 150–154.
33. BORST, J.G.G. & B. SAKMANN. 1996. Calcium influx and transmitter release in a fast CNS synapse. Nature **383:** 431–434.
34. SHEN, V. & P. PFAFFINGER. 1995. Molecular recognition and assembly sequences involved in the subfamily-specific assembly of voltage-gated K^+ channel subunit proteins. Neuron **11:** 67–76.

The Eag Family of K^+ Channels in *Drosophila* and Mammals

BARRY GANETZKY,[a,b] GAIL A. ROBERTSON,[c] GISELA F. WILSON,[b] MATTHEW C. TRUDEAU,[c] AND STEVEN A. TITUS[b]

[b]*Laboratory of Genetics and* [c]*Department of Physiology, University of Wisconsin, Madison, Wisconsin 53706, USA*

ABSTRACT: Mutations of *eag*, first identified in *Drosophila* on the basis of their leg-shaking phenotype, cause repetitive firing and enhanced transmitter release in motor neurons. The encoded EAG polypeptide is related both to voltage-gated K^+ channels and to cyclic nucleotide-gated cation channels. Homology screens identified a family of *eag*-related channel polypeptides, highly conserved from nematodes to humans, comprising three subfamilies: EAG, ELK, and ERG. When expressed in frog oocytes, EAG channels behave as voltage-dependent, outwardly rectifying K^+-selective channels. Mutations of the human *eag*-related gene (*HERG*) result in a form of cardiac arrhythmia that can lead to ventricular fibrillation and sudden death. Electrophysiological and pharmacological studies have provided evidence that HERG channels specify one component of the delayed rectifier, I_{Kr}, that contributes to the repolarization phase of cardiac action potentials. An important role for HERG channels in neuronal excitability is also suggested by the expression of these channels in brain tissue. Moreover, mutations of ERG-type channels in the *Drosophila sei* mutant cause temperature-induced convulsive seizures associated with aberrant bursting activity in the flight motor pathway. The *in vivo* function of ELK channels has not yet been established, but when these channels are expressed in frog oocytes, they display properties intermediate between those of EAG- and ERG-type channels. Coexpression of the K^+-channel β subunit encoded by *Hk* with EAG in oocytes dramatically increases current amplitude and also affects the gating and modulation of these currents. Biochemical evidence indicates a direct physical interaction between EAG and HK proteins. Overall, these studies highlight the diverse properties of the *eag* family of K^+ channels, which are likely to subserve diverse functions *in vivo*.

The isolation and molecular characterization of behavioral mutants in *Drosophila* that alter membrane excitability has proved to be a powerful method of identifying genes encoding ion channel subunits.[1] Analysis of the Shaker (*Sh*) locus, which initiated molecular studies of K^+ channels, is a case in point.[2] Once the *Sh* locus was cloned, a number of related K^+ channel genes were subsequently identified in flies and mammals on the basis of homology screens.[3,4] Although these screens gave an early indication that the diversity of K^+ channels at the molecular level was much greater than had been anticipated on the basis of electrophysiological and pharmacological studies, the homology screens were biased towards the isolation of genes that shared substantial similarity with *Sh* and therefore did not reveal the true extent of this diversity. Other K^+ channel genes only distantly related to *Sh* were likely to be missed in these screens.

[a]Address for correspondence: Barry Ganetzky, Laboratory of Genetics, 445 Henry Mall, University of Wisconsin, Madison, Wisconsin 53706. Phone: 608-262-3896; fax: 608-262-2976; e-mail: ganetzky@facstaff.wisc.edu

An alternative approach, not subject to the same limitations, was to continue to pursue the molecular analysis of other behavioral mutations in *Drosophila* whose phenotypes were consistent with K^+ channel defects. This approach led to the first isolation of a gene encoding a Ca^{2+}-activated K^+ channel (*slo*),[5] the discovery of a novel voltage-activated K^+ channel encoded by *eag* (ether à go-go),[6] and the cloning of a K^+ channel β subunit (*Hk*).[7] Each of these studies contributed to present understanding of the structural and functional diversity of K^+ channels and provided deeper insights into the physiological significance of the various K^+ channel components *in vivo*. This article focuses on studies of *eag* and on the evolutionarily conserved family of K^+ channel genes whose existence was discovered because of their homology with *eag*.

DISCOVERY AND MOLECULAR ANALYSIS OF THE EAG FAMILY OF K+ CHANNELS

The *eag* locus was originally discovered in 1969 on the basis of a mutation causing an ether-sensitive leg-shaking phenotype.[8] However, the signficance of this gene was largely overlooked until it was rediscovered in 1983 as a mutation that greatly enhanced the behavioral and electrophysiological defects of *Sh* in double mutant individuals.[9] Electrophysiological studies at the larval neuromuscular junction demonstrated that *eag* mutations cause a high frequency of spontaneous action potentials in motor axons that elicit synaptic potentials with increased amplitude and duration compared with normal larvae (FIG. 1). These results indicated that loss of *eag* activity resulted in a phenotype of neuronal hyperexcitability. This phenotype and the synergistic interaction of *eag* with *Sh* in double mutants suggested that mutations in *eag* disrupted the function of K^+ channels distinct from those altered by mutations in *Sh*.[9]

To explore this hypothesis and to elucidate the molecular nature of the *eag* phenotype, the gene was cloned on the basis of its chromosomal location.[6,10] Sequence analysis of the corresponding cDNA revealed that the encoded protein contained characteristic features

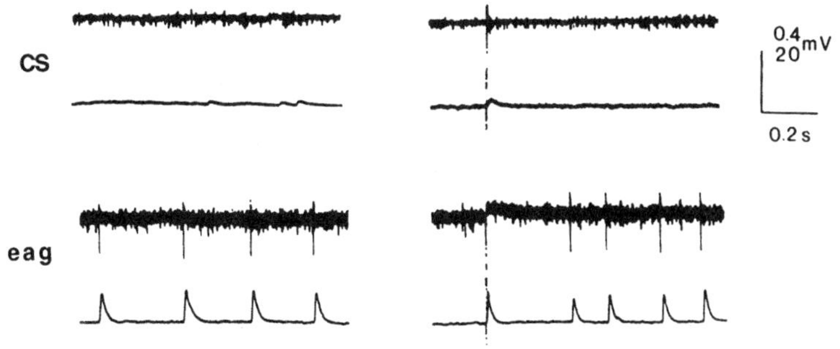

FIGURE 1. *Eag* mutants cause neuronal membrane hyperexcitability. Simultaneous nerve (**top traces** in each panel) and muscle recordings (**bottom traces** in each panel) from wildtype (CS) and *eag* larvae. Responses evoked by nerve stimulation **(right panels)** and occurring spontaneously without stimulation **(left panels)** are shown. The motor axon spikes are the most prominent units that can be detected by the extracellular recording suction electrode. At low external Ca^{2+} (0.2 mM), the evoked synaptic responses in *eag* are larger and more prolonged than normal. Similar synaptic responses occur spontaneously in *eag*, correlated with spontaneous repetitive firing of action potentials in motor axons.

previously found in members of the Sh family of K^+ channels. Most notably there are seven hydrophobic segments (six transmembrane segments plus a pore domain), each of which shows significant similarity with the corresponding segment present in Sh family members. The amino acid alignments are particularly strong in the S4 (voltage-sensor) and P (pore) segments. On the basis of these results, it was concluded that *eag* encodes a K^+ channel subunit that is a distant relative of the Sh family.[6] Nonetheless, the divergence of the Eag polypeptide from all known members of the Sh family (< 20% amino acid identity within the hydrophobic core) and the existence of novel features, such as the presence of a segment in the cytoplasmic C-terminus that is homologous with a cyclic nucleotide binding domain (cNBD), indicated that *eag* encodes a novel type of channel polypeptide.[11] In fact, the Eag polypeptide has greater similarity with cyclic nucleotide-gated cation channels than with K^+ channels in the Sh family. In addition to the presence of a cNBD, these similarities include numerous amino acid identities distributed throughout the hydrophobic core. However, expression studies in *Xenopus* oocytes (see below) demonstrate that Eag channels are voltage dependent, outwardly rectifying, and highly selective for K^+ over Na^+ ions and thus represent a novel type of voltage-gated K^+ channel.[12]

These observations suggested the possibility that Eag is the prototype of an entire new family of related K^+ channel polypeptides that parallels the Sh family. To test this idea, low stringency screens and degenerate PCR were used to isolate related sequences from *Drosophila* and mammalian tissues.[13] Sequence analysis of these cDNAs defined three distinct channel subfamilies within the Eag family: Eag, Elk (Eag-like K^+ channel), and Erg (Eag-related gene). At least one member of each subfamily has been identified in *Drosophila*, mouse, rat, and human genomes by isolation of clones and by genome database searches. Sequences corresponding to Eag and Erg have also been found in the *C. elegans* genome database, but a nematode Elk sequence has not yet been found (S. Titus and B. Ganetzky, unpublished). The general pattern revealed by sequence comparisons among the various family members is that two members of the same subfamily in different species share about 60–70% amino acid identities in the region encompassing the first membrane-spanning region (S1) through the cNBD segment. In contrast, members of two different subfamilies within the same species share about 40–50% identities across the same region. These structural relationships closely parallel those among the four subtypes of K^+ channel polypeptides in the Sh family.[3]

MUTATIONS IN HERG UNDERLIE A FORM OF CARDIAC ARRHYTHMIA

The human Erg gene (*HERG*) was mapped to a chromosome by PCR analysis of a panel of human-hamster hybrid cell lines each carrying a known subset of human chromsomes: only chromosome 7 showed complete concordance with the presence or absence of the *HERG* PCR product.[13] An autosomal dominant form of long QT (LQT) syndrome, a type of cardiac arrhythmia that can lead to ventricular fibrillation and sudden death, was mapped by family studies to the same chromosome.[14] Because LQT syndrome is associated with a defect in ventricular repolarization and no other known human K^+ channels mapped to chromosome 7, the discovery of a novel type of K^+ channel gene on this chromosome offered a plausible candidate gene for the disease. Subsequent sequence studies uncovered a variety of mutations in the *HERG* genes segregating in families affected with the chromosome 7 form of LQT syndrome, demonstrating that perturbations of HERG channels cause this disorder.[14]

CHANNEL PROPERTIES IN HETEROLOGOUS EXPRESSION SYSTEMS

Because of the extensive divergence of the Eag polypeptide from members of the Sh family of K^+ channels and its higher similarity with cyclic nucleotide–gated cation channels, it was of interest to examine the functional properties of these channels when expressed in *Xenopus* oocytes. Expression of either Eag or Meag in oocytes results in the production of voltage-dependent outward currents that activate around –40 mV and increase in amplitude at more positive potentials.[12,15] The Eag and Meag currents are similar in most respects, but only Eag currents exhibit a fast, inactivating component, which is evident at voltages above about 0 mV.[12]

Another characteristic feature of Eag, Meag, and Reag (rat Eag) currents is that activation becomes slower and more sigmoidal with increasingly negative prepulses or holding potentials.[12,16] This dependence of activation kinetics on prepulse voltage is reminiscent of the Cole-Moore shift observed for K^+ currents in squid and crayfish axons.[17,18] These results indicate that at the microscopic level, Eag channels make a series of voltage-dependent transitions between different closed states before opening. According to this model, hyperpolarizing prepulses drive the channels to more remote closed states and increase the time required for activation by subsequent depolarization.

In contrast with cyclic nucleotide–gated cation channels, both Eag and Meag channels show a high selectivity for K^+ over Na^+ ions as determined by measurements of reversal potentials in different external K^+ concentrations using a tail current protocol.[12] Similar results have also been observed for Reag.[16] It has been reported that *Drosophila* Eag channels are also permeable to Ca^{2+},[15] but this finding has been brought into question by the failure of other groups to replicate this result either for *Drosophila* Eag or mammalian Eag channels.[12,16,19] Also in contrast with cyclic nucleotide–gated channels, Eag and Meag currents do not require cyclic nucleotides for activation. In fact, no changes were observed in the properties of Meag currents when cyclic nucleotides were applied to excised macropatches.[12] A recently characterized Eag channel cloned from bovine retina has also been shown to be insensitive to cyclic nucleotides.[19] Thus, on the basis of their gating mechanism and ion selectivity, Eag-type channels behave unlike cyclic nucleotide–gated cation channels. Instead, they represent a novel class of voltage-activated K^+ channels.

Other novel features of Eag channels are their sensitivity to Mg^{2+}, H^+, and Ca^{2+} ions.[19–21] Increasing the extracellular concentration of Mg^{2+} (as well as other divalent cations) causes a striking decrease in the rate of activation of Reag (and also bovine Eag) channels in a dose- and voltage-dependent manner.[19,20] Lowering the external pH produces a similar effect. H^+ ions appear to compete with Mg^{2+} since Mg^{2+} does not slow activation at low pH.[20] In cultured 293 cells expressing Reag currents following stable transfection, induction of Ca^{2+} transients by the activation of muscarinic receptors blocked these currents.[21] In inside-out patches, Reag channels are almost completely suppressed by Ca^{2+} concentrations above 150 nM. The responses of Reag channels to voltage as well as to the extracellular and intracellular concentrations of various ions suggest that these channels may play an important integrative role in shaping neuronal firing properties under a variety of physiological conditions.

The particular K^+ currents mediated by Eag-type channels *in vivo* have not been identified with certainty. Because of their Ca^{2+} sensitivity, absence of inactivation, relative insensitivity to TEA and 4-AP, and other properties, it has been proposed that Reag channels are responsible for the M-type K^+ current found in pyramidal cells of the neocortex, hippocampus, and olfactory cortex.[21,22] However, this conclusion is still under debate, and the molecular identity of channels mediating M-currents remains uncertain.[23,24] Characterization of currents mediated by a bovine Eag-type channel expressed in HEK 293 cells reveals similarities with I_{Kx}, a non-inactivating K^+ conductance in the inner segment of rod photoreceptors, suggesting a possible molecular link.[19] This suggestion also awaits final confirmation.

When HERG channels are expressed in oocytes at levels comparable to those used for Eag or Meag, only small outward currents are detected in response to depolarizing steps.[25] Upon repolarization to negative membrane potentials, large tail currents are observed. This unusual current phenotype is conferred by a rapid inactivation that limits the amplitude of the outward currents during depolarization. During a subsequent repolarization, the channels recover from inactivation, giving rise to the marked tail current.[25,26] At higher expression levels, it is apparent that the inactivation accounts for a nonlinear I-V relation with a pronounced negative slope conductance, or inward rectification, at more positive voltages.[26] Subsequent analysis in which inactivation was removed by mutagenesis or slowed by the application of external TEA demonstrated that the mechanism of the rapid, voltage-dependent inactivation process was similar to that of C-type inactivation.[27–29]

Comparison of the properties of HERG currents with those of native cardiac currents was motivated by the connection between *HERG* mutations and LQT syndrome.[14] The pronounced inward rectification, together with a pharmacological sensitivity to the methanesulfonanilide drug E-4031, identified HERG as a component of I_{Kr},[25,26,30,31] a current important in the repolarization of the cardiac action potential.[32,33] Analysis of HERG currents during a voltage clamp command that mimics a ventricular action potential illustrates how the physiological role of these channels in cardiac cells is determined by fast inactivation: initial repolarization of the action potential facilitates recovery of HERG channels from inactivation, ensuring that HERG currents subsequently reach peak amplitude at the appropriate time to mediate terminal repolarization.[34]

A cDNA encoding an Elk-type channel cloned from mouse brain (Melk) was also recently expressed in *Xenopus* oocytes (M. Trudeau, S. Titus, B. Ganetzky & G. Robertson, in preparation). These channels mediate large outward currents that peak within 25 ms and then rapidly inactivate. Inactivation is more complete at increasingly positive voltages, resulting in a nonlinear steady state I-V relation that shows inward rectification at higher voltages. The rate of inactivation and the degree of inward rectification are intermediate to the corresponding properties of HERG and Meag (or Eag) channels. These observations suggest the distinctive characteristics of Eag family members may largely be determined by differences in inactivation properties.

A *DROSOPHILA* MUTATION AFFECTING Erg CHANNELS

Herg channels are expressed predominantly in cardiac tissue, and known mutations of the *HERG* gene cause LQT syndrome but are not associated with any reported neurological abnormalities. Nonetheless, the fact that *HERG* cDNAs were originally recovered from a hippocampus library indicates that HERG channels may have an important neural function as well. It is possible that cardiac myocytes are more sensitive than neurons to a partial loss of HERG currents and that neurological abnormalities will be manifested only in individuals homozgyous for *HERG* mutations. To explore the importance of Erg channels in neural functions in an organism more amenable to experimental manipulations, mammalian probes were used to isolate *Drosophila* cDNAs encoding Erg-type channels.[35] The amino acid sequence encoded by this cDNA from S1 through the cNBD is 45% identical with Eag and 71% identical with HERG. The transcript is expressed diffusely throughout the embryonic CNS. The chromosomal location of the gene coincides with that of the seizure (*sei*) locus, which was defined earlier by the isolation of several mutations that cause a temperature-sensitive paralytic phenotype. Paralysis at elevated temperatures in these mutants is preceded by a distinctive bout of convulsive seizure activity. Conclusive evidence that the *sei* gene encodes Erg-type channels is provided by the detection of lesions in the Erg coding sequences in four different *sei* mutants. This conclusion was also reached independently by positional cloning of the *sei* locus.[36]

To date, the most pronounced electrophysiological defect observed in *sei* mutants is a substantial increase in spontaneous bursting activity in the adult flight motor pathway at elevated temperatures.[37,38] This observation reveals that Erg channels play some role in repolarization of action potentials or in maintaining the resting potential at least in flight motor neurons. These results are of particular interest in light of the demonstration that an Erg-like current in a mammalian neuroblastoma cell line plays an important role in the spike-frequency adaptadon of these cells.[39] If Erg channels have a similar function *in vivo* in *Drosophila*, this could account for the electrophysiological abnormalities observed in *sei* mutants.

EXPANSION OF THE Eag GENE FAMILY IN MAMMALS

In *Drosophila*, each of the subfamilies within the Eag family of channels is defined by a single gene: *eag*, *elk*, and *sei* (Erg). However, in mammals genes in these subfamilies have undergone further duplication and divergence such that each subfamily contains at least two distinct members. Two different *eag* homologs have been isolated using a *Drosophila eag* probe to screen rat brain cDNA libraries (Ref. 16 and S. Titus & B. Ganetzky, unpublished). In the core region from S1 through S6, the encoded amino acid sequences are 76% identical. The Meag amino acid sequence is 99% identical with Reag1 and 72% identical with Reag2, defining the mouse sequence as a member of the Eag1 subclass.

A similar situation occurs for the Elk subfamily. Members of this subfamily have been cloned from mouse, rat, and human cDNA libraries (S. Titus & B. Ganetzky, unpublished). Sequence comparisons indicate that these genes fall into two distinct subclasses. Thus, Melk and Helk share 97% identity (S1 through the cNBD) and appear to belong to the same subclass. In contrast, Relk shares only about 63% identity with either the mouse or human genes and apparently represents a member of a second subclass.

Three different members of the Erg subfamily have been identified and characterized in rat.[40] The currents mediated by these channels when expressed in *Xenopus* oocytes each exhibit distinct kinetic properties, and the three genes are transcribed in different but overlapping sets of tissues *in vivo*. Corresponding members of the human Erg subfamily,

Herg2 and Herg3, have recently been identified and sequenced (S. Titus & B. Ganetzky, in preparation).

The existence of multiple members of each Eag subfamily expands the functional diversification of these channels in mammals. The potential for functional diversity may be greater still if different members of the same subfamily are capable of assembling into heteromultimeric channels.

ALTERNATIVE ISOFORMS OF Eag FAMILY MEMBERS

In addition to the expression of distinct proteins encoded by multiple genes within each Eag subfamily, multiple protein isoforms can also be generated from a single gene via alternative splicing and use of alternative promoters. Several examples of protein isoforms within the Eag family have been reported, and it seems likely that further investigation will uncover additional instances.

Sequence comparisons among different cDNAs for Meag identified a small, optional exon of 27 amino acids located in the extracellular linker between the S3 and S4 domains.[13] The same two splice variants have also been identified for a bovine homolog of Eag expressed in retina.[19] Differences between the channel isoforms encoded by the two splice variants in activation kinetics and Mg^{2+} sensitivity were observed, and it has been proposed that the inserted sequence between S3 and S4 modifies channel gating properties.[19]

Several isoforms have also been found for Merg.[41,42] Merg1b, the shorter form (820 amino acids), is identical with Merg1a, the longer form (1162 amino acids) from S1 through the carboxy terminus but has a novel, markedly shorter amino terminus. This amino terminus is encoded by an exon located within the intron separating exons 5 and 6 of the Merg1a mRNA. It appears that the Merg1b isoform is generated by initiation of transcription from a second promoter immediately upstream of this optional exon. The different isoforms have different patterns of expression. Whereas Merg1 is expressed in heart, brain, and testes, Merg1b is expressed preferentially in the heart. Both isoforms produce inwardly rectifying, E-4031–sensitive currents when expressed in *Xenopus* oocytes but differ in that the Merg1a currents deactivate slowly while Merg1b currents deactivate much more rapidly.[41,42] Furthermore, coexpression experiments indicate that Merg1a and Merg1b coassemble to produce heteromeric channels with deactivation kinetics that are much more rapid than for Merg1a alone and closely resemble those of the native mouse I_{Kr}. Thus, the expression of Merg1b together with the Merg1a isoforms in the mouse heart appears to be a mechanism of modifying the properties of I_{Kr} in a tissue-specific manner. Interestingly, a HERG1b isoform has been identified in human atrium and smooth muscle as well, suggesting that the electrical properties of these tissues will also be governed, in part, by expression of this isoform.

INTERACTION OF Eag CHANNELS WITH β SUBUNITS

The properties of voltage-gated K^+ channels in the Sh family are modified by the association of the pore-forming α subunits with auxiliary β subunits.[43] Experiments have demonstrated that both kinetics of channel inactivation and current amplitudes are altered when α and β subunits are coexpressed in heterologous systems.[7,44] Consequently, the formation of heteroligomeric channels *in vivo*, containing both α and β subunits, could contribute to the diversity of K^+ channel function in various neuronal types and in different subcellular compartments of a given neuron. For mammalian voltage-gated K^+ channels in the *Sh* family, coassembly with β subunits appears to be largely, if not exclusively, restricted to the K_v1 subtype of α subunits.[45–47]

To examine the possibility that voltage-gated K^+ channels in the Eag family might also be modified by coassociation with the same type of β subunits that interact with Sh channels, Eag currents were examined in *Xenopus* oocytes injected with *eag* RNA alone or in combination with *Hk* RNA.[48] Coinjected oocytes consistently display a 2–3-fold increase in Eag current amplitudes compared with oocytes injected with *eag* mRNA alone. In addition to expression levels, Eag gating kinetics are also affected by coexpression with Hk resulting in more rapid activation rate and a decrease in the rate of inactivation.[48]

Direct evidence of a physical association between Eag and Hk was obtained from immunoprecipitation experiments. Protein complexes from extracts of tsA201 cells transfected with *eag* alone, with *Hk* alone, or with constructs for both genes were immunoprecipitated with an anti-Eag antiserum. Hk protein was detected only in immunoprecipitates from cells transfected with both *eag* and *Hk*.[48] Taken together these results support the conclusion that Hk physically associates with Eag and modifies the properties of the resultant heteromeric channels. These results suggest the possibility that in *Drosophila*, Eag channels, like Sh channels, may have different functional properties in different excitable cells and in different subcellular compartments depending on their association with β subunits.

The possibility that a similar situation could occur in mammals is suggested by the observed interaction between Hk and Meag when coexpressed in oocytes. The effects of Hk on Meag closely parallel those observed on its *Drosophila* counterpart: current amplitudes are increased by about 1.5-fold, and the kinetics of activation is accelerated.[48]

Even more surprisingly, the effect of Hk is not limited to members of the Eag subfamily in mammals. Coexpression of Hk with HERG resulted in a 3–4-fold increase in the average current amplitudes compared with expression of HERG alone.[48] As in the case of Eag and Meag, Hk was also found to accelerate the activation of HERG currents. It will now be of interest to determine whether the capacity to interact with members of the Eag family in mammals is conserved in any of the mammalian Kvβ subunits and whether any of the effects observed in oocytes can be demonstrated *in vivo*.

ASSOCIATION OF Eag FAMILY MEMBERS WITH OTHER CHANNEL SUBUNITS

Channels in the Eag family have been found to associate with other proteins besides β subunits, and these associations may further contribute to the functional diversity of these channels *in vivo*. The minK K^+ channel subunit, which has been shown to assemble with various other proteins to form K^+ channels, was also found to interact with HERG.[49] Coexpression of HERG and minK in CHO cells results in a doubling of the K^+ current density and a modest increase in the voltage sensitivity of activation and inactivation of HERG channels. A direct physical association between these proteins was shown by coimmunoprecipitation experiments.

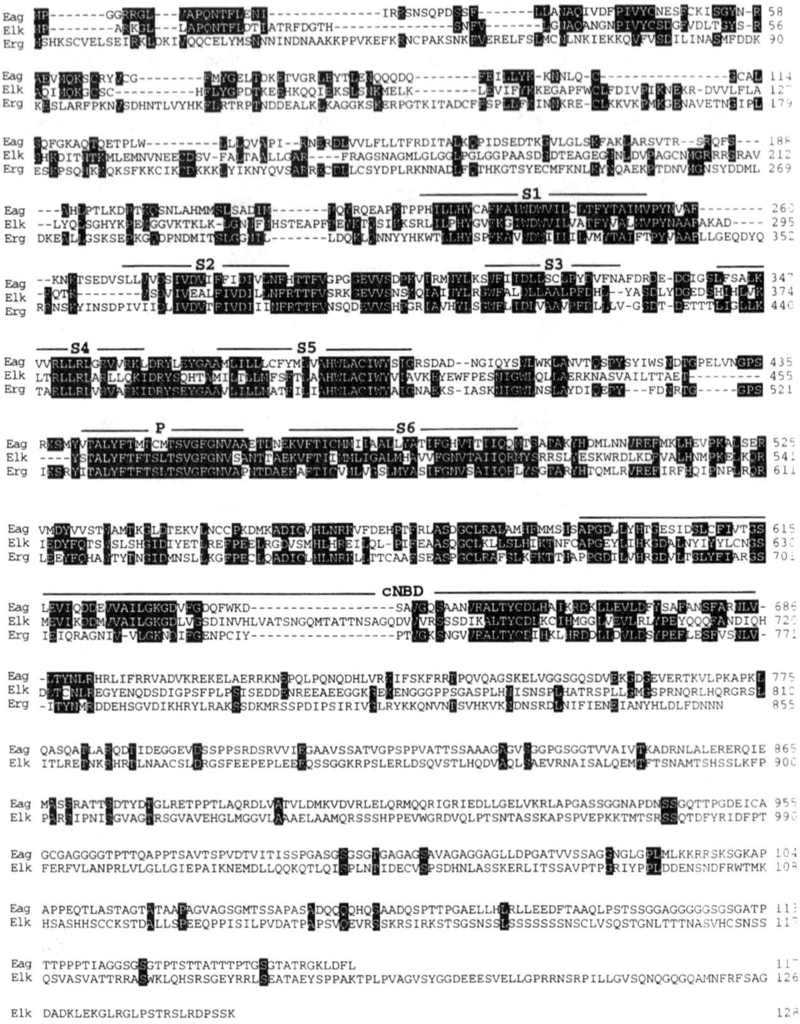

FIGURE 2. A family of channel polypeptides related to Eag. Amino acid alignment of the three members of the Eag family in *Drosophila* is shown. Identical residues are shaded in *black*. Gaps in the alignment are indicated with *dashes*. The six transmembrane segments (S1–S6), the pore domain (P), and the region of homology with a cyclic nucleotide binding domain (cNBD) are *overlined*.

A novel K^+ channel-binding protein, Slob, was recently identified based on its physical association with the *slo*-encoded Ca^{2+}-activated K^+ channel.[50] Slob also coimmunoprecipitates with Eag in transfected cells. Although Slob strongly increases the steady state open probability of Slo channels, an effect on Eag channel properties has not yet been reported.

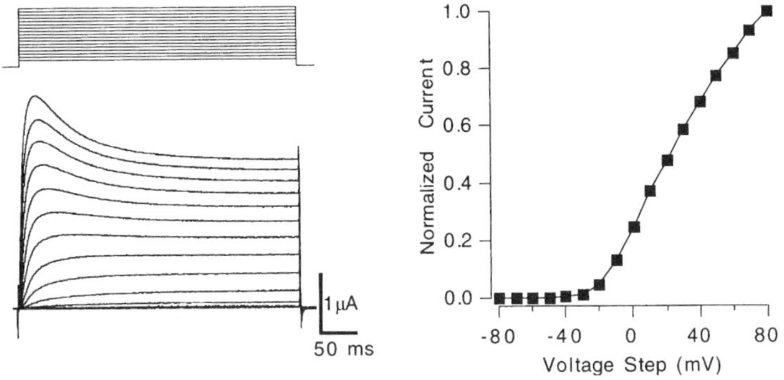

FIGURE 3. Eag channels carry a voltage-dependent outward current. Expression of Eag currents in *Xenopus* oocytes following injection of *eag* cRNA measured with a two-electrode voltage clamp. Currents **(left)** were measured at test potentials between −80 mV and +80 mV in 10-mV steps from a holding potential of −100 mV. The I/V plot **(right)** consists of normalized data from seven different oocytes. The values plotted are the mean ±SD (error bars are smaller than the symbols).

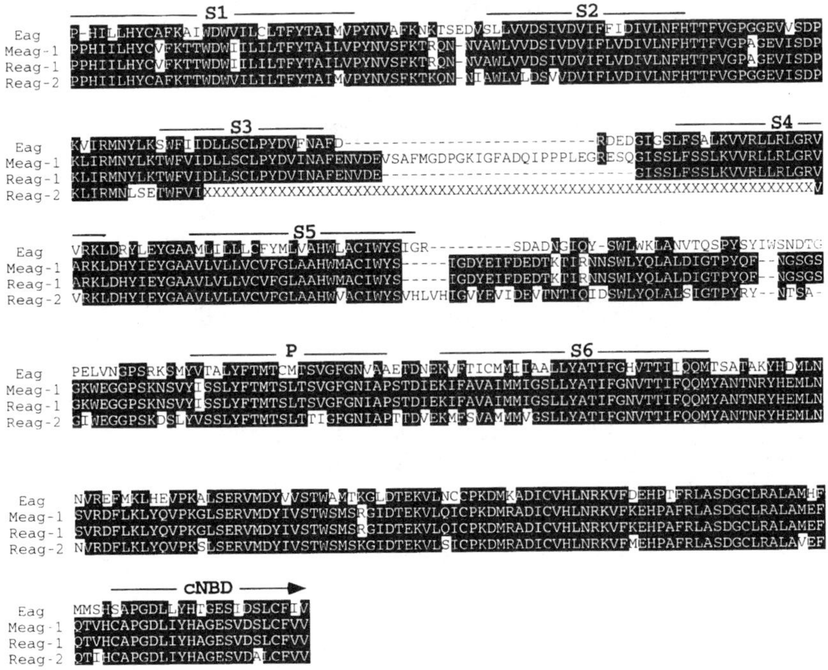

FIGURE 4. The Eag subfamily of K⁺ channel polypeptides in mammals contains at least two distinct subtypes. The polypeptide cores (S1 through the beginning of the cNBD) of two different Eag channels cloned from rat are aligned with those from *Drosophila* and mouse. Identical residues are shaded in *black*. Gaps in the alignment are indicated with *dashes*. An incompletely sequenced region of Reag-2 is indicated by the Xs. Note the high degree of identity between Meag-1 and Reag-1 and the divergence of Reag-1 and Reag-2.

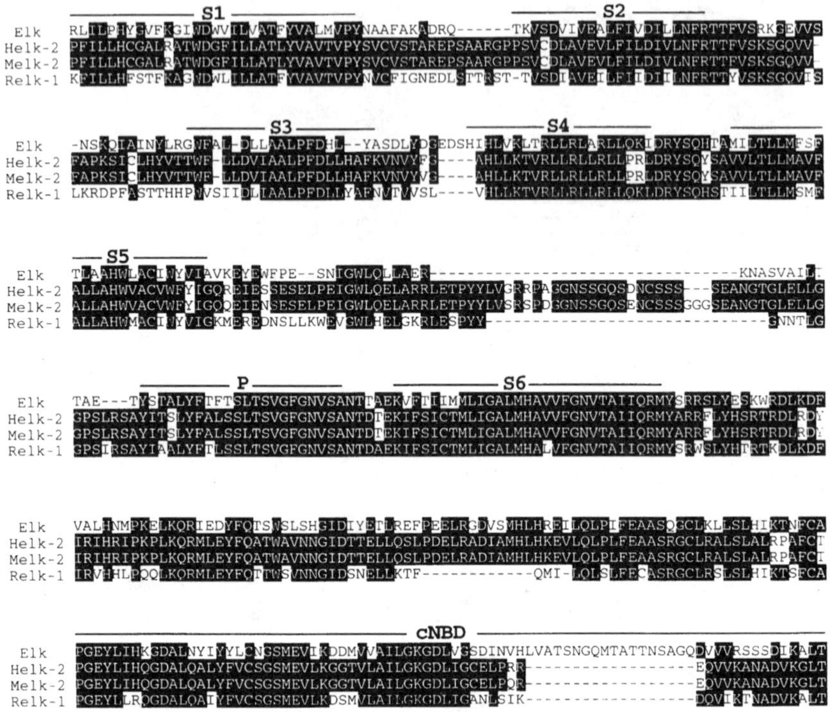

FIGURE 5. The Elk subfamily of K^+ channel polypeptides in mammals contains at least two distinct subtypes. The polypeptide cores (S1 midway through the cNBD) of Elk channels cloned from *Drosophila*, mouse, rat, and human are aligned. Identical residues are shaded in *black*. Gaps in the alignment are indicated with *dashes*. Note the high degree of identity between Helk-2 and Melk-2 and the divergence of these sequences from Relk-1.

Several observations have raised the intriguing possibility that Eag can assemble with α subunits of K^+ channels in entirely separate families to generate diverse heteromeric channels via a combinatorial mechanism. The observations on which this proposal is based include: voltage-clamp analysis of K^+ currents in larval muscles indicating that four different currents, including two voltage-activated currents and two Ca^{2+}-activated currents, are altered (but not eliminated) by various *eag* mutation[51,52]; the demonstration that modulatory effects of temperature or certain pharmacological agents on Sh currents are altered or abolished by *eag* mutation[53]; and the occurrence of novel, allele-specific phenotypic interactions in various *eag Sh* double mutant combinations.[53] The case for a combinatorial assembly of K^+ channels *in vivo* involving Eag receives support from the interaction between Eag and Sh subunits when they are coexpressed in *Xenopus* oocytes.[54] Coexpression results in more rapid inactivation of Sh currents and slower recovery from inactivation. Site-directed mutagenesis indicates that the Eag carboxyl terminus is crucial for this interaction. Despite these tantalizing results, final proof of a direct physical association between Eag and Sh subunits will probably require biochemical experiments.

CONCLUSION

The Eag family represents a distinct set of voltage-activated K$^+$ channels that has been conserved at least from nematodes to humans. The discovery and characterization of this family of K$^+$ channels has provided new insights into the structure and function of K$^+$ channels, their evolutionary history, their expanding diversity, and the molecular basis of certain genetic disorders in humans. Despite their structural similarities, Eag, Elk, and Erg channels display distinctive biophysical properties suggesting that they each subserve particular physiological functions *in vivo*. The striking electrophysiological and behavioral phenotypes displayed by *eag* and *sei* (Erg) mutations in *Drosophila* clearly demonstrate the *in vivo* significance of these channels. Association of one type of LQT sydrome with mutations of *HERG* highlight the physiological importance of the Eag family of K$^+$ channels from invertebrates to humans.

The existence of a second set of voltage-activated K$^+$ channels, comprising at least three different subfamilies, expands the known diversity of K$^+$ channels beyond that inferred from analysis of the extended Sh family. Structural and functional diversity of K$^+$ channels in the Eag family is further magnified by a variety of mechanisms including the generation of protein isoforms by alternative splicing and usage of multiple promoters, the existence in mammals of multiple genes for each subfamily, and the association with various other subunits to form heteromultimeric channel assemblies.

Further analysis of the Eag family of K$^+$ channels in *Drosophila* and mammalian systems should continue to generate novel information about the physiological roles of this family in different excitable cells and help clarify the continuing puzzle of why such an enormous diversity of K$^+$ channels has been strongly favored by evolution.

ACKNOWLEDGMENTS

This work was supported by research grants from the NIH (B.G. and G.A.R.), NSF (G.A.R.), the American Heart Association (G.A.R. and M.C.T.), and the Muscular Dystrophy Association (G.F.W.).

We thank Robert Kreber for excellent technical assistance and Blake Anson for providing FIGURE 2.

REFERENCES

1. WU, C.-F. & B. GANETZKY. 1992. Neurogenetic studies of ion channels in Drosophila. *In* Ion Chanels. Vol. 3. T. Narahashi, Ed.: 261–314. Plenum. New York.
2. PAPAZIAN, D.M., T.L. SCHWARZ, B.L. TEMPEL, L.C. TIMPE & L.Y. JAN. 1988. Ion channels in Drosophila. Annu. Rev. Physiol. **50:** 379–394.
3. SALKOFF, L., K. BAKER, A. BUTLER, M. COVARRUBIAS, M.D. PAK & A. WEI. 1992. An essential "set" of K$^+$ channels conserved in flies, mice, and humans. Trends Neurosci. **15:** 161–166.
4. CHANDY, K.G. & G.A. GUTMAN. 1995. Voltage gated K$^+$ channel genes. *In* CRC Handbook of Receptors and Channels. A.P. North, Ed.: 1–71. CRC Press. Boca Raton, FL.
5. ATKINSON, N., G. ROBERTSON & B. GANETZKY. 1991. A structural component of calcium-activated potassium channels encoded by the *Drosophila slo* locus. Science **252:** 551–555.
6. WARMKE, J.W., R. DRYSDALE & B. GANETZKY. 1991. A distinct potassium channel polypeptide encoded by the *Drosophila eag* locus. Science **252:** 1560–1562.
7. CHOUINARD, S.W., G.F. WILSON, A.K. SCHLIMGEN & B. GANETZKY. 1995. A potassium. channel β subunit related to the aldo-keto reductase superfamily is encoded by the *Drosophila* hyperkinetic locus. Proc. Natl. Acad. Sci. USA **92:** 6763–6767.

8. KAPLAN, W.D. & W.E. TROUT, III. 1969. The behavior of four neurological mutants of Drosophila. Genetics **61:** 399–409.
9. GANETZKY, B. & C.-F. WU. 1983. Neurogenetic analysis of potassium currents in *Drosophila*: Synergistic effects on neuromuscular transmission in double mutants. J. Neurogenet. **1:** 17–28.
10. DRYSDALE, R.A., J.W. WARMKE, R. KREBER & B. GANETZKY. 1991. Molecular characterization of *eag*, a gene affecting potassium channels in Drosophila. Genetics **127:** 497–505.
11. GUY, H.R., S.R. DURELL, J.W. WARMKE, R.A. DRYSDALE & B. GANETZKY. 1991. Similarities in amino acid sequences of Drosophila eag and cyclic nucleotide-gated channels. Science **254:** 730.
12. ROBERTSON, G.A., J.W. WARMKE & B. GANETZKY. 1996. Potassium currents expressed from Drosophila and mouse eag cDNAs in Xenopus oocytes. Neuropharmacology **35:** 841–850.
13. WARMKE, J.W. & B. GANETZKY. 1994. A family of potassium channel genes related to eag in *Drosophila* and mammals. Proc. Natl. Acad. Sci. USA **91:** 3438–3442.
14. CURRAN, M.E., I. SPLAWSKI, K.W. TIMOTHY, G.M. VINCENT, E.D. GREEN & M.T. KEATING. 1995. A molecular basis for cardia arrhythmia: HERG mutations cause long QT syndrome. Cell **89:** 795–803.
15. BRUGGEMAN, A., L.A. PARDO, W. STUHMER & O. PONGS. 1993. Ether a-go-go encodes a voltage-gated channel permeable to K^+ and Ca^{2+} and modulated by cAMP. Nature **365:** 445–448.
16. LUDWIG, J., H. TERLAU, F. WUNDER, A. BRUGGEMANN, L.A. PARDO, A. MARQUARDT, W. STUHMER & O. PONGS. 1994. Functional expression of a rat homologue of the voltage gated *ether à go-go* potassium channel reveals differences in selectivity and activation kinetics between the *Drosophila* channel and its mammalian counterpart. EMBO J. **13:** 4451– 4458.
17. COLE, K.S. & J.W. MOORE. 1960. Potassium ion current in the squid giant axon: Dynamic characteristic. J. Ger. Physiol. **1:** 1–14.
18. YUNG, S.W. & J.W. MOORE. 1981. Potassium ion currents in the crayfish giant axon. Dynamic characteristics. Biophys. J. **36:** 723–733.
19. FRINGS, S., N. BRULL, C. DZEJA, A ANGELE, V. HAGEN, U.B. KAUPP & A. BAUMANN. 1998. Characterization of ether-a-go-go channels present in photoreceptors reveals similarity to I_{Kx}, a K^+ current in rod inner segments. J. Gen. Physiol. **111:** 583–599.
20. TERLAU, H., J. LUDWIG, R. STEFFAN, O. PONGS, W. STUHMER & S.H. HEINEMANN. 1996. Extracellular Mg^{2+} regulates activation of rat eag potassium channel. Pflugers Arch. **432:** 301–312.
21. STANSFELD, C.E., J. ROPER, J. LUDWIG, R.M. WESELOH, S.J. MARSH, D.A. BROWN & O. PONGS. 1996. Elevation of intracellular calcium by muscarinic receptor activation induces a block of voltage-activated rat *ether à-go-go* channels in a stably transfected cell line. Proc. Natl. Acad. Sci. USA **93:** 9910–9914.
22. STANSFELD, C., J. LUDWIG, J. ROEPER, R. WESELOH, D. BROWN & O. PONGS. 1997. A physiological role for ether-à-go-go K^+ channels. Trends Neurosci. **20:** 13–14.
23. MATHIE. A. & C.S. WATKINS. 1997. Is EAG the answer to the M-current? Trends Neurosci. **20:** 14.
24. MARRION, N.V. 1997. Does R-eag contribute to the M-current? Trends Neurosci. **20:** 243.
25. TRUDEAU, M.C., J.A. WARMKE, B. GANETZKY & G.A. ROBERTSON. 1995. HERG, a human inward rectifier in the voltage-gated potassium channel gene family. Science **269:** 92–95.
26. SANGUINETTI. M.C., C. JIANG, M.E. CURRAN & M.T. KEATING. 1995. A mechanistic link between an inherited and an acquired cardiac arrhythmia: *HERG* encodes the I_{Kr} potassium channel. Cell **81:** 299–307.
27. SMITH, P.L., T. BAUKROWITZ & G. YELLEN. 1996. The inward rectification of the HERG cardiac potassium channel. Nature **379:** 833–836.
28. SCHONHERR, R. & S.H. HEINEMANN. 1996. Molecular determinants for activation and inactivation of HERG, a human inward rectifier potassium channel. J. Physiol. **493:** 635–642.
29. SPECTOR, P.S., M.E. CURRAN, A. ZOU, M.T. KEATING & M. SANGUINETTI. 1996. Fast inactivation causes rectification of the I_{Kr} channel. J. Gen. Physiol. **107:** 611–619.
30. SNYDERS, D.J. & A. CHAUDHARY. 1996. High affinity open channel block by dofetilide of HERG expressed in a human cell line. Mol. Pharrnacol. **49:** 949–955.
31. SANGUINETTI, M. & N.K. JURKIEWICZ. 1990. Two components of delayed rectifier K^+ current: Differential sensitivity to block by class III antiarrhythmic agents. J. Gen. Physiol. **96:** 195–215.
32. SHIBASAKI. T. 1987. Conductance and kinetics of delayed rectifier potassium channels in nodal cells of the rabbit heart. J. Physiol. (Lond.) **387:** 227–250.

33. Spector, P.S., M.E. Curran, M.T. Keating & M.C. Sanguinetti. 1996. Class III antiarrhythmic drugs block HERG, a human cardiac delayed rectifier K$^+$ channel. Circ. Res. **78:** 499–503.
34. Zhou, Z., Q. Gong, B. Ye, Z. Fan, J.C. Makielski, G.A. Robertson & C.T. January. 1998. Properties of HERG channels stably expressed in HEK293 cells studied at physiological temperature. Biophys. J. **74:** 230–241.
35. Titus, S.A., J.W. Warmke & B. Ganetzky. 1997. The Drosophila erg K$^+$ channel polypeptide is encoded by the seizure locus. J. Neurosci. **17:** 875–881.
36. Wang, X., E.R. Reynolds, P. Deak & L.M. Hall. 1997. The *seizure* locus encodes the *Drosophila* homolog of the HERG potassium channel. J. Neurosci. **17:** 882–890.
37. Elkins, T. & B. Ganetzky. 1990. Conduction in the giant fiber pathway in temperature-sensitive paralytic mutants in *Drosophila*. J. Neurogenet. **6:** 207–219.
38. Kasbekar, D.P., J.C. Nelson & L.M. Hall. 1987. Enhancer of seizure: A new genetic locus in *Drosophila melanogaster* defined by interactions with temperature-sensitive paralytic mutations. Genetics **116:** 423–431.
39. Chiesa, N., B. Rosati, A. Arcangeli, M. Olivotto & E. Wanke. 1997. A novel role for HERG K$^+$ channels: Spike frequency adaptation. J. Physiol. (Lond.) **501:** 313–318.
40. Shi, W., R.S. Wymore, H.-S. Wang, Z. Pan, I.S. Cohen, D. McKinnon & J.E. Dixon. 1997. Identification of two nervous system-specific members of the erg potassium channel gene family. J. Neurosci. **17:** 9423–9432.
41. London, B., M.C. Trudeau, K.P. Newton, A.K. Beyer, N.G. Copeland, D.J. Gilbert, N.A. Jenkins. C.A. Satler & G.A. Robertson. 1997. Two isoforms of the mouse *Ether-a-go-go–*related gene coassemble to form channels with properhes similar to the rapidly activating component of the cardiac delayed rectifier current. Circ. Res. **81:** 870–878.
42. Lees-Miller, J.P., C. Kondo, L. Wang & H.J. Duff. 1997. Electrophysiological characterization of an alternatively processed ERG K$^+$ channel in mouse and human hearts. Circ. Res. **81:** 719–726.
43. Isom, L.L., K.S. De Jongh & W.A. Catterall. 1994. Auxiliary subunits of voltage-gated ion channels. Neuron **12:** 1183–1194.
44. Rettig, J., S.H. Heinemann, F. Wunder, C. Lorra, D.N. Parcej, J.O. Dolly & O. Pongs. 1994. Inactivation properties of voltage-gated K$^+$ channels altered by presence of β-subunit. Nature **369:** 289–294.
45. Rhodes, K.J., S.A. Keilbaugh, N.X. Barrezueta, K.L. Lopez & J.S. Trimmer. 1995. Association and colocalization of K$^+$ channel α- and β-subunit polypeptides in rat brain. J. Neurosci. **15:** 5360–5371.
46. Sewing, S., J. Roeper & O. Pongs. 1996. Kvβ1 subunit binding specific for *Shaker*-related potassium channel α subunits. Neuron **16:** 455– 463.
47. Yu, W., J. Xu & M. Li. 1996. NAB domain is essential for the subunit assembly of both α-α and α-β complexes of Shaker-like potassium channels. Neuron, **16:** 441–453.
48. Wilson, G.F., Z. Wang, S.W. Chouinard, L.C. Griffith & B. Ganetzky. 1998. Interaction of the K channel β subunit, *Hyperkinetic*, with *eag* family members. J. Biol. Chem. **273:** 6389–6394.
49. McDonald, T.V., Z. Yu, Z. Ming, E. Palma, M.B. Meyers, K.-W. Wang, S.A.N. Goldstein & G.I. Fishman. 1997. A minK-HERG complex regulates the cardiac potassium current I_{Kr}. Nature **388:** 289–292.
50. Schopperle, W.M., M.H. Holmqvist, Y. Zhou, J. Wang, Z. Wang, L.C. Griffith, I. Keselman, F. Kusinitz, D. Dagan & I.B. Levitan. 1998. Slob, a novel protein that interacts with the slowpoke calcium-dependent potassium channel. Neuron **20:** 565–573.
51. Wu, C-F., B. Ganetzky, F.N. Haugland & A.-X. Liu. 1983. Potassium currents in *Drosophila*: Different components affected by mutations of two genes. Science **220:** 1076–1078.
52. Zhong, Y. & C.-F. Wu. 1991. Alteration of four identified K$^+$ currents in *Drosophila* muscle by mutations in *eag*. Science **252:** 1562–1564.
53. Zhong, Y. & C.-F. Wu. 1993. Modulation of different K$^+$ currents in Drosophila: A hypothetical role for the eag subunit in multimeric K$^+$ channels. J. Neurosci. **13:** 4669–4679.
54. Chen, M.-L., T. Hoshi & C.-F. Wu. 1996. Heteromultimeric interactions among K$^+$ channel subunits from Shaker and eag families in Xenopus oocytes. Neuron **17:** 535–542.

Small-Conductance Calcium-Activated Potassium Channels

CHRIS T. BOND, JAMES MAYLIE,[a] AND JOHN P. ADELMAN[b]

Vollum Institute, and [a]Department of Obstetrics and Gynecology, Oregon Health Sciences University, Portland, Oregon 97201, USA

ABSTRACT: SK channels play a fundamental role in all excitable cells. SK channels are potassium selective and are activated by an increase in the level of intracellular calcium, such as occurs during an action potential. Their activation causes membrane hyperpolarization, which inhibits cell firing and limits the firing frequency of repetetive action potentials. The intracellular calcium increase evoked by action potential firing decays slowly, allowing SK channel activation to generate a long-lasting hyperpolarization termed the *slow afterhyperpolarization* (sAHP). This spike-frequency adaptation protects the cell from the deleterious effects of continuous tetanic activity and is essential for normal neurotransmission. Slow AHPs can be classified into two groups, based on sensitivity to the bee venom toxin apamin. In general, apamin-sensitive sAHPs activate rapidly following a single action potential and decay with a time constant of approximately 150 ms. In contrast, apamin-insensitive sAHPs rise slowly and decay with a time constant of approximately 1.5 s. The basis for this kinetic difference is not yet understood. Apamin-sensitive and apamin-insensitive SK channels have recently been cloned. This chapter will compare with different classes of sAHPs, discuss the cloned SK channels and how they are gated by calcium ions, describe the molecular basis for their different pharmacologies, and review the possible role of SK channels in several pathological conditions.

SK potassium channels play a fundamentally important role in all excitable cells. They are potassium selective, voltage independent, and are activated by increases in the levels of intracellular calcium such as occur during an action potential. As the action potential decays, the membrane potential is repolarized, and internal calcium levels rise, evoking a biphasic afterhyperpolarization. The initial faster phase is due to the activation of large-conductance voltage- and calcium-activated potassium channels (BK channels), while the slower phase is due to the activation of SK channels, which are gated solely by intracellular calcium ions. As SK channels activate, they extrude potassium ions from the cell, moving the membrane to more negative potentials. The recovery of the calcium signal following an action potential is slow, permitting SK channels to generate a long-lasting hyperpolarization, the slow afterhyperpolarization (sAHP), with a time course that reflects the decay of intracellular calcium.[1,2] Thus, activation of SK channels causes membrane hyperpolarization, which inhibits cell firing. The sAHP limits burst frequency because during a train of action potentials the sAHP becomes deeper and longer lasting until the cell is no longer able to reach the action potential threshold, even though the stimulus to fire remains.[2–5] This is spike-frequency adaptation, which protects against the deleterious effects of continuous tetanic activity and is essential for normal neurotransmission.

[b]Address for correspondence: John P. Adelman, Ph.D., The Vollum Institute, L-474, Oregon Health Sciences University, 3181 S.W. Jackson Park Road, Portland, Oregon 97201-3098. Phone: 503-494-5450; fax: 503-494-4353; e-mail: adelman@ohsu.edu

SUBCLASSES OF THE SLOW AFTERHYPERPOLARIZATION PHARMACOLOGY

Two types of sAHPs have been distinguished based upon their pharmacology, regulation by transmitter-induced second messengers, and kinetics. Pharmacologically the sAHP falls into two classes, those that are blocked by the bee venom peptide toxin apamin, and those that are apamin insensitive. Apamin-sensitive sAHPs are more commonly observed, being found for instance, in hippocampal interneurons,[6] bullfrog sympathetic neurons,[7,8] and rat adrenal chromaffin cells.[9] Apamin-insensitive sAHPs are relatively rare, but have been well documented in hippocampal pyramidal neurons, where apamin application does not affect the sAHP.[2,4] The two classes of sAHPs are not mutually exclusive, as in rat and guinea pig vagal neurons,[10] cat,[11] and rat[12,13] cortical neurons, and guinea pig cholinergic nucleus basalis neurons[14] both apamin-sensitive and -insensitive AHPs are present.

REGULATION BY NEUROTRANSMITTERS

There are few examples of modulation of apamin-sensitive sAHPs by neurotransmitter-induced second messengers. In contrast, regulation of the apamin-insensitive sAHP has been well documented in hippocampal pyramidal neurons[15] and in septal cholinergic,[16] neocortical,[12] and sensorimotor[11] neurons. Many neurotransmitters act on the channels underlying the sAHP in a converging manner, exerting strong effects on neuronal excitability through its modulation. Noradrenaline, dopamine, serotonin, histamine, acetylcholine (via muscarinic receptors) and glutamate (via metabotropic receptors), as well as some neuropeptides (VIP, CRF), all suppress the apamin-insensitive sAHP. As a consequence, neuronal excitability is enhanced, spike frequency adaptation is strongly decreased, and the number of action potentials evoked by a certain depolarizing stimulus is increased. In contrast, adenosine can decrease neuronal excitability by increasing the apamin-insensitive sAHP.[15–17] Modulation of the apamin-insensitive sAHP is one of the main effector mechanisms for the ascending modulatory neurotransmitter systems controlling the functional state of the brain by setting the overall level of excitability of forebrain neurons. Such systems are most likely involved in regulating the sleep-wake cycle, arousal, attention; and in modulating sensory processing, behaviors, emotions, and cognitive functions.[18,19]

KINETICS

Apamin-sensitive sAHPs have faster kinetics than apamin-insensitive sAHPs. In some cells, such as in hippocampal interneurons,[6] guinea pig trigeminal motoneurons,[20] or septal cholinergic neurons,[16] the apamin-sensitive sAHP is maximal following an action potential and decays with a half-time on the order of hundreds of milliseconds. In contrast, apamin-insensitive AHPs such as are seen in hippocampal pyramidal neurons, have a rising phase and then decay over several seconds.[2,4] This distinction is also seen in cells that have two distinct kinetic phases to the sAHP—a relatively faster, apamin-sensitive phase, sometimes referred to as the medium, or mAHP; and a relatively slower apamin-insensitive phase, the sAHP.[10–12,21] The importance of subcellular localization and relative positioning with respect to sources of calcium for the distinct kinetics of apamin-sensitive faster AHPs and apamin-insensitive slower AHPs is emphasized by the different stimulus paradigms that effect their activation. For instance, in rat hypoglossal motoneurons,[22] cholinergic nucleus basalis neurons,[23] or basal forebrain neurons[16,24] the apamin-sensitive mAHP may be evoked by a single or a short burst of action potentials, while the slower apamin-insensitive AHP is not observed. In contrast, in these same cells long trains of action potentials evoke a long-lasting apamin-insensitive AHP.[16,22,24]

In summary, two types of sAHPs may be distinguished. One class of sAHPs is sensitive to apamin, is not regulated by neurotransmitter-induced second messengers, and is maximal following an action potential and decays over several hundred milliseconds. The other class of sAHP is apamin insensitive, is regulated by neurotransmitter-induced second messengers, and has a rising phase following an action potential, decaying over several seconds. These two classes are frequently coexpressed.

CLONES ENCODING APAMIN-SENSITIVE AND APAMIN-INSENSITIVE SK CHANNELS

We isolated and characterized three different SK channel clones from mammalian brain—SKI, SK2, and SK3.[25] The overall architecture of the SK subunits is conserved with voltage-gated potassium channels, with six membrane-spanning domains (TMs) and the N- and C-termini residing within the cell. However, the primary amino acid sequences are very different from other known potassium channels, placing the SK channels on a distinct branch of the potassium channel superfamily tree. The only notable homology with other potassium channels resides in the pore region, between the fifth and sixth TMs.[25] Among the cloned mammalian SK sequences there is a strikingly high degree of conservation. Indeed, this remarkable conservation extends evolutionarily, as the mammalian channels are very similar in structure to those from *C. elegans* and *Drosphila* (unpublished observation). There is notably less sequence conservation within the intracellular N- and C-termini. These regions also include different numbers and distributions of potential phosphorylation sites. Interestingly, even though native and cloned SK channels exhibit no voltage dependence to their gating process,[25–28] the fourth TM contains several positively charged residues (K or R), reminiscent of the voltage sensor found in voltage-gated ion channels,[29,30] suggesting a structural role for this motif. Because of the conserved topology with voltage-gated potassium channels, it is likely that four subunits assemble to form a functional channel. The site of calcium binding is not yet known; primary sequence analysis does not reveal an E-F hand,[31] C2A motif,[32] or calcium bowl sequence[33] that may mediate calcium binding. Recent studies have shown that calcium ions do not interact directly with the SK channel α subunits. Calcium gating in SK channel is due to a constitutive interaction with calmodulin and subsequent calcium-dependent conformational alterations (Xia *et al.* in press; see FIG. 1).

Heterologous expression studies show that the channels have the biophysical characteristics associated with native SK channels.[25] All three channel subtypes show the same calcium sensitivity. In macropatches, the cloned SK channels are activated by submicromolar concentrations of calcium applied to the cytoplasmic face, with half-maximal activation at 0.3 μM. Channel gating is steeply dependent upon calcium; dose-response relationships have a Hill coefficient close to 4. Macroscopic and single-channel recordings show that gating is not voltage-dependent, and the channels have unit conductances of ~10 pS in

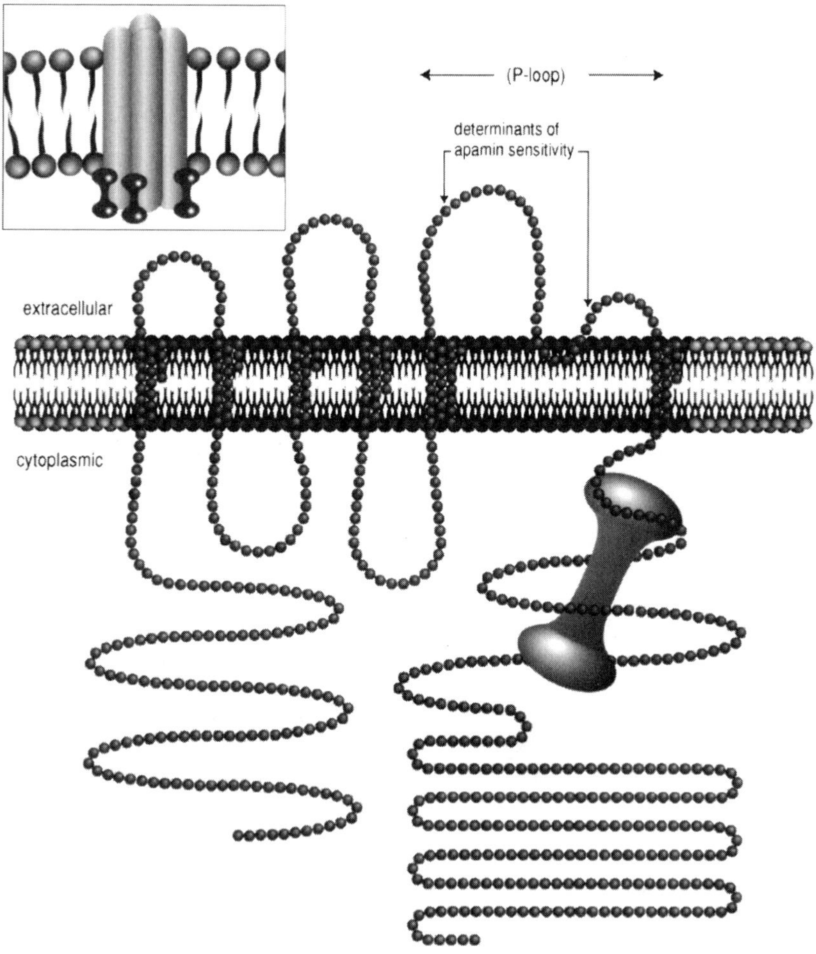

FIGURE 1. Structural representation of SK subunits. The *main panel* shows the putative subunit topology of a single SK2 subunit, with six transmembrane domains and the N- and C-termini residing within the cell. The pore region (P-loop) contains two amino acids, one on either side of the deep pore, which determine block by apamin. Calmodulin is constitutively bound to the proximal portion of the intracellular C-terminus, and within this domain, undergoes calcium-dependent interactions with the α subunit. The *inset* shows a possible model for SK channels, with a symmetrical, tetrameric stoichiometry of α subunits, each with an associated calmodulin.

symmetrical 120 mM potassium. Single-channel recordings have yielded insights into the biophysical mechanisms through which calcium ions affect SK channels.[28] Single SK2 channel activity is dependent upon calcium and is quantitatively similar to macroscopic currents. Analysis of stationary SK2 channel behavior revealed two open and three closed times, with only the longest closed time being calcium dependent, decreasing with increasing calcium concentrations. Elevated calcium concentrations also increase the relative contribution of long open times and short closures, and none of the open or closed states are altered by changing membrane voltage. Unexpectedly, the channels show two distinct gating behaviors, or modes, characterized by high or low open probability. When 1 μM calcium is applied to single SK2 channel patches, the open probability is ~0.6, but a lower open probability of ~0.05 is also observed; channels switch spontaneously and rapidly between these modes. Indeed, the frequency of mode switches and the time spent in either mode are calcium dependent. The two behaviors share many features, having similar open times and short and intermediate closed times. However, the long calcium-dependent closed time in the low open probability mode is much longer—several hundred milliseconds to seconds—than the long closed time in the high open probability mode. From these data, SK channel gating may be modeled by a scheme with four closed and two open states, which reproduces very well the single-channel data.

The pharmacological variation seen for AHPs is reflected by the cloned channels. SKI is not blocked by apamin (100 nM), while SK2 and SK3 are apamin sensitive, with IC_{50} values of 60 pM and 1 nM, respectively.[25,34] Importantly, *in situ* hybridizations in rat brain show that the distribution of SK2 and SK3 mRNA closely matches the distribution of radiolabeled apamin binding sites, and corresponds to areas in which apamin-sensitive sAHPs have been recorded.[25,35–37] SK1 is expressed in regions that have apamin-insensitive sAHPs, such as hippocampal pyramidal neurons. We have determined that the molecular basis for the difference in sensitivity to apamin is due to two residues, an aspartic acid and an asparagine, which reside on opposite sides of the deep pore and are essential for apamin sensitivity (see FIG. 1). Mutagenesis of these two residues in SKI to conform to the SK2 sequence renders SKI channels apamin sensitive. Exchange of either one of the residues found in SK2 into the sequence of SKI endows partial apamin sensitivity, while exchange of both residues renders SKI channels as sensitive to apamin as SK2 channels. SK3 channels, which have an intermediate sensitivity to apamin, contain one of the two determinant residues, consistent with the mutagenesis studies.[34] These same residues mediate differential SK channel sensitivity to d-tubocurarine[34] and are likely to be the determinants for block by other selective SK blockers. Other than pharmacology no obvious functional differences have been noted. However, the intracellular N- and C-termini of the different subunits lack the homology seen among the other domains, suggesting that they may endow functional specificity such as regulation by second messengers. The overall homology among the SK subunits suggests that they may form heteromeric channels, and this has recently been shown to occur at least for heterologously expressed subunits.[34] Since our first report of SKI, 2, and 3, no other SK channels have been cloned, nor are likely candidates yet present in the database. A closely related channel encoding an intermediate-conductance calcium-activated potassium channel (IK channel) with functional and pharmacological characteristics consistent with the Gardos channel has been reported by several groups.[38–40]

As discussed above, the AHP shows distinct time courses; relatively faster AHPs are apamin sensitive, while relatively slower AHPs are apamin insensitive. One possible mechanism for the different AHP time courses might be if the kinetics of gating were different for the channels underlying each type of AHP. However, since steady state calcium gating is not different for cloned apamin-sensitive (SK2, SK3) or apamin-insensitive (SKI) channels, it is unlikely that the kinetics of calcium activation of these channels is different. This was confirmed using a rapid, piezo-driven application of a saturating concentration of calcium, which showed that all three channels gate very rapidly, with activation time constants of <10 ms;[28] Xia *et al.*, in press). Therefore, intrinsic differences in SK channel gating do not account for the different kinetics of apamin-sensitive and apamin-insensitive AHPs, suggesting that the kinetic differences of the sAHPs reflects different rates of calcium exposure.

SK CHANNELS AND DISEASE

Apamin-sensitive SK channels have been implicated in several important physiological processes such as excitability, sleep-wake cycles, learning and memory, and digestive system activity. Application of apamin to the sensory motor portion of the inferior colliculis caused seizure activity,[41] and intracerebroventricular injections of apamin altered the circadian cycle and disrupted normal sleep patterns.[42] Rats injected with apamin prior to training paradigms showed accelerated acquisition rates and increased retention times of learned tasks compared to control animals injected after training paradigms;[43] accelerated acquisition rates correlated with increased levels of c-fos and c-jun mRNAs in the hippocampus.[44] In the periphery, apamin application to guinea pig proximal colon blocked neurotensin-induced relaxation, instead resulting in contraction.[45] More recently, a provocative study suggests that the SK3 gene may be linked to some forms of bipolar disorder.[46]

Denervated skeletal muscle[47,48] or skeletal muscle from patients with myotonic dystrophy (DM[49]) contains apamin-sensitive SK channels and radiolabeled apamin binding sites, while normal adult skeletal muscle does not. These studies suggest that apamin-sensitive SK channels may be important in these pathological conditions and that SK channels may be valuable targets for therapeutic intervention. This was dramatically illustrated by Behrens and Vergara,[50] who induced myotonic runs in the thenar muscles of a patient with myotonial congenita (MC), a disorder due to defects in a chloride channel,[51] and a patient with DM. Subsequent application of apamin was ineffective for the patient with MC; however, the myotonia was virtually abolished in the patient with DM. Therefore, even though DM is not due to a genetic lesion in one of the SK genes,[52] SK channels are central to the genesis of one of the major symptoms of DM.

CONCLUSIONS

SK channels are important regulators of neuronal excitability. The recent isolation and characterization of clones encoding SK channels present opportunities to understand the basis for the different kinetics and regulation of apamin-sensitive and apamin-insensitive sAHPs, and their roles in integrated physiological processes such as learning and memory, circadian rhythms, and sleep-wake cycles. Moreover, they may represent potential therapeutic targets for the treatment of several prominent disorders.

REFERENCES

1. Blatz, A.L. & K.L. Magleby. 1987. Calcium-activated potassium channels. Trends Neurosci. **10:** 463–467.
2. Sah, P. 1996. Ca^{2+}-activated K^+ currents in neurons: Types, physiological roles and modulation. Trends Neurosci. **4:** 150–154.
3. Madison, D.V. & R.A. Nicoll. 1984. Control of the repetitive discharge of rat CA1 pyramidal neurons *in vivo*. J. Physiol. **354:** 319–331.
4. Lancaster, B. & P.R. Adams. 1986. Calcium-dependent current generating the afterhyperpolarization of hippocampal neurons. J. Neurophys. **55:** 1268–1282.
5. Hille, B. 1992. Ionic Channels of Excitable Membranes. Sinauer Associates. Sunderland, MA.
6. Zhang, L. & C.J. McBain. 1995. Potassium conductances underlying repolarization and after-hyperpolarization in rat CA1 hippocampal interneurones. J. Physiol. **488:** 661–672.
7. Pennefather, P., B. Lancaster, P.R. Adams & R.A. Nicoll. 1985. Two distinct Ca-dependent K currents in bullfrog sympathetic ganglion cells. Proc. Natl. Acad. Sci. USA **82:** 3040–3044.
8. Goh, J.W. & P.S. Pennefather. 1987. Pharmacological and physiological properties of the after-hyperpolarization current of bullfrog ganglion neurons. J. Physiol. **394:** 315–330
9. Park, Y.B. 1994. Ion selectivity and gating of small conductance Ca^{2+}-activated K^+ channels in cultured rat adrenal chromaffin cells. J. Physiol. **481:** 555–570.
10. Sah, P. & E.M. McLachlan. 1991. Ca^{2+}-activated K^+ currents underlying the afterhyperpolarization in guinea pig vagal neurons: A role for Ca^{2+}-activated Ca^{2+} release. Neuron **7:** 257–264.
11. Schwindt, P.C., W.J. Spain & W.E. Crill. 1992. Calcium-dependent potassium currents in neurons from cat sensorimotor cortex. J. Neurophysiol. **67:** 216–226.
12. Lorenzon, N.M. & R.C. Foehring. 1992. Relationship between repetitive firing and afterhyperpolarizations in human neocortical neurons. J. Neurophysiol. **67:** 350–363.
13. Lorenzon, N.M. & R.C. Foehring. 1993. The ontogeny of repetitive firing and its modulation by norepinephrine in rat neocortical neurons. Dev. Brain. Res. **73:** 213–223,
14. Williams, S., M. Serafin, M, Muhlethaler & L. Bernheim. 1997. Distinct contributions of high- and low-voltage–activated calcium currents to afterhyperpolarizations in cholinergic nucleus basalis neurones of the guinea pig. J. Neurosci. **17:** 7307–7315.
15. Nicoll, R.A. 1988. The coupling of neurotransmitter receptors to ion channels in the brain. Science **241:** 545–551.
16. Gorelova, N. & P.B. Reiner. 1996. Role of the afterhyperpolarization in control of discharge properties of septal cholinergic neurons *in vitro*. J. Neurophysiol. **75:** 695–706.
17. Storm, J.F. 1990. Potassium currents in hippocampal pyramidal cells. Prog. Br. Res. **83:** 161–187.
18. McCormick, D.A. 1989. Cholinergic and noradrenergic modulation of thalamocortical processing. Trends Neurosci. **12:** 215–221.
19. Steriade, M. & R.W. McCarley. 1990. Brainstem Control of Wakefulness and Sleep. Plenum. New York.
20. Chandler, S.H. C.-F. Hsaio, T. Inoue & L.J. Goldberg. 1994. Electrophysiological properties of guinea pig trigeminal motoneurons recorded in vitro. J. Neurophysiol. **71:** 129–145.
21. Davies, P.J., D.R. Ireland & E.M. McLachlan. 1993. Sources of Ca2+ for different Ca2+-activated K+ conductases in neurones of the rat superior cervical ganglion. J. Physiol. **495:** 353–366.
22. Viana, F., D.A. Bayliss & A.J. Berger. 1993. Multiple potassium conductances and their role in action potential repolarization and repetitive firing behavior of neonatal rat hypoglossal motoneurons. J. Neurophysiol. **69:** 2150–2163.
23. Khateb, A., P. Fort, M. Serafin, B.E. Jones & M. Muhlethaler. 1995. Rhythmical burst induced by NMDA in guinea-pig cholinergic nucleus basalis neurones *in vitro*. J. Physiol. **487:** 623–638.
24. Matthews, R.T. & W.L. Lee. 1991. A comparison of extracellular and intracellular recordings from medial septum/diagonal band neurons *in vitro*. Neuroscience **42:** 451–462.
25. Köhler, M. *et al.* 1996. Small-conductance, calcium-activated potassium channels from mammalian brain. Science **273:** 1709–1714.

26. GRISSMER, S., R.S. LEWIS & M.D. CAHALAN. 1992. Ca2$^+$-activated K$^+$ channels in human leukemic T cells. J. Gen. Physiol. **99:** 63–84.
27. LANCASTER, B., R.A. NICOLL & D.J. PERKEL. 1991. Calcium activates two types of potassium channels in rat hippocampal neurons in culture. J. Neurosci. **11:** 23–30.
28. HIRSCHBERG, B., J. MAYLIE, J.P. ADELMAN & N.V. MARRION. 1998. Gating of recombinant small conductance Ca-activated K$^+$ channels by calcium. J. Gen. Physiol. **111:** 565–581.
29. NODA, M. *et al.* 1984. Primary structure of *Electrophorus electricus* sodium channel deduced from cDNA sequence. Nature **312:** 121–127.
30. JAN, L.Y. & Y.N. JAN. 1989. Voltage-sensitive ion channels. Cell **56:** 13–25.
31. PERSECHINI, A., N.D. MONCRIEF & R.H. KRETSINGER. 1989. The EF-hand family of calcium-modulated proteins. Trends Neurosci. **12:** 462–467.
32. SHAO, X. *et al.* 1996. Bipartite Ca^{2+}-binding motif in C_2 domains of synaptotagmin and protein kinase C. Science **273:** 248–251.
33. SCHREIBER, M. & L. SALKOFF. 1997. A novel calcium-sensing domain in the BK channel. Biophys. J. **73:** 1355–1363.
34. ISHII, T.M., J. MAYLIE & J.P. ADELMAN. 1997. Determinants of apamin and d-tubocurarine block in SK potassium channels. J. Biol. Chem. **272:** 23195–23200.
35. MOURRE, C., H. SCHMID-ANTOMARCHI, M. HUGUES & M. LAZDUNSKI. 1984. Autoradiographic localization of apamin-sensitive Ca^{2+}-dependent K$^+$ channels in rat brain. Eur. J. Pharm. **100:** 135–136.
36. MOURRE, C., J. HUGUES & M. LAZDUNSKI. 1986. Quantitative autoradiographic mapping in rat brain of the receptor of apamin, a polypeptide toxin specific for one class of Ca^{2+}-dependent K$^+$ channels. Brain Res. **382:** 239–249.
37. GELHERT, D.R. & S.L. GACHENHEIMER. 1993. Comparison of the distribution of binding sites for the potassium channel ligands [125I]apamin, [125I]charybdotoxin and [125I]iodoglyburide in the rat brain. Neuroscience **52:** 191–205.
38. ISHII, T.M., C. SILVIA, B. HIRSCHBERG, C.T. BOND, J.P. ADELMAN & J.A. MAYLIE. 1997. Human intermediate conductance calcium-activated potassium channel. Proc. Natl. Acad. Sci. USA **94:** 11651–11656.
39. JOINER, W.J., L.-Y. WANG, M.D. TANG & L.K. KACZMAREK. 1997. hSK4, a member of a novel subfamily of calcium-activated potassium channels. Proc. Natl. Acad. Sci. USA **94:** 11013–11018.
40. LOGSDON, N.J., J. KANG, J.A. TOGO, E.P. CHRISTIAN & J.A. AIYAR. 1997. A novel gene, hKCa4, encodes the calcium-activated potassium channel in human T lymphocytes. J. Biol. Chem. **272:** 32723–32726.
41. MCCOWN, T.J. & G.R. BREESE. 1990. Effects of apamin and nicotinic acetylcholine receptor antagonists on inferior collicular seizures. Eur. J. Pharm. **187:** 49–58.
42. GANDOLFO, G., H. SCHWEITZ, M. LAZDUNSKI & C. GOTTESMANN. 1996. Sleep cycle disturbances induced by apamin, a selective blocker of Ca^{2+}-activated K$^+$ channels. Brain Res. **736:** 344–347.
43. MESSIER, C. *et al.* 1991. Effect of apamin, a toxin that inhibits Ca^{2+}-dependent K$^+$ channels, on learning and memory processes. Brain Research **551:** 322–326.
44. HEURTEAUX, C., C. MESSIER, C. DESTRADE & M. LAZDUNSKI. 1993. Memory processing and apamin induce immediate early gene expression in mouse brain. Mol. Brain Res. **3:** 17–22.
45. HUGUES, M. *et al.* 1982. The Ca2+-dependent slow K+ conductance in cultured rat muscle cells: Characterization with apamin. EMBO **9:** 1039–1042.
46. CHANDY, K.G., E. FATINO, K. KALMAN, G.A. GUTMAN & J.J. GARGUS. 1997. Gene encoding neuronal calcium-activated potassium channel has polymorphic CAG repeats, a candidate role in excitotoxic neurodegeneration and maps to 22ql1-ql3, critical region for bipolar disease and schizophrenia disorder 4. Am. Soc. Human Genet. Abst.
47. BLATZ, A.L. & K.L. MAGLEBY. 1986. Single apamin-blocked Ca-activated K+ channels of small conductance in cultured rat skeletal muscle. Nature **323:** 718–720.
48. SCHMID-ANTOMARCHI, H. *et al.* 1985. The all-or-none role of innervation in expression of apamin receptor and of apamin-sensitive Ca^{2+}-activated K$^+$ channel in mammalian skeletal muscle. Proc. Natl. Acad. Sci. USA **82:** 2188–2191.
49. RENAUD, J.F. *et al.* 1986. Expression of apamin receptor in muscles of patients with myotonic muscular dystrophy. Nature **319:** 678–680.

50. Behrens, M.I., P. Jalil, A. Serani, F. Vergara & O. Alvarez. 1994. Possible role of apamin-sensitive K+ channels in myotonic dystrophy. Muscle Nerve **17:** 1264–1270.
51. George, J., A.L.Crackower, M.A. Abdalla, J.A. Hudson & G.C. Ebers. 1993. Molecular basis of Thomsen's disease (autosomal dominant myotonia congenita). Nature Genetics **3:** 305–309.
52. Harley, H.G. *et al.* 1992. Expansion of an unstable DNA region and phenotypic variation in myotonic dystrophy. Nature **355:** 545–546.

The Functional Role of Alternative Splicing of Ca^{2+}-Activated K^+ Channels in Auditory Hair Cells

E. M. C. JONES, M. GRAY-KELLER, J. J. ART, AND R. FETTIPLACE[a]

Department of Physiology, University of Wisconsin Medical School, Madison, Wisconsin 53706, USA

ABSTRACT: Turtle auditory hair cells are frequency tuned by the activity of large-conductance calcium-activated potassium (KCa) channels, the frequency range being dictated primarily by the channel kinetics. Seven alternatively spliced isoforms of the KCa channel alpha-subunit, resulting from exon insertion at two splice sites, were isolated from turtle hair cells. These, when expressed in *Xenopus* oocytes, produced KCa channels with a range of apparent calcium sensitivities and channel kinetics. However, most expressed channels were less calcium sensitive than the hair cells' native KCa channels. Coexpression of alpha-subunit with a bovine beta-subunit substantially increased the channel's calcium sensitivity while markedly slowing its kinetics, but kinetic differences between isoforms were preserved. These data suggest a molecular mechanism for hair cell frequency tuning involving differential expression of different KCa channel alpha-subunits in conjunction with an expression gradient of a regulatory beta-subunit.

The large-conductance Ca^{2+}-activated K^+ (BK_{Ca}) channel is formed from a multimeric complex of a pore-forming α-subunit and a regulatory β-subunit.[1] The channel's α-subunit, cloned from both invertebrates[2] and vertebrates,[3,4] comprises an N-terminus similar to a voltage-gated K_V channel appended to a long C-terminus that incorporates the intracellular Ca^{2+}-binding domain.[5] The α-subunit arises from a single *slo* gene that displays extensive alternative splicing in the cytoplasmic tail region,[3,4,6] but the physiological significance of the different isoforms is obscure. Additional complexity is conferred by the β-subunit, which substantially augments the Ca^{2+} sensitivity endowed by the α-subunit.[7]

A specialized deployment of BK_{Ca} channels is in the electrical tuning of auditory hair cells, where in combination with voltage-dependent Ca^{2+} channels they cause the hair cells to be maximally responsive to particular sound frequencies.[8–10] This role demands that the BK_{Ca} channels, triggered via changes in intracellular Ca^{2+}, open or close with a time course that determines the frequency to which the hair cell is tuned. In the turtle cochlea, hair cells are tuned to different frequencies, a cell's optimal frequency being correlated with the kinetic properties of its BK_{Ca} channels.[8,11,12] Such variation might be accomplished by expression in different hair cells of channel isoforms with kinetically distinct properties.[13] BK_{Ca} α-subunits have been cloned from a chick cochlear library[14] and shown to exhibit extensive alternate splicing.[15,16] Alternative splice variants have also been identified from turtle cochlear hair cells,[17] but in both chick and turtle the functional consequences of the alternative splicing patterns are unknown. Here we describe the channel

[a]Address for correspondence: Robert Fettiplace, 185 Medical Sciences Building, 1300 University Avenue, Madison, Wisconsin 53706. Phone: 608-262-9320; fax: 608-265-5512; e-mail: fettiplace@physiology.wisc.edu

properties of α-subunit variants and show that, by appropriate mixing of these variants with or without β-subunit, it is possible to achieve a range of properties commensurate with those in native hair cells.

METHODS

Auditory hair cells were dissociated from the basilar papilla of the red-eared turtle, *Trachemys scripta elegans,*[8] and alternatively spliced Ca^{2+}-activated K^+ channel transcripts were obtained by RT-PCR amplification of hair cell RNA.[17] Splice-site numbering is identical to that in Jones *et al.*[17] Full-length expression constructs were obtained by ligating overlapping PCR fragments that encode the C-terminal two-thirds of turtle *tSlo* (nts 1085–3560 in the smallest variant) with the N-terminal *cSlo*1 (nts 1–1231, a gift from P. Fuchs, differing from the same region in *tSlo* in seven mainly conservative substitutions) into the oocyte expression vector pGH19. Expression constructs are named according to the number of amino acids inserted at splice sites 1 and 2. Both strands of a 600–base pair region containing these splice sites of all expression constructs were sequenced on an ABI automated sequencer. Bovine Ca^{2+}-activated K^+ channel β-subunit (a gift from R. Swanson) was ligated into pGH19 vector for RNA production.

Oocytes were isolated from *Xenopus laevis* ovaries and maintained using standard procedures. Within 24 hours of isolation, they were injected with 10–30 ng of α-subunit cRNA with or without various amounts of bovine β-subunit cRNA. Altering the mole ratio of injected RNA for the β- to α-subunit from 2:1 to 0.1:1 produced currents with similar properties. Electrical recordings were made at 23 °C from inside-out membrane patches detached from oocytes 4 to 24 days postinjection. Electrodes were filled with a solution (composition in mM: 110 KMethyl sulfate; 2 KCl; 5 KEGTA; 10 KHEPES, pH 7.4) and had resistances of 0.5–1 MΩ for patches containing between 100 and 1000 channels. Patches were exposed to different solutions of composition (in mM): 110 KMethyl sulfate; 2 KCl; 2 K_4dibromoBAPTA (Molecular Probes, Eugene, OR); 1 dithiothreitol; 10 KHEPES, pH 7.4, with addition of $CaCl_2$ to yield free Ca^{2+} concentrations from 0.3–20 μM. Solutions with more than 20 μM free Ca^{2+} contained no dibromoBAPTA. All Ca^{2+} activities were verified with an MI-600 Ca^{2+} electrode (Microelectrodes Inc., NH). Single BK_{Ca} channel currents were recorded from inside-out patches from hair cells derived from known regions of the cochlea as previously described.[11] These patches were exposed to solutions of composition (in mM): 130 KCl; 0.5 $MgCl_2$; 1 MgATP; 2 K_4dibromoBAPTA: 5 KHEPES, pH 7.4, with different free Ca^{2+} concentrations.

RESULTS AND DISCUSSION

The Hair Cell's Native BK_{Ca} Channels

A hair cell's frequency tuning can be characterized by delivering small current steps that evoke a damped oscillation in membrane potential, reminiscent of an electrical resonance.[18] The resonant frequency derived from the period of the oscillations (FIG. 1A) ranges among cells from about 30 Hz to 600 Hz and matches the frequency to which the cell is most sensitive acoustically. The hair cells are mapped "tonotopically" along the cochlea, with their resonant frequency increasing monotonically from the apical to basal

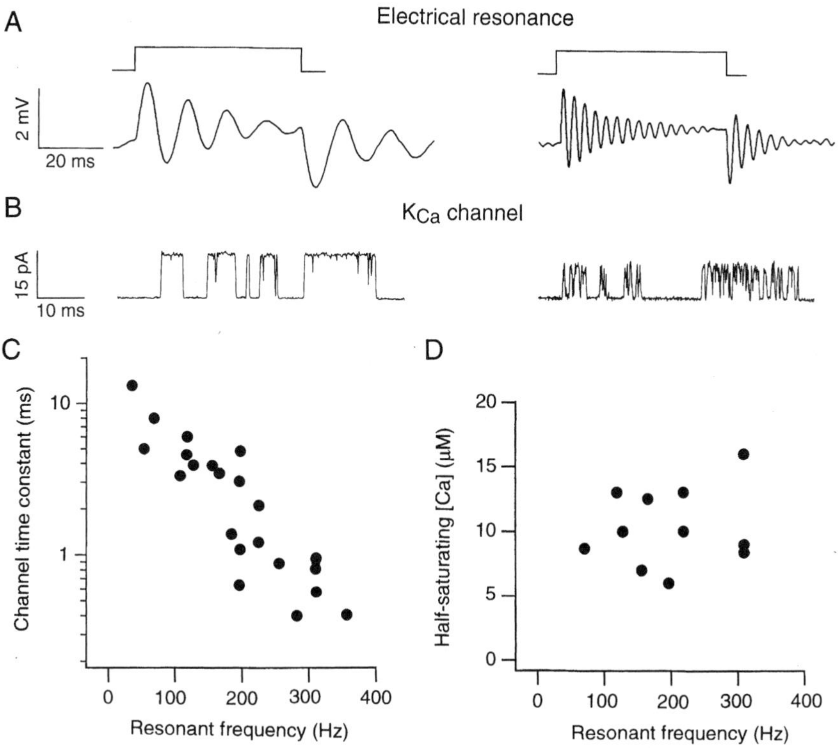

FIGURE 1. Properties of hair cell's native BK_{Ca} channels. **A.** Examples of electrical resonance in two turtle hair cells elicited by depolarizing current steps from the resting potential (−50 mV). Resonant frequencies were 75 Hz (**left**) and 309 Hz (**right**). **B.** Examples of single BK_{Ca} channels recorded from two cells with a low (**left**) and high resonant frequency. **C.** Correlation between the BK_{Ca} channel time constant and the hair cell's resonant frequency. Channel time constants derived from the relaxation of the ensemble-average current at −50 mV in 4 μM intracellular Ca^{2+}. **D.** Ca^{2+} sensitivities of BK_{Ca} channels, inferred from the half-saturating Ca^{2+} concentration at −50 mV, in hair cells of different resonant frequency.

end of the organ. The BK_{Ca} current is the major voltage-dependent current in hair cells, and its size and kinetics are correlated with resonant frequency.[8,12] Hair cells tuned to higher frequencies possess both larger and faster BK_{Ca} currents, implying that the number of BK_{Ca} channels per cell and their intrinsic properties change systematically with hair cell location.

This conclusion was confirmed by recordings of single BK_{Ca} channels, which demonstrated a similar unitary conductance of ~320 pS (in symmetrical K^+), but a systematic variation in channel kinetics in hair cells tuned to different frequencies (FIG. 1). Channel kinetics were inferred from the time constant of relaxation of ensemble-average currents at −50 mV, similar to the cell's resting potential *in vivo*. This time constant, which paralleled changes in the channel's mean open time, varied inversely with resonant frequency from

14 ms to 0.3 ms. An important observation, pertinent to the behavior of the cloned channel, was that there was no obvious trend in the channel's Ca^{2+} sensitivity with resonant frequency (FIG. 1D); the mean half-saturating Ca^{2+} was approximately 2 μM at +50 mV and 12 μM at −50 mV. One mechanism that could account for the difference in BK_{Ca} channel properties involves the differential expression of kinetically distinct variants of the K_{Ca} channel in different regions of the cochlea.[13] To test this hypothesis, we used RT-PCR to identify the channel variants present in the turtle hair cells.[17]

Properties of Cloned α-Subunits of the BK_{Ca} Channel

We initially identified six alternatively spliced cDNAs homologous to the *slo* gene that encodes the BK_{Ca} channel's α-subunit. The minimal sequence generates an 1166–amino acid protein that is 94% identical at the amino acid level to a mouse brain *Slo* form, mbr8.[3] Sequence variation occurs by insertion of exons at two splice sites lying between the eighth and ninth amphipathic α-helical segments in the extended intracellular C-terminus. These splice sites will be labeled SS1 and SS2, according to Reference 4, though additional splice sites have now been described nearer to the 5′ end.[15,16] The six variants, denoted by the number of amino acids (aa) inserted at SS1 and SS2, are 0:0, 4:0, 4:3, 4:61, 31:3, and 31:0. The cysteine-rich 61-aa insert is similar to the STREX-2 exon recently described in rat adrenal chromaffin cells, the expression of which was found to be under control of the hormone ACTH.[19] The 31-aa insert is novel and is composed of 4 aa + 27 aa, where the 4-aa sequence (**SRKR**) occurs alone in three further variants. Since we were unable to detect splicing at other potential sites, we suggest that these six variants correspond to naturally occurring combinations. We have subsequently obtained a seventh variant by amplification of hair cell RNA using a forward primer specific for **SRKR** in splice site 1, and a reverse primer downstream of splice site 2. Remarkably, this variant was smaller than the expected "minimal" sequence due to a 26-aa deletion encompassing splice site 2, and has therefore been labeled 4:−26 (GenBank accession number AF086646). Generation of the 4:−26 variant requires an additional splice site in this region, and implies a complexity of splicing reminiscent of *dSlo* at site G,[6] which is equivalent to SS2 in vertebrate *Slo*.

All isoforms, including the 4:−26 variant, produced functional Ca^{2+}-activated K^+ channels of large unitary conductance (286 ± 7 pS; mean ± 1 SEM) when expressed in *Xenopus* oocytes. The main feature distinguishing the variants was their apparent Ca^{2+} sensitivity. The half-saturating Ca^{2+} concentration at a membrane potential of −50 mV varied from 6 μM for the 4:61 to greater than 200 μM for the 4:−26. The disparities in Ca^{2+} sensitivity were also evident in the range of membrane potentials required to half-activate ($V_{1/2}$) the channel (FIG. 2C). In 12 μM Ca^{2+}, there was a 65-mV difference between the channels of greatest (4:61) and least (4:−26) sensitivity. There were also differences in kinetics among the isoforms with relaxation time constants at −50 mV ranging between 0.9 and 3.6 ms (FIG. 2D). Such kinetic differences were related to the Ca^{2+} sensitivities, with the 4:−26 isoform being the fastest and the 4:61 isoform being the slowest. The relaxation time constants and mean open times increased with Ca^{2+} concentration, and consequently, at saturating Ca^{2+} levels, the kinetic range among the different variants diminished.

The performance of the cloned α-subunit variants (FIG. 2C) suggests that the sequence around splice site 2 influences the Ca^{2+} sensitivity of the BK_{Ca} channel, which can be enhanced by large inserts and reduced by deletions in this region. Whether the sequence

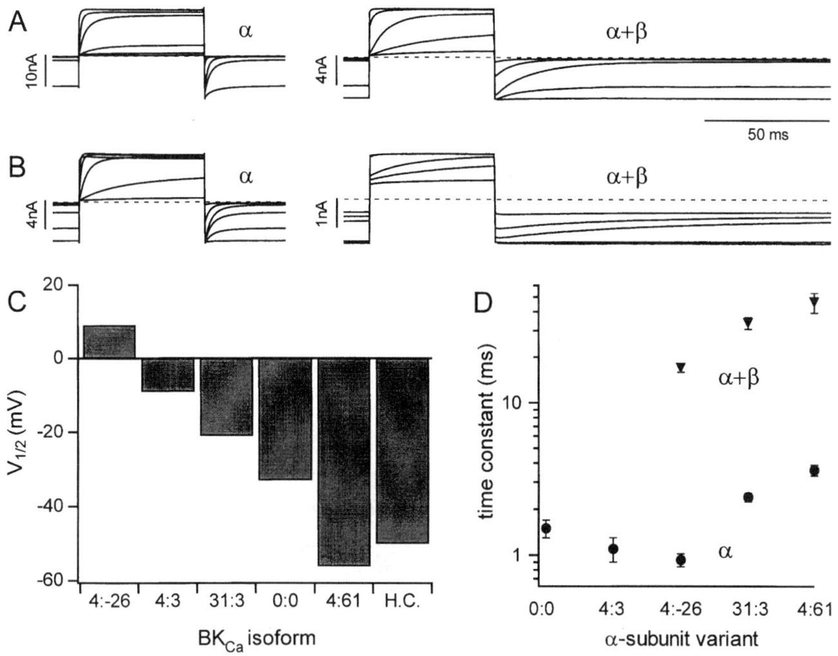

FIGURE 2. Properties of BK_{Ca} channel isoforms expressed in *Xenopus* oocytes. **A.** BK_{Ca} currents in response to voltage steps from −50 to +50 mV in a range of intracellular Ca^{2+} concentrations for the 4:−26 α subunit alone (**left**) and the 4:−26 α + β subunits (**right**). **B.** BK_{Ca} currents for 4:61 α subunit (**left**) and 4:61 α + β subunits (**right**). In both **A** and **B,** the Ca^{2+} concentrations were 0.4, 1, 2.5, 5, 20, and 300 μM (α-subunit) and 0.4, 1, 2.5, 5, and 20 μM (α- +β-subunits). **C.** The membrane potential ($V_{1/2}$) at which different channel isoforms were half-activated in 12 μM intracellular Ca^{2+} compared with the hair cell's native channel (H.C.). 12 μM is the mean Ca^{2+} concentration for half-activation of the native channel at the hair cell's resting potential of −50 mV. **D.** Channel time constants for different α-subunit isoforms (*circles*) or for α- +β-subunits (*triangles*) derived from fits to current relaxations at −50 mV. Internal Ca^{2+}, 5 μM except for 4:61 and 31:3 α- +β-subunits where more accurate measurements obtained at 2.5 μM Ca^{2+}. All α-subunit variants are specified in terms of the amino acid inserts at splice sites 1 and 2.

constitutes part of the Ca^{2+} binding site is unclear since at least one portion of the Ca^{2+} sensor is thought to lie a significant distance downstream of splice site 2 just proximal to the tenth amphipathic α-helix.[20] From a physiological perspective, all of the α-subunit variants apart from the 4:61 isoform were less Ca^{2+} sensitive at −50 mV than the hair cell's native BK_{Ca} channel. However, the kinetics lie within the correct range of hair cell values.

Coexpression of α- and β-Subunits

Coexpression of a bovine β-subunit with different α-subunit variants dramatically augmented the channel's Ca^{2+} sensitivity to the extent that it exceeded that of any of the α-subunits alone. The increased Ca^{2+} sensitivity was also manifested by a 100-mV hyperpolarizing shift in the half-activation ($V_{1/2}$) for a given Ca^{2+} concentration. The total shift in $V_{1/2}$ was similar for the two variants (4:-26 and 4:61), which were chosen since they spanned the range of Ca^{2+} sensitivities for channels composed of α-subunit alone. Thus for the 4:−26, $V_{1/2}$ in 5 μM Ca^{2+} shifted from 28 mV to −84 mV; and for the 4:61, $V_{1/2}$ shifted from −52 mV to −150 mV. This suggests that the potentiation endowed by the β-subunit may operate via a mechanism separate from that produced by exon insertion at splice site 2. A second consequence of β-subunit coexpression was a marked slowing of the channel kinetics.[21] As with the observed shifts in $V_{1/2}$, the retardation in kinetics preserved the original order of kinetic differences. Thus for the 4:−26 α-subunit the time constant at −50 mV increased from 0.9 to 17 ms; and for the 4:61, it increased from 3.6 to 45 ms, implying that the β-subunit magnified existing kinetic disparities. These observations may be explained by postulating that a major action of the β-subunit is to stabilize the channel's open state.

Two conclusions may be drawn from comparison of the properties of the cloned and native channels. First, with the exception of the 4:61 variant, the Ca^{2+} sensitivities of the α-subunits alone are lower than the native channels (FIG. 2C), which suggests that if the other α-subunits are employed in frequency tuning of the hair cell, they are likely to be complexed with a β-subunit. Second, if only those combinations with high-Ca^{2+} sensitivity are considered, then their kinetics at −50 mV range from 3.6 ms (4:61 α-subunit) to 45 ms (4:61 α- + β-subunit). While the overall kinetic range is roughly comparable for the native and cloned channels, the absolute time constants are severalfold faster for the native channels. For example, the fastest isoform, 4:−26 α alone, had a time constant of 0.9 ms compared to the fastest time constant of 0.3 ms for the native channel; the 4:61 α + β had a time constant of 45 ms compared to the slowest time constant of 14 ms for the native channel. One possible explanation for this disparity might be the existence of a specialized auxiliary subunit in hair cells that exerts greater influence on the Ca^{2+} sensitivity than on the kinetics. Nevertheless, our results clearly demonstrate that use of different splice variants of the α-subunit is likely to be a major factor generating the frequency map in the turtle cochlea. Furthermore, differential expression of different isoforms[17] may need to be supplemented by a cochlear gradient in the β-subunit with the highest concentration at the low-frequency end.

ACKNOWLEDGMENTS

This research was supported by grants to R.F from NIDCD (DC01362) and the Steenbock Fund of the University of Wisconsin. We thank M. Moscu for technical assistance.

REFERENCES

1. KACZOROWSKI, G.J., H.-G. KNAUS, R.J. LEONARD, O.B. MCMANUS & M.L. GARCIA. 1996. High-conductance calcium-activated potassium channels: Structure pharmacology and function. J. Bioenerg. Biomembr. **28:** 255–267.
2. ATKINSON, N.S., G.A. ROBERTSON & B. GANETZKY. 1991. A component of calcium-activated potassium channels encoded by the *Drosophila slo* locus. Science **253:** 551–555.
3. BUTLER, A., S. TSUNODA, D.P. MCCOBB, A. WEI & L. SALKOFF. 1993. *mSlo*, a complex mouse gene encoding "maxi" calcium-activated potassium channels. Science **261:** 221–224.

4. TSENG-CRANK, J., C.D. FOSTER, J.D. KRAUSE, R. MERTZ, N. GODINOT, T.J. DICHIARA & P.H. REINHART. 1994. Cloning, expression, and distribution of functionally distinct Ca^{2+}-activated K^+ channel isoforms from human brain. Neuron **13:** 1315–1330.
5. WEI, A., C. SOLARO, C. LINGLE & L. SALKOFF. 1994. Calcium-sensitivity of BK-type K_{Ca} channels determined by a separable domain. Neuron **13:** 671–681.
6. LAGRUTTA, A., K.-Z. SHEN, R.A. NORTH & J.P. ADELMAN. 1994. Functional differences among alternatively spliced variants of *slowpoke*, a *Drosophila* calcium-activated potassium channel. J. Biol. Chem. **269:** 20347–20351.
7. MCMANUS, O.B., L.M. HELMS, L. PALLANCK, B. GANETZKY, R. SWANSON & R.J. LEONARD. 1995. Functional role of the β-subunit of high conductance calcium-activated potassium channels. Neuron **14:** 645–650.
8. ART, J.J. & R. FETTIPLACE. 1987. Variation of membrane properties in hair cells isolated from the turtle cochlea. J. Physiol. **385:** 207–242.
9. HUDSPETH, A.J. & R.S. LEWIS. 1988. Kinetic analysis of voltage- and ion-dependent conductances in saccular hair cells of the bullfrog Rana catesbeiana. J. Physiol. **400:** 237–274.
10. FUCHS, P.A., T. NAGAI & M.G. EVANS. 1988. Electrical tuning in hair cells isolated from the chick cochlea. J. Neurosci. **8:** 2460–2467.
11. ART, J. J., WU, Y.-C. & R. FETTIPLACE. 1995. The calcium-activated potassium channels of turtle hair cells. J. Gen. Physiol. **105:** 49–72.
12. WU, Y-C., J.J. ART, M.B. GOODMAN & R. FETTIPLACE. 1995. A kinetic description of the calcium-activated potassium channel and its application to electrical tuning of hair cells. Prog. Biophys. Mol. Biol. **63:** 131–158.
13. WU, Y-C. & R. FETTIPLACE. 1996. A developmental model for generating frequency maps in the reptilian and avian cochleas. Biophys. J. **70:** 2557–2570.
14. JIANG, G., M. ZIDANIC, R.L. MICHAELS, T.H. MICHAEL, C. GRIGUER & P.A. FUCHS. 1997. *cSlo* encodes calcium-activated potassium channels in the chick's cochlea. Proc. R. Soc. Lond. B. **264:** 731–737.
15. NAVARATNAM, D.S., T.J. BELL, T.D. TIU, E.L. COHEN & J.C. OBERHOLTZER. 1997. Differential distribution of Ca^{2+}-activated K^+ channel splice variants among hair cells along the tonotopic axis of the chick cochlea. Neuron **19:** 1077–1085.
16. ROSENBLATT, K.P., Z.-P. SUN, S. HELLER & A.J. HUDSPETH. 1997. Distribution of Ca^{2+} -activated K^+ channel isoforms along the tonotopic gradient of the chicken's cochlea. Neuron **19:** 1061–1075.
17. JONES, E.M.C., C. LAUS & R. FETTIPLACE. 1998. Identification of Ca^{2+}-activated K^+ channel splice variants and their distribution in the turtle cochlea. Proc. R. Soc. Lond. B. **265:** 685–692.
18. CRAWFORD, A.C. & R. FETTIPLACE. 1981. An electrical tuning mechanism in turtle cochlear hair cells. J. Physiol. **312:** 377–412.
19. XIE, J. & D.P. MCCOBB. 1998. Control of alternative splicing of potassium channels by stress hormones. Science **280:** 443–446.
20. SCHREIBER, M. & L. SALKOFF. 1997. A novel calcium sensing domain in the BK channel. Biophys. J. **73:** 1355–1363.
21. DWORETZKY, S.I., C.G BOISSARD, J.T. LUM-REGAN, M.C. MCKAY, D.J. POST-MUNSON, J.T. TROJNACKI, C.-P. CHANG & V.K. GRIBKOFF. 1996. Phenotypic alteration of a human BK (*hSlo*) channel by *hSlo*β subunit coexpression: Changes in blocker sensitivity, activation/relaxation and inactivation kinetics and protein kinase A modulation. J. Neurosci. **16:** 4543–4550.

Structure, G Protein Activation, and Functional Relevance of the Cardiac G Protein–Gated K^+ Channel, I_{KACh}

KEVIN WICKMAN,[a] GRIGORY KRAPIVINSKY,[a] SHAWN COREY,[a] MATT KENNEDY,[a] JAN NEMEC,[b] IGOR MEDINA,[a] AND DAVID E. CLAPHAM[c,d]

Departments of [a]Cardiology and [c]Neurobiology, Howard Hughes Medical Institute, Harvard Medical School, Children's Hospital, 320 Longwood Avenue, Boston, Massachusetts 02115, USA

[b]Division of Cardiovascular Diseases, Department of Internal Medicine, Mayo Clinic, 200 First Street SW, Rochester, Minnesota 55905, USA

ABSTRACT: The muscarinic-gated atrial potassium channel I_{KACh} has been well characterized functionally, and has been an excellent model system for studying G protein/effector interactions. Complementary DNAs encoding the composite subunits of I_{KACh} have been identified, allowing direct probing of structural and functional features of the channel. Here, we highlight recent approaches taken in our laboratory to determine the oligomeric structure of native cardiac I_{KACh}, the mechanism of activation of I_{KACh} by G proteins, and the relevance of I_{KACh} to cardiac physiology.

Inwardly rectifying K^+ channels are present in both excitable and nonexcitable tissues, where they contribute to such diverse functions as setting the resting membrane potential, regulating insulin secretion, controlling heart rate, and influencing neuronal excitability.[1] Inwardly rectifying K^+ channels pass K^+ ions more effectively in the inward direction since outward flux is blocked by positively charged cytosolic substances such as Mg^{2+} and polyamines.[2–4] It should be noted, however, that it is the small amount of outward current carried by these channels that is physiologically relevant. Recent cloning efforts have revealed a large number of primary subunits that presumably form native inwardly rectifying K^+ channels, all of which share common structural features.[1] All identified subunits possess two hydrophobic domains predicted to span the membrane and surround a pore sequence that shares a high degree of similarity to other K^+ selective channels. The muscarinic-gated atrial K^+ channel, I_{KACh}, belongs to this subfamily of K^+ channels; its native structure, distribution, regulation, and physiological function have been areas of interest to our laboratory for several years.[5]

Cardiac I_{KACh} is named for the first identified agonist, but it and its neuronal relatives are also activated by a large number of neurotransmitters, including somatostatin, GABA, and adenosine.[6,7] The neurotransmitters and hormones that activate I_{KACh} all stimulate pertussis toxin–sensitive G proteins of the $G_{i/o}$ subclass, leading to the dissociation of the G protein complex into $G\beta\gamma$ and GTP-bound $G\alpha$ subunits. While many channels are modulated to a small extent as a result of G protein modulation of second messenger levels, I_{KACh} activity is rapidly increased approximately 1000-fold upon full G protein stimulation

[d]Corresponding author. Phone: 617-355-6163; fax: 617-730-0692; e-mail: clapham@rascal.med.harvard.edu

due to a direct Gβγ/channel interaction.[8–12] Other studies, which will not be discussed here, indicate that I_{KACh} activity can also be regulated by intracellular Na^+ and phosphotidyl inositol bisphosphate (PIP_2), phosphorylation, as well as membrane stretch.[13–15]

This paper is designed to provide an overview of recent studies undertaken in our laboratory to better understand the native composition of I_{KACh}, the G protein regulation of channel activity, and the physiological relevance of this conductance; for more detailed descriptions of the approaches, results, and interpretations, please consult the following recent publications.[16–18]

QUATERNARY STRUCTURE OF CARDIAC I_{KACh}

Cardiac I_{KACh} is composed of two homologous G protein–regulated inwardly rectifying K^+ (GIRK) channel subunits, termed GIRK1 and GIRK4.[19–21] GIRK1 and GIRK4, along with GIRK2 and GIRK3, comprise the mammalian G protein–coupled subfamily of inward-rectifying K^+ channel (IRK) subunits. The GIRK1-GIRK4 physical interaction was revealed initially by coimmunoprecipitation from solubilized atrial membranes.[21] Coexpression of GIRK1 with GIRK4 in oocyte and mammalian cell systems yielded channels with biophysical characteristics identical to I_{KACh}.[21] In contrast, transient transfection of mammalian cell lines with GIRK1 alone yielded no novel channel activity, despite the expression of GIRK1 protein.[21,22] In the absence of GIRK4, GIRK1 colocalizes with cytoskeletal elements.[22] Expression of GIRK2, GIRK3, or GIRK4 alone in oocyte, mammalian, or insect cells yielded G protein–activated inwardly rectifying K^+ conductances; key biophysical characteristics of these channels, such as single channel conductance and mean open time, however, were inconsistent with previously described native channels.[21,23] Together, these findings suggest that GIRK1 forms native channels together with the functional homologs GIRK2, GIRK3, and/or GIRK4, depending on which subunits are available in a given cell.

Several groups have suggested that, like voltage-gated K^+ channels and a recently crystallized and structurally related bacterial channel, eukaryotic inwardly rectifying K^+ channels are tetramers.[24–27] To date, most studies of inwardly rectifying K^+ channel stoichiometry have involved concatemeric constructs. Using this approach, GIRK1 and GIRK4 were suggested to form a tetramer in 1:1 stoichiometry.[25] Subsequently, it was shown that the GIRK1-GIRK4-GIRK1-GIRK4, rather than the GIRK1-GIRK1-GIRK4-GIRK4 tandem arrangement, produced a higher ratio of agonist-induced to basal current.[28] Conclusions from concatemer-based studies, however, are limited by the following assumptions: (1) combinations of tandemly linked subunits that yield the most current are the most representative of the native channel configuration; (2) tandemly linked subunits do not coassemble with other tandemly linked subunits; and (3) tandemly linked subunits are completely translated and are not proteolytically cleaved to release individual subunits. Unfortunately, each of these assumptions has been shown to be incorrect in certain instances.[25,29] To further address the oligomeric nature of the native channel, we purified bovine cardiac I_{KACh} to near homogeneity, and used several independent methods to estimate the quaternary structure of the complex and stoichiometry of the composite subunits.[17]

Protein complexes containing both GIRK1 and GIRK4 were purified from bovine atrial plasma membranes using a multistep scheme designed to exclude homomultimeric channels and dissociated monomers. The final product was purified to greater than 95% homogeneity. Aliquots of purified native I_{KACh} were analyzed by SDS-PAGE and silver staining or sequential immunoblotting with GIRK4- and GIRK1-specific antibodies (FIG. 1a). The predominant bands revealed by silver staining were recognized by anti-GIRK1 (54 and 56–76 kDa) and anti-GIRK4 (48 kDa) antibodies. The ratio of band intensities for GIRK1:GIRK4 in the silver-stained gels, when corrected for the molecular weight difference between the two subunits, was consistent with a molar ratio of 1:1. The high degree of similarity between GIRK1 and GIRK4 (57% overall amino acid identity) minimizes the potential inaccuracies of protein quantification due to differential sensitivities of the subunits to silver staining.

To determine the number of subunits within the native I_{KACh} complex, we extensively cross-linked the purified channel. The high degree of native channel protein purity was key to our approach since it eliminated potential nonspecific cross-linking between native

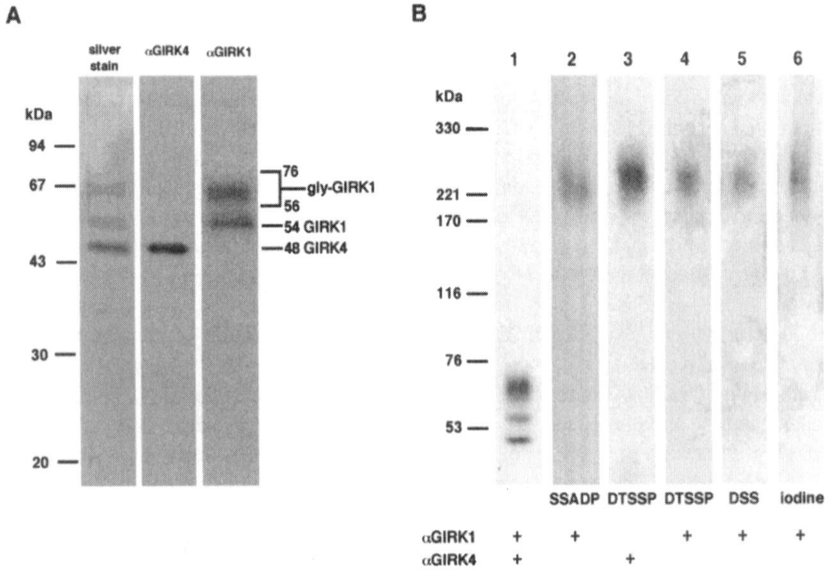

FIGURE 1. Oligomeric structure of native cardiac I_{KACh}. **(A)** Stoichiometry of GIRK1 and GIRK4 in purified native I_{KACh}. Purified native I_{KACh} was silver stained or immunoblotted with anti-GIRK4 antibodies (αGIRK4) or anti-GIRK1 antibodies (αGIRK1). GIRK1 is present in glycosylated (56–76-kDa) and unglycosylated (54-kDa) forms. Densitometry analysis of lane 1 was consistent with a 1:1 GIRK1:GIRK4 stoichiometry. **(B)** Complete cross-linking of native I_{KACh} yields a single product most consistent with tetrameric formation. Native I_{KACh} was treated with a wide variety of cross-linking reagents, and products were then analyzed by SDS-PAGE and immunoblotted. **Lane 1**, no cross-link control immunoblotted with αGIRK1 and αGIRK4 antibodies. **Lane 2**, solubilized atrial membrane proteins treated with 3 mM SSADP followed by photolysis for 1 min and immunoblotted with αGIRK1 antibodies. **Lane 3**, pure I_{KACh} treated for 1 h with 3 mM DTSSP and immunoblotted with αGIRK1 antibodies. **Lane 4**, same as lane 3, but stripped and reimmunoblotted with αGIRK4 antibodies. **Lane 5**, pure I_{KACh} treated for 1 h with 1 mM DSS and immunoblotted with αGIRK4 antibodies. **Lane 6**, pure I_{KACh} treated for 1 h with 50% saturated iodine and immunoblotted with αGIRK4 antibodies. Molecular weight markers were run with lanes 1 and 2; lanes 3–6 were derived from separate gels and aligned based on corresponding molecular weight standards.

I_{KACh} and unrelated membrane proteins. Purified I_{KACh} was first treated with the cross-linking agent dithiobis[sulfosuccinimidylpropionate]. This reaction yielded a single product, recognized by both GIRK1- and GIRK4-specific antibodies, with a molecular weight of 234 ± 7 kDa (FIG. 1b). Similar results were obtained with other cross-linking agents. By comparison, a tetramer containing two GIRK1 and two GIRK4 subunits is predicted to have a molecular weight of ~226 kDa. To test whether the native complex is actually an integral multiple of the ~235-kDa complex, atrial membranes were treated with the highly reactive cross-linking agent sulfosuccinimidyl[4-azidophenyldithio]propionate prior to channel purification. Again, a single product of 224 ± 3 kDa band was recognized by GIRK1-specific antibodies. Thus the molecular weight of the native cross-linked complex is consistent with a tetrameric structure. Furthermore, partial cross-linking of purified bovine I_{KACh} yielded four adducts representing monomers, dimers, trimers, and tetramers of GIRK subunits in various combinations. Three species of dimers, GIRK1-GIRK1, GIRK1-GIRK4, and GIRK4-GIRK4, were detected based upon interpretation of molecular weights and reactivity toward subunit-specific antibodies. The formation of GIRK1-GIRK1 and GIRK4-GIRK4 cross-linked subunits within the native I_{KACh} tetramer provided further support that the I_{KACh} tetramer is composed of two GIRK1 and two GIRK4 subunits. Interestingly, similar approaches with recombinant GIRK1 and GIRK4 have shown that both form homotetrameric complexes in heterologous expression systems.[17]

UNDERSTANDING THE MECHANISM OF I_{KACh} ACTIVATION BY Gβγ

Previous experiments demonstrated that Gβγ binds to immunopurified native I_{KACh}, as well as to the independently expressed recombinant GIRK1 and GIRK4 subunits, with comparable affinities.[10] To further localize the channel domain(s) responsible for Gβγ binding and activation, the binding of radiolabeled Gβγ to immunoprecipitated cardiac I_{KACh} was assessed in the presence of peptides corresponding to GIRK1 and GIRK4 cytoplasmic domains.[18] We speculated that peptides corresponding to or overlapping with Gβγ binding sites would interfere with the interaction between Gβγ and I_{KACh}. Some subunit domains were specifically excluded (transmembrane, pore, and extracellular domains), while some peptides were insoluble and could not be evaluated. A summary of peptide influence on Gβγ binding is shown in FIGURE 2. No peptides tested corresponding to amino-terminal cytoplasmic domains of either GIRK1 and GIRK4 had any effect on the Gβγ/I_{KACh} interaction. While a role for the GIRK1 amino-terminal cytoplasmic domain (residues 34–86) in Gβγ binding has been suggested,[8,30] we could not confirm this result as peptides corresponding to this region of GIRK1(67–83)—and a similar domain in GIRK4(66–85)—were insoluble.

Four peptides tested against the carboxyl-terminal cytoplasmic domains inhibited Gβγ binding to I_{KACh} at concentrations less than 200 μM. A GIRK1 peptide spanning residues 364–383 potently inhibited Gβγ binding to native I_{KACh}. Truncation of the first four amino acids of this peptide (364–367; MLLM) destroyed its ability to inhibit the Gβγ/channel interaction. However, when expressed in CHO cells, heteromultimeric channels containing wild-type GIRK4 and a GIRK1 subunit lacking residues 364–367 were activated by Gβγ indistinguishably from wild-type GIRK1/GIRK4 channels (data not shown). Interestingly, the GIRK1 region from 364–383 also contains a motif previously implicated in the activa-

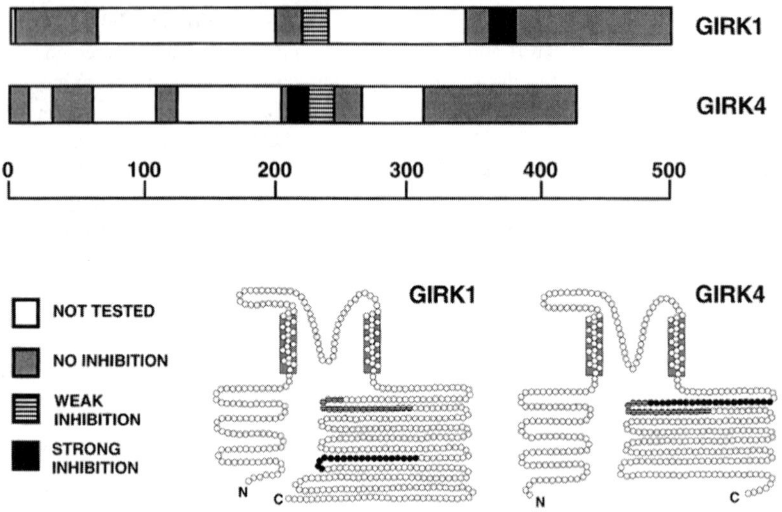

FIGURE 2. Summary of GIRK1 and GIRK4 peptide competition. The diagram displays the regions of GIRK1 and GIRK4 that were synthesized as peptides to be used for competition experiments in an *in vitro* assay for Gβγ binding to purified native atrial I_{KACh}.

tion of I_{KACh} by Gβγ. The putative Gβγ binding region (Q/NXXER) within adenylyl cyclase type 2 (AC2) has homology to GIRK1(378–382; NSKER), and the AC2-derived peptide (residues 956–982) was reported to block I_{KACh} activation by Gβγ in inside-out patches of atrial myocytes.[31] However, the AC2 and GIRK1(378–387) peptides did not inhibit Gβγ binding to I_{KACh} in our assay. Furthermore, mutation and deletion of the putative Gβγ binding motif NSKER (GIRK1 378–382) did not yield any functional differences when compared to wild-type GIRK1.

The most potent peptide inhibitor of Gβγ binding to I_{KACh} was GIRK4 (209–225), which completely inhibited Gβγ binding to the channel at concentrations of 3–5 μM. Deletion of amino acids 209–225 of the GIRK4 subunit (GIRK4Δ209–225) resulted in a subunit that did not interact strongly with GIRK1 and did not produce functional channel activity. The simplest interpretation of this result is that this region in GIRK4 is critical for both Gβγ binding and multimeric assembly. This peptide was compared to homologous regions of related GIRK subunits as well as to the more distantly related and G protein–insensitive IRK subunit, IRK1. The assumption was that the IRK1 subunit would lack critical residues involved in formation of a Gβγ binding site, while the functionally homologous GIRK subunits would all possess a comparable domain. This comparison suggested that Ser209, Phe219, or Asp224 might be involved in Gβγ binding (FIG. 3a). However, the suprising observation that the IRK1(202–218) peptide (but not the comparable GIRK1-derived peptide) potently inhibited Gβγ binding to native I_{KACh} directed us to examine the functional relevance of Cys216. Heteromultimeric channels composed of wild-type GIRK1 and GIRK4(C216T) produced functionally active channels with biophysical characteristics indistinguishable from cardiac I_{KACh}, but with a greater than 50-fold lower sen-

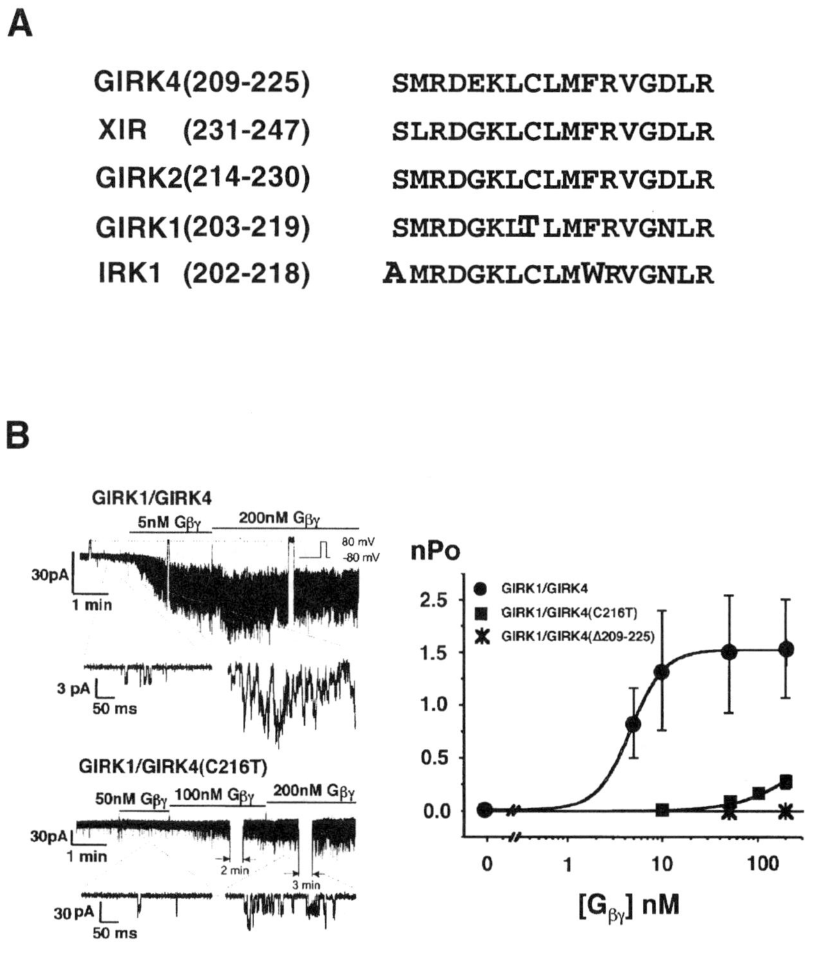

FIGURE 3. Cysteine 216 of the GIRK4(209–225) inhibitory peptide is critical for inhibition of Gβγ binding to I_{KACh}. **(A)** Comparison of the amino acid sequence of the GIRK4(209–225) peptide with homologous sequences from other inward rectifier K^+ channel subunits. Only two amino acids (Ala202 and Trp212) in the Gβγ-insensitive IRK1 differ from the corresponding region of the GIRK subunits shown. GIRK1 contained a unique Thr at position 210, which corresponded to Cys216 in GIRK4. Although not shown here, the GIRK1 peptide containing the unique Thr residue did not inhibit Gβγ binding, in contrast to the GIRK4(209–225) peptide. The IRK1(202–218) peptide did inhibit Gβγ binding. The GIRK1(203–219) peptide was 50 μM, while other peptides were 10 μM. **(B)** Functional studies with the GIRK4(C216T) mutant. Single-channel recordings from inside-out patches of CHO cells expressing either GIRK4/GIRK1 **(upper left)** and GIRK4(C216T)/GIRK1 **(lower left)**; note the much lower Gβγ-sensitivity for the C216T mutant. Single-channel amplitudes were not affected by the C216T GIRK4 mutation. The GIRK1/GIRK4(C216T) mutant channel was significantly less sensitive to Gβγ (> 50-fold) than the GIRK1/GIRK4 wild-type channel (n = 8 for mutant, n = 9 for wild type; ± SEM). Deletion of GIRK4 amino acids 209–225 led to loss of Gβγ-dependent GIRK1/GIRK4(Δ209–225) channel activity (n = 11).

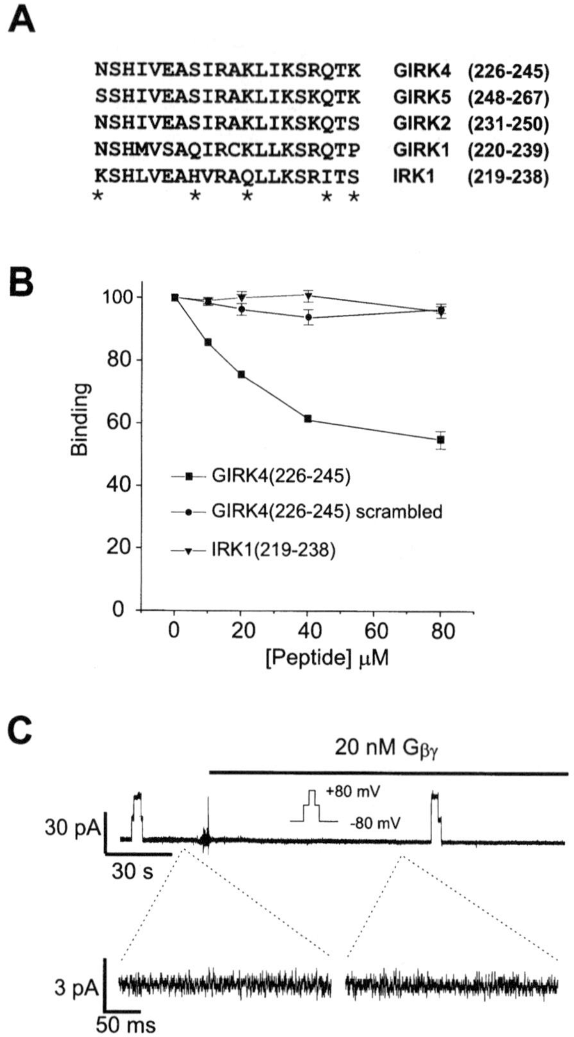

FIGURE 4. Amino acids 226–245 of GIRK4 are necessary for Gβγ binding to and activation of I_{KACh}. **(A)** Comparison of amino acids 226–245 of GIRK4 with homologous sequences from other inward rectifier K^+ channel subunits. Five amino acids of GIRK4 not conserved in IRK1 were evaluated for their potential importance for Gβγ-binding. **(B)** The IRK1 peptide homologous to the GIRK4(226–245) peptide did not inhibit Gβγ binding. This suggests that these five amino acids were crucial for the inhibitory potency of the GIRK4(226–245) peptide. The scrambled GIRK4(226–245) peptide did not inhibit Gβγ binding, demonstrating its specificity. **(C)** Single-channel recordings from inside-out patches of GIRK1/mutant GIRK4(N226K, S233H, K237Q, Q243I, and K245S) expressed in Cos-7 cells. Gβγ-induced channel activity was completely abolished (n = 21 for Cos-7 and n = 5 for CHO cells).

sitivity to activation by Gβγ (FIG. 3b). The shift in potency of Gβγ for this mutant channel implicates this region of GIRK4 in mediating the activation of I_{KACh} by G proteins.

The GIRK4(226–245) peptide also inhibited Gβγ binding to I_{KACh}. Sequence comparison between other GIRK subunits and IRK1, together with results from the peptide inhibition assay suggested that five amino acids within this domain may participate in Gβγ binding to I_{KACh} (FIG. 4a,b). Mutation of these five amino acids in GIRK4 to the corresponding amino acids in IRK1 (GIRK4[5IRK]) ablated Gβγ activation upon coexpression with wild-type GIRK1 (FIG. 4c). When coexpressed with wild-type GIRK1, however, no channel activity was observed with this mutant. Importantly, both the GIRK4(C216T) and GIRK4(5IRK) mutants were shown to be expressed at reasonable levels, to interact with GIRK1, and to promote surface membrane expression of the heterocomplex.

Gβγ binding and activation of I_{KACh} appear to involve a carboxyl-terminal GIRK4 domain near the second transmembrane domain. The comparable region in GIRK1 does not appear to be important for Gβγ binding or channel activation. Although we have shown that Gβγ binds to GIRK1, our attempts to demonstrate that this binding plays a role in Gβγ-induced channel activation have been unsuccessful to date. However, it is likely that a combined GIRK1/GIRK4 conformation determines the Gβγ binding site, and that activation will not be understood completely without high-resolution structural data.

I_{KACh} IS CRITICAL FOR EFFECTIVE HEART RATE REGULATION

Early studies suggested that vagally mediated heart rate regulation was due to ACh activation of I_{KACh}.[32–36] However, several subsequent lines of evidence have suggested that vagal regulation of heart rate, particularly at low-to-moderate levels of vagal activity, is due to changes in a cAMP-modulated cationic conductance, termed I_f, often referred to as the "pacemaker" current (reviewed in Ref. 37). In addition to I_{KACh} and I_f, L-type voltage-gated Ca^{2+} channels and the "sustained inward current" (I_{st}) have been implicated in the autonomic regulation of cardiac pacing.[38–40] Unfortunately, the lack of selective pharmacological agents for I_{KACh} and other currents relevant to cardiac pacing has made dissection of the relative contributions of these conductances *in vivo* impossible. We chose to use a genetic approach in mice to estimate the contribution of I_{KACh} to cardiac physiology and regulation.[16] Since heterologous expression studies argued that GIRK1 alone does not form functional channels, our strategy was to ablate the mouse GIRK4 gene. Heart atria is an ideal tissue for studying the functional significance of a G protein–gated K^+ channel by a genetic ablation approach since atrial I_{KACh} is formed solely by GIRK1 and GIRK4. In virtually all other tissues, there is considerable overlap in the expression patterns of the functional homologs GIRK2, GIRK3, and GIRK4.[41,42]

The first mouse GIRK4 coding sequence exon was targeted because it encodes critical functional domains in GIRK4, including both membrane-spanning domains and the pore-lining sequence. Viable GIRK4-deficient mice were obtained from two independently targeted embryonic stem cell clones; these mice appeared normal by all visual criteria. Northern analysis of atrial tissue indicated that GIRK4 mRNA was absent, GIRK1 mRNA levels were unaltered, and GIRK2 and GIRK3 mRNAs were not detectable in wild-type or GIRK4 knockout mice. Lack of atrial GIRK4 protein was demonstrated by Western blotting of atrial membrane proteins. More importantly, disruption of the GIRK4 gene eliminated the I_{KACh} conductance. Characteristic I_{KACh} activity was observed in 47/50 patches from wild-type and heterozygous mice (FIG. 5a). In contrast, no I_{KACh} activity was observed in patches from knockout mice ($n = 49$; FIG. 5b). Other cardiac inwardly rectifying K^+ channels were observed with characteristic frequencies. This finding proved that GIRK4 is required for I_{KACh} formation and argued that GIRK4 functional homologs did not compensate for the GIRK4 deficiency in heart.

Using ECG telemetry, we demonstrated that resting mean heart rate did not differ between GIRK4 knockout and wild-type mice (647.2 and 646.8 bpm, respectively; FIG. 6a). At rest, however, wild-type mice frequently exhibited episodes of mild bradycardia with rapid onset (FIG. 6b). In contrast, the heart rates of GIRK4-deficient mice did not change over time scales shorter than ~2 s. This observation suggested a specific deficit in heart rate regulation in the knockout mice. Thus, we tested a panel of pharmacological agents that selectively target signaling pathways thought to culminate in I_{KACh} activation. Methoxamine, an agonist of α_1-adrenergic receptors, causes vasoconstriction, which leads to elevated blood pressure and an increase in vagal activity through baroreflex stimulation; increased vagal activity leads to bradycardia following muscarinic receptor stimulation. CCPA, a stable adenosine analog, is thought to activate I_{KACh} via selective stimulation of sinoatrial nodal A_1 adenosine receptors.[43]

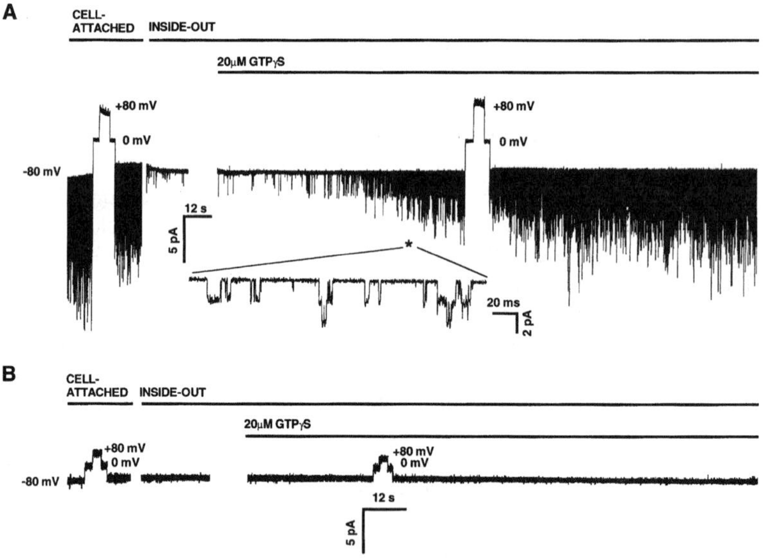

FIGURE 5. Electrophysiological analysis of mouse atrial myocytes. **(A)** Single-channel recording from a wild-type atrial myocyte. With 10 μM ACh in the pipette, I_{KACh} activity was observed in the cell-attached configuration, but declined after patch excision. Channel activity was restored upon addition of GTPγS to the bath. Recording artifacts due to addition of GTPγS to the bath have been deleted. The *asterisk* indicates the recording segment expanded in the *inset*. **(B)** Single-channel recording from a knockout atrial myocyte. No I_{KACh} activity was observed in either the cell-attached or inside-out configuration ($n = 49$). In both (A) and (B) holding potential was −80 mV with periodic voltage jumps to +80 mV to test the inward rectification of the observed channels.

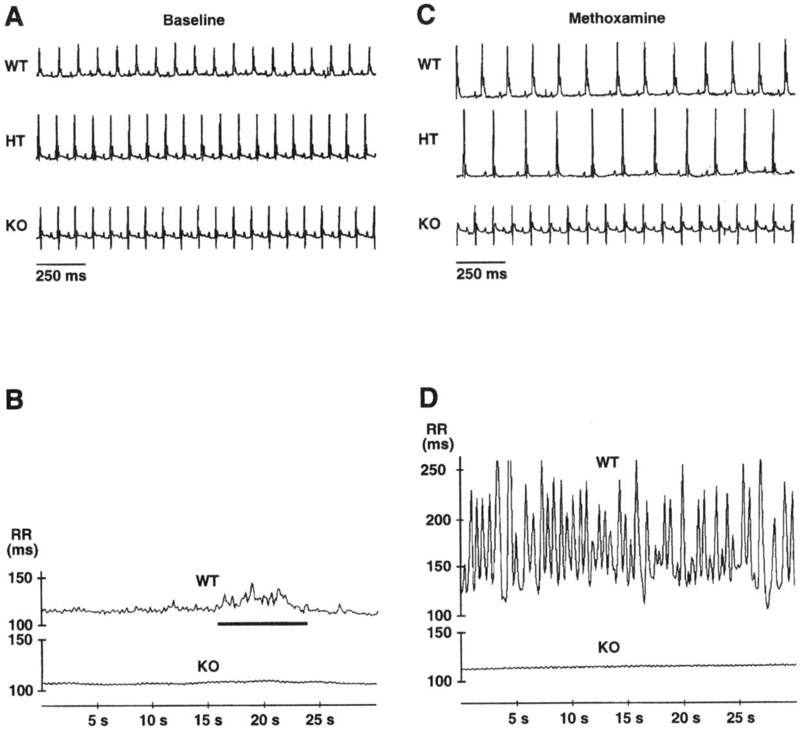

FIGURE 6. ECG and HR fluctuations in wild-type and GIRK4-deficient mice. **(A)** One-lead baseline ECG recorded simultaneously from wild-type, heterozygous, and knockout mice. **(B)** Comparison of beat-to-beat changes in RR intervals (time interval between successive R peaks on an ECG) between wild-type and knockout littermates at rest. RR intervals, the intervals between successive heart beats (ventricular depolarizations), are plotted against time (s). Note the spontaneous slowing of heart rate in the wild-type animal (*bar*). **(C)** One-lead ECG from the same mice as in (A), after methoxamine administration. Bradycardia is more pronounced in wild-type and heterozygous mice. **(D)** Comparison of beat-to-beat changes in RR intervals between wild-type and knockout mice after administration of methoxamine. Note that the RR interval fluctuations in the wild-type mouse contrast dramatically to the steady heart rate of the knockout mouse.

Shortly after methoxamine administration (usually within 2–3 minutes), wild-type mice developed bradycardia. While the heart rate of GIRK4-deficient mice also decreased after methoxamine administration, the effect was less than in wild-type siblings (564 ± 30 vs. 463 ± 44 bpm, respectively, $p < 0.01$; FIG. 6c). Following methoxamine administration, wild-type mice also exhibited a dramatic increase in heart rate variability (HRV; beat-to-beat fluctuations in heart rate); the increase in HRV seen in wild-type mice was absent in their knockout littermates (FIG. 6d). Similarly, CCPA administration caused bradycardia in both wild-type (WT) and knockout (KO) mice, with the effect being more pronounced in wild-type mice (469 ± 65 vs. 230 ± 33 bpm, KO vs. WT; $p < 0.001$; not shown).

Vagal activity and adenosine have been implicated in atrioventricular conduction properties.[43–45] Therefore, we measured the effects of GIRK4 disruption on PR interval duration (time interval between atrial and ventricular depolarization) at baseline and after methoxamine administration. The small difference in PR interval duration between wild-type and GIRK4-deficient mice was of borderline statistical significance (32.1 ± 2.6 vs. 34.9 ± 2.8 ms, KO vs. WT; $p < 0.05$). Since this difference disappeared after CCPA administration (37.7 ± 4.1 vs. 36.9 ± 3.6 ms, KO vs. WT), the importance of this finding is uncertain, but it appears that I_{KACh} is not critical for atrioventricular electrical conduction properties in mice. Although the lack of a key regulatory conductance in pacemaker tissue might have been expected to predispose the affected hearts to ectopy (heart beat initiation from regions other than the normal pacemaker tissue), there was no obvious difference in the frequency of ectopic beats between the wild-type and GIRK4 knockout mice. In over three hours of analyzed baseline data, no more than 10 ventricular ectopic beats were detected in any mouse. Apart from one wild-type mouse with frequent supraventricular ectopy (>200 ectopic beats over 3h), supraventricular ectopic beats were infrequent in mice of all genotypes.

Our findings argue that GIRK4 is indispensable for the formation and G protein regulation of cardiac I_{KACh}, and that I_{KACh} accounts for approximately half of the bradycardia induced by vagal stimulation and adenosine administration. Furthermore, it appears that the I_{KACh} branch of the parasympthetic cardiac regulatory system is absolutely required for rapid adjustments (< 2s) in heart rate, even at rest. The finding that heart rate variability differs between GIRK4 knockout and wild-type mice even at rest, when vagal activity is presumably not elevated, contrasts somewhat with previous electrophysiological data obtained from isolated sinus nodal cells, where concentrations of applied ACh too low to stimulate I_{KACh} activity nevertheless produced a negative chronotropic response.[37,46] These authors argued that I_{KACh} probably did not play a significant role in heart rate regulation at low-to-moderate levels of vagal activity. Our data indicates that the concentration of ACh released from the low-to-moderately stimulated vagus is sufficient to recruit I_{KACh} activity. We conclude that I_{KACh} is a critical effector in the parasympathetic branch of the heart rate regulatory system.

REFERENCES

1. Doupnik, C., N. Davidson & H. Lester. 1995. The inward rectifier potassium channel family. Curr. Opin. Neurol. **5:** 268–277.
2. Matsuda, H., A. Saigusa & H. Irisawa. 1987. Ohmic conductance through the inwardly rectifying K channel and blocking by internal Mg^{2+}. Nature **325:** 156–159.
3. Fakler, B. *et al.* 1994. A structural determinant of differential sensitivity of cloned inward rectifier K^+ channels to intracellular spermine. FEBS Lett. **356:** 199–203.
4. Lopatin, A., E. Makhina & C. Nichols. 1994. Potassium channel block by cytoplasmic polyamines as the mechanism of intrinsic rectification. Nature **372:** 366–369.
5. Wickman, K. & D. Clapham. 1995. Ion channel regulation by G proteins. Physiol. Rev. **75:** 865–885.
6. Hartzell, H. 1979. Adenosine receptors in frog sinus venosus: Slow inhibitory potentials produced by acetylcholine and adenine compounds. J. Physiol. **293:** 23–49.
7. North, R. 1989. Drug receptors and the inhibition of nerve cells. Br. J. Pharmacol. **98:** 13–28.
8. Huang, C. *et al.* 1995. Evidence that direct binding of Gβγ to the GIRK1 G protein-gated inwardly rectifying K^+ channel is important for channel activation. Neuron **15:** 1133–1143.
9. Inanobe, A. *et al.* 1995. G beta gamma directly binds to the carboxyl terminus of the G protein–gated muscarinic K^+ channel, GIRK1. Biochem. Biophys. Res. Comm. **212:** 1022–1028.
10. Krapivinsky, G. *et al.* 1995. Gβγ binds directly to the G protein–gated K^+ channel I_{KACh}. J. Biol. Chem. **270:** 29059–29062.

11. WICKMAN, K. *et al.* 1994. Recombinant Gβγ activates the muscarinic-gated atrial potassium channel I_{KACh}. Nature **368:** 255–257.
12. LOGOTHETIS, D.E. *et al.* 1987. The beta gamma subunits of GTP-binding proteins activate the muscarinic K$^+$ channel in heart. Nature **325:** 321–326.
13. SUI, J., J. PETIT-JACQUES & D. LOGOTHETIS. 1998. Activation of the atrial KACh channel by the betagamma subunits of G proteins or intracellular Na$^+$ ions depends on the presence of phosphatidylinositol phosphates. Proc. Natl. Acad. Sci. USA **95:** 1307–1312.
14. SUI, J., K. CHAN & D. LOGOTHETIS. 1996. Na$^+$ activation of the muscarinic K$^+$ channel by a G-protein–independent mechanism. J. Gen. Physiol. **108:** 381–391.
15. JI, S. *et al.* 1998. Mechanosensitivity of the cardiac muscarinic potassium channel. A novel property conferred by Kir3.4 subunit. J. Biol. Chem. **273:** 1324–1328.
16. WICKMAN, K. *et al.* 1998. Abnormal heart rate regulation in GIRK4 knockout mice. Neuron **20:** 103–114.
17. COREY, S. *et al.* 1998. Number and stoichiometry of subunits in the native atrial G-protein-gated K$^+$ channel, IKACh. J. Biol. Chem. **273:** 5271–5278.
18. KRAPIVINSKY, G. *et al.* 1998. Gβγ binding to GIRK4 subunit is critical for G protein–gated K$^+$ channel activation. J. Biol. Chem. **273:** 16946–16952.
19. KUBO, Y. *et al.* 1993. Primary structure and functional expression of a rat G-protein–coupled muscarinic potassium channel. Nature **364:** 802–806.
20. DASCAL, N. *et al.* 1993. Atrial G protein-activated K$^+$ channel: Expression cloning and molecular properties. Proc. Natl. Acad. Sci. U.S.A. **90:** 10235–10239.
21. KRAPIVINSKY, G. *et al.* 1995. The G-protein-gated atrial K$^+$ channel I_{KACh} is a heteromultimer of two inwardly rectifying K$^+$ channel proteins. Nature **374:** 135–141.
22. KENNEDY, M., J. NEMEC & D. CLAPHAM. 1996. Localization and interaction of epitope-tagged GIRK1 and CIR inward rectifier K$^+$ channel subunits. Neuropharmacology **35:** 831–839.
23. VELIMIROVIC, B. *et al.* 1996. The K$^+$ channel inward rectifier subunits form a channel similar to neuronal G protein–gated K$^+$ channel. FEBS Lett. **379:** 31–37.
24. DOYLE, D. *et al.* 1998. The structure of the potassium channel: Molecular basis of K$^+$ conduction and selectivity. Science **280:** 69–77.
25. SILVERMAN, S., H. LESTER & D. DOUGHERTY. 1996. Subunit stoichiometry of a heteromultimeric G protein–coupled inward-rectifier K$^+$ channel. J. Biol. Chem. **271:** 30524–30528.
26. INANOBE, A. *et al.* 1995. Immunological and physical characterization of the brain G protein–gated muscarinic potassium channel. Biochem. Biophys. Res. Commun. **217:** 1238–1244.
27. KRAPIVINSKY, G. *et al.* 1995. The cardiac inward rectifier K$^+$ channel subunit, CIR, does not comprise the ATP-sensitive K$^+$ channel, I_{KATP}. J. Biol. Chem. **270:** 28777–28779.
28. TUCKER, S., M. PESSIA & J. ADELMAN. 1996. Muscarine-gated K$^+$ channel: Subunit stoichiometry and structural domains essential for G protein stimulation. Am. J. Physiol. **271:** H379–H385.
29. MCCORMACK, K. *et al.* 1992. Tandem linkage of Shaker K$^+$ channel subunits does not ensure the stoichiometry of expressed channels. Biophys. J. **63:** 1406–1411.
30. HUANG, C., Y. JAN & L. JAN. 1997. Binding of the G protein betagamma subunit to multiple regions of G protein–gated inward-rectifying K$^+$ channels. FEBS Lett. **405:** 291–298.
31. CHEN, J. *et al.* 1995. A region of adenylyl cyclase 2 critical for regulation by G protein beta gamma subunits. Science **268:** 1166–1169.
32. LOEWI, O. & E. NAVRATIL. 1926. Uber humorale ubertragbarkeit der herznervenwirkung. Pfluegers Arch. **189:** 239–242.
33. GILES, W. & S. NOBLE. 1976. Changes in membrane currents in bullfrog atrium produced by acetylcholine. J. Physiol. **261:** 103–123.
34. GARNIER, D. *et al.* 1978. The action of acetylcholine on background conductance in frog atrial trabeculae. J. Physiol. **274:** 381–396.
35. NOMA, A. & W. TRAUTWEIN. 1978. Relaxation of the ACh-induced potassium current in the rabbit sino-atrial node. Pflugers Arch. **377:** 193–200.
36. SAKMANN, B., A. NOMA & W. TRAUTWEIN. 1983. Acetylcholine activation of single muscarinic K channels in isolated pacemaker cells of the mammalian heart. Nature **303:** 250–253.
37. DIFRANCESCO, D. 1995. The onset and autonomic regulation of cardiac pacemaker activity: Relevance of the f current. Cardiovasc. Res. **29:** 449–456.
38. NOMA, A., M. MORAD & H. IRISAWA. 1983. Does the pacemaker current generate the diastolic depolarization in the rabbit SA node cells? Pflugers Arch. **397:** 190–194.

39. Hagiwara, N., H. Irisawa & M. Kameyama. 1988. Contribution of two types of calcium currents to the pacemaker potentials of rabbit sino-atrial node cells. J. Physiol. **395:** 233–253.
40. Guo, J., K. Ono & A. Noma. 1995. A sustained inward current activated at the diastolic potential range in rabbit sino-atrial node cells. J. Physiol. (Lond.) **483:** 1–13.
41. Karschin, C. *et al.* 1996. IRK(1-3) and GIRK(1-4) inwardly rectifying K^+ channel mRNAs are differentially expressed in the adult rat brain. J. Neurosci. **16:** 3559–3570.
42. Ferrer, J. *et al.* 1995. Pancreatic islet cells express a family of inwardly rectifying K^+ channel subunits which interact to form G-protein–activated channels. J. Biol. Chem. **270:** 26086–26091.
43. Kurachi, Y., T. Nakajima & T. Sugimoto. 1986. On the mechanism of activation of muscarinic K^+ channels by adenosine in isolated atrial cells: Involvement of GTP-binding proteins. Pflugers Arch. **407:** 264–274.
44. DiMarco, J. *et al.* 1983. Adenosine: Electrophysiological effects and therapeutic use for terminating paroxysmal supraventricular tachycardia. Circulation **68:** 1254–1263.
45. Clemo, H. & L. Belardinelli. 1986. Effect of adenosine on atrioventricular conduction. I. Site and characterization of adenosine action in guinea pig atrioventricular node. Circ. Res. **59:** 427–436.
46. DiFrancesco, D. 1995. Cardiac pacemaker: 15 years of "new" interpretation. Acta Cardiologica **50:** 414–427.

Potassium Currents in Developing Neurons

ANGELES B. RIBERA[a]

Department of Physiology and Biophysics, Box C-240, University of Colorado Health Sciences Center, 4200 East Ninth Avenue, Denver, Colorado 80262, USA

ABSTRACT: In *Xenopus* spinal neurons, delayed rectifier type voltage-dependent potassium currents (I_{Kv}) are developmentally regulated. These currents play a pivotal role in maturation of the action potential from a long-duration calcium-dependent impulse to a brief sodium-dependent one. Although spinal neurons are heterogeneous, I_{Kv} undergoes a synchronized and homogenous developmental functional up-regulation across this diverse population of motor, sensory, and interneurons. This finding suggested that the diverse population of neurons expressed a common potassium channel. Thus, recent efforts have been directed towards cloning the relevant potassium channel gene. However, these molecular studies reveal an unsuspected heterogeneity in the molecular components of voltage-dependent potassium channels. Further, synchronous differentiation of I_{Kv} is achieved via heterogeneous Kv channel gene expression.

When neuronal precursors exit the cell cycle, they begin terminal differentiation. Extensive cytological and morphological differentiation occurs. A critical aspect of terminal differentiation is acquisition of electrical excitability, which underlies rapid signaling—a hallmark of the adult nervous system. Considerable evidence has accumulated over recent years indicating that excitability is acquired early by embryonic neurons and is subject to continual developmental modulation.

When neurons are synaptically interconnected, excitability is required for functional output of newly forming networks. In addition, developmental changes in excitability of individual cells will modify circuit output and resultant behaviors. Prior to the formation of synapses and establishment of sensory input, embryonic neurons are already excitable. However, it is not immediately obvious what the role of excitability is at these early stages of neuronal differentiation. Recent evidence suggests that ion channel activity, while typically associated with the rapid signaling of mature nervous systems, can also initiate intracellular signaling cascades that act over a longer time domain.[1] Further, the signaling cascades that are triggered by activity can have a profound influence on subsequent differentiation of fundamental neuronal properties such as process outgrowth, connectivity, neurotransmitter selection, and the properties of ion channels themselves.[2]

Thus, studies of excitability in embryonic neurons can be divided into two periods: one prior to synapse formation, and the second after connections are formed. The majority of our work has concerned neurons during the first period at the earliest stages of their differentiation. This is for many reasons, notably because the changes that neurons display during this period are important, and elucidating the mechanism that underlies these changes has been informative. Practical considerations have also influenced our choice of period of study. It is clear that for studies of development of electrical excitability, a key requirement of the system is electrophysiological access to neurons of interest. Primary spinal neurons of the developing *Xenopus laevis* embryo meet this essential criterion. These neurons are a

[a]Phone: 303-315-8060; fax: 303-315-8110; e-mail: Angie.Ribera@UCHSC.edu

diverse set of sensory, inter-, and motor neurons that underlie the essential swimming behavior of the embryo.[3,4] Moreover, with respect to birthdates, these neurons are a relatively synchronized population.[5,6] Secondary neurons, in contrast, are born later over a protracted period of time.[6] Primary spinal neurons can be studied during early stages prior to synapse formation both in intact preparations as well as *in vitro*. This review will briefly summarize development of electrical excitability in these neurons, since these results provide the rationale for our focus on potassium currents. Next, the development of potassium currents in these neurons will be discussed. Our recent work is aimed at identifying the molecular components of developmentally regulated potassium currents, and these studies will be summarized. The article will close with considerations of future directions.

DEVELOPMENT OF ELECTRICAL EXCITABILITY IN *XENOPUS* PRIMARY SPINAL NEURONS

Excitability is acquired by developing *Xenopus laevis* primary spinal neurons 7–8 hours after their last round of DNA synthesis[5,6] and several hours before initial signs of chemosensitivity and synapse formation.[7,8] A similar pattern has been observed for the differentiation of primary spinal neurons in culture. Typically, the neural plate is dissected from a 17 1/2–hour-old embryo. Neurons initiate neurite outgrowth 6h after plating,[9] which parallels the time of neurite outgrowth *in vivo* (22h). At this time, action potentials can be elicited from every cell with a neuronal morphology, indicating that neurite extension *in vitro* does not precede the acquisition of excitability. Thus, the initial appearance of electrical excitability occurs at similar times *in vivo* and *in vitro*.

When action potentials are first detected at the neural tube stage (22h), they are calcium dependent and of long duration. During the following day *in vivo* and *in vitro*, the impulse matures to a brief sodium-dependent spike.[9,10] This program is expressed in a cell-autonomous manner, since a neuron developing in the absence of other cells also exhibits a transient period of long-duration impulses.[11]

Calcium-dependent action potentials are spontaneously expressed in embryonic amphibian spinal neurons, and their pharmacological removal disrupts the subsequent elaboration of other phenotypes, suggesting their developmental significance.[12] The transient elevations of intracellular calcium, as assessed by calcium ion imaging with fluorescent indicators, occur most frequently at early stages of differentiation, when potassium currents are small and neurons fire long, calcium-dependent action potentials. With further development the incidence of spontaneous calcium transients declines in parallel with the conversion of the action potential to brief sodium-dependent impulses. Furthermore, early blockade of calcium influx for brief periods shows that perturbations applied during the period of greatest spontaneous calcium influx have the most profound effects on subsequent development. These data suggest that transient elevations of intracellular calcium may trigger specific sequences of differentiation.

More recently, it has been shown that the differentiation cues imparted by spontaneous transient elevations of calcium are frequency encoded.[13,14] Suppression of calcium-dependent action potentials prevents normal maturation of several key neuronal properties: expression of the neurotransmitter GABA, neurite outgrowth, and the kinetics of voltage-dependent potassium current. These properties will be referred to here as calcium-dependent phenotypes. Calcium-dependent action potentials were suppressed and then reimposed at desired frequencies. Calcium-dependent phenotypes were best restored when impulses were reimposed at their natural frequency.[14,15]

DEVELOPMENTAL REGULATION OF VOLTAGE-DEPENDENT POTASSIUM CURRENTS IN *XENOPUS* SPINAL NEURONS

Analysis of the voltage dependent currents that underlie the action potential indicates that the transition from a long-duration calcium-dependent to a sodium-dependent brief impulse is due primarily to the maturation of a delayed rectifier potassium current.[16,17] During the first day *in vitro*, calcium current shows very little change, and sodium current doubles in density during this period. Delayed rectifier potassium current, however, triples in density, and its kinetics are accelerated.[17]

Quantitative analysis using mathematical reconstruction of action potentials also indicates that the developmental changes noted for potassium current are most significant.[18] Further, changes in potassium current are not the only theoretical way to bring about the developmental change in the impulse. In fact, in mathematical reconstructions, the action potential duration is most sensitive to changes in calcium current. Thus, in addition to understanding how potassium current density is regulated during development, it will be important to determine why this current is the major determinant of impulse duration.

RNA synthesis inhibition prevents development of the mature delayed rectifier and arrests the normal maturation of the action potential. Further, a critical period of transcription is required for the differentiation of this potassium current.[19] This critical period is specific to the delayed rectifier, since the two inward currents do not show sensitivity to inhibitors of transcription when applied during the same period,[20] and a potassium A current recovers and matures following removal of transcriptional block.[19]

The appearance of neuronal morphology also requires new RNA and protein synthesis[19,21,22] Application of RNA or protein synthesis inhibitors applied before 3 or 5 h *in vitro*, respectively, completely suppresses the appearance of morphologically identifiable neurons, defining a sensitive period of transcription required for the differentiation of this phenotype. It does not appear to contain a critical period, however, since acute application of a reversible inhibitor permits neurite extension after the appropriate delay, the duration of inhibition—3 h.[19]

In sum, properties of the delayed rectifier potassium current undergo significant developmental changes during the first 48 hours of development of primary spinal neurons.

IDENTIFICATION OF THE MOLECULAR COMPONENTS OF DEVELOPMENTALLY REGULATED POTASSIUM CURRENTS

The properties of I_{Kv} determine the phenotype of the action potential in young and mature neurons.[17,18] Accordingly, a major focus of recent work has been to identify the molecular components of developmentally regulated potassium currents. These efforts have taken advantage of the advances in molecular cloning of potassium channel genes.

Voltage-dependent potassium (Kv) channels are thought to be tetramers of pore-forming α-subunits[23,24]; Kv α-subunits genes belong to four major subfamilies (Kv1, Kv2, Kv3, Kv4[25–28]) with an additional fifth family that has been found in Aplysia (Kv5)[29]. More recently, Kv6–Kv9 subfamilies have been reported.[30–32] Subunits from these subfamilies do not form functional channels when expressed by themselves. However, they affect the properties of channels when coexpressed with members of major subfamilies.

An initial way to determine the functional properties of channels encoded by the Kv genes is to express them heterologously (e.g., in *Xenopus* oocytes) and electrophysiologically characterize the resultant currents. Although the properties of homomultimeric channels are more often studied, the function of heteropolymeric channels formed from different type of Kv clones has also been examined.[33–35] The data indicate that the formation of heteropolymers yields potassium channels that are functionally distinct from those formed by association of identical subunits. Interestingly, although different Kv1 gene products can form heteropolymers, a Kv1 gene product does not form a functional channel with, for example, a Kv2, Kv3, or Kv4 gene product, suggesting that each potassium channel gene family functions independently.[36]

The biophysical properties of I_{Kv}[17] suggested that genes encoding noninactivating potassium current would contribute to the endogenous I_{Kv}. Indeed, members of the Kv1 and Kv2 subfamilies are expressed in primary spinal neurons.[37–40] However, Kv1.1 and Kv2.2 mRNAs are detected in different neuronal subpopulations. Further, heterologous expression of Kv1.1 and Kv2.2 in oocytes leads to expression of rather different sustained Kv currents (e.g., Ref. 39) in contrast to the uniform I_{Kv} of diverse spinal neurons. On the basis of whole-cell recording of voltage-dependent currents, neurons would be considered to be a homogeneous population, because subtle or no differences are noted in the whole-cell currents recorded at a given stage of development.[17] In fact, a simple scenario to account for synchronized developmental regulation of I_{Kv} is that the heterogeneous population of spinal neurons expresses a common potassium channel gene that is functionally up-regulated in a similar manner throughout the population.

Neurons in culture are heterogeneous and include motor, sensory, and interneurons.[8,9,41] Heterogeneity is revealed by the different chemosensitivity profiles that neurons acquire as they differentiate.[8] Further, a subset acquire GABA-like immunoreactivity.[2]

Our initial analyses of Kv channel gene expression were raising the possibility, however, of molecular heterogeneity in the endogenous population of functional Kv channels. This suspected molecular diversity has been recently confirmed at a functional level by overexpression of a dominant negative Kv1 subunit.[42] This subunit is expected to suppress current carried by Kv1 channels but not that of non-Kv1 channels (e.g., Kv2). Following overexpression of the dominant negative Kv1 subunit, I_{Kv} was suppressed in some, but not all, neurons. Despite the uniformity in the properties of I_{Kv} of neurons in culture, only a subset of neurons in culture. This result indicates that Kv1 channels are not expressed in all neurons. Moreover, synchronous differentiation of I_{Kv} is achieved via heterogeneous Kv channel gene expression. Examination of the contribution of Kv2 channels is currently being examined by overexpression of a dominant negative Kv2 subunit.[43]

FUTURE DIRECTIONS

A limited number of studies have examined primary neurons at stages subsequent to synapse formation, because analysis at these later stages is complicated by poor extended survival under standard culture conditions. Further, the overexpression strategies that have proved useful for study of the early period of differentiation of electrical excitability are not likely to be applicable to analyses at later stages since injected RNA and resultant exogenous proteins have half-lives that limit their duration of action. A system that holds promise for study of this later period of differentiation of electrical excitability is the zebrafish (*Danio rerio*). The swimming behavior of the early zebrafish embryo is mediated by the same set of primary neurons that do so in *Xenopus*. Large-scale mutagenesis is feasible in *Danio rerio,*[44,45] and thousands of embryonic genetic mutants have been isolated.[46,47] Zebrafish start displaying behaviors, which allow evaluation of the function of the nervous system and muscle and isolation of genetic motility mutants, as early as one day after fertilization.[48] Moreover, it is possible to record from zebrafish primary neurons in intact preparations[49] and examine directly the electrical membrane properties of primary neurons in these motility mutants.

ACKNOWLEDGMENT

Research in the author's laboratory is supported by NIH Grant NS25217 (A.B.R.).

REFERENCES

1. Spitzer, N.C. 1991. A developmental handshake: Neuronal control of ionic currents and their control of neuronal differentiation. J. Neurobiol. **22:** 659–673.
2. Spitzer, N.C., R.C. DeBaca, K.A. Allen & J. Holliday. 1993. Calcium dependence of differentiation of GABA immunoreactivity in spinal neurons. J. Comp. Neurol. **337:** 323–333.
3. Roberts, A. & J.D.W. Clarke. 1982. The neuroanatomy of an amphibian spinal cord. Philos. Trans. R. Soc. Lond. **296:** 195–212.
4. Roberts, A. & J.A. Khan. 1982. Intracellular recordings from spinal neurons during "swimming" in paralyzed amphibian embryos. Philos. Trans. R. Soc. Lond. **296:** 213–228.
5. Lamborghini, J.E. 1980. Rohon-Beard cells and other large neurons in *Xenopus* embryos originate during gastrulation. J. Comp. Neurol. **189:** 323–333.
6. Hartenstein, V. 1989. Early neurogenesis in *Xenopus*: The spatio-temporal pattern of proliferation and cell lineages in the embryonic spinal cord. Neuron **3:** 399–411.
7. Blackshaw, S.E. & A.E. Warner. 1976. Onset of acetylcholine sensitivity and endplate activity in developing myotome muscles of *Xenopus*. Nature **262:** 217–128.
8. Bixby, J.L. & N.C. Spitzer. 1984. The appearance and development of neurotransmitter sensitivity in *Xenopus* embryonic spinal neurons *in vitro*. J. Physiol. **353:** 143–155.
9. Spitzer, N.C. & J.E. Lamborghini. 1976. The development of the action potential mechanism of amphibian neurons isolated in cell culture. Proc. Natl, Acad. Sci. USA **73:** 1641–1645.
10. Baccaglini, P.I. & N.C. Spitzer. 1977. Developmental changes in the inward current of the action potential of Rohon-Beard neurons. J. Physiol. **271:** 93–117.
11. Henderson, L.P., Smith, M.A. & N.C. Spitzer. 1984. The absence of calcium blocks impulse-evoked release of acetylcholine but not *de novo* formation of functional neuromuscular synaptic contacts in culture. J. Neurosci. **4:** 3140–3150.
12. Holliday, J. & N.C. Spitzer. 1990. Spontaneous calcium influx: Roles in differentiation of spinal neurons in culture. Dev. Biol. **141:** 13–23.
13. Gu, X., E.C. Olson & N.C. Spitzer. 1994. Spontaneous neuronal calcium spikes and waves during early differentiation. J. Neurosci. **14:** 6325–6335.
14. Gu, X. & N.C. Spitzer. 1995. Distinct aspects of neuronal differentiation encoded by frequency of spontaneous Ca^{2+} transients. Nature **375:** 784–787.
15. Gu, X. & N.C. Spitzer. 1997. Breaking the code: Regulation of neuronal differentiation by spontaneous calcium transients. Dev. Neurosci. **19:** 33–41.
16. Barish, M.E. 1986. Differentiation of voltage-gated potassium current and modulation of excitability in cultured amphibian spinal neurons. J. Physiol. **375:** 229–250.
17. O'Dowd, D.K., Ribera, A.B. & N.C. Spitzer. 1988. Development of voltage-dependent calcium, sodium and potassium currents in *Xenopus* spinal neurons. J. Neurosci. **8:** 792–805.

18. Lockery, S.R. & N.C. Spitzer. 1992. Reconstruction of action potential development from whole cell currents of differentiating spinal neurons. J. Neurosci. **12:** 2268–2287.
19. Ribera, A.B. & N.C. Spitzer. 1989. A critical period of transcription required for differentiation of the action potential of spinal neurons. Neuron **2:** 1055–1062.
20. O'Dowd, D.K. 1985. Changes in membrane conductance that underlie the development of sodium-dependent action potentials in *Xenopus* spinal neurons. Ph.D. Thesis, University of California at San Diego.
21. Blair, L.A.C. 1983. The timing of protein synthesis required for the development of the sodium action potential in embryonic spinal neurons. J. Neurosci. **3:** 1430–1436.
22. O'Dowd, D.K. 1983. RNA synthesis dependence of action potential development in spinal cord neurons. Nature **303:** 619–621.
23. MacKinnon, R. 1991. Determination of the subunit stoichiometry of a voltage-activated potassium channel. Nature **350:** 232–235.
24. Liman, E.R., J. Tytgat & P. Hess. 1992. Subunit stoichiometry of a mammalian K^+ channel determined by construction of multimeric cDNAs. Neuron **9:** 861–871.
25. Butler, A., A. Wei, K. Baker & L. Salkoff. 1989. A family of putative potassium channel genes in *Drosophila*. Science **243:** 943–947.
26. Wei, A., M. Covarrubias, A. Butler, K. Baker, M. Pak & L. Salkoff. 1990. K^+ current diversity is produced by an extended gene family conserved in *Drosophila* and mouse. Science **248:** 599–603.
27. Salkoff, L., K. Baker, A. Butler, M. Covarrubias, M.D. Pak & A. Wei. 1992. An essential 'set' of K^+ channels in flies, mice and humans. Trends Neurosci. **15:** 161–166.
28. Chandy, K.G. & G.A. Gutman. 1994. Voltage-gated K^+ channel genes. *In* CRC Handbook of Receptors and Channels: 1–71. CRC Press. Boca Raton, FL.
29. Zhao, B., F. Rassendren, B.-K. Kaang, Y. Furukawa, T. Kubo & E.R. Kandel. 1994. A new class of noninactivating K^+ channels from Aplysia capable of contributing to the resting potential and firing patterns of neurons. Neuron **13:** 1205–1213.
30. Hugnot, J.P., M. Salinas, F. Lesage, E. Guillmare, J. DeWeile, C. Heurteaux, M.G. Mattei & M. Lazdunski. 1996. Kv8.1, a new neuronal potassium channel subunit with specific inhibitory properties towards Shab and Shaw channel. EMBO J. **15:** 3322–3331.
31. Castellano, A., M.D. Chiara, B. Mellströ, A. Molina, F. Monje, J.R. Naranjo & J. López-Barneo. 1997. Identification and functional characterization of a K^+ channel α subunit with regulatory properties specific to brain. J. Neurosci. **17:** 4652–4661.
32. Salinas, M., F. Duprat, C. Heurteaux, J.P. Hugnot & M. Lazdunski. 1997. New modulatory alpha subunits for mammalian Shab K^+ channels. J. Biol. Chem. **272:** 24371–24379.
33. Christie, M.J., R.A. North, P.B. Osborne, J. Douglass & J.P. Adelman. 1989. Heteropolymeric potassium channels expressed in *Xenopus* oocytes from cloned subunits. Neuron **3:** 405–411.
34. Isacoff, E.Y., Y.N. Jan & L.Y. Jan. 1990. Evidence for the formation of heteromultimeric potassium channels in *Xenopus* oocytes. Nature **345:** 530–534.
35. Ruppersberg, J.P., K.H. Schroter, B. Sakmann, M. Stocker, S. Sewing & O. Pongs. 1990. Heteromultimeric channels formed by rat brain potassium-channel proteins. Nature **345:** 535–537.
36. Covarrubias, M., A.A. Wei & L. Salkoff. 1991. Shaker, Shal, Shab and Shaw express independent K^+ current systems. Neuron **7:** 763–773.
37. Ribera, A.B. 1990. A potassium channel gene is expressed at neural induction. Neuron **5:** 691–701.
38. Ribera, A.B. & D.A. Nguyen. 1993. Primary sensory neurons express a *Shaker*-like potassium channel gene. J. Neurosci. **13:** 4988–4996.
39. Burger, C & A.B. Ribera. 1996. *Xenopus* spinal neurons express Kv2 potassium channel transcripts during embryonic development. J. Neurosci. **16:** 1412–1421.
40. Gurantz, D., A.B. Ribera & N.C. Spitzer. 1996. Temporal expression of *Shaker*- and *Shab*-like potassium channel gene expression in single embryonic spinal neurons during K^+ current development. J. Neurosci. **16:** 3287–3295.
41. Lamborghini, J.E. & A. Iles. 1985. Development of a high-affinity GABA uptake system in embryonic amphibian spinal neurons. Dev. Biol. **112:** 167–176.

42. RIBERA, A.B. 1996. Homogeneous development of electrical excitability via heterogeneous ion channel expression. J. Neurosci. **16:** 1123–1130.
43. BLAINE, J.T. & A.B. RIBERA. 1998. Heteromultimeric potassium channels formed by members of the Kv2 subfamily. J. Neurosci. In press.
44. MULLINS, M.C., M. HAMMERSCHMIDT, P. HAFFTER & C. NÜSSLEIN-VOLHARD. 1994. Large-scale mutagenesis in the zebrafish: In search of genes controlling development in a vertebrate. Curr. Biol. **4:** 189–202.
45. SOLNICA-KREZEL, L., A.F. SCHIER & W. DRIEVER. 1994. Efficient recovery of ENU-induced mutations from the zebrafish germline. Genetics **136:** 1401–1420.
46. HAFFTER, P., M. GRANATO, M. BRAND, M.C. MULLINS, M. HAMMERSCHMIDT, D.A. KANE, J. ODENTHAL, F.J.M. VAN EEDEN, Y.-J. JIANG, C.-P. HEISENBERG, R.N. KELSH, M. FURUTANI-SEIKI, E. VOGELSANG, D. BEUCHLE, U. SCHACH, C. FABIAN & C. NÜSSLEIN-VOLHARD. 1996. The identification of genes with unique and essential functions in the development of the zebrafish, *Danio rerio.* Development **123:** 1–36.
47. DRIEVER, W., L. SOLNICA-KREZEL, A.F. SCHIER, S.C.F. NEUHAUSS, J. MALICKI, D.L. STEMPLE, D.Y.R. STAINIER, F. ZWARTKRUIS, S. ABDELILAH, Z. RANGINI, J. BELAK & C. BOGGS. 1996. A genetic screen for mutations affecting embryogenesis in zebrafish. Development **123:** 37–46.
48. GRANATO, M., F.J.M. VAN EEDEN, U. SCHACH, T. TROWE, M. BRAND, M. FURUTANI-SEIKI, P. HAFFTER, M. HAMMERSCHMIDT, C.-P. HEISENBERG, Y.-J. JIANG, D.A. KANE, R.N. KELSH, M.C. MULLINS, J. ODENTHAL & C. NÜSSLEIN-VOLHARD. 1996. Genes controlling and mediating locomotion behavior of the zebrafish embryo and larva. Development **123:** 399–413.
49. RIBERA, A.B. & C. NÜSSLEIN-VOLHARD. 1998. Zebrafish touch-insensitive mutants reveal an essential role for developmental regulation of sodium current. J. Neurosci. In press.

Dysfunction of Delayed Rectifier Potassium Channels in an Inherited Cardiac Arrhythmia

MICHAEL C. SANGUINETTI[a]

Department of Medicine, Division of Cardiology, University of Utah, Salt Lake City, Utah 84112, USA

ABSTRACT: The rapid (I_{Kr}) and slow (I_{Ks}) delayed rectifier K^+ currents are key regulators of cardiac repolarization. *HERG* encodes the K_r channel, and *KVLQT1* and *hminK* encode subunits that coassemble to form K_s channels. Mutations in any one of these genes cause Romano-Ward syndrome, an autosomal dominant form of long QT syndrome (LQT). Mutations in *KVLQT1* and *HERG* are the most common cause of LQT. Not all missense mutations of *HERG* or *KVLQT1* have the same effect on K^+ channel function. Most mutations result in a dominant-negative effect, but the severity of the resulting phenotype varies widely, as judged by reduction of current induced by coexpression of wild-type and mutant subunits in heterologous expression systems. Mutations in *hminK* (S74L, D76N) reduce I_{Ks} by shifting the voltage dependence of activation and accelerating channel deactivation. A recessive form of LQT is caused by mutations in either *KVLQT1* or *hminK*. The functional consequences of mutations in delayed rectifier K^+ channel subunits are delayed cardiac repolarization, lengthened QT interval, and an increased risk of torsade de pointes and sudden death.

Potassium channels determine the resting potential of most cells and modulate the action potential duration of excitable cells. The molecular identification of K^+ channels has proceeded at a rapid rate since the *Shaker* gene was first cloned from *Drosophila*, and it is likely that there are more than 100 K^+ channel genes in the human genome. This genetic diversity is accompanied by a lesser, but impressive, array of functional diversity. Hence, it is not too surprising that mutations in K^+ channel genes can lead to a spectrum of physiological disorders including episodic ataxia with myokymia,[1–3] long QT syndrome,[4–10] Bartter's syndrome,[11] and benign familial neonatal convulsions[12,13] in humans, and *weaver* ataxia in mice.[14] The great diversity of K^+ channels and their importance in modulating cell membrane excitability makes it likely that other inherited paroxysmal disorders of neural and muscle tissue will be found to result from mutations in other K^+ channel genes. This review will highlight some of the recent findings of the physiological consequences of mutations in cardiac K^+ channel genes that cause long QT syndrome (LQTS).

LONG QT SYNDROME

LQTS is a disorder of ventricular repolarization that predisposes affected individuals to fatal cardiac arrhythmias. Delayed ventricular repolarization causes prolongation of the QT interval on the surface electrocardiogram (ECG). LQTS is most commonly caused by

[a]Address for correspondence: M.C. Sanguinetti, Ph.D., University of Utah, Eccles Institute of Human Genetics, 15 N 2030 E, Room 4220, Salt Lake City, Utah 84112-5330. Phone: 801-585-6336; fax: 801-585-3501; e-mail: mike.sanguinetti@hci.utah.edu

treatment with medications that block cardiac K^+ channels, such as class III antiarrhythmic drugs or certain antihistamines.[15] These unwanted side effects, sometimes but not always associated with overdose, are exacerbated by bradycardia, hypokalemia, or hypomagnesemia. Less commonly, LQTS is inherited as an autosomal dominant or recessive disorder. Oftentimes affected individuals are asymptomatic, but the more severely affected individuals can have intermittent syncope caused by arrhythmia. Most often, this arrhythmia is self-terminating. Sudden death can result if the ventricular arrhythmia degenerates into fibrillation.

The ventricular arrhythmia associated with LQTS is a polymorphic ventricular tachycardia called torsades de pointes, and is characterized by a sinusoidal twisting of the QRS axis around the isoelectric line of the ECG.[16] It has been hypothesized that spatial dispersion of ventricular repolarization and an alteration in the predominance of two ectopic foci is the underlying cause of this sinusoidal waveform.[17] A possible cellular mechanism is a microreentrant electrical circuit, caused by premature Ca^{2+}-dependent depolarizations during the plateau of the cardiac action potential. The most commonly used pharmacological treatment for LQTS is β-adrenergic blockers.[15] These drugs probably act by preventing the enhancement of L-type Ca^{2+} channel current caused by an increase in sympathetic tone.

Because inherited LQTS is characterized by delayed ventricular repolarization, the underlying defect was postulated to be mutations in genes encoding cardiac ion channels. Likely candidate genes were those that encoded Na^+ and L-type Ca^{2+} channels that conduct inward current, and delayed rectifier K^+ channels that determine repolarization from the plateau phase of the action potential. These predictions have proved correct, except for the Ca^{2+} channel, which has not been associated with the disorder. LQTS is inherited either as an autosomal dominant (Romano-Ward syndrome) or autosomal recessive (Jervell and Lange-Nielsen syndrome) disorder.[15] Keating and colleagues[4–8] used a candidate gene approach and positional cloning to four identified genes—*SCN5A, HERG, KVLQT1,* and *hminK*—that cause LQTS. These genes encode ion channels that modulate membrane repolarization of the cardiac myocyte. *SCN5A* encodes the cardiac sodium (I_{Na}) channel, *HERG* encodes subunits that form a rapid delayed rectifier K^+ (I_{Kr}) channel, and *KVLQT1* and *hminK* encode subunits that coassemble to form the slow delayed rectifier K^+ (I_{Ks}) channel. Gain of function mutations in *SCN5A* cause an impaired Na^+ channel inactivation, which results in a very small maintained inward current during membrane depolarization.[18] Mutations in *HERG, KVLQT1,* or *hminK* cause a decrease in delayed rectifier K^+ channel (I_{Kr} or I_{Ks}) current. The functional consequence of either a persistent inward Na^+ current or a decreased outward K^+ current during the plateau phase of the cardiac action potential is delayed repolarization and an increased QT interval. Not all individuals with mutations in these genes have syncopal episodes, indicating that additional environmental (e.g., bradycardia, hypokalemia, and medications that block K^+ channels) or genetic factors likely contribute to the pathology of this disorder.

MOLECULAR IDENTITY AND PROPERTIES OF CARDIAC DELAYED RECTIFIER K^+ CHANNELS

The delayed rectifier K^+ current (I_K) of ventricular myocytes is composed of multiple components (I_{Ks}, I_{Kr}, I_{Kur}) that can be distinguished based on their rate of activation and pharmacology. All three currents have been recorded from isolated myocytes of several species, including human.[19–21] In rate of activation I_{Ks} is < I_{Kr} < I_{Kur}. I_{Kr} is specifically blocked by methanesulfonanilide class III antiarrhythmic agents such as d-sotalol, dofetilide, and E-4031.[22–24] I_{Ks} is specifically blocked by indapamide[25] and the benzodiazepine L-735,821.[26] I_{Kur} is blocked by a low concentration (50 μM) of 4-aminopyridine (4-AP).[27] The current-voltage relationships of I_{Ks} and I_{Kur} are nearly linear. In contrast, the current-voltage relationship of I_{Kr} exhibits inward rectification due to rapid channel inactivation.

HERG SUBUNITS COASSEMBLE TO FORM THE I_{Kr} CHANNEL

HERG encodes a channel with properties nearly identical to I_{Kr} in myocytes. This gene was discovered during a high-stringency screen of a human hippocampus cDNA library[28] and encodes a protein of 1159 amino acids with a predicted molecular weight of 127 kDa. HERG has the usual voltage-gated K^+ channel topology with six transmembrane spanning regions (S1–S6), a pore region, and long cytoplasmic amino- and carboxyl-terminal regions. The biophysical properties of HERG channels expressed in heterologous systems is similar to I_{Kr} measured in cardiac myocytes.[29,30] The single-channel conductance for inward HERG currents is 12 pS, but at positive potentials the slope conductance is reduced to 5 pS,[31] and the probability of channel opening is very low. Open probability is also low after patch excision into a divalent cation-free bathing solution, indicating that rectification of HERG results from an intrinsic gating process, and is not mediated by block of channels by Mg^{2+}, polyamines, or other intracellular blocking particle.[32,33] The intrinsic rectification of HERG results from rapid voltage-dependent inactivation that occurs about 100 times faster than channel opening at potentials near 0 mV. Activation and deactivation of HERG are about 10 times slower than I_{Kr} measured in isolated cardiac myocytes. This apparent discrepancy has been explained by the discovery of an alternatively spliced variant of HERG (HERGB) that activates and deactivates faster than HERG, and at a rate similar rate to that of I_{Kr}.[34] Similar alternatively spliced variants of *merg*, the mouse homologue of *HERG*, have been described,[34,35] and, like HERGB, are specifically expressed in the heart.[34,35]

Like I_{Kr} in cardiac myocytes, heterologously expressed HERG channels are specifically blocked by methanesulfonanilide class III antiarrhythmic drugs such as d-sotalol, E-4031, MK-499, and dofetilide.[30,36,37] This finding helps to explain why drug-induced LQTS is so similar to certain forms of inherited LQTS.

KvLQT1 AND minK SUBUNITS COASSEMBLE TO FORM I_{Ks} CHANNELS

KVLQT1 was discovered by a positional cloning strategy during the intensive effort to identify the gene associated with LQT1 that mapped to chromosome 11.[5] Full-length clones have been isolated from human pancreas and heart,[38,39] mouse heart,[40] human heart,[41] and human kidney[42] cDNA libraries. The most common transcript encodes a 676–amino acid protein that is expressed most abundantly in the heart, pancreas, kidney, small intestine, and prostate.[42] When expressed in *Xenopus* oocytes or cultured mammalian cells, *KVLQT1* induced a K^+ current with biophysical properties unlike any known cardiac current. When coexpressed with minK, KvLQT1 induced a very slowly activating current that was very similar to cardiac I_{Ks}.[38,40] MinK is a small (129 amino acids in human) protein that induces an I_{Ks}-like current when heterologously expressed in oocytes, but not mammalian cells. Evidently, minK coassembles with a *Xenopus* KvLQT1 that is homologous to the human protein, and is constitutively expressed in oocytes[38] to form I_{Ks} channels. The stoichiometry for assembly of KvLQT1 and minK subunits is not known, but it has been suggested that functional I_{Ks} channels contain two minK subunits.[43] MinK may also associate with HERG to regulate its expression. Transfection of AT-1 cardiac myocytes with *minK* antisense decreased the magnitude of I_{Kr},[44] and cotransfection of CHO cells with minK increased the magnitude of HERG by about twofold.[45] Moreover, physical interaction of minK and HERG was demonstrated by coimmunoprecipitation.

MUTATIONS IN *HERG* CAUSE LQTS

A candidate gene approach was used to identify the gene associated with chromosome 7–linked LQTS (LQT2). *HERG*, *h*uman *e*ther a-go-go *r*elated *g*ene, was mapped to human chromosome 7 [28] and thus became a candidate gene for LQT2. Single-strand conformation polymorphism (SSCP) analysis of DNA from affected individuals revealed several abnormal conformers, and sequence analysis revealed many mutations, including one *de novo* mutation.[6]

Similar to other voltage-gated K^+ channels, I_{Kr} channels are likely to form by coassembly of four subunits. Therefore, it was hypothesized that mutant HERG subunits might coassemble with wild-type subunits to form heterotetramers with altered or loss of function, a dominant-negative effect.[6] We tested this hypothesis by coexpressing LQTS-associated mutant HERG subunits with wild-type HERG subunits in *Xenopus* oocytes.[46] Deletion of base pair 1261 (*Δbp1261*) results in truncation of HERG near the S1 transmembrane region. Another deletion mutant (*ΔI500-F508*) lacked nine amino acids in the S3 transmembrane region. Both mutations cause loss of function and do not interact with wild-type HERG subunits and would be expected to result in decreased I_{Kr} in myocytes by haploinsufficiency. The functional effects of Δbp1261 HERG have also been studied in transfected COS cells.[47] In these cells, Δbp1261 HERG had a variable, but significant, dominant-negative effect when cotransfected with wild-type HERG; this suggests that the N-terminal domain of HERG was required for subunit interaction in these cells.[47] In oocytes, this region is evidently not important for subunit interaction because functional HERG channels are formed from subunits lacking the majority of the N-terminal region.[33,48]

The missense mutations in *HERG* that have been characterized have a dominant-negative effect.[47] G638S caused loss of function and altered the highly conserved K^+-selective pore sequence GFG. Coexpression of wild-type and G628S HERG subunits indicated that coassembly of even a single G628S subunit in the tetrameric channel results in loss of function, a "lethal" dominant-negative effect. N470D HERG subunits were capable of coassembling and forming functional channels. Injection of oocytes with *N470D HERG* cRNA induced currents in oocytes that were smaller and deactivated more slowly than oocytes injected with *WT HERG* cRNA, and the voltage dependence of activation was shifted by −18 mV relative to control currents. Two other missense mutations in *HERG*, one in the pore region[49] and another in the carboxyl terminus,[50] were recently reported; and it is likely that many other mutations in HERG will be found that are associated with LQT2.

HERG current is paradoxically increased by an elevation of external [K^+].[29] This finding prompted a limited clinical trial to determine if K^+ supplementation might normalize QT intervals of individuals with known HERG mutations. Elevation of serum [K^+] using K^+ supplements and spironolactone caused a significant reduction of QT interval, reduced dispersion of ventricular refractoriness, and corrected abnormal T-wave morphology.[51] Because it is difficult to maintain an elevated level of serum K^+, it is unlikely that K^+ supplementation will be useful for chronic treatment of LQT2. However, these findings provide a mechanistic basis behind the correlation between low serum [K^+] and increased risk of torsade de pointes, and emphasize the need to avoid hypokalemia in individuals with LQTS.

MUTATIONS OF *KVLQT1* OR *hminK* CAUSE LQTS

The most common form of inherited LQTS is linked to chromosome 11p15.5 (LQT1). Keating and colleagues used a positional cloning approach to identify a novel gene, *KVLQT1*, at this locus.[5] Many different mutations of *KVLQT1* that cause LQT1 have been published,[5,42] and it is likely that many more will be described. Most KvLQT1 subunits containing single missense mutations (e.g., R174C, A177P, G269D, T311I, G314S, D317N, Y315S, L342F) exhibited loss of function when expressed alone or in the presence of hminK.[39,42,52] However, these subunits reduced the function of wild-type KvLQT1 subunits in coexpression experiments by a dominant-negative effect. Of all the LQT1 missense mutations studied, only one (L272F in S5 transmembrane domain) has been described that formed functional channels when expressed alone in *Xenopus* oocytes.[52] Currents induced by coexpression of L272F KvLQT1 and wild-type KvLQT1 were half-activated at a potential 10 mV more negative than wild-type KvLQT1 alone. Another KvLQT1 mutant, R555C in the carboxyl terminus, formed functional channels when coexpressed with hminK, but the resulting current density was 75% smaller than current induced by transfection of COS cells with wild-type KVLQT1 alone. In addition, currents deactivated at a faster rate, and voltage-dependent activation was shifted by +50 mV.[42]

Dominant missense mutations in *hminK* cause LQT5, another form of Romano-Ward syndrome.[8] Two mutations (S74L and D76N) have been found in hminK. Both are located in the putative cytoplasmic region of the protein, and alter the function of I_{Ks} channels formed by coassembly of KvLQT1 and hminK subunits. Both mutants shift the voltage dependence of activation to more positive potentials and increase the rate of deactivation. In addition, D76N has a strong dominant-negative effect. These alterations in channel gating would reduce I_{Ks} and lengthen action potential duration of a cardiac myocyte.

Homozygous mutations in either *KVLQT1* or *hminK* cause Jervell and Lange-Nielsen syndrome, a severe form of LQTS, and bilateral deafness.[7,53,54] Recessive mutations in *KVLQT1* include a missense mutation (W305S) and a deletion-insertion that leads to a frameshift, a premature stop codon, and truncation of the carboxyl terminus.[53] These mutations have a less severe phenotype as assessed in the oocyte expression system than the dominant mutations that cause Romano-Ward syndrome.[39,42] In contrast, a homozygous mutation in *hminK*, D76N, that causes Jervell and Lange-Nielsen syndrome[54] is the same as a heterozygous mutation associated with Romano-Ward syndrome that has a strong dominant-negative effect.[8] Most mutations of *KVLQT1* that cause autosomal dominant (Romano-Ward) LQTS have a strong dominant-negative effect. Mutations in

KVLQT1 that cause Jervell and Lange-Nielsen syndrome have a weaker dominant-negative effect, consistent with its recessive pattern of inheritance.[39,42] Dominant or recessive mutations in *hminK* can also cause Romano-Ward or Jervell and Lange-Nielsen syndromes.[8,9,54,55]

Further molecular genetic studies will lead to the identification of many more mutations in *HERG* and *KVLQT1* that cause LQTS. LQTS-associated mutations in *SCN5A* will be more rare because of the requirement that such mutations would have to cause a gain in function. Mutations in *hminK* will also be relatively rare due to the small size of the gene. It is also possible that mutations in other ion channels such as the L-type Ca^{2+} channel will be found to be associated with LQTS. Identification of all the mutations that cause LQTS will enable presymptomatic diagnosis and risk stratification, and design of gene- and mechanism-specific strategies for prevention and management of this potentially lethal disease.

REFERENCES

1. Browne, D.L., S.T. Gancher, J.G. Nutt *et al.* 1994. Episodic ataxia/myokymia syndrome is associated with point mutations in the human potassium channel gene, KCNA1. Nature Genet. **8:** 136–140.
2. Browne, D.L., E.R.P. Brunt, R.C. Griggs *et al.* 1995. Identification of two new KCNA1 mutations in episodic ataxia/myokymia families. Human Mol. Genet. **4:** 1671–1672.
3. Comu, S., M. Giuliani & V. Narayanan. 1996. Episodic ataxia and myokymia syndrome: A new mutation of potassium channel gene Kv1.1. Am. Neurolog. Assoc. **40:** 684–687.
4. Wang, Q., J. Shen, I. Splawski *et al.* 1995. *SCN5A* mutations associated with an inherited cardiac arrhythmia, long QT syndrome. Cell **80:** 805–811.
5. Wang, Q., M.E. Curran, I. Splawski *et al.* 1996. Positional cloning of a novel potassium channel gene: KVLQT1 mutations cause cardiac arrhythmias. Nature Genet. **12:** 17–23.
6. Curran, M.E., I. Splawski, K.W. Timothy *et al.* 1995. A molecular basis for cardiac arrhythmia: *HERG* mutations cause long QT syndrome. Cell **80:** 795–804.
7. Splawski, I., K.W. Timothy, G.M. Vincent *et al.* 1997. Molecular basis of the long-QT syndrome associated with deafness. N. Engl. J. Med. **336:** 1562–1567.
8. Splawski, I., M. Tristani-Firouzi, M.H. Lehmann *et al.* 1997. Mutations in the *hminK* gene cause long QT syndrome and suppress I_{Ks} function. Nature Genet. **17:** 338–340.
9. Tyson, J., L. Traneebjaerg, S. Bellman *et al.* 1997. IsK and KvLQT1: Mutation in either of the two subunits of the slow component of the delayed rectifier potassium channel can cause Jervell and Lange-Nielsen syndrome. Hum. Mol. Genet. **6:** 2179–2185.
10. Russell, M.W., M. Dick II, F.S. Collins *et al.* 1996. KVLQT1 mutations in three families with familial or sporadic long QT syndrome. Hum. Molec. Genet. **5:** 1319–1324.
11. Simon, D.B., F.E. Karet, J.R. Soriano *et al.* 1996. Genetic heterogeneity of Bartter's syndrome revealed by mutations in the K^+ channel, ROMK. Nature Genet. **14:** 152–156.
12. Biervert, C., B.C. Schroeder, C. Kubisch *et al.* 1997. A potassium channel mutation in neonatal human epilepsy. Science **279:** 403–406.
13. Charlier, C., N.A. Singh, S.G. Ryan *et al.* 1998. A pore mutation in a novel KQT-like potassium channel gene in an idiopathic epilepsy family. Nat. Genet. **18:** 6–8.
14. Patil, N., D.R. Cox, D. Bhat *et al.* 1995. A potassium channel mutation in weaver mice implicates membrane excitability in granule cell differentiation. Nature Genet. **11:** 126–129.
15. Roden, D.M., R. Lazzara, M. Rosen *et al.* 1996. Multiple mechanisms in the long QT syndrome: current knowledge, gaps, and future directions. Circulation **94:** 1996–2012.
16. Dessertenne, F. 1966. La tachycardie ventriculaire a deaux foyers opposes variables. Arch. Mal. Coeur. Vaiss. **59:** 263–272.
17. El-Sherif, N., E.B. Caref, H. Yin *et al.* 1996. The electrophysiological mechanism of ventricular arrhythmias in the long QT syndrome. Circ. Res. **79:** 474–492.

18. BENNETT, P.B., K. YAZAWA, N. MAKITA *et al.* 1995. Molecular mechanism for an inherited cardiac arrhythmia. Nature **376:** 683–685.
19. WANG, Z., B. FERMINI & S. NATTEL. 1993. Delayed rectifier outward current and repolarization in human atrial myocytes. Circ. Res. **73:** 276–285.
20. WANG, Z., B. FERMINI & S. NATTEL. 1994. Rapid and slow components of delayed rectifier currents in human atrial myocytes. Cardiovasc. Res. **28:** 1540–1546.
21. LI, G.-R., J. FENG, L. YUE *et al.* 1996. Evidence for two components of delayed rectifier K^+ current in human ventricular myocytes. Circ. Res. **78:** 689–696.
22. CARMELIET, E. 1992. Voltage- and time-dependent block of the delayed K^+ current in cardiac myocytes by dofetilide. J. Pharmacol. Exp. Ther. **262:** 809–817.
23. CARMELIET, E. 1993. Use-dependent block and use-dependent unblock of the delayed rectifier K^+ current by almokalant in rabbit ventricular myocytes. Circ. Res. **73:** 857–868.
24. SANGUINETTI, M.C. & N.K. JURKIEWICZ. 1990. Two components of cardiac delayed rectifier K^+ current: Differential sensitivity to block by class III antiarrhythmic agents. J. Gen. Physiol. **96:** 195–215.
25. TURGEON, J., P. DALEAU, P.B. BENNETT *et al.* 1994. Block of I_{Ks}, the slow component of the delayed rectifier K^+ current, by the diuretic agent indapamide in guinea pig myocytes. Circ. Res. **75:** 879–886.
26. SALATA, J.J., N.K. JURKIEWICZ, M.C. SANGUINETTI *et al.* 1996. The novel class III antiarrhythmic agent, L-735,821 is a potent and selective blocker of I_{Ks} in guinea pig ventricular myocytes. Circulation **94:** I529.
27. WANG, Z., B. FERMINI & S. NATTEL. 1993. Sustained depolarization-induced outward current in human atrial myocytes: Evidence for a novel delayed rectifier K^+ current similar to Kv1.5 cloned channel currents. Circ. Res. **73:** 1061–1076.
28. WARMKE, J.W. & B. GANETZKY. 1994. A family of potassium channel genes related to *eag* in *Drosophila* and mammals. Proc. Natl. Acad. Sci. USA **91:** 3438–3442.
29. SANGUINETTI, M.C., C. JIANG, M.E. CURRAN *et al.* 1995. A mechanistic link between an inherited and an acquired cardiac arrhythmia: *HERG* encodes the I_{Kr} potassium channel. Cell **81:** 299–307.
30. TRUDEAU, M., J.W. WARMKE, B. GANETZKY *et al.* 1995. HERG, A human inward rectifier in the voltage-gated potassium channel family. Science **269:** 92–95.
31. ZOU, A., M.E. CURRAN, M.T. KEATING *et al.* 1997. Single HERG delayed rectifier K^+ channels in Xenopus oocytes. Am. J. Physiol. **272:** H1309–H1314.
32. SMITH, P.L., T. BAUKROWITZ & G. YELLEN. 1996. The inward rectification mechanism of the HERG cardiac potassium channel. Nature **379:** 833–836.
33. SPECTOR, P.S., M.E. CURRAN, A. ZOU *et al.* 1996. Fast inactivation causes rectification of the I_{Kr} channel. J. Gen. Physiol. **107:** 611–619.
34. LEES-MILLER, J.P., C. KONDO, L. WANG *et al.* 1997. Electrophysiological characterization of an alternatively processed ERG K^+ channel in mouse and human hearts. Circ. Res. **81:** 719–726.
35. LONDON, B., M.C. TRUDEAU, K.P. NEWTON *et al.* 1997. Two isoforms of the mouse *ether-a-go-go*–related gene coassemble to form channels with properties similar to the rapidly activating component of the cardiac delayed rectifier K^+ current. Circ. Res. **81:** 870–878.
36. SPECTOR, P.S., M.E. CURRAN, M.T. KEATING *et al.* 1996. Class III antiarrhythmic drugs block HERG, a human cardiac delayed rectifier K^+ channel; open channel block by methanesulfonanilides. Circ. Res. **78:** 499–503.
37. SNYDERS, D.J. & A. CHAUDHARY. 1996. High affinity open channel block by dofetilide of *HERG* expressed in a human cell line. Molec. Pharmacol. **49:** 949–955.
38. SANGUINETTI, M.C., M.E. CURRAN, A. ZOU *et al.* 1996. Coassembly of KvLQT1 and minK (IsK) proteins to form cardiac I_{Ks} potassium channel. Nature **384:** 80–83.
39. WOLLNIK, B., B.C. SCHROEDER, C. KUBISCH *et al.* 1997. Pathophysiological mechanisms of dominant and recessive *KVLQT1* K^+ channel mutations found in inherited cardiac arrhythmias. Hum. Molec. Genet. **6:** 1943–1949.
40. BARHANIN, J., F. LESAGE, E. GUILLEMARE *et al.* 1996. KvLQT1 and IsK (minK) proteins associate to form the I_{Ks} cardiac potassium channel. Nature **384:** 78–80.
41. YANG, W.P., P.C. LEVESQUE, W.A. LITTLE *et al.* 1997. KvLQT1, a voltage-gated potassium channel responsible for human cardiac arrhythmias. Proc. Natl. Acad. Sci. USA **94:** 4017–4021.

42. CHOUABE, C., N. NEYROUD, P. GUICHNEY *et al.* 1997. Properties of KvLQT1 K^+ channel mutations in Romano-Ward and Jervell and Lange-Nielsen inherited cardiac arrhythmias. EMBO J. **16:** 5472–5479.
43. WANG, K.-W. & S.A.N. GOLDSTEIN. 1995. Subunit composition of minK potassium channels. Neuron **14:** 1303–1309.
44. YANG, T., S. KUPERSHMIDT & D. RODEN. 1995. Anti-minK antisense decreases the amplitude of the rapidly activating cardiac delayed rectifier K^+ current. Circ. Res. **77:** 1246–1253.
45. MCDONALD, T.V., Z. YU, Z. MING *et al.* 1997. A minK-HERG complex regulates the cardiac potassium current I_{Kr}. Nature **388:** 289–292.
46. SANGUINETTI, M.C., M.E. CURRAN, P.S. SPECTOR *et al.* 1996. Spectrum of HERG K^+ channel dysfunction in an inherited cardiac arrhythmia. Proc. Natl. Acad. Sci. USA **93:** 2208–2212.
47. LI, X., J. XU & M. LI. 1997. The human Δ1261 mutation of the *HERG* potassium channel results in a truncated protein that contains a subunit interaction domain and decreases the channel expression. J. Biol. Chem. **272:** 705–708.
48. SCHONHERR, R. & S.H. HEINEMANN. 1996. Molecular determinants for activation and inactivation of HERG, a human inward rectifier potassium channel. J. Physiol. **493.3:** 635–642.
49. BENSON, D.W., C.A. MACRAE, M.R. VESELY *et al.* 1996. Missense mutation in the pore region of HERG causes familial long QT syndrome. Circulation **93:** 1791–1795.
50. SATLER, C.A., E.P. WALSH, M.R. VESELY *et al.* 1996. Novel missense mutation in the cyclic nucleotide-binding domain of *HERG* causes long QT syndrome. Am. J. Med. Genet. **65:** 27–35.
51. COMPTON, S., R. LUX, M. RAMSEY *et al.* 1996. Genetically defined therapy of inherited long QT syndrome: Correction of abnormal repolarization by potassium. Circulation **94:** 1018–1022.
52. SHALABY, F.Y., P.C. LEVESQUE, W.-P. YANG *et al.* 1997. Dominant-negative *KVLQT1* mutations underlie the LQT1 form of long QT syndrome. Circulation **96:** 1733–1736.
53. NEYROUD, N., F. TESSON, I. DENJOY *et al.* 1997. A novel mutation in the potassium channel gene *KVLQT1* causes the Jervell and Lange-Nielsen cardioauditory syndrome. Nature Genet. **15:** 186–189.
54. SCHULZE-BAHR E., Q. WANG, H. WEDEKIND *et al.* 1997. KCNE1 mutations cause Jervell and Lange-Nielsen syndrome. Nature Genet. **17:** 267–268.
55. DUGGAL, P., M.R. VESELY, D. WATTANASIRICHAIGOON *et al.* 1998. Mutation of the gene for IsK associated with both Jervell and Lange-Nielsen and Romano-Ward forms of the long-QT syndrome. Circulation **97:** 142–146.

Topology of the Pore Region of an Inward Rectifier K^+ Channel, Kir2.1

C. DART,[a] M. L. LEYLAND,[a] P. J. SPENCER,[b] P. R. STANFIELD,[a,d] AND M. J. SUTCLIFFE[c]

Ion Channel Group, Departments of [a]Cell Physiology and Pharmacology, [b]Biochemistry, and [c]Chemistry, University of Leicester, PO Box 138, Leicester LE1 9HN, UK

Inwardly rectifying K^+ channels differ from voltage-gated and Ca^{2+}-activated K^+ channels by having only two membrane-spanning segments (M1 and M2) per subunit in the tetramer.[1–3] They retain, however, a short stretch of amino acids between M1 and M2 known as the H5 or P region, which in Kv channels has been shown to project into the center of the pore to form the K^+ selectivity filter.[4] Sequence similarity between the H5 region of Kir and Kv channels has led to the suggestion that the two different classes of K^+ channel may share the same basic K^+-selective pore design, possibly reflecting the pore structure of an ancestral channel.

Direct evidence that the Kir and Kv pores are similar in shape stems from binding studies using peptide neurotoxins, which bind tightly to the mouth of the pore and act as structural templates. The scorpion toxin Lq2, a potent inhibitor of both Kv and Kca channels also blocks the pore of the inward rectifier, Kir1.1.[5] The interaction surface of the toxin is the same for binding to Kir, Kv, and Kca channels, suggesting that a broadly similar pore structure exists across the different K^+ channel families.

We have produced a structural model of the pore-forming H5 region of the inward rectifier K^+ channel, Kir2.1 (residues 138–149, ETQTTIGYGFRC), based initially on an existing molecular model of the pore region of the voltage-gated channel, Kv1.3.[6] Residues whose side chains are predicted to line the channel pore were tested using cysteine-scanning mutagenesis and subsequent blockage by Ag^+ in monomers and tandem concatamers. Our results suggest that the topology of the Kir pore is similar, but not identical, to that of Kv channels. Since the completion of this study, the crystal structure of a bacterial K^+ channel, KcsA, has been published.[7] We discuss our findings in light of this new and important work.

METHODS

Details of methods are given in.[8] All experiments were carried out on channels in which the exposed Cys residue (C149) had been replaced with Ser, since we have shown this mutation prevents blockage of wild-type channels by Ag^+.[8]

[d]Corresponding author. Phone: ++116-252-3300; fax: ++116-223-1401; e-mail: prs@le.ac.uk

RESULTS

The results of cysteine scanning mutagenesis of the H5 region of Kir2.1 are summarized in FIGURE 1. Residues that, when mutated to cysteine, were inhibited by the application of 200-nM external Ag^+ were: Thr 141, Thr 142, Ile 143, Tyr 145, and Phe 147. We would predict that the side chains of these residues project into the pore. Residues that, when mutated to cysteine, remained insensitive to the application of external Ag^+ were: Thr 139, Gln 140, Gly 144, and Gly 146. Based on these results, the positions of three residues in our model (Thr 141, Ile 143, Cys 149) required refinement.

The final version of the model of the H5 region of Kir2.1 is shown in FIGURE 2. Cys 149, the residue responsible for Ag^+ block in wild-type channels, lies just above the aromatic residue Phe 147. The other aromatic, Tyr 145, also lines the pore and lies on the

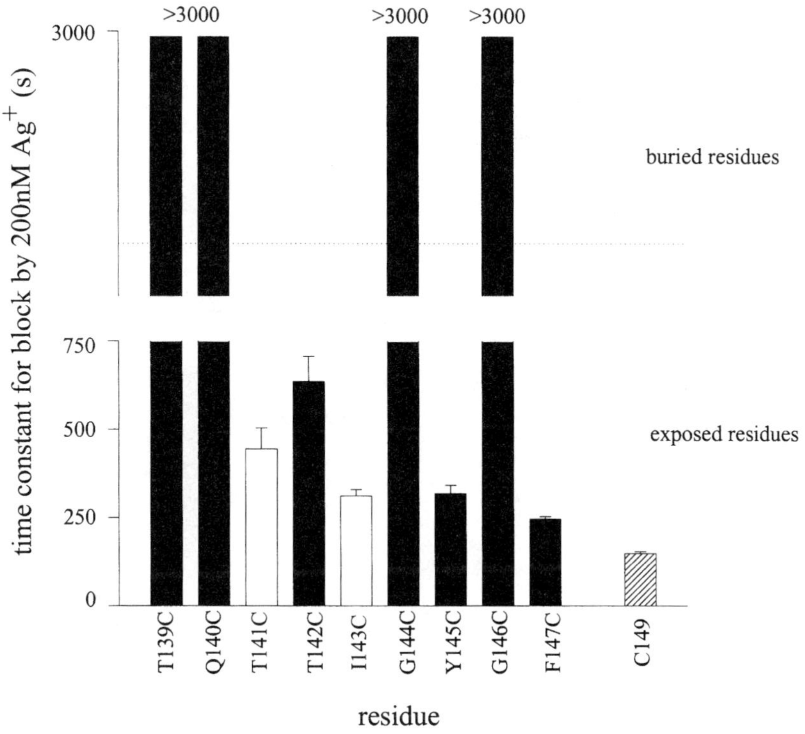

FIGURE 1. Exposed and buried residues in the H5 region of Kir2.1: results of cysteine scanning mutagenesis and agreement with the initial model. The development of block was fitted with a single exponential, whose time constant is given (mean ± SEM, n = 3–7). This time constant was >3000 s for buried residues. Except for Q140C (tetramer) and I143C (monomer), all measurements were made using dimeric constructs. It is expected that I143C would be blocked more slowly if expressed as a dimer, since the number of potential binding sites would be reduced from 4 to 2. *Filled histogram bars* indicate results that agree with the initial model; *open bars* indicate disagreement. The *cross-hatched bar* for C149 indicates the need to reposition the residue slightly.

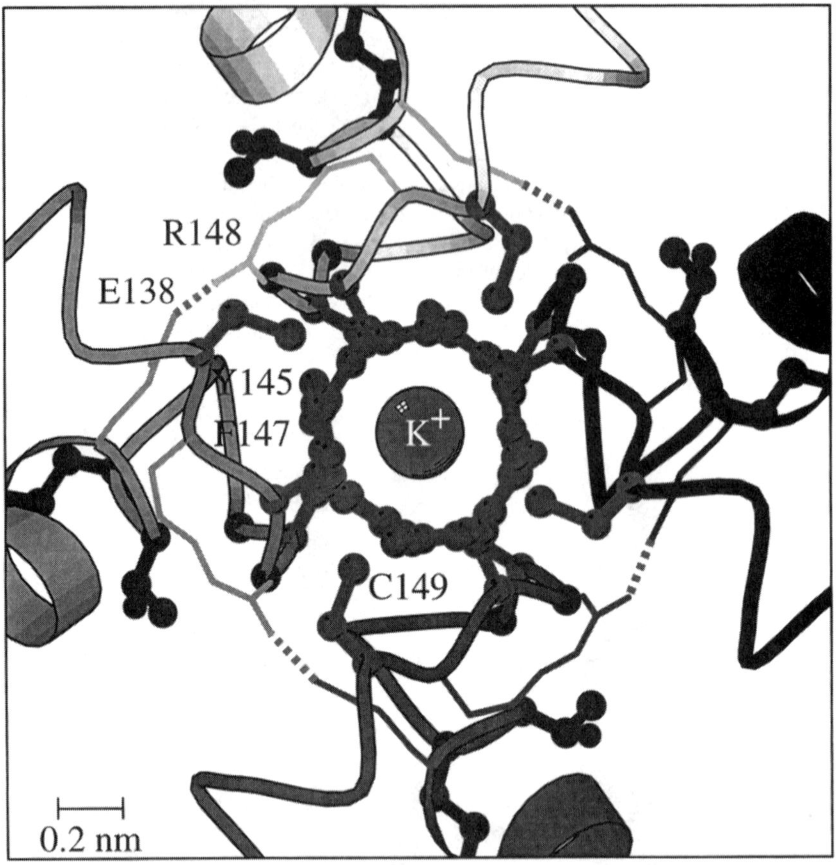

FIGURE 2. Molecular model of H5 region of Kir2.1. Schematic "ball and stick" representation as viewed from the outside of the cell. Residues found to be accessible to external Ag^+ (Tyr 145, Phe 147, and Cys 149) line the narrowest part of the pore; residues Thr141, Thr 142, and Ile 143 are not visible from this view.

other side of Phe 147 from Cys 149. Thr 141, Thr 142, and Ile 143 are positioned below Tyr 145.

DISCUSSION

The results of our experiments are in reasonable agreement with the initial version of the model, suggesting that the topology of the pore regions of Kir2.1 and Kv1.3 are similar, although not identical.

The identification of residues that line the channel pore is crucial for understanding the molecular basis of ion selectivity. In our model, Phe 147 and Tyr 145 point in toward the pore lumen and would be able to coordinate K^+ in the pore in a cage of π electrons generated at the face of their aromatic rings.[9] In addition, the carbonyl groups of Thr 142 and Gly 144 may also contribute to selectivity.

Since the completion of this study, the crystal structure of the K^+ channel from the bacterium *Streptomyces lividans* (KcsA) has been published.[7] This elegant study indicates that residues within the H5 signature sequence of KcsA orient their side chains away from the pore so that main-chain carbonyl oxygens line the selectivity filter. These carbonyl oxygen atoms are geometrically constrained so that a dehydrated K^+ ion fits precisely, but a Na^+ ion is too small to coordinate properly.

This picture of the pore is clearly different from the selectivity filter of Kir2.1 as presented here. The differences between the two models may reflect differences in the channels studied. Alternatively, side chains may orient as in the KcsA crystal structure, but mutation to the shorter side-chained Cys may allow the thiol to reorient so as to be accessible to Ag^+ in the pore. Reorientation of the thiol would, however, be expected to lead to a disruption in K^+ selectivity, which we do not observe. Additionally, Kir2.1 lacks key aromatic residues (pore helix Trp), which in KcsA act to constrain the rigid "cuff" of carbonyl oxygens within the selectivity filter. We conclude that the general mechanism proposed by Doyle *et al.* for K^+ selectivity may not apply to Kir2.1, and therefore may not be universal among K^+ channels.

REFERENCES

1. Ho, K. *et al.* 1993. Cloning and expression of an inwardly rectifying ATP-regulated potassium channel. Nature **362:** 31–38.
2. Kubo, Y. *et al.* 1993. Primary structure and functional expression of a mouse inward rectifier potassium channel. Nature **362:** 127–133.
3. Yang, J. *et al.* 1995. Determination of the subunit stoichiometry of an inwardly rectifying potassium channel. Neuron **15:** 1441–1447.
4. Heginbotham, L. *et al.* 1994. Mutations in the K^+ channel signature sequence. Biophys. J. **66:** 1061–1067.
5. Lu, Z. *et al.* 1997. Purification, characterization and synthesis of an inward-rectifier K^+ channel inhibitor from scorpion venom. Biochemistry **36:** 6936–6940.
6. Aiyar, J. *et al.* 1995. Topology of the pore-region of a K^+ channel revealed by the NMR-derived structures of scorpion toxins. Neuron **15:** 1169–1181.
7. Doyle, D.A. *et al.* 1998. The structure of the potassium channel: Molecular basis of K^+ conduction and selectivity. Science **280:** 69–77.
8. Dart, C. *et al.* 1998. The selectivity filter of a potassium channel, murine Kir2.1, investigated using scanning cysteine mutagenesis. J. Physiol. **511:** 25–32.
9. Kumpf, R.A. *et al.* 1993. A mechanism for ion selectivity in potassium channels: Computational studies of cation-π interactions. Science **261:** 1708–1710.

Virus-Mediated Modification of Cellular Excitability

DAVID C. JOHNS, EDUARDO MARBAN,[a] AND H. BRADLEY NUSS

Section of Molecular and Cellular Cardiology, 844 Ross Building, Johns Hopkins University School of Medicine, 720 North Rutland Avenue, Baltimore, Maryland 21205, USA

Ion channels in the plasma membrane play a critical role in cellular function. These proteins are the gatekeepers that control ion homeostasis and shape excitability. Excitable cells use a variety of different ion channels to fashion their hallmark electrical signal, the action potential. Advances in molecular electrophysiology have led to the identification of more ion-channel genes than there are identified membrane currents. This excess is particularly striking with potassium channels, where a wide diversity of genes is compounded by variable levels of heteromultimerization, alternative splicing, and posttranslational modification. The classic methods of studying the roles of each gene rely either on exogenous expression in frog oocytes or pharmacological manipulation of native currents. While these techniques have yielded a wealth of information concerning ion-channel structure and function, they have come up short in linking individual genes and their products to physiology and disease.

Defects in ion channels have been linked to a number of inherited diseases, including epilepsy, periodic paralysis, cystic fibrosis, and long QT syndrome.[1] In addition, changes in cellular excitability are associated with several common disease states, including drug addiction, depression, and heart failure.[2–4] Finally, an enormous number of pharmaceutical

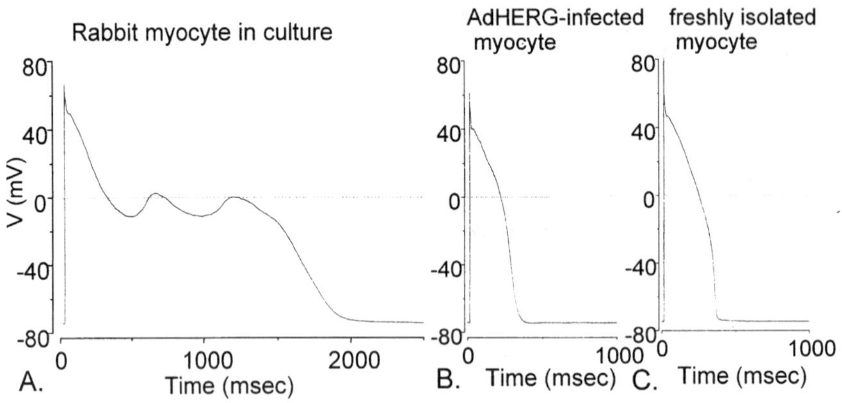

FIGURE 1. Action potentials recorded in ventricular myocytes that were isolated and kept in primary culture for 48 h are markedly prolonged and exhibit frequent and multiple early after depolarizations (**A**). Exogenous expression of *HERG* suppresses the frequency of EADs and normalizes the AP duration to close to normal values (**B,C**).

[a]Corresponding author. Phone: 410-955-2776; fax: 410-955-7953; e-mail: marban@welchlink.welch.jhu.edu

agents either directly or indirectly affect cellular excitability. This may be by design, as in the case of local anesthetic block of sodium channels, or as an unwanted side effect that limits the potential usefulness of the agent, such as erythromycin-induced block of cardiac potassium channels, leading to fatal arrhythmias.

The use of viral vectors to modify cellular excitability genetically has been previously demonstrated.[5–8] Using this approach, we were able to reverse the salient phenotypic changes that occur in an animal model of heart failure.[9] However, the reversal was often accompanied by an excessive abbreviation of the action-potential waveform, which could be arrhythmogenic or at the very least could compromise contractile function. This suggested two areas for improvement: (1) screening of other channels whose ionic-flux properties may provide a more physiological modification of the action potential waveform, and (2) tighter control of the amount of channel expressed. Here we report our progress in addressing each of these issues.

HERG SUPPRESSES EADS

The identification of mutations in the human *ether-a-go-go*-related gene (*Herg*) underlying a form of congenital long QT syndrome (LQT2) makes exogenous expression of the wild-type *Herg* gene a logical strategy for prevention of arrhythmias. *Herg* is particularly

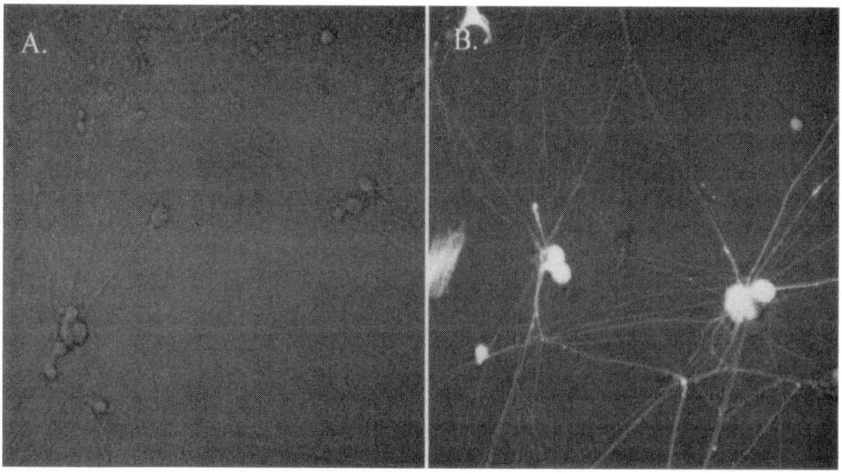

FIGURE 2 *(above and overleaf)*. Confocal images of SCG neurons infected with AdEGI-Kir2.1 and AdVgRXR in the absence **(A)** and presence **(B)** of muristerone A. The effects of a suprathreshold stimulus on a control cell do not change with the addition of barium **(C,D)**. The membrane current recorded from the same cell under these conditions using a ramp protocol from −38 mV to +22 mV over 500 ms **(E)**. The addition of barium had little or no detectable effect on the outwardly rectifying I–V curves in control cells. The response to a subthreshold stimulus on a Kir2.1-infected cell **(F)** is markedly changed by the addition of barium **(G)**; in both panels the *arrow* indicates the cessation of the applied stimulus. The I–V curve for this cell is much different upon addition of barium **(H)**. (Reprinted by permission from Johns *et al.*[12])

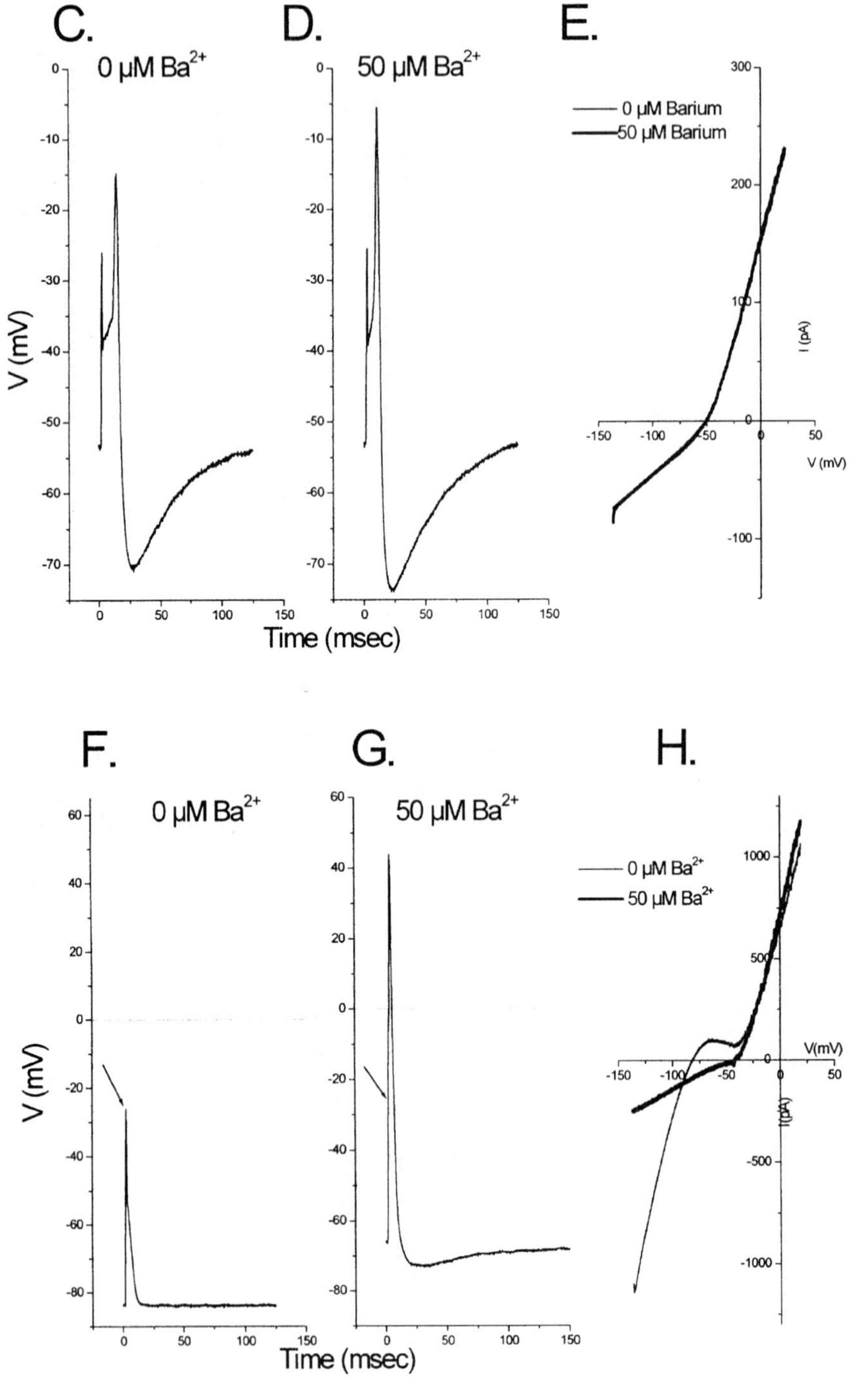

C.
0 μM Ba2+
D.
50 μM Ba2+
E.
0 μM Barium
50 μM Barium
V (mV)
I (pA)
Time (msec)
F.
0 μM Ba2+
G.
50 μM Ba2+
H.
0 μM Ba2+
50 μM Ba2+
V(mV)
I(pA)
V (mV)
Time (msec)

attractive because its ion flux is greatest during repolarization, so that it may not interfere with normal excitation. To test the effects of exogenous expression of the *Herg* channel in cardiac myocytes, we engineered an adenovirus containing the gene under the control of the Rous sarcoma virus promoter. This virus directed high-level expression of *Herg* currents when infected into CHO-K1 cells.

Changes that occur when heart cells are maintained in primary culture recapitulate many of the changes that have been noted in heart failure. In particular, there is an enhanced frequency of occurrence of spontaneous early afterdepolarizations (EADs) in cultured rabbit myocytes (FIG. 1A). EADs are the cellular triggers of long QT arrhythmias. We have used the cultured cells as a model system to test the hypothesis that *Herg* expression would suppress EADs without dramatically affecting the action-potential waveform. Indeed, there was a marked suppression in the frequency of EADs ($40 \pm 17\%$ of AdGFP infected cells vs. $9 \pm 5\%$ AdHerg infected cell: $p < 0.05$) without significant changes in the waveform or duration (FIG. 1B) as compared to freshly isolated cells (FIG. 1C). These results show that exogenous expression of *Herg* (or pharmacologic agonists of the channel) may have therapeutic value for suppressing long QT-related anhythmias, including those that cause sudden death in heart failure patients.

INDUCIBLE EXPRESSION

Control of gene expression was achieved by use of an ecdysone-regulatable promoter[10] either directly or in concert with an internal ribosome entry site (IRES).[11] To demonstrate the utility of this system we constructed viruses that express GFP alone, a GFP-Kir2.1 fusion protein, or a bicistronic message containing GFP and Kir2.1. When superior cervical ganglion (SCG) neurons were infected with these vectors, expression (as judged by green fluorescence) was only seen in the presence of the ecdysone analog muristerone A (FIG. 2A and 2B). FIGURE 2C shows an action potential elicited from a cell infected with a GFP control virus that is not changed by superfusion with 50-μM barium (Ba^{2+}), to specifically block the introduced Kir2.1 channels (FIG. 2D). Membrane voltage recordings from this cell are superimposable in the absence and presence of Ba^{2+} (FIG. 2E). FIGURE 2F–2H show a similar experiment performed on a cell that was infected with AdEGI-Kir2.1. Here a sub-threshold stimulus was given prior to application Ba^{2+}, to illustrate the effects on excitability of exogenous expression of this ion channel.

In summary the use of viral gene transfer to modify cellular excitability genetically has powerful applications in both clinical and basic science. Through logical design of the vector, both in the choice of the gene to be expressed and the cis-acting elements that control its expression, we have modified excitability of cardiac and neuronal cells in ways that are physiologically relevant.

ACKNOWLEDGMENTS

This work was supported by Tanabe Seiyaku Co. Ltd. We thank Drs. R.E. Mains and R. Marx for help with culturing SCG neurons, and Dr. B. O'Rourke for his help and advice with this work.

REFERENCES

1. Curran, M.E., I. Splawski, K.W. Timothy, G.M. Vincent, E.D. Green & M.T. Keating. 1995. A molecular basis for cardiac arrhythmia: HERG mutations cause long QT syndrome. Cell **80**(5): 795–803.
2. Nestler, E.J., M.T. Berhow & E.S. Brodkin. 1996. Molecular mechanisms of drug addiction—Adaptations in signal transduction pathways. Mol. Psychiatry. **1**(3): 190–199.
3. Sanguinetti, M.C., C. Jiang, M.E. Curran & M.T. Keating. 1995. A mechanistic link between an inherited and an acquired cardiac arrhythmia: HERG encodes the IKr potassium channel. Cell **81**(2): 299–307.
4. Hoffman, P.L. & B. Tabakoff. 1994. The role of the NMDA receptor in ethanol withdrawal. Exper. Suppl. **71:** 61–70.
5. Johns, D.C., H.B. Nuss, N. Chiamvimonvat, B.M. Ramza, E. Marban & J.H. Lawrence. 1995. Adenovirus-mediated expression of a voltage-gated potassium channel in vitro (rat cardiac myocytes) and in vivo (rat liver). A novel strategy for modifying excitability. J. Clin. Invest. **96**(2): 1152–1158.
6. Karschin, A., J. Aiyar, A. Gouin, N. Davidson & H.A. Lester. 1991. K^+ Channel expression in primary cell cultures mediated by vaccinia virus. FEBS Lett. **278**(2) 229–233.
7. Leonard, R.J., A. Karschin, S. Jayashree-Aiyar, N. Davidson, M.A. Tanouye, L. Thomas, G. Thomas & H.A. Lester. 1989. Expression of Drosophila Shaker potassium channels in mammalian cells infected with recombinant vaccinia virus. Proc. Natl. Acad. Sci. USA **86**(19): 7629–7633.
8. Ehrengruber, M.U., C.A. Doupnik, Y. Xu, J. Garvey, M.C. Jasek, H.A. Lester & N. Davidson. 1997. Activation of heteromeric G protein-gated inward rectifier channels overexpressed by adenovirus gene transfer inhibits the excitability of hippocampal neurons. Proc. Natl. Acad. Sci. USA **94:** 7070–7075.
9. Nuss, H.B., D.C. Johns, S. Kaab, G.F. Tomaselli, D. Kass, J.H. Lawrence & E. Marban. 1996. Reversal of potassium channel deficiency in cells from failing hearts by adenoviral gene transfer: A prototype for gene therapy for disorders of cardiac excitability and contractility. Gene Therapy **3**(10): 900–912.
10. No, D., T.P. Yao & R.M. Evans. 1996. Ecdysone-inducible gene expression in mammalian cells and transgenic mice. Proc. Natl. Acad. Sci. USA **93**(8): 3346–3351.
11. Dirks, W., M. Wirth & H. Hauser. 1993. Dicistronic transcription units for gene expression in mammalian cells. Gene **128**(2): 247–249.
12. Johns, D.C., R. Marx, R.E. Maïns, B. O'Rourke & E. Marbar. 1999. Inducible genetic suppression of neuronal excitability. J. Neurosci. In press.

Functional Characterization of a Cloned Human Intermediate-Conductance Ca^{2+}-Activated K^+ Channel

TINO DYHRING JØRGENSEN,[a] BO SKAANING JENSEN, DORTE STRØBÆK, PALLE CHRISTOPHERSEN, SØREN-PETER OLESEN, AND PHILIP KIÆR AHRING

NeuroSearch A/S, 26B Smedeland, DK-2600 Glostrup, Denmark

Ca^{2+}-activated K^+ channels are almost ubiquitously distributed in mammalian cells and constitute a major link between second-messenger systems and the electrical activity of the cell. Based on their electrophysiological characteristics, three major classes of Ca^{2+}-activated K^+ channels have been described: large-conductance channels (BK), small-conductance channels (SK), and intermediate-conductance channels (IK). Each type of channel shows a distinct pharmacology and expression pattern. Unlike SK and BK channels, IK is absent in excitable tissues, but its presence has been demonstrated in red blood cells,[1] B- and T-lymphocytes,[2,3] endothelial cells,[4,5] as well as cells derived from epithelia.[6,7] hIK is known to participate in the pathological dehydration of sickle cells,[8] making it a possible target in treatment of sickle cell anemia. In salt-transporting tissues IK channels are believed to play a universal role in epithelial function, and therefore pharmacological modulators of hIK might be used in the therapy of secretory disorders like diarrhea. In addition, blockade of Ca^{2+}-activated K^+ channels in T-lymphocytes has been shown to prevent secondary immune responses,[9] which demonstrate the possible relevance of IK modulators in treatment of immune disorders. In order to identify novel hIK channel modulators a high throughput fluorescence-based assay was developed.

METHODS

Cloning and Stable Expression of hIK

Cloning of hIK was based on the identification of clones in the GenBank EST database with homology to hSK1. A placenta-derived hIK cDNA was subcloned into the mammalian expression vector pNS1Z, a custom-designed derivative of pcDNA3Zeo (InVitrogen) and stably expressed into HEK 293 cells (HEK-hIK).[10]

FLIPR Experiments

A *fl*uometric *i*maging *p*late *r*eader (FLIPR) assay based on the membrane potential sensitive dye $DiBAC_4(3)$ was designed in 96 well plates seeded with HEK-hIK cells. Each

[a]Corresponding author. Phone: +45-43 43 50 10; fax: +45-43 43 59 99; e-mail: tdj@neurosearch.dk

well contained a 5-μM $DiBAC_4(3)$/FLIPR buffer solution with the following composition (mM): 145 NaCl, 1 KCl, 1 $CaCl_2$, 1 $MgCl_2$, 10 HEPES, 10 glucose (pH = 7.4). Fluorescence excitation (488 nm) was effected by an argon laser, and emission was monitored kinetically using a 510–560-nm bandpass interference filter.

RESULTS AND DISCUSSION

Patch-clamp recordings of HEK-hIK cells revealed a noninactivating potassium current exhibiting weak inward rectification with chord conductances of 11 and 30 pS at ±100 mV. The channel is blocked by classic inhibitors of the erythrocyte IK channel, such as clotrimazole and charybdotoxin, whereas it is insensitive to the selective SK channel blocker apamin.[10]

In addition to patch-clamp experiments optical recordings of membrane potential changes in stably transfected HEK-hIK cells were obtained using FLIPR. FIGURE 1 shows membrane potential measurements in HEK-hIK cells. A thapsigargin-induced release of stored calcium was used to activate hIK. Opening of hIK channels induces an efflux of K^+, which results in a cellular hyperpolarization seen as a decrease in the fluorescence intensity. Inhibition of hIK is demonstrated by addition of charybdotoxin, which reduces the hIK-mediated hyperpolarization in a dose-dependent manner. The depicted FLIPR data are in agreement with electrophysiological studies that showed an IC_{50} value for charybdotoxin of 28 ± 3 nM (n = 19). hIK channel activation could also be initiated by 1-ethyl-2-benzimidazolinone (EBIO), a compound that previously was reported to stimulate Ca^{2+}-activated K^+ currents in colonic T-84 epithelial cells[11] (FIG. 2). The EBIO-induced activation of hIK did not result from an increase in the intracellular Ca^{2+} concentration, since Fluo-3 measurements showed no effect of EBIO on the cellular Ca^{2+} level (data not

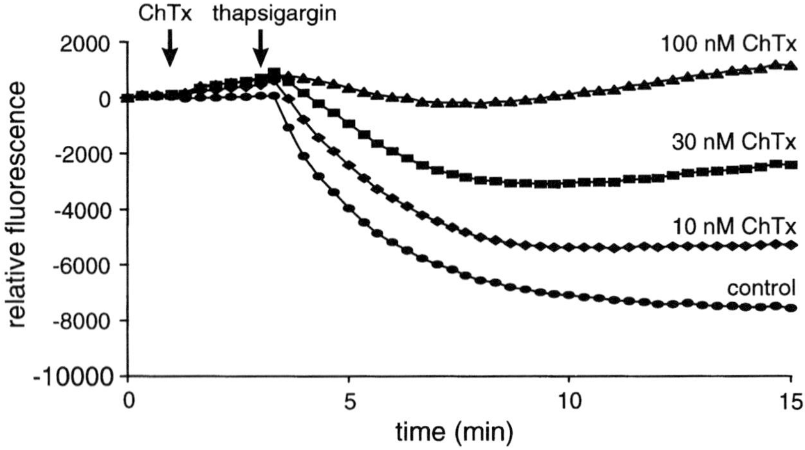

FIGURE 1. Time courses of membrane potential measurements in hIK expressing HEK 293 cells. Various concentrations of charybdotoxin (ChTx) and 0.5 μM thapsigargin were added at the times indicated by the *arrows*. Each trace is an average of data obtained from 8 wells.

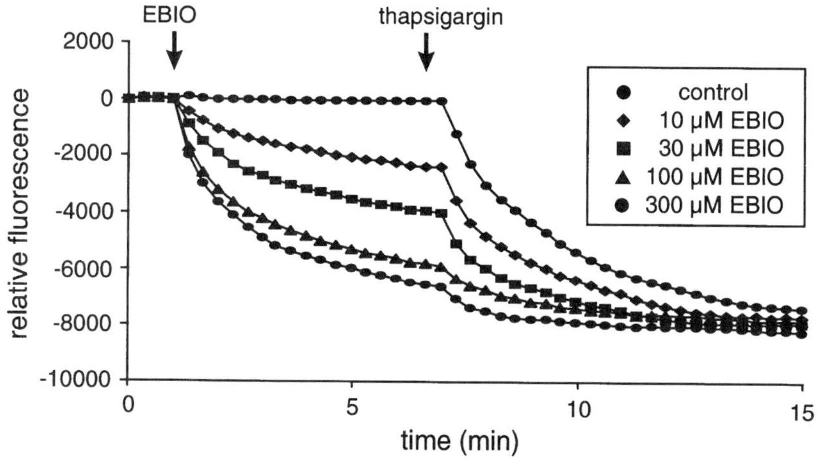

FIGURE 2. Time courses of EBIO-induced hIK channel activation. Various concentrations of EBIO and 0.5 μM thapsigargin were added at the times indicated by the *arrows*. Each trace is an average of data obtained from 8 wells.

shown). In addition, electrophysiological studies, in which intracellular Ca^{2+} was clamped to 100 nM, showed EBIO-induced hIK channel activation with an EC_{50} of 74 ± 11 μM (n = 3). As described previously, hIK may be an important molecular target in treatment of sickle cell anemia, immune disorders, and various secretory diseases. The demonstrated fluorescence-based assay provides the means of identifying novel pharmacological modulators of the hIK channel.

REFERENCES

1. Grygorczyk, R. & W. Schwarz. 1983. Properties of the Ca^{2+}-activated K^+ conductance of human red cells as revealed by the patch clamp technique. Cell Calcium **4:** 499–510.
2. Partiseti, M., D. Choquet, A. Diu & H. Korn. 1992. Differential regulation of voltage- and calcium-activated potassium channels in human B lymphocytes. J. Immunol. **148:** 3361–3368.
3. Grissmer, S., A.N. Nguyen & M.D. Cahalan. Calcium-activated potassium channels in resting and activated human T lymphocytes. J. Gen. Physiol. **102:** 601–630.
4. Sauvé, R, L. Parent, C. Simoneau & G. Roy. 1988. External ATP triggers a biphasic activation process of a calcium dependent K^+ channel in cultured bovine aortic endothelial cells. Pflügers Arch. **412:** 469–481.
5. Olesen, S.-P. & M. Bundgaard. 1993. ATP-dependent closure and reactivation of inward rectifier K^+ channels in endothelial cells. Circ. Res. **73:** 492–495.
6. Sauvé, R., C. Simoneau, R. Monette & G. Roy. 1986. Single-channel analysis of the potassium permeability in HeLa cancer cells: Evidence for a calcium-activated potassium channel of small unitary conductance. J. Membr. Biol. **92:** 269–282.
7. Devor, D.C. & R.A. Frizzell. 1993. Calcium-mediated agonists activate an inward rectified K^+-channel in colonic secretory cells. Am. J. Physiol. **265:** C1271–C1280.
8. Brugnara, C., B. Gee, C.C. Armsby, S. Kurth, M. Sakamoto, N. Rifai, S.L. Alper & O. Platt. 1996. Therapy with oral clotrimazole induces inhibition of the Gardos channel and reduction of erythrocyte dehydration in patients with sickle cell diseases. J. Clin. Invest. **97:** 1227–1234.

9. VERHEUGEN, J.A.H., F. LE DEIST, V. DEVIGNOT & H. KORN. 1997. Enhancement of calcium signaling and proliferation responses in activated human T lymphocytes. Inhibitory effects of K^+ channel block by charybdotoxin depend on the T cell activation state. Cell Calcium **21:** 1–17.
10. JENSEN, B.S., D. STRØBÆK, P. CHRISTOPHERSEN, T.D. JØRGENSEN, C. HANSEN, A. SILAHATAROGLU, S.-P. OLESEN & P.K. AHRING. 1998. Pharmacological and biophysical characterization of the cloned human intermediate-conductance Ca^{2+}-activated K^+ channel. Am. J. Physiol. **275:** C848–C856.
11. DEVOR, D.C., A.K. SINGH, R.A. FRIZZELL & R.J. BRIDGES. 1996. Modulation of Cl^- secretion by benzimidazolones. I. Direct activation of a Ca^{2+}-dependent K^+ channel. Am. J. Physiol. **271:** L775–L784.

Probing the Potassium Channel $K_v\beta1$/ $K_v1.1$ Interaction Using a Random Peptide Display Library

STEPHEN J. LOMBARDI,[a] AMY TRUONG, PAUL SPENCE, KENNETH J. RHODES, AND PHILIP G. JONES

Wyeth-Ayerst Research, CNS-Disorders, CN-8000 Princeton, New Jersey 08543-8000, USA

Voltage-gated (K_v) K^+ channels are heterooctomeric protein complexes composed of four integral membrane ion-conducting α-subunits and four tightly associated cytoplasmic β-subunits. Coexpression of certain β-subunits, for example $K_v\beta1$, with K_v1-family α-subunits in heterologous cells has revealed that the surface expression, stability, and conductance properties of the resulting channels can be dramatically altered by the presence of a β-subunit.[1,2] Analysis of the sequences of cloned β-subunits together with mutagenesis studies indicate that the amino terminal 30 amino acids of $K_v\beta1$ contain a domain that is necessary and sufficient to confer rapid inactivation kinetics upon $K_v1.1$ homomeric channels.[2] Interestingly, this N-terminal "inactivation ball" domain of $K_v\beta1$ shares primary amino acid sequence homology with the inactivation domain of the A-type K_v channels $K_v1.4$ and the *Drosophila* Shaker channel, reflecting a common role for these domains. Point mutations in the S4–S5 cytoplasmic loop of the $K_v1.4$ α-subunit, which lies near the inner mouth of the channel pore, indicate that this domain may be the acceptor site for the N-terminal inactivation ball,[3] and suggest that analysis of the interaction of these two channel domains may provide clues to the structural requirements for N-type inactivation.

RESULTS AND DISCUSSION

Here we used a bacterial peptide display library (>10^8 peptides), in which random peptides are expressed on the surface of bacteria, allowing a large array of diverse polypeptides to be screened for their ability to bind immobilized protein targets. The FliTRX random peptide display library (Invitrogen) was used to identify peptides that bound either a synthetic peptide "bait" corresponding to amino acids 313–328 of the human $K_v1.1$ α-subunit or a recombinant protein "bait" representing the N-terminal 31 amino acids of the $K_v\beta1$. 1×10^{10} cells were added to 100 μg of peptide immobilized on a 60-mm tissue culture plate. Unbound cells were removed by washing and those that bound, were harvested and grown overnight. This panning procedure was repeated up to four times. Plates were streaked following 2, 3, and 5 days of panning and the sequence of the expressed, interacting peptides determined. The sequences of the identified peptides were aligned to identify conserved residues that may be important for binding to the respective $K_v1.1$ and $K_v\beta1$ baits. Synthetic peptides containing alanine substitutions of residues thought to be

[a]Corresponding author. Phone: 732-274-4074; fax: 732-274-4020; e-mail: Lombars@war.wyeth.com

A.

wild-type Kv1.1	Q	I	L	G	Q	T	L	K	A	S	M	R	E	L	G	L	L
Round 2						T	L				M	R			G	L	
Round 3						T	L		A	S	M	R	E	L		L	
Round 5					Q	T	L	K	A	S	M	R	E	L	G	L	

B.

wild-type ß1	N	L	K	S	R	N	G	E	D	R	L	L	S	K	Q	S	S	T
Round 2					R		G	E	D	R	L							
Round 3			K	S	R	N	G	E	D	R	L	L	S					
Round 5		L	K	S	R	N	G	E	D	R	L	L	S	K	Q		S	

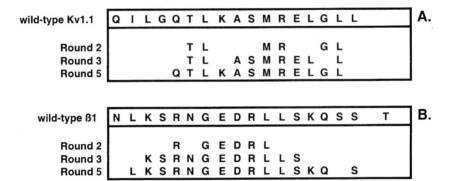

FIGURE 1. $K_v1.1$ **(Panel A)** and $K_v\beta1$ **(Panel B)** consensus sequence emergence among the pFliTrx plasmids. The $hK_v1.1$ and $hK_v\beta1$ consensus sequences derived from second, third, and fifth rounds of panning of pFliTrx clones.

important for this interaction, were analyzed in a protein interaction assay to determine their effect on $K_v1.1$(313–328) - $K_v\beta1$ binding. Briefly, a biotinylated $K_v1.1$ was incubated with recombinant full-length $K_v\beta1$ immobilized on a microtiter plate, in the presence or absence of wild-type or alanine-substituted peptides. Bound $K_v1.1$ was quantitated using an alkaline phosphatase colorimetric assay and IC_{50} values for inhibition of the protein–protein interaction determined.

Following five rounds of panning, 50 clones from each screen were isolated and the sequence of the expressed peptide determined. FIGURE 1 shows the consensus sequences obtained following 2, 3, and 5 days of panning in both screens. Interestingly, peptides which bound the $K_v1.1$ peptide showed significant sequence homology to the N-terminus of $K_v\beta1$. The sequence of all clones analyzed overlapped, identifying a core domain corresponding to residues 11–21 of $K_v\beta1$. Although this motif can be identified in the N-terminus of $K_v\beta1$, it is not present in $K_v\beta2$, which has a truncated N-terminus and cannot perform the fast inactivating functions of $K_v\beta1$. In the screen for $K_v\beta1$ binding peptides, sequences homologous to the S4–S5 loop of $K_v1.1$ were obtained, identifying a core domain represented by amino acids 317–324 and suggest that the S4–S5 loop of $K_v1.1$ forms a good acceptor for the inactivation ball domain of $K_v\beta1$.

Residues that play a key role in the interaction are likely to be found within identified peptides whose overall affinity for the bait sequence is low. Since the panning procedure is designed to enrich for peptides that bind with high affinity, we chose to isolate and examine the sequence of peptides following just 2 or 3 days of panning. The consensus sequence obtained from these early timepoints are shown in FIGURE 1, and identify a subset of amino acids in both $K_v1.1$ and $K_v\beta1$ that may play an essential and direct role in their association.

To confirm the importance of these residues for the interaction, we examined a series of synthetic mutant peptides for their ability to inhibit $K_v\beta1/K_v1.1$ binding (TABLE 1). Mutation of amino acids R20 and L21 in the $K_v\beta1$ sequence had the largest effect on binding, resulting in a 200 and a 30-fold decrease in affinity, respectively. Accordingly, these amino acids were highly represented at day 5, being present in 57 and 77% of all clones,

TABLE 1. Inhibition of Binding of hKvB1/hKv1.1 by Synthetic Peptides[a]

	Synthetic Peptides	IC50
Kvβ1 Wild-type	LKSRNGEDRLLS	9 ± 2nM
Kvβ1 K13A	LASRNGEDRLLS	9 ± 4nM
Kvβ1 S14A	LKARNGEDRLLS	10 ± 6nM
Kvβ1 R15A	LKSANGEDRLLS	25 ± 4nM
Kvβ1 N16A	LKSRAGEDRLLS	35 ± 5nM
Kvβ1 G17A	LKSRNAEDRLLS	10 ± 3nM
Kvβ1 E18A	LKSRNGADRLLS	11 ± 3nM
Kvβ1 D19A	LKSRNGEARLLS	9 ± 4nM
Kvβ1 R20A	LKSRNGEDALLS	2000 ± 3nM
Kvβ1 L21A	LKSRNGEDRALS	300 ± 5nM
Kvβ1 L22A	LKSRNGEDRLAS	10 ± 3nM
Kvβ1 S23A	LKSRNGEDRLLA	35 ± 4nM
Kv1.1 Wild-type	TLKASMRELGLL	15 ± 2nM
Kv1.1 T318A	ALKASMRELGLL	100 ± 2nM
Kv1.1 L319A	TAKASMRELGLL	17 ± 3nM
Kv1.1 S322A	TLKAAMRELGLL	2000 ± 4nM
Kv1.1 M323A	TLKASARELGLL	45 ± 2nM
Kv1.1 R324A	TLKASMAELGLL	NI
Kv1.1 E325A	TLKASMRALGLL	100 ± 3nM
Kv1.1 L326A	TLKASMREAGLL	25 ± 3nM
Kv1.1 G327A	TLKASMRELALL	18 ± 2nM
Kv1.1 L328A	TLKASMRELGAL	900 ± 5nM

[a]The dose-response for the ability of each high-affinity mutant peptide to compete for binding in a K^+ channel–specific plate assay ($n = 3$).

respectively. In $K_v\beta 1.2$ these residues are replaced by lysine and serine, respectively, and this perhaps explains the apparent lower efficacy of this β-subunit isoform in inducing rapid inactivation.

Residues R324 and L328 in the $hK_v1.1$ sequence were also found to be important for the interaction with the N-terminus of $K_v\beta 1$. The R234A mutation resulted in no inhibition, while the L328A substitution reduced the potency by 60-fold. The mutations T318A, S322A, and E325 also reduced the affinity of the peptides and interestingly, these same mutations in Shaker B reduced rapid inactivation,[3] perhaps by reducing the affinity of the interaction between Shaker channel's N-terminal inactivation domain and the S4–S5 loop.

This study demonstrates that random peptide libraries are powerful tools for the study of protein–protein interactions. While confirming that the $K_v\beta 1$ subunit is likely to exert its effects on inactivation by binding the $K_v1.1$ S4–S5 loop, we have identified residues that are likely to play key roles in the $K_v\beta 1$-mediated K_v channel inactivation.

REFERENCES

1. Rettig, J., S.H. Heinemann, F. Wunder, C. Lorra, D.N. Parcej, J.O. Dolly & O. Pongs. 1994. Inactivation properties of voltage-gated K+ channels altered by presence of β-subunit. Nature **369:** 289–294.
2. Shi, G., A.K. Kleinklaus, N.V. Marrion & J.S. Trimmer. 1994. Properties of Kv2.1 K^+ channels expressed in transfected mammalian cells. J. Biol. Chem. **269:** 23204–23211.
3. Isacoff, E.Y., Y.N. Jan & L.Y. Jan. 1991. Putative receptor for the cytoplasmic inactivation gate in the Shaker K^+ channel. Nature **353:** 86–90.

PKC Modulation of MinK Current Involves Multiple Phosphorylation Sites

C. FREDERICK LO AND RANDY NUMANN[a]

Wyeth-Ayerst Research, CN8000, Princeton, New Jersey 08543-8000, USA

Delayed-rectifier type K^+ currents are important in determining the shape and repolarization of the cardiac action potential. One component of the cardiac delay rectifier is termed I_{Ks}, and this has recently been suggested to be composed of two molecular entities: K_vLQT1 and minK.[1] Mutations have been found in the genes for both of these clones that are responsible for the most common forms of inherited long QT syndrome in man.[2] I_{Ks} is modulated in the heart by β-adrenergic receptors through a cAMP pathway and by α-adrenergic receptors through the protein kinase C (PKC) pathway. PKC activation stimulates I_{Ks} in guinea pig myocytes, but decreases I_{Ks} in mouse myocytes. When minK message is injected into *Xenopus* oocytes it coassembles with endogenous *Xenopus* K_vLQT1[1] to produce a current with properties very similar to I_{Ks}. Mouse and rat minK currents expressed this way are decreased by PKC activation, whereas guinea pig minK current is activated by PKC.[3,4] In mouse and rat minK cDNA there is a putative PKC phosphorylation site at serine 102. Amino acid sequence comparison reveals that guinea pig minK protein has an asparagine residue at the same position. Mutating the guinea pig asparagine into serine creates a guinea pig minK current that is down-regulated by PKC.[3] Conversely mutating the mouse serine 102 and neighboring residues produces a mutant mouse minK current that is up-regulated by PKC.[5] These results suggest that position 102 in minK protein is critical in determining the effect of PKC. Human minK has a serine residue at position 102. It would be predicted that PKC would down-regulate the associated current. However, we found that PKC activation can elicit a complex biphasic effect, with both up- and down-regulation, on human minK current. Here we investigate the PKC modulation of human minK current in more detail, including mutagenesis of position 102. We speculate that multiple phosphorylation sites are required to account for the complex action of PKC on the minK-K_vLQT1 channel.

MATERIALS AND METHODS

Human minK cDNA was subcloned in pSP64(polyA) vector, and mRNA made with the mMESSAGE mMACHINE kit (Ambion). Site-directed mutagenesis was performed using the Quik Change mutagenesis kit (Stratagene). Mutants were identified and confirmed by DNA sequencing. *Xenopus* oocytes were isolated and mRNA injected using conventional methods. Currents were measured using the two microelectrode voltage-clamp technique with a Turbo Tec-10c amplifier. Electrodes were filled with 3 M KCl, and the typical resistance was 1 MΩ. Holding potential was −80 mV in all experiments. Pulse/Pulsefit software (HEKA) was used for data acquisition and analysis, and a P/2 protocol

[a]Corresponding author. Phone: 732-274-4103; fax: 732-274-4004; e-mail: numannr@war.wyeth.com

was used for leakage subtraction. MinK current amplitudes were measured at the end of 5-s pulses to +30 mV repeated every 30–60 s. Pooled data are reported as mean ± SEM. Inhibitory peptides of PKC and PKA were dissolved in sterile H_2O, and injected into oocytes at a final concentration of 250 μM, assuming the volume of oocyte to be 1 μL. Oocytes were allowed to recover for 8–12 h before experiments.

RESULTS

HminK contains S102; however 50–100 nM phorbol-12,13-didecanoate (PDD) produced a biphasic effect with the current amplitude increasing during the first 20 min, and then decreasing below control levels. HminK current was 145.5 ± 23% of control current at the peak of the potentiating response and subsequently declined to 52 ± 7% of control

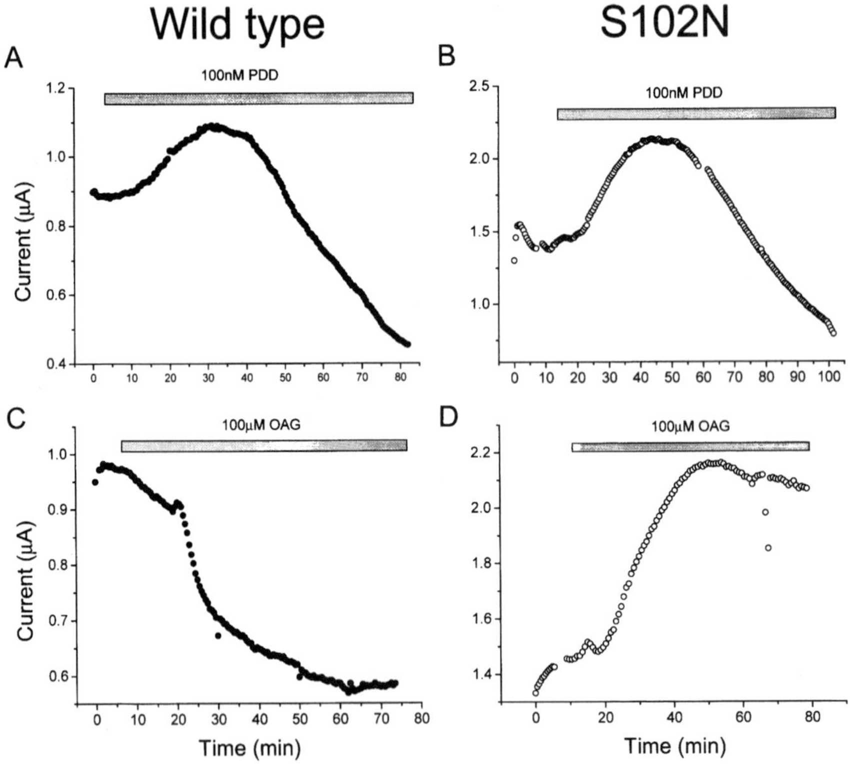

FIGURE 1. Effects of S102N mutation on PKC modulation of human minK current. **(A)** Human minK (wild-type) current response to 100 nM PDD (phorbol 12,13-didecanoate), showing the typical biphasic modulation. **(B)** 100 nM PDD also induces a biphasic effect on hminKS102N current. **(C)** Wild-type human minK current is down-regulated by 64.2 ± 6.8%, n = 5 with 100 μM OAG (1-oleoyl-2-acetyl-*sn*-glycerol). **(D)** Mutant hminKS102N current shows an increase (139 ± 18%, n = 6) after exposure to 100 μM OAG.

current after 1 h (n = 9). The current–voltage relationship of hminK current was not altered during PKC activation and the capacitance remained constant. This biphasic effect of PDD on hminK was completely eliminated by injection of a PKC-specific peptide inhibitor, while the PKA-specific peptide was without effect. Brief applications of 100 nM PDD or long applications of 5 nM PDD only elicited an increase in hminK current (134.6 ± 11.2%, n = 5). Interestingly, diacylglycerol analogs that activate PKC (100 μM 1,2-dioctanoyl-*sn*-glycerol [DOG] or 1-oleoyl-2-acetyl-*sn*-glycerol [OAG]) only produced a decrease in hminK current (59.7 ± 9.7%, n = 9). We mutated the putative PKC phosphorylation site Ser102 to Asn on hminK. The mutant's response to 100 nM PDD was biphasic, as was the wild type; however, 100 μM OAG increased the hminKS102N current amplitude (139 ± 18%, n = 6), opposite to the decrease seen in wild type (64.2 ± 6.8%, n = 5) (see FIG. 1). These results suggest that there are multiple PKC modulatory sites for the hminK current. S102 determines some of the PKC modulatory effects, while other PKC sites may reside elsewhere such as on K_VLQT1. One possible explanation of the data is that PDD and OAG/DOG activate different isoforms of PKC, and thereby produce different hminK responses mediated through different PKC sites.

REFERENCES

1. SANGUINETTI, M.C. *et al.* 1996. Coassembly of KvLQT1 and minK (IsK) proteins to form cardiac Iks potassium channel. Nature **384:** 80–83.
2. SPLAWSKI, I. *et al.* 1997. Mutations in the hminK gene cause long QT syndrome and suppress Iks function. Nat. Genet. **17:** 338–340.
3. HONORE, E. *et al.* 1991. Cloning, expression, pharmacology and regulation of a delayed rectifier K^+ channel in mouse heart. EMBO J. **10:** 2805–2811.
4. VARNUM, M.D. *et al.* 1993. The minK channel underlies the cardiac potassium current Iks and mediates species-specific responses to protein kinase C. Proc. Natl. Acad. Sci. USA **90:** 11528–11532.
5. ZHANG, J.E. *et al.* 1994. K^+ currents expressed from the guinea pig cardiac IsK protein are enhanced by activators of proteinkinase C. Proc. Natl. Acad. Sci. USA **91:** 1766–1770.

The Role of Kir2.1 in the Genesis of Native Cardiac Inward-Rectifier K+ Currents during Pre- and Postnatal Development

TOMOE Y. NAKAMURA,[a] KAREN LEE,[a] MICHAEL ARTMAN,[a,b]
BERNARDO RUDY,[b,c] AND WILLIAM A. COETZEE[a,b]

Departments of [a]Pediatrics, [b]Physiology and Neuroscience, and [c]Biochemistry, New York University Medical Center, New York 10016, USA

The prominent increase in cardiac resting potential during fetal development has been reported to be due to an increased density of the inward-rectifier K^+ current (I_{K1}). Several groups have described the existence of several K^+ channels with different unitary conductances in cardiac tissue. We recently found that a 21-pS channel was specifically reduced in rat ventricle by antisense phosphorothioate oligonucleotides (AS-oligos) targeted to the Kir2.1 transcript,[1] whereas the occurrence of K^+ channels with unitary conductances of ~8 and ~15 pS was unchanged. This suggests the possibility that the various unitary events might be functional manifestations of different channel-forming proteins, each possibly formed as homo- or heteromultimeric assembly of proteins coded by different genes. The aim of the present study is to examine the role of Kir2.1 channel proteins in the progressively increasing ventricular I_{K1} during fetal development. We chose to use mouse cardiac myocytes as a model system for our studies since the structural and functional developmental aspects of mouse embryos are increasingly well documented and also because of the obvious advantages that mouse models offer in terms of possible future gene-directed knockout/overexpression studies.

Kir2.1 mRNA EXPRESSION

As a first approach, we determined expression levels of Kir2.1 mRNA in ventricle from fetal (14 days gestational age) or adult hearts. We performed RNAse protection assays (RPA) using antisense biotinylated gel-purified riboprobe for Kir2.1 and cyclophilin (as an internal control). The probe quality was verified by the existence of distinct bands of the undigested cyclophilin and Kir2.1 probes (165 and 349 bp, respectively; FIG. 1A). In the absence of target RNA, the probes were totally digested by the RNAse mixture. More of the Kir2.1 probe was protected in the adult than in the fetal group, suggesting that transcriptional regulation of Kir2.1 occurs during fetal development. We used competitive RT-PCR (CRT-PCR) as a more exact method to quantitate the developmental changes in Kir2.1 mRNA levels (FIG. 1 B). Total RNA (5 μg) from adult (left) and 13-day fetal (right) mouse ventricles was used in a reverse transcription reaction (M-MLV RT enzyme) using a gene-specific reverse primer (25 pmole). Aliquots (5 μl) of the RT reaction were used as template in a PCR reaction (we used Amplitaq Gold to avoid nonspecific amplification). Different amounts of competitor were added (lanes 1–7, respectively, contained 3×10^{-21}, 3×10^{-20}, 1×10^{-19}, 3×10^{-19}, 1×10^{-18}, and 0 mole competitor). After staining with the ultrasensitive DNA indicator SYBR-Gold (Molecular Probes), two distinct bands were

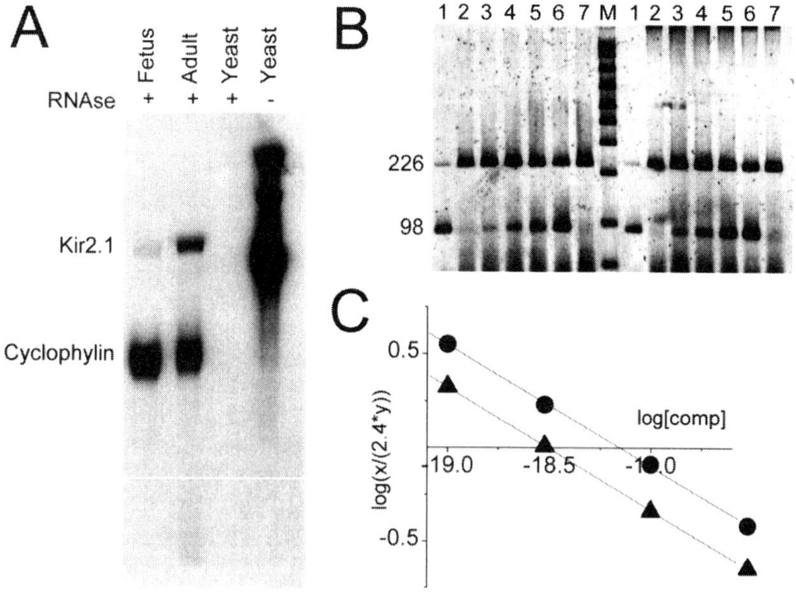

FIGURE 1. Kir2.1 mRNA expression in fetal and adult mouse heart tissue. **A:** RNAse protection assay showing more protected Kir2.1 probe in adult than in total RNA isolated from fetal heart. **B:** Competitive RT-PCR to determine Kir2.1 mRNA expression levels in total RNA isolated from adult (**left**) or fetal (**right**) hearts. The denotation of the lanes is indicated in the text. The 226-bp band represents amplification of Kir2.1 cDNA and the 98-bp band the amplification of the competitor (comp; the Kir2.1 RCP fragment with an internal deletion). The size marker is 50–2000 bp. (Bio-Rad). **C:** The ratio of Kir2.1 and competitor (comp), corrected for size difference, is plotted as a function of competitor concentration. The lines are drawn according to a linear regression fit of the data points. ● = adult heart; ▲ = fetal heart.

observed on a 3% NuSieve agarose gel: the first corresponding to the Kir2.1 target mRNA (226 bp) and the second to the competitor (98 bp). As the concentration of the competitor is increased, the ratio between these two bands is seen clearly to decrease. Correcting for the size difference between these bands, the molar ratio is plotted in panel C as a function of the competitor concentration. Interpolation of these data points (linear regression) yields values of 1.18×10^{-18} and 5.16×10^{-19} mole Kir2.1 RNA per μg of starting total RNA respectively for adult (●) and fetal (▲) heart.

ELECTROPHYSIOLOGY STUDIES

Adult ventricular myocytes and fetal myocytes were placed in short-term culture for 1–6 days or used on the day of isolation. Rod-shaped adult cells, or spontaneously beating spindle-shaped fetal myocytes, were chosen to record whole-cell or single-channel currents. Whole-cell currents, recorded with suction pipettes (1–4 MΩ), indicated a smaller Cs^+-sensitive I_{K1} current density in fetal myocytes compared to I_{K1} recorded from adult myocytes (data not shown). To determine the role of Kir2.1 proteins in I_{K1} during development, we uses AS-oligos for which we have demonstrated their efficacy and efficiency against Kir2.1.[1] For these experiments, we injected *Xenopus* oocytes with poly(A^+) RNA from heart, which led to the expression of Cs^+-sensitive K^+ currents. In different groups of experiments, oocytes were injected with poly(A^+) RNA (0.1–0.3 µg/µl) prepared for fetal (16 days) or adult heart. The oocytes were coinjected with anti-Kir2.1 AS-oligos or with control oligos (50-nl injection volume). Strong inward-rectifying K^+ currents (much larger than intrinsic currents in oocytes) were measured 3 days after injection. FIGURE 2 shows the current-voltage relationship of Ba^{2+}-sensitive currents (means ± SEM) measured from oocytes coinjected with AS-oligos or control oligos. Control oligos, with no sequence similarity with Kir2.1, had no significant effect on poly(A^+)RNA-expressed Ba^{2+}-sensitive currents in any age group. In contrast, coinjection with Kir2.1 AS-oligos led to a significantly smaller expression of Ba^{2+}-sensitive currents in the adult group, suggesting that Kir2.1 channel proteins contribute to at least one of the I_{K1} channel proteins in this age group. No such inhibition was observed in oocytes injected with fetal mouse heart poly(A^+) RNA (FIG. 2), which can be interpreted as indicating that pore-forming subunits coded by genes other than Kir2.1 may account for I_{K1} channel proteins in the fetal developmental stage. To asses this possibility further, we examined single-channel events in cell-attached or inside-out configurations[2] (at room temperature) on cardiac myocytes isolated from fetal mouse hearts (12–17 days gestational age). Recordings were made after 1–4 days of short-term culturing conditions. In the majority of the recordings made from

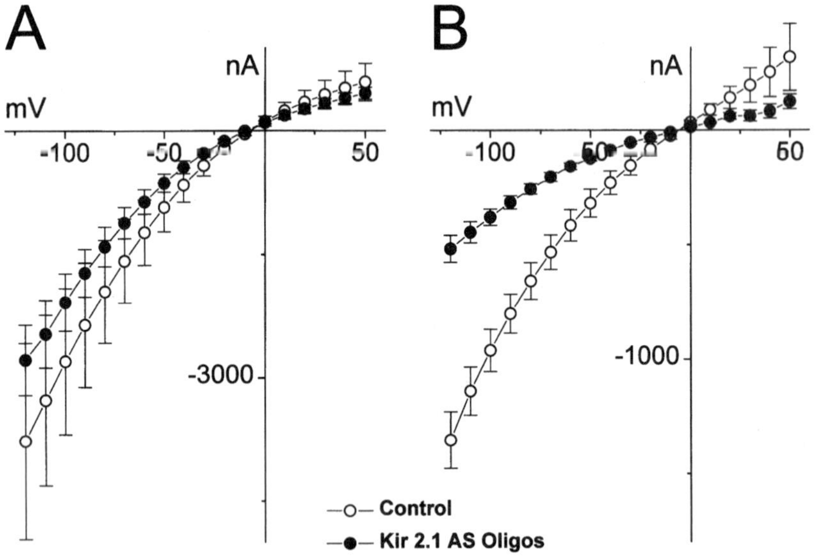

FIGURE 2. Current-voltage relationships of *Xenopus* oocytes that have been injected with poly(A^+) RNA isolated from fetal (**A**) or adult (**B**) hearts. Some oocytes with coinjected with control oligos (*open symbols*) and others with Kir2.1 AS oligos (*filled symbols*). The data points are mean ± SEM of 5–7 experiments.

fetal myocytes, at 50 mV a single open level was observed, with a mean unitary channel amplitude of 0.39 ± 0.001 pA (not shown), implying a channel conductance of ~8 pS assuming a linear current voltage relationship over this voltage range. This channel is clearly different from the 21-pS channel that we have described to be caused by Kir2.1 channel proteins,[1] suggesting that the fetal form of the I_{K1} channel is composed of alternative subunits.

SUMMARY

Our results demonstrate that (a) the Kir2.1 gene encodes a native K^+ channel protein with a 21-pS conductance; (b) this channel has an important role in the genesis of adult ventricular I_{K1}; and (c) the contribution of Kir2.1 channel proteins to I_{K1} changes during development. The lack of contribution of Kir2.1 to fetal I_{K1} channels is interesting from the point of view of possible future generation of knockout mice lacking Kir2.1, since cardiac abnormalities would not be expected to result in fetal lethality. These observations provide further support for a generalized hypothesis that different genes may code for I_{K1} channel proteins at various developmental stages. However, the effects of these AS-oligos must first be examined on *native* I_{K1} channels in cardiac myocytes before definite conclusions can be reached.

REFERENCES

1. Nakamura, T.Y., M. Artman, B. Rudy & W.A. Coetzee. 1998. Inhibition of rat ventricular I_{K1} with antisense oligonucleotides targeted to Kir2.1 mRNA. Am. J. Physiol. **274:** H892–H900.
2. Hamill, O.P., A. Marty, E. Neher, B. Sakmann & F.J. Sigworth. 1981. Improved patch-clamp techniques for high-resolution current recording from cells and cell-free membrane patches. Pflug. Arch. Eur. J. Physiol. **391**: 85–100.

Kv2.1/Kv9.3, an ATP-Dependent Delayed-Rectifier K+ Channel in Pulmonary Artery Myocytes

AMANDA J. PATEL, MICHEL LAZDUNSKI,[a] AND ERIC HONORÉ

Pharmacologie Moléculaire et Cellulaire, CNRS, 660 route des Lucioles, Sophia Antipolis, 06560 Valbonne, France

Ion channels, in particular K^+ channels, play a critical role in adaptive hypoxic mechanisms. One such adaptation is the hypoxia-induced vasoconstriction of resistance pulmonary artery (PA) smooth muscle, which leads to the redistribution of the nonoxygenated blood toward better ventilated regions of the lung[1,2] The goal of this study was to define the molecular structure of oxygen-sensitive delayed-rectifier K^+ channels, which are implicated in this process.

FIGURES 1A and 1B compare the I–V relationships of the currents recorded in freshly dissociated and primary (I) or secondary (II) cultured PA cells.[4] The characteristics of the outward currents observed in freshly dissociated cells were identical to those in primary culture. The activation voltage threshold was –50 mV, a value close to the resting membrane potential (–54 ± 4 mV, $n = 6$) (FIG. 1B). The pharmacological properties of the PA myocytes K^+ channels are illustrated in FIGURE 1C. Twenty nM charybdotoxin (CTX), 200 nM dendrotoxin (DTX), and 300 nM mast cell degranulating peptide (MCD), potent blockers of both Kv1.2 and Kv1.3, inhibited the outward current by about 20%, with the effects of DTX and CTX being nonadditive. In agreement with previous reports,[4–6] we observed that outward K^+ currents in PA smooth muscle cells are relatively resistant to tetraethylammonium (TEA), but sensitive to 4-aminopyridine (4-AP). The appetite suppressant drug dex-fenfluramine (DFF), which has been shown to induce PA vasoconstriction,[7] inhibited the outward K^+ current by 60% at a concentration of 1 mM. The sensitivity of resistance PA myocytes K^+ channels to hypoxia is illustrated in FIGURE 1D. The hypoxic inhibition of K^+ currents in the presence and in the absence of 200 nM DTX were compared (FIG. 1E). Hypoxic inhibitions (25 ± 4% and 17 ± 2% in control and in the presence of DTX, respectively) were not statistically different ($p < 0.05$) between the two experimental conditions.

A degenerate PCR-based strategy identified Kv1.2, Kv1.3, Kv2.1, and a novel K^+ channel, Kv9.3, as potential oxygen-sensitive PA K^+ channels. FIGURE 1F shows the expression of the various channels in lung, brain, conduit, and resistance pulmonary arteries by RT-PCR. Kv9.3 transfected cells did not express any exogenous channel activity. When Kv9.3 was coexpressed with Kv2.1, the activation threshold was shifted toward negative values (–50 mV), and the amplitude of the currents was greatly enhanced (FIG. 2A, 2B). Examination of the single-channel properties of Kv2.1 and Kv2.1/Kv9.3 revealed that Kv9.3 alters the single channel conductance of Kv2.1 (FIG. 2C, 2D). The activity of both Kv2.1 and Kv2.1/Kv9.3 in excised inside-out patches was sensitive to the presence of

[a]Corresponding author. Phone: 33 (0)4 93 95 77 02; fax: 33 (0)4 93 95 77 04; e-mail: patel@ipmc.cnrs.fr

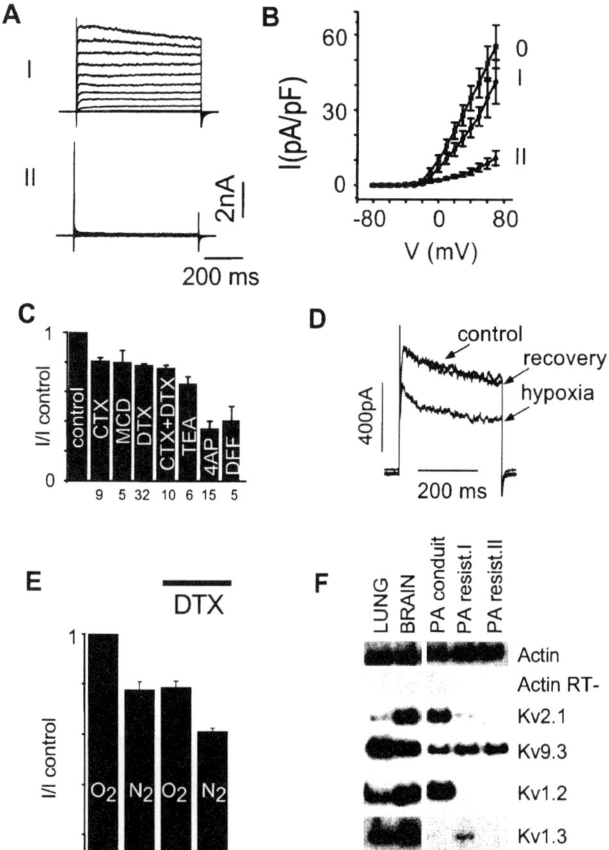

FIGURE 1. Delayed-rectifier K^+ channels in rat resistance pulmonary artery myocytes. **(A)** K^+ channel currents in PA myocytes maintained either in primary (I) or secondary (II) culture (n = 7). The holding potential was −80 mV, and the cells were depolarized from −60 mV to +40 mV by 10-mV increments. **(B)** I–V curves of freshly dissociated, primary, and secondary cultured PA resistance myocytes. **(C)** Pharmacology of K^+ channels in PA myocytes. The effects of 20 nM charybdotoxin (CTX), 300 nM mast cell degranulating factor (MCD), 200 nM dendrotoxin (DTX), 3 mM tetraethylammonium (TEA), 1 mM 4-aminopyridine (4-AP), and 1 mM dex-fenfluramine (DFF) are indicated. Numbers of cells are indicated. **(D)** Reversible hypoxic inhibition of outward K^+ current in PA myocytes. The holding potential was −80 mV and the test pulse +30 mV. **(E)** Hypoxic inhibition of outward K^+ current elicited by a test pulse at +30 mV in myocytes either in the absence (N_2) or the presence (N_2/DTX) of 200 nM DTX. **(F)** Spatial expression of pulmonary artery K^+ channel subunits. RT-PCR was performed on RNA from the tissues indicated (see **top**) with primers against actin or the K^+ channel subunits shown (see **side**).

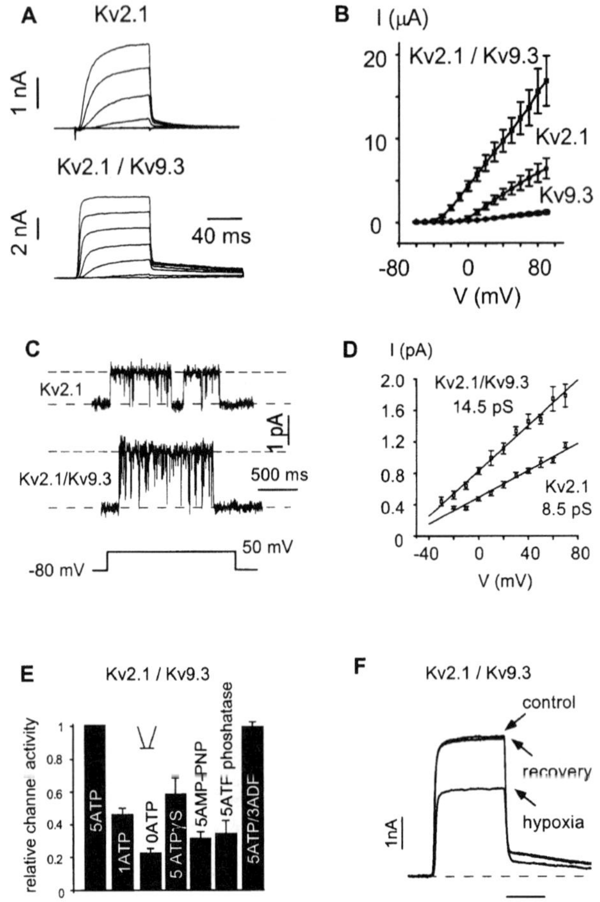

FIGURE 2. Functional expression of Kv9.3 in *Xenopus* oocytes and COS cells. (**A**) Typical recordings of Kv2.1 (**top**) and Kv2.1/Kv9.3 (**bottom**) in transfected COS cells. The holding potential was –60 mV, and cells were depolarized by 10-mV increments to 0 mV in a physiological K^+ gradient. (**B**) I–V curves of K^+ currents recorded in injected oocytes (n = 10). (**C**) Single-channel currents of Kv2.1 and Kv2.1/Kv9.3. Typical current traces of Kv2.1 channel (**top**) and Kv2.1/Kv9.3 channel (**bottom**) recorded at +50 mV in cell-attached patches under a physiological K^+ gradient from transfected COS cells. (**D**) I-V curves of Kv2.1 (n = 8) and Kv2.1/Kv9.3 (n = 8). (**E**) Effects of nucleotides (as indicated) on Kv2.1/Kv9.3 channel activity in independent inside-out patches from transfected COS cells in a physiological K^+ gradient. 100 IU/mL alkaline phosphatase was used in these experiments. The number of cells varies from 5 to 11 between the different experimental conditions. (**F**) Reversible hypoxic inhibition of Kv2.1/Kv9.3. The holding potential was –60 mV, and the test pulse was +30 mV.

internal ATP. When ATP concentration was lowered from 5 mM to 1 mM, channel activity was reduced by 55% (FIG. 2E). In a subset of transfected COS cells, Kv2.1 (21% of the cells, $n = 14$; data not shown) and Kv2.1/Kv9.3 (56% of the cells, $n = 39$) were reversibly inhibited by hypoxia (effect greater than 10% inhibition) (FIG. 2F). The mean hypoxic inhibition of the responsive cells was 28 ± 2.5% for Kv2.1/Kv9.3. Kv1.2, and Kv1.3 did not display any hypoxic sensitivity in COS cells ($n = 11$; data not shown).

Pulmonary hypertension is a disease that ultimately provokes right-heart failure and death within 2 to 5 years. It occurs in people living at high altitude and in patients suffering from chronic obstructive lung diseases such as chronic bronchitis and emphysema.[8] It has been demonstrated that anorexic agents such as aminorex fumarate and DFF have caused an epidemic of pulmonary hypertension in Europe. These compounds have been recently shown to inhibit K^+ currents in rat pulmonary artery myocytes and to cause pulmonary vasoconstriction.[7] In the present report we confirm these data. Furthermore, Kv2.1 and Kv2.1/Kv9.3 are also sensitive to DFF (IC_{50} = 300 mM; data not shown).

In conclusion our results support the idea that in PA myocytes, the Kv2.1 and Kv2.1/Kv9.3 complexes are components of the ATP-sensitive delayed-rectifier K^+ channels. The voltage activation threshold of the Kv2.1/Kv9.3 complex indicates that it may play a role in the establishment of the resting membrane potential of PA myocytes. The understanding of the molecular nature of the K^+ channels in PA myocytes will have major therapeutic significance for pathologies such as PA hypertension.

REFERENCES

1. KOZLOWSKI, R.Z. 1995. Ion channels, oxygen sensation and signal transduction in pulmonary arterial smooth muscle. Cardiovasc. Res. **30:** 318–25.
2. WEIR, E.K. & S.L. ARCHER. 1995. The mechanism of acute hypoxic pulmonary vasoconstriction: The tale of two channels. FASEB J. **9:** 183–189.
3. PATEL, A.J., M. LAZDUNSKI *et al.* 1997. Kv2.1/Kv9.3, a novel ATP-dependent delayed-rectifier K^+ channel in oxygen-sensitive pulmonary artery myocytes. EMBO J. **16:** 6615–6625.
4. ARCHER, S.L. 1996. Diversity of phenotype and function of vascular smooth muscle cells. J. Lab. Clin. Med. **127:** 524–529.
5. POST, J.M., C.H. GELBAND *et al.* 1995. $[Ca^{2+}]$i inhibition of K^+ channels in canine pulmonary artery. Novel mechanism for hypoxia-induced membrane depolarization. Circ. Res. **77:** 131–139.
6. YUAN, X.J., M.L. TOD *et al.* 1995. Hypoxic and metabolic regulation of voltage-gated K^+ channels in rat pulmonary artery smooth muscle cells. Exp. Physiol. **80:** 803–813.
7. WEIR, E.K., H.L. REEVE *et al.* 1996. Anorexic agents aminorex, fenfluramine, and dexfenfluramine inhibit potassium current in rat pulmonary vascular smooth muscle and cause pulmonary vasoconstriction. Circulation **94:** 2216–2220.
8. BARNES, P.J. & S.F. LIU. 1995. Regulation of pulmonary vascular tone. Pharmacol. Rev. **47:** 87–131.

Functional Characterization of a Novel Mutation in KCNA1 in Episodic Ataxia Type 1 Associated with Epilepsy

ALEXANDER SPAUSCHUS,[a] LOUISE EUNSON, MICHAEL G. HANNA, AND DIMITRI M. KULLMANN

University Department of Clinical Neurology, Institute of Neurology/UCL, Queen Square, London WC1N 3BG, United Kingdom

Episodic ataxia type 1 (EA1) is a rare autosomal dominant neurological disorder in which patients develop sudden episodes of ataxia precipitated by physical or emotional stress.[1] These can last seconds to minutes, and between attacks patients often have spontaneous, repetitive muscle activity (myokymia), which is not always clinically apparent. Missense point mutations in KCNA1, the gene encoding the human orthologue of Kv1.1 on chromosome 12p13, have been linked to EA1,[2–5] and the physiological properties of some mutations have been studied.[6–8]

In a large kindred with EA1 where two out of five affected family members have epilepsy we recently detected a previously undescribed point mutation in the second membrane-spanning domain of KCNA1.[9] This mutation results in a change of threonine at amino acid position 226 to arginine (T226R). Further linkage study using mismatch primer PCR revealed that patients were heterozygous for T226R, while none of the unaffected family members nor 100 control individuals carried the mutation. T226 is highly conserved in the Kv1 subfamily through different species, and the replacement by an arginine side chain is a radical exchange. We therefore expressed wild-type and mutant hKv1.1 subunits in *Xenopus laevis* oocytes to investigate the functional implications of T226R.

MATERIALS AND METHODS

Human Kv1.1 (hKv1.1) wild-type and mutant DNAs were amplified using PCR on genomic DNA from a blood sample of one of the affected family members. PCR products were subcloned into the oocyte expression vector pSGEM (courtesy of Dr. M. Hollmann, Göttingen, Germany). Wild-type and mutant clones were completely sequenced on both strands and transcribed *in vitro* using T7-RNA polymerase. *X. laevis* oocytes were defolliculated manually after collagenase treatment and injected with 0.2 to 1.4 ng of mRNA using a Nanoject automatic injector (Drummond). Injected oocytes were kept in Barth's medium at 18 °C for 3 to 7 days prior to recording in a solution containing (in mM): NaCl 115, KCl 2.5, $CaCl_2$ 1.8, HEPES 10, pH = 7.4. Two-electrode voltage-clamp recording was performed at 22 °C using a GeneClamp 500 amplifier, and data were acquired and analyzed using pClamp6 (Axon Instruments). Leak and capacitative currents were subtracted applying a −P/4 protocol.

[a]Corresponding author. Phone: +44-171-837 3611 extension 4181; fax: +44-171-278 5616; e-mail: A.Spauschus@ion.ucl.ac.uk

RESULTS

We compared voltage-activated whole-cell currents in oocytes injected with either mutant or wild-type mRNA. Oocytes injected with mutant mRNA showed currents that were significantly reduced as compared to oocytes injected with wild-type mRNA (FIG. 1A, solid bars): by 97% (3 days, $p < 0.0005$), or by 90% (7 days, $p < 0.0005$). Since

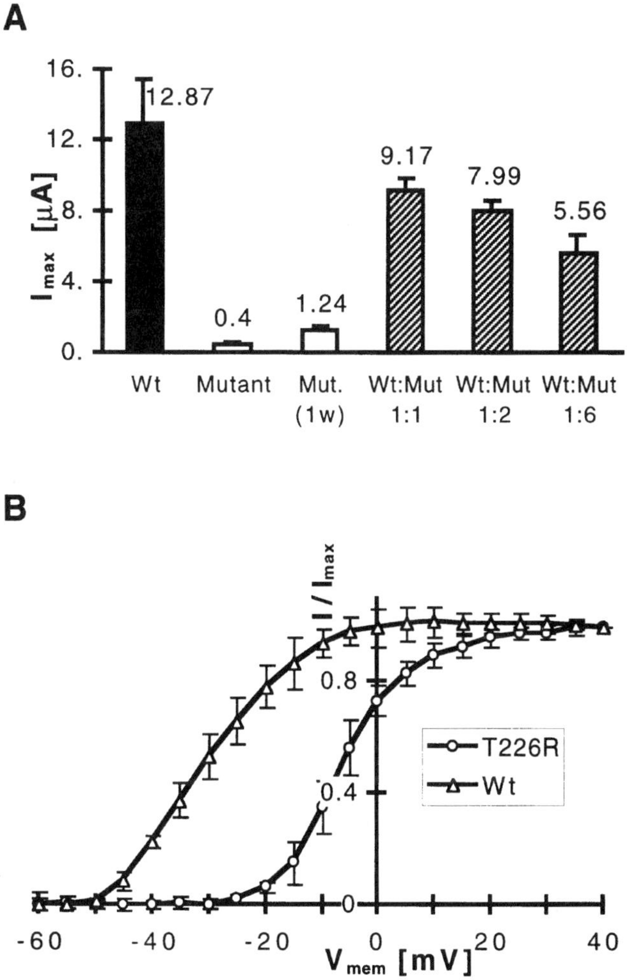

FIGURE 1. (A) Histogram showing current amplitudes recorded from oocytes injected with mRNA as indicated (Wt = wild-type; Mut = mutant; 1w = recorded after one week). Currents were measured at the end of a 350-ms pulse to +40 mV. *Error bars* indicate SD (*n* between 5 and 9). **(B)** Voltage dependence of activation. Tail currents were recorded at −50 mV, normalized, and plotted versus the preceding depolarizing potential. The tail currents were fitted with single exponentials to determine the initial current amplitudes.

heterozygous patients will express both alleles, we mimicked this situation by coinjecting constant amounts of wild-type mRNA and increasing amounts of mutant mRNA to yield ratios as indicated in FIGURE 1A (hatched bars). We found that increasing the ratio of mutant to wild-type mRNA reduced the current amplitude, consistent with a dominant negative effect of the mutant allele. Wild-type mRNA was also injected at higher concen-

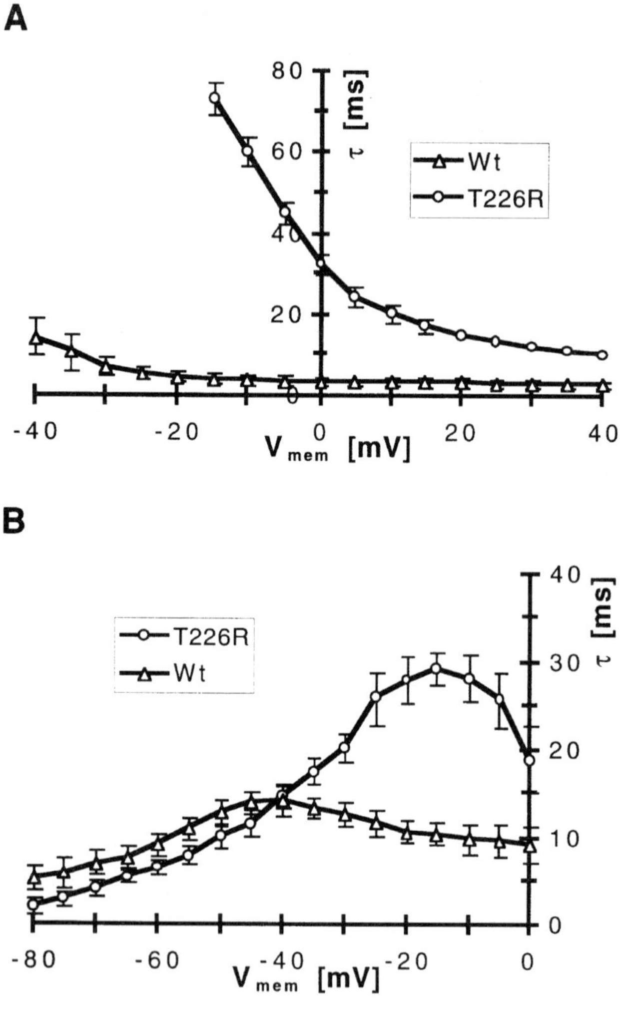

FIGURE 2. Activation **(A)** and deactivation **(B)** time constants of currents in oocytes injected with mutant (T226R) or wild-type (Wt) mRNA. Single exponentials were fitted to the rising phase of current responses A, to depolarizing voltage steps, and B, to tail currents elicited by stepping back to different test potentials after a 100-ms pulse to +40 mV. Time constants were plotted as functions of the depolarizing pulse (A), or of the postpulse tail (B) potentials.

trations on its own to exclude the possibility that reduced whole-cell currents were due to a saturating translation machinery (data not shown).

The voltage dependence of activation was examined by measuring tail currents at −50 mV. Compared with wild-type, the $V_{1/2}$ was shifted ~25 mV toward more positive potentials (Fig. 1B), while the steepness factor was reduced by 32%. Further analyses of current responses to voltage-step and tail-current protocols revealed that T226R slows the activation and deactivation at $V_{1/2}$ by a factor of ~6 and 2, respectively (FIG. 2). The functional properties of currents elicited in coinjected oocytes were between those of wild-type and mutant homomeric channels.

DISCUSSION

This is the first report on functional consequences of a mutation in the putative second transmembrane domain (S2) of KCNA1 in a family with EA1 and partial epilepsy. The physiological properties of potassium channels in EA1 were investigated previously for six different mutations,[6–8] none of which was, however, located in S2. Interestingly, three different mutations of threonine at amino acid position 226 in S2 have recently been linked to EA1.[4,5,9] This suggests an important role for T226 in the pathogenesis of CNS channelopathies.

Our results indicate that the T226R allele is translated and processed to the cell membrane, and that the mutant subunits differ significantly from wild-type subunits with respect to whole-cell current amplitudes, voltage-dependent parameters, and characteristic time constants. Two different mechanisms may underlie the clinical phenotype observed in heterozygous patients: haplotype insufficiency or a dominant negative effect of the mutant allele. Since currents through mutant channel subunits are much smaller than wild-type currents, haplotype insufficiency could underlie EA1 in the present kindred. In addition, mimicking the *in vivo* situation of heterozygous patients in coexpression studies revealed a dominance of the mutant allele over wild-type. Therefore, heterotetrameric channels as may be formed in neurons of patients should have a reduced potassium efflux during action potentials. Even a relatively small reduction in potassium current amplitude could disturb the complex equilibrium between a number of conductances in the membrane of affected neurons and account for a delay in repolarization, thereby facilitating the generation and spread of action potentials. Knockout mice deficient of Kv1.1 were recently bred and showed spontaneous seizures postnatally.[10] The high prevalence of complex partial seizures in individuals with EA1 points out that mutations in hKv1.1 may contribute to the pathogenesis of epilepsy in man.

ACKNOWLEDGMENTS

This work was supported by the Medical Research Council, Wellcome Trust, and Brain Research Trust.

REFERENCES

1. VAN DYKE, D.H. *et al.* 1975. Hereditary myokymia and periodic ataxia. J. Neurol. Sci. **25:** 109–118.
2. BROWNE, D.L. *et al.* 1994. Episodic ataxia/myokymia syndrome is associated with point mutations in the human potassium channel gene KCNA1. Nat. Genet. **8:** 136–140.

3. BROWNE, D.L. *et al.* 1995. Identification of two new KCNA1 mutations in episodic ataxia/myokymia families. Hum. Mol. Genet. **4:** 1671–1672.
4. ÇOMU, S. *et al.* 1996. Episodic ataxia and myokymia syndrome: A new mutation of potassium channel gene Kv1.1. Ann. Neurol. **40:** 684–687.
5. SCHEFFER, H. *et al.* 1998. Three novel KCNA1 mutations in episodic ataxia type 1 families. Hum. Genet. **102:** 464–466.
6. ADELMAN, J.P. *et al.* 1995. Episodic ataxia results from voltage-dependent potassium channels with altered functions. Neuron **15:** 1449–1454.
7. ZERR, P. *et al.* 1998. Episodic ataxia mutations in Kv1.1 alter potassium channel function by dominant negative effects or haploinsufficiency. J. Neurosci. **18:** 2842–2848.
8. D'ADAMO, M.C. *et al.* 1998. Episodic ataxia type-1 mutations in the hKv1.1 cytoplasmic pore region alter the gating properties of the channel. EMBO J. **17:** 1200–1207.
9. ZUBERI, S.M. *et al.* 1997. Episodic ataxia type 1 and epilepsy: Manifestations of a potassium channelopathy in a Scottish family [abstract]. Epilepsia **38** (Suppl. 3): 104.
10. SMART, S.L. *et al.* 1998. Deletion of the Kv1.1 potassium channel causes epilepsy in mice. Neuron **20:** 809–819.

Regulation of a Human Neuronal Voltage-Gated Potassium Channel (hKv1.1) by Protein Tyrosine Phosphorylation and Dephosphorylation

QIANG WANG[a]

Department of CNS Disorders, Wyeth-Ayerst Research, CN 8000, Princeton, New Jersey 08543, USA

Accumulating data has shown that voltage-gated potassium channels (Kv) are among many important targets for modulation by protein phosphorylation and dephosphorylation.[1–5] Although modulation of Kv channels by serine/threonine phosphorylation is well established,[1,3] tyrosine phosphorylation has proved to be another pivotal mechanism.[2,4,5] Increasing evidence has demonstrated that tyrosine phosphorylation and dephosphorylation regulate the functioning of a number of ion channels, including voltage-gated potassium channels. For instance, activation of an insulin receptor tyrosine kinase suppresses Kv1.3 current, a member of the Shaker-related potassium channel family.[5] The tyrosine kinase–dependent pathway is also involved in the suppression of a RAK potassium channel (equivalent to Kv1.2) induced by a G protein–coupled M1 muscarinic acetylcholine receptor.[6] In contrast, the activity of a Ca- and voltage-dependent potassium channel is increased by tyrosine phosphorylation and decreased by tyrosine dephosphorylation.[7]

It is well known that certain types of protein tyrosine kinases are highly expressed in many cell types including the central nervous system (CNS). It is thus conceivable that protein tyrosine phosphorylation may play an important role in regulation of CNS activity. In CHO cells that we have used to express hKv1.1 potassium channels, the basal level of tyrosine kinase and phosphatase activities is also high, and continuous modulation of Kv channel by those enzymes is therefore expected. To address the importance of nonreceptor tyrosine phosphorylation and dephosphorylation of Kv channels, I have tested the effects of inhibitors for both tyrosine kinases and tyrosine phosphatases on the hKv1.1 potassium channels using the standard patch-clamp whole-cell recording method.

Human Kv1.1 cDNA (kind gift of Dr. B. Tempel) was excised from pGexHG2 with BglII/EcoRI and inserted into the BamHI/EcoRI sites of expression vector to yield pWE1/Kv1.1. The construct was characterized by restriction enzyme mapping and DNA sequencing (dideoxynucleotide chain termination method). CHOK1 cells were transfected with linearized pWE1/Kv1.1 by electroporation using a single pulse at 250 V, 1180 μF in a Gibco-BRL Cell-Porator. Stably transfected cells were selected in DMEM supplemented with 10% dialyzed FCS, 1% NEAA, 1% HT, 50 μg/ml mycophenolic acid and 250 μg/ml xanthine. Selected clones were analyzed by RT-PCR for Kv1.1 mRNA expression and by Western blot analysis for protein expression. Clone 2-13 exhibited the highest frequency

[a]Phone: 732-274 4654; fax: 732-274 4020; e-mail: wangq@war.wyeth.com

and amplitude of current and was further subcloned by limiting dilution to ensure stability of line. Subclones 2-13/1-19 were used for further experiments.

Human Kv1.1 (hKv1.1) currents were activated by depolarizing pulses from −60 to +40 mV for 100 ms in 10-mV increments using the standard patch-clamp whole-cell recording method. As shown in FIGURE 1 (traces on left), genistein (30 μM), a tyrosine kinase inhibitor, significantly slowed down the channel activation kinetics from test potentials of 0 to +40 mV. The effect of genistein on channel activation was completely prevented when 1 mM sodium orthovanadate (a tyrosine phosphatase inhibitor) was applied with genistein (right-hand traces of Fig. 1). The channel activation kinetics (τ_a) were analyzed using a single exponential fit. FIGURE 2 illustrates that τ_a is voltage dependent, being reduced with increasing membrane depolarization. When measured at a test potential of 0 mV, τ_a was increased from 4.35 ± 0.55 ms (control) to 10.35 ± 2.19 ms (genistein) ($n = 8$). In addition, genistein decreased the peak current amplitude at pulses from −20 to +20 mV in a voltage-dependent manner ($n = 8$). To test if the effects of genistein on τ_a were specific, we used daidzein (30 μM), an inactive analogue of genistein, in the following experiments. In three cells, daidzein was without any effect on τ_a or the current amplitude. In contrast, a tyrosine phosphatase inhibitor sodium orthovanadate (1 mM) significantly accelerated τ_a and slightly augmented the current amplitude (FIGURE 2, right-hand panel). The effects of genistein on τ_a and current amplitude were almost completely reversed by orthovanadate ($n = 3$). Moreover, a nonspecific PKA and PKC inhibitor H-7

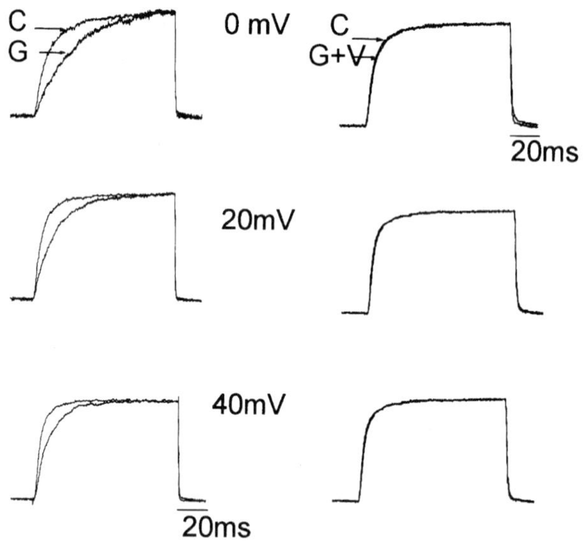

FIGURE 1. Effects of genistein and vanadate on channel activation kinetics. The **left-hand traces** show that genistein (30 μM), a tyrosine kinase inhibitor, reversibly slows down the hKv1.1 activation kinetics at varying step potentials (C stands for control and G for genistein). The effect of genistein on channel activation was prevented by sodium vanadate, an inhibitor of tyrosine phosphatase (**right-hand traces:** C, control; G+V, genistein + vanadate). Traces are scaled to the same size to gain a better comparison of the activation time course. CHO cells were all voltage clamped at −60 mV, and membrane potentials were depolarized to varying values for 100 ms.

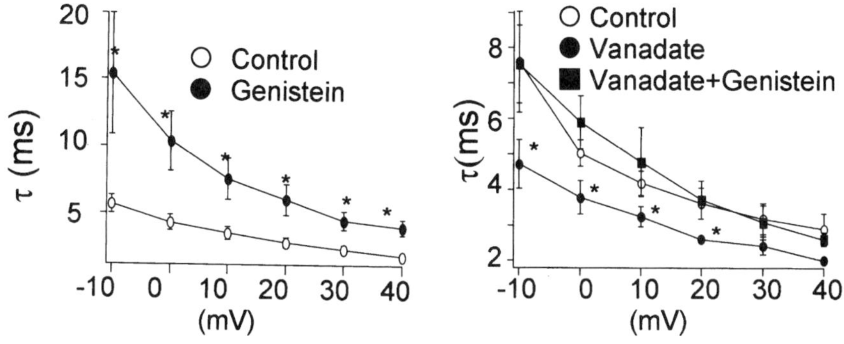

FIGURE 2. Summary of the effects of genistein (30 μM, $n = 8$) on the activation time constants (τ) of hKv1.1 from –10 to +40 mV. The activation time course was analyzed using the Pulsefit software (HEKA), and values were obtained using a monoexponential fit from the onset of the current to a steady state plateau phase. The **right-hand panel** shows that vanadate (1 mM) alone significantly accelerated channel activation. When applied together, genistein and vanadate did not change the channel activation kinetics (right-hand panel, $n = 3$). $*p < 0.05$ from paired t test.

(10 μM) did not show any effect on either τ_a or the current amplitude of hKv1.1 (data not shown). These results suggest that hKv1.1 potassium channels are subject to modulation by tyrosine phosphorylation and dephosphorylation. Tyrosine phosphorylation enhanced the channel activation and slightly increased the current amplitude. In contrast, tyrosine dephosphorylation slows down the channel activation process.

ACKNOWLEDGMENT

The author wishes to thank Drs. James Barrett and Paul McGonigle for their support and encouragement and Ms. Judith Wardwell-Swanson for providing hkv1.1 cell lines.

REFERENCES

1. Bosma, M.M., M.L. Allen, T.M. Martin & B.L. Tempel. 1993. J. Neurosci. **13:** 5242–5250.
2. Lev, S., H. Moreno, R. Martinez, P. Canoll, E. Peles, J.M. Musacchio, G.D. Plowman, B. Rudy & J. Schlessinger. 1995. Nature **376:** 737–745.
3. Murakoshi, H., G. Shi, R.H. Scannevin & J.S. Trimmer. 1997. Mol. Pharmacol. **52:** 821–828.
4. Holmes, T.C., D.A. Fadool, R. Ren & I.B. Levitan. 1996. Science **274:** 2089–2091.
5. Bowlby, M.R., D.A. Fadool, T.C. Holmes & I.B. Levitan. 1997. J. Gen. Physiol. **110:** 601–610.
6. Huang, X.-Y., A.D. Morielli & E.G. Peralta. 1993. Cell **75:** 1145–1156.
7. Prevarskaya, N.B., R.N. Skryma, P. Vacher, N. Daniel, J. Djiane & B. Dufy. 1995. J. Biol. Chem. **270:** 24292–24299.

Regulation of Firing Pattern through Modulation of Non-Sh K$^+$ Currents by Calcium/Calmodulin-Dependent Protein Kinase II in *Drosophila* Embryonic Neurons

WEI-DONG YAO AND CHUN-FANG WU[a]

Department of Biological Sciences, University of Iowa, Iowa City, Iowa 52242, USA

Calcium/calmodulin-dependent protein kinase II (CaMKII) has been implicated in cellular mechanisms underlying learning and memory. CaMKII is necessary for hippocampal long-term potentiation (LTP) and spatial memory[1–5] in vertebrate species. Inhibition of CaMKII in *Drosophila* by genetic transformation alters experience-dependent courtship behavior[6] and synaptic efficacy at neuromuscular junctions.[7] Recent evidence suggests that mutations of CaMKII may also affect nerve membrane excitability.[8,9] However, it remains unclear how different K$^+$ channels in neurons are modulated by CaMKII, which can modify spike coding.

We addressed this issue using the *Drosophila* "giant" neuron culture, a well-characterized preparation accessible for neuronal electrophysiological analyses.[10–12] The voltage-dependent outward K$^+$ currents, recorded under whole-cell voltage clamp, were drastically modified by the CaMKII-specific inhibitor, KN-93 (1 μM, FIG. 1). Within minutes of bath application, this membrane-permeable drug significantly suppressed the amplitude of both peak and steady state currents in wild-type neurons (FIG. 1, WT). Interestingly, the reduction was far more pronounced for steady state currents. Furthermore, the transient component decayed much faster following KN-93 treatments (τ_{fast} = 82 ± 19 ms vs. 42 ± 8 ms, n = 10). Inhibition of CaMKII by another antagonist, KN-62, as well as by an inducible inhibitory peptide of CaMKII in transgenic flies,[6] yielded consistent results (not shown). Similar observations have been made in *Drosophila* photoreceptor cells (A. Peretz, I. Abitbol, A. Sobko, C.-F. Wu, and B. Attali, in preparation).

The distinct effects of KN-93 on the transient and sustained components suggest that CaMKII may differentially modulate different types of K$^+$ channels. Among the genes encoding the identified K$^+$ channel subunits, *Shaker* (*Sh*) has been the best characterized physiologically.[13–15] *Sh* mutations cause ether-induced leg shaking in adult flies and alter the A-type current in muscle[13,14] and in neurons.[16] We used a null mutant to see if Sh subunits are major targets for CaMKII modulation. The mutation Sh^M appears to eliminate all alternatively spliced Sh products.[17] As shown in FIGURE 1, the remaining non-Sh currents in Sh^M neurons were affected by KN-93 in both amplitude and kinetics to a degree similar to those of WT neurons. Thus, the major effects of CaMKII modulation on K$^+$ currents do not appear to be conferred by Sh channels.

The diversity of K$^+$ channels enables neurons to generate a rich variety of firing patterns and action potential shapes.[18] Differential modulation of transient and sustained K$^+$

[a]Corresponding author. Phone: 319-335-1091; fax: 319-335-1103; e-mail: chun-fang-wu@uiowa.edu

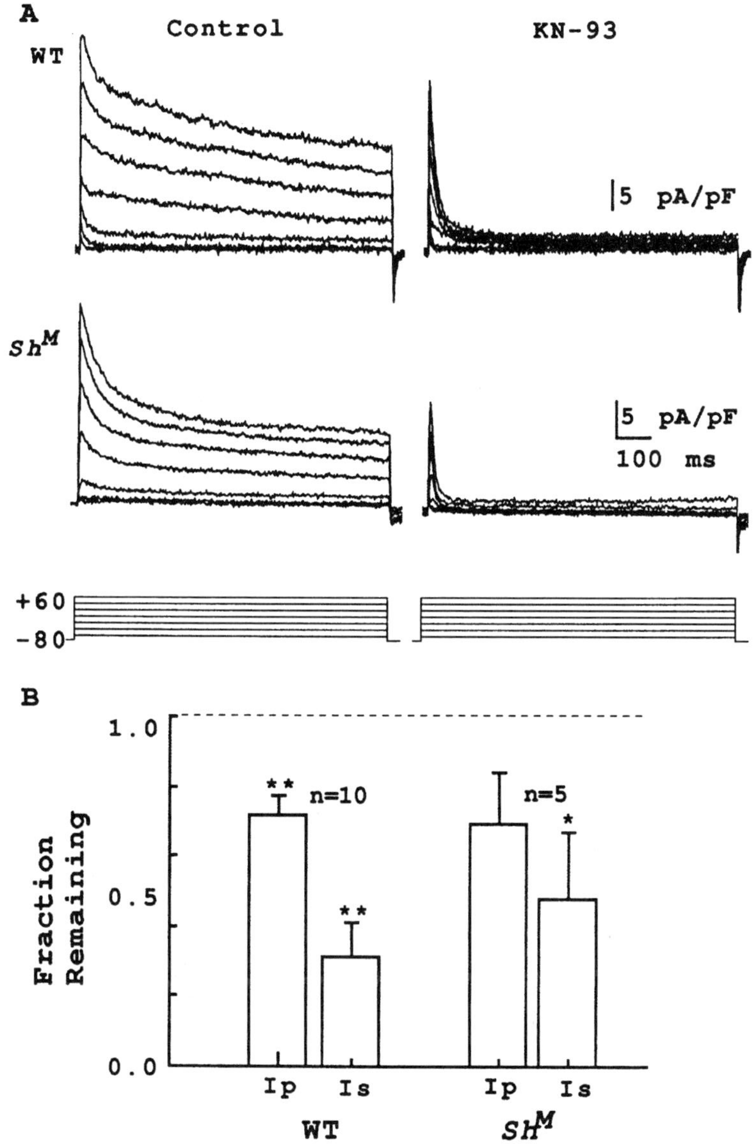

FIGURE 1. Modulation of neuronal K^+ currents by the CaMKII inhibitor KN-93. **(A)** K^+ currents were elicited by depolarization steps (1 s) from a holding potential of −80 mV to between −60 and +60 mV in 20-mV increments. TTX and Cd^{2+} were added to saline to eliminate inward Na^+ and Ca^{2+} currents and outward Ca^{2+}-activated K^+ currents. KN-93 was applied during experiments at a final concentration of 1 μM. **(B)** Following KN-93 treatment, the remaining currents (at +40 mV) were measured at the peak (I_p) and steady state (I_s), normalized to the corresponding values before treatment. *$p < 0.05$, **$p < 0.001$, one-sample two-sided *t*-test.

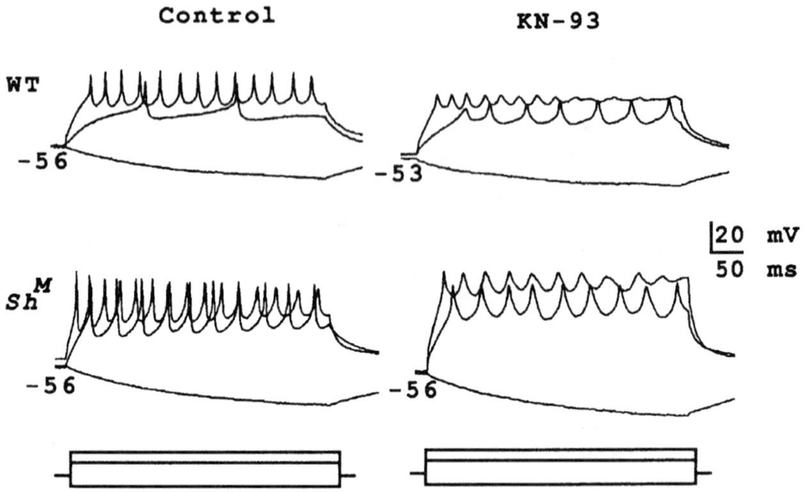

FIGURE 2. Modulation of firing patterns by the CaMKII inhibitor KN-93 in wild-type and *Sh*^{M} neurons. Action potentials were evoked by current injection (−10, +30, and +50 pA). The resting membrane potential is indicated. The final concentration of KN-93 was 1 μM.

currents by CaMKII may have significant consequences on neuronal excitability. By current clamp in neurons that fire all-or-none action potentials,[11,12] we found that KN-93 drastically altered the spike patterns of WT neurons by decreasing the amplitude and lengthening the duration of spikes (FIG. 2). Furthermore, regular spike trains were converted into irregular oscillations, especially at higher levels of current injection. Similar results were also obtained from neurons of CaMKII-inhibited transgenic flies (in preparation). This transition can be explained by weakened repolarization due to reduced amplitude coupled with accelerated inactivation kinetics of non-Sh currents. Consistently, the effects of KN-93 on firing pattern and action potential shape in *Sh*^{M} neurons were indistinguishable from those seen in WT neurons (FIG. 2).

This study supports the important role of CaMKII in the regulation of ion currents, action potential shapes, and firing patterns in neurons. Our results suggest that CaMKII exerts profound effects on certain non-Sh K^+ channel subunits, such as Eag[19,20] (in preparation). Further investigation is under way to reveal how this signaling cascade modulates different channel subunits to regulate neuronal excitability and information coding.

[NOTE ADDED IN PROOF: Similar effects of pharmacological inhibition of CaMKII on *Drosophila* photoreceptor K+ currents have appeared in literature[21] since the submission of this paper.]

REFERENCES

1. MELENKA, R.C. *et al.* 1989. An essential role for postsynaptic calmodulin and protein kinase activity in long-term potentiation. Nature **340:** 554–557.

2. MALINOW, R. *et al.* 1989. Inhibition of postsynaptic PKC or CaMKII blocks induction but not expression of LTP. Science **245:** 862–866.
3. SILVA, A.J. *et al.* 1992. Deficient hippocampal long-term potentiation in α-calcium-calmodulin kinase II mutant mice. Science **257:** 201–206.
4. SILVA, A.J. *et al.* 1992. Impaired spatial learning in α-calcium-calmodulin kinase II mutant mice. Science **257:** 206–211.
5. MAYFORD, M. *et al.* 1996. Control of memory formation through regulated expression of a CaMKII transgene. Science **274:** 1678–1682.
6. GRIFFITH, L.C. *et al.* 1993. Inhibition of calcium/calmodulin-dependent protein kinase in *Drosophila* disrupts behavioral plasticity. Neuron **10:** 501–509.
7. WANG, J. *et al.* 1994. Concomitant alterations of physiological plasticity in *Drosophila* CaM kinase II-inhibited synapses. Neuron **13:** 1373–1384.
8. BUTLER, L.S. *et al.* 1995. Limbic epilepsy in transgenic mice carrying a Ca^{2+}/calmodulin-dependent kinase II alpha-subunit mutation. Proc. Natl. Acad. Sci. USA **92:** 6852–6855.
9. GRIFFITH, L.C. *et al.* 1994. Calcium/calmodulin-dependent protein kinase II and potassium channel subunit eag similarly affect plasticity in *Drosophila.* Proc. Natl. Acad. Sci. USA **91:** 10044–10048.
10. WU, C.-F. *et al.* 1990. Giant *Drosophila* neurons differentiated from cytokinesis-arrested embryonic neuroblasts. J. Neurobiol. **21:** 499–507.
11. SAITO, M. & C.-F. WU. 1991. Expression of ion channels and mutational effects in giant *Drosophila* neurons differentiated from cell division arrested embryonic neuroblasts. J. Neurosci. **11:** 2135–2150.
12. ZHAO, M.-L. & C.-F. WU. 1997. Alterations in frequency coding and activity dependence of excitability in cultured neurons of *Drosophila* memory mutants. J. Neurosci. **17:** 2187–2199.
13. SALKOFF, L. & R. WYMAN. 1981. Genetic modification of potassium channels in *Drosophila Shaker* mutants. Nature **293:** 228–230.
14. WU, C.-F. & F. HAUGLAND. 1985. Voltage clamp analysis of membrane currents in larval muscle fibers of *Drosophila*: Alteration of potassium currents in *Shaker* mutants. J. Neurosci. **5:** 2626–2640.
15. IVERSON, L.E. & B. RUDY. 1990. The role of the divergent amino and carboxyl domains on the inactivation properties of potassium channels derived from the *Shaker* gene of *Drosophila.* J. Neurosci. **10:** 2903–2916.
16. BAKER, K. & L. SALKOFF. 1990. The *Drosophila Shaker* gene codes for a distinctive K^+ current in a subset of neurons. Neuron **2:** 129–140.
17. ZHAO, M.-L. *et al.* 1995. Functional expression of Shaker K^+ channels in cultured *Drosophila* "giant" neurons derived from Sh cDNA transformants: Distinct properties, distribution and turnover. J. Neurosci. **15:** 1406–1418.
18. RUDY, B. 1988. Diversity and ubiquity of K^+ channels. Neurosci. **25:** 729–749.
19. WARMKE, J. *et al.* 1991. A distinct potassium channel polypeptide encoded by the *Drosophila eag* gene. Science **252:** 1560–1502.
20. ZHONG, Y. & C.-F. WU. 1993. Modulation of different K^+ currents in *Drosophila*: A hypothetical role for the Eag subunit in multimeric K^+ channels. J. Neurosci. **13:** 4669–4679.
21. PERETZ, A. *et al.* 1998. A Ca^{2+}/calmodulin-dependent protein kinase modulates *Drosophila* photoreceptor K^+ currents: A role in shaping the photoreceptor potential. J. Neurosci. **18:** 9153–9162.

Expression of Kv1.2 Potassium Channels in Rat Sensory Ganglia

An Immunohistochemical Study

SHIGERU YOKOYAMA,[a] HISASHI TAKEDA, AND HARUHIRO HIGASHIDA

Department of Biophysical Genetics, Kanazawa University Graduate School of Medicine, Kanazawa 920-8640, Japan

Kv1.2, a component of voltage-gated potassium (K^+) channels, is distributed in a variety of tissues including brain, heart, and pancreas.[1] We have previously demonstrated that Kv1.2, expressed in *Xenopus* oocytes and mammalian fibroblasts, generates delayed-rectifier type K^+ currents sensitive to 4-aminopyridine and dendrotoxin.[2–5] However, it remains unclear whether such K^+ currents contribute to biological functions in the nervous system. In the present study, as an initial step to address this question, we have examined whether the Kv1.2 protein is involved in the sensory system by immunohistochemical approach.

For production of antiserum, a synthetic peptide, corresponding to the N-terminus (amino acids 1–15) of the predicted rat Kv1.2 gene product, was conjugated to keyhole limpet hemocyanin and injected into rabbits. Specific antibodies were purified with an affinity column prepared by coupling the synthetic peptide to EAH Sepharose 4B (Pharmacia).

The affinity-purified antibodies recognized ~55 kD protein on immunoblots of membranes from B82 mouse fibroblast cells transformed with rat Kv1.2 cDNA (FIG. 1, lane 1), while no monospecific band was detected in Kv3.1-transformed and parental fibroblast cells (FIG. 1, lanes 2 and 3). In the cerebellum, ~75 kD protein was detected by immunoblot analysis (FIG. 1, lane 4). The difference between ~55kD and ~75 kD proteins may be due to posttranslational modification.

In histochemical analysis, intense Kv1.2 immunoreactivity was observed in the basket cell terminal at the base of Purkinje cells (FIG. 2, A). Immunolabeling was also seen in small cell bodies in the molecular layer and relatively large cell bodies in the granule cell layer (FIG. 2, A), which may correspond to basket cells and Golgi cells, respectively. In contrast, no significant staining was detectable with antibodies preabsorbed with the antigenic peptide (FIG. 2, B). These observations, together with the data obtained by immunoblot analysis, agree well with those previously obtained with antisera against C-terminal regions.[7–9] In addition, the antibodies labeled very few cells in the hippocampal CA3 region, where Kv1.1 is abundantly expressed in this region.[9] Thus it seems likely that the antibodies are largely devoid of cross-reaction to the Kv1.1 protein.

Using this polyclonal antibody, we stained sensory ganglia of adult rats. In dorsal root ganglia, intense Kv1.2 immunoreactivity was detected in a limited number of neurons, with other neurons being weakly stained (FIG. 2, C). This tendency was more prominent in trigeminal ganglia (FIG. 2, E). Preabsorbed antibodies did not detect any significant signal

[a]Corresponding author. Phone/fax: +81-76-234-4236; e-mail: biophys@med.kanazawa-u. ac.jp

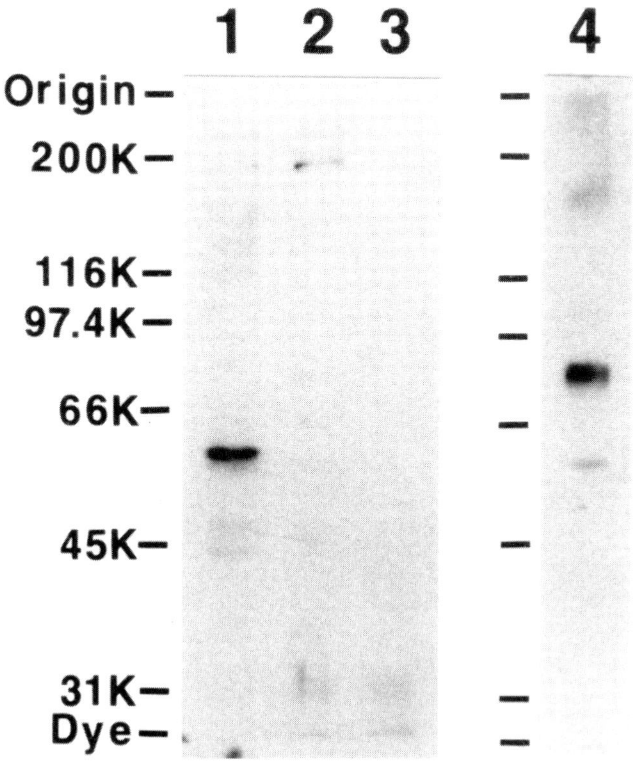

FIGURE 1. Immunoblot analysis with an anti-Kv1.2 polyclonal antibody. Membranes from CL1023, rat Kv1.2-transformed fibroblast cells[4] (**lane 1**, 5 μg protein), CL301, mouse Kv3.1a-transformed fibroblast cells[6] (**lane 2**, 5 μg protein), parental B82 fibroblast cells (**lane 3**, 5 μg protein), and rat cerebellum (**lane 4**, 10 μg protein) were electrophoresed on an SDS 9%-polyacrylamide gel and transferred to Immobilon-P membranes. The blots were probed with a polyclonal antibody raised against the N-terminal region of the Kv1.2 protein. Immunoreactive bands were visualized with enhanced chemiluminescence reagents (ECL, Amersham). Numbers at left denote relative molecular mass of standard proteins.

in both dorsal and trigeminal ganglia (FIG. 2, D, F). We also performed double-labeling experiments using anti-peripherin and RT97 anti-neurofilament monoclonal antibodies. In both dorsal and trigeminal ganglia, the cells strongly positive for Kv1.2 were more frequently observed in RT97-positive large neurons than in peripherin-positive small neurons.

These results show that the expression level of Kv1.2 is differently regulated among primary sensory neurons, and suggest that the Kv1.2 protein plays a crucial role in sensory conduction. Further studies would provide molecular correlates for native currents.

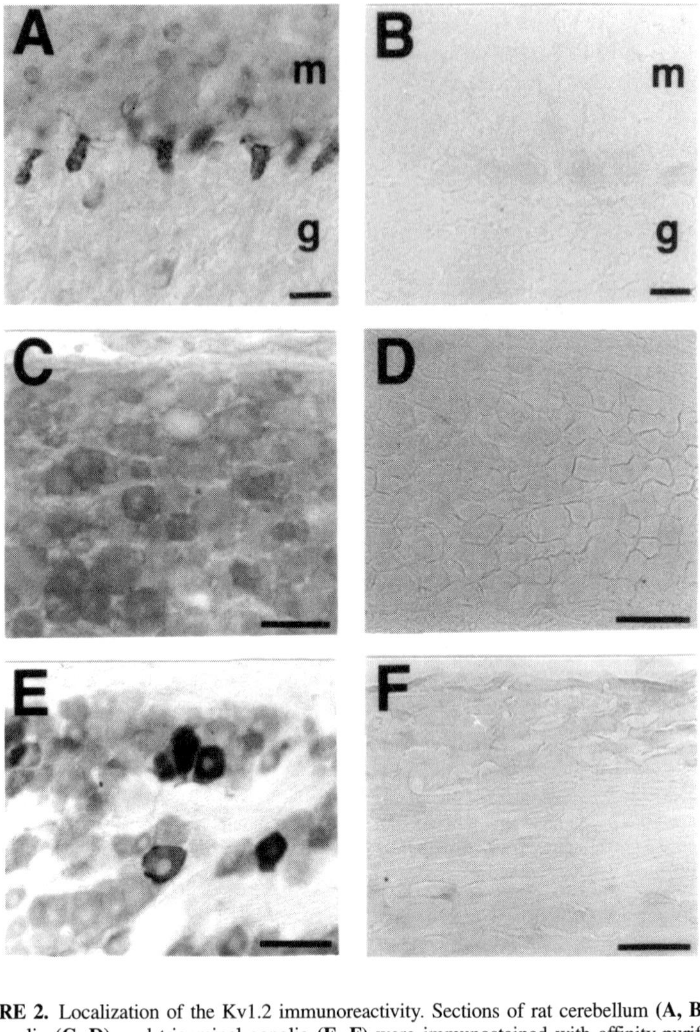

FIGURE 2. Localization of the Kv1.2 immunoreactivity. Sections of rat cerebellum **(A, B)**, dorsal root ganglia **(C, D)**, and trigeminal ganglia **(E, F)** were immunostained with affinity-purified anti-Kv1.2 polyclonal antibodies before (**A**, **C**, and **E**) and after (**B**, **D**, and **F**) preincubation with an excess of the antigenic peptide. The bound antibodies were visualized using an avidin–biotin–peroxidase complex system (Vectastain ABC Elite Kit, Vector), with 3,3′-diaminobenzidine as chromogen. Molecular layer (m) and granule cell layer (g) of cerebellar cortex are indicated. *Scale bar,* 50 μm.

ACKNOWLEDGMENT

This work was supported by a grant from the Ministry of Education, Science, Sports and Culture of Japan.

REFERENCES

1. Chandy, K.G. & G.A. Gutman. 1995. *In* Ligand- and Voltage-gated Ion Channels, R.A. North, Ed. : 1–71. CRC Press, Boca Raton.
2. Yokoyama, S., K. Imoto, T. Kawamura, H. Higashida, N. Iwabe, T. Miyata & S. Numa. 1989. FEBS Lett. **259:** 37–42.
3. Ito, Y., S. Yokoyama & H. Higashida. 1992. Proc. R. Soc. Lond. Ser. B **248:** 95–101.
4. Werkman, T. R., T. Kawamura, S. Yokoyama, H. Higashida & M. A. Rogawski. 1992. Neurosci. **50:** 935–946.
5. Yokoyama, S., T. Kawamura, Y. Ito, N. Hoshi, K.-I. Enomoto & H. Higashida. 1993. Ann. N.Y. Acad. Sci. **707:** 60–73.
6. Kawamura, T., K.-I. Enomoto, S. Yokoyama, N. Hoshi, J. Yamashita & H. Higashida. 1995. Neurosci. Lett. **197:** 164–166.
7. McNamara, N.M.C., Z.M. Muniz, G.P. Wilkin & J.O. Dolly. 1993. Neuroscience **57:** 1039–1045.
8. Sheng, M., M.-L. Tsaur, Y.N. Jan & L.Y. Jan. 1994. J. Neurosci. **14:** 2408–2417.
9. Wang, H., D.D. Kunkel, P.A. Schwartzkroin & B.L. Tempel. 1994. J. Neurosci. **14:** 4588–4599.

Effects on Ion Permeation with Hydrophobic Substitutions at a Residue in Shaker S6 That Interacts with a Signature Sequence Amino Acid

PAUL C. ZEI,[a] EVA M. OGIELSKA,[a,c] TOSHINORI HOSHI,[b] AND RICHARD W. ALDRICH[a]

[a]*Department of Molecular and Cellular Physiology, Howard Hughes Medical Institute, Stanford University, Stanford, California 94305, USA*

[b]*Department of Physiology and Biophysics, University of Iowa, Iowa City, Iowa 52242, USA*

Although potassium channels differ in their gating properties, they all select for K^+ with an extremely high fidelity and without compromising high flux rates. Based on reversal potential measurements, potassium channels display a relative selectivity sequence of $K^+ > Rb^+ > NH4^+ > Cs^+ >> Na^+$.[1–3] Potassium channels are tetramers, and each subunit contributes equally to the formation of the ion-conducting pore, which is formed along the central axis of the protein.[4–7] Voltage-gated K^+ channels have six transmembrane domains (S1–S6) in each subunit, as well as a membrane-spanning region (P-region) between S5 and S6. The P-region possesses a highly conserved sequence (TXTTXGYG) required for K^+ selectivity.[3–8] The prokaryotic K^+ channel, KscA, has only two membrane-spanning domains, but these are homologous to the S5 and S6 regions of the voltage-gated K^+ channels. The crystal structure of KscA has revealed that ion binding sites in the pore are formed by the backbone carbonyls of the signature sequence amino acids, while the S6-like helices cradle the selectivity filter and line the internal vestibule of the channel.[7] In accordance with the idea that S6 helices of voltage-gated K^+ channels also form part of the channel pore are the observations that mutations in this region can alter several pore properties.[9–14] Substitutions at position A463 in the S6 of *Shaker* are known to alter K^+ affinity, the rate of C-type inactivation, internal blocker efficacy, and the interaction of external permeant ions with channel closing.[15–18] In the KscA channel a methionine occupies position 463, and the crystal structure has revealed that this residue inhabits a region of tight protein packing.[7] In this study we examined the effect of hydrophobic substitutions of varying bulkiness at position 463 on the single-channel conductance and the relative K^+ selectivity of the *Shaker* channel.

[c]Corresponding author. Phone: 650-723-7557; fax: 650-725-4463; e-mail: ogielski@cmgm.stanford.edu

METHODS

All mutations were made in the ShakerΔ6-46 background with *N*-type inactivation removed. Mutations at 463 were introduced through site-directed mutagenesis, cRNAs were transcribed from cDNA and injected into *Xenopus* oocytes, and currents were recorded in the outside-out configuration as previously described.[15] For FIGURE 1, the internal solutions contained (in mM): 140 KCl, 2 $MgCl_2$, 11 EGTA, 1 $CaCl_2$, and 10 HEPES, and the external solutions contained (in mM): 140 NaCl, 2 KCl, 6 $MgCl_2$, and 5 HEPES (pH 7.1). For FIGURE 2, the external solutions contained (in mM): 140 XCl, 6 $MgCl_2$, and 5 HEPES (pH 7.1), where X^+ is Na^+, $NH_4{}^+$, Rb^+, or K^+.

RESULTS AND DISCUSSION

Single channel currents elicited by steps to +50 mV for the wild-type channel, (A463), and three other substitutions at this position (*G, V, I*) are shown in FIGURE 1A. The resultant amplitude histograms are plotted in FIGURE 1B. The current amplitude and hence the single-channel conductance increases with increasing size of the amino acid side chain ($G < A < V <$ I). In support of the idea that position 463 resides in a region of tight protein packing is our observation that, while the A463I mutation leads to functional channels, the A463L mutant, differing only in side-chain geometry, fails to express. In order to determine whether these substitutions also affected the selectivity of the channel, we measured the reversal potentials in biionic conditions with either K^+, $NH_4{}^+$, Rb^+, or Na^+ present in the external solution. The single-channel *i*(*V*)'s shown in FIGURE 2 indicate that although some variability is observed, the rank order is unaltered by the mutations ($K^+ \geq Rb^+ > NH_4$ $>> Na^+$). Although each of the three substitutions also affected the C-type inactivation rate of the channel (data not shown), the changes were not correlated with the bulkiness of the amino acid side chain.

The equivalent residue in the KscA channel is in direct contact with the side chain of a conserved valine in the signature sequence.[7,19] This valine (V443 in Shaker) contributes its backbone carbonyl to the formation of an ion binding site in the narrow region of the pore.[7] Substitutions at position V443 in the *Shaker* channel, can lead to the loss of K^+ selectivity, suggesting that V443 is an important component of the selectivity filter.[3] Substitutions at position A463 affected the single-channel conductance in a graded fashion by presumably altering the interactions between the two amino acids (463 and 443). In accordance with the hypothesis that a disruption of this interaction affects the integrity of an ion binding site in the pore is the observation that the A463C substitution decreases the K^+ affinity of the *Shaker* channel without affecting K^+ selectivity in physiological solutions.[17] The decrease in K^+ affinity leads to decreased repulsive interactions among ions in the pore,[20] leading to the speculation that perhaps the A463V,I substitutions increase the K^+ affinity of that site and concomitantly increase the electrostatic repulsion among ions, and thus increase the rate of ion flux through the channel. Although channel function appears to be very sensitive to structural alterations at position A463, the alanine is not well conserved in other potassium channels. Perhaps the residue at position 463 plays an important role in determining the observed variability among the pore properties of different K^+ channel subtypes.

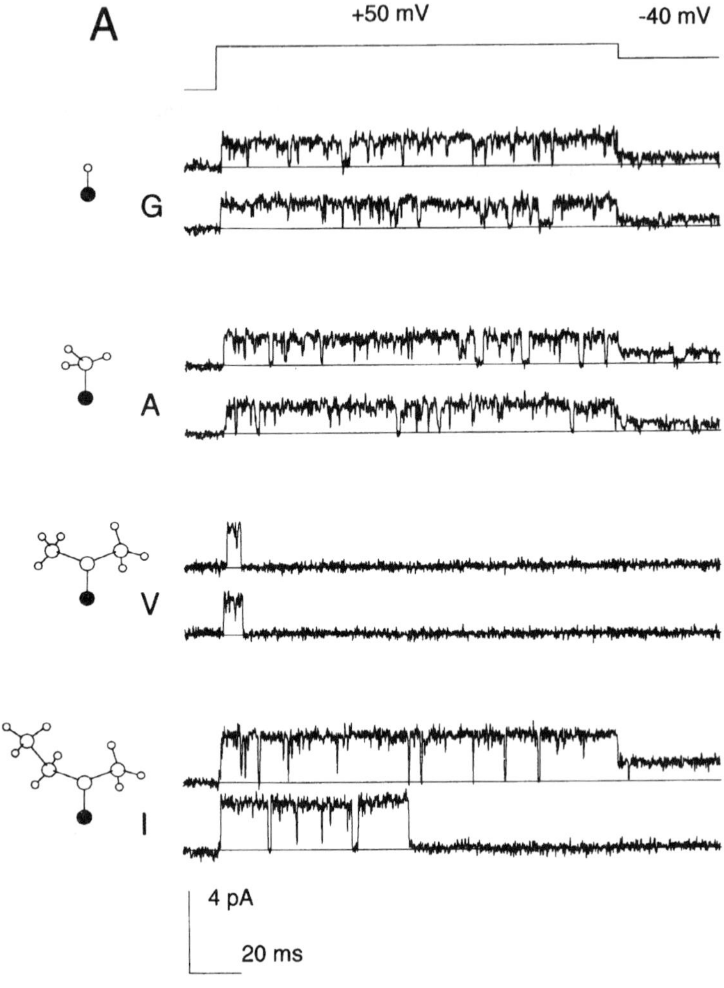

FIGURE 1 *(above and facing page).* Effect of hydrophobic substitutions at position 463 on single-channel conductance at +50 mV. **(A)** Single-channel currents elicited from *Shaker* channels containing amino acid substitutions at position 463 (from top to bottom: A463G, wild-type, A463V, and A463I, with ball-and-stick representations of the respective amino acid side chains shown).The mutations are arranged in increasing bulk size of the side chain from top to bottom.Currents were elicited with pulses to +50 mV, with a prepulse and holding potential of −90 mV, and a tail potential of −40 mV. The currents were filtered at 1.5 kHz and sampled at 20 kHz. Pulses were delivered every 4–6 s.

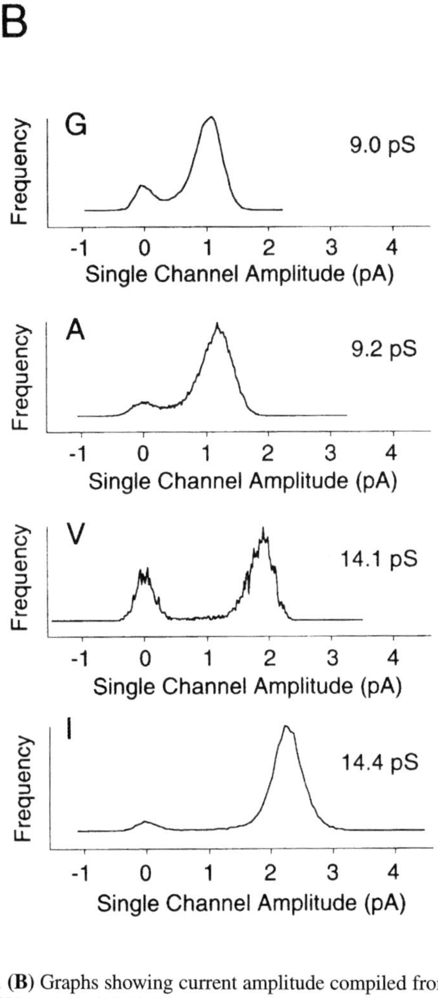

FIGURE 1 (*continued*). **(B)** Graphs showing current amplitude compiled from individual sweeps for A463G, wild-type, A463V, and A463I *Shaker* channels. Current amplitudes of every data point in a given sweep were compiled and plotted as frequency graphs. Conductances shown next to each graph were calculated by taking the peak current value at 0 mV (not shown) and +50 mV and applying the following formula: $g = (i_{50} - i_0)/50$ mV, where g is conductance, and i_{50} and i_0 are the peak-currents at 50 and 0 mV, respectively.

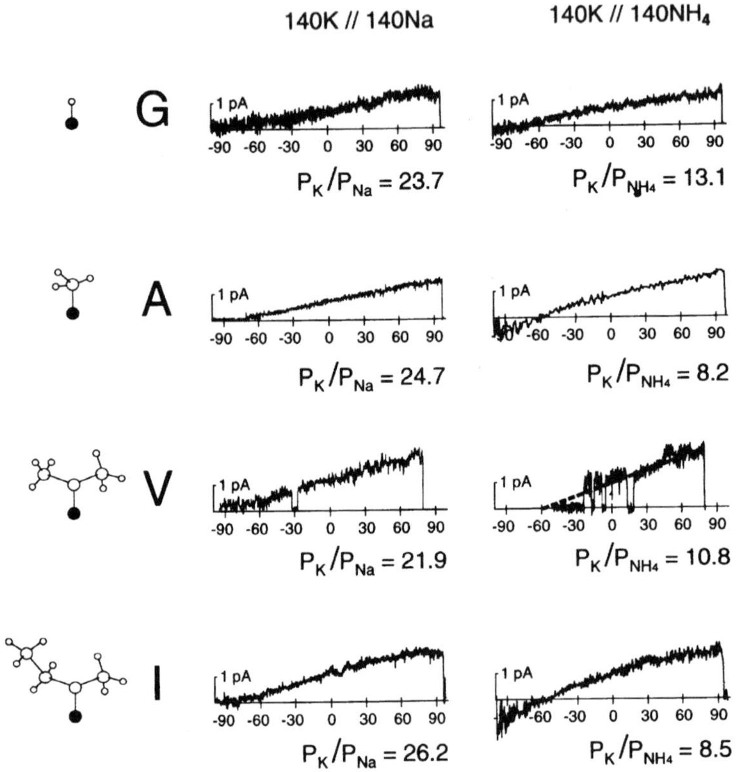

FIGURE 2 *(above and facing page).* Single-channel current–voltage [$i(V)$] relationships forA463G, wild-type, A463V, and A463I *Shaker* channels in various bi-ionic conditions, with ball-and-stick representations of the respective amino acid side chains shown. Single-channel $i(V)$'s were obtained by applying voltage ramps from 100 to −100 mV to single-channel patches. Ramp durations were generally between 10 and 50 ms. Several current traces without any openings were averaged to produce a leak template that was subtracted from each sweep.

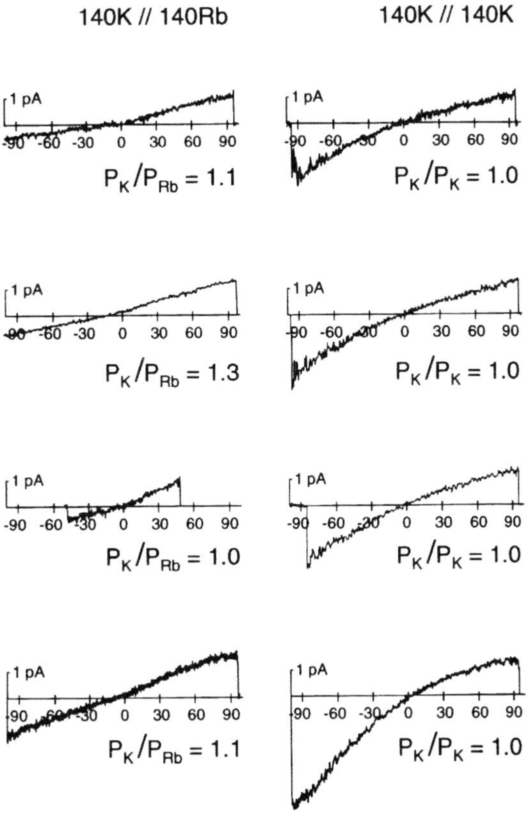

FIGURE 2 (*continued*). Open intervals were selected from sweeps that contained openings and averaged to produce a composite *i*(*V*).The selectivity ratio P_K/P_X, where X is the external cation species, is shown below each *i*(*V*).Selectivity ratios were calculated using the formula: $(P_K/P_X) = e^{-zFE_r/RT}$, where P_K and P_X are the permeabilities of K^+ and X, where X is Na^+, $NH4^+$, Rb^+, or K^+, z is the charge of the cation species ($z = 1$); E_r is the measured reversal potential, and R, T, and F have their usual meanings.

REFERENCES

1. Hille, B. 1973. J. Gen. Physiol. **61:** 669–686.
2. Blatz, A. L. & K. Magleby. 1984. J Gen. Physiol. **84:** 1–23.
3. Heginbotham, L. & R. MacKinnon. 1993. Biophys. J. **65:** 2089–2096.
4. MacKinnon, R. 1991. Nature **350:** 232–235.
5. Liman, E.R. *et al.* 1992. Neuron **9:** 861–871.
6. Kavanaugh, M.P. *et al.* 1992. Neuron **8:** 493–497.
7. Doyle, D.A. *et al.* 1998. Science **280:** 69–77.
8. Heginbotham, L. & R. MacKinnon. 1994. Science **258:** 1152–1155.
9. Choi, K.L. *et al.* 1993. Neuron **10:** 533–541.

10. LOPEZ, G. *et al.* 1994. Nature **367:** 179–182.
11. TAGLIATELA, M. *et al.* 1994. J. Biol. Chem. **269:** 13867–13873.
12. AIYAR, J. *et al.* 1994. Biophys. J. **67:** 2261–2264.
13. LIU, Y. *et al.* 1997. Neuron **19:** 175–184.
14. LIU, Y. & R.H. JOHO. 1998. Pflügers Arch. **435:** 654–661.
15. HOSHI, T. *et al.* 1991. Neuron **7:** 547–556.
16. AVDONIN, V. *et al.* 1997. J. Gen. Physiol. **109:** 169–180.
17. OGIELSKA, E.M. & R.W. ALDRICH. 1998. J. Gen. Physiol. **112:** 243–257.
18. OGIELSKA, E.M. & R.W. ALDRICH. Unpublished observations.
19. MACKINNON, R. 1998. Personal communication.
20. OGIELSKA, E.M. & R.W. ALDRICH. 1998. J. Gen. Physiol. In press.

Recent Excitement in the Ionotropic Glutamate Receptor Field

EDWARD B. ZIFF[a]

Howard Hughes Medical Institute, Department of Biochemistry, New York University Medical Center, 550 First Avenue, New York, New York 10016, USA

ABSTRACT: The synapse is a specialized cellular junction with an elaborate and highly evolved capacity for signal transduction. At excitatory synapses, the neurotransmitter glutamate is released from the presynaptic nerve terminal and stimulates several types of glutamate receptors in the postsynaptic membrane. These include the ionotropic receptors, which are glutamate-gated cation channels, and the metabotropic receptors, which are G protein–coupled seven-transmembrane receptors. The ionotropic glutamate receptors have received special attention because of growing evidence that changes in their synaptic abundance, posttranslational modification, or molecular interactions can provide long-term changes in synaptic strength. This review summarizes new information about the ionotropic glutamate receptors and relates receptor function to the organization of the postsynaptic membrane and the regulation of electrophysiologic and biochemical signaling at the synapse.

Glutamate is the major excitatory neurotransmitter of the central nervous system. Glutamate stimulates both metabotropic glutamate receptors, which are G protein-coupled receptors, and ionotropic receptors, glutamate-gated cation channels. The latter are the subject of this review. Pharmacological analysis of electrophysiologic responses to synthetic agonists has disclosed different ionotropic glutamate receptor classes. Synthetic agonists have been identified that selectively induce individual receptor types. Individual receptor types are named by their selective agonists and include the NMDA, AMPA, and kainate receptors (reviewed in Refs. 1–3).

cDNA cloning confirmed that the pharmacologically defined receptor classes correspond to different molecular receptor entities. The amino acid sequences of the subunits of the different ionotropic receptors are substantially homologous with one another, and a common topology of membrane-spanning regions, shown in FIGURE 1, applies to all of these subunits. The number of subunits that make up a functional receptor is not firmly established, but recent studies indicate that the NMDA and AMPA receptors are tetrameric.[4–6] By implication, kainate receptors are also tetrameric.

NMDA RECEPTORS

NMDA receptors are composed of two subunit types.[1] One of these, the NR1 subunit, is a ubiquitous and necessary component of functional NMDA receptor channels. There are four species of the second subunit type, NR2A-D, each encoded by a different gene.[1] The different NR2s are differentially expressed during different stages of development and in different tissues.[7,8] Recently, a NR3A subunit has been reported.[9]

Each subunit has a large extracellular N-terminal domain and four membrane (M) regions (FIG. 1). Regions M-1, -3, and -4 are conventional membrane-spanning domains.

[a]Phone: 212-263-5774; fax: 212-683-8453; e-mail: ziffe01@med.nyu.edu

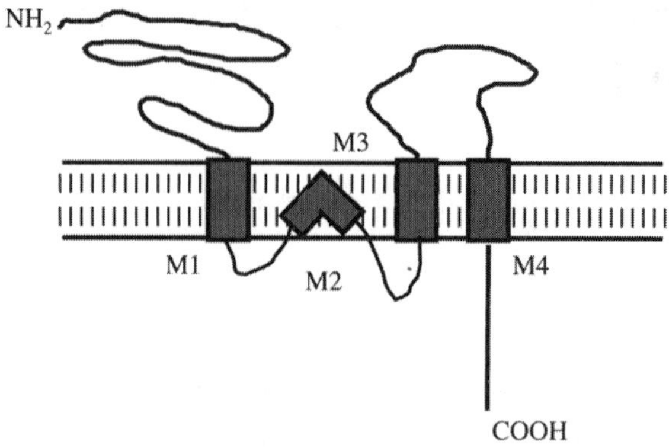

FIGURE 1. Receptor subunit topology. Subunits of ionotropic glutamate receptors have a common topology, with an extracellular N terminus, an intracellular C terminus, and four membrane regions, M1–4. M2 is thought to have a re-entrant, hairpin structure and to contribute to the lining of the channel pore. See text.

M-2 is novel and is predicted to have a membrane-reentrant hairpin structure that contributes to the receptor channel pore. The N-terminal domain, which is large, glycosylated, and extracellular, has a subdomain that is homologous to bacterial amino acid binding proteins and contributes to the agonist binding site. NMDA receptors bind two agonist ligands: glutamate and the coagonist, glycine. The NR1 subunit contains the binding site for glycine[10] and the NR2 subunit, the binding site for glutamate.[11] Modeling based on crystal structures of bacterial amino acid binding proteins indicates that each ligand binding domain is composed of residues from both the N-terminal extracellular domain and the extracellular loop that joins M3 to M4.

NMDA receptors respond to agonist more slowly than AMPA receptors, and require greater than 2 ms to open. However, they have a higher affinity for glutamate, and their currents persist longer than AMPA receptor currents.[1] The NMDA receptor admits both Na^+ and Ca^{2+} ions. At the resting potential of the cell, a Mg^{2+} ion blocks the NMDA receptor pore, but the Mg^{2+} ion is released from the pore upon cell depolarization. Therefore, opening of the channel requires binding of ligand and simultaneous depolarization of the cell.[1] The channel thus operates as a coincidence detector that admits current only when agonist binding and cell depolarization take place simultaneously.

NMDA receptors display activity-dependent current decreases of several types,[12] two of which are ligand dependent. One of these, glycine-dependent desensitization, occurs following receptor stimulation by glutamate when glycine concentrations are subsaturating, in the nanomolar range. The second, glycine-independent desensitization takes place in the presence of saturating glycine, at concentrations in the micromolar range. When glutamate binds to the receptor in the presence of low concentrations of glycine, the receptor rapidly enters a desensitized (low-conductance) state. This desensitizing transition is blocked by glycine. NMDA receptor desensitization may limit receptor currents during persistent stimulation by glutamate. Domains of NR2 that influence receptor desensitization characteristics have been defined.[13] A different activity-dependent NMDA receptor current decrease is Ca^{2+}-dependent inactivation.[14–17] Ca^{2+}-dependent inactivation may be induced by increases in intracellular [Ca^{2+}] that follow activity-dependent fluxes of Ca^{2+} through the receptor. The increase in intracellular [Ca^{2+}] triggers biochemical modifications of the receptor that decrease receptor mean opening time. It has recently been shown that Ca^{2+}-dependent inactivation results from binding of Ca^{2+}-calmodulin to the membrane proximal region of the C-terminal domain of the NR1 subunit.[18]

AMPA RECEPTORS

AMPA receptors are complexes of four subunit types, GluR1–4, which may be homomeric or heteromeric.[5] AMPA receptors account for the great majority of fast excitatory CNS synaptic transmission. AMPA receptors have a lower affinity for glutamate than NMDA receptors, and their currents are typically rapid, rising within less than 1 ms.[1] AMPA receptor channels that contain GluR1, GluR3, or GluR4 subunits, or subunit GluR2 that is encoded by an unmodified GluR2 mRNA, can admit both Ca^{2+} ions and Na^{+} ions. RNA editing of GluR2 mRNA changes the structure of the GluR2 subunit by replacing glutamine with arginine at the "Q/R site" in the pore filter region, at the apex of the M2 hairpin. AMPA receptors containing GluR2 subunits encoded by edited mRNA are impermeable to Ca^{2+} ions.[1,19,20] The effect of an edited subunit is dominant, such that inclusion of a single edited GluR2 subunit in an AMPA channel prevents Ca^{2+} entry through a receptor otherwise composed of GluR1, -3, -4, or unedited GluR2 subunits.

KAINATE RECEPTORS

Kainate receptors are composed of GluR5, -6, and -7 and KA-1 and KA-2 subunits.[1] Kainate receptors have received less attention than either AMPA or NMDA receptors. However, knockout mice lacking the GluR6 kainate receptor subunit have recently been produced and exhibit the phenotype of loss of kainate receptor currents.[21] Although normal in a number of respects, these mice show reduced locomotor activity and decreased susceptibility to kainate-induced seizures.

INTERACTION OF THE POSTSYNAPTIC DENSITY WITH NMDA RECEPTORS

Glutamate receptors at excitatory synapses are associated with a complex assemblage of structural and regulatory proteins called the postsynaptic density (PSD) (reviewed in Ref. 22). The PSD stains in electron micrographs as a dense structure that lies just below the postsynaptic membrane. The PSD is a specialization of the synaptic junction that is analagous to junctional complexes found at other types of cellular junctions (e.g., tight junctions, adherens junctions, and focal adhesions). One face of the PSD makes direct contact with the cytoplasmic domains of ion channels, while the other face contacts cytoskeletal components, such as actin filaments. Individual protein components of the PSD have been purified. These include receptor-binding proteins, cytoskeletal proteins, kineses, phosphatases, proteases, and regulatory factors.[22]

The binding of receptors to components of the PSD tethers receptors at synapses and clusters them. Receptor clustering is best understood for the NMDA receptor. C-terminal domains of NR2 and certain NR1 isoforms bind to a major PSD component, the PSD-95 protein (also called SAP90)[23,24] (FIG. 2). PSD-95 is a member of a family of structurally related proteins that includes chapsyn 110/PSD-93,[25,26] SAP102,[27] and SAP97.[28,29] The binding site within PSD-95 for the receptor C-terminus is a globular, 80–90–amino acid–long structure called a PDZ domain.[30,31] Three PDZ domains are grouped within the PSD-95 N-terminal region, and PDZ-1 and -2 are the principal attachment sites for NMDA receptor subunits. Within the PSD-95 structure, the three PDZ domains are followed by an SH3 domain and a C-terminal guanylate kinase (GK)–related domain. The GK domain binds a third factor, the GKAP protein,[32–34] which colocalizes with PSD-95. Each GKAP molecule contains multiple, tandem sites capable of GK domain association. PSD-95 also homodimerizes through the PSD-95 N-terminal leader peptide, which precedes PDZ-1. This leader contains a pair of adjacent cysteines that are thought to form disulfide bridges that link two PSD-95s.[35] The leader cysteines may also be palmitoylated.[36] Multimerization of PSD-95 is proposed to generate a subsynaptic matrix that clusters NMDA receptors. This matrix has received great attention because of its ability to cluster NMDA receptors (reviewed in Ref. 22). Thus far, clustering has been demonstrated only in heterologous cells, but is highly likely to take place also in neurons by PSD-95–dependent mechanisms.

While PDZ-1 and -2 of PSD-95 bind to NMDA receptor subunits, PDZ-3 has a different target, which may align the receptor with presynaptic structures. PSD-95 binds to the protein neuroligan via PDZ-3.[37] Neuroligan is a transmembrane protein whose N-terminal

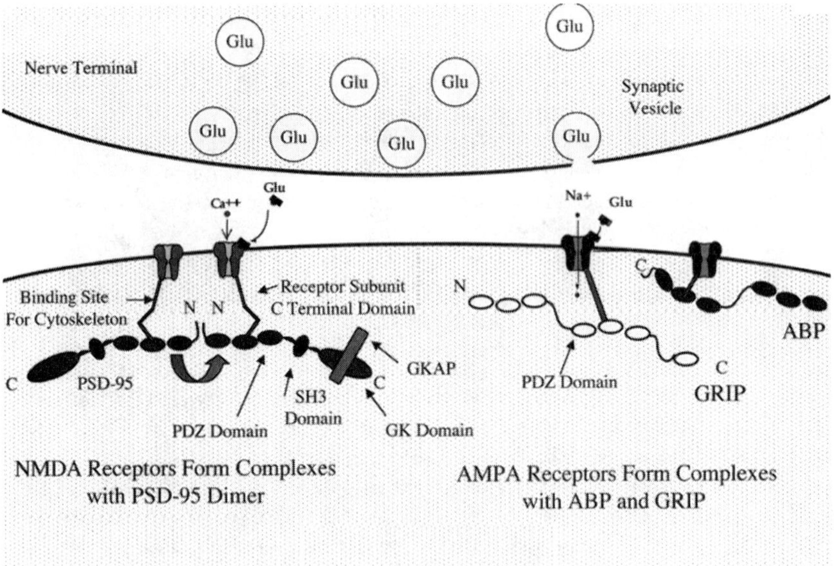

FIGURE 2. NMDA and AMPA receptors and their interactions with cytoplasmic proteins at synapses. NMDA and AMPA receptors are composed of integral membrane protein subunits whose C-terminal domains make contact with subsynaptic, cytoplasmic proteins. NMDA receptors bind to members of the PSD-95 protein family, and AMPA receptors bind to members of the ABP/GRIP family. PSD-95 binds in turn to GKAP via its C-terminal guanylate kinase related domain. PSD-95 form dimers and higher multimers. The NMDA receptor can also make contact with the cytoskeleton. Synaptic vesicles that contain glutamate are shown within the nerve terminal. See text.

extracellular domain interacts across the synaptic cleft with the presynaptic neurexin proteins. This interaction is thought to align the NMDA receptor with pre- and postsynaptic structures. PDZ-3 of PSD-95 also binds to CRIPT, an 11-kDa, cysteine-rich protein, which may act as a bridge between PSD-95 and microtubules.[38] The NMDA receptor NR1 subunit also binds directly to α-actinin-2, an actin-binding protein.[39] Binding to α-actinin-2 may enable the NMDA receptor to interact with the actin cytoskeleton.[39,40] The binding site for α-actinin-2 lies within the NR1 C-terminus, close to a binding site for Ca^{2+}-calmodulin. Calmodulin binds Ca^{2+} following activity-dependent influxes of Ca^{2+} through the receptor, and Ca^{2+}-calmodulin may displace alpha-actinin-2 from NR1, thus releasing the receptor from the cytoskeleton. Binding of calmodulin to NR 1 also decreases the mean opening time of the receptor.[40] The modulation by Ca^{2+} of NMDA receptor interactions with actin may be related to the phenomenon of rundown, a decrease in NMDA receptor currents associated with actin depolymerization.[41,42]

AMPA RECEPTOR BINDING PROTEINS

AMPA receptors also cluster at excitatory synapses. However, the mechanisms of clustering are not well established. Like NMDA receptors AMPA receptors bind to PDZ domain containing proteins. Two such proteins have been described, and both have multiple PDZ domains. These are GRIP (*g*lutamate *r*eceptor *i*nteracting *p*rotein)[43] and ABP (*A*MPA receptor *b*inding *p*rotein).[44] Both GRIP and ABP interact via PDZ domains with the C-terminal sequences of the GluR2 and GluR3 subunits. ABP has six PDZ domains, while GRIP has seven. The two proteins are highly homologous to one another, although they are encoded by different genes. The motif of the GluR2 subunit that is recognized by ABP and GRIP is the C-terminal sequence, S-V-K-I-COOH.[44] This motif differs from the S/T-X-V-COOH motif found at the C termini of NR2 subunits.[45] ABP and GRIP form homo- and heteromultimers.[44] The selective interaction of the NMDA receptors with PSD-95 and the interaction of AMPA-type glutamate receptors with GRIP/ABP depend upon the specificity of PDZ-C terminus contacts. The interaction of AMPA and NMDA receptor subunits with distinctive binding proteins may dictate receptor-specific features of receptor processing, localization, or control.

ROLE OF Tyr KINASE PATHWAYS

The NMDA receptor interacts both physically and biochemically with phosphotyrosine-mediated pathways. The NMDA receptor NR2B subunit is the major synaptic phosphotyrosine peptide of the excitatory synapse.[46–48] The receptor associates via PSD-95 with Syn-GAP, a PSD protein with a GTPase-activating domain that induces hydrolysis of GTP in complexes with the G protein Ras.[49,50] Ras, in its active, GTP-bound state transduces signals from a large number of tyrosine kinases. One of these, Src, is a nonreceptor tyrosine kinase that is abundant in the brain. A related kinase, Fyn, is also found in the PSD.[48] Syn-GAP may attentuate Src signaling through its ability to induce hydrolysis of Ras-GTP to Ras-GDP. Syn-GAP is itself regulated by Cam Kinase II, which is the most abundant serine-threonine kinase of the PSD. Phosphorylation of Syn-GAP by Cam Kinase II decreases the GTPase-stimulating activity of Syn-GAP. Because Cam Kinase II is activated by fluxes of Ca^{2+} through the NMDA receptor, NMDA receptor activity may potentiate tyrosine kinase signal transduction by inhibiting the negative effect of Syn-GAP. Src and Fyn acting through Ras may activate the MAP kinase cascade and regulate cell transcription and proliferation. By modulating Syn-GAP, the NMDA receptor may derepress MAP kinase–dependent effects. The precise role of phosphotyrosine itself in NR2B at the synapse is not yet known.

RECEPTOR DISTRIBUTION AT SYNAPSES AND SYNAPTIC PLASTICITY

The AMPA and NMDA receptors display different topologic distributions in the postsynaptic membrane. Electron microscopy of immunogold-labeled synapses has shown that NMDA receptors tend to be clustered near the center of the synapse, while AMPA receptors are distributed more uniformly across the postsynaptic membrane.[51,52] This difference may reflect association of the two receptor types with different subsynaptic structures. NMDA receptors and AMPA receptors show other differences at synapses. They are transported to synapses at different times during development (as measured for cultured primary neurons); and once installed at the synapse, they differ in their ease of extraction by detergents, with the NMDA receptor being more firmly attached.[53,54] Furthermore, a significant proportion of excitatory synapses lacks the AMPA receptor.[55] This finding is important in view of the influences that one receptor type can exert on the other receptor's function. Opening of the NMDA receptor channel requires cell depolarization, which may arise when currents flow through the AMPA receptor. Thus, NMDA receptors may be inactive at synapses that lack functional AMPA receptors, unless depolarization is provided by some other means, such as by depolarization of a neighboring synapse that does contain functional AMPA receptors.

Recently synapses have been described that lack functional AMPA receptors, but for which AMPA receptors may be activated by tetanic stimulation.[56–58] This finding has led to the formulation of the "silent synapse" hypothesis. In this hypothesis, synaptic strength is increased by recruitment of AMPA receptors to synapses that lack functional forms of the AMPA receptor, a process sometimes called "AMPAfication." The hypothesis has provoked much interest in the mechanisms that govern the abundance of functional AMPA receptor at the synapse. Silent synapses are converted to "talking" synapses by tetanic stimulation of the sort that induces long-term potentiation (LTP). Ca^{2+} fluxes through NMDA receptors following tetanic stimulation of silent synapses may induce biochemical pathways that trigger synaptic plastic changes. This may lead in turn to AMPA receptor functional activation. The mechanisms that govern AMPA receptor abundance at the synapse are not known. Recently it has been shown that the *N*-ethylmaleimide sensitive fusion protein NSF makes a specific complex with the GluR2 C-terminal domain.[59–61] NSF is a chaperonin, which modulates protein-protein interactions. Electrophysiologic studies suggest that NSF function may be required to maintain AMPA receptor currents. NSF plays an important role in dissociating SNARE protein complexes following vesicle fusion with target membranes and may mediate protein interaction with GluR2, although its precise role with the AMPA receptors is not yet known.

The rapid increase in our understanding of molecular and cell biologic features of ionotropic glutamate receptors holds great promise for the field. We may anticipate a corresponding increase in our grasp of receptor function in all respects, from the molecular to the systems levels.

ACKNOWLEDGMENTS

I thank T. Serra for assistance in preparing the manuscript, and B. Rudy for stimulating conversations. E.B.Z. is an Investigator of the Howard Hughes Medical Institute.

REFERENCES

1. Hollmann, M. & S. Heinemann. 1994. Cloned glutamate receptors. Annu. Rev. Neurosci. **17:** 31–108.
2. Michaelis, E.K. 1998. Molecular biology of glutamate receptors in the central nervous system and their role in excitotoxicity, oxidative stress and aging. Prog. Neurobiol. **54:** 369–415.
3. Ozawa, S., H. Kamiya & K. Tsuzuki. 1998. Glutamate receptors in the mammalian central nervous system. Prog. Neurobiol. **54:** 581–618.
4. Rosenmund, C., Y. Stern-Bach & C.F. Stevens. 1998. The tetrameric structure of a glutamate receptor channel [see comments]. Science **280:** 1596–1599.
5. Laube, B., J. Kuhse & H. Betz. 1998. Evidence for a tetrameric structure of recombinant NMDA receptors. J. Neurosci. **18:** 2954–2961.
6. Mano, I. & V.I. Teichberg. 1998. A tetrameric subunit stoichiometry for a glutamate receptor-channel complex. Neuroreport **9:** 327–331.
7. Monyer, H., N. Burnashev, D.J. Laurie, B. Sakmann & P.H. Seeburg. 1994. Developmental and regional expression in the rat brain and functional properties of four NMDA receptors. Neuron **12:** 529–540.
8. Laurie, D.J., I. Bartke, R. Schoepfer, K. Naujoks & P.H. Seeburg. 1997. Regional, developmental and interspecies expression of the four NMDAR2 subunits, examined using monoclonal antibodies. Brain Res. Mol. Brain Res. **51:** 23–32.
9. Das, S., Y.F. Sasaki, T. Rothe, L.S. Premkumar, M. Takasu, J.E. Crandall, P. Dikkes, D.A. Conner, P.V. Rayudu, W. Cheung, H.S. Chen, S.A. Lipton & N. Nakanishi. 1998. Increased NMDA current and spine density in mice lacking the NMDA receptor subunit NR3A. Nature **393:** 377–381.
10. Hirai, H., J. Kirsch, B. Laube, H. Betz & J. Kuhse. 1996. The glycine binding site of the N-methyl-D-aspartate receptor subunit NR1: Identification of novel determinants of co-agonist potentiation in the extracellular M3-M4 loop region. Proc. Natl. Acad. Sci. USA **93:** 6031–6036.
11. Laube, B., H. Hirai, M. Sturgess, H. Betz & J. Kuhse. 1997. Molecular determinants of agonist discrimination by NMDA receptor subunits: Analysis of the glutamate binding site on the NR2B subunit. Neuron **18:** 493–503.
12. Jones, M.V. & G.L. Westbrook. 1996. The impact of receptor desensitization on fast synaptic transmission. Trends Neurosci. **19:** 96–101.
13. Krupp, J.J., B. Vissel, S.F. Heinemann & G.L. Westbrook. 1998. N-terminal domains in the NR2 subunit control desensitization of NMDA receptors. Neuron **20:** 317–327.
14. Mayer, M.L. & G.L. Westbrook. 1985. The Action of N-methyl-D-aspartate on mouse spinal neurons in culture. J. Physiol. (Lond.) **361:** 65–90.
15. Zilberter, Y., V. Uteshev, S. Sokolova & B. Khodorov. 1991. Desensitization of N-methyl-D-aspartate receptors in neurons dissociated from adult rat hippocampus. Mol. Pharmacol. **40:** 337–341.
16. Legendre, P., C. Rosenmund & G.L. Westbrook. 1993. Inactivation of NMDA channels in cultured hippocampal neurons by intracellular calcium. J. Neurosci. **13:** 674–684.
17. Medina, I., N. Filippova, G. Barbin, Y. Ben-Ari & P. Bregestivski. 1994. Kainate-induced activation of NMDA currents via an elevation of intracellular Ca2+ in hippocampal neurons. J. Neurophysiol. **72:** 456–465.
18. Zhang, S., M.D. Ehlers, J.P. Bernhardt, C.T. Su & R.L. Huganir. 1998. Calmodulin mediates calcium-dependent inactivation of N-methyl-D-aspartate receptors. Neuron **21:** 443–453.
19. Jonas, P. & N. Burnashev. 1995. Molecular mechanisms controlling calcium entry through AMPA-type glutamate receptor channels. Neuron **15:** 987–990.
20. Sprengel, R., B. Suchanek, C. Amico, R. Brusa, N. Burnashev, A. Rozov, O. Hvalby, V. Jensen, O. Paulsen, P. Andersen, J.J. Kim, R.F. Thompson, W. Sun, L.C. Webster, S.G.

GRANT, J. EILERS, A. KONNERTH, J. LI, J.O. MCNAMARA & P.H. SEEBURG. 1998. Importance of the intracellular domain of NR2 subunits for NMDA receptor function in vivo. Cell **92**(2): 279–289.
21. MULLE, C., A. SAILER, I. PEREZ-OTANO, H. DICKINSON-ANSON, P.E. CASTILLO, I. BUREAU, C. MARON, F.H. GAGE, J.R. MANN, B. BETTLER & S.F. HEINEMANN. 1998. Altered synaptic physiology and reduced susceptibility to kainate-induced seizures in GluR6-deficient mice. Nature **392:** 601–605.
22. ZIFF, E.B. 1997. Enlightening the postsynaptic density. Neuron **19:** 1163–1174.
23. CHO, K.O., C.A. HUNT & M.B. KENNEDY. 1992. The rat brain postsynaptic density fraction contains a homolog of the Drosophila discs-large tumor suppressor protein. Neuron **9:** 929–942.
24. KISTNER, U., B.M. WENZEL, R.W. VEH, L.C. CASES, A.M. GARNER, U. APPELTAUER, B. VOSS, E.D. GUNDELFINGER & C.C. GARNER. 1993. SAP90, a rat presynaptic protein related to the product of the Drosophila tumor suppressor gene dlg-A. J. Biol. Chem. **268:** 4580–4583.
25. BRENMAN, J.E., D.S. CHAO, S.H. GEE, A.W. MCGEE, S.E. CRAVEN, D.R. SANTILLANO, Z. WU, F. HUANG, H. XIA, M.F. PETERS, S.C. FROEHNER & D.S. BREDT. 1996. Interaction of nitric oxide synthase with the postsynaptic density protein PSD-95 and alpha1-syntrophin mediated by PDZ domains. Cell **84:** 757–767.
26. KIM, E. & M. SHENG. 1996. Differential K+ channel clustering activity of PSD-95 and SAP97, two related membrane-associated putative guanylate kinases. Neuropharmacology **35:** 993–1000.
27. LAU, L.F., A. MAMMEN, M.D. EHLERS, S. KINDLER, W.J. CHUNG, C.C. GARNER & R.L. HUGANIR. 1996. Interaction of the N-methyl-D-aspartate receptor complex with a novel synapse-associated protein, SAP102. J. Biol. Chem. **271:** 21622–21628.
28. MULLER, B.M., U. KISTNER, R.W. VEH, L.C. CASES, B. BECKER, E.D. GUNDELFINGER & C.C. GARNER. 1995. Molecular characterization and spatial distribution of SAP97, a novel presynaptic protein homologous to SAP90 and the Drosophila discs-large tumor suppressor protein. J. Neurosci. 2354–2366.
29. LUE, R.A., E. BRANDIN, E.P. CHAN & D. BRANTON. 1996. Two independent domains of hDlg are sufficient for subcellular targeting: The PDZ1-2 conformational unit and an alternatively spliced domain. J. Cell Biol. **135:** 1125–1137.
30. CABRAL, J.H., C. PETOSA, M.J. SUTCLIFFE, S. RAZA, O. BYRON, F. POY, S.M. MARFATIA, A.H. CHISHTI & R.C. LIDDINGTON. 1996. Crystal structure of a PDZ domain. Nature **382:** 649–652.
31. DOYLE, D.A., A. LEE, J. LEWIS, E. KIM, M. SHENG & R. MACKINNON. 1996. Crystal structures of a complexed and peptide-free membrane protein-binding domain: Molecular basis of peptide recognition by PDZ. Cell **85:** 1067–1076.
32. KIM, E., S. NAISBITT, Y.P. HSUEH, A. RAO, A. ROTHSCHILD, A.M. CRAIG & M. SHENG. 1997. GKAP, a novel synaptic protein that interacts with the guanylate kinase-like domain of the PSD-95/SAP90 family of channel clustering molecules. J. Cell Biol. **136:** 669–678.
33. TAKEUCHI, M., Y. HATA, K. HIRAO, A. TOYODA, M. IRIE & Y. TAKAI. 1997. SAPAPs. A family of PSD-95/SAP90-associated proteins localized at postsynaptic density. J. Biol. Chem. **272:** 11943–11951.
34. SATOH, K., H. YANAI, T. SENDA, K. KOHU, T. NAKAMURA, N. OKUMURA, A. MATSUMINE, S. KOBAYASHI, K. TOYOSHIMA & T. AKIYAMA. 1997. DAP-1, a novel protein that interacts with the guanylate kinase–like domains of hDLG and PSD-95. Genes Cells **2:** 415–424.
35. HSUEH, Y.P., E. KIM & M. SHENG. 1997. Disulfide-linked head-to-head multimerization in the mechanism of ion channel clustering by PSD-95. Neuron **18:** 803–814.
36. TOPINKA, J.R. & D.S. BREDT. 1998. N-terminal palmitoylation of PSD-95 regulates association with cell membranes and interaction with K^+ channel Kv1.4. Neuron **20:** 125–134.
37. IRIE, M., Y. HATA, M. TAKEUCHI, K. ICHTCHENKO, A. TOYODA, K. HIRAO, Y. TAKAI, T.W. ROSAHL & T.C. SUDHOF. 1997. Binding of neuroligins to PSD-95. Science **277:** 1511–1515.
38. NIETHAMMER, M., J.G. VALTSCHANOFF, T.M. KAPOOR, D.W. ALLISON, T.M. WEINBERG, A.M. CRAIG & M. SHENG. 1998. CRIPT, a novel postsynaptic protein that binds to the third PDZ domain of PSD-95/SAP90. Neuron **20:** 693–707.
39. WYSZYNSKI, M., J. LIN, A. RAO, E. NIGH, A.H. BEGGS, A.M. CRAIG & M. SHENG. 1997. Competitive binding of alpha-actinin and calmodulin to the NMDA receptor. Nature **385:** 439–442.
40. EHLERS, M.D., S. ZHANG, J.P. BERNHADT & R.L. HUGANIR. 1996. Inactivation of NMDA receptors by direct interaction of calmodulin with the NR1 subunit. Cell **84:** 745–755.

41. Rosenmund, C. & G.L. Westbrook. 1993a. Calcium-induced actin depolymerization reduces NMDA channel activity. Neuron **10:** 805–814.
42. Rosenmund, C. & G.L. Westbrook. 1993b. Rundown of N-methyl-D-aspartate channels during whole-cell recording in rat hippocampal neurons: Role of Ca2+ and ATP [published erratum appears in J. Physiol. (Lond.) 1994. **475(3):** 547–548. J. Physiol. **470:** 705–729.
43. Dong, H., R.J. O'Brien, E.T. Fung, A.A. Lanahan, P.F. Worley & R.L. Huganir. 1997. GRIP: A synaptic PDZ domain–containing protein that interacts with AMPA receptors. Nature **386:** 279–284.
44. Srivastava, S., P. Osten, F.S. Vilim, L. Khatri, G.J. Inman, B. States, C. Daly, S. DeSouza, R. Abagyan, J.G. Valtschanoff, R.J. Weinberg & E.B. Ziff. 1998. Novel anchorage of GluR2/3 to the postsynaptic density by the AMPA receptor-binding protein ABP. Neuron **21**(3): 581–591.
45. Kornau, H.C., L.T. Schenker, M.B. Kennedy & P.H. Seeburg. 1995. Domain interaction between NMDA receptor subunits and the postsynaptic density protein PSD-95. Science **269:** 1737–1740.
46. Gurd, J.W. & N. Bissoon. 1985. In vivo phosphorylation of the postsynaptic density glycoprotein gp180. J. Neurochem. **45:** 1136–1140.
47. Moon, I.S., M.L. Apperson & M.B. Kennedy. 1994. The major tyrosine-phosphorylated protein in the postsynaptic density fraction is N-methyl-D-aspartate receptor subunit 2B. Proc. Natl. Acad. Sci. USA **91:** 3954–3958.
48. Suzuki, T. & N.K. Okumura. 1995. NMDA receptor subunits epsilon 1 (NR2A) and epsilon 2 (NR2B) are substrates for Fyn in the postsynaptic density fraction isolated from the rat brain. Biochem. Biophys. Res. Commun. **216:** 582–588.
49. Chen, H.J., M. Rojas-Soto, A. Oguni & M.B. Kennedy. 1998. A synaptic Ras-GTPase activating protein (p135 SynGAP) inhibited by CaM kinase II. Neuron **20:** 895–904.
50. Kim, J.H., D. Liao, L.F. Lau & R.L. Huganir. 1998. SynGAP: A synaptic RasGAP that associates with the PSD-95/SAP90 protein family. Neuron **20:** 683–691.
51. Kharaziga V.N. & R.J. Wienberg. 1997. Tangential synaptic distribution of NMDA and AMPA receptors in rat neocortex. Neurosci. Lett. **238:** 41–44.
52. Somogyi, P., G. Tamás, R. Lujan & E.H. Buhl. 1998. Salient features of synaptic organisation in the cerebral cortex. Brain Res. Rev. **27:** 113–135.
53. Allison, D.W., V.I. Gelfand, I. Spector & A.M. Craig. 1998. Role of actin in anchoring postsynaptic receptors in cultured hippocampal neurons: Differential attachment of NMDA versus AMPA receptors. J. Neurosci. **18:** 2423–2436.
54. Rao, A., E. Kim, M. Sheng & A.M. Craig. 1998. Heterogeneity in the molecular composition of excitatory postsynaptic sites during development of hippocampal neurons in culture. J. Neurosci. **18:** 1217–1229.
55. Nusser, Z., R. Lujan, G. Laube, J.D. Roberts, E. Molnar & P. Somogyi. 1998. Cell type and pathway dependence of synaptic AMPA receptor number and variability in the hippocampus. Neuron **21:** 545–559.
56. Isaac, J.T., R.A. Nicoll & R.C. Malenka. 1995. Evidence for silent synapses: Implications for the expression of LTP. Neuron **15:** 427–434.
57. Liao, D., N.A. Hessler & R. Malinow. 1995. Activation of postsynaptically silent synapses during pairing-induced LTP in CA1 region of hippocampal slice. Nature **375:** 400–404.
58. Malenka, R.C. & R.A. Nicoll. 1997. Silent synapses speak up. Neuron **19:** 473–476.
59. Nishimune, A., J.T.R. Isaac, E. Molnar, J. Noel, S.R. Nash, M. Tagaya, G.L. Collingridge, S. Nakanishi & J.M. Henley. 1998. NSF binding to GluR2 regulates synaptic transmission. Neuron **21:** 87–97.
60. Osten, P., S. Srivastava, G.J. Inman, F.S. Vilim, L. Khatri, L.M. Lee, B.A. States, S. Einheber, T.A. Milner, P.I. Hanson & E.B. Ziff. 1998. The AMPA receptor GluR2 C-Terminus can mediate a reversible, ATP-dependent interaction with NSF and a- and b- SNAPs. Neuron **21:** 99–110.
61. Song, I., S. Kamboj, J. Xia, H. Dong, D. Liao & R.L. Huganir. 1998. Interaction of the N-ethylmaleimide-sensitive factor with AMPA receptors. Neuron **21:** 393–400.

The Arrangement of Glutamate Receptors in Excitatory Synapses

YUTAKA TAKUMI,[a,b] ATSUSHI MATSUBARA,[b] ERIC RINVIK,[a] AND OLE P. OTTERSEN[a,c]

[a]*Department of Anatomy, Institute of Basic Medical Sciences, University of Oslo, POB 1105 Blindern, 0317 Oslo, Norway*

[b]*Department of Otorhinolaryngology, Hirosaki University School of Medicine, 5 Zaifu-cho, Hirosaki 036, Japan*

ABSTRACT: Electron microscopic immunogold analyses have revealed a highly differentiated arrangement of glutamate receptors at excitatory synapses in the central nervous system. Studies focused on the hippocampus and cerebellum have shown that the postsynaptic specialization is the preferential site of NMDA and AMPA receptor expression, and that the δ2 receptor is similarly concentrated at this site. In cases of colocalization (AMPA and NMDA, or AMPA and δ2) the two receptor types appear to be intermingled rather than segregated to separate parts of the membrane. The different groups of metabotropic receptor exhibit distinct distributions at the synapse: group I receptors occur in membrane domains lateral to the postsynaptic specialization; group II receptors are expressed in preterminal membranes or extrasynaptically; whereas group III receptors are found in, or close to, the presynaptic active zone consistent with their roles as autoreceptors. The differentiated distribution of glutamate receptors reflects their functional heterogeneity and explains why some receptors are activated only at high firing frequencies.

The probability that a given glutamate receptor is activated following an exocytotic event depends not only on its intrinsic properties but also on its position in the synapse.[1,2] This calls for a better understanding of how glutamate receptors are arranged in synaptic membranes and how they are targeted and anchored at their respective membrane domains. The present chapter reviews recent analyses of glutamate receptor distribution based on electron microscopic immunogold techniques—the only immunocytochemical techniques in current use that allow reliable localization of membrane proteins at the nanometer level.[3,4] A major advantage of immunogold techniques compared to the peroxidase-antiperoxidase technique is the short and well-defined distance between the epitope and the immunocytochemical signal (i.e., gold particle).[5]

STRUCTURAL FEATURES OF THE GLUTAMATE SYNAPSE

Glutamatergic synapses typically exhibit an electron-dense zone, the postsynaptic density,[6,7] just beneath the postsynaptic membrane (FIG. 1). The extent of the synapse is defined by the postsynaptic density, although it is now established (see below) that glutamate receptors also occur in a narrow zone around it (termed the *perisynaptic annulus*) and even more laterally (in *extrasynaptic* membranes). Whereas the postsynaptic spe-

[c]Corresponding author. Phone: +472-285-1270; fax: +472-285-1299; e-mail: o.p.ottersen@basalmed.uio.no

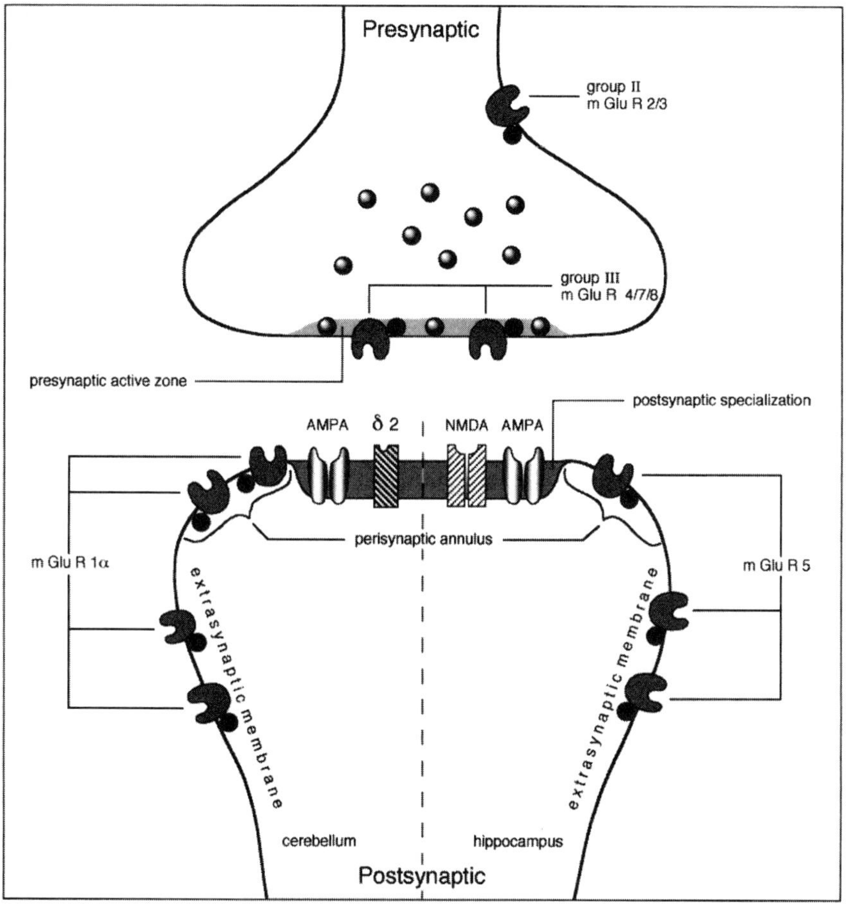

FIGURE 1. Diagram showing the arrangement of glutamate receptors at the synapse as revealed by high-resolution immunogold labeling. The drawing is not representative of all excitatory synapses since the identities and subsynaptic localization of glutamate receptors may vary considerably among brain regions and cells, and even between different glutamate synapses on the same cell (see text). Most of the data on which the diagram was based were obtained in the hippocampus (presynaptic element and right half of spine) or cerebellum (parallel fiber–Purkinje cell synapses; left half of spine). NMDA, AMPA, and δ2 receptors are concentrated in the postsynaptic specialization, whereas metabotropic receptors occur peri- or extrasynaptically or in the presynaptic element. The differential postsynaptic distribution of the group I metabotropic receptors (mGluR1α and mGluR5) is according to Luján *et al.*[2] Group III presynaptic mGluRs are found in the presynaptic active zone (shown in *gray*), while the group II presynaptic mGluRs are expressed in the preterminal axolemma. The *filled circles* associated with the mGluRs indicate G proteins. Kainate receptors and extrasynaptic ionotropic receptors are not included in the drawing.

cialization is directly apposed to the presynaptic active zone, the access of glutamate to perisynaptic and extrasynaptic receptors may be constrained by glial lamellae and their complement of glutamate transporters.[8–10] It is of importance to note that the degree of glial investment varies considerably among different synapses; thus the synaptic geometry must be included among the factors that could be envisaged to influence the activation of glutamate receptors.

COLOCALIZATION OF RECEPTORS IN THE POSTSYNAPTIC SPECIALIZATION

Three classes of glutamate receptor—NMDA, AMPA, and δ2—are concentrated at the postsynaptic specialization (FIG. 1). Postembedding immunogold analyses of the cerebral cortex revealed postsynaptic specializations that coexpress NMDA and AMPA receptors.[11] Such double-labeled synapses also occur in the substantia nigra pars reticulata (Rinvik *et al.*, unpublished), and in the hippocampus, including the stratum radiatum of CA1 (Takumi *et al.*, submitted). However, many spine synapses in the latter area (about 25%) were single labeled for NMDA receptors (ibid.). This is likely to reflect absence of AMPA receptors at these synapses, although it should be emphasized that any AMPA receptors that are not exposed at the surface of the section will be inaccessible for immunogold labeling. Thus a small pool of AMPA receptors may easily be missed even in serial section analyses. This problem is aggravated by the fact that the labeling efficiency is necessarily less than 100%.

The heterogeneous labeling for AMPA receptors in CA1 is of interest in view of physiological data suggesting that some of the glutamatergic synapses at this site are "silent" at resting membrane potentials.[12–14] It has been hypothesized that such silent synapses may be brought to express AMPA currents, by recruitment of new AMPA receptors to the postsynaptic specialization or by conversion of nonfunctional AMPA receptors into functional ones, and that this may be one of the mechanisms underlying long-term potentiation (LTP).[13,15] Notwithstanding the technical limitations mentioned above, this hypothesis is now amenable to direct testing by double immunogold labeling.

Like NMDA and AMPA receptors, the δ2 receptors are preferentially expressed in postsynaptic specializations, notably in the cerebellar cortex (FIG. 1).[16] According to recent knockout studies this receptor is required for the induction of long-term depression and for motor coordination.[17] Whereas early studies in expression systems indicated that the δ2 receptor was nonfunctional,[18,19] mutated δ2 receptors have recently been shown to sustain a large, constitutive inward current.[20] The mutation in question is believed to underlie the neurological defect in the atactic Lurcher mice and to produce the Purkinje cell degeneration that is a characteristic feature of these mice.[20]

DO DIFFERENT GLUTAMATE RECEPTORS SEGREGATE TO DISTINCT DOMAINS OF THE POSTSYNAPTIC SPECIALIZATION?

As described in the preceding paragraph a colocalization of NMDA and AMPA receptors is typical for some populations of neocortical and hippocampal synapses, whereas a colocalization of δ2 and AMPA receptors prevails in the cerebellar cortex (FIG. 1). In both cases the coexpressed receptors (visualized by two different gold particle sizes) appear to be intermingled rather than segregated to different parts of the synaptic specialization. This is particularly evident in tangentially cut synapses in which the receptor matrix can be viewed "en face" (Ref. 16 and unpublished observations).

The failure to demonstrate a distinct subcompartmentation of the postsynaptic specialization does not necessarily imply that the receptors are homogeneously distributed at this site. In particular, in synapses with a fixed and anatomically defined release site one could envisage that the receptor concentration could be tuned to the concentration of transmitter in the synaptic cleft. This hypothesis was tested in the synapse between the inner hair cells and afferent dendrites in the organ of Corti. This putative glutamatergic synapse[21,22] is equipped with a presynaptic dense body that is thought to represent the site of exocytotic release. It was found that the concentration of gold particles signaling AMPA receptors GluR 2/3 and GluR 4 was lowest in the part of the postsynaptic density facing the release site, and highest more laterally in the synapse, near the margin of the postsynaptic specialization. It was speculated that the mediolateral gradient in AMPA receptor concentration could compensate for the lateral attenuation of glutamate that occurs following vesicular release.[21]

In most synapses of the central nervous system the release site is not defined morphologically and is probably not stationary. This may explain why central synapses generally do not show the mediolateral concentration gradient of AMPA receptors typical of the inner hair cell synapses. An even distribution of AMPA receptors has been reported in postsynaptic specializations of parallel fiber-spine synapses in the cerebellum[23] as well as of auditory nerve–spherical bushy cell synapses in the anteroventral cochlear nucleus.[24] A subpopulation of striatal synapses was found to exhibit a mediolateral AMPA receptor gradient,[25] but this was less distinct than that of the inner hair cell synapse.

METABOTROPIC RECEPTORS APPEAR TO BE EXCLUDED FROM THE POSTSYNAPTIC SPECIALIZATION

The metabotropic receptors can be divided in three groups.[26] These are coupled to polyphosphoinositide hydrolysis (group I) or negatively coupled to adenylyl cyclase (groups II and III). Analyses based on pre- or postembedding immunogold procedures have shown that the different groups of metabotropic receptors are expressed in distinct membrane domains (FIG. 1), as will be discussed below. Most of the analyses have been performed in the cerebellum or hippocampus, and it remains to be seen whether the results are generally valid for all brain regions and fiber systems.

Group I receptors (mGluR 1 and 5) occur at highest concentrations in the perisynaptic annulus, i.e., the narrow zone surrounding the postsynaptic specialization.[1,2,23,27–29] Subtle differences were found between the distributions of the two receptors: whereas 50% of the immunogold particles signaling mGluR1α occurred within 60 nm of the margin of the postsynaptic density (the rest being associated with extrasynaptic membranes), the corresponding value for mGluR5 was 25%.[2] It should be noted, however, that the data were obtained from two different cell types—Purkinje cells in the case of mGluR1α and hippocampal pyramidal cells in the case of mGluR5. This complicates the interpretation of these results.

Group II receptors (mGluR2 and 3) appear to be preferentially expressed outside the synapse (FIG. 1). For example, mGluR2 immunoreactivity in the hippocampal mossy fiber system is concentrated in the preterminal axolemma,[30–32] consistent with the observation that high frequency-stimulation is required for mGluR2 activation.[30] Very high concentrations of group II receptors have been demonstrated in cerebellar Golgi cells,[33] which express mGluR2/3 immunoreactivity in all extrasynaptic membrane domains.[34,35] A recent immunogold analysis confirmed the even distribution of mGluR2 along the Golgi cell plasmalemma and its lack of association with synaptic membranes.[2]

A presynaptic localization is typical of the group III receptors (FIG. 1). mGluR4,7, and 8 have all been demonstrated in the presynaptic grid,[31,36,37] where they may function as inhibitory autoreceptors.[38] Lesion studies have revealed that different fiber systems are endowed with different types or subtypes of group III receptors.[31] Type 7a and 8 are localized in the perforant path, type 7a and 7b in the mossy fibers, and type 7a in the Schaffer collaterals. The mGluR8 in the perforant path was preferentially associated with the lateral component of this pathway, whereas the group II receptor mGluR2 predominated in the medial component.[31] The inner zone of the dentate molecular layer contains mGluR4, which probably depends on the projection from the hilar mossy cells.[31] mGluR4 has also been identified on parallel fiber terminals in the cerebellum.[39]

From the above description and relying exclusively on immunogold data, one can conclude that metabotropic receptors appear to be excluded from the postsynaptic specializations of the fiber systems that have been studied so far (FIG. 1). The virtual absence of labeling at the postsynaptic specialization cannot be explained by accessibility problems since the pre-embedding immunogold observations have been largely confirmed by postembedding procedures.[2] As the latter procedures only label epitopes expressed at the surface of the sections, they should be relatively immune to penetration artefacts. The mGluR staining at the postsynaptic specialization observed in pre-embedding peroxidase-antiperoxidase (PAP) studies may reflect diffusion of reaction product from extrasynaptic sites, although it must be recalled that the PAP technique has a sensitivity that exceeds by far that of the immunogold procedures (see Ref. 3 for discussion).

FUNCTIONAL IMPLICATIONS OF THE DIFFERENTIATED GLUTAMATE RECEPTOR DISTRIBUTION AT THE SYNAPSE

The receptors localized in the postsynaptic specialization are directly apposed to the presynaptic active site. At these receptors the peak concentration of glutamate should be largely independent of lateral diffusion or binding to glial glutamate transporters, although these factors may affect the transmitter time course in the synaptic cleft.[40] The enrichment of NMDA and AMPA receptors at the site of the postsynaptic specialization reflects their roles in mediating fast glutamatergic transmission in the CNS.

The peri- and extrasynaptic localization of the metabotropic glutamate receptors implies that their activation will be sensitive to factors that affect the lateral diffusion of glutamate.[1] Thus it is clear from modeling studies[41] that the concentration of glutamate will be much reduced over the distance required to reach the most distant metabotropic receptors, which in the case of mGluR1α in the cerebellum may be located up to 900 nm away from the synapse.[2] Increasing the amount of released glutamate would compensate for the lateral attenuation of transmitter. One would therefore expect to find a positive correlation between presynaptic firing frequency and degree of metabotropic receptor activation. This prediction has been borne out by experimental studies.[42–46]

The concentration of glutamate at extrasynaptic sites is not only a function of the amount of released transmitter but also depends on many other factors including the geometry of the synapse, the tortuosity of the extracellular space, and the density of glutamate-binding membrane molecules (neuronal and glial transporters and receptors) in the synapse and its vicinity.[41] It is notable in this regard that the degree of glial investment varies considerably among different types of synapse. In parallel fiber-spine synapses in the cerebellar cortex, the outlet of the synapse is restricted by glial lamellae[8,47] which may extend almost to the margin of the postsynaptic specialization. In these synapses it is likely that the lateral diffusion of glutamate (and activation of distant metabotropic receptors) is modified by glial glutamate transporters (GLAST, and to a lesser extent GLT1; Ref. 8), as well as by the neuronal glutamate transporter EAAT4. The latter transporter is concentrated in the lateral membrane of the spine, facing glial lamellae.[9] That glutamate transporters may serve to limit glutamate access to distant receptors would be in line with recent observations of Asztely *et al.*[48] suggesting that glial glutamate transporters restrict cross-talk between neighboring excitatory synapses. However, some types of glutamatergic synapses (such as those between cerebellar parallel fibers and dendritic stems of interneurons) show a less intimate relation to glial processes. In such synapses the activation of distant metabotropic receptors may be less dependent on diffusional constraints and glial glutamate transporters.

Many aspects of the above discussion are relevant also for a subpopulation of presynaptic metabotropic receptors, namely, those that are expressed at a distance from the active site. For example, it has been shown that the preterminal mGluR2 receptors on hippocampal mossy fibers are activated following high- but not low-frequency stimulation.[33] In contrast, the concentration of glutamate at those presynaptic metabotropic receptors that are expressed in the presynaptic grid (group III receptors; see above) should be largely independent of lateral diffusion.

GLUTAMATE RECEPTORS MAY BE TARGETED TO SELECT SUBPOPULATIONS OF GLUTAMATE SYNAPSES

The Purkinje cells receive two distinct glutamatergic inputs, the parallel and climbing fibers.[49] Both fiber systems terminate on spines or thorns that are immunopositive for GluR2/3.[16,23] However, antibodies to the δ2 receptor produce labeling exclusively of parallel fiber synapses.[16] It thus appears that neurons may be capable of targeting a given receptor to one subset of postsynaptic spines. This would be in line with the observations of Rubio and Wenthold,[50] who demonstrated that the fusiform cells of the dorsal cochlear nucleus express GluR4 and mGluR1 postsynaptic to auditory nerve terminals, but not postsynaptic to another category of glutamatergic input (the parallel fibers).

Presynaptic glutamate receptors are subject to an equally selective targeting. Notably, Shigemoto *et al.*[37] compared different terminals emanating from the same hippocampal pyramidal cells and found that the terminals expressed different densities of the group III metabotropic receptor mGluR7. The density of receptor depended on the identity of the postsynaptic target. A similar target-specific segregation of receptor has been reported in the mossy fiber system. Mossy fibers were found to express mGluR7a and 7b at contacts with mGluR1α positive neurons, but not at contacts with hippocampal pyramidal cells.[31]

Whereas several proteins involved in the anchoring of glutamate receptors have now been identified,[6,51–55] the molecular mechanisms underlying the highly selective targeting of glutamate receptors remain to be resolved. It is clear, however, that the capacity of neurons to direct glutamate receptors to select populations of glutamate synapses adds a new level of complexity that should be taken into account when attempts are made to extrapolate from the cellular level to the level of the individual synapse.

ACKNOWLEDGMENTS

This work was supported by the Norwegian Research Council, the EU Biomed program (BMH-CT96-0851), Professor Letten F. Saugstad's Fund, and the Sasakawa foundation.

REFERENCES

1. Baude, A., Z. Nusser, J.D.B. Roberts *et al.* 1993. The metabotropic glutamate receptor (mGluR1() is concentrated at perisynaptic membrane of neuronal subpopulations as detected by immunogold reaction. Neuron **11:** 771–787.
2. Luján, R., J.D.B. Roberts, R. Shigemoto *et al.* 1997. Differential plasma membrane distribution of metabotropic glutamate receptors mGluR1α, mGluR2 and mGluR5, relative to neurotransmitter release sites. J. Chem. Neuroanat. **13:** 219–241.
3. Ottersen, O.P. & A.S. Landsend. 1997. Organization of glutamate receptors at the synapse. Eur. J. Neurosci. **9:** 2219–2224.
4. Takumi, Y., L. Bergersen, A.S. Landsend *et al.* 1998. Synaptic arrangement of glutamate receptors. Prog. Brain Res. **116:** 105–122.
5. Merighi, A. 1992. Post-embedding electron microscopic immunocytochemistry. *In* Electron Microscopic Immunocytochemistry—Principles and practice. J.M. Polak & J.Y. Priestley, Eds.: 51–87. Oxford University Press. New York.
6 Kennedy, M.B. 1997. The postsynaptic density at glutamatergic synapses. Trends Neurosci. **20:** 264–268.
7. Peters, A., S.L. Palay & H.D. Webster. 1991. The fine structure of the nervous system: Neurons and their supporting cells. 3rd edit. Oxford University Press. New York.
8. Chaudhry, F.A., K.P. Lehre, M. Van Lookeren Campagne *et al.* 1995. Glutamate transporters in glial plasma membranes: Highly differentiated localizations revealed by quantitative ultrastructural immunocytochemistry. Neuron **15:** 711–720.
9. Dehnes, Y., F.A. Chaudhry, K. Ullensvang *et al.* 1998. The glutamate transporter EAAT4 in rat cerebellar Purkinje cells: A glutamate-gated chloride channel concentrated near the synapse in parts of the dendritic membrane facing astroglia. J. Neurosci. **18:** 3606–3619.
10. Rusakov, D.A. & D.M. Kullmann. 1998. Extrasynaptic glutamate diffusion in the hippocampus: Ultrastructural constraints, uptake, and receptor activation. J. Neurosci. **18:** 3158–3170.
11. Kharazia, V.N., K.D. Phend, A. Rustioni *et al.* 1996. EM colocalization of AMPA and NMDA receptor subunits at synapses in rat cerebral cortex. Neurosci. Lett. **210:** 37–40.
12. Isaac, J.T., R.A. Nicoll & R.C. Malenka. 1995. Evidence for silent synapses: Implications for the expression of LTP. Neuron **15:** 427–434.
13. Liao, D., N.A. Hessler & R. Malinow. 1995. Activation of postsynaptically silent synapses during pairing-induced LTP in CA1 region of hippocampal slice. Nature **375:** 400–404.
14. Malenka, R.C. & R.A. Nicoll. 1997. Silent synapses speak up. Neuron **19:** 473–476.
15. Kullmann, D.M. 1994. Amplitude fluctuations of dual-component EPSCs in hippocampal pyramidal cells: Implications for long-term potentiation. Neuron **12:** 1111–1120.
16. Landsend, A.S., M. Amiry-Moghaddam, A. Matsubara *et al.* 1997. Differential localization of δ glutamate receptors in the rat cerebellum: Coexpression with AMPA receptors in parallel fiber-spine synapses and absence from climbing fiber-spine synapses. J. Neurosci. **17:** 834–842.

17. KASHIWABUCHI, N., K. IKEDA, K. ARAKI *et al.* 1995. Impairment of motor coordination, Purkinje cell synapse formation, and cerebellar long-term depression in GluR δ2 mutant mice. Cell **81:** 245–252.
18. ARAKI, K., H. MEGURO, E. KUSHIYA *et al.* 1993. Selective expression of the glutamate receptor channel δ2 subunit in cerebellar Purkinje cells. Biochem. Biophys. Res. Commun. **197:** 1267–1276.
19. LOMELI, H., R. SPRENGEL, D.J. LAURIE *et al.* 1993. The rat delta-1 and delta-2 subunits extend the excitatory amino acid receptor family. FEBS Lett. **315:** 318–322.
20. ZUO, J., P.L. DE JAGER, K.A. TAKAHASHI *et al.* 1997. Neurodegeneration in Lurcher mice caused by mutation in δ2 glutamate receptor gene. Nature **388:** 769–773.
21. MATSUBARA, A., J.H. LAAKE, S. DAVANGER *et al.* 1996. Organization of AMPA receptor subunits at a glutamate synapse: A quantitative immunogold analysis of hair cell synapses in the rat organ of Corti. J. Neurosci. **16:** 4457–4467.
22. OTTERSEN, O.P., Y. TAKUMI, A. MATSUBARA *et al.* 1998. Molecular organization of a type of peripheral glutamate synapse: The afferent synapses of hair cells in the inner ear. Prog. Neurobiol. **54:** 127–148.
23. NUSSER, Z., E. MULVIHILL, P. STREIT *et al.* 1994. Subsynaptic segregation of metabotropic and ionotropic glutamate receptors as revealed by immunogold localization. Neuroscience **61:** 421–427.
24. WANG, Y.X., R.J. WENTHOLD, O.P. OTTERSEN *et al.* 1998. Endbulb synapses in the anteroventral cochlear nucleus express a specific subset of AMPA-type glutamate receptor subunits. J. Neurosci. **18:** 1148–1160.
25. BERNARD, V., P. SOMOGYI & J.P. BOLAM. 1997. Cellular, subcellular, and subsynaptic distribution of AMPA-type glutamate receptor subunits in the neostriatum of the rat. J. Neurosci. **17:** 819–833.
26. NAKANISHI, S. & M. MASU. 1994. Molecular diversity and functions of glutamate receptors. Annu. Rev. Biophys. Biomol. Struct. **23:** 319–348.
27. LUJÁN, R., Z. NUSSER, J.D.B. ROBERTS *et al.* 1996. Perisynaptic location of metabotropic glutamate receptors mGluR1 and m GluR5 on dendrites and dendritic spines in the rat hippocampus. Eur. J. Neurosci. **8:** 1488–1500.
28. NÉGYESSY, L., Z. VIDNYÁNSZKY, R. KUHN *et al.* 1997. Light and electron microscopic demonstration of mGluR5 metabotropic glutamate receptor immunoreactive neuronal elements in the rat cerebellar cortex. J. Comp. Neurol. **385:** 641–650.
29. VIDNYÁNSZKY, Z., T.J. GÖRCS, L. NÉGYESSY *et al.* 1996. Immunocytochemical visualization of the mGluR1a metabotropic glutamate receptor at synapses of corticothalamic terminals originating from area 17 of the rat. Eur. J. Neurosci. **8:** 1061–1071.
30. PETRALIA, R.S., Y.X. WANG, A.S. NIEDZIELSKI *et al.* 1996. The metabotropic glutamate receptors, mGluR2 and mGluR3, show unique postsynaptic, presynaptic and glial localizations. Neuroscience **71:** 949–976.
31. SHIGEMOTO, R., A. KINOSHITA, E. WADA *et al.* 1997. Differential presynaptic localization of metabotropic glutamate receptor subtypes in the rat hippocampus. J. Neurosci. **17:** 7503–7522.
32. YOKOI, M., K. KOBAYASHI, T. MANABE *et al.* 1996. Impairment of hippocampal mossy fiber LTD in mice lacking mGluR2. Science **273:** 645–647.
33. SCANZIANI, M., P.A. SALIN, K.E. VOGT *et al.* 1997. Use-dependent increases in glutamate concentration activate presynaptic metabotropic glutamate receptors. Nature **385:** 630–634.
34. NEKI, A., H. OHISHI, T. KANEKO *et al.* 1996. Metabotropic glutamate receptors mGluR2 and mGluR5 are expressed in two non-overlapping populations of Golgi cells in the rat cerebellum. Neuroscience **75:** 815–826.
35. OHISHI, H., R. OGAWA-MEGURO, R. SHIGEMOTO *et al.* 1994. Immunohistochemical localization of metabotropic glutamate receptors, mGluR2 and mGluR3, in rat cerebellar cortex. Neuron **13:** 55–66.
36. LI, H., H. OHISHI, A. KINOSHITA *et al.* 1997. Localization of a metabotropic glutamate receptor, mGluR7, in axon terminals of presumed nociceptive, primary afferent fibers in the superficial layers of the spinal dorsal horn: An electron microscope study in the rat. Neurosci. Lett. **223:** 153–156.

37. Shigemoto, R., A. Kulik, J.D.B. Roberts *et al.* 1996. Target-cell-specific concentration of a metabotropic glutamate receptor in the presynaptic active zone. Nature **381:** 523–525.
38. Conn, P.J., D.G. Winder & R.W. Gereau IV. 1994. Regulation of neuronal ciruits and animal behavior by metabotropic glutamate receptors. *In* The Metabotropic Glutamate Receptors. P.J. Conn & J. Patel, Eds.: 195–229. Humana Press. Totowa, NJ.
39. Mateos, J.M., J. Azkue, R. Sarria *et al.* 1998. Localization of the mGlu4a metabotropic glutamate receptor in rat cerebellar cortex. Histochem. Cell Biol. **109:** 135–139.
40. Clements, J.D. 1996. Transmitter timecourse in the synaptic cleft: Its role in central synaptic function. Trends Neurosci. **19:** 163–171.
41. Barbour, B. & M. Häusser. 1997. Intersynaptic diffusion of neurotransmitter. Trends Neurosci. **20:** 377–384.
42. Bashir, Z.I., Z.A. Bortolotto, C.H. Davies *et al.* 1993. Induction of LTP in the hippocampus needs synaptic activation of glutamate metabotropic receptors. Nature **363:** 347–350.
43. Batchelor, A.M. & J. Garthwaite. 1997. Frequency detection and temporally dispersed synaptic signal association through a metabotropic glutamate receptor pathway. Nature **385:** 74–77.
44. Charpak, S. & B.H. Gähwiler. 1991. Glutamate mediates a slow synaptic response in hippocampal slice cultures. Proc. R. Soc. Lond. B. Biol. Sci. **243:** 221–226.
45. McCormick, D.A. & M. Von Krosigk. 1992. Corticothalamic activation modulates thalamic firing through glutamate "metabotropic" receptors. Proc. Natl. Acad. Sci. USA **89:** 2774–2778.
46. Miles, R. & J.C. Poncer. 1993. Metabotropic glutamate receptors mediate a post-tetanic excitation of guinea-pig hippocampal inhibitory neurones. J. Physiol. (Lond.) **463:** 461–473.
47. Palay, S.L. & V. Chan-Palay. 1974. Cerebellar cortex: Cytology and organization. Springer-Verlag. New York.
48. Asztely, F., G. Erdemli & D.M. Kullmann. 1997. Extrasynaptic glutamate spillover in the hippocampus: Dependence on temperature and the role of active glutamate uptake. Neuron **18:** 281–293.
49. Ottersen, O.P., N. Zhang & F. Walberg. 1992. Metabolic compartmentation of glutamate and glutamine: Morphological evidence obtained by quantitative immunocytochemistry in rat cerebellum. Neuroscience **46:** 519–534.
50. Rubio, M.E. & R.J. Wenthold. 1997. Glutamate receptors are selectively targeted to postsynaptic sites in neurons. Neuron **18:** 939–950.
51. Brakeman, P.R., A.A. Lanahan, R. O'Brien *et al.* 1997. Homer: A protein that selectively binds metabotropic glutamate receptors. Nature **386:** 284–288.
52. Dong, H., R.J. O'Brien, E.T. Fung *et al.* 1997. GRIP: A synaptic PDZ domain-containing protein that interacts with AMPA receptors. Nature **386:** 279–284.
53. Hsueh, Y.P., E. Kim & M. Sheng. 1997. Disulfide-linked head-to-head multimerization in the mechanism of ion channel clustering by PSD-95. Neuron **18:** 803–814.
54. Kim, E., K.O. Cho, A. Rothschild *et al.* 1996. Heteromultimerization and NMDA receptor-clustering activity of Chapsyn-110, a member of the PSD-95 family of proteins. Neuron **17:** 103–113.
55. Kornau, H.C., L.T. Schenker, M.B. Kennedy *et al.* 1995. Domain interaction between NMDA receptor subunits and the postsynaptic density protein PSD-95. Science **269:** 1737–1740.

Glutamate Receptor Anchoring Proteins and the Molecular Organization of Excitatory Synapses

MORGAN SHENG[a] AND DANIEL T. PAK

Howard Hughes Medical Institute and Department of Neurobiology, Massachussets General Hospital and Harvard Medical School, Boston, Massachusetts 02114, USA

ABSTRACT: Ionotropic glutamate receptors are concentrated at postsynaptic sites in excitatory synapses. The cytoplasmic C-terminal tail of certain glutamate receptor subunits interact with specific PDZ domain–containing proteins. NMDA receptor NR2 subunits bind to the PSD-95 family of proteins, whereas AMPA receptor subunits GluR2/3 bind to GRIP. These interactions may underlie the clustering, targeting, and immobilization of the glutamate receptors at postsynaptic sites. By virtue of their multiple protein-binding domains (e.g., three PDZs in PSD-95 and seven PDZs in GRIP), PSD-95 and GRIP can function as multivalent proteins that organize a specific cytoskeletal and signaling complex associated with each class of glutamate receptor. The network of protein-protein interactions mediated by these abundant PDZ proteins is likely to contribute significantly to the molecular scaffold of the postsynaptic density.

In recent years it has become apparent that membrane ion channels and receptors do not diffuse freely by themselves on the neuronal cell surface, but rather are localized at specific subcellular sites and associated with specific intracellular proteins.[1] Several potential functions of interactions between intracellular proteins and receptors and ion channels have emerged, and can be grouped as follows: (1) aggregation or clustering of the receptor/channel; (2) targeting of the receptor/channel to specific membrane domains; (3) coupling of the receptor/channel to intracellular signaling pathways; and (4) immobilization of the receptor/channel by anchoring to the cortical cytoskeleton. A single class of receptor/channel binding protein can be involved in all four activities, as exemplified by the PDZ domain containing proteins that interact with the NMDA and AMPA subclasses of ionotropic glutamate receptors.

INTERACTION BETWEEN NMDA RECEPTORS AND PSD-95/SAP90

The interaction between NMDA receptors and PSD-95 (also known as SAP90) was the first identified case of specific binding between the cytoplasmic C-terminal tail of an ionotropic receptor and a PDZ domain containing protein.[2,3] Concomitantly, an analogous interaction was described between the C-terminus of Shaker-type K+ channels and PSD-95/SAP90.[4] The C-terminal sequences of NMDA receptor NR2 subunits (last four amino acids -ESDV) and of Shaker-type K+ channels (-ETDV) are almost identical, and both C-termini bind to the same domains of PSD-95 (namely, PDZ1 and PDZ2). These findings were influential because they revealed for the first time that PDZ domains were modular

[a]Corresponding author: Morgan Sheng, HHMI (Wellman 423), Massachusetts General Hospital, 50 Blossom Street, Boston, MA 02114. Phone: 617-724-2800; fax: 617-724-2805; e-mail: sheng@helix.mgh.harvard.edu

protein interaction domains that bound to specific short peptide sequences at the very C-termini of interacting proteins.[2–5] Soon afterwards, the interaction between a PDZ domain from PSD-95 and its cognate C-terminal peptide was visualized crystallographically at 2-angstrom resolution.[6]

PSD-95 was originally purified from the postsynaptic density (PSD) fraction of brain,[7] and the protein is exquisitely localized to synaptic sites, predominantly though not exclusively on the postsynaptic side.[2,4,8–12] Evidence for an *in vivo* association between PSD-95 and NMDA receptors came from double-label colocalization studies in neurons[2,9,10,13] and from coimmunoprecipitation of NMDA receptors and PSD-95 family proteins from brain extracts.[9,14] To complicate matters somewhat, PSD-95 is merely the prototype of a family of proteins that to date contains four members in mammals. Except for one protein, SAP97,[15] the other members of the PSD-95 family are predominantly postsynaptic and seem to behave *in vitro* identically to PSD-95 itself.[14,16,17]

CLUSTERING FUNCTION OF PSD-95 FAMILY PROTEINS

A remarkable property of PSD-95 is its ability to induce the clustering of its binding partners (such as Shaker K+ channels and NMDA receptors) when coexpressed in heterologous cells.[4,17,18] This behavior is reminiscent of the ability of rapsyn to cluster nicotinic acetylcholine receptors in cultured nonmuscle cells.[19] PSD-95, however, does not form clusters on its own as rapsyn does; instead, the coexpression of the membrane receptor/channel and PSD-95 induces the coclustering of both proteins. The aggregating activity of PSD-95 suggests that it may play a role in concentrating its binding partners in a microdomain of the membrane. The coclustering phenomenon is dependent on the interaction between PSD-95 and the C-terminal PDZ-binding motif of the ion channels. PSD-95 or its relative chapsyn-110 can cluster NR2 subunits in COS-7 cells in the absence of NR1.[20] They can also cluster NR1 when NR2 subunits are coexpressed, but they cannot cluster NR1 directly since the common splice variants of NR1 do not interact with PSD-95. The mechanism of membrane protein clustering by PSD-95 depends on a head-to-head multimerization of PSD-95 that is mediated by the conserved N-terminal regions of PSD-95 family proteins.[18] A pair of cysteines in the N-terminal region of PSD-95 are also critical for its clustering activity. These cysteines appear to be the sites of palmitoylation and/or disulfide bond formation.[18,21]

SYNAPTIC TARGETING/LOCALIZATION FUNCTION OF PSD FAMILY PROTEINS

In *Drosophila*, there is a single known homolog of the PSD-95 family of proteins called Discs large (Dlg).[22] Dlg has the same domain organization as PSD-95, and its three PDZ domains have binding specificities that are indistinguishable *in vitro* from its mammalian counterparts. The Dlg protein is concentrated in *Drosophila* neuromuscular junctions (NMJs), which are glutamatergic synapses.[23,24] Moreover, they colocalize there with Shaker K+ channels and with glutamate receptors. Dlg mutant flies have abnormal NMJ morphology.[25] More significantly, Shaker K+ channels are no longer localized to the synapse in Dlg loss-of-function mutants.[23] In a complementary approach, it has been shown that the C-terminal tail of Shaker (which binds to Dlg *in vitro*) can confer upon a heterologous protein the ability to target to the neuromuscular synapse in wild-type but not in *dlg* mutant flies.[24] Together, these genetic studies in *Drosophila* provide strong evidence that the interaction between cytoplasmic tails and PSD-95 family proteins is important for the localization of specific ion channels at synaptic sites. The function of Dlg in glutamate receptor localization at the fly NMJ remains to be resolved; however, so far there appears not to be a direct binding between glutamate receptors and Dlg in *Drosophila* analogous to that between NMDA receptors and PSD-95 in mammals.

COUPLING OF NMDA RECEPTORS TO INTRACELLULAR SIGNALING PATHWAYS MEDIATED BY PSD-95

In addition to clustering and subcellular targeting of the receptor/channel, PSD-95 family proteins appear to play a role in coupling the membrane proteins to intracellular signaling pathways. This is a function that might be predicted from the multivalent domain organization of PSD-95 and its relatives. Since the discovery of PDZ-mediated binding of PSD-95 to NMDA receptors and Shaker K+ channels (which are integral membrane proteins), several intracellular proteins have also been identified that interact specifically with the PDZ domains of PSD-95 (FIG. 1). The neuronal nitric oxide synthase enzyme (nNOS) binds to PDZ2 of PSD-95. Interestingly, rather than via the typical C-terminus binding mechanism, this association occurs by a PDZ-PDZ interaction involving the single PDZ domain at the N-terminus of nNOS.[26] The interaction of nNOS with PSD-95 would presumably bring nNOS into the close vicinity of the NMDA receptor at postsynaptic sites.

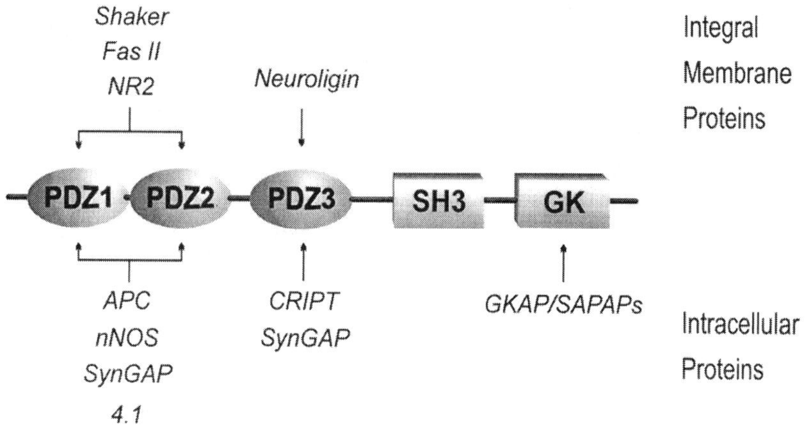

FIGURE 1. Summary of interactions with PSD-95 family proteins. Domain organization of the prototypical PSD-95 is shown with interacting membrane proteins above and intracellular proteins below their respective interaction domains. Because of the highly similar binding characteristics of PDZ1 and PDZ2, these domains are grouped into one functional unit. Not all of the interactions shown have been demonstrated to occur with the same PSD-95 family member, nor is it fully known which interactions can exist concurrently. See text for details.

Since nNOS is a Ca^{2+}-calmodulin–regulated enzyme, and NMDA receptors are Ca^{2+}-permeable channels, their physical approximation could neatly explain the specific coupling of NMDA receptor activation and nNOS stimulation that is observed in neuronal cells (reviewed in Refs. 27, 28). However, to confirm this attractive model, it will be important to show directly that PSD-95 can form a ternary complex with NMDA receptors and nNOS and mediate functional coupling between receptor and enzyme.

Perhaps a more surprising molecule that interacts with PSD-95 is the recently discovered SynGAP, a novel Ras GTPase activating protein (RasGAP) that is specifically and abundantly localized at postsynaptic sites in neurons.[29,30] SynGAP can bind to all three PDZ domains of PSD-95 via its C-terminus (last four amino acids: -QTRV), and its GAP activity can be inhibited by phosphorylation by Ca^{2+}/CaM–dependent protein kinase II, an abundant enzyme of the PSD. The presence of SynGAP in the PSD-95–associated protein complex suggests that Ras signaling pathways may operate at the postsynaptic density, perhaps in response to NMDA receptor activation and Ca^{2+} influx, or resulting from stimulation of postsynaptic receptor tyrosine kinases. SynGAP would be expected to inhibit Ras signaling by stimulating the GTPase activity of Ras. In this regard, it is interesting that the *Drosophila* PSD-95 homolog *dlg* was originally identified as a tumor suppressor gene, and was proposed to play a role in guanine nucleotide metabolism.[22] Another tumor suppressor gene product, adenomatosis polyposis coli (APC), has also been shown to bind to the PDZ domains of PSD-95 and to be localized at synapses.[31] In the Wnt signaling pathway, APC is believed to regulate the levels of armadillo/β-catenin,[32] but its function at synapses is unclear. Whatever the role of SynGAP and APC at the synapse, it is clear that we have only obtained the first glimpse of the complex signal transduction complex (or "transducisome") built on the PSD-95 scaffold at postsynaptic sites.

In addition to its three N-terminal PDZ domains, the PSD-95 family proteins also contain an SH3 domain and a guanylate kinase-like (GK) domain in their C-terminal region (FIG. 1). SH3 domains are well-known modular domains that bind to proline-rich peptides, and are frequently found in signaling proteins.[33] Binding partners for the SH3 domain of PSD-95 presumably exist but are presently unknown. Recent evidence also implicate GK domains as sites of protein-protein interaction. A family of abundant PSD proteins (termed GKAP, or SAPAPs) have been identified that bind to the GK domains of all members of the PSD-95 family.[34–36] Protein binding may well be the primary function of these GK domains, since they appear not to have catalytic activity.[37] The functions of GKAP are unknown, but it is likely that GKAP in turn binds to other synaptic proteins, thereby extending the size and complexity of the postsynaptic machinery attached to the cytoplasmic tail of NMDA receptors.

Recent studies have provided genetic evidence for the functional significance of the interaction between NMDA receptors and PSD-95 *in vivo*. Knockout mice deleted for the C-terminal cytoplasmic regions of individual NR2 subunits exhibited phenotypes resembling those of the null mutations of these genes, indicating the functional importance of the C-terminal region of NR2 proteins.[38] It is not clear, however, that the entire phenotype is attributable to a loss of interaction with PSD-95 and PSD-95–associated proteins. Only a few (perhaps as few as four) amino acids are required at the very C-terminus of NR2 subunits to bind to the PDZ domains of PSD-95, whereas hundreds of amino acids were deleted in the C-terminal truncations in the mutant mice. Interestingly, the synaptic localization of the C-terminal mutant NMDA receptors was inferred to be normal, based on electrophysiological and calcium imaging experiments,[38] but this interpretation needs to be reexamined more directly.

ORGANIZATION OF A HETEROGENEOUS MEMBRANE PROTEIN CLUSTER BY PSD-95

In the previous section, we emphasized the concept of a large signaling complex attached to the cytoplasmic tail of NMDA receptors. This model emerged from the identification of the intracellular signaling proteins that bind to PSD-95. Perhaps an equally important idea is that PSD-95 also organizes a functionally interconnected set of proteins in the plane of the membrane. Since the identification of NR2 subunits and Shaker K+ channels as ligands for PSD-95, other integral membrane proteins have been shown to interact with the PDZs of PSD-95 family proteins. These include other ion channels such as inward rectifying K+ channels,[39,40] a plasma membrane Ca^{2+} pump PMCA4b,[41] and cell adhesion molecules such as FasII and neuroligin.[24,42,43] All these membrane proteins have C-terminal sequences that mediate their binding to the various PDZ domains of PSD-95. Since PSD-95 can cluster Shaker channels and NMDA receptors, it should also be able to cluster these more recently discovered binding partners, and, more significantly, to organize heterogeneous clusters in which two or more membrane protein ligands are mixed together in the same cluster. This principle been directly confirmed for Shaker K+ channels and FasII.[42] In addition, genetic evidence in *Drosophila* indicates that an interaction with Dlg is important for the synaptic localization of FasII as well as for Shaker.[24,42]

The significance of a heterogeneous membrane protein cluster organized by a common protein such as PSD-95 can be viewed in different ways. First, such a mechanism allows the possibility of physically bringing together distinct membrane proteins with connected functions. The coclustering of NMDA receptors and the Ca^{2+} pump PMCA4b may provide an example of this. NMDA receptors allow influx of Ca^{2+}, and this has to be extruded from the postsynaptic compartment sooner or later by plasma membrane Ca^{2+} ATPases, some of which are localized in dendritic spines. The coincident binding of both NR2 and PMCA4b proteins to PSD-95 provides a simple mechanism for localizing them at the same subcellular site, namely, the postsynaptic specialization. Second, coclustering of two different membrane proteins by PSD-95 in theory allows one membrane protein to direct the localization of the other. For instance, Südhof and colleagues[43] have postulated that a postsynaptic concentration of neuroligin, a cell surface receptor for neurexins, could result from interactions with neurexins presented by the presynaptic terminal, and that this could be the primary event in positioning the PSD-95–associated complex. Because neuroligins bind preferentially to PDZ3 of PSD-95, the initial localization of neuroligin could then "recruit" PSD-95 and NMDA receptors (which bind to PDZ1/2) into the developing postsynaptic site. This conceptually attractive model remains to be tested experimentally. In conclusion, however, it is clear that many different membrane proteins structurally unrelated to NMDA receptors have evolved the appropriate C-terminal sequence to allow interaction with PSD-95 PDZ domains. Thus, multiple classes of membrane proteins probably use a common PSD-95–based mechanism to achieve localization and concentration at postsynaptic sites.

ANCHORING OF NMDA RECEPTORS TO CYTOSKELETON

The specific concentration of NMDA receptors at postsynaptic sites suggests that NMDA receptors are anchored to the cytoskeleton at the postsynaptic membrane. Consistent with tight cytoskeletal association is the fact that NMDA receptors are core components of the PSD, being extremely insoluble in all but strongly denaturing detergents such as SDS. How are NMDA receptors attached to the cytoskeleton? One mechanism is likely to involve PSD-95 family proteins, which are also insoluble components of the PSD. SAP97, a relative of PSD-95, has been shown to bind *in vitro* to protein 4.1, an actin-binding protein of the ERM family.[44,45] It remains to be determined if other members of the PSD-95 family can bind to ERM proteins, and if ERM proteins are present in the PSD. Filamentous actin is certainly abundant postsynaptically, especially in dendritic spines (Ref. 46 and references therein).

More recently, a postsynaptic protein called CRIPT has been identified that binds specifically to the third PDZ domain of PSD-95 family members.[12] When coexpressed with PSD-95 in heterologous cells, CRIPT causes the redistribution of PSD-95 to microtubules. By itself, CRIPT has a reorganizing effect on microtubule structure. These findings suggest that CRIPT might link the PSD-95–associated membrane complex to a tubulin-based cytoskeleton at postsynaptic sites. The model is especially engaging since CRIPT binds to PDZ3, whereas NMDA receptors bind to PDZ1/2. A problem with this hypothesis is that although tubulin is probably a major component of PSDs, microtubules are believed to be absent or rare in dendritic spines (where most excitatory synapses are situated). It is possible that CRIPT links PSD-95 to a tubulin cytoskeleton in the PSD that is distinct from conventional microtubules.

With linkages to the cytoskeleton in addition to direct associations with membrane proteins and intracellular signaling molecules, the myriad of interactions mediated by the PDZ domains of PSD-95 is getting increasingly complex. One might wonder how just three PDZ domains can accommodate this great variety of protein-protein interactions. It should be remembered, however, that PSD-95 probably exists as a multimer *in vivo*, and, indeed, these PSD-95 multimers may be further cross-linked into a lattice by interaction with proteins such as GKAP.[34] Thus the PSD-95 based transducisome need not necessarily be scaffolded on a PSD-95 monomer, but is more likely organized in a complex three-dimensional array around a scaffold of interlinked molecules of PSD-95 and related proteins.

Further complicating the intracellular interactions of NMDA receptors is the fact that only NR2 subunits directly associate with the PSD-95 complex. The essential NR1 subunit of NMDA receptors (at least in its common splice variations) does not interact with PSD-95, but instead binds to a distinct set of proteins (FIG. 2). A membrane proximal region found in all NR1 splice variants (termed C0) is the binding site for α-actinin, an actin cross-linking protein.[9,47] Such an interaction could provide an alternate connection to the cytoskeleton for NMDA receptors. Binding of α-actinin to NR1 is competitive with Ca^{2+}/calmodulin, raising the possibility that this mode of cytoskeletal attachment is regulated by activity of NMDA receptors themselves. Since Ca^{2+}/calmodulin inhibits the NMDA receptor channel,[48] α-actinin may play a role in the mechanism of Ca^{2+}-dependent inactivation of NMDA receptors. Consistent with an involvement of α-actinin in NMDA receptor regulation is the dependence of NMDA receptor function on the actin cytoskeleton.[49]

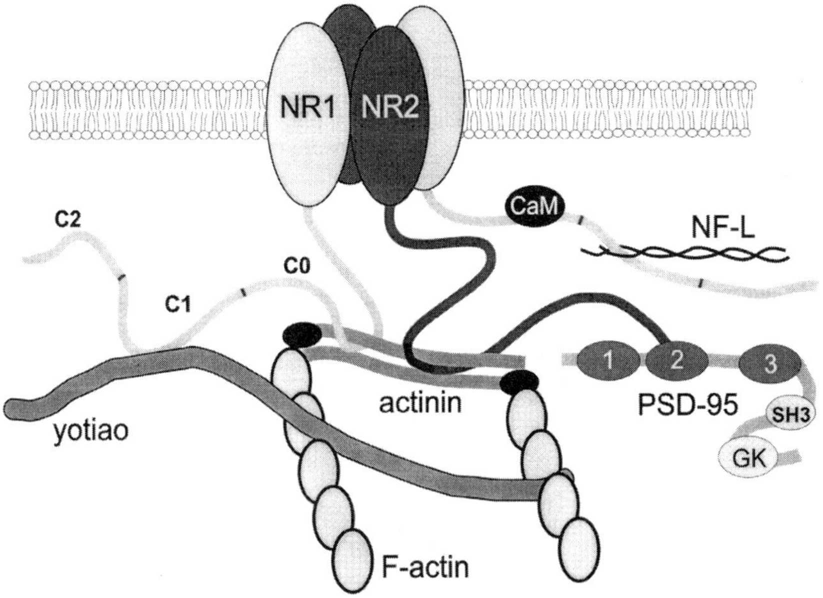

FIGURE 2. Multiple mechanisms for anchoring NMDA receptors. The NMDA receptor, here depicted as a heteromultimer of NR1 and NR2 subunits within the cell membrane, is anchored to various structural elements via its C-terminal tails. NR1 interacts through its invariant C0 exon to the actin-binding protein α-actinin, which is competitively inhibited by calmodulin (CaM) binding to the same C0 region. The alternatively spliced C1 exon of NR1 binds both to the putative cytoskeletal-associated protein yotiao and to the neurofilament subunit NF-L. NR2 subunits interact with PSD-95, as shown in FIGURE 1, and the NR2B subtype can also bind to α-actinin. PDZ domains of PSD-95 are labeled 1, 2, and 3.

Most NR1 variants contain an alternatively spliced exon-segment (known as C1), which harbors the major sites of phosphorylation by PKC.[50] The presence of the C1 segment, and its phosphorylation state, regulate the distribution of NR1 proteins expressed in heterologous cells, suggesting yet another possible interaction with cytoskeletal components.[51] Recent studies have identified two proteins that bind directly to the C1 exon-segment of NR1: NF-L (a neurofilament chain[52]) and yotiao (a novel coiled-coil protein named for Chinese fried dough strips[53]). Yotiao is not only enriched in PSD fractions in the brain, but is also highly concentrated at the postsynaptic side of the NMJ in muscle. Thus, yotiao may be a novel cytoskeletal protein of both neuromuscular and neuronal synapses. The functional significance of NF-L, yotiao, and α-actinin interactions with NMDA receptors remains to be examined *in vivo*.

INTERACTION OF AMPA RECEPTORS WITH GRIP

Notable absentees from the list of proteins that interact with PSD-95 have been the subunits of the AMPA receptors (GluR1 through GluR4), despite the fact that AMPA receptors are localized postsynaptically in glutamatergic synapses with NMDA receptors. Recently GluR2 and GluR3 subunits have been found to bind to a novel protein (termed GRIP) that has seven PDZ domains.[54,55] GluR2 and GluR3 have the same C-terminal sequence (-SVKI), and they bind to PDZ4 and PDZ5 of GRIP. Overexpression of an interfering GluR2 C-terminal peptide inhibited synaptic localization of AMPA receptors, suggesting that GRIP binding is important for synaptic targeting of AMPA receptors.[54] In other respects, the biochemical and functional characterization of GRIP lags behind PSD-95. Nevertheless, the identification of a distinct multivalent PDZ protein as the probable anchoring protein for AMPA receptors points to the division of glutamatergic postsynaptic sites into NMDA receptor and AMPA receptor microdomains. GRIP and PSD-95 will presumably organize distinct sets of proteins associated with the AMPA receptor and the NMDA receptor, respectively. This could mediate differential coupling of these receptors to intracellular signaling pathways and cytoskeletal components, and allow for differential regulation of the density of these receptors at the synapse.[55] The separation of AMPA and NMDA receptor–associated proteins may occur even at a macroscopic level, such that there is anatomical segregation of these different classes of glutamate receptor to distinct regions within the synapse.[56,57]

CONCLUSIONS

The PDZ domain–containing proteins (PSD-95 and GRIP) that bind to the cytoplasmic tails of glutamate receptors appear to play a central role in the molecular organization of the postsynaptic specialization. As multivalent scaffold proteins, PSD-95 and GRIP can cluster together the receptors with other membrane proteins, and couple the receptors to intracelluar signaling and cytoskeletal networks. Analogous PDZ-based signaling complexes are also being characterized at other types of cell junctions, such as epithelial cell junctions,[58,59] and in other specialized signal transduction organelles, such as the rhabdomeres of *Drosophila* photoreceptors.[60] These analogies emphasize that synapses are signaling devices specialized for neuronal communication, but utilizing the same fundamental molecular and cell biological principles as found in other systems.

ACKNOWLEDGMENT

M.S. is an Assistant Investigator of the Howard Hughes Medical Institute.

REFERENCES

1. Sheng, M. & E. Kim. 1996. Ion channel associated proteins. Curr. Opin. Neurobiol. **6:** 602–608.
2. Kornau, H.-C., L.T. Schenker, M.B. Kennedy & P.H. Seeburg. 1995. Domain interaction between NMDA receptor subunits and the postsynaptic density protein PSD-95. Science **269:** 1737–1740.
3. Niethammer, M., E. Kim & M. Sheng. 1996. Interaction between the C terminus of NMDA receptor subunits and multiple members of the PSD-95 family of membrane-associated guanylate kinases. J. Neurosci. **16:** 2157–2163.
4. Kim, E., M. Niethammer, A. Rothschild, Y.N. Jan & M. Sheng. 1995. Clustering of shaker-type K^+ channels by interaction with a family of membrane-associated guanylate kinases. Nature **378:** 85–88.

5. Sheng, M. 1996. PDZs and receptor/channel clustering: Rounding up the latest suspects. Neuron **17:** 575–578.
6. Doyle, D.A., A. Lee, J. Lewis, E. Kim, M. Sheng & R. MacKinnon. 1996. Crystal structures of a complexed and peptide-free membrane protein-binding domain: Molecular basis of peptide recognition by PDZ. Cell **85:** 1067–1076.
7. Cho, K.-O., C.A. Hunt & M.B. Kennedy. 1992. The rat brain postsynaptic density fraction contains a homolog of the drosophila discs-large tumor suppressor protein. Neuron **9:** 929–942.
8. Kistner, U., B.M. Wenzel, R.W. Veh, C. Cases-Langhoff, A.M. Garner, U. Appeltauer, B. Voss, E.D. Gundelfinger & C.C. Garner. 1993. SAP90, a rat presynaptic protein related to the product of the *Drosophila* tumor suppressor gene *dlg*-A. J. Biol. Chem. **268:** 4580–4583.
9. Wyszynski, M., J. Lin, A. Rao, E. Nigh, A.H. Beggs, A.M. Craig & M. Sheng. 1997. Competitive binding of α-actinin and calmodulin to the NMDA receptor. Nature **385:** 439–442.
10. Rao, A., E. Kim, M. Sheng & A.M. Craig. 1998. Heterogeneity in the molecular composition of excitatory postsynaptic sites during development of hippocampal neurons in culture. J. Neurosci. **18:** 1217–1229.
11. Hunt, C.A., L.J. Schenker & M.B. Kennedy. 1996. PSD-95 is associated with the postsynaptic density and not with the presynaptic membrane at forebrain synapses. J. Neurosci. **16:** 1380–1388.
12. Niethammer, M., J.G. Valtschanoff, T.M. Kapoor, D.W. Allison, R.J. Weinberg, A.M. Craig & M. Sheng. 1998. CRIPT, a novel postsynaptic protein that binds to the third PDZ domain of PSD-95/SAP90. Neuron **20:** 693–707.
13. Rao, A. & A.M. Craig. 1997. Activity regulates the synaptic localization of the NMDA receptor in hippocampal neurons. Neuron **19:** 801–812.
14. Müller, B.M., U. Kistner, S. Kindler, W.J. Chung, S. Kuhlendahl, L.F. Lau, R.W. Veh, R.L. Huganir, E.D. Gundelfinger & C.C. Garner. 1996. SAP102, a novel postsynaptic protein that interacts with the cytoplasmic tail of the NMDA receptor subunit NR2B. Neuron **17:** 255–265.
15. Müller, B.M., U. Kistner, R.W. Veh, C. Cases-Langhoff, B. Becker, E.D. Gundelfinger & C.C. Garner. 1995. Molecular characterization and spatial distribution of SAP97, a novel presynaptic protein homologous to SAP90 and the *Drosophila* discs-large tumor suppressor protein. J. Neurosci. **15:** 2354–2366.
16. Brenman, J.E., K.S. Christopherson, S.E. Craven, A.W. McGee & D.S. Bredt. 1996. Cloning and characterization of postsynaptic density 93, a nitric oxide synthase interacting protein. J. Neurosci. **16:** 7407–7415.
17. Kim, E. & M. Sheng. 1996. Differential K^+ channel clustering activity of PSD-95 and SAP97, two related membrane-associated putative guanylate kinases. Neuropharmacology. **35:** 993–1000.
18. Hsueh, Y.-P., E. Kim & M. Sheng. 1997. Disulfide-linked head-to-head multimerization in the mechanism of ion channel clustering by PSD-95. Neuron **18:** 803–814.
19. Phillips, W.D., C. Kopta, P. Blount, P.D. Gardner, J.H. Steinbach & J.P. Merlie. 1991. ACh receptor-rich membrane domains organized in fibroblasts by recombinant 43-kildalton protein. Science **251:** 568–570.
20. Kim, E., K.-O. Cho, A. Rothschild & M. Sheng. 1996. Heteromultimerization and NMDA receptor-clustering activity of chapsyn-110, a member of the PSD-95 family of proteins. Neuron **17:** 103–113.
21. Topinka, J.R. & D.S. Bredt. 1998. N-terminal palmitoylation of PSD-95 regulates association with cell membranes and interaction with K+ channel Kv1.4. Neuron **20:** 125–134.
22. Woods, D.F. & P.J. Bryant. 1991. The discs-large tumor suppressor gene of drosophila encodes a guanylate kinase homolog localized at septate junctions. Cell **66:** 451–464.
23. Tejedor, F.J., A. Bokhari, O. Rogero, M. Gorczyca, J. Zhang, E. Kim, M. Sheng & V. Budnik. 1997. Essential role for *dlg* in synaptic clustering of shaker K^+ channels *in vivo*. J. Neurosci. **17:** 152–159.
24. Zito, K., R.D. Fetter, C.S. Goodman & E.Y. Isacoff. 1997. Synaptic clustering of fasciclin II and shaker: Essential targeting sequences and role of dlg. Neuron **19:** 1007–1016.
25. Lahey, T., M. Gorczyca, X.-X. Jia & V. Budnik. 1994. The *Drosophila* tumor suppressor gene *dlg* is required for normal synaptic bouton structure. Neuron **13:** 823–835.

26. Brenman, J.E., D.S. Chao, S.H. Gee, A.W. McGee, S.E. Craven, D.R. Santillano, Z. Wu, F. Huang, H. Xia, M.F. Peters, S.C. Froehner & D.S. Bredt. 1996. Interaction of nitric oxide synthase with the postsynaptic density protein PSD-95 and α1-syntrophin mediated by PDZ domains. Cell **84:** 757–767.

27. Schuman, E.M. & D.V. Madison. 1994. Nitric oxide and synaptic function. Annu. Rev. Neurosci. **17:** 153–183.

28. Jaffrey, S.R. & S.H. Snyder. 1995. Nitric oxide: A neural messenger. Annu. Rev. Cell Dev. Biol. **11:** 417–440.

29. Kim, J.H., D. Liao, L.-F. Lau & R.L. Huganir. 1998. SynGAP: A synaptic RasGAP that associates with the PSD-95/SAP90 protein family. Neuron **20:** 683–691.

30. Chen, H.J., M. Rojas-Sota, A. Oguni & M.B. Kennedy. 1998. A synaptic RasGTPase activating protein inhibited by CaM Kinase II. Neuron **20:** 895–904.

31. Matsumine, A., A. Ogai, T. Senda, N. Okumura, K. Satoh, G.-H. Baeg, T. Kawahara, S. Kobayashi, M. Okada, K. Toyoshima & T. Akiyama. 1996. Binding of APC to the human homolog of the *Drosophila* discs large tumor suppressor protein. Science **272:** 1020–1023.

32. Willert, K. & R. Nusse. 1998. β-catenin: A key mediator of Wnt signaling. Curr. Opin. Genet. Dev. **8:** 95–102.

33. Pawson, T. 1995. Protein modules and signalling networks. Nature **373:** 573–580.

34. Kim, E., S. Naisbitt, Y.-P. Hsueh, A. Rao, A. Rothschild, A.M. Craig & M. Sheng. 1997. GKAP, a novel synaptic protein that interacts with the guanylate kinase-like domain of the PSD-95/SAP90 family of channel clustering molecules. J. Cell Biol. **136:** 669–678.

35. Naisbitt, S., E. Kim, R.J. Weinberg, A. Rao, F.-C. Yang, A.M. Craig & M. Sheng. 1997. Characterization of guanylate kinase-associated protein, a postsynaptic density protein at excitatory synapses that interacts directly with posysynaptic density-95/synapse-associated protein 90. J. Neurosci. **17:** 5687–5696.

36. Takeuchi, M., Y. Hata, K. Hirao, A. Toyoda, M. Irie & Y. Takai. 1997. SAPAPs. A family of PSD-95/SAP90-associated proteins localized at postsynaptic density. J. Biol. Chem. **272:** 11943–11951.

37. Kistner, U., C.C. Garner & M. Linial. 1995. Nucleotide binding by the synapse associated protein SAP90. FEBS Lett. **359:** 159–163.

38. Sprengel, R., B. Suchanek, C. Amico, R. Brusa, N. Burnasheve, A. Rozov, O. Hvalby, V. Jensen, O. Paulsen, P. Andersen, J.J. Kim, R.F. Thompson, W. Sun, L.C. Webster, S.G.N. Grant, J. Eilers, A. Konnerth, J. Li, J.O. McNamara & P.H. Seeburg. 1998. Importance of the intracellular domain of NR2 subunits for NMDA receptor function in vivo. Cell **92:** 279–289.

39. Cohen, N.A., J.E. Brenman, S. Snyder & D.S. Bredt. 1996. Binding of the inward rectifier K^+ channel Kir 2.3 to PSD-95 is regulated by protein kinase A phosphorylation. Neuron **17:** 759–767.

40. Horio, Y., H. Hibino, A. Inanobe, M. Yamada, M. Ishii, Y. Tada, E. Satoh, Y. Hata, Y. Takai & Y. Kurachi. 1997. Clustering and enhanced activity of an inwardly rectifying potassium channel, Kir4.1, by an anchoring protein, PSD-95/SAP90. J. Biol. Chem. **272:** 12885–12888.

41. Kim, E., S.J. DeMarco, S.M. Marfatia, A.H. Chishti, M. Sheng & E.E. Strehler. 1998. Plasma membrane Ca2+ ATPase isoform 4b binds to membrane-associated guanylate kinase (MAGUK) proteins via their PDZ (PSD-95/Dlg/ZO-1) domains. J. Biol. Chem. **273:** 1591–1595.

42. Thomas, U., E. Kim, S. Kuhlendahl, Y. Ho Koh, E.D. Gundelfinger, M. Sheng, C.C. Garner & V. Budnik. 1997. Synaptic clustering of the cell adhesion molecule fasciclin II by discs-large and its role in the regulation of presynaptic structure. Neuron **19:** 787–799.

43. Irie, M., Y. Hata, M. Takeuchi, K. Ichtchenko, A. Toyoda, K. Hirao, Y. Takai, T.W. Rosahl & T.C. Südhof. 1997. Binding of neuroligins to PSD-95. Science **277:** 1511–1515.

44. Lue, R.A., E. Brandin, E.P. Chan & D. Branton. 1996. Two independent domains of hDlg are sufficient for subcellular targeting: The PDZ1-2 conformational unit and an alternatively spliced domain. J. Cell Biol. **135:** 1125–1137.

45. Marfatia, S.M., J.H.M. Cabral, L. Lin, C. Hough, P.J. Bryant, L. Stolz & A.H. Chishti. 1996. Modular organization of the PDZ domains in the human discs-large protein suggests a mechanism for coupling PDZ domain-binding proteins to ATP and the membrane cytoskeleton. J. Cell Biol. **135:** 753–766.

46. KAECH, S., M. FISCHER, T. DOLL & A. MATUS. 1997. Isoform specificity in the relationship of actin to dendritic spines. J. Neurosci. **17:** 9565–9572.
47. WYSZYNSKI, M., V. KHARAZIA, R. SHANGHVI, A. RAO, A.H. BEGGS, A.M. CRAIG, R. WEINBERG & M. SHENG. 1998. Differential regional expression and ultrastructural localization of α-actinin-2, a putative NMDA receptor-anchoring protein, in rat brain. J. Neurosci. **18:** 1383–1392.
48. EHLERS, M.D., S. ZHANG, J.P. BERNHARDT & R.L. HUGANIR. 1996. Inactivation of NMDA receptors by direct interaction of calmodulin with the NR1 subunit. Cell **84:** 745–755.
49. ROSENMUND, C. & G.L. WESTBROOK. 1993. Calcium-induced actin depolymerization reduces NMDA channel activity. Neuron **10:** 805–814.
50. TINGLEY, W.G., K.W. ROCHE, A.K. THOMPSON & R.L. HUGANIR. 1993. Regulation of NMDA receptor phosphorylation by alternative splicing of the C-terminal domain. Nature **364:** 70–73.
51. EHLERS, M.D., W.G. TINGLEY & R.L. HUGANIR. 1995. Regulated subcellular distribution of the NR1 subunit of the NMDA receptor. Science **269:** 1734–1737.
52. EHLERS, M.D., E.T. FUNG, R.J. O'BRIEN & R.L. HUGANIR. 1998. Splice variant-specific interaction of the NMDA receptor subunit NR1 with neuronal intermediate filaments. J. Neurosci. **18:** 720–730.
53. LIN, J.W., M. WYSZYNSKI, R. MADHAVAN, R. SEALOCK, J.U. KIM & M. SHENG. 1998. Yotiao, a novel protein of neuromuscular junction and brain that interacts with specific splice variants of NMDA receptor subunit NR1. J. Neurosci. **18:** 2017–2027.
54. DONG, H., R.J. O'BRIEN, E.T. FUNG, A.A. LANAHAN, P.F. WORLEY & R.L. HUGANIR. 1997. GRIP: A synaptic PDZ domain-containing protein that interacts with AMPA receptors. Nature **386:** 279–284.
55. SHENG, M. 1997. Glutamate receptors put in their place. Nature **386:** 221–223.
56. KHARAZIA, V.N., K.D. PHEND, A. RUSTIONI & R.J. WEINBERG. 1996. EM colocalization of AMPA and NMDA receptor subunits at synapses in rat cerebral cortex. Neurosci. Lett. **210:** 37–40.
57. BERNARD, V., P. SOMOGYI & J.P. BOLAM. 1997. Cellular, subcellular, and subsynaptic distribution of AMPA-type glutamate receptor subunits in the neostriatum of the rat. J. Neurosci. **17:** 819–833.
58. KIM, S.K. 1995. Tight junctions, membrane-associated guanylate kinases and cell signaling. Curr. Opin. Cell Biol. **7:** 641–649.
59. ANDERSON, J.M., M.S. BALDA & A.S. FANNING. 1993. The structure and regulation of tight junctions. Curr. Opin. Cell Biol. **5:** 772–778.
60. TSUNODA, S., J. SIERRALTA, Y. SUN, R. BODNER, E. SUZUKI, A. BECKER, M. SOCOLICH & C.S. ZUKER. 1997. A multivalent PDZ-domain protein assembles signalling complexes in a G-protein-coupled cascade. Nature **388:** 243–249.

Mice with Genetically Modified NMDA and AMPA Receptors

ROLF SPRENGEL[a] AND FRANK N. SINGLE

Department of Molecular Neurobiology, Max-Planck-Institute for Medical Research, Jahnstrasse 29, 69120 Heidelberg, Germany

ABSTRACT: This manuscript summarizes mouse mutants for ionotropic glutamate receptors that were generated by different laboratories to analyze the function of the NMDA and AMPA receptors in the mouse. Thus, NMDA receptor mutant mice that were generated by the "knock-in" technology demonstrate that the NR1 and the NR2B subunits participate in the formation of NMDA receptors that are involved in vital functions like breathing and suckling of a newborn mouse. Mice that lack NR2A,-2C, and -2D subunits were described to be viable and have been used to study the role of NMDA receptors in adult mice. The depletion of the GluR-B subunit revealed an NMDA receptor–independent form of long-term potentiation (LTP). This AMPA receptor–mediated LTP at CA3/CA1 synapses was also observed in mice that carry an editing-deficient GluR-B allele even though these mice die prematurely after heavy epileptic seizures. In other mutants, the intracellular COOH-terminal domain of the NMDA receptor was truncated; and when compared to NMDA receptor "knock-out" mice, a functional knock-out of the NMDA receptor was observed. However, in the synapses of $NR2^{\Delta C/\Delta C}$ mutants, gatable NMDA receptors were synaptically activated, indicating that the knock-out phenotypes mediated by the COOH-terminally truncated NMDA receptors appear to reflect defective intracellular signaling.

In the vertebrate brain, rapid excitatory neurotransmission is mediated by ionotropic glutamate receptors of the NMDA and non-NMDA (AMPA/kainate) receptor families.[1,2] The AMPA receptors contribute the fast component of excitatory postsynaptic currents, whereas the slow component is mediated by NMDA receptor channels,[3] which might basically work as coincidence detectors of pre- and postsynaptic activity, since the gating of the integral ion channel requires simultaneous presynaptic glutamate release and depolarization of the postsynaptic membrane.[2–4] Depolarization occurs primarily by the activation of synaptically colocalized AMPA receptors. Coincidence detection by the NMDA receptor rests on its voltage-dependent channel block by extracellular Mg^{2+}. This block regulates the Ca^{2+} influx through the highly Ca^{2+}-permeable ion channel. Controlled Ca^{2+} influx through this channel is thought to be essential for activity-dependent synaptic modulation (reviewed in Ref. 4). Furthermore, excessive Ca^{2+} influx through NMDA receptors has pathophysiological consequences, including epileptiform activities and neurodegeneration.[5] By contrast, AMPA receptors in most central neurons display low divalent to monovalent ion permeability ratios (reviewed in Ref. 6).

Both types of glutamate receptors are presumably tetrameric ion channels,[7] very likely composed of at least two different subunits. All subunits of the ionotropic glutamate

[a]Corresponding author: Dr. Rolf Sprengel, Max Planck Institut für medizinische Forschung, Jahnstrasse 29, D-69120 Heidelberg, Germany. Phone: 49-6221-486101; fax: 49-6221-486110; e-mail: sprengel@mpimf-heidelberg.mpg.de

receptors are sequence related. They contain, in brief, an NH_2-terminal signal peptide, followed by a large extracellular domain, four hydrophobic regions (M1 to M4) and an intracellular COOH-terminal domain. The hydrophobic region M2 seems to form a hinge region in the channel pore; COOH-terminal amino acid sequences were shown to mediate interactions with intracellular proteins (for review see Refs. 8, 9).

For the formation of most NMDA receptor channels two distantly sequence-related subunits, the NR1 subunit[10] and one of the four NR2 subunits (NR2A to 2D)[11–15] are necessary. The NR2 subunits impart distinct gating and ion conductance properties on the respective NR1/NR2 receptor subtypes.[16,17] NR1 is in the vast majority of central neurons throughout all developmental stages, whereas the NR2 subunits are expressed in distinct spatial and temporal patterns. Prenatal NMDARs contain NR2B or NR2D, whereas NR2A and NR2C are expressed only after birth, the former predominantly in the forebrain and the latter mainly in cerebellar granule cells.[18–20]

For the AMPA receptor channels four subunits with selective expression patterns and different splice forms (GluR-A, -B, -C, and -D) have been described.[21–23] In principal neurons the AMPA receptors are described as being impermeable for Ca^{2+} ions, whereas in many interneurons, such as dentate gyrus basket cells, Ca^{2+}-permeable AMPA receptors can be detected.[24] The low permeability for divalent Ca^{2+} ions is achieved by a surplus of the AMPAR subunit GluR-B over other AMPAR subunits. GluR-B confers low Ca^{2+} permeability on heteromeric AMPA receptors because it carries an arginine (R) residue in its pore-forming segment M2,[25,26] in a position occupied by glutamine (Q) in the other AMPAR subunits. The critical arginine residue in the Q/R site of GluR-B is not encoded on the gene, but an arginine codon (CIG) is created at the pre-mRNA stage by adenosine deamination within the exonic glutamine codon CAG[27] with participation of intronically localized sequences ("ECS element")[28] by a mechanism termed *RNA editing*.

MOUSE MUTANTS THAT LACK NMDA RECEPTOR SUBUNITS

NR1 Subunit (ζ1)

The knock-out technology was used to investigate the biological importance of different NMDA receptor subtypes. Mouse mutants constructed to lack the NR1 subunit, which participates in all NMDA receptor subtypes, died 8–15 h after birth, indicating a vital neonatal function for the NMDA receptor. Although the NMDA receptor has been implicated in several aspects of neurodevelopment, overall neuroanatomy of $NR1^{-/-}$ mice appeared normal. Pathological evidence suggested that respiratory failure was the ultimate cause of death.[29] Li and colleagues[30] used the $NR1^{-/-}$ mice to demonstrate that whisker-related patterns in the trigeminal nuclei of the mutant mice were impaired, compared to their wild-type littermates, and showed that in the $NR1^{-/-}$ mice pathfinding, initial targeting, and crude topographic projection of trigeminal axons in the brainstem were unaffected, but that whisker-specific patches failed to form. Their results provided further evidence of the involvement of the NMDA receptor in the formation of sensory-related neural patterns in the mammalian brain, and suggested that the formation of this pattern is activity dependent. In a follow-up study[31] ectopic expression of a transgene of an NR1 splice variant rescued neonatally fatal $NR1^{-/-}$ mice, although the average life span was dependent on the level of the transgene expression. If the NR1 transgene was expressed at high levels, the sensory periphery-related patterns were normal along both the trigeminal and dorsal column pathways; if expression was low, the patterns were absent in the trigeminal pathway, again indicating that NMDA receptor–mediated neural activity plays a critical role in pattern formation along the ascending somatosensory pathways.

A similar approach—but now using the Cre-recombination system[32]—was described to create mice in which the deletion of the NR1 gene was restricted to hippocampal CA1 region in the mouse brain.[33,34] These mice developed normally but were impaired in the hippocampal representation of space, which was the first direct evidence that NMDA receptor–mediated synaptic plasticity is necessary for the proper representation of space in the CA1 region of the hippocampus.[35]

NR2B Subunit [ε1]

Perinatal lethality was also found for NR2B$^{-/-}$ mutants,[36] arguing that during CNS development, the NR1/NR2B subtype is the important NMDA receptor subtype. In NR2B$^{-/-}$ mice, breathing was normal; but the suckling response was missing, and NR2B$^{-/-}$ mice starved to death soon after birth. As in the NR1$^{-/-}$ mice, the formation of the whisker-related neuronal barrelette structure and the clustering of primary sensory afferent terminals in the brainstem trigeminal nucleus were disturbed. In addition, in the hippocampus of the mutant mice, synaptic NMDA responses and long-term depression (LTD) were abolished. These results suggest that the NR2B subunit is essential for both neuronal pattern formation and synaptic plasticity.

Mice lacking the NR2A, NR2C, and NR2D subtypes were found to be viable and were used for studies on NMDA receptor function in adult mice (see below).

NR2A Subunit [ε2]

The targeted disruption of the NR2A subunit gene resulted in significant reduction of the NMDA receptor channel current and LTP at the hippocampal CA1 synapses. The NR2A$^{-/-}$ mice also showed a moderate deficiency in spatial learning, thus supporting the notion obtained from the NR1 mutants (see above) that the NMDA receptor–dependent synaptic plasticity at synapses between CA3 and CA1 pyramidal cells of the hippocampus might be the cellular basis of certain forms of learning.[37,38]

Ito used in his study NR2A$^{-/-}$ and NR2B$^{-/-}$ mice to show that NMDA receptors with different subunit compositions function within a single hippocampal CA3 pyramidal cell in a synapse-selective manner. Thus the depletion of the NR2A subunit selectively reduced NMDA receptor excitatory postsynaptic currents (EPSCs) and LTP in the commissural/associational CA3 synapse without significantly affecting stimulations at the fimbrial-CA3 synapse; whereas the NR2B depletion diminished NMDA EPSCs and LTP in the fimbrial/CA3 synapse with no appreciable functional modifications of the commissural/associational CA3 synapse.[39]

NR2C Subunit [ε3]

Gene-manipulated mice lacking the NMDA receptor subunit NR2C, which is restricted mostly to granular cells of the cerebellum, showed that the presence of NR2C reduces NMDA receptor–mediated currents in mossy fiber/granule cell synapses and might increase the non-NMDA component of the synaptic current.[40] Despite these changes, the $NR2C^{-/-}$ mice were normal in motor coordination tests and did not show any measurable phenotype.[40–42] However, mice with motor coordination deficits were generated when $NR2A^{-/-}$ and $NR2C^{-/-}$ mice were crossed to produce $NR2A^{-/-}/NR2C^{-/-}$ genotypes. These mutants express no NMDA receptors in the adult cerebellum, and NMDA receptor–mediated components of EPSCs in granule cells were abolished. Phenotypically the mice managed simple coordinated tasks, such as staying on a stationary or slowly rotating rod, but failed more challenging tasks, such as staying on a quickly rotating rod, demonstrating that the NMDA receptors play an active role in motor coordination.[41]

MOUSE MUTANTS THAT LACK AMPA RECEPTOR SUBUNITS

For the AMPA receptor, GluR-B is the only subunit the expression of which is abolished up to now.

The GluR-B Subunit [GluR-2]

The $GluR\text{-}B^{-/-}$ mice exhibited increased mortality, and those survivors showed reduced exploration and impaired motor coordination. Acute brain slice preparations of these mice showed a twofold increase in LTP at CA3/CA1 synapses, demonstrating that the high increase in Ca^{2+} permeability of AMPA receptors at these synapses can induce plasticity changes similar to plasticity changes mediated by NMDA receptors.[43]

MOUSE MUTANTS THAT EXPRESS MODIFIED GLUTAMATE RECEPTORS

NMDA Receptors

Site-selective recombination in mouse embryonic stem cells can be used to abolish the expression of genes and can be used to manipulate exonic sequences to alter single amino acids or whole protein domains in a gene product. We have tried several approaches to genetically manipulate different subunits of the family of the ionotropic glutamate receptors.

Point Mutations in the NR1 Subunit

A single asparagine residue (N598) in the M2 segment of the NR1 subunit, which determines the Ca^{2+} permeability of the NMDA receptor, was changed to a glutamine.[44] Electrophysiological recordings from nucleated patches of hippocampal CA1 pyramidal cells in acute brain slices from $NR1^{Q/Q}$ mice, which carry the point mutation in both alleles, revealed the expected reduction in Ca^{2+} permeability and voltage-dependent Mg^{2+} blockade. This change had a severe effect on the function of the NMDA receptor, and $NR1^{Q/Q}$ mice die at P_0 like the $NR1^{-/-}$ mice, demonstrating that Ca^{2+} permeability is an essential function of the NMDA receptor.

NR2 Subunits with Truncated COOH Terminus

By replacing the last translated exons of the mouse NR2A, -2B, and -2C genes by gene targeting, we generated mice that express NMDA receptors without the large intracellular COOH-terminal domain of any one of three NR2 subunits.[42] Phenotypically these $NR2^{\Delta C/\Delta C}$ mice resemble those mice that are deficient in that particular subunit (see above). Thus, mice expressing the NR2B subunit in a COOH-terminally truncated form ($NR2B^{\Delta C/\Delta C}$ mice) die perinatally, and $NR2A^{\Delta C/\Delta C}$ mice are viable but exhibit impaired synaptic plasticity and contextual memory, indicating that the deletion of the COOH terminus induced a functional knock-out of the receptor. This functional NMDA receptor knock-out was not due to a lack of functional NMDA receptor channels but appeared to reflect defective intracellular signaling of activated NMDA receptors. The study with these mice clearly demonstrated that the intracellularly located COOH termini of the NR2 subunits are indispensable for the physiological functionality of the respective NMDA receptor subtypes.[42] The controlled gating of the integral Ca^{2+} ion–preferring channel appears to be only one facet of the elaborate functional design of this receptor. A critical facet is contributed by intracellular receptor domains, which may mediate interaction of the ion channel with components transducing the synaptically evoked Ca^{2+} signal.

AMPA RECEPTORS

Mutations in the GluR-B Subunit Gene

As mentioned in the introduction, within the heteromeric assembly of AMPA receptors the GluR-B subunit determines the low permeability for Ca^{2+} ions of AMPA receptors. A single arginine residue (R586) of the M2 segment was identified as critical residue. This arginine residue is not gene encoded, but its codon is generated at the level of the GluR-B precursor RNA by site-selective deamination of an adenosine of an exonic CAG glutamine codon (for review see Ref. 45). In genomes of mice we deleted intronic sequences that are necessary for the deamination of the GluR-B pre-mRNA at codon position Q586.[46] Heterozygous mice that carried this editing-incompetent GluR-B allele (GluR-$B^{\Delta ECS}$) synthesized GluR-B(Q586) subunits and therefore carried in principal neurons AMPA receptors with increased Ca^{2+} permeability. Like the GluR-$B^{-/-}$ mice, the GluR-$B^{\Delta ECS}$ mice show an NMDA receptor–independent form of LTP in hippocampal Schaffer collateral-CA1 pyramidal cell connections.[47] In contrast to the GluR-$B^{-/-}$ mice, the GluR-$B^{\Delta ECS}$ mice developed a severe phenotype with epileptic seizures and died by three weeks of age. As far as we know, this difference is best explained by an altered macroscopic conductance of the AMPA receptors on pyramidal cells of both mutants, with low conductance for the GluR-$B^{-/-}$ and higher conductance for the GluR-$B^{\Delta ECS}$ mutants,[48] which still express a substantial amount of the GluR-B(Q) subunit. Nevertheless the GluR-$B^{-/-}$ and the GluR-$B^{\Delta ECS}$

mice both demonstrate the importance of the AMPA receptors with low Ca^{2+} permeability in the pyramidal cells of the hippocampus.

CONCLUDING REMARKS

Gene targeting technology has been used successfully by several groups to delete and to manipulate the genes for several members of the AMPA and NMDA receptor family. In this way, a crude picture of the functional importance of these subunits has emerged, and some studies have helped redefine our thinking about the interactions of these ion channels. However, a conclusive correlation between receptor function and phenotype of the mouse cannot be drawn from these types of gene-manipulated mice. First, in most of the cases described, the introduced mutation was expressed in too many areas of the brain; and second, the mutation could develop its phenotype during development, leading to alterations in the mature nervous system. Therefore, a phenotype can result from these alterations during development or might be caused by synergistic activities of different brain regions where the mutation is expressed. Thus, our future studies will make use of inducible gene switches[32,49] to create defined changes of ionotropic glutamate receptors in specific brain regions. This should help in discriminating between developmental deficits in brain function and the acute effects of the mutated subunit.

ACKNOWLEDGMENTS

Many thanks to Rossella Brusa, Carla Amico, Bettina Suchanek, Frank Zimmermann, Jasna Jerecic, Nail Burnashev, Oivind Hvalby, Andrei Rozov, Daniel Zamanillo, Miyoko Higuchi, and Annette Herold.

This work was done in the laboratories of P. H. Seeburg and funded, in part, by grants to P. H. Seeburg from HFSP, the Volkswagenstiftung, and the German Chemical Society, and an unrestricted grant from Bristol-Meyers Squibb.

REFERENCES

1. Collingridge, G.L. & R.A. Lester. 1989. Excitatory amino acid receptors in the vertebrate central nervous system. Pharmacol. Rev. **41:** 143–210.
2. Mayer, M.L. & G.L. Westbrook. 1987. The physiology of excitatory amino acids in the vertebrate central nervous system. Prog. Neurobiol. **28:** 197–276.
3. Stern, P., P. Behe, R. Schoepfer & D. Colquhoun. 1992. Single-channel conductances of NMDA receptors expressed from cloned cDNAs: Comparison with native receptors. Proc. R. Soc. Lond. Biol. **250:** 271–277.
4. Bliss, T.V. & G.L. Collingridge. 1993. A synaptic model of memory: Long-term potentiation in the hippocampus. Nature **361:** 31–39.
5. Choi, D.W. & S.M. Rothman. 1990. The role of glutamate neurotoxicity in hypoxic-ischemic neuronal death. Annu. Rev. Neurosci. **13:** 171–182.
6. Jonas, P. & N. Burnashev. 1995. Molecular mechanisms controlling calcium entry through AMPA-type glutamate receptor channels. Neuron **15:** 987–990.
7. Rosenmund, C., B.Y. Stern & C.F. Stevens. 1998. The tetrameric structure of a glutamate receptor channel. Science **280:** 1596–1599.
8. Kornau, H.C., P.H. Seeburg & M.B. Kennedy. 1997. Interaction of ion channels and receptors with PDZ domain proteins. Curr. Opin. Neurobiol. **7:** 368–373.

9. TSUNODA, S. *et al.* 1997. A multivalent PDZ-domain protein assembles signalling complexes in a G-protein-coupled cascade. Nature **388:** 243–249.
10. MORIYOSHI, K. *et al.* 1991. Molecular cloning and characterization of the rat NMDA receptor. Nature **354:** 31–37.
11. MONYER, H. *et al.* 1992. Heteromeric NMDA receptors: Molecular and functional distinction of subtypes. Science **256:** 1217–1221.
12. MEGURO, H. *et al.* 1992. Functional characterization of a heteromeric NMDA receptor channel expressed from cloned cDNAs. Nature **357:** 70–74.
13. KUTSUWADA, T. *et al.* 1992. Molecular diversity of the NMDA receptor channel. Nature **358:** 36–41.
14. IKEDA, K. *et al.* 1992. Cloning and expression of the epsilon 4 subunit of the NMDA receptor channel. FEBS Lett. **313:** 34–38.
15. ISHII, T. *et al.* 1993. Molecular characterization of the family of the *N*-methyl-D-aspartate receptor subunits. J. Biol. Chem. **268:** 2836–2843.
16. HOLLMANN, M. & S. HEINEMANN. 1994. Cloned glutamate receptors. Annu. Rev. Neurosci. **17:** 31–108.
17. NAKANISHI, S. & M. MASU. 1994. Molecular diversity and functions of glutamate receptors. Annu. Rev. Biophys. Biomol. Struct. **23:** 319–348.
18. WATANABE, M., Y. INOUE, K. SAKIMURA & M. MISHINA. 1993. Distinct distributions of five *N*-methyl-D-aspartate receptor channel subunit messenger RNAs in the forebrain. J. Comp. Neurol. **338:** 377–390.
19. AKAZAWA, C., R. SHIGEMOTO, Y. BESSHO, S. NAKANISHI & N. MIZUNO. 1994. Differential expression of five *N*-methyl-D-aspartate receptor subunit mRNAs in the cerebellum of developing and adult rats. J. Comp. Neurol. **347:** 150–160.
20. MONYER, H., N. BURNASHEV, D.J. LAURIE, B. SAKMANN & P.H. SEEBURG. 1994. Developmental and regional expression in the rat brain and functional properties of four NMDA receptors. Neuron **12:** 529–540.
21. KEINÄNEN, K. *et al.* 1990. A family of AMPA-selective glutamate receptors. Science **249:** 556–560.
22. HOLLMANN, M., A. O'SHEA-GREENFIELD, S.W. ROGERS & S. HEINEMANN. 1989. Cloning by functional expression of a member of the glutamate receptor family. Nature **342:** 643–648.
23. SOMMER, B. *et al.* 1990. Flip and flop: A cell-specific functional switch in glutamate-operated channels of the CNS. Science **249:** 1580–1585.
24. GEIGER, J.R. *et al.* 1995. Relative abundance of subunit mRNAs determines gating and Ca^{2+} permeability of AMPA receptors in principal neurons and interneurons in rat CNS. Neuron **15:** 193–204.
25. HUME, R.I., R. DINGLEDINE & S.F. HEINEMANN. 1991. Identification of a site in glutamate receptor subunits that controls calcium permeability. Science **253:** 1028–1031.
26. BURNASHEV, N., H. MONYER, P.H. SEEBURG & B. SAKMANN. 1992. Divalent ion permeability of AMPA receptor channels is dominated by the edited form of a single subunit. Neuron **8:** 189–198.
27. SOMMER, B., M. KÖHLER, R. SPRENGEL & P.H. SEEBURG. 1991. RNA editing in brain controls a determinant of ion flow in glutamate-gated channels. Cell **67:** 11–19.
28. HIGUCHI, M. *et al.* 1993. RNA editing of AMPA receptor subunit GluR-B: A base-paired intron-exon structure determines position and efficiency. Cell **75:** 1361–1370.
29. FORREST, D. *et al.* 1994. Targeted disruption of NMDA receptor 1 gene abolishes NMDA response and results in neonatal death. Neuron **13:** 325–338.
30. LI, Y., R.S. ERZURUMLU, C. CHEN, S. JHAVERI & S. TONEGAWA. 1994. Whisker-related neuronal patterns fail to develop in the trigeminal brainstem nuclei of NMDAR1 knockout mice. Cell **76:** 427–437.
31. IWASATO, T. *et al.* 1997. NMDA receptor-dependent refinement of somatotopic maps. Neuron **19:** 1201–1210.
32. GU, H., J.D. MARTH, P.C. ORBAN, H. MOSSMANN & K. RAJEWSKY. 1994. Deletion of a DNA polymerase beta gene segment in T cells using cell type-specific gene targeting. Science **265:** 103–106.
33. TSIEN, J.Z., P.T. HUERTA & S. TONEGAWA. 1996. The essential role of hippocampal CA1 NMDA receptor-dependent synaptic plasticity in spatial memory. Cell **87:** 1327–1338.

34. TSIEN, J.Z. *et al.* 1996. Subregion- and cell type-restricted gene knockout in mouse brain. Cell **87:** 1317–1326.
35. MCHUGH, T.J., K.I. BLUM, J.Z. TSIEN, S. TONEGAWA & M.A. WILSON. 1996. Impaired hippocampal representation of space in CA1-specific NMDAR1 knockout mice. Cell **87:** 1339–1349.
36. KUTSUWADA, T. *et al.* 1996. Impairment of suckling response, trigeminal neuronal pattern formation, and hippocampal LTD in NMDA receptor epsilon 2 subunit mutant mice. Neuron **16:** 333–344.
37. SAKIMURA, K. *et al.* 1995. Reduced hippocampal LTP and spatial learning in mice lacking NMDA receptor epsilon 1 subunit. Nature **373:** 151–155.
38. KIYAMA, Y. *et al.* 1998. Increased thresholds for long-term potentiation and contextual learning in mice lacking the NMDA-type glutamate receptor 1 subunit. J. Neurosci. **18:** 6723–6739.
39. ITO, I. *et al.* 1997. Synapse-selective impairment of NMDA receptor functions in mice lacking NMDA receptor epsilon 1 or epsilon 2 subunit. J. Physiol. (Lond.) **500**(2)**:** 401–408.
40. EBRALIDZE, A.K., D.J. ROSSI, S. TONEGAWA & N.T. SLATER. 1996. Modification of NMDA receptor channels and synaptic transmission by targeted disruption of the NR2C gene. J. Neurosci. **16:** 5014–5025.
41. KADOTANI, H. *et al.* 1996. Motor discoordination results from combined gene disruption of the NMDA receptor NR2A and NR2C subunits, but not from single disruption of the NR2A or NR2C subunit. J. Neurosci. **16:** 7859–7867.
42. SPRENGEL, R. *et al.* 1998. Importance of the intracellular domain of NR2 subunits for NMDA receptor function *in vivo*. Cell **92:** 279–289.
43. JIA, Z. *et al.* 1996. Enhanced LTP in mice deficient in the AMPA receptor GluR2. Neuron **17:** 945–956.
44. SINGLE, F.N. *et al.* 1997. Effects of targeted point mutations at the N-site of the NMDA receptor channel M2 domain in mice. Soc. Neurosci. Abstr. No. 451.5.
45. SEEBURG, P.H., M. HIGUCHI & R. SPRENGEL. 1998. RNA editing of brain glutamate receptor channels: Mechanism and physiology. Mol. Brain Res. **26:** 217–229.
46. BRUSA, R. *et al.* 1995. Early-onset epilepsy and postnatal lethality associated with an editing-deficient GluR-B allele in mice. Science **270:** 1677–1680.
47. FELDMEYER, D. *et al.* 1999. Neurological dysfunctions in mice expressing different levels of the Q/R site-unedited AMPA receptor subunit GluR-B. Nature Neurosci. In press.
48. SWANSON, G.T., D. FELDMEYER, M. KANEDA & S.G. CULL-CANDY. 1996. Effect of RNA editing and subunit co-assembly single-channel properties of recombinant kainate receptors. J. Physiol. (Lond.) **492:** 129–142.
49. GOSSEN, M., A.L. BONIN, S. FREUNDLIEB & H. BUJARD. 1994. Inducible gene expression systems for higher eukaryotic cells. Curr. Opin. Biotechnol. **5:** 516–520.

GluRδ2 and the Development and Death of Cerebellar Purkinje Neurons in Lurcher Mice

NATHANIEL HEINTZ[a] AND PHILIP L. DE JAGER

Laboratory of Molecular Biology, Howard Hughes Medical Institute, The Rockefeller University, 1230 York Avenue, New York, New York 10021, USA

ABSTRACT: Lurcher (*Lc*) is a spontaneous, semidominant mouse neurological mutation. Heterozygous lurcher mice (*Lc*/+) display ataxia due to a selective, cell-autonomous, apoptotic death of 90% of cerebellar Purkinje cells during postnatal development. Homozygous lurcher mice (*Lc*/*Lc*) die shortly after birth due to massive loss of mid- and hindbrain neurons during late embryogenesis. We identified the mutations responsible for neurodegeneration in two independent *Lc* alleles as identical G-to-A transitions that change a highly conserved alanine to a threonine residue in transmembrane domain III of the mouse δ2 glutamate receptor gene (GluRE2). *Lc*/+ Purkinje cells displayed a very high membrane conductance and a depolarized resting potential, indicating the presence of a large, constitutive inward current. Expression of the mutant GluRδ2^{Lc} protein in *Xenopus* oocytes confirmed these results, demonstrating that lurcher is an inherited neurodegenerative disorder resulting from a gain-of-function mutation in a glutamate receptor gene. Further characterization of GluRδ2 signaling and the activation of apoptotic death in *Lc* Purkinje cells have begun to yield mechanistic insights into this neurodegenerative disease, and to highlight its relationship to neuronal loss following ischemia.

The ability of a particular brain structure to perform its function rests on highly specialized properties of particular neurons and glia resident in that structure, and in their ability to integrate into the complex circuitry of the adult brain. As part of a broad-based effort to understand the differentiation of particular cell types in the mammalian cerebellum, we became interested in the lurcher (Lc) mutant mouse, because of its dramatic and rather specific defect in the final stages of cerebellar Purkinje cell development.[1] Using a positional cloning strategy, we have identified the lurcher phenotype as resulting from a gain-of-function mutation in the orphan glutamate receptor GluRδ2.[2] In this review, we summarize the identification of the lurcher mutation, the impact of this mutation on GluRδ2 function *in vivo* and *in vitro*, and the insights we have gained from this work concerning the function of GluRδ2 in Purkinje cells and the effects of the mutant allele (GluRδ2^{Lc}) on cerebellar Purkinje neurons.

THE LURCHER PHENOTYPE

Lc is a semidominant mouse mutation first described by Philips as an ataxic mouse strain with gross cerebellar abnormalities.[3] Heterozygous *Lc* mice develop ataxia during the second postnatal week due to the loss of cerebellar Purkinje.[1] Homozygous lurcher animals are much more severely affected, dying at birth due to massive loss of neurons in the hindbrain and brainstem.[4,5] Analysis of mouse chimeras created by fusion of wild-type

[a]Phone: 212-327-7956; fax: 212-327-7878; email: heintz@rockvax.rockefeller.edu

and heterozygous *Lc* (*Lc*/+) embryos established that the *Lc* gene acts cell-autonomously in cerebellar Purkinje cells *in vivo*.[6] These studies also established that the death of other cerebellar cell types in *Lc*/+ animals is a consequence of Purkinje cell loss and thus not a direct effect of the *Lc* mutation.[6,7] *Lc*/+ Purkinje cells do not die when crossed onto a genetic background (staggerer/staggerer mice—*sg/sg*) that prevents their terminal differentiation, demonstrating a requirement for maturation of Purkinje cells prior to their degeneration as a consequence of lurcher gene action.[8] The mode of cell death in *Lc*/+ Purkinje cells has been extensively characterized by two laboratories.[9,10] Both studies report features characteristic of apoptosis, although a definitive genetic demonstration that Purkinje cell death in these animals is programmed has not yet been reported.

While the precise phenotype of *Lc* mice does not correspond to any known human disease, several features of *Lc* animals are reminiscent of human neurodegenerative disorders: the phenotypic sensitivity to gene dosage, the focal death of specific neuronal populations followed by much more widespread neuronal loss, and the requirement of neuronal maturation for the onset of cell death have all been observed in human disease. Neither the cell-autonomous nature of a mutation nor the mode of cell death can be definitively established in cases of human neurodegeneration, but suggestive evidence supporting each of these points has been obtained. Given these considerations and the obvious experimental advantages to studying a spontaneous neurologic mutant mouse strain, it seemed evident that careful studies of the pathogenic process in *Lc*/+ mice could contribute to our knowledge of human neurodegeneration.

POSITIONAL CLONING OF THE *Lc* MUTATION

To begin a molecular dissection of the *Lc* phenotype, we first sought to identify the *Lc* mutation by traditional positional cloning. Thus, a 504-animal intersubspecific backcross using the original *Lc* mutation was generated by Norman and colleagues[11] and used to map the *Lc* mutation to a very small segment of mouse chromosome 6. Further refinement of this map demonstrated that the critical genetic interval carrying the *Lc* mutation was a 110–kb chromosomal segment flanked by polymorphic marker *D6Rck353* on the centromeric side and polymorphic marker *D6Rck357* on the telomeric side.[12] A search for genes within this critical 110-kb genomic DNA segment revealed the presence of three exons of the ionotropic glutamate receptor delta-2 (GluRδ2) gene within this critical interval. Furthermore, we demonstrated that the GluRδ2 gene extends over approximately 800 kb of genomic DNA, completely encompassing the 110-kb *D6Rck353–D6Rck357* segment. Analysis of GluRδ2 exons from the critical genetic interval revealed a single missense mutation that results in the substitution of a threonine residue for an alanine residue in the third transmembrane domain of GluRδ2 (FIG. 1a and b).[2] This point mutation was present in both the original *Lc* allele and in a recently isolated allele (Lc^J) that arose on an inbred genetic background, providing conclusive evidence that the mutation is responsible for the lurcher phenotype. The position of the mutation in the third transmembrane span of GluRδ2 is intriguing, since it highlights a nine–amino acid domain that is highly conserved among all ionotropic glutamate receptors, including those present in *Caenorhabtidis elegans* and *Drosophila melanogaster*.

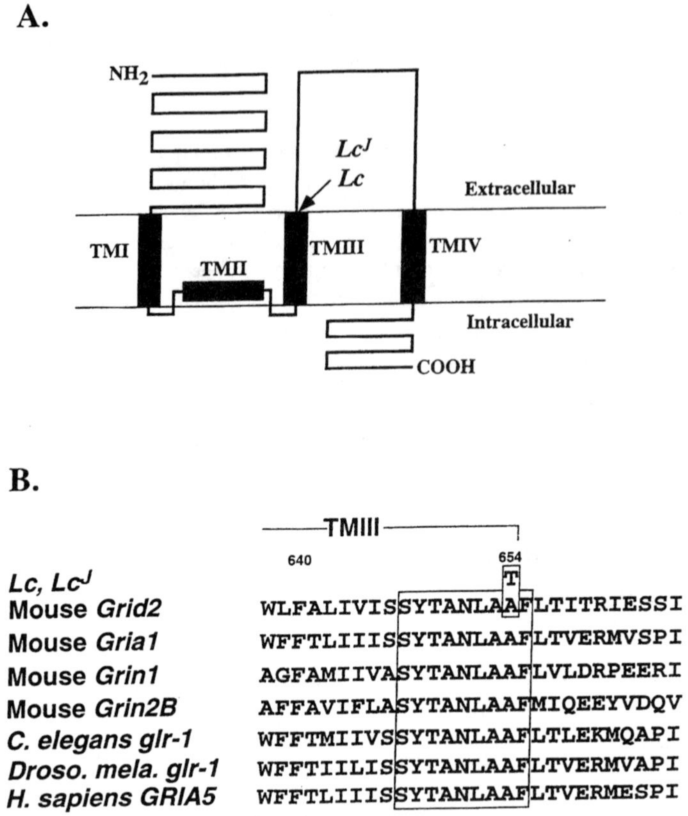

FIGURE 1. The *Lc* mutation occurs in an evolutionarily conserved domain of *Grid2*. **(A)** A diagram of the topology of ionotropic glutamate receptors. The amino acid change caused by the *Lc* point mutation takes place in the third transmembrane domain (TMIII), near the extracellular surface. **(B)** Conservation of the affected Ala residue in members of the ionotropic glutamate receptor family from different species. The *Lc* and Lc^J mutations result in an Ala-to-Thr substitution at a.a. position 654 (*small box*) of mouse *Grid2* in a highly conserved domain of nine amino acids (*large box*) in TMIII. Six representative members of the ionotropic glutamate receptor family are used to demonstrate the conservation of this domain through evolution: mouse *Gria1* (genbank accession no. X57497); mouse *Grin1* (D10028); mouse *Grin2B* (D10651); *C. elegans glr-1* (U34661); *D. melanogaster GluR-1* (M97192); and human *Gria5* (L19058). The significance of this extensive conservation throughout the evolution of ionotropic glutamate receptors is unknown. (Reprinted by permission from Zuo *et al.*[2])

Lc IS A GAIN-OF-FUNCTION MUTATION

The identification of the *Lc* mutation as a missense mutation in the GluRδ2 gene (the $GluR\delta2^{Lc}$ allele), taken together with the semidominant nature of the lurcher phenotype, immediately suggested that it might be a gain-of-function mutation. Definitive proof that this hypothesis is correct came from phenotypic comparison of $GluRd2^{Lc}$ with a null mutation (the *GluRδ2*$^{-/-}$ allele) in the *GluRd2* gene generated by gene targeting.[13] While both *Lc* alleles of *GluRd2* result in ataxia and loss of motor learning, the *GluRd2*$^{-/-}$ allele is recessive and does not result in either Purkinje cell degeneration in the cerebellum or perinatal death when carried in a homozygous state. These phenotypic differences are important for two reasons. First, when considering the mechanisms resulting in Purkinje cell death in lurcher animals, it is important to realize that the properties of the *GluRδ2*^{Lc} receptor may not accurately reflect the role of the wild-type GluRδ2 molecule *in vivo*. Second, the dose-dependent effect of the *GluRδ2*^{Lc} allele is readily explained because functional ionotropic glutamate receptors are thought to be either tetramers or pentamers.[14] Thus, if the presence of wild-type subunits in the mixed channels of *Lc*/+ animals can mitigate the effects of the mutant subunits, then the phenotypic severity of this gain-of-function allele will be lessened in heterozygous animals.

GluRδ2 IS AN "ORPHAN" IONOTROPIC GLUTAMATE RECEPTOR

The delta family (GluRδ1 and GluRδ2) of ionotropic glutamate receptors was isolated using low-stringency hybridization and degenerate reverse transcriptase–PCR to identify additional members of the ionotropic glutamate receptor superfamily.[15–17] Sequence similarity comparisons clearly indicate that these two genes are members of the ionotropic glutamate receptor family, since they share approximately 20–30% identity with both NMDA and AMPA/kainate receptors; in addition, the predicted membrane topology of the delta receptors is the same as that proposed for other family members.[15–17] However, sequence comparisons also indicate that the GluRδ1 and GluRδ2 genes are much more highly related to one another than they are to either the NMDA or AMPA/kainate subfamilies, suggesting that they might comprise a separate functional subclass of this superfamily of receptors (FIG. 2). This suggestion is supported by functional analysis of the delta family of receptors.

Ionotropic glutamate receptors are ligand-gated ion channels that can function either as homo- or heteromultimers. The different family members have been extensively investigated after expression in cultured cells or *Xenopus laevis* oocytes, revealing significant differences in their responses to agonist binding and in their physiological properties.[18] In contrast to other members of this gene family, neither GluRδ1 nor GluRδ2 have been shown to bind glutamate or to display ion channel activity alone or in combination with other members of the family.[15,16] Furthermore, immunoprecipitation studies of GluRδ2 from extracts of cerebellum failed to reveal interacting proteins with the correct stoichiometry to be considered as candidates for additional subunits of this receptor.[19] The delta receptors, therefore, remain "orphan" receptors because there are no data that identifies them as being responsive to glutamate or other known agonists of ionotropic glutamate receptors. As a result, the ion channel activities of the wild-type GluRδ1 and GluRδ2 receptors remain unknown.

In spite of the lack of functional data concerning the properties of GluRδ2, a great deal is known concerning its pattern of expression *in vivo*. GluRδ2 is expressed at high levels in cerebellar Purkinje cells, and at lower levels in some brainstem neurons.[15,16] In general, the cells affected by the lurcher mutation correspond to those known to express the receptor, as expected from the cell-autonomous action of the *Lc* mutation.[6] Detailed *in situ* hybridization and immunocytochemical localization studies have indicated that GluRδ2 is

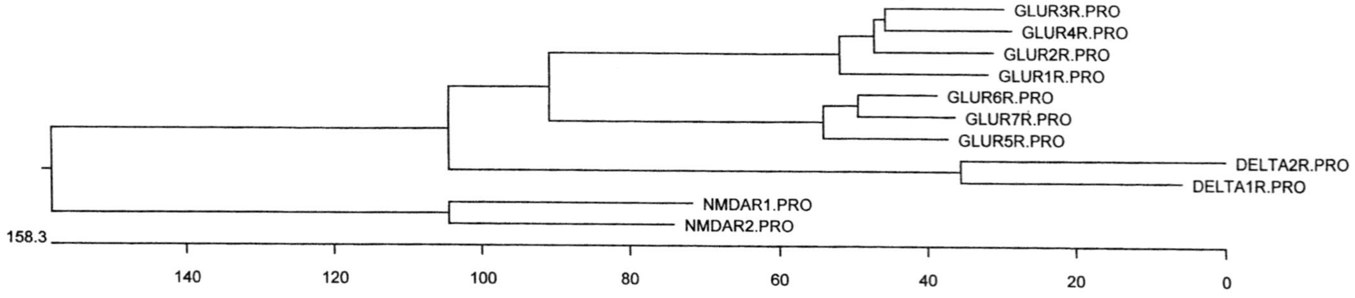

FIGURE 2. Dendrogram showing evolutionary relationships between mouse glutamate receptor proteins. GluRδ1 and GluRδ2 comprise a separate branch, indicating that they may comprise a separate subfamily. Scale is evolutionary divergence in millions of years.

expressed as early as embryonic day 15 in cerebellar Purkinje cells, and that its level of expression in this cell type increases significantly after birth.[20] One of the most interesting properties of GluRδ2 is its specific subcellular localization; it is found only in the postsynaptic density of those Purkinje cell dendritic spines that make contact with the parallel fibers of granule cells in the molecular layer of the cerebellar cortex.[21,22] This contrasts with the localization of other ionotropic glutamate receptors in Purkinje cells, which can be found both at the parallel fiber synapse and at the climbing fiber synapse—the other major excitatory input to Purkinje cells.[21,22]

A LARGE, CONSTITUTIVE INWARD CONDUCTANCE IS OBSERVED IN *Lc*/+ PURKINJE CELLS

To investigate the physiological effects of the *Lc* mutation, and to gain insight into its mode of action, Purkinje cells in thin slices of cerebellar vermis from postnatal day 10 and 11 mutant and wild-type animals were analyzed electrophysiologically.[2] *Lc*/+ Purkinje cells exhibited a dramatic and severe physiological phenotype: when compared to wild-type Purkinje cells, *Lc*/+ Purkinje cells required a holding current of larger magnitude to clamp the neuron at −70 mV. In addition, measurements of the initial currents and membrane conductances in the affected cells were much greater, and their resting potential was elevated.[2] In addition, these abnormal currents could be largely eliminated by the substitution of N-methyl-D-glucamine (NMDG), a relatively large organic cation, for most of the Na^+ in the external saline, demonstrating that the *Lc*-specific conductance is selective and not the result of poor *Lc*/+ membrane integrity or leakage at the pipette/membrane interface. Furthermore, the large reduction in holding-current magnitude and decrease in membrane conductance caused by reducing the external Na^+ concentration strongly suggested that Na^+ is a major current carrier of the *Lc*-specific, constitutive, inward current. All of these results were obtained in the presence of tetrodotoxin, a potent voltage-gated Na^+ channel blocker that should eliminate the release of glutamate evoked by any action potentials arising spontaneously within the slice preparation; this suggests that the constitutively active current in *Lc*/+ Purkinje cells does not depend on activation by ambient neurotransmitters.

GluRδ2^{Lc} ENCODES A CONSTITUTIVELY ACTIVE HOMOMERIC CHANNEL

To examine the properties of the GluRδ2^{Lc} channel, the cRNAs coding for the wild-type and mutant alleles of GluRδ2 were assayed for electrophysiological activity in *Xenopus laevis* oocytes (FIG. 3). Consistent with previous reports,[15,16] oocytes expressing wild-type GluRδ2 were not significantly different from their uninjected counterparts in resting potential observed either in the absence or presence of NMDG. In contrast, injection of GluRδ2^{Lc} cRNA, which was prepared by changing a single nucleotide from G to A at position 1960 in the full-length cDNA sequence to recreate the *GluRδ2^{Lc}* allele, produced a dramatic depolarization in the resting potential. This depolarization could be completely reversed by replacement of external Na^+ with NMDG.[2] Since these changes in resting potential and whole-cell conductance were observed in the absence of any ligand, these

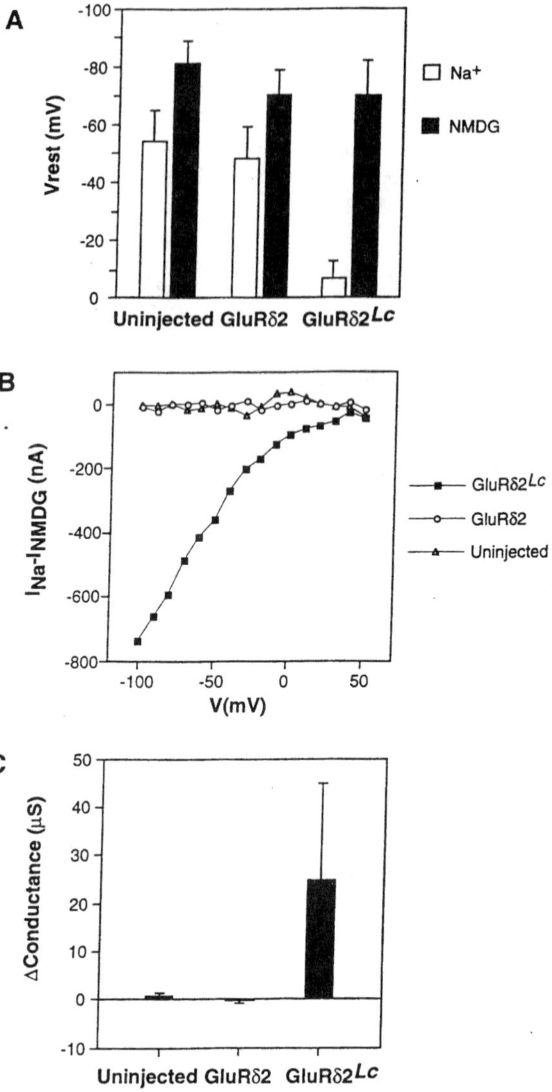

FIGURE 3. Physiological characterization of *Grid2*^{Lc} in oocytes. A large, constitutive conductance is observed in *X. laevis* oocytes injected with *Grid2*^{Lc}. *Error bars* in (A) and (C) represent SEM of 2 electrode voltage measurements from 4 uninjected oocytes, 5 oocytes injected with wild-type *Grid2* cRNA, and 15 oocytes injected with *Grid2*^{Lc} cRNA. **(A)** Average resting membrane potentials (Vrest) of uninjected oocytes and oocytes injected with *Grid2* and *Grid2*^{Lc} cRNAs in external Na^+ bath (*open bars*) and in external NMDG bath (*filled bars*). **(B)** Current-voltage relationships of two representative oocytes before and after NMDG substitution. *Open squares* (Na^+) and *triangles* (NMDG) are from the oocyte injected with *Grid2*^{Lc} cRNA. **(C)** Average changes in whole-cell conductance at −60 mV membrane potential before and after NMDG substitution. (Reprinted by permission from Zue *et al.*[2])

measurements demonstrated that cells injected with the mutant GluRδ2^{Lc} expressed a large, constitutive conductance under physiological conditions.

The discovery that the expression of GluRδ2^{Lc} in *Xenopus* oocytes results in the formation of a large, constitutively active conductance is important for two reasons. First, and most important, the fact that the currents observed in *Lc*/+ Purkinje cells and GluRδ2^{Lc} expressing *Xenopus laevis* oocytes display the same basic properties strongly suggests that the major electrophysiological phenotype of *Lc*/+ Purkinje cells results from the direct action of the mutant allele. Second, the formation of a constitutively active channel by expression of the *GluRδ2^{Lc}* allele in oocytes provides the first evidence that wild-type GluRδ2 subunits are competent to form homomeric channels. This result thus suggests that the failure to observe channel activity of wild-type GluRδ2 in injected oocytes or transfected mammalian cells may reflect the inability to properly activate these channels rather than an inherent lack of channel-forming properties in the wild-type protein. The inability to properly inactivate these channels could result from our ignorance concerning the relevant ligand, from a requirement for an additional subunit to participate in gating, or a combination of these two possibilities.

NEURODEGENERATION IN *Lc* MICE IS SIMILAR TO DELAYED CELL DEATH FOLLOWING ISCHEMIA

Since the original demonstration that brain lesions can occur in animals exposed to glutamate,[23] excitotoxic cell death due to prolonged exposure to this neurotransmitter has been thought to play a role in neurodegeneration *in vivo*.[24,25] Recent evidence implicating glutamate toxicity in delayed neuronal death following ischemia[26] and aberrant processing of astrocytic glutamate transporter mRNAs in sporadic cases of amyotrophic lateral sclerosis (ALS)[27] has supported this general hypothesis. The demonstration that neuronal death in lurcher mice results from the constitutive activation of an ionotropic glutamate receptor provided the first genetic proof that this type of pathway can be the primary cause of neurodegeneration *in vivo*.[2] Obviously, this raises important issues concerning the mechanisms of neuronal cell death in these disorders, and suggests that *Lc* mice can provide an important experimental system in which to address them.

The prevailing model for glutamate toxicity has been developed through *in vitro* studies using glutamate to elicit excitotoxic cell death in cultured neurons; in this system, it has been shown that Ca^{2+} influx into cells through the NMDA and/or AMPA receptors is a critical step in initiating neuronal death and that the mechanism of death is necrosis.[26,28,29] *In vivo*, the situation appears to be more complex. Although our knowledge of the precise mechanisms involved in widespread neuronal death following ischemia is still quite primitive, a synthesis of the available data suggests that the immediate necrotic death of neurons at the site of an ischemic lesion may be very closely related to excitotoxic cell death in culture.[26] On the other hand, the delayed neuronal death that is found in hypoperfused areas near the site of ischemia appears to occur through the activation of ionotropic glutamate receptors and of an apoptotic pathway.[26,30–32] This is consistent with genetic results demonstrating a role for the apoptotic machinery in ischemic cell death *in vivo*.[33–35] Given the mechanistic distinctions between these two forms of cell death, it is of obvious importance to study the cell death pathway in any case of neurodegeneration where activation of glutamate receptors is thought to play an important role.

Studies of Purkinje cell loss in *Lc*/+ animals has provided strong evidence that these cells are dying *in vivo* through an apoptotic mechanism. Thus, at both the light and electron microscopic levels, the morphology of dying *Lc*/+ Purkinje cells is typical of neurons undergoing apoptotic death. Furthermore, several genes that are known to be expressed during the programmed death of neurons are expressed in *Lc*/+ Purkinje cells prior to their death *in vivo*, and the nuclei of *Lc*/+ Purkinje cells contain nicked DNA ends during the death process.[9] These results are certainly consistent with the activation of an apoptotic process in response to the GluRδ2^{Lc} stimulus, suggesting that neurodegeneration in this case is very closely related to the delayed death of neurons following ischemia, and that it does not resemble the necrotic process observed in studies of cultured neurons exposed to excess glutamate.

THE SPECIFICITY AND TIMING OF PURKINJE CELL DEATH IN *Lc*/+ ANIMALS

The discovery of the *GluRδ2*^{Lc} allele provided an opportunity to understand the lurcher phenotype in the context of the very detailed knowledge of the expression and location of the GluRδ2 protein *in vivo*. The fact that those cell populations first dying in the cerebellum of *Lc*/+ animals, or in the brainstem of *Lc/Lc* embryos, are those known to express the *GluRδ2* gene[15,16] might have been predicted from the genetic demonstration that the lurcher phenotype is cell autonomous.[6] However, the timing of cell death in lurcher animals is not so readily explained. The *GluRδ2* gene is first expressed at embryonic day 15 in cerebellar Purkinje cells, and the protein accumulates to quite high levels in individual Purkinje cells by birth.[20] Yet, there is no evidence of apoptosis in this cell population until approximately postnatal day 8, when degenerating Purkinje cells are first observed. Neuronal loss occurs quickly thereafter, and 95% of Purkinje cells are dead by postnatal day 25.[1] On the other hand, in (*Lc*, *sg* / +, *sg*) double mutants, Purkinje cells do not die postnatally.[8] While neither the requirement for Purkinje cell maturation nor the timing of Purkinje cell death in *Lc*/+ mice is yet understood, a possible explanation for these results has been provided by the demonstration that the GluRδ2 protein is redistributed within Purkinje cells from a homogeneous distribution throughout the dendritic arbor to the postsynaptic density of the parallel fiber synapse at approximately the same time that Purkinje cell death commences.[1,22] These results suggest that localization of the GluRδ2^{Lc} channels to the synapse might be important for the activation of apoptotic death in *Lc*/+ Purkinje cells and raise the general issue of the importance of subcellular localization to the generation of aberrant signals that initiate neurodegeneration *in vivo*.

WHY DO NEURONS DIE IN RESPONSE TO CONSTITUTIVE ACTIVATION OF *GluRδ2*^{Lc}?

In trying to formulate a working hypothesis for the mechanisms involved in Purkinje cell death in *Lc*/+ mice, we have considered two very different models of neurodegeneration. The first is based on the demonstrated involvement of Ca^{2+} in excitotoxic cell death in cultured neurons. *In vitro* results suggest that excitotoxic cell death in culture, even when induced slowly by low concentrations of the excitotoxic agent, occurs through necrosis.[26,29] These data suggest that the excitotoxic pathway may be quite distinct from the delayed apoptotic death of neurons in ischemia or death of *Lc*/+ Purkinje cells. However, it seems quite possible that necrosis and apoptosis simply represent different cellular responses to the same primary metabolic event. In this case, it is not the nature of the signal emanating from the constitutively activated receptor that dictates the choice of pathway, but the quantitative impact it has upon the cell. Were this to be the case, the primary signal in both forms of cell death would be elevated intracellular Ca^{2+} levels (FIG. 4). While a mechanism by which elevated intracellular Ca^{2+} levels would activate apoptosis has not been established, the robust signaling between the synapse and cell nucleus through cAMP response element binding protein (CREB) activation[14] provides a precedent for the type of mechanism that may be relevant in this form of neurodegeneration. A key prediction of this model is that it is the ion flux through the activated channel per se that is the initiating event in neurodegeneration.

The second model we are considering derives from our previous attempts to organize facts concerning the general properties of neurons and the characteristics of neurodegeneration into a framework that is analogous to the one developed for the action of oncogenes in cell transformation.[36–38] Incorporated into this idea are three key elements: the involvement of neurodegenerative disease genes in the activation of inappropriate signal transduction events, the integration of these aberrant metabolic signals by intracellular mechanisms akin to cell cycle checkpoints, and the activation of programmed cell death in response to these signals as the sole effector pathway downstream from these metabolic integrators (FIG. 4). The central tenet of this hypothesis is that all cells contain a mechanism for integrating signal transduction events with internal metabolic information and that pro-

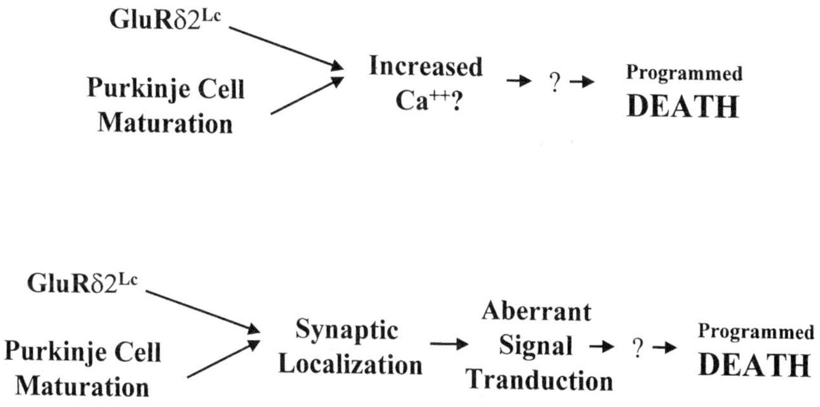

FIGURE 4. Two models of the initiation of apoptosis by ionotropic glutamate receptors. We propose two models to explain the manner in which an apoptotic process is initiated in the context of delayed cell death following an ischemic insult. In model A (**top panel**), the abnormal Ca^{2+} influx plays a major role by interacting with and activating signaling molecules that ultimately initiate the neuronal death program. In model B (**lower panel**), Ca^{2+} may play a role in activating the various signaling modalities of a particular ionotropic glutamate receptor complex. A critical step in this model is maturation of the *GluRδ2*^{Lc} signaling complex and its localization to the synapse. It is the persistent synaptic signaling from this receptor complex that is responsible for the initiation of apoptosis.

grammed cell death is the dominant effector pathway that is activated by this mechanism in postmitotic neurons in response to aberrant stimuli. Strong evidence that activation of programmed neuronal death plays a role in both delayed cell loss following ischemia[26] and Purkinje cell death in *Lc*/+ mice[9] has been obtained. Furthermore, the existence of cell death "checkpoints" as natural regulatory mechanisms for the initiation of apoptotic death is now established. The critical issue, therefore, is the nature of the stimulus that results in activation of the cell death pathway in response to ionotropic glutamate receptor activation. According to this hypothesis, one might predict that the signal transducing capability of ionotropic glutamate receptors, as well as their role as ligand-gated ion channels, may be crucial for activation of the cell death pathway. It is becoming increasingly apparent that large complexes of proteins involved in various modalities of signaling—such as α-actinin,[39] calmodulin,[20,26] glutamate receptor interacting protein (GRIP),[40] PSD-95,[41,42] and SAP102[43]—are assembled onto the C-terminal cytoplasmic domain of ionotropic glutamate receptors; and evidence is now accumulating that these complexes are critical for the *in vivo* functions of these molecules.[44] Furthermore, the neuronal isoform of nitric oxide synthase, which interacts with the C-terminal domain of the NR1 receptor, has been implicated in the initiation of programmed cell death following activation of NR1-containing ionotropic glutamate receptors.[45,46] While the coincident ion flux through the channel may be important in signal generation, its proposed role would be to act through the C-terminal signaling components associated with the receptor. Thus, a critical role for the C-terminus that is independent from its possible involvement in gating the channel is proposed.

To distinguish between these possibilities, or to determine their relative contribution in a specific neurodegenerative disorder, will require an understanding of the precise signals emanating from the mutant receptor, their integration by the cell, and the molecular mechanism responsible for activation of the programmed death pathway. While *in vitro* systems will be very useful in illustrating the menu of possible pathways that should be considered in trying to understand neurodegeneration, a clear definition of the molecular mechanisms participating *in vivo* will require genetic analysis of these events in the context of the intact CNS. It is toward this end that we have directed our research into Purkinje cell death in lurcher mutant mice.

ACKNOWLEDGMENTS

We would like to thank Dr. J. Zuo for his contributions to the cloning and characterization of the *Lc* mutation; Drs. L. Feng, D. J. Norman, and W. Jiang for their contribution to the cloning and histopathological analysis of the Lc mutation; as well as K. Takahashi and Dr. D. Linden for their physiological characterization of *Lc* Purkinje cells in slices of cerebellum. P.L.D. is supported by National Institutes of Health / National Institute of General Medical Sciences (NIH/NIGMS) Grant GMO7739, and N.H. is supported by the Howard Hughes Medical Institute.

REFERENCES

1. Caddy, K.W. & T.J. Biscoe. 1979. Structural and quantitative studies on the normal C3H and lurcher mutant mouse [review]. Philos. Trans. R. Soc. Lond. B. Biol. Sci. **287:** 167–201.

2. Zuo, J., P.L. De Jager, K.A. Takahashi, W. Jiang, D.J. Linden & N. Heintz. 1997. Neurodegeneration in lurcher mice caused by a mutation in the δ2 glutamate receptor gene. Nature 388: 769–773.
3. Philips, R.J.S. 1960. "Lurcher," a new gene in linkage group XI of the house mouse. J. Genet. **57:** 35–42.
4. Cheng, S. & N. Heintz. 1997. Massive loss of mid- and hindbrain neurons during embryonic development of homozygous lurcher mice. J. Neurosci. **17:** 2400–2407.
5. Resibois, A., L. Cuvelier & A.M. Goffinet. 1997. Abnormalities in the cerebellum and brainstem in homozygous lurcher mice. Neuroscience **80:** 175–190.
6. Wetts, R. & K. Herrup. 1982a. Cerebellar Purkinje cells are descended from a small number of progenitors committed during early development: Quantitative analysis of lurcher chimeric mice. J. Neurosci. **2:** 1494–1498.
7. Wetts, R. & K. Herrup. 1982. Interaction of granule, Purkinje and inferior olivary neurons in lurcher chimeric mice. I. Qualitative studies. J. Embryol. Exp. Morphol. **68:** 87–98.
8. Messer, A., B. Eisenberg & J. Plummer. 1991. The lurcher cerebellar mutant phenotype is not expressed on a staggerer mutant background. J. Neurosci. **11:** 2295–2302.
9. Norman, D.J., L. Feng, S.S. Cheng, J. Gubbay, E. Chan & N. Heintz. 1995. The lurcher gene induces apoptotic death in cerebellar Purkinje cells. Development **121:** 1183–1193.
10. Wullner, U., P.A. Loschmann, M. Weller & T. Klockgether. 1995. Apoptotic cell death in the cerebellum of mutant weaver and lurcher mice. Neurosci. Lett. 200(**2**): 109–112.
11. Norman, D.J., Fletcher, C. & N. Heintz. 1991. Genetic mapping of the lurcher locus on mouse chromosome 6 using an intersubspecific backcross. Genomics **9:** 147–153.
12. De Jager, P.L., J. Zuo & N. Heintz. 1997a. An ~1.2 Mb BAC contig refines the genetic and physical maps of the lurcher locus on mouse chromsome 6. Genome Res. **7:** 736–746.
13. Kashiwabuchi, N., K. Ikeda, K. Araki, T. Hirano, K. Shibuki, C. Takayama, Y. Inoue, T. Kutsuwada, T. Yagi, Y. Kang, S. Aizawa & M. Mishina. 1995. Impairment of motor coordination, Purkinje cell synapse formation, and cerebellar long-term depression in GluRδ2 mutant mice. Cell **81:** 245–252.
14. Laube, B., J. Kuhse & H. Betz. 1998. Evidence for a tetrameric structure of recombinant NMDA receptors. J. Neurosci. **18:** 2954–2961.
15. Araki, K., H. Meguro, E. Kushiya, C. Takayama, Y. Inoue & M. Mishina. 1993. Selective expression of the glutamate receptor channel δ2 subunit in cerebellar Purkinje cells. Bioch. Biophys. Res. Commun. **197:** 1267–1276.
16. Lomeli, H., R. Sprengel, D.J. Laurie, G. Köhr, A. Herb, P.H. Seeburg & W. Wisden. 1993. The rat delta-1 and delta-2 subunits extend the excitatory amino acid receptor family. FEBS Lett. **315:** 318–322.
17. Yamazaki, M., K. Araki, A. Shibata & M. Mishina. 1992. Molecular cloning of a cDNA encoding a novel member of the mouse glutamate receptor channel family. Biochem. Biophys. Res. Commun. **183:** 886–892.
18. Hollmann, M. & S. Heinemann. 1994. Cloned glutamate receptors. Annu. Rev. Neurosci. **17:** 31–108.
19. Mayat, E., R.S. Petralia, Y.-X. Wang & R.J. Wenthold. 1995. Immunoprecipitation, immunoblotting, and immunocytochemistry studies suggest that gluatmate receptor δ subunits form novel postsynaptic receptor complexes. J. Neurosci. **15:** 2533–2546.
20. Takayama, C., S. Nakagawa, M. Watanabe, H. Kurihara, M. Mishina & Y. Inoue. 1996. Developmental changes in expression and distribution of the glutamate receptor channel δ2 subunit according to the Purkinje cell maturation. Dev. Brain Res. **92:** 147–155.
21. Landsend, A.S., M. Amiry-Moghaddam, A. Matsubara, L. Bergersen, S.-I. Usami, R.J. Wenthold & O.P. Ottersen. 1997. Differential localization of δ glutamate receptors in the rat cerebellum: Coexpression with AMPA receptors in parallel fiber-spine synapses and absence from climbing fiber-spine synapses. J. Neurosci. **15:** 834–842.
22. Zhao., H.-M., R.J. Wenthold,Y.-X. Wang & R.S. Petralia. 1997. δ-glutamate receptors are differentially distributed at parallel and climbing fiber synapses on Purkinje cells. J. Neurochem. **68:** 1041–1052.
23. Olney, J.W. 1969. Brain lesions, obesity, and other disturbances in mice treated with monosodium glutamate. Science **164:** 719–721.
24. Choi, D.W. 1988. Calcium-mediated neurotoxicity: relationship to specific channel types and role in ischemic damage. Trends Neurosci.**11:** 465–469.

25. ROTHMAN, S.M. & J.W. OLNEY. 1987. Excitotoxicity and the NMDA receptors. Trends Neurosci. **10:** 299–302.
26. CHOI, D.W. 1996. Ischemia-induced neuronal apoptosis. Curr. Opin. Neurobiol. **6:** 667–672.
27. LIN, C.L., L.A. BRISTOL, L. JIN, M. DYKES-HOBERG, T. CRAWFORD, L. CLAWSON & J.D. ROTHSTEIN. 1998. Aberrant RNA processing in a neurodegenerative disease: The cause for absent EAAT2, a glutamate transporter, in amyotrophic lateral sclerosis. Neuron **20:** 589–602.
28. PELLIGRINI-GIAMPETRO, D.E., J.A. GORTER, M.V.L. BENNETT & R.S. ZUKIN. 1997. The GluR2 (GluR-B) hypothesis: Ca^{2+}-permeable AMPA receptors in neurological disorders. Trends Neurosci. **10:** 464–470.
29. SZATKOWSKI, M. & D. ATTWELL. 1994. Triggering and execution of neuronal death in brain ischemia: Two phases of glutamate release by different mechanisms. Trends Neurosci. **17:** 359–365.
30. DU, C., R. HU, C.A. CSERNANSKY, C.Y. HSU & D.W. CHOI. 1996. Very delayed infarction after mild focal cerebral ischemia: A role for apoptosis? J. Cereb. Blood Flow Metab. **16:** 195–201.
31. LINNIK, M.D., R.H. ZOBRIST & M.D. HATFIELD. 1993. Evidence supporting a role for programmed cell death in focal cerebral ischemia in rats. Stroke **24:** 2002–2009.
32. TOMINAGA, T., S. KURE, K. NARISAWA & T. YOSHIMOTO. 1993. Endonuclease activation following focal ischemic injury in the rat brain. Brain Res. **608:** 21–26.
33. CRUMRINE, R.C., A.L. THOMAS & P.F. MORGAN. 1994. Attenuation of p53 expression protects against focal ischemic damage in transgenic mice. J. Cereb. Blood Flow Metab. 14(**6**): 887–891.
34. MARTINOU, J.-C., M. DUBOIS-DAUPHIN, J.K. STAPLE, I. RODRIGUEZ, H. FRANKOWSKI, M. MISSOTTEN, P. ALBERTINI, D. TALABOT, S. CATSICA, C. PIETRA & J. HUARTE. 1994. Overexpression of Bcl-2 in transgenic mice protects neurons from naturally occurring cell death and experimental ischemia. Neuron **13:** 1017–1030.
35. PARSADANIAN, A.S., Y. CHENG, C.R. KELLER-PECK, D.M. HOLTZMAN & W.D. SNIDER. 1998. Bcl-xL is an antiapoptotic regulator for postnatal CNS neurons. J. Neurosci. **18:** 1009–1019.
36. HEINTZ, N. 1993. Cell death and the cell cycle: A relationship between transformation and neurodegeneration? Trends Biochem. Sci. **18:** 157–159.
37. HEINTZ, N. 1996. Ataxia telangiectasia: Cell signaling, cell death and the cell cycle. Curr. Opin. Neurol. **9:** 137–140.
38. HEINTZ, N. & H. ZOGHBI. 1997. Alpha-Synuclein--a link between Parkinson and Alzheimer diseases? Nat. Genet. **16:** 325–327.
39. WYSZYNSKI, M., J. LIN, A. RAO, E. NIGH, A.H. BEGGS, A.M. CRAIG & M. SHENG. 1997. Competitive binding of α-actinin and calmodulin to the NMDA receptor. Nature **385:** 439–442.
40. DONG, H., R.J. O'BRIEN, E.T. FUNG, A.A. LANAHAN, P.F. WORLEY & R. HUGANIR. 1997. GRIP: A synaptic PDZ domain-containing protein that interacts with AMPA receptors. Nature **386:** 279–284.
41. KORNAU, H.-C., L.T. SCHENKER, M.B. KENNEDY & P.H. SEEBURG. 1995. Domain interaction between NMDA receptor subunits and the postsynaptic density protein PSD-95. Science **269:** 1737–1740.
42. NIETHAMMER, M., E. KIM & M. SHENG. 1996. Interaction between the C-terminus of NMDA receptor subunits and multiple members of the PSD-95 family of membrane-associated guanylate kinases. J. Neurosci. **16:** 2157–2163.
43. MÜLLER, B.M., U. KISTNER, S. KINDLER, W.J. CHUNG, S. KUHLENDAHL, S.D. FENSTER, L-F. LAU, R.W. VEH, R.L. HUGANIR, E.D. GUNDELFINGER & C.C. GARNER. 1996. SAP102, a novel postsynaptic protein that interacts with NMDA receptor complexes *in vivo*. Neuron **17:** 255–265.
44. SPRENGEL, R., B. SUCHANEK, C. AMICO, R. BRUSA, N. BURNASHEV, A. ROZOV, O. HVALBY, V. JENSEN, O. PAULSEN, P. ANDERSEN, J.J. KIM, R.F. THOMPSON, W. SUN, L.C. WEBSTER, S.G. GRANT, J. EILERS, A. KONNERTH, J. LI, J.O. MCNAMARA & P.H. SEEBURG. 1998. Importance of the intracellular domain of NR2 subunits for NMDA receptor function *in vivo*. Cell **92:** 279–289.
45. LEIST, M., C. VOLBRACHT, S. KUHNLE, E. FAVA, E. FERRANDO-MAY & P. NICOTERA. 1997. Caspase-mediated apoptosis in neuronal excitotoxicity triggered by nitric oxide. Mol. Med. **3:** 750–764.
46. AYATA, C., G. AYATA, H. HARA, R.T. MATTHEWS, M.F. BEAL, R.J. FERRANTE, M. ENDRES, A. KIM, R.H. CHRISTIE, C. WAEBER, P.L. HUANG, B.T. HYMAN & M.A. MOSKOWITZ. 1997. Mechanisms of reduced striatal excitotoxicity in Type I nitric oxide synthase knock-out mice. J. Neurosci. **17:** 6908–6917.

Expression Mechanisms Underlying NMDA Receptor–Dependent Long-Term Potentiation

R. A. NICOLL[a,b,c] AND R. C. MALENKA[b,d]

Departments of [a]Cellular and Molecular Pharmacology, [b]Physiology, and [d]Psychiatry, University of California, San Francisco, California 94143-0450, USA

ABSTRACT: Long-term potentiation (LTP) is currently the best available cellular model for learning and memory in the mammalian brain. In the CA1 region of the hippocampus, as well as in many other areas of the CNS, its induction requires a rise in postsynaptic Ca^{2+} via activation of NMDA receptors. What happens after the rise in postsynaptic Ca^{2+} is less clear. This paper summarizes experiments performed over the last decade in slice preparations that address the site of expression of LTP. While a large number of laboratories have contributed importantly to this issue, this review will rely primarily on experiments performed in the authors' laboratory. The experiments to be discussed can be broadly divided into two groups: those designed to determine if an increase in glutamate release occurs during LTP and those designed to determine if a change in postsynaptic sensitivity to glutamate occurs during LTP. Experiments in the first category include the analysis of dual-component excitatory postsynaptic currents (EPSCs), paired-pulse facilitation, saturating release probability, the use of MK-801 to measure release probability, and glial glutamate transporter currents to measure directly the synaptic release of glutamate. Experiments in the second category include analysis of miniature EPSC amplitudes, measurements of synaptic potency, the consequences of loading cells with the constitutively activated form of CaM kinase II, and the evidence that during LTP postsynaptically silent synapses become functional. We will argue that, while numerous experiments fail to support a presynaptic expression mechanism, many experiments *do* point to a postsynaptic expression mechanism. The decrease in synaptic failures during LTP, the only generally accepted experimental result that supports a presynaptic expression mechanism, can be explained by postsynaptically silent synapses. Future directions for research in this field include activity-dependent targeting of glutamate receptors and the functional consequences of phosphorylation of AMPA receptors.

Long-term potentiation (LTP) incorporates many of the features expected for a form of synaptic modification appropriate for learning and memory, such as a Hebbian associative property. However, it is important to emphasize that the actual linkage of this cellular phenomenon to behavior is very limited. The induction of LTP requires the near-simultaneous coincidence of synaptic activity and adequate postsynaptic depolarization. We now know that the NMDA receptor provides the molecular basis for this coincidence detection. In order for the NMDA receptor channel to conduct, glutamate must bind to the receptor and the postsynaptic membrane must be depolarized. The basis for this voltage dependence is a voltage-dependent block of the ion channel by extracellular Mg^{2+}. The NMDA receptor channel is known to be highly permeable to Ca^{2+} as well as to monovalent cations, and therefore activation of this channel will lead to an influx of Ca^{2+} into the

[c]Address for correspondence: Roger A. Nicoll, Department of Cellular and Molecular Pharmacology, University of California, San Francisco, California 94143. Phone: 415-476-2018; fax: 415-476-5292; e-mail: nicoll@phy.ucsf.edu

postsynaptic spine. It is the rise in Ca^{2+} that serves as the trigger for LTP. The events that are initiated by Ca^{2+} and ultimately lead to an increased synaptic strength are the subject of this review. It is important to keep in mind that these processes must all occur within the time scale of approximately 10 to 20 seconds.[1,2]

LTP has been the subject of numerous reviews.[3–8] The purpose of this paper is not to provide a comprehensive review of the LTP field but rather to summarize some of the experiments carried out in the authors' labs that specifically address important issues related to the expression mechanism of LTP.

EXPERIMENTS DESIGNED TO TEST WHETHER GLUTAMATE RELEASE INCREASES DURING LONG-TERM POTENTIATION

In this section we will summarize experiments, which have spanned the last 10 years, that specifically address whether the release of glutamate increases during LTP. The general approach in these experiments has been to examine various parameters of synaptic transmission that have been unambiguously documented by many laboratories as changing in highly predictable ways in response to known changes in transmitter release, and to determine if these parameters similarly change during LTP. The brief answer is that this approach, in our hands, has failed to detect any evidence that transmitter release changes during LTP. The detailed experimental evidence follows.

Relative Change in the AMPA and NMDA Component of the EPSP/C

This is the first approach we took to address the expression mechanisms of LTP. We argued that if there were a change in the amount of glutamate released, the two receptors, which are colocalized at excitatory synapses, would sense this increase, and there would be a parallel increase in both receptor-mediated responses. On the other hand if LTP were due to a selective postsynaptic enhancement in the responsiveness of AMPARs, only the AMPA component would increase. We[9,10] found that well-established presynaptic manipulations (in this paper the term *well-established presynaptic manipulations* will refer to manipulations that include changing release probability by changing Ca/Mg ratios, applying presynaptic inhibitory transmitter agonists, and posttetanic potentiation), including increasing the number of active synapses by increasing the stimulus strength, did cause a parallel increase in the two synaptic receptor mediated-components.

In marked contrast LTP was found to affect primarily the AMPA receptor component. Given that changes in the probability of transmitter release, as well as changes in the number of activated synapses, cause a parallel increase in the two components, such a selective augmentation of the AMPAR excitatory postsynaptic current (EPSC) argues strongly for a postsynaptic modification of AMPARs. However, based on a variety of experimental protocols, there has been a great deal of controversy concerning the effects of LTP on the two components, ranging from changes primarily associated with the AMPAR component,[11–15] to equal changes in the two components.[16] While it seems clear that under some conditions the NMDAR component can change during LTP, the fact that under some conditions the enhancement appears to be essentially restricted to the AMPAR component argues strongly for a selective postsynaptic change in AMPAR function.

Paired-Pulse Facilitation

Paired-pulse facilitation (PPF) is generally agreed to be of presynaptic origin and is modified by well-established presynaptic manipulations that change the probability of transmitter release. This does not change when the number of activated synapses is changed as a result of changing the stimulus strength. There is a general consensus that PPF is not altered during LTP (see Refs. 3–8).[17] The fact that all well-established presynaptic manipulations that change the probability of transmitter release change PPF, but LTP causes no change, indicates that LTP cannot be due to a simple change in the probability of transmitter release. An increase in the number of functional synapses is not excluded by these findings, although these new synapses would have to have, on average, the same amount of PPF as those synapses monitored prior to LTP.

It might be argued that the PPF causes only a modest change in presynaptic function and that a more dramatic perturbation of release would be altered by LTP. Indeed, it has recently been reported that in the neocortex LTP causes a redistribution of synaptic strength within a burst of stimuli.[18] We have therefore reinvestigated this issue by applying prolonged trains of stimuli that result in stable vesicular depletion.[19] Contrary to the results in the neocortex, we found that there was a scaling of the synaptic responses within the burst, with no evidence of redistribution. This finding is of considerable importance because it indicates that in the hippocampus the fidelity of synaptic transmission is preserved during LTP. The simplest way to maintain fidelity when changing the gain of synaptic transmission is to place the plasticity in the postsynaptic cell so that it cannot interact with short-term alterations in release probability.

Saturating Release Probability

If LTP were due to an increase in the probability of transmitter release, then it should be possible to occlude LTP by increasing the release probability to its maximum. We have examined this prediction by applying the K channel blocker 4-aminopyridine (4-AP).[20] This markedly increased synaptic transmission and other manipulations that increase the probability of release, such as high Ca and Sp-cAMPS, fail to further increase synaptic transmission, confirming that the probability of release is close to saturation. When we compared the magnitude of saturated LTP induced by low-frequency stimulation paired with depolarization of the membrane to approximately 0 mV (which we will refer to as *pairing*) in control slices to those bathed in 4-AP, no difference could be detected. In addition, when the effects of 4-AP were compared between a pathway expressing saturating levels of LTP to a naive pathway, no difference in the 4-AP–induced enhancement could be seen. These results indicate that a simple increase in the probability of release cannot explain the enhancement seen with LTP.

Analysis of the Probability of Transmitter Release with MK-801

MK-801 is an essentially irreversible, use-dependent antagonist of NMDA receptors. In the presence of MK-801 there is a progressive decrease in the size of the NMDAR EPSC, and the rate of this decrease is directly related to the probability of transmitter release.[21,22]

We therefore decided to use MK-801 to see if an increase in the probability of transmitter release could be detected during LTP. Two independent pathways were monitored, and LTP was induced on one pathway by pairing. After establishing that the LTP was stable, the AMPARs were blocked with CNQX, stimulation was stopped, and the membrane potential shifted to +30 mV. MK-801 was applied for 15 min, and then synaptic stimulation in the two pathways was resumed. No difference was detected in the decay of the NMDAR EPSC between the two pathways. Similar negative results have been obtained by Malinow (personal communication), although Kullmann *et al.*[13] have reported a difference between the two pathways. In their study a tetanus was used to induce LTP, rather than pairing. To explain the lack of difference when LTP is induced by pairing it was proposed that a substantial fraction of the NMDAR EPSC may be generated by spillover from synapses on neighboring cells, a situation that would mask any increase in transmitter release due to the LTP that was generated in the recorded cells. Further studies will be required to settle this issue.

Monitoring Transmitter Release with Glial Glutamate Transporter Currents

Part of the difficulty in resolving the controversy over the site of expression of LTP might be due to the indirect nature of many of the approaches used and the assumptions that underlie these approaches. The ability to monitor the release of glutamate directly at the synapse would circumvent many of these potential problems. Glial cells, which ensheath synapses,[23] respond to synaptically released glutamate by activation of electrogenic transporters,[24–26] and thus these transporter currents may reflect the amount of transmitter released. We have, therefore, quantified the sensitivity of these glial transporter currents to well-established presynaptic manipulations and have used this assay system to monitor the level of glutamate release during LTP in the CA1 region of the hippocampus.[27]

Glial transporter currents were isolated pharmacologically by blocking ionotropic glutamate receptors with kynurenate. We first established that the transporter current was sensitive to well-established presynaptic manipulations that change glutamate release and quantified the sensitivity of this assay by using PPF. The magnitude of PPF depends upon the interval between stimuli and therefore provides a manipulation by which the probability of transmitter release can be changed quantitatively. A comparison of the degree of facilitation of the glial transporter current to the field EPSP at different intervals showed a close congruence between the two parameters, indicating that glial transporter currents provide a very accurate assay for the synaptic release of glutamate.

An analysis of glial cell responses during LTP is complicated by the fact that the conditions required to measure glial transporter currents (i.e., the blockade of postsynaptic glutamate receptors) preclude monitoring LTP. To circumvent this problem we took advantage of the observation that the nonselective ionotropic glutamate receptor antagonist kynurenate can be rapidly washed from the preparation. In this experiment the responses to two independent inputs were recorded in a glial cell in the presence of kynurenate. After a stable baseline was obtained, kynurenate was washed from the preparation so that the simultaneously recorded synaptic field potential responses in the immediate vicinity of the recorded glial cell could be monitored. Following stabilization of the field potential response LTP was induced in one of the pathways and monitored for approximately 10 minutes. Kynurenate was then reapplied to the preparation. The size of the glial response

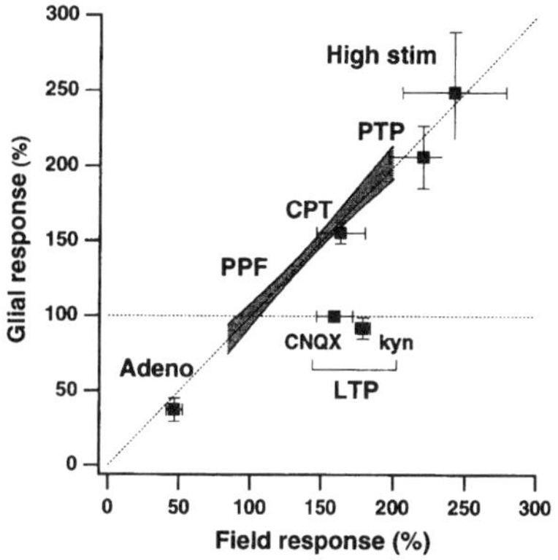

FIGURE 1. Summary diagram of glial recording experiments. Figure shows the correlation between field response and glial response for the manipulations described in the text. Manipulations known to affect release probability (Adenosine, CPT, PPF, and PTP) and number of release sites (increased stimulation strength) fall onto the line of identity. In contrast, the two experiments where LTP was induced show an unchanged glial response.

in the tetanized pathway was compared to that in the control pathway. No difference was detected ($-8 \pm 7\%$), despite an average LTP in the field EPSPs of $179 \pm 5\%$. To ensure that the LTP had not decayed back to baseline while recording the glial cell responses, we again washed out the kynurenate and confirmed that the LTP was stable. A summary of all the experiments including manipulations in transmitter release and LTP is shown in FIGURE 1. Failure to detect any increase in glutamate release, as monitored by glial transporter currents, has also been reported recently by Diamond, *et al.*[28] These findings strongly support a postsynaptic expression mechanism for LTP.

To summarize the results thus far, we have designed a number of experiments to determine if we could detect an increase in the release of transmitter during LTP. Although a variety of approaches were used, no evidence could be found for an increase in transmitter release, either as a consequence of an increase in the probability of release or in the number of releasing synapses.

EXPERIMENTS DESIGNED TO TEST WHETHER THE SENSITIVITY AND/OR NUMBER OF AMPA RECEPTORS INCREASES DURING LONG-TERM POTENTIATION

Miniature Excitatory Postsynaptic Currents

Analysis of the size and frequency of miniature synaptic events at the neuromuscular junction has proved very effective in defining the site responsible for a change in synaptic strength. The classical interpretation is that a change in miniature EPSC (mEPSC) size reflects a postsynaptic change, whereas a change in frequency reflects a presynaptic change. This approach has limitations in the CNS because pyramidal cells receive thousands of excitatory synapses that will all release quanta spontaneously, and only a small percentage of synapses can be electrically stimulated to induce LTP. We have used two approaches to circumvent this problem. First, the effect of brief applications of NMDA[29] and strong repetitive depolarizing voltage pulses[30–32] was examined. In the second approach we replaced Ca with Sr, which causes the asynchronous release of quanta from activated synapses.[33]

The rationale behind the first approach was to potentiate as many synapses impinging onto the cell as possible. Brief application of NMDA can cause an enhancement in evoked EPSCs,[34] and this is accompanied by a substantial increase in the amplitude of mEPSCs.[29] In addition, repeated activation of Ca channels by depolarizing voltage pulses causes a very robust increase in the amplitude of mEPSCs,[31] and this effect appears to be mediated by CaM kinase II.[32] These changes were also associated with a modest increase in the apparent frequency of mEPSCs. To a considerable extent this increase in frequency is likely to be due to small events that, prior to the potentiation, failed to reach the threshold for detection. These results unequivocally demonstrate that a rise in intracellular Ca can result in an increase in the amplitude of mEPSCs, which classically indicates that the sensitivity and/or number of postsynaptic receptors has increased. One limitation to these studies is the uncertainty that the potentiation induced by these procedures is identical to LTP. For instance, the potentiation induced by voltage pulses is transient and does not appear to occlude with LTP.[30,35]

The rationale in the second approach was to devise a way of looking specifically at mEPSCs arising from the small subset of electrically stimulated synapses. In this way we could look for changes in mEPSCs associated with standard pairing induced LTP. To accomplish this we replaced Ca with Sr.[33] It has been known for some time that, while Sr fails to substitute for Ca in the synchronous release of quanta, asynchronous release is dramatically enhanced. This then allows one to collect and analyze those quanta that are released specifically from the stimulated synapses that impinge on a pyramidal cell. In these experiments we used two pathways, and quantal size was compared between the control pathway and the one expressing LTP. Using this approach, we found that there was a clear increase in quantal size during LTP.

Synaptic Potency

When conditions are optimized for activating a single synaptic site, the response fluctuates between no response (failures) and responses. If one subtracts out the failures and then averages together the successes, the average quantal size is obtained, and this param-

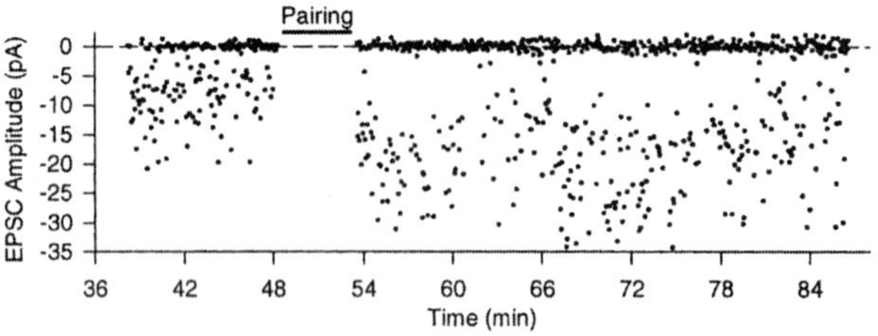

FIGURE 2. Example of LTP that was monitored with perforated patch recording and single-axon stimulation and that was associated with an increase in potency and no decrease in failure rate. Individual response amplitudes during the course of an experiment. Time 0 (not shown) was the time at which a 10-GΩ seal was established. Note that the gap between successes and failures increases following pairing, indicating that the potency has increased.

eter has been referred to as *potency*.[36] We have carried out a study on LTP using minimal synaptic stimulation and tests for single-fiber and single-synapse activation.[37] We have found that during LTP the potency invariably increases, although the magnitude of the increase varies across cells. FIGURE 2 shows an example in which there was no change in failures, but a large increase in potency. However, it should be pointed out that this result conflicts with another study that concluded that there was no change in potency following LTP.[36] The reason for this difference is not apparent.

CaM Kinase II and Long-Term Potentiation

A great deal of evidence has been obtained indicating that the Ca-sensitive kinase, CaM kinase II, which is present in high concentrations in the postsynaptic density, plays an important role in LTP.[38] We have loaded cells with a constitutively active form of CaM kinase II and found that it enhances EPSCs, as would be expected if this enzyme were involved in LTP.[39] In addition we found that responses to iontophoretically applied AMPA were also enhanced to a similar magnitude. If this postsynaptic enhancement by CaM kinase II is identical to LTP, synapses expressing saturating levels of LTP should no longer be enhanced by CaM kinase II. We have therefore compared in the same cell the enhancing effects of CaM kinase II on two independent pathways, one expressing saturating levels of LTP and the other serving as a control. CaM kinase II had its usual enhancing action on the control pathway, but failed to enhance the pathway expressing LTP. These findings indicate that the enhancement seen with LTP and CaM kinase II share the same mechanism. Since we also demonstrated that CaM kinase II enhances responses to exogenous AMPA, these findings also support a postsynaptic expression mechanism for LTP.

Postsynaptically Silent Synapses

While the evidence reviewed thus far overwhelmingly supports a postsynaptic expression mechanism for LTP, there is one widely accepted finding that seems incompatible with a postsynaptic expression mechanism. Minimal stimulation experiments have shown that LTP is typically, although not invariably, associated with a decrease in synaptic failures[37,40] (see also Refs. 1–8). Classically a change in failures has been interpreted as a change in the probability of transmitter release and thus as indicating that LTP is associated with an increase in this probability. Since all other experiments that have addressed the site of expression of LTP have pointed decisively to a postsynaptic expression mechanism, we wondered if there might be a postsynaptic explanation for the change in failures seen with LTP. For instance, what if some synapses lacked functional AMPA receptors. In this case a failure could be due to the failure to detect released transmitter, rather than the failure to release transmitter. We therefore searched for the possible existence of synapses that had the normal compliment of NMDA receptors but had no detectable AMPA receptors[41] (FIG. 3). This was done by decreasing the stimulus strength to just below the threshold for activating any response, while holding the membrane at −70 to −80 mV. We then shifted the membrane to +30 to +40 mV, which would remove the Mg block of the NMDA receptor, and in approximately half the cases we were able to clearly discern an NMDA receptor–mediated response. We were also able to show that stimuli that gave no

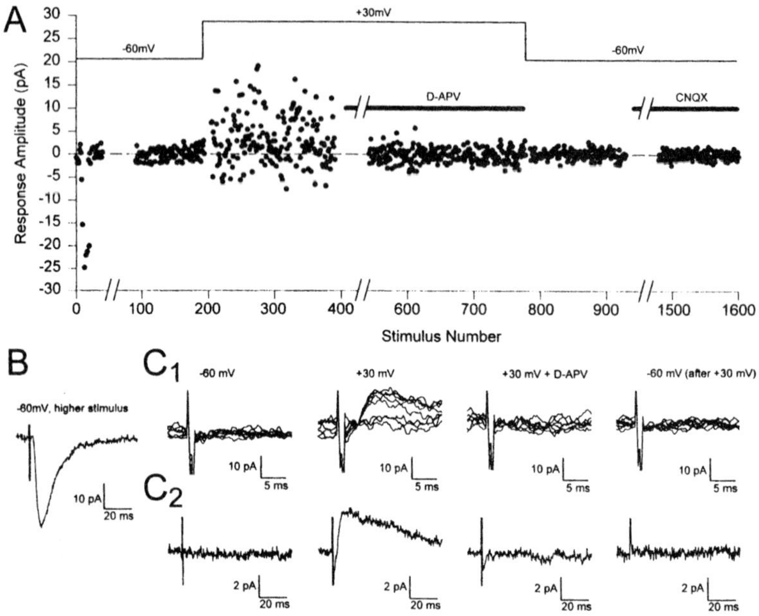

FIGURE 3. Example of an experiment demonstrating the existence of silent synapses. **(A)** Graph of an experiment to illustrate the standard experimental protocol. The cell was held at −60 mV, and after obtaining a small EPSC, stimulus intensity was decreased so that no EPSCs were detected for 100 consecutive stimuli. The cell was then depolarized to +30 mV, and stimulation now evoked responses that were completely blocked by the application of D-APV (25 μM), indicating that they were mediated by NMDARs. The cell was then returned to −60 mV, where again no EPSCs could be detected, as evidenced by the lack of effect of CNQX (10 μM), which was applied at the end of this experiment. **(B)** Sample of EPSC (average of 10 responses) recorded at the beginning of the experiment. **(C)** Examples of 8 consecutive traces **(C_1)** or the average of 100 consecutive responses **(C_2;** including the traces in C_1) taken at the indicated times during the course of the experiment. For all experiments in this study, once stimulation was commenced, it was maintained at the same rate throughout the entire experiment.

evidence of an AMPA EPSC in control conditions evoked AMPA EPSCs after the induction of LTP by pairing (FIG. 4). Thus it appears that a fraction of synapses in the CA1 region have NMDA receptors but not functional AMPA receptors, and that LTP-inducing stimuli can switch on these silent synapses. Similar results in the CA1 region,[12,14] as well as in a number of other preparations[42–44] have been observed by others.

CONCLUSION

This review has focused on experiments performed over the past decade in the authors' laboratories that have specifically addressed the site of expression of NMDA receptor–dependent LTP. Since this is a highly controversial field, we decided to summarize in one manuscript the evidence that has led us to the view that LTP is expressed postsynaptically.

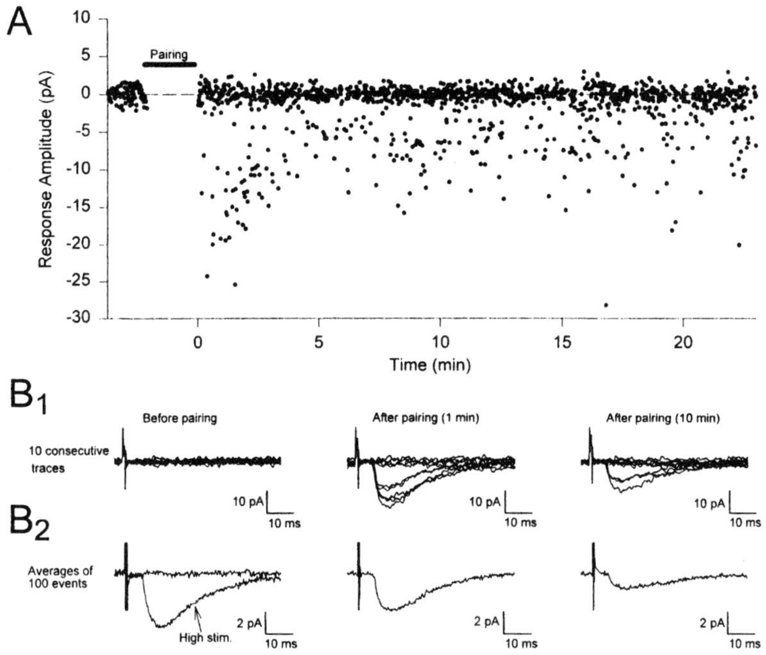

FIGURE 4. A silent synapse can exhibit LTP. (**A**) Graph of an experiment in which stimulus intensity was reduced so that no EPSCs were elicited at −60 mV. While continuing afferent stimulation, the cell was depolarized to −10 mV (pairing; *bar*) When the cell was returned to −60 mV, clear EPSCs could now be elicited, and these remained for the duration of the recording. (**B**) Examples of 10 consecutive traces (**B_1**) or the average of 100 consecutive traces (**B_2**; including traces in B_1).

Our experiments can be divided into two groups: those that were designed to determine if there is any evidence to support a presynaptic site of expression, and those that were designed to determine if there is any evidence for a change in the postsynaptic responsiveness of AMPA receptors. The strategy underlying the experiments in the first category was to compare the effects of well-established presynaptic manipulations to LTP on a variety of tests. These included paired pulse facilitation, saturating the probability of transmitter release, the use of MK-801 to measure the probability of release, and the use of glial glutamate transporter currents to measure directly the synaptic release of glutamate. None of these experiments gave any evidence that LTP is associated with an increase in the release of glutamate. Experiments in the second category included analysis of mEPSC amplitude, synaptic potency, and the effects of applying constitutively active CaM kinase II. All of these experiments yielded results consistent with a postsynaptic expression mechanism. The one universally accepted finding that seems incompatible with the experiments summarized above is the decrease in synaptic failures that occurs during LTP. However, we and others have found evidence suggesting that a portion of excitatory synapses in the CA1 region may be postsynaptically silent—that is, they lack functional AMPA receptors, but do contain a full compliment of NMDA receptors. Furthermore, these silent synapses can be "switched on" during LTP, thus providing a postsynaptic explanation for a decrease in failures during LTP. The mechanisms involved in controlling the trafficking and stabilization of AMPA receptors at the synapse remain to be elucidated.

REFERENCES

1. Gustafsson, B., F. Asztely, E. Hanse & H. Wigström. 1989. Onset characteristics of long-term potentiation in the guinea-pig hippocampal CA1 region *in vitro*. Eur. J. Neurosci. **1:** 382–394.
2. Petersen, C.C.H., R.C. Malenka, R.A. Nicoll & J.J. Hopfield. 1998. All-or-none potentiation at CA3-CA1 synapses. Proc. Natl. Acad. Sci. USA **95:** 4732–4737.
3. Bliss, T.V.P. & M.A. Lynch. 1988. Long-term potentiation of synaptic transmission in the hippocampus: Properties and mechanisms. *In* Long-Term Potentiation: From Biophysics to Behavior. P.W. Landfield & S.A. Deadwyler, Eds.: 3–72. Liss. New York.
4. Gustafsson, B. & H. Wigström. 1988. Physiological mechanisms underlying long-term potentiation. Trends Neurosci. **11:** 156–162.
5. Kullmann, D.M. & S.A. Siegelbaum. 1995. The site of expression of NMDA receptor-dependent LTP: New fuel for an old fire. Neuron **15:** 997–1002.
6. Larkman, A.U. & J.J.B. Jack. 1995. Synaptic plasticity: Hippocampal LTP. Curr. Opin. Neurobiol. **5:** 324–334.
7. Madison, D.V., R.C. Malenka & R.A. Nicoll. 1991. Mechanisms underlying long-term potentiation of synaptic transmission. Annu. Rev. Neurosci. **14:** 379–397.
8. Nicoll, R.A. & R.C. Malenka. 1995. Contrasting properties of two forms of long-term potentiation in the hippocampus. Nature **377:** 115–118.
9. Kauer, J.A., R.C. Malenka & R.A. Nicoll. 1988. A persistent postsynaptic modification mediates long-term potentiation in the hippocampus. Neuron **1:** 911–917.
10. Perkel, D.J. & R.A. Nicoll. 1993. Evidence for all-or-none regulation of neurotransmitter release: Implications for long-term potentiation. J. Physiol. **471:** 481–500.
11. Asztely, F., H. Wigström & B. Gustafsson. 1992. The relative contribution of NMDA receptor channels in the expression of long-term potentiation in the hippocampal CA1 region. Eur. J. Neurosci. **4:** 681–690.
12. Durand, G.M., Y. Kovalchuk & A. Konnerth. 1996. Long-term potentiation and functional synapse induction in developing hippocampus. Nature **381:** 71–75.
13. Kullmann, D.M., G. Erdemli & F. Asztely. 1996. LTP of AMPA and NMDA receptor-mediated signals: Evidence for presynaptic expression and extrasynaptic glutamate spill-over. Neuron **17:** 461–474.
14. Liao, D., N.A. Hessler & R. Malinow. 1995. Activation of postsynaptically silent synapses during pairing-induced LTP in CA1 region of hippocampal slice. Nature **375:** 400–404.
15. Muller, D., M. Joly & G. Lynch. 1988. Contributions of quisqualate and NMDA receptors to the induction and expression of LTP. Science **242:** 1694–1697.
16. Clark, K.A. & G.L. Collingridge. 1995. Synaptic potentiation of dual-component excitatory postsynaptic currents in the rat hippocampus. J. Physiol. (Lond.) **482:** 39–52.
17. Manabe, T., D.J.A. Wyllie, D.J. Perkel & R.A. Nicoll. 1993. Modulation of synaptic transmission and long-term potentiation: Effects on paired pulse facilitation and EPSC variance in the CA1 region of the hippocampus. J. Neurophysiol. **70:** 1451–1459.
18. Markram, H. & M. Tsodyks. 1996. Redistribution of synaptic efficacy between neocortical pyramidal neurons. Nature **382:** 807–810.
19. Selig, D.K., R.A. Nicoll & R.C. Malenka. 1999. Hippocampal long-term potentiation preserves the fidelity of postsynaptic responses to presynaptic bursts. J. Neurosci. In press.
20. Hjelmstad, G.O., R.A. Nicoll & R.C. Malenka. 1997. Synaptic refractory period provides a measure of probability of release in the hippocampus. Neuron **19:** 1309–1318.
21. Hessler, N.A., A.M. Shirke & R. Malinow. 1993. The probability of transmitter release at a mammalian central synapse. Nature **366:** 569–572.
22. Rosenmund, C., J.D. Clements & G.L. Westbrook. 1993. Nonuniform probability of glutamate release at a hippocampal synapse. Science **262:** 754–757.

23. SPACEK, J. 1983. Three-dimensional analysis of dendritic spines. III Glial sheath. Anat. Embryol. **167:** 289–310.
24. BERGLES, D.E., F.A. DZUBAY & C.E. JAHR. 1997. Glutamate transporter currents in Bergmann glial cells following the time course of extrasynaptic glutamate. Neuron **19:** 14821–14825.
25. BERGLES, D.E. & C.E. JAHR. 1997. Synaptic activation of glutamate transporters in hippocampal astrocytes. Neuron **19:** 1297–1308.
26. SARANTIS, M., L. BALLERINI, M. EDWARDS, B. MILLER, A. SILVER & D. ATTWELL. 1993. Effects of the glutamate uptake blocker L-*trans*-PDC on isolated salamander retinal glia and isolated rat neurones. J. Physiol. **459:** 246P.
27. LÜSCHER, C., R.C. MALENKA & R.A. NICOLL. 1998. Monitoring glutamate release during LTP with glial transporter currents. Neuron **21:** 435–441.
28. DIAMOND, J.S., D.E. BERGLES & C.E. JAHR. 1998. Glutamate release monitored with estrocyte transporter current during LTP. Neuron **21:** 425–433.
29. MANABE, T., P. RENNER & R.A. NICOLL. 1992. Postsynaptic contribution to long-term potentiation revealed by the analysis of miniature synaptic currents. Nature **355:** 50–55.
30. KULLMANN, D.M., D.J. PERKEL, T. MANABE & R.A. NICOLL. 1992. Ca^{2+} entry via postsynaptic voltage-sensitive Ca^{2+} channels can transiently potentiate excitatory synaptic transmission in the hippocampus. Neuron **9:** 1175–1183.
31. WYLLIE, D.J.A., T. MANABE & R.A. NICOLL. 1994. A rise in postsynaptic Ca^{2+} potentiates miniature excitatory postsynaptic currents and AMPA responses in hippocampal neurons. Neuron **12:** 127–138.
32. WYLLIE, D.J.A. & R.A. NICOLL. 1994. A role for protein kinases and phosphatases in the Ca^{2+}-induced enhancement of AMPA receptor-mediated synaptic responses in the hippocampus. Neuron **13:** 635–643.
33. OLIET, S.H.R., R.C. MALENKA & R.A. NICOLL. 1996. Bidirectional control of quantal size by synaptic activity in the hippocampus. Science **271:** 1294–1297.
34. KAUER, J.A., R.C. MALENKA & R.A. NICOLL. 1988. NMDA application potentiates synaptic transmission in the hippocampus. Nature **334:** 250–252.
35. CHEN, H.-X., E. HANSE, M. PANANCEAU & B. GUSTAFSSON. 1998. Distinct expressions for synaptic potentiation induced by calcium through voltage-gated calcium and *N*-methyl-D-aspartate receptor channels in the hippocampal CA1 region. Neuroscience **86:** 415–422.
36. STEVENS, C.F. & Y. WANG. 1994. Changes in reliability of synaptic function as a mechanism for plasticity. Nature **371:** 704–707.
37. ISAAC, J.T.R., G.O. HJELMSTAD, R.A. NICOLL & R.C. MALENKA. 1996. Long-term potentiation at single fiber inputs to hippocampal CA1 pyramidal cells. Proc. Natl. Acad. Sci. USA **93:** 8710–8715.
38. LISMAN, J. 1994. The CaM kinase II hypothesis for the storage of synaptic memory. Trends Neurosci. **17:** 406–412.
39. LLEDO, P.-M., G.O. HJELMSTAD, S. MUKHERJI, T.R. SODERLING, R.C. MALENKA & R.A. NICOLL. 1995. Calcium/calmodulin-dependent kinase II and long-term potentiation enhance synaptic transmission by the same mechanism. Proc. Natl. Acad. Sci. USA **92:** 11175–11179.
40. KULLMANN, D.M. & R.A. NICOLL. 1992. Long-term potentiation is associated with increases in both quantal content and quantal amplitude. Nature **357:** 240–244.
41. ISAAC, J.T.R., R.A. NICOLL & R.C. MALENKA. 1995. Evidence for silent synapses: Implications for the expression of LTP. Neuron **15:** 427–434.
42. ISAAC, J.T.R., M.C. CRAIR, R.A. NICOLL & R.C. MALENKA. 1997. Silent synapses during development of thalamocortical inputs. Neuron **18:** 269–280.
43. LI, P. & M. ZHUO. 1998. Silent glutamatergic synapses and nociception in mammalian spinal cord. Nature **393:** 695–698.
44. WU, G., R. MALINOW & H.T. CLINE. 1996. Maturation of a central glutamatergic synapse. Science **274:** 972–976.

Activation of *N*-Methyl-D-Aspartate Receptors Reverses Desensitization of Metabotropic Glutamate Receptor, mGluR5, in Native and Recombinant Systems

S. ALAGARSAMY,[a] S. T. ROUSE,[a] R. W. GEREAU IV,[b] S. F. HEINEMANN,[b] Y. SMITH,[c] AND P. J. CONN[a,d]

[a]*Department of Pharmacology, Division of Neuroscience, Emory University School of Medicine, Atlanta, Georgia 30322, USA*

[b]*Molecular Neurobiology Laboratory, Salk Institute for Biological Studies, La Jolla, California 92037, USA*

[c]*Yerkes Regional Primate Research Center, Emory University School of Medicine, Atlanta, Georgia 30322, USA*

Glutamate, the primary excitatory neurotransmitter in the central nervous system, elicits fast excitatory synaptic responses by activation of glutamate-gated cation channels termed ionotropic glutamate receptors (iGluRs).[1] Glutamate also modulates synaptic transmission and cell excitability by activation of G protein–coupled receptors, termed metabotropic glutamate receptors (mGluRs).[2]

The NMDA receptor is unique among the ionotropic glutamate receptors in its critical role in formation of new memory and other forms of synaptic plasticity.[1] In addition, excessive activation of NMDA receptors leads to excitotoxicity that may participate in cell death associated with a variety of pathological conditions, including stroke, seizure activity, and certain neurodegenerative disorders.[1] Interestingly, one of the most prominent effects of mGluR activation in a number brain regions is an enhancement of agonist-evoked currents through the NMDA receptor cation channel.[3] In hippocampus and other cortical regions, this response is likely mediated by mGluR5.[3] Recent studies in hippocampal and cortical slices suggest that activation of NMDA receptors can also potentiate responses to activation of mGluR5.[4,5] This positive feedback reciprocal regulation between mGluR5 and NMDA receptors may play an important role in regulating NMDA receptor-dependent forms of synaptic plasticity and could contribute to pathological responses to NMDA receptor activation. However, little is known about the mechanisms by which NMDA receptor activation potentiates mGluR5-mediated responses.

Recent studies suggest that mGluR5 undergoes a rapid agonist-induced desensitization that is mediated by activation of protein kinase C (PKC) and PKC-mediated phosphorylation of identified serine residues on mGluR5.[6] This, coupled with previous findings that NMDA receptor activation can lead to activation of calcium-dependent serine/threonine phosphatases,[7] led us to postulate that mGluR5 and NMDA receptor subunits are colocal-

[d]Address for correspondence: P. Jeffrey Conn, Rollins Research Building, Department of Pharmacology, Emory University, 1510 Clifton Road, Atlanta, Georgia 30322. Phone: 404-727-5617; fax: 404-727-0365; e-mail: jconn@pharm.emory.edu

ized at specific postsynaptic sites and that activation of NMDA receptors reverses agonist-induced desensitization of mGluR5 by activating a phosphatase and inducing dephosphorylation of this receptor. We now report a series of studies in recombinant and native systems that provide direct support for this hypothesis.

Consistent with previous studies, the mGluR agonist 1S,3R-ACPD activates a calcium-dependent chloride current in *Xenopus* oocytes expressing mGluR5 and NMDA NRl-1a/NR2B (FIG. 1). FIGURE 1 shows representative traces to the current response of an oocyte injected with mGluR5, NRl-la and NR2B RNA to two sequential applications of lS,3R-ACPD, with an interval between 1S,3R-ACPD applications of 2–3 minutes. Thus, the first agonist application elicited a robust response, whereas there was little or no discernible current response to the second 1S,3R-ACPD application with this short interval between agonist applications.

To test the hypothesis that NMDA receptor activation reduces mGluR5 desensitization, NMDA was applied during the interval between lS,3R-ACPD applications. In oocytes in which NMDA was applied for a period of 2 minutes between lS,3R-ACPD applications, the second application of 1S,3R-ACPD elicited a robust response that was similar in magnitude to the response to the first 1S,3R-ACPD application (FIG. 1B). The response to NMDA was clearly mediated by NMDA receptor activation, since it was inhibited by 100 μM AP5 (FIG. 1C), a competitive NMDA receptor antagonist, or by adding Mg^{2+} (1.2 mM) to the extracellular perfusion solution (10.8 ± 5.6% of first current). Phosphatases were also involved in the NMDA effects since the phosphatase inhibitor, sodium orthovanadate (1 mM), and a more selective inhibitor of phosphatase 2B, cyclosporin-A (500 μM), blocked the effect of NMDA on mGluR5-mediated responses (FIG. lE).

The preceding studies suggest that NMDA reverses desensitization of mGluR5 by a mechanism that involves activation of a phosphatase in *Xenopus* oocytes. To determine whether NMDA has a similar action in a native preparation, we determined the effect of NMDA on the DHPG-induced phosphoinositide hydrolysis in cortical slices. Consistent with a previous report,[5] NMDA (10 μM) evoked a robust potentiation of agonist-induced [^{3}H]-inositol phosphate accumulation in rat cortical slices (FIG. 2C) if this is due to reversal of PKC-induced desensitization of mGluR5, a selective PKC inhibitor should also potentiate the response by eliminating PKC-induced desensitization, and should prevent any further potentiation by NMDA. As shown in FIGURE 2B, bisindolylmaleimide I (BIS) (10 μM), a selective inhibitor of PKC induced a potentiation of the response of DHPG that was similar to that induced by NMDA. Furthermore, BIS prevented any further potentiating effect of NMDA. In addition, the potentiating effect of NMDA was completely blocked by the nonselective phosphatase inhibitor sodium orthovanadate and by the phosphatase 1b-selective inhibitor cyclosporin-A (FIG. 2C). In contrast, okadaic acid had no effect on the response to NMDA at concentrations that selectively inhibit phosphatase 1 and 2A (data not shown). Additionally, we directly tested the hypothesis that NMDA induces dephosphorylation of mGluR5 in cortical neurons. Neurons were incubated with ^{32}P-orthophosphate and treated with agonists to induce changes in phosphorylation. mGluR5 was then isolated by immunoprecipitation, and radioactivity in mGluR5 was determined by autoradiography. DHPG (0.5 mM) induced a robust phosphorylation of mGluR5. NMDA (10 μM) had little or no effect on mGluR5 phosphorylation when added alone at this concentration. However, NMDA markedly reduced DHPG-induced mGluR5 phosphorylation (data not shown).

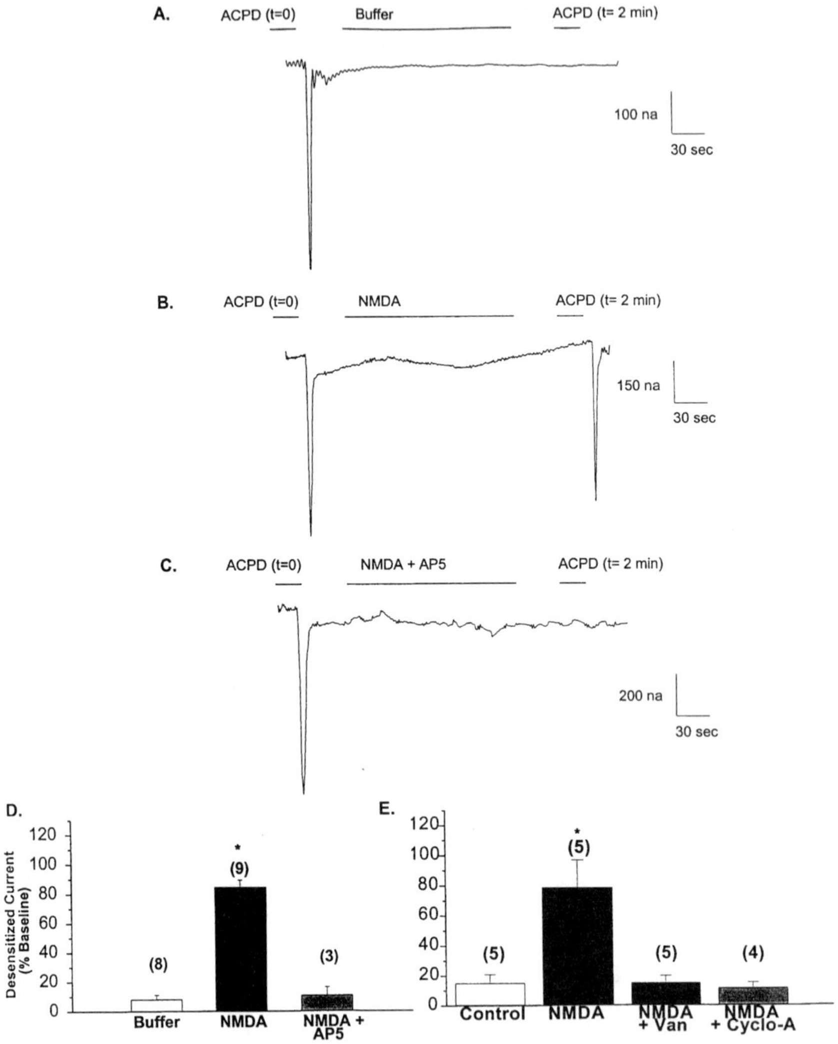

FIGURE 1. NMDA receptor activation reduces mGluR5 desensitization in *Xenopus* oocytes. **(A)** Representative traces of two sequential applications of 1S,3R-ACPD with a 2-min interval between agonist application. Note the marked desensitization to the second 1S,3R-ACPD application. **(B)** NMDA application during the 2-min interval between 1S,3R-ACPD applications reverses mGluR5a desensitization. **(C)** The NMDA receptor antagonist AP5 blocks this reversal of desensitization. **(D)** Mean data ± SEM from (*N*) experiments of the peak amplitude of second 1S,3R-ACPD-evoked current/peak amplitude of first 1S,3R-ACPD-evoked current × 100. (*$p < 0.01$; Student's *t*-test). **(E)** Phosphatase inhibitors block the NMDA-induced reversal of desensitization. Mean ± SEM of (*N*) experiments using phosphatase inhibitors, sodium orthovanadate (VAN) or cyclosporin-A (Cyclo-A). Data are calculated as previously described (*$p < 0.01$).

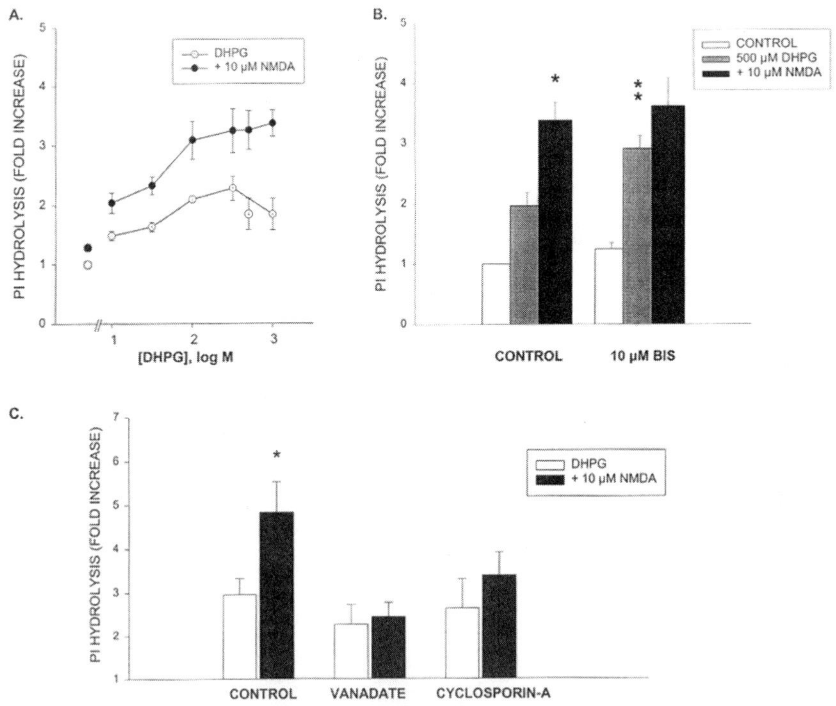

FIGURE 2. **(A)** DHPG dose dependently increases phosphoinositide hydrolysis, and NMDA increases this response in adult rat cortex. Data are represented as percent of no drug control and are means ± SEM of 3 independent experiments, each done in triplicate. **(B)** PKC inhibitor bisinolmaleimide (BIS) potentiates the DHPG-induced phosphoinositide hydrolysis and prevents NMDA from potentiating this response. Data are calculated as described for panel A and are the means ± SEM of 3 experiments, each done in triplicate. All groups, except BIS control, are statistically different from untreated control, $p < 0.01$. **(C)** Phosphatase inhibitors block the NMDA-induced potentiation of phosphoinositide hydrolysis. The nonspecific and 2B specific inhibitors block the NMDA response. Data represent 3–7 independent experiments done in triplicate and are calculated as previously described ($*p < 0.05$).

Finally, we performed immunocytochemistry studies at the light and electron microscopic levels using antibodies specific for mGluR5 and NR1 to determine whether these receptors are colocalized at individual postsynaptic targets. Analysis at the light microscopic level revealed a very similar pattern of immunoreactivity for mGluR5 and NR1 in neocortex. Immunoreactivity for both proteins was abundant in the cell bodies and neuropil of all cortical areas. Pyramidal cell bodies as well as their apical dendrites were filled with immunoreactivity for both proteins, and it appeared that mGluR5 and NR1 may be expressed in the same population of cortical cells. At the electron microscopic level, immunoreactivity for both the mGluR5 (gold particles) and NR1 (DAB reaction product) was indeed colocalized in a large number of cortical pyramidal cell bodies, as well as proximal and distal dendrites and spines that received asymmetric synaptic inputs (data not shown). Taken together these studies provide converging lines of evidence that activation of NMDA receptors potentiates mGluR5 responses by activation of a serine/threonine protein phosphatase and dephosphorylation of mGluR5.

REFERENCES

1. HOLLMANN, M. & S. HEINEMANN. 1994. Cloned glutamate receptors. Annu. Rev. Neurosci. **17:** 31–108.
2. CONN, P.J. & J.P. PIN. 1997. Pharmacology and functions of metabotropic glutamate receptors. Annu. Rev. Pharmacol. Toxicol. **37:** 205–237.
3. FITZJOHN, S.M. *et al.* 1996. Activation of group I mGluRs potentiates NMDA responses in rat hippocampal slices. Neurosci. Lett. **203:** 211–213.
4. LÜTHI, A. *et al.* 1994. Potentiation of a metabotropic glutamatergic response following NMDA receptor activation in rat hippocampus. Pflügers Arch. **427:** 197.
5. CHALLISS, R.A.J. *et al.* 1994. Modulatory effects of NMDA on phosphoinositide responses evoked by the metabotropic glutamate receptor agonist 1S, 3R-ACPD in neonatal rat cerebral cortex. Br. J. Pharmacol. **112:** 231–239.
6. GEREAU, R.W. & S.F. HEINEMANN. 1998. Role of protein kinase C phosphorylation in rapid desensitization of metabotropic glutamate receptor 5. Neuron **20:** 143–151.
7. WANG, J.H. & A. STELZER. 1994. Inhibition of phosphatase 2B prevents expression of hippocampal long-term potentiation. Neuroreport **5:** 2377–2380.

Distribution of Group III mGluRs in Rat Basal Ganglia with Subtype-Specific Antibodies

STEFANIA RISSO BRADLEY,[a,d] DAVID G. STANDAERT,[b] ALLAN I. LEVEY,[c] AND P. JEFFREY CONN[a]

[a]*Department of Pharmacology, Emory University School of Medicine, Atlanta, Georgia 30322, USA*

[b]*Neurology Service, Massachusetts General Hospital and Harvard Medical School, Boston, Massachusetts 02114, USA*

[c]*Department of Neurology, Emory University School of Medicine, Atlanta, Georgia 30322, USA*

Parkinson's disease (PD) is a common basal ganglia neurodegenerative disorder resulting in disabling motor impairment (tremor, rigidity, and bradykinesia). Loss of nigrostriatal dopamine neurons results in a series of neurophysiological changes that lead to overactivity of the globus pallidus (GP; the main output nucleus of the basal ganglia) and consequent "shutdown" of thalamocortical structures, to produce the symptoms of PD. While therapies have traditionally utilized dopamine replacement strategies, this approach eventually fails in most patients. Exciting advances in understanding the molecular pharmacology of basal ganglia and recent findings about abundant localization of metabotropic glutamate receptors (mGluRs) in basal ganglia have provided the foundation for this study to examine the localization of two group III mGluR subtypes proteins, mGluR4a and mGluR7a, at crucial sites within basal ganglia circuits.

Glutamate is the principal excitatory neurotransmitter in the brain, and is present at many synapses along the basal ganglia circuits. It is now clear that the physiological effects of glutamate are mediated by ligand-gated cation channels, known as ionotropic glutamate receptors (iGluRs), and by G-protein-linked receptors, referred to as mGluRs. By activating mGluRs, glutamate can modulate transmission and neuronal excitability at the same synapses at which it elicits fast excitatory synaptic responses.[1] To date, eight mGluR subtypes have been identified by molecular cloning, and these receptors can be placed into three groups based on sequence homology, coupling to second messenger systems, and pharmacological profiles.[1]

Previous studies suggest that presynaptic group II and group III mGluRs play important roles in regulating excitatory[2] and inhibitory[3] transmission in the striatum (STR). However, the localization and physiological roles of mGluRs in other basal ganglia structures are not known. We produced and characterized polyclonal antibodies that specifically react with the C-terminus of mGluR4a and mGluR7a.[4,5] Confocal laser microscopic analysis of mGluR4a immunoreactivity showed intense staining for mGluR4a in fibers in GP, whereas the STR is virtually devoid of mGluR4a immunoreactivity (FIG. 1A). In contrast, mGluR7a immunoreactivity was strong in STR (FIG. 1C, left). Notable immunoreactivity was also

[d]Address for correspondence: Stefania Risso Bradley, Rollins Research Building, Department of Pharmacology, Emory University, Room 5160, 1510 Clifton Road, Atlanta, Georgia 30322. Phone: 404-727-0352; fax: 404-727-0365; e-mail: sbrad02@emory.edu

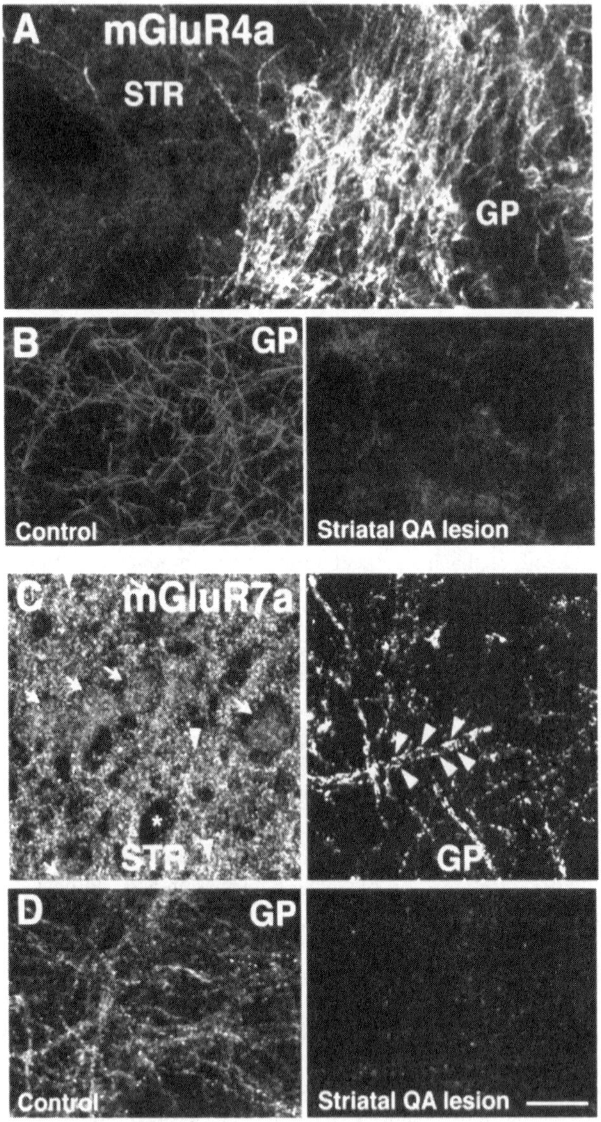

FIGURE 1. mGluR4a immunoreactivity in basal ganglia. **Panel A** shows intense localization of mGluR4a in fibers in the rat globus pallidus (GP). On the other hand, the striatum (STR) is practically depleted of mGluR4a immunoreactivity. **Panel B** shows mGluR4a staining in GP, contralateral **(left)** and ipsilateral **(right)** to a quinolate lesion of the striatum. *Scale bar*: A, 50 µm; B, 100 µm. **Panel C** shows confocal laser microscopy of mGluR7a immunoreactivity in STR **(left)** and GP **(right)**. Cell bodies in STR show no staining (*asterisk*). *Arrows* point to cells that show diffuse mGluR7a immunoreactivity, while *arrowheads* point to surrounding neuropil staining. *Arrowheads* in GP indicate intense staining along fibers. **Panel D** shows mGluR7a staining in GP, contralateral **(left)** and ipsilateral **(right)** to a quinolate lesion of the striatum. *Scale bar*: C, D, 20 µm.

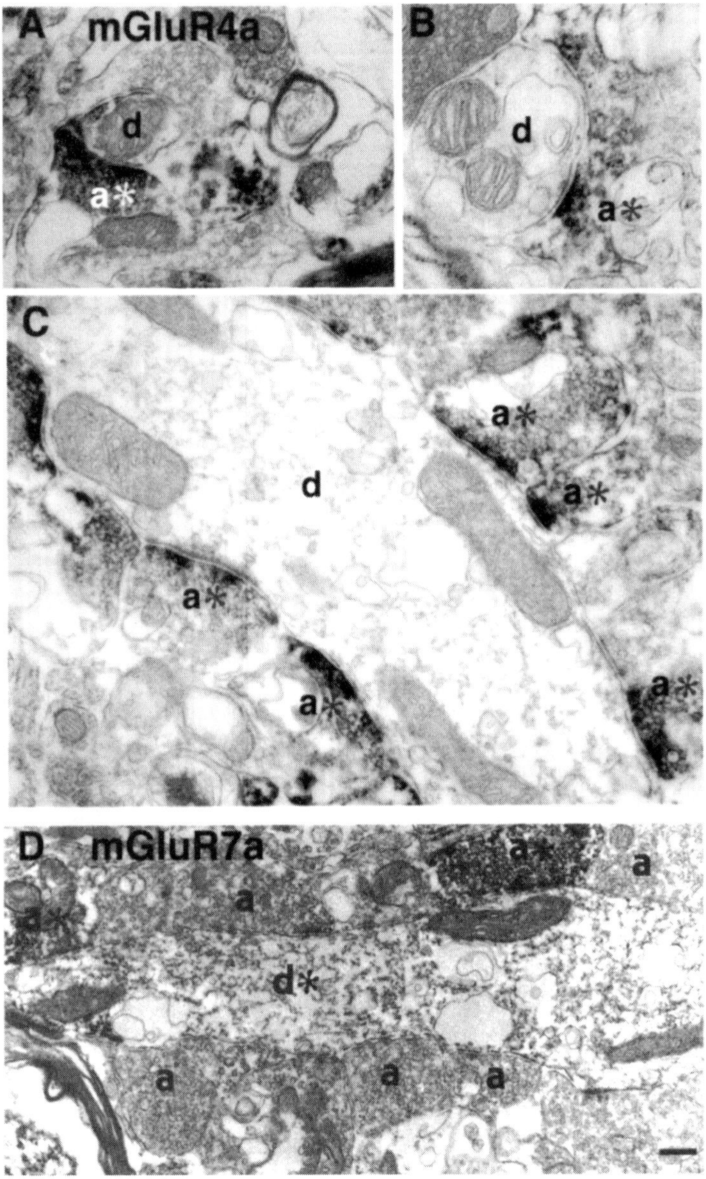

FIGURE 2. Electron micrographs demonstrating presynaptic mGluR4a immunoreactivity in the GP. Examples of mGluR4a-immunoreactive axon terminals (a*) synapsing with dendrites (d) of cells in GP (**Panels A, B, C**). **Panel D** shows pre- and postsynaptic localization of mGluR7a in GP: two labeled axons (a*) and a labeled dendrite (d*). *Scale bar*: A, B, C: 400 nm; D, 3.5 μm.

present in GP (FIG. 1C, right), and substantia nigra pars reticulata (SNr: not shown). In both of these regions mGluR7a immunoreactivity was localized along fibers. Omission of primary antibodies abolished both mGluR4a and mGluR7a immunoreactivity (not shown).

Double labeling with either mGluR4a or mGluR7a antibodies and the presynaptic marker SVE showed that mGluR4a or mGluR7a and SVE immunoreactivity are clearly colocalized along the outside of dendrites in the GP (not shown), suggesting a presynaptic localization of these mGluRs in GP. Consistent with this, FIGURE 1B and 1D showed mGluR4a and mGluR7a staining in the GP, contralateral (control) and ipsilateral to a quinolinate lesion of STR. Quinolinate lesioning of the projecting neurons to GP from STR induced a marked decrease in both mGluR4a and mGluR7a immunoreactivity in ipsilateral (FIG. 1B, 1D right) but not contralateral (FIG. 1B, 1D left) GP. To further test the hypothesis that mGluR4a and mGluR7a are presynaptically localized in GP, we performed immunocytochemistry with analysis at the electron microscopy (EM) level. EM analysis revealed that mGluR4a immunoreactivity was prominently on axon terminals, which form symmetric synapses with dendrites (FIG. 2A, 2B, 2C). Although the majority of dendrites were not labeled, a small percentage contained localized patches of reaction product (not shown). In contrast, mGluR7a seems to be present in both pre- and postsynaptic sites. FIGURE 2D shows examples of labeled axons and dendrite.

In conclusion, lesion studies, along with double labeling and immunocytochemistry at the electron microscopy level, revealed that, in the globus pallidus, mGluR4a is mainly localized in presynaptic sites in striatopallidal projections. mGluR7a also appears to be associated with presynaptic terminals of corticostriatal and striatopallidal projections. It is possible that presynaptically localized mGluR4a and mGluR7a could serve as heteroreceptors involved in regulating GABA release from striatopallidal and striato–SNr terminals. If so, these mGluR's could provide novel targets for new therapeutic agents that could be useful in treating Parkinson's disease and other disorders of basal ganglia function.

REFERENCES

1. CONN, P.J. & J.-P. PIN. 1997. Pharmacology and functions of metabotropic glutamate receptors. Ann. Rev. Pharmacol. Toxicol. **37:** 205–237.
2. PISANI, A. *et al.* 1997. Activation of group III metabotropic glutamate receptors depresses glutamatergic transmission at corticostriatal synapse. Neuropharmacol. **36:** 845–851.
3. STEFANI, A. *et al.* 1994. Activation of metabotropic glutamate receptors inhibits calcium currents and GABA-mediated synaptic potentials in striatal neurons. J. Neurosci. **14:** 6734–6743.
4. BRADLEY, S.R. *et al.* 1999. Immunohystochemical localization of subtype 4a metabotropic glutamate receptors in the rat and mouse basal ganglia. J. Comp. Neurol. In press.
5. BRADLEY, S.R. *et al.* 1998. Distribution and developmental regulation of metabotropic glutamate receptors in the rat brain. J. Neurochem. **71:** 636–645.

Characterization, Expression, and Distribution of GRIP Protein

HUALING DONG,[a] PEISU ZHANG, DEZHI LIAO, AND RICHARD L. HUGANIR

The Howard Hughes Medical Institute, Department of Neuroscience, The Johns Hopkins University School of Medicine, 725 North Wolfe Street, Baltimore, Maryland 21205, USA

Glutamate receptors are the major receptors in excitatory synapses. They also play important roles in neuronal development, excitotoxicity, and synaptic plasticity. Glutamate receptors are highly enriched as clusters on the postsynaptic membrane. The mechanisms underlying the sorting, targeting, anchoring, and clustering of glutamate receptors have been unclear. Recently we have isolated a glutamate receptor interacting protein (GRIP) that directly binds to the GluR2 and three subunits of the AMPA receptors. GRIP may be involved in the mechanisms underlying the targeting and clustering of AMPA receptors. Here we characterize the expression and distribution of GRIP protein in brain (FIG.1). GRIP is expressed in many neurons in brain and is present in a somatodendritic staining pattern in cerebral cortex, hippocampus, and cerebellum (FIGS. 2, 3). Immunostaining of cultured hippocampal neurons shows that GRIP colocalizes with both AMPA receptors and GABAergic synapses, suggesting that GRIP might be involved in the function of both excitatory and inhibitory synapses (FIG. 4).

DISCUSSION

Spatial localization and clustering of ion channels and neurotransmitter receptors are necessary for neurons to fire and propagate action potentials and for normal synaptic transmission.[1–4] Recent studies suggest that localization may be due to specific interactions between synaptic membrane proteins and a variety of anchoring or clustering proteins.[5–9] Individual neurons receive both glutamate-mediated excitatory synaptic input and GABA-mediated inhibitory synaptic input, with the corresponding expression of appropriate neurotransmitter receptors at each synaptic junction.[1,8] Thus, neurons must have some means of routing excitatory neurotransmitter receptors to excitatory synapses, and inhibitory neurotransmitter receptors to inhibitory synapses. Molecules that interact with the receptor complexes might play a role in these targeting processes.

AMPA glutamate receptors mediate the majority of the rapid excitatory synaptic transmissions in the central nervous system and play a role in neurodevelopment, excitotoxicity, and synaptic plasticity underlying learning and memory. AMPA receptors are heteromeric complexes of four homologous subunits (GluR1–4) that differentially combine to form a variety of AMPA receptor subtypes. These subunits are thought to have a large extracellular aminoterminal domain, three transmembrane domains, and an intracelular carboxyterminal domain. AMPA receptors are localized to excitatory synapses and

[a]Corresponding author. Phone: 410-955-4052; fax: 410-955-4857; e-mail: Hdong@ welchlink.jhu.edu

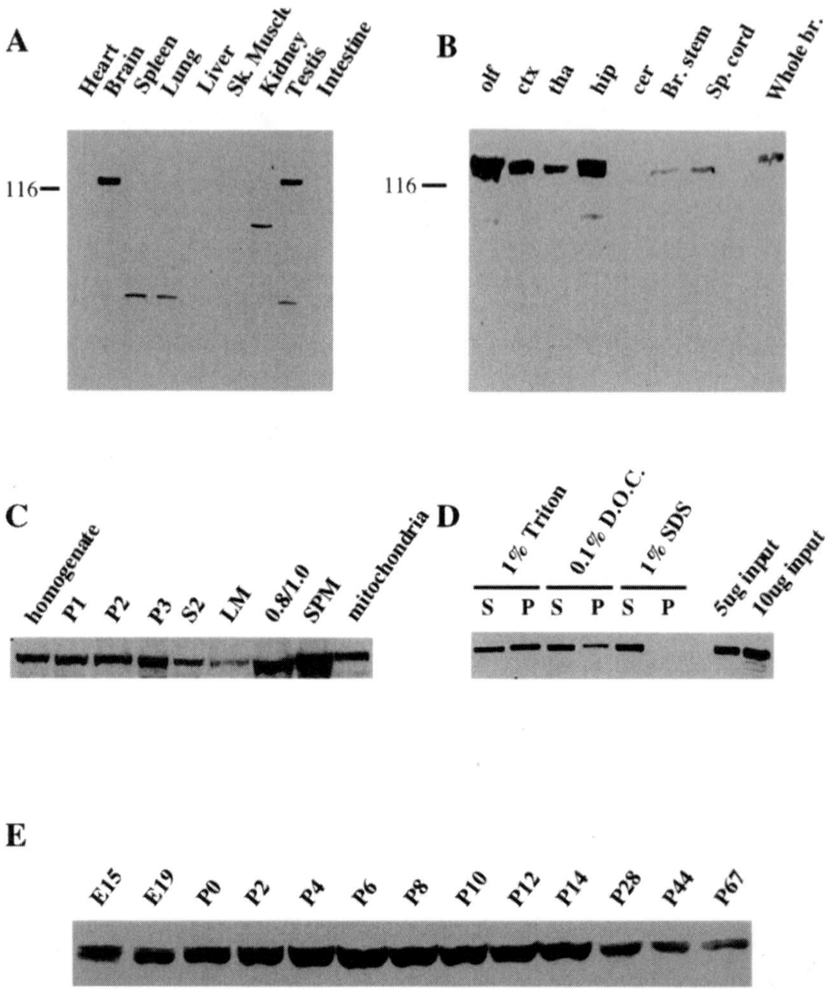

FIGURE 1. Characterization of GRIP protein. **(A)** Multiple tissue distribution of GRIP. GRIP is expressed as a 135-kD doublet in brain and testis. **(B)** Distribution of GRIP in subregions of the brain. GRIP is enriched in olfactory bulb, cortex, and hippocampus. **(C)** Subcellular fractionation of GRIP. GRIP is expressed in different cytosolic and membrane fractions of tissue lysates, but is highly enriched in synaptic plasmic membrane. **(D)** Solubility of GRIP in different detergents. GRIP is partially solubilized by 1% Triton X-100, and is solubilized very well in 0.1% deoxycholate and 1% SDS. **(E)** Developmental profile of GRIP expression. Compared with GluR1 expression, which continues to increase until adult, GRIP expression appears in earlier stages (as early as E10; data not shown), and exhibits increases through early development, peaks at about postnatal day 6–8, and then decreases slowly until adult.

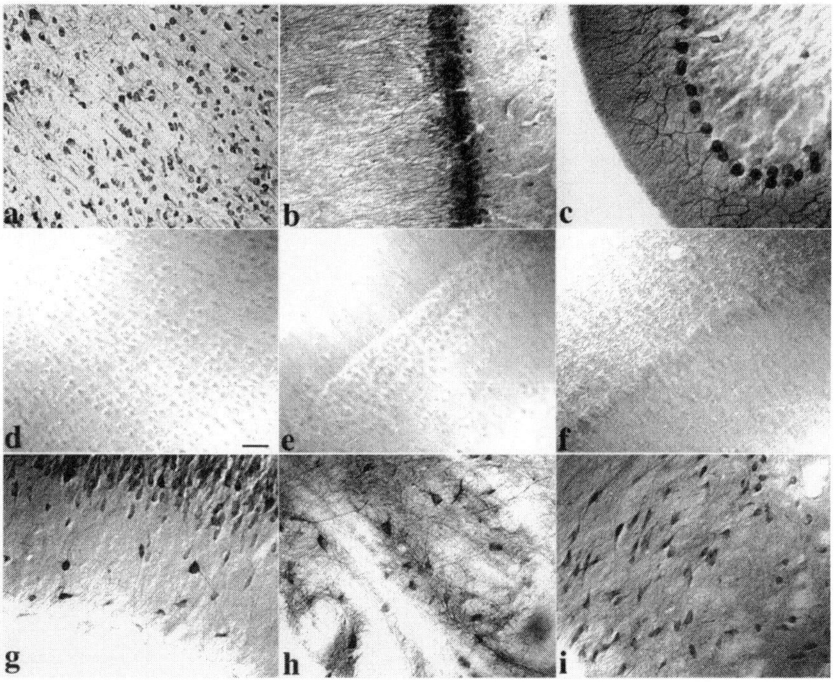

FIGURE 2. Immunohistochemical localization of GRIP in rat brain. Coronal rat brain sections were immunostained with GRIP antibody and visualized by DAB. GRIP immunoreactivity was observed in cerebral cortex **(a)**, hippocampal formation **(b)**, and cerebellum **(c)**, with neglectable staining in corresponding controls (antibody preincubated with immunogen peptide, **d–f**). **(a)** Immunostaining of cell soma and apical dendrites of pyramidal neurons in the cerebral cortex. Staining of basal dendrites is also observed. **(b)** Labeling of hippocampus CA1 pyramidal cells in a somatodendritic manner. Note scattered interneuron immunoreactivity and neuropil staining in stratum oriens and stratum radiatum. **(c)** Staining of Purkinje neurons extending from cell soma throughout major dendritic arborizations. **(g)** Strong staining of scattered inhibitory interneurons in stratum oriens of CA3 area. GRIP immunoreactivity is also observed in many other areas such as striatum **(h)** and substantia nigra **(i)**. *Calibration bar*: 40 μm.

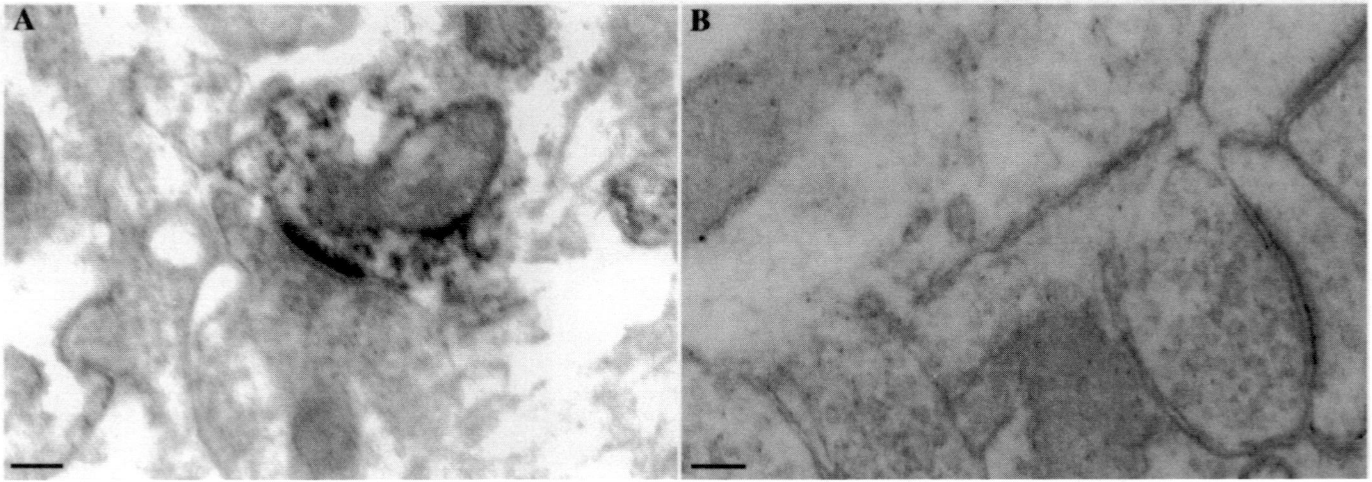

FIGURE 3. Immunoelectromicroscopy localization of GRIP protein in cortex. Coronal rat brain sections were immunostained with GRIP antibody and visualized by DAB. Cortex regions were further subjected to electromicroscopy analysis. GRIP protein shows an enriched expression in postsynaptic density areas (**A**), compared to unstained areas (**B**). *Calibration*: 0.12 μm.

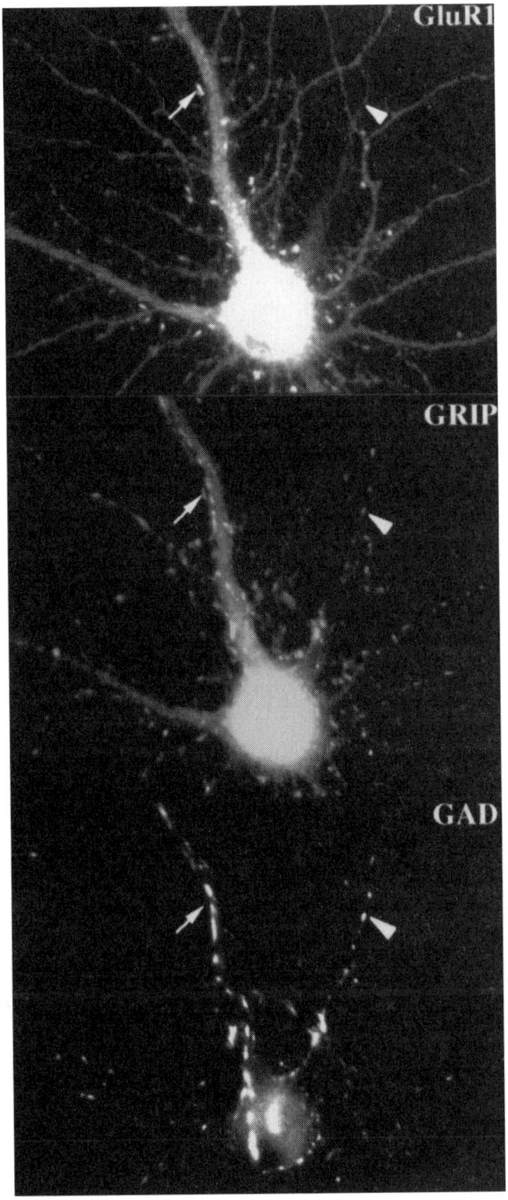

FIGURE 4. GRIP colocalizes at excitatory synapses with GluR1 and at inhibitory synapses with glutamate amino decarboxylase (GAD). Hippocampal neurons about 3 weeks in culture were triple stained with cy3-labeled anti-GluR1, Fluo-Ex-labeled anti-GRIP, and monoclonal anti-GAD. GRIP is distributed in puncta that colocalize with clusters of GluR1 at dendritic spines and dendritic shaft. GRIP also colocalizes with GAD as elongated clusters in dendritic shafts.

are not found on adjacent inhibitory synapses that are enriched in GABA receptors.[8] The targeting of neurotransmitter receptors, such as AMPA receptors, and ion channels to synapses is essential for efficient synaptic transmission. The mechanisms underlying the targeting and mobilization of AMPA receptors on the postsynaptic membrane has not been known. It has been suggested that cytoplasmic proteins that interact with the cytoplasmic domains of membrane receptors play a role in the process of receptor targeting and localization.[6,7] We have recently identified a PDZ domain-containing protein, GRIP, that specifically interacts with the C-termini of GluR 2 and 3 subunits of AMPA receptors.[5] GRIP has seven PDZ domains and appears to serve as an adaptor protein that links AMPA receptors to other cellular proteins. GRIP may thus mediate association of AMPA receptors with various signaling pathways and may play a critical role in clustering AMPA receptors at excitatory central synapses.[9,10]

In this report we use different biochemical and immunohistochemical approaches to characterize GRIP at the protein level, and to study its expression and distribution pattern in both hippocampal cultures and in rat brain (see Figs. 1 and 2). Our results confirm the idea that GRIP is a neuronal protein that is enriched at synaptic sites, and is also expressed in a large variety of brain regions (see Fig. 3). GRIP colocalizes with both the glutamatergic excitatory synapses and GABAergic inhibitory synapses (see Fig. 4), suggesting that it may play a role in both excitatory and inhibitory synaptic transmission.

REFERENCES

1. Ehlers, M.D. *et al.* 1996. Synaptic targeting of glutamate receptors. Curr. Opin. Cell Biol. **8:** 484–489.
2. Gomperts, S.N. 1996. Clustering membrane proteins: It's all coming together with the PSD-95/SAP90 protein family. Cell **84:** 659–662.
3. Sheng, M. *et al.* 1996. PDZs and receptor/channel clustering: Rounding up the latest suspects. Neuron **17:** 575–578.
4. Sheng, M. *et al.* 1996. Ion channel associated proteins. Curr. Opin. Neurobiol. **6:** 602–608.
5. Dong, H.-L. *et al.* 1997. GRIP: A synaptic PDZ domain-containing protein that interacts with AMPA receptors. Nature **386:** 279–284.
6. Kim, E. *et al.* 1995. Clustering of shaker-type K^+ channels by interaction with a family of membrane-associated guanylate kineses. Nature **378:** 85–88.
7. Kornau, H.-C. 1995. Domain interaction between NMDA receptor subunits and the postsynaptic density protein PSD-95. Science **269:** 1737–1740.
8. Craig, A.M. *et al.* 1994. Selective clustering of glutamate and -aminobutyric acid receptors opposite terminals releasing the corresponding neurotransmitters. Proc. Natl. Acad. Sci. USA **91:** 12373–12377.
9. Brenman, J.E. *et al.* 1996. Interaction of nitric oxide synthase with the postsynaptic density protein PSD-95 and alphal-syntrophin mediated by PDZ domains. Cell **84:** 757–767.
10. Tsunoda, S. *et al.* 1997. A multivalent PDZ-domain protein assembles signalling complexes in a G-protein-coupled cascade. Nature **388:** 243–249.

GluR2 Antisense Knockdown Produces Seizure Behavior and Hippocampal Neurodegeneration during a Critical Window

LINDA K. FRIEDMAN[a,b,c] AND JANA VELÍSKOVÁ[b]

[a]*Department of Neuroscience, New Jersey Neuroscience Institute/Seton Hall University, Edison, New Jersey 08818, USA*

[b]*Departments of Neuroscience and Neurology, Albert Einstein College of Medicine, Bronx, New York 10461, USA*

Overaccumulation of intracellular Ca^{2+} may contribute to delayed hippocampal cell death.[1–3] AMPA (α-amino-3-hydroxy-5-methyl-4-isoxazole-propionic acid)–type glutamate receptors lacking the GluR2 subunit are highly permeable to Ca^{2+}.[4–6] Formation of Ca^{2+}-permeable AMPA receptors in principle neurons may contribute to the delayed neurodegenerative process (GluR2 hypothesis). In adult rats, kainic acid–induced status epilepticus reduces GluR2 subunit expression prior to neurodegeneration of hippocampal CA3 neurons.[7–9] In young rats, which are highly prone to seizures but resistant to kainic acid seizure–induced brain damage,[10–11] GluR2 mRNA and protein expression remain constant in CA3 neurons and may contribute to their survival.[12] Interestingly, GluR2 hippocampal mRNA expression peaks in development, then subsequently declines,[13–14] suggesting that the level of GluR2 expression may be pertinent to normal brain function during a critical stage in development. In keeping with this, Brusa *et al.*[15] elegantly showed that knockout of the GluR2 RNA editing mechanism (but not the overall level of GluR2 mRNA expressed) increases AMPA-mediated Ca^{2+} influx throughout the brain, and induces a delayed postnatal epilepsy that is fatal by postnatal (P) day 20 and selective CA3 hippocampal neurodegeneration. Thus, editing of the GluR2 subunit appears essential for normal brain function in development.

Antisense *in vivo* knockdown can be powerful to examine particular roles of single receptor subunits by selectively inhibiting the translation, stability, or transport of sense mRNAs within a specific brain region.[16] Here we decreased the expression of GluR2 by unilateral microinfusion of GluR2 antisense oligodeoxynucleotides (AS-ODNs) into the dorsal hippocampus of young and adult rats to determine whether reduced GluR2 subunit synthesis may influence the viability of hippocampal neurons to seizure-induced damage. Unilateral knockdown of hippocampal GluR2 subunit expression induced age-dependent seizure-like behavioral manifestations, altered EEG recordings, and reduced the survival of CA3 neurons in young rats during a critical window when GluR2 expression peaks in development and glutamatergic inputs are maturing.

[c]Address for correspondence: Dr. Linda K. Friedman, NJ Neuroscience Institute, Seton Hall University, 65 James Street, Edison, NJ 08820. Phone: 732-321-7950; fax: 718-430-8899; e-mail: lfriedma@aol.com

PROCEDURES

Sprague-Dawley male albino rats at three ages (P8, 15–18 g; P13, 25–30 g, $n = 5–10$; P60, 175–250 g, $n = 10$) were used to administer ODNs and obtain ipsilateral hippocampal EEG recordings with stereotaxically cannula/bipolar electrode assemblies that were implanted into the right hippocampus. EEG recordings were obtained daily between 10:00–3:00 PM and 4:00–8:00 PM to correlate with behavioral changes. Animals were sacrificed following 4 days of twice daily treatment of Tris buffer, antisense (AS), sense (S), or missense (NS, scrambled) ODN microinfusions. Two sets of phosphorothiolated 21-mers targeted to the GluR2 subunit were synthesized (5′-CCTGTTTTATGGGGACT-GATT-3′ and 5′-AACCATTTTATCCACTTCACT-3′) and microinfused (0.25 μl/2 min). Histological analysis of the dorsal hippocampus was carried out in fixed coronal brain sections (29 μm) stained with thionin at the three ages.

RESULTS AND DISCUSSION

At P8, GluR2 knockdown did not affect the behavior, EEG recording, or morphology of the CA3 hippocampus (TABLE 1 and FIG. 1A). At P13, GluR2 knockdown resulted in seizure-like behavioral manifestations; approximately half exhibited automatisms by 48 h. In 36% of rat pups, there were jerking and stumbling movements, ball curling, head nods and shaking, frequent wet dog shakes, or quadrilateral forelimb jerking movements with backward jumping, occasional wild running, and loss of postural control. This behavior resembled automatisms described in rat pups treated with kainic acid at similar ages.[10,17–18] Following GluR2 knockdown, the EEG revealed occasional paroxysmal activity; high-frequency and high-amplitude discharges were associated with long episodes of vigorous and continuous scratching followed by immobility. Tremor was observed in only two of nine pups, suggesting that automatisms may not be epileptic or that conditions still need to be further characterized. In the control animals tested (sense/nonsense/tris/no treatment), there were no jerking movements or paroxysmal activity found in the EEG; normal scratching movement artifacts were short lasting. GluR2 knockdown resulted in specific degeneration of CA3a neurons distant from the medial intrahippocampal cannula microinfusion site (TABLE 1 and FIG. 1B). Loss of CA3 neurons was observed over a 1.8–2.4-mm distance. In adult rats, GluR2 knockdown did not affect behavior, the ipsilateral hippocampal EEG or CA3 morphology being similar to the P8 age group (TABLE 1 and FIG. 1C).

TABLE 1. Quantitation of Behavioral Manifestations per Hour after GluR2 Hippocampal Knockdown

Age	Scratching	Wet Dog Shakes	Bilateral Jerks	n
P8	0	0	0	5
P13	$133 \pm 35^*$	3.9 ± 1.9	$5 \pm 1.5^*$	9
P60	2.4 ± 0.5	1.6 ± 0.5	0	5

NOTE: Rats were infused with GluR2 AS-ODNs for 4 days into the right hippocampus. The number of behavioral events per hour were measured. Values expressed represent the mean ± SEM at three ages. $^*p < 0.01$, two-way ANOVA.

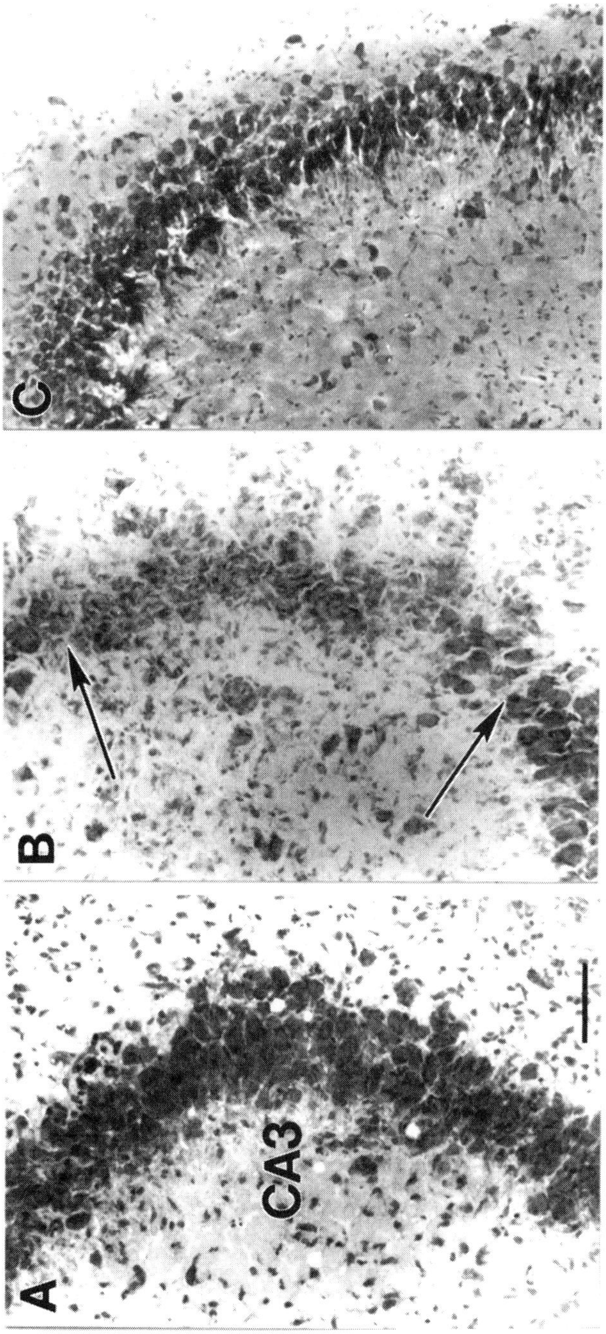

FIGURE 1. Photomicrographs depicting GluR2 knockdown age-dependent induced cell loss of CA3 hippocampal neurons. (**A**) At P8, thionin-stained section from GluR2 AS-ODN treated pup shows no apparent cell loss. (**B**) At P13, Nissl stain shows marked loss of CA3a pyramidal neurons (between *arrows*), particularly after the bend. (**C**) At P60, thionin staining shows normal cytoarchitechture and no detectable cell loss by the knockdown procedure. *Scale* = 50 µm.

GluR2 protein was selectively decreased by the GluR2-ODN treatment at the three ages, suggesting that down-regulation of GluR2 expression is not acutely needed for survival of hippocampal neurons when they have maturated.

The present study introduces the novel finding that unilateral hippocampal GluR2 antisense knockdown induces spontaneous seizure-like behavior, occasional paroxysmal activity in EEG recording traces, and specific CA3 hippocampal damage during the third postnatal week, when rat pups are highly prone to seizures but relatively resistant to damage.[10–11] This is also a time in development that corresponds to a peak in GluR2 expression.[13–14] The antisense *in vivo* GluR2 knockdown approach produces seizure-like manifestations at a similar age to the genetic GluR2 editing–incompetent mice that developed spontaneous seizure behavior. Although obvious alterations in hippocampal morphology or EEG activity were not reported in GluR2 knockout mutants (at P30), during the second-to-third postnatal week mutants had marked decreases in body weight, increased mortality, and enhancement of long-term potentiation (LTP).[19] Similar to *in vivo* knockdown experiments, in adulthood, GluR2 knockout mutants did not show obvious alterations in behavior or weight. Differences in developmental compensatory mechanisms may exist in the GluR2 mutants compared with GluR2 editing-deficient mice or our *in vivo* GluR2 knockdown rat pups. GluR2 antisense knockdown did not affect behavior or morphology in the P8 and adult rat groups, suggesting that maturation of glutamatergic afferents plays a role in the age-specific effect. Previous studies and present findings suggest that the GluR2 subunit has certain functions at different ages. It appears that the level of GluR2 subunit expression may be involved in the control of seizures and synaptic plasticity during a critical window in development of high seizure susceptibility and maturation of excitatory and inhibitory connections.

ACKNOWLEDGMENTS

This work was supported by the Epilepsy Foundation of America and March of Dimes (to L.K.F.) and Epilepsy Foundation of America (to J.V.). The authors thank Dr. Solomon L. Moshé for helpful comments and for making these experiments possible.

REFERENCES

1. Choi, D. 1988. Glutamate neurotoxicity and diseases of the nervous system. Neuron **1:** 623–634.
2. Siesjo, B.K. & F. Bengtsson. 1989. Ca^{2+} fluxes, Ca^{2+} antagonists, and Ca^{2+}-mediated pathology in brain ischemia, hypoglycemia, and spreading depression. J. Cereb. Blood Flow Metab. **9:** 127–140.
3. Randall, R.D. & S.A. Thayer. 1992. Glutamate-calcium transient triggers delayed calcium overload and neurotoxicity in rat hippocampal neurons. J. Neurosci. **12:** 1882–1895.
4. Hollman, M., M. Hartley & S. Heinemann. 1991. Ca^{2+} permeability of KA-AMPA-gated glutamate receptor channels depends on subunit composition. Science **253:** 1028–1031.
5. Verdoorn, T.A., N. Burnashev, H. Monyer, P.H. Seeburg B. Sakmann. 1991. Structural determinants of ion flow through recombinant glutamate receptor channels. Science **252:** 1715–1718.
6. Burnashev, N., A. Khodorova, P. Jonas, P.J. Helm, W. Wisden, H. Monyer, P.H. Seeburg & B. Sakmann. 1992. Calcium-permeable AMPA-kainate receptors in fusiform cerebellar glial cells. Science **25:** 1566–1570.

7. Pollard, H., H.J. Moreau, Y. Ben-Ari & M. Khrestchatisky. 1993. Alterations of the GluR-B AMPA receptor subunit flip/flop expression in kainate-induced epilepsy and ischemia. Neuroscience **57:** 545–554.
8. Friedman, L.K., D.E. Pellegrini-Giampietro, E.F. Sperber, M.V. L. Bennett, S.L. Moshé & R.S. Zukin. 1994. Kainate-induced status epilepticus alters glutamate and $GABA_A$ receptor gene expression in adult rat hippocampus: An *in situ* hybridization study. J. Neurosci. **14:** 2697–2707.
9. Friedman, L.K. 1998. Selective reduction of GluR2 protein in adult hippocampal CA3 neurons following status epilepticus but prior to cell loss. Hippocampus **8:** 511–525.
10. Albala, B.J., S.L. Moshé & R. Okada. 1984. Kainic-acid-induced seizures: A developmental study. Dev. Brain. Res. **13:** 139–148.
11. Nitecka, L., E. Tremblay, G. Charton, J.P. Bouillot, M.L. Berger & Y. Ben-Ari. 1984. Maturation of kainic acid seizure brain damage syndrome in the rat. II. Histopathological sequelae. Neuroscience **13:** 1073–1094.
12. Friedman, L.K., E. Sperber, S.L. Moshé, M.V.L. Bennett & R.S. Zukin. 1997. Developmental regulation of glutamate and $GABA_A$ receptor gene expression in rat hippocampus following kainate-induced status epilepticus. Dev. Neurosci. **19:** 529–542.
13. Monyer, H., P.H. Seeburg & W. Wisden. 1991. Glutamate-operated channels: Developmentally early and mature forms arise by alternative splicing. Neuron **6:** 799–810.
14. Standley, S., G. Tocco, M.F. Tourigny, G. Massicotte, R.F. Thompson & M. Baudry. 1995. Developmental changes in α-amino-3-hydroxy-5-methyl-4-isoxazole propionate receptor properties and expression in the rat hippocampal formation. Neuroscience **67:** 881–892.
15. Brusa, R., F. Zimmermann, D.S. Koh, D. Feldmeyer, P. Gass, P.H. Seeburg & R. Sprengel. 1995. Early onset epilepsy and postnatal lethality associated with an editing-deficient GluR-B allele in mice. Science **270:** 1677–1680.
16. Uhlmann, E. & A. Peyman. 1990. Antisense oligonucleotides: A new therapeutic principle. Chem. Rev. **90:** 544–584.
17. Cherubini, E., M.R. De Feo, O. Mecarelli & G.F. Ricci. 1983. Behavioral and electrographic patterns induced by systemic administration of kainic acid in developing rats. Dev. Brain Res. **9:** 69–77.
18. Tremblay, E., L. Nitecka, M.L. Berger & Y. Ben-Ari. 1984. Maturation of kainic acid seizure-brain damage syndrome in the rat: Clinical, electrographic and metabolic observations. Neuroscience **13:** 1015–1072.
19. Jia, Z., A. Nadia, P. Miu, Z. Xiong, J. Henderson, R. Geriai, F.A. Taverna, A. Velumian, J. Mac Donald, P. Carlen, W. Abrominov-Newerly & J. Roder. 1996. Enhanced LTP in mice deficient in the AMPA receptor GluR2. Neuron **17:** 945–956.

Activation of Kainate Receptors on Rat Sensory Neurons Evokes Action Potential Firing and May Modulate Transmitter Release

C. JUSTIN LEE,[a] H. S. ENGELMAN, AND A. B. MacDERMOTT

Department of Physiology and Cellular Biophysics and Center for Neurobiology and Behavior, Columbia University, 630 West 168th Street, New York, New York 10032, USA

Evidence for functional presynaptic kainate receptors in the cortex and spinal cord has been slowly accumulating over the last 20 years. Recently, glutamate release from CA1 hippocampal pyramidal neurons was shown to be transiently enhanced, then depressed by selective activation of presynaptic kainate receptors.[1] GABA release is also depressed by activation of presynaptic kainate receptors.[2,3] *In vivo* expression of kainate receptors by nociceptive primary afferents was suggested by recordings from isolated rat dorsal-root preparations[4,5] and acutely dissociated dorsal root ganglia (DRG) neurons.[6] However, the role of kainate receptors in the regulation of synaptic transmission in the superficial dorsal horn where nociceptive primary afferents terminate remains uncertain. Using microisland cocultures of DRG and dorsal horn neurons and acutely prepared spinal cord slices, we have begun to investigate the possible presence of presynaptic kainate receptors and their function in regulating synaptic transmission onto dorsal horn neurons. We demonstrate here that pharmacologically defined kainate receptors are expressed by DRG neurons grown in coculture with dorsal horn neurons, that kainate receptor activation of DRG neurons elicits a burst of action potential firing, and that application of kainate to the spinal cord slice increases the frequency of spontaneous postsynaptic currents (sPSCs) recorded from dorsal horn neurons.

DRG and dorsal horn neurons were harvested from E16 embryos, dissociated, and plated on top of previously prepared astrocytes growing on collagen-coated microislands. Cultures were grown for at least a week before recording. DRG neuron recordings were made using perforated patch technique with 25 μg/mL gramicidin and (mM) 75 K_2SO_4, 10 KCl, 0.1 $CaCl_2$, 10 HEPES, pH 7.2. Spinal cord slices were acutely prepared from post-natal day 7–10 rats. Slices were 400 μm thick and lamina II was visually identified. Whole-cell patch recordings were made using (mM) 140 Cs-methanesulfonate, 10 HEPES, 10 NaCl, 5 EGTA, 0.5 $CaCl_2$, pH 7.2.

Application of 100 μM kainate evoked an inward current from DRG neurons in coculture voltage clamped at −60 mV. As shown in FIGURE 1A, the inward current had both transient and steady-state components, as originally described for kainate currents in DRG neurons by Huettner.[6] CNQX (250 μM), a non-NMDA antagonist, mostly blocked the kainate-evoked current. It was completely blocked by applying 3 μM SYM 2081, a selec-

[a]Address for correspondence: C. Justin Lee, Department of Physiology, Columbia University, 630 West 168th Street, New York, New York 10032. Phone: 212/305-3817; fax: 212/305-3723; e-mail: cjl2@columbia.edu

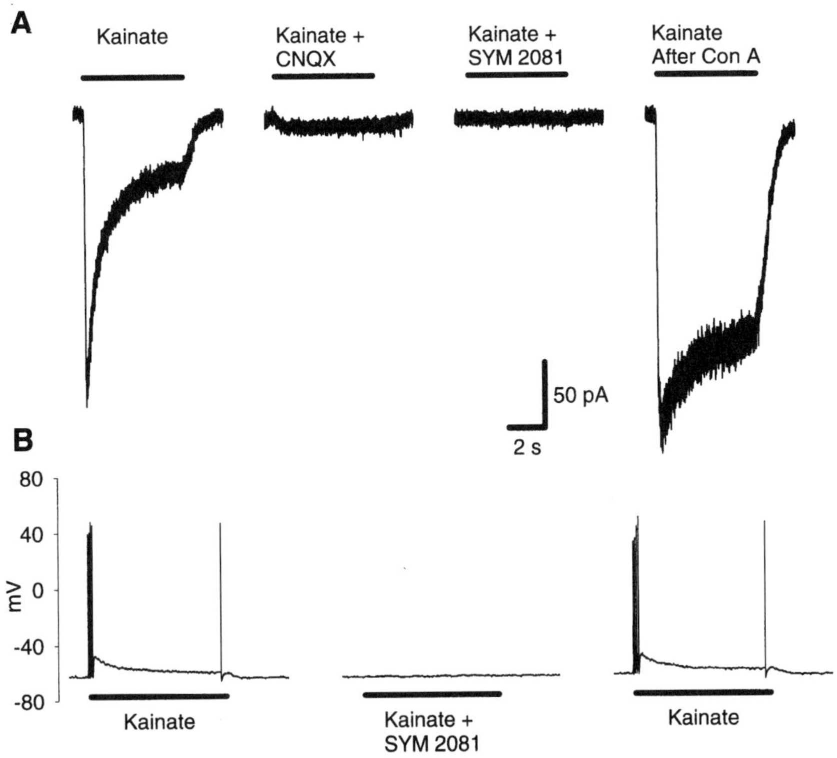

FIGURE 1. Kainate evokes whole-cell currents **(A)** and action potential firing **(B)** in cultured DRG neurons by activating kainate receptors. **(A)** 5-s application of 100 μM kainate evoked a desensitizing current, which could be mostly blocked by 250 μM CNQX, completely blocked by 3 μM SYM 2081, and enhanced by 5-min pretreatment with 300 μg/ml Con A. Holding potential: −60 mV. **(B)** Under current clamp, a 5-s application of 100 μM kainate elicited a burst of action potentials at the onset of the application, which is completely and reversibly blocked by 3 μM SYM 2081.

tive kainate receptor desensitizing agonist, for 1 minute before and during kainate application. These data indicate that the kainate-induced current was mediated exclusively by kainate receptors. To further confirm receptor identity, 300 μg /mL of Concanavalin A (ConA) was added to the bath for 5 min before kainate application, and the subsequent kainate-induced current was greatly enhanced (FIG. 1A). ConA has been shown to selectively block desensitization of kainate receptors.[7] Thus these data provide strong evidence for the expression of kainate receptors by DRG neurons grown in coculture with dorsal horn neurons. Under current clamp conditions, a 5-s application of 100-μM kainate to a DRG neuron elicited a burst of action potential firing, an effect that was reversibly blocked by SYM 2081 (FIG. 1B). These data suggest that activation of DRG neurons through kainate receptors is potentially able to drive action potential-evoked transmitter release from DRG central and peripheral terminals.

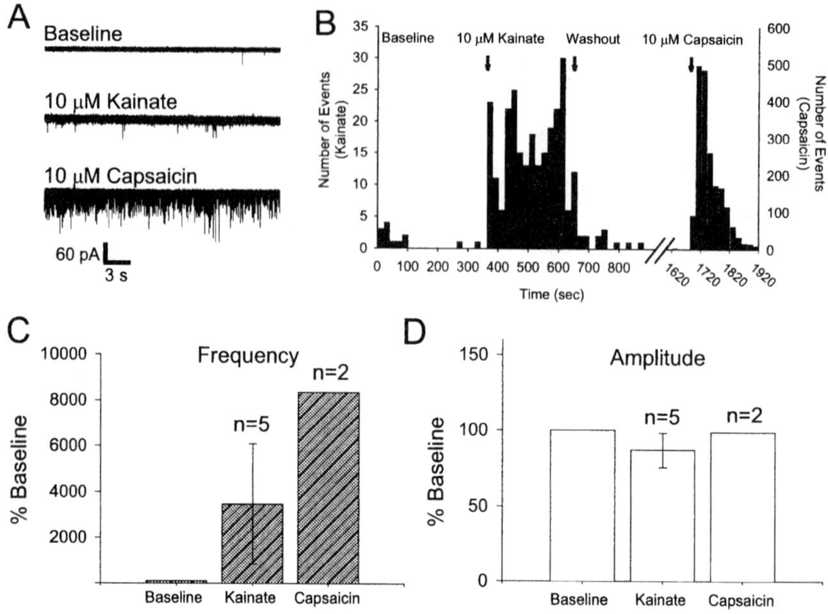

FIGURE 2. Kainate and capsaicin induce an increase in the frequency of spontaneous dorsal horn neuron PSCs in the presence of TTX. **(A)** Representative traces are shown from recordings in normal bath, 10 μM kainate, and 10 μM capsaicin from a single dorsal horn neuron held at −70 mV. Recordings are from a transverse spinal cord slice taken from a P8 rat. **(B)** For the same neuron as in **(A)**, the number of PSCs is plotted versus time (*bars* represent the number of events over 20-s intervals). Note difference in scale for kainate and capsaicin. **(C)** Average PSC frequency was calculated over 4–5 min under baseline conditions. Changes in frequency in the presence of kainate and capsaicin determined over 4–5 min were calculated as a percent of baseline. The average percent of baseline was then computed for all cells tested and plotted (for kainate, $n = 5$, for capsaicin, $n = 2$). **(D)** PSC amplitudes were measured over the same periods as for frequency and expressed as percent of baseline.

We have begun to investigate the effect of kainate on spontaneous release of transmitter onto lamina II neurons in the postnatal rat dorsal horn. This region of the spinal cord receives inputs from cutaneous nociceptive primary afferents that express Ca^{2+}-permeable capsaicin receptors on their central terminals.[8] Capsaicin was used to confirm that the dorsal horn neurons under study with kainate received input from nociceptors by recording both the kainate- and capsaicin-induced change in frequency of spontaneous PSCs from dorsal horn neurons. Bath application of 10 μM kainate to a lamina II neuron in the presence of 0.5 μM TTX increased the sPSC frequency 17-fold, suggesting the presence of presynaptic receptors sensitive to kainate (FIG. 2A and 2B). After washout of kainate, sPSC frequency decreased back to baseline levels. Subsequent bath application of 10 μM capsaicin increased sPSC frequency 163-fold. This change in sPSC frequency in response to capsaicin indicates that this neuron received innervation from primary afferents. sPSC frequency was expressed as a percentage of baseline and averaged across cells (FIG. 2C).

Both kainate and capsaicin caused dramatic and significant, though widely variable, increases in sPSC frequency with no significant change in amplitude (FIG. 2D).

In these studies, kainate receptors were pharmacologically identified on cultured DRG neurons. Activation of kainate receptors elicited a burst of action potential firing, raising the possibility that if kainate receptors are expressed presynaptically on DRG terminals, they could initiate action potential firing and transmitter release when activated by glutamate. The kainate-induced increase in frequency of sPSCs recorded from lamina II neurons in spinal cord slices supports our hypothesis of a role for presynaptic kainate-sensitive receptors in the regulation of synaptic transmission in the nociceptive pathway.

REFERENCES

1. CHITTAJALLU, R. *et al.* 1996. Regulation of glutamate release by presynaptic kainate receptors in the hippocampus. Nature **379:** 78–81.
2. CLARKE, V.R.J. *et al.* 1997. A hippocampal GluR5 kainate receptor regulating inhibitory synaptic transmission. Nature **389:** 559–603.
3. RODRIGUEZ-MORENO, A. *et al.* 1997. Kainate receptors presynaptically downregulate GABAergic inhibition in the rat hippocampus. Neuron **19:** 893–901.
4. DAVIES, J. *et al.* 1979. Excitatory amino acid receptors and synaptic excitation in the mammalian central nervous system. J. Physiol. (Paris) **75:** 641–654.
5. AGRAWAL, S.G. & R.H. EVANS. 1986. The primary afferent depolarizing action of kainate in the rat. Br. J. Pharmacol. **87:** 345–355.
6. HUETTNER, J.E. 1990. Glutamate receptor channels in rat DRG neurons: Activation by kainate and quisqualate and blockade of desensitization by Con A. Neuron **5:** 255–266.
7. PARTIN, K.M. *et al.* 1993. Selective modulation of desensitization at AMPA versus kainate receptors by cyclothiazide and concanavalin A. Neuron **11:** 1069–1082.
8. ENGELMAN, H. *et al.* The distribution of neurons expressing calcium-permeable AMPA receptors in the superficial laminae of the spinal cord dorsal horn. J. Neurosci. In press.

An Immunocytochemical Assay for Activity-Dependent Redistribution of Glutamate Receptors from the Postsynaptic Plasma Membrane

DMITRI V. LISSIN,[a,b] ROBERT C. MALENKA,[a,c] AND MARK VON ZASTROW[a,d]

Departments of [a]Psychiatry, [c]Physiology and [d]Cellular and Molecular Pharmacology, University of California, San Francisco, California 94143, USA

Two major subtypes of ionotropic glutamate receptors, AMPA and NMDA, are present at the majority of excitatory synapses. Recent evidence suggests that the localization of individual receptors may be differentially regulated at individual synapses in response to synaptic activity.[1,2] While it is suggested that regulation of receptor distribution among individual postsynaptic specializations of the plasma membrane plays an important role in mediating synaptic plasticity, direct experimental study of this hypothesis has been limited by a lack of methods to clearly identify receptors present at the cell surface. We have developed a novel method of directly visualizing receptor localization at individual postsynaptic membrane specializations, and distinguishing receptors present on the plasma membrane from those present in intracellular membranes.[1] Here we describe our use of this method to study differential effects of synaptic activity on the surface expression of epitope-tagged GluR1 (an AMPA subunit) and NR1 (an NMDA subunit)[3] in individual postsynaptic specializations of the plasma membrane.

METHODS

Two-week-old dissociated hippocampal cultures[4] were transfected with adenovirus[1,5] encoding epitope-tagged GluR1 or NR1. Cultures were fixed 48 h later, incubated with anti-Flag mouse monoclonal antibody (M1, Kodak) or anti-HA monoclonal (BAbCO) to specifically label receptors exposed at the cell surface. Antibody-labeled receptors were detected using HRP-conjugated antimouse IgG followed by Fluorescein Tyramide Signal Amplification (NEN™), after which specimens were permeabilized and incubated with rabbit antisynaptophysin to label presynaptic membranes followed by Cy3-conjugated antirabbit secondary antibody. Localization of receptors and synaptophysin in the same specimens was then performed by dual-label fluorescence microscopy.

[b]Present address and address for correspondence: Dmitri Lissin, M.D., AIRS, Inc., 455 Market Street, Suite 1850, San Francisco, California 94105. Phone: 415-817-8931; fax: 415-817-8999; e-mail: dmitri.lissin@airs-usa.com

RESULTS AND DISCUSSION

Epitope-tagged GluR1 and NR1 subunits were concentrated in clusters at the cell surface. Some synapses were receptor-negative (FIG. 1A). Most synapses defined by synaptophysin immunoreactivity contained GluR1. Quantitative analysis of randomly selected fields ($n = 20$) (FIG. 1B) demonstrated that approximately 90% of the Flag-GluR1 and HA-NR1 surface receptor clusters were localized at synapses. The majority of synapses expressed Flag-GluR1 and HA-NR1 (FIG. 1C). The remaining synapses were likely to be inhibitory, since approximately 20% of the synaptophysin puncta colocalized with the GABA synthetic enzyme, GAD65, identified by staining with Gad6 specific antibody (FIG. 2D).

Increasing network activity using the GABA-A receptor antagonist picrotoxin caused pronounced reduction in the number of synaptic membranes containing detectable amounts of GluR1 immunoreactivity, suggesting an activity-dependent down regulation of AMPA receptors in postsynaptic sites (FIG. 2A). Quantitative analysis of multiple fields ($n = 20$) from several experiments (FIG. 2B) confirmed these observations. Picrotoxin treatment had no discernible effect on the surface expression of HA-NR1 at synapses, sug-

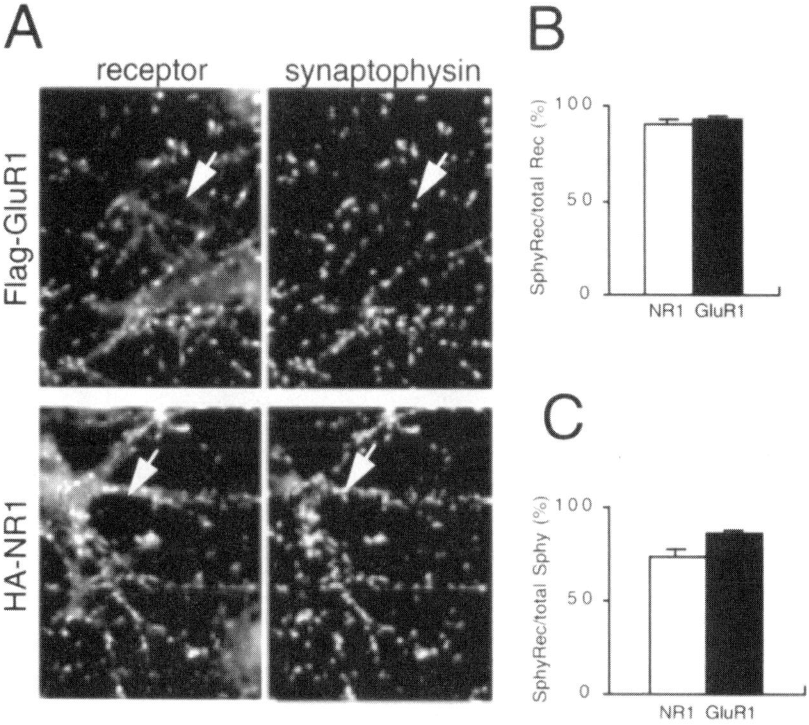

FIGURE 1. Epitope-tagged GluR1 and NR1 are targeted to synaptic membranes. *Error bars* represent SEM. (Adapted with permission from Lissin *et al.*[1])

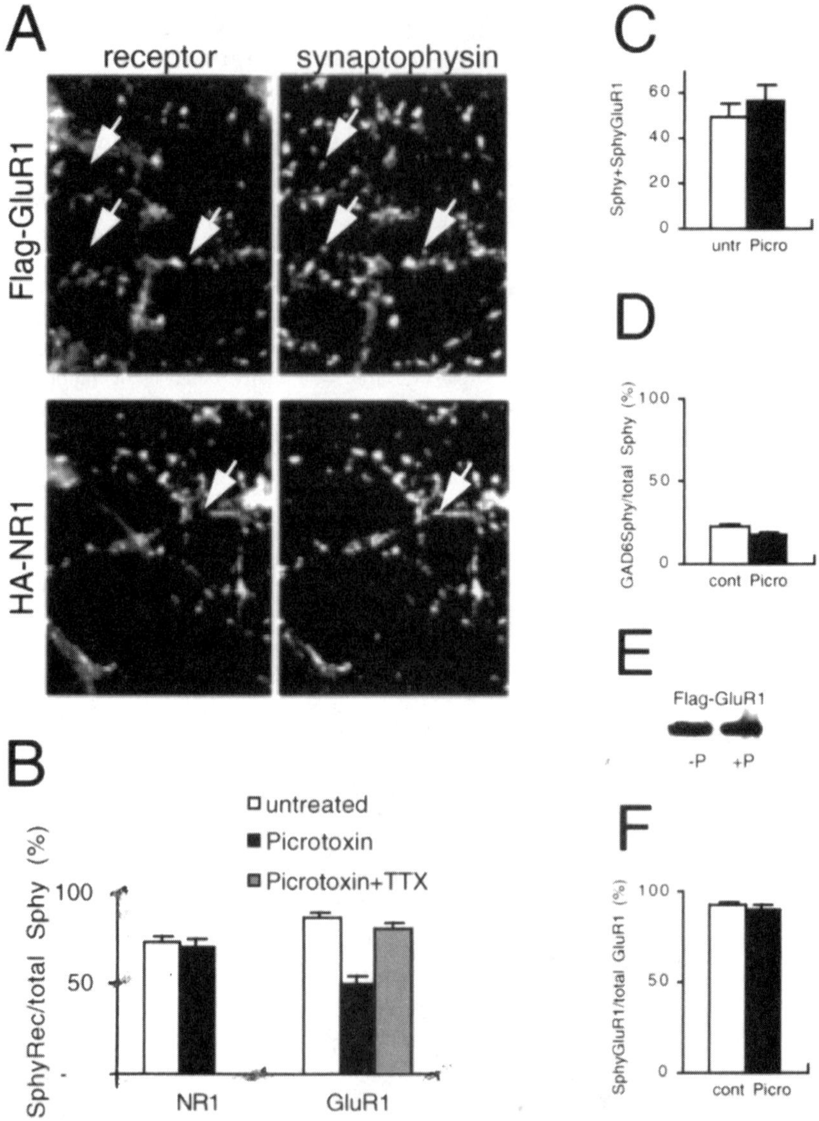

FIGURE 2. Increasing activity causes a decrease in synaptic membrane clusters of Flag-GluR1 but not HA-NR1. *Arrows* demonstrate receptor-negative synapses. (Adapted with permission from Lissin *et al.*[1])

gesting that the activity-dependent down-regulation observed under these conditions is specific for AMPA receptors. The effect of picrotoxin on GluR1 receptor localization was completely prevented by concomitant application of TTX, confirming that the synaptic down-regulation of GluR1-containing receptors is activity-dependent (FIG. 2B).

The effect of picrotoxin on the surface expression of GluR1 appeared to be highly specific. The total number of synapses was the same in control and picrotoxin-treated cultures (FIG. 2C), as was the percentage of inhibitory synapses (FIG. 2D). Furthermore, Western blotting of culture extracts revealed no effect of the picrotoxin treatment on the total number of receptors present in cells (FIG. 2E). Despite this pronounced down-regulation, the vast majority (~90%) of the Flag-GluR1 clusters observed in the picrotoxin-treated cultures were located at synapses (FIG. 2F).

The observed down-regulation of synaptic GluR1 immunoreactivity may be mediated by an activity-dependent signal that reduces or prevents the insertion of new AMPARs, or by regulated removal of existing AMPARs from the synaptic plasma membrane. Such mechanisms might account for the formation of silent synapses observed electrophysiologically.[6]

REFERENCES

1. LISSIN, D.V. *et al.* 1998. Activity differentially regulates the surface expression of synaptic AMPA and NMDA glutamate receptors. Proc. Natl. Acad. Sci. USA **12:** 7097–7102.
2. RAO, A. & A.M. CRAIG. 1997. Activity regulates the synaptic localization of the NMDA receptor in neurons. Neuron **19:** 801–812.
3. HOLLMANN, M. & S. HEINEMANN. 1994. Cloned glutamate receptors. Annu. Rev. Neurosci. **17:** 31–108.
4. BREWER, G.J., J.R. TORRICELLI, E.K. EVEGE & P.J. PRICE. 1993. Optimized survival of hippocampal neurons in B27-supplemented Neurobasal, a new serum-free medium combination. J. Neurosci. Res. **35:** 567–576.
5. HARDY, S. *et al.* 1997. Construction of adenovirus vectors through Cre-lox recombination. J. Virol. **71:** 1842–1849.
6. ISAAC, J.T.R., R.A. NICOLL & R.C. MALENKA. 1995. Evidence for silent synapses: Implications for the expression of LTP. Neuron **15:** 427–434.

Activation of PKC Disrupts Presynaptic Inhibition by Group II and Group III Metabotropic Glutamate Receptors and Uncouples the Receptor from GTP-Binding Proteins

THOMAS A. MACEK,[a,c] HERVÉ SCHAFFHAUSER,[b] AND P. JEFFREY CONN[a,b]

[a]*Program in Molecular Therapeutics and Toxicology, Emory University School of Medicine, Atlanta, Georgia 30322, USA*

[b]*Department of Pharmacology, Emory University School of Medicine, Atlanta, Georgia 30322, USA*

Throughout the central nervous system, cell excitability and fast synaptic transmission are modulated by the activation of a family of G protein–coupled receptors termed metabotropic glutamate receptors (mGluRs). Eight mGluR subtypes have been cloned and have been classified into three major groups. Group I mGluRs (mGluR1 and mGluR5) couple primarily to phosphoinositide hydrolysis, whereas group II (mGluR2 and mGluR3) and group III (mGluRs 4,6,7, and 8) couple to inhibition of adenylyl cyclase in expression systems.[1]

One of the primary functions of mGluRs seen throughout the central nervous system (CNS) is to serve as presynaptic receptors involved in reducing transmission at glutamatergic synapses. A diversity exists within the CNS as to which mGluR subtypes serve this role at a given synapse. For example, different complements of mGluR subtypes serve this role at each of the three major excitatory synapses in the hippocampal formation. We now report that activation of protein kinase C (PKC) inhibits the function of presynaptic mGluR's at three major hippocampal synapses. At the Schaffer collateral (SC)-CA1 synapse, this effect can be elicited by activation of A3 adenosine receptors. Furthermore, we provide biochemical evidence that suggests that PKC elicits this effect by inhibiting coupling of mGluR's to GTP-binding proteins. Regulation of presynaptic mGluRs by PKC could play a critical role in fine-tuning transmission at glutamatergic synapses.

RESULTS

Previous studies revealed that a group II mGluR inhibits transmission at the medial perforant path-dentate gyrus (MPP-DG) synapse.[1] Group II mGluR function at the corticostriatal synapse can be inhibited by activation of PKC.[2] Therefore, we performed a series of studies to test the hypothesis that activation of PKC would also inhibit group II mGluR

[c]Address for correspondence: Thomas A. Macek, Rollins Research Building, Department of Pharmacology, Emory University, 1510 Clifton Road, Atlanta, Georgia 30322. Phone: 404-727-5617; fax: 404-727-0365; e-mail: tmacek@emory.edu

function at the MPP-DG synapse. DCG-IV (3 μM), a group II-selective mGluR agonist, reduced fEPSPs at this synapse (FIG. 1). Activation of PKC by the PKC-activating phorbol ester phorbol 12,13-dibutyrate (PDBu) dramatically inhibited this inhibitory action of DCG-IV (FIG. 1). In contrast, a non-PKC-activating phorbol ester, 4α-phorbol had no effect on the inhibitory action of DCG-IV (FIG. 1).

Group III mGluRs also inhibit transmission at the MPP-DG, SC-CA1, and lateral perforant path-dentate gyrus (LPP-DG) synapses.[1] The group III mGluR that is present at the LPP-DG synapse is thought to be of a different molecular subtype than the group III mGluR expressed at MPP-DG and SC- CA1 synapses. Interestingly, activation of PKC by PDBu reduced the inhibitory actions of the group III mGluR agonist L-AP4 at each of these synapses (FIG. 1B), whereas 4α-phorbol had no effect on the inhibition by L-AP4. These data suggest that PKC also inhibits the presynaptic actions of at least two group III mGluR subtypes.

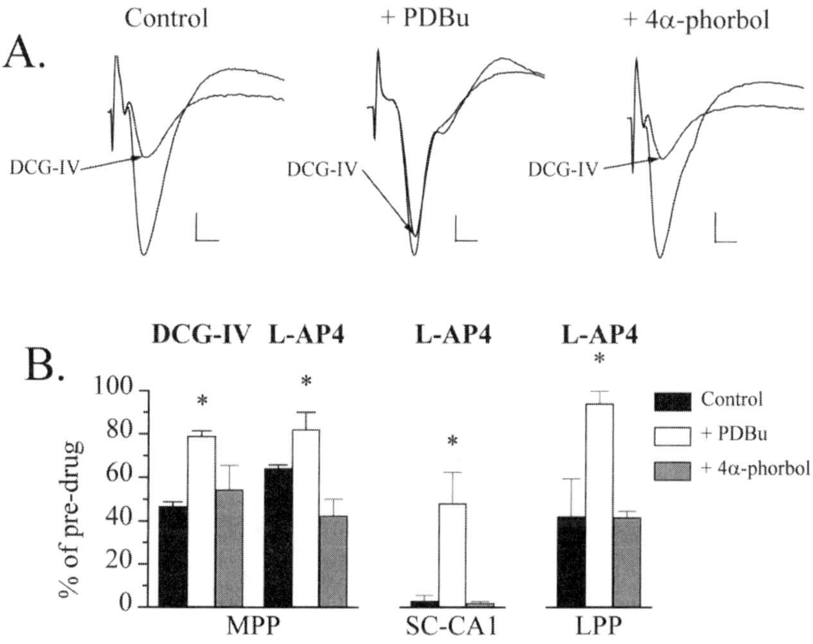

FIGURE 1. Activation of PKC disrupts presynaptic inhibition by mGluRs at three hippocampal synapses. **(A)** Representative fEPSP traces showing that the inhibitory effect of DCG-IV (3 μM) at the MPP synapse is attenuated when preceded by PDBu (10 μM), an activator of PKC. The inactive analog, 4α-phorbol (10 μM) had no effect on DCG-IV-mediated inhibition (*scale bars* = 0.2 mV, 2.5 mS). **(B)** Graph summary showing the mean effect (± SEM) of mGluR agonists on synaptic transmission at MPP (DCG-IV, 3 μM; L-AP4, 3 mM), SC-CA1 (L-AP4, 1 mM), and LPP (L-AP4, 20 μM) in the presence and absence of phorbol esters. PDBu (10 μM) reduced mGluR-mediated inhibition, and the inactive analog 4α-phorbol (10 μM) had no effect on mGluR-mediated inhibition at all three hippocampal synapses (n = 3–5 for each experiment; one-way ANOVA; *$p < 0.05$).

These studies suggest that PKC-induced regulation of presynaptic mGluR function is a relatively widespread phenomenon that occurs at multiple glutamatergic synapses and is not restricted to actions on group II mGluRs at the corticostriatal synapse. It is not clear, however, whether this response can only be elicited with PKC activators, or whether it is a physiologically relevant response that can be elicited by agonists of receptors coupled to activation of phosphoinositide hydrolysis (and PKC). Thus, we determined the effect of A3 adenosine receptor activation on the response to L-AP4 at the SC-CA1 synapse. Interestingly, the A_3-selective adenosine receptor agonist, Cl-IB-MECA, inhibited the response to L-AP4 at this synapse (FIG. 2). The effect of Cl-IB-MECA was inhibited by application of an A_3 adenosine-selective antagonist, MRS1191, and by the cell-permeable PKC inhibitor Bisindolylmaleimide I HCl (Bis)(1 μM), suggesting that this inhibition is mediated by activation of PKC (FIG. 2).

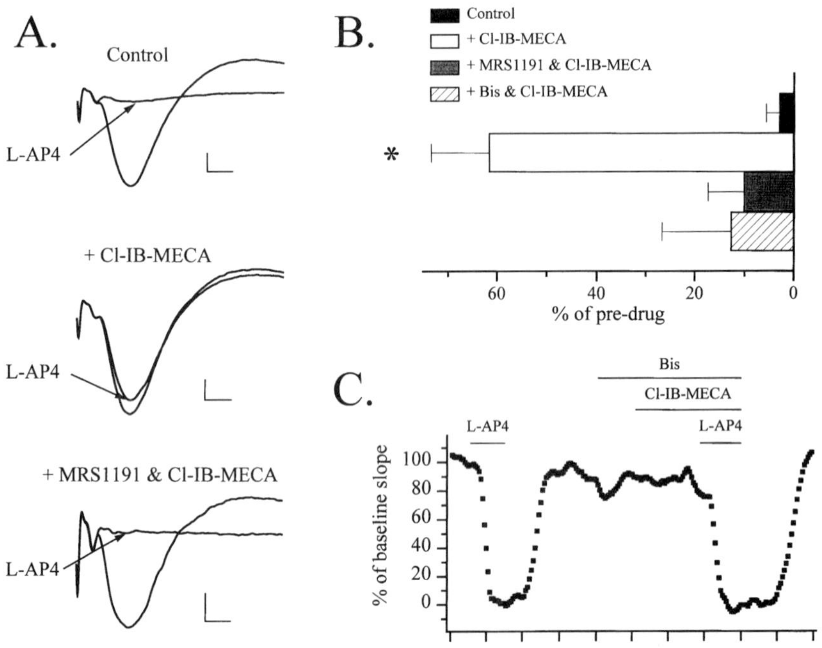

FIGURE 2. A_3 adenosine receptor activation reduces group III mGluR-mediated inhibition at the SC-CA1 synapse by a PKC-mediated mechanism. **(A)** Representative fEPSP traces showing the effect of a 20-minute application of Cl-IB-MECA (2-chloro-N^6-(3-iodobenzyl)-adenosine-5-*N*-methyluronamide; 1 μM) on the inhibitory actions of L-AP4 at the SC-CA1 synapse. Application of MRS1191 (3-ethyl-5-benzyl-2-methyl-4-phenylethynyl-6-phenyl-1,4-(±)-dihydropyridine-3,5-dicarboxylate; 10 μM) for 10 minutes prior to Cl-IB-MECA blocked the effect of Cl-IB-MECA on L-AP4-induced inhibition (*scale bars* = 0.2 mV, 2.5 mS). **(B)** Meaned (± SEM) data depicting the effect of A_3 adenosine agonists and antagonists on L-AP4-induced inhibition. A_3 adenosine-mediated inhibition of mGluR function was disrupted when preceded by Bis (1 μM), a selective inhibitor of PKC. (n = 3–5 for each experiment; one-way ANOVA; $^*p < 0.05$). **(C)** Time course showing that application of Bis (1 μM) blocks the effect of Cl-IB-MECA (1 μM) on L-AP4-induced inhibition. *Axis ticks* represent 10-minute intervals.

It has been suggested that PKC-induced inhibition of presynaptic receptor function at some other synapses is mediated by phosphorylation of a calcium channel subunit.[3] Evidence suggests, however, that group III mGluR function at the SC-CA1 synapse may be mediated by a calcium-independent mechanism. Therefore, we postulated that PKC effects on presynaptic mGluR function may be mediated by actions at the level of the receptor or GTP-binding protein. To test this hypothesis, we used a biochemical system in which receptor activation by agonist application is determined by measuring binding of the nonhydrolyzable [^{35}S]-GTPγS to GTP binding proteins. Since this system isolates the receptor and G protein from other downstream effectors, any PKC-induced effect on [^{35}S]-GTPγS binding would be mediated by actions at the level of the receptor or the GTP-binding protein. Membranes were prepared from cortical synaptosomes under conditions in which the cortical slices were first incubated for 30 minutes in either vehicle alone, PDBu, or 4α-phorbol. The effects of DCG-IV (10 μM) and L-AP4 (30 μM) on [^{35}S]-GTPγS binding were significantly reduced in synaptosomes prepared from slices incubated in PDBu (10 μM) for 30 minutes prior to synaptosome preparation (58 ± 4% of control for DCG-IV and 43 ± 8% of control for L-AP4). [^{35}S]-GTPγS binding in membranes prepared from cortical slices exposed to 4α-phorbol were not different from control. These data suggest that the inhibitory actions of PKC on mGluR function may be due to an uncoupling of the mGluR from the GTP- binding protein.

In summary, these data, coupled with previous findings, suggest that activation of PKC inhibits function of multiple presynaptic mGluR subtypes at different glutamatergic synapses, and that this effect may be partially mediated by inhibition of coupling of mGluRs to GTP-binding proteins. Furthermore, this response can be elicited by activation of receptors that couple to activation of phosphoinositide hydrolysis (i.e., A_3 adenosine receptors). This effect of PKC could play a critical role in fine-tuning transmission at glutamatergic synapses. Selective agonists or antagonists of receptors that activate PKC could provide novel therapeutic targets for drug development that could be used to regulate transmission.

REFERENCES

1. CONN, P.J. & J.P. PIN. 1997. Pharmacology and functions of metabotropic glutamate receptors. Annu. Rev. Pharmacol. Toxicol. **37:** 205–237.
2. SWARTZ, K.J. *et al.* 1993. Protein kinase C modulates glutamate receptor inhibition of Ca^{2+} channels and synaptic transmission. Nature **361:** 165–168.
3. ZAMPONI, G.W. *et al.* 1997. Crosstalk between G proteins and protein kinase C mediated by the calcium channel α_1 subunit. Nature **385:** 442–446.

AMPA Receptor Forms a Biochemically Functional Complex with NSF and α- and β-SNAPs

PAVEL OSTEN[a] AND EDWARD B. ZIFF

Howard Hughes Medical Institute, Department of Biochemistry, New York University Medical Center, 550 First Avenue, New York, New York 10016, USA

The α-amino-3-hydroxy-5-methyl-4-isoxazolepropionate (AMPA) glutamate receptor is proposed to be a complex of four subunits, termed GluR1 through GluR4. Each subunit is a transmembrane protein with an extracellular N-terminus, four membrane domains, and an intracellularly located C-terminus. Recently, several GluR2 subunit interacting proteins were identified by yeast two-hybrid screening, using the GluR2 C-terminal peptide as a bait. From these, two novel PDZ-domain proteins, glutamate receptor interacting protein (GRIP)[1] and AMPA receptor binding protein (ABP),[2] have been found to bind at the VKI-COOH motif of the end of the GluR2 C-terminus through a PDZ domain class II interaction.[2] It has been proposed that both GRIP and ABP anchor the AMPA receptor in the postsynaptic density membrane. In addition, *N*-ethylmaleimide–sensitive fusion protein (NSF) also has been found to bind at the GluR2 C-terminus, residues 844–853, in the yeast two-hybrid assays.[3,4] NSF is an ATPase that is known to disrupt a multiprotein SNARE complex formed by syntaxin, SNAP25, and synaptobrevin. The SNARE complex has been shown to mediate various membrane fusion events, including Golgi transport and exocytosis of synaptic vesicles. The chaperone-like function of NSF is proposed to occur after each fusion cycle during a step of protein sorting and recycling.

Our data demonstrate that NSF, together with α- and β-SNAPs form a protein complex with the AMPA receptor, which can be isolated by immunoprecipitation of the receptor from hippocampal triton lysate using antibody binding to the extracellular N-terminal portion of the GluR2 subunit under conditions of inhibited ATP hydrolysis (FIG. 1, lane 1). This complex is effectively dissociated in the presence of ATP and Mg (FIG. 1, lane 2) and cannot be isolated when the AMPA receptor is precipitated by GluR2-specific antibody binding at the last 13 amino acids of the GluR2 C-terminus (FIG. 1, lanes 4 and 5). This last result suggests a direct competition of binding between the NSF-SNAP complex and the C-terminal targeted antibody, and implies that proteins binding at the end of the C-terminus, such as GRIP and ABP, may be excluded from the interaction with GluR2 by the NSF-SNAP complex. In addition, we have established that the binding ratio of NSF:SNAP is essentially the same in the NSF-SNAP complex with GluR2 and with the t-SNARE syntaxin, suggesting that the same multiprotein NSF-SNAP complex is involved in binding to the AMPA receptor and the SNARE complex.[3]

Our work thus suggests that one function of the GluR2 C-terminus is to recruit the chaperone NSF-SNAP complex in an ATP reversible binding with the AMPA receptor, which may functionally resemble the interaction of NSF and SNAP with the membrane fusion complex SNARE. Interestingly, postsynaptic injection of a peptide derived from the

[a]Corresponding author. Phone: 212-263-5936; fax: 212-683-8453; e-mail: ostenp01@mcrcr6.med.nyu.edu

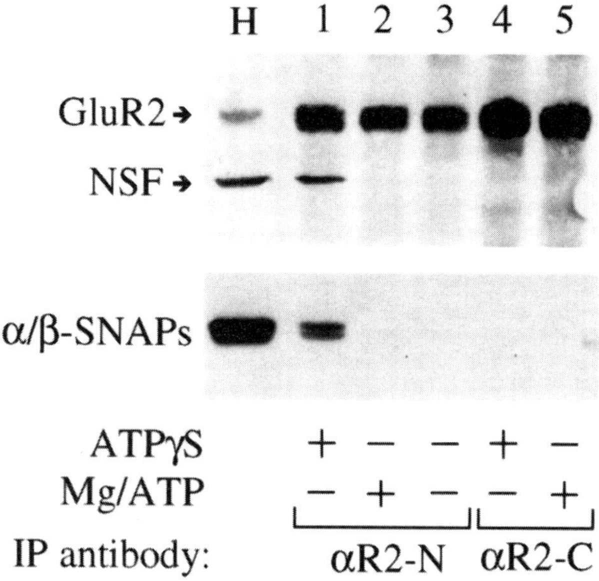

FIGURE 1. GluR2-NSF-SNAP complex can be coimmunoprecipitated using the N-terminal targeted αR2-N antibody under conditions of inhibited ATP-hydrolysis. Hippocampal triton extract prepared either in the general IP buffer lacking any nucleotides (25 mM Hepes-NaOH, pH 7.4; 150 mM NaCl; and 1 mM DTT; **lane 3**), or in the ATPγS-IP buffer (0.5 mM ATPγS and 2 mM EDTA; **lanes 1 and 4**), or ATP-IP buffer (0.5 mM ATP and 2 mM $MgCl_2$; **lanes 2 and 5**), was precipitated with anti-R2-N mAb **(lanes 1–3);** or anti-R2/3-C terminus Ab **(lanes 4 and 5).** The precipitates were resolved by SDS-PAGE and immunoblotted. H = hippocampal extract, 50 μg, used as a Western blot marker for precipitated proteins in **lanes 1–5**.

binding region of NSF at the GluR2 C-terminus results in about 40% inhibition of the AMPA receptor–mediated current in hippocampal slices.[4] The electrophysiological effect of the peptide is abolished by a change in a single amino acid, Asn851.[4] This result well agrees with our finding that a double mutation of Asn851 and Pro852 to Ala results in approximately 70% inhibition of NSF binding to the GluR2 C-terminus *in vitro*.[3] Thus our results, when combined with those of Nishimune and colleagues, can be interpreted to propose a model for the GluR2-NSF-SNAP interaction in which the NSF-SNAP function is required for maintaining the AMPA receptor current (FIG. 2). According to this model, NSF and SNAP may mediate a chaperone-like priming of the AMPA receptors during a continuous process of receptor recycling between the postsynaptic membrane and a cytoplasmic pool. Similarly to what has been proposed for NSF function at the SNARE complex, the interaction of NSF and SNAP with the AMPA receptor could involve disruption of multiprotein complexes, such as those formed between the membrane-inserted receptor and the proteins of the postsynaptic density GRIP and ABP. NSF-driven disassembly of

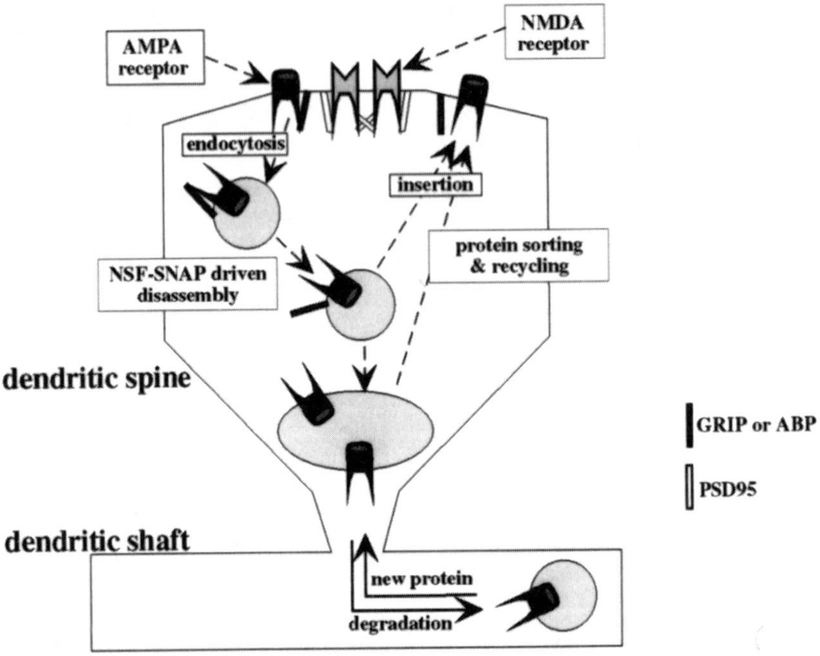

FIGURE 2. Model of the GluR2-NSF-SNAP complex function. According to this model NSF-SNAP function may be required for disassembly of the GluR2-GRIP and/or GluR2-ABP complexes after endocytosis of the receptor from postsynaptic membrane. This step would allow for protein sorting of the components of these complexes and their recycling back to the postsynaptic membrane. Breakdown of NSF function would result in a block of the reinsertion pathway, while the endocytotic part of the cycle would not be affected. This would lead to a decreased number of membrane-inserted receptors and subsequently to a decrease of the AMPA receptor-mediated current.

these complexes could be required for proper sorting of these proteins prior to a new cycle of insertion and anchoring.

REFERENCES

1. Dong, H. *et al.* 1997. GRIP: A synaptic PDZ domain-containing protein that interacts with AMPA receptors. Nature **386:** 279–284.
2. Srivastava, *et al.* 1998. Novel anchorage of GluR2/3 to the postsynaptic density by the AMPA receptor binding protein, ABP. Neuron **21:** 581–591.
3. Osten, P. *et al.* 1998. The AMPA receptor GluR2 C-terminus can mediate a reversible, ATP-dependent interaction with NSF and α- and β-SNAPs. Neuron. **21:** 99–110.
4. Nishimune, A. *et al.* 1998. NSF binding to GluR2 regulates synaptic transmission. Neuron **21:** 87–97.

ABP: A Novel AMPA Receptor Binding Protein

S. SRIVASTAVA[a] AND E. B. ZIFF

Howard Hughes Medical Institute, Department of Biochemistry, New York University Medical Center, 550 First Avenue, New York, New York 10016, USA

INTRODUCTION

Two types of ionotropic receptors that are glutamate-gated cation channels present at excitatory synapses are characterized by their selective agonists *N*-methyl-D-aspartate (NMDA) and α-amino-3-hydroxy-5-methyl-4-isoxazole propionic acid (AMPA). AMPA receptors, which are formed from GluR1–4 (GluR A–D) subunits,[1] mediate fast excitatory transmission, while the NMDA receptors, which are assembled from the NR1 and NR2A–D subunits, admit Ca^{2+} ions and may mediate plasticity of central excitatory synapses.[2] NMDA receptors are anchored to the postsynaptic density (PSD) by interaction with PSD-95.[3–5] The PDZ domains of PSD-95 interact with the C-terminus motif S/TXV of the NMDA receptor.[3] PDZ domains are globular structures, which are known to mediate protein–protein interactions.[6,7] Recently, a new PDZ protein, glutamate receptor interacting protein (GRIP),[8] which binds to GluR2 and GluR3 subunits, has been reported to have a similar anchoring function for AMPA receptors. Here we review our studies of the cloning of a novel PSD protein, ABP (AMPA receptor binding protein), which also binds to the C-termini of the GluR2 and GluR3 subunits via PDZ domains.[9]

RESULTS

To identify the proteins that interact with the GluR2 C-terminus, we used the last 50 amino acids of GluR2 as bait to screen an adult rat brain library using the two hybrid yeast system.[10] We isolated two different proteins containing PDZ domains. One was the previously characterized protein GRIP, and the other was the novel protein ABP. By screening an adult rat brain library we isolated the full-length ABP clone of 5422 base pairs with an open reading frame of 822 amino acids. ABP and GRIP have similar structural organizations. The first six PDZ domains of both proteins are arranged in two clusters, with three PDZ domains in each cluster (PDZ domain sets S_I and S_{II}), followed by two linker regions, referred to as linker 1 (L_1) and linker 2 (L_2) (FIG. 1).

We analyzed ABP protein by Western blotting adult rat brain extract with an affinity-purified polyclonal antiserum directed to an ABP peptide in linker 2, whose sequence is not present in GRIP. The antiserum recognized three bands, migrating at 130, 120, and 98 kDa. The recognition of all three bands was competed by antigen peptide but not by the control peptide. The band at 98 kDa comigrated with the protein encoded by ABP cDNA transfected in 293T cells. The three isoforms were detected in cortex, hippocampus, thalamus, cerebellum, and brain stem, but not in lung or liver. We found that all three isoforms

[a]Address for correspondence: Sapna Srivastava, Room 397, Medical Science Building, Department of Biochemistry, New York University Medical Center, 550 First Avenue, New York, NY 10016. Phone: 212-532-9858; fax: 212-683-8453; e-mail: srivas01@mcrcr6.med.nyu.edu

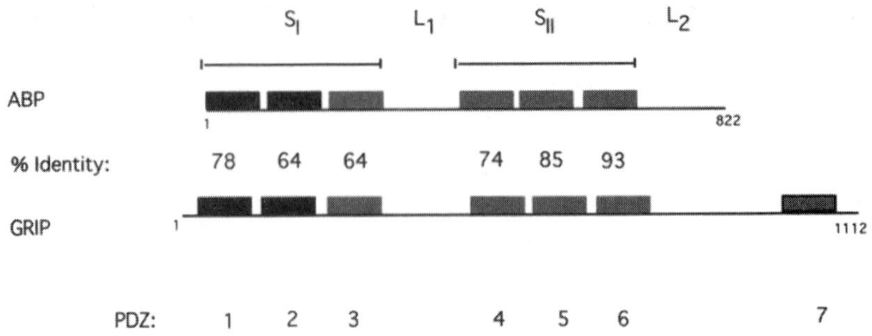

FIGURE 1. Structure of ABP protein and relationship to GRIP. Comparison of the structures of the ABP and GRIP proteins includes the extent of sequence homology between corresponding PDZ domains. PDZ domains with a His at the αB1 residue position are shown as *black boxes*, and those with a hydrophobic residue (Met or Leu) as *gray boxes*. PDZ 7 of GRIP has a Cys residue at αB1. The positions of PDZ domain sets, S_I and S_{II}, and linkers, L_1 and L_2, are indicated.

of ABP were enriched in the synaptosomal and PSD fractions, along with GluR2 in the subcellular fractions made from the forebrain. The expression of the 98-kDa form was developmentally regulated, increasing from P9 to adult. The other two forms were expressed at constant levels from P2.

The spatial distribution of ABP in the brain was studied by immunocytochemistry.[9] Strong staining in the hippocampus could be seen in the pyramidal neurons of CA2, interneurons in the stratum radiatum of Ammon's horn, and the granular layer of the dentate gyrus. In neocortex, layer III pyramidal neurons, large nonpyramidal neurons, and a few neurons in layer I showed strong immunoreactivity. In cerebellum, stellate cells in the molecular layer stained intensely in contrast to weak staining in Purkinje and granule cells. Several other parts of the brain also showed considerable immunoreactivity. At the level of electron microscopy, immunogold-labeled sections of adult rat brain were examined. Maximum staining was observed at the postsynaptic membrane and the PSD. A much weaker stain could be seen extending into the cytoplasm. On quantitation of gold particles labeling 60 randomly selected synapses from three rats, the highest concentration was seen within 15–30 nm inside the postsynaptic membrane, extending ~100 nm into the cytoplasm.

We next determined the ability of ABP to bind AMPA and NMDA receptor subunits. Using yeast interaction assays, we established that ABP binds selectively to the GluR2 and three C-termini (Fig. 2), but not to GluR1, 4, or the NMDA receptor subunit NR2A C-terminus. We confirmed the specificity of ABP-GluR2 interaction by coexpressing these proteins in 293T cells. ABP bound to GluR2, but not to GluR1. We then identified the binding sites on GluR2 and on ABP. We found that the last 13 aa of the GluR2 C-terminus, which are identical in the GluR3 C-terminus, were sufficient to bind ABP. PDZ domains 3, 5, and 6 of ABP were capable of binding GluR2, but not capable of binding NR2A. PDZ5 bound most strongly (Fig. 2).

Two classes of PDZ domains have been defined based on residue 1 of α-helix-B (αB1) in the hydrophobic pocket that determines the binding specificity of the interacting peptide.[11] Class I PDZ domains as found in the PSD-95 family have a histidine at the αB1

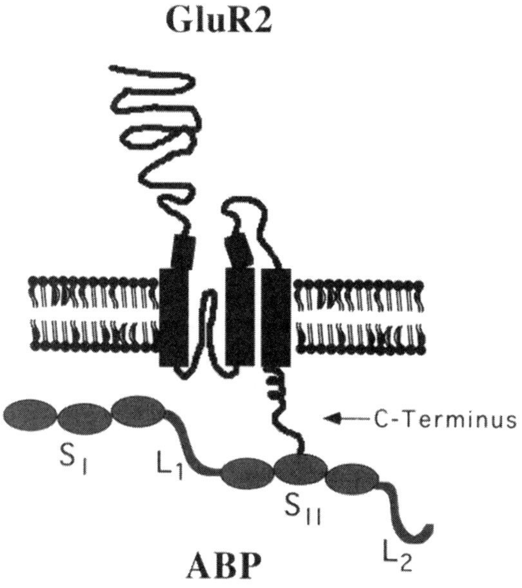

FIGURE 2. Model: structure of ABP and its binding to GluR2. GluR2 is shown in *black* with the extracellular N-terminus, four transmembrane domains, and an intracellular C-terminus. ABP is shown in *gray.* The intracellular GluR2 C-terminus binds most strongly to the 5th PDZ domain of ABP.

position, which forms a hydrogen bond with Ser or Thr at the –2 position of the carboxy terminus.[7,11] In the class II PDZ domains the αB1 residue is hydrophobic. Sequence alignment showed that PDZ1 and two of ABP were class I domains (αB1 is His), while PDZ3, 4, 5, and 6 are class II (αB1 are, respectively, Leu, Met, Met and Leu) (FIG. 1). Thus all of the PDZ domains of ABP that bound to GluR2 were class II domains.

To analyze the binding specificity further, we mutated GluR2 C-terminal amino acids one at a time to Ala, and coexpressed the point mutants with ABP in 293T cells. Wild-type GluR2 and the mutant K861A, in which Lys at position –1 was changed to Ala, bound to ABP. Mutants V860A and I862A, in which the 0 and –2 amino acids were, respectively, changed to Ala, did not bind GluR2. We concluded that the 0 and –2 aa positions of GluR2 are critical for binding ABP, as found previously for the NMDA receptor interaction with PSD-95.[3]

PSD-95 and its family members can multimerize.[12] We asked whether ABP and GRIP could also multimerize. Using the yeast interaction assay we found that S_{II} was sufficient for homo- or heteromultimerization. To confirm the interaction we coexpressed ABP with GRIP or PSD-95 in 293T cells and assayed for complexes by coimmune precipitation. GRIP bound to ABP but not to PSD-95. We also expressed ABP and GRIP in COS7 cells and looked at the distribution by immunofluorescence. When expressed by themselves, they showed a distribution of tiny microaggregates, and in a few cells a large accumulation could be seen in some intracellular organelles. When ABP and GRIP were coexpressed, there was a dramatic redistribution of both the proteins in large plaquelike aggregates. This redistribution suggested that ABP and GRIP can homo- or heteromultimerize to form large scaffolds.

SUMMARY

We review the cloning of a novel AMPA receptor binding protein (ABP) that interacts with GluR2/3 and is homologous to GRIP. ABP is enriched in the PSD with GluR2 and is localized to the PSD by EM. ABP binds GluR2 via the C-terminal VXI motif through a Class I PDZ interaction. ABP and GRIP can also homo- and heteromultimerize. Thus, ABP and GRIP may be involved in AMPA receptor regulation and localization, by linking it to other cytoskeletal or signaling molecules. We suggest that the ABP/GRIP and PSD-95 families form distinct scaffolds that anchor, respectively, AMPA and NMDA receptors. We are currently investigating proteins that bind ABP and that may regulate the AMPA receptor.

ACKNOWLEDGMENTS

This work was supported by the Howard Hughes Medical Institute. S.S. was supported by NIH Grant AG13620 (to E.B.Z); E.B.Z. is an Investigator of the Howard Hughes Medical Institute.

REFERENCES

1. HOLLMANN, M. & S. HEINEMANN. 1994. Cloned glutamate receptors. Annu. Rev. Neurosci. **17:** 31–108.
2. MCBAIN, C.J. & M.L. MAYER. 1994. N-methyl-D-aspartic acid receptor structure and function. Physiol. Rev. **74:** 723–760.
3. KORNAU, H.C. *et al.* Domain interaction between NMDA receptor subunits and the postsynaptic density protein PSD-95. Science **269:** 1737–1740.
4. LAU, L.F. *et al.* 1996. Interaction of the N-methyl-D-aspartate receptor complex with a novel synapse-associated protein, SAP102. J. Biol. Chem. **271:** 21622–21628.
5. NIETHAMMER, M. *et al.* 1996. Interaction between the C terminus of NMDA receptor subunits and multiple members of the PSD-95 family of membrane-associated guanylate kinases. J. Neurosci. **16:** 2157–2163.
6. CABRAL, H.C. *et al.* 1996. Crystal structure of a PDZ domain. Nature **382:** 649–652.
7. DOYLE, D.A. *et al.* 1996. Crystal structures of a complexed and peptide-free membrane protein-binding domain: Molecular basis of peptide recognition by PDZ. Cell **85:** 1067–1076.
8. DONG, H. *et al.* 1997. GRIP: A synaptic PDZ domain-containing protein that interacts with AMPA receptors. Nature **386:** 279–284.
9. SRIVASTAVA, S. *et al.* 1998. Novel anchorage of CluR2/3 to the postsynaptic density by the AMPA receptor binding protein, ABP. Neuron **21:** 581–591.
10. FIELDS, S. & O. SONG. 1989. A novel genetic system to detect protein-protein interactions. Nature **340:** 245–246.
11. SONGYANG, Z. *et al.* 1997. Recognition of unique carboxyl-terminal motifs by distinct PDZ domains. Science **275:** 73–77.
12. HSUEH, Y. *et al.* 1997. Disulfide-linked head-to-head multimerization in the mechanism of ion channel clustering by PSD-95. Neuron **18:** 803–814.

Molecular Diversity of Neuronal Nicotinic Acetylcholine Receptors

DANIEL S. McGEHEE[a]

Department of Anesthesia and Critical Care, The University of Chicago, 5841 South Maryland Avenue, MC 4028, Chicago, Illinois 60637, USA

ABSTRACT: The potent behavioral and cognitive effects of nicotine highlight the physiological importance of nicotinic acetylcholine receptors (nAChRs). These receptors are part of the superfamily of neurotransmitter-gated ion channels that are responsible for rapid intercellular communication. Molecular cloning of the protein subunits that make up these receptors has led to greater understanding of the pharmacology and physiology of nAChRs. This review outlines our current understanding of the molecular constituents of these receptors and some of the recent studies of the structural determinants of receptor function.

Nicotine exerts potent behavioral effects primarily through interactions with neuronal nicotinic acetylcholine receptors (nAChRs).[1,2] These receptors are part of a superfamily of neurotransmitter-gated ion channels that mediate rapid intercellular communication. Our current understanding of the structural, molecular, and biophysical properties of these multisubunit proteins has arisen largely from studies of the nicotinic acetylcholine receptor channels in vertebrate muscle cells and *Torpedo* electrocytes (for reviews see Refs. 3–5). Molecular cloning of homologous nAChR subunits expressed by central and peripheral neurons opened the door to investigation of the molecular physiology of this diverse family of receptors. This review will outline our current understanding of the molecular constituents of neuronal nAChRs and discuss some of the recent studies of the structural determinants of receptor function. The reader is also referred to other reviews of the molecular and functional properties of nAChRs.[6–13]

IDENTIFICATION OF A FAMILY OF nAChR SUBUNITS

The molecular cloning of a neuronal nAChR subunit-encoding gene by Heinemann, Patrick and their colleagues ushered in a new era of nAChR research.[14] Subsequently, multiple "α-type" subunit genes were identified,[15] which has since emerged as the pattern for most ligand-gated ion channel gene families.[5] As the nicotinic receptor subunit gene family was revealed, a total of 11 subunit genes and two subfamilies were identified, including eight α subunits (named α2, α3, α4, α5, α6, α7, α8, α9) and three β subunits (named β2, β3, β4) (reviewed in Refs. 9, 11, 13; see also Refs. 16–19).

The α subunits are highly homologous to the α1 subunit of the muscle nAChR, including the vicinal cysteine residues that are known to be important for ACh binding (Cys 192, Cys 193; FIGURE 1).[4] These subunits also resemble the muscle α1 in both the ligand-binding and proposed transmembrane-spanning domains. In particular, they contain a tyrosine residue that aligns with the α1 Tyr 190, previously demonstrated to be an important determi-

[a]Phone: 773-834-0790; fax: 773-702-4791; e-mail: dmcgehee@midway.uchicago.edu

nant of agonist affinity (Fig. 1).[4] Finally, the high degree of sequence identity between the muscle and neuronal α subunits in the first two transmembrane-spanning regions (TM1 and TM2) was consistent with the similarity in the muscle and neuronal nAChR conductance and ion permeability profiles, since these regions are thought to be involved in lining the muscle nAChR channel pore.

Identification of a second type of neuronal nAChR subunit by Patrick, Heinemann, and colleagues, alternately referred to as the "β" or "non-α," led to a flood of physiological studies that have outlined the properties of neuronal nAChR channels. The β nomenclature did not result from homology with the muscle-type β subunit, rather it is due to the non-α nature of the N-terminal domain. The neuronal nAChR β subunits are distinguished by the absence of the ACh-binding cysteine residues. However, this does not imply that neuronal β subunits do not contribute to ligand binding profiles.

Physiological studies of heterologous expression of α2, α3, or α4 with either β2 or β4 subunit genes in *Xenopus* oocytes have demonstrated that the combination of any of these α/β pairs is sufficient to reconstitute an ACh-gated cationic conductance of appropriate pharmacology for a neuronal nAChR (i.e., nicotine activated, αBgTx insensitive). Despite the effort of several labs, using a battery of molecular techniques, neuronal homologues of the muscle type AChR β, γ, δ, or ε subunits have not been identified. Early studies suggested that the formation of nAChR channels required coexpression of both α and β subunit genes. The exception to this rule emerged with the isolation and expression of neuronal α7, α8, and α9 subunit genes, as each was shown to form functional channels when expressed alone. Another unique property of α7, α8, and α9 is that currents carried

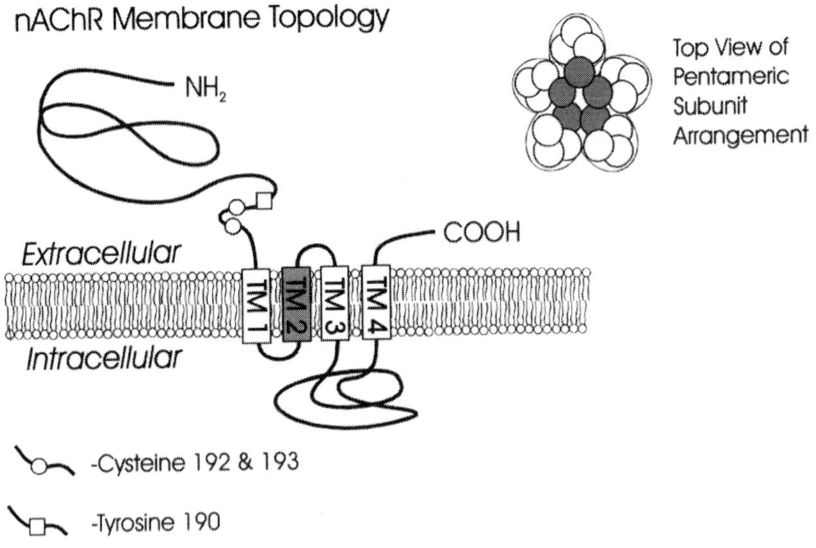

FIGURE 1. Putative membrane topology of nAChR subunits. The symbols denote the residues that are known to contribute to ligand binding. TM2 is the reported to include residues that line the aqueous pore within the ion channel structure. TM = transmembrane domains.

by these homomeric channels are blocked by nanomolar concentrations of αBgTx, unlike the α/β subunit combinations.[17,19–22]

Although the search for additional α- and β-like subunit genes has been extensive, some studies suggest that the nAChR family tree may not be complete. The pattern of subunit expression revealed by *in situ* hybridization still leaves some neuroanatomical regions apparently lacking an appropriate α or β partner. In addition, the array of functional characteristics of native nAChRs is still poorly matched to the channel characteristics obtained in heterologous expression studies to date. While the complexity of the known nAChR combinations is considerable, there is ample precedent in other ligand-gated channel families to expect that additional neuronal nAChR subunits may still be awaiting discovery (e.g., see discussion in Ref. 23).

This is the repertoire of known subunits available for the assembly of neuronal nAChRs. It is clear that a huge number of subunit combinations are theoretically possible—even if we assume restrictions on subunit association and assembly. As discussed below, it is unlikely that nAChR subunit diversity is simply an "evolutionary hangover," bereft of functional consequences. The convergence of evidence from multiple fields indicates that diversity in the molecular components of nAChRs provides an important means of attaining functional diversity of these receptors *in vivo*.

THE PHYSIOLOGICAL IMPORTANCE OF nAChR SUBUNIT DIVERSITY

The physiological importance of expressing more than one α-type nAChR subunit was first demonstrated in analyses of α3 and α4 subunit expression (reviewed in Refs. 9, 10, 13, 16). In particular, immunoprecipitation studies of brain nAChRs, extended by *in situ* hybridization, Northern blot, and RNAase protection assays demonstrated that the α4 subunit was widely expressed in the CNS. Interestingly, the expression pattern of the α4 gene was closely correlated with that of nicotine binding sites, but did not overlap with either α3 or αBgTx sites. In addition, the expression of α4 was considerably more abundant in brain than in autonomic ganglia. Following up earlier biochemical and immunological studies, complimentary results emerged from Berg and Lindstrom and their colleagues that provided compelling evidence for prominent expression of α3 in the PNS (e.g., Ref. 24). This comparison suggested distinct contributions of the α3 and α4 subunits to nAChRs in the CNS vs. PNS. In addition, expression of α3 and α4 in *Xenopus* oocytes yielded functionally distinguishable nAChR channels when expressed with the same β subunit (e.g., Ref. 25). These studies provided the first indication that the expression of different subunit genes produces a rich array of biophysically and pharmacologically distinct nAChRs.

Although expression of functional nAChRs in *Xenopus* oocytes has both technical and practical limitations, it is still the predominant system used to study varied subunit expression. As a result, most of the discussion below will focus on the properties of nAChRs expressed in oocytes. However, several recent reports have utilized stably transfected mammalian cell lines to examine nAChR properties (e.g., Refs. 26, 27).

nAChR CHANNELS FORMED BY EXPRESSION OF α/β SUBUNIT PAIRS AND BY α SUBUNITS ALONE

The α subunits, α2, α3, and α4 form distinct ACh-activated channels when expressed in combination with either β2 or β4 in *Xenopus* oocytes.[25,28–35] In contrast, the expression of α5 or α6, in combination with any β subunit, or expression of β3 with any of the α's, does not produce ACh-activated current[16,36] (M. Ballivet, unpublished). More recently, some of these "orphan" subunits have been found to combine with α/β pairs, which has expanded the array of known functional properties of nAChRs.

As stated previously, the α7, α8, or α9 subunits are unique in that solo expression of these genes alone supports robust ACh-activated currents that are blocked by nanomolar concentrations of αBgTx.[17,19–22] Early efforts to express α7 with other α's or β's produced ACh currents that are pharmacologically indistinguishable from those expressed when oocytes were injected with α7 alone (β2, β3, β4, and muscle βγδ tested with chicken α7;[20] α3, α5, β2, and β4 were tested with the rat α7[22]). One exception to this observation is that α7 and α8, which have a high degree of sequence identity, will form heteromeric complexes with pharmacology very similar to α7 homomers.[21] Note that recent efforts have suggested possible combinations of α7 with other subunits (see below).

SUBUNIT STOICHIOMETRY

Relatively little is known about the regulation of assembly and insertion of neuronal nAChRs into the plasma membrane. The neuronal nAChR is likely to be a pentamer with a subunit stoichiometry of at least two α subunits.[37,38] There is as yet no information on how many alternative α/β stoichiometries may exist or if there is any flexibility in the ordering of subunits around the ion channel (e.g., ααβββ vs. αβαβα). Such flexibility could explain the observation that coexpression of a single set of α and β subunits often produces nAChRs with multiple conductances and opening kinetics (e.g., Refs. 31,34). Another possibility is that the differences reflect a medley of postranslational modifications of one combination of subunits.

BIOPHYSICAL CHARACTERISTICS OF NaChRs EXPRESSED IN OOCYTES

An overview of the biophysical characteristics of the various α/β combinations expressed in oocytes yields some valuable insights into the functional contribution of the α- vs. β-type nAChR subunits. The single-channel conductances for the α/β pairwise combinations range between 5–45 pS, with most between 15–20 pS. The channel conductances are quite similar for the α4β2, α3β2, and α3β4 channels, which could reflect the high degree of amino acid identity in the pore lining TM2 domains of all of these subunits.[4]

Ca^{2+} Permeability of nAChRs Expressed in Oocytes

The relative permeability to Ca^{2+} of a ligand-gated ion channel is an important determinant of its physiological impact. Vernino *et al.*[39] showed that α3β4 channels permeated enough Ca^{2+} to activate endogenous Ca^{2+}-activated Cl^- conductances in oocytes, and that external [Ca^{2+}] can modulate neuronal nAChR function, independent of increases in intracellular [Ca^{2+}]. Using a superfused oocyte preparation, Dani and colleagues determined the Ca^{2+}:Na^+ permeability ratio (P_{Ca}/P_{Na}) for the muscle $\alpha_2\beta\gamma\delta$ receptors and neuronal α3β4 receptors to be 0.2 and 1.1, respectively.[40] Other α/β combinations have Ca^{2+} permeabilities similar to α3β4 channels (J. Dani, personal communication). While Ca^{2+} permeation through these channels is physiologically important, α7 homomeric channels in oocytes are extremely permeable to Ca^{2+}, with a P_{Ca}/P_{Na} close to 20.[22]

Desensitization Characteristics of nAChRs Expressed in Oocytes

Different subunit combinations in the oocyte also have clear differences in the rate and voltage dependence of nAChR desensitization.[30,32] The α3β4 currents desensitize relatively slowly (τ_d of 2 and 20 s),[29,30,32,35] whereas α3β2, α7α8 heteromeric, and α7 or α8 homomeric nAChRs desensitize very rapidly (e.g., α8: $\tau_d = 0.1$ s with 1 mM nicotine).[21] Desensitization of nAChR currents is a complex phenomenon, which can confound comparisons between oocyte expression and nAChR currents in neurons. However, the rapid decay of the α7 and α8 currents are qualitatively similar to the αBgTx-sensitive currents measured in ciliary ganglion[41] and hippocampal neurons.[42,43]

Agonist/Antagonist Sensitivity of nAChRs Expressed in Oocytes

The agonist rank-order potencies vary markedly among the different nAChR subunit combinations. One of the most profound examples is the comparison of rat α3β4 and α3β2 receptors, which display a tenfold difference in the ACh sensitivity (α3β4 > α3β2). Cytisine is a potent agonist for all β4-containing receptors but is the least effective agonist for β2-containing receptors. In fact, cytisine can antagonize the interaction of ACh with β2-containing nAChRs.[44] Similarly, whereas αBgTx is a potent blocker of α7 and α8 homooligomeric channels at nanomolar concentrations, coexpressed α/β combinations are blocked only by thousandfold higher concentrations of this toxin. However, another component of cobra venom, known as neuronal BgTx (nBgTx), differentially inhibits distinct nAChR subunit combinations.[45] Papke and colleagues[46] report that nBgTx block of α3β2 receptors is prolonged due to the high-affinity binding of this toxin. nBgTx inhibits other nAChR subtypes only if it is coapplied with the agonist at micromolar concentrations, and the inhibition reverses rapidly with removal of the toxin. These responses in oocyte expression systems can be used to implicate the contribution of specific subunits to responses *in vivo*. In addition, these studies provide evidence that both the α and β subunits contribute to nAChR agonist and antagonist pharmacology.

Given the extent of sequence similarity of the chick and rat nAChR subunits, the differences in the channels from these species are surprising. As mentioned above, rat α3β2 is nearly tenfold less sensitive to ACh than rat α3β4, whereas the opposite is true for the chick homologues. The agonist rank order of α7 homomers is also quite different between rat and chick clones, since the latter is insensitive to the agonist 1,1 dimethyl 4-phenylpiperazinium (DMPP) up to 3 mM[21] compared with robust responses of rat α7 to DMPP at 100 μM.[22]

nAChR FUNCTIONAL DOMAINS REVEALED BY MUTAGENESIS ANALYSES

The nAChR subunits that form homomeric channels provide unique opportunities to examine and identify functional domains within an α subunit—i.e., regions of the molecule involved in single-channel conductance, channel open time, agonist/antagonist binding, Ca^{2+} permeability, and receptor subunit assembly. Changeux, Bertrand, and colleagues have examined the effects of a series of mutations of functional domains of homomeric α7 channels. Mutation of a single site (Leu 247 to Thr) in the putative channel domain (TM2) of α7 yields a higher conductance channel (80 pS) that desensitizes slowly and is *activated* by classical antagonists such as hexamethonium and d-tubocurarine.[29,35] These results suggest that changes in conductance, as well as the forward and reverse rates of channel opening, can occur with mutations in a presumed channel domain. An alternative possibility is that one or more of the desensitized states of the channel are now in an open conformation. Combined substitution and addition of amino acid residues near the cytoplasmic end of TM2 of α7 with residues found in the TM2 region of the anion-selective glycine receptor α1 subunit converted the ion selectivity from cationic to anionic.[47] Another mutation also near the cytoplasmic end of TM2—in the position referred to as the cytoplasmic ring (Glu 237 to Ala)—abolishes the Ca^{2+} permeability without significantly affecting other characteristics of the ACh responses. The Ca^{2+} permeability, the apparent affinity for ACh, and the rates of activation and desensitization were also altered by mutations of two adjacent amino acids (Leu 254, Leu 255) close to the extracellular end of TM2.[48]

α3β2 and α2β2 receptors expressed in oocytes have dramatically different pharmacology profiles, particularly in their sensitivity to nBgTx and the ratio of responses to nicotine and ACh. Luetje *et al.*[49,50] have exploited these differences to determine the portions of the extracellular domain of α2 and α3 that are critical in ligand-receptor interactions. A series of chimeric α subunits, with portions of α3 replaced by the corresponding regions of α2 (and vice versa) were coexpressed with β2. Testing for nBgTx sensitivity as well as the ratio of nicotine/ACh responses revealed seven amino acids located near the conserved residues of the ligand binding domain that contribute to NBT sensitivity (Thr143, Tyr184, Lys185, His186, Ile188, Gln198, Ser203). In particular, mutation of Thr143 to the α2 residue lysine results in 1000-fold lower nBgTx sensitivity. Since Thr143 is part of a consensus sequence for glycosylation at Asn141, they conclude that this glycosylation is required for nBgTx sensitivity.[50]

The region of the β subunit that participates in ligand interactions has been identified using chimeric β2/β4 subunits expressed with α3.[46] These studies demonstrated that the N-terminal 121 amino acids of the β2 subunit are sufficient to regulate the kinetics of nBgTx block as well as the partial agonist properties of cytisine on α3β2 receptors.

FUNCTIONAL ROLES FOR "ORPHAN" SUBUNITS: COEXPRESSION OF THREE nAChr SUBUNITS

While pairwise expression of nAChR subunits produces a variety of physiologically and pharmacologically distinct receptors, there are very few examples where the properties of nAChR channels measured in neurons actually match those of the combinations expressed in oocytes. An explanation that has gained considerable support from recent immunoprecipitation studies is that native nAChRs may be composed of more than one type of α and/or β subunit.[51–53]

Several recent studies have addressed the possibility of "triplex" subunit expression with intriguing results. Triplex coexpression of subunits has been exploited to determine the role of the α5 subunit, which was a puzzle due to the lack of functional nAChR formation with α5/β2 or α5/β4 expression, despite the abundance of α5 throughout the nervous system. When α5 is coexpressed with other α and β subunits, the properties of the resultant channels are substantially altered. Thus, when α5 mRNA is coinjected along with α4 and β2, the channels expressed have a 100-fold lower sensitivity to ACh than those expressed with injection of α4β2 alone. Furthermore, the α5α4β2 complex has a conductance more than twice that of the α4β2 complex.[36] Participation of the α5 subunit in the resultant nAChRs was confirmed by site-directed mutagenesis of α5 to provide a specific reactive site and hence an assayable α5-specific tag. These results indicate that α5 can combine with other nAChR subunits to form functional nAChRs and that inclusion of α5 in these channels contributes novel functional features to the nAChR complexes formed.

Conroy and Berg[52] have recently reported that immunoprecipitation of nAChRs from chick brain supports the combination of α5 with α4 and β2 subunits *in vivo*. Related experiments have shown that α5 can be immunoprecipitated with complexes containing α3 and β4 from chick ciliary ganglion neurons.[51] Taken together, these findings indicate that the α5 subunit can contribute to a variety of native nAChR subtypes.

Role and colleagues recently found that the combined expression of α7 with α5 and β4 mRNAs produces functional nAChRs. The functional contribution of α5 was demonstrated by the same reporter mutation described above.[54] It is intriguing that immunoprecipitation analyses indicate that α7 does not associate with any known subunits, other than the highly homologous α8 in chick.[53,55] However, differences in the sensitivity of autonomic ganglion neurons to α7-selective antagonists MLA and αBgTx support a potential role for α7-containing heteromeric complexes *in vivo*.[56]

Given that the α6 and α3 sequences are 75% identical, it is indeed puzzling that the coexpression of α6 with β subunits does not yield functional channels.[10,11,13] A recent report by Gerzanich and colleagues[57] has provided a glimmer of hope for a functional role of this subunit. Expression in oocytes of either rat or chick α6 with the human β4 subunit results in the expression of nicotine-gated currents with a pharmacology that is quite similar to α3β4 channels.[57] Why α6 forms functional channels with the human β4 and not with the subunits from the same species may indicate intrinsic properties of α6 that regulate assembly and expression. These results highlight the need for progress in mammalian expression systems, which will hopefully identify the structural determinants of nAChR receptor assembly.

The last "orphan" to find a conceptual home is the β3 subunit, which was found to coimmunoprecipitate with α4, β2, and β4 subunits from striatum and cerebellum. These findings were extended using heterologous expression of all four of these subunits in COS cells, where the elimination of any one of the subunits altered the ligand-binding characteristics of these receptors.[58] Additional evidence that β3 is a functional nAChR subunit has come from oocyte expression of β3 with α3 and β4. In these experiments, a reported mutation in the β3 sequence enabled independent verification that β3 is part of the receptor complex.[59]

COMPOSITION AND FUNCTION OF NATIVE nAChRs USING MOLECULAR AND GENETIC APPROACHES

Determining the function of specific nAChR subunits *in vivo* has relied primarily upon correlation of the pharmacological sensitivities with those described for known subunit combinations expressed in oocytes. Molecular and genetic approaches, such as antisense oligonucleotide-mediated subunit deletion and transgenic technology, are providing important insights into the functional contribution of specific nAChR subunits. Role and colleagues have used antisense treatment to identify the subunits that contribute to both pre- and postsynaptic nAChRs.[11,60,61] Recent investigations by Yu and Role[56] into the subunit composition of nAChRs in autonomic neurons support functional roles for some of the triplet subunit combinations described above as well as a number of other novel combinations.

Changeux and colleagues have investigated the contribution of the β2 subunit to the behavioral effects of nicotine using genetically altered mice with a null mutation in the β2 subunit gene (reviewed in Ref. 65). These mice do not express high-affinity nicotine binding sites in the CNS, and electrophysiological recording from brain slices reveals that thalamic neurons from these mice do not respond to nicotine application. Finally, behavioral tests have demonstrated that nicotine does not augment associative memory, as it does in nonmutant siblings.

Hippocampal neurons express rapidly desensitizing ACh-activated currents that are similar to the α7 currents in oocytes. Recent studies of knockout mice that lack a functional α7 subunit gene have provided additional support for α7 involvement in hippocampal nAChRs. These mice do not express detectable α7 mRNA or high-affinity [^{125}I]-BGT sites. Hippocampal neurons isolated from these mice lack rapidly desensitizing, methyllycaconitine-sensitive, nicotinic currents.[63] It will be exciting to learn of other changes in the nicotine sensitivity of these animals.

REASONS FOR THE DIVERSITY IN nAChRs

The findings described above, combined with the wide range of functional properties of native nAChRs, demonstrates an impressive array of nAChR channels. Differences in subunit composition, pattern of expression, and cellular localization as well as finer distinctions in conductance, kinetics, ion permeability, and transmitter affinity suggest that nAChRs in the CNS and PNS are indeed diverse. Some of the advantages resulting from this diversity are outlined below.

Role of Diversity in nAChR Expression in Subcellular Domains

A convergence of anatomical, biochemical, and electrophysiological approaches have determined that a predominant role for nAChRs in the CNS is to modify neurotransmitter release due to their expression on presynaptic structures. In addition to the evidence that presynaptic nAChRs can alter transmitter release, there is an intriguing diversity in the pharmacology of these receptors that seems to correlate with the type of transmitter whose release is enhanced.[6,45,64–67] The targeting of specific proteins to axonal vs. somatic domains is not well understood; however, the addresses for such targeting are generally thought to be determined by specific sequence domains.[68] In view of the importance of presynaptic nAChRs in the regulation of transmitter release, the ability to selectively ship the appropriate nAChR subtype to a terminal vs. somatic membrane is essential. Thus, differences in the primary structure of nAChR subunits may have evolved to allow specific targeting.

Role of nAChR Diversity in Synaptic Development and Efficacy

Comparison of nAChRs at different stages of development and with different types of synaptic input reveal alterations in channel conductance, kinetics, pharmacology, and distribution (for reviews see Refs. 11, 13, 69). Innervation of autonomic neurons increases ACh sensitivity[70–72] as well as the number and apparent clustering of surface nAChRs *in vivo* and *in vitro*.[73–75] Specific nAChR subunit expression levels also increase concurrent with innervation and target contact in autonomic ganglia and interfering with presynaptic input inhibits the increase in nAChR subunit gene expression.[76–78]

Increasing the number of nAChRs enhances excitability and the reliability of synaptic transmission, but the advantage of expressing multiple nAChR channel subtypes is less clear. After innervation, autonomic neurons up-regulate expression of larger conductance channels, with longer openings, which are segregated into high-density clusters.[74] Computer modeling has indicated that "postinnervation" type nAChRs can enhance the probability that a given input will cause a suprathreshold depolarization (D. McGehee, L. Role & A. Brussaard, in preparation). This analysis supports one rationale for multiple nAChR subtypes: expression of specific nAChR subtypes can fine-tune the efficacy of synaptic transmission to optimize neuronal excitability.

Role of nAChR Diversity in Short-Term Changes in nAChR Function

The subunit composition of nAChRs may also contribute to differences in the susceptibility to modulation in channel function. Modulation of neuronal nAChRs in peripheral autonomic neurons by both A and C kinase has been studied at the macroscopic current level, with some corroborative single-channel data (reviewed in Ref. 11). Peptide neurotransmitters have been implicated as primary messengers due to their presence in fibers contacting these neurons, consistent with a physiological, rather than strictly pharmacological, role for nAChR modulation. In support of nAChR subtype–specific modulation, recent studies have shown that individual nAChRs have different sensitivity to A kinase in on-cell patch recordings with external application of kinase activators and inhibitors.[79]

CONCLUSION

The widespread pattern of expression and the rich diversity of nAChRs correlate with the potent behavioral effects of nicotine. Chronic administration of this drug has profound cognitive effects; but as we increase our understanding of the CNS effects of nicotine, the mechanistic questions seem to multiply. Further investigations into the molecular basis of nAChR function are required if we are to understand how these similar molecular entities can influence so many aspects of behavior and experience.

ACKNOWLEDGMENTS

This work was supported by NIH Grant Number NS35090 and a grant from the Brain Research Foundation of Chicago. I want to thank Drs. L. Role, P. Devay, and J. Dani for their helpful suggestions, along with J. Genzen and K. Oppen for their assistance with the manuscript preparation.

REFERENCES

1. Balfour, D.J. & K.O. Fagerstrom. 1996. Pharmacology of nicotine and its therapeutic use in smoking cessation and neurodegenerative disorders. Pharmacol. & Ther. **72**(1): 51–81.
2. Dani, J.A. & S. Heinemann. 1996. Molecular and cellular aspects of nicotine abuse. Neuron **16**(5): 905–908.
3. Galzi, J.-L., F. Revah, A. Bessis & J.-P. Changeux. 1991. Functional architecture of the nicotinic acetylcholine receptor: from electric organ to brain. Annu. Rev. Pharmacol. **31:** 37–72.
4. Karlin, A. 1993. Structure of nicotinic acetylcholine receptors. Curr. Opin. Neurobiol. **3:** 299–309.
5. Karlin, A. & M.H. Akabas. 1996. Toward a structural basis for the function of nicotinic acetylcholine receptors and their cousins. Neuron **15**(6): 1231–1244.
6. Albuquerque, E.X., M. Alkondon, E.F. Pereira, N.G. Castro, A. Schrattenholz, C.T. Barbosa, R. Bonfante-Cabarcas, Y. Aracava, H.M. Eisenberg & A. Maelicke. 1997. Properties of neuronal nicotinic acetylcholine receptors: Pharmacological characterization and modulation of synaptic function. J. Pharmacol. Exp. Ther. **280**(3): 1117–1136.
7. Bertrand, D. & J-P. Changeux. 1995. Nicotinic receptor: an allosteric protein specialized for intercellular communication. Semin. Neurosci. **7:** 75–90.
8. Colquhoun, L.M. & J.W. Patrick. 1997. Pharmacology of neuronal nicotinic acetylcholine receptor subtypes. Adv. Pharmacol. **39:** 191–220.
9. Heinemann, S., J. Boulter, E. Deneris, J. Connolly, R. Duvousin, R. Papke & J. Patrick. 1990. The brain nicotinic acetylcholine receptor gene family. Prog. Brain Res. **86:** 195–203.
10. Lindstrom, J., R. Schoepfer, W.G. Conroy & P. Whiting. 1990. Structural and functional heterogeneity of nicotinic receptors. In The Biology of Nicotine Dependence, CIBA Foundation Symposium, Vol.152. G. Bock & J. Marsh, Eds. 23–52. Wiley & Sons. New York.
11. McGehee, D.S. & L.W. Role. 1995. Physiological diversity of nicotinic acetylcholine receptors expressed by vertebrate neurons. Annu. Rev. Physiol. **57:** 521–546.
12. Role, L.W. & D.K. Berg. 1996. Nicotinic receptors in the development and modulation of CNS synapses. Neuron. **16**(6): 1077–1085.
13. Sargent, P.B. 1993. The diversity of neuronal nicotinic acetylcholine receptors. Annu. Rev. Neurosci. **16:** 403–433.
14. Boulter, J., K. Evans, D. Goldman, G. Martin, D. Treco, S. Heinemann & J. Patrick. 1986. Isolation of a cDNA clone coding for a possible neural nicotinic acetylcholine receptor α-subunit. Nature **319:** 368–374.
15. Boulter, J., J. Connolly, E. Deneris, S. Goldman, S. Heinemann & J. Patrick. 1987. Functional expression of two neuronal nicotinic acetylcholine receptors from cDNA clones identifies a gene family. Proc. Natl. Acad. Sci. USA **84:** 7763–7767.
16. Couturier, S., L. Erkman, S. Valera, D. Rungger, S. Bertrand, J. Boulter, M. Ballivet & D. Bertrand. 1990. α5, α3, and non-α3: Three clustered avian genes encoding neuronal nicotinic acetylcholine receptor–related subunits. J. Biol. Chem. **265:** 17560–17567.
17. Elgoyhen, A.B., D.S. Johnson, J. Boulter, D.E. Vetter, S. Heinemann. 1994. Alpha 9: an acetylcholine receptor with novel pharmacological properties expressed in rat cochlear hair cells. Cell 79(**4**)**:** 705–715.
18. Nef, P., C. Oneyser, C. Alliod, S. Couturier & M. Ballivet. 1988. Genes expressed in the brain define three distinct neuronal nicotinic acetylcholine receptors. EMBO J. **7:** 595–601.
19. Schoepfer, R., W.G. Conroy, P. Whiting, M. Gore & J. Lindstrom. 1990. Brain α-bungarotoxin binding protein cDNAs and MAbs reveal subtypes of this branch of the ligand-gated ion channel gene superfamily. Neuron **5:** 35–48.
20. Couturier, S., D. Bertrand, J. M. Matter, M.C. Hernandez, S. Bertrand, N. Millar, S. Valera, T. Barkas & M. Ballivet. 1990b. A neuronal nicotinic acetylcholine receptor subunit (α7) is developmentally regulated and forms a homooligomeric channel blocked by α-BgTx. Neuron **5:** 847–856.
21. Gerzanich, V., R. Anand & J. Lindstrom. 1994. Homomers of α8 and α7 subunits of nicotinic receptors exhibit similar channel but contrasting binding site properties. Mol. Pharmacol. **45:** 212–220.

22. SEGUELA, P., J. WADICHE, K. DINELEY-MILLER, J.A. DANI & J.W. PATRICK. 1993. Molecular cloning, functional properties and distribution of rat brain α7: A nicotinic cation channel highly permeable to calcium. J. Neurosci. **13:** 596–604.
23. DUVOSIN, R.M., E.S. DENERIS, J. PATRICK & S. HEINEMANN. 1989. The functional diversity of the neuronal nicotinic acetylcholine receptors is increased by a novel subunit: β4. Neuron **3:** 487–496.
24. BOYD, R.T., M.H. JACOB, S. COUTURIER, M. BALLIVET & D.K. BERG. 1988. Expression and regulation of neuronal acetylcholine receptor mRNA in chick ciliary ganglia. Neuron **1:** 495–502.
25. LUETJE, C.W. & J. PATRICK. 1991. Both α- and β-subunits contribute to the agonist sensitivity of neuronal nicotinic acetylcholine receptors. J. Neurosci. **11:** 837–845.
26. GOPALAKRISHNAN, M., B. BUISSON, E. TOUMA, T. GIORDANO, J.E. CAMPBELL, I.C. HU, D. DONNELLY-ROBERTS, S.P. ARNERIC, D. BERTRAND & J.P. SULLIVAN. 1995. Stable expression and pharmacological properties of the human alpha 7 nicotinic acetylcholine receptor. Eur. J. Pharmacol. **290(3):** 237–246.
27. ROTHHUT, B., S.J. ROMANO, S. VIJAYARAGHAVAN & D.K. BERG. 1996. Post-translational regulation of neuronal acetylcholine receptors stably expressed in a mouse fibroblast cell line. J. Neurobiol. **29**(1): 115–225.
28. BALLIVET, M., P. NEF, S. COUTURIER, D. RUNGGER, C.R. BADER, D. BERTRAND & E. COOPER. 1988. Electrophysiology of a chick neuronal nicotinic acetylcholine receptor expressed in Xenopus oocytes after cDNA injection. Neuron **1:** 847–852.
29. BERTRAND, D., A. DEVILLERS-THIERY, F. REVAH, J.H. GALZI, N. HUSSY, C. MULLE, S. BERTRAND, M. BALLIVET & J.P. CHANGEUX. 1992. Unconventional pharmacology of a neuronal nicotinic receptor mutated in the channel domain. Proc. Natl. Acad. Sci. USA **89:** 1261–1265.
30. CACHELIN, A.B. & R. JAGGI. 1991. β subunits determine the time course of desensitization in rat α3 neuronal nicotinic acetylcholine receptors. Pfluegers Archiv. Eur. J. Physiol. **419:** 579–582.
31. CHARNET, P., C. LABARCA, B.N. COHEN, N. DAVIDSON, H.A. LESTER & G. PILAR. 1992. Pharmacological and kinetic properties of α4β2 neuronal nicotinic acetylcholine receptors expressed in Xenopus oocytes. J. Physiol. **450:** 375–394.
32. GROSS, A., M. BALLIVET, D. RUNGGER & D. BERTRAND. 1991. Neuronal nicotinic acetylcholine receptors expressed in Xenopus oocytes: Role of the α subunit in agonist sensitivity and desensitization. Pfluegers Arch. Eur. J. Physiol. **419:** 545–551.
33. LUETJE, C.W., K. WADA, S. ROGERS, S.N. ABRAMSON, K. TSUJI, S. HEINEMANN & J. PATRICK. 1990a. Neurotoxins distinguish between different neuronal nicotinic acetylcholine receptor subunit combinations. J. Neurochem. **55:** 632–640.
34. PAPKE, R.L., J. BOULTER, J. PATRICK & S. HEINEMANN. 1989. Single channel currents of rat neuronal nicotinic acetylcholine receptors expressed in Xenopus oocytes. Neuron **3:** 589–596.
35. REVAH, F., D. BERTRAND, J.L. GAIZI, A. DEVILLERS-THIERY, C. MULLE, N. HUSSY, S. BERTRAND, M. BALLIVET & J.P. CHANGEUX. 1991. Mutations in the channel domain alter desensitization of a neuronal nicotinic receptor. Nature **353:** 846–849.
36. RAMIREZ-LATORRE, J., C.R.YU, X. QU, F. PERIN, A. KARLIN & L. ROLE. 1996. Functional contributions of alpha5 subunit to neuronal acetylcholine receptor channels. Nature **380**(6572): 347–351.
37. ANAND, R., W.G. CONROY, R. SCHOEPFER, P. WHITING & J. LINDSTROM. 1991. Neuronal nicotinic acetylcholine receptors expressed in Xenopus oocytes have a pentameric quaternary structure. J. Biol. Chem. **266:** 11192–11198.
38. COOPER, E., S. COUTURIER & M. BALLIVET. 1991. Pentameric structure and subunit stoichiometry of a neuronal actylcholine receptor. Nature **350:** 235–238.
39. VERNINO, S., M. AMADOR, C.W. LUETJE, J. PATRICK & J.A. DANI. 1992. Calcium modulation and high calcium permeability of neuronal nicotinic acetylcholine receptors. Neuron **8:** 127–134.
40. COSTA, A.C.S., J.W. PATRICK & J.A. DANI. 1994. Improved technique for studying ion channels expressed in Xenopus oocytes, including fast superfusion. Biophys. J. **67:** 1–7.
41. ZHANG, Z.W., S. VIJAYARAGHAVAN & D.K. BERG. 1994. Neuronal acetylcholine receptors that bind α-bungarotoxin with high affinity function as ligand-gated ion channels. Neuron **12:** 167–177.

42. Alkondon, M. & E.X. Albuquerque. 1993. Diversity of nicotinic acetylcholine receptors in rat hippocampal neurons. I. Pharmacological and functional evidence for distinct structural subtypes. J. Pharmacol. Exp. Ther. **265:** 1455–1473.
43. Frazier, C.J., Y.D. Rollins, C. R. Breese, S. Leonard, R. Freedman & T.V. Dunwiddie. 1998. Acetylcholine activates an α-bungarotoxin–sensitive nicotinic current in rat hippocampal interneurons, but not pyramidal cells. J. Neurosci. **18:** 1187–1195.
44. Papke, R.L. & S.F. Heinemann. 1993. Partial agonist properties of cytisine on neuronal nicotinic receptors containing the β2 subunit. Mol. Pharm. **45:** 142–149.
45. Guo, J.Z., T.L. Tredway & V.A. Chiappinelli. 1998. Glutamate and GABA release are enhanced by different subtypes of presynaptic nicotinic receptors in the lateral geniculate nucleus. J. Neurosci. **18**(6): 1963–1969.
46. Papke, R.L., R.M. Duvoisin & S.F. Heinemann. 1993. The amino terminal half of the nicotinic β-subunit extracellular domain regulates the kinetics of inhibition by neuronal bungarotoxin. Proc. R. Soc. Lond. Biol. Ser. B. Sci. **252**(1334): 141–148.
47. Galzi, J.-L., A. Devillers-Thiery, N. Hussy, S. Bertrand, J.-P. Changeux & D. Bertrand. 1992. Mutations in the channel domain of a neuronal nicotinic receptor convert ion selectivity from cationic to anionic. Nature **359:** 500–505.
48. Bertrand, D., J.H. Galzi, A. Devillers-Thiery, S. Bertrand & J.P. Changeux. 1993. Mutations at two distinct sites within the channel domain M2 alter calcium permeability of neuronal α7 nicotinic receptor. Proc. Natl. Acad. Sci. USA **90:** 6971–6975.
49. Luetje, C.W., M. Piattoni & J. Patrick. 1993. Mapping of ligand binding sites of neuronal nicotinic acetylcholine receptors using chimeric alpha subunits. Mol. Pharmacol. **44**(3): 657–666.
50. Luetje, C.W., F.N. Maddox & S.C. Harvey. 1998. Glycosylation within the cysteine loop and six residues near conserved Cys192/Cys193 are determinants of neuronal bungarotoxin sensitivity on the neuronal nicotinic receptor alpha3 subunit. Mol. Pharmacol. **53**(6): 1112–1119.
51. Conroy, W.G. & D.K. Berg. 1995. Neurons can maintain multiple classes of nicotinic acetylcholine receptors distinguished by different subunit compositions. J. Biol. Chem. **270:** 4424–4431.
52. Conroy, W.G. & D.K. Berg. 1998. Nicotinic receptor subtypes in the developing chick brain: Appearance of a species containing the alpha4, beta2, and alpha5 gene products. Mol. Pharmacol. **53**(3): 392–401.
53. Whiting, P.J., R. Schoepfer, W.G. Conroy, M.J. Gore, K.T. Keyser, S. Shimanski, F. Esch, J.M. Lindstrom. 1991. Expression of nicotinic acetylcholine receptor subtypes in brain and retina. Mol. Brain Res. **10:** 61–70.
54. Crabtree, G., J. Ramirez-Latorre, L.W. Role. 1997. Assembly and Ca^{2+} regulation of neuronal nicotinic receptors including the α7 and α5 subunits. Soc. Neurosci. Abstr. **23**(1): 391.
55. Chen D. & J.W. Patrick. 1997. The α-bungarotoxin–binding nicotinic acetylcholine receptor from rat brain contains only the α7 subunit. J. Biol. Chem. **272:** 24024–24029.
56. Yu, C.R. & L.W. Role. 1998. Functional contribution of the α7 subunit to multiple subtypes of nicotinic receptors in embryonic chick sympathetic neurones. J. Physiol. **509:** 651–665.
57. Gerzanich, V., A. Kuryatov, R. Anand & J. Lindstrom. 1998. "Orphan" alpha6 nicotinic AchR subunit can form a functional heteromeric acetylcholine receptor. Mol. Pharmacol. **51**(2): 320–327.
58. Forsayeth, J.R. & E. Kobrin. 1997. Formation of oligomers containing the β3 and β4 subunits of the rat nicotinic receptor. J. Neurosci. **17:** 1531–1538.
59. Groot-Kormelink, P.J., W.H.M.L. Luyten, D. Colquhoun & L.G. Sivilotti. 1998. A reporter mutation approach shows incorporation of the "orphan" subunit 3 into a functional nicotinic receptor. J. Biol. Chem. **273:** 15317–15320.
60. Brussaard, A.B., X. Yang, J.P. Doyle, S. Huck & L.W. Role. 1994. Developmental regulation of multiple nicotinic AChR channel subtypes in embryonic chick habenula neurons: Contributions of both the α2 and α4 subunit genes. Pfuegers Archiv. Eur. J. Physiol. **429**(1): 27–43.
61. Listerud, M., A.B. Brussard, P. Devay, D.R. Colman & L.W. Role. 1991. Functional contribution of neuronal AChR subunits by antisense oligonucleotides. Science **254:** 1518–1521.
62. Picciotto, M.R., M. Zoli, V. Zachariou & J.-P. Changeux. 1997. Contribution of nicotinic acetylcholine receptors containing the β2-subunit to the behavioural effects of nicotine. Biochem. Soc. Trans. **25**(3): 824–829.

63. Orr-Urtreger, A., F. M. Göldner, M. Saeki, I. Lorenzo, L. Goldberg, M. De Biasi, J. A. Dani, J.W. Patrick & A.L. Beaudet. 1997. Mice deficient in the α7 neuronal nicotinic acetylcholine receptor lack α-bungarotoxin binding sites and hippocampal fast nicotinic currents. J. Neurosci. **17:** 9165–9171.
64. Clarke, P.B.S., G.S. Hamill, N.S. Nadi, D.M. Jacobowitz & A. Pert. 1986. 3H-nicotine– and 125I-alpha-bungarotoxin–labeled nicotinic receptors in the interpeduncular nucleus of rats. II. Effects of habenular deafferentation. J. Comp. Neurol. **251:** 407–413.
65. Gray, R., A.S. Rajan, K.A. Radcliffe, M. Yakehiro & J.A. Dani. 1996. Hippocampal synaptic transmission enhanced by low concentrations of nicotine. Nature **383**(6602): 713–716.
66. McGehee, DS. & L.W. Role. 1996. Presynaptic ionotropic receptors. Curr. Opin. Neurobiol. **6**(3): 342–349.
67. Wonnacott, S. 1997. Presynaptic nicotinic ACh receptors. Trends Neurosci. **20**(2): 92–98.
68. Kelly, R.B. & E. Grote. 1993. Protein targeting in the neuron. Annu. Rev. Neurosci. **16:** 95–127.
69. Berg, D.K., R.T. Boyd, S.W. Halvorsen, L.S. Higgins, M.H. Jacob, J.F. Margiotta. 1989. Regulating the number and function of neuronal acetylcholine receptors. Trends Neurosci. **12:** 16–21.
70. Gardette, R., M.D. Listerud, A.B. Brussaard & L.W. Role. 1991. Developmental changes in transmitter sensitivity and synaptic transmission in embryonic chicken sympathetic neurons innervated *in vitro*. Dev. Biol. **147:** 83–95.
71. Jacob, M.H. 1991. Acetylcholine receptor expression in developing chick ciliary ganglion neurons. J. Neurosci. **11:** 1701–1712.
72. Margiotta, J.F. & D. Gurantz. 1989. Changes in the number, function, and regulation of nicotinic acetylcholine receptors during neuronal development. Dev. Biol. **135:** 326–339.
73. Engisch, K.L. & G.D. Fischbach. 1992. The development of ACH- and GABA-activated currents in embryonic chick ciliary ganglion neurons in the absence of innervation *in vivo*. J. Neurosci. 1992, **12:** 1115–1125.
74. Moss, B.L. & L.W. Role. 1993. Enhanced ACh sensitivity is accompanied by changes in ACh receptor channel properties and segregation of ACh receptor subtypes on sympathetic neurons during innervation in vivo. J. Neurosci. **13:** 13–28.
75. Sargent, P.B. & D.Z. Pang. 1989. Acetylcholine receptor–like molecules are found in both synaptic and extrasynaptic clusters on the surface of neurons in the frog cardiac ganglion. J. Neurosci. **9:** 1062–1072.
76. Corriveau, R.A. & D.K. Berg. 1993. Coexpression of multiple acetylcholine receptor genes in neurons: Quantification of transcripts during development. J. Neurosci. **13**(6): 2662–2671.
77. Devay, P., X. Qu & L.W. Role. 1994. Regulation of nAChR subunit gene expression relative to the development of pre- and postsynaptic projections of embryonic chick sympathetic neurons. Dev. Biol. **162:** 56–70.
78. Levey, M.S., C. Brumwell, S. Dryer & M. Jacob. 1995. Innervation and target tissue interactions differentially regulate acetylcholine receptor subunit transcript levels in developing neurons in situ. Neuron. **14**(1): 153–162.
79. Tan, W., C. Du, S.A. Siegelbaum & L.W. Role. 1998. Modulation of nicotinic AChR channels by prostaglandin E2 in chick sympathetic ganglion neurons. J. Neurophysiol. **79**(2): 870–878.

Heteromeric Complexes of α5 and/or α7 Subunits

Effects of Calcium and Potential Role in Nicotine-Induced Presynaptic Facilitation

R. GIROD,[a] G. CRABTREE,[a] G. ERNSTROM,[a] J. RAMIREZ-LATORRE,[b] D. McGEHEE,[c] J. TURNER,[a] AND L. ROLE[a,d]

[a]*Department of Anatomy and Cell Biology in the Center for Neurobiology and Behavior, Columbia University College of Physicians and Surgeons, P.I. Annex #807, 1051 Riverside Drive, New York, New York 10032, USA*

[b]*Department of Pharmacology, Temple University, Philadelphia, Pennsylvania 19122, USA*

[c]*Department of Anesthetics and Critical Care, University of Chicago, Chicago, Illinois 60637, USA*

ABSTRACT: Nicotine alters a broad spectrum of behaviors, including attention, arousal, anxiety, and memory. The cellular physiology of nicotine is comparably diverse: nicotine interacts with an array of ionotropic receptors whose gating can lead to direct depolarization of neurons or to an indirect modulation of neuronal excitability by presynaptic facilitation. Furthermore, as many laboratories have shown, the α- and β-type subunits that comprise neuronal nicotinic acetylcholine receptors (nAChRs) are encoded by multiple, homologous genes, yielding at least seven α and three β subunits, distinct in primary sequence. nAChRs that differ in subunit composition differ in pharmacology, conductance, and kinetics as well as in their permeability to and modulation by calcium. We will first discuss recent studies on the biophysics of a special (peculiar?) subset of nAChRs, focusing on heteromeric nAChRs comprised of α4β2 ± α5 or α7 ± β2 and α5. These nAChR channel subtypes are potently and differentially modulated by changes in intracellular calcium ([Ca]). Thus, the P_o, τ_o, and desensitization kinetics of α4β2 channels are altered by changes in $[Ca]_{int}$ from 0 to 50 μM; nAChRs that include the α5 subunit are oppositely regulated. Mutagenesis of specific residues within the M1 to M2 domain of α4, β2, and α5 suggest a possible Ca binding "pocket." The assembly of functional nAChRs that include α5 and/or α7 and the potential role of these novel heteromeric complexes in presynaptic facilitation will also be presented.

Nicotine alters a multiplicity of behavioral and cognitive processes in the CNS through its interaction with the neuronal nicotinic acetylcholine receptors (nAChRs), receptors for which acetylcholine (ACh) is the natural ligand. Nicotine has been shown to enhance working memory, to enhance attention, to increase arousal and alertness, to decrease anxiety and—last but not least—to reinforce its own self-administration. (For recent reviews, see Refs. 1,2). The breadth of nicotine's actions reflects the widespread distribution and variety of nAChRs expressed throughout the mammalian CNS. The cloning and spatial mapping of several homologous genes encoding the nAChR subunits suggest that different combinations of these homologous subunits coassemble in distinct brain

[d]Corresponding author.

regions. These nAChR subtypes have distinct pharmacology and biophysical profiles. Thus, neuronal responses to exogenous nicotine or to endogenous acetylcholine are also likely to be influenced by the particular nAChR subtype(s) expressed.

Following their activation by nicotinic agonists, nAChRs desensitize and recover from desensitization with kinetics that also depend on receptor composition.[3–7] The net effect of sustained exposure to nicotine, such as that seen during intake of nicotine through smoking, will depend on the balance between active and desensitized nAChR subtypes in specific areas of the brain (see review in Ref. 8).

The following discussion briefly reviews the current knowledge of the pharmacology and biophysics of some major nAChR subtypes encountered in the vertebrate brain. We then focus on results from our lab, where heterologous expression studies reveal differential modulation of nAChR subtypes by changes in internal calcium. Recent work suggests that presynaptic nAChRs may contribute to the effects of nicotine and endogenous ACh *in vivo* (see review in Ref. 9). As such, we conclude with a very speculative discussion of how distinct subtypes of nAChRs may differentially modulate synaptic transmission.

BIOPHYSICAL AND PHARMACOLOGICAL CHARACTERISTICS OF NEURONAL nAChRs

Heinemann, Patrick, and their colleagues introduced the first neuronal sibling of the ligand-gated ion channel gene family, with the isolation of a gene encoding the α nAChR subunit.[10] Since this report, 11 homologous genes encoding neuronal nAChR receptor subunits have been identified in rat and chick. Eight are designated as α-type (α2-α9), as they contain a cysteine pair reminiscent of that found within the n-terminal of muscle α1 subunit (i.e., cysteines 192, 193). Three subunits, designated β-type (β2–β4), do not contain the cysteine pair (for reviews, see Refs. 11,12).

Nicotinic Receptor Channels Expressed in Heterologous Systems

Heterologous expression studies have been conducted to investigate which combination(s) of nAChR subunits can assemble to form functional nAChRs. Of course, the underlying goal of these studies is to determine the composition of native nAChRs by identifying subunit combinations that recapitulate the biophysical and pharmacological profile of native receptors. (For reviews, see Refs. 13–18.)

Work in heterologous systems has revealed that functional receptor complexes can be formed from α2, α3, and α4 in association with either β2 or β4.[14–16,18] Three of the neuronal α-type subunits yield functional nAChRs when expressed singly: α7, α8, and α9.[19–23] Heterologous expression studies of known subunit combinations have also established that nAChRs can be distinguished from one another based on biophysical and/or pharmacological criteria. In particular, homomeric α7, α8, or α9 nAChR channels are strikingly different from their heteromeric counterparts. Expression of either α7-, α8-, or α9-encoding RNA generates ACh-evoked currents sensitive to nanomolar concentrations of α-bungarotoxin (αBgTx), a toxin that is without effect on neuronal α/

β-type nAChRs.[19–21] Homomeric α7 receptors are also selectively inhibited by MLA and activated by choline.[24–26]

Other striking features of α7 and α8 homomers are their relatively high permeability to Ca^{2+} (see below) and their rapid desensitization in response to high concentrations of nicotinic agonists[27,28] (for review see Ref. 16). However, recent studies reveal that α7 receptors desensitize very slowly in response to sustained application of a low concentration of ACh or nicotine.[6] Hence, the kinetics of agonist delivery as well as the local concentration of agonist are important determinants of the overall role of nAChRs in synaptic transmission.

Are Native nAChRs Composed of More Than Two Subunit Types?

Heterologous expression studies have yielded the bulk of information currently available on the function and pharmacology of nAChRs subtypes and constitute the essential groundwork for ongoing work on the composition of native nAChRs. Nevertheless, the biophysical and pharmacological profiles of most recombinant nAChRs studied to date do not match well with those of native nAChRs. The rule seems to be that heterologous systems poorly recapitulate native nAChR (see recent review in Ref. 29). These discrepancies may reflect posttranslational modifications unique to the heterologous systems or that native nAChRs include more than two types of nAChR subunit.

Several labs (including ours) have started to address this issue by testing combinations of more than one α or β subunit in heterologous systems. The first "nontraditional" nAChR combinations we have examined tested the hypothesis that α5 coassembles with other α and β subunits to form functional, more apparently "*in vivo*-like" nAChR complexes. Previous biochemical studies suggested that α5 *could* coassemble with other nAChR subunits,[15,30] but the coexpression of α5 with any of the other α or β genes in heterologous systems did not yield agonist-gated currents.[31–33] Subsequent studies reveal that if α5 is coexpressed with both another α *and* β subunit, α5 can participate in functional channels. Similar results have been reported for other "silent" subunits such as α6 and β3.[29,33–36]

Inclusion of the α5 subunit in a nAChR complex decreases apparent affinity for ACh (but has relatively little effect on the apparent affinity or potency of nicotine) and dramatically alters the modulation of the receptor complex by Ca^{2+} (see below and Refs. 80,81). This observation supports the idea that the specific subtypes of nAChRs, despite overall low levels, may subserve important and distinct functions in regulating synaptic transmission (see below and Ref. 37 for review).

Another subject of recent controversy is the subunit composition of αBgTx-binding sites in the brain. Although everyone agrees that most if not all αBgTx sites on neurons include the α7 subunit,[38] it is the contention of some that all α7-containing nAChRs are homomeric. We have found that heteromeric combinations of α7 (e.g., with α5 and β2) can be expressed in heterologous systems.[81] In the chick visual system, α7/α8 heteromers, perhaps in combination with as of yet unidentified subunits, are expressed in native neurons (see Ref. 15). The existence of *native* nAChRs that are heteromeric complexes including α7 is further supported by experiments demonstrating that native α7 nAChRs are physiologically and pharmacologically distinct from heterologously expressed α7-homomeric channels.[25,26,39–45]

REGULATION OF nAChRs' OPENING PROBABILITY BY CHANGES IN INTRACELLULAR CALCIUM

In addition to differing in the characteristics of their agonist interactions, nAChR subtypes show a complex regulation by Ca^{2+}, which can both permeate and modulate the open channel.[6,46–55]

Studies in *Xenopus* oocytes demonstrate that the relative permeability for Ca^{2+} versus Na^{+} ($P_{Ca/Na}$) of neuronal heteromeric α/β nAChRs ranges from 1 to 1.5, compared with less than 0.2 for muscle but nearly 20 for α7-homomeric complexes.[47–50,52] Approximately 5% of the total current through α/β combinations is carried by Ca^{2+}, compared to 2% via muscle nAChRs.[50] Ca^{2+} permeation through NMDA receptor channels is about 12%[52] and is estimated at about 10% for α7-homomeric nAChRs. α7-homomeric nAChRs may be an important determinant of intracellular Ca^{2+} in neurons, as they constitute a major route of Ca^{2+} entry at membrane potentials equal to or hyperpolarized from rest potential (see, e.g., Refs. 15,52,56).

A number of studies have highlighted the effects of divalent cations on nAChR channel conductance and opening probability.[46,52,57,58] Extracellular Ca^{+2} decreases the single-channel conductance of most nAChRs examined to date. Other divalent cations, including Ba^{+2} and Sr^{+2}, also decrease nAChR conductance, and recent studies suggest that such effects may be subtype specific.[53,54,59,60]

In adrenal chromaffin cells, which express α3, α5, β4, and/or β2, external Ca^{+2} increases the opening frequency of nAChRs fourfold. The burst length is increased by a factor of 1.7.[61] In rat medial habenula, the presence of Ca^{+2} did not change the open time of the channel,[46] suggesting that the closing rate α did not change, but rather that the opening rate β increased by a factor of 15. Other reports, however, suggest that habenula neurons express multiple nAChR subtypes, so it is possible that effects of external calcium on the opening and/or closing rate of a particular nAChR subtype may have been overlooked.[62,63]

Recent Studies on the Mechanism of nAChR Gating by Internal Ca^{2+}

Although the locus of calcium-dependent modulation of nAChR gating is not known, the short cytoplasmic loop between the M1 and M2 loop appears to be well suited to coordinate Ca^{+2} ion binding. In particular, our preliminary studies toward identifying sites related to nAChR modulation by changes in internal Ca^{2+} suggest that one or more cysteine residues within the M1-M2 cytoplasmic loop are functionally "close" to a Ca^{+2} regulatory site. Thus, chemical modification of such cysteine residues with MTS reagents mimics the effects of increased $[Ca]_{int}$, whereas site-directed mutagenesis of the M1-M2 cysteines to serine residues ablates nAChR modulation by internal Ca^{2+} (see below).

Because of the inherent complexity of dissecting the effects and sites of calcium modulation in native nAChR channels, we have pursued this issue by studying specific subunit combinations using heterologous expression in *Xenopus* oocytes.[33] Comparison of recombinant α4β2 and α4α5β2 nAChRs reveals distinct effects of altering intracellular and extracellular Ca. Increasing calcium on the cytoplasmic face of excised patches of α4β2 nAChRs (from nominally zero to approximately 1 μM Ca) markedly increases opening probability and open duration while decreasing the interburst interval.[80,81] In contrast, preliminary studies of α5 + α4β2 nAChRs reveal that identical changes in cytoplasmic calcium have precisely the opposite effect. Thus the inclusion (or lack thereof) of the α5 subunit critically regulates the extent of nAChR activation. As these changes in calcium are likely to occur within activated synaptic terminals, one might predict that the inclusion

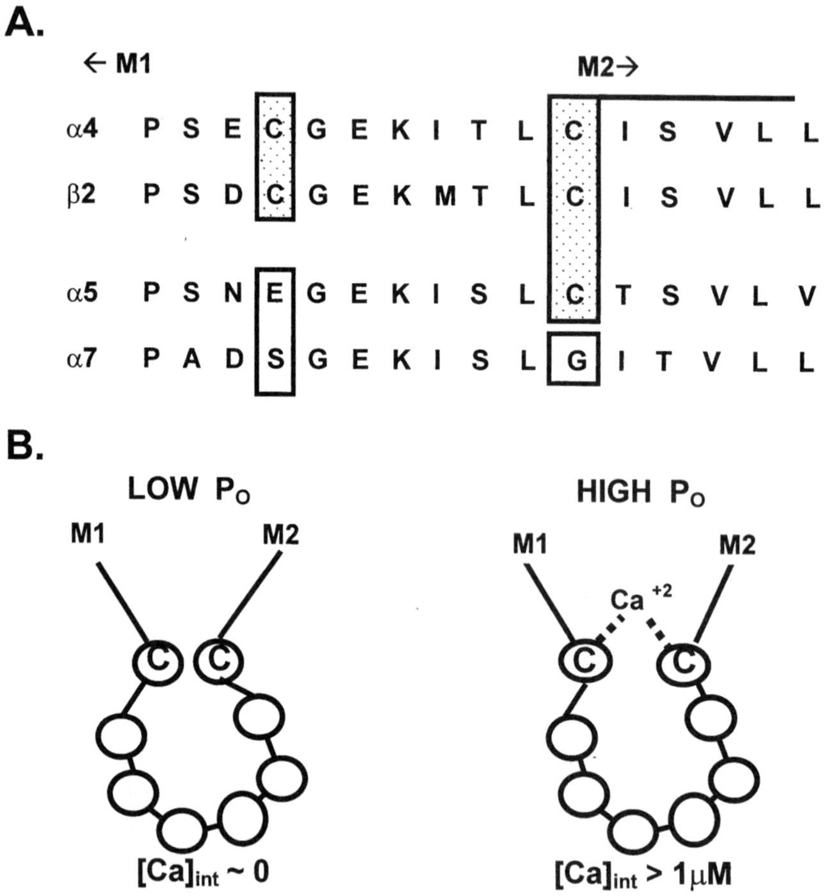

FIGURE 1. Hypothetical mechanism for modulation of nAChR opening probability by internal calcium. **A:** Sequence domains of chick α4, β2, α5, and α7 nAChRs that comprise the proposed cytoplasmic loop between M1-M2. Note the upstream and downstream cysteine residues, shared by α4 and β2 (as well as α3 and β4; not shown) but not by α5 and α7. **B:** Sheer fantasy for a possible mechanism whereby increases in internal calcium concentration stabilize the open state of α4β2-type complexes. The presence of both the upstream and downstream cysteine residues in the M1-M2 loop of each subunit within the complex is putatively required for the increased P_0.

of α5 in presynaptic nAChR complexes would be an important determinant of both the duration and extent of synaptic facilitation by nicotine.

The differential modulation of α4β2 compared to α5α4β2 nAChR complexes appears to be due to the presence or absence of two cysteine residues within the cytoplasmic loop between M1-M2. Whereas these two cysteine moieties, separated by six amino acid residues, are conserved in α2, α3, α4, β2, and β4, one or both cysteines are absent from the α5 and α7 sequences. To test the role of the M1-M2 cysteines, we examined the sensitivity of α4-, β2-, and α5-containing receptors to modulation by Ca^{2+} following site-directed mutagenesis of the cysteine residues flanking the M1-M2, region as noted in FIGURE 1. Mutation of the upstream cysteine residue in α4 and expression of mutant α4 cRNA with wild-type (WT) β2 in *Xenopus* oocytes yields ACh-gated currents similar in time course, Ca^{2+} sensitivity, and Ca^{2+} permeability to WT α5α4β2 nAChR complexes. The nature of this effect is dependent on the amino acid substituted for the cysteine residue. Thus the currents elicited by gating of α4(cysteine → glycine)β2 complexes closely resemble α7-homomeric channels, in rapid activation and inactivation kinetics and with respect to enhanced calcium permeability. These preliminary studies are thus consistent with our hypothesis that the first intracellular loop of neuronal nAChRs is critical to both the permeability of nAChRs to Ca^{2+} and the susceptibility of these receptors to modulation by $[Ca]_{int}$. Based on this hypothesis we would predict that the P_o of WT α7 nAChRs would *not* be modulated by internal Ca^{2+}, due to the absence of either cysteine residue in the M1-M2 region of α7.

Additional support for the functional significance of the M1-M2 cytoplasmic loop in Ca^{2+}-nAChR interactions comes from a recent report describing a point mutation in this region of the α4 subunit.[64] This mutation diminishes the Ca^{2+} permeability of α4β2 nAChRs, and the defect is restored by inclusion of the α5 subunit in the complex. Further studies directed at dissecting molecular mechanisms of calcium permeability and calcium modulation of α5- and α7-containing complexes should prove to be of great interest.

MODULATION OF SYNAPTIC TRANSMISSION BY PRESYNAPTIC nAChRs

Despite the fascination of trying to dissect out the functional contribution of particular nAChR subunits and to microdissect the physiology of individual domains, these questions pale by comparison to attempts to understand how nAChRs participate in the development, aging, and plasticity of the CNS. By analogy with the well-established role of nAChRs in autonomic ganglia, CNS nAChRs are likely to mediate fast excitatory synaptic transmission in the CNS. Nevertheless, the documented examples of direct nicotine receptor–mediated synapses within the brain are strikingly few in number. The limited success of such demonstrations does *not*, however, mean that nicotinic synaptic transmission is not important in the CNS. The technical challenges of such studies are considerable due to the highly diffuse nature of cholinergic projections.

In addition to the ongoing efforts to document direct nAChR-mediated synaptic transmissions in the CNS, many studies indicate an important contribution of CNS nAChRs to the modulatory control of transmitter release (for reviews, see Refs. 9,16,18,37). This idea emerged from the work of Wonnacott and collaborators, who demonstrated that nicotinic agonists enhance dopamine release from synaptosomes purified from the rat striatum. Biochemical and electrophysiological studies have since confirmed that nicotine modulates the release of most classical neurotransmitters—including GABA, norepinephrine, serotonin, acetylcholine, and glutamate. Both the targeting of nAChRs to presynaptic terminals and the role of nAChRs in modulating CNS transmission have been further characterized in studies of nAChR axonal transport, nicotine-induced facilitation of both evoked and spontaneous synaptic transmission, and even direct recording of nAChR channel on presynaptic terminals.[40,41,65,66]

Recent Studies on the Subunit Composition of Presynaptic nAChR Receptors

Several studies have begun the sure-to-be arduous task of trying to determine the subunit composition of the small percentage of nAChRs targeted to pre- as opposed to postsynaptic domains. Not surprisingly, this work has revealed that presynaptic nAChRs, like those expressed in somato-dendritic domains, include a variety of nAChR subtypes and that the diverse array of nAChR complexes expressed vary among the brain regions and neurotransmitter systems studied (see reviews in Refs. 9,37). For example, recent work suggests that the facilitation of dopamine secretion in the striatum is mediated by presynaptic nAChR complexes containing the α3, β2, and perhaps α5 subunits.[67–70] Nicotinic autoreceptors controlling norepinenphrine secretion in hippocampus are similar heteromeric complexes.[67,71,72] In contrast, the presynaptic nAChRs that potentiate glutamate transmission include α7, whether one records in the hippocampus,[41] in the olfactory bulb,[73] or at synapses between medial habenula (MHN) and interpeduncular neurons (IPN).[40] Thus at each of these sites, synaptic facilitation by nicotine is blocked by αBgTx and/or by functional deletion of α7 subunits,[40,41,74,75] (also Girod, McGehee & Role, unpublished).

There are now several examples of brain regions where distinct nAChR subtypes appear to be segregated to distinct afferent synapses. For example, release of GABA from interneurons within the IPN is modulated by activation of presynaptic nAChRs that do not include α7.[76] A similar situation, where GABA and glutamate transmission within the same brain region are under the control of both α7-containing and α/β type nAChRs, has been reported in the lateral geniculate nucleus.[43] Given the high permeability of α7 receptors to calcium, we speculate that their inclusion in presynaptic receptors will significantly alter the duration and extent of presynaptic facilitation by nicotine (see below).

Speculations on the Impact of nAChR Subtypes on Nicotine-Induced Synaptic Facilitation

We like to think that the presence of nAChRs on axons and axon terminals and the diversity of these presynaptic nAChRs reflect an important role of specific nAChR *subtypes* in the modulation of synaptic transmission *in vivo*. The following discussion is a highly speculative view of the subunit composition of presynaptic nAChRs and what we might expect of their impact on physiological responses to nicotine and ACh.

Before starting this discussion, we must recall important differences in the way nicotinic agonists are delivered to their targets. Clearly, the temporal profile and concentration of ACh delivered by stimulation of cholinergic inputs will be quite different from those of nicotine delivered by inhalation (i.e., smoking). Action potential–evoked secretion of ACh at axo-axonic synapses results in an essentially instantaneous (milliseconds) rise and rapid decline (tens of milliseconds) of the local concentration of ACh (>1mM at the peak of the synaptic release). The high turnover rate of acetylcholinesterase both limits the diffusion of ACh and rapidly decreases the synaptic ACh concentration to resting levels. Thus, activation of presynaptic nAChRs will be terminated within milliseconds, and significant desensitization may occur only during periods of high-frequency activation of the cholinergic input. In contrast, delivery of nicotine by smoking results in fairly slow (~1–2 s) pulses of nicotine superimposed on a very gradual (hours) buildup of low concentrations of agonist (0.3–1 μM at the "peak" of the puffs!)[77] throughout the brain. The half-life of nicotine *in vivo* is ~2 hours,[78] so that the CNS synapses are likely to be exposed to sustained concentrations of low-level nicotine. Under these conditions, the extent of receptor desensitization may be significant, depending on the subtypes of nAChRs expressed.[79] Clearly, a circumspect discussion of "how nicotine works" must consider that the effects of nicotine, at low but sustained concentrations (*or*, during smoking per se, at somewhat higher pulsatile concentrations), will be superimposed on the pre- and postsynaptic actions of endogenous ACh. Of course, consideration of nicotine-ACh interactions is extremely complex and even more speculative, as relatively little is known of the firing patterns of cholinergic projections within the CNS, and axo-axonic cholinergic synapses have largely evaded detailed study (but see abstract by Lubin *et al.*, this volume).

Despite these myriad limitations, we forge ahead to our proposed model of how nAChR composition might influence "how nicotine works" at presynaptic sites. For simplicity, we assume that cholinergic projections are inactive so that the local concentration of ACh is low (less than 1 μM). Let us consider the effect of the first cigarette of the day, resulting in the pulsatile administration of nicotine at a presynaptic terminal studded with α4β2-type receptors. Transmitter release elicited by the first nicotine pulse will be relatively small because the terminal is "at rest," and hence the intraterminal Ca^{2+} is initially in the nanomolar range, yielding a low α4β2 P_o. However, repeated pulsatile delivery of nicotine will result in a gradual increase in release as intracellular calcium rises, thereby increasing the P_o of the α4β2 nAChRs. As long as intraterminal calcium remains elevated and receptor desensitization is negligible, the response to subsequent nicotine applications (i.e., inhalations) will continue to increase. Finally, in view of the slow turnover of nicotine, release remains elevated above precigarette levels and declines slowly during the "intercigarette interval." The sustained exposure to nicotine, however, is predicted to elicit some desensitization of α4β2-type nAChRs, so that subsequent exposure of the synapse to nicotine or ACh may be less efficacious.

In contrast, if α5 is a component of the presynaptic nAChRs, we predict that initial facilitation will be more robust, because of the higher calcium permeability of the α4α5β2 receptor complex. We would also predict that facilitation will decay faster and the response to a second pulse will be smaller because of the decrease in P_o of α4α5β2 nAChRs with increasing intraterminal calcium and the greater susceptibility of α4α5β2 receptors to desensitization.

Finally, we would predict a very different profile of nicotine-induced synaptic facilitation at axonal terminals expressing α7-containing nAChRs. Although α7 nAChRs have very fast desensitization kinetics when exposed to high agonist concentration, even sustained exposure to low levels elicits very little desensitization.[6] Furthermore, since α7-containing receptors are highly permeable to calcium, robust and protracted potentiation of neurotransmitter release is predicted throughout the first, and with subsequent, nicotine administrations.

Although it is admittedly speculative, we propose that the composition of presynaptic nAChRs should profoundly influence the temporal profile and extent of synaptic facilitation by nicotine and/or by cholinergic inputs. Likewise, the relative levels of synaptic facilitation at two different inputs to a common postsynaptic neuron will depend on the subtypes of presynaptic nAChRs expressed. In this regard, we would predict that the profile of (for example, GABA vs. glutamate) release from convergent terminals expressing $\alpha4\beta2$ vs. $\alpha7$-type receptors could be differentially regulated—essentially "fine-tuned"—depending on the level of activity of cholinergic afferents to this site. As more refined tools for the dissection of nAChR composition emerge, the role of particular nAChR subunits in the modulation of CNS synapses will be more amenable to experimental scrutiny.

ACKNOWLEDGMENTS

Work presented in this paper was supported by Grants NS22061 and DA09366 to L.W.R.

REFERENCES

1. Schroder, H., E. Giacobini, A. Wevers, C. Birtsch & U. Schutz. 1995. Nicotinic receptors in Alzheimer's disease. Brain Imag. Nicotine Tob. Smok. :73–93.
2. Vidal, C. 1996. Nicotinic receptors in the brain. Molecular biology, function, and therapeutics. Mol. Chem. Neuropathol. **28:** 3–11.
3. Cachelin, A.B. & R. Jaggi. 1991. Beta subunits determine the time course of desensitization in rat alpha 3 neuronal nicotinic acetylcholine receptors. Pflugers Arch. **419:** 579–582.
4. Gross, A., M. Ballivet, D. Rungger & D. Bertrand. 1991. Neuronal nicotinic acetylcholine receptors expressed in *Xenopus* oocytes: Role of the alpha subunit in agonist sensitivity and desensitization. Pflugers Arch. **419:** 545–551.
5. Hsu, K.S., C. Huang, C.H. Yang & P.W. Gean. 1995. Presynaptic D_2 dopaminergic receptors mediate inhibition of excitatory synaptic transmission in rat neostriatum. Brain Res. **690:** 264–268.
6. Fenster, C.P., M.F. Rains, B. Noerager, M.W. Quick & R.A. Lester. 1997. Influence of subunit composition on desensitization of neuronal acetylcholine receptors at low concentrations of nicotine. J. Neurosci. **17:** 5747–5759.
7. Vibat, C.R., J.A. Lasalde, M.G. McNamee & E.L. Ochoa. 1995. Differential desensitization properties of rat neuronal nicotinic acetylcholine receptor subunit combinations expressed in *Xenopus laevis* oocytes. Cell Mol. Neurobiol. **15:** 411–425.
8. Dani, J.A. & S. Heinemann. 1996. Molecular and cellular aspects of nicotine abuse. Neuron **16:** 905–908.
9. Wonnacott, S. 1997. Presynaptic nicotinic ACh receptors. Trends Neurosci. **20:** 92–98.
10. Boulter, J., K. Evans. D. Goldman, G. Martin, D. Treco, S. Heinemann & J. Patrick. 1986. Isolation of a cDNA clone coding for a possible neural nicotinic acetylcholine receptor a-subunit. Nature **319:** 368–374.
11. Le Novere, N. & P. Changeux. 1995. Molecular evolution of the nicotinic acetylcholine receptor: An example of multigene family in excitable cells. J. Mol. Evol. **40:** 155–172.
12. Ortells, M.O. & G.C. Lunt. 1995. Evolutionary history of the ligand-gated ion-channel superfamily of receptors. TINS **18:** 121–127.
13. Patrick, J., P. Sequela, S. Vernino, M. Amador, C. Luetje & J.A. Dani. 1993. Functional diversity of neuronal nicotinic acetylcholine receptors. Prog. Brain Res. **98:** 113–120.

14. SARGENT, P.B. 1993. The diversity of neuronal nicotinic acetylcholine receptors. Annu. Rev. Neurosci. **16:** 403–433.

15. LINDSTROM, J., R. ANAND, X. PENG, V. GERZANICH, F. WANG & Y. LI. 1995. Neuronal nicotinic receptor subtypes. Ann. N.Y. Acad. Sci. **757:** 100–116.

16. MCGEHEE, D.S. & L.W. ROLE. 1995. Physiological diversity of nicotinic acetylcholine receptors expressed by vertebrate neurons. Annu. Rev. Physiol. **57:** 521–546.

17. CHANGEUX, J.P., A. BESSIS, J.P. BOURGEOIS, P.J. CORRINGER, A. DEVILLERS-THIERY, J.L. EISELE, M. KERSZBERG, C. LENA, N. LE NOVERE, M. PICCIOTTO & M. ZOLI. 1996. Nicotinic receptors and brain plasticity. Cold Spring Harbor Symp. Quant. Biol. **61:** 343–362.

18. COLQUHOUN, L.M. & J.W. PATRICK. 1997. Pharmacology of neuronal nicotinic acetylcholine receptor subtypes. Adv. Pharmacol. **39:** 191–220.

19. COUTURIER, S., D. BERTRAND, J.M. MATTER, M.C. HERNANDEZ, S. BERTRAND, N. MILLAR, S. VALERA, T. BARKAS & M. BALLIVET. 1990. A neuronal nicotinic acetylcholine receptor subunit (alpha 7) is developmentally regulated and forms a homo-oligomeric channel blocked by alpha-BTX. Neuron **5:** 847–856.

20. SCHOEPFER, R., W.G. CONROY, P. WHITING, M. GORE & J. LINDSTROM. 1990. Brain alpha-bungarotoxin binding protein cDNAs and mAbs reveal subtypes of this branch of the ligand-gated ion channel gene superfamily. Neuron **5:** 35–48.

21. SEGUELA, P., J. WADICHE, K. DINELEY-MILLER, J. DANI & J.W. PATRICK. 1993. Molecular cloning, functional properties, and distribution of rat a7: A nicotinic cation channel highly permeable to calcium. J. Neurosci. **13:** 596–604.

22. ELGOYHEN, A.B., D.S. JOHNSON. J. BOULTER, D.E. VETTER & S. HEINEMANN. 1994. Alpha 9: An acetylcholine receptor with novel pharmacological properties expressed in rat cochlear hair cells. Cell **79:** 705–715.

23. GERZANICH, V., R. ANAND & J. LINDSTROM. 1994. Homomers of a8 and a7 subunits of nicotinic receptors exhibit similar channel but contrasting binding site properties. Mol. Pharmacol. **45:** 212–220.

24. PAPKE, R.L., M. BENCHERIF & P. LIPPIELLO. 1996. An evaluation of neuronal nicotinic acetylcholine receptor activation by quaternary nitrogen compounds indicates that choline is selective for the alpha 7 subtype. Neurosci. Lett. **213:** 201–204.

25. ALKONDON, M., E.F. PEREIRA, C.T. BARBOSA & E.X. ALBUQUERQUE. 1997. Neuronal nicotinic acetylcholine receptor activation modulates gamma-aminobutyric acid release from CA1 neurons of rat hippocampal slices. J. Pharmacol. Exp. Ther. **283:** 1396–1411.

26. ALKONDON, M., E.F. PEREIRA, W.S. CORTES, A. MAELICKE & E.X. ALBUQUERQUE. 1997. Choline is a selective agonist of alpha7 nicotinic acetylcholine receptors in the rat brain neurons. Eur. J. Neurosci. **9:** 2734–2742.

27. ALKONDON, M. & E.X. ALBUQUERQUE. 1991. Initial characterization of the nicotinic acetylcholine receptors in rat hippocampal neurons. J. Recept. Res. **11:** 1001–1021.

28. ALKONDON, M. & E.X. ALBUQUERQUE. 1993. Diversity of nicotinic acetylcholine receptors in rat hippocampal neurons. I. Pharmacological and functional evidence for distinct structural subtypes. J. Pharmacol. Exp. Ther. **265:** 1455–1473.

29. SIVILOTTI, L.G., D.K. MCNEIL, T.M. LEWIS. M.A. NASSAR, R. SCHOEPFER & D. COLQUHOUN. 1997. Recombinant nicotinic receptors, expressed in *Xenopus* oocytes, do not resemble native rat sympathetic ganglion receptors in single-channel behaviour. J. Physiol. (Lond.) **500:** 123–138.

30. CONROY, W.G., A.B. VERNALLIS & D.K. BERG. 1992. The α5 gene product assembles with multiple acetylcholine receptor subunits to form distinctive receptor subtypes in brain. Neuron **9:** 679–691.

31. BOULTER, J., J. CONNOLLY, E. DENERIS, S. GOLDMAN, S. HEINEMANN & J. PATRICK. 1987. Functional expression of two neuronal nicotinic acetylcholine receptors from cDNA clones identifies a gene family. Proc. Natl. Acad. Sci. USA **84:** 7763–7767.

32. WADA, E., D. MCKINNON, S. HEINEMANN, J. PATRICLK & L.W. SWANSON. 1990. The distribution of mRNA encoded by a new member of the neuronal nicotinic acetylcholine receptor gene family (alpha 5) in the rat central nervous system. Brain Res. **526:** 45–53.

33. RAMIREZ-LATORRE, J., C.R. YU, X. QU, F. PERIN, A. KARLIN & L. ROLE. 1996. Functional contributions of alpha5 subunit to neuronal acetylcholine receptor channels. Nature **380:** 347–351.

34. Wang, D.X. & L.G. Abood. 1996. Expression and characterization of the rat alpha 4 beta 2 neuronal nicotinic cholinergic receptor in baculovirus-infected insect cells. J. Neurosci, Res. **44:** 350–354.
35. Forsayeth, J.R. & E. Kobrin. 1997. Formation of oligomers containing the beta3 and beta4 subunits of the rat nicotinic receptor. J. Neurosci. **17:** 1531–1538.
36. Gerzanich, V., A. Kuryatov, R. Anand & J. Lindstrom. 1997. "Orphan" alpha6 nicotinic AChR subunit can form a functional heteromeric acetylcholine receptor. Mol. Pharmacol. **51:** 320–327.
37. Role, L.W. & D.K. Berg. 1996. Nicotinic receptors in the development and modulation of CNS synapses. Neuron **16:** 1077–1085.
38. Chen, D. & J.W. Patrick. 1997. The alpha-bungarotoxin-binding nicotinic acetylcholine receptor from rat brain contains only the alpha7 subunit. J. Biol. Chem. **272:** 24024–24029.
39. Gerzanich, V., R. Anand & J. Lindstrom. 1994. Homomers of $\alpha 8$ and $\alpha 7$ subunits of nicotinic receptors exhibit similar channel but contrasting binding sites properties. Mol. Pharmacol. **45:** 212–220.
40. McGehee, D., M. Heath, S. Gelber, P. Deva & L.W. Role. 1995. Nicotine enhancement of fast excitatory synaptic transmission in CNS by presynaptic receptors. Science **269:** 1692–1697.
41. Gray, R., A.S. Rajan, K.A. Radcliffe, M. Yakehiro & J.A. Dani. 1996. Hippocampal synaptic transmission enhanced by low concentrations of nicotine [see comments]. Nature **383:** 713–716.
42. Liang, S.D. & S. Vizi. 1997. Positive feedback modulation of acetylcholine release from isolated rat superior cervical ganglion. J. Pharmacol. Exp. Ther. **280:** 650–655.
43. Guo, J.Z., T.L. Tredway & V.A. Chiappinelli. 1998. Glutamate and GABA release are enhanced by different subtypes of presynaptic nicotinic receptors in the lateral geniculate nucleus. J. Neurosci. **18:** 1963–1969.
44. Yu, C.R. & L.W. Role. 1998. Functional contribution of the alpha5 subunit to neuronal nicotinic channels expressed by chick sympathetic ganglion neurones. J. Physiol. (Lond.) **509:** 667–681.
45. Yu, C.R. & L.W. Role. 1998. Functional contribution of the alpha7 subunit to multiple subtypes of nicotinic receptors in embryonic chick sympathetic neurones. J. Physiol. (Lond.) **509:** 651–665.
46. Mulle, C., D. Choquet, H. Korn & J.P. Changeux. 1992. Calcium influx through nicotinic receptor in rat central neurons: Its relevance to cellular regulation. Neuron **8:** 135–143.
47. Vernino, S., M. Amador, C.W. Luetje, J. Patrick & J.A. Dani. 1992. Calcium modulation and high calcium permeability of neuronal nicotinic acetylcholine receptors. Neuron **8:** 127–134.
48. Bertrand, D., J.H. Galzi, A. Devillers-Thiery, S. Bertrand & J.P. Changeux. 1993. Mutations at two distinct sites within the channel domain M2 alter calcium permeability of neuronal $\alpha 7$ nicotinic receptor. Proc. Natl. Acad. Sci. USA **90:** 6971–6975.
49. Seguela, P., J. Wadiche, K. Dineley-Miller, J.A. Dani & J.W. Patrick. 1993. Molecular cloning, functional properties, and distribution of rat brain alpha 7: A nicotinic cation channel highly permeable to calcium. Neurosci. **13:** 596–604.
50. Vernino, S., M. Rogers, K.A. Radcliff & J.A. Dani. 1994. Quantitative measurement of calcium flux through muscle and neuronal nicotinic acetylcholine receptors. J. Neurosci. **14:** 5514–5524.
51. Albuquerque, E.X., E.F. Pereira, N.G. Castro, M. Alkondon, S. Reinhardt, H. Schroder & A. Maelicke. 1995. Nicotinic receptor function in the mammalian central nervous system. Ann. N.Y. Acad. Sci. **757:** 48–72.
52. Rogers, M. & J.A. Dani. 1995. Comparison of quantitative calcium flux through NMDA, ATP, and ACh receptor channels. Biophys. J. **68:** 501–506.
53. Zwart, R., R.G. Van Kleef, J.M. Milikan, M. Oortgiesen & H.P. Vijverberg. 1995. Potentiation and inhibition of subtypes of neuronal nicotinic acetylcholine receptors by Pb2+. Eur. J. Pharmacol. **291:** 399–406.
54. Galzi, J.L., S. Bertrand, P.J. Corringer, J.P. Changeux & D. Bertrand. 1996. Identification of calcium binding sites that regulate potentiation of a neuronal nicotinic acetylcholine receptor. Embo J. **15:** 5824–5832.

55. WILKIE, G.I., P. HUTSON, J.P. SULLIVAN & S. WONNACOTT. 1996. Pharmacological characterization of a nicotinic autoreceptor in rat hippocampal synaptosomes. Neurochem. Res. **21:** 1141–1148.
56. CASTRO, N.G. & E.X. ALBUQUERQUE. 1995. The α-bungarotoxin-sensitive hippocampal nicotinic receptor channel has a high calcium permeability. Biophys. J. **68:** 516–524.
57. ZWART, R., M. OORTGIESEN & H.P. VIJVERBERG. 1992. The nitromethylene heterocycle 1-(pyridin-3-yl-methyl)-2-nitromethylene-imidazolidine distinguishes mammalian from insect nicotinic receptor subtypes. Eur. J. Pharmacol. **228:** 165–169.
58. NEUHAUS, R. & A.B. CACHELIN. 1990. Changes in the conductance of the neuronal nicotinic acetylcholine receptor channel induced by magnesium. Proc. R. Soc. Lond. **B241:** 78–84.
59. SANDS, S.B., A.C. COSTA & J.W. PATRICK. 1993. Barium permeability of neuronal nicotinic receptor alpha 7 expressed in *Xenopus* oocytes. Biophys. J. **65:** 2614–2621.
60. NUTTER, T.J. & D.J. ADAMS. 1995. Monovalent and divalent cation permeability and block of neuronal nicotinic receptor channels in rat parasympathetic ganglia. J. Gen. Physiol. **105:** 701–723.
61. AMADOR, M. & J.A. DANI. 1995. Mechanism for modulation of nicotinic acetylcholine receptors that can influence synaptic transmission. J. Neurosci. **15:** 4525–4532.
62. CONNOLLY, J.G., A.J. GIBB & D. COLQUHOUN. 1995. Heterogeneity of neuronal nicotinic acetylcholine receptors in thin slices of rat medial habenula. J. Physiol. (Lond.) **484:** 87–105.
63. BRUSSAARD, A.B., X. YANG, J.P. DOYLE, S. HUCK & L.W. ROLE. 1994. Developmental regulation of multiple nicotinic AChR channel subtypes in embryonic chick habenula neurons: Contributions of both the alpha 2 and alpha 4 subunit genes. Pflugers Arch. **429:** 27–43.
64. KURYATOV, A., V. GERZANICH, M. NELSON, F. OLALE & J. LINDSTROM. 1997. Mutation causing autosomal dominant nocturnal frontal lobe epilepsy alters Ca^{2+} permeability, conductance, and gating of human alpha4beta2 nicotinic acetylcholine receptors. J. Neurosci. **17:** 9035–9047.
65. SARGENT, P.B., S.H. PIKE, D.B. NADEL & J.M. LINDSTROM. 1989. Nicotinic acetylcholine receptor-like molecules in the retina, retinotectal pathway, and optic tectum of the frog. J. Neurosci. **9:** 565–573.
66. COGGAN, J.S., J. PAYSAN, W.G. CONROY & D.K. BERG. 1997. Direct recording of nicotinic responses in presynaptic nerve terminals. J. Neurosci. **17:** 5798–5806.
67. CLARKE, P.B. & M. REUBEN. 1996. Release of [^{3}H]-noradrenaline from rat hippocampal synaptosomes by nicotine: Mediation by different nicotinic receptor subtypes from striatal [^{3}H]-dopamine release. Br. J. Pharmacol. **117:** 595–606.
68. SOLIAKOV, L. & S. WONNACOTT. 1996. Voltage-sensitive Ca^{2+} channels involved in nicotinic receptor-mediated [^{3}H]dopamine release from rat striatal synaptosomes. J. Neurochem. **67:** 163–170.
69. SACAAN, A.I., J.L. DUNLOP & G.K. LLOYD. 1995. Pharmacological characterization of neuronal acetylcholine gated ion channel receptor-mediated hippocampal norepinephrine and striatal dopamine release from rat brain slices. Pharmacol. Exp. Ther. **274:** 224–230.
70. KAISER, S.A., L. SOLIAKOV, S.C. HARVEY, C.W. LUETJE & S. WONNACOTT. 1998. Differential inhibition by alpha-conotoxin-MII of the nicotinic stimulation of [^{3}H]dopamine release from rat striatal synaptosomes and slices. J. Neurochem. **70:** 1069–1076.
71. KISS, J.P., H. SERSHEN, A. LAJTHA & E.S. VIZI. 1996. Inhibition of neuronal nitric oxide synthase potentiates the dimethylphenylpiperazinium-evoked carrier-mediated release of noradrenaline from rat hippocampal slices. Neurosci. Lett. **215:** 115–118.
72. SERSHEN, H., A. BALLA, A. LAJTHA, & E.S. VIZI. 1997. Characterization of nicotinic receptors involved in the release of noradrenaline from the hippocampus. Neuroscience **77:** 121–130.
73. ALKONDON, M., E.S. ROCHA, A. MAELICKE & E.X. ALBUQUERQUE. 1996. Diversity of nicotinic acetylcholine receptors in rat brain. V. Alpha Bungarotoxin-sensitive nicotinic receptors in olfactory bulb neurons and presynaptic modulation of glutamate release. J. Pharmacol. Exp. Ther. **278:** 1460–1471.
74. SARGENT, P.B. & D.Z. PANG. 1989. Acetylcholine receptor-like molecules are found in both synaptic and extrasynaptic clusters on the surface of neurons in the frog cardiac ganglion. Neurosci. **9:** 1062–1072.

75. ALKONDON, M., E.F. PEREIRA & E.X. ALBUQUERQUE. 1996. Mapping the location of functional nicotinic and gamma-aminobutyric acids receptors on hippocampal neurons. J. Pharmacol. Exp. Ther. **279:** 1491–1506.
76. LENA, C., J.-P. CHANGEUX & C. MULLE. 1993. Evidence for "preterminal" nicotinic receptors on GABAergic axons in the rat interpeduncular nucleus. J. Neurosci. **13:** 2680–2688.
77. BENOWITZ, N.L. 1990. Pharmacokinetic considerations in understanding nicotine dependence. CIBA Found. Symp. **152:** 186–200.
78. BENOWITZ, N.L., H. PORCHET & P.D. JACOB. 1989. Nicotine dependence and tolerance in man: Pharmacokinetic and pharmacodynamic investigations. Prog. Brain Res. **79:** 279–287.
79. PIDOPLICHKO, V.I., M. DEBIASI, J.T. WILLIAMS & J.A. DANI. 1997. Nicotine activates and desensitizes midbrain dopamine neurons. Nature **390:** 401–404.
80. YU, C., J. RAMIREZ-LATORRE & L. ROLE. 1996. Ca^{2+} modulation of $\alpha4\beta2$ and $\alpha4\alpha5\beta2$ combinations of neuronal nicotinic receptors. Soc. Neurosci. Abstr.
81. CRABTREE, G. *et al.* 1997. Assembly and Ca^{2+} regulation of neuronal nicotinic receptors including the $\alpha7$ and $\alpha5$ subunits. Soc. Neurosci. Abstr.

Nicotinic Modulation of Glutamate and GABA Synaptic Transmission in Hippocampal Neurons

KRISTOFER A. RADCLIFFE, JANET L. FISHER, RICHARD GRAY, AND JOHN A. DANI[a]

Division of Neuroscience, Baylor College of Medicine, Houston, Texas 77030-3498, USA

ABSTRACT: Although the hippocampus expresses nicotinic acetylcholine receptors (nAChRs) and receives cholinergic innervation, the functional roles of these receptors are not completely understood. Our results indicated that presynaptic nAChRs mediated a calcium influx that enhanced the release of both glutamate and GABA. Fura-2 detection of calcium in single mossy fiber presynaptic terminals indicated that nAChRs directly mediated a calcium influx. In hippocampal neurons in primary culture, both spontaneous vesicular release and evoked release of glutamate and GABA were enhanced by nicotine. The nicotinic current displayed rapid desensitization kinetics, and the response to nicotine was inhibited by α-bungarotoxin and methyllcaconitine, suggesting that nAChRs containing the α7 subunit mediated the effect. Modulation of synaptic activity by presynaptic calcium influx may represent a physiological role of acetylcholine in the brain, as well as a mechanism of action of nicotine.

The nicotinic acetylcholine receptors (nAChRs) are members of the ligand-gated ion channel superfamily that includes $GABA_A$, glycine, and 5-HT_3 serotonin receptors.[1] Nicotinic AChRs are responsible for fast excitatory transmission at the mammalian neuromuscular junction and in the peripheral nervous system. In the central nervous system (CNS), however, the roles of nicotinic cholinergic receptors can be subtle and varied. Nicotinic mechanisms commonly modulate the release of neurotransmitters, but only on rare occasions have they been observed to mediate fast synaptic transmission in the CNS (for reviews see Refs. 2–4).

Neuronal Nicotinic Acetylcholine Receptor Subunits

Both neuronal and neuromuscular nAChRs are pentameric in structure.[5] The neuromuscular receptors are composed of a combination of four different subunits, α1, β1, δ and γ (fetal) or ε (adult). Messenger RNA for eight α (α2–α9) and three β (β2–4) subunits are expressed in neurons and form neuronal nAChRs. The neuronal α subunits are homologous in structure to the α1 muscle subunit, and share the two consecutive N-terminal cysteines that contribute to the acetylcholine binding site (see Ref. 6). Although the β subunits are not closely related structurally to the β1 subunit, β2 can substitute for β1 to form muscle-type nAChRs in the oocyte expression system.[7] Additionally, the β subunits are grouped together because they lack the N-terminal cysteines that characterize the α subunits.

[a]Address for correspondence: John A. Dani, Ph.D., Division of Neuroscience, Baylor College of Medicine, One Baylor Plaza, Houston, Texas 77030-3498. Phone: 713-798-3710; fax: 713-798-3946; e-mail: jdani@bcm.tmc.edu

The α subunits expressed in neurons form two distinct groups based on functional and pharmacological properties.[2] The α2, α3, α4, and α6 subunits require β subunits to form functional nAChRs in expression systems and produce receptors that are not inhibited by α-bungarotoxin (α-BGT).[7–12] The α4 and β2 subunits are the most highly expressed and probably contribute to the majority of CNS nAChRs.[13] In contrast, in expression systems the α7, α8, and α9 subunits form functional homomeric channels that are inhibited by α-BGT.[14–17] The α7 subunit is the most highly expressed of these α subunits, and its expression correlates well with the α-BGT binding sites identified in the rat brain. The α8 subunit has been found in chicken but not in mammals. The α9 subunit has been located in rat cochlea but is not distributed in the nervous system.[3]

Functional Properties of Neuronal Nicotinic Acetylcholine Receptors

The neuronal nAChRs differ from the muscle receptor not only in their subunit structure, but also in their functional properties. One of the most important functional characteristics of the neuronal nAChRs is their higher Ca^{2+} permeability compared to the muscle receptor.[18–20] Of all the nAChRs tested to date, the α7 homomeric channels have the highest permeability to Ca^{2+} relative to Na^+, comparable to that of the NMDA-activated glutamate receptor.[15,21,22]

In whole-cell recordings from brain slice and in primary tissue culture, several different types of responses to nicotine or acetylcholine have been observed. In hippocampal neurons, the predominant response is an α-BGT– and methyllycaconitine (MLA)–sensitive current with rapid activation and desensitization kinetics.[23,24] These currents correlated with α7 expression, suggesting that the nAChRs contain the α7 subunit.[25] In many brain regions, however, the majority of the nAChRs have characteristics similar to α4β2 receptors. It can be difficult to identify the subunit composition of nAChRs functionally because the properties of native receptors often do not match well with the properties of expressed recombinant receptors. This result may be due to the wide variety of possible subunit combinations, posttranslational processing, interactions with endogenous cellular components, or contributions from subunits that have not yet been cloned.

Functional Roles of Neuronal nAChRs

Neuronal nAChRs are important in both the development and modulation of synapses. Activation of nAChRs enhances cell proliferation[26] and contributes to the stabilization of neurite outgrowth.[22,27] Nicotinic agonists have varied effects on brain function, including alterations of cognitive function.[28] In addition, some nicotinic receptor agonists have anxiolytic or analgesic activity.[29,30] Activation of nAChRs has been shown to increase the release of a wide variety of neurotransmitters, including GABA, glutamate, serotonin, dopamine, acetylcholine, and norepinephrine (see Ref. 4). Thus, cholinergic inputs may serve to modulate the strength and fidelity of many different kinds of synaptic connections.

NICOTINE INFLUENCES GLUTAMATERGIC TRANSMISSION

In the mammalian CNS, most fast excitatory neurotransmission is mediated by glutamate. Activity-dependent changes of the properties of hippocampal glutamatergic synapses are believed to provide the mechanism for several forms of learning and memory[26] (for reviews see Refs. 31, 32). There is increasing evidence that cholinergic neurons can play a role in some forms of learning and memory and that cholinergic pathways may interact with glutamatergic transmission.[28,33–36]

Nicotine has been shown to increase glutamate release from several types of neurons.[37–39] Furthermore, the hippocampus receives extensive cholinergic innervation and exhibits a high level of nAChR expression.[15,40,41] To determine whether nicotine could also modulate hippocampal glutamatergic transmission, we examined the effect of low concentrations of nicotine on spontaneous glutamate release from hippocampal neurons in slice and primary culture.[24]

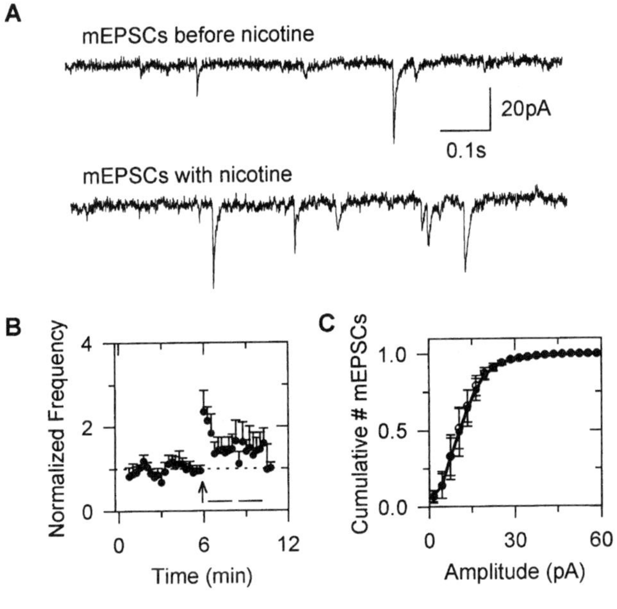

FIGURE 1. Nicotine enhances the frequency but not the amplitude of mEPSCs recorded from CA3 cells in rat hippocampal slice. **(A)** Representative traces of mEPSCs recorded before and during application of nicotine. Nicotine was injected locally onto the proximal dendrites of the voltage-clamped CA3 neuron. **(B)** Nicotine increased the frequency of mEPSCs (6 trials from 4 neurons, ± SEM). The frequency was normalized to the frequency before nicotine for each cell *(circles)*. The *dashed line* indicates the continued presence of nicotine in the area of the dendrites following the initial 5-s injection from a pipette containing 20 μM nicotine in bath solution. **(C)** Nicotine did not affect the amplitude of the mEPSCs. The amplitude distributions before and during nicotine applications overlap. Peak amplitudes of the events were combined into 3-pA bins. The total number of events was normalized to 1. Hippocampal slices were prepared from adult rats (20–63 days).[24] The external bath solution was composed of (in mM): 149 NaCl, 2.5 KCl, 2 $CaCl_2$, 2 $MgCl_2$, 26 $NaHCO_3$, 1 dextrose, 1 μM TTX, and 200 μM $CdCl_2$. The bath was bubbled with 95% O_2, 5% CO_2. The puffer pipette solution containing 20 μM nicotine had the same composition except that 10 mM HEPES acted as the buffer (pH = 7.3). The internal solution contained (in mM): 150 $CsCH_3SO_3$, 5 NaCl, 0.2 EGTA, 2–5 MgATP, 0–2 Na_2ATP, 0–0.3 Na_3GTP, 10 HEPES, with pH = 7.3. (Adapted from Gray *et al.*[24] by permission).

A Low Concentration of Nicotine Increased mEPSC Frequency in Hippocampal Slice

Pyramidal neurons from the CA3 region were voltage clamped at –60mV in whole-cell configuration. Spontaneous miniature excitatory postsynaptic currents (mEPSCs) were observed in these neurons (FIG. 1A). To record only spontaneous vesicular glutamate release, TTX (1μM) and Cd^{2+} (200μM) were included in the bath to block voltage-gated Na^{+} and Ca^{2+} channels, respectively. Nicotine was applied to the area surrounding the proximal dendrites with a pressure-ejection pipette. Nicotine (20μM) was applied in a brief puff (5s), and the nicotine-containing solution then mixed around the dendritic processes. The actual concentration of nicotine reaching the dendrites is not known but was certainly much less than 20μM. Application of nicotine to the dendrites increased the frequency of the spontaneous events over two-fold (FIG. 1B), but the amplitudes of the mEPSCs were not affected (FIG. 1C). The increase in mEPSC frequency was observed in four of four neurons examined. There was some variability in the time required for nicotine to reach the dendrites, and the time course for the increase in mEPSC frequency varied among the neurons. After an initial increase, the frequency declined toward the baseline level as the nicotine was slowly washed out from the slice.

Presynaptic nAChRs Mediated a Direct Ca^{2+} Influx into Mossy Fiber Terminals

Nicotine increased the frequency, but not the amplitude, of the mEPSCs. This result suggested a presynaptic site of action, with nicotine increasing the release of neurotransmitter, rather than altering the number or responsiveness of postsynaptic glutamate receptors. Because nAChRs are known to be permeable to Ca^{2+}, we examined whether nicotine increased intracellular Ca^{2+} in presynaptic terminals of the mossy fibers, which synapse onto the CA3 neurons. Slices were bathed with an external solution containing Fura-2 acetoxymethyl ester for 10–15 min. The presynaptic terminals took up this dye much more efficiently than the rest of the slice (FIG. 2A). Recording the fluorescent signal from a single mossy fiber terminal, we found that application of 20μM nicotine caused a decrease in the fluorescence at 380nm, indicating an increase in the concentration of intracellular Ca^{2+} (FIG. 2B).

FIGURE 2. Nicotine increases intracellular Ca^{2+} in mossy fiber presynaptic terminals. **(A)** Image of mossy fiber presynaptic terminals loaded with Fura-2. The central, bright fluorescent area is a mossy fiber presynaptic terminal, 4 μM in diameter. **(B)** Nicotine injection for 5 s *(arrow)* decreased the relative fluorescence (ΔF/F). The decreased fluorescence at 380 nm indicated an increase in intraterminal Ca^{2+}. A puff of control solution without nicotine had no effect. **(C)** Presynaptic Ca^{2+} influx was induced through voltage-gated Ca^{2+} channels by electrically stimulating the mossy fibers. The Ca^{2+} influx indicated by the sharp decrease in ΔF/F was the same in control solution and after adding 100 μM APV and 20 μM CNQX to block ionotropic glutamate receptors. Adding 1 μM TTX and 500 μM Cd^{2+} inhibited Na^{+} channels and voltage-gated Ca^{2+} channels and blocked the stimulus-induced fluorescence change. Each ΔF/F trace is the average of >11 individual traces. **(D)** Nicotine application (5-s puff, *arrow head*) induced the same ΔF/F before (control) and after inhibiting ionotropic glutamate receptors with 100 μM APV and 20 μM CNQX. Preincubation of the slice with 100 nM α-BGT inhibited the nicotine-induced ΔF/F. The traces were constructed from three 5-s records each separated by a 10-s gap (gaps not shown). **(E)** Ca^{2+} signal induced by nicotinic receptors was similar in magnitude to stimulus-evoked influx via voltage-gated Ca^{2+} channels. Mossy fibers were electrically stimulated at 0.1 Hz. A 2-s puffer application of nicotine increased Ca^{2+} in the presynaptic terminal (decreased ΔF/F) by about the same amount as electrical stimulation. The trace was constructed from 18 2-s records each separated by a 10-s gap (gaps not shown). (Reprinted from Gray *et al.*[24] by permission.)

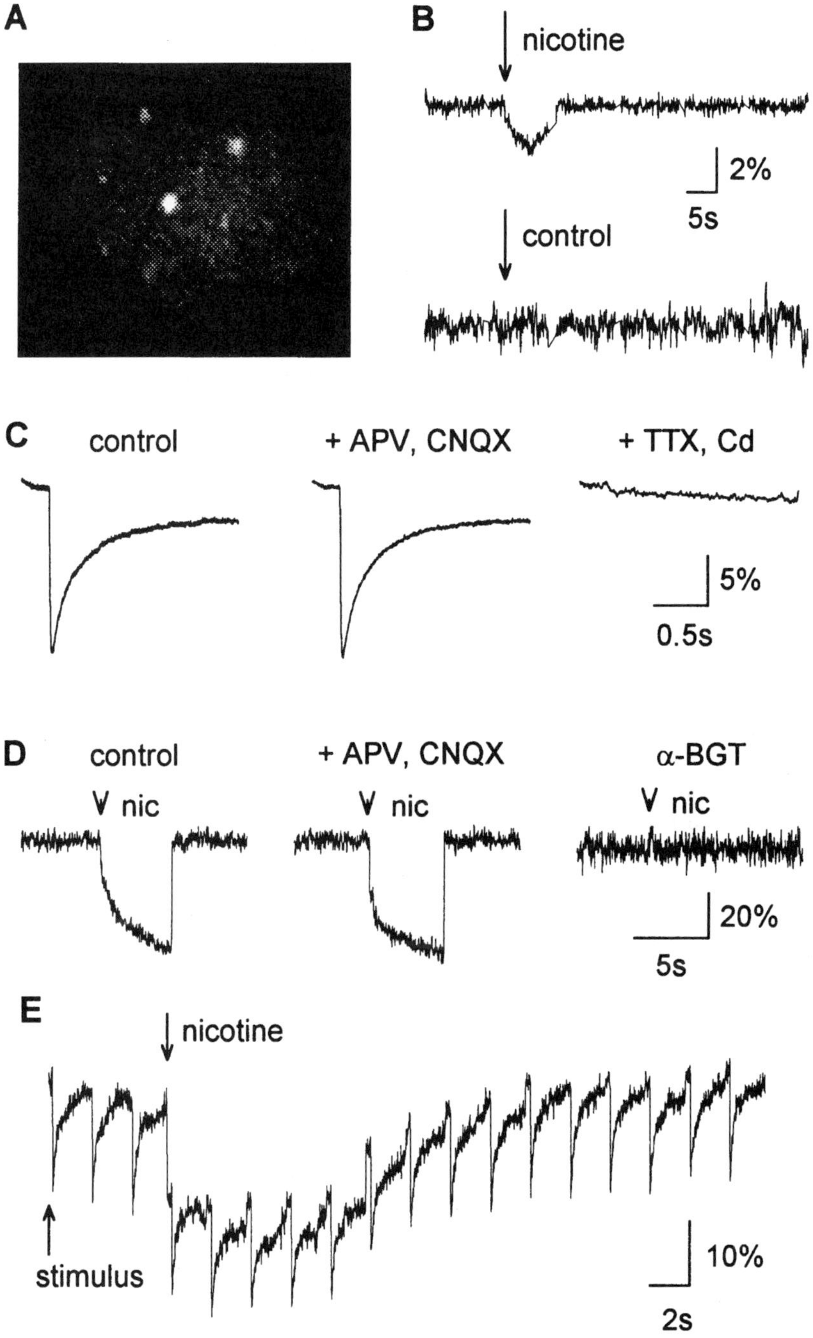
A
B
nicotine
2%
5s
control
C
control
+ APV, CNQX
+ TTX, Cd
5%
0.5s
D
control
nic
+ APV, CNQX
nic
α-BGT
nic
20%
5s
E
nicotine
stimulus
10%
2s

To determine whether the Fura-2 signal was influenced by activation of ionotropic glutamate receptors, we stimulated the dentate region to increase glutamate release from presynaptic terminals. This stimulation activated voltage-gated Ca^{2+} channels, increasing intracellular Ca^{2+} in the mossy fiber terminals (FIG. 2C). The fluorescent signal was not affected by CNQX (20μM) and APV (100μM) (FIG, 2C). Therefore, the Ca^{2+} signal was not due in any part to the activation of ionotropic glutamate receptors. The Ca^{2+} signal in response to nicotine application was also not reduced by inhibitors of glutamate receptors (FIG. 2D). The response to nicotine was prevented by α-BGT, suggesting that α7-containing nAChRs were responsible for initiating the Ca^{2+} signal (FIG. 2D). These results strongly support the conclusion that presynaptic nAChRs serve to initiate the nicotine-induced Ca^{2+} signals.

We then wanted to determine whether the magnitude of the Ca^{2+} signal initiated by nAChRs was physiologically meaningful. Influx of Ca^{2+} through voltage-gated channels normally mediates the release of neurotransmitter in response to an action potential. We compared the presynaptic Ca^{2+} signal mediated by voltage-gated Ca^{2+} channels to the Ca^{2+} signal mediated by nAChRs. Voltage-gated Ca^{2+} channels were activated by electrical stimulation of the incoming mossy fibers, and nAChRs were activated by local pressure injection of 20μM nicotine. Similar levels of Ca^{2+} arose from both routes (FIG. 2E). When averaged over the whole mossy fiber presynaptic terminal, Ca^{2+} signals initiated by nAChRs can be comparable to Ca^{2+} entry through voltage-gated Ca^{2+} channels.

A Low Concentration of Nicotine Increased the Frequency of mEPSCs in Hippocampal Cultures

To allow better control of the time course and effective concentration of the nicotine applications, we also examined the effect of nicotine on spontaneous mEPSCs in rat hippocampal neurons in primary tissue culture. Neurons were obtained from whole hippocampi of postnatal rats (1–3 days) and cultured on microislands to allow the neurons to form autaptic connections.[42,43] Cells were used for experiments between 10 and 30 days in culture.

FIGURE 3. Nicotine acting through α-BGT–sensitive receptors increased the frequency but not the amplitude of mEPSCs in hippocampal neurons in primary culture. **(A)** Representative traces of mEPSCs recorded before and during a 2-min application of 0.5 μM nicotine. **(B)** Nicotine (0.5 μM) applied for 2 min *(solid bar)* increased mEPSC frequency (n = 8 out of 13). For each neuron, mEPSC frequency was normalized to the frequency before nicotine application *(dashed line)*. **(C)** The cumulative amplitude distribution of mEPSCs was the same before nicotine *(open circles)* and during nicotine application *(closed circles)*. Events were combined in bins of 3 pA, and the total number of events in each condition was normalized to 1. **(D)** Nicotine did not increase the frequency of mEPSCs in a Ca^{2+}-free external solution (n = 14). **(E)** With 10 of the same neurons as in D, when external solution containing 1 mM Ca^{2+} was added, nicotine increased the frequency of mEPSCs (n = 7). **(F)** When neurons were pretreated with 50 nM α-BGT, nicotine had no effect on the mEPSC frequency (n = 6). The external solution was composed of (in mM): 150 NaCl, 2.5 KCl, 1 $CaCl_2$, 1 $MgCl_2$, 10 glucose, 10 HEPES, 0–200 μM $CdCl_2$, 20–100 μM picrotoxin, 0.5–1 μM TTX, 10 μM CNQX, with pH = 7.3. The internal solution contained (in mM): 150 $CsCH_3SO_3$, 5 NaCl, 0.2 EGTA, 2–5 MgATP, 0–2 Na_2ATP, 0–0.3 Na_3GTP, 10 HEPES, with pH = 7.3. Standard patch-clamp or perforated patch techniques were used. The mEPSCs were recorded from cultured rat hippocampal neurons voltage-clamped at –50 mV. (Adapted from Gray *et al.*[24] by permission.)

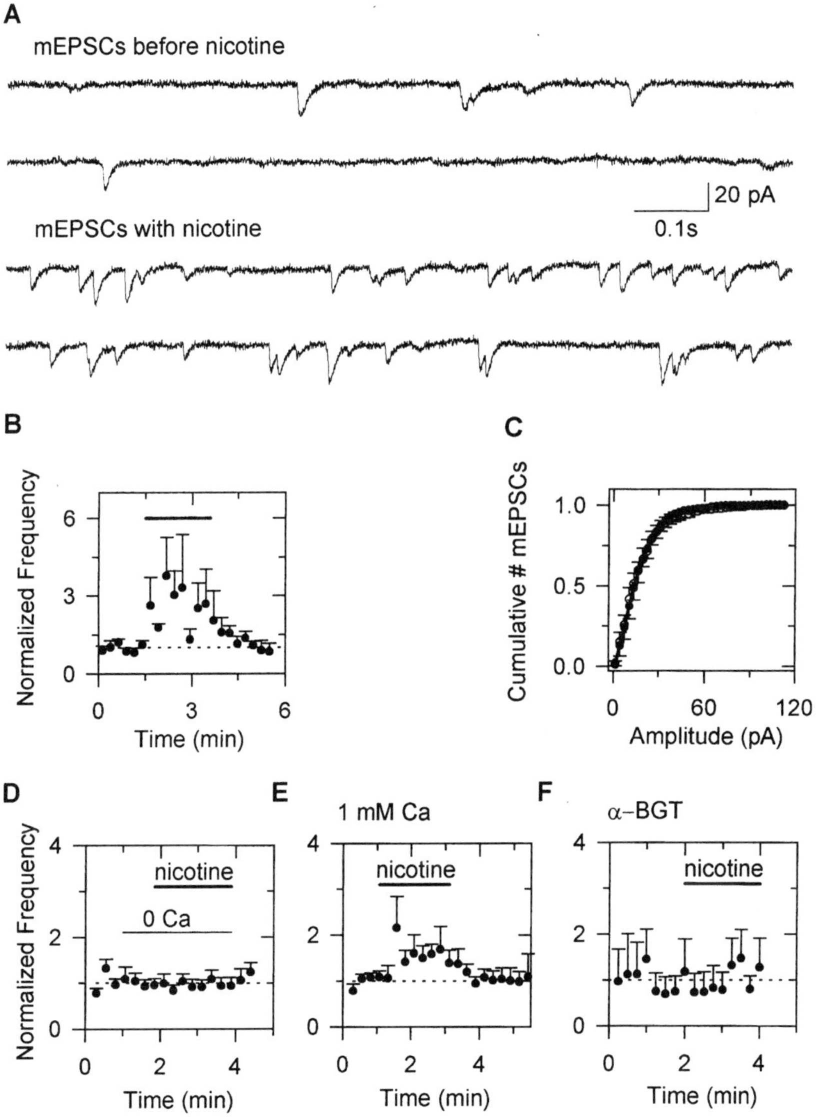
A
mEPSCs before nicotine
20 pA
0.1s
mEPSCs with nicotine
B
Normalized Frequency
Time (min)
C
Cumulative # mEPSCs
Amplitude (pA)
D
nicotine
0 Ca
Normalized Frequency
Time (min)
E
1 mM Ca
nicotine
Time (min)
F
α-BGT
nicotine
Time (min)

Spontaneous mEPSCs were observed in neurons voltage-clamped at –50 mV (FIG. 3A). These events could be completely prevented by inclusion of 10 μM CNQX or 1 mM kynurenic acid (+1 mM Mg^{2+}) in the extracellular solution, indicating that they arose from the activation of glutamate receptors. Similar to the observation in slice, 0.5 μM nicotine increased the frequency, but not the amplitude, of mEPSCs in 8 of 13 neurons examined (FIG. 3B and 3C). The frequency of mEPSCs was highly variable, but in general the activity increased in response to nicotine and then began to return toward the baseline even in the continued presence of nicotine. The effect of nicotine was dependent on the presence of extracellular Ca^{2+} (FIG. 3D and 3E) and was prevented by pretreatment with 50 nM α-BGT (n = 5 of 6, FIG. 3F), suggesting the response was mediated mainly through α-7–containing nAChRs.

NICOTINE INFLUENCES GABAERGIC NEUROTRANSMISSION

GABA is the principal inhibitory neurotransmitter in the mammalian CNS. GABAergic neurons in the hippocampus are known to express nAChRs, and are generally more responsive to nicotine than glutamatergic hippocampal neurons.[44–46] Nicotine has been shown to increase GABA release from a variety of neurons.[47–50] It also was recently shown that high concentrations of acetylcholine increase the activity of GABAergic CA1 hippocampal neurons.[44] We further investigated nicotinic modulation of GABAergic neurotransmission in hippocampal neurons by analyzing spontaneous and evoked GABA currents in cultured neurons. In addition, we examined the effects of both high and low concentrations of nicotine. The low concentrations of nicotine were comparable to the levels experienced by smokers. In addition, high concentrations of nicotinic agonist were briefly applied to mimic the physiological situation in the synaptic cleft of cholinergic afferents to the hippocampus.

A Low Concentration of Nicotine Increased the Frequency of GABAergic mIPSCs

Hippocampal neurons were grown on microislands to limit the synaptic connections between neurons and to encourage the formation of autapses. Neurons were voltage-clamped at –20 mV in whole-cell configuration. TTX (0.5 μM) and CNQX (10 μM) were included in the extracellular solution to allow observation of spontaneous GABA-mediated events. GABAergic miniature inhibitory postsynaptic currents (mIPSCs) were generally observed at a relatively low frequency in the absence of nicotine. Low concentrations of nicotine (0.5 μM) increased the frequency of GABAergic mIPSCs (FIG. 4A) in a manner similar to that seen with glutamatergic mEPSCs. During a 3–4 min nicotine application, the frequency of spontaneous events increased 80 ± 17% (n = 3). However, the amplitudes of the mIPSCs were not affected by nicotine (FIG. 4B). Since TTX did not prevent the response to nicotine, it is likely that the increased vesicular release of GABA occurred though activation of presynaptic nAChRs.

High Concentrations of Nicotine Increase the Frequency of mIPSCs

To mimic experimentally the high agonist concentrations expected at cholinergic synapses, brief applications of saturating concentrations of agonist were used to synchro-

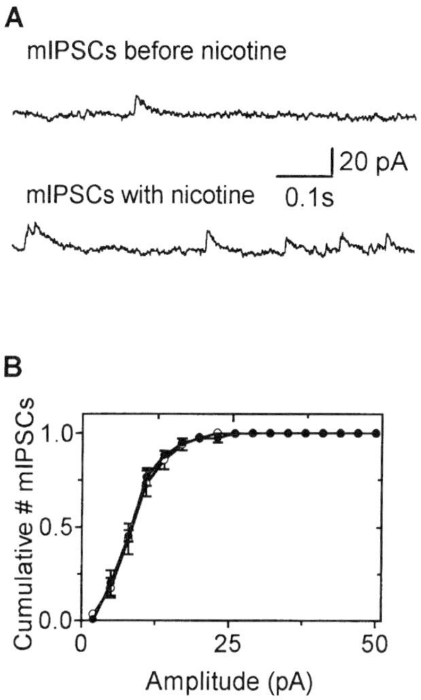

FIGURE 4. A low concentration of nicotine increased the frequency of mIPSCs, but did not change the amplitude. **(A)** Representative traces of mIPSCs before and during nicotine (0.5 μM) application. Neurons were voltage-clamped at –20 mV. **(B)** Cumulative amplitude histograms show no change in mIPSC amplitude before *(open circles)* or during *(closed circles)* nicotine application (n = 3). Data not previously published.

nously activate nAChRs. Because of the large degree of desensitization caused by these high concentrations, we waited 8 seconds for recovery between agonist applications. This 8-second interval was chosen to maximize the amount of nicotinic current during the time of the five agonist applications.

Miniature IPSCs were monitored from neurons voltage-clamped at –20 to 0 mV. Because the internal solution contained a solution low in Cl^-, at these holding potentials the anionic mIPSCs were recorded as upward current deflections, while the cationic mEPSCs were downward. For most experiments, CNQX (10 μM) and APV (50 μM) were included in the external bath solution to prevent activation of glutamate receptors. In some experiments, however, glutamate receptor inhibitors were not included and both outward mIPSCs and inward mEPSCs could be observed at the same time (FIG. 5A).

Brief applications (200 ms) of high concentrations of nicotine (0.5 mM) or acetylcholine (3 mM) caused an inward, rapidly desensitizing nAChR current at a holding potential of –50 mV (FIG. 5A, middle). The frequency of mIPSCs increased immediately after the agonist application, and remained elevated for several minutes before returning to the original baseline frequency (FIG. 5B and 5C, n = 4 of 5 cells). Interestingly, for the neuron shown in FIGURE 5A, the frequency of mIPSCs increased (FIG. 5B, filled circles), while the

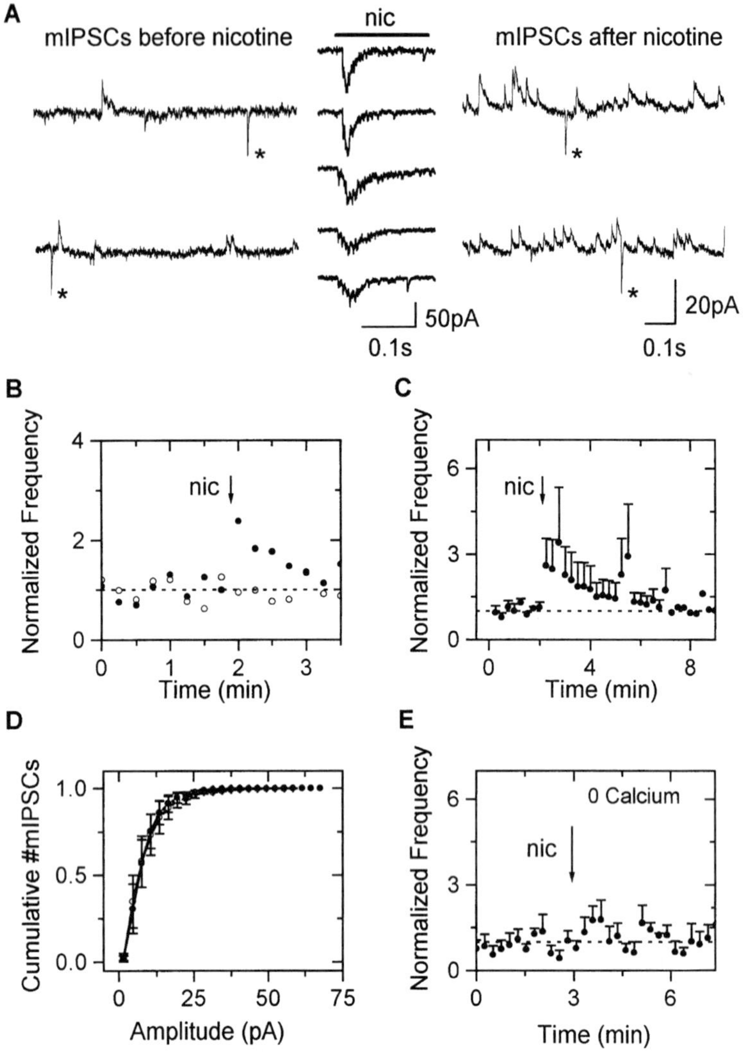

FIGURE 5. Brief repetitive activation of nAChRs by high concentrations of nicotine increased GABAergic mIPSC frequency without affecting the amplitude. **(A)** Traces of mIPSCs are shown before *(left)* and after *(right)* 5 rapid nicotine applications (500 µM, 200 ms; *middle* records). The neuron was voltage-clamped at –20 mV; so the mIPSCs are upward deflections and mEPSCs (*) are downward. For agonist applications, the cell was briefly stepped to a holding potential of –50 mV to increase the inward nicotinic currents. This voltage step had no effect on the mIPSC frequency in the absence of agonist application (n = 4). **(B)** The nicotinic currents increased the frequency of mIPSCs *(filled circles)* in this experiment, but did not alter the frequency of mEPSCs *(open circles)*. **(C)** In 4 out of 5 neurons (SEM), activation of nAChRs (as shown in A) increased mIPSC frequency. **(D)** Cumulative amplitude distributions show no change in mIPSC amplitudes following nicotine application (n = 4). **(E)** In external solutions containing ≤ 50 µM Ca^{2+}, no increase in mIPSC frequency was observed (n = 4, SEM). Data not previously published.

frequency of mEPSCs was unchanged (FIG. 5B, open circles). As was seen with lower concentrations of nicotine, the amplitudes of the mIPSCs were unaffected (FIG. 5D). To determine the role of Ca^{2+} influx in the response to nicotine, experiments were performed with external solutions containing ≤ 50μM calcium. In the nominal absence of Ca^{2+}, nicotine did not increase mIPSC frequency (FIG. 5E, $n = 4$ of 4 cells).

Evoked GABAergic IPSCs

One important role of the cholinergic inputs may be to enhance the response of a presynaptic terminal to an action potential, increasing the amount of neurotransmitter released and thereby increasing the probability of synaptic transmission. Therefore, we examined the effect of nicotine on the properties of evoked (e) GABAergic IPSCs.

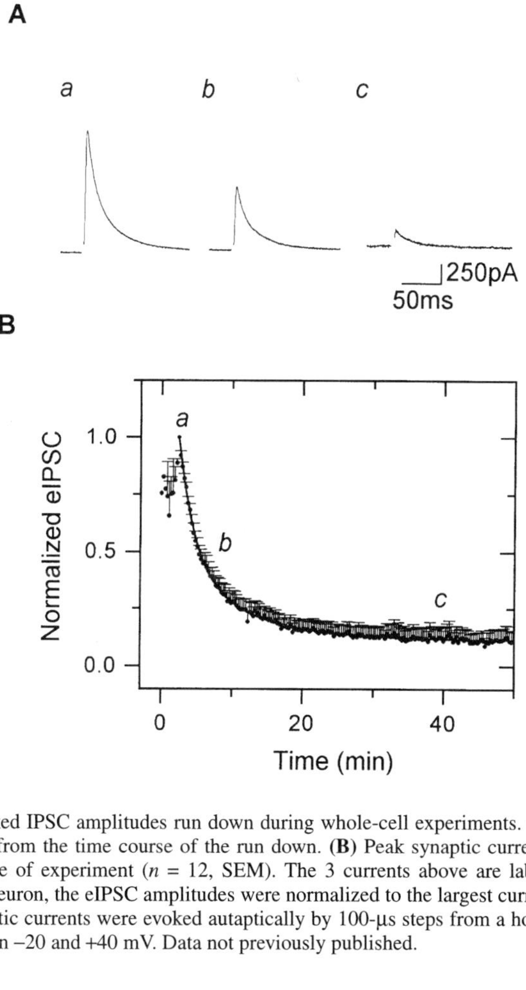

FIGURE 6. Evoked IPSC amplitudes run down during whole-cell experiments. **(A)** Three evoked IPSCs are shown from the time course of the run down. **(B)** Peak synaptic current amplitudes are plotted versus time of experiment (n = 12, SEM). The 3 currents above are labeled on the time course. For each neuron, the eIPSC amplitudes were normalized to the largest current obtained. The GABAergic synaptic currents were evoked autaptically by 100-μs steps from a holding potential of –60 mV to between –20 and +40 mV. Data not previously published.

GABAergic IPSCs were evoked from single neurons forming autaptic connections or by stimulation of a neuron synaptically paired to one other neuron. Postsynaptic neurons were voltage-clamped in the whole-cell configuration. Pairs of stimuli separated by 150ms were given to the presynaptic neuron. Individual pairs of eIPSCs were separated by 16–60s. The study was complicated by rundown of the eIPSCs over time (FIG. 6). To control for the occurrence of rundown, the time course of the decrease in amplitude was fit with the sum of two exponentials with time constants (and relative contributions) of 2.22min (62%) and 9.14min (38%). The decrease in amplitude was not accompanied by a change in the cell input resistance ($n = 3$). In subsequent experiments, we corrected for eIPSC rundown by adjusting the amplitude and then subtracting the exponential fit shown in FIGURE 6.

Low Concentrations of Nicotine Increase the Amplitude of Evoked IPSCs

GABAergic IPSCs were evoked using stimulation of neuron pairs. Postsynaptic neurons were voltage-clamped in the whole-cell configuration while pairs of stimuli were given to the presynaptic neuron. Prior to nicotine application, several synaptic currents were evoked in control solutions. Rundown of the current amplitude during this period was always observed. Continuous application of 0.1 µM nicotine for 1–10 min increased the peak amplitude of the eIPSCs (FIG. 7A). The average amplitude of the first eIPSC of the pair increased 73.5 ± 28.6% ($n = 6$ neurons) in the presence of nicotine (2min application). Enhancement was observed in 9 of 12 neurons in response to 0.1 µM nicotine using a variety of application protocols. MLA (5nM) prevented the increase by nicotine ($n = 3$), indicating that the response was mediated through α7-containing receptors (FIG. 7B and 7D). Picrotoxin (100µM) completely blocked the evoked currents ($n = 4$), confirming that they were $GABA_A$-receptor mediated (FIG. 7C and 7D).

High Concentrations of Nicotine Increase the Amplitude of Evoked IPSCs

We also examined the effects of high concentrations of nicotine on eIPSCs. These experiments were conducted on individual neurons forming autaptic connections. Voltage stimulations were given in pairs (150ms apart) every 16 seconds. After several stimulations under control conditions, two applications of 500 µM nicotine (arrow heads, FIG. 8A) were given between pairs of eIPSCs. The nicotine applications were 8 seconds apart, and the second application was given 1–2 s before the next evoked stimulus. Nicotine increased the peak amplitude of the eIPSCs in 10 of 33 neurons examined. The amplitude of the first current of the evoked pair increased 50.5% ($p < 0.004$, paired t-test) in response to nicotine. In most cases, the amplitude of the eIPSC increased immediately following the first nicotine application, then rose more gradually to a plateau with continued applications (FIG. 8B, 8C, and 8D). The eIPSC amplitude returned to baseline levels during or up to several minutes following the end of the nicotine applications.

High Concentrations of Nicotine Reduced the Paired-Pulse Ratio of Evoked IPSCs

The relationship between the amplitude of the first and second eIPSC of a pair can provide clues to the mechanism of action underlying changes in synaptic transmission.[51] The paired-pulse ratio (PPR, amplitude of the second eIPSC/amplitude of the first eIPSC) is

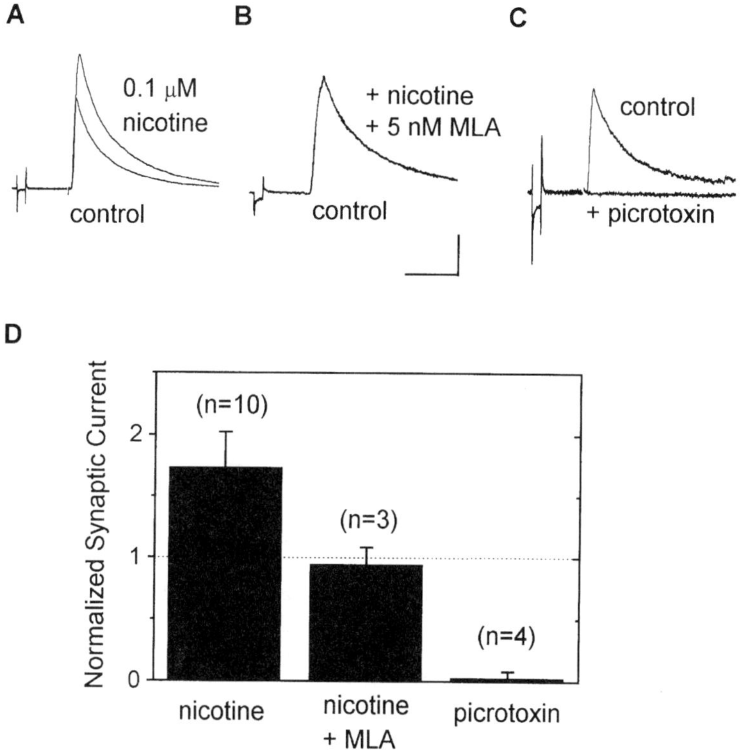

FIGURE 7. A low concentration of nicotine increased the amplitude of evoked GABAergic IPSCs. **(A)** Application of nicotine (1 μM) enhanced eIPSC amplitude compared to synaptic currents obtained prior to nicotine application (control). **(B)** Evoked IPSCs recorded during an incubation in MLA (5 nM) were not enhanced by a 2-min application of nicotine. The two traces overlie. **(C)** The eIPSCs were completely inhibited by picrotoxin (100 μM), indicating the currents were GABAergic. Changes in series resistance were monitored with brief small hyperpolarizations shown at the beginning of each trace. **(D)** Amplitudes of eIPSCs were normalized to the amplitude obtained under control conditions (*dotted line* at 1; SEM). *Vertical scale bar* represents 500 pA for A, 200 pA for B, and 125 pA for C. *Horizontal scale bar* = 50 ms. Data not previously published.

reasoned to be sensitive to the baseline probability of neurotransmitter release. All GABAergic neurons we studied displayed paired-pulse depression (PPR < 1) prior to nicotine application (FIG. 9). Although nicotine typically increased the amplitude of both the first and second eIPSC, the first eIPSC always increased more than the second. As a result, the paired pulse ratio (eIPSC2/eIPSC1) decreased following nicotine application, from 0.60 ± 0.05 before, to 0.52 ± 0.05 after ($p < 0.002$, paired t-test). This result suggests that nicotine increased the amplitude of the eIPSC by increasing the amount of neurotransmitter released, rather than by increasing the responsiveness of the postsynaptic receptors to the same amount of neurotransmitter.

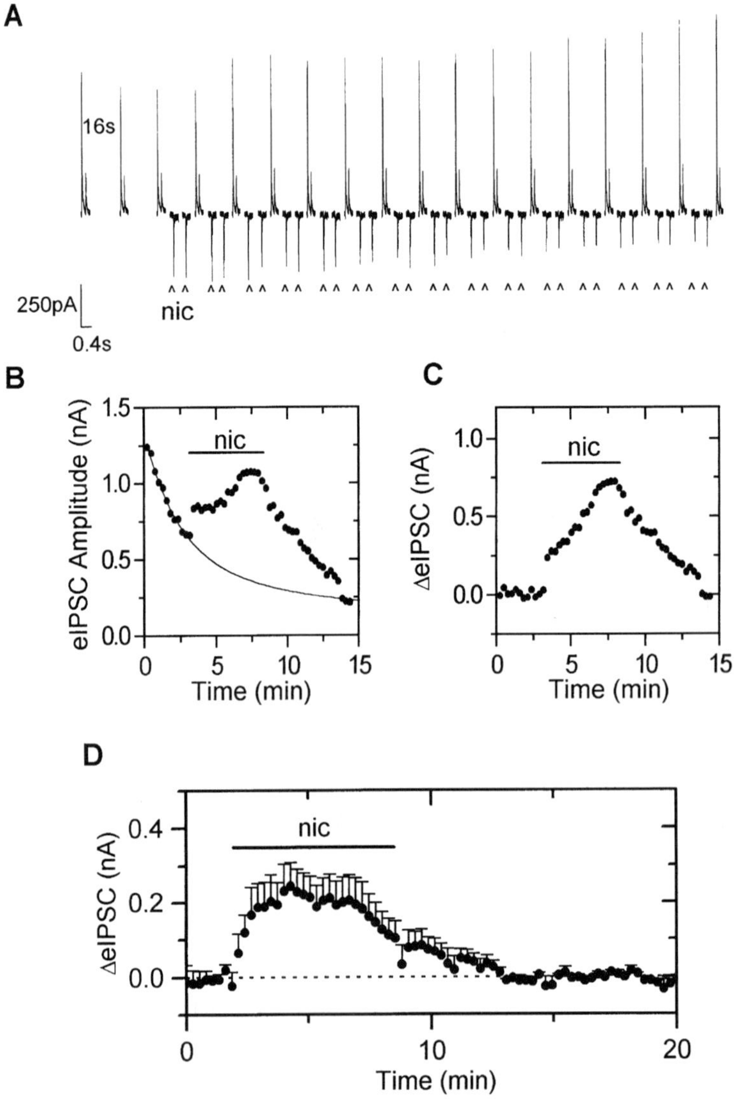
A
16s
250pA
0.4s
nic
B
eIPSC Amplitude (nA)
1.5
1.0
0.5
0.0
nic
0
5
10
15
Time (min)
C
ΔeIPSC (nA)
1.0
0.5
0.0
nic
0
5
10
15
Time (min)
D
ΔeIPSC (nA)
0.4
0.2
0.0
nic
0
10
20
Time (min)

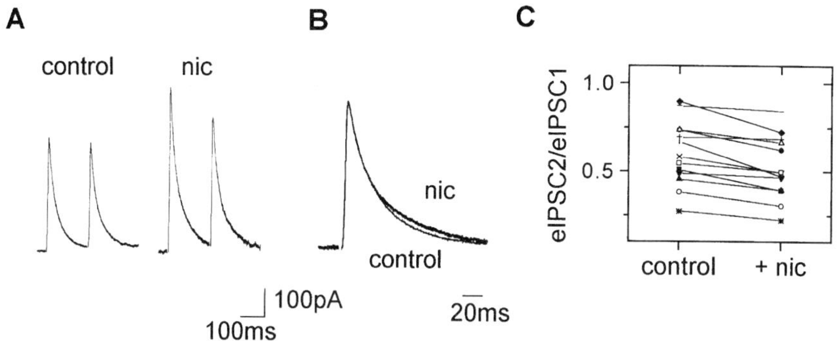

FIGURE 9. Nicotine decreased the paired-pulse ratio of evoked IPSCs. **(A)** Pairs of eIPSC are shown before (control) and during (nic) the interleaved synchronous activation of nAChR currents. Each trace shown is the average of four individual traces. Both eIPSCs of the pair are increased, but the first is increased more than the second. **(B)** The synaptic currents shown in A were normalized and overlaid to verify that the kinetics of the eIPSCs were similar after enhancement by nicotine. **(C)** For each of the experiments in which nicotine enhanced the IPSC amplitude, the ratio of the amplitude of the second eIPSC in the pair to that of the first eIPSC (paired-pulse ratio) decreased following nicotine application. The paired-pulse ratio (PPR) was determined for the period immediately preceding nicotine application (control) and for the period during the peak of the amplitude enhancement by nicotine (+nic). Five successive pairs for each condition were averaged for each cell. Data not previously published.

FIGURE 8. Brief, rapid applications of a high concentration of nicotine increased the amplitude of evoked GABAergic IPSCs. **(A)** Evoked autaptic IPSCs are shown to increase following synchronous activation of nAChRs (indicated by ∧). Paired inhibitory synaptic currents were evoked from autapses of cultured hippocampal neurons. Every 16 seconds, pairs of depolarizing steps (100 μs in duration, 150 ms apart) from a holding potential of –60 mV or –70 mV to –20 mV were delivered. Two applications of 500 μM nicotine 8 s apart were given in the interval between the stimulus pairs. **(B)** Peak amplitudes for the first evoked synaptic current in each pair measured from the cell shown in A were enhanced during nicotine applications *(bar)*. The exponential decay (rundown) of the eIPSCs as shown in FIG. 6 was scaled to the data observed. **(C)** The exponential decay was subtracted from the amplitude data to yield the corrected form of the data expressed as ΔeIPSC. **(D)** Averaged data showing the time course of the increase in eIPSCs as a consequence of nAChR currents (induced between the eIPSCs during the *solid bar*). An increase in IPSC amplitude was observed in 10 of 33 neurons examined. Data were aligned to the start of nicotine applications prior to averaging. The time course of IPSC rundown was subtracted from all data. Between 5 and 56 nicotine applications were used for these experiments. The *solid line* indicates the time for delivery of 56 nicotine applications. Data not previously published.

DISCUSSION

Summary

These results show that activation of hippocampal nAChRs can increase the release of glutamate and GABA. In each case, the effect of nicotine was dependent on the presence of extracellular Ca^{2+} and occurred through activation of presynaptic, α7-containing nAChRs. This conclusion is consistent with previous data indicating that α7-containing receptors make up the predominant population of nAChRs in the hippocampus.[23,25] In slice preparations, application of nicotine increased the Ca^{2+} influx into the presynaptic terminals of the mossy fibers.[24] When averaged over the whole presynaptic terminal, the magnitude of the Ca^{2+} signal induced by nAChRs was comparable to that via voltage-gated Ca^{2+} channels in response to an action potential. The frequency of spontaneous mEPSCs and mIPSCs increased, but their amplitudes were unchanged, again suggesting a presynaptic mechanism of action. Overall, there was little difference between the effects of nicotine on glutamatergic or GABAergic transmission. However, we observed a positive response to nicotine (increased frequency of spontaneous neurotransmitter release) more frequently with cultured GABAergic neurons than with glutamatergic neurons. This result is consistent with evidence indicating that nAChR expression is higher on inhibitory interneurons than on pyramidal neurons in the hippocampus.[45,46,52]

Nicotine also increased the amplitude of evoked IPSCs. This indicates that activation of nAChRs can modulate the responsiveness of the presynaptic terminal to a subsequent action potential. The influx of Ca^{2+} through voltage-gated channels in response to an action potential increases the probability of releasing a vesicle or two of neurotransmitter per presynaptic active zone. However, the presynaptic action potential does not guarantee release[53] (but see Ref. 54). Therefore, activation of presynaptic nAChRs just prior to the arrival of an action potential can increase the probability of successful synaptic transmission because the Ca^{2+} entry via the two pathways integrates in the presynaptic terminal. Consequently, properly timed nicotinic activity can increase the fidelity of particular synaptic events.

We also showed the nicotinic stimulation can alter the relationship between closely spaced incoming presynaptic activity and outgoing postsynaptic signaling. The relationship between the amplitudes of evoked pairs of IPSCs changed as a result of nAChR activity. This result has implications for decoding the information contained in a train of action potentials. The same information arriving along the same GABAergic afferents will be processed differently depending on the arrival of properly timed nicotinic activity that converges onto the GABAergic terminals. Work underway in our laboratory suggests similar mechanisms operate at glutamatergic synapses.[39]

Role of Cholinergic Inputs in the CNS

In many regions of the mammalian CNS, one role of nAChRs is to modulate the release of various neurotransmitters. Activation of nAChRs has been shown to increase release of virtually every major neurotransmitter, including GABA and glutamate (for review see Ref. 4). Activity of cholinergic inputs could increase the strength of some synapses, increasing the probability that action potentials result in successful synaptic transmission. Because cholinergic neurons can make broad, diffuse synaptic connections, alterations in their activity could alter neural information processing in many areas of the brain. For example, mutations in neuronal nAChR subunits have been associated with the occurrence of schizophrenia[55] and at least one form of epilepsy.[56] Selective degeneration of some cholinergic pathways has been observed in patients suffering from Alzheimer's disease.[57] Despite its addictive properties, nicotine has been proposed as a therapeutic agent for several neurological disorders, including schizophrenia, Alzheimer's disease, and Tourette's syndrome.[49]

Direct mediation of fast nicotinic cholinergic synaptic transmission has only rarely been observed thus far in the CNS.[58,59] It is likely, however, that these types of mechanisms will be uncovered in other areas of the brain.

Potential Effects of Nicotine from Tobacco

Tobacco use has been estimated to be the largest single cause of premature death in developed countries.[60] Nicotine is widely accepted as the addictive agent from cigarettes, and the addictive properties are likely to be mediated through nAChRs.[61] Self-administration of nicotine is reduced in mice that do not express the nAChR β2 subunit, supporting this hypothesis.[62] Furthermore, there are nicotinic effects on cognitive function, including increases in arousal, attention, and learning and memory.[28] Our results show that even the relatively low concentrations of nicotine achieved by smokers can effectively increase synaptic activity in hippocampal neurons. Longer exposure to nicotine, however, will cause desensitization of nAChRs.[63] A multitude of psychopharmacological factors will contribute to addiction, but the ability of nicotine to influence hippocampal synaptic transmission may contribute to some of nicotine's cognitive influences. These mechanisms also might participate in the association of contextual memory with nicotine use, such that a person can experience cravings during particular environment cues even years after having quit smoking.

REFERENCES

1. Barnard, E.A. 1992. Receptor classes and the transmitter-gated ion channels. Trends Biochem. Sci. **17:** 368–374.
2. Role, L.W. & D.K. Berg. 1996. Nicotinic receptors in the development and modulation of CNS synapses. Neuron **16:** 1077–1085.
3. Colquhoun, L.M. & J.W. Patrick. 1997. Pharmacology of neuronal nicotinic acetylcholine receptor subtypes. Adv. Pharmacol. **39:** 191–220.
4. Wonnacott, S. 1997. Presynaptic nicotinic ACh receptors. Trends Neurosci. **20:** 92–98.
5. Cooper, E., S. Couturier & M. Ballivet. 1991. Pentameric structure and subunit stoichiometry of a neuronal nicotinic acetylcholine receptor. Nature **350:** 235–238.
6. Karlin, A. & M.H. Akabas. 1995. Toward a structural basis for the function of nicotinic acetylcholine receptors and their cousins. Neuron **15:** 1231–1244.
7. Deneris, E.S., J. Connolly, J. Boulter, E. Wada, K. Wada, L.W. Swanson, J. Patrick & S. Heinemann. 1988. Primary structure and expression of beta 2: A novel subunit of neuronal nicotinic acetylcholine receptors. Neuron **1:** 45–54.
8. Boulter, J., K. Evans, D. Goldman, G. Martin, D. Treco, S. Heinemann & J. Patrick. 1986. Isolation of a cDNA clone coding for a possible neural nicotinic acetylcholine receptor alpha-subunit. Nature **319:** 368–374.

9. GOLDMAN, D., E. DENERIS, W. LUYTEN, A. KOCHHAR, J. PATRICK & S. HEINEMANN. 1987. Members of a nicotinic acetylcholine receptor gene family are expressed in different regions of the mammalian central nervous system. Cell **48:** 965–973.
10. WADA, K., M. BALLIVET, J. BOULTER, J. CONNOLLY, E. WADA, E.S. DENERIS, L.W. SWANSON, S. HEINEMANN & J. PATRICK. 1988. Functional expression of a new pharmacological subtype of brain nicotinic acetylcholine receptor. Science **240:** 330–334.
11. DUVOISIN, R.M., E.S. DENERIS, J. PATRICK & S. HEINEMANN. 1989. The functional diversity of the neuronal nicotinic acetylcholine receptors is increased by a novel subunit: β4. Neuron **3:** 487–496.
12. GERZANICH, V., A. KURYATOV, R. ANAND & J. LINDSTROM. 1997. "Orphan" alpha6 nicotinic AChR subunit can form a functional heteromeric acetylcholine receptor. Mol. Pharmacol. **51:** 320–327.
13. SARGENT, P.B. 1993. The diversity of neuronal nicotinic acetylcholine receptors. Annu. Rev. Neurosci. **16:** 403–443.
14. COUTURIER, S., D. BERTRAND, J.-M. MATTER, M.-C. HERNANDEZ, S. BERTRAND, N. MILLAR, S. VALERA, T. BARKAS & M. BALLIVET. 1990. A neuronal nicotinic acetylcholine receptor subunit (α7) is developmentally regulated and forms a homo-oligomeric channel blocked by α-BTX. Neuron **5:** 847–856.
15. SEGUELA, P., J. WADICHE, K. DINELEY-MILLER, J.A. DANI & J.W. PATRICK. 1993. Molecular cloning, functional properties, and distribution of rat brain alpha 7: A nicotinic cation channel highly permeable to calcium. J. Neurosci. **13:** 596–604.
16. GERZANICH, V., R. ANAND & J. LINDSTROM. 1994. Homomers of alpha 8 and alpha 7 subunits of nicotinic receptors exhibit similar channel but contrasting binding site properties. Mol. Pharmacol. **45:** 212–220.
17. ELGOYHEN, A.B., D.S. JOHNSON, J. BOULTER, D.E. VETTER & S. HEINEMANN. 1994. Alpha 9: An acetylcholine receptor with novel pharmacological properties expressed in rat cochlear hair cells. Cell **79:** 705–715.
18. DECKER, E.R. & J.A. DANI. 1990. Calcium permeability of the nicotinic acetylcholine receptor: The single-channel calcium influx is significant. J. Neurosci. **10:** 3413–3420.
19. VERNINO, S., M. AMADOR, C.W. LUETJE, J. PATRICK & J.A. DANI. 1992. Calcium modulation and high calcium permeability of neuronal nicotinic acetylcholine receptors. Neuron **8:** 127–134.
20. VERNINO, S., M. ROGERS, K.A. RADCLIFFE & J.A. DANI. 1994. Quantitative measurement of calcium flux through muscle and neuronal nicotinic acetylcholine receptors. J. Neurosci. **14:** 5514–5524.
21. CASTRO, N.G. & E.X. ALBEQUERQUE. 1995. alpha-Bungarotoxin-sensitive hippocampal nicotinic receptor channel has a high calcium permeability. Biophys. J. **68:** 516–524.
22. PUGH, P.C. & D.K. BERG. 1994. Neuronal acetylcholine receptors that bind alpha-bungarotoxin mediate neurite retraction in a calcium-dependent manner. J. Neurosci. **14:** 889–896.
23. ALKONDON, M. & E.X. ALBUQUERQUE. 1993. Diversity of nicotinic acetylcholine receptors in rat hippocampal neurons. I. Pharmacological and functional evidence for distinct structural subtypes. J. Pharmacol. Exp. Ther. **265:** 1455–1473.
24. GRAY, R., A.S. RAJAN, K.A. RADCLIFFE, M. YAKEHIRO & J.A. DANI. 1996. Hippocampal synaptic transmission enhanced by low concentrations of nicotine. Nature **383:** 713–716.
25. ZAREI, M.M., K.A. RADCLIFFE, D. CHEN, J.W. PATRICK & J.A. DANI. 1998 Distribution of nicotinic acetylcholine α7 subunits on cultured hippocampal neurons. Neurosci. In press.
26. QUIK, M., J. CHAN & J. PATRICK. 1994. alpha-Bungarotoxin blocks the nicotinic receptor mediated increase in cell number in a neuroendocrine cell line. Brain Res. **655:** 161–167.
27. ZHENG, J.Q., M. FELDER, J.A. CONNOR & M.M. POO. 1994. Turning of nerve growth cones induced by neurotransmitters. Nature **368:** 140–144.
28. LEVIN, E.D. 1992. Nicotinic systems and cognitive function. Psychopharmacology (Berl.) **108:** 417–431.
29. POMERLEAU, O.F. 1986. Nicotine as a psychoactive drug: Anxiety and pain reduction. Psychopharmacol.Bull. **22:** 865–869.
30. BADIO, B. & J.W. DALY. 1994. Epibatidine, a potent analgetic and nicotinic agonist. Mol. Pharmacol. **45:** 563–569.
31. MADISON, D.V., R.C. MALENKA & R.A. NICOLL. 1991. Mechanisms underlying long-term potentiation of synaptic transmission. Annu. Rev. Neurosci. **14:** 379–397.

32. GODA, Y. & C.F. STEVENS. 1996. Synaptic plasticity: The basis of particular types of learning. Curr. Biol. **6:** 375–378.
33. OHNO, M., T. YAMAMOTO & S. WATANABE. 1993. Blockade of hippocampal nicotinic receptors impairs working memory but not reference memory in rats. Pharmacol. Biochem. Behav. **45:** 89–93.
34. MAURICE, T., M. HIRAMATSU, T. KAMEYAMA, T. HASEGAWA & T. NABESHIMA. 1994. Behavioral evidence for a modulating role of sigma ligands in memory processes. II. Reversion of carbon monoxide-induced amnesia. Brain Res. **647:** 57–64.
35. AIGNER, T.G. 1995. Pharmacology of memory: Cholinergic-glutamatergic interactions. Curr. Opin. Neurobiol. **5:** 155–160.
36. FELIX, R. & E.D. LEVIN. 1997. Nicotinic antagonist administration into the ventral hippocampus and spatial working memory in rats. Neurosci. **81:** 1009–1017.
37. VIDAL, C. & J.-P. CHANGEUX. 1993. Nicotinic and muscarinic modulations of excitatory synaptic transmission in the rat prefrontal cortex *in vitro*. Neurosci. **56:** 23–32.
38. MCGEHEE, D.S., M.J. HEATH, S. GELBER, P. DEVAY & L. ROLE. 1995. Nicotine enhancement of fast excitatory synaptic transmission in CNS by presynaptic receptors. Science **22:** 1692–1696.
39. RADCLIFFE, K.A. & J.A. DANI. 1998. Nicotinic stimulation produces multiple forms of increased glutamatergic synaptic transmission. J. Neurosci. **18:** 7075–7083.
40. WADA, E., K. WADA, J. BOULTER, E. DENERIS, S. HEINEMANN, J. PATRICK & L.W. SWANSON. 1989. Distribution of alpha 2, alpha 3, alpha 4, and beta 2 neuronal nicotinic receptor subunit mRNAs in the central nervous system: A hybridization histochemical study in the rat. J. Comp. Neurol. **284:** 314–335.
41. WOOLF, N.J. 1991. Cholinergic systems in mammalian brain and spinal cord. Prog. Neurobiol. **37:** 475–524.
42. BEKKERS, J.M. & C.F. STEVENS. 1991. Excitatory and inhibitory autaptic currents in isolated hippocampal neurons maintained in cell culture. Proc. Natl. Acad. Sci. USA **88:** 7834–7838.
43. ZAREI, M.M. & J.A. DANI. 1995. Structural basis for explaining open-channel blockade of the NMDA receptor. J. Neurosci. **15:** 1446–1454.
44. ALKONDON, M., E.F. PEREIRA, C.T. BARBOSA & E.X. ALBUQUERQUE. 1997. Neuronal nicotinic acetylcholine receptor activation modulates gamma-aminobutyric acid release from CA1 neurons of rat hippocampal slices. J. Pharmacol. Exp. Ther. **283:** 1396–1411.
45. JONES, S. & J.L. YAKEL. 1997. Functional nicotinic ACh receptors on interneurones in the rat hippocampus. J. Physiol. **504:** 603–610.
46. FRAZIER, C.J., Y.D. ROLLINS, C.R. BREESE, S. LEONARD, R. FREEDMAN & T.V. DUNWIDDIE. 1998. Acetylcholine activates an α-Bungarotoxin–sensitive nicotinic current in rat hippocampal interneurons, but not pyramidal cells. J. Neurosci. **18:** 1187–1195.
47. LENA, C., J.P. CHANGEUX & C. MULLE. 1993. Evidence for "preterminal" nicotinic receptors on GABAergic axons in the rat interpeduncular nucleus. J. Neurosci. **13:** 2680–2688.
48. MCMAHON, L.L., K.W. YOON & V.A. CHIAPPINELLI. 1994. Nicotinic receptor activation facilitates GABAergic neurotransmission in the avian lateral spiriform nucleus. Neurosci. **59:** 689–698.
49. LENA, C. & J.P. CHANGEUX. 1997. Pathological mutations of nicotinic receptors and nicotine-based therapies for brain disorders. Curr. Opin. Neurobiol. **7:** 674–682.
50. BERTOLINO, M., K.J. KELLAR, S. VICINI & R.A. GILLIS. 1997. Nicotinic receptor mediates spontaneous GABA release in the rat dorsal motor nucleus of the vagus. Neurosci. **79:** 671–681.
51. MENNERICK, S. & C.F. ZORUMSKI. 1995. Paired-pulse modulation of fast excitatory synaptic currents in microcultures of rat hippocampal neurons. J. Physiol. (Lond.) **488:** 85–101.
52. FREEDMAN, R., C. WETMORE, I. STROMBERG, S. LEONARD & L. OLSON. 1993. Alpha-bungarotoxin binding to hippocampal interneurons: Immunocytochemical characterization and effects on growth factor expression. J. Neurosci. **13:** 1965–1975.
53. HUANG, E.P. & C.F. STEVENS. 1997. Estimating the distribution of synaptic reliabilities. J. Neurophysiol. **78:** 2870–2880.
54. HARDINGHAM, N.R. & A.U. LARKMAN. 1998. Rapid report: The reliability of excitatory synaptic transmission in slices of rat visual cortex *in vitro* is temperature dependent. J. Physiol. (Lond.) **507:** 249–256.
55. FREEDMAN, R., H. COON, M. MYLES-WORSLEY, A. ORR-URTREGER, A. OLINCY, A. DAVIS, M. POLYMEROPOULOS, J. HOLIK, J. HOPKINS, M. HOFF, J. ROSENTHAL, M.C. WALDO, F. REIMHERR, P.

WENDER, J. YAW, D.A. YOUNG, C.R. BREESE, C. ADAMS, D. PATTERSON, L.E. ADLER, L. KRUGLYAK, S. LEONARD & W. BYERLEY. 1997. Linkage of a neurophysiological deficit in schizophrenia to a chromosome 15 locus. Proc. Natl. Acad. Sci. USA **94:** 587–592.

56. STEINLEIN, O.K., J.C. MULLEY, P. PROPPING, R.H. WALLACE, H.A. PHILLIPS, G.R. SUTHERLAND, I.E. SCHEFFER & S.F. BERKOVIC. 1995. A missense mutation in the neuronal nicotinic acetylcholine receptor alpha 4 subunit is associated with autosomal dominant nocturnal frontal lobe epilepsy. Nature Genet. **11:** 201–203.
57. KASA, P., Z. RAKONCZAY & K. GULYA. 1997. The cholinergic system in Alzheimer's disease. Prog. Neurobiol. **52:** 511–535.
58. ZHANG, M., Y.T. WANG, D.M. VYAS, R.S. NEUMAN & D. BIEGER. 1993. Nicotinic cholinoceptor–mediated excitatory postsynaptic potentials in rat nucleus ambiguus. Exp. Brain Res. **96:** 83–88.
59. ROERIG, B., D.A. NELSON & L.C. KATZ. 1997. Fast synaptic signaling by nicotinic acetylcholine and serotonin 5-HT_3 receptors in developing visual cortex. J. Neurosci. **17:** 8353–8362.
60. PETO, R., A.D. LOPEZ, J. BOREHAM, M. THUN & C.J. HEATH. 1992. Mortality from tobacco in developed countries: Indirect estimation from national vital statistics. Lancet **339:** 1268–1278.
61. DANI, J.A. & S. HEINEMANN. 1996. Molecular and cellular aspects of nicotine abuse. Neuron **16:** 905–908.
62. PICCIOTTO, M.R., M. ZOLI, R. RIMONDINI, C. LENA, L.M. MARUBIO, E.M. PICH, K. FUXE & J.P. CHANGEUX. 1998. Acetylcholine receptors containing the beta2 subunit are involved in the reinforcing properties of nicotine. Nature **391:** 173–177.
63. PIDOPLICHKO, V.I., M. DEBIASI, J.T. WILLIAMS & J.A. DANI. 1997. Nicotine activates and desensitizes midbrain dopamine neurons. Nature **390:** 401–404.

The Role of β2-Subunit–Containing Nicotinic Acetylcholine Receptors in the Brain Explored with a Mutant Mouse

CLÉMENT LÉNA AND JEAN-PIERRE CHANGEUX[a]

Laboratoire de Neurobiologie Moléculaire - CNRS UA D1284, Institut Pasteur, 28, rue du Dr. Roux, 75724 Paris Cédex 15, France

ABSTRACT: Neuronal nicotinic receptors comprise a family of pentameric oligomers made up of a combination of 10 different subunits. The β2 subunit has the widest pattern of expression in the brain and is thus likely to form a significant fraction of neuronal nicotinic receptors. Using mice lacking the β2 subunit, we have shown that nAChRs containing this subunit are responsible for most of the high-affinity binding sites for nicotine, cytisine, and epibatidine in the brain. Functional receptors containing the β2-subunit are found in the somatodendritic compartment as well as the axonal compartment of neurons. We have examined the contribution of these receptors to the effects of nicotine on the mesolimbic DA system, which mediates the reinforcing properties of many addictive drugs (including nicotine). Submicromolar doses of nicotine, corresponding to the concentrations of nicotine *in vivo* in self-administration paradigms, increased the firing rate of dopaminergic neurons *in vitro* in normal mice but not in mice lacking the β2 subunit. Consistently, systemic injection of nicotine induced an increase in extracellular dopamine in normal mice but not in mutant mice, and nicotine self-administration was reduced or suppressed in mutant mice. These data support the view that the β2-containing receptors are involved in mediating the reinforcing properties of nicotine.

The nicotinic acetylcholine receptors (nAChRs) are present in most brain regions, and the pharmacological effects of nicotine on behavior as well as the diseases associated with defects in various nAChR genes (in particular, point mutations) support an important functional significance of these receptors in the central nervous system.[1,2] The molecular and cellular diversity of the nAChRs have led to the functional dissection of these receptors in mutant mice altered in a single nicotinic subunit.

nAChRs in neurons are formed by the combination of various subunits (see, e.g., Refs. 2, 3). To date, 10 "neuronal" nicotinic subunits have been identified in mammals (α2–α7, α9, β2–β4). These subunits assemble following combinatorial rules to form functional oligomers (most likely pentamers). The α7 and α9 subunits are members of a first family of subunits able to form homoligomers when expressed in the *Xenopus* oocyte. The eight other subunits compose another distinct family of nAChRs.[4] In particular, none of them is able to form functional receptors when expressed alone in the oocyte. The α2–α4, α6 subunits indeed associate into functional heterooligomers together with the β2 or β4 subunits (references in Refs. 2, 5). The α5 subunit may associate with α3β2/4 and α4β2 and thus forms heterooligomers with three different subunits. The β3 subunit may also be included

[a]Corresponding author. Phone: 33-145 68 88 05; fax: 33-145 68 88 36; e-mail: changeux@pasteur.fr

in heteromers with other subunits following a pattern similar to the α5 subunit.[6,7] In summary, the β2 and β4 subunits critically contribute to the formation of functional nAChRs from the second family.

The pattern of expression of these various neuronal subunits strikingly differs from one to the other. The β2 subunit is expressed in most if not all neurons, while the expression of the β4 subunit is restricted to a few nuclei.[8–10] Therefore, inactivation of the β2-subunit gene in mutant animals is expected to result in an important reduction in the number of functional nAChRs in the brain.

INACTIVATION OF THE β2 NICOTINIC-SUBUNIT GENE

The β2-subunit gene has been disrupted by introducing a construct containing a reporter gene and a selection gene covering the initial methionine and the rest of the first exon. Following homologous recombination, viable mutant mice have been generated.[11] These mutant mice survive, feed, and mate normally. The brains of the animals exhibit normal organization and morphology. No expression of the β2 subunit could be demonstrated in the β2-/- mutant mice by *in situ* hybridization or by immunocytochemistry. On the other hand, the expression of the other nicotinic subunits examined by *in situ* hybridization was unchanged. Measurements of acetylcholinesterase activity and muscarinic M1 and M2 ligand binding failed to reveal any significant change in the cholinergic system.[12] Little, if any, compensatory process in the development of the cholinergic system could be demonstrated. The removal of the β2-subunit is thus expected to produce discrete deficits directly related to the function of β2-containing nAChRs oligomers.[13]

The distribution of the nAChRs in the brain of β2-/- mice was then analyzed using nicotinic ligand–binding experiments.[12] The pattern of [^{125}I]α-bungarotoxin binding, which corresponds to α7-containing nAChRs,[14] was not modified in the mutant mice. On the other hand, the binding of [^{3}H]-nicotine, [^{3}H]-cytisine, and [^{3}H]-epibatidine dramatically decreased in most brain regions. No high-affinity [^{3}H]-nicotine binding was observed in any brain region. On the other hand, [^{3}H]epibatidine binding activity persisted in a few brain regions (habenulo-interpeduncular system, dorsal medulla oblongata, dorsal cortex of the inferior colliculus and a few other regions with faint labeling). This pattern largely overlapped the pattern of expression of the β4 subunit observed with oligomers (Ref. 10, but see Ref. 9 for experiments with riboprobes); it may possibly correspond to the binding activity of β4-containing nAChRs. Furthermore, [^{3}H]-cytisine was found to label only a subset of the nuclei that bind epibatidine with high affinity in the mutant mice, indicating that the nAChRs that do not contain β2 are heterogeneous. A heterogeneity was also noted in electrophysiological recordings, the nicotinic responses exhibiting faster desensitization in the dorsal interpeduncular nucleus (which binds both [^{3}H]-epibatidine and [^{3}H]-cytisine with high affinity) than in the medial part of the medial habenula (which binds only [^{3}H]-epibatidine with high affinity). The diversity of nAChRs subunits expressed in these regions (references in Ref. 6) suggests that the heterogeneity of binding activity and electrophysiological properties may actually correspond to various combinations of nAChR subunits.

FUNCTIONAL CHARACTERIZATION OF β2-CONTAINING NICOTINIC RECEPTORS

The contribution of β2-containing nAChRs to neuronal function was then investigated in electrophysiological experiments with the patch-clamp technique in slices of mouse brain. These nAChRs are found both in the somato-dendritic compartment and in the axonal compartment of various types of neurons. In neurons from regions that are strongly labeled by [^{3}H]-nicotine, such as the anterior thalamus or the mesencephalic dopaminergic nuclei, cytisine behaves as a poor agonist compared to nicotine.[11,12,15,16] The responses observed so far with this particular pharmacological specificity are absent in neurons from β2-/- mutant mice, indicating that β2-containing nAChRs mediate this response in accordance with previous data in the *Xenopus* oocyte.[17,18] Residual nicotinic responses in mutant mice have been recorded in regions exhibiting residual labeling by [^{3}H]-epibatidine, and these β2-independent nAChRs had a different pharmacology, cytisine being as efficient as nicotine.[12]

β2-containing nAChRs are also present on nerve terminals.[15] nAChR activation facilitates the release of GABA in slices of sensory thalamus from normal mice but not from mutant mice. In normal mice, micromolar concentrations of nicotinic agonists reduce the number of failures in the synaptic release of GABA following electrical stimulation and increase the frequency of GABAergic spontaneous synaptic events. This last effect is also observed when the propagation of action potentials is prevented by the application of tetrodotoxin, or when the GABAergic terminals have been separated from the presynaptic cell soma with a razor blade. These data demonstrate that nicotinic agonists facilitate the release of GABA by activating nAChRs present at the level of nerve terminals. The nicotine-elicited increase in spontaneous release was altered when the extracellular calcium concentration was changed. The replacement of 90% of the extracellular calcium ions by magnesium ions suppresses the increase in frequency of miniature synaptic currents. On the other hand, the duration of this effect following the removal of nicotinic agonist increased when the extracellular calcium concentration was raised. Presynaptic β2-containing nAChRs thus facilitate neurotransmitter release by eliciting an influx of calcium in the nerve terminals. Such effect has also been observed for α7-containing nAChRs in glutamatergic terminals.[19,20] Interestingly, nicotinic responses disappeared in knockout mice from GABAergic neurons in the *reticularis* thalamus as well as in their terminal field in the ventrobasal thalamus,[15] indicating that β2-containing nAChRs may be present in the somato-dendritic and presynaptic compartments of a same class of neurons. On the other hand, preterminal nicotinic receptors, which enhance neurotransmitter release by triggering an action potential in the nerve terminal,[21] are still functional in the interpeduncular nucleus in β2-/- mutant mice (C. Léna, unpublished observation). The subunit composition of the nAChRs may thus differ in the terminal and preterminal axonal compartments.

BEHAVIORAL ANALYSIS OF THE β2-/- MUTANT MICE

The contribution of the β2-containing nAChRs to brain-defined functions was then examined in behavioral tests. Passive avoidance procedures are widely used to measure cognitive effects of pharmacological agents active in the central nervous system. The β2-/- mutant mice tested for the retention of an avoidance task showed marked differences compared to nonmutant siblings.[11] The test utilized in a standard manner for pharmacological assay of nicotine sensitivity consists of the measurement of the delay by a mouse placed in a well-lit compartment in entering an adjacent dark chamber. In the sample run, the mouse is given an electric shock upon entry into the dark chamber, and vehicle or nicotine are immediately injected into the mouse. After 24 hours, retention of the electric shock memory was assessed by measuring the delay in entering the dark chamber. Low doses of nicotine (0.01 mg/kg ip) consistently increased the delay in wild-type animals but was completely ineffective in β2-/- mice. β2-containing nAChRs are thus a major pharmacological target of low nicotine concentration in passive avoidance learning.

Preliminary experiments indicate a slight modification of the learning behavior that takes place in the absence of nicotine. The mice spontaneously show longer retention latencies compared to their littermate siblings. Altogether, these experiments support the notion that β2-containing nAChRs mediate endogenous effects of acetylcholine in the brain under defined behavioral conditions.

The hypothesis has been formulated and formalized[22–24] that neuronal circuits involved in learning by selection include evaluation circuits that treat the reinforcing signals in the central nervous system. Moreover, reinforcers in general are viewed as signals possessing an "incentive value" and as a consequence modulate memory acquisition (references in Ref. 25). The addictive drugs act as reinforcers since they promote their own intake. Nicotine is a widely used drug of abuse, and laboratory animals as well as humans self-administer nicotine (review in Ref. 26). Accumulating evidence indicates that nicotine self-administration is related to the activation of the mesocorticolimbic system similarly to other drugs of abuse[27,28] (review in Ref. 26). The systemic injection of nicotine produces an increase in the release of striatal dopamine, possibly as a consequence of activation of both somatodendritic and presynaptic nAChRs in dopaminergic neurons. However, recent evidence suggests that the increase in dopamine release *in vivo* is more likely to be mediated by activation of nAChRs within the mesencephalic nuclei rather than in their terminal field.[29,30] Consistently, nicotine self-administration is strongly reduced by the local injection of a nicotinic antagonist at the level of the mesencephalic dopaminergic neurons.[31]

We have tested the effects of nicotine in the dopaminergic system both by electrophysiology and microdialysis experiments, and in a paradigm of nicotine self-administration in wild-type and β2-/- mutant mice.[16] The concentrations of nicotine in the arterial blood of smokers during cigarette consumption are in the submicromolar range (see, e.g., Ref. 32). In agreement with parallel experiments,[33] we found that 500 nM nicotine significantly increased the discharge frequency in the mesencephalic dopaminergic neurons recorded *in vitro* in normal mice. This effect disappears in β2-/- mutant mice.[16] Consistent with these findings, microdialysis experiments in the striatum showed that nicotine produces a dose-dependent increase in the extracellular dopamine concentration in normal mice but not in mutant mice.[16] Nicotine self-administration was tested in normal and in mutant mice. The animals were first trained for a couple of days to self-administer cocaine. This compound experimentally facilitates the acquisition of self-administration behavior in mice and serves as an internal control to demonstrate that mutant mice do normally learn this behavior. Following the days of training, cocaine was switched to nicotine in the self-administration system. Normal mice exhibited a sustained behavior of nicotine intake, while β2-/- mice exhibited a progressive extinction of nicotine self-administration behavior. β2-containing nAChRs thus contribute to the reinforcing properties of nicotine.

In summary, β2-containing nAChRs are widely expressed at the brain level and at the neuronal level. They contribute to a variety of physiological and behavioral processes. They are activated by low doses of nicotine *in vivo* and are likely to contribute to a significant fraction of the effects of nicotine, in particular the modulation of learning and the activation of the mesolimbic system in relation to nicotine addiction. The involvement of these nAChRs in endogeneous cholinergic transmission remains to be explored, and future work may show how these receptors participate in the fine tuning of cognition and behavior.

ACKNOWLEDGMENTS

This work was supported by the Collège de France, the Centre National de la Recherche Scientifique, the Association Francaise contre la Myopathie, the Council for Tobacco Research; Biotech/Biomed contracts from the Commission of the European Communities; grants from the Human Frontiers Science Program, the Institut Pasteur; and a grant from the Association France Alzheimer for C.L.

REFERENCES

1. CHANGEUX, J.P., A. BESSIS, J.P. BOURGEOIS, P.J. CORRINGER, A. DEVILLERS-THIERY, J.L. EISELE, M. KERSZBERG, C. LÉNA, N. LE NOVERE, M. PICCIOTTO & M. ZOLI. 1996. Nicotinic receptors and brain plasticity. Cold Spring Harbor Symp. Quant. Biol. **61:** 343–362.
2. LÉNA, C. & J.P. CHANGEUX. 1997. Pathological mutations of nicotinic receptors and nicotine-based therapies for brain disorders. Curr. Opin. Neurobiol. **7**(5): 674–682.
3. LINDSTROM, J. 1996. Neuronal nicotinic acetylcholine receptors. Ion Channels **4:** 377–450.
4. LE NOVÈRE, N. & J.P. CHANGEUX. 1995. Molecular evolution of the nicotinic acetylcholine receptor: An example of multigene family in excitable cells. J. Mol. Evol. **40**(2): 155–172.
5. FUCILE, S., J.M. MATTER, L. ERKMAN, D. RAGOZZINO, B. BARABINO, F. GRASSI, S. ALEMA, M. BALLIVET & F. EUSEBI. 1998. The neuronal alpha6 subunit forms functional heteromeric acetylcholine receptors in human transfected cells. Eur. J. Neurosci. **10**(1): 172–178.
6. LE NOVÈRE, N., M. ZOLI & J.P. CHANGEUX. 1996. Neuronal nicotinic receptor alpha6 subunit mRNA is selectively concentrated in catecholaminergic nuclei of the rat brain. Eur. J. Neurosci. **8**(11): 2428–2439.
7. ROGERS, M., L.M. COLQUHOUN, J.W. PATRICK & J. DANI. 1997. Calcium flux through predominantly independent purinergic ATP and nicotinic acetylcholine receptors. J. Neurophysiol. **77**(3): 1407–1417.
8. WADA, E., K. WADA, J. BOULTER, E. DENERIS, S. HEINEMANN, J. PATRICK & L.W. SWANSON. 1989. Distribution of alpha2, alpha3, alpha4, and beta2 neuronal nicotinic subunit mRNAs in the central nervous system: A hybridization histochemical study in rat. J. Comp. Neurol. **284:** 314–335.
9. DINELEY-MILLER, K. & J. PATRICK. 1992. Gene transcripts for the nicotinic acetylcholine receptor subunit, beta4, are distributed in multiple areas of the rat central nervous system. Mol. Brain Res. **16**(3–4): 339–344.
10. ZOLI, M., N. LE NOVÈRE, J.J. HILL & J.P. CHANGEUX. 1995. Developmental regulation of nicotinic ACh receptor subunit mRNAs in the rat central and peripheral nervous systems. J. Neurosci. **15:** 1912–1939.
11. PICCIOTTO, M.R., M. ZOLI, C. LÉNA, A. BESSIS, Y. LALLEMAND, N. LE NOVÈRE, P. VINCENT, E. MERLO PICH, P. BRÛLET & J.P. CHANGEUX. 1995. Abnormal avoidance learning in mice lacking functional high-affinity nicotine receptor in the brain. Nature **374**(6517): 65–67.
12. ZOLI, M., C. LÉNA, M.R. PICCIOTTO & J.P. CHANGEUX. 1998. Identification of four classes of brain nicotinic receptors using beta2 mutant mice. J. Neurosci. **18**(12): 4461–4472.

13. Picciotto, M.R., M. Zoli, V. Zachariou & J.P. Changeux. 1997. Contribution of nicotinic acetylcholine receptors containing the beta2-subunit to the behavioural effects of nicotine. Biochem. Soc. Trans. **25**(3): 824–829.
14. Orr-Urtreger, A., F.M. Goldner, M. Saeki, I. Lorenzo, L. Goldberg, M. De Biasi, J.A. Dani, J.W. Patrick & A. Beaudet. 1997. Mice deficient in the alpha7 neuronal nicotinic acetylcholine receptor lack alpha-bungarotoxin binding sites and hippocampal fast nicotinic currents. J. Neurosci. **17**(23): 9165–9171.
15. Léna, C. & J.P. Changeux. 1997. Role of Ca^{2+} ions in nicotinic facilitation of GABA release in mouse thalamus. J. Neurosci. **17**(2): 576–585.
16. Picciotto, M.R., M. Zoli, R. Rimondini, C. Léna, L.M. Marubio, E. Merlo Pich, K. Fuxe & J.P. Changeux. 1998. Acetylcholine receptors containing the beta2 subunit are involved in the reinforcing properties of nicotine. Nature **391**(6663): 173–177.
17. Luetje, C.W. & J. Patrick. 1991. Both alpha- and beta-subunits contribute to the agonist sensitivity of neuronal nicotinic acetylcholine receptors. J. Neurosci. **11**: 837–845.
18. Papke, R.L. & S.F. Heineman. 1994. Partial agonist properties of cytisine on neuronal nicotinic receptors containing the beta 2 subunit. Mol. Pharmacol. **45**(1): 509–531.
19. McGehee, D.S., M.J. Heath, S. Gelber, P. Devay & L.W. Role. 1995. Nicotine enhancement of fast excitatory synaptic transmission in CNS by presynaptic receptors. Science **269**(5231): 1692–1696.
20. Gray, R., A.S. Rajan, K.A. Radcliffe, M. Yakehiro & J.A. Dani. 1996. Hippocampal synaptic transmission enhanced by low concentrations of nicotine. Nature **383**(6602): 713–716.
21. Léna, C., J.P. Changeux & C. Mulle. 1993. Evidence for "preterminal" nicotinic receptors on GABAergic axons in the rat interpeduncular nucleus. J. Neurosci. **13**(6): 2680–2688.
22. Dehaene, S. & J.P. Changeux. 1989. Neuronal models of cognitive functions. Cognition **33:** 63–109.
23. Dehaene, S. & J.P. Changeux. 1991. The Wisconsin Card Sorting Test: Theoretical analysis and modeling in a neuronal network. Cerebral Cortex **1**(1): 62–79.
24. Dehaene, S. & J.P. Changeux. 1997. A hierarchical neuronal network for planning behavior. Proc. Natl. Acad. Sci. USA **94:** 13293–13298.
25. White, N.M. 1996. Addictive drugs as reinforcers: Multiple partial actions on memory systems. Addiction **91**(7): 921–949.
26. Rose, J.E. & W.A. Corrigall. 1997. Nicotine self-administration in animals and humans—similarities and differences. Psychopharmacology **130**(1): 28–40.
27. Pontieri, F.E., G. Tanda, F. Orzi & G. Di Chiara. 1996. Effects of nicotine on the nucleus accumbens and similarity to those of addictive drugs. Nature **382**(6588): 255–257.
28. Merlo Pich, E., S.R. Pagliusi, M. Tessari, D. Talabot-Ayer, R. Hooft van Huijsduijnen & C. Chiamulera. 1997. Common neural substrates for the addictive properties of nicotine and cocaine. Science **275**(5296): 83–86.
29. Benwell, M.E., D.J. Balfour & H.M. Lucchi. 1993. Influence of tetrodotoxin and calcium on changes in extracellular dopamine levels evoked by systemic nicotine. Psychopharmacology **112**(4): 467–474.
30. Nisell, M., G.G. Nomikos & T.H. Svensson. 1994. Systemic nicotine-induced dopamine release in the rat nucleus accumbens is regulated by nicotinic receptors in the ventral tegmental area. Synapse **16**(1): 36–44.
31. Corrigall, W.A., K.M. Coen & K.L. Adamson. 1994. Self-administered nicotine activates the mesolimbic dopamine system through the ventral tegmental area. Brain Res. **653**(1–2): 278–284.
32. Henningfield, J., K. Miyasato & D. Jasinski. 1983. Cigarette smokers self-administer intravenous nicotine. Pharmacol. Biochem. Behav. **19**(5): 887–890.
33. Pidoplichko, V., M. DeBiasi, J.T. Williams & J.A. Dani. 1997. Nicotine activates and desensitizes midbrain dopamine neurons. Nature **390**(6658): 401–404.

Development of a Novel Class of Subtype-Selective Nicotinic Receptor Antagonist: Pyridine-*N*–Substituted Nicotine Analogs

LINDA P. DWOSKIN,[a] LINCOLN H. WILKINS, JAMES R. PAULY, AND PETER A. CROOKS

Department of Pharmaceutical Sciences, College of Pharmacy, University of Kentucky, Rose Street, Lexington, Kentucky 40536-0082, USA

In spite of the extensive diversity in neuronal nicotinic receptors, only a limited number of tools with receptor subtype selectivity are available to study the pharmacology of native receptors. Primarily, efforts have been directed toward development of novel nicotinic receptor *agonists* as targets for drug discovery (see Ref. 1). A limited number of nicotinic receptor antagonists are available, but most of these are noncompetitive open-channel blockers (e.g., mecamylamine).[2] Dihydro-β-erthyroidine (DHβE) is a commonly used competitive antagonist that blocks the agonist binding site of neuronal nicotinic receptors, but has no inherent receptor subtype-selectivity.[3–5] Neuronal-bungarotoxin (*n*-BTX) may be a selective, competitive antagonist at α3β2 receptors as evidenced by potent antagonism of nicotinic responses in oocytes expressing these receptors; however, while *n*-BTX is inactive at α3β4 and α2β2 receptors, it has an intermediate potency at α4β2 receptors.[6] More recently, α3β2 receptor subtype selectivity has been reported for the competitive antagonist, α-conatoxin MII.[7] However, limited availability and high cost of these toxins limit their usefulness. The focus of the present work is the development of subtype-selective nicotinic receptor antagonists via facile pyridine-*N* alkylation of the nicotine (NIC) molecule, which will provide useful tools for elucidating structural/functional diversity of native nicotinic receptors in brain.

A series of quaternary, pyridine-*N*-*n*-alkyl nicotinium analogs with alkyl substituents ranging from C_1 to C_{12} were synthesized and tested for their ability to inhibit NIC-evoked [^{3}H]dopamine ([^{3}H]DA) release from [^{3}H]DA-preloaded striatal slices using the methods of Dwoskin and Zahniser,[8] to inhibit [^{3}H]NIC binding to rat striatal membranes using the methods of Crooks *et al.*,[5] and to inhibit NIC-evoked $^{86}Rb^+$ efflux from rat striatal synaptosomes using the methods of Marks *et al.*[9] Results indicate that increasing the alkyl chain length from C_1 to C_9 directly correlated with increasing potency to inhibit NIC-evoked [^{3}H]dopamine release, purportedly mediated by the α3β2 receptor subtype. The C_{10} analog, *N*-*n*-decylnicotinium iodide (NDNI), did not inhibit NIC-evoked [^{3}H]DA release. Schild analysis with the C_8 analog, *N*-*n*-octylnicotinium iodide (NONI) indicated a competitive interaction with this nicotinic receptor site and the potential involvement of more than one subtype. Assays for [^{3}H]NIC binding to striatal membranes and $^{86}Rb^+$ efflux from striatal synaptosomes were used to assess interaction of these analogs with the α4β2 nicotinic receptor subtype. All analogs displaced [^{3}H]NIC binding; however, no correlation was found between alkyl chain length and binding affinity. Both NDNI and NONI competitively displaced [^{3}H]NIC binding with Ki values of 60 nM and 20 μM, respec-

[a]Corresponding author. Phone: 606-257-4743; fax: 606-257-7564; e-mail: ldwoskin@pop.uky.edu

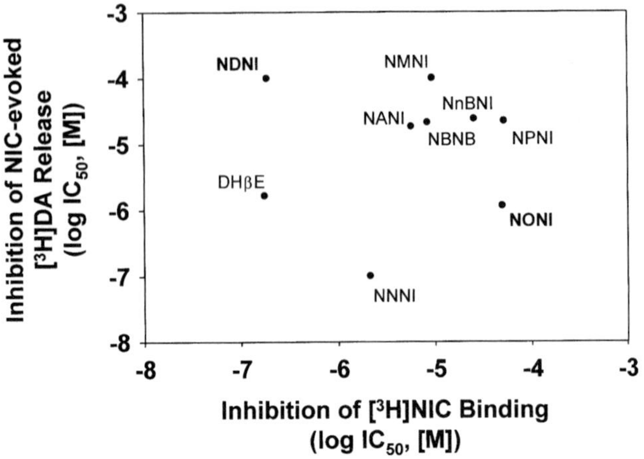

FIGURE 1. Lack of correlation of DHβE and pyridine-*N*–substituted nicotine analog-induced inhibition of nicotine-evoked [^{3}H]dopamine release and displacement of [^{3}H]nicotine binding in rat striatum.

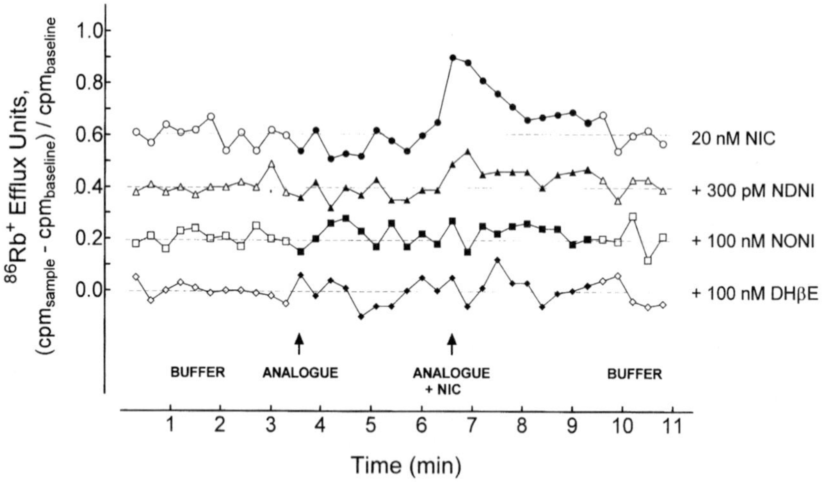

FIGURE 2. Representative data illustrating NDNI, NONI, and DHβE inhibition of nicotine-evoked $^{86}Rb^+$ efflux from rat striatal synaptosomes. Striatal synaptosomes were superfused with buffer for 3 min, followed by 3-min superfusion in the absence or presence of NDNI (300 pM), NONI (100 nM), or DHβE (100 nM), followed by 3-min superfusion with nicotine (20 nM) added to the buffer in the absence and presence of antagonist. *Filled symbols* indicate samples collected during the period of drug exposure. Data are expressed as $^{86}Rb^+$ efflux units, which represent cpms collected in superfusate samples after subtracting basal efflux and expressing this value as a percentage of basal efflux. Each curve has been offset by 0.2 units from the curve below for ease of presentation.

tively. FIGURE 1 illustrates the lack of correlation between analog-induced displacement of [^{3}H]NIC binding and inhibition of NIC-evoked [^{3}H]DA release, suggesting that different nicotinic receptor subtypes are involved in these two assays. FIGURE 1 also illustrates that NONI was the most potent, and at the same time the most efficacious antagonist in the [^{3}H]DA release assay, while it was the least potent in the [^{3}H]NIC binding assay. Thus, a more than 100-fold greater concentration of NONI was required to displace [^{3}H]NIC binding compared to inhibition of [^{3}H]DA release, suggesting a selective interaction with the nicotinic $\alpha3\beta2$ subtype. Furthermore, NDNI was most potent in the binding assay and was completely ineffective in inhibiting NIC-evoked [^{3}H]DA release, suggesting selectivity for the $\alpha4\beta2$ subtype, purportedly the NIC binding site. Interestingly, DHβE was equipotent in inhibiting NIC-evoked [^{3}H]DA release and displacing [^{3}H]NIC binding, indicating a lack of nicotinic receptor subtype selectivity. NDNI and NONI were also examined in the NIC-evoked $^{86}Rb^+$ efflux assay, and Ki values were 30 pM and 85 nM, respectively. FIGURE 2 illustrates representative results of maximally effective concentrations of NDNI, NONI, and DHβE to inhibit NIC (20 μM) -evoked $^{86}Rb^+$ efflux. As in the [^{3}H]NIC binding assay, NDNI inhibited NIC-evoked $^{86}Rb^+$ efflux at concentrations 3–4 orders of magnitude higher than did NONI. Taken together, these results support a selective interaction for NDNI with $\alpha4\beta2$ nicotinic receptors, and for NONI with $\alpha3\beta2$ receptors. These subtype-selective nicotinic receptor antagonists will undoubtedly be valuable tools for probing the consequences of activating different subtypes of nicotinic receptor.

ACKNOWLEDGMENT

This work was supported by Public Health Service Grant DA 10934 from the National Institute on Drug Abuse and the Tobacco and Health Research Institute (Lexington, KY).

REFERENCES

1. HOLIDAY, M.W., M.J. DART & J.K. LYNCH. 1997. Neuronal nicotinic acetylcholine receptors as targets for drug discovery. J. Med. Chem. **40:** 4169–4193.
2. PENG, O., V. GERZANICH, R. ANAND, P.J. WHITING & J. LINDSTROM. 1994. Nicotine-induced increase in the neuronal nicotinic receptors results from a decrease in rate of receptor turnover. Mol. Pharmacol. **46:** 523–530.
3. GRADY, S.R., M.J. MARKS, S. WONNACOTT & A.C. COLLINS. 1992. Characterization of nicotinic receptor mediated [^{3}H]dopamine release from synaptosomes prepared from mouse striatum. J. Neurochem. **59:** 848–856.
4. CROOKS, P.A., A. RAVARD, L.H. WILKINS, L.H. TENG, S.T. BUXTON & L.P. DWOSKIN. 1995. Inhibition of nicotine-evoked [^{3}H]dopamine release by pyridino N-substituted nicotine analogues: A new class of nicotinic antagonists. Drug Dev. Res. **36:** 91–102.
5. TENG, L.H., P.A. CROOKS, S.T. BUXTON & L.P. DWOSKIN. 1997. Nicotinic receptor mediation of *S*(-)nornicotine-evoked [^{3}H]overflow from rat striatal slices preloaded with [^{3}H]dopamine. J. Pharmacol. Exp. Ther. **283:** 778–787.
6. LUETJE, C.W., M. PIATTONI & J. PATRICK. 1993. Mapping of ligand binding sites of neuronal nicotinic acetylcholine receptors using chimeric alpha subunits. Mol. Pharmacol. **44:** 657–666.
7. CARTIER, G.E., D. YOSHIKAMI, W.R. GRAY, S. LUO, B.M. OLIVERA & J.M. MCINTOSH. 1996. A new α-conotoxin which targets $\alpha3\beta2$ nicotinic acetylcholine receptors. J. Biol. Chem. **271:** 7522–7528.
8. DWOSKIN, L.P. & N.R. ZAHNISER. 1986. Robust modulation of [^{3}H]dopamine release from rat striatal slices by D-2 dopamine receptors. J. Pharmacol. Exp. Ther. **239:** 442–453.
9. MARKS, M.J., A.E. BULLOCK & A.C. COLLINS. 1995. Sodium channel blockers partially inhibit nicotine-stimulated $^{86}Rb^+$ efflux from mouse brain synaptosomes. J. Pharmacol. Exp. Ther. **274:** 833–841.

Desensitization of Nicotinic Receptors in the Central Nervous System

C. P. FENSTER, J. H. HICKS, M. L. BECKMAN, P. J. O. COVERNTON, M. W. QUICK, AND R. A. J. LESTER[a]

Department of Neurobiology, University of Alabama at Birmingham, 1719 Sixth Avenue South, Birmingham, Alabama 35294, USA

Chronic exposure to nicotine renders some CNS neuronal nicotinic acetylcholine receptors (nAChRs) "permanently" desensitized.[1,2] These long-lived inactive receptor states are likely to underlie the development of tolerance to nicotine and may be responsible for some of the more long-term addictive properties of this drug.[3,4] An understanding of nicotine dependency would benefit from knowledge of the neuronal factors that control desensitization of nAChRs.

During recordings of nAChR channel activity in excised outside-out patches from medial habenula (MHb) neurons, there is a time-dependent increase in the rate of onset of receptor desensitization, assessed from the decay of macroscopic currents produced by brief pulses of nicotine.[5] This alteration in desensitization precedes a complete rundown in channel activity.[5,6] FIGURE 1A shows that after rundown of channel activity in an outside-out patch, a partial recovery of channel function is possible, provided that nAChRs are not exposed to nicotine for a prolonged period. These data are consistent with the idea that rundown of channel activity results from a build-up of nAChRs in a nonfunctional desensitized state; this may occur because recovery from desensitization is slowed due to the loss of certain factors upon patch formation. Knowledge of the factors that control how nAChRs exit from the desensitized states of the receptor should lead to an understanding of how neurons control the level of nAChR activity. Because receptors at synapses are only likely to be exposed to neurotransmitter very transiently, the physiological importance of these slowly reached desensitized states is unknown. However, because desensitized receptor conformations have high affinities for agonist, these states will become occupied during the continuous presence of drug, for example, as occurs during smoking. Furthermore, because certain nAChRs appear to become "permanently" inactivated during chronic exposure to nicotine, we hypothesize that nAChRs become trapped in a desensitized state(s)[1,2] due to a long-term biochemical alteration.

Chronic nicotine, in addition to causing a long-term inactivation of nAChRs, causes an up-regulation in the number of high-affinity [^{3}H]nicotine ($\alpha4\beta2$ nAChR subunit-containing) binding sites.[3,4,7] In accordance with these observations, we have shown that tobacco-related levels of nicotine ($\approx$30–300 nM) will preferentially interact with the desensitized state of $\alpha4$-containing nAChRs: in *Xenopus* oocytes expressing rat $\alpha4\beta2$ nAChRs, nicotine opens channels with an EC_{50} of ≈ 10 μM, whereas the IC_{50} for inducing receptor desensitization is ≈ 60 nM. After a 30 min exposure to 300 nM nicotine, receptors recover from desensitization with a relatively slow time course.[8] Consistent with the idea that nAChRs could become trapped in the desensitized state, the rate of recovery from desensitization is strongly influenced by

[a]Corresponding author. Phone: 205-934-4483; fax: 205-934-6571; e-mail: rlester@nrc.uab.edu

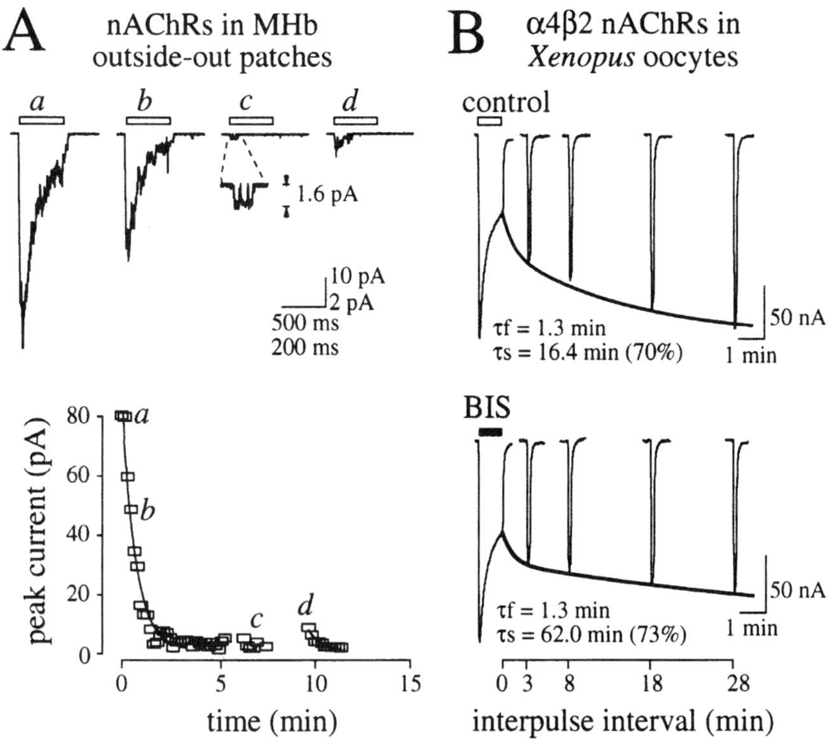

FIGURE 1. Regulation of recovery from desensitization **(A)** Time-dependent rundown in nAChR channel activity in MHb cell outside-out patches (nicotine 100 μM; 500 ms; 10-s intervals). Example response (**upper traces**) and a plot of the peak current versus time (**lower**) are shown. Immediately following patch excision (*a,b*) a large current due to the simultaneous opening of ≈50 channels could be evoked (1.6-pA single-channel current, *inset*). After 5-min recording (*c*) few or no channel openings could be detected. After ≈1-min period with no stimulation by nicotine, some recovery of channel activity was observed (*d*). **(B)** Recovery from desensitization of α4β2 receptors in *Xenopus* oocytes is slowed by inhibition of PKC. Desensitization was induced by a 1-min application of nicotine in control (**upper traces**) or after injection of BIS into the oocyte (final concentration ≈100 μM). Recovery from desensitization was followed by further brief (≈5-s) pulses of nicotine at various interpulse intervals. The *solid lines* are double exponential fits to the time course of recovery, and the respective time constants are indicated.

other factors. For example, in nominally free extracellular Ca^{2+}, although the onset of receptor desensitization is similar to control (1.8 mM Ca^{2+}), recovery is dramatically slowed.[8] We have also observed that the desensitized state of α4β2 receptors is influenced by protein kinase/phosphatase activity. Inhibition of protein kinase C by bisindolylmaleimide (BIS) and other agents, causes a dramatic slowing of recovery from desensitization (FIG. 1B). Thus, these data are consistent with the idea that both the "permanently" inactive state of α4β2 nAChRs entered during chronic nicotine treatment and rundown of channel activity in

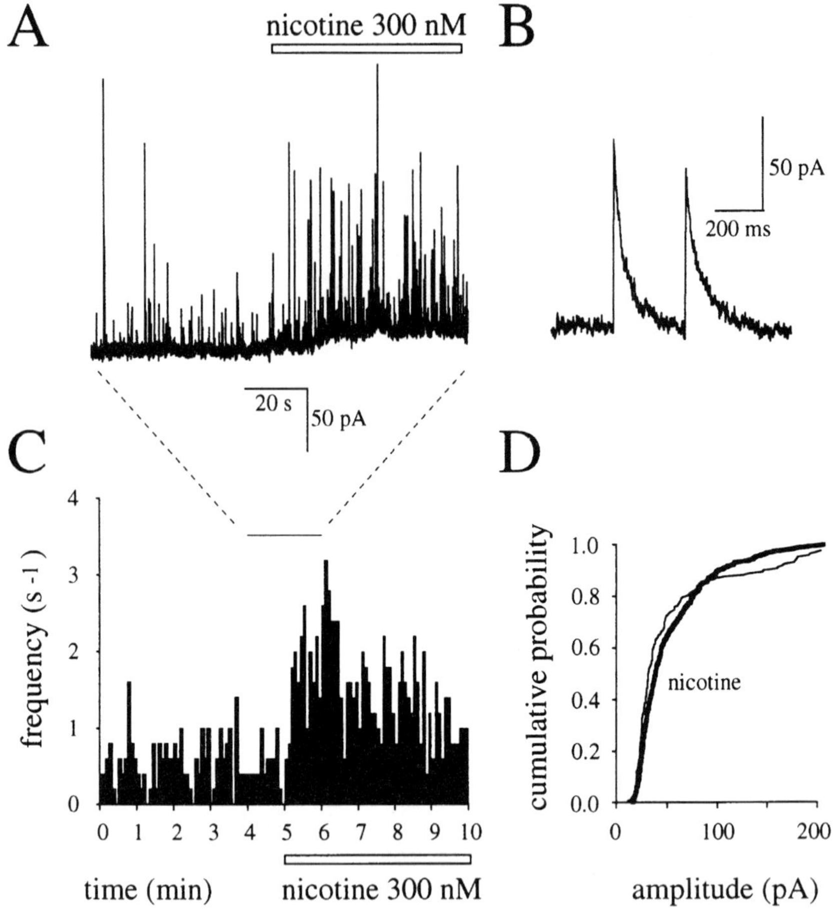

FIGURE 2. Desensitization of presynaptic nAChRs? **(A)** Spontaneous GABA postsynaptic currents (sPSCs) recorded from an interpeducular (IPN) neuron in a thin brain slice, before and during bath application of nicotine (300 nM). **(B)** sPSCs are shown on an expanded time scale. **(C)** Plot of the change in frequency of sPSCs during nicotine exposure. **(D)** Cumulative frequency graphs of sPSC amplitudes before and during nicotine.

patches may be due to dephosphorylation of nAChRs at a PKC site (see Ref. 9). The "permanently" inactive state may be the trigger for nAChR upregulation.[2]

Enhancement of neurotransmitter release by presynaptically localized nAChRs, in particular dopamine in the nucleus accumbens, is thought to be responsible for the euphoric/positive reinforcing qualities of nicotine.[10,11] In brain slices, we find that the increased frequency of spontaneous GABA postsynaptic currents, caused by continuous perfusion with low concentrations of nicotine (100–300 nM), can in some cells be very transient (FIG. 2), implying that tolerance to the rewarding aspects of tobacco abuse could be due to desensitization of presynaptic receptors. If desensitization controls up-regulation of nAChRs during chronic exposure to nicotine,[4] then these data predict that presynaptic nAChRs should be up-regulated in number.[10]

In summary, our data suggest that regulation of the desensitized states of nAChRs localized at presynaptic and postsynaptic sites will be of critical importance for control of receptor function during both physiological and nonphysiological conditions.

ACKNOWLEDGMENT

This work was supported by Public Health Service Grant NINDS NS31669.

REFERENCES

1. LUKAS, R.J. 1991. Effects of chronic nicotinic ligand exposure on functional activity of nicotinic acetylcholine receptors expressed by cells of the PC12 rat pheochromocytoma or the TE671/RD human clonal line. J. Neurochem. **56:** 1134–1145.
2. PENG, X., V. GERZANICH, R. ANAND, P.J. WHITING & J. LINSTROM. 1994. Nicotine-induced increase in neuronal receptors results from a decrease in the rate of receptor turnover. Mol. Pharmacol. **46:** 523–530.
3. MARKS, M.J., J.B. BURCH & A.C. COLLINS. 1983. Effects of chronic nicotine infusion on tolerance development and nicotinic receptors. J. Pharmacol. Exp. Ther. 226: 817-825.
4. SCHWARTZ, R.D. & K.J. KELLAR. 1985. *In vivo* regulation of [^{3}H]acetylcholine recognition sites in brain by nicotinic cholinergic drugs. J. Neurochem. **45:** 427–433.
5. LESTER, R.A.J. & J.A. DANI. 1994. Time-dependent changes in central nicotinic acetylcholine channel kinetics in excised patches. Neuropharmacology **33:** 27–34.
6. MULLE, C. & J.P. CHANGEUX. 1990. A novel type of nicotinic receptor in the rat central nervous system characterized by patch-clamp techniques. J. Neurosci. **10:** 169–175.
7. FLORES, C.M., S.W. ROGERS, L.A. PABREZA, B.B. WOLFE & K.J. KELLAR. 1992. A subtype of nicotinic cholinergic receptor in rat brain is composed of α4 and β2 subunits and is up-regulated by chronic nicotine treatment. J. Pharmacol. Exp. Ther. **41:** 31–37.
8. FENSTER, C.P., M.F. RAINS, B. NOERAGER, M.W. QUICK & R.A.J. LESTER. 1997. Influence of subunit composition on desensitization of neuronal acetylcholine receptors at low concentrations of nicotine. J. Neurosci. **17:** 5747–5759.
9. EILERS, H., E. SCHAEFFER, P.E. BICKLER & J.R. FORSAYETH. 1997. Functional deactivation of the major neuronal nicotinic receptor caused by nicotine and a protein kinase C-dependent mechanism. Mol. Pharmacol. **52:** 1105–1112.
10. WONNACOTT, S. 1997. Presynaptic nicotinic ACh receptors. Trends Neurosci. **20:** 92–98.
11. DANI, J.A. & S. HEINEMANN. 1996. Molecular and cellular aspects of nicotine abuse. Neuron **16:** 905–908.

Methyllycaconitine-, α-Bungarotoxin–Sensitive Neuronal Nicotinic Receptor Operates Slow Calcium Signal in Skeletal Muscle End Plate

IKUKO KIMURA[a] AND KATSUYA DEZAKI

Department of Chemical Pharmacology, Faculty of Pharmaceutical Sciences, Toyama Medical and Pharmaceutical University, 2630 Sugitani, Toyama 930-0194, Japan

Activation of neuronal-type nicotinic acetylcholine receptors (N-nAChRs) via a transmitter elicits elementary cation currents through the channels that lead to the entry of much Ca^{2+} into cells.[1–3] Long-lasting noncontractile slow Ca^{2+} (RAMIC: *r*eceptor-*a*ctivity *m*odulating *i*ntracellular *c*alcium) is generated in mouse skeletal muscles via N-nAChR,[4,5] in addition to the usual muscle-type (M-) nAChR-operated fast contractile Ca^{2+} mobilization. The sensitivities of N- and M-type nicotinic agonists and antagonists to RAMIC were compared with the fast contractile Ca^{2+} signals.

RESULTS

ACh (0.1–3 mM), (-)-nicotine (10 mM), and 1,1-dimethyl-4-phenyl-piperazinium (10 mM), locally applied onto the end plate region, elicited biphasic elevation of $[Ca^{2+}]_i$ (fast contractile Ca^{2+} transients and slow noncontractile Ca^{2+} transients measured as aequorin luminescence) in mouse diaphragm muscle cells.[6] The peak amplitude of slow Ca^{2+} transients was concentration-dependently increased by ACh (0.1–3 mM, FIG. 1A). Fast Ca^{2+} transients were generated at a lower concentration (10 μM) of ACh, and the peak amplitude reached a maximum response at approximately 30 μM.

Bath application of ACh (0.3–3 μM) to isolated single skeletal (flexor digitorum brevis) muscle cells of mice using a perfusion system elicited long-lasting elevation of $[Ca^{2+}]_i$ (measured as fluo-3 fluorescence) localized at the end plate region. Choline (0.3–3 mM) also elicited the slow Ca^{2+} signal; its maximum response was equivalent to that of ACh (FIG. 1B).

Lower concentrations of (+)-tubocurarine (70–100 nM), pancuronium (50–70 nM), dihydro-β-erythroidine (1 μM) and α-bungarotoxin (α-BgTx, 60–90 nM), which affected neither the fast Ca^{2+} transients nor twitch tension, depressed the nerve-stimulated slow Ca^{2+} signal in the presence of anti-cholinesterase neostigmine (0.3 μM). Moreover, N-nAChR antagonists methyllycaconitine (MLA, 1 μM) and hexamethonium (Hex, 0.1 mM) significantly depressed the slow Ca^{2+} signal.[7]

[a]Corresponding author. Phone: +81-764-34-2281; fax: +81-764-34-5045; e-mail: ikukokim@ms.toyama-mpu.ac.jp

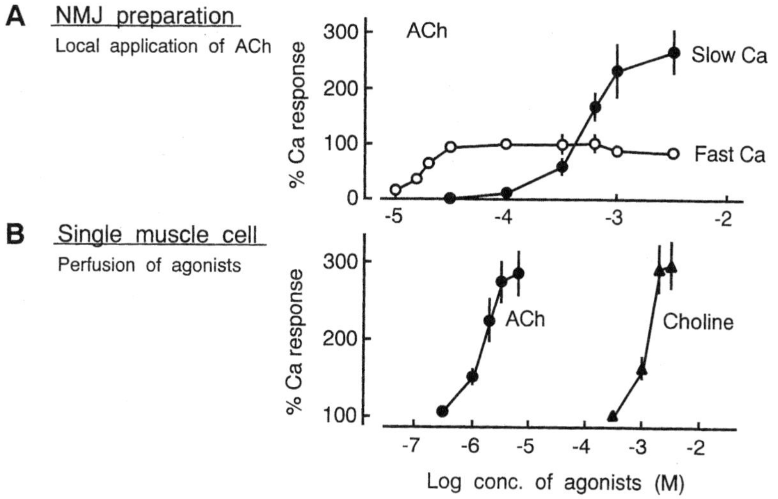

FIGURE 1. RAMIC signal and agonist features. **A.** Log concentration-response curves of ACh for the peak amplitudes of fast and slow Ca^{2+} (RAMIC) signals (measured as aequorin luminescence) elicited by ACh locally applied onto the end plate region of mouse diaphragm muscle preparations. Data (means ± SEM) are expressed as percentages of ACh (0.1 mM)–induced fast Ca^{2+} transients. **B.** Choline as a full agonist for RAMIC signal measured as fluo-3 fluorescence at mouse flexor digitorum brevis muscle cells. Data (means ± SEM) are expressed as percentages of resting fluorescence intensity.

DISCUSSION

The present study indicates that noncontractile slow Ca^{2+} (RAMIC) signal has low sensitivity to ACh, whereas it is more potently inhibited by nAChR antagonists than the fast Ca^{2+} signal associated with the usual M-nAChR–operated excitation-contraction coupling. RAMIC may be operated by N-nAChR because choline, an agonist at some N-nAChR subtypes, acts as full agonist and the N-nAChR antagonists MLA and Hex selectively depress RAMIC. Thus, the N-nAChR subtype may postsynaptically operate the slow Ca^{2+} response.

The RAMIC signal depresses the contractile Ca^{2+} mobilization elicited by paired-pulse nerve stimulation via protein kinase–C activation,[8] suggesting that one of physiological roles of RAMIC via the highly Ca^{2+} -permeable N-nAChR subtype may stabilize the usual M-nAChR in refractory states and protect it from overexcitation at the motor end plate.

In conclusion, our present study strongly suggests that postsynaptic MLA-, α-BgTx–sensitive N-nAChR subtype operates RAMIC signal at the neuromuscular junction.

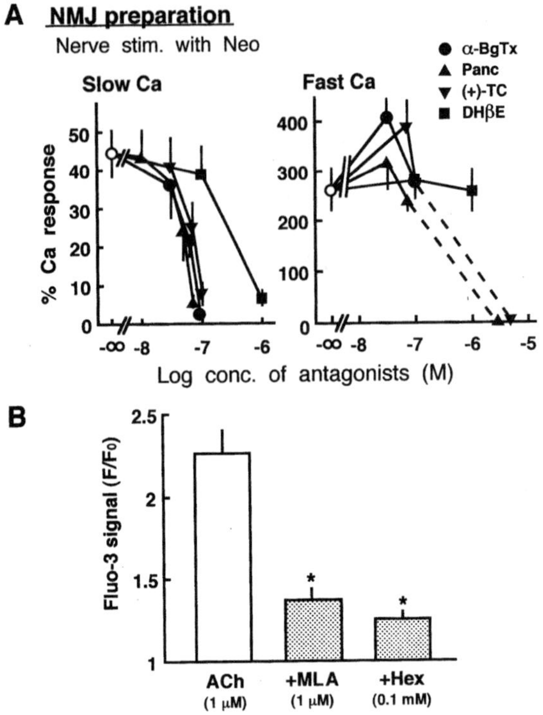

FIGURE 2. RAMIC signal and antagonist features. **A.** Log concentration-response curves of nAChR antagonists on the peak amplitudes of the fast Ca^{2+} transients and the RAMIC signals. Data (means ± SEM) are expressed as percentages of the control response obtained before the bath application of antagonists. **B.** The MLA- or hexamethonium-induced depression of RAMIC signal elicited by ACh perfusion. The fluorescence intensity was quantitated at the end plate and expressed as a ratio (F/F_0; means ± SEM) of peak response (F) to resting intensity (F_0). $^*p < 0.01$; significantly different from ACh alone by Student's *t*-test.

REFERENCES

1. Mulle, C. *et al.* 1992. Calcium influx through nicotinic receptor in rat central neurons: Its relevance to cellular regulation. Neuron **8:** 135–143.
2. Seguela, P. *et al.* 1993. Molecular cloning, functional properties, and distribution of rat brain α_7: A nicotinic cation channel highly permeable to calcium. J. Neurosci. **13:** 596–604.
3. Vernino, S. *et al.* 1992. Calcium modulation and high calcium permeability of neuronal nicotinic acetylcholine receptors. Neuron **8:** 127–134.
4. Kimura, I. *et al.* 1990. Postsynaptic nicotinic ACh receptor-operated Ca^{2+} transients with neostigmine in phrenic nerve-diaphragm muscles of mice. Brain Res. **507:** 309–311.
5. Kimura, I. *et al.* 1994. Monoclonal antibody to β2 subunit of neuronal nicotinic receptor depresses the postjunctional non-contractile Ca^{2+} mobilization in the mouse diaphragm muscle. Neurosci. Lett. **180:** 101–104.
6. Dezaki, K. & Kimura, I. 1998. Acetylcholine sensitivity of biphasic Ca^{2+} mobilization induced by nicotinic receptor activation at the mouse skeletal muscle endplate. Br. J. Pharmacol. **123:** 1418–1424.

7. Kimura, I. 1998. Calcium-dependent desensitizing function of the postsynaptic neuronal-type nicotinic acetylcholine receptors at the neuromuscular junction. Pharmacol. Ther. **77:** 183–202.
8. Kimura, I. *et al.* 1995. Postsynaptic nicotinic receptor desensitized by non-contractile Ca^{2+} mobilization via protein kinase-C activation at the mouse neuromuscular junction. Br. J. Pharmacol. **114:** 461–467.

Ultrastructural Immunolocalization of the α7 nAChR Subunit in Guinea Pig Medial Prefrontal Cortex

MONA LUBIN,[a] ALEV ERISIR, AND CHIYE AOKI

Center for Neural Science, New York University, 4 Washington Place, Room 809, New York, New York 10003, USA

The alpha 7 nicotinic cholinergic receptor subunit (α7 nAChR) is one of 11 genes for neuronal nicotinic receptor subunits cloned to date (see Refs. 1 and 2 for reviews). Shown to form functional homomeric receptor-channels when expressed in *Xenopus* oocytes, currents recorded from the α7 nAChR subunit are fast, rapidly desensitize, and show high permeability to calcium and nanomolar affinity to an antagonist derived from the snake toxin α-bungarotoxin (see Ref. 3 and references therein). The functions of the α7 nAChR subunit vary considerably among brain regions and cell types. For example, evidence for α7-mediated presynaptic enhancement of both glutamate and GABA release has been found in the hippocampus[4] and elsewhere,[5,6] while enhancement of firing via a postsynaptic mechanism has been found in the hippocampus only in interneurons.[7] In pyramidal neurons of the somatosensory cortex, α7 nAChR selectively enhances thalamocortical synaptic potentials,[8] suggesting that thalamocortical terminals preferentially contain the subunit or that pyramidal neurons preferentially locate the α7 nAChR subunit apposing thalamic input. Since labeling of axons may not be detectable by light microscopy due to limited resolution, and since no previous ultrastructural studies have been performed in cortex or have indicated a synaptic location for the α7 nAChR subunit elsewhere,[9] we sought to determine the pre- versus postsynaptic locations of the α7 nAChR subunit in the cerebral cortex, and their inclusion at glutamatergic versus GABAergic synapses by using dual immunocytochemistry at the electron microscopic (EM) level. Cortical tissue was therefore immunolabeled for the α7 nAChR subunit with HRP-DAB (for maximum detectability) and/or immunogold (for more precise localization) and was analyzed along with tissue dually immunolabeled for the α7 nAChR subunit and GABA.

FIGURE 1. Immunoreactivity for the α7 nAChR subunit in perikarya and dendrites occurs within the cytoplasm and at the plasma membrane. **(A)** Silver-intensified gold (SIG) particles (*arrowheads*) mark α7 nAChR subunit immunoreactivity within the perikaryal cytoplasm (cyt) of a pyramidal neuron, adjacent to the smooth endoplasmic reticulum (ser) and at the nuclear envelope (NE). A symmetric synaptic junction is formed by a labeled terminal (T) onto the soma of this neuron. **(B)** Concentrated HRP-DAB labeling (*open arrows*) is shown within the perikaryal cytoplasm (Cyt) of a pyramidal neuron apposed to an unlabeled terminal (UT). **(C)** An SIG-labeled dendrite (D_1) shows α7 nAChR subunit immunoreactivity within the shaft and along the membrane. A smaller-caliber dendrite in cross-section (D_2) demonstrates a similar labeling pattern. **(D)** A dendrite is dually immunolabeled for GABA by postembedding gold (*arrowheads*) and for α7 nAChR subunit by HRP-DAB (*open arrows*). An apposing terminal (T) is GABA positive. *Bar* = 500 nm for (A), 200 nm for (B) and (D), and 1 μm for (C).

[a]Corresponding author. Phone: 212-998-3611; fax: 212-995-4011; e-mail: mona@cns.nyu.edu

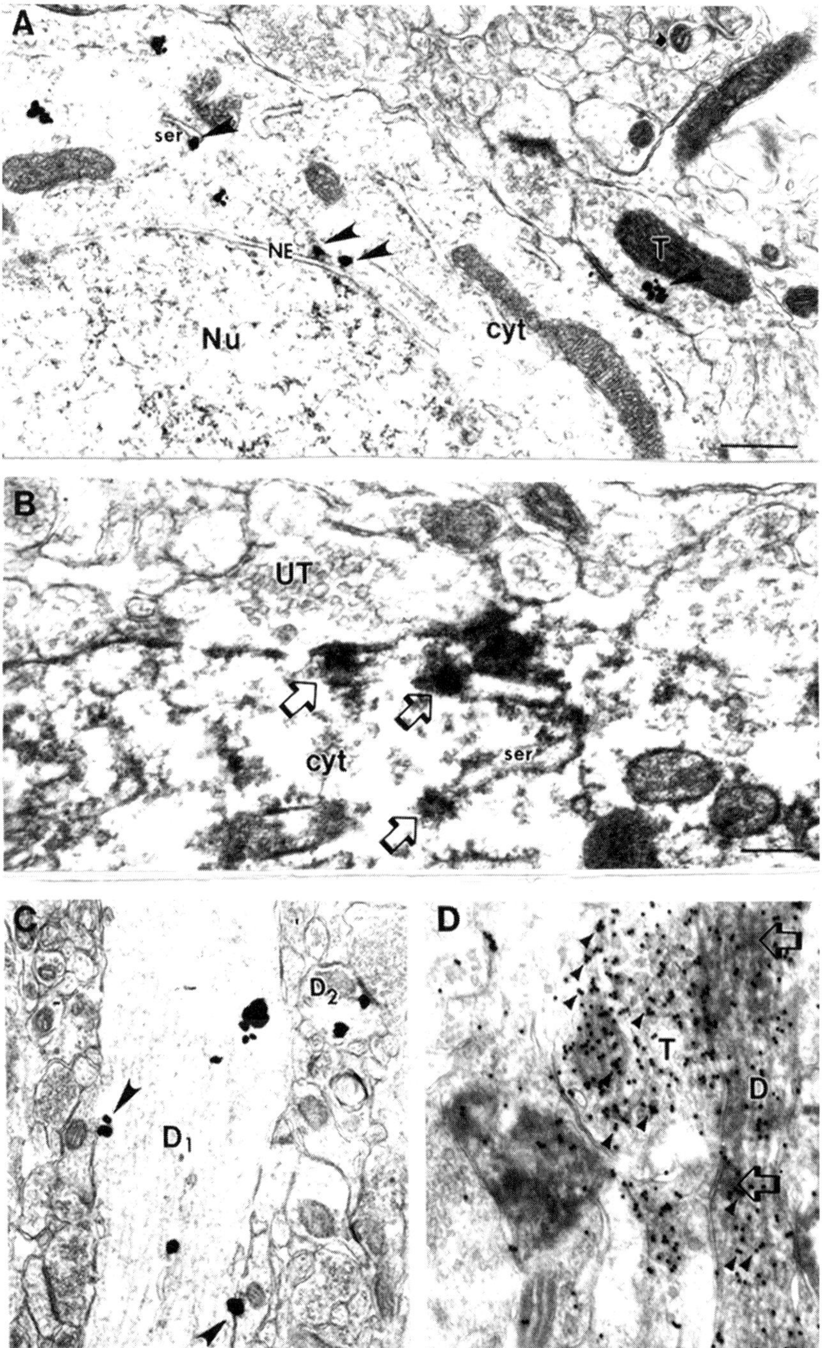
A
ser
NE
Nu
cyt
T
B
UT
cyt
ser
C
D2
D1
D
T
D

METHODS

The primary monoclonal antibody (mAB 306) (Research Biochemicals International), used to immunolabel the α7 nAChR subunit, has been shown to recognize an intracellular domain of the molecule.[1] Adult female Hartley guinea pigs were deeply anesthetized and transcardially perfused with aldehydes. Vibratome sections were collected to include the medial prefrontal cortex and processed for single labeling, either by ABC/HRP-DAB or by preembedding silver intensified gold (SIG). Tissue was dually labeled by HRP-DAB for the α7 nAChR subunit and then by postembedding gold for either the α7 nAChR subunit antibody or for a monoclonal antibody to GABA (Sigma).

RESULTS

Light microscopy (LM) revealed α7 nAChR subunit immunoreactivity in pyramidal and nonpyramidal neurons throughout layers II through VI of the medial prefrontal cortex of the guinea pig. Immunolabeling in axons and spines was not evident at this magnification. Upon EM examination, the prominent diffuse somatic labeling seen by LM was revealed as concentrated near the smooth endoplasmic reticulum (FIG. 1A, 1B), the nuclear envelope (FIG. 1A), and Golgi apparatus, and along the cytoplasmic surface of plasma membranes, often apposing terminals (FIG. 1B). Within dendrites, α7 nAChR subunit immunoreactivity was found in varying degrees within shafts and consisted of both intracellular and plasmalemmal sites (FIG. 1C). Dendritic labeling at the plasma membrane was found at sites receiving symmetric-type synaptic inputs from labeled or unlabeled terminals as well as at sites of terminal appositions, that is, those without morphologically identifiable synaptic junctions. Dual immunoelectron microscopy, where postembedding gold labeling for GABA was combined with preembedding HRP-DAB labeling for the α7 nAChR subunit, confirmed the presence of the α7 nAChR subunit on GABAergic-immunoreactive dendritic shafts (FIG. 1D) and somata. Within spines, labeling appeared clustered over the postsynaptic density (FIG. 2), the spine neck (FIG. 2A), and at the base of spines. Labeling often appeared both pre- and postsynaptically at axospinous junctions. Within terminals at axospinous junctions (which presumably are glutamatergic), immunoreactivity for the α7 nAChR subunit was found at sites near as well as away from the synaptic junction (FIG. 2). Also within axons, α7 nAChR subunit immunoreactivity occurred in terminals forming symmetric-type synaptic junctions with somata (FIG. 1A) and dendrites, suggesting that these are GABAergic.

DISCUSSION

The anatomical evidence just described supports the physiological evidence garnered to date suggesting that, as in other brain regions,[4–6,10] in cortex, the α7 nAChR subunit modulates excitatory synaptic transmission by enhancing the release of glutamate. Specific electrophysiological evidence of α7 nAChR subunit participation in the cortex, specifically at thalamocortical synapses in the rat somatosensory cortex,[8] has indeed been found. However, in addition to its location in terminals forming asymmetric-type synaptic junctions (presumably excitatory, and thus glutamatergic), the preceding data suggest

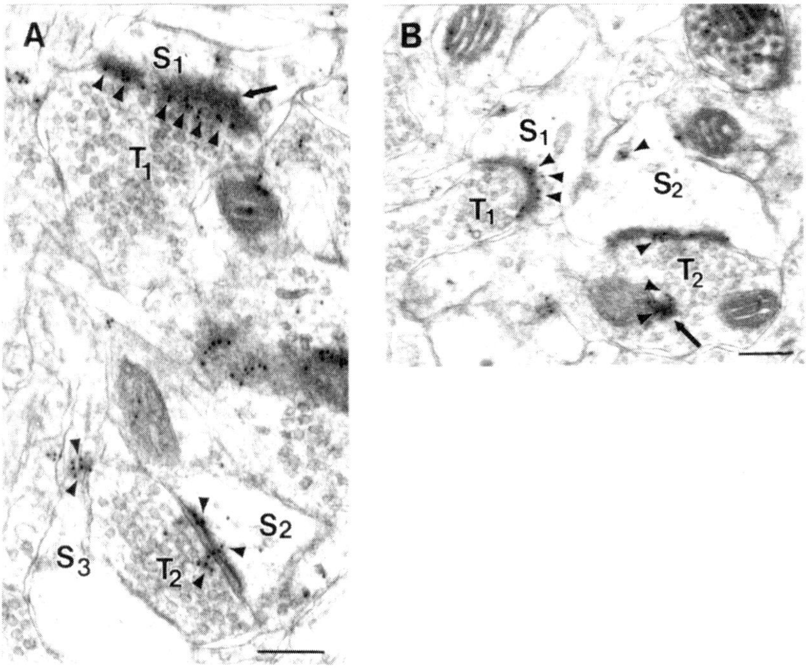

FIGURE 2. The α7 nAChR subunit occurs at axospinous junctions both pre- and postsynaptically. Tissue was immunolabeled by both HRP-DAB (*arrows*) and postembedded gold (*arrowheads*). Presynaptic labeling appears within terminals (T) at the synaptic junction (T_1, T_2 in **panel A**) and away from the junction (T_2 in **B**). The postsynaptic density is labeled in spines (S_2 in **panel A** and S_1 in **panel B**); however, it also appears within the spine neck (S_3 in **panel A**) and elsewhere within the spine head (S_2 in **panel B**). Note that at times, both pre- and postsynaptic labeling occurs at a single axospinous junction (see T_2/S_2 in **panel A**). *Bar* = 200 nm for both panels.

additional roles. First, the subunit's presence in cortical spines suggests a postsynaptic role in modulating glutamatergic synaptic transmission, possibly by depolarizing the spine to engage voltage-dependent calcium channels and/or opening NMDA receptor-channels. Second, the presence of the α7 nAChR subunit in synaptic locations on GABAergic cell dendritic shafts, suggests that the receptor subunit may also play a role in modulating the firing rate of cortical interneurons. Moreover, the α7 nAChR subunit may play a dual role of modulating inhibitory synaptic transmission since terminals forming symmetric-type synaptic junctions (presumably inhibitory, and thus GABAergic) were also immunoreactive for the α7 nAChR subunit. Thus, the α7 nAChR subunit may have multiple roles that will depend on its location within certain brain regions, cell types, and subcellular locations within those cells. It will be interesting to learn if the functions suggested by the discovery of these cortical sites of α7 nAChR subunit immunoreactivity are corroborated by future physiological investigations.

ACKNOWLEDGMENT

Funded by NIMH Training Grant MH 19524 and NIH-EY8055, NSF-RCD 92-53750, and HFSP RG-1693 to C. Aoki.

REFERENCES

1. LINDSTROM, J. 1996. Neuronal nicotinic acetylcholine receptors. Ion Channels **4:** 377–450.
2. ROLE, L.W. & D.K. BERG. 1996. Nicotinic receptors in the development and modulation of CNS synapses. Neuron **16**(6): 1077–1085.
3. SEGUELA, P., J. WADICHE, K. DINELEY-MILLER, J.A. DANI & J.W. PATRICK. 1993. Molecular cloning, functional properties, and distribution of rat brain alpha 7: A nicotinic cation channel highly permeable to calcium. J Neurosci. **13**(2): 596–604.
4. GRAY, R., A.S. RAJAN, K.A. RADCLIFFE, M. YAKEHIRO & J.A. DANI. 1996. Hippocampal synaptic transmission enhanced by low concentrations of nicotine [see comments]. Nature. **383**(6602): 713–716.
5. MCGEHEE, D.S., M.J. HEATH, S. GELBER, P. DEVAY & L.W. ROLE. 1995. Nicotine enhancement of fast excitatory synaptic transmission in CNS by presynaptic receptors [see comments]. Science **269**(5231): 1692–1696.
6. GUO, J.Z., T.L. TREDWAY & V.A. CHIAPPINELLI. 1998. Glutamate and GABA release are enhanced by different subtypes of presynaptic nicotinic receptors in the lateral geniculate nucleus. J. Neurosci. **18**(6): 1963–1969.
7. FRAZIER, C.J., Y.D. ROLLINS, C.R. BREESE, S. LEONARD, R. FREEDMAN & T.V. DUNWIDDIE. 1998. Acetylcholine activates an alpha-bungarotoxin-sensitive nicotinic current in rat hippocampal interneurons, but not pyramidal cells. J. Neurosci. **18**(4): 1187–1195.
8. GIL, Z., B.W. CONNORS & Y. AMITAI. 1997. Differential regulation of neocortical synapses by neuromodulators and activity. Neuron **19**(3): 679–686.
9. CARUNCHO, H.J., A. GUIDOTTI, J. LINDSTROM, E. COSTA & C. PESOLD. 1997. Subcellular localization of the alpha 7 nicotinic receptor in rat cerebellar granule cell layer. Neuroreport **8**(6): 1431–1433.
10. ORR-URTREGER, A., F.M. GOLDNER, M. SAEKI, I. LORENZO, L. GOLDBERG, M. DE BIASI, J.A. DANI, J.W. PATRICK & A.L. BEAUDET. 1997. Mice deficient in the alpha7 neuronal nicotinic acetylcholine receptor lack alpha-bungarotoxin binding sites and hippocampal fast nicotinic currents. J. Neurosci. **17**(23): 9165–9171.

Nicotinic Receptor Subunit mRNA Expression in Dopaminergic Neurons of the Rat Brain

FRÉDÉRIC SGARD,[a,b] ERIC CHARPANTIER,[a] PASCAL BARNÉOUD,[c] AND FRANÇOIS BESNARD[a]

Departments of [a]Genomic Biology and [c]Neurosciences, Synthélabo Recherche, 10 rue des Carrières, 92500 Rueil Malmaison, France

Nicotinic agonists have been shown to increase dopamine release both *in vivo* and *in vitro*.[1] In particular, modulation of dopaminergic neurotransmission in the mesolimbic system is thought to be responsible for the addictive properties of nicotine.[2] However, the composition of the native nicotinic acetylcholine receptors (nAChRs) involved remains unclear.

In situ hybridization studies have shown the presence of most of the known neuronal nAChR subunit mRNA (α3–6 and β2–3) in the substantia nigra (SN) and ventral tegmental area (VTA).[3] In addition, studies with β2 subunit knockout mice suggest that β2-containing nAChRs play an important role in nicotine-evoked dopamine release,[4] and pharmacological analyses indicate that at least two different receptor subtypes, one containing an α3/β2 subunit interface, may be involved.[5] However, only the α6 subunit has been recently unambiguously localized on dopaminergic neurons.[6]

In order to determine the nAChR subunits specifically expressed in dopaminergic neurons, we have used the RT-PCR technique to analyze nAChR mRNA expression in rat SN and VTA following total unilateral lesion of the dopaminergic system induced by injection of 6-hydroxydopamine (6-OHDA).

RESULTS

The results obtained from sham-operated animals show the presence of α3-7 and β2-4 subunit mRNAs in the SN and α2–7 and β2–4 subunit mRNAs in the VTA (TABLE 1). These results were identical for both hemispheres, demonstrating that the injection procedure does not affect nAChR mRNA expression. Identical results were also obtained with contralateral (uninjected) hemispheres of 6-OHDA–treated animals (FIG. 1 and TABLE 1). The presence of β4 and α7 subunit mRNA in the SN and VTA and of α2 subunit in the VTA, which had not been previously reported from *in situ* hybridization studies, probably reflects the greater sensitivity of the RT-PCR technique, suggesting that these subunits are expressed at a low level in these brain regions. Thus, recent electrophysiological studies indicate the presence of α7 receptors only within a minority of VTA dopaminergic neurons,[7] and pharmacological analyses using the nicotinic agonist cytisine also suggest that β4-containing nAChRs are involved in the facilitation of dopamine release.[1]

In contrast to the nonlesioned side, the 6-OHDA lesion resulted in a complete absence of amplification product for α3, α5, α6, and β4 subunits in the SN and for α2, α3, α5, α6,

[b]Corresponding author. Phone: + 33-141 39 13 85; fax: +33-141 39 13 04.

TABLE 1. RT-PCR Results Obtained for SN and VTA of 6-OHDA– and Sham-Injected Animals

	Brain Region	Localization	*n*	α2	α3	α4	α5	α6	α7	β2	β3	β4
6-OHDA	SN	Ipsilateral	5	–	–	+	–	–	+	+	+	–
		Contralateral	5	–	+	+	+	+	+	+	+	+
	VTA	Ipsilateral	5	–	–	+	–	–	–	+	+	–
		Contralateral	5	+	+	+	+	+	+	+	+	+
Sham	SN	Ipsilateral	5	–	+	+	+	+	+	+	+	+
		Contralateral	5	–	+	+	+	+	+	+	+	+
	VTA	Ipsilateral	5	+	+	+	+	+	+	+	+	+
		Contralateral	5	+	+	+	+	+	+	+	+	+

NOTE: Results obtained by RT-PCR for the different regions studied. *n* = number of animals tested. + corresponds to positive amplifications (5 samples out of 5 except for β3, for which 4 positive amplifications out of 5 samples were detected). – corresponds to absence of signal for all samples tested, as visualized on ethidium bromide–stained agarose gel after RT-PCR amplification.

α7, and β4 subunits in the VTA (FIG. 1 and TABLE 1). These results corroborate recent studies that have shown the probable presence of α3- and α6-containing nAChRs in dopaminergic neurons.[5,6] In addition, they demonstrate that α5, α7, and β4 subunits are exclusively present in dopaminergic neurons of the SN and VTA, and therefore suggest that nAChRs mediating dopamine release may also contain these subunits.

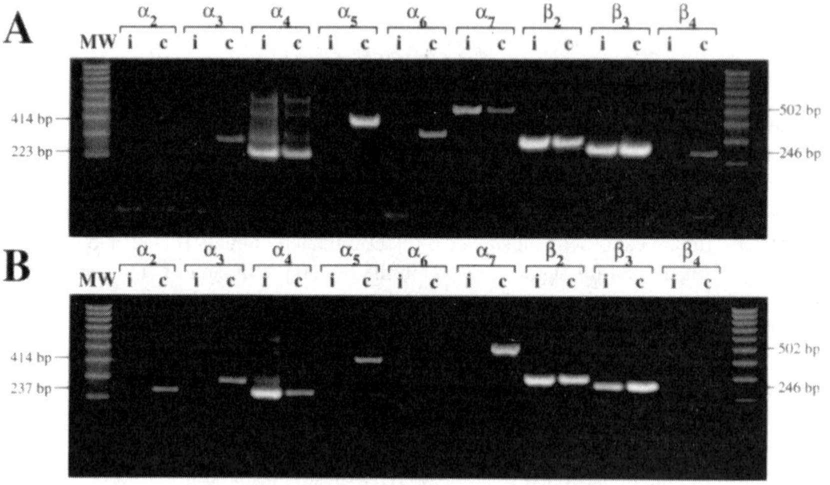

FIGURE 1. Analysis of PCR products obtained from SN (**A**) and VTA (**B**) RNA of a 6-OHDA–injected rat. Results from injected ipsilateral (i) and noninjected contralateral (c) hemispheres are presented. Subunit-specific primers used for each amplification are indicated above each lane. MW indicates molecular weight marker.

α4, α7 (in the SN), β2, and β3 subunit mRNAs were still detectable following 6-OHDA–induced lesion. While we cannot exclude the possibility that these subunits are present on dopaminergic neurons (a recent study has demonstrated the probable presence of β2-subunit–containing nAChRs in such neurons[4]), these results indicate that these subunits are expressed in nondopaminergic neurons such as GABAergic neurons and interneurons of the SN and VTA. Indeed, nAChRs thought to be composed of α4/β2 and α7 subunits have been shown to modulate GABA release in hippocampal neurons.[8] In addition, our data also indicate the presence of β3 subunit mRNA in those nondopaminergic neurons, which has been shown to associate with α4β2β4 subunits in the cerebellum.[9]

The molecular nature of native nAChRs involved in the modulation of dopaminergic neurotransmission is an object of considerable interest, particularly in relation with the known addictive properties of nicotine.[2] Using the RT-PCR methodology, we have identified the different neuronal nicotinic acetylcholine receptor subunits expressed selectively in the dopaminergic neurons of the SN and VTA and provided direct evidence for the heterogeneity of nicotinic receptor subtypes in dopaminergic neurons. The molecular characterization of these receptors will help us to understand the relative contribution of different nAChRs in the modulation of the physiological processes taking place in midbrain dopaminergic systems and thus contribute to the development of novel agents for the treatment of pathologies affecting these systems.

REFERENCES

1. Wonnacott, S. 1997. Presynaptic nicotinic ACh receptors. Trends Neurosci. **20:** 92–98.
2. Dani, J.A. *et al.* 1996. Molecular and cellular aspects of nicotine abuse. Neuron **16:** 905–908.
3. LeNovere, N. *et al.* 1996. Neuronal nicotinic receptor α6 subunit mRNA is selectively concentrated in catecholaminergic nuclei of the rat brain. Eur. J. Neurosci. **8:** 2428–2439.
4. Picciotto, M.R. *et al.* 1998. Acetylcholine receptors containing the β2 subunit are involved in the reinforcing properties of nicotine. Nature **391:** 173–177.
5. Kaiser, S.A. *et al.* 1998. Differential inhibition by α-conotoxin-MII of the nicotinic stimulation of [^{3}H]dopamine release from rat striatal synaptosomes and slices. J. Neurochem. **70:** 1069–1076.
6. Goldner, F.M. *et al.* 1997. Immunohistochemical localization of the nicotinic acetylcholine receptor subunit α6 to dopaminergic neurons in the substantia nigra and ventral tegmental area. Neuroreport **8:** 2739–2742.
7. Pidoplichko, V.I. *et al.* 1997. Nicotine activates and desensitizes midbrain dopamine neurons. Nature **390:** 401–404.
8. Alkondon, M. *et al.* 1997. Neuronal nicotinic acetylcholine receptor activation modulates γ-aminobutyric acid release from CA1 neurons of rat hippocampal slices. J. Pharmacol. Exp. Ther. **283:** 1396–1411.
9. Forsayeth, J.R. *et al.* 1997. Formation of oligomers containing the β3 and β4 subunits of the rat nicotinic receptor. J. Neurosci. **17:** 1531–1538.

Physostigmine and Atropine Potentiate and Inhibit Neuronal α4β4 Nicotinic Receptors

R. ZWART, R. G. D. M. VAN KLEEF, AND H. P. M. VIJVERBERG

Research Institute of Toxicology, Utrecht University, P.O. Box 80.176, NL-3508 TD Utrecht, the Netherlands

The muscarinic acetylcholine receptor antagonist atropine is able to potentiate and to inhibit neuronal α4β4 nicotinic acetylcholine receptor–mediated responses (nAChRs) by competitive and noncompetitive mechanisms.[1] The concentration-dependent effects of atropine on inward currents evoked by a low ACh concentration result in a bell-shaped concentration-effect curve (FIG. 1A). The potentiating effect of atropine is surmounted by elevating the concentration of ACh, indicating that the potentiation is due to a competitive interaction of atropine with the nAChR. The inhibitory effect of atropine is predominantly due to noncompetitive, voltage-dependent ion channel block. Steady state effects of ACh and atropine are accounted for by a model for combined receptor occupation and channel block (FIG. 2), in which atropine acts on two distinct sites. The first is associated with competitive potentiation, which occurs when one of the agonist recognition sites of the receptor is occupied by ACh and the other agonist recognition site is occupied by atropine. The second site is associated with noncompetitive ion channel block.[1]

The acetylcholinesterase inhibitor physostigmine exerts agonistic and antagonistic effects on nAChR. Physostigmine induces nAChR-mediated single-channel openings in primary cultured rat hippocampal neurons, which are thought to be due to binding of physostigmine to a site on the nAChR distinct from the ACh recognition site. Interestingly, no physostigmine-induced whole-cell currents are observed in vertebrate cells.[2] Recently, we have observed physostigmine-induced nAChR-mediated whole-cell ion currents in insect neurons. In addition, physostigmine inhibits ACh-induced ion current in these cells. Block

FIGURE 1. **(A)** Effects of atropine on α4β4 nAChR-mediated ion currents. **Left:** Effects of various concentrations of atropine on ion currents induced by 1 μM ACh. *Horizontal bars*, application of ACh and atropine. At concentrations of 0.1–3 μM, atropine potentiates the ACh-induced ion current, whereas at higher atropine concentrations the initial potentiation is followed by inhibition of ACh-induced ion current. **Right:** Concentration dependence of the steady state effects of atropine on α4β4 nAChR-mediated ion current. Steady state effects were determined as the steady current level in the presence of atropine relative to the current level just before atropine application. Values are mean ± SD for four oocytes. *Bell-shaped curve*, fit of the model (see FIG. 2) combining receptor occupation and ion channel block by atropine to the data. **(B)** Effects of physostigmine on α4β4 nAChR-mediated ion currents. **Left:** Effects of various concentrations of physostigmine on ion currents induced by 1 μM ACh. *Horizontal bars*, application of ACh and physostigmine. At concentrations of 0.1–10 μM, physostigmine potentiates the ACh-induced ion current, whereas at higher physostigmine concentrations the initial potentiation is followed by inhibition of ACh-induced ion current. **Right:** Concentration dependence of the steady state effects of physostigmine on α4β4 nAChR-mediated ion current. Steady state effects were determined as the steady current level in the presence of physostigmine relative to the current level just before physostigmine application. Values are mean ± SD for three oocytes. *Bell-shaped curve*, fit of the receptor occupation model (see FIG. 2) without ion channel block by physostigmine to the data.

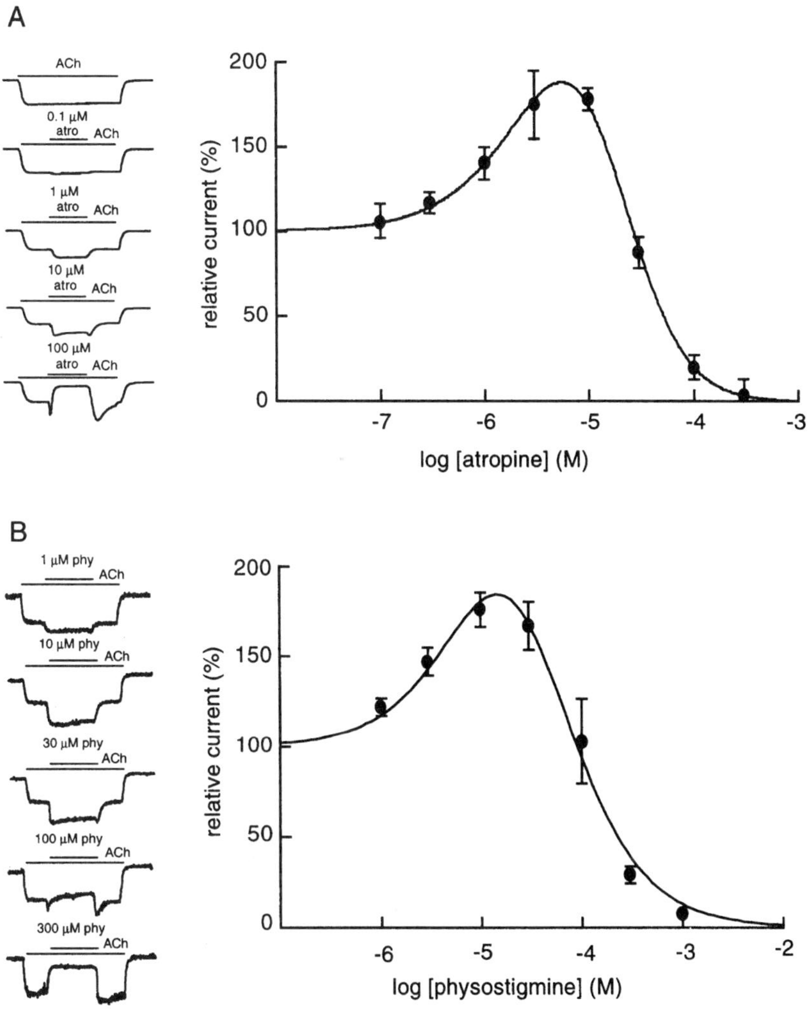
A
ACh
0.1 µM atro ACh
1 µM atro ACh
10 µM atro ACh
100 µM atro ACh
relative current (%)
200
150
100
50
0
-7
-6
-5
-4
-3
log [atropine] (M)
B
1 µM phy ACh
10 µM phy ACh
30 µM phy ACh
100 µM phy ACh
300 µM phy ACh
relative current (%)
200
150
100
50
0
-6
-5
-4
-3
-2
log [physostigmine] (M)

of physostigmine-induced ion currents by competitive nAChR antagonists suggests that, unlike in vertebrate cells, physostigmine directly interacts with the agonist recognition site of insect nAChR.[3] In the present study we have investigated effects of physostigmine on rat α4β4 nAChRs expressed in *Xenopus* oocytes, and have compared these effects to those of atropine.

In oocytes voltage clamped at −40 mV, ion currents were evoked by superfusion with external solution containing a low concentration of 1 μM ACh. Superfusion of oocytes with physostigmine alone did not induce detectable inward currents, indicating that physostigmine is not an agonist of α4β4 nAChR. However, potentiation of ACh-induced ion currents by 1–10 μM physostigmine is observed, whereas higher concentrations of physostigmine initially

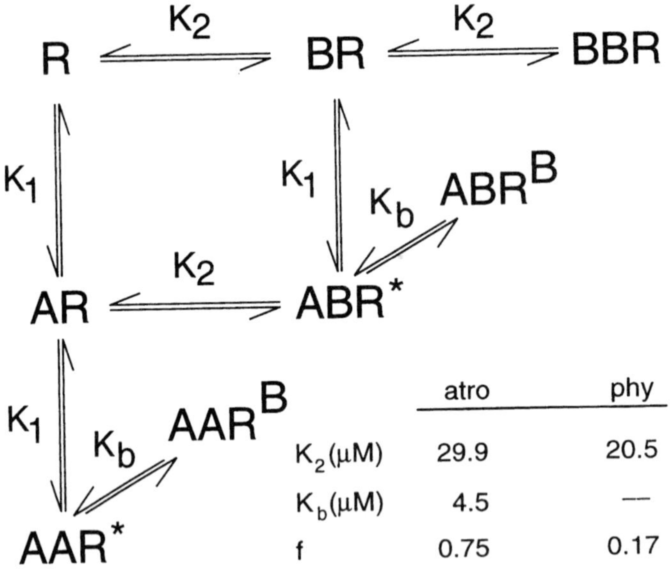

	atro	phy
K_2(μM)	29.9	20.5
K_b(μM)	4.5	—
f	0.75	0.17

FIGURE 2. Eight-state model to explain the equilibrium effects of atropine and physostigmine on ACh-induced ion current mediated by the α4β4 nAChR. R, A, and B denote the free receptor, ACh, and atropine or physostigmine, respectively. Occupation of two agonist recognition sites by ACh (AAR*), as well as occupation of one site by ACh and a second site by atropine or physostigmine (ABR*), result in activated receptor states. The conformational change from the activated states (marked by *asterisks*) to the open states is omitted. Occupation of all agonist recognition sites by atropine or physostigmine (BBR) does not lead to an activated receptor, and open ion channels may be blocked by atropine (AAR^B and ABR^B). The ion current amplitude as a function of steady state occupancy of the AAR* and ABR* states combined with ion channel block is described by

$$I(x_A, x_B) \propto [(c_A^2 + 2fc_Ac_B)/(1 + c_A + c_B)^2] \cdot [1/(1 + x_B/K_b)] \ ,$$

where x_A and x_B are the ACh and atropine or physostigmine concentrations, respectively; c_A represents x_A/K_1 ($K_1 = 16.9\ \mu M^1$); and c_B represents the normalized atropine or physostigmine concentration (x_B/K_2); f is a factor to account for a difference in efficacy of ACh and atropine or physostigmine; and K_b is the apparent affinity of atropine for ion channel block. All ion current amplitudes were normalized to control amplitudes in the absence of atropine or physostigmine ($I(x_A, 0) = [c_A/(1 + c_A)]^2$) and multiplied by 100 to obtain percentages as described by:

$$I(x_A, x_B) = 100 \cdot \{[(1 + c_A)/c_A]^2 \cdot [(c_A^2 + 2fc_Ac_B)/(1 + c_A + c_B)^2]\} \cdot [1/(1 + x_B/K_b)] \ .$$

potentiate and successively inhibit the ACh-induced ion currents. The potentiating effect of physostigmine is surmounted by elevating the ACh concentration (not shown), indicating a competitive interaction of physostigmine with the nAChR. The steady state effects of physostigmine on ion currents induced by the low concentration of ACh result in a bell-shaped concentration-effect curve (FIG. 1B), similar to that obtained for atropine (FIG. 1A). The main difference is that the inhibitory part of the physostigmine curve has a more shallow slope than that of the atropine curve. This suggests that ion channel block is less pronounced for physostigmine than for atropine. The physostigmine data could be fitted adequately by the atropine model, omitting open-channel block. Curve fitting yielded estimates for the apparent affinity of physostigmine for the agonist recognition site (K_2) of 20.5 μM, and for the efficacy factor (f) of 0.17 (FIG. 2). The small value of the factor f is in agreement with short open times of physostigmine-induced single-channel events as compared to ACh-evoked single-channel events.[2] However, the present results do allow for "agonistic" effects of physostigmine on α4β4 nAChRs only in the presence of low concentrations of ACh. Whether and to what extent physostigmine is able to cause open-channel block is currently being investigated.

It is concluded that our model to account for the mechanisms of potentiation and inhibition of α4β4 nAChRs by atropine may also account for the effects of physostigmine. The model predicts that interaction of physostigmine with the agonist recognition sites on the α4β4 nAChR causes potentiation of ACh-induced inward current at low agonist concentrations and mainly competitive inhibition at high agonist concentrations.

ACKNOWLEDGMENT

This work was supported by the Netherlands Organization for Scientific Research (N.W.O. #903.42.011).

REFERENCES

1. ZWART. R. & H.P.M. VIJVERBERG. 1997. Potentiation and inhibition of neuronal nicotinic receptors by atropine: Competitive and noncompetitive effects. Mol. Pharmacol. **52:** 886–895.
2. PEREIRA. E.F., S. REINHARDT-MAELICKE, A. SCHRATTENHOLZ, A. MAELICKE, & E.X. ALBUQUERQUE. 1993. Identification and functional characterization of a new agonist site on nicotinic acetylcholine receptors of cultured hippocampal neurons. J. Pharmacol. Exp. Ther. **265:** 1474–1491.
3. VAN DEN BEUKEL, I., R.G.D.M. VAN KLEEF, R. ZWART & M. OORTGIESEN. 1998. Physostigmine and acetylcholine differentially activate nicotinic receptor subpopulations in *Locusta migratoria* neurons. Brain Res. **789:** 263–273.

The Long Cytoplasmic Loop of the α3 Subunit Targets Specific nAChR Subtypes to Synapses on Neurons *in Vivo*

BRIAN M. WILLIAMS,[a] MURALI KRISHNA TEMBURNI,[a] SONIA BERTRAND,[b] DANIEL BERTRAND,[b] AND MICHELE H. JACOB[a,c]

[a]*Department of Neuroscience, Tufts University, School of Medicine, 136 Harrison Avenue, Boston, Massachusetts 02111, USA*

[b]*Department of Physiology, University of Geneva, 1 Rue Michel-Servet, 1211 Geneva 4, Switzerland*

Different types of neurotransmitter receptors are active within a single neuron and must be trafficked to discrete synaptic regions for proper function. Chick parasympathetic ciliary ganglion (CG) neurons express two classes of nicotinic cholinergic receptors: acetylcholine receptors (nAChRs) and α-bungarotoxin receptors (Bgt-nAChRs). Both receptor types mediate excitatory synaptic transmission, but differ in subunit composition and synaptic distribution.[1–4] The nAChRs, which contain α3 and α5 subunits, are concentrated in the specialized postsynaptic membrane; whereas the Bgt-nAChRs, composed of α7 subunits, are excluded from the synapse, being localized perisynaptically. Molecular interactions that mediate the accumulation of nAChRs at neuronal synapses are undefined.

We tested the hypothesis that particular domains of neuronal nAChR subunits specify their *in vivo* localization. Subunits are segregated among the two cholinergic receptor classes, and do not coassemble.[3] The nAChR and Bgt-nAChR subunits are highly homologous and have a stereotyped structure with extracellular N- and C-termini, four transmembrane regions (TMs), and a long cytoplasmic loop between TM3 and TM4.[5] This large loop region shows the greatest divergence in sequence and length between the individual subunits. The function of this domain is undefined.

We tested the ability of the long cytoplasmic loop to influence the localization of nAChRs and Bgt-nAChRs relative to synapses on CG neurons. We generated chimeric subunits by replacing the long cytoplasmic loop of α7 with the α3 or α5 cytoplasmic loop between TM3 and TM4. The N-terminus up to TM2 regulates subunit assembly and α7 does not coassemble with nAChR subunits, as demonstrated for endogenous subunits in CG neurons and for chimeric α7 subunits coexpressed with wild-type (WT) nAChR subunits in *Xenopus laevis* oocytes (this study; Refs. 3,6,7). Thus, a difference in the distribution of chimeric α7 as compared to WT α7 on the infected CG neuron surface would be due to the added α3 or α5 cytoplasmic loop. It cannot be explained by assembly with an endogenous subunit that can target to the synapse. To distinguish the exogenous subunits, a sequence encoding the myc proto-oncogene tag (10 amino acids) was added to the end of the C-terminus, being readily accessible for external surface labeling.[8]

Chimeric subunits were overexpressed in CG neurons developing *in vivo* by using avian-specific replication competent retroviral vector variants of the B envelope subgroup

[c]Corresponding author. Phone: 617-636-2429; fax: 617-636-2413: e-mail: mjacob@opal.tufts.edu

(RCASBP-B).[9,10] RCASBP-B was injected into the neural tube of Stage 9–10 chick embryos *in ovo* to infect neural crest precursors that give rise to CG neurons. The viral vector stably integrates into the genome of dividing cells and leads to high levels of exogenous gene expression in the progeny throughout development. Using this injection protocol, almost all CG neurons and very few brain regions were infected. For all receptor subunit constructs, age-matched chick embryos were injected with equivalent viral titers (5×10^7 to 10^9).[10]

A key aspect of this study is the previous demonstration that nAChRs and Bgt-nAChRs have restricted and largely exclusive surface localizations on normal developing CG neurons.[1,2] Again, α3 and α5 containing nAChRs are concentrated in the postsynaptic membrane. In contrast, α7 Bgt-nAChRs are excluded from the synapse, and are present perisynaptically on the surface membrane of dendrites that emerge from the postsynaptic cell in the region of innervation. Synapses with even light Bgt labeling are rare, representing 6% of the synapses examined (n = 158) over a range of developmental ages from embryos to adults (this study; Ref. 1). Synapses are distinguished by the presence of a parallel arrangement and thickening of the pre- and postsynaptic membranes, a widened synaptic cleft, an enhanced postsynaptic density, and an accumulation of synaptic vesicles adjacent to the presynaptic membrane.

The α7/α3 chimeric subunit is preferentially localized to the postsynaptic membrane, demonstrating that the long cytoplasmic loop of α3 is sufficient to target extrasynaptic α7 to the synapse (FIG. 1). The ultrastructural distribution of the chimeric subunits was established in the CG at stages when all of the neurons are functionally innervated by using an anti-myc monoclonal antibody (9E10), followed by a biotinylated antimouse antibody and a strepavidin-biotinylated horseradish peroxidase complex to achieve maximum sensitivity. Reaction product lines the postsynaptic membrane outer surface and often fills the synaptic cleft in the α7/α3 infected neurons. In contrast, overexpression of α7/α5 chimeric subunit or myc-tagged WT α7 (as a negative control) does not result in myc labeling of the synapse (FIG. 1). In α7/α3 infected neurons, 90% of the synapses (n = 163) were labeled, whereas only 9% and 12% of the synapses (n = 65 and n = 57) had any reaction product in α7/α5 and α7 infected neurons, respectively.

All three exogenous myc-tagged proteins have a perisynaptic distribution, being expressed on the dendrite surface membrane (FIG. 1 bottom). Higher levels of perisynaptic labeling are present on neurons overexpressing WT α7 and α7/α5 as compared to α7/α3. These results demonstrate that overexpression and myc-tagging do not cause a change in the distribution of WT α7. In all cases, chick embryonic development and the formation of CG synaptic specializations, in particular the length of the postsynaptic membrane opposite an active zone, appear normal.

Measurements of surface expression using the anti-myc antibody followed by a radiolabeled secondary antibody show that α7/α3 levels arc 1.8-fold lower than exogenous α7 and 1.6-fold lower than α7/α5 levels on acutely dissociated CG neurons. The number of surface sites was normalized to total protein and to the proportion of infected neurons, which was similar for the three constructs, as established by immunolabeling for viral *gag*. The predominantly synaptic localization of α7/α3, even though there are fewer total surface sites as compared to exogenous α7 and α7/α5, demonstrates that the α3 cytoplasmic loop is specifically and preferentially targeted to the specialized postsynaptic membrane *in vivo*.

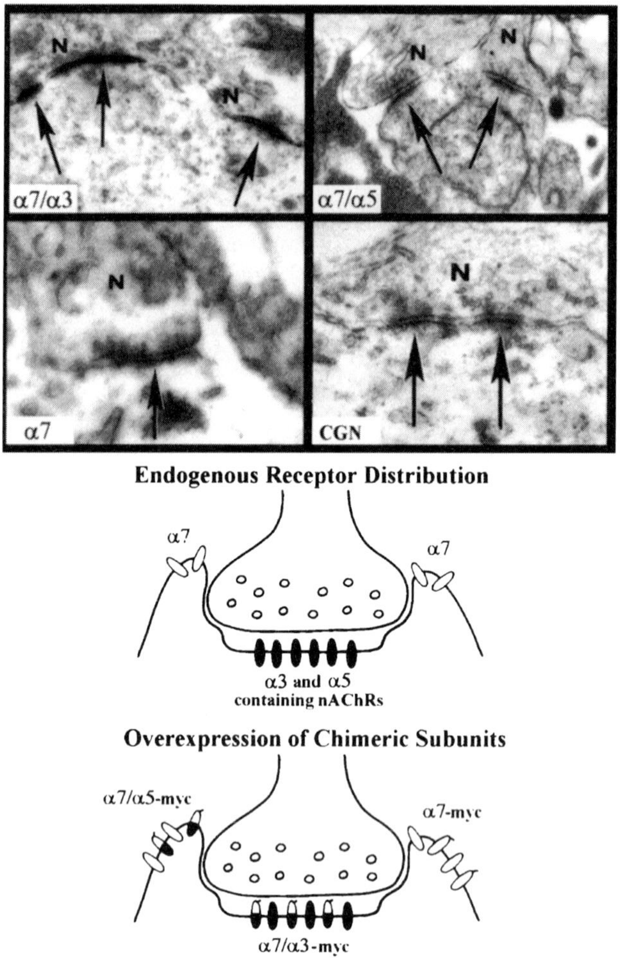

FIGURE 1. Electron micrographs and schematic representations demonstrating the distribution of overexpressed chimeric subunits in relation to synapses on ciliary ganglion neurons developing *in vivo*. **Top:** The myc-tagged subunits were detected by the binding of an anti-myc monoclonal antibody to the surface of the neurons followed by horseradish peroxidase immunolocalization. The α7/α3 chimeric subunit is concentrated in the specialized postsynaptic membrane (*arrow*). In contrast, synaptic membranes (*arrows*) are not labeled on the surface of neurons expressing α7/α5 or exogenous α7. Control uninfected CG neurons (CGN) were processed in parallel and are included to demonstrate background levels of myc immunolabeling at the synapse and the inherent density associated with pre- and postsynaptic membrane specializations (*arrow*). N = presynaptic nerve terminal. **Bottom:** Schematic representation of the ultrastructural localization of the chimeric subunits and endogenous nAChRs and Bgt-nAChRs in synaptic and perisynaptic surface regions on infected and control ciliary ganglion neurons, respectively.

We show that myc-tagged α7/α3, α7/α5, and WT α7 all form functional channels, as established by reconstitution studies in *Xenopus laevis* oocytes and two-electrode voltage clamps. Myc-tagging α7 has no apparent effect on the maximal ACh-evoked current. However, peak ACh current is reduced with both α7/α3 and α7/α5 as compared to WT α7. Importantly, the peak ACh current per oocyte closely resembles the relative level of surface expression as measured by radiolabeled α-Bgt binding. At ACh concentrations close to the EC_{50} for these receptors, there is no significant difference in the time course of the ACh response, suggesting normal agonist-induced gating. The low levels of α7/α5 and α7/α3 in comparison to WT α7 on the oocyte surface further suggest that the cytoplasmic loop plays an important role in influencing surface expression as well as spatial distribution of nAChRs.

This study provides novel mechanistic insights into neuronal synaptogenesis *in situ*. We show that the long cytoplasmic loop of α3, but not α5, can target nAChRs to the synapse during normal *in vivo* synapse formation. This is the first demonstration that a particular domain of one subunit plays an essential role in targeting receptor subtypes to the synapse in intact animals.

The sorting of nAChRs and Bgt-nAChRs seems to be mediated by molecular interactions that cause α3-containing nAChRs to be targeted to or selectively retained in the synapse. Since the biophysical and pharmacological properties of neuronal nicotinic cholinergic receptors are exquisitely sensitive to subunit composition,[11] the spatial segregation of receptor subtypes is likely to result in the formation of functionally specialized synaptic microdomains, as suggested by microiontophoretic electrophysiological analyses of innervated sympathetic neurons.[12] In comparison to nAChRs, Bgt-nAChRs have faster kinetics of activation and desensitization, lower affinity for ACh, and higher permeability to Ca^{2+}.[4,13] Segregating nAChRs and Bgt-nAChRs to synaptic and perisynaptic (dendritic) surface regions, respectively, is likely to influence synaptic firing properties and establish distinct spatial and temporal patterns of Ca^{2+} influx that may target different downstream signaling events in these local domains.[14,15]

Our study suggests a difference in the mechanisms that govern assembly of interneuronal synapses as compared to the neuromuscular junction (nmj) in vertebrates. Localization to the nmj seems to occur via the interaction of any of the muscle-type nAChR subunits with rapsyn, an important synapse organizing molecule,[16–18] suggesting that all of the subunits share a redundant targeting function that may optimize formation of the characteristic high-density clusters. In contrast, our work suggests that the intracellular domain of a particular subunit is responsible for targeting specific nAChR subtypes to the interneuronal synapse. Our *in vivo* study on neuronal nAChRs provides support for the *in vitro* work on the glutamate receptor family[19,20] and the glycine receptor,[21] and suggests a general principle of neuronal synapse differentiation. In neurons, which receive a far greater number and variety of synaptic connections as compared to skeletal muscle, only a limited number of highly specific molecular interactions are required for the formidable task of sorting the multiple neurotransmitter receptor subtypes to the appropriate synaptic regions within a single cell.

REFERENCES

1. Jacob, M.H. & D.K. Berg. 1983. The ultrastructural localization of alpha-bungarotoxin binding sites in relation to synapses on chick ciliary ganglion neurons. J. Neurosci. **3:** 260–271.

2. Jacob, M.H., J.M. Lindstrom & D.K. Berg. 1986. Surface and intracellular distribution of a putative neuronal nicotinic acetylcholine receptor. J. Cell Biol. **103:** 205–214.
3. Conroy, W.G. & D.K. Berg. 1995. Neurons can maintain multiple classes of nicotinic acetylcholine receptors distinguished by different subunit compositions. J. Biol. Chem. **270:** 4424–4431.
4. Zhang, Z.W., J.S. Coggan & D.K. Berg. 1996. Synaptic currents generated by neuronal acetylcholine receptors sensitive to alpha-bungarotoxin. Neuron **17:** 1231–1240.
5. Lindstrom, J., R. Anand, V. Gerzanich, X. Peng, F. Wang & G. Wells. 1996. Structure and function of neuronal nicotinic acetylcholine receptors. Prog. Brain Res. **109:** 125–137.
6. Vicente-Agullo, F., J.C. Rovira, A. Campos-Caro, C. Rodiguez-Ferrer, J.J. Ballesta, S. Sala, F. Sala & M. Criado. 1996. Acetylcholine receptor subunit homomer formation requires compatibility between amino acid residues of the M1 and M2 transmembrane segments. FEBS Lett. **399:** 83–86.
7. Eisele, J.L., S. Bertrand, J.L. Galzi, A. Devillers-Thiery, J.P. Changuex & D. Bertrand. 1993. Chimaeric nicotinic-serotonergic receptor combines distinct ligand binding and channel specificities. Nature **366:** 479–483.
8. Squinto, S.P., T.H. Aldrich, R.M. Lindsay, D.M. Morrissey, N. Panayotatos, S.M. Bianco, M.E. Furth & G.D. Yancopoulos. 1990. Identification of functional receptors for ciliary neurotrophic factor on neuronal cell lines and primary neurons. Neuron **5:** 757–766.
9. Homburger, S.A. & D.M. Fekete. 1996. High efficiency gene transfer into the embryonic chicken CNS using B-subgroup retroviruses. Dev. Dyn. **206:** 112–120.
10. Morgan, B.A. & D.M. Fekete. 1996. Manipulating gene expression with replication-competent retroviruses. Methods Cell Biol. **51:** 185–218.
11. McGehee, D.S. & L.W. Role. 1995. Physiological diversity of nicotinic acetylcholine receptors expressed by vertebrate neurons. Annu. Rev. Physiol. **57:** 521–546.
12. Moss, B.L. & L.W. Role. 1993. Enhanced ACh sensitivity is accompanied by changes in ACh receptor channel properties and segregation of ACh receptor subtypes on sympathetic neurons during innervation in vivo. J. Neurosci. **13:** 13–28.
13. Ullian, E.M., J.M. McIntosh & P.B. Sargent. 1997. Rapid synaptic transmission in the avian ciliary ganglion is mediated by two distinct classes of nicotinic receptors. J. Neurosci. **17:** 7210–7219.
14. Rathouz, M.M., S. Vijayaraghavan & D.K. Berg. 1995. Acetylcholine differentially affects intracellular calcium via nicotinic and muscarinic receptors on the same population of neurons. J. Biol. Chem. **270:** 14366–14375.
15. Vijayaraghavan, S., B. Huang, E.M. Blumenthal & D. K. Berg. 1995. Arachidonic acid as a possible negative feedback inhibitor of nicotinic acetylcholine receptors on neurons. J. Neurosci. **15:** 3679–3687.
16. Maimone, M.M. & J.P. Merlie. 1993. Interaction of the 43 kd postsynaptic protein with all subunits of the muscle nicotinic acetylcholine receptor. Neuron **11:** 53–66.
17. Yu, X.M. & Z.W. Hall. 1994. The role of the cytoplasmic domains of individual subunits of the acetylcholine receptor in 43 kDa protein-induced clustering in COS cells. J. Neurosci. **14:** 785–795.
18. Apel, E.D., D.J. Glass, L.M. Moscoso, G.D. Yancopoulos & J.R. Sanes. 1997. Rapsyn is required for MuSK signaling and recruits synaptic components to a MuSK-containing scaffold. Neuron **18:** 623–635.
19. Kornau, H.C., P.H. Seeburg & M.B. Kennedy. 1997. Interaction of ion channels and receptors with PDZ domain proteins. Curr. Opin. Neurobiol. **7:** 368–373.
20. Sheng, M. & M. Wyszynski. 1997. Ion channel targeting in neurons. Bioessays **19:** 847–853.
21. Kirsch, J., I. Wolters, A. Triller & H. Betz. 1993. Gephyrin antisense oligonucleotides prevent glycine receptor clustering in spinal neurons. Nature **366:** 745–748.

Molecular and Functional Diversity of the Expanding GABA-A Receptor Gene Family

PAUL J. WHITING,[a] TIMOTHY P. BONNERT, RUTH M. McKERNAN, SOPHIE FARRAR, BEATRICE LE BOURDELLÈS, ROBERT P. HEAVENS, DAVID W. SMITH, LOUISE HEWSON, MICHAEL R. RIGBY, DALIP J. S. SIRINATHSINGHJI, SALLY A. THOMPSON, AND KEITH A. WAFFORD

Neuroscience Research Centre, Merck Sharp & Dohme Research Laboratories, Eastwick Road, Harlow, Essex CM20 2QR, England, UK

ABSTRACT: Fast inhibitory neurotransmission in the mammalian CNS is mediated primarily by the neurotransmitter γ-aminobutyric acid (GABA), which, upon binding to its receptor, leads to opening of the intrinsic ion channel, allowing chloride to enter the cell. Over the past 10 years it has become clear that a family of GABA-A receptor subtypes exists, generated through the coassembly of polypeptides selected from α1–α6, β1–β3, γ1–γ3, δ, ε, and π to form what is most likely a pentomeric macromolecule. The gene transcripts, and indeed the polypeptides, show distinct patterns of temporal and spatial expression, such that the GABA-A receptor subtypes have a defined localization that presumably reflects their physiological role. A picture is beginning to emerge of the properties conferred to receptor subtypes by the different subunits; these include different functional properties, differential modulation by protein kinases, and the targeting to different membrane compartments. These properties presumably underlie the different physiological roles of the various receptor subtypes. Recently we have identified a further member of the GABA-A receptor gene family, which we have termed θ, which appears to be most closely related to the β subunits. The structure, function, and distribution of θ-containing receptors, and receptors containing the recently reported ε subunit, are described.

Inhibitory neurotransmission within the mammalian CNS is mediated primarily by γ-aminobutyric acid (GABA). Indeed, 20–30% of all synapses within the mammalian brain are thought to be GABAergic. GABA mediates its effects by activating two types of receptors, GABA-A receptors and GABA-B receptors. The latter is a G protein–coupled receptor[1] that is structurally related to the metabotropic glutamate receptor; it will not be discussed further here. GABA-A receptors are ligand-gated ion channels, the binding of GABA to specific sites on the receptor leading to conformational changes resulting in the opening of an intrinsic ion channel and, in most neurons, the flux of chloride into the cell. This leads to a hyperpolarization of the cell membrane and an increase in the inhibitory tone. GABA-A receptors are the site of action of a number of clinically and pharmacologically important drugs, all of which are modulator agents acting at distinct binding sites on the receptor to allosterically potentiate or inhibit the action of GABA.[2–5] Benzodiazepines are a class of such drugs, which are very widely prescribed as anxiolytics, sedatives (hypnotics), and anticonvulsants. Diverse classes of clinically prescribed anesthetics including barbiturates,[6] the volatile anesthetics such as halothane,[7] and agents such as etomidate[8] are thought to mediate their effect partially, if not completely, through the GABA-A receptor.

[a]Corresponding author. Phone: 1279-440535; fax: 1279-440712; e-mail: paul_whiting@merck.com

Similarly, the well-known behavioral manifestations of alcohol are mediated, at least in part, through the GABA-A receptor.[9] Thus it is obvious from the above statements that an understanding of the structure and function of the GABA-A receptor family is important not only because of their central role in the function of the CNS, but also because of their demonstrated utility as targets for therapeutic intervention.

The tools of molecular biology have revealed that the GABA-A receptor, like most receptors and ion channels, exists as a gene family. From the initial cloning of cDNAs encoding the α1 and β1 subunits,[10] low-stringency hybridization cDNA library screening, degenerate polymerase chain reaction (PCR), and sequence searching of expressed sequence tag (EST) and genomic DNA sequence databases have lead to the identification of a large number of subunits that make up the GABA-A receptor gene family. FIGURE 1 shows the gene family represented as a dendrogram so as to show the relative sequence identities of the deduced amino acid sequences of all the family members. The polypep-

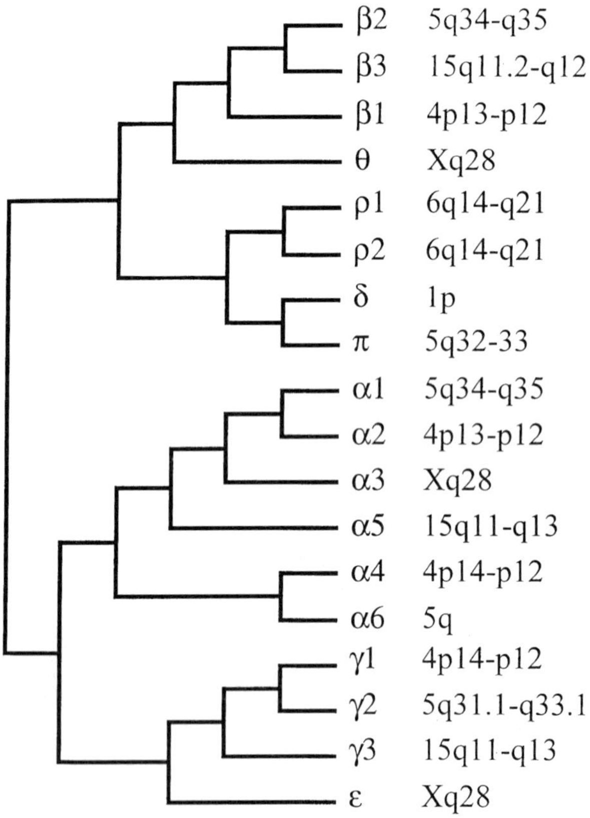

FIGURE 1. The human GABA-A receptor gene family. The dendrogram indicates the homologies between the deduced amino acid sequences of the subunits. The human chromosome assignment of each gene is also indicated.

tides are 450–627 amino acids in length, and all share the same putative structural features: a signal peptide, a large extracellular domain, three transmembrane domains, a large cytoplasmic domain followed by a fourth transmembrane domain so as to leave the C-terminal tail extracellular, as shown in FIGURE 2. The other motif that is conserved within the gene family is the so-called Cys-loop, two cysteines separated by 13 amino acids, located in the extracellular domain. All of these features are also found within the nicotinic acetylcholine receptor gene family,[11] the strychnine-sensitive glycine receptor gene family,[12] and the 5-HT3 receptor,[13] all of which show significant sequence homology with each other and GABA-A receptor subunits, indicating the existence of a "supergene" family of ligand-gated ion channels. FIGURE 2 shows diagrammatically the regions of amino acid sequence homology conserved between different GABA-A receptor subunits. There are regions of sequence homology within the extracellular domain and the putative transmembrane domains. The region between TM3 and TM4, the large cytoplasmic loop, shows little or no sequence homology between the subunits, suggesting that this domain can tolerate many changes without affecting any possible functional role, or that this domain has undergone significant divergence so as to subserve specific functions for each subunit.

Currently there are 18 known members of the mammalian GABA-A receptor family, shown in FIGURE 1. The inclusion of ρ1 and ρ2 in the GABA-A family is debatable. At the level of amino acid sequence identity they clearly fit. However, their GABA site pharmacology differs from that of other GABA-A receptors,[14] and they appear to function quite well as homooligomers,[15] again unlike other GABA-A receptors. This has led some to suggest that ρ1 and ρ2 be considered GABA-C receptors. The most recently identified members of the gene family are π, ε, and θ. All were identified through searches of DNA databases. The role of the π subunit is currently unclear;[16] it is appears expressed outside the CNS, in tissues such as the uterus. The ε and θ subunits will be discussed in more detail below.

FIGURE 1 also indicates the human chromosomal localization of the GABA-A receptor subunit genes. Intriguingly, the genes are distributed as clusters throughout the genome, on chromosomes 4, 5, 15, and X. The curious exception is the δ subunit, which is located by

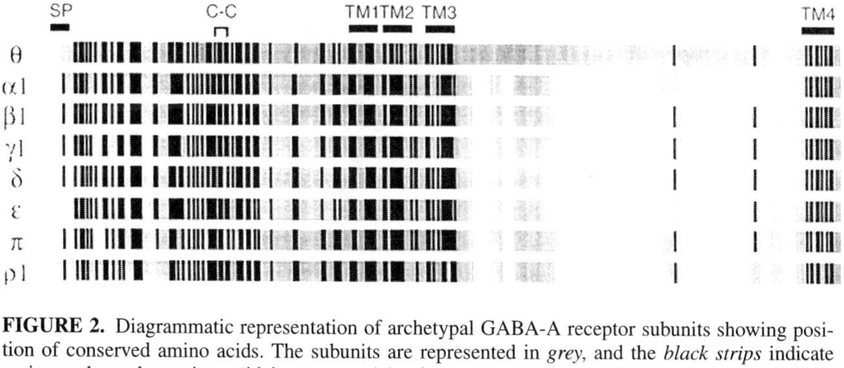

FIGURE 2. Diagrammatic representation of archetypal GABA-A receptor subunits showing position of conserved amino acids. The subunits are represented in *grey*, and the *black strips* indicate regions where the amino acid is conserved in six or more subunits. The breaks in the domain between TM3 and TM4 for all except the θ subunit represent the much greater length of θ in this domain. SP, signal peptide; C-C, cysteine loop; TM1–4, transmembrane regions 1–4.

itself on chromosome 1. The biological significance of this gene clustering, if indeed there is any, is currently unknown.

It can be demonstrated that *in vitro* (e.g., in a *Xenopus* oocyte or transfected mammalian cell) coexpression of an α, a β, and a γ subunit is sufficient to recapitulate the properties of a native GABA-A receptor.[17] Such a prototypical GABA-A receptor is represented in FIGURE 3. It is thought that αβγ receptors have the stoichiometry $(\alpha)_2\ (\beta)_2\ (\gamma)_1$,[18–19] although $(\alpha)_2\ (\beta)_1\ (\gamma)_2$ is also possible.[20] The subunit order depicted in FIGURE 3 is based upon the $(\alpha)_2\ (\beta)_2\ (\gamma)_1$ stoichiometry and is consistent with data indicating that the GABA site is located between the α and β subunits[21,22] and data indicating that the benzodiazepine site is located between the α and γ subunits.[23] FIGURE 3 also indicates the proposed localization of binding sites for modulatory agents, referred to earlier, which are known to act at the GABA-A receptor. Conclusions are all based upon site-directed mutagenesis studies, and thus need to be interpreted in light of the possibility that amino acids identified may not be directly at the binding site but could be affecting the modulation of the compound at a site distant from the binding site. Loreclezole and etomidate have been shown to act at a site on the β subunit,[8,24,25] the effects of volatile anesthetics and alcohols is dependent upon amino acids in TM2 and TM3,[7] the inhibitory action of zinc is dependent upon certain histidine residues thought to be in or around the pore of the channel,[26–28] and the inhibitory action of furosemide is dependent upon amino acid residues around TM1 of the α subunits (Whiting *et al.*, unpublished). The localization of binding sites for barbiturates, steroids, avermectins, and propofol are currently unknown.

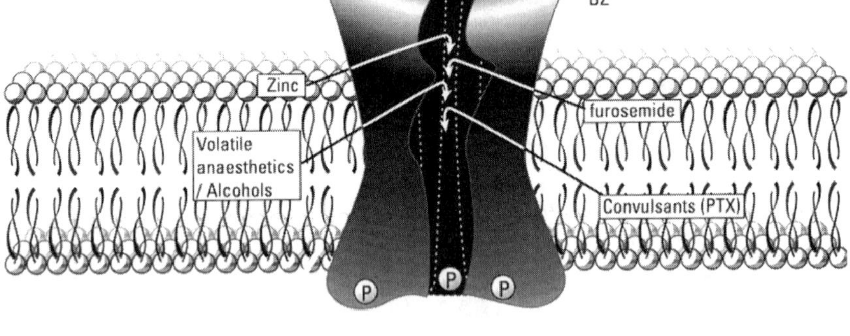

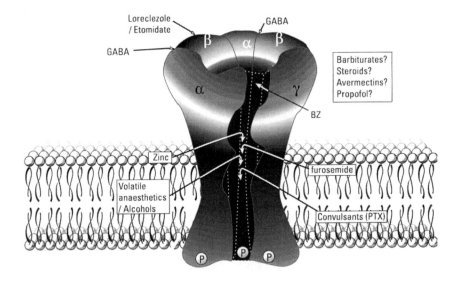

FIGURE 3. Diagrammatic representation of a prototypic GABA-A receptor showing the putative localization of the binding sites of the agonist and various modulatory agents. BZ, benzodiazepine; PTX, picrotoxin. The locations of binding sites for barbiturates, steroids, avermectins, and propofol remain unclear. The *P in a circle* represents the phosphorylation sites located on the cytoplasmic domain between TM3 and TM4 of the α, β, and γ subunits.

The major issue of how receptor subunits coassemble *in vivo* to form heterooligomeric subtypes is one that has been addressed by numerous labs, using subunit-specific antibodies and immunoprecipitation methodologies. As discussed above, there is little argument that the GABA-A receptor, like other members of the ligand-gated ion channel "super family," is a pentamer. There is, however, considerable debate over what five subunits make up the receptor subtypes. FIGURE 4 is a pie chart representing what is known about the composition of the *major* receptor subtypes in the rat brain. There is evidence that receptors can contain more than one α subunit,[29–32] more than one β subunit,[33] or more than one γ subunit.[34] The subunit combination α1β2γ2 is the most abundant, while α2β3γ2 and α2βxγ2 (βx meaning β1, 2, or 3) are the next most prevalent subtypes (see Ref. 29).[35] Receptors containing α5 are prevalent in the hippocampus,[36] while α6-containing receptors are expressed almost exclusively within the cerebellar granule cells, where they coassemble with a β subunit and either γ or δ.[29,32] Receptors containing α4 constitute more minor receptor subtypes and are expressed primarily in the thalamus and the hippocampus.[37,38] Essentially nothing is currently known about the subunit composition of native ε (see below) or γ3-containing receptors. There is considerable debate over whether receptors exist that contain, for instance, both α1 and α6 subunits.[32,39–41] Determining the composition of native receptor subtypes, especially the less abundant ones, is undoubtedly a difficult task, and one that will continue to occupy laboratories for the foreseeable future. The two members of the gene family to be identified most recently are the ε subunit[42–45] and the θ subunit. While they are "low-abundance" subunits (see below and FIGURE 4), it is not clear why their identification has eluded the now-standard molecular biology methods of low-stringency hybridization cDNA library screening or degenerate PCR, their identification being through searches of the Genbank nucleotide sequence databases. At the time of writing there are over 30 ε subunit EST sequences in Genbank, though this is obviously

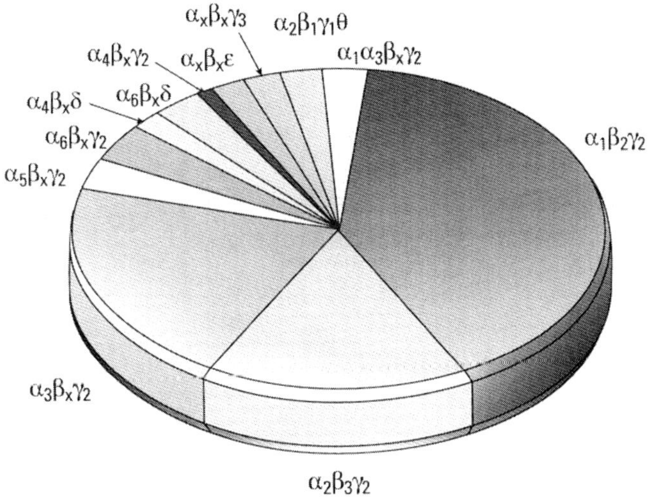

FIGURE 4. Pie chart model of the subunit composition of the major GABA-A receptor subtypes in rat brain. βx indicates β1, 2, or 3.

considerably more than there was two or three years ago. Additionally, the ε subunit is something of a special case, being abundantly expressed in many peripheral tissues as an incompletely spliced mRNA, with appropriately spliced mRNA being found in the brain and to a much lesser degree in heart tissue.[42] The ε EST sequences were derived primarily from cDNA libraries made from RNA extracted from peripheral tissues, where the incompletely spliced ε mRNA is relatively more abundantly expressed. The ε subunit is most closely related to the γ subunits (FIGS. 1 and 5). Within the brain the expression of the ε subunit mRNA is one of the most restricted of all members of this gene family, with the major sites of expression restricted to the hypothalamic region and the hippocampus.[42] A sequence polymorphism exists within the ε subunit at amino acid residue 102, which is a serine in the deduced amino acid sequence of the cDNAs we have cloned and also in the deduced amino acid sequence of cDNA sequence published by Wilke *et al.*,[39] but is an alanine in the deduced amino acid sequence published by Davies *et al.*[43] and also by Garret *et al.*[45] This is due to a T (serine) to G (alanine) transition at the first base of the codon. Coexpression in oocytes of either forms of ε with an α1 and β1 subunit has no obvious effect upon the functional properties of the receptor.

In functional studies, expressing ε-containing receptors in *Xenopus* oocytes or HEK293 cells, it has been demonstrated by two groups[42,43] that ε can substitute for a γ subunit; that is, coexpression of ε with an α and a β subunit leads to the formation of a receptor with novel properties, such as decreased sensitivity to zinc and increased rate of desensitization,[42] which differ from those of receptors formed from α and β subunits alone. However, there is some discrepancy over the response of ε containing receptors to allosteric modulators such as steroids, propofol, and pentobarbital. While we[42] have found that ε-containing receptors respond to these agents in a manner not significantly different to other GABA-A receptors—i.e., are allosterically potentiated, Davies *et al.*[43] found that the coexpression of ε with an α and β subunit lead to the formation of GABA-A receptors that are insensitive to these agents. The reason for the difference in findings is not entirely clear at this moment, but clearly whether or not native ε-containing receptors are sensitive

α1	β1	γ1	δ	ε	ρ1	π	
36.7	50.5	40.1	42.0	39.2	39.5	40.2	θ
	42.7	52.7	36.4	44.7	39.5	36.7	α1
		44.3	45.9	37.6	44.9	47.8	β1
			43.2	48.5	41.0	39.0	γ1
				38.8	42.5	44.3	δ
					38.3	36.0	ε
						41.4	ρ1

FIGURE 5. Deduced percent amino acid sequence identity between archetypal GABA-A receptor subunits.

to these agents has important implications for our understanding of the biological role of this subtype *in vivo*.

The most recent member of the GABA-A receptor gene family is a subunit we have christened θ (theta). This was identified as Genbank submission U47334, a short cDNA sequence identified through exon trap studies of human chromosome Xq28.[46] Cloning of the full-length cDNA sequence indicates that the deduced amino acid sequence is 627 residues, making it the largest member of the GABA-A receptor gene family, and, indeed, only two residues shorter than the deduced amino acid sequence of the nicotinic acetylcholine receptor α4 subunit, the largest known subunit of the ligand-gated ion channel superfamily. From FIGURE 1 it can be seen that θ is most similar to the β subunits. It has 50% deduced amino acid sequence identity with the β1 subunit (FIG. 5). Indeed, the original annotation of the U47334 Genbank submission stated that the deduced amino acid sequence was most similar to chick β4. However, θ is sufficiently different from the human β1–β3 deduced amino acid sequences that it should not be classified as β4. Preliminary data also suggests that, like the ε subunit, θ has a very limited pattern of expression in the CNS, restricted primarily to monoaminergic neurons; and has the subunit composition, in rat striatum, of α2β1γ1θ (see FIG. 4; Whiting *et al.*, unpublished). Its functional role remains to be determined.

An interesting question is whether there are any further members of the GABA-A receptor gene family yet to be identified. The last three to be identified, π, ε, and θ, were all found through searches of the nucleotide sequence databases. Thus one would predict that any others will also be identified through this approach. One could also predict that they will be of very low abundance; if not, they probably would already have been identified. Within the next five years the human genome sequence will be completed, and we will know the full extent of the gene family. It will take many more years of effort by many researchers to understand the structure, function, and physiological role of these subtypes; indeed, this will be the greatest challenge.

REFERENCES

1. KAUPMANN, K., K. HUGGEL, J. HEID *et al.* 1997. Expression cloning of $GABA_B$ receptors uncovers similarity to metabotropic glutamate receptors. Nature **386:** 239–246.
2. WHITING, P.J., R.M. MCKERNAN & K.A. WAFFORD. 1995. Structure and pharmacology of vertebrate GABA-A receptor subtypes. Int. Rev. Neurobiol. **38:** 95–138.
3. LAMBERT J.J., D. BELILLI, C. HILL-VENNING & J.A. PETERS. 1995. Neurosteroids and GABA-A receptor function. Trends Pharmacol. Sci. **16:** 295–303.
4. DARLISON, M.G. & B.E. ALBRECHT. 1995. GABA-A receptors; which, where and why? Semin. Neurosci. **7:** 115–126.
5. SIEGHART, W. 1995. Structure and pharmacology of gamma-aminobutyric acidA receptor subtypes. Pharmacol. Rev. **47:** 181–234.
6. THOMPSON, S. A., P.J. WHITING & K.A. WAFFORD. 1996. Alpha subunits influence the action of pentobarbital on recombinant GABA-A receptors. Br. J. Pharmacol. **117:** 521–527.
7. MIHIC, S.J., Q. YE, M.J. WICK *et al.* 1997. Sites of alcohol and volatile anaesthetic action on GABA(A) and glycine receptors. Nature **389:** 385–389.
8. BELLELLI, D., J.J. LAMBERT, J.A. PETERS, K.A. WAFFORD & P.J. WHITING. 1997. The interaction of the general anesthetic etomidate with the γ-aminobutyric acid type A receptor is influenced by a single amino acid. Proc. Natl. Acad. Sci. USA **94:** 11031–11036.
9. WAFFORD, K.A., D.M. BURNETT, N.J. LEIDENHEIMER *et al.* 1991. Ethanol sensitivity of the GABAA receptor expressed in Xenopus oocytes requires 8 amino acids contained in the gamma 2L subunit. Neuron. **7:** 27–33.

10. Schofield, P.R., M.G. Darlison, N. Fujita *et al.* 1987. Sequence and functional expression of the GABA A receptor shows a ligand-gated receptor super-family. Nature **328:** 221–227.
11. Lindstrom, J., R. Anand, X. Peng, V. Gerzanich *et al.* 1995. Neuronal nicotinic receptor subtypes. Ann. N.Y. Acad. Sci. **757:** 100–116.
12. Vannier, C. & A. Triller. 1997. Biology of the postsynaptic glycine receptor. Int. Rev. Cytol. **176:** 201–244.
13. Maricq, A.V., A.S. Peterson, A.J. Brake *et al.* 1991. Primary structure and functional expression of the 5HT3 receptor, a serotonin-gated ion channel. Science **254:** 432–437.
14. Johnston, G.A. 1996. GABAc receptors: Relatively simple transmitter-gated ion channels? Trends Pharmacol. Sci. **17:** 319–323.
15. Cutting, G.R., L. Lu, B.F. O'Hara *et al.* 1991. Cloning of the gamma-aminobutyric acid (GABA) rho 1 cDNA: A GABA receptor subunit highly expressed in the retina. Proc. Natl. Acad. Sci. USA **88:** 2673–2677.
16. Hedblom, E. & E.F. Kirkness. 1997. A novel class of GABAA receptor subunit in tissues of the reproductive system. J. Biol. Chem. **272:** 15346–15350.
17. Pritchett, D.B., H. Sontheimer, B.H. Shivers *et al.* 1989. Importance of a novel GABA-A receptor subunit for benzodiazepine pharmacology. Nature **338:** 582–585.
18. Chang, Y., R. Wang, S. Barot & D.S. Weiss. 1996. Stoichiometry of a recombinant GABAA receptor. J. Neurosci. **16:** 5415–5424.
19. Tretter, V., N. Ehya, K. Fuchs & W. Sieghart. 1997. Stoichiometry and assembly of a recombinant GABAA receptor subtype. J. Neurosci. **17:** 2728–2737.
20. Backus, K.H., M. Arigoni, U. Drescher *et al.* 1993. Stoichiometry of a recombinant GABA receptor deduced from mutation-induced rectification. NeuroReport **5:** 285–288.
21. Smith, G.B. & R.W Olsen. 1994. Identification of a [3H]muscimol photoaffinity substrate in the bovine gamma-aminobutyric acidA receptor alpha subunit. J. Biol. Chem. **269:** 20380–20387.
22. Amin, J. & D. S. Weiss. 1993. $GABA_A$ receptor needs two homologous domains of the β subunit for activation by GAB_A but not by pentobarbital. Nature **366:** 565–569.
23. Sigel, E. & A. Buhr. 1997. The benzodiazepine binding site of GABAA receptors. Trends Pharmacol. Sci. **18:** 425–429.
24. Wingrove, P. B., K.A. Wafford, S.A. Thompson & P.J. Whiting. 1997. Identification of key amino acids in the γ subunit of the GABA-A receptor which contribute to the benzodiazepine binding site. Mol. Pharmacol. **52:** 874–881.
25. Wingrove, P.B., K.A. Wafford, C. Bain & P.J. Whiting. 1994. The modulatory action of loreclezole at the γ-aminobutyric acid type A receptor is determined by a single amino acid in the β2 and β3 subunit. Proc. Natl. Acad. Sci. USA **91:** 4569–4573.
26. Horenstein, J. & M.H. Akabas. 1998. Location of a high affinity Zn2+ binding site in the channel of alpha1beta1 gamma-aminobutyric acidA receptors. Mol. Pharmacol. **53:** 870–877.
27. Wooltorton, J.R., B.J. McDonald, S.J. Moss & T.G.Smart. 1997. Identification of a Zn2+ binding site on the murine GABAA receptor complex: dependence on the second transmembrane domain of beta subunits. J Physiol. (Lond.) **505:** 633–640.
28. Fisher J.L. & R.L. Macdonald. 1998 The role of an alpha subtype M2-M3 His in regulating inhibition of GABAA receptor current by zinc and other divalent cations. J. Neurosci. **18:** 2944–2953.
29. McKernan, R.M. & P.J. Whiting. 1996. Which GABA-A receptor subtypes really occur in the brain? Trends Neurosci. **19:** 139–143.
30. Duggan M.J., S. Pollard & F.A. Stephenson. 1991. Immunoaffinity purification of GABAA receptor alpha-subunit iso-oligomers. Demonstration of receptor populations containing alpha 1 alpha 2, alpha 1 alpha 3, and alpha 2 alpha 3 subunit pairs. J. Biol. Chem. **266:** 24778–24784.
31. McKernan R.M., K. Quirk, R. Prince *et. al.* 1991. GABA-A receptor subtypes immunopurified from rat brain with alpha subunit specific antibodies have unique pharmacological properties. Neuron **7:** 667–676.
32. Jechlinger, M., R. Pelz, V. Tretter *et al.* 1998. Subunit composition and quantitative importance of hetero-oligomeric receptors: GABAA receptors containing alpha6 subunits. J. Neurosci. **18:** 2449–2457.
33. Li, M. & A.L. De Blas. 1997. Coexistence of two beta subunit isoforms in the same gamma-aminobutyric acid type A receptor. J. Biol. Chem. **272:** 16564–16569.

34. QUIRK K., N.P. GILLARD, C.I. RAGAN *et al.* 1994. γ-aminobutyric acid type A receptors in the rat brain can contain both γ2 and γ3 subunits, but γ1 does not exist in combination with another γ subunit. Mol. Pharmacol. **45:** 1061–1070.
35. BENKE, D., J.M. FRITSCHY, A. TRZECIAK *et al.* 1994. Distribution, prevalence, and drug binding profile of gamma-aminobutyric acid type A receptor subtypes differing in the beta-subunit variant. J. Biol. Chem. **269:** 27100–27107.
36. QUIRK, K., P. BLURTON, S. FLETCHER *et al.* 1996. [3H]L-655,708, a novel ligand selective for the benzodiazepine site of GABAA receptors which contain the alpha 5 subunit. Neuropharmacology **35:** 1331–1335.
37. WISDEN W., A. HERB, H. WIELAND *et al.* 1991 Cloning, pharmacological characteristics and expression pattern of the rat GABAA receptor alpha 4 subunit. FEBS Lett. **289:** 227–230.
38. KHAN, Z.U., A. GUTIERREZ, A.K. MEHTA *et al.* 1996. The alpha 4 subunit of the GABAA receptors from rat brain and retina. Neuropharmacol. **35:** 1315–1322.
39. KHAN Z.U., A. GUTIERREZ & A.L. DE BLAS. 1996. The alpha 1 and alpha 6 subunits can coexist in the same cerebellar GABAA receptor maintaining their individual benzodiazepine-binding specificities. J. Neurochem. **66:** 685–691.
40. QUIRK K., N. P. GILLARD, C.I. RAGAN *et al.* 1994. Model of subunit composition of gamma-aminobutyric acid A receptor subtypes expressed in rat cerebellum with respect to their alpha and gamma/delta subunits. J. Biol. Chem. **269:** 16020–16028.
41. POLLARD, S., C.L. THOMPSON & F.A. STEPHENSON. 1995. Quantitative characterization of alpha 6 and alpha 1 alpha 6 subunit-containing native gamma-aminobutyric acidA receptors of adult rat cerebellum demonstrates two alpha subunits per receptor oligomer. J. Biol. Chem. **270:** 21285–21290.
42. WHITING, P. J., G. MCALLISTER, D. VASILATIS *et al.* 1997. Neuronal restricted RNA splicing regulates the expression of a novel GABA-A receptor subunit conferring atypical functional properties. J. Neurosci. **17:** 5027–5037.
43. DAVIES, P.A., M.C. HANNA, T.G. HALES & E.F. KIRKNESS. 1997. Insensitivity to anaesthetic agents conferred by a class of GABAA receptor subunit. Nature **385:** 820–823.
44. WILKE, K., R. GAUL, S.M. KLAUCK & A. POUSTKA. 1997. A gene in human chromosome band Xq28 (GABRE) defines a putative new subunit class of the GABAA neurotransmitter receptor. Genomics **45:** 1–10.
45. GARRET, M., L. BASCLES, E. BOUE-GRABOT *et al.* 1997. An mRNA encoding a putative GABA-gated chloride channel is expressed in the human cardiac conduction system. J. Neurochem. **68:** 1382–1389.
46. LEVIN, M.L., A. CHATTERJEE, A. PRAGLIOLA *et al.* 1996. A comparative transcription map of the murine bare patches (Bpa) and striated (Str) critical regions and human Xq28. Genome Res. **6:** 465–477.

Activity-Dependent Regulation of $GABA_A$ Receptors

SILKE PENSCHUCK, JACQUES PAYSAN,[a] OLIVIA GIORGETTA, AND JEAN-MARC FRITSCHY[b]

Institute of Pharmacology, University of Zurich, CH - 8057 Zurich, Switzerland

ABSTRACT: $GABA_A$ receptor heterogeneity is based on the combinatorial assembly of a large family of subunits into distinct receptor subtypes. A neuron-specific expression pattern of receptor subtypes has been demonstrated in adult rat brain, which can be reproduced *in vitro* in primary neuron cultures. This suggests that genetic programs established during ontogeny govern the expression of γ-aminobutyric acid ($GABA_A$) receptor subtypes. Activity-dependent mechanisms nevertheless modulate on a short-term basis the cell surface expression of $GABA_A$ receptors, as demonstrated in cultured hippocampal neurons upon blockade of synaptic transmission or application of brain-derived neurotrophic factor. Preliminary evidence points to changes in protein phosphorylation as a mechanism underlying short-term activity-dependent regulation of $GABA_A$ receptors. *In vivo*, chronic pharmacological modulation of neuronal activity during development, while having marked effects on the rate of cortical growth, failed to influence the expression of $GABA_A$ receptor subtypes, suggesting that additional factors are involved.

NEURON-SPECIFIC EXPRESSION OF $GABA_A$ RECEPTOR SUBTYPES

Fast synaptic inhibition in the CNS is mediated predominantly by γ-aminobutyric acid ($GABA_A$) receptors, which form GABA-gated heteromeric chloride ion channels. The drug-induced modulation of $GABA_A$ receptors contributes to the regulation of anxiety, vigilance, memory, epileptogenic activity, and muscle tension. Based on a repertoire of subunits encoded by at least 18 genes (α1–6, β1–3, γ1–3, δ, ε, π, ρ1–3)[1–5] $GABA_A$ receptors are prime examples of receptor heterogeneity in the CNS. The major $GABA_A$ receptor subtypes are made up of an α, β, and the γ2 subunit. They are best distinguished by the type of α subunit, which determines their pharmacological profile and distribution in the CNS.[3,6,7] The diverse receptor subtypes are expressed in specific neurons, thereby adapting GABAergic transmission to the requirements of the respective neuronal circuits.

The subunit combination α1β2γ2 constitutes the major $GABA_A$ receptor subtype in the brain, as demonstrated by immunoprecipitation and immunohistochemical localization of the three subunits in the same neurons.[6,8–10] With regard to identified neurons, high levels of α1β2γ2 receptors are expressed in numerous populations of GABAergic neurons at all levels of the neuraxis. In particular, interneurons in cerebral cortex and hippocampus, and GABAergic neurons in the cerebellum, brain stem reticular formation, pallidum, substantia nigra, and basal forebrain are intensely immunoreactive for the α1 subunit.[9,11–13] The α1β2γ2 subunit combination has also been allocated to non-GABAergic neurons, such as

[a]Present address: Institute of Physiology, University of Tübingen, D - 72074 Tübingen, Germany.

[b]Address for correspondence: PD Dr. Jean-Marc Fritschy, Institute of Pharmacology, University of Zurich, Winterthurerstrasse 190, CH - 8057 Zurich, Switzerland. Phone: (41-1) 635 5926; fax: (41-1) 635 5708; e-mail: fritschy@pharma.unizh.ch

olfactory bulb mitral cells, in association with the α3 subunit, and relay neurons in the thalamus, in combination with the δ subunit. Only a few regions are devoid of α1 subunit staining, notably the granule cell layer of the olfactory bulb, the reticular nucleus of the thalamus, and motoneurons.[6]

Receptors containing the α2 subunit are most abundant in regions where the α1 subunit is absent or expressed at low levels,[6,14,15] such as striatum, hippocampal formation, and olfactory bulb. As for the α2 subunit, receptors containing the α3 subunit are frequently distributed in regions expressing only low or moderate levels of the α1 subunit, including the lateral septum, the reticular nucleus of the thalamus, and several brainstem nuclei.[6,11,14] The coexpression of α2 or α3 subunits with the β3 and γ2 subunits is evident on the cellular level—e.g., in hippocampal pyramidal cells (α2β3γ2) and in cholinergic neurons of the basal forebrain (α3β3γ2).

The allocation of $GABA_A$ receptor subtypes to defined populations of neurons suggests that expression of a particular type of $GABA_A$ receptor is part of the neurochemical phenotype of neurons. As tested in primary cultures for several types of neurons (cerebellar granule cells, neocortical neurons, hippocampal neurons) the expression of $GABA_A$ receptor subtypes is determined relatively early during development and is not dependent on the formation of normal synaptic circuits. Therefore, cerebellar granule cells isolated from 5-day-old rats and maintained for 6–8 days *in vitro* (DIV) express high levels of the same six subunits (α1, α6, β2,3, γ2, δ) as they do *in vivo*. Furthermore, these subunits are clustered at synaptic sites on dendrites, even though the majority of these synapses are formed between granule cells, which are excitatory neurons.[16] When these cells are isolated from newborn rat pups, they express only moderate levels of certain $GABA_A$ receptor subunits,[17–19] indicating that the expression program seen in older cells is established shortly after birth. Hippocampal neurons dissociated from E17 rat embryos develop in culture in two morphologically distinct populations, corresponding to pyramidal cells and interneurons.[20] Most strikingly, pyramidal-like cells express the α2/β2,3/γ2 subunit IR, as do pyramidal cells *in vivo*, whereas interneuron-like cells express a conspicuous α1/β2,3/γ2 subunit staining, as *in vivo*. In double-staining experiments, all α1 subunit–positive neurons are also positive for glutamic acid decarboxylase, indicating that they are GABAergic (Fritschy, unpublished).

The cell-type specificity and maturation of $GABA_A$ receptor subtypes *in vitro* are therefore largely similar to those *in vivo*, indicating that they are driven by genetic programs already determined in the late embryonic or early postnatal brain. The functional significance of $GABA_A$ receptor diversity is, however, largely unknown. One possibility is that it provides a substrate for posttranslational modification—e.g., by protein phosphorylation mechanisms,[21–24] to adjust GABAergic inhibition to the requirements of neuronal circuits under physiological and pathophysiological conditions. To test this hypothesis, we have investigated *in vitro* and *in vivo* the role of neuronal activity on the expression levels of $GABA_A$ receptor subtypes in hippocampus and neocortical neurons.

REGULATION OF $GABA_A$ RECEPTOR SUBUNIT EXPRESSION *IN VITRO*

Role of Neuronal Activity

To investigate the role of neuronal activity on the expression of $GABA_A$ receptor subunits, mature hippocampal neuron cultures (14–21 DIV) were treated either with tetrodotoxin (TTX; 5 μM) to block synaptic activity nonselectively, or with the NMDA receptor antagonist MK-801 (50 μM) to block the major source of synaptic Ca^{2+} influx. Both substances resulted in an approximately 30% loss of the α1, α2, and γ2 subunit immunoreactivity (IR), as measured by densitometric analysis after 1- and 12-hour treatment. Qualitatively, the effects of TTX and of MK-801 appeared largely similar, with the $GABA_A$ receptor subunit staining disappearing from the cell surface (FIG. 1). The rapid onset of this effect argues against a change in gene expression. Rather, these results suggest that cell surface expression of $GABA_A$ receptor subtypes can be regulated by neuronal activity in cultured hippocampal neurons. Furthermore, the observation that $GABA_A$ receptor subunit staining is reduced to the same extent by selective blockade of NMDA receptors with MK-801 as by overall blockade of synaptic transmission with TTX points

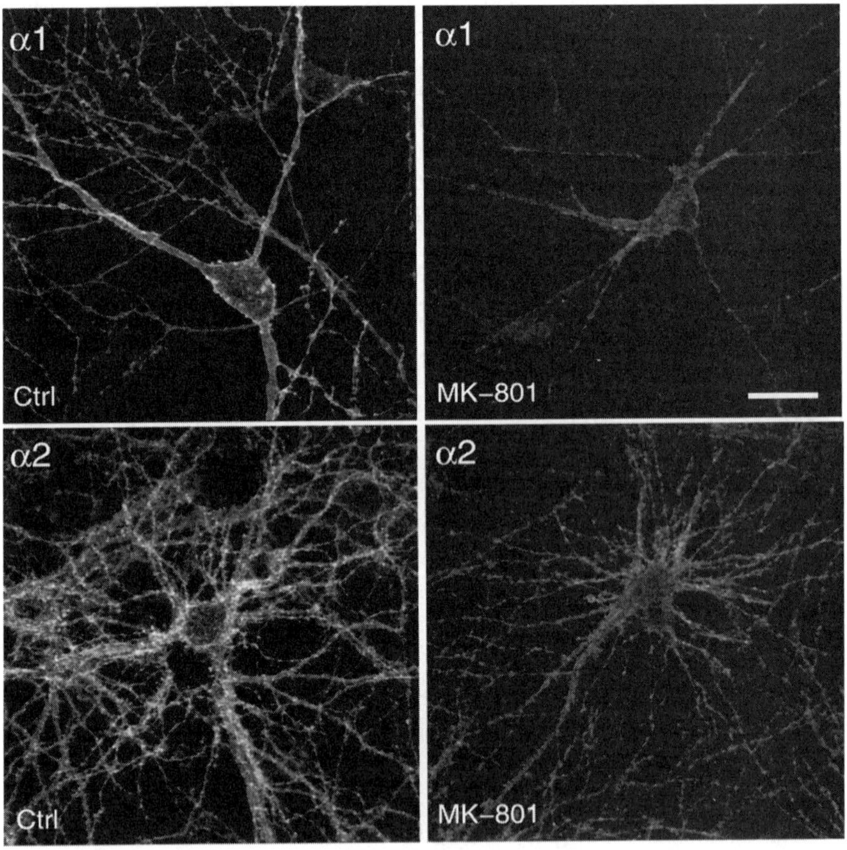

FIGURE 1. Activity-dependent modulation of $GABA_A$ receptor α1 and α2 subunit IR in cultured hippocampal neurons, as visualized by immunofluorescence staining with subunit-specific antisera. Following 12-hour incubation with the NMDA receptor antagonist MK-801, a profound reduction of staining intensity was evident in both types of neurons. Each panel depicts the superposition of 8 consecutive optical images from confocal laser scanning microscopy spaced by 0.3 μm. *Scale bar*, 25 μm.

to a role for Ca^{2+}-dependent mechanisms in the regulation of $GABA_A$ receptor surface expression.

Role of BDNF

The neurotrophin brain-derived neurotrophic factor (BDNF), which is a candidate factor mediating short-term effects of neuronal activity in the hippocampus, has been shown to acutely reduce GABAergic transmission in pyramidal cells via a postsynaptic mechanism.[25] The effects of BDNF on the expression of $GABA_A$ receptors were assessed in cultured hippocampal neurons upon chronic treatment with 20 ng/ml BDNF once a day (starting at 16 DIV) for five consecutive days. This treatment schedule resulted in a marked decrease (40–50%) of $GABA_A$ receptor subunit staining in both $\alpha 1$ and $\alpha 2$ subunit–positive neurons. The $\alpha 1$, $\alpha 2$, and $\gamma 2$ subunit IR outlining neurites and somata had largely disappeared, and the residual staining appeared uniformly distributed intracellularly, as seen by confocal laser scanning microscopy. No overt difference in neuronal morphology was apparent between BDNF-treated and control cells, indicating that the change of $GABA_A$ receptor subunit IR was not due to a nonspecific effect of BDNF on cell differentiation.

To determine whether acute BDNF exposure was sufficient to produce this reduction of $GABA_A$ receptor cell surface IR, 14–21 DIV cultures were treated once with 20 ng/ml BDNF. The changes in $GABA_A$ receptor subunit staining intensity were quantified 1–24 hours later by densitometric analysis. A substantial decrease of IR was already detected after 1-hour exposure. The decrease reached a maximum after 3–12 hours, and subsided thereafter to return to control levels after 24 hours. This second phase indicates that the effect of acute BDNF treatment is reversible, due either to a desensitization process or to the degradation of BDNF in the culture dish. As observed after chronic, 5-day BDNF exposure, the reduction in staining intensity was apparently due to a redistribution of the $GABA_A$ receptor subunits, with loss of membrane staining and increased intracellular IR. The specificity of the acute effects of BDNF on $GABA_A$ receptor subunit staining was assessed by cotreatment of 20 DIV neurons with BDNF and 200 nM K252, a protein kinase inhibitor with preference for trk-kinases.[26] No change in staining intensity was detected after this treatment for 1 hour, indicating that binding of BDNF to trkB is required for reducing cell surface expression of $GABA_A$ receptor subunits.[20]

BDNF-Induced Dephosphorylation of $GABA_A$ Receptor Subunits

To investigate whether protein phosphorylation mechanisms contribute to the effects of BDNF on $GABA_A$ receptor cell surface expression, hippocampal neurons were treated with BDNF in the presence of 80 pM cypermethrin, an inhibitor of the phosphatase calcineurin.[27] Calcineurin is the major serine/threonine phosphatase expressed by neurons and has been implicated in the regulation of ion channel function.[28] Immunohistochemical analysis revealed that this treatment for 1 or 12 hours completely prevented the decrease of $GABA_A$ receptor subunit IR induced by BDNF alone. Addition of cypermethrin in the absence of BDNF had no effect on $GABA_A$ receptor subunit IR.[20]

Functional Significance

These results demonstrate that blockade of synaptic transmission or of NMDA receptor–mediated Ca^{2+} influx reduces $GABA_A$ receptor expression in cultured neurons. The same effect can be achieved by exposure to BDNF, dependent on trk receptor and on calcineurin activity. The reduction was apparently due to a redistribution of $GABA_A$ receptor subunits, with loss of membrane staining and increased intracellular immunoreactivity. Since all subunits were affected to the same degree, it is likely that functional receptor complexes, rather than isolated subunit proteins, are removed from the neuronal membrane. These results suggest that posttranslational mechanisms, including protein phosphorylation,[29,30] play a major role in regulating the cell surface expression and thereby the function of $GABA_A$ receptors in hippocampal neurons. Such an effect might account for the decreased amplitude of IPSCs induced by BDNF in adult CA1 pyramidal neurons.[25] Thus, $GABA_A$ receptors themselves appear to be a major target for the short-term modulation of inhibitory synaptic transmission by BDNF.

Recently, it was shown that insulin induces a rapid recruitment of $GABA_A$ receptor subunits to the plasma membrane of hippocampal neurons through a process requiring tyrosine kinase activity.[31] Thus, insulin elicits in hippocampal neurons a response opposite to that of BDNF, indicating that rapid signaling through trks can either increase or decrease the surface expression of $GABA_A$ receptors, depending on the intracellular pathway involved.

The regulation of $GABA_A$ receptor membrane expression by growth factors such as BDNF and insulin[31] might reflect a general mechanism regulating neurotransmitter receptor distribution during activity-dependent plasticity. Recruitment of neurotransmitter receptors to synaptic sites has been proposed as a potential mechanism of long-term synaptic enhancement.[32] Growth factor–induced phosphorylation of receptors might enable a rapid, activity-dependent insertion of new receptors into the synaptic membrane. Consistent with these ideas, BDNF increases the phosphorylation of the NMDA receptor NR1 subunit.[33] At the same time, a BDNF-induced decrease in the number of $GABA_A$ receptors at postsynaptic sites could contribute to enhance synaptic transmission. A similar mechanism might also be operative under pathophysiological conditions and account for the reduced $GABA_A$ receptor–mediated responses after epileptic seizures,[34] which are known to cause dramatic increases in BDNF expression.[35,36] While direct evidence for receptor recruitment/removal during synaptic plasticity is still lacking, the finding that neurotrophic factors modulate cell-surface expression of neurotransmitter receptors adds plausibility to this hypothesis.

REGULATION OF $GABA_A$ RECEPTOR SUBUNIT EXPRESSION *IN VIVO*

Area-Specific Expression in Developing Neocortex

$GABA_A$ receptors containing the α1 and α5 subunits display a reciprocal area-specific distribution in developing neocortex. This pattern is most apparent in layers III-IV of primary visual (V1) and somatosensory (S1) cortex and is best visualized by immunohistochemistry at postnatal day 7 (P7) (FIGS. 2, 3). The somatotopic organization is strikingly

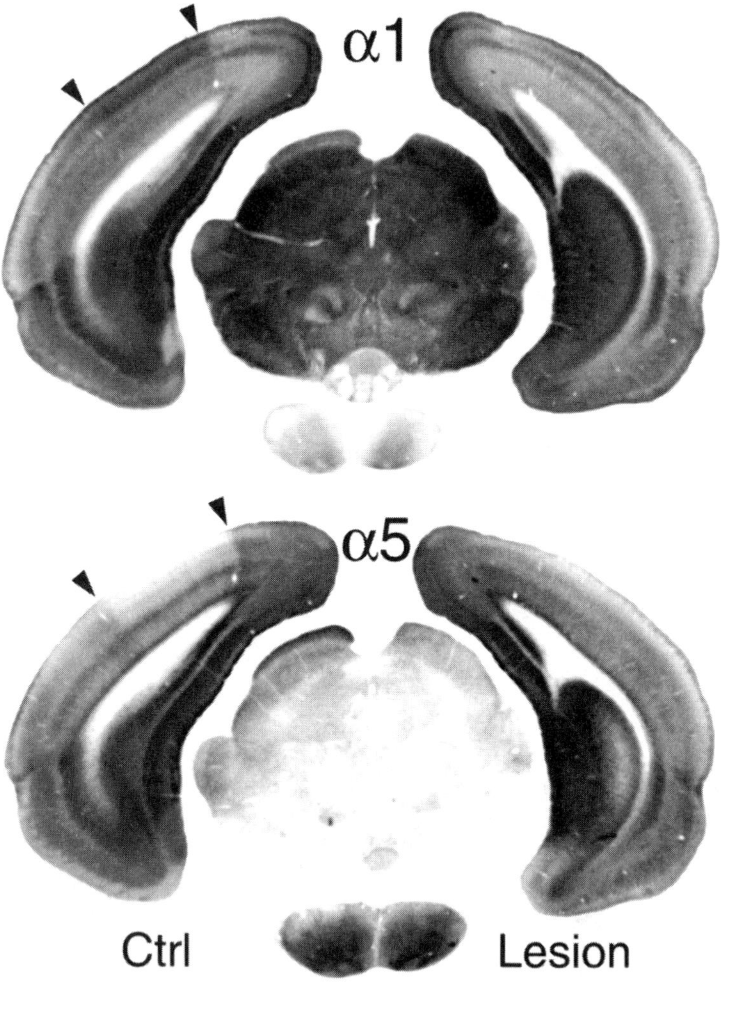

FIGURE 2. Effect of an electrolytic lesion of the LGN on the area-specific distribution of the α1 and α5 subunit IR in V1, as seen at P7 by immunoperoxidase staining with subunit-specific antisera. On the control (Ctrl), unlesioned side, the boundaries of V1 (*arrowheads*) were clearly demarcated by either an intense α1 subunit IR, or a lack of α5 subunit IR, in the superficial layers. On the lesioned side, the increased α5 subunit IR in V1 resulted in the disappearance of the boundaries of V1, whereas the α1 subunit IR was clearly reduced.

revealed in S1, where distinct patches of intense α1 subunit IR correspond to the barrels, the cortical representations of individual whiskers of the rat snout. The pattern of α1 subunit IR in layers III-IV precisely reflects the histochemical distribution of thalamocortical afferents in V1 and S1 at P7.[37,38] In striking contrast, the α5 subunit IR displays a reciprocal pattern of distribution, being almost completely absent in the regions of high α1 sub-

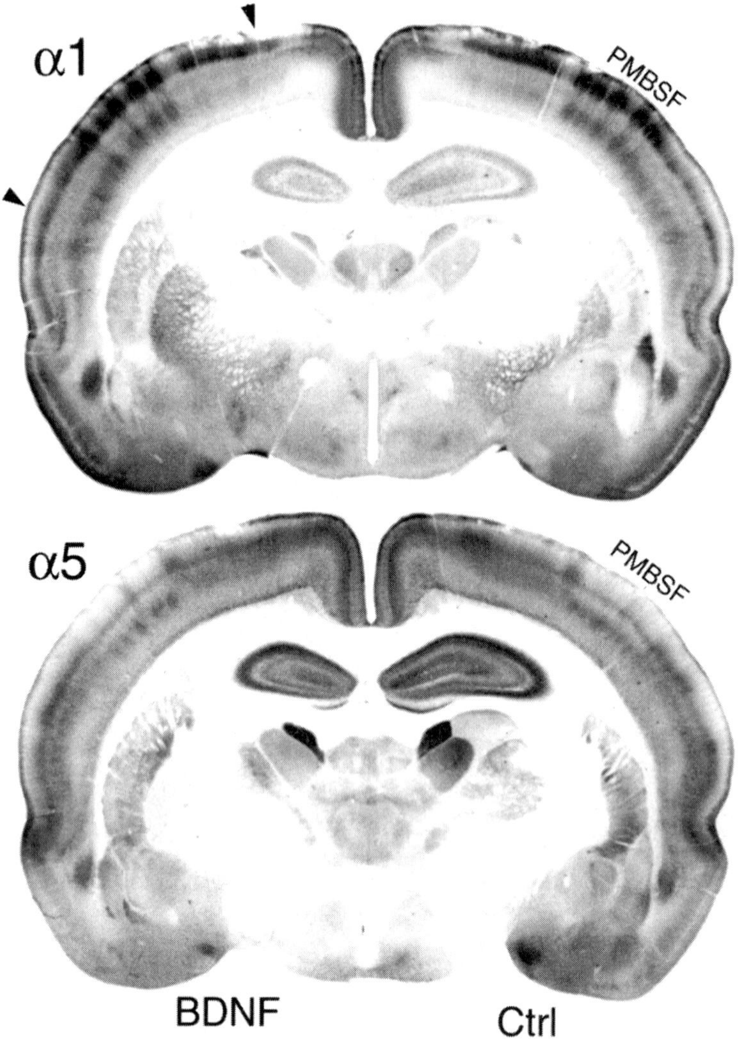

FIGURE 3. Effects of local, chronic application of BDNF released from Elvax polymers placed at birth on the neocortex on expression of the $GABA_A$ receptor α1 and α5 subunit IR. The areal and laminar distribution of the α1 and α5 subunits were visualized at P7 in transverse sections processed for immunoperoxidase staining with subunit-specific antisera. *Arrowheads* in the BDNF-treated hemisphere (**left**) point to the location of the implant, which entirely covered S1. Note the opposite expression pattern of the α1 and α5 subunits in layers III-IV of S1, with the α1 subunit IR being most intense in the barrels and the α5 subunit IR lacking in the corresponding sector. No difference in the distribution and staining intensity of these subunits is apparent in the control hemisphere (**right**), indicating that chronic BDNF treatment did not affect the expression of $GABA_A$ receptors.

unit expression (layers III-IV of S1 and V1) but enriched in association areas lacking the α1 subunit. This complementary expression pattern, which emerges from very low levels of both the α1 and α5 subunit IR at birth in the neocortex, suggests a reciprocal regulation of α1 and α5 subunit expression by thalamocortical afferents in V1 and S1 during the first postnatal week.[39]

Effect of Thalamic Lesions

To directly determine the role of thalamic input on GABA$_A$ receptor subunit expression in V1, unilateral electrolytic ablations of the lateral geniculate nucleus (LGN) were performed in neonates, and the pattern of α1 and α5 subunit staining was analyzed immunohistochemically in V1 at P7, with the contralateral side serving as control.[39] Ablation of the LGN resulted in a profound loss of α1 subunit staining in layers I and III-IV of V1, which normally receive most of the innervation from the LGN, whereas the α5 subunit IR was increased (FIG. 2). The lesion-induced up-regulation of α5 subunit expression resulted in a laminar distribution in V1 similar to that in the adjacent association areas receiving only sparse thalamic innervation. At P7, the areal boundaries of V1 normally formed by the α5 subunit IR were therefore no longer apparent on the lesioned side (FIG. 2). The area-specific expression of GABA$_A$ receptor subtypes results therefore from interactions between an intrinsic program operative in the neocortex, which is seen on the lesioned side, and extrinsic factors provided by the thalamic innervation, which selectively modify the expression of the α1 and α5 subunits in primary sensory areas.

Effect of Chronic Modulation of Neuronal Activity on GABA$_A$ Receptor Expression

To search for these factors, the effects of chronic manipulation of neuronal activity were investigated in S1. In particular, since NMDA receptor–dependent mechanisms and BDNF have been implicated in synaptic plasticity and in the formation of topographic maps during development, we analyzed the effect of chronic, topical delivery of MK-801 and of BDNF on the expression of GABA$_A$ receptor subunits. The drugs were loaded in Elvax polymers[40,41] that were implanted unilaterally over the parietal cortex of newborn rats. After 7 days, alterations in GABA$_A$ receptor expression were investigated immunohistochemically in S1, using the contralateral, untreated side as control. No change in the laminar distribution or staining intensity of the α1 and α5 subunits could be detected at P7 (FIG. 3). Thus, the intense α1 subunit IR clearly delineating barrels was indistinguishable in control and treated hemispheres. On adjacent sections, the corresponding barrels lacked the α5 subunit IR on both sides of the brain (FIG. 3). The characteristic laminar distribution of these subunits in the other layers of cortex was also not affected by these pharmacological manipulations. In control experiments, release of ^{3}H-MK-801 from the Elvax polymers was found to yield pharmacologically relevant concentrations in the underlying cortical tissue during the entire treatment period (Penschuck *et al.*, submitted for publication).

These results contrast with the effects of unilateral thalamic lesions, which result in a lack of differentiation of barrels and in marked changes of the distribution of GABA$_A$ receptor subunits in the affected hemisphere.[39] Topical blockade of NMDA receptors or release of BDNF, expected to profoundly influence the overall level of neuronal activity in neocortex (see below), are by themselves not sufficient to reproduce the effects of cortical deafferentation, suggesting that additional factors are involved.

Effect of Chronic Modulation of Neuronal Activity on Cortical Growth

The growth rate of cortical tissue during development correlates with neuronal activity. To provide an independent index of the effects of MK-801 and BDNF, the cross-sectional area on the main barrels in the posteromedial barrel subfield (PMBSF) was measured in drug-treated and control hemispheres, using tangential sections stained for the $\alpha 1$ subunit. In this plane, the entire barrel field can be visualized, allowing the identification and measurement of corresponding barrels on the treated and control sides (FIG. 4). Following drug treatment, surface quantification revealed on average a $16 \pm 0.7\%$ (mean $\pm$ SD; $n = 11$) increase in cross-sectional area of $\alpha 1$ subunit–positive barrels (sum of B1-B3, C1-C3, and D1-D3) in the MK-801–treated hemisphere, whereas BDNF exposure caused an opposite change of the same magnitude ($-16 \pm 1.8\%$; FIG. 4). These differences were statistically significant in paired *t*-tests, in spite of the large interindividual variation of the size of barrels.[42] In control experiments, rats treated with saline-impregnated Elvax were analyzed to verify that the changes of barrel size resulted from the substances released, and not from the surgical procedure or implantation of the Elvax polymers. No difference in the size of barrels between control and saline-treated hemisphere was detected. To determine whether the changes of barrel size were reflected in the entire PMBSF, the total surface area (including barrels and interbarrel septa) of the PMBSF was measured separately. In both drug treatment paradigms, the size of the PMBSF was altered, being enlarged by about 13 $\pm$ 2% following MK-801 treatment, and decreased by $11 \pm 4\%$ following BDNF treatment (FIG. 4). These values are within the same range as those observed for individual barrels, indicating that the pharmacological manipulations uniformly influenced the growth of entire PMBSF (Penschuck *et al.*, submitted for publication).

These results confirm that the level of neuronal activity influences the growth of cortical modules during development. While the changes in size of barrels were small, only minor anatomical changes were detected in experiments inducing pronounced changes in the size of functional barrels.[43] Furthermore, the effects of chronic drug treatment were measured at a single time point, and it cannot be ruled out that larger effects could have been detected earlier (or later).

Most strikingly, the changes in barrel size induced by BDNF and MK-801 are the opposite to those expected from the pharmacological profile of these drugs. By antagonizing NMDA receptor–mediated transmission, MK-801 was expected to lead to a decrease in barrel size. Conversely, BDNF is known to potentiate excitatory neurotransmission,[44] and was expected to increase the size of barrels. In light of the findings of Riddle *et al.*[45,46] the present results suggest that MK-801 actually increases, and BDNF decreases, the overall level of activity in developing neocortex. This hypothesis is plausible, considering that GABAergic interneurons, notably those in layer IV, play a key role in controlling cortical activity. By blocking NMDA receptors on these neurons, MK-801 might release pyramidal neurons in the supragranular layers from inhibition. Conversely, since these GABAergic neurons express high levels of trkB receptors, they are likely to be a preferential target of BDNF. By activating these cells, BDNF would enhance inhibitory activity in the cor-

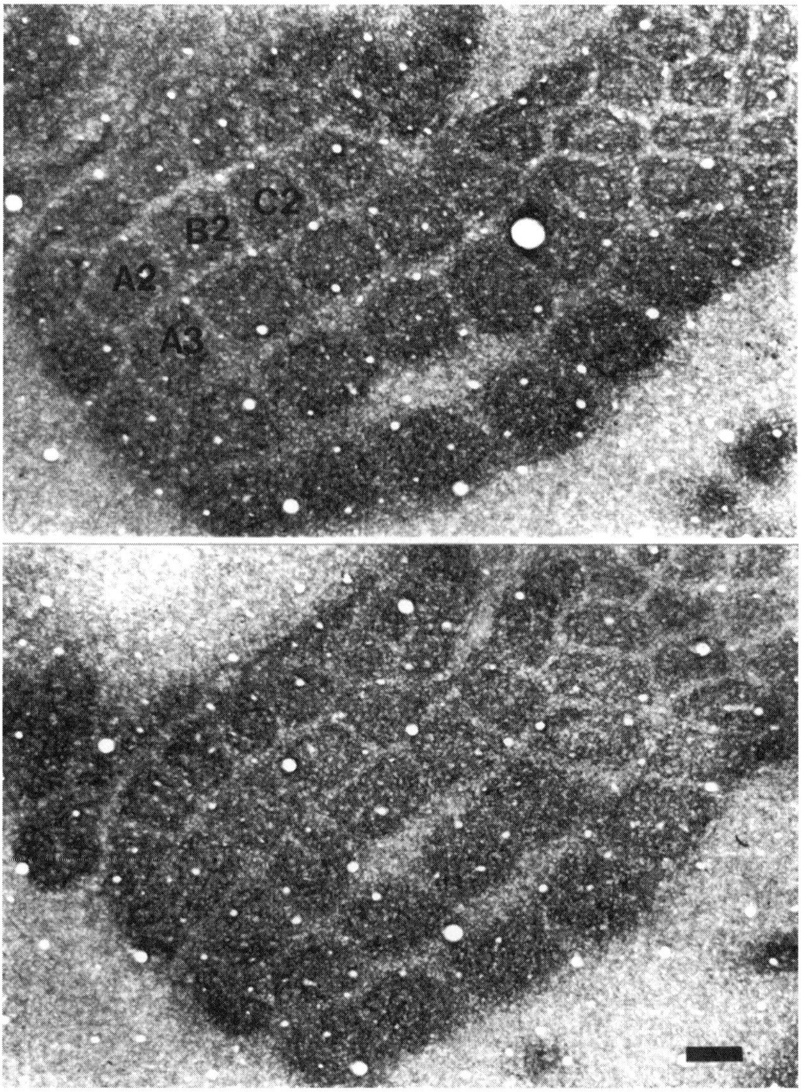

FIGURE 4. Reduction of the cross-sectional area of barrels in PMBSF induced by chronic local application of BDNF. Barrels were visualized in tangential sections processed for immunoperoxidase staining with an antiserum recognizing the $GABA_A$ receptor α1 subunit. Compared to the control, untreated hemisphere (**top panel**), a distinct reduction in barrel size is evident in the BDNF-treated hemisphere (**bottom panel**) from the same animal. *Scale bar,* 200 μm.

tex.[47] Additional support for this theory is provided by the fact that general blockade of synaptic activity by TTX or of glutamatergic activity (both NMDA and AMPA receptors) by AP5 does not result in an increase of barrel size.[40,48] In addition, an inhibitory effect of BDNF has been demonstrated *in vivo* by optical imaging, with focal application of BDNF causing a rapid decrease of the area of barrel representations in adult rat S1.[49] While the present hypothesis remains to be verified, it would imply that GABAergic neurons play a major role in mediating the effect of activity-dependent mechanisms on cortical maturation.

CONCLUSIONS

The neuron-specific expression of $GABA_A$ receptor subtypes in adult mammalian brain is the result of complex interactions between genetic programs determined during embryogenesis, before the period of synapse formation, and of epigenetic effects, which include posttranslational regulation mechanisms such as protein phosphorylation. Our studies suggest that activity-dependent mechanisms modulate short-term regulation of $GABA_A$ receptor function, but do not play a preeminent role in the establishment of neuron-specific expression patterns of $GABA_A$ receptor subtypes in developing brain.

REFERENCES

1. Davies, P.A., M.C. Hanna, T.G. Hales & E.F. Kirkness. 1997. Insensitivity to anaesthetic agents conferred by a class of $GABA_A$ receptor subunit. Nature **385:** 820–823.
2. MacDonald, R.L. & R.W. Olsen. 1994. $GABA_A$ receptor channels. Annu. Rev. Neurosci. **17:** 569–602.
3. Mohler, H., D. Benke & J.M. Fritschy. 1997. The $GABA_A$ receptors. *In* Molecular and Cellular Targets for Anti-Epileptic Drugs. G. Avanzini, G. Regesta, P. Tanganelli & M. Avoli, Eds.: 39–53. John Libbey & Co. Ltd. London.
4. Mohler, H., J.M. Fritschy, B. Luscher, U. Rudolph, J. Benson & D. Benke. 1996. The $GABA_A$-receptors: From subunits to diverse functions. *In* Ion channels, Vol. 4. T. Narahashi, Ed.: 89–113. Plenum Press. New York.
5. Whiting, P.J., G. McAllister, D. Vasilatis, T.P. Bonnert, R.P. Heavens, D.W. Smith, L. Hewson, R. O'Donnell, M.R. Rigby, D.J.S. Sirinathsinghji, G. Marshall, S.A. Thompson & K.A. Wafford. 1997. Neuronally restricted RNA splicing regulates the expression of a novel $GABA_A$ receptor subunit conferring atypical functional properties. J. Neurosci. **17:** 5027–5037.
6. Fritschy, J.M. & H. Mohler. 1995. $GABA_A$-receptor heterogeneity in the adult rat brain: Differential regional and cellular distribution of seven major subunits. J. Comp. Neurol. **359:** 154–194.
7. McKernan, R.M. & P.J. Whiting. 1996. Which $GABA_A$ receptor subtypes really occur in the brain? Trends Neurosci. **19:** 139–143.
8. Benke, D., S. Mertens, A. Trzeciak, D. Gillessen & H. Mohler. 1991. $GABA_A$ receptors display association of γ2-subunit with α1- and β2/3 subunits. J. Biol. Chem. **266:** 4478–4483.
9. Fritschy, J.M., D. Benke, S. Mertens, W.H. Oertel, T. Bachi & H. Mohler. 1992. Five subtypes of type A γ-aminobutyric acid receptors identified in neurons by double and triple immunofluorescence staining with subunit-specific antibodies. Proc. Natl. Acad. Sci. USA **89:** 6726–6730.
10. Somogyi, P., J.M. Fritschy, D. Benke, J.D.B. Roberts & W. Sieghart. 1996. The γ2 subunit of the $GABA_A$-receptor is concentrated in synaptic junctions containing the α1 and β2/3 subunits in hippocampus, cerebellum and globus pallidus. Neuropharmacology **35:** 1425–1444.

11. GAO, B. & J.M. FRITSCHY. 1994. Selective allocation of GABA$_A$-receptors containing the α1-subunit to neurochemically distinct subpopulations of hippocampal interneurons. Eur. J. Neurosci. **6:** 837–853.
12. GAO, B., J.M. FRITSCHY, D. BENKE & H. MOHLER. 1993. Neuron-specific expression of GABA$_A$ receptor subtypes: Differential associations of the α1- and α3-subunits with serotonergic and GABAergic neurons. Neuroscience **54:** 881–892.
13. GAO, B., J.P. HORNUNG & J.M. FRITSCHY. 1995. Identification of distinct GABA$_A$-receptor subtypes in cholinergic and parvalbumin-positive neurons of the rat and marmoset medial-septum-diagonal band complex. Neuroscience **65:** 101–117.
14. BENKE, D., J.M. FRITSCHY, A. TRZECIAK, W. BANNWARTH & H. MOHLER. 1994. Distribution, prevalence and drug-binding profile of GABA$_A$-receptors subtypes differing in β-subunit isoform. J. Biol. Chem. **269:** 27100–27107.
15. MARKSITZER, R., D. BENKE, J.M. FRITSCHY & H. MOHLER. 1993. GABA$_A$-receptors: Drug binding profile and distribution of receptors containing the α2-subunit in situ. J. Recept. Res. **13:** 467–477.
16. GAO, B. & J.M. FRITSCHY. 1995. Cerebellar granule cells in vitro recapitulate the in vivo pattern of GABA$_A$-receptor subunit expression. Dev. Brain Res. **88:** 1–16.
17. BEATTIE, C.E. & R.E. SIEGEL. 1993. Developmental cues modulate GABA$_A$ receptor subunit mRNA expression in cultured cerebellar granule neurons. J. Neurosci. **13:** 1784–1792.
18. BEHRINGER, K.A., L.M. GAULT & R.E. SIEGEL. 1996. Differential regulation of GABA receptor subunit mRNAs in rat cerebellar granule neurons—importance of environmental cues. J. Neurochem. **66:** 1347–1353.
19. GAULT, L.M. & R.E. SIEGEL. 1997. Expression of the GABA$_A$ receptor δ subunit is selectively modulated by depolarization in cultured rat cerebellar granule neurons. J. Neurosci. **17:** 2391–2399.
20. PENSCHUCK, S. *et al.* 1997. Soc. Neurosci. Abstr. 23, 45.
21. LEIDENHEIMER, N.J., S.J. MCQUILKIN, L.D. HAHNER, P. WHITING & R.A. HARRIS. 1992. Activation of protein kinase C selectively inhibits the γ-aminobutyric acid$_A$ receptor: Role of desensitization. Mol. Pharmacol. **41:** 1116–1123.
22. MOSS, S.J., G.H. GORRIE, A. AMATO & T.G. SMART. 1995. Modulation of GABA$_A$ receptors by tyrosine phosphorylation. Nature **377:** 344–348.
23. MOSS, S.J. & T.G. SMART. 1996. Modulation of amino-acid gated ion channels by protein phosphorylation. Int. Rev. Neurobiol. **39:** 1–52.
24. WHITING, P., R.M. MCKERNAN & L.L. IVERSEN. 1990. Another mechanism for creating diversity in γ-aminobutyrate type A receptors: RNA splicing directs expression of two forms of γ2-subunit, one of which contains a protein kinase C phosphorylation site. Proc. Natl. Acad. Sci. USA **87:** 9966–9970.
25. TANAKA, T., H. SAITO & N. MATSUKI. 1997. Inhibition of GABA$_A$ synaptic responses by brain-derived neurotrophic factor (BDNF) in rat hippocampus. J. Neurosci. **17:** 2959–2966.
26. KOIZUMI, S., M.L. CONTRERAS, Y. MATSUDA, T. HAMA, P. LAZAROVICI & G. GUROFF. 1988. K-252a: A specific inhibitor of the action of nerve growth factor on PC12 cells. J. Neurosci. **8:** 715–721.
27. ENAN, E. & F. MATSUMURA. 1992. Specific inhibition of calcineurin by type II synthetic pyrethroid insecticides. Biochem. Pharmacol. **43:** 1777–1784.
28. YAKEL, J.L. 1997. Calcineurin regulation of synaptic function—From ion channels to transmitter release and gene transcription. Trends Pharmacol. Sci. **18:** 124–134.
29. HUGANIR, R.L. & P. GREENGARD. 1990. Regulation of neurotransmitter receptor desensitization by protein phosphorylation. Neuron **5:** 555–567.
30. SMART, T.G. 1997. Regulation of excitatory and inhibitory neurotransmitter-gated ion channels by protein phosphorylation. Curr. Opin. Neurobiol. **7:** 358–367.
31. WAN, Q., Z.G. XIONG, H.Y. MAN, C.A. ACKERLEY, J. BRAUNTON, W.Y. LU, L.E. BECKER, J.F. MACDONALD & Y.T. WANG. 1997. Recruiment of functional GABA$_A$ receptors to postsynaptic domains by insulin. Nature **388:** 686–690.
32. MALENKA, R.C. & R.A. NICOLL. 1997. Silent synapses speak up. Neuron **19:** 473–476.
33. SUEN, P.C., K. WU, E.S. LEVINE, H.T.J. MOUNT, J.L. XU, S.Y. LIN & I.B. BLACK. 1997. Brain-derived neurotrophic factor rapidly enhances phosphorylation of the postsynaptic N-methyl-D-aspartate receptor subunit 1. Proc. Natl. Acad. Sci. USA **94:** 8191–9195.

34. Isokawa, M. 1996. Decrement of $GABA_A$ receptor–mediated inhibitory postsynaptic currents in dentate granule cells in epileptic hippocampus. J. Neurophysiol. **75:** 1901–1908.
35. Ernfors, P., J. Bengzon, Z. Kokaia, H. Persson & O. Lindvall. 1991. Increased levels of messenger RNAs for neurotrophic factors in the brain during kindling epileptogenesis. Neuron **7:** 165–176.
36. Gall, C.M., J.C. Lauterborn, K.M. Guthrie & C.T. Stinis. 1997. Seizures and the regulation of neurotrophic factor expression: Associations with structural plasticity in epilepsy. Adv. Neurol. **72:** 9–24.
37. Broide, R.S., R.T. Robertson & F.M. Leslie. 1996. Regulation of α7 nicotinic acetylcholine receptors in the developing rat somatosensory cortex by thalamocortical afferents. J. Neurosci. **16:** 2956–2971.
38. Schlaggar, B.L. & D.D.M. O'Leary. 1994. Early development of the somatotopic map and barrel patterning in rat somatosensory cortex. J. Comp. Neurol. **346:** 80–96.
39. Paysan, J., A. Kossel, J. Bolz & J.M. Fritschy. 1997. Area-specific regulation of γ-aminobutyric acid A receptor subtypes by thalamic afferents in developing rat neocortex. Proc. Natl. Acad. Sci. USA **94:** 6995–7000.
40. Schlaggar, B.L., K. Fox & D.D.M. O'Leary. 1993. Postsynaptic control of plasticity in developing somatosensory cortex. Nature **364:** 623–626.
41. Simon, D.K., G.T. Prusky, D.D.M. O'Leary & M. Constantine-Paton. 1992. N-methyl-D-aspartate receptor antagonist disrupts the formation of a mammalian neural map. Proc. Natl. Acad. Sci. USA **89:** 10593–10597.
42. Riddle, D.R. & D. Purves. 1995. Individual variation and lateral asymmetry of the rat primary somatosensory cortex. J. Neurosci. **15:** 4184–4195.
43. Fox, K. 1992. A critical period for experience-dependent synaptic plasticity in rat barrel cortex. J. Neurosci. **12:** 1825–1838.
44. Carmignoto, G., T. Pizzorusso, S. Tia & S. Vicini. 1997. Brain-derived neurotrophic factor and nerve growth factor potentiate excitatory synaptic transmission in the rat visual cortex. J. Physiol. **498:** 153–164.
45. Riddle, D.R., G. Gutierrez, D. Zheng, L.E. White, A. Richards & D. Purves. 1993. Differential metabolic and electrical activity in the somatic sensory cortex of juvenile and adult rats. J. Neurosci. **13:** 4193–4213.
46. Riddle, D., A. Richards, F. Zsuppan & D. Purves. 1992. Growth of the rat somatic sensory cortex and its constituent parts during postnatal development. J. Neurosci. **12:** 3509–3524.
47. Thoenen, H. 1995. Neurotrophins and neuronal plasticity. Science **270:** 593–598.
48. Chiaia, N.L., S.E. Fisch, W.R. Bauer, C.A. Bennett-Clarke & R.W. Rhoades. 1992. Postnatal blockade of cortical activity by tetrodotoxin does not disrupt the formation of vibrissa-related patterns in the rat's somatosensory cortex. Dev. Brain Res. **66:** 244–250.
49. Prakash, N., S. Cohen-Cory & R.D. Frostig. 1996. Rapid and opposite effects of BDNF and NGF on the functional organization of the adult cortex in vivo. Nature **381:** 702–706.

Structure and Functions of Inhibitory and Excitatory Glycine Receptors

HEINRICH BETZ,[a] JOCHEN KUHSE, VOLKER SCHMIEDEN,[b] BODO LAUBE, JOACHIM KIRSCH,[c] AND ROBERT J. HARVEY[d]

Max-Planck-Institut für Hirnforschung, Abteilung Neurochemie, Deutschordenstrasse 46, 60528 Frankfurt am Main, Germany

ABSTRACT: The strychnine-sensitive glycine receptor (GlyR) is a pentameric chloride channel protein that exists in several developmentally and regionally regulated isoforms in the CNS. These result from the differential expression of four genes encoding different variants (α1–α4) of the ligand-binding subunit of the GlyR. Their assembly with the structural β subunit is governed by "assembly cassettes" within the extracellular domains of these proteins and creates chloride channels of distinct conductance properties. GlyR gating is potentiated by Zn^{2+}, a metal ion co-released with different neurotransmitters. Site-directed mutagenesis has unraveled major determinants of agonist binding and Zn^{2+} potentiation. During development, glycine receptors mediate excitation that results in Ca^{2+} influx and neurotransmitter release. Ca^{2+} influx triggered by the activation of embryonic GlyRs is required for the synaptic localization of the GlyR and its anchoring protein gephyrin. In the adult, mutations in GlyR subunit genes result in motor disorders. The *spastic* and *spasmodic* phenotypes in mouse as well as human hereditary startle disease will be discussed.

Glycine is a major inhibitory neurotransmitter in adult spinal cord and brain stem, where it controls both motor and sensory pathways.[1] After its release from the presynaptic terminals of glycinergic interneurons, glycine causes an increase in chloride conductance by activating postsynaptic glycine receptors (GlyRs). This hyperpolarizing action of glycine is selectively antagonized by strychnine, the most potent GlyR antagonist known.[2] In embryonic neurons, glycine application induces a strong depolarizing response that can cause Ca^{2+} influx via voltage-dependent calcium channels and thus trigger neurotransmitter release.[3,4] This excitatory action of glycine is also mediated by strychnine-sensitive GlyR channels and results from a more positive equilibrium potential of chloride in differentiated neurons. Due to a decrease of intracellular chloride upon further development, the character of GlyR-mediated membrane currents inverts around birth from excitatory to inhibitory.[5] As discussed below, the excitatory GlyR response seen in embryonic neurons is pivotal for the accumulation of GlyRs at the developing postsynaptic membrane. In addition, it has been postulated to be important in early neocortical development.[6]

[a]Address for correspondence: Dr. Heinrich Betz, MPIH, Department of Neurochemistry, Deutschordenstr. 46, D-60528, Frankfurt/Main, Germany. Phone: +49-69-96769-220; fax: +49-69-96769-441; e-mail: neurochemie@mpih-frankfurt.mpg.de

[b]Present address: Universitätsklinikum Charité, Humboldt-Universität Berlin, Abteilung Neurophysiologie, Tucholskystr. 2, 10117 Berlin, Germany.

[c]Present address: Abteilung Anatomie, Universität Ulm, Albert-Einstein Allee 11, 89069 Ulm, Germany.

[d]Present address: Department of Pharmacology, The School of Pharmacy, 29-39 Brunswick Square, London WC1N 1AX, United Kingdom.

STRUCTURE, ASSEMBLY AND DIVERSITY OF GLyRs

The GlyR has been purified from rat, porcine, and mouse spinal cord by affinity chromatography on aminostrychnine-agarose columns.[7–9] Purified GlyR preparations contain two glycosylated integral membrane proteins of 48 kDa (α) and 58 kDa (β) and an associated peripheral membrane protein of 93 kDa,[10] named gephyrin.[11] The primary structures of GlyR α and β subunits deduced by molecular cloning[12,13] show significant sequence and structural similarity to nicotinic acetylcholine receptor (nAChR), γ-aminobutyric acid type A ($GABA_A$) receptor, and serotonin type 3 ($5HT_3$) receptor proteins.[14] Sequence conservation is particularly high in a conserved cysteine motif in the large N-terminal extracellular domain and in the four hydrophobic membrane-spanning regions (M1–M4).

The quaternary structure of GlyRs has been analyzed using cross-linking and sedimentation techniques.[15] GlyRs purified from adult rat spinal cord are pentameric proteins that contain three α and two β subunits. This stoichiometry closely resembles that of nAChR and $GABA_A$ receptors, which also contain five membrane-spanning subunits. Mutational analysis in combination with heterologous expression in *Xenopus* oocytes and voltage clamp recording has shown that this subunit stoichiometry is determined by regions of sequence divergence within the first half of the N-terminal domain, the so-called assembly boxes, of the GlyR β subunit.[16] Apparently, heteromeric GlyRs originate by a sequential assembly pathway that involves an initial α/β heterodimerization step.

Heterogeneity of the GlyR was first revealed biochemically by the identification of a neonatal GlyR isoform in rat spinal cord, whose α subunit differs in strychnine binding affinity, molecular weight (49 kDa), and immunological properties from the adult 48-kDa polypeptide.[17] GlyR diversity was confirmed further by oocyte expression studies[18] and molecular cloning methods. Initially, peptide sequences derived from affinity-purified adult rat spinal cord GlyRs were used to isolate cDNAs for the adult 48-kDa (α1)[12] and 58-kDa (β) subunits.[13] Subsequently, cDNA clones corresponding to two novel GlyR α subunits, the embryonic α2 and the adult α3 protein, were cloned by homology screening.[19–22] A partial mouse genomic sequence encoding part of a fourth α subunit[23] recently allowed the isolation of full-length α4 subunit cDNAs (Harvey *et al.*, in preparation). *In situ* hybridization showed that the different GlyR subunit genes exhibit unique spatial and temporal expression patterns in spinal cord, brain stem, and some higher brain regions.[22,24–28] GlyR α2 subunit transcripts predominate in the embryonic and neonatal brain and spinal cord, but are replaced postnatally by the α1 or α3 subunit mRNAs. Notably, transcripts for the GlyR β subunit are widely expressed even in brain regions that lack α1, α2, and α3 mRNAs.[24,25] Since the α4 subunit mRNA is not abundant in brain,[23] these findings may indicate that either additional GlyR α subunits remain to be identified, or that the β subunit forms part of another receptor complex.[25] Further diversity of GlyRs arises from alternative splicing; splice variants of the α1,[29] α2,[22] and β subunits[30] have been identified so far. A rat α2 subunit variant (α2*) has also been described[31] that, in contrast to the human or rat α1, α2, and α3 subunits, forms channels of low strychnine affinity and may represent the neonatal GlyR isoform described above. Biochemical analysis and single-channel recordings suggest a homooligoneric structure for these neonatal GlyRs.

THE LIGAND BINDING DOMAIN

Initial evidence that the GlyR ligand binding site resides on α subunits came from photoaffinity labeling experiments using the GlyR antagonist strychnine. Peptide mapping of

[^{3}H]strychnine-labeled GlyR preparations revealed covalent incorporation of this antagonist between amino acids 170 and 220 of the N-terminal domain of the rat GlyR α1 subunit.[32,33] This was confirmed by functional expression studies using cloned GlyR cDNAs. In both *Xenopus laevis* oocytes and mammalian cells, GlyR α subunits form homooligomeric chloride channels that are gated by micromolar concentrations of glycine, taurine, and β-alanine.[34,35] The GlyR β subunit alone does not form functional GlyRs,[36,37] but its incorporation into heteromeric receptors alters several functional aspects of the ion channel (see below). By comparing the GlyRs generated from different α subunits[21,31,34,38,39] in combination with site-directed mutagenesis, several discontinuous domains of the α subunit extracellular domain were shown to contribute to ligand binding. In particular, residue G167 of the α2 subunit (equivalent to G160 in the α1 subunit) was shown to be an important determinant of glycine and strychnine binding,[31] and the two neighboring residues (F159 and Y161 in the α1 subunit) were found to be crucial for agonist selectivity and antagonist efficacy.[39] Another domain in the GlyR α1 subunit, encompassing K200 and Y202, has been demonstrated to also contribute to the strychnine binding site,[40] whereas substitution of residues I111 and A212 alters the potency of the glycinergic agonists β-alanine and taurine.[38]

Naturally occurring mutations of the GlyR α1 subunit (see below) have revealed additional determinants of agonist binding. In the mouse mutant *spasmodic*, a single substitution (A52S) results in a modest reduction of glycine affinity, but does not affect strychnine binding.[41,42] Point mutations in the human GlyR α1 subunit gene that underlie hereditary hyperekplexia have uncovered domains that may link agonist binding and channel gating. Heterologous expression of mutants R271L, R271Q, K276E, or Y279C (residues found in the M2-M3 loop) results in GlyRs that exhibit a decreased sensitivity to glycine and a loss of β-alanine and taurine responses.[43–47] Some of these mutations (R271L/Q and K276E) have been shown to reduce the single-channel conductance and/or the open-channel probability of the expressed GlyRs,[43,46,48] implying that the M2-M3 loop is vital for coupling signal transduction and ligand binding. Mutation of I244N within segment M1 also reduces channel gating,[45] but additionally impairs the efficiency of GlyR expression. Taken together, these data point to a multisite ligand binding/signal transduction mechanism that involves distant segments of the large extracellular domain and residues between transmembrane segments M2 and M3.

Zn^{2+}, A PHYSIOLOGICAL REGULATOR OF CHANNEL FUNCTION?

The divalent cation Zn^{2+} is stored in the synaptic vesicles of different neuronal populations and co-released with the transmitter upon stimulation. Changes in synaptic Zn^{2+} concentrations may therefore play important modulatory roles in synaptic transmission. Electrophysiological data indicate that Zn^{2+} exhibits biphasic effects on both native GlyRs in rat spinal cord neurons and on recombinantly expressed homooligomeric and heteromeric GlyRs.[49,50] At low concentrations (nanomolar and low micromolar) Zn^{2+} potentiates glycine-induced currents, whereas at high micromolar concentrations Zn^{2+} decreases the glycine response.[49,50] Dose-response analysis suggests that both the potentiating and inhibitory effects of Zn^{2+} result from changes in apparent agonist affinity. Using chimeric GlyR subunit cDNA constructs, Laube and colleagues[50] revealed that the positive and negative modulatory effects of Zn^{2+} are mediated by different regions of α subunits, and that the determinants of the potentiating Zn^{2+} binding site are localized between amino acids 74–86 of the rat GlyR α1 subunit. This Zn^{2+} modulation of GlyRs is of potential physiological importance, since GlyRs are known to be expressed in regions that accumulate Zn^{2+} presynaptically—e.g., the developing hippocampus.

SYNAPTIC CLUSTERING BY THE ANCHORING PROTEIN GEPHYRIN

GlyRs are densely clustered within postsynaptic specializations in spinal cord neurons. This ordered arrangement is thought to be mediated by gephyrin, a peripheral membrane protein of 93 kDa that copurifies with GlyRs.[10] Gephyrin is located at the cytoplasmic face of postsynaptic specializations containing GlyRs[51] and binds with high affinity to polymerized tubulin.[52] This has led to the proposal that gephyrin acts as a GlyR-cytoskeleton linker protein. Indeed, compounds that disrupt the integrity of microtubules (e.g., demecolcine) and microfilaments (e.g., cytochalsin D) affect the size and packing density of gephyrin and GlyR clusters in these cultures.[53] These cytoskeletal structures appear to operate antagonistically: microtubules condense GlyR clusters, while microfilaments disperse them. Gephyrin binds to GlyRs via an 18–amino acid motif that lies within the large intracellular loop of the β subunit.[54]

Molecular cloning has elucidated several different isoforms of gephyrin that result from alternative splicing of four distinct exons.[11] Northern blot analysis,[11] *in situ* hybridization,[55] and immunocytochemical studies[56,57] indicate a widespread expression of gephyrin in embryonic and adult rat brain and spinal cord as well as in other tissues. Antisense oligonucleotide treatment of cultured embryonic spinal cord neurons indicates that gephyrin is required for the correct targeting of GlyRs to postsynaptic specializations.[58] Also the addition of strychnine or L-type Ca^{2+} channel blockers to the culture medium has shown that the activation of embryonic GlyRs, resulting in Ca^{2+} influx,[3] is crucial for the formation of gephyrin and GlyR clusters at developing postsynaptic sites.[59] The latter observation assigns an important role in synaptogenesis to the GlyR-mediated excitatory responses seen during embryonic development and suggests that functional verification constitutes an essential step in the proper targeting of postsynaptic receptors to developing synaptic connections.

MOLECULAR PATHOLOGY

In several mammalian species, defects in glycinergic neurotransmission have been implicated in complex motor disorders characterized by hypertonia and an exaggerated startle reflex (reviewed in Ref. 60). Recently, mutations in GlyR subunit genes have been identified in mouse as well as in human startle disease, hereditary hyperekplexia.

The gene responsible for the recessive mouse mutant *spasmodic* ($Glra^{spd}$) is located on mouse chromosome 11[61] at 29.0 cm, a region that exhibits synteny to human chromosome 5q31.3, where the human GlyR α1 subunit gene (GLRA1) has been mapped.[62,63] Homozygous *spasmodic* mice appear normal at rest, but around postnatal day 14 acquire an exaggerated acoustic startle reflex: when subjected to loud noises or handling, animals show rigidity, tremor, and an impaired righting reflex. This phenotype is caused by an alanine-to-serine conversion at position 52 (A52S) of the extracellular domain of the α1 subunit, which lowers the agonist affinity of GlyRs containing the mutant subunit.[41,42] The mouse mutant *oscillator* also carries a mutation in the *Glra1* gene. At three weeks of age, homozygous *oscillator* mice show prolonged periods of rapid tremor, producing extreme rigor and stiffness, and normally die around this time. This is due to a complete loss of the adult α1 GlyR[64] and results from a microdeletion of seven nucleotides within exon 8 of *Glra1*.[65] The *spastic* mutation *($Glrb^{spa}$)* in contrast maps to mouse chromosome 3 at 38.5 cM.[66–68] Homozygous *spastic* mice have a phenotype similar to that of *spasmodic* animals; at 14 days of age they suffer from muscle spasms, rapid tremor, stiffness of posture, and difficulty in righting.[7] However, unlike GlyR levels in *spasmodic* mice, GlyR levels in *spastic* homozygotes are drastically reduced.[7,69] The *spastic* phenotype has been shown[67,68] to be due to the insertion of a LINE-1 transposable element into intron 5 of the mouse GlyR β subunit gene *(Glrb),* which induces aberrant splicing with "skipping" of exons 4 and/or 5 and thus reduces the level of full-length β subunit transcripts. Introduction of a transgene encoding the rat GlyR β subunit into the $Glrb^{spa}$ genetic background rescued the *spastic* phenotype, thus confirming its causal link with the LINE-1 element insertion in *Glrb.*[70] To date, no mutations have been identified in the mouse α2, α3, or α4 subunit genes, which map to chromosome X at 71.5 cM,[71] chromosome 8 at 25.0 cM,[72] and chromosome X at 56.0 cM,[23] respectively.

Hereditary hyperekplexia or startle disease is a human autosomal neurological disorder whose symptoms closely resemble sublethal strychnine poisoning (reviewed in Refs. 60, 73). Affected individuals show an exaggerated startle reaction in response to sudden stimuli, such as noise, light, or touch, which can result in general rigidity triggering loss of posture and unprotected falling. In some babies and young infants, the stiff baby syndrome is seen, an excessive startle reaction involving strong muscle spasms that may result in apnea and death. These symptoms normally ameliorate with age, and in most cases adults experience only a mild acoustic startle reaction. Genetic linkage analysis of two large families mapped hyperekplexia to human chromosome 5q32, the GLRA1 locus,[74] and identified two point mutations at the same position (R271Q or R271L) of the short extracellular loop linking segments M2 and M3.[63] Subsequently, further mutations of GLRA1 have been discovered that cause both dominant and recessive forms of startle disease. These include other substitutions in the M2-M3 loop (K276E and Y279C)[75,76] and within segments M1 (I244N)[77] and M2 (Q266H).[78] In addition, a sporadic case of a recessive form of hyperekplexia has been found that results from a deletion of the first six exons of GLRA1.[79] A recent report[80] has also uncovered families with hyperekplexia-like syndromes that do not have mutations in GLRA1, suggesting that mutations in other genes involved in glycinergic neurotransmission might also cause hyperekplexia. Candidates include the human GlyR β subunit locus (GLRB) which has been mapped to human chromosome 4q32[81] and the α2, α3, and α4 subunit genes, which have been localized to human chromosomes Xp21.2–p22.1 (GLRA2),[20] 4q33–q34 (GLRA3; Nikolic & Becker, unpublished data), and Xq21–q22 (GLRA4; Harvey & Betz, unpublished data).

PERSPECTIVES

The past years have seen considerable progress in our understanding of GlyR structure, function, and pathology. At present, major questions concern the functions of GlyRs during development and the mechanisms that regulate the synaptic targeting of these membrane proteins. In addition, the role of GlyR mutations in motor disorders like hyperekplexia will have to be resolved further. Finally, compounds that potentiate GlyR function in a fashion similar to Zn^{2+} have considerable promise as novel anticonvulsive, analgesic, and muscle relaxant drugs.

ACKNOWLEDGMENTS

The work from our laboratory summarized in this article is supported by Deutsche Forschungemeinschaft, EC Program BIOMED 2, and Fonds der Chemischen Industrie.

We thank M. Baier and H. Reitz for assistance with the preparation of this manuscript.

REFERENCES

1. Aprison, M.H. 1990. The discovery of the neurotransmitter role of glycine. *In* Glycine Neurotransmission. O.P. Ottersen & J. Storm-Mathiesen, Eds.: 1–23. John Wiley & Sons. New York.
2. Becker, C.-M. 1992. Convulsants acting at the inhibitory glycine receptor. *In* Handbook of Experimental Pharmacology, Vol. 102. H. Herken & F. Hucho, Eds.: 539–575. Springer Verlag. Berlin-Heidelberg.
3. Boehm, S., R.J. Harvey, A. von Holst, H. Rohrer & H. Betz. 1997. Glycine receptors in cultured chick sympathetic neurons are excitatory and trigger neurotransmitter release. J. Physiol. **504:** 683–694.
4. Reichling, D.B., A. Kyrozis, J. Wang & A.B. MacDermott. 1994. Mechanisms of GABA and glycine depolarization-induced calcium transients in rat dorsal horn neurons. J. Physiol. **476:** 411–421.
5. Wang, J., D.B. Reichling, A. Kyrozis & A.B. MacDermott. 1994. Developmental loss of GABA- and glycine-induced depolarization and Ca^{2+} transients in embryonic rat dorsal horn neurons in culture. Eur. J. Neurosci. **6:** 1275–1280.
6. Flint, A.C., X. Liu & A.R. Kriegstein. 1998. Nonsynaptic glycine receptor activation during early neocortical development. Neuron **20:** 43–53.
7. Becker, C.-M., I. Hermans-Borgmeyer, B. Schmitt & H. Betz. 1986. The glycine receptor deficiency of the mutant mouse *spastic*: Evidence for normal glycine receptor structure and localization. J. Neurosci. **6:** 1358–1364.
8. Graham, D., F. Pfeiffer & H. Betz. 1983. Photoaffinity-labelling of the glycine receptor of rat spinal cord. Eur. J. Biochem. **131:** 519–525.
9. Pfeiffer, F., D. Graham & H. Betz. 1982. Purification by affinity chromatography of the glycine receptor of rat spinal cord. J. Biol. Chem. **257:** 9389–9393.
10. Schmitt, B., P. Knaus, C.-M. Becker & H. Betz. 1987. The Mr 93,000 polypeptide of the postsynaptic glycine receptor complex is a peripheral membrane protein. Biochemistry **26:** 805–811.
11. Prior, P., B. Schmitt, G. Grenningloh, I. Pribilla, G. Multhaup, K. Beyreuther, Y. Maulet, P. Werner, D. Langosch, J. Kirsch & H. Betz. 1992. Primary structure and alternative splice variants of gephyrin, a putative glycine receptor-tubulin linker protein. Neuron **8:** 1161–1170.
12. Grenningloh, G., A. Rienitz, B. Schmitt, C. Methfessel, M. Zensen, K. Beyreuther, E.D. Gundelfinger & H. Betz. 1987. The strychnine-binding subunit of the glycine receptor shows homology with nicotinic acetylcholine receptors. Nature **328:** 215–220.
13. Grenningloh, G., I. Pribilla, P. Prior, G. Multhaup, K. Beyreuther, O. Taleb & H. Betz. 1990. Cloning and expression of the 58 kd β subunit of the inhibitory glycine receptor. Neuron **4:** 963–970.
14. Betz, H. 1990. Ligand-gated ion channels in the brain: The amino acid receptor superfamily. Neuron **5:** 383–392.
15. Langosch, D., L. Thomas & H. Betz. 1988. Conserved quaternary structure of ligand gated ion channels: The postsynaptic glycine receptor is a pentamer. Proc. Natl. Acad. Sci. USA **85:** 7394–7398.

16. KUHSE, H., B. LAUBE, D. MAGALEI & H. BETZ. 1993. Assembly of the inhibitory glycine receptor: Identification of amino-acid sequence motifs governing subunit stoichiometry. Neuron **11:** 1049–1056.
17. BECKER, C.M., W. HOCH & H. BETZ. 1988. Glycine receptor heterogeneity in rat spinal cord during postnatal development. EMBO J. **7:** 3717–3726.
18. AKAGI, H. & R. MILEDI. 1988. Heterogeneity of glycine receptors and their messenger RNAs in rat brain and spinal cord. Science **242:** 270–273.
19. AKAGI, H., K. HIRAI & F. HISHINUMA. 1991. Cloning of a glycine receptor subtype expressed in rat brain and spinal cord during a specific period of neuronal development. FEBS Lett. **281:** 160–166.
20. GRENNINGLOH, G., V. SCHMIEDEN, P.R. SCHOFIELD, P.H. SEEBURG, T. SIDDIQUE, T.K. MOHANDAS, C.-M. BECKER & H. BETZ. 1990. Alpha subunit variants of the human glycine receptor: Primary structures, functional expression and chromosomal localisation of the corresponding genes. EMBO J. **9:** 771–776.
21. KUHSE, J., V. SCHMIEDEN & H. BETZ. 1990. Identification and functional expression of a novel ligand binding subunit of the inhibitory glycine receptor. J. Biol. Chem. **265:** 22317–22320.
22. KUHSE, J., A. KURYATOV, Y. MAULET, M.L. MALOSIO, V. SCHMIEDEN & H. BETZ. 1991. Alternative splicing generates two isoforms of the α2 subunit of the inhibitory glycine receptor. FEBS Lett. **283:** 73–77.
23. MATZENBACH, B., Y. MAULET, L. SEFTON, B. COURTIER, P. AVNER, J.-L. GUÇNET & H. BETZ. 1994. Structural analysis of mouse glycine receptor α subunit genes: identification and chromosomal localization of a novel variant, α4. J. Biol. Chem. **269:** 2607–2612.
24. FUJITA, M., K. SATO, M. SATO, T. INOUE, T. KOZUKA & M. TOHYAMA. 1991. Regional distribution of the cells expressing glycine receptor β subunit mRNA in the rat brain. Brain Res. **560:** 23–37.
25. MALOSIO, M.L., B. MARQUÈZE-POUEY, J. KUHSE & H. BETZ. 1991. Widespread expression of glycine receptor subunit mRNAs in the adult and developing rat brain. EMBO J. **10:** 2401–2409
26. SATO, K., J.H. ZHANG, T. SAIKA, M. SATO, K. TADA & M. TOHYAMA. 1991. Localization of glycine receptor α subunit mRNA-containing neurons in the rat brain: An analysis using *in situ* hybridization histochemistry. Neuroscience **43:** 381–395.
27. SATO, K., H. KIYAMA & M. TOHYAMA. 1992. Regional distribution of cells expressing glycine receptor α2 subunit mRNA in the rat brain. Brain Res. **590:** 95–108.
28. WATANABE, E. & H. AKAGI. 1995. Distribution patterns of mRNAs encoding glycine receptor channels in the developing rat spinal cord. Neurosci. Res. **23:** 377–382.
29. MALOSIO, M.L., G. GRENNINGLOH, J. KUHSE, V. SCHMIEDEN, B. SCHMITT, P. PRIOR & H. BETZ. 1991. Alternative splicing generates two variants of the α subunit of the inhibitory glycine receptor. J. Biol. Chem. **266:** 2048–2053.
30. HECK, S., R. ENZ, C. RICHTER-LANDSBERG & D.H. BLOHM. 1997. Expression and mRNA splicing of glycine receptor subunits and gephyrin during neuronal differentiation of P19 cells *in vitro*, studied by RT-PCR and immunocytochemistry. Dev. Brain Res. **98:** 211–220.
31. KUHSE, J., V. SCHMIEDEN & H. BETZ. 1990. A single amino acid exchange alters the pharmacology of neonatal rat glycine receptor subunit. Neuron **5:** 867–873.
32. GRAHAM, D., F. PFEIFFER & H. BETZ. 1981. UV light-induced cross-linking of strychnine to the glycine receptor of rat spinal cord membranes. Biochem. Biophys. Res. Commun. **102:**1330–1335.
33. RUIZ-GOMEZ, A., E. MORATO, M. GARCIA-CALVO, VALDIVIESO & F. MAYOR, JR. 1990. Localization of the strychnine binding site on the 48-kilodalton subunit of the glycine receptor. Biochemistry **29:** 7033–7040.
34. SCHMIEDEN, V., G. GRENNINGLOH, P.R. SCHOFIELD & H. BETZ. 1989. Functional expression in *Xenopus* oocytes of the strychnine binding 48 kd subunit of the glycine receptor. EMBO J. **8:** 695–700.
35. SONTHEIMER, H., C.-M. BECKER, D.B. PRITCHETT, P.R. SCHOFIELD, G. GRENNINGLOH, H. KETTENMANN, H. BETZ & P.H. SEEBURG. 1989. Functional chloride channels by mammalian cell expression of rat glycine receptor subunit. Neuron **2:** 1491–1497.
36. BORMANN, J., N. RUNDSTRÖM, H. BETZ & D. LANGOSCH. 1993. Residues within transmembrane segment M2 determine chloride conductance of glycine receptor homo- and hetero-oligomers. EMBO J. **12:** 3729–3737.

37. PRIBILLA, I., T. TAKAGI, D. LANGOSCH, J. BORMANN & H. BETZ. 1992. The atypical M2 segment of the β subunit confers picrotoxinin resistance to inhibitory glycine receptor channels. EMBO J. **11:** 4305–4311.
38. SCHMIEDEN, V., J. KUHSE & H. BETZ. 1992. Agonist pharmacology of neonatal and adult glycine receptor α-subunits: Identification of amino acid residues involved in taurine activation. EMBO J. **11:** 2025–2032.
39. SCHMIEDEN, V., J. KUHSE & H. BETZ. 1993. Mutation of glycine receptor subunit creates β-alanine receptor responsive to GABA. Science **262:** 256–258.
40. VANDENBERG, R.J., C.R. FRENCH, P.H. BARRY, J. SHINE & P.R. SCHOFIELD. 1992. Antagonism of ligand-gated ion channel receptors: Two domains of the glycine receptor α subunit form the strychnine-binding site. Proc. Natl. Acad. Sci. USA **89:** 1765–1769.
41. RYAN, S.G., M.S. BUCKWALTER, J.W. LYNCH, C.A. HANDFORD, L. SEGURA, R. SHIANG, J.J. WASMUTH, S.A. CAMPER, P. SCHOFIELD & P. O'CONNELL. 1994. A missense mutation in the gene encoding the α subunit of the inhibitory glycine receptor in the *spasmodic* mouse. Nature Genet. **7:** 131–135.
42. SAUL, B., V. SCHMIEDEN, C. KLING, C. MÜLHARDT, P. GASS, J. KUHSE & C.-M. BECKER. 1994. Point mutation of glycine receptor α subunit in the *spasmodic* mouse affects agonist responses. FEBS Lett. **350:** 71–76.
43. LANGOSCH, D., B. LAUBE, N. RUNDSTRÖM, V. SCHMIEDEN, J. BORMANN & H. BETZ. 1994. Decreased agonist affinity and chloride conductance of mutant glycine receptors associated with human hereditary hyperekplexia. EMBO J. **13:** 4223–4228.
44. LAUBE, B., D. LANGOSCH, H. BETZ & V. SCHMIEDEN. 1995. Hyperekplexia mutations of the glycine receptor unmask the inhibitory subsite for β-amino-acids. Neuroreport **6:** 897–900.
45. LYNCH, J.W., S. RAJENDRA, K.D. PIERCE, C.A. HANDFORD, P.H. BARRY & P.R. SCHOFIELD. 1997. Identification of intracellular and extracellular domains mediating signal transduction in the inhibitory glycine receptor. EMBO J. **16:** 110–120.
46. RAJENDRA, S., J.W. LYNCH, K.D. PIERCE, C.R. FRENCH, P.H. BARRY & P.R. SCHOFIELD. 1994. Startle disease mutations reduce the agonist sensitivity of the human inhibitory glycine receptor. J. Biol. Chem. **269:** 18739–18742.
47. RAJENDRA, S., J.W. LYNCH, K.D. PIERCE, C.R. FRENCH, P.H. BARRY & P.R. SCHOFIELD. 1995. Mutation of an arginine residue in the human glycine receptor transforms β-alanine and taurine from agonists into competitive antagonists. Neuron **14:** 169–175.
48. LEWIS, T.M., L.G. SIVILOTTI, D. COLQUHOUN, R.M. GARDINER, R. SCHOEPFER & M. REES. 1998. Properties of human glycine receptors containing the hyperekplexia mutation α(K276E), expressed in *Xenopus* oocytes. J. Physiol. **507:** 25–40.
49. BLOOMENTHAL, A.B., E. GOLDWATER, D.B. PRITCHETT & N.L. HARRISON. 1994. Biphasic modulation of the strychnine-sensitive glycine receptor by Zn^{2+}. Mol. Pharmacol. **46:** 1156–1159.
50. LAUBE, B., J. KUHSE, N. RUNDSTRÖM, J. KIRSCH, V. SCHMIEDEN & H. BETZ. 1995. Modulation by zinc ions of native rat and recombinant human inhibitory glycine receptors. J. Physiol. **483:** 613–619.
51. TRILLER, A., F. CLUZEAUD, F. PFEIFFER, H. BETZ & H. KORN. 1985. Distribution of glycine receptors at central synapses of the rat spinal cord. J. Cell Biol. **101:** 683–688.
52. KIRSCH, J., D. LANGOSCH, P. PRIOR, U.Z. LITTAUER, B. SCHMITT & H. BETZ. 1991. The 93 kDa glycine receptor–associated protein binds to tubulin. J. Biol. Chem. **266:** 22242–22245.
53. KIRSCH, J. & H. BETZ. 1995. The postsynaptic localization of the glycine receptor–associated gephyrin is regulated by the cytoskeleton. J. Neurosci. **15:** 4148–4156.
54. MEYER, G., J. KIRSCH, H. BETZ & D. LANGOSCH. 1995. Identification of a gephyrin binding motif on the glycine receptor β subunit. Neuron **15:** 563–572.
55. KIRSCH, J., M.L. MALOSIO, I. WOLTERS & H. BETZ. 1993. Distribution of gephyrin transcripts in the adult and developing rat brain. Eur. J. Neurosci. **5:** 1109–1117.
56. ARAKI, T., M. YAMANO, T. MURAKAMI, A. WANAKA, H. BETZ & M. TOHYAMA. 1988. Localization of glycine receptors in the rat central nervous system: An immunocytochemical analysis using monoclonal antibody. Neuroscience **25:** 613–624.
57. KIRSCH, J. & H. BETZ. 1993. Widespread expression of gephyrin, a putative glycine receptor-tubulin linker protein, in rat brain. Brain Res. **621:** 301–310.
58. KIRSCH, J., I. WOLTERS, A. TRILLER & H. BETZ. 1993. Gephyrin antisense oligonucleotides prevent glycine receptor clustering in spinal neurons. Nature **366:** 745–748.

59. KIRSCH, J. & H. BETZ. 1998. Glycine-receptor activation is required for receptor clustering in spinal neurons. Nature **392:** 717–720.
60. BECKER, C.-M. 1995. Glycine receptors: Molecular heterogeneity and implications for disease. Neuroscientist **1:** 130–141.
61. LANE, P.W., A.L. GANSER, A.L. KERNER & W.F. WHITE. 1987. *Spasmodic*, a mutation on chromosome 11 in the mouse. J. Hered. **78:** 353–356.
62. BAKER, E., G.R. SUTHERLAND & P.R. SCHOFIELD. 1994. Localization of the glycine receptor α subunit gene (GLRA1) to chromosome 5q32 by FISH. Genomics **40:** 396–400.
63. SHIANG, R., S.G. RYAN, Y.-Z. ZHU, A.F. HAHN, P. O'CONNELL & J.J. WASMUTH. 1993. Mutations in the α subunit of the inhibitory glycine receptor cause the dominant neurologic disorder hyperekplexia. Nature Genet. **5:** 351–358.
64. KLING, C., M. KOCH, B. SAUL & C.-M. BECKER. 1997. The frameshift mutation *oscillator* ($Glra1^{spd\text{-}ot}$) produces a complete loss of glycine receptor α-polypeptide in mouse central nervous system. Neuroscience **78:** 411–417.
65. BUCKWALTER, M.S., S.A. COOK, M.T. DAVISSON, W.F. WHITE & S. CAMPER. 1994. A frameshift mutation in the mouse α glycine receptor gene (*Glra1*) results in progressive neurological symptoms and juvenile death. Hum. Mol. Genet. **3:** 2025–2030.
66. EICHER, E.M. & P. LANE. 1980. Assignment of LG XV1 to chromosome 3 in the mouse. J. Hered. **71:** 315–318.
67. KINGSMORE, S.F., B. GIROS, D. SUH, M. BIENIARZ, M.G. CARON & M.F. SELDIN. 1994. Glycine receptor β-subunit gene mutation in *spastic* mice associated with LINE-1 element insertion. Nature Genet. **7:** 136–142.
68. MÜLHARDT, C., M. FISCHER, P. GASS, D. SIMON CHAZOTTES, J.-L. GUENET, J. KUHSE, H. BETZ & C.-M. BECKER. 1994. The *spastic* mouse: aberrant splicing of glycine receptor β subunit mRNA caused by intronic insertion of L1 element. Neuron **13:** 1003–1015.
69. WHITE, W.F. & A.H. HELLER. 1982. Glycine receptor alteration in the mutant mouse *spastic*. Nature **298:** 655–657.
70. HARTENSTEIN, B., J. SCHENKEL, J. KUHSE, B. BESENBECK, C. KLING, C.-M. BECKER, H. BETZ & H. WEIHER. 1996. Low level expression of glycine receptor β subunit transgene is sufficient for phenotype correction in *spastic* mice. EMBO J. **15:** 1275–1282.
71. DERRY, J.M. & P.J. BARNARD. 1991. Mapping of the glycine receptor α2-subunit gene and the $GABA_A$ receptor α3-subunit gene on the mouse X chromosome. Genomics **10:** 593–597.
72. KINGSMORE, S.F., D. SUH & M.F. SELDIN. 1994. Genetic mapping of the glycine receptor α3 subunit on mouse chromosome 8. Mamm. Genome **5:** 831–832.
73. RAJENDRA, S. & P.R. SCHOFIELD. 1995. Molecular mechanisms of inherited startle syndromes. Trends Neurosci. **18:** 80–82.
74. RYAN, S.G., S.L. SHERMAN, J.C. TERRY, R.S. SPARKES, M. TORRES & R.W. MACKEY. 1992. Startle disease, or hyperekplexia: Response to clonazepam and assignment of the gene (STHE) to chromosome 5q by linkage analysis. Ann. Neurol. **31:** 663–668.
75. ELMSLIE, F.V., S.M. HUTCHINGS, V. SPENCER, A. CURTIS, T. COVANIS, R.M. GARDINER & M. REES. 1996. Analysis of GLRA1 in hereditary and sporadic hyperekplexia: A novel mutation in a family co-segregating for hyperekplexia and spastic paraparesis. J. Med. Genet. **33:** 435–436.
76. SHIANG, R., S.G. RYAN, Y.-Z. ZHU, T.J. FIELDER, R.J. ALLEN, A. FRYER, S. YAMASHITA, P. O'CONNELL & J.J. WASMUTH. 1995. Mutational analysis of familial and sporadic hyperekplexia. Ann. Neurol. **38:** 85–91.
77. REES, M.I., M. ANDREW, S. JAWAD & M.J. OWEN. 1994. Evidence for recessive as well as dominant forms of startle disease (hyperekplexia) caused by mutations in the α subunit of the inhibitory glycine receptor. Hum. Mol. Genet. **3:** 2175–2179.
78. MILANI, N., L. DALPRE, A. PRETE, R. DEL ZANINI & L. LARIZZA. 1996. A novel mutation (Gln266 His) in the α subunit of the inhibitory glycine receptor gene (GLRA1) in hereditary hyperekplexia. Am. J. Hum. Genet. **58:** 420–422.
79. BRUNE, W., R.G. WEBER, B. SAUL, M. VON KNEBEL DOEBERITZ, C. GROND-GINSBACH, K. KELLERMANN, H.-M. MEINCK & C.-M. BECKER. 1996. A *GLRA1* null mutation in recessive hyperekplexia challenges the functional role of glycine receptors. Am. J. Hum. Genet. **58:** 989–997.
80. VERGOUWE, M.N., M.A.J. TIJSSEN, R. SHIANG, J.G. VAN DIJK, S.A. SHAHWAN, R.A OPHOFF & R.R. FRANTS. 1997. Hyperekplexia-like syndromes without mutations in the GLRA1 gene. Clin. Neurol. Neurosurg. **99:** 172–178.

81. Handford, C.A., J.W. Lynch, E. Baker, G.C. Webb, J.H. Ford, G.R. Sutherland & P.R. Schofield. 1996. The human glycine receptor β subunit: Primary structure, functional characterisation and chromosomal localisation of the human and murine genes. Mol. Brain Res. **25:** 211–219.

Changes in $GABA_A$ Receptor–Mediated Synaptic Transmission in Oxytocin Neurons during Female Reproduction: Plasticity in a Neuroendocrine Context

ARJEN B. BRUSSAARD[a] AND KAREL S. KITS

Membrane Physiology Section, Research Institute Neuroscience (RIN)—Vrije Universiteit Amsterdam (VUA), De Boelelaan 1087, 1081 HV Amsterdam, the Netherlands

Magnocellular neurons located in the dorsomedial region of the supraoptic nucleus (SON) in female rats secrete systemic oxytocin during parturition and lactation. One of the key players in the regulation of these cells is the inhibitory input mediated via postsynaptic $GABA_A$ receptors.[1–3] Several neuronal plasticity phenomena occur in the SON during each cycle of female reproductive activity. These include (1) complex alterations in neuron–glia interactions;[4] (2) an increase in the number of GABAergic release sites and the appearance of multiple synapses;[5] and (3) changes in the cellular mRNA content of oxytocin neurons, encoding two α subunits of the $GABA_A$ receptor, leading to an α1-to-α2 subunit switch in expression around the time of parturition.[2,3] We have recently reported that the latter alteration in α subunit expression affects the ion channel gating properties of the $GABA_A$ receptor in these neurons.[2] In addition, we found that this change in α subunit expression correlates with a change in $GABA_A$ receptor potentiation by the neurosteroid 3α-OH-DHP (allopregnanolone).

We describe here some of the physiological consequences of changes in GABAergic synaptic innervation of oxytocin neurons during the first female reproductive cycle. To this end, *in situ* patch-clamp recordings were made from dorsomedial SONs of female rats at all relevant stages of reproduction. In the analysis of the data three central hypotheses were followed. First, as previously suggested,[5] pregnancy may increase the number of GABAergic release sites available per individual oxytocin neuron. If true, one would predict that the average interval between miniature sIPSCs would be reduced. The data shown in FIGURE 1A indicate that is the case: at late pregnancy the interval between sIPSCs is significantly reduced. Second, an increasing number of GABAergic release sites during the female reproductive active period[5] may lead to enhanced insertion of postsynaptic receptors, which in turn may give rise to alterations in receptor density under individual synaptic boutons. FIGURE 1B shows that robust changes in sIPSC amplitude do not occur in parallel to the change in sIPSC intervals during pregnancy. Third, synaptic plasticity in the SON includes the endogenous regulation of $GABA_A$ receptor expression in the oxytocin neurons leading to changes in the subunit composition of these receptors.[2] Two distinct subtypes of $GABA_A$ receptor were found to be expressed, each yielding a characteristic synaptic current decay and sensitivity to 3α-OH-DHP potentiation. In late pregnancy we found that an α1 subunit–dominated and allopregnanolone-sensitive

[a]Corresponding author. Phone: 31-20-444-7098; fax: 31-20-444-7123; email: brssrd@bio.vu.nl

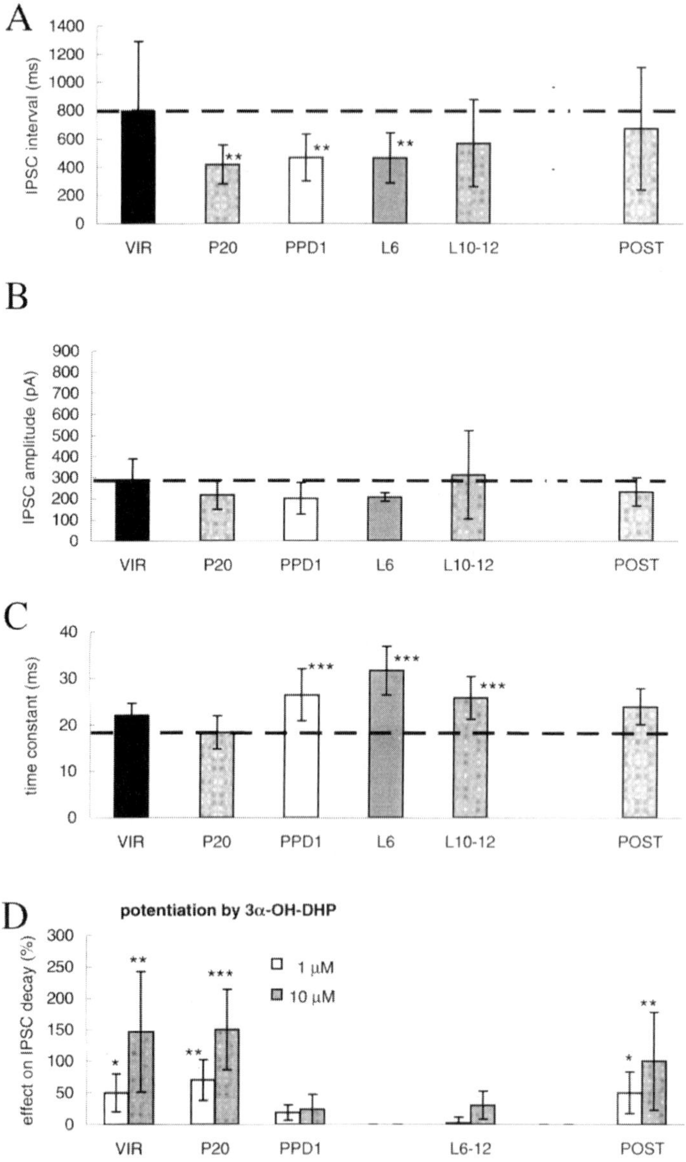

FIGURE 1. Monoquantal sIPSC properties in oxytocin neurons at various stages of female reproduction: VIR (virgins), P20 (pregnancy, day 20), PPD1 (postparturition, day 1), L6 (lactation, day 6), and PL (postlactation, 6 weeks). **(A)** Intervals; **(B)** amplitudes; **(C)** monoexponential decay time constant; and **(D)** neurosteroid (allopregnanolone) potentiation of sIPSCs (mean and standard deviations). $GABA_A$ receptor activity was recorded at −70 mV. Comparisons were made to levels indicated by *broken lines* (or in FIG. 1D: the 0% effect level): *, **, and *** indicate a $p < 0.05$, 0.01, and 0.001, respectively.

$GABA_A$ receptor subtype (P20, FIG. 1D) gives rise to a relatively fast monoexponential synaptic current decay (18 ± 4 ms, P20, Fig. 1C). In contrast, two days later, at parturution, an α2-subunit-dominated $GABA_A$ receptor subtype was found yielding a significantly slower synaptic current decay (26 ± 6 ms, PPD1; FIG. 1C), plus a largely reduced sensitivity to 3α-OH-DHP (PPD1; FIG. 1D).

This hypothesis leads to three specific research questions: (1) Is the $GABA_A$ receptor subtype previously described[2] as occurring during late pregnancy (P20) also found prior to pregnancy? (2) Is the $GABA_A$ receptor subtype observed at parturition (PPD1; Ref. 2) also present during the subsequent lactation period? (3) Is there complete recovery of sIPSC properties during the postlactation period (PL)? The data of FIGURE 1C and 1D indicate that all these questions can be answered positively: We found no noticeable alteration either in sIPSC decay kinetics or in neurosteroid regulation from virginity (VIR) to late pregnancy (P20). Second, the synaptic current decay of sIPSCs remained significantly slower during the first and the second week of the lactation period (L6–12) Third, after six weeks of the postlactation period (PL), both sIPSC decay and its neurosteroid regulation returned to prepregnancy levels.

Measurements of the monoquantal sIPSC's properties provide the key parameters that together determine the impact of the fast synaptic inhibition of the dorsomedial SON neurons. We show here that, in oxytocin neurons, changes in three out of four sIPSC properties (interval, decay kinetics, and neurosteroid regulation) occur during particular stages of female reproduction. In addition, we found that these changes are cyclical and recover to prepregnancy levels upon the end of lactation. The postsynaptic receptor plasticity studied here is highly temporal and reminiscent of developmental subunit switching of postsynaptic nicotinic ACh receptors in the neuromuscular synapse.[6] Also in the CNS, subunit switching in ligand-gated receptor expression has previously been reported to occur during neonatal development in a period of enhanced synaptogenesis.[7,8] We claim here that mature neuroendocrine neurons are also capable of reshaping their postsynaptic ligand-gated receptor subunit expression profile, in order to shift between distinct modes of synaptic transmission.

REFERENCES

1. BRUSSAARD, A.B., K.S. KITS, R.E. BAKER, W.P.A. WILLEMS, J.W. LEYTING-VERMEULEN, P. VOORN, A.B. SMIT & A.E. HERBISON. 1997. Plasticity in fast synaptic inhibition of adult oxytocin neurons caused by switch in $GABA_A$ receptor subunit expression. Neuron **19:** 1103–1114.
2. BRUSSAARD, A.B., K.S. KITS & T.A. DE VLIEGER. 1996. Postsynaptic mechanism of depression of GABAergic synapses by oxytocin in the supraoptic nucleus of immature rat. J. Physiol. (Lond.) **497:** 495–507.
3. FENELON, V.S. & A.E. HERBISON. 1996. Plasticity in $GABA_A$ receptor subunit mRNA expression by hypothalamic magnocellular neurons in the adult rat. J. Neurosci. **16:** 4872–4880.
4. HATTON, G.I. 1997. Function-related plasticity in hypothalamus. Annu. Rev. Neurosci. **20:** 375–397.
5. GIES, U. & D.T. THEODOSIS. 1994. Synaptic plasticity in the rat supraoptic nucleus during lactation involves GABA innervation and oxytocin neurons: A quantitative immunocytochemical analysis. J. Neurosci. **14:** 2861–2869.
6. SAKMANN, B. & H. BRENNER. 1978. Change in synaptic channel gating during neuromuscular development. Nature **276:** 401–402.
7. FRITSCHY, J.-M., J. PAYSAN, A. ENNA & H. MOHLER. 1994. Switch in the expression of rat GABA(A)-receptor subtypes during postnatal development: An immunohistochemical study. J. Neurosci. **14:** 5302–5324.

8. Laurie, D.J., W. Wisden, & P.H. Seeburg. 1992. The Distribution of 13 $GABA_A$ receptor subunit messenger RNAs in the rat brain, III: Embryonic and postnatal development. J. Neurosci. **12:** 4151–4172.

Structure-Function Relationships of the Human Glycine Receptor: Insights from Hyperekplexia Mutations

TREVOR M. LEWIS[a] AND PETER R. SCHOFIELD

Garvan Institute of Medical Research, 384 Victoria Street, Darlinghurst, Sydney NSW 2010, Australia

The glycine receptor (GlyR) is a member of the ligand-gated ion channel (LGIC) superfamily, and the major inhibitory receptor in the spinal cord and brain stem. Missense mutations in the gene encoding the α1 subunit of the human GlyR have been linked to the rare neurological disease hyperekplexia (or familial startle disease), where affected individuals present an exaggerated startle reflex to unexpected sensory stimuli. This is similar to the behavior produced by mild poisoning with the plant alkaloid strychnine, a competitive antagonist for the GlyR. Six missense mutations have been identified by genetic analysis; four in the extracellular loop between transmembrane domains M2–M3 (R271L, R271Q, K276E, Y279C), one in the intracellular M1–M2 loop (I244N), and one in the pore-lining M2 transmembrane domain (Q266H).[1] By investigating the dysfunction of recombinant receptors containing hyperekplexia mutations, information about the normal function of these regions of the receptor protein have been deduced.

The qualitative description of the hyperekplexia mutations in the M1–M2 loop and M2–M3 loop suggested that these regions were important in the activation of the receptor. An alanine scan, where each residue in turn is replaced by an alanine, systematically probed the function of these regions.[2] Each mutant α1 receptor was expressed as a homomer in transiently transfected HEK293 cells and glycine-activated currents recorded with the whole-cell patch-clamp technique. Four residues (R271, K276, V277, Y279), which included the three hyperekplexia mutation positions in the M2–M3 loop, showed a "full" disruption phenotype when mutated to alanine; an increase in the EC_{50} for glycine and the conversion of the agonists β-alanine and taurine to competitive antagonists. Other residues in both the M2–M3 loop (L274, S278, K281) and the M1–M2 loop (W243, I244, M246) showed a "partial" disruption; an increase in the EC_{50} for glycine, β-alanine and taurine, and the conversion of β-alanine and taurine to classic partial agonists. Mutations at other positions in the M1–M2 loop and M2–M3 loop showed little or no change in the concentration–response curves compared to wild-type. The observed changes in the efficacy of β-alanine and taurine (conversions to antagonists or partial agonists) suggested that the main effect of these mutations was upon the gating of the receptor; affecting the time spent in the open state once the ligand is bound. Direct evidence for a change in the gating of the receptor was obtained from single-channel analysis of the hyperekplexia mutation K276E[3] (see FIG. 1). The α1(K276E) subunit was expressed as a heteromer with the β subunit, in *Xenopus* oocytes, and glycine-activated whole-cell currents were recorded. The concentra-

[a]Corresponding author. Phone: +61-2-9295 8290; fax: +61-2-9295 8281; e-mail: t.lewis@garvan.unsw.edu.au

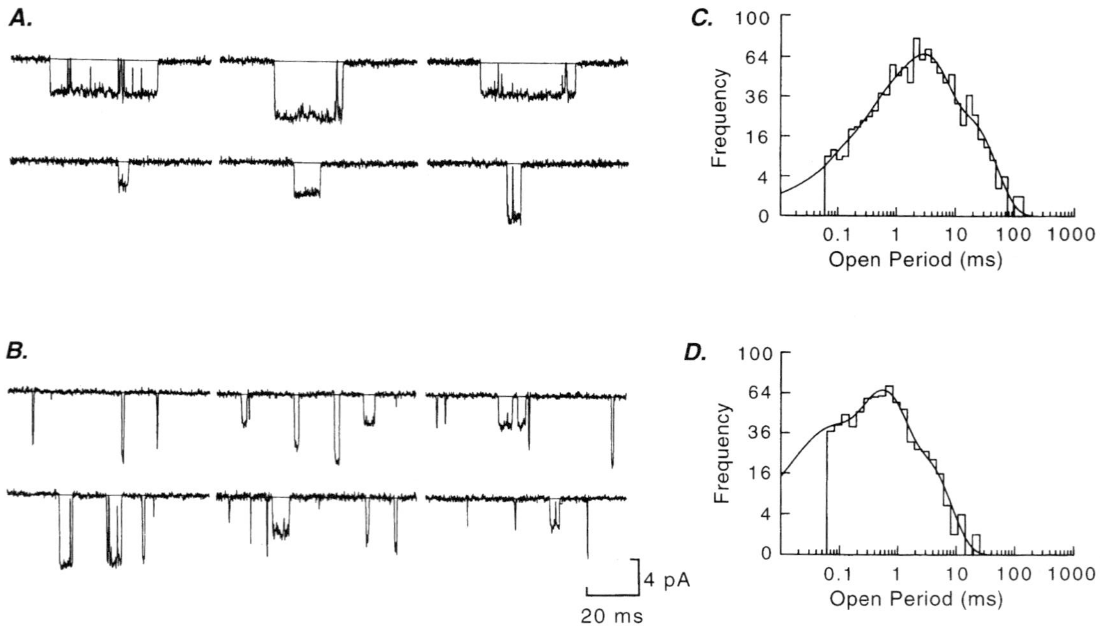

FIGURE 1. Examples of single-channel recordings from α1β (*A*) and α1(K276E)β (***B***) glycine receptors. Recordings were obtained from outside-out patches (held at −100 mV; 3 kHz filter) in response to the application of glycine [10 μM for α1β and 100 μM for α1(K276E)β receptors]. Both wild-type and mutant receptors opened to a similar range of conductance levels. Analysis of open period distributions for α1β (***C***) and α1(K276E)β (***D***) were both fitted with a mixture of three exponential densities and displayed as the log duration of the open times against the frequency (plotted on a square root scale). Open periods for α1(K276E)β are shorter than those for α1β. The fitted time constants and relative areas (%) are for α1β: 0.476 ms, 9.5%; 2.54 ms, 62.6%; 13.7 ms, 27.9%; and for α1(K276E)β: 0.042 ms, 21.9%; 0.464 ms, 55.1%; 2.29 ms, 23.0%.

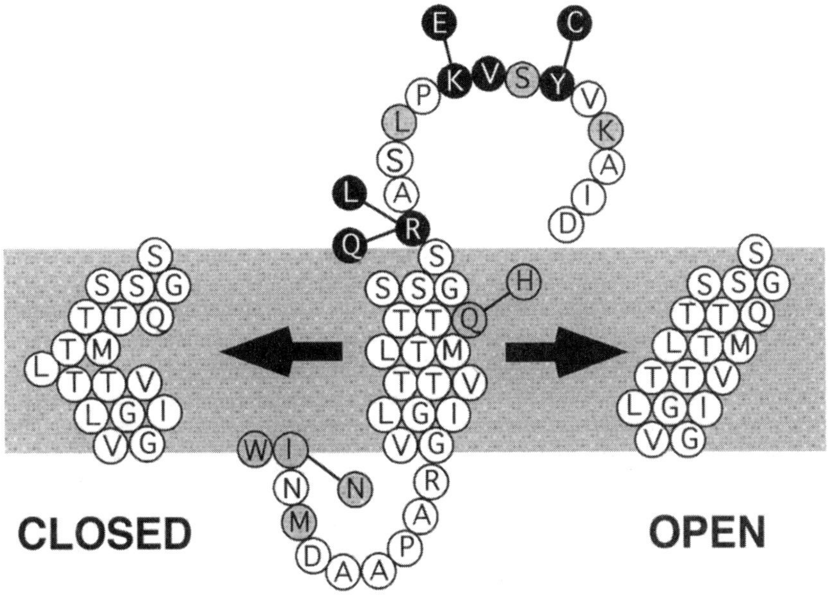

FIGURE 2. Model of the M2 domain and flanking M1–M2 intracellular loop and M2–M3 extracellular loop, illustrating the putative open (*extended*) and closed (*kinked*) conformations of the M2 domains of the GlyR α1 subunit. The hyperekplexia mutations are shown joined to the wild-type sequence by stems. From the results of the alanine scan, residues shaded in *black* show a "full" disruption phenotype, where β-alanine and taurine are converted to competitive antagonists; and residues shaded in *gray* show a "partial" disruption, phenotype, where β-alanine and taurine are converted to partial agonists. The M1–M2 loop and M2–M3 loop are postulated to act like "hinges" to allow the allosteric conformational change or gating of the ion channel to occur.

tion-response curves, fitted with the empirical Hill equation, showed a 10-fold decrease in I_{max}, a 29-fold increase in EC_{50}, and a decrease in the Hill coefficient for α1(K276E)β compared to wild-type α1β. The K_B for the competitive antagonist strychnine was shown to be unchanged, suggesting that there had been no dramatic change in the structure of the agonist binding site. Fitting of a simple sequential mechanism, which involves the binding of three molecules of agonist followed by a conformation change to the open state, further indicated that a change in channel gating is able to account entirely for the change in the concentration–response curve seen with α1(K276E)β. Single-channel recordings showed that there was no change in the unitary conductance of α1(K276E)β compared with wild-type; however, analysis of open periods showed that they were very much shorter for α1(K276E)β. The overall mean open period for wild-type was 6.87 ± 1.43 ms, compared to 0.82 ± 0.09 ms for α1(K276E)β. This directly indicates, in terms of the sequential mechanism just described, that the mutation has impaired the gating of the receptor.

From the results of alanine scanning the M1–M2 and M2–M3 loops, and the specific analysis of the α1(K276E) mutation, it is clear that these regions are important in the signal transduction of ligand binding to gating of the ion channel pore. Given the relationship of these loops flanking the pore-lining M2 domain, it is attractive to propose that they act as "hinges" for a conformational change in the M2 (FIG. 2), in a manner that may be similar to that proposed from electron microscope images of the *Torpedo* acetylcholine receptor.[4]

REFERENCES

1. VAFA, B. & P.R. SCHOFIELD. 1998. Identification and functional analysis of heritable mutations in the glycine, γ-aminobutyric acid type A and nicotinic acetylcholine receptors provide new insights into the structure and function of the ligand-gated ion channel receptor superfamily. Int. Rev. Neurobiol. **42:** 285–332.
2. LYNCH, J.W., S. RAJENDRA, K.D. PIERCE, C.A. HANDFORD, P.H. BARRY & P.R. SCHOFIELD. 1997. Identification of intracellular and extracellular domains mediating signal transduction in the inhibitory glycine receptor chloride channel. EMBO J. **16:** 110–120.
3. LEWIS, T.M., L.G. SIVILOTTI, D. COLQUHOUN, R.M. GARDINER, R. SCHOEPFER & M. REES. 1998. Properties of human glycine receptors containing the hyperekplexia mutation α1(K276E), expressed in *Xenopus* oocytes. J. Physiol. **507:** 25–40.
4. UNWIN, N. 1995. Acetylcholine receptor channel imaged in the open state. Nature **373:** 37–43.

Regulation of Glycine Transport in Cultured Müller Cells by Ca^{2+}/Calmodulin-Dependent Enzymes

ANA MARÍA LÓPEZ-COLOMÉ[a,b,c] AND ANA GADEA[a]

[a]*Instituto de Fisiología Celular, Departamento de Neurociencias, UNAM, Apartado Postal 70-253, 04510, México, D.F., México*

[b]*Facultad de Medicina, Departamento de Bioquímica, UNAM, 04510, México, D.F., México*

The amino acid glycine (gly) is a classic inhibitory neurotransmitter in the spinal cord, brain stem, and retina, where its effects are mediated by the gly receptor, a ligand-gated chloride channel competitively antagonized by strychnine.[1] Gly also modulates excitatory neurotransmission as an obligatory coagonist of glutamate (glu) at *N*-methyl-D-aspartate (NMDA) receptors through a strychnine-insensitive binding site.[2]

The termination of chemical neurotransmission in the CNS involves the rapid removal of neurotransmitter from synapses by reuptake either into the presynaptic terminal or into surrounding glia through specific transport systems. Drugs that block the action of the transporters involved in the termination of neurotransmission can modulate neural function by increasing the duration of neurotransmitter action.[3]

Although the reuptake process is subject to physiological regulation, very little information is currently available regarding the regulatory possibilities of these proteins by second messengers. Arachidonic acid, which may be released via phospholipase A_2, has been shown to inhibit several sodium-coupled uptake systems, including those for glycine[4] and glutamate;[5] also, the modulation of glutamate transporters by PKC,[6] as well as the regulation by cAMP of GABA transporters, have been demonstrated.[7]

To date, two different glycine transporters have been cloned. GLYT1, which seems to colocalize mainly with NMDA receptors,[8] and GLYT2, which colocalizes with the inhibitory glycine receptors.[9] Recent results from our laboratory have demonstrated the presence of a gly transport system in Müller cells showing two components, one of high affinity for gly and one of lower affinity (FIG. 1). The high-affinity system was identified as GLYT1, inhibited by sarcosine. Since the gly transport system in these cells could participate in the modulation of glu excitatory transmission in the vertical pathways of the retina, the regulation of gly transport in confluent monolayer cultures of Müller cells from 7-day-old chick embryos was studied.

RESULTS

Uptake was measured at 37 °C in Krebs-Ringer Bicarbonate buffer containing [^{3}H]-gly/gly 1:25,000. Incubation of Müller cells with the PKC activators phorbol-12-

[c]Corresponding author. Phone: 525-622-5617; fax: 525-622-5607; e-mail: acolome@ifcsun1.ifisiol.unam.mx

myristate-13-acetate (0.1, 0.5, and 1 μM), 1,2-dioctanoyl-rac-glycerol 100 μg/mL, or the PKC inhibitors staurosporine (100 nM), H7 (50 μM) and polymyxine B (400 μM) had no effect on transport. Neither did the PLC inhibitor neomicin (110 μM), the AC activator forskolin (2.5 μM), nor the AC selective inhibitor SQ-22536 (10 and 100 μM). Incubation of the cells with the calmodulin inhibitors trifluoperazine (25 μM) and W-7 (10 μM) led to a 74% and 40% decrease in the transport, respectively. R-24571 (10 μM), inhibitor of Ca^{2+}/calmodulin-stimulated enzymes, and the selective inhibitor of CaMPK II (Ca^{2+}/calmodulin-dependent protein kinase II), KN-62 (20 μM), inhibited glycine transport 82% and 46%, respectively (FIG. 2).

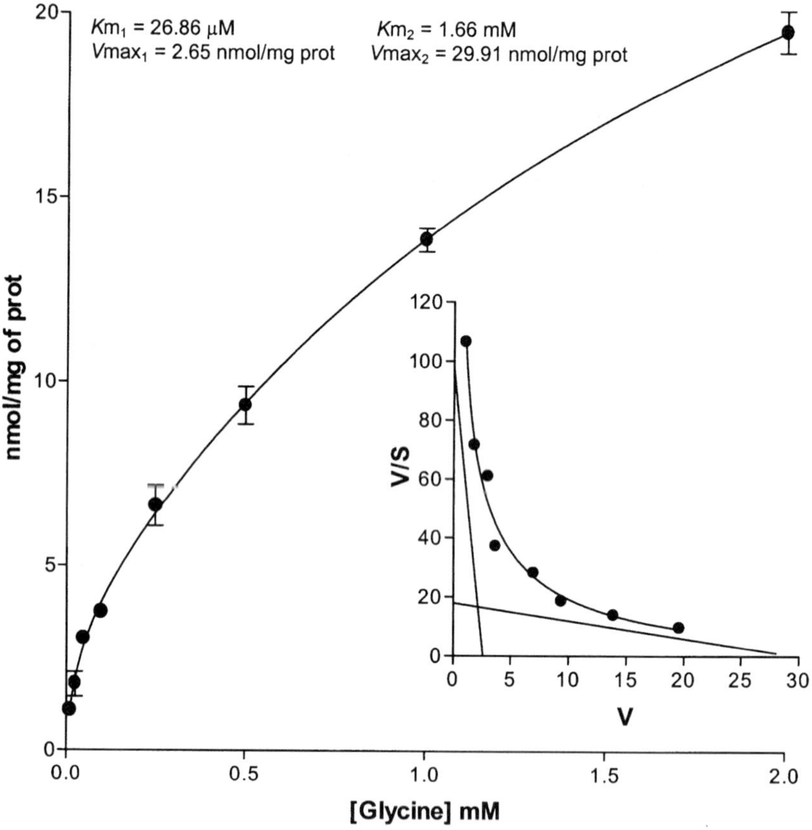

FIGURE 1. Kinetics of glycine uptake in Müller cells. Müller cells grown to confluence were incubated for 10 min in the presence of glycine at concentrations ranging from 0.01 mM to 2 mM ([^{3}H]-Gly/Gly 1:5000). Scatchard analysis is described in the *inset* of the figure. Values given are the mean ± standard error of two to four triplicate determinations.

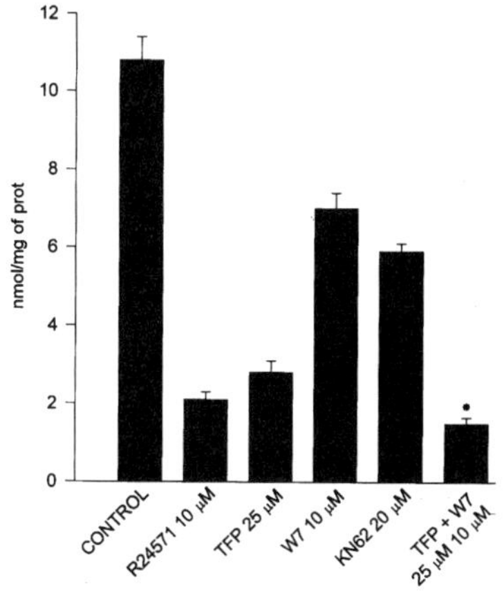

FIGURE 2. Müller cells were preincubated with the indicated compounds for 20 min in KRB medium. The medium was removed, and the transport assay was then carried out for 10 min in KRB with 1 mM glycine ([^{3}H]-Gly/Gly 1:25000). Values are expressed as the mean ± standard error of two experiments performed in triplicate. All values were significantly different from control ($p < 0.001$, Student's *t*-test). * Significantly different from TFP ($p < 0.01$, Student's *t*-test).

CONCLUSIONS

The gly transport system in Müller cells, as a contrast with those for other neuroactive amino acids such as glutamate[6] and GABA,[7] is not regulated through phosphorylation by PKC. Changes in cAMP concentration, shown to modulate GABA transport,[10] do not affect glycine transport in these cells.

The results from this work strongly suggest that the gly transport system, present in Müller cells in primary culture, is regulated by Ca^{2+}/calmodulin-dependent enzymes, one of them being Ca^{2+}/calmodulin-dependent protein kinase II (CaMPK II).

In the retina, Müller cells are the most abundant glial cell type. These cells span the entire width of the retina from the inner limiting membrane to the outer nuclear layer. Anatomic localization of these cells, ensheathing excitatory synapses, has suggested their participation in the modulation of neurotransmission. The regulation of glycine transport into Müller radial glia could be of physiological relevance, since gly concentration at the synaptic cleft influences the activity of the NMDA receptors involved in glu excitatory transmission at the vertical pathways of the retina.

REFERENCES

1. Aprison, M.H. 1990. The discovery of the neurotransmitter role of glycine. *In* Glycine Neurotransmission. O.P. Ottersen & J. Storm-Mathisen, Eds.: 1–23. Wiley. New York.
2. Fletcher, E.J. *et al.* 1990. Involvement of glycine in excitatory neurotransmission. *In* Glycine Neurotransmission. O.P. Ottersen & J. Storm-Mathisen, Eds: 193–218. Wiley. New York.
3. Kanner, B.I. 1994. Sodium-coupled neurotransmitter transport: Structure, function and regulation. J. Exp. Biol. **196:** 237–249.

4. ZAFRA, F. *et al.* 1990. Arachidonic acid inhibits glycine transport in cultured glial cells. Biochem. J. **271:** 237–242.
5. BARBOUR, B. 1989. Modulaton of glutamate transporters by arachidonic acid. Nature **342:** 918–920.
6. CASADO, M.A. *et al.* 1993. Phosphorylation and modulation of brain glutamate transporters by protein kinase C. J. Biol. Chem. **268**(36): 27313–27317.
7. GOMEZA, J.M. *et al.* 1991. Inhibition of high-affinity γ-aminobutyric acid uptake in primary astrocyte cultures by phorbol esters and phospholipase C. Biochem. J. **275:** 435–439.
8. SMITH, K.E. *et al.* 1992. Cloning and expression of a glycine transporter reveal colocalization with NMDA receptors. Neuron **8:** 927–936.
9. JURSKY, F. & N. NELSON. 1995. Localization of glycine neurotransmitter transporter (GLYT2) reveals correlation with the distribution of glycine receptor. J. Neurochem. **64**(3): 1026–1033.
10. GOMEZA, J. *et al.* 1994. Cellular distribution and regulation by cAMP of the GABA transporter (GAT-1) mRNA. Brain Res. Mol. Brain Res. **21**(1–2): 150–156.

Processing of $GABA_BR1$ in Heterologous Expression Systems

J. MOSBACHER,[a,c] K. KAUPMANN,[a] V. SCHULER,[a] D. RISTIG,[a] K. STRUCKMEYER,[a] T. PFAFF,[b] A. KARSCHIN,[b] M. F. POZZA,[a] AND B. BETTLER[a]

[a]*Novartis Pharma AG, CH-4002 Basel, Switzerland*

[b]*MPI f. Biophysical Chemistry, Am Fassberg 11, D-37070 Göttingen, Germany*

The recently cloned $GABA_B$ receptors belong to the family of neuronal G protein–coupled receptors (GPCRs) and modulate via $G_{i/o}$ protein K^+ and Ca^{2+} channels at pre- and postsynaptic sites (reviewed in Ref. 1). Although the binding pharmacology is similar for native and cloned receptor, it remains difficult to achieve a robust functional coupling of the cloned receptor to the effectors of native $GABA_B$ receptors.[2] We analyzed the cellular localization of a GFP-tagged $GABA_BR1$ and the coupling of the cloned receptor to potassium channels. Our results may in retrospect explain why functional expression cloning of $GABA_BR1$ was unsuccessful.

RESULTS

After cloning the rat $GABA_BR1$, hybridization screening of brain libraries resulted in the subsequent cloning of the human $GABA_BR1$. Similar to the rat receptor, the human counterpart exists in two splice variants, R1a and R1b.[2] Human and rat receptors have >98% sequence similarity and do not show pharmacological differences. Both splice variants have the same molecular weight as the two native $GABA_B$ proteins detected by [^{125}I]CGP71872 photoaffinity labeling of cortical cell membranes. The antagonist CGP71872 acts on both, pre- and postsynaptic $GABA_B$ receptors, as shown from functional studies (M. F. Pozza, unpublished). This suggests that the cloned receptor represents the most abundant $GABA_B$ receptor in rat and human brain. In support of this suggestion, the rank order of agonist affinity and the antagonist affinities are similar for recombinant and native $GABA_B$ receptors.[3]

When coexpressed with Kir3.0 inwardly rectifying potassium channels (formerly GIRK) in HEK293 cells, $GABA_BR1$ occasionally up-regulated the inward current (I_{Kir}). However, agonist and antagonist affinities measured in these cells by current up-modulation are comparable to those from studies with native receptors.[4] Even though the approximately twofold up-regulation of basal I_{Kir} by $GABA_BR1$ activation is comparable to the up-regulation by other GPCRs, only less than 10% of the putative transfected cells showed a positive receptor-channel coupling, a rate that is significantly lower than that for other $G_{i/o}$-coupled receptors (A. Karschin, unpublished observations). Thus, $GABA_BR1$ showed a rather large variation in the capability of activating the coexpressed Kir3.0 channels than a generally low coupling efficacy in every cell.

[c]Corresponding author. Phone: +41-61-32-44249; fax: +41-61-32-45474; e-mail: Johannes.Mosbacher@pharma.novartis.com

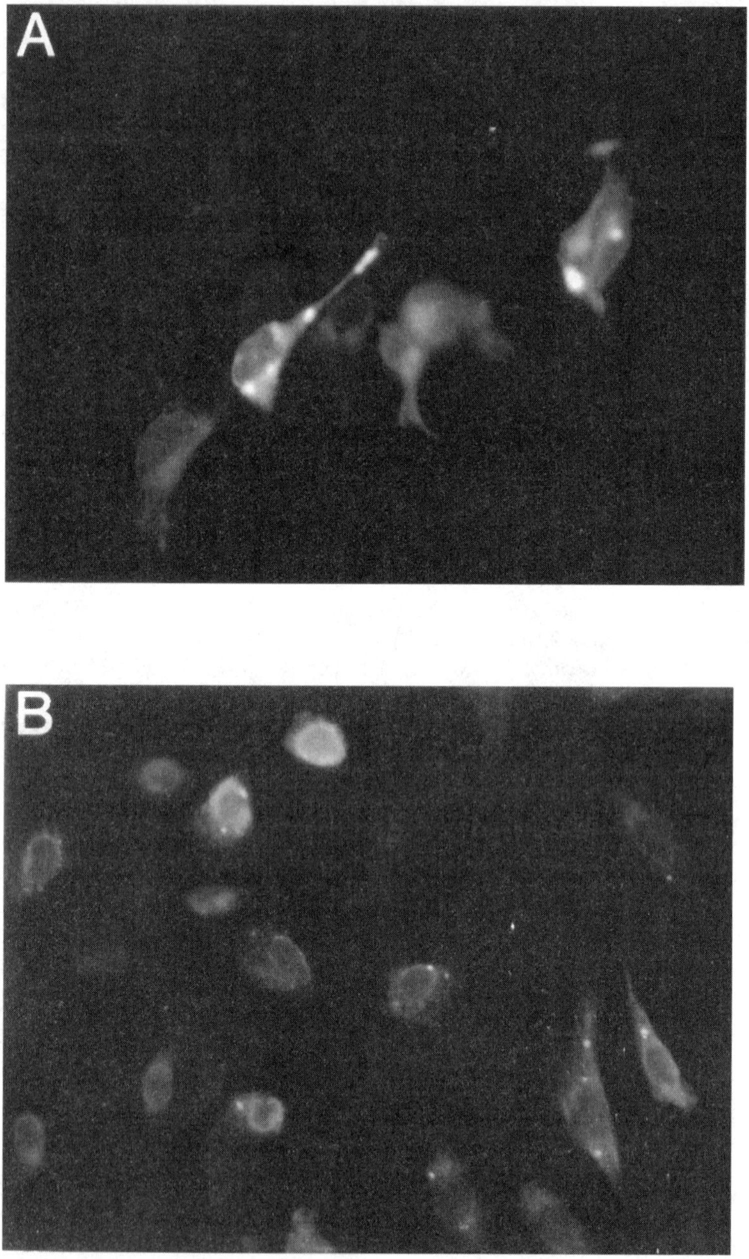
A
B

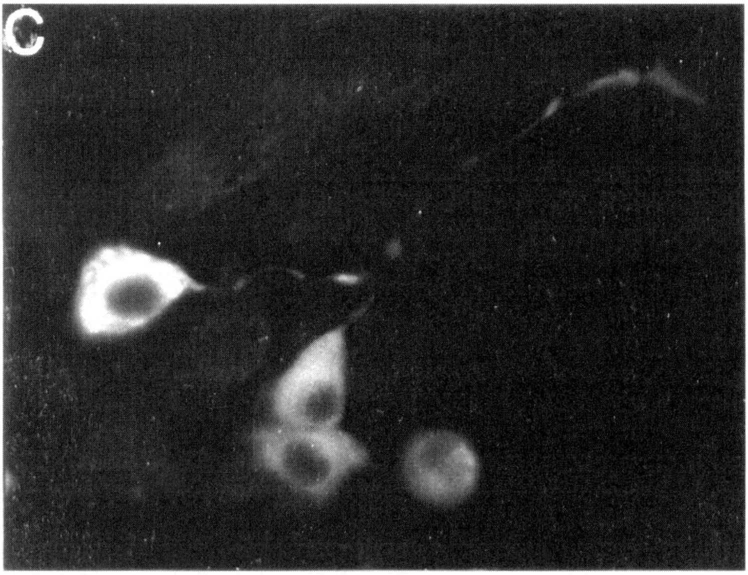

FIGURE 1 *(above and facing page).* Cellular localization of GABA$_B$R1 monitored by immunocytochemistry and GFP tagging. **(A)** Rat GABA$_B$R1a was inserted into pEGFP-N1 (Clontech, CA), and the fusion protein was expressed in HEK293 cells. Binding affinities of the fusion protein were tested and were similar to wild-type GABA$_B$R1 receptor. GFP fluorescence was measured in living cells 3–5 days posttransfection. Note the high-density fluorescence in intracellular organelles. **(B)** Fixed and permeabilized CCL-39 cells with stable expression of hGABA$_B$R1a were incubated with an antibody against a C-terminal part of GABA$_B$R1 and a FITC-labeled secondary antibody. Fluorescence could be detected throughout the cytoplasm and in intracellular spots, but only weakly at the cellular membrane. **(C)** GABA$_B$R1a/GFP-fusion protein expressed in NG108-15 cells. The fusion protein migrated into dendritic processes at day 3 after differentiation of cells.

It has been reported that an unequilibrated expression ratio of receptor to G protein may cause a poor coupling of the GPCR to its effectors.[5] However, cotransfection of different G protein subtypes with GABA$_B$R1 did not result in a more robust functional coupling. Therefore, we questioned which other mechanisms may underlie the large variation in functional coupling of GABA$_B$R1. Results from recent studies suggested that membrane targeting and compartmentalization determine the interaction between specific proteins. Therefore, we studied the cellular localization of the transfected GABA$_B$R1 protein.

When HEK293 cells were transiently transfected with an GABA$_B$R1/eGFP-fusion protein, most of the GFP fluorescence can be attributed to the intracellular membranes (FIG. 1A). Transiently transfected cells may overexpress the cloned receptor. However, also in cells stable transfected with GABA$_B$R1 the majority of GABA$_B$ receptor proteins were localized in intracellular membranes (FIG. 1B).

Since it is possible that GABA$_B$R1 needs a neuron-specific cofactor for surface expression, we transfected NG108-15 mouse neuroblastoma × glioma cells with the GABA$_B$/eGFP fusion protein. As shown in FIGURE 1C, the protein accumulated 3 days after cell differentiation in the dendritic processes as expected for a synaptic protein. We were unable to determine whether the protein was localized intracellularly or whether it was correctly inserted into the cell membrane. However, when we looked at the modulation of endogenous high voltage–activated Ca^{2+} channels or cotransfected Kir3.0 channels by $GABA_BR1$ in NG108-15 cells, we did not detect functional coupling of the $GABA_BR1$ to these channels.

DISCUSSION

The studies with NG108-15 cells indicate that the targeting of the receptor might not be sufficient to result in robust coupling of the cloned $GABA_BR1$ to its effector molecules. Furthermore, $GABA_BR1$ is glycosylated when heterologously expressed, which indicates that the protein does not remain in the endoplasmic reticulum.

Recently, the cloning of an auxiliary protein was reported that controls transport and glycosylation of a GPCR.[6] This finding may stimulate the search for other auxiliary proteins that influence the function of $GABA_B$. Since a strong positive coupling to Kir 3.0 channels is detectable in some transfected cells, this cofactor is also present in heterologous expression systems—at least under certain conditions.

[**Note added in proof:** We and others have cloned $GABA_BR2$, a receptor subtype that promotes surface expression of $GABA_BR1$ by forming a heteromeric complex. The heteromeric receptor shows robust coupling to Kir3.0 channels. (Kaupmann, K. *et al.* 1998. $GABA_B$ receptor subtypes assemble into functional heteromeric complexes. Nature **396:** 683–687. See also two other related articles in the same volume.)]

REFERENCES

1. Bettler, B., K. Kaupmann & N.G. Bowery. 1998. $GABA_B$ receptors—Drugs meet clones. Curr. Opin. Neurobiol. **8:** 345–350.
2. Kaupmann, K. *et al.* 1997. Structure, pharmacology and chromosomal localization of $GABA_B$ receptors. Soc. Neurosci. Abstr. **23:** 954.
3. Kaupmann, K. *et al.* 1997. Expression cloning of $GABA_B$ receptors uncovers similarity to metabotropic glutamate receptors. Nature **386:** 239–246.
4. Kaupmann, K. *et al.* 1998. Human $GABA_B$ receptors are differentially expressed and regulate inwardly rectifying K^+ channels. Proc. Natl. Acad. Sci. USA **95:** 14991–14996.
5. Kenakin, T. 1997. Differences between natural and recombinant G-protein coupled receptor systems with varying receptor/G-protein stoichiometry. Trends Pharmacol. Sci. **18:** 456–464.
6. McLatchie, L.M. *et al.* 1998. RAMPs regulate the transport and ligand specificity of the calcitonin-receptor-like receptor. Nature **393:** 333–339.

Postsynaptic Colocalization of Gephyrin and $GABA_A$ Receptors

MARCO SASSOÈ-POGNETTO,[a,d] MAURIZIO GIUSTETTO,[a] PATRIZIA PANZANELLI,[a] DARIO CANTINO,[a] JOACHIM KIRSCH,[b] AND JEAN-MARC FRITSCHY[c]

[a]*Department of Anatomy, Pharmacology and Forensic Medicine, University of Turin, Corso Massimo d'Azeglio 52, I-10126 Turin, Italy*

[b]*Max-Planck-Institut für Hirnforschung, Frankfurt am Main, Germany*

[c]*Institute of Pharmacology, University of Zürich, Zürich, Switzerland*

The high concentration of neurotransmitter receptors in the postsynaptic membrane is achieved by the interaction of receptor polypeptides with anchoring and clustering proteins. One of these proteins is gephyrin, which was originally copurified with the glycine receptor (GlyR) from mammalian spinal cord.[1] Gephyrin acts as a linker protein that

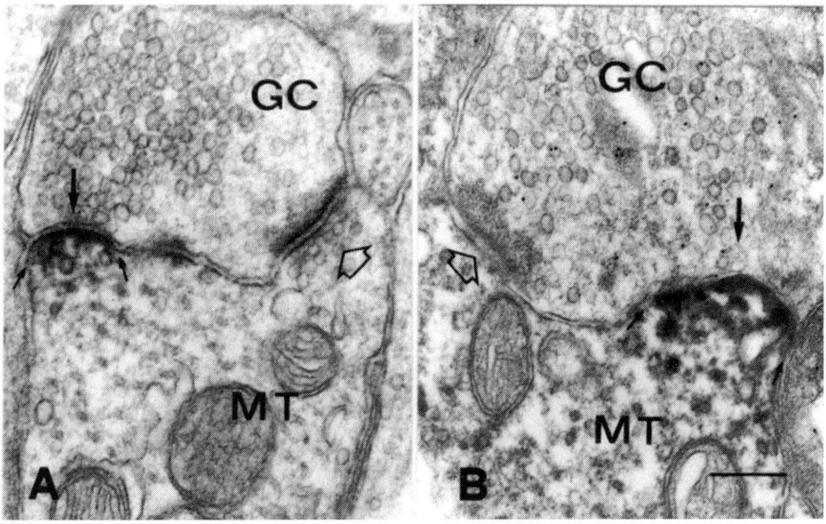

FIGURE 1. Synaptic localization of gephyrin in the olfactory bulb. **(A)** and **(B)** show two pairs of reciprocal dendrodendritic synapses between granule (GC) and mitral/tufted (MT) cells. The granule-to-mitral/tufted synapses (*arrows*) are symmetric and labeled by the antibody against gephyrin (*small arrows*). The mitral/tufted-to-granule synapses (*hollow arrows*) are asymmetric and not labeled. In **(B),** preembedding immunostaining with mAb 7a was combined with postembedding immunogold staining with an antiserum to GABA. The granule cell spine, which is presynaptic to gephyrin, is GABA-positive, whereas the mitral/tufted dendrite is not. *Scale bar* = 0.3 μm. (Reprinted from Giustetto *et al.*[7] with permission.)

[d]Corresponding author: Phone: +39-11-6707-725; fax: +39-11-6707-732; e-mail: marco.sassoe@unito.it

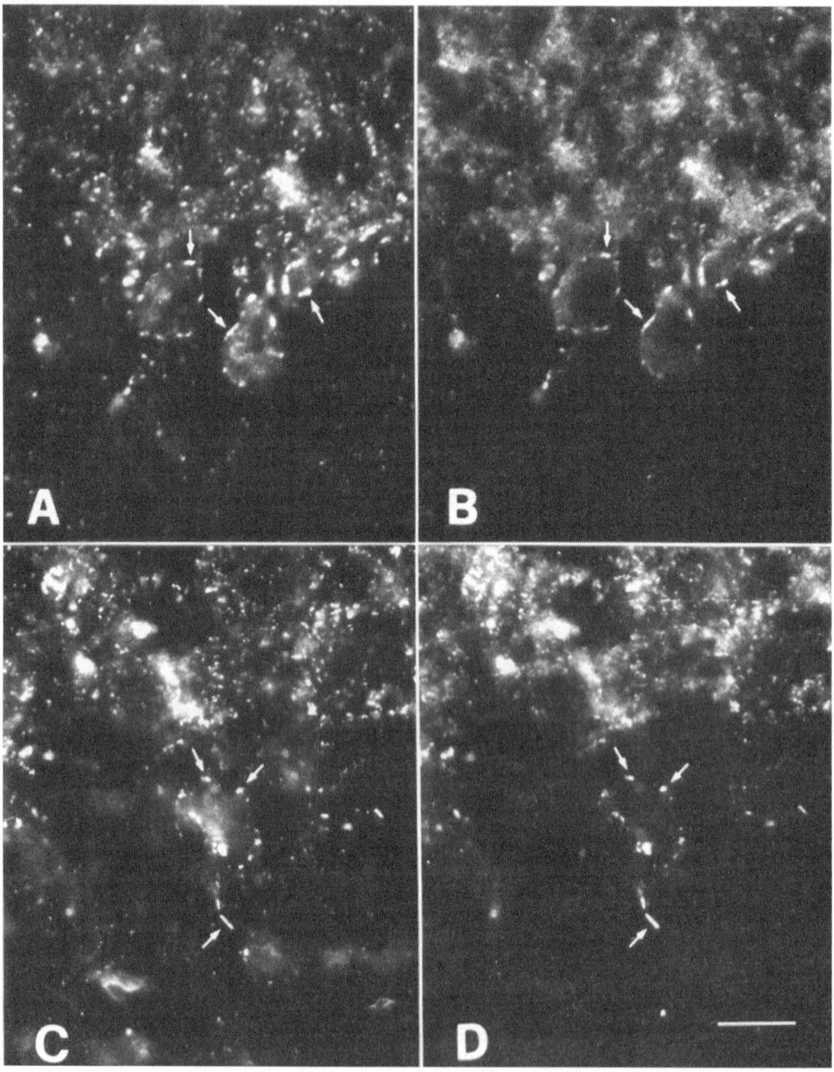

FIGURE 2. Colocalization of gephyrin and $GABA_A$ receptors. Gephyrin **(A)**, **(C)** colocalizes with the α1 **(B)** and γ2 **(D)** subunits of the $GABA_A$ receptor. Some of the double-labeled puncta surround the mitral cells (*arrows*). *Scale bar* = 12 μm. (Reprinted from Giustetto *et al.*[7] with permission.)

connects the GlyRs to the cytoskeleton, thus promoting the clustering of GlyRs at postsynaptic sites.[2,3] There is increasing evidence that gephyrin is not uniquely associated with glycinergic synapses. As a matter of fact, gephyrin is widely distributed throughout the brain and spinal cord and is present at many locations where no GlyRs have been detected.[4] In this paper, we show that gephyrin colocalizes with GABA$_A$ receptors at synapses that lack GlyRs.

We studied the distribution of gephyrin in the rat olfactory bulb, using a monoclonal antibody (mAb 7a).[5] Mitral/tufted cells, that are the output neurons of the olfactory bulb, form reciprocal dendrodendritic synapses with a class of inhibitory interneurons, the granule cells. In these bidirectional contacts, the mitral/tufted-to-granule synapses are glutamatergic, whereas the granule-to-mitral/tufted synapses are GABAergic.[6] When gephyrin immunoreactivity was investigated in the electron microscope, it was found to be located exclusively at the granule-to-mitral/tufted synapses, where it was concentrated at the cytoplasmic side of the postsynaptic membrane (FIG. 1A and 1B). Immunogold labeling for GABA was distributed over the granule cell profiles, which were presynaptic to gephyrin (FIG. 1B). The gephyrin-positive synapses were not labeled by the GlyR-specific monoclonal antibody 4a,[5] suggesting that glycine does not function as a neurotransmitter in these synapses. Taken together, these data indicate that gephyrin is clustered at GABAergic, but not glutamatergic synapses in the olfactory bulb.[7] Significantly, Craig *et al.*,[8] who have studied the distribution of gephyrin on cultured hippocampal neurons, also found that gephyrin clusters were associated selectively with GABAergic synapses.

To assess the colocalization of gephyrin with GABA$_A$ receptors, mAb 7a was used in combination with polyclonal antibodies directed against different GABA$_A$ receptor subunits ($\alpha 1$, $\alpha 2$, $\alpha 3$, $\gamma 2$). These antibodies produced a punctate immunoreactivity, and electron microscopy showed that each punctum corresponds to a synaptic site.[7] By using double-immunofluorescence, we could demonstrate that in the olfactory bulb gephyrin colocalizes extensively with the $\alpha 1$ and $\gamma 2$ subunits of the GABA$_A$ receptor (FIG. 2).

More recently, we have started to investigate the distribution of gephyrin in other brain regions, and we found that in the rat cerebellar cortex gephyrin also colocalizes with GABA$_A$ receptors containing the $\alpha 1$ and $\gamma 2$ subunits (not shown). In the retina, however, gephyrin only scarcely colocalizes with the $\alpha 1$ subunit of the GABA$_A$ receptor, but colocalizes extensively with the $\alpha 2$ subunit.[9] Therefore, gephyrin may have the ability to interact with GABA$_A$ receptors of different subunit combinations. The existence of different splice variants may confer specificity to the binding of gephyrin to different receptor subtypes.

The results revised here indicate that gephyrin is not exclusively associated with GlyRs, but can also be found at distinct GABAergic synapses. Thus, they raise the possibility that gephyrin is involved in anchoring certain GABA$_A$ receptor subtypes in the postsynaptic membrane. Since it appears that gephyrin is not present at every GABAergic synapse, distinct mechanisms may be involved in the clustering of different GABA$_A$ receptor subtypes. Such mechanisms could contribute to provide the local specificity to target distinct receptor subtypes to their appropriate synapses.

REFERENCES

1. PFEIFFER, F., D. GRAHAM & H. BETZ. 1982. Purification by affinity chromatography of the glycine receptor of the rat spinal cord. J. Biol. Chem. **257:** 9389–9393.
2. KIRSCH, J., I. WOLTERS, A. TRILLER & H. BETZ. 1993. Gephyrin antisense oligonucleotides prevent glycine receptor clustering in spinal neurons. Nature **366:** 745–748.

3. Kirsch, J., G. Meyer & H. Betz. 1996. Synaptic targeting of ionotropic neurotransmitter receptors. Mol. Cell. Neurosci. **8:** 93–98.
4. Kirsch, J. & H. Betz. 1993. Widespread expression of gephyrin, a putative glycine receptor-tubulin linker protein, in rat brain. Brain Res. **621:** 301–310.
5. Pfeiffer, F., R. Simler, G. Grenningloh & H. Betz. 1984. Monoclonal antibodies and peptide mapping reveal structural similarities between the subunits of the glycine receptor of rat spinal cord. Proc. Natl. Acad. Sci. USA **81:** 7224–7227.
6. Shepherd, G.M. & C.A. Greer. 1998. Olfactory bulb. *In* The Synaptic Organization of the Brain, 4th ed., G.M. Shepherd, Ed. : 159–203. Oxford Univ. Press. New York.
7. Giustetto, M., J. Kirsch, J.-M. Fritschy, D. Cantino & M. Sassoè-Pognetto. 1998. Localization of the clustering protein gephyrin at GABAergic synapses in the main olfactory bulb of the rat. J. Comp. Neurol. **395:** 231–244.
8. Craig, A.M., G. Banker, W.R. Chang, M.E. Mcgrath & A.S. Serpinskaya. 1996. Clustering of gephyrin at GABAergic but not glutamatergic synapses in cultured rat hippocampal neurons. J. Neurosci. **16:** 3166–3177.
9. Sassoè-Pognetto, M., J. Kirsch, U. Grünert, U. Greferath, J.-M. Fritschy, H. Möhler, H. Betz & H. Wässle. 1995. Colocalization of gephyrin and $GABA_A$-receptor subunits in the rat retina. J. Comp. Neurol. **357:** 1–14.

Structural Requirements for the Interaction of Unsaturated Free Fatty Acids with Recombinant Human $GABA_A$ Receptor Complexes

MICHAEL-ROBIN WITT,[a,c] CLAUS FOG POULSEN,[a] BIRTHE LÜKENSMEJER,[a] SVEND ERIK WESTH-HANSEN,[a] JUNICHI NABEKURA,[b] NORIO AKAIKE,[b] AND MOGENS NIELSEN[a]

[a]*Research Institute of Biological Psychiatry, Sct. Hans Hospital, Roskilde, Denmark*

[b]*Department of Neurophysiology, Kyushu University School of Medicine, Fukuoka, Japan*

The major inhibitory neurotransmitter in the mammalian brain, γ-amino-butyric acid (GABA), elicits some of its effects by gating chloride ion channels that are an integral part of the postsynaptic $GABA_A$/chloride channel receptor complex located on the neuronal membrane. Unsaturated free fatty acids (FFA) interact with integral membrane proteins such as neurotransmitter receptor-ion channel complexes.[1] Agonist benzodiazepine receptor binding *in vitro* is markedly increased by the addition of unsaturated fatty acids—e.g., oleic acid, arachidonic acid (AA), or docosahexaenoic acid (DHA)—both in rat brain preparations[2] and in recombinant human $GABA_A$ receptor complexes expressed in the baculovirus/Sf-9 insect cell expression system.[3] Furthermore it has been shown that agonist binding to the $GABA_A$ receptor in brain membrane preparations is increased by unsaturated fatty acids.[4,5] Recent electrophysiological studies on recombinant human $GABA_A$ receptor complexes assembled in a $\alpha_1\beta_2\gamma_{2S}$ subunit composition have shown a marked effect of AA and DHA on GABA-induced whole-cell chloride ion currents.[6] Extending these findings, the present data show an effect of FFA on ^{3}H-muscimol binding to $GABA_A$ receptor complexes containing $\alpha_1\beta_2\gamma_{2S}$, $\alpha_2\beta_2\gamma_{2S}$, $\alpha_3\beta_2\gamma_{2S}$, and $\alpha_5\beta_2\gamma_{2S}$ subunits. Electrophysiologically it is found that the addition of AA (10^{-5} M) to $GABA_A$ receptors consisting of $\alpha_3\beta_2\gamma_{2S}$ subunits leads to an approximately twofold increase in the whole-cell chloride ion current induced by a submaximal concentration of GABA.

METHODS

$GABA_A$ receptor complexes were produced in Sf-9 insect cells by infection with baculovirus containing various human cDNA of $GABA_A$ receptor subunits α_1, α_2, α_3, or α_5 in combination with β_2 or β_3 as well as with γ_{2S} (for details, see references 7 and 8). Electrophysiological measurements of GABA-induced whole-cell currents in Sf-9 cells containing recombinant $GABA_A$ receptor complexes were done as described previously.[6,8]

RESULTS AND DISCUSSION

TABLE 1 shows that seven structurally different FFAs markedly increase specific ^{3}H-muscimol binding to $GABA_A$ receptor complexes and that complexes consisting of

[c]Corresponding author. Phone: +45 4633 4971; fax: +45 4633 4367; e-mail: michael-robin.witt@shh.hosp.dk

TABLE 1. Effect of Free Fatty Acids (10^{-4} and 10^{-5} M) on ^{3}H-Muscimol (10 nM) Binding to Recombinant Human $GABA_A$ Receptors Expressed in Sf-9 Cells

	$\alpha_1\beta_2\gamma_{2S}$		$\alpha_2\beta_2\gamma_{2S}$		$\alpha_3\beta_2\gamma_{2S}$		$\alpha_5\beta_2\gamma_{2S}$	
	10^{-4} M	10^{-5} M	10^{-4} M	10^{-5} M	10^{-4} M	10^{-5} M	10^{-4} M	10^{-5} M
Arachidonic acid	193 ± 28	123 ± 9	246 ± 38	114 ± 8	135 ± 4	115 ± 4	145 ± 4	109 ± 5
Oleic acid	200 ± 27	156 ± 15	258 ± 27	142 ± 10	147 ± 8	106 ± 3	138 ± 16	111 ± 18
Linoleic acid	187 ± 23	122 ± 7	208 ± 16	118 ± 11	148 ± 11	112 ± 7	143 ± 10	113 ± 6
Docosahexaenoic acid	174 ± 22	119 ± 12	184 ± 25	109 ± 6	135 ± 2	119 ± 18	140 ± 6	110 ± 5
Docosapentaenoic acid	189 ± 20	119 ± 16	186 ± 25	110 ± 5	132 ± 3	110 ± 10	138 ± 5	109 ± 5
Docosatetraenoic acid	183 ± 19	123 ± 10	179 ± 26	125 ± 16	133 ± 8	115 ± 9	142 ± 11	108 ± 8
Docosatrienoic acid	187 ± 30	118 ± 8	181 ± 28	108 ± 3	130 ± 9	107 ± 6	148 ± 7	106 ± 3

NOTE: Values are mean ± SD of 3–4 determinations.

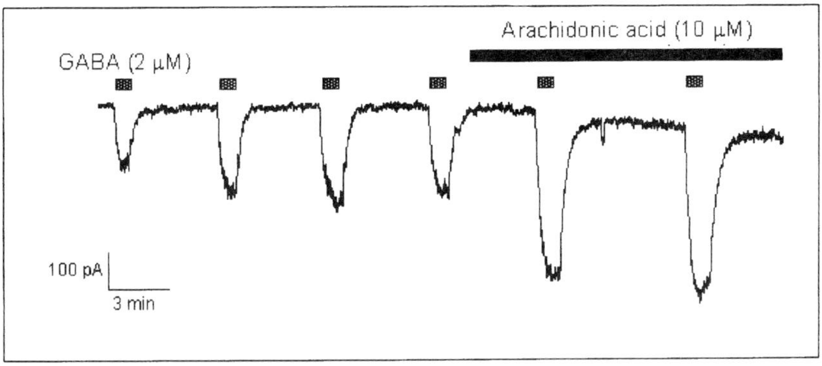

FIGURE 1. Arachidonic acid–induced enhancement of GABA-gated chloride current in recombinant human $GABA_A$ receptor complexes consisting of $\alpha_3\beta_2\gamma_{2S}$ subunits.

$\alpha_1\beta_2\gamma_{2S}$ and $\alpha_2\beta_2\gamma_{2S}$ subunits apparently had a higher increase in binding induced by fatty acids (approx. 200% of control) than receptor complexes consisting of $\alpha_3\beta_2\gamma_{2S}$ and $\alpha_5\beta_2\gamma_{2S}$ (approx. 140% of control binding). Preliminary studies indicate that the observed increase in specific binding is due to either an increase in the number of binding sites (B_{max} value) or a combination of an increase in the B_{max} value together with an increase in binding affinity (decrease in K_D value). It should be noted that since the actual concentration of FFAs in the binding assay is unknown, the indicated values (10^{-4}M and 10^{-5}M) should be regarded as nominal values.

Electrophysiologically it is found that AA (10^{-5}M), dissolved in physiological bath buffer and applied by the Y-tube system, markedly increased (approximately twofold) whole-cell chloride ion currents induced by submaximal concentrations of GABA (2×10^{-6} M) in $GABA_A$ receptor complexes consisting of $\alpha_3\beta_2\gamma_{2S}$ subunits (FIG. 1). The present data show that unsaturated fatty acids influence $GABA_A$ receptor complex binding and function in recombinant human receptors. Previously we have found that unsaturated FFAs increase agonist benzodiazepine receptor binding[2] as well as agonist $GABA_A$ receptor binding in rat brain membrane preparations.[4] The binding data indicate a small difference in the potency of FFAs among $GABA_A$ receptor complexes (stronger effects on complexes containing α_1 and α_2 subunits; TABLE 1). Saturation ^{3}H-muscimol binding experiments will clarify whether the effect of FFA is mediated by an effect on the number of receptors and/or binding affinity. The electrophysiological findings indicate a strong potentiation of GABA responses by application of AA to Sf-9 insect cells expressing $GABA_A$ receptor complexes containing $\alpha_3\beta_2\gamma_{2S}$ subunits. Previously, using almost identical experimental conditions on Sf-9 cells expressing $\alpha_1\beta_2\gamma_{2S}$ receptor complexes, it has been shown that AA and DHA potentiate GABA-induced whole-cell currents by approximately 15% at a FFA concentration of 10^{-6}M, while an application of 10^{-5}M FFA led to a strong inhibition of GABA-gated chloride currents.[6] The findings indicate that the effect of FFA on GABA responses is dependent on the subunit present in $GABA_A$ receptor complexes.

REFERENCES

1. Ordway, R.W., J.J. Singer & J.V. Walsh. 1991. Trends Neurosci. **14:** 96–100.
2. Witt, M.R. & M. Nielsen. 1994. J. Neurochem. **62:** 1432–1439.
3. Witt, M.R., M. Nielsen & S.E. Westh-Hansen. 1996. J. Neurochem. **67:** 2141–2145.
4. Nielsen, M., M.R. Witt & H. Thøgersen. 1988. Eur. J. Pharmacol. **146:** 349–353.
5. Koenig, I.A. & I.L. Martin. 1992. Biochem. Pharmacol. **44:** 11–15.
6. Nabekura, J., K. Noguchi, M.R. Witt, M. Nielsen & N. Akaike. 1998. J. Biol. Chem. **273:** 11056–11061.
7. Westh-Hansen, S.E., M.R Witt, P. Rasmussen, J. Nabekura, K. Noguchi, N. Akaike & M. Nielsen. 1997. Eur. J. Pharmacol. **329:** 253–257.
8. Ai, J., K. Dekermendjian, X. Wang, M. Nielsen & M.R.Witt. 1997. Drug Dev. Res. **41:** 99–107.

5-HT Receptor Knockout Mice: Pharmacological Tools or Models of Psychiatric Disorders

K. SCEARCE-LEVIE, J.-P. CHEN, E. GARDNER, AND R. HEN[a]

Center for Neurobiology and Behavior, Columbia University, 722 West 168th Street, New York, New York 10032, USA

ABSTRACT: The molecular diversity of cloned serotonin receptor subtypes in the brain makes it difficult to understand the specific modulatory roles played by different receptors. In order to understand the role of the 5-HT$_{1B}$ receptor subtype in behavior and neuropsychiatric disorders, we have been studying genetic knockout mice lacking the 5-HT$_{1B}$ receptor. The 5-HT$_{1B}$ knockout mice show evidence of increased aggression and impulsivity, behavioral patterns that are also associated with reduced 5-HT function. They also show reduced or absent locomotor stimulation to some serotoninergic drugs, indicating that the locomotor effects of these drugs require the 5-HT$_{1B}$ receptor. However, in some cases, data obtained with knockout mice conflicts with the pharmacological data. The 5-HT$_{1B}$ receptor knockout mice show a phenotype of increased vulnerability to drugs of abuse such as cocaine. However, pharmacological studies suggest that 5-HT$_{1B}$ stimulation enhances the effects of cocaine, while 5-HT$_{1B}$ blockade can attenuate some of the effects of cocaine. Compensations that enhance dopamine function appear to be responsible for the drug-vulnerable phenotype of 5-HT$_{1B}$ receptor knockout mice. By studying these compensations and changes in neural function, we can learn more about the fundamental mechanisms underlying addiction. The 5-HT$_{1B}$ knockout mice should be considered a model for the disease state of vulnerability to drugs of abuse, rather than a direct pharmacological model of 5-HT$_{1B}$ receptor function.

Serotonin (5-hydroxytryptamine, or 5-HT) is a modulatory neurotransmitter that acts throughout the brain. It has been suggested to play a role in a number of behaviors, including sleep, feeding, reward, locomotion, and mood. Yet it has proved quite difficult to understand how specific changes in serotonin neurotransmission affect specific behaviors or neurological functions. This task is made even more challenging by the molecular cloning of more than 14 5-HT receptor subtypes, each with its own expression pattern, coupling mechanism, and pharmacological profile.[1] This tremendous diversity increases the modulatory power of 5-HT, since different receptor subtypes are stimulated in different neural circuits.

With the exception of the 5-HT$_3$ receptor, which is coupled to a ligand-gated ion channel, all of the 5-HT receptor subtypes belong to the family of seven-transmembrane spanning, G protein–coupled receptors. These 5-HT receptor subtypes can be further subdivided into distinct families, based on structure and function. The 5-HT1 family of receptors, including the 1A, 1B, 1D, 1E, and 1F receptors, is negatively coupled to adenylyl cyclase. The 5-HT2 family includes the 2A, 2B, and 2C, and stimulates phospholipase C. The 5-HT$_4$, 5-HT$_6$, and 5-HT$_7$ receptors all simulate adenylyl cyclase, but a lack of homology in their amino acid sequences suggests that they belong to distinct families.

[a]Corresponding author. Phone: 212-543-5569; fax: 212-543-6059; e-mail: rh95@columbia.edu

Finally, the 5-HT_{5A} and 5-HT_{5B} receptors seem to belong to the same family, based on their amino acid sequences, but their effector systems remain unknown.

FOCUS ON THE 5-HT_{1B} RECEPTOR

The 5-HT_{1B} receptor is a particularly interesting research target for several reasons. First, the size of the 5-HT1 family suggests an evolutionary need for finely controlled serotoninergic signaling in several distinct neuronal populations. However, the large number of closely related receptors makes it difficult to dissect out the role of specific receptors in this neural modulation. The pattern of expression of the 5-HT_{1B} receptor is particularly intriguing. The localization of 5-HT_{1B} receptor mRNA shows that the receptor is produced in neurons in the raphé nuclei, striatum, cerebellum, hippocampus, entorhinal and cingulate cortical regions, subthalamic nuclei, and the nucleus accumbens.[2,3] However, autoradiographic studies demonstrate that the protein has a different expression pattern that is highest in the projection zones of the neurons that produce the mRNA: substantia nigra, globus pallidus, dorsal subiculum, and superior colliculi.[4,5] This suggests that the receptor is transported to axon terminals in regions that receive serotoninergic projections.[6,7] Although the receptor always seems to be expressed on terminals, it can be expressed either as an autoreceptor on serotoninergic projections or as a heteroreceptor modulating the release of other neurotransmitters.

The 5-HT_{1B} receptor generally has an inhibitory effect on neuronal activity. Specifically, stimulation of the 5-HT_{1B} autoreceptor has been shown to inhibit 5-HT release in rat cortex.[8] Similarly, 5-HT_{1B} heteroreceptor stimulation also inhibits release of acetylcholine in rat hippocampus. In the substantia nigra pars reticulata and ventral tegmental area (VTA), 5-HT_{1B} agonists can inhibit the release of GABA from axon terminals.[9–11] However, it should be noted that when considered in the broader neural circuitry, the net effects of 5-HT_{1B} activation can be stimulatory. For instance, since GABA is an inhibitory neurotransmitter, its release from terminals in the VTA inhibits dopamine (DA) neurotransmission. But stimulation of 5-HT_{1B} receptors on those GABAergic terminals inhibits GABA release, and can indirectly stimulate DA release by removing the GABA inhibition. Benloucif and colleagues demonstrated this by showing that RU24969, a 5-$HT_{1B/1A}$ agonist, can increase extracellular DA release in rat striatum in a dose-dependent manner.[12]

To better understand the role of the 5-HT_{1B} receptor in behavior and neurotransmission, we have used homologous recombination to create a line of knockout mice that completely lacks functional expression of the 5-HT_{1B} receptor gene.[13] Briefly, a cassette containing the neomycin-resistance gene was inserted so that it disrupted most of the single-exon coding sequence for the 5-HT_{1B} receptor gene. Chimeric males that demonstrated germline transmission of this mutation were bred with wild-type female 129/Sv mice to produce offspring heterozygous for the 5-HT_{1B} receptor gene. These were then bred together to generate a line of homozygous 5-HT_{1B} knockout mice lacking both copies of the 5-HT_{1B} gene. The functional deletion of the receptor was confirmed by autoradiographic studies showing an absence of specific binding of the 5-HT_{1B} radioligand [^{125}I]-iodocyanopindolol in brain tissue from knockout mice.[13]

This knockout has few obvious developmental or behavioral abnormalities. The mice develop at a normal rate, feed normally, mature normally, breed and nurture offspring normally, and die at a normal age (unpublished observations and Ref. 13). In general, their behavior is similar to wild type, except when the mice are presented with an environmental or pharmacological challenge (for a review of phenotypic differences, see Ref. 14). We have therefore used these mice, in combination with traditional pharmacological methods, to explore the role of the 5-HT_{1B} receptor in various behaviors.

BEHAVIORAL PHARMACOLOGY OF THE 5-HT_{1B} RECEPTOR

Classically, behavioral pharmacology only examines the effects of acute blockade or stimulation of a receptor. However, chronic blockade may reveal additional information about the function of the receptor. For instance, some receptors may play a critical role in the maintenance of neural homeostasis. But changes in homeostatic regulation may not become apparent unless the receptor is absent from the neural circuitry for a long time. Similarly, long-term activation of certain receptors may cause long-lasting changes in gene expression through the induction of certain transcription factors. An acute pharmacological manipulation may not be able to detectably affect longer-term changes in gene expression.

A knockout mouse lacking a particular receptor allows us to look at the long-term effects of receptor absence, including changes in homeostatic regulation and gene expression. Since genetic defects and diseases also may cause long-term changes in homeostasis or gene transcription, knockout mice are appropriate models for genetic diseases. However, in some cases, the behavioral consequences of genetic knockouts are strikingly different from the results suggested by traditional behavioral pharmacology. Therefore, a combination of behavioral pharmacology and behavioral genetics is most powerful for understanding how the 5-HT_{1B} receptor affects various behaviors.

Locomotion

Although stimulation of the 5-HT_{1B} receptor increases locomotion, the receptor does not seem to be required for normal locomotor behavior. Acute or chronic administration of GR127935 (an antagonist of $5\text{-HT}_{1B/1D}$ receptors, with partial 1B agonist activity) does not affect locomotion (unpublished observations). However, when introduced into a novel environment, the 5-HT_{1B} knockout mice (KO) display more locomotion and exploration than wild-type mice (WT).[15] With repeated exposure to this environment, the locomotor levels eventually return to normal, indicating that the increased locomotion may reflect an increased response to novelty.

In contrast, the $5\text{-HT}_{1B/1A}$ receptor agonist RU24969 appears to increase locomotor activity in rats and mice via a 5-HT_{1B}-dependent mechanism. The locomotor stimulatory effects of RU24969 are attenuated by the $5\text{-HT}_{1B/1A}$ antagonists (–)-pindolol and (–)-propanolol, but not by other 5-HT antagonists, metergoline or methysergide.[16] Similarly, the $5\text{-HT}_{1B/1A}$ agonist CP94253 also increases locomotion in rodents.[17] These effects are probably not due to the stimulation of other 5-HT receptors, since stimulation of 1A receptors[18] and 2A/2C receptors[19] decreases locomotion. We have confirmed that RU24969 increases locomotion by stimulating the 5-HT_{1B} receptor by demonstrating the KO mice do not show increased locomotion in response to this drug[13] (FIG. 1). Similarly, pretreatment of WT mice with GR127935 can block the locomotor effects of RU24969.[20]

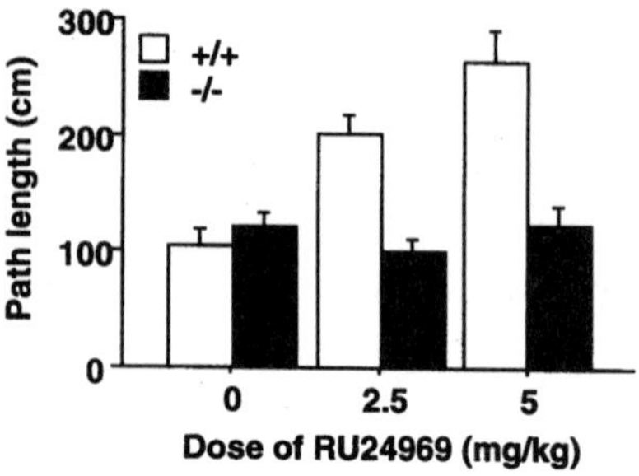

FIGURE 1. Locomotor response to acute ip injection of 5-HT$_{1A/1B}$ agonist RU24969 (0, 2.5, or 5 mg/kg), as measured by videotracking. Each point represents mean path length (in cm) + SEM covered in 5 minutes and averaged over 1 hour for 10 to 12 mice per group. Locomotor stimulation is seen in WT (+/+, *white bars*), but not KO (–/–, *black bars*) mice.

The increased locomotion in rodents caused by MDMA (3,4-methylenedioxy-N-methamphetamine, commonly known as ecstasy) has also been suggested to be mediated by the 5-HT$_{1B}$ receptor. MDMA causes release of 5-HT from terminals and thus acts as an indirect agonist of most 5-HT receptors. However, the pattern of locomotion, which includes a decrease in exploratory behavior, is qualitatively similar to the effects of RU24969, suggesting that similar neural mechanisms are involved in both cases.[21] Furthermore, repeated administration of RU24969, but not 5-HT$_{1A}$ or 5-HT$_{2A/2C}$ agonists, reduces the behavioral response to MDMA, indicating the development of behavioral cross-tolerance via possible desensitization of 5-HT$_{1B}$ receptors.[22] Neither (–)-pindolol nor (–)-propanolol can antagonize MDMA-induced hyperactivity.[21,23] Taken together, these experiments point to 5-HT$_{1B}$ receptor stimulation as the primary mechanism underlying the hyperlocomotor effects of both RU24969 and MDMA. We tested the role of the 5-HT$_{1B}$ receptor in the effects of MDMA by assessing the response of KO mice to the drug.[24] Their locomotor response to MDMA was greatly attenuated, confirming that the 5-HT$_{1B}$ receptor is necessary for the locomotor effects of MDMA (FIG. 2). However, the drug decreased exploration equally in both WT and KO mice, indicating that MDMA affects exploratory activity independently of the 5-HT$_{1B}$ receptor. Pretreatment of WT mice with GR127935 before MDMA administration mimicked the KO response to MDMA, blocking the increase in locomotion, but not the decrease in exploration.[24]

In summary, we can conclude that blockade or removal of the 5-HT$_{1B}$ receptor has little effect on baseline locomotor behavior, but stimulation of the receptor by direct or indirect agonists can specifically increase locomotion. Although certain nonspecific 5-HT$_{1B}$ agonists also decrease exploration, this effect persists in KO mice and WT mice pretreated with 5-HT$_{1B}$ antagonists, suggesting that the 5-HT$_{1B}$ receptor does not mediate the effects of these drugs on exploratory behavior.

Aggression

There are numerous links between 5-HT levels and aggression. In primates, a central 5-HT deficit has been associated with suicidal behavior and impulsive violence.[25] Particularly aggressive mouse strains have lower brain 5-HT levels.[26] There are even several correlations between low 5-HT levels and aggression in humans. Violent offenders,[27] arsonists,[28] and people who have committed suicide violently[29,30] all show decreased levels of 5-hydroxyindoleacetic acid (5-HIAA, a 5-HT metabolite) in their cerebrospinal fluid. Decreases in dietary tryptophan (a precursor for 5-HT) increase predatory mouse-killing behavior in rats, while increases in dietary tryptophan inhibit this behavior.[31] Specific serotonin reuptake inhibitors (SSRIs), which increase synaptic 5-HT levels, decrease aggressive behavior in isolated male mice.[32,33] Specific stimulation of 5-HT_1 receptor subtypes is also able to reduce aggression in rodents. Drugs that can act as 5-HT_{1B} agonists, including RU24969, TFMPP, and eltoprazine, reduce aggression in mice and rats without inducing general sedation.[34,35]

Given this background, it was exciting to note that the most striking behavioral phenotype of the 5-HT_{1B} KO mice is increased aggression. When an isolated male mouse is confronted with an unfamiliar intruder mouse in his home cage (resident-intruder paradigm), KO mice attack the intruder more quickly, more often, and more intensely than WT controls.[13,36] Lactating female KO mice will also attack an unfamiliar male more quickly and violently.[37] Even when they are stably housed with littermates, there is more evidence of aggression among the KO mice (unpublished observations). This fits well with the suggestion that 5-HT_{1B} receptor stimulation can reduce aggression (see above, and Ref. 38 for review). With this receptor removed, some of the circuitry modulating aggression may be able to run unchecked. It is notable, however, that administration of the 5-HT_1 antagonist, eltoprazine, can significantly reduce aggressive behavior in KO as well as WT mice. MDMA, which can increase feelings of emotional affinity in humans, also decreases aggression in both KO and WT mice.[37] This suggests that while the 5-HT_{1B} receptor contributes to aggression, it is not the sole 5-HT receptor subtype modulating this behavior. In particular, 5-HT_{1A} receptor activation can also influence aggressive behavior.

Feeding

In general, drugs that increase 5-HT availability, such as agonists, SSRIs, and the 5-HT precursor tryptophan, decrease feeding behavior.[39] Most notably, fenfluramine, a 5-HT releaser, was marketed as an appetite suppressant until reported cardiovascular side effects drove it off the market. Fenfluramine's anorectic effects may be mediated by the 5-HT_{1B} receptor, as indicated by the ability of cyanopindolol, a 5-$HT_{1A/1B}$ antagonist, to attenuate the hypophagic effects of the drug in rats.[40] Similarly, a number of drugs that activate 5-HT_1 receptors, including CP94253,[17] RU24969, m-CPP, and TFMPP, all reduce food intake in rats.[41] This is probably due to 1B activation, since 5-HT_{1A} receptor agonists increase food consumption in rats.[42] Serotonin receptors in the paraventricular nucleus of the hypothalamus may play an important role in this response. Microinjection of indirect 5-HT agonists fluoxetine and fenfluramine into the paraventricular nucleus has been shown to have hypophagic effect.[43–45] Moreover, injection of the 5-$HT_{1A/1B}$ agonist RU24969 into this nucleus can also elicit hypophagia without affecting locomotor activity.[46]

In agreement with these results, 5-HT_{1B} KO mice did not show any anorectic response to systemic fenfluramine treatment, although WT controls did.[47] In addition, although fenfluramine induces the expression of the immediate-early gene c-fos in the paraventricular nucleus of WT mice, the induction of c-fos was greatly attenuated in KO mice. This indicates that the 5-HT_{1B} receptor is required for both the anorectic effects of fenfluramine, as well as its physiological effects in the paraventricular nucleus.

Impulsivity

There have been some suggestions, that, when considered together, the types of behaviors induced by 5-HT_{1B} receptor stimulation can be classed as an overall decrease in impulsive behavior. In most cases, stimulation of this receptor seems to inhibit behaviors related to fundamental drives like feeding, aggression, or sexual behavior. This fits with the suggestion that increased 5-HT levels can inhibit impulsive behavior in both risky and appetitive situations.[48] The type of aggression in humans related to low 5-HT levels has been specifically characterized as impulsive aggression. For instance, impulsive violent offenders had low levels of 5-HIAA in cerebrospinal fluid, while people who premeditated their acts had higher levels of 5-HIAA.[49] Similarly, low levels of 5-HIAA were found only in suicide victims who performed violent suicides.[30] Although it is difficult to determine which specific 5-HT receptor subtype may underlie a behavior as generalized as "impulsivity," the 5-HT_{1B} receptor is an intriguing candidate since stimulation inhibits aggression, feeding, and sexual behavior (for a review, see Ref. 50).

The decreased attack latency found in KO mice in all aggression tests suggests that a general increase in impulsivity may contribute to this behavior. Indeed, there are several other suggestions that the KO mice may be more impulsive, or at least less inhibited when interacting with novel objects or environments.[50] They will explore a novel environment more.[15] When learning an operant task, they tend to press a lever more and make stimulus discriminations more quickly. In the elevated plus maze, KO mice appear to spend less time exhibiting exploratory behavior in the center of the plus, suggesting that they spend less time making decisions about which arm to enter.[51] Pilot studies of sexual behavior suggest that KO males will mount females faster and more often (Saudou, unpublished observation). In a study of maternal behavior, KO mothers would retrieve pups placed in a far corner of the cage more quickly than WT controls.[51] It has therefore been suggested that the KO mice may represent a model of an impulsive behavior pattern.[13,50]

CAN THE 5-HT_{1B} RECEPTOR MODULATE THE EFFECTS OF ADDICTIVE DRUGS?

Although few researchers had directly assessed the role of the 5-HT_{1B} receptor in the response to addictive and abused drugs, the pattern of expression of the receptor suggested it might be able to modulate the response to these drugs. Cameron and Williams suggested that 5-HT_{1B} stimulation may be able to indirectly facilitate or enhance dopamine (DA) neurotransmission from the VTA and substantia nigra to striatal and cortical regions.[9] Since this DA pathway appears to be the main pathway stimulated by addictive drugs,[52] we reasoned that 5-HT_{1B} stimulation might enhance some of the effects of certain abused drugs. Indeed, other researchers have found that RU24969 can increase the rewarding effects of DA agonists in a self-administration paradigm.[53,54] Similarly, 5-HT_{1B} agonists can partially substitute for cocaine on a drug-discrimination test.[55]

This reasoning suggested that 5-HT_{1B} KO mice would be less sensitive to the behavioral effects of drugs of abuse. In fact, for ecstasy (MDMA), this did appear to be the case. As mentioned above, the KO mice showed much less locomotor stimulation in response to this drug (see FIG. 2).

The actions of cocaine and amphetamine are similar to MDMA, but are thought to be primarily via the DA system. Cocaine inhibits reuptake of DA, 5-HT, and norepinephrine (NE), and thereby increases synaptic availability of those neurotransmitters. Most research has pointed to the importance of increased extracellular DA in the addictive and stimulatory effects of cocaine. However, it is important to remember that cocaine's affinity for the 5-HT transporter is actually greater than its affinity for the DA transporter.[56] Moreover, drugs that block the DA transporter but have little serotoninergic activity, like mazindol or GBR12909, are not readily self-administered by animals and are not abused by humans.[57] Furthermore, mice lacking the DA transporter still self-administer cocaine and amphetamine.[58] Therefore, the increased extracellular 5-HT caused by cocaine may enhance the stimulatory or addictive effects of the drug. This evidence, along with the attenuated response to MDMA in 5-HT_{1B} KO mice, suggests that 5-HT_{1B} receptor stimulation may increase the effects of drugs of abuse.

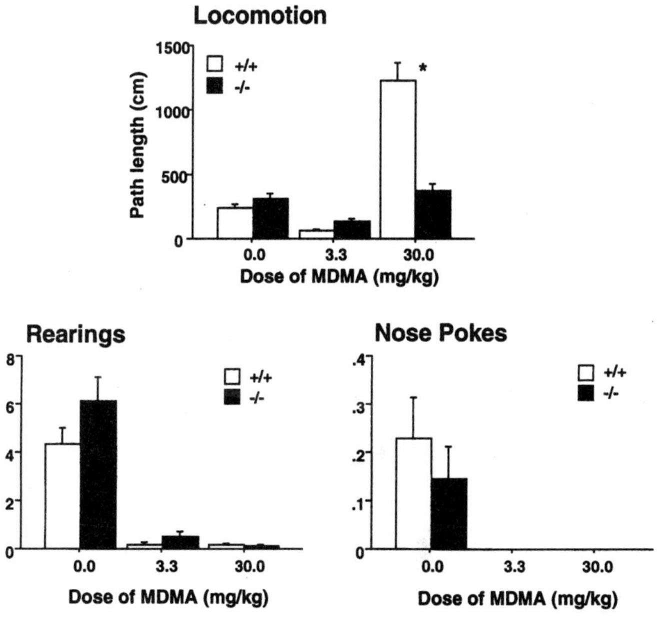

FIGURE 2. Top, locomotor response to acute ip injection of MDMA (0, 3.3, or 10 mg/kg), $n = 10$ mice per group. Each *bar* represents mean path length (in cm) + SEM covered in 5 minutes and averaged over 30 minutes. Locomotor stimulation is seen in WT (+/+, *white bars*), but not KO (-/-, *black bars*) mice. **Bottom**, exploratory responses of rearing and nose pokes are blocked by MDMA in both WT and KO mice. Each *bar* is mean number of rearings or nose pokes occurring in 5 minutes, averaged over 30 minute test session.

We therefore expected that mice lacking the 5-HT$_{1B}$ receptor would have attenuated behavioral responses to psychostimulants like cocaine. However, cocaine actually stimulates locomotion and exploration in 5-HT$_{1B}$ KO mice much more than in WT controls (FIG. 3).[59] This enhanced locomotor response is apparent over a range of doses, as well as after repeated administrations of the drug, suggesting that the increased sensitivity in KO mice cannot be explained by a simple shift in the dose-response curve for cocaine. The KO mice also appear to be more sensitive to the reinforcing effects of cocaine. In an intravenous self-administration paradigm, they acquire self-administration behavior more quickly than WT mice.[60] Once they are stably self-administering, their behavior is the same as WT on a fixed-ratio reinforcement schedule. However, when the animals are switched to a progressive ratio schedule, the KO mice appear to be significantly more motivated to obtain the cocaine reinforcer (FIG. 4). In this test, each subsequent reinforcer requires progressively more lever presses. The dependent measure is the "break point," or the number of reinforcers an animal obtains before it stops lever pressing. The KO mice will obtain 9 reinforcers, which require 25 lever presses, while the WT stop pressing after only 4 reinforcers, or 9 lever presses.[59] When food is the reinforcer, however, there is no difference between WT and KO behavior. This suggests that cocaine is specifically more rewarding in KO mice.

It is not clear, however, that this increased sensitivity to the locomotor and reinforcing effects of cocaine in KO mice is a direct result of the absence of the 5-HT$_{1B}$ receptor. Behavioral pharmacology in rats suggests that 5-HT$_{1B}$ stimulation enhances the effects of cocaine, and that antagonism of the receptor can reverse this enhancement.[54] Indeed, pretreatment of WT mice with the 5-HT$_{1B/1D}$ antagonist GR127935 actually decreases the locomotor effects of cocaine—the opposite effect of a genetic KO of the receptor.[20] Furthermore, GR127935 alone does not affect cocaine self-administration behavior of either WT or KO mice.[20] Since KO mice are missing the receptor throughout development, numerous neurochemical changes may have occurred to compensate for the developmental absence of the receptor.

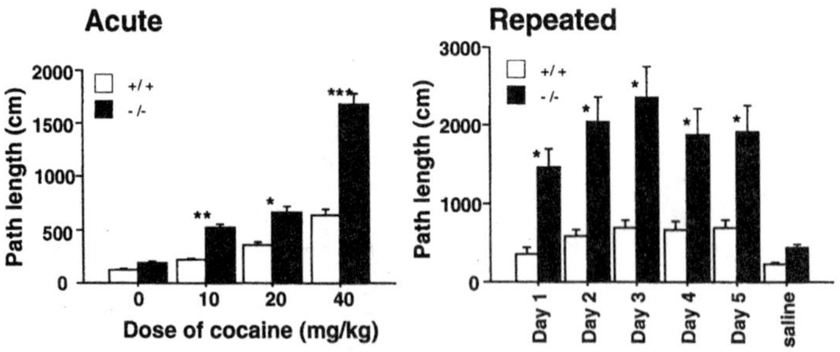

FIGURE 3. Left, locomotor response to acute ip injection of cocaine (0, 10, 20, or 40 mg/kg), n = 12 mice per group. **Right**, locomotor response to repeated injections of 20 mg/kg cocaine, n = 16 mice per group. Each *bar* represents mean path length (in cm) + SEM covered in 5 minutes and averaged over 1 hour. In all treatment conditions, KO mice (*black bars*) showed significantly more locomotion than WT (*white bars*): *($p < 0.05$), **($p < 0.01$), or *** ($p < 0.001$).

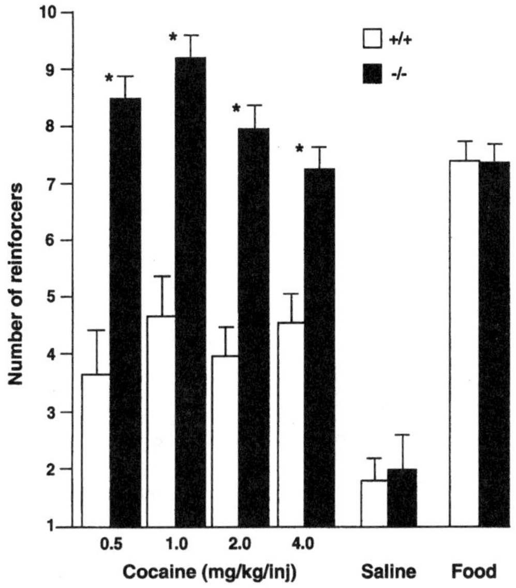

FIGURE 4. Self-administration behavior. KO mice (*black bars*) obtain more cocaine reinforcements, but not food or saline, under a PR reinforcement schedule. Each *bar* represents the number of reinforcements expressed as mean + SEM for 10 KO mice and 5 WT (cocaine experiment) or 7 WT and 8 KO (food experiment). * indicates a significant ($p < 0.05$) difference between genotypes.

The behavioral response of KO mice to other, more specific, DA agonists suggests that developmental compensations may have enhanced the function of the DA system in the KO mice. They show increased locomotor responses to the DA releasers amphetamine and methylphenidate (ritalin), as well as to the specific DA reuptake inhibitor GBR12909 (FIG. 5). Since these drugs are more specific for DA than cocaine is, it is difficult to explain the increased response to these drugs in 5-HT_{1B} KO mice without postulating some change in DA function.

Although there is no indication of enhanced basal levels of tissue DA (as shown by HPLC) or extracellular DA (as shown by microdialysis), there is some evidence for increased DA function. When a moderate dose of cocaine (20 mg/kg) is given, there is a slight increase in DA release in the nucleus accumbens of KO mice relative to WT (FIG. 6). Postsynaptically, D1 DA receptor levels are increased in the striatum of KO mice, as indicated by increased binding of the D1 antagonist [^{3}H]-SCH23390, as well as by increased D1 mRNA.[61] Interestingly, mRNA for dynorphin and substance P, neuropeptides expressed in the same striatal neurons that express D1 receptors, are also increased in KO mice. In contrast, there is no evidence of changes in the D2 receptor or its associated neuropeptide, enkephalin, in the striatum.[61]

A simple model of how DA neurotransmission may be perturbed in the KO mouse is posited in FIGURE 7. In the normal circuitry, activation of 5-HT_{1B} receptors on GABAergic terminals inhibits GABA release, which indirectly stimulates DA release in projections from the VTA and substantia nigra to the striatum. When an antagonist acutely blocks the

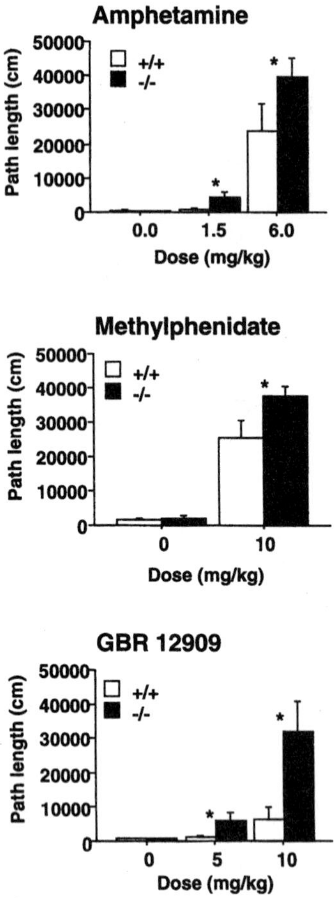

FIGURE 5. Locomotor response to acute ip injections of indirect DA agonists: amphetamine, 1.5 or 6 mg/kg, $n = 10$ mice per group (**top**); methylphenidate, 10 mg/kg, $n = 8$ mice per group (**middle**); and GBR 12909, 5 or 10 mg/kg, $n = 10$ mice per group (**bottom**). Each *bar* is the average total locomotion + SEM over 1 hour. * indicates a significant ($p < 0.05$) difference between genotypes.

5-HT_{1B} receptor, the GABA input is enhanced, and, consequently, the DA output is reduced. Initially, the knockout may have had a similar neurochemical effect. But it is possible that the chronic reduction in DA neurotransmission is deleterious, so the organism must compensate by boosting DA transmission. Increased DA function may maintain homeostasis in normal baseline conditions, but when drugs further heighten DA levels, behavioral changes can be observed. Since a modulatory element of the circuitry, the 5-HT_{1B} receptor, is missing, the organism is less able to buffer the large increases in DA caused by drugs like cocaine, and therefore it has a heightened response to the drug.

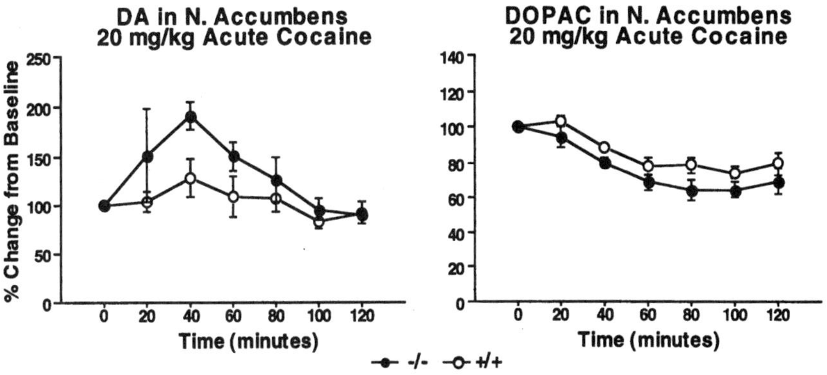

FIGURE 6. *In vivo* microdialysis measurements of extracellular changes in DA and DOPAC changes in the NAc in response to acute ip injection of 20 mg/kg cocaine. Each *point* is the percent baseline ± SEM for 3–4 mice per group.

A mechanism behind these compensations is suggested by studies of gene expression in the KO mice. After repeated injections of cocaine or other psychostimulants, the level of the ΔFosB transcription factor is greatly increased in the striatum and nucleus accumbens of normal rodents.[62] ΔFosB can dimerize with Jun proteins to form a persistent activator protein–1 (AP-1) complex. This persistent AP-1 complex can last for weeks and regulates the transcription of genes with an AP-1 recognition site in their promotor regions.[62] Interestingly, drug-naïve KO mice have elevated levels of ΔFosB and persistent AP-1 complex, approaching the levels found in WT mice that have received repeated drug treatments.[59] Since ΔFosB can regulate the expression of many other genes, the elevated levels of this transcription factor may be responsible for some of the compensations seen in the KO mice.

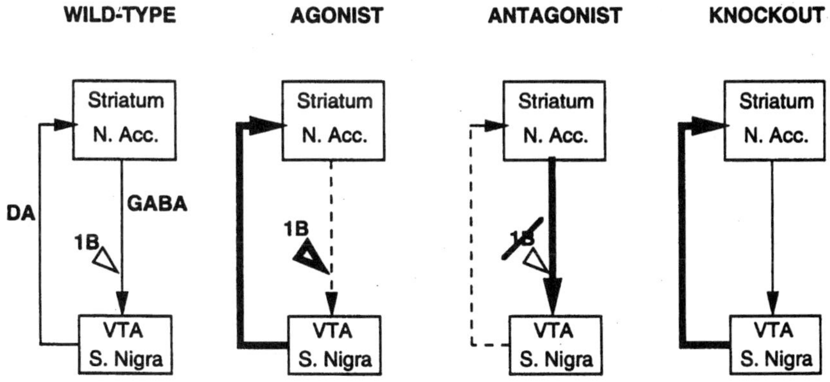

FIGURE 7. Proposed model of 5-HT1B modulation of GABA and DA neurotransmission, comparing the acute effects of pharmacological manipulations and the chronic compensations after genetic manipulations.

A MODEL OF GENETIC VULNERABILITY TO DRUGS OF ABUSE

It is notable that the ΔFosB levels found in KO mice are similar to the levels found in normal mice that have been sensitized to cocaine by repeated injections. In several other respects, drug-naïve KO mice resemble sensitized mice: increased locomotor response to cocaine, increased self-administration, increased DA release after cocaine injection, increased DA function, and an altered pattern of gene expression. The KO mice, therefore, could be considered "presensitized" to drugs, or in a state of enhanced vulnerability to drugs of abuse. We would predict that the 5-HT_{1B} KO mice would also be more responsive to other drugs of abuse. Indeed, they drink more alcohol than WT controls.[63] By studying the changes in gene expression and neural transmission that have caused this phenotype, we can gain valuable insights into the neurochemistry underlying individual vulnerability to drugs of abuse in humans. Although it can be difficult to extrapolate from animal models to human disorders, a recent report has linked mutations in the 5-HT_{1B} receptor to antisocial alcoholism in two human populations.[64]

It is important to consider constitutive KO mice as a model of a disease state—in this case, increased vulnerability to drugs of abuse—rather than as a pharmacological model. The fact that pharmacological studies of the role of the 5-HT_{1B} receptor in drug abuse yield the opposite result of the genetic KO emphasizes the dangers of assuming that KO mice mimic pharmacological antagonism. However, when the genetic KO is used in concert with traditional pharmacology, it provides a powerful method for understanding the role of a specific receptor in complex behaviors and neural circuitry. Pharmacology provides a way to assess how stimulation or blockade of a specific receptor affects a specific response at a single moment of time. Genetic knockouts help explore how that receptor interacts with other molecules to modulate neurotransmission, maintain homeostasis, and control complex behaviors.

REFERENCES

1. Hoyer, D., D.E. Clarke, J.R. Fozard, P.R. Hartig, G.R. Martin, E.J. Mylecharane, P.R. Saxena & P.P. Humphrey. 1994. International Union of Pharmacology classification of receptors for 5-hydroxytryptamine. Pharmacol. Rev. **46:** 157–203.
2. Adham, N., P. Romanienko, P. Hartig, R.L. Weinshank & T. Branchek. 1992. The rat 5-hydroxytryptamine 1B receptor is the species homologue of the human 5-hydroxytryptamine 1DB. Mol. Pharm. **41:** 1–7.
3. Maroteaux, L., F. Saudou, N. Amlaiky, U. Boschert, J.-L. Plassat & R. Hen. 1992. The mouse 5-HT1B receptor: Cloning, functional expression and localization in the motor control centers. Proc. Natl. Acad. Sci. USA **89:** 3020–3024.
4. Boulenguez, P., J. Chauveau, L. Segu, A. Morel, J. Lanoir & M. Delaage. 1991. A new 5-hydroxy-indole derivative with preferential affinity for 5-HT1B binding sites. Eur. J. Pharm. **194:** 91–98.
5. Palacios, J. M., A. Raurich, G. Mengod, S. D. Hurt & R. Cortés. 1996. Autoradiographic analysis of 5-HT receptor subtypes labeled by 3H-5-CT. Behav. Brain Res. **295:** 271–274.
6. Boschert, U., D.A. Amara, L. Segu & R. Hen. 1994. The mouse 5-HT1B receptor is localized predominantly on axon terminals. Neuroscience **58:** 167–182.
7. Ghavami, A., K. Stark, M. Jareb, L. Segu & R. Hen. 1999. Differential addressing of 5-HT1A and 5-HT1B receptors in epithelial cells and neurons. J. Cell Sci. **112:** 967–976.
8. Engel, G., M. Gothert, D. Hoyer, E. Schlicker & K. Hillenbrand. 1986. Identity of inhibitory presynaptic 5-hydroxytryptamine (5-HT) autoreceptors in the rat brain cortex with 5-HT1B binding sites. Naunyn-Schmiedebergs Arch. Pharmakol. **332:** 1–7.

9. CAMERON, D.L. & J.T. WILLIAMS. 1994. Cocaine inhibits GABA release in the VTA through endogenous 5-HT. J. Neurosci. **14:** 6763–6767.
10. JOHNSON, S.W., N.B. MERCURI & R.A. NORTH. 1992. 5-HT1B receptors block the GABA B synaptic potential in rat dopamine neurons. J. Neurosci. **12:** 2000–2006.
11. STANFORD, I.M. & M.G. LACEY. 1996. Differential actions of serotonin, mediated by 5-HT1B and 5-HT2C receptors on GABA-mediated synaptic input to rat substantia nigra pars reticulata neurons in vitro. J. Neurosci. **16:** 7566–7573.
12. BENLOUCIF, S., M.J. KEEGAN & M.P. GALLOWAY. 1993. Serotonin-facilitated dopamine release in vivo: Pharmacological characterization. J. Pharmacol. Exp. Ther. **265:** 373–377.
13. SAUDOU, F., D.A. AMARA, A. DIERICH, M. LE MEUR, S. RAMBOZ, L. SEGU, M.C. BUHOT & R. HEN. 1994. Enhanced aggressive behavior in mice lacking 5-HT1B receptor. Science **265:** 1875–1878.
14. STARK, K.L. & R. HEN. 1999. 5-HT1B knockout mice. Int. J. Neuropsychopharmacol. **2:** in press.
15. ZHAUNG, X., C. GROSS, L. SANTARELLI, V. COMPAN & R. HEN. 1999. Altered emotional states in knockout mice lacking the 5-HT1A or 5-HT1B receptors. Submitted for publication.
16. TRICKLEBANK, M.D., D.N. MIDDLEMISS & J. NEILL. 1986. Pharmacological analysis of the behavioral and thermoregulatory effects of the putative 5-HT1 receptor agonist, RU24969, in the rat. Neuropharmacology **25:** 877–886.
17. KOE, B.K., J.A. NIELSEN, J.E. MACOR & J. HEYM. 1992. Biochemical and behavioral studies of the 5-HT1B receptor agonist, CP-94,253. Drug Devel. Res. **26:** 241–250.
18. MITTMAN, S.M. & M.A. GEYER. 1989. Effects of 5-HT1A agonists on locomotor and investigatory behaviors in rats differ from those of hallucinogens. Psychopharmacology **98:** 183–188.
19. WING, L.L., G.S. TAPSON & M.A. GEYER. 1990. 5-HT2 mediation of acute behavioral effects of hallucinogens in rats. Psychopharmacology **100:** 417–425.
20. CASTANON, N., K. SCEARCE-LEVIE, J.J. LUCAS, B.A. ROCHA & R. HEN. 1999. Modulation of the effects of cocaine on behavioral reactivity and striatal c-fos expression by pharmacological manipulations of 5-HT1B receptors. Submitted for publication.
21. REMPEL, N., C.W. CALLAWAY & M.A. GEYER. 1993. The 5-HT1B receptor activation mimics behavioral effects of presynaptic serotonin release. Neuropsychopharmacology **8:** 201–212.
22. CALLAWAY, C.W. & M.A. GEYER. 1992. Tolerance and cross-tolerance to the activating effects of 3,4-methylenedioxymethamphetamine and a 5-hydroxytryptamine1B agonist. J. Pharmacol. Exp. Ther. **263:** 318–326.
23. CALLAWAY, C.W., N. REMPEL, R.Y. PENG & M.A. GEYER. 1992. Serotonin 5-HT1-like receptors mediate hyperactivity in rats induced by MDMA. Neuropsychopharmacology **7:** 113–127.
24. SCEARCE-LEVIE, K.A., S.S. VISHWANATHAN & R. HEN. 1999. Locomotor response to MDMA is attenuated in knockout mice lacking the 5-HT1B receptor. Psychopharmacology **141:** 154–161.
25. HIGLEY, J.D., P.T. MEHLMAN, D.M. TAUB, S.B. HIGLEY, S.J. SUOMI, J.H. VICKERS & M. LINNOILA. 1992. Cerebrospinal fluid monoamine and adrenal correlates of aggression in free-ranging rhesus monkeys. Arch. Gen. Psychiatry **49:** 436–441.
26. MAAS, J.W. 1962. Neurochemical differences between two strains of mice. Science **137:** 621–625.
27. BROWN, G.L., F.K. GOODWIN, J.C. BALLENGER, P.F. GOYER & L.F. MAJOR. 1979. Aggression in humans correlates with cerebrospinal fluid amine metabolites. Psych. Res. **1:** 131–139.
28. VIRKKUNEN, M., A. NUUTILA, F.K. GOODWIN & M. LINNOILA. 1987. Cerebrospinal fluid monoamine metabolite levels in male arsonists. Arch. Gen. Psych. **44:** 241–247.
29. COCCARO, E.F. 1989. Central serotonin and impulsive aggression Brit. J. Psychol. **155**(Suppl.): 52–62.
30. MANN, J.J., V. ARANGO, P.M. MARZUK, S. THECCANANT & D.J. REIS. 1989. Evidence of the 5-HT hypothesis of suicide. A review of post-mortem studies. Brit. J. Psych. **155:** (Suppl.): 7–14.
31. GIBBONS, J.L., G.A. BARR, W.H. BRIDGER & S.F. LEIBOWITZ. 1979. Manipulations of dietary tryptophan: Effects on mouse killing and brain serotonin in the rat. Brain Res. **169:** 139–153.
32. OGREN, S.-O., A.-C. HOLM, A.L. RENYI & S.B. ROSS. 1980. Anti-aggressive effect of zimelidine in isolated mice. Acta Pharmacol. Toxicol. **47:** 71–74.
33. OLIVIER, B., J. MOS, J.A.M. VAN DER HEYDEN & J. HARTOG. 1989. Serotonergic modulation of social interactions in isolated male mice. Psychopharmacology **97:** 154–156.
34. OLIVIER, B. & J. MOS. 1992. Rodent models of aggressive behavior and serotonergic drugs. Prog. Neurophychopharmacol. & Biol. Psychiatry **16:** 847–870.

35. SANCHEZ, C., J. ARNT, J. HYTTEL & E.K. MOLTZEN. 1993. The role of serotonergic mechanisms in inhibition of isolation-induced aggression in male mice. Psychopharmacology. **110:** 53–59.
36. CASTANON, N., S. RAMBOZ, F. SAUDOU & R. HEN. 1997. Behavioral consequences of 5-HT1B receptor gene deletion. *In* Serotoninergic Neurons and 5-HT Receptors in the CNS, H.G. Baumgarten & M. Gothert, Eds.: 351–365. Springer. Berlin.
37. RAMBOZ, S., N. CASTANON & R. HEN. 1999. Territorial and maternal aggression—inhibitory roles of the 5-HT1B and 5-HT1A receptors. Submitted for publication.
38. OLIVIER, B., J. MOS, R. VAN OORSCHOT & R. HEN. 1995. Serotonin receptors and animal models of aggressive behavior. Pharmacopsychiatry **28:** 80–90.
39. CHOPIN, P., C. MORET & M. BRILEY. 1994. Neuropharmacology of 5-HT1B/1D receptor ligands. Pharmacol. Ther. **62:** 385–405.
40. GRIGNASCHI, G. & R. SAMININ. 1992. Role of 5-HT receptors in the effect of d-fenfluramine on feeding patterns in the rat. Eur. J. Pharm. **212:** 287–289.
41. GARATTINI, S., T. MENNINI & R. SAMANIN. 1989. Reduction of food intake by manipulation of central serotonin. Current experimental results. Brit. J. Psych. **8:** 41–51.
42. VICKERS, S.P., P.G. CLIFRON & C.T. DOURISH. 1996. Behavioral evidence that d-fenfluramine–induced anorexia in the rat is not mediated by the 5-HT1A receptor subtype. Psychopharmacology **125:** 168–175.
43. LEIBOWITZ, S.F., G.F. WEISS & J.S. SUH. 1990. Medial hypothalamic nuclei mediate serotonin's inhibitory effect on feeding behavior. Pharmacol. Biochem. Behav. **37:** 735–742.
44. WEISS, G.F., N. ROGACKI, A. FUEG, D. BUCHEN & S.F. LEIBOWITZ. 1990. Impact of hypothalamic D-norfenfluramine and peripheral D-fenfluramine injection of macronutrient intake in the rat. Brain Res. Bull. **25:** 849–859.
45. WEISS, G.F., N. ROGACKI, A. FUEG, J.S. SUH, D.T. WONG & S.F. LEIBOWITZ. 1991. Effect of hypothalamic and peripheral fluoxetine injection on natural patterns of macronutrient intake in the rat. Psychopharmacology **105:** 467–476.
46. HUTSON, P.H., T.P. DONOHOW & G. CURZON. 1988. Infusion of the 5-hydroxytryptamine agonists RU24969 and TFMPP into the paraventricular nucleus of the hypothalamus causeshypophagia. Psychopharmacology 93–100.
47. LUCAS, J.J., A. YAMAMOTO, K.S. SCEARCE-LEVIE, F. SAUDOU & R. HEN. 1998. Absence of fenfluramine-induced anorexia and reduced c-fos induction in the hypothalamus and central amygdaloid complex of serotonin 1B receptor knock-out mice. J. Neurosci. **18:** 5537–5544.
48. SOUBRIÉ, P. 1986. Reconciling the role of central serotonin neurons in human and animal behavior. Behav. Brain Sci. **9:** 319–364.
49. VIRKKUNEN, M., R. RAWLINGS, R. TOKOLA, R.E. POLAND, A. GUIDOTTI, C. NEMEROFF, G. BISSETTE, K. KALOGERAS, S.L. KARONEN & M. LINNOILA. 1994. CSF biochemistries, glucose metabolism, and diurnal activity rhythms in alcoholic, violent offenders, fire setters, and healthy volunteers. Arch. Gen. Psych. **51:** 20–27.
50. BRUNNER, D. & H. HEN. 1997. Insights into the neurobiology of impulsive behavior from serotonin receptor knockout mice. Ann. N.Y. Acad. Sci. **836:** 81–105.
51. BRUNNER, D., M.C. BUHOT, R. HEN & M. HOFER. 1999. Anxiety, motor activation and maternal-infant interactions in 5-HT1B knockout mice. Behav. Neurosci. In press.
52. KOOB, G.F. 1992. Drugs of abuse: Anatomy, pharmacology and function of reward pathways. TIPS **13:** 177–184.
53. PARSONS, L.H., F. WEISS & G.F. KOOB. 1996. Serotonin1b receptor stimulation enhances dopamine-mediated reinforcement. Psychopharmacology **128:** 150–160.
54. PARSONS, L.H., F. WEISS & G.F. KOOB. 1998. Serotonin1B receptor stimulation enhances cocaine reinforcement. J. Neurosci. **18:** 10078–10089.
55. CALLAHAN, P.M. & K.A. CUNNINGHAM. 1995. Modulation of the discriminative stimulus properties of cocaine by 5-HT1B and 5-HT2C receptors. J. Pharmacol. Exp. Ther. **274:** 1414–1424.
56. RITZ, M.C. & M.J. KUHAR. 1989. Relationship between self-administration of amphetamine and monoamine receptors in brain: comparison with cocaine. J. Pharmacol. Exp. Ther. **248:** 1010–1017.
57. MANSBACH, R.S. & R.L. BALSTER. 1993. Effects of mazindol on behavior maintained or occasioned by cocaine. Drug Alcohol Depend. **31:** 183–191.

58. ROCHA, B.A., G. FUMAGALLI, R.R. GAINETDINOV, S.R. JONES, R. ATOR, B. GIROS, G.W. MILLER & M.G. CARON. 1998. Cocaine self-administration in dopamine-transporter knockout mice. Nature Neurosci. **1:** 132–137.
59. ROCHA, B.A., K.S. SCEARCE-LEVIE, J.J. LUCAS, N. HIROI, N. CASTANON, J.C. CRABBE, E.J. NESTLER & R. HEN. 1998. Increased vulnerability to cocaine in mice lacking the serotonin 1B receptor. Nature **393:** 175–178.
60. ROCHA, B., R. ATOR, M. EMMETT-OGLESBY & R. HEN. 1997. Intravenous cocaine self-administration in mice lacking 5-HT1B receptors. Pharmacol. Biochem. Behav. **57:** 407–412.
61. SCEARCE, K., S. KASSIR, J. LUCAS, N. CASTANON, L. SEGU, V. ARANGO & R. HEN. 1997. Dopaminergic compensations in knockout mice lacking the serotonin1B receptor. Presented at the 27th Annual Meeting of the Society for Neuroscience. New Orleans, LA.
62. HOPE, B.T., H.E. NYE, M.B. KELZ, D.W. SELF, M.J. IADAROLA, Y. NAKABEPPU, R.S. DUMAN & E.J. NESTLER. 1994. Induction of a long-lasting AP-1 complex composed of altered Fos-like proteins in brain by chronic cocaine and other chronic treatments. Neuron **13:** 1235–1244.
63. CRABBE, J.C., T.J. PHILLIPS, D.J. FELLER, R. HEN, C.D. WENGER, C.N. LESSOV & G.L. SCHAFER. 1996. Elevated alcohol consumption in null mutant mice lacking 5-HT1B serotonin receptors. Nature Genet. **14:** 98–101.
64. LAPPALAINEN, J., J.C. LONG, M. EGGERT, N. OZAKI, R.W. ROBIN, G.W. BROWN, H. NAUKKARINEN, M. VIRKKUNEN, M. LINNOILA & D. GOLDMAN. 1998. Linkage of antisocial alcoholism to the serotonin 5-HT1B receptor gene in 2 populations. Arch. Gen. Psych. **55:** 989–994.

Functional and Molecular Diversity of Purinergic Ion Channel Receptors

A. B. MacKENZIE,[a] A. SURPRENANT, AND R. A. NORTH

Institute of Molecular Physiology, University of Sheffield, Alfred Denny Building, Western Bank, Sheffield S10 2TN, England, UK

ABSTRACT: P2X receptors are membrane ion channels gated by extracellular adenosine 5′-triphosphate (ATP); nucleotides also activate a family of seven transmembrane G protein–coupled receptors (P2Y). P2X receptors are widely expressed on mammalian cells, where they can be broadly differentiated into three groups. The first group is almost equally well activated by ATP and its analog αβmethyleneATP (αβmeATP), whereas a second group is not activated by αβmeATP. A third-group type of receptor (termed P2Z) is distinguished by the fact that the channel opening is followed by cell permeabilization and lysis if the agonist application is continued for more than a few seconds. Seven cDNAs have been cloned that encode P2X receptor subunits. When expressed individually in heterologous systems, P2X1 and P2X3 subunits form channels activated by ATP or αβmeATP; whereas P2X2, P2X4, and P2X5 form channels activated by ATP but not αβmeATP. P2X6 receptors do not express readily, and P2X7 receptors correspond closely in their properties to P2Z. Further phenotypes can be produced when two subunits are coexpressed, indicating heteromultimerization. This chapter compares the properties of the native P2X receptors with those of the cloned and expressed subunits.

Transmission between cells by ATP was first proposed by Burnstock (1972); he suggested that it mediates nonadrenergic, noncholinergic transmission to the smooth muscle of the gastrointestinal tract.[1] Receptors activated by extracellular ATP were subsequently termed *P2 receptors* and were subdivided into two major classes; P2X and P2Y receptors.[2] This classification was based on the selectivity of certain agonists, but it is now known to be coincident with the molecular structure of the receptors and has been retained. P2X receptors are ligand-gated ion channels; P2Y receptors are heptahelical receptors coupled to G proteins. The first demonstration that ATP could activate cation channels without the involvement of diffusible second messengers was made for embryonic skeletal muscle and sensory and spinal cord neurons[3–5] (see Refs. 6,7 for review of later work). It was generally thought that this channel might belong to the nicotinic superfamily of ligand-gated channels, but cloning of the complementary DNAs in 1994 showed that ATP-gated cation channels are structurally distinct from other channel-forming proteins (reviewed in Refs. 8–12). To date, cDNAs for seven P2X receptor subunits have been identified: these are $P2X_1$,[13–14] $P2X_2$,[15] $P2X_3$,[16–18] $P2X_4$,[19–25] $P2X_5$,[26,27] $P2X_6$,[26,28,29] and $P2X_7$.[30,31] The aim of this review is to relate this molecular diversity to the known functional diversity of these channels.

[a]Corresponding author. Fax: 44 114 222 2360; e-mail: A.MacKenzie@Sheffield.ac.uk

FUNCTIONAL DIVERSITY

Native Receptors

ATP-gated ion channels have been studied in a broad range of cell types including smooth muscle, cardiac muscle, platelets, antigen-presenting immune cells, epithelial cells, and peripheral and central neurons. Native receptors have been classified into three major groups based on the relative effectiveness of ATP analogs, and biophysical characteristics of the ionic current. Selective antagonism is the best way to discriminate among receptors in native tissues, but the limited arsenal of compounds currently available has hampered this approach. Antagonists currently used include suramin, pyridoxal-5-phosphate and its 6-azo-2′,4′-disulfonic acid derivative (PPADS), P^1,P^5-bis(5′-inosyl)pentaphosphate, and 2′(or 3′)-(2,4,6-trinitrophenyl)-ATP (TNP-ATP).

The first of the three groups of native receptors is represented by responses of smooth muscle,[32–38] cardiac muscle,[39] platelets,[40] and sensory neurons of nodose and dorsal root ganglia[3,4,41–47] (FIG. 1). In these tissues, αβmethyleneATP (αβmeATP) is about equipotent with ATP itself, and concentrations producing half-maximal effects (EC_{50}) are in the range 1–3 μM. The currents evoked by ATP (or αβmeATP) show inward rectification, but they fall into two subgroups depending on the degree to which they decline during the continued agonist application. In smooth muscle and platelets such desensitization is profound; the time constant of desensitization (τ) is typically 100–300 ms (FIG. 1) but is largely independent of the concentration of agonist.[35] The currents in neonatal sensory neurons also desensitize strongly, but this is usually best fit by the sum of two exponentials and is strongly dependent on the agonist concentration (fast component $\tau = 50$ ms; slow component $\tau = 1$ s). In contrast, currents evoked by αβmeATP (or ATP) in adult rat sensory neurons or guinea pig celiac ganglion cells show little or no desensitization during applications of several seconds.[41] Currents evoked at these αβmeATP-sensitive receptors in dissociated mesenteric smooth muscle cells (desensitizing)[48] and rat nodose ganglion neurons (nondesensitizing)[49] and are very sensitive to inhibition by TNP-ATP (IC_{50} about 1 nM).

The second group of tissues are those where the action of ATP is not mimicked by αβmeATP. This αβmeATP-insensitive receptor is expressed by a diverse range of cells including NGF-differentiated PC12 cells,[50–53] superior cervical ganglion neurons,[41] submucosal enteric neurons,[54] outer hair cells of the cochlea,[55–57] nucleus tractus solitarius neurons,[58] ciliary ganglion,[59] and epithelial cells including the submandibular gland[19] and $LLC\text{-}PK_1$[60] cells. The effective concentrations of ATP at these receptors tend to be higher than those at the first class, with EC_{50}s typically 3–30 μM. The current is nonselective for cations, and shows strong inward rectification. These currents do not desensitize much during applications of several seconds, and repeated applications evoke currents with relatively constant peak amplitude (FIG. 1). Within this group, the receptors can be further distinguished by their sensitivity to suramin and PPADS; those in the rat submandibular gland are relatively insensitive.[19] TNP-ATP is a weak antagonist ($IC_{50} > 1$ μM) at this class of receptor in nodose ganglion neurons, even though the same cells express a current activated by αβmeATP that is highly sensitive to block by TNP-ATP.[48]

The third main phenotype is membrane permeabilization in response to ATP. The P2Z receptor was first described in rat mast cells,[61] where high concentrations of ATP lead to

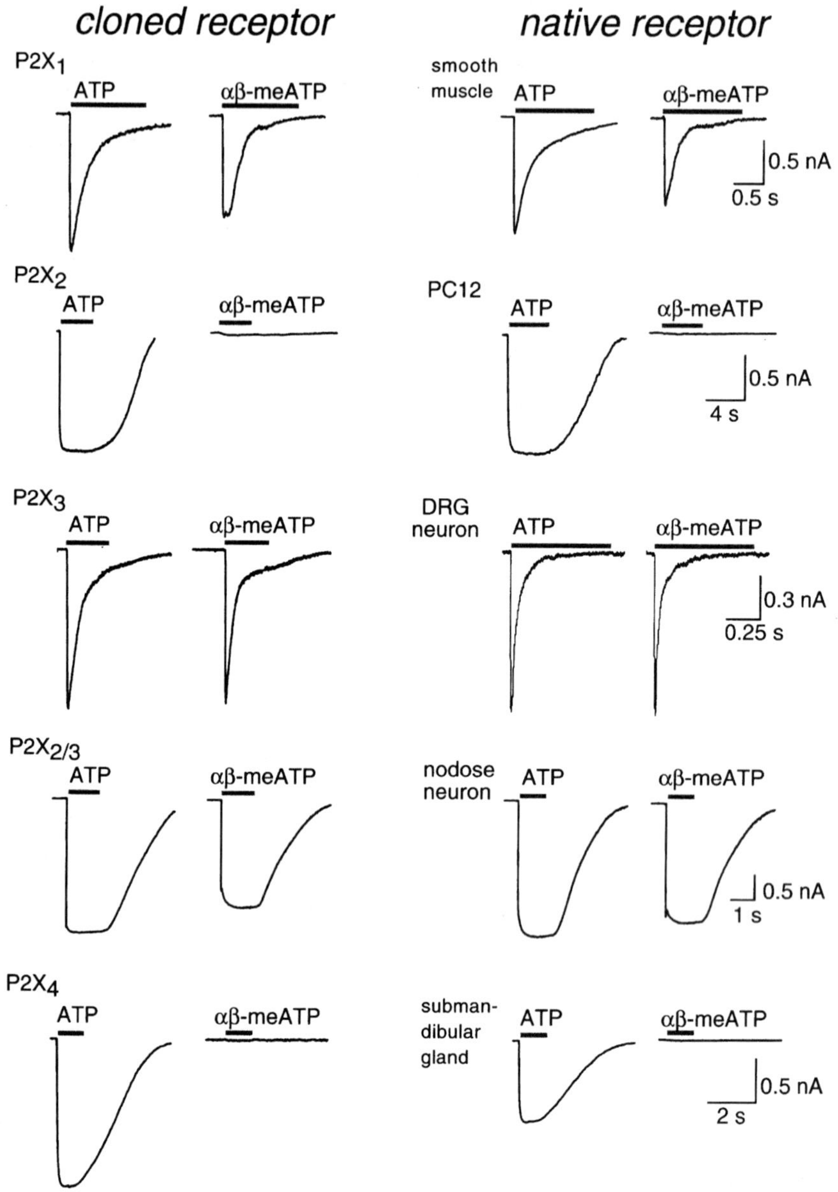

FIGURE 1. Inward currents in response to brief applications of ATP or αβmeATP (as indicated by *bars* above traces) recorded from HEK293 cells transfected with $P2X_1$, $P2X_2$, $P2X_3$, $P2X_4$, or $P2X_2$/$P2X_3$ receptors (*left-hand traces*) or from native tissues (*right-hand traces*). Recordings were made at holding potentials of −70 mV except in the case of dissociated DRG neurons, where holding potential was −40 mV. Richard Evans kindly provided unpublished recordings from DRG neuron; other traces are unpublished records from our laboratory. For the most part, functional properties of cloned receptors resemble closely those from native tissue.

membrane permeabilization. A similar P2Z receptor has been described in several antigen-presenting immune cells, studied primarily in macrophages and microglial cells.[62–64] Activation of P2Z receptors usually has been studied by following the ATP-induced uptake into cells of dyes such as ethidium bromide,[65] propidium iodide, or the cyanine dye YO-PRO-1 iodide (Molecular Probes); these cations (molecular weight 314, 414, and 375, respectively) enter the cell and can readily be visualized when they intercalate nucleic acid. The two characteristic features of this action are (i) that 2′,3′-(4-benzoyl)-benzyl-ATP (BzATP) is the most potent agonist (typical EC_{50} 30–100 μM; whereas ATP acts at some 10-fold higher concentrations); and (ii) that responses are greatly potentiated by reducing the concentration of extracellular divalent cations. This latter observation has been widely interpreted to indicate that ATP^{4-} rather than Mg-ATP is the effective agonist. The dye uptake is much reduced at reduced temperature, and the most effective compound to block the dye uptake and cell permeabilization is the isoquinoline KN-62.[66,67] KN-62 was introduced as a calmodulin kinase type II antagonist, but the effectiveness of the analog KN-04 at P2Z receptors indicates that this action is unrelated to calmodulin kinase type II:[66] it also blocks ATP-induced ethidium uptake in several human and mouse leucocyte cells lines.[66]

These two features—BzATP more potent than ATP, and potentiation by reduced divalent ion concentration—also characterize an action of ATP that has been studied electrophysiologically in human B lymphocytes,[68] microglial cells,[69,70] and smooth muscle cells of toad stomach[71] and human saphenous vein.[72] In some[70,72] but not other[68,71] of these tissues, dye uptake or cell lysis was also observed. The currents induced by ATP (or BzATP) in these cases show little desensitization and little rectification; this is in distinction to the first two classes of current described above (i.e. αβmeATP sensitive and αβmeATP insensitive), which show marked inward rectification.

The functional roles attributed to P2Z receptor activation include elimination of cells from the immune system via a fas- and perforin-independent pathway,[73,74] release of the cytokine Il-1β,[75,76] activation of the transcription factor NF-κB,[77] cell-cell fusion in the macrophage-like J774 cell line,[78] and destruction of intracellular mycobacteria infecting human macrophages.[79]

Cloned Receptors

Heterologous expression of each of the single cDNAs in *Xenopus* oocytes or mammalian cell lines results in ATP-induced cation-selective currents; the exception is the $P2X_6$ receptor, for which currents have been reported in only a small fraction of transfections into HEK293 cells.[26] The pharmacological and biophysical properties allow the receptors to be classed into three groups, which correspond in broad terms to the three phenotypes described for the native receptors (FIG. 1).

The first group comprises the $P2X_1$ and $P2X_3$ receptors; cells expressing these cDNAs are activated by αβmeATP as well as ATP, and half-maximal concentrations are close to 1 μM.[13,16–18] Desensitization is also profound for cells expressing through $P2X_1$ receptors and $P2X_3$ receptors: in the first case the desensitization has a time constant of about 100 to 300 ms, and in the second case the decay is best fit by two exponentials (τ of about 50 ms and 1 s). The unitary conductance of cloned $P2X_1$ receptors[80] (18 pS) is comparable to that observed in smooth muscle (20 pS)[37,38] under similar ionic conditions. ATP-induced currents in cells expressing $P2X_1$ or $P2X_3$ receptors are very sensitive to inhibition by TNP-ATP (IC_{50} about 1 nM).[81] In all respects examined to date, the properties of $P2X_1$ receptors correspond to those observed in smooth muscle of the vas deferens; this is consistent with, but by no means proves, the hypothesis that the receptor in that tissue is a homomeric $P2X_1$ receptor.

The second group comprises currents in cells expressing $P2X_2$, $P2X_4$, $P2X_5$, and $P2X_6$ receptors. In these cases, ATP activates the channel with an EC_{50} of about 10 μM, but αβmeATP is ineffective even at 100–300 μM.[15,19,20,24,26,82] These receptors can be further subdivided according to antagonist sensitivity: $P2X_2$ and $P2X_5$ receptors are reversibly inhibited by PPADS and suramin (<30 μM), while $P2X_4$ and $P2X_6$ receptors are not blocked by this concentration. The single-channel properties[80] of the cloned $P2X_2$ receptor are comparable to those of the native receptor of PC12 cells.[53] The functional properties of the cloned $P2X_4$ receptors correspond quite well to those recorded from submandibular gland cells, particularly in respect of their insensitivity to suramin and PPADS.[19] There is an interesting difference in sensitivity to PPADS and suramin between the rat and human $P2X_4$ receptor (see below).[25] $P2X_2$ and $P2X_4$ receptors are relatively insensitive to TNP-ATP.[81]

Alternative splicing of some of the P2X receptor primary transcripts occurs.[83–87] A shortened form of the $P2X_2$ receptor has been expressed ($P2X_{2b}$; it lacks 69 amino acids in the intracellular C-terminal domain; FIG. 2) and was found to desensitize rather more than the wild-type $P2X_2$ receptor.[85,86] Other forms for which splice variants have been detected (e.g., human $P2X_4$[87] and $P2X_5$: Genbank U49395/6) either do not form functional ATP-gated channels or have not been tested.

Further distinct phenotypes can be provided by coexpression of some of the cloned receptors. The first example was the coexpression of $P2X_2$ and $P2X_3$ receptors, which leads

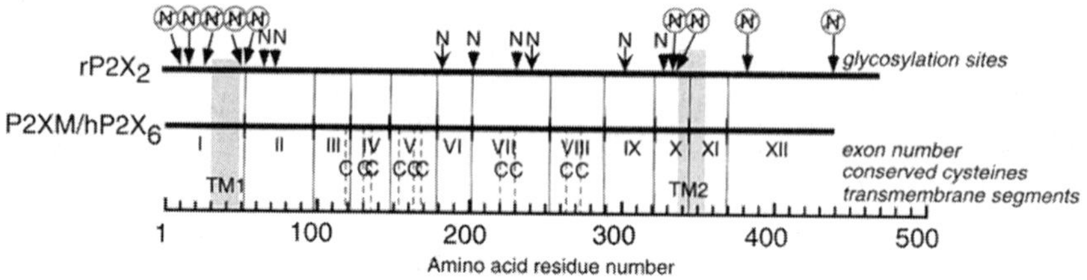

FIGURE 2. Schematic representation of some of the molecular features of P2X receptors. *Top line:* rat $P2X_2$ receptor, to indicate sites that are glycosylated (N) either in the native receptor (*arrowheads*) or when introduced artificially (*triangles*), and sites that are not glycosylated when introduced artificially. The site corresponding to N182 is also glycosylated in the $rP2X_1$ receptor. *Bottom line*: exons of the P2XM receptor. This is 80% identical to the rat $P2X_6$ receptor, and is considered to be the human $P2X_6$ receptor. Intron/exon boundaries at homologous positions have also been mapped for the $hP2X_7$ receptor that has an additional exon XIII (G. Buell, unpublished observations). Several of the possible splice sites have also been described in other receptors. For example, the $P2X_2$ receptor missing exon XII desensitizes more quickly than the full-length $P2X_2$ and is at least as abundant as the longer form. The cDNA RP-2 isolated from apoptotic thymocytes corresponds to $P2X_1$ exons VII and up (with some 38 bases of intron at its 5′ end). The human $P2X_5$ receptor cDNA has been isolated only in forms that lack either exon X or both exon III and exon X. For references see text.

to the assembly of an ATP-gated channel that can be activated by αβmeATP, but that does not desensitize;[17] these properties could not be accounted for by the simple summation of $P2X_2$ and $P2X_3$ homomeric channels (FIG. 1). Further evidence for functional heteropolymerization has recently been provided by using the antagonist TNP-ATP. The IC_{50} at the $P2X_3$ receptor is about 1 nM, and that at the $P2X_2$ receptor is more than 1 μM. In cells that have been transfected with both receptor cDNAs the currents induced by αβmeATP are also very sensitive to inhibition by TNP-ATP (IC_{50} about 1 nM), although the lack of desensitization indicates that at least one $P2X_2$ subunit must have been incorporated. The properties of the channels formed by coexpression of $P2X_2$ and $P2X_3$ subunits closely reproduce those seen in the adult sensory neurons of the dorsal root and nodose ganglia (activated by αβmeATP, show little desensitization, blocked by nM TNP-ATP).[17,48]

Baculovirus expression of $P2X_2$ and $P2X_3$ receptor subunits with distinct C-terminal epitopes has provided direct evidence for physical association between these subunits; each of the pair of proteins could be immunoprecipitated with antibodies specific for an epitope attached to the other.[88] These experiments are consistent with the finding of extensive coexpression of $P2X_2$ and $P2X_3$ receptor subunits in the nodose ganglion, as detected by immunohistochemistry.[89]

There is also evidence that the $P2X_4$ and $P2X_6$ receptor can heteropolymerize,[90] which would be consistent with their very widespread colocalization through the nervous system.[26] Thus, coexpression of the two subunits produced channels with some properties distinct from either alone, and direct coimmunoprecipitation of the two proteins was also demonstrated.[90] It remains unclear to what extent this combination can account for the functional responses exhibited by many central neurons.

The third distinct type of functional response exhibited by a cloned receptor is that observed with the $P2X_7$ receptor. Electrophysiological experiments in HEK293 cells transfected with the rat or human $P2X_7$ receptor show that brief application (1–2 s) of ATP activates an inward cation current.[30,31] Key properties of this current are (i) little rectification, (ii) little desensitization, (iii) strong potentiation by reducing the concentration of extracellular calcium and/or magnesium, and (iv) a relative insensitivity to ATP (EC_{50} about 300 μM) as compared to BzATP (EC_{50} 3 –30 μM). It was particularly the last two of these properties that suggested that the $P2X_7$ receptor might correspond to the P2Z receptor;[30] this interpretation was strongly supported by the findings that activation of $P2X_7$ receptors allowed the uptake of YO-PRO1 into transfected HEK293 cells, and that more prolonged exposure to agonists (10–30 s) led to cytolysis.

There are differences between the properties of the rat and human $P2X_7$ receptors: activation of the human receptor requires higher agonist concentrations, for the human receptor the inward current declines more quickly when agonist is removed, and YO-PRO-1 uptake is less for the human than rat receptor under otherwise comparable conditions.[31] KN-62 blocks both the cation current and the ethidium uptake in HEK293 cells expressing the human $P2X_7$ receptor, but has little or no effect in the case of the rat receptor.[67] Similar species differences have been observed for the native receptors (see above), but there is close correspondence between the actions of ATP on the HEK293 cells expressing the human $P2X_7$ receptor and on human macrophages.

On the other hand, agonist-induced cell permeabilization (to YO-PRO1) and lysis have not been uniform findings in cells expressing $P2X_7$ receptors,[91] just as some native cells with electrophysiological phenotypes resembling $P2X_7$ also show no permeabilization (see above). This raises the concern that ancillary proteins found in some but not other expression systems might also play a role in conferring this aspect of the phenotype, or that some cells might contain substances that inhibit permeabilization. The finding that extracellular calmidazolium (100 nM) almost completely inhibits the current induced by ATP (or BzATP) in HEK293 cells expressing the $P2X_7$ receptor but has no effect on their uptake of YO-PRO-1 indicates that the two properties can be distinguished. Clearly, further information is needed regarding the mechanism of formation of the large pores that allow YO-PRO-1 uptake, both biophysically and pharmacologically.

MOLECULAR DIVERSITY

The seven P2X receptor subunits range from 379 to 595 amino acids in length, and these amino acids are 35 to 48% identical between any two subunits. The conservation is dispersed throughout the sequence, but striking among the fully conserved residues are 10 cysteines, 13 glycines, and 7 lysines. There is no detectable sequence homology with other known ion channels or ATP binding proteins.

Assignment of Membrane Sidedness

The disposition of the P2X receptor subunit in the membrane has been assigned from a range of experimental results. First, the lack of signal sequence is consistent with the N-terminus being on the cytoplasmic side of the membrane. There are two hydrophobic regions sufficiently long to span the membrane (residues 30–50, and 335–355 in $P2X_2$); this suggests that the region between 50 and 335 lies outside of the cell. Native and artificially introduced glycosylation sites were used to test this topological model.[92,93] The $P2X_2$ receptor is glycosylated at each of its three N-linked consensus sites (N182, N239, N298) when expressed in *Xenopus* oocytes[92] or HEK293 cells.[93] Site-directed mutagenesis of each asparagine to serine[92] or glutamine[93] resulted in a $P2X_2$ receptor that was not glycosylated (Δ3N-$P2X_2$); receptors lacking any one of the sites functioned normally, but cells expressing receptors lacking two or three of the asparagines showed no response to ATP.[92] Consensus sequences (N-X-S/T) were introduced into the Δ3N-$P2X_2$ receptor, and Western blotting showed that these were glycosylated when introduced at positions 62, 66, 206, 300, and 324, but not at positions 9, 16, 22 or 28 or 328, 335, 381, 434 (FIG. 2).[92,93] These results are consistent with an intracellular N-terminus (1–30), extracellular domain (50–325), and intracellular C-terminus (355–472).

Further evidence that the N- and C-termini lie on the same side of the membrane comes from studies in which fully functional channels are formed by expressing concatenated cDNAs that encode dimeric (tandem) subunits. In the case of $P2X_2$-$P2X_2$ constructs,[92] incorporation of both halves of the dimer was indicated by observing the effects of a point mutation (T336C) on channel block by a methanethiosulphonate derivative (see below); in the case of the $P2X_2$-$P2X_3$ dimeric constructs,[93] the incorporation of both subunits into the functional channel was indicated by the unique phenotype of the $P2X_2$/$P2X_3$ heteromer (this is activated by $\alpha\beta$meATP, but is nondesensitizing; see above).

Residues Contributing to the Pore

Rassendren and coworkers[94] made point mutations of the $P2X_2$ receptor in which individual residues (V316 to T354) were substituted by cysteine. All 38 constructs yielded functional ATP-gated channels when expressed in HEK293 cells. Exposure of the cysteine side chain to the aqueous solution was then assessed by determining the effect of extracellular application of methanethiosulfonates (MTS) on the currents evoked by ATP. Three methanethiosulfonates were used: these were the positively charged ethylamine derivative (MTSEA), which is small enough to permeate the P2X receptor channel (permeability 0.70 that of sodium);[94] the larger ethyltrimethylammonium derivative (MTSET), which was almost impermeant (0.14 relative to sodium); and the negatively charged ethylsulphonate derivative (MTSES).

For three positions situated at the outer (extracellular) edge of the second transmembrane domain (I328, N333, and T336), the introduction of cysteine resulted in ATP-evoked currents that were blocked by all three methanethiosulfonates. The effectiveness of MTSES implies that these positions are not significantly sensing the membrane electric field. Three possible interpretations of this result were considered; namely, that the attachment of the MTS moiety (i) inhibits the binding of ATP, (ii) interferes with the conformational change needed for channel opening, or (iii) extends into, and blocks, the ion conducting pore. The third interpretation was favored by the observation that, in the case of T336C, the ATP-induced outward currents were blocked more rapidly than inward currents; in other words, the attachment of the MTS moiety introduced new inward rectification into the ion conduction.

For two positions (I338 and D349), the cysteine-containing channels were blocked by MTSEA, but not by MTSET or MTSES. This would be consistent with these side chains being further along the conducting ion path, accessible only to the permeant MTSEA. D349C was unique among the substitutions conferring sensitivity to MTS in that the block required prior channel opening. This most likely indicates that the external MTSEA is reaching the cysteine side chain only by passing through the open channel; in other words, D349 is located internal to the "gate." There was no consistent pattern of "hits" with MTS analogs that one might expect from organized secondary structure; indeed, the relatively small number of substitutions that conferred MTS sensitivity strongly suggests that further regions of the molecule also contribute.

Egan, Voigt, and coworkers[95] used MTSEA and silver as cysteine modifying agents and also found residues in and around the second transmembrane domain that reduced the action of ATP. Coapplication of extracellular silver and ATP irreversibly altered current at 15 of the 25 positions. Silver is a small ion, with a crystal ionic radius (1.26 Å) similar to that of potassium (1.33 Å), and might be expected to penetrate any water-accessible crevice of the protein: for none of the positions was there evidence that the modification resulted from the silver thiolate lying in the ion conducting path. MTSEA produced a pattern of "hits" distinct from those seen with silver. However, as also reported by Rassendren and colleagues,[94] the effects of MTSEA were quite variable from cell to cell, and large numbers of experiments would be required to obtain a reliable interpretation. The variability in the action of MTSEA might be related to the fact that it can enter the cell, either in its nonpolarized form or through open cation channels, and therefore act from the inside. Efforts to limit this complication by "scavenging" the MTSEA with intracellular free cysteine go some way to reducing this complication[92] but are also fraught with difficulty; such intracellular cysteine may also pass out through the open channel and bind to MTSEA in the vestibule even when applied from the outside. Conclusions about the "sidedness" of specific residues based on such an approach must be made with caution.

Ligand Binding Sites

The $P2X_4$ receptor is relatively insensitive to block by extracellular PPADS and suramin;[19] however, when the receptor is provided with a lysine residue in place of the glutamate at position 249 ($P2X_4$-E249K), then PPADS causes very slowly reversible inhibition of current. This substitution was chosen because a lysine occupies this position in wild-type $P2X_2$ and $P2X_1$ receptors, and because the slow onset and offset of block by PPADS suggested Schiff base formation between its aldehyde moiety and a receptor lysine. Consistent with this interpretation, the reciprocal mutation in the $P2X_2$ receptor (K246E) does not prevent inhibition by PPADS, but it makes it rapidly reversible. The human $P2X_4$ receptor is more sensitive to inhibition by PPADS (IC_{50} about 30 μM) and suramin than its rat homolog (IC_{50} > 500 μM).[25] Exchange of part of the extracellular domain (82 to 183) between the rat and human homolog transferred the PPADS sensitivity, but suramin sensitivity was determined by the region of the molecule preceding residue 82 (particularly Lys^{78}). Taken together, these results suggest that several parts of the extracellular domain contribute to the PPADS and suramin binding sites, and that some of the conserved lysine residues play important roles.

Subunit Stoichiometry

Four approaches have been used to deduce the stoichiometry of the P2X receptor. In the first, most of the extracellular domain of the $P2X_2$ receptor (K53 to K308) was expressed in *E. coli* and purified by an attached hexahistidine tag.[96] The purified protein was refolded in an excess of sodium sulfite to prevent disulfide formation between the conserved cysteines, and further purified by gel filtration chromatography. The resulting protein (29 kDa) was photoaffinity labeled with [α-^{32}P]ATP, and this could be inhibited by suramin (1 μM). The labeled protein had a molecular weight of about 140 kDa, suggesting that it formed as a stable tetramer.

The second approach used cross-linking with bifunctional derivatives of PPADS (i.e., two PPADS moieties joined by a spacer).[97] $P2X_1$ receptors were N-terminal tagged, expressed in *Xenopus* oocytes, solubilized with digitonin, purified on nickel-agarose beads, and resolved by SDS-PAGE. Addition of the PPADS derivative caused the protein to run as dimers and trimers; the yield of trimers was increased when the length of the spacer was increased, and a similar result was obtained with 3,3′-dithiobis(sulfosuccimidylproprionate), in which the reactive groups are 12 Å apart. The third approach, also reported by the Schmalzing laboratory,[97] was to use blue native PAGE to resolve radioiodinated proteins from the plasma membrane of oocytes. After solubilization with *n*-octylglucoside, only trimeric and hexameric forms of the protein were observed; under identical conditions, the nicotinic acetylcholine receptor was clearly resolved as a pentamer.

The fourth kind of approach is functional. $P2X_2$ receptors incorporating a FLAG epitope were expressed in oocytes. Membrane currents were recorded in response to maximal concentrations of ATP, and the binding of ^{125}I-anti-FLAG antibody was then measured in the same cells. For many oocytes, at various expression levels, the relation between current and binding sites was consistent with three to four subunits per channel; more accurate estimates were not possible in view of assumptions about unitary currents and maximum open probability. Results consistent with these were obtained by measuring the inhibition by MTSET of ATP-induced currents in oocytes expressing concatenated cDNAs. There was no inhibition in dimers, trimers, and tetramers of wild-type $P2X_2$ receptors, but the inhibition increased stepwise when the T336C mutation was introduced into progressively more of the concatenated domains (R. Stoop and R. A. North, unpublished observations).

CONCLUSIONS

The functional diversity of P2X receptors in native cells is rather well explained by the properties of cloned receptors, either alone or in combination. However, there are discrepancies that lead one to question whether further P2X receptors, or other ancillary proteins, significantly affect channel function. The coexpression of $P2X_4$ and $P2X_6$ subunits seems not to account fully for the responses of some central neurons that are activated by αβmeATP and readily blocked by antagonists. $P2X_7$ receptors reproduce well the P2Z receptor of macrophages, but the properties assigned to "P2Z receptor" activation in other native cells (in particular whether dye uptake and/or permeabilization occurs) is variable and is not perfectly reproduced by heterologous expression of $P2X_7$ receptors. One can expect that it will be the experimental pursuit of these discrepancies that will be most rewarding in furthering our understanding of the physiological role of these purinergic ion channel receptors.

REFERENCES

1. BURNSTOCK, G. 1972. Purinergic nerves. Pharmacol. Rev. **24:** 509–581.
2. BURNSTOCK, G. & C. KENNEDY. 1985. Is there a basis for distinguishing two types of P2 purinoceptor? Gen. Pharmacol. **16:** 433–440.
3. KRISHTAL, O.A., S.M. MARCHENKO & V.I. PIDOPLICHKO. 1983. Receptor for ATP in the membrane of mammalian sensory neurones. Neurosci. Lett. **35:** 41–45.
4. JAHR, C.E. & T.M. JESSELL. 1983. ATP excites a subpopulation of rat dorsal horn neurones. Nature **304:** 730–733.
5. KOLB, H.A. & M.J. WAKELAM. 1983. Transmitter-like action of ATP on patched membranes of cultured myoblasts and myotubes. Nature **303**: 621–623.
6. BEAN, B.P. 1992. Pharmacology and electrophysiology of ATP-activated ion channels. Trends Pharmacol. Sci. **13:** 87–90.
7. EVANS, R.J. & A. SURPRENANT. 1996. P2X receptors in autonomic and sensory neurons. Semin. Neurosci. **8:** 217–223.
8. NORTH, R.A. 1996. P2X receptors: A third major class of ligand gated ion channels. *In* P2 Purinoceptors: Localization, Function and Transduction Mechanisms. Ciba Foundation Symposium: 91–109. John Wiley & Sons. Chichester, England.
9. SURPRENANT, A., G. BUELL & R.A. NORTH. 1995. P_{2X} receptor brings new structure to ligand-gated ion channels. Trends Neurosci. **18:** 224–229.
10. NORTH, R.A. 1996. P2X purinoceptor plethora. Semin. Neurosci. **8:** 187–194.

11. North, R.A. 1996. Families of ion channels with two hydrophobic segments. Curr. Opin. Cell Biol. **8:** 474–483.
12. North, R.A. & E.A. Barnard. 1997. Nucleotide receptors. Curr. Opin. Neurobiol. 7: 346–357.
13. Valera, S., N. Hussy, R.J. Evans *et al.* 1994. A new class of ligand-gated ion channel defined by P2X receptor for extracellular ATP. Nature **371:** 516–519.
14. Valera, S., F. Talabot, R.J. Evans *et al.* 1995. Characterization and chromosomal localization of a human P2X receptor from the urinary bladder. Recept. Channel **3:** 283–289.
15. Brake, A.J., M.J. Wagenbach & D. Julius. 1994. New structural motif for ligand-gated ion channels defined by an ionotropic ATP receptor. Nature **371:** 519–523.
16. Chen, C-C., A.N. Akopian, L. Sivilotti *et al.* 1995. A P2X purinoceptor expressed by a subset of sensory neurons. Nature **377:** 428–431.
17. Lewis, C., S. Neidhart, C. Holy *et al.* 1995. Co-expression of P_{2X2} and P_{2X3} receptor subunits can account for ATP-gated currents in sensory neurons. Nature **377:** 432–435.
18. Garcia-Guzman, M., W. Stuhmer & F. Soto. 1997. Molecular characterization and pharmacological properties of the human P2X3 purinoceptor Mol. Brain Res. **47:** 59–66.
19. Buell, G., C. Lewis, G. Collo *et al.* 1996. An antagonist-insensitive P2X receptor expressed in epithelia and brain. EMBO J. **15:** 55–62.
20. Seguela, P., A. Haghighi, J.-J. Soghomonian *et al.* 1996. A novel neuronal P2X ATP receptor ion channel with widespread distribution in the brain. J. Neurosci. **16:** 448–455.
21. Wang, C-Z., N. Namba, T. Gonoi *et al.* 1996. Cloning and pharmacological characterization of a fourth P2X receptor subtype widely expressed in brain and peripheral tissues including various endocrine tissues. Biochem. Biophys. Res. Commun. **220:** 196–202.
22. Xuenong, B., Y. Zhang, M. Nassar *et al.* 1995. A P2X purinoceptor cDNA conferring a novel pharmacological profile. FEBS Lett. **375:** 129–133.
23. Genbank submission AF08975.
24. Soto, F., M. Garcia-Guzman, J.M. Gomez-Hernandez *et al.* 1996. P2X4: An ATP-activated ionotropic receptor cloned from rat brain. Proc. Natl. Acad. Sci. USA **93:** 3684–3688.
25. Garcia-Guzman, M., F. Soto, J. Manuel *et al.* 1996. Characterization of recombinant human P2X4 receptor reveals pharmacological differences to the rat homologue. Mol. Pharmacol. **51:** 109–118.
26. Collo, G., R.A. North, E. Kawashima *et al.* 1996. Cloning of $P2X_5$ and $P2X_6$ receptors and the distribution and properties of an extended family of ATP-gated ion channels. J. Neurosci. **16:** 2495–2507.
27. Garcia-Guzman, M., F. Soto, B. Laube *et al.* 1996. Molecular cloning and functional expression of a novel rat heart P2X purinoceptor. FEBS Lett. **388:** 123–127.
28. Soto, F., M. Garcia-Guzman, C. Karschin *et al.* 1996. Cloning and tissue distribution of a novel P2X receptor from rat brain. Biochem. Biophys. Res. Commun. **223:** 456–460.
29. Urano, T., H. Nishimori, H.-J. Han, *et al.* 1997. Cloning of P2XM, a novel human P2X receptor gene regulated by p53. Cancer Res. **57:** 3281–3287.
30. Surprenant, A., F. Rassendren, E. Kawashima *et al.* 1996. The cytolytic P2Z receptor for extracellular ATP identified as a P2X receptor (P2X7). Science **272:** 735–738.
31. Rassendren, F., G.N. Buell, C. Virginio *et al.* 1997. The permeabilizing ATP receptor, P2X7. Cloning and expression of a human cDNA. J. Biol. Chem. **272:** 5482–5486.
32. Khakh, B.S., P.P.A. Humphrey & A. Surprenant. 1995. A study on P2X purinoceptors mediating the electrophysiological and contractile effects of purine nucleotides in rat vas deferens. Br. J. Pharmacol. **115:** 177–185.
33. Evans, R.J. & A. Surprenant. 1992. Vasoconstriction of guinea-pig submuscosal arterioles following sympathetic nerve stimulation is mediated by release of ATP. Br. J. Pharmacol. **106:** 242–249.
34. Benham, C.D. & R.W. Tsien. 1987. A novel receptor-operated Ca^{2+} permeable channel activated by ATP in smooth muscle. Nature **328:** 275–278.
35. Evans, R.J. & C. Kennedy. 1994. Characterization of P2 purinoceptors in the smooth muscle of the rat tail artery: A comparison between contractile and electrophysiological responses. Br. J. Pharmacol. **113:** 853–860.
36. Inoue, R. & A.F. Brading. 1990. The properties of the ATP-induced depolarization and current in single cells isolated from the guinea-pig urinary bladder. Br. J. Pharmacol. **100:** 619–625.

37. Friel, D. 1988. An ATP-sensitive conductance in single smooth muscle cells from the rat vas deferens. J. Physiol. **401:** 361–380.
38. Nakazawa, K. & N. Matsuki. 1987. Adenosine triphosphate–activated inward current in isolated smooth muscle cells from rat vas deferens. Pfluegers Arch. Eur. J. Physiol. **409:** 644–646.
39. Parker, K.E. & A. Scarpa. 1995. An ATP-activated nonselective cation channel in guinea pig ventricular myocytes. Am. J. Physiol. **269:** H789–H797.
40. MacKenzie, A.B., M.P. Mahaut-Smith & S.O. Sage. 1996. Activation of receptor-operated cation channels via P2X1 not P2T purinoceptors in human platelets. J. Biol. Chem. **271:** 2879–2881.
41. Khakh, B.S., P.P.A. Humphrey & A. Surprenant. 1995. Electrophysiological properties of P2-purinoceptors in rat superior cervical, nodose and guinea-pig coeliac neurones. J. Physiol. **484:** 385–396.
42. Krishtal, O.A., S.M. Marchenko & A.G. Obukhov. 1988. Receptors for ATP in rat sensory neurones: The structure-function relationship for ligands. Br. J. Pharmacol. **95:** 1057–1062.
43. Li, C., R.W. Peoples, Z. Li *et al.* 1993. Zn^{2+} potentiates excitatory action of ATP on mammalian neurons. Proc. Natl. Acad. Sci. USA **90:** 8264–8267.
44. Krishtal, O.A., S.M. Marchenko & A.G. Obukhov. 1988. Cationic channels activated by extracellular ATP in rat sensory neurons. Neuroscience **3:** 995–1000.
45. Bean, B.P., C. Williams & P.W. Ceelen. 1990. ATP activated channels in rat and bullfrog sensory neurons: Current-voltage relationship and single channel behaviour. J. Neurosci. **10:** 11–19.
46. Bean, B.P. 1990. ATP-activated channels in rat and bullfrog sensory neurons: Concentration dependence and kinetics. J. Neurosci. **10:** 1–10.
47. Robertson, S.J., M.G. Rae, E.G. Rowan *et al.* 1996. Characterization of a P2X purinoceptor in cultured neurones of the rat dorsal root ganglia. Br. J. Pharmacol. **118:** 951–956.
48. Lewis, C.J., A. Surprenant & R.J. Evans. 1998. 2′,3′-O-(2,4,6-trinitrophenyl)adenosine 5′-triphosphate (TNP-ATP)—A nanomolar affinity antagonist at rat mesenteric artery P2X receptor ion channels. Br. J. Pharmacol. **124**(7): 1463–1466.
49. Thomas, S., C. Virginio, R.A. North *et al.* 1998. The antagonist trinitrophenyl-ATP reveals co-existence of distinct P2X receptor channels in rat nodose neurones. J. Physiol. **509:** 411–417.
50. Nakazwa, K., K. Inoue, K. Fujimori *et al.* 1991. Effects of ATP antagonists on purinoceptor-operated inward currents in rat phaeochromocytoma cells. Pfluegers Arch. Eur. J. Physiol. **418:** 214–219.
51. Nawazawa, K., K. Fuimori, A. Takanaka & K. Inoue. 1990. An ATP activated conductance in phaeochromocytoma cells and its suppression by extracellular calcium. J. Physiol. **428:** 257–272.
52. Neuhaus, R., B.F.X. Reber & H. Reuter. 1991. Regulation of bradykinin- and ATP-activated Ca2+ permeable channels in rat pheochromocytoma (PC12) cells. J. Neurosci. **11:** 3984–3990.
53. Nakazawa, K. & P. Hess. 1993. Block by calcium of ATP-activated Ca^{2+} permeable channels in pheochromocytoma cells. J. Gen. Physiol. **101:** 377–392.
54. Barajas-Lopez, C., R. Espinosa-Luna & V. Gerzanich. 1994. ATP closes a potassium and opens a cationic conductance through different receptors in neurons of guinea pig submucous plexus. J. Pharmacol. Exp. Ther. **58:** 1396–1402.
55. Nakagawa, T., N. Akaike, T. Kimitsuki *et al.* 1990. ATP-induced current in isolated outer hair cells of guinea pig cochlea. J. Neurophysiol. **63:** 1068–1074.
56. Glowatzki, E., J.P. Ruppersberg, H.P. Zenner *et al.* 1997. Mechanically and ATP-induced currents of mouse outer hair cells are independent and differentially blocked by d-tubocurarine. Neuropharmacology **36:** 1269–1275.
57. Raybould, N.P. & G.D. Housley. 1997. Variation in expression of the outer hair cell P2X receptor conductance along the guinea-pig cochlea. J. Physiol. **498:** 717–727.
58. Ueno, S., N. Harata & K. Inoue. 1992. ATP-gated current in dissociated rat nucleus solitarii neurons. J. Neurophysiol. **68:** 778–785.
59. Sun, X-P. & E.F. Stanley. 1996. An ATP activated, ligand-gated ion channel on a cholinergic presynaptic nerve terminal. Proc. Natl. Acad. Sci. USA **93:** 1859–1863.

60. FILIPOVIC, D.M., O.A. ADEBANJO & M. ZAIDI. 1998. Functional and molecular evidence for P2X receptors in LLC-PK1 cells. Am. J. Physiol. **274:** F1070–F1077.
61. COCKCROFT, S. & B.D. GOMPERTS. 1979. ATP induces nucleotide permeability in rat mast cells. Nature **279:** 541–542.
62. FERRARI, D., M. VILLALBA, P. CHIOZZI *et al.* 1996. Mouse microglial cells express a plasma membrane pore gated by extracellular ATP. J. Immunol. **156:** 1531–1539.
63. DI VIRGILIO, F., D. FERRARI, P. CHIOZZI *et al.* 1996. Purinoceptor function in the immune system. Drug Dev. Res. **39:** 319–329.
64. DI VIRGILIO, F. 1995. The P2Z receptor: An intriguing role in immunity, inflammation and cell death. Immunol. Today **16:** 524–528.
65. STEINBERG, T.H., A.S. NEWMAN, J.A. SWANSON *et al.* 1987. ATP^{4-} permeabilizes the plasma membrane of mouse macrophages to fluorescent dyes. J. Biol. Chem. **262:** 8884–8888.
66. GARGETT, C.E. & S. WILEY. 1997. The isoquinoline derivative KN-62 a potent antagonist of the P2Z-receptor of human lymphocytes. Br. J. Pharmacol. **120:** 1483–1490.
67. HUMPHREYS, B.D., C. VIRGINIO, A. SURPRENANT *et al.* 1998. Isoquinolines as antagonists of the $P2X_7$ nucleotide receptor: High selectivity for the human versus rat receptor homologues. Mol. Pharmacol. **54:** 22–32.
68. MARKWARDT, F., M. LOHN, T. BOHM *et al.* 1997. Purinoceptor operated cationic channels in human B lymphocytes. J. Physiol. **498:** 143–151.
69. HAAS, S., J. BROCKHAUS, A. VEHKHRATSKY *et al.* 1996. ATP-induced membrane currents in ameboid microglia acutely isolated from mouse brain slices. Neuroscience **75:** 257–261.
70. CHESSELL, I.P., A.D. MICHEL & P.P.A. HUMPHREY. 1997. Properties of the pore-forming $P2X_7$ purinoceptor in mouse NTW8 microglial cells. Br. J. Pharmacol. **121**(7): 1429–1437.
71. UGAR, M., R.M. DRUMMOND, H. ZOU *et al.* 1997. An ATP-gated cation channel with some P2Z-like characteristics in gastric smooth muscle cells of toad. J. Physiol. **498:** 427–442.
72. CARIO-TOUMANIANTZ, C., G. LOIRAND, A. LADOUX *et al.* 1998. $P2X_7$ receptor activation-induced contraction and lysis in human saphenous vein. Circ. Res. **83:** 196–203.
73. APASOV, S., M. KOSHIBA, F. REDGELD *et al.* 1995. Role of extracellular ATP and P1 and P2 classes of purinergic receptors in T-cell development and cytotoxic lymphocyte effector functions. Immunol. Rev. **146:** 5–19.
74. BLANCHARD, D.K., S. WEI, C. DUAN *et al.* 1995. Role of extracellular adenosine triphosphate in the cytotoxic T-lymphocyte-mediated lysis of antigen presenting cells. Blood **85:** 3173–3182.
75. FERRARI, D., P. CHIOZZI, S. FALZONI *et al.* 1997. Purinergic modulation of interleukin 1β release from microglial cells stimulated with bacterial endotoxin. J. Exp. Med. **185:** 1–4.
76. PERREGAUX, D. & C.A. GABEL. 1994. Interleukin-1β maturation and release in response to ATP and nigericin. J. Biol. Chem. **269:** 15195–15203.
77. FERRARI, D., S. WESSELBORG, M.K.A. BAUER *et al.* 1997. Extracellular ATP activates transcription factor NF-κB through the P2Z purinoceptor by selectively targeting NF-dB p65 (RelA). J. Cell Biol. **139:** 1635–1643.
78. CHIOZZI, P., J.M. SANZ, D. FERRARI *et al.* 1997. Spontaneous cell fusion in macrophage cultures expressing high levels of the P2Z/P2X7 receptor. J. Cell Biol. **138:** 697–706.
79. LAMMAS, D.A., C. STOBER, C.J. HARVEY *et al.* 1997. ATP-induced killing of mycobacteria by human macrophages is mediated by purinergic P2Z(P2X7) receptor. Immunity **7**(3): 433–444.
80. EVANS, R.J. 1996. Single channel properties of ATP-gated cation channels (P2X receptors) heterologously expressed in Chinese hamster ovary cells. Neurosci. Letts. **212:** 212–214.
81. VIRGINIO, C., G. ROBERTSON, A. SURPRENANT *et al.* 1998. Trinitrophenyl-substituted nucleotides are potent antagonists selective for $P2X_1$, $P2X_3$, and heteromeric $P2X_{2/3}$ receptors. Mol. Pharmacol. **53:** 969–973.
82. EVANS, R.J., C. LEWIS, G. BUELL *et al.* 1995. Pharmacological characterization of heterologously expressed ATP-gated cation channels (P_{2X}-purinoceptors). Mol. Pharmacol. **48:** 178–183.
83. PARKER, M.S., M.L. LARROQUE, J.M. CAMPBELL *et al.* 1998. Novel variant of the P2X2 ATP receptor from the guinea pig organ of Corti. Hear. Res. **121:** 62–70.
84. SALIH, S.G., G.D. HOUSLEY, L.D. BURTON *et al.* 1998. $P2X_2$ receptor subunit expression in a subpopulation of cochlear type I spiral ganglion neurones. Neuroreport **9:** 279–282.
85. BRANDLE, U., P. SPIELMANNS, R. OSTEROTH *et al.* 1997. Desensitization of the $P2X_2$ receptor controlled by alternative splicing. FEBS Lett. **404:** 294–298.

86. SIMON, J., E.J. KIDD, F.M. SMITH *et al.* 1997. Localization and functional expression of splice variants of the $P2X_2$ receptor. Mol. Pharmacol. **52:** 237–248.
87. DHULIPALA, P.D.K., Y.-X. WANG & M.I. KOTLIKOFF. 1998. The human $P2X_4$ receptor gene is alternatively spliced. Gene **207:** 259–266.
88. RADFORD, K., C. VIRGINIO, A. SURPRENANT *et al.* 1997. Baculovirus expression provides direct evidence for heteromeric assembly of $P2X_2$ and $P2X_3$ receptors. J. Neurosci. **17:** 6529–6533.
89. VULCHANOVA, L., M.S. RIEDL, S.J. SHUSTER *et al.* 1998. Immunohistochemical study of the $P2X_2$ and $P2X_3$ receptor subunits in rat and monkey sensory neurons and their central terminals. Neuropharmacology **36:** 1229–1242.
90. LÊ, K.-T., K. BABINSKI & P. SÉGUÉLA. 1998. Central $P2X_4$ and $P2X_6$ channel subunits coassemble into a novel heteromeric ATP receptor. J. Neurosci. **18**(18): 7152–7159.
91. PETROU, S., M. UGUR, R.M. DUMMOND *et al.* 1997. $P2X_7$ purinoceptor expression in Xenopus oocytes is not sufficient to produce a pore-forming P2Z-like phenotype. FEBS Lett. **411:** 339–345.
92. NEWBOLT, A., R. STOOP, C. VIRGINIO *et al.* 1998. Membrane topology of an ATP-gated ion channel (P2X receptor). J. Biol. Chem. **273:** 15177–15182.
93. TORRES, G.E., T.M. EGAN & M.M. VOIGT. 1998. Topological analysis of the ATP-gated ionotropic P2X2 receptor subunit. FEBS Lett. **425:** 19–23.
94. RASSENDREN, F., G. BUELL, A. NEWBOLT *et al.* 1997. Identification of amino acid residues contributing to the pore of a P2X receptor. EMBO J. **16:** 3446–3454.
95. EGAN, T.M., W.R. HAINES & M.M. VOIGT. 1998. A domain contributing to the ion channel of ATP-gated P2X receptors identified by the substituted cysteine accessibility method. J. Neurosci. **18**(7): 2530–2539.
96. KIM, M., O.-J. YOO & S. CHOE. 1997. Molecular assembly of the extracellular domain of $P2X_2$, an ATP gated ion channel. Biochem. Biophys. Res. Comm. **240:** 618–622.
97. NICKE, A., H.G. BAUMERT, J. RETTINGER *et al.* 1998. $P2X_1$ and $P2X_3$ receptors form stable trimers: A novel structural motif of ligand gated ion channels. EMBO J. **17:** 3016–3028.

Cyclic Nucleotide–Gated Channels

Molecular Mechanisms of Activation

MARIE-CHRISTINE BROILLET[a,b] AND STUART FIRESTEIN[c]

[a]*Institut de Pharmacologie et de Toxicologie, Université de Lausanne, CH-1005 Lausanne, Switzerland*

[c]*Department of Biological Sciences, Columbia University, New York, New York 10027, USA*

ABSTRACT: Activation of cyclic nucleotide–gated (CNG) channels represents the final step in the transduction pathways in both vision and olfaction. Over the past several years, CNG channels have been found in a variety of other cell types where they might fulfill various physiological functions. The olfactory and photoreceptor CNG channels rely on the binding of at least two molecules of cAMP or cGMP at intracellular sites on the channel protein to open a nonspecific cation conductance with a significant permeability to Ca ions. A series of elegant experiments with cloned channels and chimeric constructs has revealed significant information regarding the binding and gating reactions that lead to CNG channel activation. These recent studies have identified several regions as well as specific amino acid residues distributed on the retinal or the olfactory CNG channel subunits that play a key role in channel regulation. In this review, we will focus on these specific molecular sites of activation and modulation of CNG channels.

Activation of cyclic nucleotide–gated (CNG) channels represents the final step in the transduction pathways in both vision and olfaction.[1] Over the past several years, CNG channels have been found in a variety of other cell types including kidney, testis, and heart (for review see Refs. 2 and 3), where they may fulfill various physiological functions. More recently, these channels have been found elsewhere in the nervous system[4–6] and have been implicated in processes as diverse as synaptic modulation and axon outgrowth in animals ranging from the nematode to mammals.[7,8] The olfactory and photoreceptor CNG channels rely on the binding of at least two molecules of cAMP or cGMP at intracellular sites on the channel protein to open a nonspecific cation conductance that also has a significant permeability to Ca ions.[1] Thus their activation by the ubiquitous cyclic nucleotide second messengers can lead not only to depolarization but also to Ca influx.

Native CNG channels are constructed from different but highly homologous subunits. The olfactory CNG channel is, for example, constructed of at least three subunits variously called α or CNG2 (original designation OCNC1[9]), β or CNG5 (original designation OCNC2[10,11]), and CNG4.3[12] (for nomenclature of CNG channel subunits see Ref. 13). These subunits, in an unknown stoichiometry, probably form a heterotetrameric structure.[14,15] The different subunits are similar in structure to voltage-gated K^+ channels except that they possess a cyclic nucleotide binding site on the intracellular C-terminal tail, and have no apparent voltage sensitivity.[1] In heterologous expression systems, the CNG1 (retinal) or the CNG2 (olfactory) α subunits can form functional homomeric channels,

[b]Corresponding author. Phone: 0041 21 692 5370; fax: 0041 21 692 5355; e-mail: Marie-Christine.Broillet@ipharm.unil.ch)

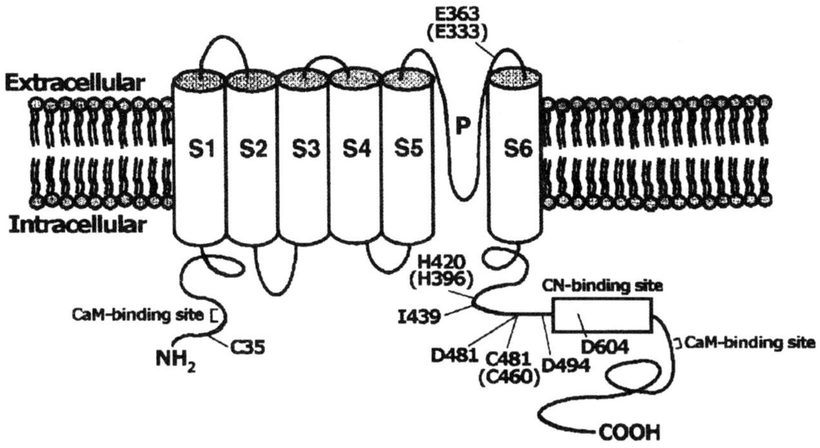

FIGURE 1. Hypothetical model of the two-dimensional architecture of a cyclic nucleotide–gated channel subunit. S1-S6 are the putative transmembrane domains; P is the putative pore region. The cyclic nucleotide (CN) binding site is defined by homology to the sequences of cAMP and cGMP binding proteins. The position of several important functional regions are labeled. The amino acid positions in the olfactory channel are shown in *parenthesis*.

although their biophysical properties are different from the native channels.[9,16] The CNG4 (retinal) and the CNG5 (olfactory) β subunits cannot be activated by cyclic nucleotides, although the olfactory CNG5, for example, form functional homomeric channels.[10,11,17] Heteromeric olfactory CNG2-5-4.3 channels behave very much like the native channel.[12]

A series of elegant experiments with cloned channels and chimeric constructs has revealed significant information regarding the binding and gating reactions that lead to CNG channel activation.[1,15,18–26] These studies have identified several regions as well as specific residues distributed throughout the approximately 500 amino acids of the CNG1 (retinal) or the CNG2 (olfactory) α subunit proteins that play a key role in channel regulation (FIG. 1). In this review, we will concentrate on activation and modulation of the CNG channels at these specific molecular sites.

THE CYCLIC NUCLEOTIDE BINDING SITE

Since the demonstration by Fesenko *et al.* (1985) and Nakamura *et al.* (1987)[27,28] that CNG channels could be directly activated by intracellular cyclic nucleotides (CN), a highly conserved stretch of approximately 120 amino acids homologous to the cyclic nucleotide binding domains of other proteins like the cAMP or cGMP protein kinases has been identified. It consists of a short amino-terminal α helix preceding an eight-stranded antiparallel β roll that is followed by two α helices.[1] A structural model of the retinal CNG channel CN binding site has also been constructed.[29]

The retinal and olfactory CNG channels share a high degree of sequence similarity (over 80% amino acids identity) in the CN binding sites, but they have very different cyclic nucleotide selectivities. cGMP is a more potent and effective agonist of the retinal channel,[27] while cAMP and cGMP have very similar effects on the olfactory channel.[30] Varnum *et al.* (1995)[31] investigated the molecular mechanism for ligand discrimination of CNG channels. They found that the retinal photoreceptors', and olfactory neurons' CNG channels were differentially activated by ligands that varied only in their purine ring structure. The nucleotide selectivity of the retinal bovine CNG channel (cGMP > cIMP >> cAMP) was significantly altered by neutralization of a single aspartic acid residue (D604) in the CN binding domain (cGMP > or = cAMP > cIMP). Substitution by a nonpolar residue at this position inverted agonist selectivity (cAMP >> cIMP > or = cGMP). These effects resulted from an alteration in the relative ability of the agonists to promote the allosteric conformational change associated with channel activation, not from a modification in their initial binding affinity. These authors proposed a general mechanism for guanine nucleotide discrimination, in common with that observed in high-affinity GTP binding proteins, involving the formation of a pair of hydrogen bonds between the aspartic acid side chain and N1 and N2 of the guanine ring. This amino acid residue (D604) appears to play a critical role in the selective activation of the retinal CNG channel by cGMP.

The functional effects of each ligand binding event have always been difficult to assess because ligands continuously bind and unbind at each site. Furthermore, in retinal rod photoreceptors the low cytoplasmic concentration of cyclic GMP means that CNG channels exist primarily in partially liganded states. Ruiz *et al.* (1997)[32] studied single CNG channel behavior with the use of a photoaffinity analogue of cGMP that tethered cGMP moieties covalently to their binding sites to show that single retinal CNG channels could be effectively locked in four distinct ligand-bound states. Their results indicated that CNG channels open more than they would spontaneously when two ligands were bound (approximately 1% of the maximum current), open significantly more with three ligands bound (approximately 33%), and open maximally with four ligands bound. In each ligand-bound state, channels opened to two or three different conductance states. These findings placed strong constraints on the activation mechanism of CNG channels.

Recently, this issue was investigated further by Liu *et al.* (1998).[15] These authors observed the effects of individual binding events on channel activation by studying CNG channels containing one, two, three, or four functional CN binding sites. They found that the binding of a single ligand significantly increased channel opening, although four ligands were required for full channel activation. Their data were inconsistent with models in which the four subunits activated in a single concerted step (Monod-Wyman-Changeux model) or in four independent steps (Hodgkin-Huxley model). Instead, the four subunits of the channel might associate and activate as two independent channel dimers.

THE C-LINKER REGION

The cyclic nucleotide binding site is connected to the last transmembrane segment of the channel by a chain of approximately 90 amino acids known as the C-linker. Recent evidence suggests that the N-terminal region of the channel as well as the C-linker region influence the apparent agonist affinity and efficacy with which the cyclic nucleotides open the channel. The C-linker region has been implicated by several studies as being critical in the gating reaction that leads to channel activation subsequent to cyclic nucleotide binding.

As a first attempt toward understanding the channel gating process, Gordon *et al.* (1995)[20] have studied the mechanism of potentiation of expressed rod CNG channels by Ni^{2+}. They found that coordination binding of Ni^{2+} between histidine residues (H420) on adjacent channel subunits occurred when the channels were open. Mutation of H420 to lysine completely eliminated the potentiation by Ni^{2+} but did not markedly alter the apparent cGMP affinity of the channel, indicating that the introduction of positive charge at the $Ni(^{2+})$ binding site was not sufficient to produce potentiation. Deletion or mutation of most of the other histidines present in the channel did not diminish Ni^{2+} potentiation. These authors also examined the role of subunit interactions in Ni^{2+} potentiation by generating heteromultimeric channels using dimers of the rod CNG channel. Injection of single heterodimers (wt/H420Q or H420Q/wt) in which one subunit contained H420 and the other did not resulted in channels that were not potentiated by Ni^{2+}. However, coinjection of both heterodimers into *Xenopus* oocytes resulted in channels that exhibited potentiation. The H420 residues probably occurred predominantly in nonadjacent subunits when each heterodimer was injected individually; but when the two heterodimers were coinjected, the H420 residues could occur in adjacent subunits as well. Their results suggest that the mechanism of Ni^{2+} potentiation involves intersubunit coordination of Ni^{2+} by H420. Based on the preferential binding of Ni^{2+} to open channels, they suggest that alignment of H420 residues of neighboring subunits into the $Ni(^{2+})$ coordinating position may be associated with channel opening. These authors also identified the corresponding histidine residue on olfactory neurons (H396) to have an opposite effect, an inhibition, upon Ni^{2+} binding.[21] Thus, this particular C-linker region of the channel probably undergoes a movement during the opening transition.[20,21]

The cone and olfactory CNG channels also differ considerably in cyclic nucleotide affinity and efficacy. Recently, Zong *et al.* (1998)[33] found that three amino acids in the C-linker region are major determinants of gating in these CNG channels. Indeed, the replacement of three amino acids in the cone C-linker by the corresponding amino acids of the olfactory channel (I439V, D481A, and D494S) profoundly enhanced the cAMP efficacy and increased the affinities for cAMP and cGMP. Unlike the wild-type cone channel, the mutated channel exhibited similar (olfactory) single-channel kinetics for both cGMP and cAMP, explaining the increase in cAMP efficacy. Therefore, the identified amino acids appear to be major determinants of CNG channel gating.

Among the other key amino acid residues of the C-terminal end of the channel are a series of intracellularly located cysteine residues in or near the cyclic nucleotide binding site (see Ref. 13 for review). These residues are thought to affect the gating reaction either through subunit-subunit interactions or within single subunits.[34,35] At least one of those cysteines has been proposed as a site that undergoes redox modulation by reactive nitrogen species that are downstream of the gaseous messenger nitric oxide (NO).[36] Especially notable among these is the nitrosonium ion, NO^+, which participates in nitrosothiol reactions with free SH groups on unpaired cysteine residues forming a nitrosylated protein.[37] The formation of a nitrosothiol through this chemical reaction has now been shown to occur *in vivo* in an array of proteins[37,38] and may represent an alternative pathway for protein modulation analogous to phosphorylation. The regulation of protein by NO has been proposed to play a critical role in many processes such as blood pressure regulation, host defense, and neurotransmission.[39] Ion channel regulation has also been postulated to occur by this direct NO pathway.[40–43] Among ion channels, only the CNG channel has been shown to be directly activated by NO; in other channels normal activation and inactivation parameters are altered. Biochemical experiments allowed Broillet *et al.* (1996)[36] to find that the cysteine in position C460 in the α (CNG2) subunit of the rat olfactory CNG channel is the critical residue in the reaction that leads to channel activation by NO. This particular cysteine residue is located within the C-linker region just N-terminal to the CN binding site.

Brown *et al.* (1998)[26] have probed the structural changes that occur during channel activation by using SH-modifying reagents on the α subunit of the bovine retinal channel. Treatment with these reagents dramatically potentiated the channel's response to both cAMP and cGMP. This potentiation was abolished by conversion of the cysteine residue C481 to a nonreactive alanine residue. Potentiation occurred more rapidly in the presence of saturating cGMP, indicating that this region of the channel is more accessible when the channel is open. C481 is located in the C-linker region between the S6 transmembrane domain and CN binding site and corresponds to the C460 residue of the olfactory channel that is the NO target site. These results suggest that this region of the channel undergoes significant movement during the activation process and is critical for coupling ligand binding to pore opening. Brown *et al.* (1998) also claim, however, that potentiation is not mediated by the recently reported interaction between the amino- and carboxy-terminal regions of the α subunit because deletion of the entire amino-terminal domain had little effect on potentiation by SH-modifying reagents (see N-C INTERACTION section below).

THE PORE REGION

In voltage-gated ion channels and in the homologous CNG channels, the loop between the S5 and S6 transmembrane segments (P region) is thought to form the lining of the pore. To investigate the structure and the role in gating of the P region of the bovine retinal CNG channel, Sun *et al.* (1996)[25] determined the accessibility of 11 cysteine-substituted P-region residues to small, charged sulfhydryl reagents applied to the inside and outside of membrane patches in the open and closed states of the channel. The results they obtained suggest that the P region forms a loop that extends toward the central axis of the channel, analogous to the L3 loop of bacterial porin channels. Furthermore, the P region, in addition to forming the ion selectivity filter, functions as the channel gate, the structure of which changes when the channel opens. The pore region of the channel controls both the single-channel conductance and the pore diameter of the channel.[18] The determination of how ions permeate the channel and how the ionic selectivity occur is of great physiological importance because ion permeation is responsible for generating membrane depolarization, which is critical for electrical signaling; but also, the CNG channels are permeable to Ca^{2+}, which is an important element in the activation of intracellular targets. Balasubramanian *et al.* (1997)[44] demonstrated that the permeation properties of the olfactory CNG channels are significantly different from those of photoreceptor CNG channels. Their results further indicated that Na^+ currents through these channels did not obey the independence principle and showed saturation kinetics with K(m)s in the range of 100–150 mM. They also displayed a lack of voltage dependence of conductance in asymmetric solutions, which suggested that ion binding sites were situated midway along the channel pore.

Wells *et al.* (1997)[45] have developed a two-site, Eyring rate theory model of ionic permeation for CNG channels. The parameters of the model were optimized by simultaneously fitting current-voltage (IV) data sets from excised photoreceptor patches in electrolyte solutions containing one or more of the following ions: Na^+, Ca^{2+}, Mg^{2+}, and K^+. The model accounted well for the shape of the IV relations, the binding affinity for Na^+, the reversal potential values with single-sided additions of calcium or magnesium and biionic KCl, and the $K_{1/2}$ and voltage dependence for divalent block from the cytoplasmic side of the channel. The differences between the predicted $K_{1/2}$s for extracellular block by Ca^{2+} and Mg^{2+} and the values obtained from heterologous expression of only the α subunit of the channel suggest that the β subunit or a cell-specific factor affects the interaction of divalent cations at the external but not the internal face of the channel. The model predicts concentration-dependent permeability ratios with single-sided addition of calcium and magnesium and anomalous mole fraction effects under a limited set of conditions for both monovalent and divalent cations. Calcium and magnesium are predicted to carry 21% and 10%, respectively, of the total current in the retinal rod cell at –60 mV.

In addition to permeating the CNG channel, Ca^{2+} also profoundly blocks the current flow carried by monovalent cations through the channels,[1] similar to the behavior observed in voltage-activated Ca^{2+}-channels.[46] Supposedly, this behavior implies the high-affinity binding of Ca^{2+} to a single acidic amino acid residue located in the pore of the channel (E363 for the rod CNG channel and E333 for the catfish olfactory CNG channel).[22] Root *et al.* (1993)[22] also implicated this particular glutamate residue as being important in the external rapid proton block of CNG channels, another characteristic that the CNG channels share with Ca^{2+} channels. Recently, Bucossi *et al.* (1997)[47] did an extensive analysis of single-channel properties of CNG channels composed of the α and β subunits from bovine rods expressed in *Xenopus laevis* oocytes. At least three types of channels with different properties were observed. Indeed, CNG channels had different conductive levels, leading to the existence of multiple open states in homomeric channels and to the flickering behavior in heteromeric channels, and further demonstrated that the pore is an essential part of the gating of CNG channels. One type of channel had well-resolved, multiple conductive levels at negative voltages, but not at positive voltages. The other two types of channels were characterized by flickering openings and two different conductances. The α subunit of CNG channels had a well-defined conductance of about 28 pS, but multiple conductive levels are observed in mutant channels E363D and T364M. The conductance of these open states is modulated by protons and by the membrane voltage. The relative probability of occupying any of these open states is independent of the cGMP concentration, but depends on extracellular protons. The open probability in the presence of saturating cGMP in the wild-type and mutants E363D, T364M, and E363G depends also on temperature, indicating that the thermodynamics of the transition between the closed and open state is also affected by mutations in the pore region.

Gavazzo *et al.* (1997)[48] have examined the modulation by internal protons of the α subunit of the bovine olfactory and retinal CNG channels. Increasing internal proton concentrations caused a partial blockage of the single-channel current, consistent with protonation of a single acidic site with a pK1 of 4.5–4.7, both in rod and in olfactory CNG channels. Channel gating properties were also affected by internal protons. The open probability at low cyclic nucleotide concentrations was greatly increased by lowering pHi, and the increase was larger when channels were activated by cAMP than by cGMP. Therefore, internal protons affected both channel permeation and gating properties, causing a reduction in single-channel current and an increase in open probability. These effects are likely to be caused by different titratable groups on the channel.

THE CALCIUM-CALMODULIN BINDING SITE

A characteristic of CNG channels is their Ca^{2+} permeability. At physiological extracellular Ca^{2+} concentrations, Ca^{2+} represents a significant fraction of the current passing through CNG channels. Ca^{2+} is very important for both adaptation and excitation of photoreceptors and olfactory neurons, as it controls the activity of several signaling enzymes including the CNG channels themselves (for reviews see Refs. 30 and 49). Ca^{2+}/calmodulin (CaM) attenuates the activity of rod and olfactory CNG channels by increasing their apparent $K_{1/2}$ for cGMP and cAMP (for review see Ref. 50). This modulation is believed to serve as one of the several Ca^{2+}-mediated feedback mechanisms that terminate the electrical response and set the sensitivity of the photoreceptor cells and olfactory neurons (for reviews see Refs. 50 and 51).

Weitz *et al.* (1998)[52] and Grunwald *et al.* (1998)[53] have examined the mechanism of CaM modulation using electrophysiological and biochemical techniques and surface plasmon resonance spectroscopy. They have identified a domain on the β subunit of the rod cGMP-gated cation channel that mediates inhibition by CaM. Using heteromeric channels, consisting of retinal α and β subunits that display a high CaM sensitivity ($EC_{50} \leq 5$ nM) similar to the native channel, they identified two unconventional CaM binding sites (CaM1 and CaM2), one in each of the N- and the C-terminal regions of the β subunit. Ca^{2+} cooperatively stimulates binding of CaM to these sites exactly within the range of $[Ca^{2+}]$ occurring during a light response. Deletion of the N-terminal CaM1 site results in channels that are no longer CaM sensitive, whereas deletion of CaM2 has only minor effects. These results indicate that CaM controls the activity of the rod cGMP-gated ion channel by decreasing the apparent cGMP affinity through an unconventional binding site in the N-terminus of the β subunit. Müller *et al.* (1998)[54] showed that phosphorylation by PKC of a serine residue near the CaM binding site sensitized the olfactory CNG channel to cAMP and thereby extended the range of CaM modulation.

N-C INTERACTION

Varnum *et al.* (1996)[34] have shown that an intramolecular protein-protein interaction between the amino-terminal domain and the carboxyl-terminal ligand-binding domain of the rat olfactory CNG channel exerts an autoexcitatory effect on channel activation. CaM, which modulates CNG channel activity during odorant adaptation, blocked this interaction. A specific deletion within the amino-terminal domain disrupted the interdomain interaction *in vitro* and altered the gating properties and CaM sensitivity of expressed channels. Thus, the amino-terminal domain might promote channel opening by directly interacting with the carboxyl-terminal gating machinery; CaM regulated channel activity by targeting this interaction.

Cysteine residues again seem to be playing an important role in CNG channel activation; indeed, Gordon *et al.* (1997),[35] working on the rod α subunit of the CNG channel, have shown that the N-terminal and the C-terminal regions of each channel subunit interacted. This interaction involved precise cysteine residues and the formation of a disulfide bond between Cys 35 (N-terminal) and Cys 481 (C-linker region). The corresponding cysteine residue located in the C-terminal region corresponds exactly to the NO target site in the olfactory CNG channel (Cys 460), confirming that this amino acid—which is highly conserved among the different cloned CNG channels—plays a very important role in channel gating and in the potentiation of cyclic nucleotide action.

CONCLUSION

The activation of CNG channels is a complex process comprising the initial ligand binding and a consecutive allosteric transition from a closed to an open configuration. The recent studies on cyclic nucleotide–gated channels have demonstrated the importance of different regions of the protein in activation and modulation of this class of ion channels.

Not only is the direct activation by cyclic nucleotides controlled at the CN-binding site where cyclic nucleotide recognition and channel activation take place, but particular amino acid residues located elsewhere on the protein also modulate the affinity and efficacy of cAMP or cGMP. Allosteric changes in the CN binding site and movement of certain regions of the channel molecule, like the C-linker region, seem to be of fundamental importance in the control of the gating mechanisms of the channel. The pore region is responsible for ionic selectivity and for the control of physiological Ca^{2+} permeation, whereas the N-teminal and C-terminal tails of the channel take part in channel modulation. CNG channels are becoming an interesting channel model for understanding how the various functions of a protein can be distributed throughout the molecule.

CNG channels first identified in rods and olfactory receptors are now known to be distributed throughout the different cells of the body and may have important roles in such functions as cell motility, secretion, development, and neural plasticity. Future work will reveal new functions for CNG channels as well as more information about their structure, functions, and modulation.

ACKNOWLEDGMENTS

We gratefully thank Olivier Randin for assistance in producing FIGURE 1. M.-C.B. and S.F. are supported by grants from the NIH, ONR, and Fonds National Suisse de la Recherche.

REFERENCES

1. ZAGOTTA, W.N. & S.A. SIEGELBAUM. 1996. Structure and function of cyclic nucleotide-gated channels. Annu. Rev. Neurosci. **19:** 235–263.
2. KAUPP, U.B. 1991. The cyclic nucleotide-gated channels of vertebrate photoreceptors and olfactory epithelium. TINS **14:** 150–157.
3. YAU, K.-W. 1994. Cyclic nucleotide-gated channels: An expanding new family of ion channels. Proc. Natl. Acad. Sci. USA **91:** 3481–3483.
4. BRADLEY, J., Y. ZHANG, R. BAKIN, H.A. LESTER, G.V. RONNET & K. ZINN. 1997. Functional expression of the heteromeric "olfactory" cyclic nucleotide-gated channel in the hippocampus: A potential effector of synaptic plasticity in brain neurons. J. Neurosci. **17:** 1993–2005.

5. KINGSTON, P.A., F. ZUFALL & C.J. BARNSTABLE. 1996. Rat hippocampal neurons express genes for both rod retinal and olfactory cyclic nucleotide-gated channels: Novel targets for cAMP/cGMP function. Proc. Natl. Acad. Sci. USA **93:** 10440–10445.
6. WEI, J.Y., D.S. ROY, L. LECONTE & C.J. BARNSTABLE. 1998. Molecular and pharmacological analysis of cyclic nucleotide-gated channel function in the central nervous system. Prog. Neurobiol. **56**(1): 37–64.
7. COBURN, C.M. & C.I. BARGMANN. 1996. A putative cyclic nucleotide-gated channel is required for sensory development and function in C. elegans. Neuron **17:** 695–706.
8. ZUFALL, F., G.M. SHEPHERD & C.J. BARNSTABLE. 1997. Cyclic nucleotide gated channels as regulators of CNS development and plasticity. Curr. Opin. Neurobiol. **7:** 404–412.
9. DHALLAN, R.S., K.W. YAU, K.A. SCHRADER & R.R. REED. 1990. Primary structure and functional expression of a cyclic nucleotide–activated channel from olfactory neurons. Nature **347:** 184–187.
10. BRADLEY, J., J. LI, N. DAVIDSON, H.A. LESTER & K. ZINN. 1994. Heteromeric olfactory cyclic nucleotide-gated channels: A subunit that confers increased sensitivity to cAMP. Proc. Natl. Acad. Sci. USA **91:** 8890–8894.
11. LIMAN, E.R. & L.B. BUCK. 1994. A second subunit of the olfactory cyclic nucleotide–gated channel confers high sensitivity to cAMP. Neuron **13:** 611–621.
12. SAUTTER, A., X. ZONG, F. HOFMANN & M. BIEL. 1998. An isoform of the rod photoreceptor cyclic nucleotide–gated channel β subunit expressed in olfactory neurons. Proc. Natl. Acad. Sci. USA **95:** 4696–4701.
13. BIEL, M., X. ZONG & F. HOFMANN. 1996. Cyclic nucleotide-gated cation channels: Molecular diversity, structure and cellular functions. Trends Cardiovasc. Med. **6:** 274–280.
14. LIU, D.T., G.R. TIBBS & S.A. SIEGELBAUM. 1996. Subunit stoichiometry of cyclic nucleotide–gated channels and effects of subunit order on channel function. Neuron **16:** 983–990.
15. LIU, D.T., G.R. TIBBS, P. PAOLETTI & S.A. SIEGELBAUM. 1998. Constraining ligand-binding site stoichiometry suggests that a cyclic nucleotide-gated channel is composed of two functional dimers. Neuron **21:** 235–248.
16. KAUPP, U.B., T. NIIDOME, T. TANABE, S. TERADA, W. BONIGK, W. STUHMER, N.J. COOK, K. KANGAWA, H. MATSUO, T. HIROSE, T. MIYATA & S. NUMA. 1989. Primary structure and functional expression from complementary DNA of the rod photoreceptor cyclic GMP-gated channel. Nature **342:** 762–766.
17. BROILLET, M.-C. & S. FIRESTEIN. 1997. β Subunits of the olfactory cyclic nucleotide–gated channel form a nitric oxide activated Ca^{2+} channel. Neuron **18:** 951–958.
18. GOULDING, E.H., G.R. TIBBS, D. LIU & S.A. SIEGELBAUM. 1993. Role of H5 domain in determining pore diameter and ion permeation through cyclic nucleotide–gated channels. Nature **364**(6432): 61–64.
19. GOULDING, E.H., G.R. TIBBS & S.A. SIEGELBAUM. 1994. Molecular mechanism of cyclic nucleotide-gated channel activation. Nature **372:** 369–374.
20. GORDON, S.E. & W.N. ZAGOTTA. 1995. A histidine residue associated with the gate of the cyclic nucleotide-activated channels in rod photoreceptors. Neuron **14:** 177–183.
21. GORDON, S.E. & W.N. ZAGOTTA. 1995. Localization of regions affecting an allosteric transition in cyclic nucleotide-activated channels. Neuron **14:** 857–864.
22. ROOT, M. & R. MACKINNON. 1993. Identification of an external divalent cation–binding site in the pore of a cGMP-activated channel. Neuron **11:** 459–466.
23. ROOT, M.J. & R. MACKINNON. 1994. Two identical noninteracting sites on an ion channel revealed by proton transfer. Science **265:** 1852–1856.
24. PARK, C.S. & R. MACKINNON. 1995. Divalent cation selectivity in a cyclic nucleotide–gated ion channel. Biochemistry **34:** 13328–13333.
25. SUN, Z., M.H. AKABAS, E.H. GOULDING, A. KARLIN & S.A. SIEGELBAUM. 1996. Exposure of residues in the cyclic nucleotide–gated channel pore: P-region structure and function in gating. Neuron **16:** 141–149.
26. BROWN, L.A., S.D. SNOW & T.L. HALEY. 1998. Movement of gating machinery during the activation of rod cyclic nucleotide–gated channels. Biophys. J. **75:** 825–833.
27. FESENKO, E.E., S.S. KOLESNIKOV & A.L. LYUBARSKY. 1985. Induction by cyclic GMP of cationic conductance in plasma membrane of retinal rod outer segment. Nature **313:** 310–313.

28. NAKAMURA, T. & G.H. GOLD. 1987. A cyclic-nucleotide gated conductance in olfactory receptor cilia. Nature **325:** 442–444.
29. KUMAR, V.D. & I.T. WEBER. 1992. Molecular model of the cyclic GMP-binding domain of the cyclic GMP–gated ion channel. Biochemistry **31:** 4643–4649.
30. ZUFALL, F., S. FIRESTEIN & G.M. SHEPHERD. 1994. Cyclic nucleotide–gated ion channels and sensory transduction in olfactory receptor neurons. Annu. Rev. Biophys. Biomol. Struct. **23:** 577–607.
31. VARNUM, M.D., K.D. BLACK & W.N. ZAGOTTA. 1995. Molecular mechanism for ligand discrimination of cyclic nucleotide–gated channels. Neuron **15:** 619–625.
32. RUIZ, M.L. & J.W. KARPEN. 1997. Single cyclic nucleotide–gated channels locked in different ligand-bound states. Nature **389:** 389–392.
33. ZONG, X., H. ZUCKER, F. HOFMANN & M. BIEL. 1998. Three amino acids in the C-linker are major determinants of gating in cyclic nucleotide–gated channels. EMBO J. **17:** 353–362.
34. VARNUM, M.D. & W.N. ZAGOTTA. 1996. Subunit interactions in the activation of cyclic nucleotide-gated ion channels. Biophys. J. **70:** 2667–2679.
35. GORDON, S.E., M.D. VARNUM & W.N. ZAGOTTA. 1997. Direct interaction between amino- and carboxyl-terminal domains of cyclic nucleotide–gated channels. Neuron **19:** 431–441.
36. BROILLET, M.-C. & S. FIRESTEIN. 1996. Direct activation of the olfactory cyclic nucleotide–gated channel through modification of sulfhydryl groups by NO compounds. Neuron **16:** 377–385.
37. BUTLER, A.R., F.W. FLITNEY & D.L.H. WILLIAMS. 1995. NO, nitrosonium ions, nitroxide ions, nitrosothiols, and iron-nitrosyls in biology: A chemist's perspective. TIPS **16:** 18–22.
38. LANDER, H.M. 1997. An essential role for free radicals and derived species in signal transduction. FASEB J. **11:** 118–124.
39. MONCADA, S. & A. HIGGS. 1993. The L-arginine-nitric oxide pathway. N. Eng. J. Med. **329**(27): 2002–2012.
40. LEI, S.Z., Z.H. PAN, S.K. AGGARWAL, H.S.V. CHEN, J. HARTMAN, N.J. SUCHER & S.A. LIPTON. 1992. Effect of nitric oxide production on the redox modulatory site of the NMDA receptor-channel complex. Neuron **8:** 1087–1099.
41. BOLOTINA, V.M., S. NAJIBI, J. PALACINO, P. PAGANO & R.A. COHEN. 1994. Nitric oxide directly activates calcium-dependent potassium channels in vascular smooth muscle. Nature **368:** 850–853.
42. LI, Z., M.W. CHAPLEAU, J.N. BATES, K. BIELEFELDT, H.-C. LEE & F.M. ABBOUD. 1998. Nitric oxide as an autocrine regulator of sodium currents in baroreceptor neurons. Neuron **20:** 1039–1049.
43. XU, L., J.P. EU, G. MEISSNER & J.S. STAMLER. 1998. Activation of the cardiac calcium release channel (ryanodine receptor) by poly-S-nitrosylation. Science **279:** 234–236.
44. BALASUBRAMANIAN, S., J.W. LYNCH & P.H. BARRY. 1997. Concentration dependence of sodium permeation and sodium ion interactions in the cyclic AMP–gated channels of mammalian olfactory receptor neurons. J. Membr. Biol. **159**(1): 41–52.
45. WELLS, G.B. & J.C. TANAKA. 1997. Ion selectivity predictions from a two-site permeation model for the cyclic nucleotide–gated channel of retinal rod cells. Biophys. J. **72**(1): 127–140.
46. ALMERS, W. & E.W. MCCLESKEY. 1984. Non-selective conductance in calcium channels of frog muscle: Calcium selectivity in a single-file pore. J. Physiol. **353:** 585–608.
47. BUCOSSI, G., M. NIZARRI & V. TORRE. 1997. Single-channel properties of ionic channels gated by cyclic nucleotides. Biophys. J. **72:** 1165–1181.
48. GAVAZZO, P., C. PICCO & A. MENINI. 1997. Mechanism of modulation by internal protons of cyclic nucleotide–gated channels cloned from sensory cells. Proc. R. Soc. Lond. B. Biol. Sci. **264**(1385): 1157–1165.
49. SHEPHERD, G.M. 1994. Discrimination of molecular signals by the olfactory receptor neuron. Neuron **13:** 771–790.
50. MOLDAY, R.S. 1996. Calmodulin regulation of cyclic nucleotide–gated channels. Curr. Opin. Neurobiol. **6**(4): 445–452.
51. FINN, J.T. & K.W. YAU. 1996. Cyclic nucleotide–gated channels: An extended family with diverse functions. Annu. Rev. Physiol. **58:** 395–426.
52. WEITZ, D., M. ZOCHE, F. MULLER, M. BEYERMANN, H.G. KORSCHEN, U.B. KAUPP & K.W. KOCH. 1998. Calmodulin controls the rod photoreceptor CNG channel through an unconventional binding site in the N-terminus of the beta-subunit. EMBO J. **17**(8): 2273–2284.

53. GRUNWALD, M.E., W.P. YU, H.H. YU & K.W. YAU. 1998. Identification of a domain on the beta-subunit of the rod cGMP-gated cation channel that mediates inhibition by calcium-calmodulin. J. Biol. Chem. **273**(15): 9148–9157.
54. MULLER, F., W. BONIGK, F. SESTI & S. FRINGS. 1998. Phosphorylation of mammalian olfactory cyclic nucleotide-gated channels increases ligand sensitivity. J. Neurosci. **18**(1): 164–173.

The HCN Gene Family: Molecular Basis of the Hyperpolarization-Activated Pacemaker Channels

BINA SANTORO[a] AND GARETH R. TIBBS[b,c]

[a]Center for Neurobiology and Behavior and [b]Department of Anesthesiology, Columbia University, 722 West 168th Street, New York, New York 10032, USA

ABSTRACT: The molecular basis of the hyperpolarization-activated cation channels that underlie the anomalous rectifying current variously termed I_h, I_q, or I_f is discussed. On the basis of the expression patterns and biophysical properties of the newly cloned HCN ion channels, an initial attempt at defining the identity and subunit composition of channels underlying native I_h is undertaken. By comparing the sequences of HCN channels to other members of the K channel superfamily, we discuss how channel opening may be coupled to membrane hyperpolarization and to direct binding of cyclic nucleotide. Finally, we consider some of the questions in cardiovascular physiology and neurobiology that can be addressed as a result of the demonstration that I_h is encoded by the HCN gene family.

Periodic activity in biological systems, such as the rhythmic firing of neuronal networks, the autonomous beating of the heart, and the steady cycle of circadian rhythms, is generated over a remarkable spectrum of frequencies—with periods ranging from milliseconds to days. However, the molecular bases of the oscillators that generate those rhythms are only now coming to light. One example of this is the recent cloning of clock genes and the description of the rhythmic interplay of transcription and translation that appears to serve as the primary circadian pacemaker.[1] In contrast to this low-frequency, nuclear-cytoplasmic timekeeper that regulates homeostatic events, faster rhythms depend on the biophysical properties of distinct classes of ion channels in the excitable membranes of specialized pacemaking nerve and muscle cells. Over the past 20 years, one channel in particular—a hyperpolarization-activated cation channel termed I_h, I_q, or I_f, which we shall consistently refer to as I_h—has been implicated as a primary component of rhythmic firing in many loci in heart and brain.[2–4]

The focus of this paper is to describe the molecular basis of the I_h channels and to consider some of the questions in cardiovascular physiology, neurobiology, and channels biophysics that can be addressed as a result of the recent demonstration that I_h is encoded by the HCN gene family.[5–7]

THE PHYSIOLOGICAL SIGNIFICANCE OF RHYTHMIC FIRING PATTERNS AND THE ROLE OF I_h

I_h was first described in sinoatrial node cells of the heart, but has since been identified in cardiac Purkinje fibers, atrial and ventricular muscle, and both peripheral and central neurons (for references, see TABLE 2). Native I_h channels share a number of striking char-

[c]Corresponding author. Phone: 212-543-5259; fax: 212-795-7997; e-mail: GRT1@columbia.edu

acteristics that underlie their ability to serve as crucial components of rhythm generators. First, unlike most other members of the voltage-gated channel superfamily, I_h activates upon hyperpolarization, not depolarization. Second, I_h displays only a weak selectivity for K over Na; as a result, the reversal potential of these channels lies at –35 mV, so they carry an inward current at the hyperpolarized potentials where they open. It is the generation of this depolarizing current that serves to drive the membrane potential of a cell back towards threshold, thereby maintaining rhythmic firing. Third, transmitters and hormones acting through second messenger systems can elicit profound shifts in the voltage dependence of I_h activation. Thus, the binding of cAMP to the channel shifts the activation curve of I_h to more depolarized potentials, altering how rapidly and completely the channels activate upon repolarization, which, in turn, alters the amplitude of the inward current and the rate at which the cell depolarizes.[2,3]

In the heart, I_h contributes to the pacemaker potential that drives the rhythmic firing and beating of the atria and ventricles. Thus, the increase in heart rate in response to β-adrenergic agonists (FIG. 1) and the slowing of the heart rate during vagal stimulation are mediated by the ability of cAMP to directly modulate the activation of the pacemaker current.[2]

Why is I_h present in many neurons? To answer this, it is instructive to consider what roles rhythmicity may serve in the central nervous system (CNS). While rhythmic activity in some lower brainstem regions—such as those involved in the generation of respiratory rhythms—are obviously important, it has been postulated that, at the other end of the cognitive spectrum, the rhythmic patterns observed in electroencephalograms (EEGs) reflect

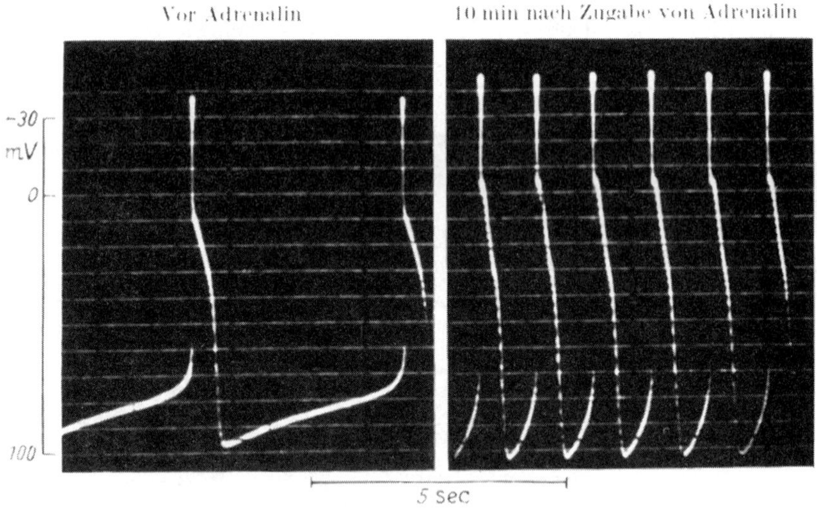

FIGURE 1. Effect of adrenaline on the pacemaker action potential in sheep Purkinje fibers. **Left panel:** before the application of adrenaline. **Right panel:** 10 minutes after the application of 10 µg/ml adrenaline (Reprinted from Otsuka[69] by permission.)

underlying mechanisms used in encoding and controlling information flow in the CNS.[4,8–10] For example, synchronization of the activity of neuronal populations in higher cortical regions, through an endogenous 40-Hz oscillation, may serve to bind together the separate analyzed components of a perceptual representation into a coherent representation of the external world.[9] Perhaps the best-studied example of the generation of rhythmic and tonic activity at the cellular level, and of the contribution of this to higher-order neural function, is the thalamus.

Isolated thalamic neurons display a range of complex firing patterns such as tonic firing, slow δ oscillations, and the waxing and waning of spindle activity—bursts of a few action potentials separated by long periods of quiescence. Remarkably, it has been demonstrated that neurons will switch between these modes in response to small changes in the resting potential, due to modulatory transmitter actions, and this behavior is dependent on I_h.[8,11,12] The physiological relevance of such behavior is suggested from the observation that similar patterns dominate the EEG during separate phases of sleep.[4,8–10,13] This correlation between cellular physiology and general CNS function is important because the thalamus serves as a primary path for sensory input to, and motor output from, the cerebral cortex. Thus, it is thought that the various firing patterns of thalamic neurons control access to and from the cortex during sleep.[4,8,10,13] Indeed, it is possible that the rhythmic firing of thalamocortical neurons may even be the drive that leads to synchronization of the entire cerebral cortex and thus may be a key element in the development and maintenance of behavioral states of sleep and wakefulness. Not surprisingly, perhaps, disturbance of rhythm generation is observed in major CNS pathologies, such as epilepsy, although causal relationships are unclear.[4,9]

Several classes of interneurons have been shown to exhibit spontaneous firing activity, providing a tonic inhibition to principal neurons, which can be modulated by noradrenalin and other neurotransmitters.[14] I_h currents are present in these cells, and contribute to the regulation of firing frequency, as well as to the regulation of excitability. Interestingly, different classes of interneurons, which do not have spontaneous activity, do not appear to express notable I_h-type currents.[15] Thus, it is possible that I_h might also act as a pacemaker conductance in the generation of oscillatory activity in local networks (see also Ref. 16).

While the role of I_h in such pacemaking processes may be the most well known function of these channels, the channels also help determine other electrical properties of non-pacemaking neurons. I_h currents contribute to the facilitation of repetitive activity, most importantly in sensory pathways such as the auditory system, where the number and timing of incoming action potentials must be accurately transmitted even at high input rates to ensure fidelity in stimulus coding.[17] Also, I_h currents produce rebound depolarization after hyperpolarizing responses—for example, in the rod photoresponse to bright light[18,19]—and can generate rebound action potentials following inhibitory inputs.[20] Finally, I_h channels have been shown to impact the shape and propagation of subthreshold voltage potentials in dendritic trees, and thus to regulate the integration of synaptic inputs.[21]

Despite the important roles that have been conceptualized for I_h in the heart and the CNS, the absence of either specific pharmacological tools that can target the I_h channels or knowledge of their molecular identity has severely hampered rigorous investigation of the contribution of these channels to cardiovascular physiology and neurobiology.

THE HCN GENE FAMILY

In spite of the extensive physiological characterization of the native currents, the search for the molecular basis of I_h has been a long and frustrating venture. Nonetheless, the hypothesis that I_h would be a member of the voltage-gated K channel superfamily, and a distant cousin of the cyclic nucleotide–gated (CNG) channels, contributed to all three approaches that recently identified the gene family that encodes I_h channels.

The first member of the gene family encoding I_h was identified somewhat serendipitously following a protein-protein interaction screen, using the SH3 domain of the neural form of Src tyrosine kinase as bait.[5] Several lines of evidence suggested that the interacting gene, mBCNG-1 (now termed mHCN1), was a candidate gene for I_h. First, the deduced amino acid sequence of mBCNG-1 revealed it to be a member of the superfamily of voltage-gated K channels, but with an unusual pore (FIG 2). This was consistent with the weak selectivity for K over Na that is a property of all native I_h currents. Second, the carboxy terminus includes a conserved cyclic nucleotide–binding (CNB) domain, which accorded with the observation of direct modulation of native pacemaker currents by cyclic nucleotide (FIG. 2). Third, both mBCNG-1 mRNA and protein were expressed in brain regions where I_h had been identified (see EXPRESSION PATTERNS OF THE HCN GENES). Fourth, we obtained partial clones for three additional members of the HCN family from mouse (mBCNG-2,3,4) and orthologs of two of these channels from human (hBCNG-1 and 2). In addition to showing widespread distribution in the CNS, two of these clones were also expressed in the heart. The assumption that mBCNG-1 encoded a brain-specific form of I_h was confirmed upon heterologous expression in *Xenopus* oocytes and by the demonstration that mBCNG-1 formed a hyperpolarization-activated channel that was weakly selective for K over Na and was blocked by cesium.[6]

A second line of research used the EST database as a source for cDNA sequences containing CNB domains, similar to the site present in CNG channels. This screening resulted in the identification of a fragment of a second member of the newly identified channel family, which was named HAC1 (now termed mHCN2). After cloning the full-length cDNA of HAC1 and cDNAs encoding two related genes (HAC2 and HAC3) from a mouse brain cDNA library, expression in HEK293 cells revealed that HAC1 also encoded an I_h type current.[7] Sequence comparison demonstrates that HAC1 is the full-length counterpart to the partial clone mBCNG-2, reported by Santoro and colleagues.[6] The correspondence between the other two cDNAs identified by Ludwig *et al.* and those identified by Santoro *et al.* are shown in TABLE 1.

Together, the studies of Santoro *et al.* and Ludwig *et al.* revealed the existence of products of at least four separate genes encoding mammalian I_h channels. Given the overlap between the genes identified in these two studies, a unified nomenclature has been proposed that includes all known members of this new mammalian gene family (TABLE 1; see also Ref. 22). In this nomenclature, HCN stands for Hyperpolarization-activated, Cyclic Nucleotide–sensitive, Cation Non-selective. At present, the fourth member of the mouse gene family is not cloned to full length, and only two human orthologs (hHCN1 and hHCN2) have been identified. However, cloning of mHCN4 is nearly complete, and it seems likely that hHCN3 and hHCN4 will be identified soon.

In a third approach, Kaupp and colleagues designed degenerate oligonucleotide primers based on conserved CNG channel sequences and used these to amplify cDNA fragments from a sea urchin sperm library.[23] Full-length cloning identified a polypeptide,

SPIH (FIG. 2), that forms a channel with characteristics that are similar to mammalian I_h channels.

FIGURE 3A presents the percent similarity between each of the identified HCN genes and SPIH. The four mammalian channels are very closely related to each other, with an overall similarity in the protein sequences of 56–60% and a similarity within the core sequence (domains S1 through the CNB domain) of 80–90%. The overall protein sequence similarity between the SPIH and the HCN gene products is 38–40%, with a similarity of 52–54% in the core region of the proteins. At present, there is little evidence that the HCN family includes any other immediate members; however, the presence of closely related or more distantly related members cannot be ruled out. A database search for the presence of HCN-related genes identified a further homologue in the tobacco budworm, *Heliothis virescens*, (accession AJO12664, here called HVIH). The overall protein sequence similarity of this gene to the mammalian genes is 46–49% (51% to the sea urchin gene), and the similarity in the core region is 58–61% to the HCN proteins (65% to SPIH). The inferred phylogenetic relationship between the HCN1-4, SPIH, and HVIH genes is presented in FIGURE 3B.

Although mHCN1, mHCN2, and SPIH show relatively high sequence similarity, the biophysical properties of channels formed upon homomeric expression display some unique and revealing properties. The possible molecular bases for the observed properties of these channels will be discussed later in this chapter.

The phylogenetic relationship between the HCN gene family (represented by mHCN1 and SPIH) and other representative members of the K channel superfamily is presented as a dendrogram in FIGURE 3C. From this it is clear that mHCN1, its homologues, and SPIH form a new subfamily within the K channel superfamily. Not surprisingly, the HCN and SPIH channels are most closely related to other K channels that contain a CNB domain, such as the EAG-related channels, the CNG channels, and the plant inward rectifiers related to KAT1. Nonetheless, HCN and SPIH form a new branch within this group. A comparison of the hydrophobic core region (domains S1 through S6) of HCN1 and SPIH with the corresponding region in other CNB site–containing channels shows that the highest sequence similarity occurs with the EAG and HERG proteins, albeit this amounts to only ~22%. The lowest similarity is to bRET1 (17%). However, if the CNB regions of these channels are compared, the binding domain in HCN is found to be most similar to the CNB site of bRET1 (28%), while the corresponding domains of HERG or KAT1 are less similar to HCN (22% or 15%, respectively). This observation is congruous with the fact that HCN channels display a voltage-dependent inward rectification reminiscent of HERG or KAT1, but a cyclic nucleotide regulation that is not seen in any other channels except the CNG channels. A complete sequence alignment between the HCN1, SPIH, HERG, KAT1, and bRET1 proteins is presented in FIGURE 2.

It is interesting to speculate on the evolutionary relationship among the various members of the K channel superfamily. It has been suggested that the fusion between an ancestral K channel and an ancestral cyclic nucleotide binding site might have occurred before the evolutionary separation of plants and animals.[24] Interestingly, among all channels bearing a CNB site, the HCN binding site shows the closest homology to binding sites present in protein kinases and most remarkably to the yeast cAMP-dependent protein kinase (25% similarity). At the same time, the hydrophobic core region of HCN1 retains a relatively high similarity to that of voltage-gated K channels, Shaker in particular (14%). Thus, the HCN genes might be a closer representation of the genealogical link that existed

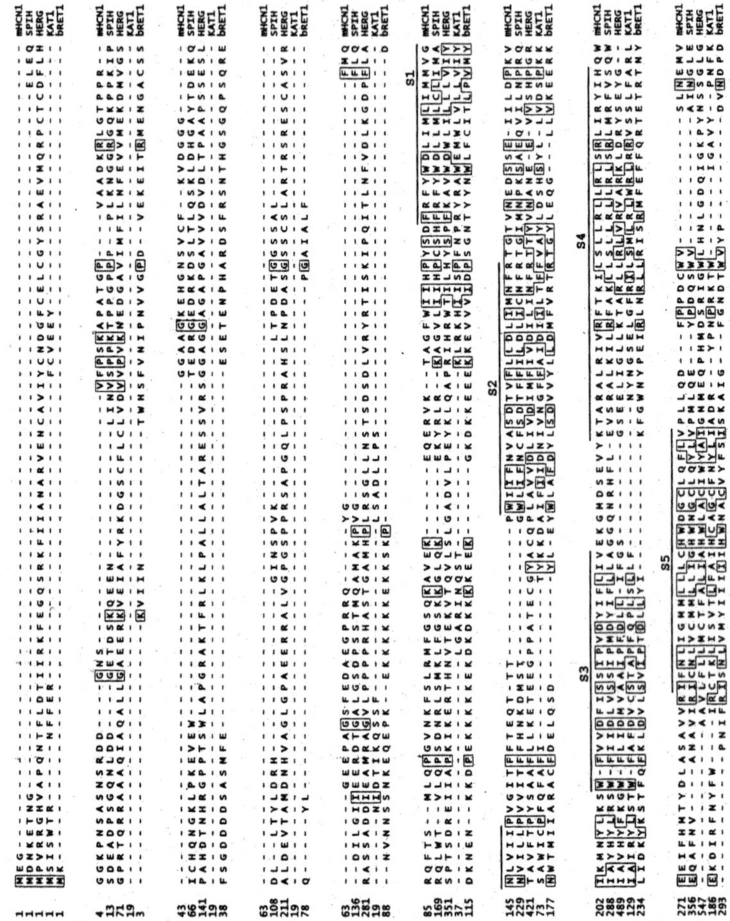
mHCN1
SPIH
HERG
KAT1
bRET1
S1
S2
S3
S4
S5

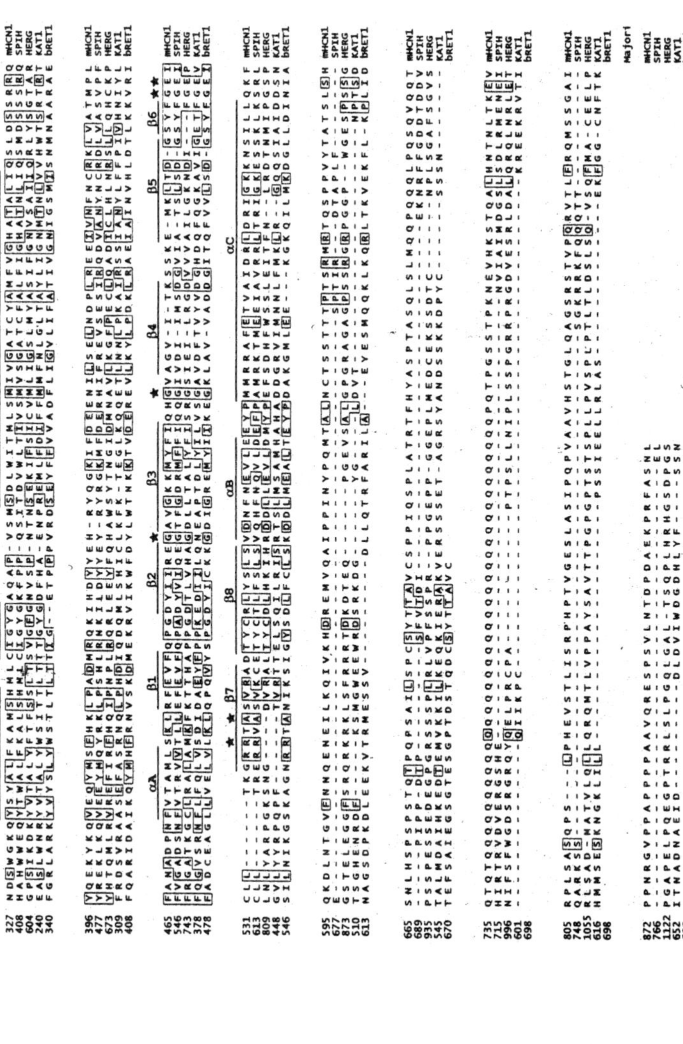

FIGURE 2. Amino acid sequence alignment of the mouse HCN1 protein (mHCN1),[5] sea urchin SPIH protein,[23] human ERG protein (HERG)[24] Arabidopsis KAT1 protein,[70] and α-subunit of the bovine retinal CNG-channel (bRET1).[71] The putative transmembrane domains (S1 through S6), the pore helix (PH), the selectivity filter (SF),[50] and the three α-helices (A–C) and eight β-strands (1–8) comprising the cyclic nucleotide binding site are indicated. *Asterisks* mark the six critical residues that are conserved in all functional CNB sites of the protein kinase A type.[54,55]

TABLE 1. Classification and Tissue Distribution of HCN Gene Family Members

Nomenclature			Tissue Distribution									
Consensus	Santoro *et al.*	Ludwig *et al.*	Br	He	Mu	Lu	Li	Ki	Sp	Pa	Te	P1
mHCN1	mBCNG-1	HAC2	+++	-	-	–	-	–	–	nd	–	nd
mHCN2	mBCNG-2	HAC1	++++	+++	–	–	-	–	–	nd	–	nd
mHCN3	mBCNG-4	HAC3	++	–	–	–	?	?	–	nd	–	nd
mHCN4	mBCNG-3	–	++	++	++	+	-	-	–	nd	–	nd
hHCN1	hBCNG-1	–	+++	–	+	–	–	–	nd	+/–	nd	–
hHCN2	hBCNG-2	–	++++	+++	–	–	–	–	nd	-	nd	–

NOTE: Data were collected from Northern blot analysis of polyA+ RNA isolated from the following tissues[5–7]: brain (Br), heart (He), skeletal muscle (Mu), lung (Lu), liver (Li), kidney (Ki), spleen (Sp), pancreas (Pa), testis (Te) and placenta (P1); nd, not determined.

between voltage-gated and CNG channels, than any of the channels known so far. The presence of an HCN-related gene in the budworm, a member of the phylum Arthropoda, would also predict that the primordial HCN gene appeared before the separation of Schizocela (Protostoma) and Enterocela (Deuterostoma). Somewhat surprisingly, no HCN-related genes are present in the *Caenorhabditis elegans* genome, while both CNG and EAG-related genes are clearly represented.[25] Further data will be needed to assess whether HCN genes were lost during nematode evolution, as happened with Na channels,[25] or else arose after the branching off of the Pseudocoelomata, and perhaps concomitantly with the appearance of the Coelomata.

EXPRESSION PATTERNS OF THE HCN GENES

TABLE 1 summarizes the tissue-specific expression of HCN1-4 mRNA as determined by Northern Blot analysis (data summarized from Refs. 5–7). This analysis reveals that all four genes are expressed in the brain, with HCN2 being the most abundant mRNA species. HCN2 and HCN4 are also expressed in the heart. Lower levels of expression of HCN1 and HCN4 are also detected in several other tissues, such as skeletal muscle, lung, and pancreas; and the expression pattern might be broadened by the presence of alternatively spliced transcripts. Indeed, a splice variant of HCN3 is present in liver and kidney,[6] although the significance of this transcript remains to be established. Moreover, the physiological relevance of HCN gene expression in nonexcitable tissues is unclear and might be due, in part, to contaminating components of the peripheral and autonomic nervous system.

Northern blot, *in situ* hybridization, and immunohistochemical approaches have been used to obtain a more detailed analysis of the distribution of HCN1 and HCN2. These data show that HCN2 is widely expressed throughout the brain with prominent labeling of tha-

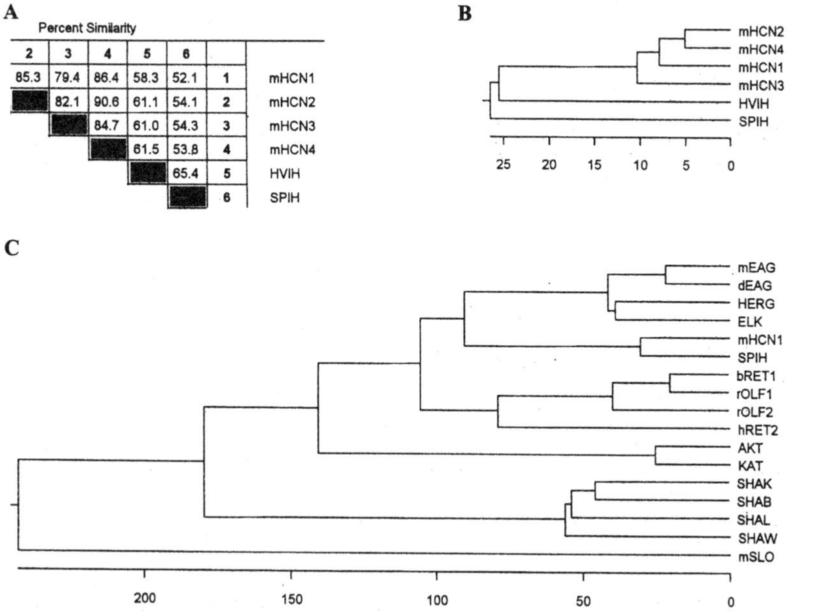

FIGURE 3. Amino acid sequence similarity between ion channel proteins. All alignments were generated using the MegAlign program from DNASTAR. **Panel A:** Percent similarity between mouse HCN1-4, SPIH (*S. purpuratus* I_h), and HVIH (*H. virescens* I_h) proteins. Similarities were calculated from an alignment of the region included between transmembrane domain S1 and the CNB site. **Panel B:** Phylogenetic tree deduced from sequence analysis in panel A. **Panel C:** Phylogenetic tree of representative members of the K channel superfamily. The dendrogram was derived from an alignment of the hydrophobic core of the proteins, including transmembrane domains S1 through S6. The following sequences were used: mouse HCN1 protein (mHCN1), sea urchin SPIH protein (SPIH), mouse Eag protein (mEAG),[24] *Drosophila* Eag protein (dEAG),[72] human Erg protein (HERG), *Drosophila* Elk protein (ELK),[24] *Arabidopsis* AKT1 protein (AKT)[73] *Arabidopsis* KAT1 protein (KAT), α-subunit of bovine retinal CNG-channel (bRET-1), α-subunit of rat olfactory CNG-channel (rOLF-1)[74] β-subunit of human retinal CNG-channel (hRET-2)[75] β-subunit of rat olfactory CNG-channel (rOLF-2),[76] *Drosophila* Shaker protein (SHAK),[77] *Drosophila* Shab protein (SHAB),[78] *Drosophila* Shaw protein (SHAW)[78] *Drosophila* Shal protein (SHAL),[78] and mouse slo protein (mSLO).[79]

lamic and brainstem nuclei.[6,7] In the heart, HCN2 transcripts can be detected in ventricle, atrium, and sino-atrial node preparations.[6,7] Thus, HCN2 gene products appear to be a predominant and very widespread species in both these tissues.

In contrast, HCN1 expression shows a more restricted distribution pattern. HCN1 transcripts are particularly abundant in cortical structures—namely, the neocortex, hippocampus and cerebellum. Within these structures, *in situ* hybridization shows a marked labeling of the cell bodies of layer V pyramidal neurons, and a comparably high labeling of the cell body layer of CA1 pyramidal neurons.[5,7] In the cerebellum, labeling appears to be particularly prominent in the Purkinje cell body layer. Immunohistochemical staining is consistent with these mRNA distribution patterns, but further reveals that the HCN1 protein shows discrete subcellular localization patterns that vary depending on the cell type. Thus, HCN1 protein appears to be localized to the apical dendrites of the cortical pyramidal neurons, with a dense staining of the terminal dendritic plexus (FIG. 4A). In contrast, a very distinct staining pattern is observed in the cerebellar cortex. Here, the distinctive pattern of antibody labeling indicates that HCN1 protein is predominantly localized in the terminals of basket cells that innervate the Purkinje cells (FIG. 4C). Thus, HCN1 protein can be found either postsynaptically (apical dendrites) or presynaptically (axon terminals) depending on the cell type. The presence of a high density of HCN channels at these criti-

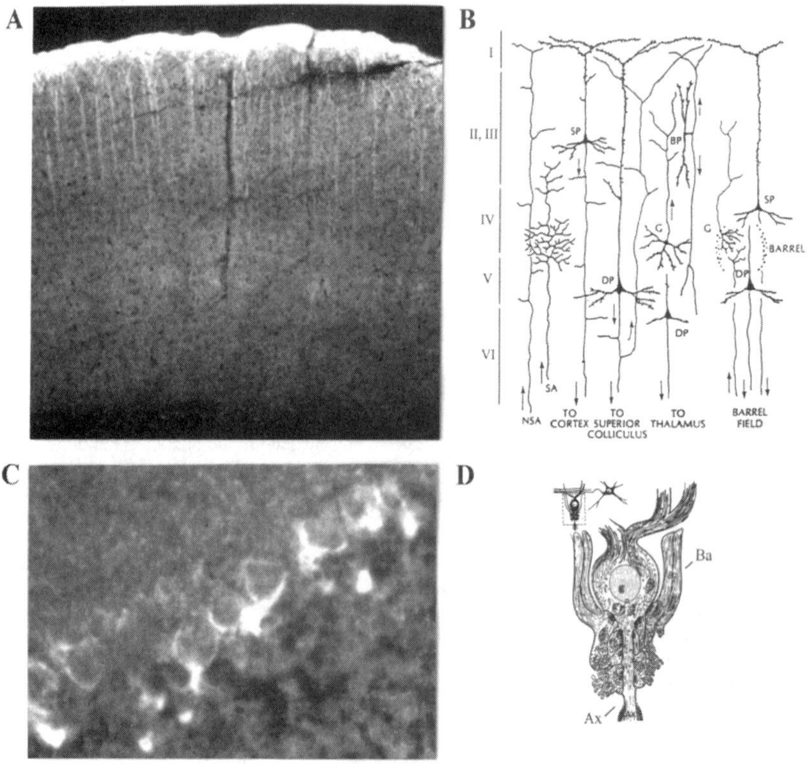

FIGURE 4. Immunohistochemical analysis of mHCN1 distribution in the brain. **Panel A:** Polyclonal antisera generated against the mouse HCN1 protein (αq1 and αq2)[5] specifically label the apical dendrites of cortical pyramidal neurons. **Panel B:** Schematic representation of neuronal elements in the neocortex; NSA, nonspecific afferents; SA, specific afferents; DP, deep pyramidal neurons; SP, superficial pyramidal neurons; G, granule cells; BP, bipolar cell. (Modified from G.M. Shepherd, *The Synaptic Organization of the Brain*, second edit., Oxford University Press, New York, 1979; now in fourth edition.) **Panel C:** Labeling of the cerebellar basket cell nerve terminals by the anti-HCN1 sera (αq1 and αq2).[5] **Panel D:** Diagram illustrating the ultrastructure of the basket synapse (Ba) on the initial segment of the Purkinje cell axon (Ax). (Modified from J.C. Eccles, M. Ito & J. Szentagothai, *The Cerebellum as a Neuronal Machine*, Springer-Verlag, New York, 1967. Panels A and C reproduced from Santoro *et al.*;[5] copyright 1997 National Academy of Sciences, USA.)

cal locations in projection neurons and in interneurons is likely to have profound physiological consequences, and provides a unique opportunity for an in-depth study of the properties and roles of native I_h currents at anatomically defined loci in the brain.

A COMPARISON OF HETEROLOGOUSLY EXPRESSED HCN CHANNELS WITH NATIVE I_h

Upon heterologous expression, both mHCN1 and mHCN2 channels give rise to hyperpolarization-activated currents, but these channels display different voltage-dependent kinetics.[6,7] FIGURE 5 shows a comparison of whole-cell voltage-clamp currents recorded from HEK293 cells transfected with either mHCN1 or mHCN2 cDNA. The characteristics of the mHCN1 currents obtained under these conditions are essentially identical to the properties originally reported for this clone upon heterologous expression in *Xenopus* oocytes,[6] demonstrating that the difference in kinetics between mHCN1 and mHCN2 is not a consequence of the different expression systems used in the earlier studies, but reflects intrinsic difference in the channels. From the records shown in FIGURE 5, it is clear that mHCN2 activates and deactivates considerably more slowly across the entire voltage range when compared to mHCN1. Indeed, a comparison of the time constants of exponential fits to the activation curves for each of these channels at their $V_{1/2}$ reveals that mHCN2 activates considerably slower than mHCN1 (τ at $V_{1/2}$ was 3740 ms for mHCN2 compared to τ at $V_{1/2}$ of 170 to 350 ms for mHCN1; recorded at 22–25°C in HEK293 cells or *Xenopus* oocytes; see FIG. 5; TABLE 2; (Refs. 6; 7). Significantly, mHCN2 was also shown to be strongly regulated by the direct binding of cAMP ($V_{1/2}$ shifted by up to 25 mV in the depolarizing direction), while the channel encoded by mHCN1 shows only feeble direct modulation by cAMP (see TABLE 2; Refs. 6, 7).

Studies of hyperpolarization-activated currents in numerous tissues have revealed that where conduction and permeation properties are almost invariant, considerable divergence is observed in the gating behavior of native I_h channels (see TABLE 2, and references therein). It is possible that the highly variant gating properties may arise either from the assembly of channels from alternative gene products (or alternatively spliced transcripts); from the interaction of channels with other subunits or auxiliary proteins; or as a consequence of posttranslational modification events or the influence of cytoplasmic components such as cAMP and Ca ions. Here we now ask: how do the properties of the mHCN1 and mHCN2 expressed channels compare to native currents? Specifically, do the heterologously expressed channels display the defining permeation properties of native I_h channels, and do the activation properties of the expressed channels correspond to native channels? Based on this comparison of biophysical properties, coupled with the preliminary distribution patterns of the HCN gene products discussed earlier, we can start to make tentative assignments as to the subunit identities of the native channels and to speculate on the possible contribution of posttranslational modifications to the observed properties of native I_h channels.

How do the permeation and conduction properties of the HCN gene products compare with native I_h channels? Independent of cell type, native I_h is invariably reported to be weakly selective for K over Na with permeability ratios generally in the range of 3:1 to 4:1. Thus, the inward current through I_h is largely carried by Na. Native I_h is impermeant to anions. Nonetheless, native I_h displays an unusual dependency on both external K and

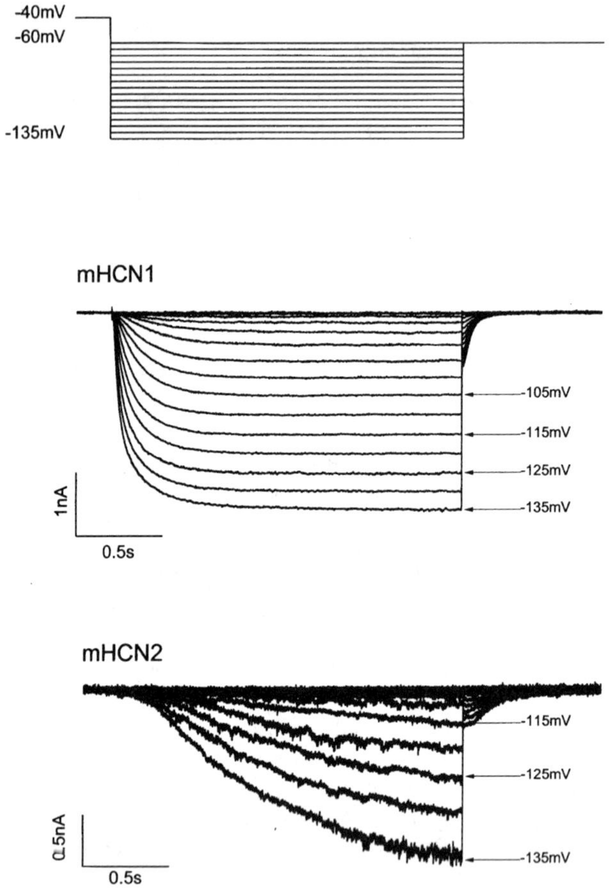

FIGURE 5. Representative recordings of mHCN1 and mHCN2 following expression in HEK293 cells. Records are whole-cell voltage-clamp currents obtained from HEK293 cells (Edge Biosystems) that had been transfected (Superfect, Qiagen) with either mHCN1 or mHCN2 subcloned into pTracer-cmv2 (Invitrogen). The composition of the intracellular and extracellular solutions in mM were 140 KC1, 1 EGTA, 10 HEPES, pH7.4 with KOH and 110 NaC1, 30 KC1, 1CaC1$_2$, 10 HEPES, pH 7.4 with NaOH, respectively. Five-MOhm borosilicate electrodes (Kimax 51, Kimble Products) were coated with sylgard to reduce capacitative artifacts. Data were acquired using an Axopatch 200B integrating patch clamp amplifier in the resistive mode. The holding potential was –40mV. Voltage clamp protocols were applied using a P/10 protocol to subtract uncompensated linear leak and capacitance, and data were digitized (Pulse software, Heka Instruments; ITC-16 interface Instrutech Corp.; Mac Centris 650 computer) at 5 KHz following low-pass filtering at 2.5 kHz with an 8-pole Bessel filter (Frequency Devices). The midpoint potential for mHCN2 shown here is somewhat more negative than reported by Ludwig *et al.*[7] This may reflect the failure of activation to reach steady state during the hyperpolarizing test voltage. However, it may also suggest a modulatory role for Mg ions that were absent in these experiments but present in the work of Ludwig and colleagues.

external C1. Thus, the complete removal of external K or the replacement of external C1 by large organic ions results in an almost complete loss of inward current. The K ion dependency has been attributed to loss of conduction due to tight Na binding in the pore rather than to alterations in gating. The basis of the C1 ion dependency is as yet unestablished. Finally, all native I_h isoforms display a similar profile with regard to channel blockers. Thus, micromolar concentrations of a number of bradycardic organic molecules, such as ZD7288, block these channels, while submillimolar concentrations of external Cs block the inward current but potentiate the outward current through native I_h channels. I_h channels are relatively insensitive to Ba ions, which block most other forms of K channels.[3]

Both mHCN1 and mHCN2 have been shown to express cation-selective channels that display a K:Na permeability ratio between 3:1 and 4:1 depending on the ionic conditions, and both channels are blocked by external Cs but not Ba.[6,7] Moreover, mHCN1 conducts little if any inward current in the absence of external K, it is blocked by ZD 7288, and low concentrations of external Cs potentiate the outward current (G.R. Tibbs & S.A. Siegelbaum, unpublished observations). Thus, the HCN gene products appear to encode channels that fully recapitulate the basic selectivity and conduction properties of native I_h channels.

A comparison of the basic gating parameters of a number of native I_h channels is summarized in TABLE 2. As discussed above, the gating properties of I_h are very sensitive to the ionic environment. Thus, it is possible that some of the observed differences arise from variations in the recording conditions. However, as differences in gating are seen even when similar approaches are used to study channels in different populations of cells (see references for TABLE 2), other explanations are necessary.

I_h activation and deactivation are nearly always described as sigmoidal (Ref. 3 and references for TABLE 2). However, following an often pronounced lag, the time course of activation is very variable. This is particularly evident if the later phases of activation are fit by exponential functions. Although a single exponential term is sometimes adequate, often two exponentials are required (TABLE 2). Thus, hippocampal pyramidal neurons display fast single-exponential activation (τ at $V_{1/2}$ is 18 to 62 ms), while other cell types exhibit biexponential and relatively slow activation–for example, dorsal lateral and medial geniculate thalamic neurons and sino-atrial cells (τ_f at $V_{1/2}$ is 230 to 500 ms and τ_s at $V_{1/2}$ is 1100 to 3000 ms). In yet other cell types, such as superior colliculus projecting (SCP) or dorsal root ganglion (DRG) neurons, fast biexponential behavior can be seen, with one kinetic component resembling the hippocampal current (τ_f at $V_{1/2}$ is 38–67 ms) in conjunction with a second component that is in the range of the fast component seen in thalamic and sino-atrial cells (τ_f at $V_{1/2}$ is 154–481 ms). In neurons from the ventrobasal thalamus, a very slowly activating isoform of I_h has recently been detected (τ_f is 800 ms, and τ_s is 7500 ms).[26]

Clearly, there are two possible interpretations for this type of behavior. First, the two components represent distinct fast and slow I_h channels within the same cell. Alternatively, these data may simply reflect a complex activation path for a single population of channels. Evidence that the multiexponential behavior reflects different channel isoforms comes from the work of Mark Fishman and colleagues. They report that in zebrafish myocytes I_h displays two kinetic components with time constants of 280 ms and 1600 ms,[27] very similar to the parameters for mammalian cardiac preparations (see TABLE 2). Interestingly, they have identified a mutant zebrafish, *smo*, that has a selective loss of the predominant, fast component while the minor, slow component is unaffected.

TABLE 2. Comparison of the Activation Properties of Heterologously Expressed Channels and Native Isoforms of I_h

	$V_{1/2}$	Time Constants at $V_{1/2}$		cAMP Modulation		
	(mV)	τ_f (ms)	τ_s (ms)	shift (mV)	$K_{1/2}$ (μM)	References
Clone						
mHCN1	–100	47*	--	+2	--	6[a]
mHCN2	–103	720*	--	+26	0.5	7[a]
Cell type						
Sino atrial node	–75	360–500	1560–2400	+11	0.2	31,34,84,85,89,90,91
Purkinje fibers	–85[b]	--	800–4000	--	--	32,81,83
Right atrium	–81	--	1900	--	--	100
Ventricular myocytes	–140[b]	--	5000	--	--	32, 36,37,38
Thalamic relay neurons	–75	230–250	1100–3000	+7	--	11,12,35,86,94,98,101[c]
Cortical SCP neurons	–81	67*	481*	--	--	20,96,97,102
Hippocampal pyramidal neurons	–95	18–62*	--	+2	--	21,80,82,95
Amygdala pyramidal neurons	–89	62*	--	--	--	99
Cerebellar Purkinje cells	–85	86	--	--	--	88
Hippocampal SOA interneurons	–84	530*	--	--	--	14
MNTB neurons	–76	230*	1250*	+14	--	17
Spiral ganglion neurons	–100[b]	92*	600*	+3	--	87,93
DRG sensory neurons	–88[b]	38*	154*	--	--	43,92
Nodose/trigeminal ganglion neurons	–70	--	--	+12	--	46

TABLE 2. Comparison of the Activation Properties of Heterologously Expressed Channels and Native Isoforms of I_h (*continued*)

NOTE: Heterologous current data were taken from Santoro *et al.*[6] or Ludwig *et al.*[7] or are unpublished observations from Tibbs, Santoro and Siegelbaum. For determination of $V_{1/2}$ and time constants of native currents only data obtained from intracellular recording, perforated patch or whole cell patch clamp studies were included. Data describing cAMP modulation of native currents also included recordings obtained from excised inside-out patch clamp studies. Time constants were either measured at 35–37 °C or, if marked by an *, were normalized to that temperature assuming a Q_{10} of 4 for I_h gating.[21,80,81] MNTB, medial nucleus trapezoid body; SOA, stratum oriens-alveus; SCP, superior colliculus projecting; DRG, dorsal root ganglion. --, data not determined.

[a]G.R. Tibbs & S.A. Siegelbaum, unpublished data.

[b]$V_{1/2}$ reported for I_h in these cells show substantial variability; the value indicated here approximates the middle of the range.

[c]Drs. Anita Lüthi and David McCormick, personal communication.

TABLE 2 also shows that native I_h currents display a very variable behavior with respect to the midpoint potential for channel activation (ranging from –75 mV to –100 mV, with the most negative value at –130 mV). Taken together with the zebrafish data, these results argue that there are multiple native isoforms of I_h and that these have divergent voltage sensitivity and kinetic properties.

Clearly, the likelihood of multiple channel isoforms underlying the whole-cell current of any particular neuron does not exclude the possibility of complex kinetics for individual channel types. Although in the original reports by Santoro *et al.* and Ludwig *et al.* it was concluded that activation of both mHCN1 and mHCN2 were well approximated by single-exponential functions following a lag, it is possible that very slow components of mHCN1 or mHCN2 activation could have been overlooked, as these would become well resolved only with very long voltage steps or with recordings where the current density is large enough that a small slow component can be readily resolved. There is some evidence from experiments such as that shown in FIGURE 5 that two exponentials are required to adequately fit the time course of mHCN1 at the more extreme hyperpolarized potentials (G.R. Tibbs & S.A. Siegelbaum, unpublished results). Resolution of this issue is of considerable importance, but it is probable that the multiexponential behavior of native whole-cell currents reflects both complex activation of each I_h isoform and multiple isoforms in some cell types.

While it is still premature to draw strong conclusions about the identity of the subunits underlying the channels of particular cell types, it is pertinent to note the following general observations. Thus, the properties of the native I_h current recorded from hippocampal pyramidal cells (TABLE 2) are almost indistinguishable from the properties reported for mHCN1 (FIG. 5).[6] Moreover, dendritic recordings performed at an increasing distance from the cell body of pyramidal neurons showed a significant increase in I_h current density, which parallels the increasing intensity of HCN1-antibody labeling observed along the dendritic arbor of pyramidal neurons in the neocortex (FIG. 4; Refs. 5, 21, 103).

As discussed earlier, HCN2 appears to be the major isoform of the HCN channel family expressed in the adult heart and appears to be ubiquitously expressed in the CNS, with particular abundance in thalamic structures. Interestingly, mHCN2 activates slowly, as does I_h of cardiac cells and thalamic neurons. The slower kinetic component of native I_h in those cells may arise from a very slow, and as yet undetected, gating process in mHCN2 channels. Alternatively, this behavior may represent the expression of another channel, such as HCN4, which is also present in the heart[6] and thalamus (B. Santoro, G. Shumyatsky & S.A. Siegelbaum; unpublished results). The observation that the channel encoded by mHCN2 is strongly regulated by the direct binding of cAMP—as is also seen for both the thalamic and cardiac currents—is a further indication that this subunit contributes to the native currents in those cells.

However, there is an important difference between the heterologously expressed mHCN2 channel, and the current recorded in the nonventricular cells of the heart and the thalamic relay neurons. Notably, the midpoint of activation for the heterologously expressed current is between –100 and –125mV, whereas the nonventricular cardiac channel and the thalamic currents typically have midpoint potentials around –65 to –75mV. If mHCN2 really makes a major contribution to the native currents of these cells, what is the basis for this difference in the voltage sensitivity? First, all the data on native channels reported in TABLE 2 (except for data describing cAMP modulation) were acquired using intracellular recording, perforated patch clamp, or whole-cell patch clamp with ATP and often GTP in the pipette. In contrast, all the heterologous data were obtained in excised-patch configurations or in whole-cell patch without added nucleotides. Second, upon patch excision from sino-atrial node cells, the midpoint potential for I_h activation shifts from –75mV to –125mV.[28–30] Thus, the currents expressed in these cells appear to be strongly modulated by cytoplasmic elements, perhaps requiring the presence of ATP and/or GTP. In part, this can be explained by the demonstrated influence of direct cAMP binding to these channels (see TABLE 2). However, the largest depolarizing shift in the $V_{1/2}$ that has been reported upon exposure of native channels to cAMP in inside-out excised patches is only approximately 25 mV (see TABLE 2), insufficient to account for the 50-mV hyperpolarizing shift seen following patch excision. These data suggest that other modulatory processes influence the channels. Indeed, it has been demonstrated that both phosphorylation[31–33] and changes in cytoplasmic Ca concentrations[34,35] can result in modulation of the channel, including depolarizing shifts in the $V_{1/2}$ for sino-atrial node and thalamic isoforms of I_h.

DEVELOPMENTAL CHANGES IN I_h EXPRESSION

I_h currents display not only a broad and very specific pattern of spatial expression across different cell types, but also a regulated pattern of temporal (or developmental) expression within a particular cell type. It has been shown that in ventricular myocytes I_h currents are present very early in development, and persist throughout adult life. However, the activation threshold of the current varies greatly depending on the developmental stage of the tissue. Whereas in the newborn rat ventricle I_h begins to activate at potentials negative to –70 mV, in the adult ventricle I_h can be measured only at potentials negative to approximately –115 mV (TABLE 2).[36–38] This shift parallels a reduction in the spontaneous pacemaker activity of these cells. Interestingly, in the hypertrophied or failing heart, ventricular I_h currents have been shown to activate in a physiological voltage range, an observation that might be linked to the increased predisposition to arrythmias present in these conditions[39,40] Adult ventricular myocytes also resume spontaneous contractility when cultured in conditions leading to cell "dedifferentiation," and this transition is again accompanied by a positive shift in the I_h activation curve.[41] This phenomenon may be of clinical importance since different chronic cardiac pathologies, including fibrillation, cause structural changes of the myocardium that have been interpreted as a "dedifferentiation" rather than degeneration.[42] Thus, changes in the activation curve of ventricular I_h appear to be dynamically regulated, and can be modeled in *in vitro* preparations.

The observed shifts in the activation curve of I_h currents may be due to regulated switches in gene expression (including alternative splicing), which would result in a different subunit composition of the underlying channels, or posttanslational modifications of preexisting channels including, but perhaps not limited to, protein phosphorylation or dephosphorylation. Preliminary observations suggest a significant modulation of HCN gene expression during development (B. Santoro & S.A. Siegelbaum, unpublished observations). However, the observation that the activation threshold of I_h in ventricular myocytes can be rapidly shifted by the application of phosphatase and kinase inhibitors[32] suggests an important contribution of posttranslational modifications as well.

I_h currents have also been shown to contribute to different aspects of nervous system development—for example, the organization of neuronal networks. Synchronized electrical activity is a determining factor during synaptic circuitry formation. In the developing hippocampus, at a time when the connections between dentate gyrus granule cells and CA3 neurons have not been established, a network of electrically coupled hilar interneurons provides the excitatory input to the CA1 and CA3 subfields. This network generates a spontaneous oscillating activity, whose frequency is regulated by an inwardly rectifying cationic conductance similar to I_h.[16]

I_h is also found in the growth cones of developing neurites.[43] Interestingly, the properties of I_h currents measured at the tip of the growing neurite or in the soma of the cell appear to be different, again reflecting a developmental regulation of I_h activity during fiber outgrowth. Hyperpolarization-activated cationic conductances are found in mature nerves, both myelinated[44] and unmyelinated,[45] where they appear to regulate excitability in conditions of high-frequency impulse conduction, in both normal and pathological conditions. Thus, the modulation of I_h activity in primary afferent neurons might contribute to sensitization to painful stimuli during the inflammatory response,[46] and an abnormal inward rectification of peripheral axons has been associated with nerve degeneration in diabetic neuropathy.[47]

The tools made available by the identification of the molecular components underlying I_h channels will undoubtedly allow a deeper understanding of the causal relationships existing between I_h current expression and the developmental and functional physiology of many cell types in the near future.

POSSIBLE MOLECULAR BASES FOR THE LOW ION SELECTIVITY, INWARD RECTIFICATION, AND CYCLIC NUCLEOTIDE MODULATION OF HCN-ENCODED CHANNELS

Having established the molecular identity of the proteins that underlie the pacemaker currents, it is now possible to address the compelling questions surrounding the unusual biophysical properties of I_h. What is the mechanism by which the HCN channels activate upon hyperpolarization rather than depolarization? What is the basis of the unusual permeation properties of the pacemaker channels, and how are these properties related to the gating machinery? How does cyclic nucleotide binding interact with the voltage-dependent gating apparatus to allow the dual regulation seen in both native and cloned channels?

Clearly, one instructive approach to understanding the basic mechanistic properties of I_h comes from studies on related channels. Thus, the HCN channels have the same basic body plan as the other members of the K channel superfamily—notably, six transmembrane domains and cytoplasmic N and C termini (as indicated in FIG. 2). Furthermore, it seems reasonable to assume that the pore will be formed from the re-entrant P loop plus the S6 segment, as demonstrated for other members of the K channel superfamily,[48,49] and this will adopt the general architectural features of the pore of the recently crystallized bacterial Kcsa channel[50] (as indicated on FIG. 2). Moreover, it is likely that the HCN-encoded channels will conform to the tetrameric arrangement of both the six-[51,52] and the two-[50] transmembrane-segment K channels. But what predictions can we make regarding the distinctive ion selectivity and gating properties of I_h channels?

With regard to activation gating, three close cousins of the HCN channels—notably, HERG, the CNG channels, and the plant inward rectifiers (KAT1 and AKT1)—are instructive. Like the HCN channels, all of these members of the K channel superfamily contain a CNB domain in the C terminus (see FIG. 2). Furthermore, elements of the activation mechanism of these channels resemble different aspects of HCN channel gating. Thus, the opening of CNG and HCN channels is promoted by the direct binding of cAMP or cGMP.[6,7,53] Both KAT1 and HERG show inward rectification that resembles that of the HCN channels.

Consistent with the evidence that activation of both heterologously expressed HCN channels and native I_h channels are altered by the direct binding of cyclic nucleotides, the six residues that have been identified as being strictly conserved in all functional CNB pockets of the type found in CNG channels[54,55] are also conserved in the HCN channels (see FIG. 2)

Although there is insufficient information to draw conclusions about the structures in HCN channels that couple cyclic nucleotide binding to opening based simply on sequence comparison with the CNG channels, studies on these channels point to domains that may be important in the cyclic nucleotide modulation of the HCN channels. Thus, mutagenesis studies have demonstrated that the CNB domain of CNG channels binds ligand and determines the energy a bound agonist can provide to open the channel.[53,56–58] However, it seems rather implausible that the striking difference in cyclic nucleotide efficacy, observed between HCN1 and HCN2 channels, could be explained by the modest sequence differences observed in the CNB site of the respective proteins.[6,7] Other studies have demonstrated that the region that links S6 to the CNB site appears to be involved in coupling the binding energy to opening,[59,60] while the amino terminus of CNG channels appears to influence the energy difference that has to be overcome upon activation–that is, how easy or hard the channel is to open.[57,58,61] Finally, the actual opening and closing of the gate appears to involve movements in the pore (J. Liu & S.A. Siegelbaum, personal communication) as seen in voltage-gated K channels.[50,62] Thus, the biophysical basis of the differential cAMP sensitivities of mHCN1 and mHCN2 may well reside outside of the CNB domain itself.

Is the voltage-dependent activation of the HCN channels mechanistically related to the activation process of either KAT1 or HERG? To consider this, it is helpful to consider a simple schematic representation that illustrates how two different domains, S4 and the pore, control the key processes of activation and inactivation in outward-rectifying, voltage-dependent K channels.

$$\text{CLOSED} \xleftrightarrow{\text{S4}} \text{OPEN} \xleftrightarrow{\text{PORE}} \text{INACTIVATED} \tag{1}$$

Ordinarily, a channel occupies the left-most closed states at hyperpolarized potentials. Upon depolarization, the S4 helix senses the change in the membrane electric field, moves outward, and triggers the conformational change to the open state. A second process that is often observed in voltage-gated channels is inactivation, a closing of the channel that follows opening at depolarized voltages and is thought to involve either block of the internal mouth of the pore by an N-terminal peptide (N-type inactivation) and/or a rearrangement of the pore involving the P region and S6 segment (C-type inactivation).[63,64]

Experiments from a number of laboratories have demonstrated that transient opening of HERG upon hyperpolarization is not due to a radical departure from the gating scheme outlined above, but is simply a consequence of alterations in the kinetics of the opening and inactivation pathways.[65,66] In HERG, activation is slow but inactivation is fast. As a result, little outward current is observed on depolarization because the channels inactivate essentially as soon as they open. However, on hyperpolarization the channels quickly recover from inactivation but then remain in the open state for a considerable time as the closing reaction to the resting state is slow. Although this gives rise to a current that is superficially similar to I_h, such kinetic changes alone are clearly insufficient to account for the behavior of I_h. Thus, the current carried by I_h does not decay on hyperpolarization (FIG. 5), whereas HERG relaxes back to a closed state.

However, the gating of I_h can in principle be explained by a relatively simple modification of the model. Miller and Aldrich[67] reported that a combination of three point mutations in the S4 voltage-sensing motif of Shaker K channels shifted the midpoint of activation to extreme hyperpolarized potentials (see FIG. 6). As inactivation remained intact, the channel occupied the right-most, nonconducting, inactivated state at the resting potential. However, upon hyperpolarization the channel recovered from inactivation (similar to HERG) but remained open (reminiscent of I_h) as it was no longer possible for the activation gates to close.[67]

Compelling evidence that supports the recovery from inactivation model for SPIH (and thus HCN) gating comes from the studies of Gauss *et al.*[23] In their original report, they demonstrated that upon moderate hyperpolarization, SPIH activates similar to its mammalian counterparts, the HCN channels. However, when hyperpolarized to more negative potentials, SPIH first opens and then closes. Although HCN channels do not shut off even upon hyperpolarization to –200 mV, this discrepancy could simply reflect that the midpoint of the activation curve for SPIH is shifted to less extreme negative voltages compared to that of the HCN channels. Interestingly, the S4 segment of both SPIH and the HCN channels contains an unusually large number of positive charges—nine in the mammalian channels—in the canonical motif of one charge every third residue. This is higher than most outward-rectifying K channels or the related inward rectifiers, KAT1 or HERG. Moreover, SPIH has one less charge in S4 than the HCN channels (see FIG. 2 and 6), which could be consistent with a less negatively shifted activation curve.

Although attractive, there are clearly alternative hypotheses to explain the activation of HCN channels, one of which is raised by recent studies suggesting that movement of S4 is coupled to activation in KAT1. Like HCN, KAT1 activates upon hyperpolarization, and it follows a sigmoidal time course with no evidence of channel closure at very negative potentials. Moreover, deactivation of KAT1 is accompanied by a small rising hook (i.e., the current increases and then decreases), as is observed when HCN channels close at depolarized potentials. Zei and Aldrich have individually mutated several residues in the KAT1 S4 motif and looked at the effect on gating.[68] Surprisingly, neutralization of R177 resulted in a +54-mV shift in open probability versus voltage relationship, while neutralization of the adjacent residue, R176, shifted the curve by +89 mV (the positions of these residues in the KAT1 S4 are indicated in FIG. 6). This shift is in the same absolute direction as is elicited by similar substitutions in outwardly rectifying channels such as Shaker, but here it results in the channel opening more readily. These data suggest that in both Shaker and KAT1, the movement of the S4 is coupled to the channel gate but with the opposite polarity—that is, the activation gate is closed when the S4 is in its outermost configuration at positive voltages in KAT1[68] but open in Shaker and related channels.

Given that SPIH and KAT1 appear to have distinct gating behaviors, it is curious to note a small but potentially significant similarity in the sequences of the S4 motifs of all three inward rectifiers, HCN, SPIH, and KAT1. Notably, in the middle of S4 at a site that would be anticipated to be a positively charged basic residue, there is a serine in all of these channels. Interestingly, this serine is flanked by the three residues that Miller and Aldrich[67] mutated to make their inwardly rectifying Shaker channel (FIG. 6).

Perhaps one of the most remarkable aspects of the HCN amino acid sequence is to be found in the pore. While the HCN channels all retain the basic hallmarks of the K channel family—in particular, the GYG motif that constitutes much of the selectivity filter and contributes to two of the K ion binding sites[50]—there are numerous unusual substitutions (FIG. 2). Some of the more striking examples are the introduction of a hydrophobic residue in HCN1 (alanine) and a positively charged residue in HCN2, HCN4, and SPIH (arginine and lysine, respectively) following the GYG triplet in place of the negative charge found in most K-selective pores. Further substitutions include a cysteine in place of the conserved threonine at –2 from the GYG triplet, and the presence of two more positively charged residues at –5 (histidine) and –9 (lysine). It is fair to speculate that these and other unusual substitutions will probably account for the altered selectivity observed for the HCN channel proteins and, perhaps, the unusual gating properties. Given that the crystal structure of a K channel pore has now been solved, the mechanistic implications of the pore substitutions are likely to be resolved soon.

SUMMARY

Although the pacemaker channels are widely distributed in the CNS and have been implicated in numerous physiological processes, a detailed analysis of the contribution of these channels to physiology has been severely hampered by the lack of both specific inhibitors and knowledge of their molecular identity. Thus, while the involvement of I_h in the generation of rhythmic activities in thalamic neurons has been fairly well characterized, the physiological consequences of the thalamic firing patterns are still highly speculative. For example, is slow-wave firing of the thalamus the driving rhythm that underlies the generation of the unconscious state, or is this an epiphenomenon? Do the rhythms observed in EEG recordings actually originate in the hypothalamus or lower brain stem

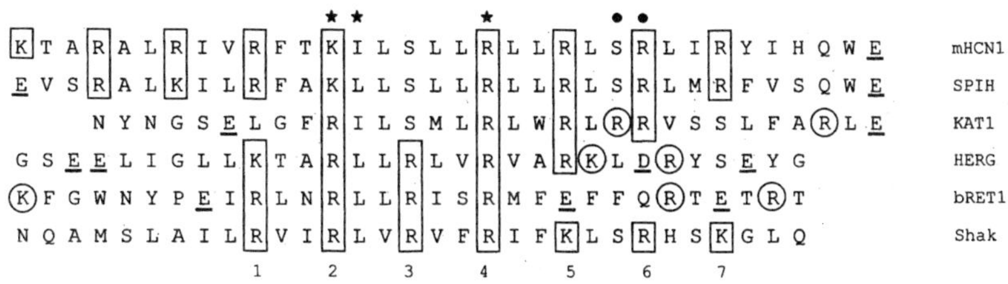
K T A R A L R I V R F T K I L S L L R L L R L S R L I R Y I H Q W E mHCN1
E V S R A L K I L R F A K L L S L L R L L R L S R L M R F V S Q W E SPIH
N Y N G S E L G F R I L S M L R L W R L R R V S S L F A R L E KAT1
G S E E L I G L L K T A R L L R L V R V A R K L D R Y S E Y G HERG
K F G W N Y P E I R L N R L L R I S R M F E F F Q R T E T R T bRET1
N Q A M S L A I L R V I R L V R V F R I F K L S R H S K G L Q Shak
1 2 3 4 5 6 7

FIGURE 6. Alignment of the S4 domain region in mHCN1, SPIH, KAT1, HERG, bRET1, and Shaker (Shak) proteins. The S4 domain is conventionally assigned as comprising the seven positively charged residues indicated in the Shaker protein. Positively charged residues (K or R) located outside of the triad repeats are *circled.* Negatively charged residues (E or D) are *underlined. Asterisks* mark residues that were mutated in the Shaker protein in the Miller and Aldrich study.[67] *Circles* mark residues that were mutated in KAT1, in the study of Zei and Aldrich.[68]

regions? Is the cyclic nucleotide–dependent modulation of I_h critical for changes in behavioral states during the sleep-wake cycle? Is I_h involved in the rhythmic firing patterns in cortical structures that are perhaps used to bind inputs from different sensory modalities?

By determining the biophysical properties and cellular distribution of each member of the HCN gene family and establishing the properties of the homomeric and heteromeric channels formed from the HCN gene products, we will gain a new insight into the molecular diversity of these rhythm-generating proteins. Furthermore, by utilizing subtype RNA probes and subtype-specific antibodies, it will be possible to accurately map the distribution of I_h both in cell types where function appears clear and in novel locations. Combining this information with the generation of spatially and temporally targeted knockouts of members of the HCN gene family, it will finally be possible to establish whether there is a causal involvement of I_h in rhythm generation and what the physiological and behavioral consequences of those rhythms are. Indeed, it is realistic to believe that with the cloning of this channel family we will soon be in a position to directly test some of the critical ideas underlying the role that pattern generation in areas such as the thalamus have in sensory processing and the generation of the conscious state. Finally, it is to be anticipated that the knowledge of the channel proteins should permit a rapid advance in the design of I_h-specific pharmacological agents and perhaps even subtype-selective ligands that will allow selective intervention within the CNS or the cardiovascular system.

ACKNOWLEDGMENTS

We would like to thank Steve Siegelbaum and Eric Kandel for their support, and Rich Robinson, David McCormick, and Anita Lüthi for sharing unpublished data and their thoughts. This work was supported in part by Grant S98-23 from The Whitehall Foundation (to G.R.T.).

REFERENCES

1. Reppert, S.M. 1998. Neuron **21:** 1–4.
2. DiFrancesco, D. 1993. Annu. Rev. Physiol. **55:** 455–472.
3. Pape, H.C. 1996. Annu. Rev. Physiol. **58:** 299–327.
4. McCormick, D.A. & T. Bal. 1997. Annu. Rev. Neurosci. **20:** 185–215.
5. Santoro, B., S.G. Grant, D. Bartsch & E.R. Kandel. 1997. Proc. Natl. Acad. Scl. USA **94:** 14815–14820.
6. Santoro, B., D.T. Liu, H. Yao, D. Bartsch, E.R. Kandel, S.A. Siegelbaum & G.R. Tibbs. 1998. Cell **93:** 717–729.
7. Ludwig, A., X. Zong, M. Jeglitsch, F. Hofmann & M. Biel. 1998. Nature **393:** 587–591.
8. Steriade, M., D.A. McCormick & T.J. Sejnowski. 1993. Science **262:** 679–685.
9. Singer, W. & C.M. Gray. 1995. Annu. Rev. Neurosci. **18:** 555–586.
10. Pare, D. & R. Llinas. 1995. Neuropsychologia **33:** 1155–1168.
11. McCormick, D.A. & H.C. Pape. 1990. J. Physiol. (Lond.) **431:** 319–342.
12. McCormick, D.A. & H.C. Pape. 1990. J. Physiol. (Lond.) **431:** 291–318.
13. McCormick, D.A. & T. Bal. 1994. Curr. Opin. Neurobiol. **4:** 550–556.
14. Maccaferri, G. & C.J. McBain. 1996. J. Physiol. (Lond.) **497:** 119–130.
15. Williams, S. D.D. Samulack, C. Beaulieu & J.C. Lacaille. 1994. J. Neurophysiol. **71:** 2217–2235.
16. Strata, F., M. Atzori, M. Molnar, G. Ugolini, F. Tempia & E. Cherubini. 1997. J. Neurosci. **17:** 1435–1446.

17. BANKS, M.I., R.A., PEARCE, & P.H. SMITH. 1993. J. Neurophysiol. **70:** 1420–1432.
18. FAIN, G.L., F.N. QUANDT, B.L., BASTIAN & H.M. GERSCHENFELD. 1978. Nature **272:** 466–469.
19. WOLLMUTH, L.P. & B. HILLE. 1992. J. Gen. Physiol. **100:** 749–765.
20. SOLOMON, J.S. & J.M. NERBONNE. 1993. J. Physiol. (Lond.) **462:** 393–420.
21. MAGEE, J.C. 1998. J. Neurosci. **18:** 7613–7624.
22. CLAPHAM, D.E. 1998. Neuron **21:** 5–7.
23. GAUSS, R., R. SEIFERT, & U.B. KAUPP. 1998. Nature **393:** 583–587.
24. WARMKE, J.W. & B. GANETZKY. 1994. Proc. Natl. Acad. Sci. USA **91:** 3438–3442.
25. BARGMANN, C. I. 1998. Science **282:** 2028–2033.
26. WILLIAMS, S.R., J.P. TURNER, S.W. HUGHES & V. CRUNELLI. 1997. J. Physiol. (Lond.) **505:** 727–747.
27. BAKER, K., K.S. WARREN, G. YELLEN & M.C. FISHMAN. 1997. Proc. Natl. Acad. Sci. USA **94:** 4554–4559.
28. DIFRANCESCO, D. & P. TORTORA. 1991. Nature **351:** 145–147.
29. DIFRANCESCO, D. & M. MANGONI. 1994. J. Physiol. **474:** 473–482.
30. BOIS, P., B. RENAUDON, M. BARUSCOTTI, J. LENFANT & D. DIFRANCESCO. 1997. J. Physiol. (Lond.) **501:** 565–571.
31. ACCILI, E.A., G. REDAELLI & D. DIFRANCESCO. 1997. J. Physiol. (Lond.) **500:** 643–651.
32. YU, H., F. CHANG & I.S. COHEN. 1995. J. Physiol. (Lond.) **485:** 469–483.
33. WU, J.Y. & I.S. COHEN. 1997. Pflügers Arch. Eur. J. Physiol. **434:** 509–514.
34. HAGIWARA, N. & H. IRISAWA. 1989. J. Physiol. (Lond.) **409:** 121–141.
35. LÜTHI, A. & D.A. MCCORMICK. 1998. Neuron **20:** 553–563.
36. YU, H., F. CHANG, & I.S. COHEN. 1993. Circ. Res. **72:** 232–236.
37. ROBINSON, R.B., H. YU, F. CHANG, & I.S. COHEN. 1997. Pflügers Arch. Eur. J. Physiol. **433:** 533–535.
38. Ranjan, R., N. Chiamvimonvat, N.V. Thakor, G.F. Tomaselli & E. Marban. 1998. Biophys. J. **74:** 1850–1863.
39. CERBAI, E., M. BARBIERI & A. MUGELLI. 1996. Circulation **94:** 1674–1681.
40. CERBAI, E., R. PINO, F. PORCIATTI, G. SANI, M. TOSCANO, M. MACCHERINI, G. GIUNTI & A. MUGELLI. 1997. Circulation **95:** 568–571.
41. FARES, N., P. BOIS, J. LENFANT & D. POTREAU. 1998. J. Physiol. (Lond.) **506:** 73–82.
42. AUSMA, J., M. WIJFFELS, G. VAN EYS, M. KOIDE, F. RAMAEKERS, M. ALLESSIE & M. BORGERS. 1997. Am. J. Pathol. **151:** 985–997.
43. WANG, Z., R.J. VAN DEN BERG & D.L. YPEY. 1997. J. Neurophysiol. **78:** 177–186.
44. BAKER, M., H. BOSTOCK, P. GRAFE & P. MARTIUS. 1987. J. Physiol. (Lond.) **383:** 45–67.
45. GRAFE, P., S. QUASTHOFF, J. GROSSKREUTZ, & C. ALZHEIMER. 1997. J. Neurophysiol. **77:** 421–426.
46. INGRAM, S.L. & J.T. WILLIAMS. 1996. J. Physiol. (Lond.) **492:** 97–106.
47. HORN, S., S. QUASTHOFF, P. GRAFE, H. BOSTOCK, R. RENNER & B. SCHRANK. 1996. Muscle & Nerve **19:** 1268–1275.
48. MACKINNON, R. & C. MILLER. 1989. Science **245:** 1382–1385.
49. GOULDING, E.H., G.R. TIBBS, D. LIU & S.A. SIEGELBAUM. 1993. Nature **364:** 61–64.
50. DOYLE, D.A., J.M. CABRAL, R.A. PFUETZNER, A. KUO, J.M. GULBIS, S.L. COHEN, B.T. CHAIT & R. MACKINNON. 1998. Science **280:** 69–77.
51. MACKINNON, R. 1991. Nature **350:** 232–235.
52. LIU, D.T., G.R. TIBBS & S.A. SIEGELBAUM. 1996. Neuron **16:** 983–990.
53. ZAGOTTA, W.N. & S.A. SIEGELBAUM. 1996. Annu. Rev. of Neurosci. **19:** 235–263.
54. SHABB, J.B. & J.D. CORBIN. 1992. J. Biol. Chem. **267:** 5723–5726.
55. TIBBS, G.R., D.T. LIU, B.G. LEYPOLD & S.A. SIEGELBAUM. 1998. J. Biol. Chem. **273:** 4497–4505.
56. VARNUM, M.D., K.D. BLACK & W.N. ZAGOTTA. 1995. Neuron **15:** 619–625.
57. GOULDING, E.H., G.R. TIBBS, & S.A. SIEGELBAUM. 1994. Nature **372:** 369–374.
58. GORDON, S.E. & W.N. ZAGOTTA. 1995. Neuron **14:** 857–864.
59. GORDON, S.E. & W.N. ZAGOTTA. 1995. Neuron **14:** 177–183.
60. PAOLETTI, P., E.C. YOUNG & S.A. SIEGELBAUM. 1999. J. Gen. Physiol. **113:** 17–33.
61. TIBBS, G.R., E.H. GOULDING & S.A. SIEGELBAUM. 1997. Nature **386:** 612–615.
62. LIU, Y., M. HOLMGREN, M.E. JURMAN & G. YELLEN. 1997. Neuron **19:** 175–184.
63. ZAGOTTA, W.N., T. HOSHI, & R.W. ALDRICH. 1990. Science **250:** 568–571.
64. HOSHI, T., W.N. ZAGOTTA, & R.W. ALDRICH. 1991. Neuron **7:** 547–556.

65. Trudeau, M.C., J.W. Warmke, B. Ganetzky & G.A. Robertson. 1995. Science **269:** 92–95.
66. Smith, P.L., T. Baukrowitz, & G. Yellen. 1996. Nature **379:** 833–836.
67. Miller, A.G. & R.W. Aldrich. 1996. Neuron **16:** 853–858.
68. Zei, P.C. & R. Aldrich. 1998. J. Gen. Physiol. **112:** 679–713.
69. Otsuka, M. 1958. Pflügers Archiv. Eur. J. Physiol. **266:** 512–517.
70. Anderson, J.A., S.S. Hupkrikar, L.V. Kochian, W.J. Lucas & R.F. Gaber. 1992. Proc. Natl. Acad. Sci. USA **89:** 3736–3740.
71. Kaupp, U.B., T. Niidome, T. Tanabe, S. Terada, W. Bonigk, W. Stuhmer, N.J. Cook, K. Kangawa, H. Matsuo, T. Hirose, *et al* 1989. Nature **342:** 762–766.
72. Warmke, J., R. Drysdale & B. Ganetzky. 1991. Science **252:** 1560–1562.
73. Sentenac, H., N. Bonneaud, M. Minet, F. Lacroute, J.M. Salmon, F. Gaymard & C. Grignon. 1992. Science **256:** 663–665.
74. Dhallan, R.S., K.W. Yau, K.A. Schrader & R.R. Reed. 1990. Nature **347:** 184–187.
75. Chen, T.Y., Y.W. Peng, R.S. Dhallan, B. Ahamed, R.R. Reed & K.W. Yau. 1993. Nature **362:** 764–767.
76. Liman, E.R. & L.B. Buck. 1994. Neuron **13:** 611–621.
77. Papazian, D.M., T.L. Schwarz, B.L. Tempel, Y.N. Jan & L.Y. Jan. 1987. Science **237:** 749–753.
78. Wei, A., M. Covarrubias, A. Butler, K. Baker, M. Pak & L. Salkoff. 1990. Science **248:** 599–603.
79. Pallanck, L. & B. Ganetzky. 1994. Hum. Mol. Gen. **3:** 1239–1243.
80. Halliwell, J.V. & P.R. Adams. 1982. Brain Res. **250:** 71–92.
81. Hart, G. 1983. J. Physiol. (Lond.) **337:** 401–416.
82. Maccaferri, G., M. Mangoni, A. Lazzari, & D. DiFrancesco. 1993. J. Neurophysiol. **69:** 2129–2136.
83. DiFrancesco, D. 1984. J. Physiol. **348:** 341–367.
84. DiFrancesco, D. 1987. European Heart Journal, **8** (Suppl. L): 19–23.
85. Accili, E.A., R.B. Robinson & D. DiFrancesco. 1997. Am. J. Physiol. **272:** H1549–H1552.
86. Budde, T., G. Biella, T. Munsch & H.C. Pape. 1997. J. Physiol. (Lond.) **503:** 79–85.
87. Chen, C. 1997. Heart Res. **110:** 179–190.
88. Crepel, F. & J. Penit-Soria. 1986. J. Physiol. (Lond.) **372:** 1–23.
89. Frace, A.M. F. Maruoka & A. Noma. 1992. J. Physiol. (Lond.) **453:** 307–318.
90. Liu, Z.W., A.R. Zou, S.S. Demir, J.W. Clark & R.D. Nathan. 1996. J. Mol. Cell. Cardiol. **28:** 2523–2535.
91. Maruoka, F., Y. Nakashima, M. Takano, K. Ono & A. Noma. 1994. J. Physiol. (Lond.) **477:** 423–435.
92. Mayer, M.L. & G.L. Westbrook. 1983. J. Physiol. (Lond.) **340:** 19–45.
93. Mo, Z.L. & R.L. Davis. 1997. J. Neurophysiol. **78:** 3019–3027.
94. Pape, H.C. & D.A. McCormick. 1989. Nature **340:** 715–718.
95. Pedarzani, P. & J.F. Storm. 1995. Proc. Natl. Acad. Sci. USA **92:** 11716–11720.
96. Solomon, J.S. & J.M. Nerbonne. 1993. J. Physiol. (Lond.) **469:** 291–313.
97. Solomon, J.S., J.F. Doyle, A. Burkhalter & J.M. Nerbonne. 1993. J. Neurosci. **13:** 5082–5091.
98. Soltesz, I., S. Lightowler, N. Leresche, D. Jassik-Gerschenfeld, C.E. Pollard & V. Crunelli. 1991. J. Physiol. (Lond.) **441:** 175–197.
99. Womble, M.D. & H.C. Moises. 1993. J. Neurophysiol. **70:** 2056–2065.
100. Zhou, Z. & S.L. Lipsius. 1992. J. Physiol. (Lond.) **453:** 503–523.
101. Pape, H.C. & R. Mager. 1992. Neuron **9:** 441–448.
102. Spain, W.J., P.C. Schwindt & W.E. Crill. 1991. J. Physiol. (Lond.) **434:** 609–626.
103. Stuart, G. & N. Spruston. 1998. J. Neurosci. **18:** 3501–3510.

Ca^{2+}-Mediated Up-Regulation of I_h in the Thalamus

How Cell-Intrinsic Ionic Currents May Shape Network Activity

ANITA LÜTHI[a] AND DAVID A. McCORMICK

Yale University School of Medicine, Section of Neurobiology, 333 Cedar Street, New Haven, Connecticut 06510, USA

Hyperpolarization-activated mixed cation currents (I_h) are widely expressed throughout the central and peripheral nervous systems as well as in cardiac tissue.[1] The h-current activates slowly upon membrane hyperpolarization below −50 to −60 mV and does not inactivate.[1,2] The ionic channels carrying I_h have recently been molecularly identified, and are structurally related to cyclic nucleotide–gated channels.[3] (See note added in proof.) Traditionally, I_h has been implied in the control of resting membrane potential levels and firing properties of single cells in the CNS as well as neuronal rhythmogenesis ("pacemaker" activity).[1,2] These roles are complemented by an exquisite sensitivity of I_h to neurotransmitter receptor activation, mediated through the activation of G proteins, adenylate cyclase, and protein kinase C.[1] In contrast, modulation of I_h by changes in intracellular Ca^{2+} concentration has been proposed, but has remained controversial for almost 10 years (see references in Ref. 1).

The h-current is expressed in thalamocortical (TC) cells that participate in spindle oscillations related to early periods of sleep.[4] Spindle oscillations are generated as a bidirectional synaptic interaction between TC cells and thalamic reticular neurons and display a strong periodicity of once every 5–20 s.[4–6] Interestingly, this periodicity can be largely abolished by reducing I_h in TC cells, suggesting that activation of this current contributes to the temporal development of these synchronized thalamocortical oscillations.[5,7] The involvement of I_h in spindle waves provides a model system for investigating the physiological factors, including Ca^{2+}, contributing to the generation of slow periodicities in coordinated network activity.

METHODS

Intracellular current and voltage-clamp recordings were performed *in vitro* in slices of lateral dorsal geniculate nucleus from 2-month-old ferrets, as described previously.[6] For flash photolysis experiments, cells were perfused with DM-Nitrophen (70–100 mM, 60–75% Ca^{2+}- loaded) present in the microelectrode and subsequently exposed to a UV-flash delivered by a Xenon Flashlamp (Rapp Optoelektronik, Hamburg), as described previously.[8]

[a]Corresponding author. Phone: 203-737-5217; fax: 203-785-5263; e-mail: luthi@biomed.med.yale.edu

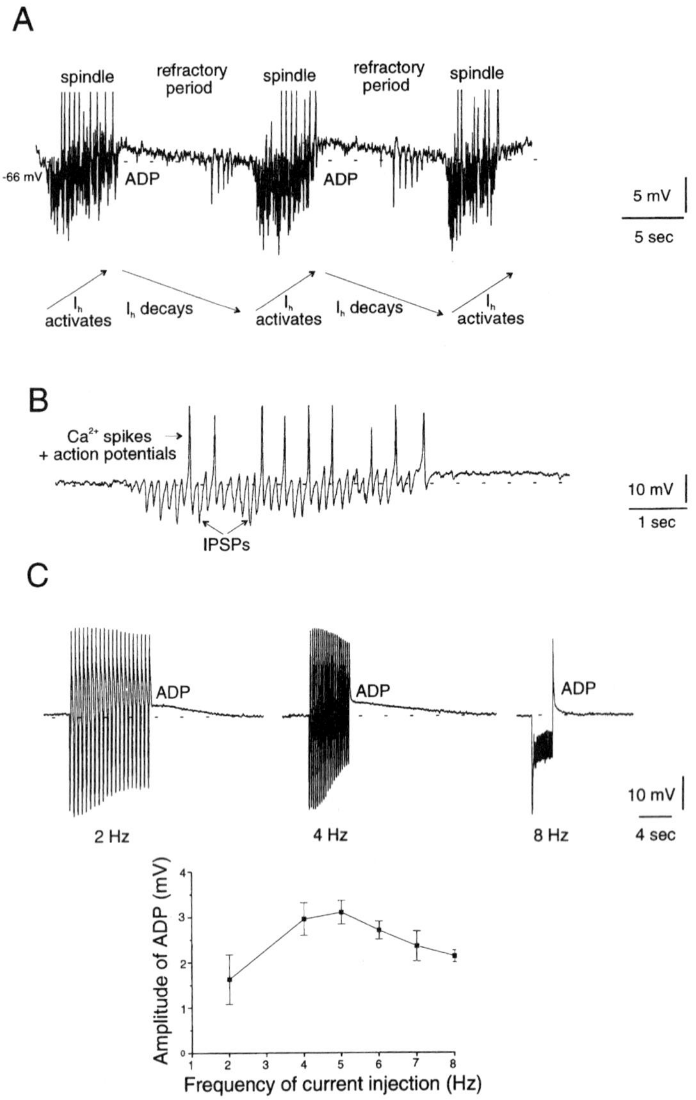

FIGURE 1. **(A)** Spindle oscillations in thalamocortical cells *in vitro* are characterized by oscillatory periods followed by silent periods associated with an afterdepolarization (ADP). This periodicity is associated with recurring activation of I_h. **(B)** Expanded trace showing the electrophysiological events during a spindle oscillation. **(C)** Injection of repetitive hyperpolarizing current pulses (−400-pA, 120-ms pulse duration) generates an ADP, and this shows a bell-shaped dependence on frequency.

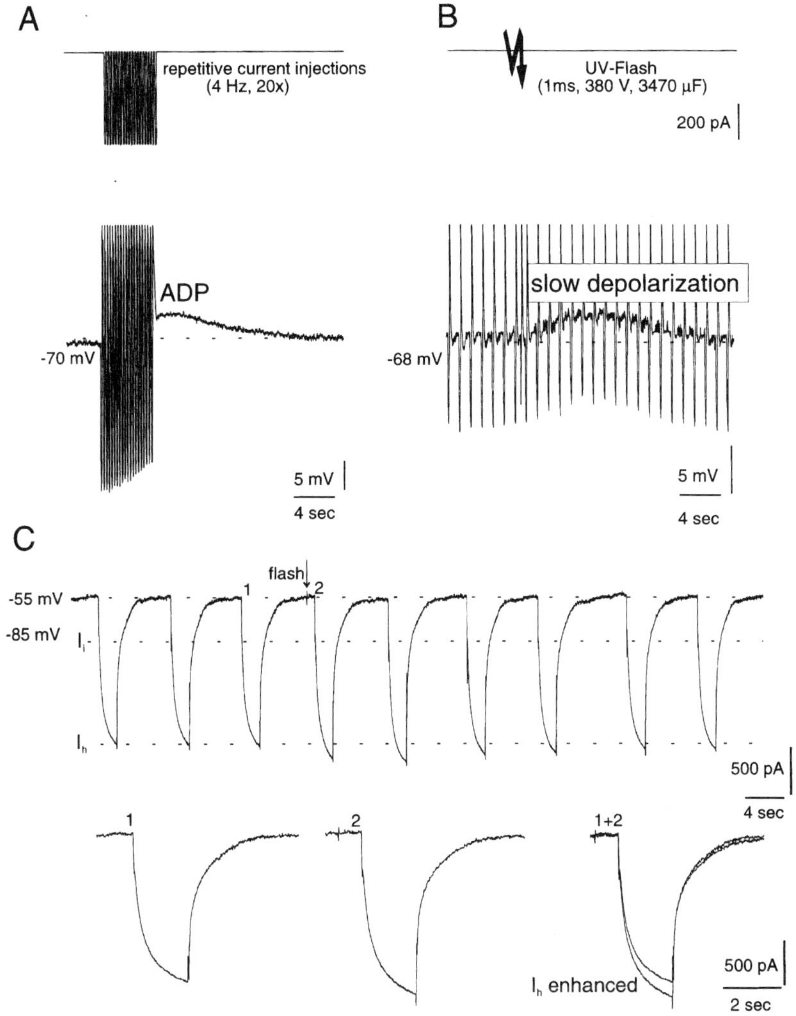

FIGURE 2. (**A**) Repetitive hyperpolarizing current injections evoke an ADP. (**B**) Release of Ca^{2+} through the application of a single UV-flash in a cell perfused with a caged Ca^{2+} chelator (DM-Nitrophen, 70 mM, 60% Ca^{2+}-loaded) also evokes a slow depolarization. Short hyperpolarizing current pulses (120 ms, −150 pA) were applied throughout the recording to monitor cellular input resistance. For (**A**) and (**B**) **top traces** represent current, **bottom traces** represent membrane voltage. (**C**) In voltage-clamp, a UV-flash induces a transient enhancement of I_h. (Part B adapted from Lüthi & McCormick.[8])

RESULTS

FIGURE 1 shows an intracellular recording of a TC cell participating in spontaneous spindle oscillations. The prominent periodic recurrence of oscillatory followed by silent phases is evident in FIGURE 1A. The silent period is associated with a slowly decaying afterdepolarization (ADP) resulting from activation of I_h,[5,7,8] while each oscillation consists of barrages of IPSPs and rebound bursts including Ca^{2+} spikes and action potentials (FIG. 1B).[4,6] To investigate which of these components may contribute to an ADP caused by up-regulation of I_h, repetitive hyperpolarizing current injections at various frequencies were delivered[5] (FIG. 1C). The number and duration of the pulses was kept unchanged to ensure a constant level of membrane hyperpolarization, while the frequency was varied from 2 Hz to 8 Hz. ADPs of maximal amplitude were evoked at frequencies of 4–5 Hz, whereas lower or higher frequencies resulted in a decreased size of the ADP (n = 3–5 per point). This bell-shaped dependence of ADP amplitude on the frequency of current injections shows that the occurrence of rebound bursts in addition to membrane hyperpolarization and the close temporal succession of these two factors is required for the development of the ADP. These findings extend and complement earlier data demonstrating that Ca^{2+} entry during repetitive current injections is a triggering event in the generation of the ADP and may thus contribute to the activation of I_h.[8]

To obtain direct evidence for a role of Ca^{2+} entry in the development of the ADP, flash photolysis of caged Ca^{2+} was used to rapidly elevate the intracellular Ca^{2+} concentration in TC cells (FIG. 2). Application of a UV-flash following perfusion of TC cells for at least 50 min resulted in a slow membrane depolarization, the time course of which was reminiscent of the ADP induced by repetitive current injections (FIG. 2A, 2B). In voltage-clamp, I_h responses evoked by step hyperpolarizations from −55 mV to −85 mV were transiently enhanced in amplitude following flash release of caged Ca^{2+}, thus demonstrating that indeed increases in $[Ca^{2+}]_i$ facilitate the activation of I_h.

DISCUSSION

Our results provide evidence for an involvement of Ca^{2+} in up-regulating the hyperpolarization-activated cation current I_h in TC cells and show its physiological role in thalamic network activity.[5,7,8] Thus, the entry of Ca^{2+} together with IPSPs during spindling activity in TC cells leads to a gradual membrane depolarization owing to the activation and up-regulation of I_h.[8] This depolarization compromises the participation of TC cells in the network activity and is therefore responsible for the cessation of spindle waves (FIG. 1A). The h-current and its modulation by Ca^{2+} provide an example by which activity-dependent second messenger production and modulation of ionic channels in neurons may contribute to shaping the temporal development of network oscillations. Whether Ca^{2+} acts directly on h-channels or via production of second-messenger compounds and how it affects the biophysical properties of these channels remain to be investigated.

[NOTE ADDED IN PROOF: The ionic channels underlying hyperpolarization-activated cation currents have now been cloned. For further references, see Santoro & Tibbs, this volume.]

REFERENCES

1. PAPE, H.-C. 1996. Queer current and pacemaker: The hyperpolarization-activated cation current in neurons. Annu. Rev. Physiol. **58:** 299–327.
2. MCCORMICK, D.A. & H.-C. PAPE. 1990. Properties of a hyperpolarization-activated cation current and its role in rhythmic oscillation in thalamic relay neurones. J. Physiol. (Lond.) **431:** 291–318.
3. FINN, J.T., M.E. GRUNWALD & K.-W. YAU. 1996. Cyclic nucleotide-gated ion channels: An extended family with diverse functions. Annu. Rev. Physiol. **58:** 395–426.
4. MCCORMICK, D.A. & BAL, T. 1997. Sleep and arousal: Thalamocortical mechanisms. Annu. Rev. Neurosci. **20:** 185–215.
5. BAL, T. & MCCORMICK, D.A. 1996. What stops synchronized thalamocortical oscillations? Neuron **17:** 297–308.
6. VON KROSIGK, M., T. BAL & D.A. MCCORMICK. 1993. Cellular mechanisms of a synchronized oscillation in the thalamus. Science **261:** 361–364.
7. LÜTHI, A., T. BAL & D.A. MCCORMICK. 1998. Periodicity of thalamic oscillations is abolished by ZD7288, a blocker of I_h. J. Neurophysiol. **79:** 3284–3289.
8. LÜTHI, A. & D.A. MCCORMICK. 1998. Periodicity of thalamic synchronized oscillations: the role of Ca^{2+}-mediated upregulation of I_h. Neuron **20:** 553–563.

Index of Contributors

Adelman, J.P., 370–378
Ahring, P.K., 423–426
Akaike, N., 697–700
Alagarsamy, S., 515–525
Aldrich, R.W., 458–464
Alonso, A., 84–87
Amarillo, Y., 233–285, 304–343
Aoki, C., 628–632
Art, J.J., 379–385
Artman, M., 434–437

Bähring, R., 344–355
Barnéoud, P., 633–635
Bean, B.P., 93–96
Beckman, M.L., 620–623
Benham, C.D., 224–227
Berg, U., 77–79
Berger, M., 344–355
Berrow, N.S., 160–174
Bertrand, S., 640–644
Bertrand, D., 640–644
Besnard, F., 633–635
Bettler, B., 689–692
Betz, H., 667–676
Bond, C.T., 370–378
Bonneau, L.J., 51–66
Bonnert, T.P., 645–653
Bradley, S.R., 531–534
Broillet, M.-C., 730–741
Brucker, C., 77–79
Brussaard, A.B., 677–680
Bulling, A., 77–79
Burgess, D.L., 199–212
Butler, A., 286–303

Campbell, D.B., 217–219
Cantí, C., 160–174
Cantino, D., 693–696
Catterall, W.A., 144–159
Champigny, G., 67–76
Changeux, J-P., 611–616
Charpantier, E., 633–635
Chen, Y., 80–83
Chen, J.-P., 701–715
Chiu, J., 233–285
Chow, A., 233–285, 304–343
Christophersen, P., 423–426
Clapham, D.E., 386–398
Clare, J.J., 80–83, 88–92
Coetzee, W.A., 233–285, 434–437
Conn, P.J., 515–525, 531–534, 554–557
Cook, N., 213–216
Corey, S., 386–398
Covernton, P.J.O., 620–623
Crabtree, G., 578–590
Cribbs, L.L., 131–143
Crooks, P.A., 617–619

Dale, T.J., 80–83
Dani, J.A., 591–610
Dart, C., 414–417
De Jager, P.L., 502–514
de Weille, J.R., 67–76
Dezaki, K., 624–627
Dolphin, A.C., 160–174
Dong, H., 535–540
Dubel, S.J., 118–130
Dwoskin, L.P., 617–619

Emson, P.C., 88–92
Engelman, H.S., 546–549
Engisch, K.L., 213–216
Erisir, A., 304–343, 628–632
Ernstrom, G., 578–590
Eunson, L., 442–446

Farrar, S., 645–653
Fenster, C.P., 620–623
Fettiplace, R., 379–385
Firestein, S., 730–741
Fisher, J.L., 591–610
Friedman, L.K., 541–545
Fritschy, J.-M., 654–666, 693–696
Fureman, B.E., 217–219

Gadea, A., 685–688

Ganetzky, B., 356–369
Gardner, E., 701–715
Ge, S., 228–232
Gereau IV, R.W., 515–525
Giese, K.P., 344–355
Giorgetta, O., 654–666
Girod, R., 578–590
Giustetto, M., 693–696
Goldin, A.L., 38–50
Goldman-Rakic, P. S., 13–26
Gratzl, M., 77–79
Gray, R., 591–610
Gray-Keller, M., 379–385

Hanna, M.G., 442–446
Hanrahan, C.J., 51–66
Harvey, R.J., 667–676
Heavens, R.P., 645–653
Hećimović, H., 220–223
Heinemann, S.F., 515–525
Heintz, N., 502–514
Hen, R., 701–715
Hernandez-Cruz, A., 304–343
Hernandez-Pineda, R., 304–343
Hess, E.J., 217–219
Heurteaux, C., 67–76
Hewson, L., 645–653
Hicks, J.H., 620–623
Higashida, H., 454–457
Higuchi, M., 27–37
Honoré, E., 438–441
Hoshi, T., 458–464
Huganir, R.L., 535–540

Jacob, M.H., 640–644
Jensen, B.S., 423–426
Jerecic, J., 27–37
Johns, D.C., 418–423
Jones, P.G., 427–430
Jones, E.M.C., 379–385
Jørgensen, T.D., 423–426
Jung, H.-Y, 97–101

Karschin, A., 689–692
Kask, K., 27–37
Kaupmann, K., 689–692
Kelly, J.S., 220–223
Kennedy, M., 386–398
Kimura, I., 624–627
Kirsch, J., 667–676, 693–696
Kits, K.S., 677–680
Kolhekar, R., 27–37
Krapivinsky, G., 386–398
Krestel, H., 27–37
Krüth, U., 27–37
Kuhse, J., 667–676
Kullmann, D.M., 442–446
Kunkel, M.T., 286–303
Kuroda, C., 228–232

Lau, D., 233–285, 304–343
Laube, B., 667–676
Lazdunski, M., 67–76, 438–441
Le Bourdellès, B., 645–653
Lee, C.J., 546–549
Lee, K., 434–437
Lee, J.-H., 131–143
Leicher, T., 344–355
Léna, C., 611–616
Leonard, C., 304–343
Lester, R.A.J., 620–623
Levey, A.L., 531–534
Lewis, T.M., 681–684
Leyland, M.L., 414–417
Liao, D., 535–540
Lingueglia, E., 67–76
Lissin, D.V., 550–553
Lo, C.F., 431–433
Lombardi, S.J., 427–430
López-Colomé, A.M., 685–688
Lubin, M., 628–632
Lükensmejer, B., 697–700
Lüthi, A., 765–769

MacDermott, A.B., 546–549
Macek, T.A., 554–557
McCormack, T., 233–285
McCormick, D.A., 765–769
McGehee, D.S., 565–577, 578–590
MacKenzie, A.B., 716–729
McKernan, R.M., 645–653
Magistretti, J., 84–87
Malenka, R.C., 515–525, 550–553
Marban, E., 418–423

Matsubara, A., 474–482
Mayerhofer, A., 77–79
Maylie, J., 370–378
Meadows, H.J., 224–227
Medina, I., 386–398
Mickus, T., 97–101
Moreno Davila, H., 102–117, 233–285, 304–343
Mosbacher, J., 689–692

Nabekura, J., 697–700
Nadal, M.S., 233–285, 304–343
Nakamura, T.Y., 434–437
Namkung, Y., 175–198
Nemec, J., 386–398
Nicol, R.A., 515–525
Nielsen, M., 697–700
Niesen, C.E., 228–232
Noebels, J.L., 199–212
North, R.A., 716–729
Nowycky, M.C., 213–216
Numann, R., 431–433
Nuss, H.B., 418–423

Ogielska, E.M., 458–464
Olesen, S.-P., 423–426
Osten, P., 558–560
Ottersen, O.P., 474–482
Ozaita, A., 233–285, 304–343

Page, K.M., 160–174
Pak, D.T., 483–493
Palladino, M.J., 51–66
Panzanelli, P., 693–696
Patel, A.J., 438–441
Pauly, J.R., 617–619
Paysan, J., 654–666
Peakman, T.C., 80–83
Penschuck, S., 654–666
Perez-Reyes, E., 131–143
Peters, J., 228–232
Pfaff, T., 689–692
Piedras-Renterìa, E.S., 175–198
Pongs, O., 344–355
Poulsen, C.F., 697–700
Pountney, D., 233–285
Pozza, M.F., 689–692
Pradhan, S.M., 228–232

Quick, M.W., 620–623

Radcliffe, K.A., 591–610
Raman, I. M., 93–96
Ramirez-Latorre, J., 578–590
Reenan, R.A., 51–66
Rhodes, K.J., 427–430
Ribera, A.B., 399–405
Rich, M.M., 213–216
Rigby, M.R., 645–653
Rinvik, E., 474–482
Ristig, D., 689–692
Robertson, G.A., 356–369
Roeper, J., 344–355
Role, L., 578–590
Rouse, S.T., 515–525
Rudy, B., 1–12, 233–285, 304–343, 434–437

Saganich, M., 233–285, 304–343
Salkoff, L., 286–303
Sanguinetti, M.C., 406–413
Santoro, B., 741–764
Sassoè-Pognetto, M., 693–696
Scearce-Levie, K., 701–715
Schaffhauser, H., 554–557
Schmieden, V., 667–676
Schofield, P.R., 681–684
Schuler, V., 689–692
Seeburg, P.H., 27–37
Sgard, F., 633–635
Sheng, M., 483–493
Shin, H.-S., 175–198
Silva, A. J., 344–355
Single, F., 27–37, 494–501
Sirinathsinghji, D.J.S., 645–653
Smith, Y., 515–525
Smith, D.W., 645–653
Smith, S.S., 175–198
Snutch, T.P., 118–130
Spauschus, A., 442–446
Spence, P., 427–430
Spencer, P.J., 414–417
Sprengel, R., 27–37, 494–501
Spruston, N., 97–101

Srivastava, S., 561–564
Standaert, D.G., 531–534
Stanfield, P.R., 414–417
Stea, A., 118–130
Stephens, G.J., 160–174
Storck, T., 27–37
Storm, J.F., 344–355
Strøbæk, D., 423–426
Struckmeyer, K., 689–692
Surprenant, A., 716–729
Sutcliffe, M.J., 414–417

Takeda, H., 454–457
Takumi, Y., 474–482
Temburni, M.K., 640–644
Thompson, S.A., 645–653
Tibbs, G.R., 741–764
Titus, S.A., 356–369
Trudeau, M.C., 356–369
Truong, A., 427–430
Tsien, R.W., 175–198
Turner, J., 578–590

Van Kleef, R.G.D.M., 636–639
Vega-Saenz de Miera, E., 233–285, 304–343
Velísková, J., 541–545
Vijverberg, H.P.M., 636–639
von Zastrow, M., 550–553

Wafford., K.A., 645–653
Waldmann, R., 67–76
Wang, Z.-W., 286–303
Wang, Q., 447–449
Wei, A., 286–303
Westh-Hansen, S.E., 697–700
Whitaker, W.R.J., 88–92
Whiting, P.J., 645–653
Wickman, K., 386–398
Wilkins, L.H., 617–619
Williams, B.M., 640–644
Wilson, G.F., 356–369
Witt, M.-R., 697–700
Wray, D., 344–355
Wu, C.-F., 450–453

Xie, X., 80–83

Yao, W.-D., 450–453
Yokoyama, S., 454–457

Zei, P.C., 458–464
Zhang, P., 535–540
Ziff, E.B., 465–473, 561–564, 558–560
Zwart, R., 636–639